ELECTROMAGNETIC RADIATION

Frequency (Hz)

10^8 10^{12} 10^{14} 10^{16} 10^{18} 10^{20}

Wavelength (m)

10^{-1} 10^{-3} 10^{-5} 10^{-7} 10^{-9} 10^{-11}

RADIO MICROWAVE INFRARED |VIS| UV X-RAY γ-RAY

Energy (kJ/mol)

10^{-3} 10^{-1} 2×10^2 10^4 10^7

VISIBLE LIGHT

Color

Red Orange Yellow Green Blue Violet

Wavelength (nm)

800 700 600 500 400

UNDERGRADUATE INSTRUMENTAL ANALYSIS

Sixth Edition

UNDERGRADUATE INSTRUMENTAL ANALYSIS

Sixth Edition

James W. Robinson
Louisiana State University
Baton Rouge, Louisiana, U.S.A.

Eileen M. Skelly Frame
Rensselaer Polytechnic Institute
Troy, New York, U.S.A.

George M. Frame II
New York State Department of Health
Albany, New York, U.S.A.

MARCEL DEKKER   NEW YORK

Permission for the publication herein of Sadtler Spectra has been granted by Bio-Rad Laboratories, Informatics Division

Although great care has been taken to provide accurate and current information, neither the author(s) nor the publisher, nor anyone else associated with this publication, shall be liable for any loss, damage, or liability directly or indirectly caused or alleged to be caused by this book. The material contained herein is not intended to provide specific advice or recommendations for any specific situation.

Trademark notice: Product or corporate names may be trademarks or registered trademarks and are used only for identification and explanation without intent to infringe.

Library of Congress Cataloging-in-Publication Data
A catalog record for this book is available from the Library of Congress.

ISBN: 0-8247-5359-3

This book is printed on acid-free paper.

Headquarters
Marcel Dekker, 270 Madison Avenue, New York, NY 10016, U.S.A.
tel: 212-696-9000; fax: 212-685-4540

Distribution and Customer Service
Marcel Dekker, Cimarron Road, Monticello, New York 12701, U.S.A.
tel: 800-228-1160; fax: 845-796-1772

World Wide Web
http://www.dekker.com

Current printing (last digit):

10 9 8 7 6 5 4 3 2 1

PRINTED IN THE UNITED STATES OF AMERICA

Preface to the Sixth Edition

Analytical chemistry today is almost entirely instrumental analytical chemistry and it is performed by many scientists and engineers who are not chemists. Analytical instrumentation is crucial to research in molecular biology, medicine, geology, food science, materials science, and many other fields. While it is true that it is no longer necessary to have almost artistic skills to obtain accurate and precise analytical results using instrumentation, the instruments should not be considered "black boxes" by those using them. The well-known phrase "garbage in, garbage out" holds true for analytical instrumentation as well as computers. We hope this book serves to provide users of analytical instrumentation with an understanding of their instruments.

In keeping with the earlier editions of this text, the book is designed for teaching undergraduates and those with no analytical chemistry background how modern analytical instrumentation works and what the uses and limitations of analytical instrumentation are. Mathematics is kept to a minimum. No background in calculus, physics, or physical chemistry is required. All major fields of modern instrumentation are covered, including applications of each type of instrumental technique. Each chapter includes discussion of the fundamental principles underlying each technique, detailed descriptions of the instrumentation and a large number of applications. Each chapter includes an updated bibliography and new problems and most chapters have suggested experiments appropriate to the technique.

This edition has been completely rewritten, revised, and expanded. To achieve this, the previous approach of having each chapter be self-contained has been abandoned; repetition has been reduced to a minimum so that more topics could be covered in more detail. The topics of chromatography and mass spectrometry have been greatly expanded when compared with the 5th edition to better reflect the predominance of chromatography and mass spectrometry instrumentation in modern laboratories. The equally important topic of NMR has been refocused on FTNMR and expanded to included ^{13}C and 2D NMR spectral interpretation.

A unique feature of this text is the combination of instrumental analysis with organic spectral interpretation (IR, NMR, and MS). The NMR, IR, and MS are all new, courtesy of Bio-Rad Laboratories, Informatics Division (IR), Aldrich Chemical Company (NMR), and one of the authors (MS), and were obtained on modern instruments, to reflect what students will encounter in modern laboratories. The use of spreadsheets for performing calculations has been introduced with examples. Reflecting the ubiquitous nature of the

Internet, we have included large numbers of instrument manufacturers' websites, which contain extensive resources for interested students.

Sampling, sample handling, and storage and sample preparation methods are extensively covered, and modern methods such as accelerated solvent extraction, solid phase microextraction (SPME), and microwave techniques are included. The NMR chapter is focused on the current technique of FTNMR. Proton, ^{13}C, 2D NMR are covered including spectral interpretation. The NMR spectra presented are from a 300 MHz NMR instrument. Instrumentation, the analysis of liquids and solids, and applications of NMR are discussed in detail. A section on hyphenated NMR techniques is included, along with an expanded section on MRI. The IR instrumentation section is focused on FTIR instrumentation. Absorption, emission, and reflectance spectroscopy are discussed, as is FTIR microscopy. Near-IR instrumentation and applications are presented. Coverage of Raman spectroscopy includes resonance Raman, surface-enhanced Raman, and Raman microscopy. Chemical imaging is described. The section on IR spectral interpretation has been greatly expanded and all new spectra are presented. UV and visible spectroscopy includes innovations such as flowthrough sample holders and fiber optic probes. UV absorption spectral interpretation for organic molecules is covered in depth. Applications described include spectrophotometric titrations and spectroelectrochemistry. Nephelometry, turbidimetry, fluorescence, and phosphorescence are described in detail, including instrumentation and applications. The techniques of thermomechanical analysis and dynamic mechanical analysis have been added to the chapter on Thermal Analysis.

All major modern atomic absorption and emission techniques and instrumentation are covered. Appendices with FAAS and GFAAS conditions have been added, and a new appendix with up-to-date limits of detection for all the atomic spectroscopic techniques is included. Chemical speciation using hyphenated chromatographic-atomic emission spectroscopy is described as is a novel microwave induced plasma emission instrument for particle characterization.

Mass spectrometry has been expanded to two chapters and covers both organic and inorganic MS instrumentation and applications. GC-MS and LC-MS, along with MS^n instruments are described, along with modern ionization methods such as electrospray and MALDI. Organic mass spectral interpretation is covered with many examples and new spectra. Organic and inorganic applications focus on speciation using GC-MS, LC-MS, hyphenated ICP-MS, with emphasis on proteomics, biomolecules, and species of environmental interest.

In contrast to earlier editions, the subject of chromatographic separations and instrumentation is expanded from one to three chapters containing more than twice as much text, illustrations, exercises, and problems. The first of these chapters covers the nature of the chromatographic process. A minimum of complex formulas is introduced; instead, extensive description and analogy are employed to give the student an intuitive grasp of the mechanisms giving rise to separation, resolution, and detection of separated components. In line with current practice, GC is treated as a method almost completely employing open tubular (capillary) columns. New developments such as SPME injection, low bleed stationary phases, compound-selective detectors, and especially interfacing to spectrometric detectors and the use of computerized data processing have been added. So-called "hyphenated techniques" such as GC-MS, GC-IR, and comprehensive GC-GC are included, and new sections on retention indices, derivitization to improve detectability and volatility, and analysis of gases in air or water, have been added. The requirements for implementing and instrumentalizing HPLC are developed from the preceding discussion of GC. The student gains an appreciation of the difficulties that caused this instrumentation to lag behind GC. Once overcome, liquid chromatographic instrumentation is seen to have wider applicability, especially in the burgeoning subdisciplines of proteomics and genomics. Detailed discussions of instrumental design and the operation of new detectors have been added. Major sections on the design and operation of HPLC interfaces to mass spectrometers (ESI and APCI) have been added, and examples of their use in protein or peptide sequencing and identification are included. Tables of amino acid structures and nomenclature have been included so that students can follow these descriptions. Separate sections on ion chromatography, affinity chromatography, size exclusion chromatography, and supercritical fluid chromatography have been expanded. Planar and capillary electrophoresis are described in detail in this chapter, despite not being strictly defined as chromatographic methods. Examples are given of the use of capillary electrophoresis with fluorescence-derivitization detection to gene sequencing in genomics, or of 2D slab-gel, isoelectric focusing/SDS-PAGE electrophoresis for protein peptide mapping in proteomics. New appendices provide links to websites providing examples of thousands of chromatographic separations, and encourage the student to learn how to utilize the resources of commercial column vendors to find a solution to particular separation or measurement problems.

James W. Robinson
Eileen M. Skelly Frame
George M. Frame II

Acknowledgments

The following people are gratefully acknowledged for their assistance in the successful completion of this edition. They provided diagrams, photographs, applications notes, spectra, chromatograms and many helpful comments and suggestions regarding the text. Thanks are due, in totally random order, to: Pam Decker, Restek; Jodi Dorfler, Alltech; Roger Blaine, TA Instruments; Jim Ferrara and John Flavell, ThermoHaake; Andy Rodman, PerkinElmer Instruments; Gwen Boone, Doug Shrader, Pat Grant, Laima Baltusis and Dan Steele, Varian, Inc.; Didier Arniaud, Lisa Goldstone, Tina Harville, Phillipe Hunault, Patrick Chapon and Phil Shymanski, JobinYvon/Horiba; Tiger Pitts, ThermoElemental; John Sotera and Phil Bennett, Leeman Labs, Inc.; Steve Sauerbrunn, Mettler Toledo, Inc.; Mark Mabry, Bob Coel, and Ed Oliver, ThermoNicolet; Lara Pryde, ThermoOrion; David Coler, PANalytical, Inc.; Marty Palkovic, ThermoARL; Dale Gedcke, ORTEC(Ametek); Bob Anderhalt, Edax, Inc. (Ametek); Mike Hurt, HHT; John Martin, RigakuMSC; Dr. Peter Codella, Dr. Elizabeth Williams, and Dr. Woodfin Ligon, GE Global Research and Development; Dr. Tom Dulski, Carpenter Technologies; Eric Francis, Dionex Corporation; Dr. Ales Medek, Pfizer Central Research and Development; Dr. James Roberts, Lehigh University; Harry Xie, Bruker Optics; Pat Wilkinson and James Beier, Bruker BioSpin; Professor F. X. Webster, State University of New York College of Environmental Science and Forestry; Toshiyuki Suzuki, Yukihiko Takamatsu, Morio Kyono and Hisao Takahara, Yokogawa Electric Company; Dr. A. Horiba, Atsuro Okada, Juichiro Ukon, Mike Pohl and Paul Dinh, Horiba, Ltd.; Pat Palumbo, Bill Strzynski and Dr. Jack Cochran, LECO Corporation; John Moulder, Physical Electronics USA, Inc.; Dr. Cedric Powell, NIST; Jill Thomas and Michael Monko, Supelco; Michael Garriques, Phenomenex, Inc.; A. Audino, SGE; Alan D. Jones, Mallinckrodt Baker, Inc.; Dr. Ronald Starcher, BURLE Electro-Optics, Inc.; Merrill Loechner, Milestone, Inc.; Dr. Mike Collins, CEM Corporation; Kerry Scoggins, SGE, Inc.; Dr. Julian Phillips and Wendy Weise, Thermo Electron Corporation; Professor Gary Siudzak, Scripps Research Institute Center for Mass Spectrometry; Dr. Robert Kobelski, Centers for Disease Control, Atlanta, GA; Professor David Hercules, Vanderbilt University; Dr. S.E. Stein, NIST Mass Spec Data Center; Volker Thomsen, NITON Corp.; Michael Fry, Dr. Chip Cody and Patricia Corkum, JEOL, Inc.; Chuck Douthitt, Thermo Electron; Giulia Orsanigo and Danielle Hawthorne, PerkinElmer Life and Analytical Sciences; Nancy Fernandes, Newport Corporation; Jackie Lathos-Markham and Joseph Dorsheimer, Thermo Electron Corp.; Ralph Obenauf, SPEX CertiPrep, Inc.; David Weber, Rensselaer Polytechnic Institute.

Special thanks go to Professor Milton Orchin, University of Cincinnati, for permission to use material from his and the late H.H. Jaffé's classic text, *Theory and Applications of Ultraviolet Spectroscopy*, Wiley, New York, 1962. The cheerful cooperation and invaluable assistance of Ms. Marie Scandone, Bio-Rad Informatics Division, in providing the Bio-Rad infrared spectra used in Chapter 4 and of Mr. Chris Wozniak and Mr. Chris Lein, Aldrich Chemical Co., for providing the Aldrich NMR spectra in Chapter 3 deserve a very special thank you. Herk Alberry, Mike Farrell and Danny Dirico of Albany Advanced Imaging, Albany, NY are thanked for the MRI images and technical assistance they provided for Chapter 3. To Dr. Christian Bock, Alfred-Wegener-Institute for Polar and Marine Research, Bremerhaven, Germany, for the use of his MRI images in Chp.3, Vielen Dank. Ms. Julie Powers, Toshiba America Medical Systems, is gratefully acknowledged for the medical MRI system photos used in Chapter 3. Special thanks go to Dr. Frank Dorman, Restek, for providing their EZ-Chrom Chromatography Simulation Program, used to create several of the Figures in Chapter 12.

The authors wish to thank the following colleagues for their helpful comments and suggestions: Professor Ronald Bailey, Rensselaer Polytechnic Institute; Professor Peter Griffiths, University of Idaho; and Professor Julian Tyson, University of Massachussets, Amherst. Professor Emeritus Robert Gale, LSU, who coauthored Chapter 15 for the 5th edition is acknowledged for his substantive contributions to and review of that Chapter for this edition.

The image on the book cover is a false-color Fourier Transform Infrared (FTIR) image of a cranefly wing, showing the location of carboxylic acid functional groups and their relative concentrations. The image is courtesy of PerkinElmer Life and Analytical Sciences (www.perkinelmer.com).

Contents

Preface to the Sixth Edition . . . . *iii*
Acknowledgments . . . . *vii*
About the Authors . . . . *xix*

1. Concepts of Instrumental Analytical Chemistry *1*
1.1. Introduction: What is Analytical Chemistry? *1*
1.2. The Analytical Approach *3*
    1.2.1. Defining the Problem *4*
    1.2.2. Designing the Analytical Method *14*
    1.2.3. Sampling *15*
    1.2.4. Storage of Samples *20*
1.3. Basic Statistics and Data Handling *21*
    1.3.1. Significant Figures *21*
    1.3.2. Accuracy and Precision *24*
    1.3.3. Types of Errors *25*
    1.3.4. Definitions for Statistics *31*
    1.3.5. Quantifying Random Error *32*
    1.3.6. Rejection of Results *39*
1.4. Sample Preparation *40*
    1.4.1. Acid Dissolution and Digestion *40*
    1.4.2. Fusions *43*
    1.4.3. Dry Ashing and Combustion *44*
    1.4.4. Extraction *44*
1.5. Performing the Measurement *51*
    1.5.1. Signals and Noise *52*
    1.5.2. Plotting Calibration Curves *56*
1.6. Assessing the Data *57*
    1.6.1. Limit of Detection *58*
    1.6.2. Limit of Quantitation *59*
Bibliography *60*
Problems *60*

2. Introduction to Spectroscopy *65*
2.1. The Interaction Between Electromagnetic Radiation and Matter *65*
    2.1.1. What is Electromagnetic Radiation? *65*
    2.1.2. How does Electromagnetic Radiation Interact with Matter? *67*

2.2. Atoms and Atomic Spectroscopy *72*
2.3. Molecules and Molecular Spectroscopy *74*
    2.3.1.   Rotational Transitions in Molecules *74*
    2.3.2.   Vibrational Transitions in Molecules *75*
    2.3.3.   Electronic Transitions in Molecules *76*
2.4. Absorption Laws *76*
    2.4.1.   Deviations from Beer's Law *80*
2.5. Methods of Calibration *81*
    2.5.1.   Calibration with Standards *81*
    2.5.2.   Method of Standard Additions *84*
    2.5.3.   Internal Standard Calibration *87*
    2.5.4.   Errors Associated with Beer's Law Relationships *90*
2.6. Optical Systems Used in Spectroscopy *93*
    2.6.1.   Radiation Sources *95*
    2.6.2.   Wavelength Selection Devices *95*
    2.6.3.   Optical Slits *103*
    2.6.4.   Detectors *104*
    2.6.5.   Single-Beam and Double-Beam Optics *105*
    2.6.6.   Dispersive Optical Layouts *107*
    2.6.7.   Fourier Transform Spectrometers *108*
2.7. Spectroscopic Technique and Instrument Nomenclature *111*
Bibliography *111*
Suggested Experiments *112*
Problems *113*

3. Nuclear Magnetic Resonance Spectroscopy *117*
3.1. Introduction *117*
    3.1.1.   Properties of Nuclei *118*
    3.1.2.   Quantization of ^{1}H Nuclei in a Magnetic Field *119*
    3.1.3.   Width of Absorption Lines *125*
3.2. The FTNMR Experiment *128*
3.3. Chemical Shifts *130*
3.4. Spin–Spin Coupling *135*
3.5. Instrumentation *148*
    3.5.1.   Sample Holder *149*
    3.5.2.   Sample Probe *150*
    3.5.3.   Magnet *151*
    3.5.4.   RF Generation and Detection *152*
    3.5.5.   Signal Integrator and Computer *153*
    3.5.6.   Wide-Line Benchtop NMR Spectrometers and Portable
            NMR Spectrometers *154*
3.6. Analytical Applications of NMR *154*
    3.6.1.   Samples and Sample Preparation for NMR *154*
    3.6.2.   Qualitative Analyses: Molecular Structure Determination *155*
    3.6.3.   Interpretation of Proton Spectra *161*
    3.6.4. ^{13}C NMR *173*
    3.6.5.   2D NMR *180*
    3.6.6.   Qualitative Analyses: Other Applications *184*
    3.6.7.   Quantitative Analyses *190*
3.7. Hyphenated NMR Techniques *194*
3.8. NMR Imaging and MRI *195*

3.9. Limitations of NMR *200*
Bibliography *200*
Spectral Databases *201*
Suggested Experiments *202*
Problems *203*

4. Infrared Spectroscopy *213*
4.1. Absorption of IR Radiation by Molecules *214*
    4.1.1. Dipole Moments in Molecules *214*
    4.1.2. Types of Vibrations in Molecules *217*
    4.1.3. Vibrational Motion *219*
4.2. IR Instrumentation *225*
    4.2.1. Radiation Sources *225*
    4.2.2. Monochromators and Interferometers *230*
    4.2.3. Detectors *236*
    4.2.4. Detector Response Time *241*
4.3. Sampling Techniques *242*
    4.3.1. Techniques for Transmission (Absorption) Measurements *242*
    4.3.2. Background Correction in Transmission Measurements *248*
    4.3.3. Techniques for Reflectance and Emission Measurements *249*
4.4. FTIR Microscopy *254*
4.5. Nondispersive IR Systems *258*
4.6. Analytical Applications of IR Spectroscopy *259*
    4.6.1. Qualitative Analyses and Structural Determination by Mid-IR
          Absorption Spectroscopy *261*
    4.6.2. Quantitative Analyses by IR Spectrometry *281*
4.7. Near IR Spectroscopy *285*
    4.7.1. Instrumentation *286*
    4.7.2. NIR Vibrational Bands and Spectral Interpretation *287*
    4.7.3. Sampling Techniques for NIR Spectroscopy *287*
    4.7.4. Applications of NIR Spectroscopy *288*
4.8. Raman Spectroscopy *290*
    4.8.1. Principles of Raman Scattering *291*
    4.8.2. Raman Instrumentation *293*
    4.8.3. Applications of Raman Spectroscopy *298*
    4.8.4. The Resonance Raman Effect *301*
    4.8.5. Surface-Enhanced Raman Spectroscopy (SERS) *302*
    4.8.6. Raman Microscopy *302*
4.9. Chemical Imaging Using NIR, IR, and
      Raman Spectroscopy *306*
Bibliography *308*
Spectral Databases *309*
Suggested Experiments *309*
Problems *310*

5. Visible and Ultraviolet Molecular Spectroscopy *317*
5.1. Introduction *317*
    5.1.1. Electronic Excitation in Molecules *319*
    5.1.2. Absorption by Molecules *323*
    5.1.3. Molar Absorptivity *325*

      5.1.4.   The Shape of UV Absorption Curves *326*
      5.1.5.   Solvents for UV/VIS Spectroscopy *328*
5.2. Instrumentation *329*
      5.2.1.   Optical System *329*
      5.2.2.   Radiation Sources *330*
      5.2.3.   Monochromators *332*
      5.2.4.   Detectors *333*
      5.2.5.   Sample Holders *341*
5.3. UV Absorption Spectra of Molecules *345*
      5.3.1.   Solvent Effects on UV Spectra *345*
5.4. UV Spectra and the Structure of Organic Molecules *348*
      5.4.1.   Conjugated Diene Systems *348*
      5.4.2.   Conjugated Ketone Systems *352*
      5.4.3.   Substitution of Benzene Rings *355*
5.5. Analytical Applications *356*
      5.5.1.   Qualitative Structural Analysis .. *356*
      5.5.2.   Quantitative Analysis *357*
      5.5.3.   Multicomponent Determinations *361*
      5.5.4.   Other Applications *362*
5.6. Accuracy and Precision in UV/VIS Absorption Spectrometry *364*
5.7. Nephelometry and Turbidimetry *364*
5.8. Molecular Emission Spectrometry *366*
      5.8.1.   Fluorescence and Phosphorescence *366*
      5.8.2.   Relationship Between Fluorescence Intensity
            and Concentration *368*
5.9. Instrumentation for Luminescence Measurements *370*
      5.9.1.   Wavelength Selection Devices *371*
      5.9.2.   Radiation Sources *371*
      5.9.3.   Detectors *373*
      5.9.4.   Sample Cells *373*
5.10. Analytical Applications of Luminescence *374*
      5.10.1. Advantages of Fluorescence
            and Phosphorescence *376*
      5.10.2. Disadvantages of Fluorescence
            and Phosphorescence *376*
Bibliography *377*
Suggested Experiments *378*
Problems *379*

6. Atomic Absorption Spectrometry *385*
6.1. Absorption of Radiant Energy by Atoms *385*
      6.1.1.   Spectral Linewidth *388*
      6.1.2.   Degree of Radiant Energy Absorption *389*
6.2. Instrumentation *389*
      6.2.1.   Radiation Sources *390*
      6.2.2.   Atomizers *393*
      6.2.3.   Spectrometer Optics *399*
      6.2.4.   Detectors *401*
      6.2.5.   Modulation *401*
6.3. The Atomization Process *402*
      6.3.1.   Flame Atomization *402*
      6.3.2.   Graphite Furnace Atomization *408*

6.4. Interferences in AAS *409*
   6.4.1. Nonspectral Interferences *410*
   6.4.2. Spectral Interferences *417*
6.5. Analytical Applications of AAS *424*
   6.5.1. Qualitative Analysis *424*
   6.5.2. Quantitative Analysis *425*
   6.5.3. Analysis of Samples *428*
Bibliography *433*
Suggested Experiments *434*
Problems *436*
Appendix 6.1 .... *438*
Appendix 6.2 .... *445*

7. Atomic Emission Spectroscopy **449**
7.1. Flame Atomic Emission Spectroscopy *450*
   7.1.1. Instrumentation for Flame OES *451*
   7.1.2. Interferences *455*
   7.1.3. Analytical Applications of Flame OES *458*
7.2. Atomic Optical Emission Spectroscopy *462*
   7.2.1. Instrumentation for Emission Spectroscopy *463*
   7.2.2. Interferences in Arc and Spark Emission Spectroscopy *476*
   7.2.3. Applications of Arc and Spark Emission Spectroscopy *479*
7.3. Plasma Emission Spectroscopy *483*
   7.3.1. Instrumentation for Plasma Emission Spectrometry *483*
   7.3.2. Interferences and Calibration in Plasma Emission Spectrometry *497*
   7.3.3. Applications of ICP and DCP Atomic Emission Spectroscopy *503*
   7.3.4. Chemical Speciation with Hyphenated Instruments *505*
7.4. Glow Discharge Emission Spectrometry *506*
   7.4.1. DC and RF GD Sources *506*
   7.4.2. Applications of GD Atomic Emission Spectrometry *507*
7.5. Particle Characterization Using a Helium MIP System *509*
7.6. Atomic Fluorescence Spectrometry (AFS) *516*
   7.6.1. Instrumentation for AFS *517*
   7.6.2. Interferences in AFS *519*
   7.6.3. Applications of AFS *520*
7.7. Commercial Atomic Emission Systems *521*
   7.7.1. Arc and Spark Systems *521*
   7.7.2. ICP and DCP Systems *522*
   7.7.3. GD Systems *522*
   7.7.4. AFS Systems *522*
7.8. Atomic Emission Literature and Resources *522*
7.9. Comparison of Atomic Spectroscopic and ICP-MS Techniques *523*
Bibliography *523*
Suggested Experiments *524*
Problems *527*
Appendix 7.1 .... *529*
Appendix 7.2 .... *531*

8. X-Ray Spectroscopy **535**
8.1. Origin of X-Ray Spectra *535*
   8.1.1. Energy Levels in Atoms *535*

8.1.2.   Moseley's Law *542*
8.1.3.   X-Ray Methods *542*
8.2. Instrumentation *547*
8.2.1.   X-Ray Source *548*
8.2.2.   Collimators *552*
8.2.3.   Filters *554*
8.2.4.   WDXRF Spectrometers *555*
8.2.5.   Sample Holders *565*
8.2.6.   Simultaneous WDXRF Spectrometers *568*
8.2.7.   EDXRF Spectrometers *568*
8.3. Analytical Applications of X-Rays *572*
8.3.1.   X-Ray Absorption *573*
8.3.2.   X-Ray Diffraction *576*
8.3.3.   X-Ray Fluorescence (XRF) *585*
8.3.4.   Electron Probe Microanalysis *593*
Bibliography *594*
Suggested Experiments *595*
Problems *597*
Appendix 8.1 .... *602*
Appendix 8.2 .... *607*

9. Mass Spectrometry I: Principles and Instrumentation *613*
9.1. Principles of MS *613*
9.1.1.   Resolving Power and Resolution of a Mass Spectrometer *619*
9.2. Instrumentation *620*
9.2.1.   Sample Input Systems *621*
9.2.2.   Ionization Sources *622*
9.2.3.   Mass Analyzers *633*
9.2.4.   Detectors *644*
Bibliography *648*
Problems *648*

**10. Mass Spectrometry II: Spectral Interpretation
     and Applications** *651*
10.1. Interpretation of Mass Spectra: Structural Determination of
      Simple Molecules *652*
10.1.1. The Molecular Ion and Fragmentation Patterns *654*
10.1.2. The Nitrogen Rule *656*
10.1.3. Molecular Formulae and Isotopic Abundances *658*
10.1.4. Compounds with Heteroatoms *662*
10.1.5. Halogen Isotopic Clusters *663*
10.1.6. Rings Plus Double Bonds *666*
10.1.7. Common Mass Losses on Fragmentation *667*
10.2. Mass Spectral Interpretation: Some Examples *667*
10.2.1. Mass Spectra of Hydrocarbons *670*
10.2.2. Mass Spectra of Other Organic Compound Classes *676*
10.2.3. Compounds Containing Heteroatoms *684*
10.3. Applications of Molecular MS *687*
10.3.1. High-Resolution Mass Spectrometry *687*
10.3.2. Quantitative Analysis of Compounds and Mixtures *689*
10.3.3. Protein-Sequencing Analysis (Proteomics) *691*
10.3.4. Gas Analysis *692*

10.3.5. Environmental Applications *693*
10.3.6. Other Applications of Molecular MS *693*
10.3.7. Limitations of Molecular MS *694*
10.4. Atomic MS *694*
10.4.1. Inductively Coupled Plasma Mass
Spectrometry (ICP-MS) *695*
10.4.2. Applications of Atomic MS *697*
10.4.3. Interferences in Atomic MS *704*
10.4.4. Instrumental Approaches to Eliminating Interferences *708*
10.4.5. Limitations of Atomic MS *709*
Bibliography *710*
Problems *711*
Appendix 10.1 . . . . *719*

11. Principles of Chromatography *721*
11.1. Introduction to Chromatography *721*
11.2. What is the Chromatographic Process? *722*
11.3. Chromatography in More than One Dimension *724*
11.4. Visualization of the Chromatographic Process at the Molecular
Level: Analogy to "People on a Moving Belt Slideway" *725*
11.5. A Digression on the Central Role of Silicon–Oxygen Compounds
in Chromatography *728*
11.6. The Basic Equations Describing Chromatographic Separations *731*
11.7. How do Column Variables Affect Efficiency (Plate Height)? *734*
11.8. Practical Optimization of Chromatographic Separations *736*
11.9. Extra-Column Band Broadening Effects *737*
11.10. Qualitative Chromatography—Analyte Identification *739*
11.11. Quantitative Measurements in Chromatography *740*
11.11.1. Peak Areas or Peak Heights—Which are Best? *740*
11.11.2. Calibration with an External Standard *741*
11.11.3. Calibration with an Internal Standard *742*
11.12. Examples of Chromatographic Calculations *743*
Bibliography *746*
Problems *746*

12. Gas Chromatography *749*
12.1. Historical Development of GC—the First
Chromatographic Instrumentation *749*
12.2. Advances in GC Leading to Present-Day Instrumentation *750*
12.3. GC Instrument Component Design (Injectors) *753*
12.3.1. Syringes *753*
12.3.2. Autosamplers *753*
12.3.3. Solid Phase Microextraction (SPME) *754*
12.3.4. Split Injections *755*
12.3.5. Splitless Injections *755*
12.4. GC Instrument Component Design (The Column) *757*
12.4.1. Column Stationary Phase *757*
12.4.2. Selecting a Stationary Phase for an Application *760*
12.4.3. Effects of Mobile Phase Choice and Flow Parameters *760*
12.5. GC Instrument Operation (Column Dimensions and Elution Values) *762*
12.6. GC Instrument Operation (Column Temperature and Elution Values) *765*

12.7. GC Instrument Component Design (Detectors) *769*
   12.7.1.   The Thermal Conductivity Detector (TCD) *771*
   12.7.2.   Flame Ionization Detector (FID) *773*
   12.7.3.   The Electron Capture Detector (ECD) *774*
   12.7.4.   The Electrolytic Conductivity Detector (ELCD) *776*
   12.7.5.   The Sulfur–Phosphorous Flame Photometric Detector (SP-FPD) *777*
   12.7.6.   The Sulfur Chemiluminescence Detector (SCD) *777*
   12.7.7.   The Nitrogen–Phosphorous Detector (NPD) *777*
   12.7.8.   The Photoionization Detector (PID) *778*
   12.7.9.   The Helium Ionization Detector (HID) *779*
   12.7.10. The Atomic Emission Detector (AED) *780*
12.8. Hyphenated GC Techniques (GC-MS; GC-IR; GC-GC; or 2D-GC) *780*
   12.8.1.   Gas Chromatography-Mass Spectrometry (GC-MS) *781*
   12.8.2.   Gas Chromatography-IR Spectrometry (GC-IR) *784*
   12.8.3.   Comprehensive 2D-Gas Chromatography (GC-GC or GC^2) *785*
12.9. Retention Indices (A Generalization of Relative R_t Information) *786*
12.10. The Scope of GC Analyses ..   *788*
   12.10.1. GC Behavior of Organic Compound Classes *789*
   12.10.2. Derivitization of Difficult Analytes to Improve GC
             Elution Behavior *789*
   12.10.3. Gas Analysis by GC *790*
   12.10.4. Limitations of Gas Chromatography *791*
Bibliography *792*
Problems *792*
Appendix 12.1 .... *795*

13. Chromatography with Liquid Mobile Phases *797*
13.1. High-Performance Liquid Chromatography (HPLC) *797*
   13.1.1.   The HPLC Column and Stationary Phases *798*
   13.1.2.   Effects on Separation of Composition of the Mobile Phase *804*
   13.1.3.   Design and Operation of an HPLC Instrument *805*
   13.1.4.   HPLC Detector Design and Operation *809*
   13.1.5.   Derivatization in HPLC *821*
   13.1.6.   Hyphenated Techniques in HPLC *824*
   13.1.7.   Applications of HPLC *829*
13.2. Chromatography of Ions Dissolved in Liquids *835*
   13.2.1.   Ion Chromatography *839*
13.3. Affinity Chromatography *843*
13.4. Size Exclusion Chromatography (SEC) *845*
13.5. Supercritical Fluid Chromatography (SFC) *848*
   13.5.1.   Operating Conditions *848*
   13.5.2.   Effect of Pressure *848*
   13.5.3.   Stationary Phases *849*
   13.5.4.   Mobile Phases *849*
   13.5.5.   Detectors *849*
   13.5.6.   SFC vs. Other Column Methods *849*
   13.5.7.   Applications *850*
13.6. Electrophoresis *850*
   13.6.1.   Capillary Zone Electrophoresis (CZE) *851*
   13.6.2.   Sample Injection in CZE *857*
   13.6.3.   Detection in CZE *858*
   13.6.4.   Modes of CE *859*
   13.6.5.   Capillary Electrochromatography (CEC) *863*

13.7. Planar Chromatography and Planar Electrophoresis 866
    13.7.1. Thin Layer Chromatography (TLC) 866
    13.7.2. Planar Capillary Electrophoresis on Slab Gels 869
Bibliography 871
Problems and Exercises 871
Appendix 13.1 875

14. Surface Analysis 877
14.1. Introduction 877
14.2. Electron Spectroscopy Techniques 879
    14.2.1. Electron Spectroscopy for Chemical Analysis (ESCA) or X-ray
        Photoelectron Spectroscopy (XPS) 880
    14.2.2. Auger Electron Spectroscopy (AES) 897
14.3. Ion Scattering Spectroscopy 906
14.4. Secondary Ion Mass Spectrometry (SIMS) 908
    14.4.1. Instrumentation for SIMS 909
    14.4.2. Analytical Applications of SIMS 911
14.5. Electron Microprobe (Electron Probe Microanalysis) 914
Bibliography 915
Problems 916

15. Electroanalytical Chemistry 919
15.1. Fundamentals of Electrochemistry 920
15.2. Electrochemical Cells 921
    15.2.1. Line Notation for Cells and Half-Cells 924
    15.2.2. Standard Reduction Potentials 925
    15.2.3. Sign Conventions 928
    15.2.4. The Nernst Equation 928
    15.2.5. Activity Series 930
    15.2.6. Reference Electrodes 931
    15.2.7. The Electrical Double Layer 933
15.3. Electroanalytical Methods 934
    15.3.1. Potentiometry 935
    15.3.2. Coulometry 961
    15.3.3. Conductometric Analysis 969
    15.3.4. Polarography 976
    15.3.5. Voltammetry 989
15.4. LC Detectors 994
    15.4.1. Voltammetric Detection 994
    15.4.2. Conductometric Detection 995
Bibliography 997
Suggested Experiments 998
Problems 999
Appendix 15.1 1001

16. Thermal Analysis 1003
16.1. Thermogravimetry 1004
    16.1.1. TGA Instrumentation 1007
    16.1.2. Analytical Applications of Thermogravimetry 1010
    16.1.3. Derivative Thermogravimetry 1017
    16.1.4. Sources of Error in Thermogravimetry 1019

16.2. Differential Thermal Analysis *1020*
    16.2.1. DTA Instrumentation *1021*
    16.2.2. Analytical Applications of DTA *1023*
16.3. Differential Scanning Calorimetry *1026*
    16.3.1. DSC Instrumentation *1026*
    16.3.2. Applications of DSC *1028*
16.4. Hyphenated Techniques *1031*
    16.4.1. Hyphenated Thermal Methods *1031*
    16.4.2. Evolved Gas Analysis *1031*
16.5. Thermometric Titrimetry *1036*
    16.5.1. Applications of Thermometric Titrimetry *1037*
16.6. Direct Injection Enthalpimetry *1038*
16.7. Thermomechanical Analysis and Dynamic Mechanical Analysis *1039*
    16.7.1. Instrumentation *1040*
    16.7.2. Applications of TMA and DMA *1043*
16.8. Summary *1048*
Bibliography *1049*
Suggested Experiments *1049*
Problems *1050*

Acronyms Index *1055*
Index *1061*

About the Authors

JAMES W. ROBINSON is Professor Emeritus of Chemistry, Louisiana State University, Baton Rouge, Louisiana. A Fellow of the Royal Chemical Society, he is the author of over 200 professional papers and book chapters and several books including *Atomic Absorption Spectroscopy* and *Atomic Spectroscopy*. He was Executive Editor of *Spectroscopy Letters* and the *Journal of Environmental Science and Health* (both titles, Marcel Dekker, Inc.) and the *Handbook of Spectroscopy* and the *Practical Handbook of Spectroscopy* (both titles, CRC Press). He received the B.Sc. (1949), Ph.D. (1952), and D.Sc. (1978) degrees from the University of Birmingham, England.

EILEEN M. SKELLY FRAME is Clinical Assistant Professor and Visiting Research Professor, Rensselaer Polytechnic Institute, Troy, New York. Dr. Skelly Frame has extensive practical experience in the use of instrumental analysis to characterize a wide variety of substances, from biological samples and cosmetics to high temperature superconductors, polymers, metals, and alloys. Her industrial career includes supervisory roles at GE Corporate Research and Development, Stauffer Chemical Corporate R&D, and the Research Triangle Institute. She is a member of the American Chemical Society, the Society for Applied Spectroscopy, and the American Society for Testing and Materials. Dr. Skelly Frame received the B.S. degree in chemistry from Drexel University, Philadelphia, Pennsylvania, and the Ph.D. in analytical chemistry from Louisiana State University, Baton Rouge.

GEORGE M. FRAME II is Scientific Director, Chemical Biomonitoring Section of the Wadsworth Laboratory, New York State Department of Health, Albany. He has a wide range of experience in the field and has worked at the GE Corporate R&D Center, Pfizer Central Research, the U.S. Coast Guard R&D Center, the Maine Medical Center, and the USAF Biomedical Sciences Corps. He is an American Chemical Society member. Dr. Frame received the B.A. degree in chemistry from Harvard College, Cambridge, Massachusetts, and the Ph.D. degree in analytical chemistry from Rutgers University, New Brunswick, New Jersey.

1

Concepts of Instrumental Analytical Chemistry

1.1. INTRODUCTION: WHAT IS ANALYTICAL CHEMISTRY?

Perhaps the most functional definition of analytical chemistry is that it is "the qualitative and quantitative characterization of matter". The word "characterization" is used in a very broad sense. It may mean the identification of the chemical compounds or elements present in a sample to answer questions such as "Is there any vitamin E in this shampoo as indicated on the label?" or "Is this white tablet an aspirin tablet?" or "Is this piece of metal iron or nickel?" This type of characterization, to tell us *what* is present is called qualitative analysis. **Qualitative analysis** is the identification of one or more chemical species present in a material. Characterization may also mean the determination of how much of a particular compound or element is present in a sample, to answer questions such as "How much acetylsalicylic acid is in this aspirin tablet?" or "How much nickel is in this steel?" This determination of *how much* of a species is present in a sample is called quantitative analysis. **Quantitative analysis** is the determination of the exact amount of a chemical species present in a sample. The chemical species may be an element, compound, or ion. The compound may be organic or inorganic. Characterization can refer to the entire sample (*bulk analysis*), such as the elemental composition of a piece of steel, or to the surface of a sample (*surface analysis*), such as the identification of the composition and thickness of the oxide layer that forms on the surface of most metals exposed to air and water. The characterization of a material may go beyond chemical analysis to include structural determination of materials, the measurement of physical properties of a material, and the measurement of physical chemistry parameters like reaction kinetics. Examples of such measurements are the degree to which a polymer is crystalline as opposed to amorphous, the temperature at which a material loses its water of hydration, how long it takes for antacid "Brand A" to neutralize stomach acid, and how fast a pesticide degrades in sunlight. These diverse applications make analytical chemistry one of the broadest in scope of all scientific disciplines. Analytical chemistry is critical to our understanding of biochemistry, medicinal chemistry, geochemistry, environmental science, atmospheric chemistry, the behavior of materials such as polymers, metal alloys, and ceramics, and many other scientific disciplines.

For many years, analytical chemistry relied on chemical reactions to identify and determine the components present in a sample. These types of classical methods, often called "wet chemical methods", usually required that a part of the sample be taken, dissolved in a suitable solvent if necessary and the desired reaction carried out. The most

1

important analytical fields based on this approach were volumetric and gravimetric analyses. Acid–base titrations, oxidation–reduction titrations, and gravimetric determinations, such as the determination of silver by precipitation as silver chloride are all examples of wet chemical analyses. These types of analyses require a high degree of skill and attention to detail on the part of the analyst if accurate and precise results are to be obtained. They are also time consuming and the demands of today's high-throughput pharmaceutical development labs and industrial quality control labs often do not permit the use of such time-consuming methods for routine analysis. In addition, it may be necessary to analyze samples without destroying them. Examples include evaluation of valuable artwork to determine if a painting is really by a famous "Old Master" or is a modern forgery, as well as in forensic analysis, where the evidence may need to be preserved. For these types of analyses, **nondestructive analysis** methods are needed, and wet chemical analysis will not do the job. Wet chemical analysis is still used in specialized areas of analysis, but many of the volumetric methods have been transferred to automated instruments. Classical analysis and instrumental analysis are similar in many respects, such as in the need for proper sampling, sample preparation, assessment of accuracy and precision, and proper record keeping. Some of the topics discussed briefly in this chapter are covered at greater length in more general texts on analytical chemistry and quantitative analysis. Several of these types of texts are listed in the bibliography.

Most analyses today are carried out with specially designed electronic instruments controlled by computers. These instruments make use of the interaction of electromagnetic radiation and matter, or of some physical property of matter, to characterize the sample being analyzed. Often these instruments have automated sample introduction, automated data processing, and even automated sample preparation. To understand how the instrumentation operates and what information it can provide requires knowledge of chemistry, physics, mathematics, and engineering. The fundamentals of common analytical instruments and how measurements are performed with these instruments are the subjects of the following chapters on specific instrumental techniques. The analytical chemist must not only know and understand analytical chemistry and instrumentation, but must also be able to serve as a problem solver to colleagues in other scientific areas. This means that the analytical chemist may need to understand materials science, metallurgy, biology, pharmacology, agricultural science, food science, geology, and other fields. The field of analytical chemistry is advancing rapidly. To keep up with the advances, the analytical chemist must understand the fundamentals of common analytical techniques, their capabilities, and their shortcomings. The analytical chemist must understand the problem to be solved, select the appropriate technique or techniques to use, design the analytical experiment to provide relevant data, and ensure that the data obtained are valid. Merely providing data to other scientists is not enough; the analytical chemist must be able to interpret the data, and communicate the meaning of the results, together with the accuracy and precision (the reliability) of the data, to scientists who will use the data. In addition to understanding the scientific problem, the modern analytical chemist often must also consider factors such as time limitations and cost limitations in providing an analysis. Whether one is working for a government regulatory agency, a hospital, a private company, or a university, analytical data must be legally defensible. It must be of known, documented quality. Record keeping, especially computer record keeping, assessing accuracy and precision, statistical handling of data, documenting, and ensuring that the data meet the applicable technical standards are especially critical aspects of the job of modern analytical chemists.

Analytical chemistry uses many specialized terms that may be new to you. The definitions of the terms, usually shown in boldface, must be learned. The units used in this text are, for the most part, the units of the Système International d'Unités (SI system). The SI system is used around the world by scientists and engineers. The tables inside the textbook covers give the primary units of measurement in the SI system. A comprehensive list of SI units, SI derived units and definitions, as well as non-SI units may be found at the US National Institute for Standards and Technology website at http://physics.nist.gov.

Many analytical results are expressed as the concentration of the measured substance in a certain amount of sample. The measured substance is called the **analyte**. Commonly used concentration units include molarity (moles of substance per liter of solution), weight percent (grams of substance per gram of sample $\times$ 100%), and units for trace levels of substances. One *part per million* (ppm) by weight is one microgram of analyte in a gram of sample, that is, 1×10^{-6} g analyte/g sample. One *part per billion* (ppb) by weight is one nanogram of element in a gram of sample or 1×10^{-9} g analyte/g sample. For many elements, the technique known as inductively coupled plasma mass spectrometry (ICP-MS), can detect *parts per trillion* of the element, that is, picograms of element per gram of sample (1×10^{-12} g analyte/g sample). To give you a feeling for these quantities, a million seconds is $\sim$12 days (11.57 days, to be exact). One part per million in units of seconds would be one second in 12 days. A part per billion in units of seconds would be 1 s in $\sim$32 years, and one part per trillion is one second in 32,000 years. Today, lawmakers set environmental levels of allowed chemicals in air and water based on measurements of compounds and elements at part per trillion levels because instrumental methods can detect part per trillion levels of analytes. It is the analytical chemist who is responsible for generating the data that these lawmakers rely on. A table of commonly encountered constants, multiplication factors, and their prefixes is found inside the textbook cover. The student should become familiar with these prefixes, since they will be used throughout the text.

1.2. THE ANALYTICAL APPROACH

A major personal care products manufacturer receives a phone call from an outraged customer whose hair has turned green after using their "new, improved shampoo". The US Coast Guard arrives at the scene of an oil spill in a harbor and finds two ship captains blaming each other for the spill. A plastics company that sells bottles to a water company bottling "pure crystal clear spring water" discovers that the 100,000 new empty bottles it is ready to ship are slightly yellow in color instead of crystal clear. A new, contagious disease breaks out and people are dying of flu-like symptoms. What caused the problem? How can it be prevented in the future? Who is at fault? Can a vaccine or drug treatment be developed quickly? These sorts of problems and many more occur daily around the world, in industry, in medicine, and in the environment. A key figure in the solution of these types of problems is the analytical chemist. The analytical chemist is first and foremost a problem solver and to do that, must understand the analytical approach, the fundamentals of common analytical techniques, their uses, and their limitations.

The approach used by analytical chemists to solve problems may include the following steps:

1. Defining the problem and designing the analytical method
2. Sampling and sample storage
3. Sample preparation
4. Performing the measurement

5. Assessing the data
6. Method validation
7. Documentation

General sample preparation will be discussed in this chapter, but instrument-specific sample preparation is included in the appropriate chapter on each technique. Method validation and documentation will not be covered as the focus of this text is on instrumentation. The text by Christian cited in the bibliography has an excellent introduction to validation and documentation for the interested student.

Although the steps in solving analytical problems usually follow the order listed above, knowledge of basic statistics is useful not just for handling the data and method validation but is required for proper sampling and selection of an analytical method. The statistics and definitions needed to understand what is meant by **accuracy**, **precision**, **error**, and so on are covered in Section 1.3. Students not familiar with these terms and concepts may want to read Section 1.3 at this point. Steps (1) and (2) are covered in this section, while steps (3) through (5) are discussed in the sections following Section 1.3.

1.2.1. Defining the Problem

The analytical chemist must find out what information needs to be known about the sample, material, or process being studied, how accurate and precise the analytical information must be, how much material or sample is available for study, and if the sample must be analyzed without destroying it. Is the sample organic or inorganic? Is it a pure material or a mixture? Does the customer want a bulk analysis or information about a particular fraction of the sample, such as the surface? Does the customer need to know if the sample is homogeneous or heterogeneous with respect to a given analyte? Does the customer need elemental information or information about the chemical species (ionic or molecular, particular oxidation states) present in the sample? The answers to such questions will guide the analyst in choosing the analytical method. Of course, sometimes the answers to some of the questions may be part of the problem. If the sample is an unknown material, the analyst must find out if it is organic or inorganic, pure or a mixture, as part of solving the problem. The analyte is the substance to be measured; everything else in the sample is called the **matrix**. Of course, there may be more than one analyte in a given sample. The terms *analysis* and *analyze* are applied to the sample under study, as in "this water was analyzed for nitrate ion" or "an analysis of the contaminated soil was performed". Water and soil are the samples being analyzed. The terms *determine* and *determination* are applied to the measurement of the analyte in the sample, as in "nitrate ion was determined in the water sample", "a determination of lead in blood was made because the symptoms indicated lead poisoning", or "an analysis of the soil was performed and cyanide levels were determined". Nitrate ion, lead, and cyanide are the analytes being determined. Other components in the sample matrix may interfere with the measurement of the analyte; such components are called **interferences**.

A sample may be **homogeneous**, that is, it has the same chemical composition everywhere within the sample. Plain vanilla pudding, a pure milk chocolate bar, and salt water are examples of homogeneous materials. Many samples are **heterogeneous**; the composition *varies* from region to region within the sample. Vanilla pudding with raisins in it and a chocolate bar with whole almonds in it are heterogeneous; you can see the composition difference. In most real samples, the heterogeneity may not be visible to the human eye. The variation in composition can be *random* or it can be *segregated* into regions of distinctly different compositions.

A significant part of defining the problem is the decision between performing a qualitative analysis and a quantitative analysis. Often the problem is first tackled with a qualitative analysis, followed by a quantitative analysis for specific analytes. The analyst needs to communicate with the customer who is requesting the analysis. Two-way communication is important, to be certain that the problem to be solved is understood and to be sure that the customer understands the capabilities and limitations of the analysis.

1.2.1.1. *Qualitative Analysis*

Qualitative analysis is the branch of analytical chemistry that is concerned with questions such as "What makes this water smell bad?", "Is there gold in this rock sample?", "Is this sparkling stone a diamond or cubic zirconia?", "Is this plastic item made of polyvinyl chloride, polyethylene or polycarbonate?", or "What is this white powder?"

Some methods for qualitative analysis are nondestructive, that is, they provide information about what is in the sample without destroying the sample. These are often the best techniques to begin with, because the sample can be used for subsequent analyses if necessary. To identify what elements are present in a sample nondestructively, a *qualitative elemental analysis* method such as X-ray fluorescence spectroscopy (XRF) can be used. Modern XRF instruments, discussed in Chapter 8, can identify all elements from sodium to uranium, and some instruments can measure elements from beryllium to uranium. The sample is usually not harmed by XRF analysis. For example, XRF could easily distinguish a diamond from cubic zirconia. Diamond is, of course, a crystalline form of carbon; most XRF instruments would see no elemental signal from the carbon in a diamond but would see a strong signal from the element zirconium in cubic zirconia, a crystalline compound of zirconium and oxygen. *Qualitative molecular analysis* will tell us what molecules are present in a material. The nondestructive identification of molecular compounds present in a sample can often be accomplished by the use of nuclear magnetic resonance (NMR) spectroscopy, discussed in Chapter 3, or by infrared (IR) spectroscopy, discussed in Chapter 4. IR spectroscopy can provide information about organic functional groups present in samples, such as alcohols, ketones, carboxylic acids, amines, thioethers, and many others. If the sample is a pure compound such as acetylsalicylic acid (the active ingredient in aspirin), the IR spectrum may be able to identify the compound exactly, because the IR spectrum for a compound is unique, like a fingerprint. Qualitative identification of polymers for recycling can be done using IR spectroscopy, for example. NMR gives us detailed information about the types of protons, carbon, and other atoms in organic compounds and how the atoms are connected. NMR can provide the chemical structure of a compound without destroying it.

Many methods used for qualitative analysis are destructive, that is, the sample is consumed during the analysis or must be chemically altered in order to be analyzed. The most sensitive and comprehensive elemental analysis methods for inorganic analysis are inductively coupled plasma atomic emission spectrometry (ICP-OES or ICP-AES), discussed in Chapter 7, and ICP-MS, discussed in Chapters 9 and 10. These techniques can identify almost all the elements in the periodic table, even when only trace amounts are present, but often require that the sample be in the form of a solution. If the sample is a rock or a piece of glass or a piece of biological tissue, the sample usually must be dissolved in some way to provide a solution for analysis. We will see how this is done later in the chapter. The analyst can determine accurately what elements are present, but information about the molecules in the sample is often lost in the sample preparation process. The advantage of ICP-OES and ICP-MS is that they are very sensitive; concentrations at or below 1 ppb of most elements can be detected using these methods.

If the sample is organic, that is, composed primarily of carbon and hydrogen, qualitative analysis can provide chemical and structural information to permit identification of the compound. The IR spectrum will provide identification of the class of compound, for example, ketone, acid, ether, and so on. NMR spectroscopy and mass spectrometry (MS), as we shall see in the appropriate chapters, provide detailed structural information, often including the molecular weight of the compound. Use of IR, NMR, and MS, combined with quantitative elemental analysis to accurately determine the percentage of carbon, hydrogen, oxygen, and other elements, is the usual process by which analytical chemists identify organic compounds. This approach is required to identify new compounds synthesized by pharmaceutical chemists, for example. In a simple example, elemental analysis of an unknown organic compound might provide an **empirical formula** of C_2H_5. An empirical formula is the simplest whole number ratio of the atoms of each element present in a molecule. For any given compound, the empirical formula may or may not coincide with the **molecular formula**. A molecular formula contains the total number of atoms of each element in a single molecule of the compound. The results from IR, NMR, and MS might lead the analytical chemist to the molecular formula, C_4H_{10}, and would indicate which of the two different structures shown below was our sample.

$$CH_3\text{-}CH_2\text{-}CH_2\text{-}CH_3 \qquad \qquad \begin{array}{c} H \\ | \\ H_3C-C-CH_3 \\ | \\ CH_3 \end{array}$$

n-butane   2-methyl-propane (or iso-butane)

These two structures are two different compounds with the same molecular formula. They are called **isomers**. Elemental analysis cannot distinguish between these isomers, but NMR and MS usually can distinguish isomers. Another example of a more difficult qualitative analysis problem is the case of the simple sugar, erythrose. The empirical formula determined by elemental analysis is CH_2O. The molecular formula, $C_4H_8O_4$, and some of the structure can be obtained from IR, NMR, and MS, but we cannot tell from these techniques which of the two possible isomers shown in Fig. 1.1 is our sample.

Mirror Plane
↓

$$\begin{array}{c} CHO \\ | \\ H-C-OH \\ | \\ H-C-OH \\ | \\ CH_2OH \end{array} \qquad \qquad \begin{array}{c} CHO \\ | \\ HO-C-H \\ | \\ HO-C-H \\ | \\ CH_2OH \end{array}$$

D-(−)-Erythrose   L-(+)-Erythrose

Figure 1.1   Isomers of erythrose.

These two erythrose molecules are **chiral**, that is, they are nonsuperimposable mirror-image isomers, called **enantiomers**. (Imagine sliding the molecule on the left *in the plane of the paper*, through the "mirror plane" indicated by the arrow, over the molecule on the right. The OH groups will not be on top of each other. Imagine turning the left molecule *in the plane of the paper* upside down and then sliding it to the right; now the OH groups are lined up, but the CHO and CH_2OH groups are not. That is what is meant by nonsuperimposable. You can do whatever you like to the two molecules except remove them from the plane of paper; no matter how you move them, they will not be superimposable.) They have the same molecular formula, $C_4H_8O_4$, the same IR spectrum, the same mass spectrum, and the same NMR spectrum, and many of the same physical properties such as boiling point and refractive index. Such chiral compounds can be distinguished from each other by interaction with something else that possesses chirality or by interaction with plane-polarized light. Chiral compounds will interact differently with other chiral molecules, and this interaction forms the basis of chiral chromatography. Chiral chromatography (Chapter 13) can be used to separate the two erythrose compounds shown. Chiral compounds also differ in their behavior toward plane-polarized light, and the technique of polarimetry can be used to distinguish them. One of the erythrose enantiomers rotates plane-polarized light to the right (clockwise); this compound is *dextrorotatory*, and is given the symbol (+) as part of its name. The other enantiomer rotates the plane of polarization to the left (counterclockwise); this compound is *levorotatory*, and is given the symbol (−) in its name. Such compounds are said to be optically active. Chiral compounds are very important because biochemical reactions are selective for only one of the two structures and only one of the two enantiomers is biologically active. Biochemists, pharmaceutical chemists, and medicinal chemists are very interested in the identification, synthesis, and separation of only the biologically active compound. The letters D and L in the name of the sugar refer to the position of the alcohol group on the carbon closest to the bottom primary alcohol. There is no relationship between the D and L configuration and the direction of rotation of plane polarized light. Figure 1.2 shows the simplest sugar, glyceraldehyde. It also has two enantiomers, one D and one L, but the D enantiomer of glyceraldehyde rotates light in the opposite direction from D-erythrose.

If organic compounds occur in mixtures, separation of the mixture often must be done before the individual components can be identified. Techniques such as gas chromatography, liquid chromatography, and capillary electrophoresis are often used to separate mixtures of organic compounds prior to identification of the components. These methods are discussed in Chapters 11–13.

Table 1.1 list some common commercially available instrumental methods of analysis and summarizes their usefulness for qualitative elemental or molecular analysis. Table 1.2

Figure 1.2   Enantiomers of glyceraldehyde.

Table 1.1 Instrumental Methods of Analysis

Method	Qualitative		Quantitative	
	Elemental	Molecular	Elemental	Molecular
Atomic absorption spectrometry	No	No	Yes	No
Atomic emission spectrometry	Yes	No	Yes	No
Capillary electrophoresis	Yes	Yes	Yes	Yes
Electrochemistry	Yes	Yes	Yes	Yes
Gas chromatography	No	Yes	No	Yes
ICP-mass spectrometry	Yes	No	Yes	No
Infrared spectroscopy	No	Yes	No	Yes
Ion chromatography	Yes	Yes	Yes	Yes
Liquid chromatography	No	Yes	No	Yes
Mass spectrometry	Yes	Yes	Yes	Yes
Nuclear magnetic resonance	No	Yes	No	Yes
Raman spectroscopy	No	Yes	No	Yes
Thermal analysis	No	Yes	No	Yes
UV/VIS spectrophotometry	Yes	Yes	Yes	Yes
UV absorption	No	Yes	No	Yes
UV fluorescence	No	Yes	No	Yes
X-ray absorption	Yes	No	Yes	No
X-ray diffraction	No	Yes	No	Yes
X-ray fluorescence	Yes	No	Yes	No

gives a very brief summary of the use of the methods. Analyte concentrations that can be determined by common methods of instrumental analysis are presented in Table 1.3. The concentration of analyte that can be determined in real samples will depend on the sample and on the instrument, but Table 1.3 gives some indication of the sensitivity and working range of methods.

1.2.1.2. Quantitative Analysis

When qualitative analysis is completed, the next question is often "How much of each or any component is present?" or "Exactly how much gold is this rock?" or "How much of the organochlorine pesticide dieldrin is in this drinking water?" The determination of how much is quantitative analysis. Analytical chemists express how much in a variety of ways, but often in terms of *concentration*, the amount of analyte in a given amount of sample. Concentration is an expression of the quantity of analyte in a given volume or mass of sample. Common concentration units include molarity, defined as moles of analyte per liter of sample and symbolized by M or mol/L; percent by weight, defined as grams of analyte per gram of sample $\times 100\%$, symbolized as % or %w/w; parts per million, defined as micrograms of analyte per gram of sample (ppm, $\mu g/g$); and others. For dilute aqueous solutions, one milliliter of solution has a mass of one gram (because the density of water is 1 g/mL), so solution concentrations are often expressed in terms of volume. A part per million of analyte in dilute aqueous solution is equal to one microgram per milliliter of solution ($\mu g/mL$), for example.

The first quantitative analytical fields to be developed were for quantitative elemental analysis, which revealed how much of each element was present in a sample. These early techniques were not instrumental methods, for the most part, but relied on chemical reactions, physical separations, and weighing of products (gravimetry), titrations

Table 1.2 Principal Applications of Instrumental Methods of Analysis

Molecular Analysis

Nuclear magnetic resonance spectroscopy (NMR)

Qualitative analysis: NMR is one of the most powerful methods available for determining the structure of molecules. It identifies the number and type of protons and carbon atoms in organic molecules, e.g., distinguishes among aromatic, aliphatic, alcohols, aldehydes, etc. Most importantly, it also reveals the positions of the nuclei in the molecule relative to each other. For example, NMR will distinguish between CH_3—CH_2—CH_2OH and CH_3—$CHOH$—CH_3. It does not provide the molecular weight of the compound. NMR is also applied to compounds containing heteroatoms such as sulfur, nitrogen, fluorine, phosphorus, and silicon.

Quantitative analysis: NMR is useful at % concentration levels, but trace levels (ppm) are becoming attainable with reasonable accuracy.

Infrared spectroscopy (IR) and Raman spectroscopy

Qualitative analysis: IR readily identifies organic functional groups present in molecules including groups containing heteroatoms—O, S, N, Si, halides. The IR spectrum is a fingerprint for a given compound, making it a very useful qualitative method. IR spectroscopy cannot be done on aqueous solutions. It does not give the molecular weight of the compound or structural information. Raman spectroscopy complements IR spectroscopy and is useful for aqueous samples.

Quantitative analysis: IR is used routinely for the quantitative analysis of organic compounds, particularly at % concentration levels. It is used mostly for liquid samples. The related field of Raman spectroscopy complements IR. Application of IR spectroscopy to gas samples is limited by lack of sensitivity.

Ultraviolet (UV) absorption spectroscopy

Qualitative analysis: UV absorption can be used for identifying functional groups and the structures of molecules containing unsaturated bonds (π electrons), such as

and aromatics and lone-pair electrons, such as those in pyridine:

lone pair

It does not indicate molecular weight or give useful information on saturated bonds (σ bonds). NMR and IR have almost entirely replaced UV absorption spectroscopy for organic compound identification.

Quantitative analysis: UV absorption is used routinely for the quantitative determination of unsaturated compounds such as those found in natural products. The method is subject to spectral overlap and therefore interference from other compounds in the sample.

UV fluorescence

Qualitative analysis: UV fluorescence is used for the determination of unsaturated compounds, particularly aromatics. It does not indicate molecular weight, but gives some indication of the functional groups present. It is much more sensitive than UV absorption.

(continued)

Table 1.2 Continued

Quantitative analysis: UV fluorescence is a very sensitive method of analysis (10^{-8} g/g or 10 ppb), but it is subject to many kinds of interference, both from quenching effects and from spectral overlap from other compounds.

UV and visible (UV/VIS) spectrophotometry
Qualitative analysis: Organic or inorganic reagents are used for specific tests for many elements or compounds by forming a compound that absorbs at specific wavelengths. The products may or may not be colored. If the compounds are colored, analysis may be carried out visually (colorimetric analysis by eye) but use of a spectrometer is more accurate.
Quantitative analysis: Sensitive and selective methods have been developed for most elements and many functional groups. It is used extensively in routine analysis of water, food, beverages, industrial products, etc.

X-ray diffraction (XRD)
Qualitative analysis: XRD is used for the measurement of crystal lattice dimensions and to identify the structure and composition of all types of crystalline inorganic and organic materials.
Quantitative analysis: XRD is used for the determination of percent crystallinity in polymers, the composition of mixtures, mixed crystals, soils, and natural products.

X-ray absorption spectroscopy
Qualitative analysis: X-ray absorption reveals the contours and location of high atomic weight elements in the presence of low atomic weight matrixes or holes in the interior of solid samples (voids). Examples are bone locations in the human body, the contents of closed suitcases, old paintings hidden under new painting on a canvas, and voids in welded joints and opaque solid objects.

Organic mass spectrometry (MS)
Qualitative analysis: MS can be used to identify the molecular weight of organic and inorganic compounds, from very small molecules to large polymers and biological molecules ($>100,000$ Da). MS is a powerful tool in the determination of the structure of organic compounds. Fragmentation patterns can reveal the presence of substructure units within the molecule.
Quantitative analysis: MS is used extensively for the quantitative determination of the organic components of liquid and gas samples. Solid samples can be analyzed using laser ablation.

Thermal analysis (TA)
Qualitative analysis: TA is used to identify inorganic and some organic compounds using very small quantities of sample. It is also used to identify phase changes, chemical changes on heating, heats of fusion, melting points, boiling points, drying processes, decomposition processes, and the purity of compounds.
Quantitative analysis: Thermal analysis can be used for the quantitative determination of the components of an inorganic sample, particularly at high concentration levels.

Gas chromatography (GC)
Qualitative analysis: GC can be used to separate the components of complex mixtures of gases or of volatile compounds. By comparison with known standards, it can identify components based on retention time.
Quantitative analysis: Gas chromatography is an accurate method for quantitative analysis based on the area of the peak and comparison with standards. It is used extensively in organic, environmental, clinical and industrial analysis. GC with MS detection (GC-MS) is a routine and powerful tool for quantitative analysis of organic compounds in environmental and biological samples.

Liquid chromatography (LC, HPLC)
Qualitative analysis: LC is used for the identification of components of liquid mixtures, including polar compounds, ions, high molecular weight components and thermally unstable compounds. Identification is based on retention time and comparison with standards.

(*continued*)

Table 1.2   Continued

Quantitative analysis: LC is used for the quantitative determination of components in mixtures, especially for high molecular weight or thermally unstable compounds. It is particularly useful for separating complicated mixtures such as natural products derived from plants or animals and biological samples such as urine and blood. Ion chromatography is used routinely in water analysis. LC with MS detection (LC-MS) is a routine and powerful tool for quantitative analysis of organic compounds in environmental and biological samples.

Capillary electrophoresis (CE)

Qualitative analysis: Used for the separation and identification of ions and neutral molecules in mixtures. Can be used for ions in aqueous solution and for organic ions.

Quantitative analysis: Quantitative determination of ions can be accomplished following separation, as in IC and LC.

Elemental Analysis

Atomic emission spectrometry (AES, OES)

Qualitative analysis: AES is an almost comprehensive methods for qualitative elemental analysis for metals, metalloids, and nonmetals with the exception of some of the permanent gases. Its sensitivity range is great, varying from parts per billion to percent levels. Many elements can be detected simultaneously. Spectral overlap is the major limitation.

Quantitative analysis: AES is used extensively for the quantitative determination of elements in concentrations from % levels down to ppb. Liquids, slurries, and solids can be analyzed using the appropriate atomization source.

Flame photometry (flame atomic emission spectrometry)

Qualitative analysis: Flame photometry is particularly useful for the determination of alkali metals and alkaline-earth metals. It provides the basis for flame tests used in qualitative analysis schemes.

Quantitative analysis: Flame photometry is used for the quantitative determination of alkaline metals and alkaline-earth metals in blood, serum, and urine in clinical laboratories. It provides much simpler spectra than those found in other types of atomic emission spectrometry, but its sensitivity is much reduced.

Atomic absorption spectrometry

Qualitative analysis: Atomic absorption spectrometry is not used routinely for qualitative analysis, since with most instruments it is only possible to test for one element at a time.

Quantitative analysis: Atomic absorption spectrometry is a very accurate and sensitive method for the quantitative determination of metals and metalloids down to absolute amounts as low as picograms for some elements. It cannot be used directly for the determination of nonmetals.

X-ray fluorescence (XRF)

Qualitative analysis: X-ray fluorescence is useful for elements with atomic numbers greater than 4, including metals and nonmetals. For qualitative analysis, no sample preparation is required and the method is generally nondestructive.

Quantitative analysis: XRF is used extensively for quantitative determination of elements in alloys and mineral samples, particularly of elements with high atomic weights. Sample preparation is complex for quantitative analysis.

Inorganic mass spectrometry (MS)

Qualitative analysis: Inorganic MS can identify elements, isotopes and polyatomic ions in solutions and solid samples.

Quantitative analysis: Inorganic MS can determine elements at ppt concentrations or below. Inorganic MS is used for simultaneous multielement analysis for metals and nonmetals. Inorganic MS provides the isotope distribution of the elements. Special mass spectrometers are used for accurate isotope ratio measurements used in geology and geochemistry.

Table 1.3 Analytical Concentration Ranges for Common Instrumental Methods

Technique	Destructive	Ultratrace (<1 ppm)	Trace (1 ppm– 0.1%)	Minor (0.1–10%)	Major (>10%)
X-ray diffraction	No	No	No	Yes	Yes
Nuclear magnetic resonance	No	No	Yes	Yes	Yes
X-ray fluorescence	No	No	Yes	Yes	Yes
Infrared spectroscopy	No	No	Yes	Yes	Yes
Raman spectroscopy	No	No	Yes	Yes	Yes
UV/VIS spectrometry	No	No	Yes	Yes	Yes
Colorimetry	No	Yes	Yes	Yes	No
Molecular fluorescence spectrometry	No	Yes	Yes	Yes	Yes
Atomic absorption spectrometry	Yes	Yes	Yes	Yes	No
Atomic emission spectrometry	Yes	Yes	Yes	Yes	Yes
Atomic fluorescence spectrometry	Yes	Yes	Yes	No	No
ICP-mass spectrometry	Yes	Yes	Yes	Yes	No
Organic mass spectrometry	Yes	Yes	Yes	Yes	Yes
GC-MS	Yes	Yes	Yes	Yes	Yes
LC-MS	Yes	Yes	Yes	Yes	Yes
Potentiometry	No	Yes	Yes	Yes	Yes
Voltammetry	No	Yes	Yes	Yes	Yes
Gas chromatography	May be	Yes	Yes	Yes	Yes
High performance liquid chromatography	May be	Yes	Yes	Yes	Yes
Ion chromatography	May be	Yes	Yes	Yes	Yes
Capillary electrophoresis	No	Yes	Yes	Yes	Yes
Thermal analysis	Yes	No	No	Yes	Yes

Note: The destructive nature of the instrumental method is characterized. A sample may be destroyed by a non-destructive instrumental method, depending on the sample preparation required. The chromatographic techniques may be destructive or nondestructive, depending on the type of detector employed. The nondestructive detectors generally limit sensitivity to "trace". Molecular fluorescence is not destructive if the molecule is inherently fluorescent. It may be if the molecule requires derivatization. A method with "yes" for ultratrace and "no" for major concentrations reflects linear working range. Such methods can measure "majors" if the sample is diluted sufficiently.

(titrimetry or volumetric analysis), or production of colored products with visual esti-mation of the amount of color produced (colorimetry). Using these methods, it was found, for example, that dry sodium chloride, NaCl, always contained 39.33% Na and 60.67% Cl. The atomic theory was founded on early quantitative results such as this, as were the concept of valency and the determination of atomic weights. Today, quantitative inorganic elemental analysis is performed by atomic absorption spectrometry (AAS), atomic emission spectrometry of many sorts, inorganic mass spectrometry such as ICP-MS, XRF, ion chromatography, and other techniques discussed in detail in later chapters.

In a similar fashion, quantitative elemental analysis for carbon, hydrogen, nitrogen, and oxygen enabled the chemist to determine the empirical formulas of organic com-pounds. An empirical formula is the simplest whole number ratio of the atoms of each element present in a molecule. For any given compound, the empirical formula may or may not coincide with the molecular formula. A molecular formula contains the total number of atoms of each element in a single molecule of the compound. For example, ethylene and cyclohexane have the same empirical formula, CH_2, but molecular formulas of C_2H_4 and C_6H_{12}, respectively. The empirical formula of many sugars is CH_2O, but the molecular formulas differ greatly. The molecular formula of glucose is $C_6H_{12}O_6$, fructose is $C_6H_{12}O_6$, erythrose is $C_4H_8O_4$, and glyceraldehyde is $C_3H_6O_3$. An example of a molecule whose empirical formula is the same as the molecular formula is tetrahydrofuran (THF), an important organic solvent. The molecular formula for THF is C_4H_8O; there is only one oxygen atom, so there can be no smaller whole number ratio of the atoms. Therefore, C_4H_8O is also the empirical formula of THF.

Empirical formulas of organic compounds were derived mainly from combustion analysis, where the organic compound is heated in oxygen to convert all of the carbon to CO_2 and all of the hydrogen to H_2O. The CO_2 and H_2O were collected and weighed or the volume of the gas was determined by displacement of liquid in a measuring device. To distinguish between butane, C_4H_{10}, which contains 82.76% C and 17.24% H and pentane, C_5H_{12}, which contains 83.33% C and 16.66% H required great skill using manual combustion analysis. Today, automated analyzers based on combustion are used for quantitative elemental analysis for C, H, N, O, S, and the halogens in organic compounds. These analyzers measure the evolved species by gas chromatography (GC), IR, or other techniques. These automated analyzers require only microgram amounts of sample and a few minutes to provide data and empirical formulas that used to take hours of skilled analytical work. Quantitative elemental analysis cannot distinguish between isomers, which are compounds with the same molecular formula but different structures. Glucose and fructose have the same molecular formula, but glucose is a sugar with an aldehyde group in its structure, while fructose is a sugar with a ketone group in its structure. They cannot be distinguished by elemental analysis, but are easily distinguished by their IR and NMR spectra.

Quantitative molecular analysis has become increasingly important as the fields of environmental science, polymer chemistry, biochemistry, pharmaceutical chemistry, natural products chemistry, and medicinal chemistry have grown explosively in the past 10 years. Techniques such as GC, liquid chromatography or high-performance liquid chromatography (LC or HPLC), capillary electrophoresis (CE), MS, fluorescence spectrometry, IR, and X-ray diffraction (XRD) are used to determine the amount of specific compounds, either pure or in mixtures. These techniques have become highly automated and extremely sensitive, so that only micrograms or milligrams of sample are needed in most cases. The chromatography techniques, which can separate mixtures, have been "coupled" to techniques like MS, which can identify and quantitatively measure the components in a mixture. Such techniques, like GC-MS and LC-MS, are called hyphenated techniques. Many hyphenated instruments are commercially available. These types of instruments for use in the pharmaceutical industry have been designed to process samples in very large batches in a completely automated fashion. The instruments will analyze the samples, store the data in computer files, "pattern-match" the spectra to identify the compounds, and calculate the concentrations of the compounds in the samples. Even then, more than one instrument is required to keep up with the need for characterization of potential drug candidates. As an example, one research department in a major pharmaceutical company bought its first LC-MS in 1989. By 1998, that one department had more than 40 LC-MS instruments running on a daily basis to support drug metabolism studies alone.

Instrumental methods differ in their ability to do quantitative analysis; some methods are more *sensitive* than others. That is, some methods can detect smaller amounts of a given analyte than other methods. Some methods are useful for wide ranges of analyte concentrations; other methods have very limited ranges. We will discuss the reasons for this in the chapters on the individual techniques, but Table 1.3 shows the approximate useful concentration ranges for common instrumental techniques. Table 1.3 is meant to serve as a guide; the actual sensitivity and useful concentration range (also called the working range) of a technique for a specific analysis will depend on many factors.

1.2.2.   Designing the Analytical Method

Once the problem has been defined, an analytical procedure, or method, must be designed to solve the problem. The analytical chemist may have to design the method to meet certain goals, such as achieving a specified accuracy and precision, using only a limited amount of sample, or performing the analysis within a given cost limit or "turnaround time". Turnaround time is the time elapsed from receipt of a sample in the lab to delivery of the results to the person who requested the analysis. This length of time may need to be very short for clinical chemistry laboratories providing support to hospital emergency rooms, for example. A common goal for modern analytical procedures is that they are "green chemistry" processes, that is, that the solvents used are of low toxicity or biodegradable, that waste is minimized, and that chemicals used in the analysis are recycled when possible.

Designing a good analytical method requires knowing how to obtain a *representative sample* of the material to be analyzed, how to store or preserve the sample until analysis, and how to prepare the sample for analysis. The analyst must also know how to evaluate possible *interferences* and *errors* in the analysis and how to assess the accuracy and precision of the analysis. These topics will be discussed subsequently and specific interferences for given instrumental methods are discussed in the following chapters.

There are many analytical procedures and methods that have been developed and published for a wide variety of analytes in many different matrices. These methods may be found in the chemical literature, in journals such as Analytical Chemistry, The Analyst, Analytical and Bioanalytical Chemistry (formerly Fresenius' Journal of Analytical Chemistry), Talanta, and in journals which focus on specific analytical techniques, such as Applied Spectroscopy, Journal of Separation Science (formerly Journal of High Resolution Chromatography), Journal of the American Society for Mass Spectrometry, Thermochimica Acta, and many others. Compilations of "standard" methods or "official" methods have been published by government agencies such as the US Environmental Protection Agency (EPA) and private standards organizations such as the American Association of Official Analytical Chemists (AOAC), the American Society for Testing and Materials (ASTM), and the American Public Health Association (APHA), among others. Similar organizations and official methods exist in many other countries. These standard methods are methods that have been tested by many laboratories and have been found to be reproducible, with known accuracy and precision. The bibliography lists several of these books on analytical methods. It is always a good idea to check the chemical literature first, so that you don't waste time designing a procedure that already exists.

If there are no methods available, then the analytical chemist must develop a method to perform the analysis. For very challenging problems, this may mean inventing entirely new analytical instruments or modifying existing instruments to handle the task.

The design of the method also requires the analyst to consider how the method will be shown to be accurate and precise. This requires knowledge of how we assess accuracy and precision, discussed in Section 1.3. The analyst must evaluate interferences. Interference is anything that (1) gives a response other than the analyte itself or (2) that changes the response of the analyte. Interferences may be other compounds or elements present in the sample, or that form on degradation of the sample. Interfering compounds or elements may respond directly in the instrumental measurement to give a false analyte signal, or they may affect the response of the analyte indirectly by enhancing or suppressing the analyte signal. Examples will be given in the chapters for each instrumental technique. The analyst must demonstrate that the method is reliable and *robust*. These will be covered in greater detail in Sections 1.5 and 1.6.

There are some fundamental features that should be part of every good analytical method. The method should require that a **blank** be prepared and analyzed. A blank is used to ascertain and correct for certain interferences in the analysis. In many cases, more than one type of blank is needed. One type of blank solution may be just the pure solvent used for the sample solutions. This will ensure that no analyte is present in the solvent and allows the analyst to set the baseline or the "zero point" in many analyses. A **reagent blank** may be needed; this blank contains all of the reagents used to prepare the sample but does not contain the sample itself. Again, this assures the analyst that none of the reagents themselves contribute analyte to the final reported value of analyte in the sample. Sometimes a **matrix blank** is needed; this is a blank that is similar in chemical composition to the sample but without the analyte. It may be necessary to use such a blank to correct for an over-lapping spectral line from the matrix in atomic emission spectrometry, for example.

All instrumental analytical methods except coulometry (Chapter 15) require **calibration standards**, which have known concentrations of the analyte present in them. These calibration standards are used to establish the relationship between the analytical signal being measured by the instrument and the concentration of the analyte. Once this relationship is established, unknown samples can be measured and the analyte concentrations determined. Analytical methods should require some sort of **reference standard** or check standard. This is also a standard of known composition with a known concentration of the analyte. This check standard is not one of the calibration standards and should be from a different lot of material than the calibration standards. It is run as a sample to confirm that the calibration is correct and to assess the accuracy and precision of the analysis. Reference standard materials are available from government and private sources in many countries. Government sources include the National Institute of Standards and Technology (NIST) in the US, the National Research Council of Canada (NRCC), and the Laboratory of the Government Chemist in the UK.

1.2.3. Sampling

The most important single step in an analysis is collecting the sample of the material to be analyzed. Real materials are usually not homogeneous, so the sample must be chosen carefully to be **representative** of the real material. A representative sample is one that reflects the true value and distribution of the analyte in the original material. If the sample is not taken properly, no matter how excellent the analytical method or how expert the analyst, the result obtained will not provide a reliable characterization of the material. Other scientists, law enforcement officials, and medical professionals often collect samples for analysis, sometimes with no training in how to take a proper sample. The analytical chemist ideally would be part of the team that discusses collection of samples before they are taken, but in reality, samples often "show up" in the lab. It is important that the analyst talks with the sample collector before doing any analyses; if the sample has been contaminated or improperly stored, the analysis will be not only a waste of time, but can also lead to erroneous conclusions. In clinical chemistry analysis, this could lead to a misdiagnosis of a disease condition; in forensic analysis, this could lead to a serious miscarriage of justice, for example.

The amount of sample taken must be sufficient for all analyses to be carried out in duplicate or triplicate, if possible. Of course, if only a small quantity of sample is available, as may be the case for forensic samples from a crime scene or rocks brought back from the moon, the analyst must do the best job possible with what is provided.

A good example of the problems encountered in sampling real materials is collecting a sample of a metal or metal alloy. When a molten metal solidifies, the first portion of solid to form tends to be the most pure (remember freezing point depression from your general chemistry class?). The last portion to solidify is the most impure and is generally located in the center or *core* of the solidified metal. It is important to bear this in mind when sampling solid metals. A sample is often ground from a representative cross-section of the solid, or a hole is drilled through a suitable location and the drillings mixed and used as the sample.

Samples have to be collected using some type of collection tool and put into some type of container. These tools and containers can often contaminate the sample. For example, stainless steel needles can add traces of metals to blood or serum samples. Metal spatulas, scissors and drill bits, glass pipets, filter paper, and plastic and rubber tubing can add unwanted inorganic and organic contaminants to samples. To avoid iron, nickel and chromium contamination from steel, some implements like tongs and tweezers can be purchased with platinum or gold tips.

The discussion of sampling which follows refers to the traditional process of collecting a sample at one location (often called "collection in the field") and transporting the sample to the laboratory at a different location. Today it is often possible to analyze samples *in situ* or during the production of the material (on-line or process analysis) with suitable instrumental probes, completely eliminating the need for "collecting" a sample. Examples of on-line analysis will be discussed in later chapters.

The process of sampling requires several steps, especially when sampling bulk materials such as coal, metal ore, soil, grain and tank cars of oil or chemicals. First a *gross representative sample* is gathered from the *lot*. The lot is the total amount of material available. Portions of the gross sample should be taken from various locations within the lot, to ensure that the gross sample is representative. For very large lots of solid material such as coal or ore, the *long pile and alternate shovel* method can be used. The material is formed into a long rectangular pile. It is then separated into two piles by shoveling material first to one side and then to the other, creating two piles. One pile is set aside. The remaining pile may be reduced in size by repeating the process, until a sample of a size to be sent to the laboratory remains. The *cone and quarter* method is also used to collect a gross sample of solid materials. The sample is made into a circular pile and mixed well. It is then separated into quadrants. A second pile is made up of two opposite quadrants, and the remainder of the first pile discarded. This process is shown in Fig. 1.3. This process can be repeated until a sample of a suitable size for analysis is obtained. This sample

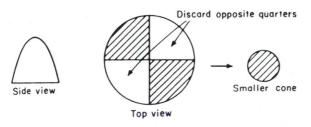

Figure 1.3   The cone and quarter method of sampling bulk materials.

can still be very large. Ferroalloys, for example, are highly segregated (i.e., inhomogeneous) materials; it is not uncommon for the amount required for a representative sample of alloy in pieces about 2 in. in diameter to be one ton (0.9 Mg) of material from the lot of alloy.

A computer program that generates random numbers can choose the sampling locations and is very useful for environmental and agricultural sampling. If the lot is a field of corn, for example, the field can be divided into a grid, with each grid division given a number. The computer program can pick the random grid divisions to be sampled. Then a smaller, homogeneous *laboratory sample* is prepared from the gross composite sample. If the sample is segregated (i.e., highly inhomogeneous), the representative sample must be a composite sample that reflects each region and its relative amount. This is often not known, resulting in the requirement for very large samples. The smaller laboratory sample may be obtained by several methods, but must be representative of the lot and large enough to provide sufficient material for all the necessary analyses. After the laboratory sample is selected, it is usually split into even smaller *test portions*. Multiple small test portions of the laboratory sample are often taken for replicate analyses and for analysis by more than one technique. The term **aliquot** is used to refer to a quantitative amount of a *dissolved* test portion; for example, a 0.100 g test portion of sodium chloride may be dissolved in water in a volumetric flask to form 100.0 mL of test solution. Three 10.0 mL aliquots may be taken with a volumetric pipet for triplicate analysis for chloride using an ion selective electrode, for example.

As the total amount of the sample is reduced, it should be broken down to successively smaller pieces by grinding, milling, chopping, or cutting. The one ton sample of ferroalloy, for example, must be crushed, ground, and sieved many times. During the process, the sample size is reduced using a sample splitter called a riffle. After all this and then a final drying step, a 1 lb (454 g) sample remains. The sample must be mixed well during this entire process to ensure that it remains representative of the original. The grinding equipment used must not contaminate the sample. For example, boron carbide and tungsten carbide are often used in grinding samples because they are very hard materials, harder than most samples. However, they can contribute boron or tungsten to the ground sample, so would not be used if boron or tungsten must be measured at low concentrations. Zirconium oxide ball mills can contribute Zr and Hf to a sample. Stainless steel grinders are a source of Fe, Cr, and Ni. Some cutting devices use organic fluids as lubricants; these must be removed from the sample before analysis.

It is also possible for the grinding or milling step to cause erroneously low results for some analytes. Malleable metals like gold may adhere to the grinding or milling surface and be removed from the sample in the process, an undesirable effect. An example of sampling a segregated material with a problematic component like gold is illustrated in Fig. 1.4. The rectangular piece at the top is a hypothetical piece of gold-bearing quartz. The gold is represented as the dark flecks. You can see that the gold appears in bands within the quartz, separated by bands of pure quartz (the white area). If the rock is crushed to Size I, the gold particles have not been liberated from the quartz; some pieces have gold flecks and many large pieces are pure quartz. At this size, it is difficult to remove a sample of the rock pieces and expect it to be representative. If the rock is crushed to a smaller size, Size II, it is evident that a representative small sample can be obtained. If the rock is crushed to Size III, the gold particles are freed from the quartz matrix. If this sample could be mixed perfectly, a smaller sample could be taken to represent the whole than the sample needed at Size II. (Why would this be desirable? The smaller the analytical sample, the less gold is used up by analysis. This is an important consideration with valuable analytes and valuable samples.) But the gold particles and the quartz particles have different densities,

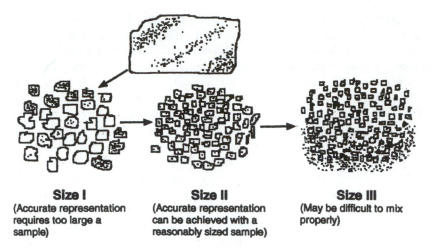

Size I
(Accurate representation
requires too large a
sample)

Size II
(Accurate representation
can be achieved with a
reasonably sized sample)

Size III
(May be difficult to mix
properly)

Figure 1.4   Sampling of a segregated material with a problematic component like gold. Extracted, with permission, from Dulski, T.R., Copyright ASTM International, 100 Barr Harbor Drive, West Conshohocken, PA 19428.

different shapes, and will be difficult to mix well. As mentioned, gold is soft and malleable. If it is broken out of the quartz it may become embedded in the grinder or smeared onto surfaces in the grinding equipment, so some gold may actually be lost from the sample particles ground to Size III. Size II will give a more representative sample than either of the other sizes.

Sampling procedures for industrial materials, environmental samples, and biological samples are often agreed upon, or *standardized*, by industry, government, and professional societies. Standard sampling procedures help to ensure that the samples analyzed are representative and are not contaminated or changed during the sampling process. Standard sampling procedures for many materials can be found in the Annual Book of ASTM Standards, for example. Sampling procedures for soil, water, and air are established by the US EPA in the United States, and similar government organizations in other countries. Procedures for sampling of water and wastewater can be found in Standards Methods for the Analysis of Water and Wastewater; the AOAC publishes procedures for food products. The bibliography provides some examples of these publications. A good analytical chemist will consult the literature before sampling an unfamiliar material. Some general guidelines for sampling different classes of materials are discussed here.

1.2.3.1.   Gas Samples

Gas samples are generally considered homogeneous, but gas mixtures may separate into layers of differing density. Samples that have been collected and allowed to settle will need to be stirred before a portion is taken for analysis. Gas samples can be taken at a single point in time (called a **grab** sample) or can be collected over a period of time or from different locations to provide an *average* or **composite** sample. Gas samples can be collected using gas-tight syringes, balloons, plastic bags, or containers made of metal or glass that can be evacuated. Sampling of toxic, flammable, or corrosive gases should be done with great care using appropriate safety equipment.

The containers used to collect the samples must not contaminate the sample with analyte. Plastic bags and balloons may leach volatile organic compounds into the gas sample, while glass may adsorb components of the sample onto the surface of the glass.

Certain components of gas samples, such as organic vapors in air, may be collected by pulling the air through activated charcoal. The organic gases are adsorbed onto the charcoal, while the majority of the air (oxygen, nitrogen, etc.) passes through. This has the advantage of *preconcentrating* the analytes of interest and reducing the physical size of the sample. Many liters of air can be pulled through an activated charcoal bed that is no bigger than a ballpoint pen. It is much easier to transport the analytes trapped on the charcoal to the laboratory than to transport hundreds of liters of air. The process of trapping an analyte out of the gas phase is called "scrubbing". Scrubbing a gas sample can also be done by bubbling the gas through a liquid that will absorb the analytes of interest.

Gas samples may contain particles of solid material that need to be removed by filtration. The filter material must be chosen so that it does not adsorb analytes or add contaminants to the gas. Filters are available which will remove particles as small as 0.2 μm in diameter from a gas stream.

1.2.3.2. Liquid Samples

Liquid samples can also be collected as grab samples or as composite samples. Sampling liquids can be quite difficult; it is not always as straightforward as "pouring some" out of a bottle or dipping a bucket into a fluid. Only a few comments with respect to general sampling of liquids can be made here. It is usual to stir liquid samples adequately to obtain a representative sample; however, there may be occasions when stirring is not desired. If the analyst is only interested in identifying an oily layer floating on water, stirring the sample is not needed; the oily layer may be pulled off with a pipet or an eyedropper, for example. Samples must be collected at locations remote from sources of contamination if a representative sample is desired. For example, if a sample of "normal" river water is desired, the sample should be collected away from riverbanks, floating froth, oil, and discharges from industrial and municipal waste treatment sites. Sampling of rivers, lakes, and similar bodies of water may require samples from different depths or different distances from shore. Such samples may be analyzed individually or blended to obtain an average composition.

Liquid samples may contain particles of solid material that need to be removed by filtration or centrifugation. The filter material must be chosen so that it does not adsorb analytes or contaminate the liquid. Some samples that are mostly liquid contain suspended solid material; orange juice and liquid antacids are examples. In these types of samples, the liquid and its associated solids may need to be sampled for analysis without removing the solids. It may be difficult to obtain a representative sample from these suspensions; a standard sampling procedure is needed to ensure that results can be compared from one day to the next. Liquid samples may consist of more than one layer because they contain two or more **immiscible** liquids. Examples include samples of oil and water from an oil spill at sea, oil and vinegar salad dressing, or cream at the top of a bottle of milk. The layers may need to be *emulsified* to provide a representative sample, but it may be more useful to sample each layer separately.

Sampling of hot molten materials such as metals, alloys, and glasses is a form of liquid sampling, but one requiring very specialized equipment and techniques.

1.2.3.3. Solid Samples

Solid samples are often the most difficult to sample because they are usually less homogeneous than gases or liquids. Large amounts of solid sample cannot be conveniently "stirred up". Moreover, unlike the situation with fluids, there are no diffusion or convection currents in solids to ensure mixing. Solids must often be ground or drilled or crushed into smaller particles to homogenize the sample. There are many types of manual and automated grinders and crushers available; the choice depends on the hardness of the material to be ground. Soft materials also pose a challenge in grinding because they often just deform instead of being reduced in size. Polymer pellets may be ground in an electric coffee grinder with a small amount of liquid nitrogen added to the grinder. The liquid nitrogen freezes the polymer, making the pellets brittle and capable of being easily powdered. Other soft solids such as foods can be handled the same way. Many solid materials must be oven-dried before sampling to remove adsorbed water in order to obtain a representative sample. There are numerous published standard methods for sampling solid materials such as cement, textiles, food, soil, ceramics, and other materials. Examples of the wide variety of analytical pulverizing, grinding, and blending equipment available can be found at the SpexCertiprep website at www.spexcsp.com.

1.2.4. Storage of Samples

When samples cannot be analyzed immediately, they must be stored. The composition of a sample may change during storage because of reactions with air, light, or interaction with the container material. The container used for collection and storage of the sample and the storage conditions must be chosen to minimize changes in the sample.

Plastic containers may leach organic components such as plasticizers and monomers into a sample. Plastic containers may also introduce trace metal impurities such as Cu, Mn, or Pt from the catalysts used to make the polymer or elements such as Si, Ti, Sb, Br, and P from inorganic fillers and flame-retardants. Glass surfaces both adsorb and release trace levels of ionic species, which can dramatically change the trace element and trace ion concentrations in solutions. It has been observed that trace metals will "plate out" of solution along strain lines in glass. Such strain lines are not reproducible from one container to another; therefore the loss of trace metals cannot be estimated accurately for one container by measuring the loss in a similar but different container. All containers require appropriate cleaning before use. Containers for organic samples are usually washed in solvent, while containers for samples for trace metals analysis are soaked in acid and then in deionized water.

Precautions such as freezing biological and environmental samples or displacing the air in a container by an inert gas will often extend the storage life of a sample. Samples should not be stored any longer than is absolutely necessary prior to analysis and should not be stored under conditions of high heat or high humidity. Some samples require storage in the dark to avoid photolytic (light-induced) changes in composition; water samples to be analyzed for silver are a good example. Such samples must be stored in dark plastic bottles to avoid the photolytic formation of colloidal silver, which will precipitate out of the sample. Many samples for environmental analysis require the addition of preservatives or adjustment of pH to prevent the sample from deteriorating. Water samples for trace metals determinations must be acidified with high purity nitric acid to keep the trace metals in solution, for example. Blood samples often require collection in tubes containing an anticoagulant to keep the blood sample fluid, but the anticoagulant must not interfere in the analysis. For example, a sample collected to measure a patient's sodium level cannot be collected in a tube that contains the sodium salt of ethylenediamminetetraacetic acid (EDTA) as the anticoagulant. Other biological samples may need to be collected in sterile containers.

Sample containers must be labeled accurately and in such a way that the label does not deteriorate on storage; do not use water-soluble marking pen on samples to be put in a freezer, for example. The label should clearly identify the sample and any hazards associated with the sample. Many analytical laboratories have computer-based sample tracking systems that generate adhesive bar coded labels for samples, exactly like the bar codes used on retail items in stores. These computer-based systems are called Laboratory Information Management Systems (LIMS) and catalog and track not only the samples but also the analytical data generated on the samples.

As a student, you should get into the habit of labeling all your containers in the laboratory with your name, date, and the contents. There is nothing worse than finding four beakers of colorless liquid on the lab bench and not knowing which one is yours or what is in the other beakers! That situation would be a serious safety violation in an industrial laboratory. Academic labs in the US are now required to follow the same safety regulations followed in industry and something as simple as beakers with "stuff" in them but no proper labels can result in large monetary fines for your school. The cost of chemical waste disposal is very high and it is not legal to dispose of unidentified chemicals, so unlabeled containers are a very expensive problem. The material must be analyzed to identify it just so that it can be disposed of properly.

How many samples do we need to collect for a given analysis? How large must the sample be to insure that it is representative? These types of questions can be answered by statistics. We also need to have a basic knowledge of statistics to understand the limitations in the other steps in method development, so we will now briefly introduce the statistical concepts and calculations used by analytical chemists.

1.3. BASIC STATISTICS AND DATA HANDLING

In order to design the correct experiment to answer the analytical question being asked, statistics is needed to select the size of the sample required, the number of samples, and the number of measurements that must be performed to obtain the needed accuracy and precision in the results generated by the experiment. Statistics is also used to express the uncertainty in measured values, so that the users of the data understand the limitations associated with results.

1.3.1. Significant Figures

The result of an analytical measurement is a number with associated units such as 50.1% iron, 10 ppm parathion (a pesticide), or 25 mg isopropanol/L. To deal with numbers that result from measurements and calculations involving these numbers, the concept of **significant figures** must be understood. All measurements have uncertainty in them, so the results should be reported as a number that tells us about the uncertainty. The numbers reported from a measurement should be all of the digits known with certainty plus the first uncertain digit. These digits, with uncertainty only in the last digit of the number, are called significant figures. This gives a result that reflects the precision of the measurement. For example, the number 50.1% means that the percentage is closer to 50.1 than to 50.2 or 50.0, but it does not mean that the percentage is exactly 50.1. In short, we are sure of the "50" part of the number, but there is some uncertainty in the last figure reported. If we were to analyze two samples containing 50.08% and 50.12% of a component by using an instrument accurate to 0.1%, we would not be able to distinguish the difference in the compositions of the samples, but would report them both as 50.1%.

The number 50.1 has three significant figures (5, 0, 1). Since the measurement is no better than 0.1%, the last digit in 50.1 is uncertain by at least ± 1. The last significant figure reported should reflect the precision of the measurement. There is no point in reporting any more figures because they would have no meaning, even though they might be obtainable mathematically. For example, scientific calculators generally display eight or more digits; that does not mean that all of the displayed digits are meaningful.

The reporting of figures implies that all the numbers are significant and only the last number is in doubt, even if that number is zero. For example, 1.21×10^6, which has three significant figures (1, 2, 1) implies that the number is closer to 1.21×10^6 than to 1.22×10^6 or 1.20×10^6. But writing 1,210,000, with seven significant figures, implies that the number is closer to 1,210,000 than to 1,210,001 or 1,209,999. As discussed below, terminal zeros can be confusing because they may not be significant if they are only used to show the decimal place. Furthermore, the number 50.10 implies 10× greater accuracy than 50.1. A zero can be a significant figure or it can be used to show the decimal place. If a zero occurs between two nonzero significant figures, it is significant, as in 12,067, which has five significant figures. In a number such as 0.024 the two initial zeros just show the decimal place. Initial zeros are never significant and the number 0.024 has only two significant figures, the 2 and the 4. One way to confirm that the zeros are only placeholders is to write the number in scientific notation, 2.4×10^{-2}, which clearly has only two significant figures. If a zero is written after a decimal point, it is significant; for example, 24.0 would have three significant figures. It is important to write a zero at the end of a number (a terminal zero) only if it is significant. For example, in a number like 54,300 or 1,210,000, the zeros at the end of each number may or may not be significant; it is impossible to tell by looking at the number. In cases like these, scientific notation should be used to indicate exactly how many figures are significant. The number 5.4300×10^4 has five significant figures; 5.430×10^4 has four significant figures; 5.43×10^4 has three significant figures. To be clear about the number 1,210,000, it should be written in scientific notation: 1.210000×10^6 is the unambiguous way to show that there are seven significant figures.

Numbers that represent discrete objects have no uncertainty. If five measurements were made, and we calculate the average by adding the five results and dividing by 5, the number 5 has no effect on the number of significant figures in the answer. There is no uncertainty in the number of measurements made, so the 5 can be considered to have an infinite number of significant figures.

The following are some rules that should be observed when reporting results.

1. In enumerating data, report all significant figures, such that only the last figure is uncertain.
2. Reject all other figures, rounding off in the process. That is, if a number such as 1.325178 must be reported to four significant figures, only the first five figures, 1.3251, should be considered. If the fifth figure is greater than 5, increase the fourth figure by one and drop the fifth figure. If the fifth figure is less than 5, the fourth figure remains unchanged and the fifth number is dropped. If the fifth figure is 5 and the fourth figure is odd, increase the fourth figure by one and drop the 5. If the fifth figure is 5 and the fourth figure is even, it is not increased when the 5 is dropped. Table 1.4 shows some examples of rounding to four significant figures using these rules.
3. In reporting results obtained by addition and subtraction, the figures in each number are significant only as far as the first uncertain figure of any one of the numbers to be added or subtracted. The result of an addition or subtraction should have the same absolute uncertainty as the number in the calculation with the largest absolute uncertainty. The number with the largest absolute uncertainty is the one with the fewest significant figures. For example, the sum of

Table 1.4   Rounding Off to Four Significant Figures

Number	Four significant figures
1.37286	1.373
1.37243	1.372
1.3735	1.374
1.3725	1.372
1.37251[a]	1.373

[a]The number 0.00051 is greater than 0.0005, even though the last figure is not significant; hence the fourth figure is increased by one.

the numbers in the set 21.1, 3.216, and 0.062 is reliable to the first decimal point, because 21.1 is uncertain in the tenths place. Therefore the sum is only known to the tenths place. One approach is to round off the other numbers to the tenths place prior to addition (or subtraction). The sum of the rounded-off numbers $(21.1 + 3.2 + 0.1)$ is 24.4. A second approach is to carry one more figure than the least significant figure and round off at the end. Using this approach, the sum of $(21.1 + 3.21 + 0.06) = 24.37$, which rounds off to 24.4 in the tenths place. The approach used should be consistent. When adding or subtracting numbers with exponents, such as numbers written in scientific notation, the numbers should be adjusted so that the exponents are all the same. For example, to add $3.25 \times 10^{-2} + 3 \times 10^{-6}$, express 3×10^{-6} as 0.0003×10^{-2}. The sum is then 3.2503×10^{-2}, which is rounded off to 3.25×10^{-2}.

4. For multiplication and division, the number of significant figures in the answer should be no greater than that of the term with the least number of significant figures. In the case of multiplication and division, the answer will have a relative uncertainty of the same order of magnitude as the number with the largest relative uncertainty. Once again, the terms can be rounded off before calculation, but it is preferable to carry one more figure and round off at the end. For example, $(1.236 \times 3.1 \times 0.18721 \times 2.36) = 1.692860653$, according to a scientific calculator that allows for 10 digits to be expressed. The figures being multiplied have four, two, five, and three significant figures, respectively. The term 3.1 has the least number of significant figures, two. The answer is therefore rounded off to two significant figures, and would be reported as 1.7. If the terms are rounded to two significant figures before multiplication, the product $(1.2 \times 3.1 \times 0.19 \times 2.4) = 1.69$, which is rounded off to 1.7, so that only two significant figures are reported.

5. The characteristic of a **logarithm** indicates an order of magnitude; it has no uncertainty and does not affect the number of significant figures in a calculation. The **mantissa** should contain no more significant figures than are in the original number. For example, the number 12.7 has a logarithm of 1.1038. The mantissa is rounded off to three significant figures and log 12.7 is reported as 1.104. The pH of a solution equals the negative logarithm of the hydrogen ion concentration. Therefore in the following calculation, $pH = -\log(3.42 \times 10^{-2}) = 1.4659 = 1.466$. The mantissa has three significant figures, as did the original number 3.42.

6. If several analyses have been obtained for a particular sample (**replicate analysis**), it should be noted at what point there is doubt in the significant numbers of the

result. The final answer should be reported accordingly, and we will see how this is done in the next section. For example, given the triplicate results 11.32, 11.35, and 11.32, there is no doubt about 11.3, but there is uncertainty in the hundreds place (the fourth figure). The average should be reported as 11.33, with four significant figures [i.e., $(11.32 + 11.35 + 11.32) \div 3$]. Remember that the number 3 (the divisor) indicates the exact number of measurements and has no uncertainty, so it does not affect the number of significant figures in the answer.

If a calculation involves a combination of multiplication (or division) and addition (or subtraction), the steps must be treated separately. When using a calculator or spreadsheet program, the best approach is to keep all significant figures throughout the calculation and round off at the end.

Why place all this emphasis on significant figures? When you know the limitations of a measurement in terms of its uncertainty, you can design an analytical method efficiently. If you can only read the absorbance of light in a spectrometric measurement to three significant figures (a typical value), it is a waste of time to weigh the sample to five significant figures.

1.3.2. Accuracy and Precision

It is very important to understand the definitions of accuracy and precision and to recognize the difference between precision and accuracy. Accuracy is a measure of how close a measured analytical result is to the true answer. For most analytical work, the "true answer" is not usually known. We often work with an "accepted" true value or "accepted reference value". Accuracy is evaluated by analyzing known, standard samples. The US NIST (formerly the National Bureau of Standards) in Washington, D.C., has well-characterized **standard reference materials** of many types that can serve as the known sample. Similar **certified reference materials** are available from government standards agencies in Canada, the UK, Europe, and other countries, as well as from a wide variety of commercial standards suppliers. Another way of assessing accuracy is to *spike* a sample with a known amount of the pure analyte. The sample is analyzed and the amount of added analyte *recovered* is reported. A spike recovery of 100% would indicate that all of the added analyte was measured by the analytical method, for example. Accuracy is documented by reporting the difference between the measured value and the true value with the appropriate **confidence level**, or by reporting the spike recovery as a percentage of added analyte.

Precision is a measure of how close replicate results on the same sample are to each other. A common analogy used to envision the difference between accuracy and precision is to imagine a bull's-eye target used by an archer. If all the arrows hit in the bull's-eye, the archer is both accurate (has hit the center) and precise (all the arrows are close together). If the archer puts all the arrows into the target close together (a "tight shot group") but to the upper left of the bull's-eye, the archer is precise but not accurate. If the arrows hit the target in many locations—top, bottom, center, left, and right of the center—the archer is neither precise nor accurate. The difference between precision and accuracy is illustrated in Table 1.5. There are several ways to express precision mathematically; Table 1.5 uses **standard deviation** (to be defined shortly) as a measure of precision. A superficial examination of the results provided by Analyst 2 could be misleading. It is very easy to be deceived by the closeness of the answers into believing that the results are accurate. The closeness, expressed as the standard deviation, shows that the results of Analyst 2 are *precise*, and not that the analysis will result in obtaining the true

Table 1.5 Replicate Determinations of Analyte in a Sample[a]

	% Analyte		
	Analyst 1[b]	Analyst 2[c]	Analyst 3[d]
	10.0	8.1	13.0
	10.2	8.0	10.2
	10.0	8.3	10.3
	10.2	8.2	11.1
	10.1	8.0	13.1
	10.1	8.0	9.3
Average (%)	10.1	8.1	11.2
Absolute error[e]	0.0	2.0	1.1
Standard deviation	0.089	0.13	1.57

[a]Accepted true answer is $10.1 \pm 0.2\%$ (obtained independently).
[b]Results are precise and accurate.
[c]Results are precise but inaccurate.
[d]Results are imprecise and inaccurate.
[e]Absolute error = |true value − measured value|.

answer. The latter must be discovered by an independent method, such as having Analyst 2 analyze a sample of known composition. The accepted true value for the determination is $10.1 \pm 0.2\%$ according to the table footnote, so the determination by Analyst 2 is not accurate. Analyst 3 is both inaccurate and imprecise. It is very unlikely that an imprecise determination will be accurate. Precision is required for accuracy, but does not guarantee accuracy.

It is important for students to realize that the inability to obtain the correct answer does not necessarily mean that the analyst uses poor laboratory techniques or is a poor chemist. Many causes contribute to poor accuracy and precision, some of which we will discuss in this chapter as well as in later chapters. Careful documentation of analytical procedures, instrument operating conditions, calculations, and final results are crucial in helping the analyst recognize and eliminate errors in analysis.

The quantitative analysis of any particular sample should generate results that are precise and accurate. The results should be reproducible, reliable, and truly representative of the sample. Unfortunately, some degree of error is always involved in analytical determinations, as discussed in Section 1.3.3.

For analytical results to be most useful, it is important to be aware of the reliability of the results. To do this it is necessary to understand the sources of error and to be able to recognize when they can be eliminated and when they cannot. Error is the difference between the true result (or accepted true result) and the measured result. If the error in an analysis is large, serious consequences may result. A patient may undergo expensive and even dangerous medical treatment based on an incorrect laboratory result or an industrial company may implement costly and incorrect modifications to a plant or process because of an analytical error. There are numerous sources of error and several types of errors, some of which are described here.

1.3.3. Types of Errors

There are two principal types of error in analysis: **determinate** or **systematic** error and **indeterminate** or **random** error.

1.3.3.1. Determinate Error

Broadly speaking, **determinate errors** are caused by faults in the analytical procedure or the instruments used in the analysis. The name determinate error implies that the cause of this type of error may be found out and then either avoided or corrected. Determinate errors are *systematic errors*; that is, they are not random. A particular determinate error may cause the analytical results produced by the method to be always too high; another determinate error may render all results too low. Sometimes the error is *constant*; all answers are too high (or too low) by the same amount. If the true results for three samples are 25, 20, and 30 mg/L of analyte, but the measured (or determined) results are 35, 30, and 40 mg/L, respectively, the analysis has a *constant error* of 10 mg/L. Since these results are all too high, the constant error is positive; a constant negative error of 10 mg/L would result in the three measured results being 15, 10, and 20 mg/L of analyte, respectively. Sometimes the determinate error is proportional to the true result, giving rise to **proportional errors**. For example, if the measured results for the same three earlier samples are 27.5, 22.0, and 33.0 mg/L analyte, respectively, the measured results are too high by 10% of the true answer. This error varies in proportion to the true value. Other determinate errors may be variable in both sign and magnitude, such as the change in the volume of a solution as the temperature changes. Although this variation can be positive or negative, it can be identified and accounted for. Determinate errors can be additive or they can be multiplicative. It depends on the error and how it enters into the calculation of the final result.

If you look again at the results in Table 1.5 for Analyst 2, the results produced by this analyst for the repetitive analysis of a single sample agree closely with each other, indicating high precision. However, the results are all too low (and therefore inaccurate), given that Table 1.5 states the true value of the sample to be 10.1 ± 0.2% analyte. There is a negative determinate error in the results from Analyst 2. This determinate error could be the result of an incorrectly calibrated balance. If the balance is set so that the zero point is actually 0.5 g too high, all masses determined with this balance will be 0.5 g too high. If this balance was used to weigh out the potassium chloride used to make the potassium standard solution used in the clinical laboratory, the standard concentration will be erroneously high, and all of the results obtained using this standard will be erroneously low. The error is reported as the absolute error, the absolute value of the difference between the true and measured values. However, there is not enough information provided to know if this is a constant or a proportional error. It can be seen that close agreement between results (i.e., high precision) does not rule out the presence of a determinate error.

Determinate errors arise from some faulty step in the analytical process. The faulty step is repeated every time the determination is performed. Whether a sample is analyzed 5 times or 50 times, the results may all agree with each other (good precision) but differ widely from the true answer (poor accuracy). An example is given in Table 1.6. Although the replicate results are close to each other, that tells us nothing about their accuracy. We can see from the true value given in Table 1.6 that the experimental results are too high; there is a determinate error in the procedure. An analyst or doctor examining the measured analytical results in Table 1.6 might be deceived into believing that the close agreement among the replicate measurements indicates high accuracy and that the results are close to the true potassium concentration. (Potassium in adult human serum has a normal range of 3.5–5.3 mmol potassium/L serum. Assume that the true value given is for this particular patient.)

In the example in Table 1.6, the true value was 4.0 mmol potassium/L and the average measured result was 5.2 mmol potassium/L in the patient's serum. However,

Table 1.6 Potassium Concentration in a Single Serum Sample

	Measured value (mmol/L)	True value[a] (mmol/L)
	5.2	4.0
	5.1	
	5.3	
	5.1	
	5.1	
Average	5.2	4.0

[a]Normal range for potassium in serum: 3.5–5.3 mmol/L.

the analyst and the doctor do not know the true value of an unknown serum sample. The measured result is in the normal range for adult human serum potassium concentrations, so neither the analyst nor the doctor is likely to be suspicious of the results.

If a faulty analytical procedure is used to analyze five different patients' serum samples and the results shown in Table 1.7 are obtained, it can be seen that in all cases the error is +1.2 mmol/L. This indicates a constant, positive determinate error. As you can see, this faulty procedure would result in one patient being misdiagnosed with a false high serum K level and a patient with a truly low serum K level being misdiagnosed as "normal".

An analyst working at a different hospital with different instrumentation obtains the results shown in Table 1.8. Examination of these analytical results shows they are all ~20% greater than the true answer. The error is *proportional* to the true concentration of the analyte. Such information as to the nature of the error is useful in the diagnosis of the source of the determinate error.

Systematic error is under the control of the analyst. It is the analyst's responsibility to recognize and correct for these *systematic errors* that cause results to be *biased*, that is, offset in the average measured value from the true value. How are determinate errors identified and corrected? Two methods are commonly used to identify the existence of systematic errors. One is to analyze the sample by a completely different analytical procedure that is known to involve no systematic errors. Such methods are often called "standard methods"; they have been evaluated extensively by many laboratories and shown to be accurate and precise. If the results from the two analytical methods agree, it is reasonable to assume that both analytical procedures are free of determinate errors. The second method is to run several analyses of a reference material of known,

Table 1.7 Potassium Concentrations in Patients' Serum

Patient	Measured value[a] (mmol/L)	True value (mmol/L)
A	5.3	4.1
B	4.8	3.6
C	6.3	5.1
D	5.0	3.8
E	4.1	2.9

[a]Constant error of +1.2 mmol/L.

Table 1.8 Potassium Concentration in Serum

Patient	Measured value (mmol/L)	True value[a] (mmol/L)
A	5.8	4.8
B	4.3	3.6
C	7.4	6.2
D	3.5	2.9
E	6.6	5.5

[a]Results indicate a positive proportional error of 20% of the true value.

accepted concentration of analyte. The difference between the known (true) concentration and that measured by analysis should reveal the error. If the results of analysis of a known reference standard are consistently high (or consistently low), then a determinate error is involved in the method. The cause of the error must be identified and either eliminated or controlled if the analytical procedure is to give accurate results. In the earlier example of potassium in serum, standard serum samples with certified concentrations of potassium are available for clinical laboratories. Many clinical and analytical laboratories participate in proficiency testing programs, where "unknown" standard samples are sent to the laboratory on a regular basis. The results of these samples are sent to the government or professional agency running the program. The unknowns are of course known to the agency that sent the test samples; the laboratory receives a report on the accuracy and precision of its performance.

Determinate errors can arise from uncalibrated balances, improperly calibrated volumetric flasks or pipettes, malfunctioning instrumentation, impure chemicals, incorrect analytical procedures or techniques, and analyst error.

Analyst error. The person performing the analysis causes these errors. They may be the result of inexperience, insufficient training, or being "in a hurry". An analyst may use the instrument incorrectly, perhaps by placing the sample in the instrument incorrectly each time or setting the instrument to the wrong conditions for analysis. Consistently misreading a meniscus in a volumetric flask as high (or low) and improper use of pipettes, such as "blowing out" the liquid from a volumetric pipette, are common analyst errors. Some other analyst-related errors are (1) *carelessness*, which is not as common as is generally believed; (2) *transcription errors*, that is, copying the wrong information into a lab notebook or onto a label; and (3) *calculation errors*. Proper training, experience, and attention to detail on the part of the analyst can correct these types of errors.

Reagents and instrumentation. Contaminated or decomposed reagents can cause determinate errors. Impurities in the reagents may interfere with the determination of the analyte, especially at the ppm level or below. Prepared reagents may also be improperly labeled. The suspect reagent may be tested for purity using a known procedure or the analysis should be redone using a different set of reagents and the results compared.

Numerous errors involving instrumentation are possible, including incorrect instrument alignment, incorrect wavelength settings, incorrect reading of values, and incorrect settings of the readout (i.e., zero signal should read zero). Any variation in proper instrument settings can lead to errors. These problems can be eliminated by a systematic procedure to check the instrument settings and operation before use. Such procedures are called standard operating procedures (SOPs) in many labs. There should be a written SOP for each instrument and each analytical method used in the laboratory.

In instrumental analysis, electrical line voltage fluctuations are a particular problem. This is especially true for automated instruments running unattended overnight. Instruments are often calibrated during the day, when electrical power is in high demand. At night, when power demand is lower, line voltage may increase substantially, completely changing the relationship between concentration of analyte and measured signal. Regulated power supplies are highly recommended for analytical instruments. The procedure for unattended analysis should include sufficient calibration checks during the analytical run to identify such problems. Many instruments are now equipped with software that can check the measured value of a standard and automatically recalibrate the instrument if that standard falls outside specified limits.

Analytical method. The most serious errors are those in the method itself. Examples of method errors include (1) incomplete reaction for chemical methods, (2) unexpected interferences from the sample itself or reagents used, (3) having the analyte in the wrong oxidation state for the measurement, (4) loss of analyte during sample preparation by volatilization or precipitation, and (5) an error in calculation based on incorrect assumptions in the procedure (errors can evolve from assignment of an incorrect formula or molecular weight to the sample). Most analytical chemists developing a method check all the compounds likely to be present in the sample to see if they interfere with the determination of the analyte; unlikely interferences may not have been checked. Once a valid method is developed, an SOP for the method should be written so that it is performed the same way every time it is run.

Contamination. Contamination of samples by external sources can be a serious source of error and may be extremely variable. An excellent example of how serious this can be has been documented in the analysis of samples for polychlorinated biphenyls (PCBs). PCBs are synthetic mixtures of organochlorine compounds that were first manufactured in 1929 and have become of concern as significant environmental pollutants. It has been demonstrated that samples archived since 1914, before PCBs were manufactured, picked up measurable amounts of PCBs in a few hours just sitting in a modern laboratory (Erickson). Aluminum levels in the dust in a normal laboratory are so high that dust prohibits the determination of low ppb levels of aluminum in samples. A special dust-free "clean lab" or "clean bench" with a filter to remove small dust particles may be required, similar to the clean rooms needed in the semiconductor industry, for determination of traces of aluminum, silicon, and other common elements such as iron. When **trace** (<ppm level) or **ultratrace** (<ppb level) organic and inorganic analysis is required, the laboratory environment can be a significant source of contamination.

Another major source of contamination in an analysis can be the analyst. It depends on what kind of analytes are being measured, but when trace or ultratrace levels of elements or molecules are being determined, the analyst can be a part of the analytical problem. Many personal care items, such as hand creams, shampoos, powders, and cosmetics, contain significant amounts of chemicals that may be analytes. The problem can be severe for volatile organic compounds in aftershave, perfume, and many other scented products and for silicone polymers, used in many health and beauty products. Powdered gloves may contain a variety of trace elements and should not be used by analysts performing trace element determinations. Hair, skin, and clothing can shed cells or fibers that can contaminate a sample.

Having detected the presence of a determinate error, the next step is to find its source. Practical experience of the analytical method or first-hand observation of the analyst using the procedure is invaluable. Much time can be wasted in an office guessing at the source of the trouble. Unexpected errors can be discovered only in the laboratory. A little data is worth a lot of discussion (Robinson's Law).

1.3.3.2. Indeterminate Error

After all the determinate errors of an analytical procedure have been detected and eliminated, the analytical method is still subject to random or indeterminate error arising from inherent limitations in making physical measurements. Each error may be positive or negative, and the magnitude of each error will vary. Indeterminate errors are not constant or biased. They are random in nature and are the cause of slight variations in results of replicate samples made by the same analyst under the same conditions.

Sources of random error include the limitations of reading balances, scales such as rulers or dials, and electrical "noise" in instruments. For example, a balance that is capable of measuring only to 0.001 g cannot distinguish between two samples with masses of 1.0151 and 1.0149 g. In one case the measured mass is low, in the other case it is high. These random errors cause variation in results, some of which may be too high and some too low, as we see for Analyst 1 in Table 1.5. The average of the replicate determinations is accurate, but each individual determination may vary slightly from the true value. Indeterminate errors arise from sources that cannot be corrected, avoided, or even identified, in some cases. All analytical procedures are subject to indeterminate error. However, because indeterminate error is random, the errors will follow a random distribution. This distribution can be understood using the laws of probability and basic statistics. The extent of indeterminate error can be calculated mathematically.

Let us suppose that an analytical procedure has been developed in which there is no determinate error. If an infinite number of analyses of a single sample were carried out using this procedure, the distribution of numerical results would be shaped like a symmetrical bell (Fig. 1.5). This bell-shaped curve is called the **normal** or **Gaussian distribution**. The frequency of occurrence of any given measured value when only indeterminate error occurs is represented graphically by a plot such as Fig. 1.5.

If only indeterminate errors were involved, the most frequently occurring result would be the true result, that is, the result at the maximum of the curve would be the true answer. In practice it is not possible to make an infinite number of analyses of a single sample. At best, only a few analyses can be carried out, and frequently only one analysis of a particular sample is possible. We can, however, use our knowledge of statistics to determine how reliable these results are. The basis of statistical calculations

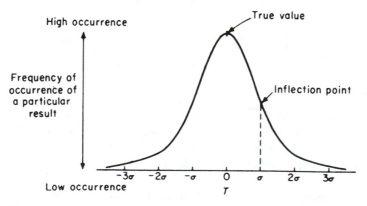

Figure 1.5 A normal or Gaussian distribution of results when only indeterminate error is present. The value that occurs with most frequency is the true value (T) or mean value, while the spread of the distribution is expressed in units of standard deviation from the mean, symbolized by σ. The larger the random error, the broader the distribution will be.

is outlined below. Statisticians differentiate between the values obtained from a finite number of measurements, N, and the values obtained from an infinite number of measurements, so we need to define these statistical terms.

1.3.4. Definitions for Statistics

True value T: the true or accepted value; also symbolized by x_t.

Observed value x_i: a single value measured by experiment.

Sample mean $\bar{x}$: the arithmetic mean of a finite number of observations, that is,

$$\bar{x} = \frac{\sum_{i=1}^{N} x_i}{N} = \frac{(x_1 + x_2 + x_3 + \cdots + x_N)}{N} \tag{1.1}$$

where N is the number of observations and $\sum x_i$ is the sum of all the individual values x_i.

Population mean μ: the limit as N approaches infinity of the sample mean, that is,

$$\mu = \lim_{N \to \infty} \sum_{i=1}^{N} \frac{x_i}{N} \tag{1.2}$$

In the absence of systematic error, the population mean μ equals the true value T of the quantity being measured.

Error E: the difference between the true value T and either a single observed value x_i or the sample mean of the observed values, $\bar{x}$; error may be positive or negative,

$$E = x_i - T \quad \text{or} \quad \bar{x} - x_t \tag{1.3}$$

The total error is the sum of all the systematic and random errors.

Absolute error: the absolute value of E, and can be defined for a single value or for the sample mean,

$$E_{abs} = |x_i - T| \quad \text{or} \quad |\bar{x} - x_t| \tag{1.4}$$

Relative error: the absolute error divided by the true value; it is often expressed as a percent by multiplying by 100,

$$E_{rel} = \frac{E_{abs}}{x_t} \quad \text{or} \quad \%E_{rel} = \frac{E_{abs}}{x_t} \times 100 \tag{1.5}$$

Absolute deviation d_i: the absolute value of the difference between the observed value x_i and the sample mean $\bar{x}$,

$$d_i = |x_i - \bar{x}| \tag{1.6}$$

Relative deviation D: the absolute deviation d_i divided by the mean $\bar{x}$,

$$D = \frac{d_i}{\bar{x}} \tag{1.7}$$

Percent relative deviation: the relative deviation multiplied by 100,

$$D\,(\%) = \frac{d_i \times 100\%}{\bar{x}} = D \times 100 \tag{1.8}$$

Sample standard deviation s: for a finite number of observations N, the sample standard deviation is defined as

$$s = \sqrt{\frac{\sum_{i=1}^{N} d_i^2}{N-1}} = \sqrt{\frac{\sum_{i=1}^{N} (x_i - \bar{x})^2}{N-1}} \tag{1.9}$$

Standard deviation of the mean s_m: the standard deviation associated with the mean of a data set consisting of N measurements,

$$s_m = \frac{s}{\sqrt{N}} \tag{1.10}$$

Population standard deviation σ: for an infinite number of measurements,

$$\sigma = \sqrt{\lim_{N \to \infty} \frac{\sum_{i=1}^{N} (x_i - \mu)^2}{N}} \tag{1.11}$$

Percent relative standard deviation % RSD,

$$\% \text{ RSD} = \frac{s}{\bar{x}} \times 100 \tag{1.12}$$

Variance σ^2 or s^2: the square of the population standard deviation σ or the sample standard deviation s.

1.3.5. Quantifying Random Error

If the systematic errors have been eliminated, the measured value will still be distributed about the true value owing to random error. For a given set of measurements, how close is the average value of our measurements to the true value? This is where the Gaussian distribution and statistics are used.

The Gaussian distribution curve assumes that an infinite number of measurements of x_i have been made. The maximum of the Gaussian curve occurs at $x = \mu$, the true value of the parameter we are measuring. So, for an infinite number of measurements, the population mean is the true value x_t. We assume that any measurements we make are a subset of the Gaussian distribution. As the number of measurements, N, increases, the difference between $\bar{x}$ and μ tends toward zero. For N greater than 20 to 30 or so, the sample mean rapidly approaches the population mean. For 25 or more replicate measurements, the true value is approximated very well by the experimental mean value. Unfortunately, even 20 measurements of a real sample are not usually possible. Statistics allows us to express the random error associated with the difference between the population mean μ and the mean of a small subset of the population, $\bar{x}$. The random error for the mean of a small subset is equal to $\bar{x} - \mu$.

The area under any portion of the Gaussian distribution, for example, between a value x_1 and a value x_2, corresponds to the fraction of the measurements which will yield a measured value of x between and including these two values. The spread of the Gaussian distribution, that is, the width of the bell-shaped curve, is expressed in terms of the population standard deviation σ. The standard deviation σ coincides with the point of inflection of the curve as can be seen in Fig. 1.5. The curve is symmetrical, so we have two inflection points, one on each side of the maximum. The x-axis relates the area under the curve to the standard deviation. For $\pm\sigma$ on either side of the maximum, 68.3% of the area under the curve lies in this range of x. This means that 68.3% of all measurements of x_i will fall within the range $x = \mu \pm \sigma$. About 95.5% of the area under the curve lies between $x = \mu \pm 2\sigma$ and 99.7% of the area lies between $x = \mu \pm 3\sigma$.

The precision of analytical results is usually stated in terms of the standard deviation σ. As just explained, in the absence of determinate error, 68.3% of all results can be expected to fall within $\pm\sigma$ of the true value, 95.5% of the results will fall within $\pm 2\sigma$ of the true answer, and 99.7% of our results will fall within $\pm 3\sigma$ of the true value if we perform enough measurements (more than 20 or so replicates). It is common practice to report analytical results with the mean value and the standard deviation expressed, thereby giving an indication of the precision of the results.

Let us suppose that we know by previous extensive testing that the standard deviation σ of a given analytical procedure for the determination of Si in an aluminum alloy is 0.1% w/w. This is the standard deviation in concentration units (0.1 g Si/100 g Al alloy) not the relative standard deviation. We know that there are no determinate errors in the analysis. Also, when we analyze a particular sample using this method, and perform sufficient replicate analyses, we obtain an average result of 19.6% Si by weight. We can now report that the analytical result is 19.6 ± 0.1% Si with the understanding that 68.3% of all results will fall in this range. We could also report that the result is 19.6 ± 0.2% (where 0.2% = 2σ). A report stating that an analysis of a sample indicated 19.6% Si and that 2σ for the method is 0.2% means that we are 95.5% certain that the true answer is 19.6 ± 0.2% Si. We say that the **confidence level** (CL) of the measurement is 95.5%, with the understanding that there are no determinate errors and that we have performed a sufficient number of replicates to know both μ and σ. Notice that the more certain we are that the answer falls within the range given, the bigger the range actually is. It is important when reporting data with both mean and precision information that the analyst tells the customer what precision is reported. If Analyst A reported 19.6 ± 0.1% Si and Analyst B reported 19.6 ± 0.2% Si, without telling the customer what the 0.1 and 0.2% mean, the customer might think incorrectly that Analyst A is more precise than Analyst B.

However, we are usually dealing with a small, finite subset of measurements, not 20 or more; so the standard deviation that should be reported is the sample standard deviation s. For a small finite data set, the sample standard deviation s differs from σ in two respects. Look at the equations given in the definitions. The equation for sample standard deviation s contains the sample mean, not the population mean, and uses $N - 1$ measurements instead of N, the total number of measurements. The term $N - 1$ is called the **degrees of freedom**.

Let us go through an example of calculating some of these statistical parameters. The equations are simple enough that the values can be calculated manually, although the calculations can be tedious for large values of N. Most scientific handheld calculators have programs that calculate mean, sample standard deviation s (sometimes marked σ_{N-1} on a calculator button), and σ (sometimes marked σ_N on a button). You should learn how to use these programs on your calculator. In addition, the calculations can be set up in a spreadsheet in programs like Microsoft Excel®.

You have measured mercury in eight representative samples of biological tissue from herons (which eat fish that may be contaminated with mercury) using cold vapor

AAS (which is discussed in Chapter 6). The values in column 2 of the table below were obtained.

Sample	Hg content (ppb)	$(x_i - \bar{x})^2$
1	5.34	0.010
2	5.37	0.005
3	5.44	0.000
4	5.22	0.048
5	5.84	0.160
6	5.67	0.053
7	5.27	0.029
8	5.33	0.012

The mean is calculated using Eq. (1.1):

$$\bar{x} = \frac{5.34 + 5.37 + 5.44 + 5.22 + 5.84 + 5.67 + 5.27 + 5.33}{8}$$

$$= \frac{43.48}{8} = 5.44 \text{ ppb Hg}$$

Since we have only eight measurements, we will calculate s, the sample standard deviation, using Eq. (1.9). The standard deviation s is calculated manually in steps, with the intermediate values for $(x_i - \bar{x})^2$ shown in the table preceding in the third column. The sum of the $(x_i - \bar{x})^2$ values $= 0.317$, therefore:

$$s = \sqrt{\frac{0.317}{8 - 1}} = 0.213$$

You could report the average concentration of Hg in the heron tissue as 5.44 ± 0.21 ppb Hg and indicate in your report that 0.21 equals $1s$. The Hg concentration can be reported in terms of $2s$, $3s$, and so on, just as for the population standard deviation. Analytical results published without such precision data lose much of their meaning. They indicate only the result obtained and not the reliability of the answer.

Software programs called spreadsheets are extremely useful for performing the repetitive calculations used by analysts, displaying each step in the calculation, tabulating data, and presenting data graphically. A variety of commercial software programs are available. The example given here uses the spreadsheet program Microsoft Excel®. The other commercial programs have similar capabilities but the instructions will differ for each. The following example assumes some familiarity with using Microsoft programs. If more fundamental directions are needed, the texts by Christian or Harris listed in the bibliography have excellent instructions and examples.

When an Excel spreadsheet is opened, the page consists of blank cells arranged in rows and columns. Each cell is identified by its column letter and row number. The individual mercury concentrations from the earlier example can be typed into separate cells, as can text such as column headings. In the sample spreadsheet page shown, text is typed into cells A1, B1, A12, and A13, while the sample numbers and data are put in as shown. The data points are in cells B3 through B10. The width of the columns can be varied to fit the contents.

	A	B	C	D
1	Sample	Hg conc. (ppb)		
2				
3	1	5.34		
4	2	5.37		
5	3	5.44		
6	4	5.22		
7	5	5.84		
8	6	5.67		
9	7	5.27		
10	8	5.33		
11				
12	Mean			
13	Std. Dev.			

It is possible to write mathematical formulas and insert them into the spreadsheet to perform calculations, but Excel has many functions already built into the program. By clicking on an empty cell and then on f_x on the toolbar, these functions can be accessed. This opens the Paste Function window. Select *Statistical* in the *Function category* on the left side of the window and a list of *Function names* appears on the right side of the window. Click on the cell B12, then select AVERAGE from the Function name list. Click OK, and type in (B3:B10) in the active box. The average (mean) will appear in cell B12. Alternatively, in cell B12, you can type =AVERAGE(B3:B10) and the mean will be calculated. The standard deviation can be calculated by selecting STDEV from the Function name list and clicking on cell B13 or by typing =STDEV(B3:B10) into cell B13.

	A	B	C	D
1	Sample	Hg conc. (ppb)		
2				
3	1	5.34		
4	2	5.37		
5	3	5.44		
6	4	5.22		
7	5	5.84		
8	6	5.67		
9	7	5.27		
10	8	5.33		

	A	B	C	D
11				
12	Mean	5.435		
13	Std. Dev.	0.212804		
14				

The values calculated by Excel are shown in cells B12 and B13. Of course they must be rounded off to the correct number of significant figures, but are the same as the results obtained manually. Learning to use spreadsheets can save time and permit the data to be stored, processed, and presented in a variety of formats. The spreadsheets can be made part of the electronic lab notebooks that are common in industry and government laboratories as well as in many college laboratories.

The value obtained for σ is an estimate of the precision of the method. If an analyst sets up a new analytical procedure and carries out 20 determinations of a standard sample, the precision obtained is called the **short-term precision** of the method. This is the optimum value of σ because it was obtained from analyses run at the same time by the same analyst, using the same instrumentation and the same chemicals and reagents. In practice the short-term precision data may be too optimistic. Routine analyses may be carried out for many years in a lab, such as the determination of Na and K in serum in a hospital laboratory. Different analysts, different chemicals and reagents, and even different instrumentation may be used. The analysis of a standard sample should be carried out on a regular basis (daily, weekly, etc.) and these results compiled on a regular basis. Over several months or a year, the **long-term precision** of the method can be calculated from these compiled results. This is a more realistic measure of the reliability of the analytical results obtained on a continuing basis from that laboratory.

There are a variety of terms used to discuss analytical methods that are related to the precision of the method. The **repeatability** of the method is the short-term precision of the method under the same operating conditions (the interassay precision). **Reproducibility** refers to the ability of multiple laboratories to obtain the same results on a given sample and is determined from collaborative studies. **Ruggedness** is the degree of reproducibility of the results obtained by one laboratory under a variety of conditions (similar to the long-term precision). Repeatability, reproducibility, and ruggedness are all expressed in terms of the standard deviation (or relative standard deviation) obtained experimentally. **Robustness** is another term for the **reliability** of the method, that is, its accuracy and precision, under small changes in conditions. These changes can be in operating conditions such as laboratory room temperature, sample variables such as concentration and pH, reagent and standard stability, and so on.

1.3.5.1. Confidence Limits

It is impossible to determine μ and σ from a limited set of measurements. We can use statistics to express the probability that the true value μ lies within a certain range of the measured average mean $\bar{x}$. That probability is called a **confidence level** (CL) and is usually expressed as a percentage (e.g., the CL is 95%). The term **confidence limit** refers to the extremes of the **confidence interval** (the range) about $\bar{x}$ within which μ is expected to fall at a given CL.

When s is a good approximation for σ, we can state a CL or confidence limits for our results based on the Gaussian distribution. The CL is a statement of how close the sample mean lies to the population mean. For a single measurement we let $s = \sigma$. The CL is then the certainty that $\mu = x \pm z\sigma$. For example, if $z = 1$, we are 68.3% confident that x lies within $\pm\sigma$ of the true value; if we set $z = 2$, we are 95.5% confident that x lies within $\pm 2\sigma$ of the true value. For N measurements, the CL for $\mu = \bar{x} \pm zs_m$.

In most cases, s is not a good estimate of σ because we have not made enough replicate analyses. In this case, the CL is calculated using a statistical probability parameter, *Student's t*. The parameter t is defined as $t = (x - \mu)/s$ and the CL for $\mu = \bar{x} \pm ts/\sqrt{N}$. An abbreviated set of t values is given in Table 1.9; complete tables can be found in mathematics handbooks or statistics books.

As an example of how to use Table 1.9 and CLs, assume that we have made five replicate determinations of the pesticide DDT in a water sample using GC. The five results are given in the spreadsheet below, along with the mean and standard deviation. How do we report our results so that we are 95% confident that we have reported the true value?

	A	B	C	D
1	Replicate sample	DDT conc. (ppb)		
2				
3	1	1.3		
4	2	1.5		
5	3	1.4		
6	4	1.4		
7	5	1.3		
8				
9	Mean	1.4		
10	Std. Dev.	0.08		

Table 1.9   Student's t Values

Degrees of freedom ($N - 1$)	t Value for confidence limit		
	90%	95%	99%
1	6.31	12.7	63.7
2	2.92	4.30	9.92
3	2.35	3.18	5.84
4	2.13	2.78	4.60
5	2.02	2.57	4.03
6	1.94	2.45	3.71
7	1.90	2.36	3.50
8	1.86	2.31	3.36
9	1.83	2.26	3.25
10	1.81	2.23	3.17
∞	1.64	1.96	2.58

The number of determinations is five, so there are four degrees of freedom. The t value for the 95% CL with $N - 1 = 4$ is 2.78, according to Table 1.9. Therefore:

$$95\% \text{ CL} = \bar{x} \pm \frac{ts}{\sqrt{N}} = 1.4 \pm \frac{2.78 \times 0.08}{5^{1/2}} = 1.4 \pm 0.1$$

We are 95% confident that the true value of the DDT concentration in the water sample is 1.4 ± 0.1 ppb, assuming no determinate error is present.

The Student's t value can be used to test for systematic error (bias) by comparing means of different determinations. The CL equation is rewritten as:

$$\pm t = (\bar{x} - \mu)\frac{\sqrt{N}}{s} \tag{1.13}$$

By using a known, valid method, μ is determined for a known sample, such as a certified reference material. The values of $\bar{x}$ and s are determined for the known sample using the new method (new instrument, new analyst, etc.). A value of t is calculated for a given CL and the appropriate degrees of freedom. If the calculated t value exceeds the value of t given in Table 1.9 for that CL and degrees of freedom, then a significant difference exists between the results obtained by the two methods, indicating a systematic error. Using the DDT data above, assume that we know that the true value of the DDT concentration is 1.38 ppb. A new analyst runs the five determinations and obtains a mean of 1.20 ppb and a standard deviation of 0.13. If we calculate t using the previous equation, we get $\pm t = (1.20 - 1.38)(5)^{1/2}/0.13 = -3.09$. At 95% CL, the absolute value of this calculated t is larger than the tabulated t for 4 degrees of freedom found in Table 1.9. Therefore a determinate error exists in the procedure as performed by the new analyst.

1.3.5.2. Variance

Variance is defined as the square of the standard deviation, σ^2. Variance is often preferred as a measure of precision because variances from m independent sources of random error are additive. The standard deviations themselves are not additive. The use of variance allows us to calculate the random error in the answer from mathematical calculations involving several numbers, each of which has its own associated random error. The total variance is the sum of the individual variances:

$$\sigma_{\text{tot}}^2 = \sum_{i=1}^{m} \sigma_i^2 \tag{1.14}$$

For addition and subtraction, the absolute variance, σ^2, is additive. For multiplication and division, the relative variances are additive, where the relative variance is just the square of the standard deviation divided by the mean, $\sigma_{\text{rel}}^2 = (\sigma/\bar{x})^2$.

The square of the standard deviation, σ^2, can also be used to determine if sets of data from two methods (analysts, instruments, etc.) are statistically significantly different from each other in terms of their precision. In this test, the variances of two sets of results are compared. The variance σ_2^2 of one set of results is calculated and compared with the variance σ_1^2 of earlier results, or the variance of a new method is compared with that of a standard method. The test is called the F-test. The ratio of the variances of the two sets of numbers is called the F-function:

$$F = \frac{\sigma_1^2}{\sigma_2^2} \tag{1.15}$$

where $\sigma_1^2 > \sigma_2^2$ (i.e., the ratio should be greater than 1). Each variance has its associated degrees of freedom, $(N-1)_1$ and $(N-1)_2$. Tables of F values are found in mathematics handbooks or statistics books. An abbreviated table for the 95% CL is given in Table 1.10. If the calculated value of F is larger than the tabulated value of F for the appropriate degrees of freedom and CL, then there is a significant difference between the two sets of data.

1.3.6. Rejection of Results

When a set of replicate results is obtained it may be the case that one of the results appears to be "out of line"; such a result is called an **outlier**. While it is tempting to discard data that does not "look good" in order to make the analysis seem more precise, it is never a good practice unless there is justification for discarding the result. If it is known that an error was made, such as spillage of the sample, use of the wrong size pipet, incorrect dilution, or allowing the sample to boil to dryness when it should not have done so, the result should be rejected and not used in any compilation of results. In practice, if something of this sort is suspected, a good analyst will discard the sample and start over if possible.

There are a variety of statistical tests that have been used to decide if a data point should be rejected, as well as some "rules of thumb". The range chosen to guide the decision will limit all of these tests and guidelines. A large range will retain possibly erroneous results, while a very small range will reject valid data points. It is important to note that the outlier must be either the highest value in the set of data or the lowest value in the set. A value in the middle of a data set cannot be discarded unless the analyst knows that an error was made.

One rule of thumb is that if the outlier value is greater than $\pm 4\sigma$ from the mean, it should be rejected. When calculating the mean and standard deviation, an outlier result should *not be* included in the calculation. After the calculation, the suspected result should be examined to see if it is more than 4σ from the mean. If it is outside this limit, it should be ignored; if it is within this limit, the value for σ should be recalculated with this result included in the calculation. It is not permissible to reject more than one result on this basis. A suspected result should not be included in calculating σ. If it is included, it will automatically fall within 4σ because such a calculation includes this number. Other reference sources recommend an even smaller range for rejection, such as a 2.5σ limit.

A statistical test called the Q-test can be used effectively for small data sets (see the reference by Rorabacher) to determine if a given data point should be rejected. The

Table 1.10 F Values at 95% CL

$(N-1)_2$	$(N-1)_1$					
	2	4	6	8	10	20
2	19.0	19.2	19.3	19.4	19.4	19.4
4	6.94	6.39	6.16	6.04	5.96	5.80
6	5.14	4.53	4.28	4.15	4.06	3.87
8	4.46	3.84	3.58	3.44	3.35	3.15
10	4.10	3.48	3.22	3.07	2.98	2.77
20	3.49	2.87	2.60	2.45	2.35	2.12

Q-test at the 90% CL is typically used. The data is arranged in order of increasing value. The range of the data is calculated, that is, the lowest value is subtracted from the highest value. The range is $x_n - x_1$. Then the "gap" is calculated, where the gap is defined as the difference between the suspect value and the nearest value, $x_n - x_{n-1}$. The Q ratio is defined as $Q = \text{gap/range}$. Using a table of Q values, if Q observed $>Q$ tabulated, the suspect value is discarded. The Q-test cannot be used if all but one data point is the same in a set. For example, if triplicate results are 1.5, 1.5, and 3.0, you cannot discard the 3.0 value using statistics. It ultimately falls to the analyst to make the decision about rejecting data, but it should not be done lightly. Having covered the basic statistics needed, we return to the discussion of solving the analytical problem.

1.4. SAMPLE PREPARATION

Few samples in the real world can be analyzed without some chemical or physical preparation. The aim of all sample preparation is to provide the analyte of interest in the physical form required by the instrument, free of interfering substances, and in the concentration range required by the instrument. For many instruments, a solution of analyte in organic solvent or water is required. We have already discussed some of the sample preparation steps that may be needed. Solid samples may need to be crushed or ground, or they may need to be washed with water, acid, or solvent to remove surface contamination. Liquid samples with more than one phase may need to be extracted or separated. Filtration or centrifugation may be required.

If the physical form of the sample is different from the physical form required by the analytical instrument, more elaborate sample preparation is required. Samples may need to be dissolved to form a solution or pressed into pellets or cast into thin films or cut and polished smooth. The type of sample preparation needed depends on the nature of the sample, the analytical technique chosen, the analyte to be measured, and the problem to be solved. Most samples are not homogeneous. Many samples contain components that interfere with the determination of the analyte. A wide variety of approaches to sample preparation has been developed to deal with these problems in real samples. Only a brief overview of some of the more common sample preparation techniques is presented. More details are found in the chapters on each instrumental method.

Note: None of the sample preparation methods described here should be attempted without approval, written instructions, and close supervision by your professor or laboratory instructor. The methods described present many potential hazards. Many methods use concentrated acids, flammable solvents, and/or high temperatures and high pressures. Reactions can generate harmful gases. The potential for "runaway reactions" and even explosions exists with preparation of real samples. Sample preparation should be performed in a laboratory fume hood for safety. Goggles, lab coats or aprons, and gloves resistant to the chemicals in use should be worn at all times in the laboratory.

1.4.1. Acid Dissolution and Digestion

Metals, alloys, ores, geological samples, ceramics, and glass react with concentrated acids and this approach is commonly used for dissolving such samples. Organic materials can be decomposed (digested or "wet ashed") using concentrated acids to remove the carbonaceous material and solubilize the trace elements in samples such as biological tissues, foods, and plastics. A sample is generally weighed into an open beaker, concentrated acid is added, and the beaker heated on a hot plate until the solid material dissolves. Dissolution often is much faster if the sample can be heated at pressures greater than atmospheric pressure. The boiling point of the solvent is raised at elevated pressure, allowing the sample and solvent to be heated to higher temperatures than can be attained at atmospheric pressure. This can be done in a sealed vessel, which also has the advantage of not allowing volatile elements to escape from the sample. Special stainless steel high-pressure vessels, called "bombs", are available for acid dissolution and for the combustion of organic samples under oxygen. While these vessels do speed up the dissolution, they operate at pressures of hundreds of atmospheres and can be very dangerous if not operated properly. Another sealed vessel digestion technique uses **microwave digestion**. This technique uses sealed sample vessels made of polymer, which are heated in a specially designed laboratory microwave oven. (NEVER use a kitchen-type microwave oven for sample preparations. The electronics in kitchen-type units are not protected from corrosive fumes, arcing can occur, and the microwave source, the magnetron, can easily overheat and burn out.) The sealed vessel microwave digestion approach keeps volatile elements in solution, prevents external contaminants from falling into the sample, and is much faster than digestion on a hot plate in an open beaker. Microwave energy efficiently heats solutions of polar molecules (such as water) and ions (aqueous mineral acids) and samples that contain polar molecules and/or ions. In addition, the sealed vessel results in increased pressure and increased boiling point. A commercial analytical microwave digestion system for sealed vessel digestion of multiple samples is shown in Fig. 1.6.

The acids commonly used to dissolve or digest samples are hydrochloric acid (HCl), nitric acid (HNO_3), and sulfuric acid (H_2SO_4). These acids may be used alone or in combination. The choice of acid or acid mix depends on the sample to be dissolved and the analytes to be measured. The purity of the acid must be chosen to match the level of analyte to be determined. Very high purity acids for work at ppb or lower levels of elements are commercially available, but are much more expensive than standard reagent grade acid. For special applications, perchloric acid ($HClO_4$) or hydrofluoric acid (HF) may be required. A student should never use $HClO_4$ or HF without specific training from an experienced analytical chemist and only then under close supervision. While a mixture of HNO_3 and $HClO_4$ is extremely efficient for wet ashing organic

Figure 1.6 A commercial analytical microwave digestion system for sealed vessel digestion of multiple samples. Courtesy of CEM Corporation, Matthews, NC (www.cem.com).

materials, $HClO_4$ presents a very serious explosion hazard. Specially designed fume hoods are required to prevent $HClO_4$ vapors from forming explosive metal perchlorate salts in the hood ducts, and reactions of hot $HClO_4$ with organic compounds can result in violent explosive decompositions. A blast shield should be used and the organic sample must first be heated with HNO_3 alone to destroy easily oxidized material before the $HClO_4$ is added. Concentrated HF is used for dissolving silica-based glass and many refractory metals such as tungsten, but it is extremely dangerous to work with. It causes severe and extremely painful deep tissue burns that do not hurt immediately upon exposure. However, delay in treatment for HF burns can result in serious medical problems and even death from contact with relatively small amounts of acid.

HCl is the most commonly used non-oxidizing acid for dissolving metals, alloys, and many inorganic materials. HCl dissolves many materials by forming stable chloride complexes with the dissolving cations. There are two major limitations to the universal use of HCl for dissolution. Some elements may be lost as volatile chlorides; examples of volatile chlorides include arsenic, antimony, selenium, and germanium. Some chlorides are not soluble in water; the most common insoluble chloride is silver chloride, but mercurous chloride, cuprous chloride, BiOCl, and $AuCl_3$ are not soluble, while $PbCl_2$ and TlCl are only partially soluble. A 3:1 mixture of HCl and HNO_3 is called *aqua regia*, and has the ability to dissolve gold, platinum, and palladium. The mixture is also very useful for stainless steels and many specialty alloys.

HNO_3 is an oxidizing acid; it has the ability to convert the solutes to higher oxidation states. It can be used alone for dissolving a number of elements, including nickel, copper, silver, and zinc. The problem with the use of HNO_3 by itself is that it often forms an insoluble oxide layer on the surface of the sample that prevents continued dissolution. For this reason, it is often used in combination with HCl, H_2SO_4, or HF. A mixture of HNO_3 and H_2SO_4 or HNO_3 and $HClO_4$ can be used to destroy the organic material in an organic sample, by converting the carbon and hydrogen to CO_2 and H_2O when the sample is heated in the acid mixture. The trace metals in the sample are left in solution. This use of acids to destroy organic matter is called wet ashing or digestion, as has been noted. H_2SO_4 is a strong oxidizing acid and is very useful in the digestion of organic samples. Its main drawback is that it forms a number of insoluble or sparingly soluble sulfate salts.

HF is a non-oxiziding, complexing acid like HCl. Its most important attribute is that it dissolves silica-based substances like glass and many minerals. All or most of the silicon is volatized on heating with sufficient HF. Glass beakers and flasks cannot be used to hold or store even dilute HF. Teflon or other polymer labware and bottles are required. Commercial "heatable" Teflon beakers with graphite bottoms are available for use on hot plates. HF is used in acid mixtures to dissolve many refractory elements and minerals by forming fluoride complexes; such elements include tungsten, titanium, niobium and tantalum, among others. Some elements can be lost as volatile fluorides (e.g., Si, B, As, Ge, and Se). There are a number of insoluble fluoride compounds, including most of the alkaline earth elements (Ca, Mg, Ba, and Sr) and the rare earth elements (lanthanides). Table 1.11 gives examples of some typical acid digestions.

Some bases, such as sodium hydroxide and tetramethylammonium hydroxide, are used for sample dissolution, as are some reagents that are not acids or bases, like hydrogen peroxide. The chemical literature contains sample dissolution procedures for virtually every type of material known and should be consulted. For elements and inorganic compounds, the CRC Handbook of Chemistry and Physics gives guidelines for dissolution in the tables of physical properties of inorganic compounds.

Table 1.11 Common Acid Dissolutions of Metals, Alloys, and Materials for Inorganic Compositional Analysis

Material[a]	Total volume of reagent (mL)	Reagent (vol:vol)
Elements		
Copper metal	20	1:1 HNO_3/H_2O
Gold metal	30	3:1 HCl/HNO_3
Iron metal	20	1:1 HCl/H_2O
Titanium metal	20	H_2SO_4, 3–5 drops HNO_3
Zinc metal	20	HCl
Zirconium metal	15	HF
Alloys		
Copper alloys	30	1:1 HNO_3/H_2O
Low alloy steels	20	3:1 HCl/HNO_3
Stainless steels	30	1:1 HNO_3/HCl
Titanium alloys	100	1:1 HCl/H_2O, 3–5 drops HNO_3
Zinc alloys	30	1:1 HCl/H_2O, dropwise HNO_3
Zirconium alloys	40	1:1 H_2SO_4/H_2O, 2 mL HF dropwise
Other materials		
Borosilicate glass	12	10 mL HF + 2 mL 1:1 H_2SO_4/H_2O
Dolomite	40	1:1 HCl/H_2O
Gypsum	50	1:1 HCl/H_2O
Portland cement	20	HCl + g NH_4Cl
Silicate minerals	30	10 mL HF + 20 mL HNO_3
Titanium dioxide	15	HF
Zinc oxide	15	1:1 HCl/H_2O

Note: "Dropwise" means add drop by drop until dissolution is complete.
Source: Extracted, with permission, from Dulski. Copyright ASTM International, 100 Barr Harbor Drive, West Conshohocken, PA 19428.
[a] A 1 g test portion is used; warm to complete reaction.

1.4.2. Fusions

Heating a finely powdered solid sample with a finely powdered salt at high temperatures until the mixture melts is called a fusion or molten salt fusion. The reacted and cooled melt is leached with water or dilute acid to dissolve the analytes for determination of elements by atomic spectroscopy or ICP-MS. Often, the molten fusion mixture is poured into a flat bottomed mold and allowed to cool. The resulting glassy disk is used for quantitative XRF measurements. Molten salt fusions are useful for the dissolution of silica-containing minerals, glass, ceramics, ores, human bone, and many difficultly soluble materials like carbides and borides. The salts used (called "fluxes") include sodium carbonate, borax (sodium tetraborate), lithium metaborate, and sodium peroxide. The fusions are carried out over a burner or in a muffle furnace in crucibles of the appropriate material. Depending on the flux used and the analytes to be measured, crucibles may be made of platinum, nickel, zirconium, porcelain, quartz, or glassy carbon. Automated "fluxers" are available that will fuse up to six samples at once and pour the melts into XRF molds or into beakers, for laboratories that perform large numbers of fusions. The drawback of fusion is that the salts used as fluxes can introduce many trace element contaminants into the sample, the crucible material itself may contaminate the sample, and the elements present in the flux itself cannot be analytes in the sample. Fusion cannot be used for boron determinations if the flux is borax or lithium metaborate, for example. Platinum crucibles cannot be used if trace levels of platinum catalyst are to be determined. Table 1.12 gives examples of typical fusions employed for materials.

1.4.3. Dry Ashing and Combustion

To analyze organic compounds or substances for the inorganic elements present, it is often necessary to remove the organic material. Wet ashing with concentrated acids has been mentioned as one way of doing this. The other approach is "dry ashing", that is, ignition of the organic material in air or oxygen. The organic components react to form gaseous carbon dioxide and water vapor, leaving the inorganic components behind as solid oxides. Ashing is often done in a crucible or evaporating dish of platinum or fused silica in a muffle furnace. Volatile elements will be lost even at relatively low temperatures; dry ashing cannot be used for the determination of mercury, arsenic, cadmium, and a number of other metals of environmental and biological interest for this reason. Oxygen bomb combustions can be performed in a high-pressure steel vessel very similar to a bomb calorimeter. One gram or less of organic material is ignited electrically in a pure oxygen atmosphere with a small amount of absorbing solution such as water or dilute acid. The organic components form carbon dioxide and water and the elements of interest dissolve in the absorbing solution. Combustion in oxygen at atmospheric pressure can be done in a glass apparatus called a *Schöniger flask*. The limitation to this technique is sample size; no more than 10 mg sample can be burned. However, the technique is used to obtain aqueous solutions of sulfur, phosphorus, and the halogens from organic compounds containing these heteroatoms. These elements can then be determined by ion selective potentiometry, ion chromatography, or other methods.

1.4.4. Extraction

The sample preparation techniques earlier discussed are used for inorganic samples or for the determination of inorganic components in organic materials by removing the organic matrix. Obviously, they cannot be used if we want to determine organic analytes. The most common approach for organic analytes is to extract the analytes out of the sample matrix using a suitable solvent. Solvents are chosen with the polarity of the analyte in mind, since "like dissolves like". That is, polar solvents dissolve polar compounds, while nonpolar

Table 1.12   Molten Salt Fusions of Materials

Material[a]	Dissolution procedure
Bauxite	2 g Na_2CO_3; Pt c&l
Corundum	3 g Na_2CO_3 + 1 g H_3BO_3; Pt c&l
Iron Ores	5 g Na_2O_2 + 5 g Na_2CO_3; Zr c&l
Niobium alloys	10 g $K_2S_2O_7$; fused silica crucible
Silicate minerals	10 g 1:1 Na_2CO_3:$Na_2B_4O_7$; Pt c&l
Tin ores	10 g Na_2O_2 + 5 g NaOH; Zr c&l
Titanium ores	7 g NaOH + 3 g Na_2O_2; Zr c&l
Tungsten ores	8 g 1:1 Na_2CO_3/K_2CO_3; Pt c&l

Note: C&l = crucible and lid.
Source: Extracted, with permission, from Dulski. Copyright ASTM International, 100 Barr Harbor Drive, West Conshohocken, PA 19428.
[a]A 1 g test portion is used.

solvents dissolve nonpolar compounds. Common extraction solvents include hexane, methylene chloride, methyisobutyl ketone (MIBK), and xylene.

1.4.4.1. *Solvent Extraction*

Solvent extraction is based on preferential solubility of an analyte in one of two immiscible phases. There are two common situations that are encountered in analysis: extraction of an organic analyte from a solid phase, such as soil, into an organic solvent for subsequent analysis, and extraction of an analyte from one liquid phase into a second immiscible liquid phase, such as extraction of polychlorinated biphenyls from water into an organic solvent for subsequent analysis.

The liquid–liquid extraction is similar to what happens when you shake oil and vinegar together for salad dressing. If you pour the oil and vinegar into a bottle carefully, you will have two separate clear layers, because the oil and vinegar are not soluble in each other (they are *immiscible*). You shake the bottle of oil and vinegar vigorously and the "liquid" gets cloudy and you no longer see the two separate phases; on standing, the two immiscible phases separate into clear liquids again. Our two immiscible solvents will be called solvent 1 and solvent 2. The analyte, which is a solute in one of the phases, say solvent 2, will distribute itself between the two phases on vigorous shaking. After allowing the phases to separate, the ratio of the concentration of analyte in the two phases is approximately a constant, K_D:

$$K_D = \frac{[A]_1}{[A]_2} \tag{1.16}$$

K_D is called the distribution coefficient and the concentrations of A, the analyte, are shown in solvent 1 and solvent 2. A large value of K_D means that the analyte will be more soluble in solvent 1 than in solvent 2. If the distribution coefficient is large enough, most of the analyte can be extracted quantitatively out of solvent 2 into solvent 1. The liquid containing the analyte and the extracting solvent are usually placed into a separatory funnel, shaken manually, and the desired liquid phase drawn off into a separate container. The advantages of solvent extraction are to remove the analyte from a more complex matrix, to extract the analyte into a solvent more compatible with the analytical instrument to be used, and to preconcentrate the analyte. For example, organic analytes can be extracted from water using solvents such as hexane. Typically, 1 L of water is extracted with 10–50 mL of hexane. Not only is the analyte extracted, but it is also now more concentrated in the hexane than it was in the water.

The analyte % extracted from solvent 2 into solvent 1 can be expressed as:

$$\%E = \frac{[A]_1 V_1}{[A]_2 V_2 + [A]_1 V_1} \times 100 \tag{1.17}$$

where $\%E$ is the percent of analyte extracted into solvent 1, the concentration of analyte in each solvent is expressed in molarity; V_1 and V_2 are the volumes of solvents 1 and 2, respectively. The percent extracted is also related to K_D:

$$\%E = \frac{100 K_D}{K_D + (V_2/V_1)} \tag{1.18}$$

The percent extracted can be increased by increasing the volume of solvent 1, but it is more common to use a relatively small volume of extracting solvent and repeat the extraction more than once. The multiple volumes of solvent 1 are combined for analysis. Multiple small extractions are more efficient than one large extraction.

Liquid–liquid extraction is used extensively in environmental analysis to extract and concentrate organic compounds from aqueous samples. Examples include the extraction of pesticides, PCBs, and petroleum hydrocarbons from water samples. Extraction is also used in the determination of fat in milk. Liquid–liquid extraction can be used to separate organometallic complexes from the matrix in clinical chemistry samples such as urine. For example, heavy metals in urine can be extracted as organometallic complexes for determination of the metals by flame AAS. The chelating agent and a solvent such as MIBK are added to a pH-adjusted urine sample in a separatory flask. After shaking and being allowed to stand, the organic solvent layer now contains the heavy metals, which have been separated from the salts, proteins, and other components of the urine matrix. In addition to now having a "clean" sample, the metals have been extracted into a smaller volume of solvent, increasing the sensitivity of the analysis. An added benefit is that the use of the organic solvent itself further increases the sensitivity of flame AAS measurement (as discussed in Chapter 6).

Extraction of organic analytes such as pesticides, PCBs, and fats from solid samples such as food, soil, plants, and similar materials can be done using a Soxhlet extractor. A Soxhlet extractor consists of a round bottom flask fitted with a glass sample/siphon chamber in the neck of the flask. On top of the sample chamber is a standard water-cooled condenser. The solid sample is placed in a cellulose or fiberglass porous holder, called a thimble; the solvent is placed in the round bottom flask. Using a heating mantle around the flask, the solvent is vaporized, condensed, and drips or washes back down over the sample. Soluble analytes are extracted and then siphoned back into the round bottom flask. This is a continuous extraction process as long as heat is applied. The extracted analyte concentrates in the round bottom flask.

As you can imagine, performing these extractions manually is time consuming and can be hard work (try shaking a 1 L separatory funnel full of liquid for 20 min and imagine having to do this all day!). There are several instrumental advances in solvent extraction that have made extraction a more efficient process. These advances generally use sealed vessels under elevated pressure to improve extraction efficiency and are classified as pressurized fluid (or pressurized solvent) extraction methods. One approach is the Accelerated Solvent Extraction system, ASE®, from Dionex (www.dionex.com). This technique is used for extracting solid and semisolid samples, such as food, with liquid solvents. The technique is shown schematically in Fig. 1.7. ASE uses conventional solvents and mixtures of solvents at elevated temperature and pressure to increase the efficiency of the extraction process. Increased temperature, up to 200°C compared with the 70–80°C normal boiling points of common solvents, accelerates the extraction rate while elevated pressure keeps the solvents liquid at temperatures above their normal boiling points, enabling safe and rapid extractions. Extraction times for many samples can be cut from hours using a conventional approach to minutes, and the amount of solvent used is greatly reduced. Table 1.13 presents a comparison of the use of a commercial ASE system with conventional Soxhlet extraction. Dozens of application examples, ranging from fat in chocolate through environmental and industrial applications can be found at the Dionex website. The US EPA has recognized ASE and other instruments that use pressure and temperature to accelerate extraction of samples for environmental analysis by issuing US EPA Method 3545A (SW-846 series) for Pressurized Fluid Extraction of samples. The method can be found at www.usepa.gov.

A second approach also using high pressure and temperature is that of microwave assisted extraction. The sample is heated with the extraction solvent in a sealed vessel by microwave energy, as was described for microwave digestion. The temperature can be raised to about 150°C with the already described advantages of high temperature and

Load sample into cell.

↓

Fill cell with solvent.

↓

Heat and pressurize cell.

↓

Hold sample at pressure and temperature.

↓

Pump clean solvent into sample cell.

↓

Purge solvent from cell with N_2 gas.

↓

Extract ready for analysis.

Solvent

Pump

N_2

Oven

Extraction Cell

Vent

Collection Vial

Figure 1.7   Accelerated solvent extraction technique. Courtesy of Dionex Corp., (www.dionex.com).

high pressure. One limitation of microwave assisted extraction is that some solvents are "transparent" to microwave radiation and do not heat; pure nonpolar solvents such as the normal alkanes are examples of such transparent solvents. Several instrument companies manufacture microwave extraction systems. Milestone Inc. (www.milestonesci.com) has numerous applications on their website as well as video CDs of microwave extraction, ashing, and digestion available. CEM (www.cem.com) also has applications notes available for microwave extraction. An example of improved performance from a microwave extraction system vs. conventional extraction is shown in Table 1.14.

The third instrumental approach is the use of supercritical fluid extraction (SFE). A **supercritical fluid** is a substance at a temperature and pressure above the critical point for the substance. (You may want to review phase diagrams and the critical point on the phase diagram in your general chemistry text.) Supercritical fluids are more dense and viscous than the gas phase of the substance but not as dense and viscous as the liquid phase. The relatively high density (compared with the gas phase) of a supercritical fluid allows these fluids to dissolve nonvolatile organic molecules. Carbon dioxide, CO_2, has a critical temperature of 31.3°C and a critical pressure of 72.9 atm; this temperature and pressure are readily attainable, making supercritical CO_2 easy to form. Supercritical CO_2 dissolves many organic compounds, so it can replace a variety of common solvents; supercritical

Table 1.13   Comparison of Soxhlet Extraction with Accelerated Solvent Extraction

Extraction method	Average solvent used per sample (mL)	Average extraction time per sample	Average cost per sample (US $)
Manual Soxhlet	200–500	4–48 h	27
Automated Soxhlet	50–100	1–4 h	16
Accelerated solvent extraction	15–40	12–18 min	14

Source: Information presented in this table courtesy of Dionex (www.dionex.com).

Table 1.14 Comparison of Microwave Assisted Extraction with Conventional Solvent Extraction for Herbicides in Soil Samples

Extraction method	Time (min)	Volume of solvent (mL)	% Recovery
Separatory funnel	15	25	42–47
Soxhlet	90	40	51–52
Microwave extraction	10 (90°C)	20	66–78

Source: Data in table courtesy of Milestone Inc. (www.milestonesci.com).

CO_2 is used widely as a solvent for extraction. The advantages of using supercritical CO_2 include its low toxicity, low cost, nonflammability, and ease of disposal. Once the extraction is complete and the pressure returns to atmospheric pressure, the carbon dioxide immediately changes to a gas and escapes from the opened extraction vessel. The pure extracted analytes are left behind. Automated SFE instruments can extract multiple samples at once at temperatures up to 150°C and pressures up to 10,000 psi (psi means pounds per square inch and is not an SI unit; 14.70 psi = 1 atm). SFE instrument descriptions and applications from one manufacturer, Isco, Inc., can be found at their website (www.isco.com). The SFE methods have been developed for extraction of analytes from environmental, agricultural, food and beverage, polymer and pharmaceutical samples, among other matrices.

1.4.4.2. Solid Phase Extraction (SPE)

In **solid phase extraction** (SPE) the "extractant" is not an organic liquid, but a solid phase material. Organic compounds are chemically bonded to a solid substrate such as silica beads or polymer beads. The bonded organic layer interacts with organic analytes in the sample solution and extracts them from the sample solution as it is poured through a bed of the solid extractant. The excess solution is allowed to drain away, and interfering compounds are washed off the extractant bed with a solution that does not remove the target analytes. The extracted organic analytes are then **eluted** from the solid phase extractant by passing a suitable organic solvent through the bed. The interactions that cause the analytes to be extracted are those intermolecular attractive forces you learned about in general chemistry: van der Waals attractions, dipole–dipole interactions, and electrostatic attractions.

The types of organic compounds that can be bonded to a solid substrate vary widely. They can be hydrophobic nonpolar molecules such as C_8 and C_{18} hydrocarbon chains, chains with highly polar functional groups such as cyano ($-C\equiv N$), amine ($-NH_2$), and hydroxyl ($-OH$) groups, and with ionizable groups like sulfonic acid anions ($-SO_3^-$) and carboxylic acid anions ($-CO_2^-$), to extract a wide variety of analytes. The term **sorbent** is used for the solid phase extractant. Commercial SPE cartridges have the sorbent packed into a polymer syringe body or disposable polymer pipet tip. Figure 1.8(a) shows commercial plastic syringe-type cartridges with a variety of sorbents, including several of those just mentioned, while Fig. 1.8(b) shows a schematic of how the sorbent is packed and held in the cartridge. These are used only once, preventing cross-contamination of samples and allowing the cleanup of extremely small sample volumes (down to 1 μL), such as those encountered in clinical chemistry samples. Specialized sorbents have been developed for the preparation of urine, blood, and plasma samples for drugs of abuse, for example. Automated SPE systems that can process hundreds of samples simultaneously are now in use in the pharmaceutical and biotechnology industries.

screen
glass fiber filter
sorbent
glass fiber filter
screen

(a)   (b)

Figure 1.8   (a) Commercial plastic syringe-type cartridges and (b) schematic of sorbent packing. Courtesy of J.T. Baker, a division of Mallinckrodt Baker, Inc., (www.jtbaker.com).

SPE is used widely for the cleanup and concentration of analytes for analysis using LC, HPLC, and LC-MS, discussed in Chapter 13. As you will see, the phases used in HPLC for the separation of compounds are in many cases identical to the bonded solid phase extractants described here. Detailed examples and applications notes are available from a number of SPE equipment suppliers: J.T. Baker (www.jtbaker.com), Supelco (www.sigma-aldrich.com/supelco), and Phenomenex (www.phenomenex.com) are a few of the companies that supply these products.

1.4.4.3.   Solid Phase Microextraction (SPME)

Solid phase microextraction (SPME, pronounced "spee-mee" by some users) is a sampling technique developed first for analysis by GC; the use of SPME for GC and related applications is discussed in greater detail in Chapter 12, Section 12.3. The solid phase in this case is a coated fiber of fused silica. The coatings used may be liquid polymers like poly(dimethylsiloxane) (PDMS), which is a silicone polymer. Solid sorbents or combinations of both solid and liquid polymers are also used. Figure 1.9(a) shows a commercial SPME unit with the coated fiber inserted into a sample vial: the coated fiber tip is shown in Fig. 1.9(b). No extracting solvent is used when the sample is analyzed by GC. The coated fiber is exposed to a liquid or gas sample or to the vapor over a liquid or solid sample in a sealed vial (this is called sampling the **headspace**) for a period of time. Analyte is adsorbed by the solid coating or absorbed by the liquid coating on the fiber and then thermally desorbed by insertion into the heated injection port of the gas chromatograph. The process is shown schematically in Fig. 1.10.

Unlike solvent extraction, the entire amount of analyte is not extracted. The amount of analyte extracted by the coated fiber is proportional to the concentration of analyte in the sample. This will be true if equilibrium between the fiber and the sample is achieved or before equilibrium is achieved if the sampling conditions are controlled carefully. SPME sampling and desorption can be used for qualitative and quantitative analyses.

(a)   (b)

Figure 1.9 (a) A commercial SPME unit with the coated fiber (b) inserted into a sample vial. Reprinted with permission of Supelco, Bellefonte, PA 16823, USA (www.sigma-aldrich.com).

Quantitative analysis using external calibration, internal standard calibration, and the method of standard additions are all possible with SPME. Calibration is discussed in Section 1.5.2 and at greater length in Chapter 2.

SPME sampling is used for a wide variety of analytes, including environmental pollutants, volatiles from botanical samples (e.g., used to identify tobacco species),

Figure 1.10 Schematic of the SPME process. Reprinted with permission of Supelco, Bellefonte, PA 16823, USA (www.sigma-aldrich.com).

explosives, and chemical agent residues. Gasoline and other accelerants in the headspace over fire debris can be sampled with SPME to determine whether arson may have caused the fire. As little as 0.01 μL of gasoline can be detected. Gas samples such as indoor air and breath have been sampled using SPME. Liquid samples analyzed by either immersion of the fiber into the sample or sampling of the headspace vapor include water, wine, fruit juice, blood, milk, coffee, urine, and saliva. Headspace samplings of the vapors from solids include cheese, plants, fruits, polymers, pharmaceuticals, and biological tissue. These examples and many other applications examples are available in pdf format and on CD from Supelco at www.sigma-aldrich.com/supelco. SPME probes that are the size of a ballpoint pen (Fig. 1.11) are available for field sampling (e.g., see www.fieldforensics.com or www.sigma-aldrich.com/supelco). These can be capped and taken to an on-site mobile lab or transported back to a conventional laboratory for analysis.

While SPME started as a solvent-free extraction system for GC analysis, it can now also be used to introduce samples into an HPLC apparatus. A new SPME–HPLC interface, Fig. 1.12, allows the use of an SPME fiber to sample nonvolatile analytes such as nonionic surfactants in water, and elute the analyte into the solvent mobile phase used for the HPLC analysis. The sampling process and elution are shown schematically in Fig. 1.10. HPLC and its applications are covered in Chapter 13.

1.5. PERFORMING THE MEASUREMENT

To determine an analyte using an instrumental method of analysis, we must establish the relationship between the magnitude of the physical parameter being measured and

Figure 1.11 An SPME probe the size of a ballpoint pen. Reprinted with permission of Supelco, Bellefonte, PA 16823, USA (www.sigma-aldrich.com).

Figure 1.12 A new SPME–HPLC interface. Reprinted with permission of Supelco, Bellefonte, PA 16823, USA (www.sigma-aldrich.com).

the amount of analyte present in the sample undergoing the measurement. In most analyses, the measurement is made and then a calculation is performed to convert the result of the measurement into the amount of analyte present in the original sample. The calculation accounts for the amount of sample taken and dilutions required in the process of sample preparation and measurement.

In an instrumental method of analysis, a **detector** is a device that records a change in the system that is related to the magnitude of the physical parameter being measured. We say that the detector records a **signal**. If the detector is properly designed and operated, the signal from the detector can be related to the amount of analyte present in the sample through a process called **calibration**. Before we discuss calibration, we need to understand a little about what the detector is recording. A detector can measure physical, chemical, or electrical changes or signals, depending on its design. A **transducer** is a detector that converts nonelectrical signals to electrical signals (and vice versa). There are transducers used in spectroscopic instruments that convert photons of light into an electrical current, for example. Another term used in place of detector or transducer is **sensor**. The operation of specific detectors is covered in the chapters on the different instrumental methods (Chapters 3–16).

1.5.1. Signals and Noise

Instrumental analysis uses electronic equipment to provide chemical information about the sample. Older instruments used vacuum tubes and output devices like strip chart recorders, while modern instruments use semiconductor technology and computers to control the instrument, collect signals, process and report data. Fundamentals of modern instrument electronics are covered in the text by Malmstadt et al. listed in the bibliography.

All instruments measure some chemical or physical characteristic of the sample, such as how much light is absorbed by the sample at a given wavelength, the mass-to-charge ratio of an ion produced from the sample, or the change in conductivity of a wire as the sample passes over it. A detector of some type makes the measurement and the detector response is converted to an electrical signal. The electrical signal should be directly related to the chemical or physical property being measured and that should be related to the amount of analyte present. Ideally, the signal would represent only information related to the analyte. When no analyte is present, there should be no signal. For example, in Fig. 1.13 we are looking at signals from a spectrometer that is measuring the amount of light emitted by a sample at a given wavelength. The three traces in Fig. 1.13 show a **peak**, which is the signal at the emission wavelength. The response on either side of the peak is called the **baseline**.

An ideal signal for intensity of light emitted by the analyte vs. wavelength would be a smooth baseline when no light is emitted and a smooth peak at the emission wavelength, as shown in Fig. 1.13(a). In this case, when the instrument does not detect the analyte, there is no signal, represented by the flat baseline. When the instrument does detect the analyte, the signal increases. When the instrument no longer detects the analyte, the signal decreases back to the baseline. In this case the entire signal (the peak) is attributed to the analyte. In practice, however, the recorded signal and baseline are seldom smooth, but include random signals called **noise**. All measured signals contain the desired information about the analyte and undesirable information, including noise. Noise can originate from small fluctuations of various types in the instrumentation, in the power provided to the instrument, from external sources such as TV and radio stations, other instruments nearby, building vibrations, electrical motors and similar sources, and even from fundamental quantum effects that cannot be eliminated. Provided that the signal is sufficiently greater than the noise, it is not difficult to make a reliable measurement of the desired analyte signal. The **signal-to-noise ratio**, S/N, is a useful quantity for comparing analytical methods or instruments.

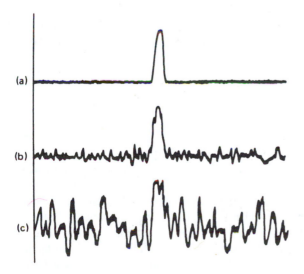

Figure 1.13 Signal vs. wavelength with different noise levels: (a) no detectable noise, (b) moderate noise, and (c) high noise.

In Fig. 1.13(b) the noise level is increased over that in Fig. 1.13(a) and is superimposed on the signal. The value of the signal is less certain as a result of the noise. The measurement is less precise, but the signal is clearly discernible. In Fig. 1.13(c) the noise level is as great as the signal, and it is virtually impossible to make a meaningful measurement of the latter. It is very important to be able to separate data-containing signals from noise. When the signal is very weak, as it might be for trace amounts of analyte, the noise must be reduced or the signal enhanced. Noise is random in nature; it can be either positive or negative at any given point, as can be seen in Fig. 1.13(b) and (c). Because noise is random, it can be treated statistically.

If we consider the signal S to be our measured value, the noise N is the variation in the measured value when repeat measurements of the same sample are made. That is, the noise can be defined as the standard deviation s of repeat measurements; the signal S is then the average value of the measurement, $\bar{x}$. While making the repeat measurements at the peak maximum would provide the best estimate of the signal-to-noise ratio at the exact point we want to measure, there are some difficulties associated with this approach. One is that the noise measured at the peak maximum may be hard to detect, as a small variation in a large signal is more difficult to measure than a large variation in a small signal. A second problem is that the signal-to-noise ratio will be dependent on the size of the signal if measured at the peak maximum. In practice, the noise is often measured along the baseline where the signal should be zero, not at the peak maximum. An easy way to measure the noise in Fig. 1.13(b) is to use a ruler to measure the maximum amplitude of the noise somewhere away from the signal along the baseline. The magnitude of the noise is independent of the magnitude of the signal along the baseline. The signal magnitude is measured from the middle of the baseline to the middle of the "noisy" peak. The middle of the baseline has to be estimated by the analyst. The effect of noise on the relative error of the measurement decreases as the signal increases.

$$\frac{S}{N} = \frac{\bar{x}}{s} = \frac{\text{mean}}{\text{std. dev.}} \tag{1.19}$$

There are several types of noise encountered in instrumental measurements. The first is **white noise**, the random noise seen in Fig. 1.13(b) and (c). White noise can be due to the random motions of charge carriers such as electrons; the random motion results in voltage fluctuations. This type of white noise is called *thermal noise*. Cooling the detector and other components in an instrument can reduce thermal noise. A second type of white noise is *shot noise*, which occurs when charge carriers cross a junction in an electric circuit. **Drift** or **flicker noise** is the second major type of instrumental noise. It is inversely proportional to the frequency of the signal being measured and is most significant for low frequencies. The origin of drift or flicker noise is not well understood. The third type of instrumental noise is that due to the surroundings of the instrument, such as the line noise due to the power lines to the instrument or building vibrations. Some of this type of noise is frequency dependent and may occur at discrete frequencies.

Improvement in S/N requires that the signal be different from the noise in some way. Most differences can be expressed in terms of time correlation or frequency. To increase S/N, either the noise must be reduced or the signal enhanced, or both must occur. There are a variety of hardware and software approaches to reduce noise in instruments. External sources of noise can be eliminated or reduced by proper grounding and shielding of instrument circuits and placement of instruments away from drafts, other instruments, and sources of vibration. The intrinsic noise can be reduced using a variety of electronic hardware such as lock-in amplifiers, filters, and signal modulators. Signals can be enhanced by a variety of computer software programs to perform signal averaging, Fourier transformation, filtering, and smoothing. Many of these software programs can be (and are) applied after the data has been collected. Many of the hardware methods for reducing noise are now being replaced by computer software methods, since even simple instruments now put out data in digital form. The discussion of analog to digital conversion and details of methods for S/N enhancement are beyond the scope of the text; the references by Enke, Malmstadt et al. and Skoog et al. can be consulted for details.

Signal averaging is one way to improve S/N; repetitive measurements are made and averaged. In this instance advantage is taken of the fact that noise is random but the signal is additive. If a signal, such as that shown in Fig. 1.13(b), is measured twice and the results added together, the signal will be twice as intense as the first measurement of the signal. If n measurements are made and added together, the signal will be n times as intense as the first signal. Because noise is random, it may be positive or negative at any point. If n measurements are added together, the noise increases only as the square root of n, or $n^{1/2}$. Since S increases by a factor of n, and N increases by $n^{1/2}$, S/N increases by $n/n^{1/2} = n^{1/2}$. Averaging multiple signal measurements will improve the signal-to-noise ratio by a factor of $n^{1/2}$ as shown in Fig. 1.14. Therefore to improve the S/N ratio by 10, about 100 measurements must be made and averaged. The disadvantage to signal averaging is the time required to make many measurements. Some instrumental methods lend themselves to rapid scanning, but others such as chromatography do not.

Instruments that use Fourier transform (FT) spectroscopy, introduced in Chapter 2, collect the entire signal, an interferogram, in a second or two. Hundreds of measurements can be made, stored by a computer and averaged by an FT instrument very quickly, greatly improving the signal-to-noise ratio using this approach. The FT approach discriminates signal from noise based on frequency. An FT is a mathematical transformation that converts a variable measured as a function of time, $f(t)$, into a function of reciprocal time, $f(1/t)$. A function of reciprocal time is also a function of frequency. The FT permits the removal of noise that differs in frequency from the signal and also permits enhancement of frequencies associated with the signal. FT and a related algorithm, the fast Fourier transform (FFT) are now available as part of many data handling software

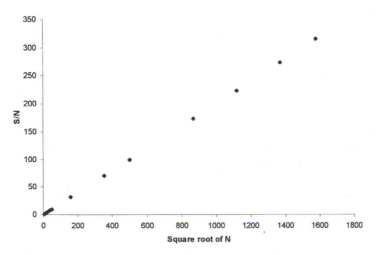

Figure 1.14 Improvement in the signal-to-noise ratio by averaging multiple signal measurements.

packages. (FFT is an algorithm for efficiently computing FT.) The use of FT is required to process data from experiments based on interferometry, such as FT-IR spectroscopy (Chapter 4), from pulsed NMR experiments (FT-NMR, Chapter 3), and the signal pulses from cyclotron resonance MS experiments (FT-MS, Chapter 9).

The signal-to-noise ratio is a limiting factor in making analytical measurements. It is theoretically possible to measure any signal in any noise level, provided the measurements are taken over a long enough period of time or repeated enough times to accumulate the signal. In practice, there is usually a limitation on the number of measurements that can be made, owing to factors such as time, cost, amount of sample, and the inherent limitations of the instrument. If many measurements are made over too long a period of time, other sources of error may occur in the measurement and these types of errors may not be random.

1.5.2. Plotting Calibration Curves

Calibration is the process of establishing the relationship between the instrument signal we measure and known concentrations of the analyte. Samples with known concentrations or amounts of analyte used to establish this signal–concentration relationship are called "calibration standards". Calibration approaches are covered in detail in Chapter 2 but it suffices to say that usually a series of known calibration standards is analyzed and a signal is measured for each. Calibration standards with a range of concentrations are prepared and measured. The magnitude of the signal for each standard is plotted against the concentration of the analyte in the standard and the equation that relates the signal and the concentration is determined. Once this relationship, called a **calibration curve**, is established, it is possible to calculate the concentration of the analyte in an unknown sample by measuring its signal and using the equation that has been determined. The calibration curve is a plot of the sets of data points (x_i, y_i) where x is the concentration or amount of analyte in the standards and y is the signal for each standard. Many calibration curves are in fact straight-line relationships or have a linear region, where the concentration of analyte is related to the signal according to the equation $y = mx + b$. The slope of the line is m and the intercept of the line with the y-axis is b. Properly plotting the calibration curve is critical to obtaining accurate results. We have already learned that all measurements have uncertainty associated with them. The calibration curve also has uncertainty associated with it. While using graph paper and a ruler and trying to draw the best straight line through the data points manually can be used to plot calibration curves, it is much more accurate to use the statistical programs on a calculator, in a spreadsheet, or in an instrument's computer data system to plot calibration curves and determine the equation that best fits the data. The use of graphing calculators, computerized data handling, and spreadsheets also permits the fitting of nonlinear responses to higher order equations accurately.

We will assume that the errors in the measured signal, y, are significantly greater than any errors in the known concentration, x. In this case, the best-fit straight line is the one that minimizes vertical deviations from the line, that is, the line for which the sum of the squares of the deviations of the points from the line is the minimum. This fitting technique is called the **method of least squares**. For a given x_i, y_i, and the line $y = mx + b$, the vertical deviation of y_i from $y = (y_i - y)$. Each point has a similar deviation, so the sum of the squares of the deviations is:

$$D = \sum (y_i - y)^2 = \sum [y_i - (mx_i + b)]^2 \tag{1.20}$$

To obtain the slope m and the intercept b of the best-fit line requires the use of calculus and the details will not be covered in this text. The results obtained are:

$$m = \frac{n \sum x_i y_i - \sum x_i \sum y_i}{n \sum x_i^2 - \left(\sum x_i\right)^2} \tag{1.21}$$

$$b = \bar{y} - m\bar{x} \tag{1.22}$$

Expressions for the uncertainty in the measured value, y, in the slope, and in the intercept are similar to the expressions for the standard deviation [Eq. (1.9)]. Two degrees of freedom are lost in the expression for the uncertainty in y, because both the slope and intercept have been defined.

$$s_y = \sqrt{\frac{\sum_{i=1}^{N} d^2}{N - 2}} = \sqrt{\frac{\sum_{i=1}^{N} (y_i - (mx + b))^2}{N - 2}} \tag{1.23}$$

$$s_m = \sqrt{\frac{s_y^2}{\sum (\bar{x} - x_i)^2}} \tag{1.24}$$

$$s_b = s_y \sqrt{\frac{1}{N - (\sum x_i)^2 / \sum x_i^2}} \tag{1.25}$$

Equation (1.23) defines the uncertainty in y, s_y; Eq. (1.24) gives the uncertainty in the slope, s_m, and Eq. (1.25) gives the uncertainty in the intercept, s_b. The uncertainty in x_i is calculated using all the associated variances, as has been discussed in Section 1.3.5. The use of a spreadsheet program such as Microsoft Excel® to plot a calibration curve eliminates the need to do manual calculations for least squares and the associated statistics. The Excel program permits the calculation of the best-fit equation using linear least squares regression (as well as many nonlinear curve fitting routines). The program also calculates the **correlation coefficient**, r, for the line. A correlation coefficient with a value of 1 means that there is a direct relationship between x and y; the fit of the data to a straight line is perfect. A correlation coefficient of 0 means that x and y are completely independent. The range of the correlation coefficient is from 1 to -1. Most linear calibration curves should have a correlation coefficient of 0.99 or greater. Statistics programs usually calculate the square of the correlation coefficient, r^2, which is a more realistic measure of the goodness-of-fit of the data. It is always a good idea to plot the data graphically so that it can be looked at, to ensure that the statistical calculations are not misleading about the goodness-of-fit.

Modern computerized analytical instruments have quantitative analysis programs that allow the analyst to specify the calibration standard concentrations, select the curve-fitting mode, and calculate the results of the samples from the calibration curve equation. Many of these programs will rerun outlier standards and samples automatically, flag suspect data, compute precision and recovery of spikes, track reference standards for quality control, and perform many other functions that used to be done manually by the analyst.

1.6. ASSESSING THE DATA

A good analytical method should be both precise and accurate; that is, it should be reliable or robust. A robust analytical method is one that gives precise and accurate results even if small changes are made in the method parameters. The robustness of a method is assessed by varying the parameters in the analysis such as temperature, pH, reaction time, and so on, and observing the effect of these changes on the results obtained. The **specificity** of a method refers to the ability of the method to determine the analyte accurately in the presence of interferences. Method development should include checking the response of the method to other chemicals known to be in the sample, to possible degradation products, and to closely related compounds or elements. Ideally, the analytical method would be specific for only the analyte of interest. It is possible that analytical methods published in the literature may appear to be valid by a compensation of errors; that is, although the results appear accurate, the method may have involved errors that balanced each other out. When the method is used in another laboratory, the errors may differ and not compensate for each other. A net error in the procedure may result. This is an example of a method that is not reliable or robust. It is always prudent to run known reference samples when employing a method from the literature to evaluate its reliability.

The **sensitivity** of an analytical method can be defined as the slope of the calibration curve, that is, as the ratio of change in the instrument response with a change in the analyte concentration. Other definitions are also used. In AAS, sensitivity is defined as the concentration of analyte that produces an absorbance of 0.0044 (an absorption of 1%), for example. When the term sensitivity is used, it should be defined.

Once the relationship between the signal and analyte concentration (i.e., the calibration curve) has been established, the linear working range of the method can be determined. The **range** is that interval between (and including) the lowest and highest analyte concentrations that have been demonstrated to be determined with the accuracy, precision, and linearity of the method. Linear working ranges vary greatly among instrumental methods, and may depend on the instrument design and the detector used, among other factors. Some instruments, such as a gas chromatograph with an electron capture detector or an atomic absorption spectrometer, have short linear ranges of one to two orders of magnitude. Other instruments, like ICP atomic emission spectrometers, may have linear ranges of five orders of magnitude, while mass spectrometers may be linear over nine orders of magnitude. All results should fall within the linear range of the method. This may require dilution of samples with analyte concentrations that are higher than the highest calibration standard in order to bring the signal into the known linear response region. Extrapolating beyond the highest standard analyzed is very dangerous because many signal–concentration relationships become nonlinear at high levels of analyte. Extrapolating below the lowest standard analyzed is also very dangerous because of the risk of falling below the limit of quantitation. If samples fall below the lowest calibration standard, they may need to be concentrated to bring them into the working range.

1.6.1. Limit of Detection

All measurements have noise and the magnitude of the noise limits the amount of analyte that can be detected and measured. As the concentration of analyte decreases, the signal decreases. At some point, the signal can no longer be distinguished from the noise, so the analyte can no longer be "detected". Because of noise, it is not possible to say that there is *no* analyte present; it is only possible to establish a **detection limit**. Detection is the ability to discern a weak signal in the presence of background noise, so reducing the noise will permit the detection of smaller concentrations of analyte. The **limit of detection** (LOD) is the lowest concentration of analyte in a sample that can be detected. Detected does not mean that this concentration can be measured quantitatively; it only specifies whether an analyte is above or below a certain value, the LOD. One common approach to establishing the LOD is to measure a suitable number of replicates (8–10 replicates is common) of an appropriate blank or low concentration standard and determine the standard deviation of the blank signal. The blank measures only the background or baseline noise. The LOD is then considered to be the concentration of analyte that gives a signal that is equal to two or three times the standard deviation of the blank. This is equivalent to defining the LOD as that concentration at which the S/N ratio $= 2$ at 95% CL or 3 at 99% CL.

$$\text{LOD} = \bar{x}_{\text{blank}} \pm 2\sigma_{\text{blank}} \quad \text{or} \quad = \bar{x}_{\text{blank}} \pm 3\sigma_{\text{blank}} \tag{1.26}$$

The use of 2σ results in an LOD with a 95% CL, that is, there is a 5% risk of a false positive or false negative. The use of 3σ is now more common and often specified by regulatory methods such as those of the US EPA; it results in an LOD with a 99% CL.

There are other approaches used for calculating LODs; it is important to specify exactly how an LOD has been determined in an analytical method. The calculated LOD should be validated by analyzing standards whose concentrations are below, at, and above the LOD. Any "results" from samples that fall below the established detection limit for the method are reported as "not detected" or as "<LOD". They should not be reported as numerical values except as "<the numerical value of the LOD"; for example, <0.5 ppb if the LOD is 0.5 ppb for this analysis.

1.6.2. Limit of Quantitation

The precision of an analysis at or near the detection limit is usually poor compared with the precision at higher concentrations. This makes the uncertainty in the detection limit and in concentrations slightly above the detection limit also high. For this reason, many regulatory agencies define another limit, the **limit of quantitation** (LOQ), which is higher than the LOD and should have better precision.

The LOQ is the lowest concentration of analyte in a sample that can be determined quantitatively with a given accuracy and precision using the stated method. The LOQ is usually defined as that concentration equivalent to a signal-to-noise ratio of 10/1. The LOQ can also be determined from the standard deviation of the blank; the LOQ is 10× the standard deviation of the blank, expressed in concentration units. The LOQ is stated with the appropriate accuracy and precision and should be validated by running standards at concentrations that can confirm the ability of the method to determine analyte with the required accuracy and precision at the LOQ.

Analytical results that fall between the LOD and the LOQ should be reported as "detected but not quantifiable". These results are only estimates of the amount of analyte present, since by definition, they cannot be determined quantitatively.

Note: In this chapter and in the following chapters, the websites for many instrument manufacturers, government agencies, and so on are given. While there are many useful technical notes, applications examples, tutorials, and some very useful photographs and videos on commercial websites, it should be understood by the student that these publications and tutorials are in most cases not peer-reviewed. The information should be treated accordingly. In addition, the student will find that there are many on-line forums (message boards, chat rooms) dedicated to analytical chemistry and to specific techniques in analytical chemistry. While these can be valuable, not all of the information provided on these types of sites is accurate and students are encouraged to use the peer-reviewed literature as the first source of answers to your questions.

BIBLIOGRAPHY

American Society for Testing and Materials *2003 Annual Book of ASTM Standards*; ASTM International: West Conshohocken, PA, 2003.

APHA, AWWA, WPCF, *Standard Methods for the Examination of Water and Wastewater*, 18th ed.; Greenberg, A.E., Clesceri, L.S., and Eaton, A.D., Eds.; American Public Health Association: Washington, D.C., 1992.

Christian, G.D. *Analytical Chemistry*, 6th ed.; John Wiley and Sons, Inc.: Hoboken, NJ, 2004.

CRC Standard Math Tables; CRC Press: Boca Raton, FL (any edition).

Diamond, D.; Hanratty, V. *Spreadsheet Applications in Chemistry using Microsoft Excel*; John Wiley and Sons, Inc.: New York, 1997.

Dulski, T.R. *A Manual for the Chemical Analysis of Metals*; ASTM International: West Conshohocken, PA, 1996.

Enke, C.G. *The Art and Science of Chemical Analysis*; John Wiley and Sons, Inc.: New York, 2001.

Erickson, M.D. *Analytical Chemistry of PCBs*, 2nd ed.; CRC Press: Boca Raton, FL, 1997.

Harris, D.C. *Quantitative Chemical Analysis*, 5th ed.; W.H. Freeman and Company: New York, 1999.

Horwitz, W., Ed. *Official Methods of Analysis of the Association of Official Analytical Chemists*, 13th ed.; Association of Official Analytical Chemists: Washington, D.C., 1980.

Keith, L.H., Ed. *Principles of Environmental Sampling*; American Chemical Society: Washington, D.C., 1988.

Malmstadt, H.V.; Enke, C.G.; Crouch, S.R. *Microcomputers and Electronic Instrumentation: Making the Right Connections*; American Chemical Society: Washington, D.C., 1994.

Mark, H.; Workman, J. *Statistics in Spectroscopy*; Academic Press, Inc.: San Diego, CA, 1991.

Rorabacher, D.B. *Anal. Chem.* **1991**, *63*, 139.

Simpson, N.J.K., Ed. *Solid-Phase Extraction: Principles, Techniques and Applications*; Marcel Dekker, Inc.: New York, 2000.

Skoog, D.A.; Holler, J.A.; Nieman, T.A. *Principles of Instrumental Analysis*, 5th ed.; Saunders College Publishing; Harcourt, Brace and Company: Orlando, FL, 1998.

Tyson, J. *Analysis. What Analytical Chemists Do*; Royal Society of Chemistry: London, 1988.

US Environmental Protection Agency. *Methods for the Chemical Analysis of Water and Wastes*, EPA-600/4-79-020, Environmental Monitoring and Support Laboratory: Cincinnati, OH, March 1983.

US Environmental Protection Agency. *Test Methods for Evaluating Solid Waste-Physical/Chemical Methods*, 3rd ed.; SW-846; Office of Solid Waste and Emergency Response: Washington, D.C., 1986. (most recent version available at www.epa.gov).

Wercinski, S.A.S., Ed. *Solid Phase Microextraction: A Practical Guide*; Marcel Dekker, Inc.: New York, 1999.

PROBLEMS

1.1 (a) Define determinate error and give two examples of determinate errors.

(b) In preparing a sample solution for analysis, the pipet used actually delivered 4.92 mL instead of the 5.00 mL it was supposed to deliver. Would this cause determinate or indeterminate error in the analysis of this sample?

1.2 (a) Define precision.

(b) Do determinate errors affect precision?

1.3 (a) Define accuracy.

(b) How can the accuracy of an analytical procedure be determined?

1.4 (a) What is the statistical definition of sigma (σ)?

(b) What percentage of measurements should fall within $\pm 2\sigma$ of the true value for a data set with no determinate error, assuming a Gaussian distribution of random error?

1.5　Calculate the standard deviation of the following set of measured values:
3.15, 3.21, 3.18, 3.30, 3.25, 3.13, 3.24, 3.41, 3.13, 3.42, 3.19

1.6　The true mass of a glass bead is 0.1026 g. A student takes four measurements of the mass of the bead on an analytical balance and obtains the following results: 0.1021 g, 0.1025 g, 0.1019 g, and 0.1023 g. Calculate the mean, the average deviation, the standard deviation, the percentage relative standard deviation, the absolute error of the mean, and the relative error of the mean.

1.7　How many significant figures are there in each of the following numbers?
3.216, 32.1, 30, 3×10^6, 3.21×10^6, $321{,}000$

1.8　Round off the following additions to the proper number of significant figures:

3.2	1.9632	1.0×10^6
0.135	0.0013	1.321×10^6
3.12	1.0	1.13216×10^6
0.028	0.0234	4.32×10^6
Total　6.483	2.9879	7.77316×10^6

1.9　The following data set represents the results of replicate measurements of lead, expressed as ppm Pb. What are (a) the arithmetic mean, (b) the standard deviation, and (c) the 95% confidence limits of the data?
2.13, 2.51, 2.15, 2.17, 2.09, 2.12, 2.17, 2.09, 2.11, 2.12
(d) Do any of the data points seem "out of line" with the rest of the data? Are any point(s) outside the 4σ "rule of thumb" for rejecting suspect data? Should the suspect data be ignored in the calculation?

1.10　(a)　Explain the importance of good sampling.
　　　(b)　Give three examples of precautions that should be taken in sample storage.

1.11　(a)　Illustrate the difference between precision and accuracy.
　　　(b)　Do indeterminate errors affect precision or accuracy?

1.12　The results in Problem 1.9 were obtained for the lead content of a food sample. The recommended upper limit for lead in this food is 2.5 ppm Pb. (a) Are the results greater than 2.5 ppm Pb with 95% confidence? (b) If the regulatory level is decreased to allow no more than 2.00 ppm Pb, is the Pb content of the food greater than 2.00 ppm with 95% confidence?

1.13　The determination of Cu in human serum is a useful diagnostic test for several medical conditions. One such condition is Wilson's disease, in which the serum Cu concentration is lowered from normal levels and urine Cu concentration is elevated. The result of a single copper determination on a patient's serum was 0.58 ppm. The standard deviation σ for the method is 0.09 ppm. If the serum copper level is less than 0.70 ppm Cu, treatment should be started. Based on this one result, should the doctor begin treatment of the patient for low serum copper? Support your answer statistically. If the doctor were unsure of the significance of the analytical result, how would the doctor obtain further information?

1.14　Add the following concentrations and express the answer with the correct units and number of significant figures:
3.25×10^{-2} M, 5.01×10^{-4} M, 8×10^{-6} M

1.15 With what confidence can an analytical chemist report data using σ as the degree of uncertainty?

1.16 The mean of eight replicate blood glucose determinations is 74.4 mg glucose/100 mL blood. The sample standard deviation is 1.8 mg glucose/100 mL blood. Calculate the 95% and the 99% confidence limits for the glucose concentration.

1.17 An analysis was reported as 10.0 with $\sigma = 0.1$. What is the probability of a result occurring within (a) ± 0.3 or (b) ± 0.2 of 10.0?

1.18 Name the types of noise that are frequency dependent. Which types of noise can be reduced by decreasing the temperature of the measurement?

1.19 The following measurements were obtained on a noisy instrument:
1.22, 1.43, 1.57, 1.11, 1.89, 1.02, 1.53, 1.74, 1.83, 1.62
(a) What is the signal-to-noise ratio, assuming that the noise is random? (b) How many measurements must be averaged to increase the S/N to 100?

1.20 Estimate the S/N ratio in Fig. 1.13(b) and (c). What is the improvement in S/N from (c) to (b)? If the S/N value in (c) is the result of one measurement, how many measurements must be made and averaged to achieve the S/N value in (b)?

1.21 The tungsten content of a reference ore sample was measured both by X-ray fluorescence (XRF) spectrometry, the standard method, and by inductively coupled plasma-atomic emission spectrometry (ICP). The results as weight percent tungsten are given in the following table. Are the results of the two methods significantly different at the 95% confidence level? Is there any bias in the ICP method? (*Hint*: The standard method can be considered to have a mean $= \mu$.)

Replicate number	XRF	ICP
1	3.07	2.92
2	2.98	2.94
3	2.99	3.02
4	3.05	3.00
5	3.01	2.99
6	3.01	2.97

1.22 Assuming that the results of many analyses of the same sample present a Gaussian distribution, what part of the curve defines the standard deviation σ?

1.23 A liquid sample is stored in a clear glass bottle for the determination of trace metals at the ppm level. What factors can cause the results to be (a) too low or (b) too high?

1.24 Round off the following numbers according to the rules of significant figures: (a) $10.2 \div 3$, (b) $10.0 \div 3.967$, (c) $11.05 \div 4.1 \times 10^{-3}$, (d) $11.059 \div 4.254$, and (e) $11.00 \div 4.03$.

1.25 Nitrate ion in potable water can be determined by measuring the absorbance of the water at 220 nm in a UV/VIS spectrometer. The absorbance is proportional to the concentration of nitrate ion. The method is described in Standard Methods for Analysis of Water and Wastewater listed in the bibliography. Calibrations standards were prepared and their absorbances measured. The results are given in the following table. (*Note*: the absorbance of the blank, the 0.0 mg nitrate/L "standard" was set to 0.000 in this experiment. A nonzero blank value is subtracted from all the standards before the calibration curve is plotted and the equation calculated.)

Nitrate ion (mg/L)	Absorbance
0.00	0.000
1.00	0.042
2.00	0.080
5.00	0.198
7.00	0.281

(a) Make an $x-y$ graph showing the experimental data with absorbance plotted on the y-axis. (b) Determine the equation of the least squares line through the data points, with y as absorbance and x as nitrate ion concentration. (c) Calculate the uncertainty in the slope and intercept. (d) If you have done this using a statistics program or spreadsheet, what is the value of the correlation coefficient for the line?

1.26 (a) Name the instrumental methods that can be used for elemental qualitative analysis.

(b) Name the instrumental methods that are used for elemental quantitative analysis.

1.27 (a) Name the instrumental methods that can be used for molecular organic functional group identification.

(b) What instrumental methods provide molecular structural information, that is, indicate which functional groups are next to each other in an organic molecule?

1.28 What instrumental methods are best for quantitative analysis of (a) complex mixtures, (b) simple mixtures, or (c) pure compounds?

1.29 What instrumental methods can provide measurements of the molecular weight of a molecule?

1.30 What is the purpose of a blank in an analysis? What is the purpose of a reference material or reference standard in an analysis?

1.31 Equation (1.26) shows how to calculate the LOD of a method at both the 95% and 99% confidence levels. You have measured the blank for a determination of arsenic in food samples by hydride-generation atomic fluorescence spectrometry. The blank values are:

0.23 ppb, 0.14 ppb, 0.16 ppb, 0.28 ppb, 0.18 ppb, 0.09 ppb, 0.10 ppb, 0.20 ppb, 0.15 ppb, and 0.21 ppb As

What is the LOD at (a) 95% CL and (b) 99% CL?

1.32 Based on your answer to problem 1.31, what are the respective method LOQs for As at the two confidence levels?

2

Introduction to Spectroscopy

2.1. THE INTERACTION BETWEEN ELECTROMAGNETIC RADIATION AND MATTER

We know from our observation of rainbows that visible *light* (white light) is composed of a **continuum** of colors from violet to red. If a beam of white light is passed through a beaker of water, it remains white. If potassium permanganate is added to the water, the white light appears purple after it passes through the solution. The permanganate solution allows the red and blue components of white light to pass through but absorbs the other colors from the original beam of light. This is one example of the interaction of **electromagnetic radiation**, or light, with *matter*. In this case, the electromagnetic radiation is visible light and we can see the effect of absorption of some of the light with our eyes. However, interactions between electromagnetic radiation and matter take place in many ways and over a wide range of *radiant energies*. Most of these interactions are not visible to the human eye, but can be measured with suitable instruments.

Interaction of electromagnetic radiation and matter is not haphazard, but follows well-documented rules with respect to the wavelengths of light absorbed or emitted and the extent of absorption or emission. The subject of **spectroscopy** is the study of the interaction of electromagnetic radiation and matter.

2.1.1. What is Electromagnetic Radiation?

The nature of electromagnetic radiation baffled scientists for many years. At times light appears to behave like a wave; at other times it behaves as though it were composed of small particles. While we now understand the "wave–particle duality" of all matter, including electromagnetic radiation, in terms of quantum mechanics, it is still convenient to consider electromagnetic radiation as having the properties of waves in many cases.

Light waves can be represented as oscillating perpendicular electric and magnetic fields. The fields are at right angles to each other and to the direction of propagation of the light. The oscillations are sinusoidal in shape, as shown in Fig. 2.1. We can easily and accurately measure the **wavelength** λ defined as the crest-to-crest distance between two successive maxima. The standard unit of wavelength is the SI unit of length, the meter (m), but smaller units such as the centimeter (cm), micrometer (μm), and nanometer (nm) are commonly used. The **amplitude** of the wave is defined as the maximum of the vector from the origin to a point displacement of the oscillation. An example of the electric field portion of a light wave propagating along only one axis is shown in Fig. 2.1. Such a wave, confined to one plane, is called **plane-polarized light**. The wave shown represents only a single

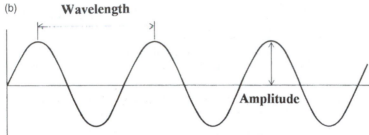

Figure 2.1 (a) A plane-polarized light wave in the x-direction, showing the mutually perpendicular electric and magnetic field components. (b) The wavelength and amplitude of a wave.

wavelength, λ. Light of only one wavelength is called **monochromatic light**. Light that consists of more than one wavelength is called **polychromatic** light. White light is an example of polychromatic light.

The **frequency** ν of a wave is the number of crests passing a fixed point per second. One crest-to-crest oscillation of a wave is called a cycle. The common unit of frequency is the hertz (Hz) or inverse second (s^{-1}); an older term for frequency is the cycle per second (cps). One hertz equals one cycle per second.

The wavelength of light, λ, is related to its frequency, ν by the equation:

$$c = \lambda\nu \tag{2.1}$$

where c is the speed of light in a vacuum, 2.997×10^8 m/s, ν is the frequency of the light in inverse seconds (Hz), and λ is the wavelength in meters. In a vacuum, the speed of light is a maximum and does not depend on the wavelength. The frequency of light is determined by the source and does not vary. When light passes through materials other than vacuum, its speed is decreased. Because the frequency cannot change, the wavelength must decrease. If we calculate the speed of light in air, it only differs by a very small amount from the speed of light in vacuum; in general, we use 3.00×10^8 m/s (to three significant figures) for the speed of light in air or vacuum.

In some cases it is more convenient to consider light as a stream of particles. We call particles of light **photons**. Photons are characterized by their energy, E. The energy of a photon is related to the frequency of light by the equation:

$$E = h\nu \tag{2.2}$$

where E is the energy in joules (J), h is Planck's constant, 6.626×10^{-34} J s, and ν is the frequency in inverse seconds (Hz). From Eqs. (2.1) and (2.2) we can deduce that

$$E = \frac{hc}{\lambda} \tag{2.3}$$

We can see from the relationships in Eqs. (2.2) and (2.3) that the energy of electromagnetic radiation is directly proportional to its frequency and inversely proportional to its wavelength. Electromagnetic radiation ranges from very low energy (long wavelength, low frequency) radiation, like radiowaves and microwaves, to very high energy (short wavelength, high frequency) radiation, like X-rays. The major regions of the electromagnetic spectrum of interest to us as analytical chemists are shown in Fig. 2.2. It is clear from this figure that visible light, that portion of the electromagnetic spectrum to which the human eye responds, is only a very small portion of all radiant energy. Table 2.1 presents some common units and symbols for various types of electromagnetic radiation.

2.1.2. How does Electromagnetic Radiation Interact with Matter?

Spectroscopy is the study of the interaction of radiant energy (light) with matter. We know from quantum mechanics that energy is really just a form of matter, and that all matter exhibits the properties of both waves and particles. However, matter composed of molecules, atoms, or ions, which exists as solid or liquid or gas, exhibits primarily the properties of particles. Spectroscopy studies the interaction of light with matter defined as materials composed of molecules or atoms or ions.

In a gas, atoms or molecules are widely separated from each other; in liquids and solids, the atoms or molecules are closely associated. In solids, the atoms or molecules may be arranged in a highly ordered array, called a **crystal**, as they are in many minerals, or they may be randomly arranged, or **amorphous**, as they are in many plastics. Whatever their physical state or arrangement atoms, molecules, and ions are in constant motion. For molecules, many types of motion are involved. Molecules can rotate, vibrate, and translate

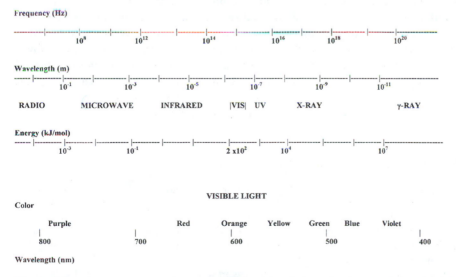

Figure 2.2 The electromagnetic spectrum. The visible light region is expanded to show the colors associated with wavelength ranges.

Table 2.1 Common Wavelength Symbols and Units for Electromagnetic Radiation

Unit	Symbol	Length (m)	Type of radiation
Angstrom	Å	10^{-10}	X-ray
Nanometer	nm	10^{-9}	UV, visible
Micrometer	μm	10^{-6}	IR
Millimeter	mm	10^{-3}	IR
Centimeter	cm	10^{-2}	Microwave
Meter	m	1	Radio

(move from place to place in space). Interaction with radiant energy can affect these molecular motions. Molecules that absorb IR radiation vibrate with greater amplitude; interaction with UV or visible light can move bonding electrons to higher energy levels in molecules. A change in any form of motion or electron energy level involves a change in the energy of the molecule. Such a change in energy is called a **transition**; we have the possibility of vibrational transitions, rotational transitions, electronic transitions, and so on in molecules. We have some of the same kinds of motion in atoms and ions; atoms can move in space, and their electrons can move between energy levels, but atoms or monoatomic ions cannot rotate or vibrate. The chemical nature of matter (its composition), its physical state, and the arrangement of the atoms or molecules in the physical state with respect to each other affect the way in which any given material interacts with electromagnetic radiation. Table 2.2 lists some of the important types of transitions studied by spectroscopy. We will cover these techniques in detail in later chapters. There are literally hundreds of types of transitions and types of spectroscopy used to investigate matter. Only a few of the most common types of spectroscopy will be covered in this text.

When light strikes a sample of matter, the light may be absorbed by the sample, transmitted through the sample, reflected off the surface of the sample, or scattered by the sample. Samples can also emit light after absorbing incident light; such a process is called **luminescence**. There are different kinds of luminescence, called **fluorescence** or **phosphorescence** depending on the specific process that occurs; these are discussed in detail in Chapter 5. Emission of light may also be caused by processes other than absorption of light. There are spectroscopic methods based on all of these interactions. Table 2.3 summarizes the major types of interaction of light with matter and gives examples of the common spectroscopic techniques based on these interactions. For the moment, we will focus on the absorption, transmission, and emission of light by matter.

Table 2.2 Some Types of Transitions Studied by Spectroscopy

Type of transition	Spectroscopic method	Wavelength range
Spin of nuclei in a magnetic field	NMR spectroscopy	0.5–10 m
Rotation and vibration of molecules	Raman and IR spectroscopy	0.8–300 μm
Bonding electron energy, valence electron energy	UV/VIS spectroscopy	180–800 nm
Core electron energy	X-ray spectroscopy	0.1–100 Å

Note: This is a very limited list of the types of transitions and spectroscopic methods in current use.

Table 2.3 Some Interactions of Light and Matter

Interaction	Radiation measured	Spectroscopic method
Absorption and transmission	Incident light, I_0	Atomic absorption
	Transmitted light, I	Molecular absorption
Absorption then emission	Emitted light, I'	Atomic fluorescence
		Molecular fluorescence
		Molecular phosphorescence
Scattering	Scattered light, I_s	Turbidimetry
		Nephelometry
		Raman
Reflection	Reflected light, I_R	Attenuated total reflection
	or relative	Diffuse reflection IR (the term
	reflected I_R	*reflectance* is also used for these
		methods)
Emission	Emitted light, I_e	Atomic emission
		Molecular emission
		Chemiluminescence

If we pass white light (i.e., visible light) through blue glass, the emerging light is blue. The glass has absorbed the other colors, such as red and yellow. We can confirm this absorption by shining red light through the blue glass. If the absorption is strong enough, all of the red light is absorbed; no light emerges from the glass and it appears black. How can this be explained?

The interaction of electromagnetic radiation and matter conforms to well-established quantum mechanical laws. Atoms, ions, and molecules exist only in certain discrete states with specific energies. The same quantum mechanical laws dictate that a change in state requires the absorption or emission of energy, ΔE, exactly equal to the difference in energy between the initial and final states. We say that the energy states are quantized. A change in state (change in energy) can be expressed as:

$$\Delta E = E_{\text{final}} - E_{\text{initial}} = h\nu \tag{2.4}$$

Since we know that $c = \lambda\nu$, then:

$$\Delta E = h\nu = \frac{hc}{\lambda} \tag{2.5}$$

These equations tell us that matter can absorb or emit radiation when a transition between two states occurs, but it can absorb or emit only the specific frequencies or wavelengths that correspond to the exact difference in energy between two states in which the matter can exist. Absorption of radiation increases the energy of the absorbing species (i.e., $E_{\text{final}} > E_{\text{initial}}$). Emission of radiation decreases the energy of the emitting species (i.e., $E_{\text{final}} < E_{\text{initial}}$). So the quantity ΔE can have either a positive sign or a negative sign, but when using ΔE to find the wavelength or the frequency of radiation involved in a transition, only the absolute value of ΔE is used. Wavelength, frequency, and the speed of light are always positive in sign.

A specific molecule, such as hexane, or a specific atom, such as mercury, can absorb or emit only certain frequencies of radiation. All hexane molecules will absorb and emit light with the same frequencies, but these frequencies will differ from those absorbed or emitted by a different molecule, such as benzene. All mercury atoms will absorb the same frequencies of incident light, but these will differ from the frequencies of light absorbed by atoms of lead or copper. Not only are the frequencies unique, but also the degree to which the frequency is absorbed or emitted is unique to a species. This degree of absorption or emission results in light of a given **intensity**. The uniqueness of the frequencies and amount of each frequency absorbed and emitted by a given chemical species is the basis for the use of spectroscopy for identification of chemicals. We call the set of frequencies and the associated intensities at which a species absorbs or emits its **spectrum**.

The lowest energy state of a molecule or atom is called the **ground state**. All higher energy states are called **excited states**. Generally at room temperature molecules and atoms exist in the ground state.

If we think about our example of blue glass, and its ability to absorb red and yellow light, we can deduce a simple picture of the energy states in the blue glass. We will assume that the glass is in its ground state before we shine any light through it since we have performed this experiment at room temperature. We will call the ground state energy E_1. If the glass is capable of absorbing red light, there must be an excited state such that the difference in energy between the ground state and this excited state is equivalent to the energy of a wavelength of red light. If we look at Fig. 2.2, we can choose a representative wavelength in the red region of the visible spectrum, such as 653 nm. If a wavelength $\lambda = 653$ nm is absorbed by the glass, we can calculate the frequency of this light by rearranging Eq. (2.1):

$$\nu = \frac{c}{\lambda} = \frac{2.997 \times 10^8 \, \text{m/s}}{(653 \, \text{nm})(10^{-9} \, \text{m/nm})}$$

$$\nu = 4.59 \times 10^{14} \, \text{s}^{-1}$$

From the frequency, we are able to calculate the difference in energy between the ground state and this excited state, which we will call E_2:

$$\Delta E = E_2 - E_1 = h\nu$$

$$\Delta E = (6.626 \times 10^{-34} \, \text{J s})(4.59 \times 10^{14} \, \text{s}^{-1})$$

$$\Delta E = 3.05 \times 10^{-19} \, \text{J}$$

So there is one excited state with an energy that is 3.05×10^{-19} J higher than the ground state in the glass. We do not know the exact energy of the ground state itself. The glass also absorbs yellow light, so we can pick a representative wavelength of yellow light, such as 575 nm, and repeat the preceding calculation. The frequency of light corresponding to a wavelength of 575 nm is 5.21×10^{14} Hz, so there must be an excited state E_3 such that:

$$\Delta E = E_3 - E_1 = h\nu$$

$$\Delta E = (6.626 \times 10^{-34} \, \text{J s})(5.21 \times 10^{14} \, \text{s}^{-1})$$

$$\Delta E = 3.45 \times 10^{-19} \, \text{J}$$

We can now construct a simplified energy diagram for the blue glass, such as the one shown in Fig. 2.3.

The diagram shows two excited states, one corresponding to the ability of the glass to absorb light with a wavelength of 653 nm and one corresponding to the ability of the glass to absorb light with a wavelength of 575 nm. Because the glass does not absorb blue light (it transmits the blue light portion of white light), there would be no energy states with a difference in energy equal to any frequency of blue light. This diagram is very oversimplified.

E_3 — 2nd Excited State
$E_3 = 3.45 \times 10^{-19}$ J

E_2 — 1st Excited State
$E_2 = 3.05 \times 10^{-19}$ J

653 nm   575 nm

E_1 — Ground State
E_1 defined = 0

Figure 2.3 Simplified energy diagram for the absorption of visible light by blue glass. Two possible excited energy states are shown, one corresponding to the absorption of 653 nm (red) light and a higher state corresponding to the absorption of 575 nm (yellow) light. If the ground state energy is defined as $E = 0$, the relative energies of the excited states can be determined.

"Red", "yellow", and "blue" light span a range of wavelengths, as can be seen from Fig. 2.2. There are actually many different energy levels associated with the transitions occurring in glass. Absorption of red light occurs from 620 to 750 nm, yellow light from 450 to 495 nm, and so on. So what is the molecular reason for this "broadband" absorption observed in spectroscopic experiments with visible light? The absorption of visible light by glass is due to excitation of bonding electrons in the molecules; in other words, due to electronic transitions. Electronic transitions require more energy than rotational or vibrational transitions. For a molecule, the relative energy of transitions is rotational < vibrational < electronic. A more realistic energy level diagram for glass (and for molecules in general) is presented in Fig. 2.4. For every electronic state E_n there are many associated rotational and vibrational sublevels. Each sublevel has a slightly different energy, with the result that a transition from one energy level E_n to a higher energy level is not a single energy but a range of closely spaced energies, because the electron can end up in any one of the many sublevels. For this reason, absorption of red light occurs over a closely spaced range of wavelengths in molecules.

Excited states are energetically unfavorable; the molecule or atom wants to return to the lowest energy ground state by giving up energy, often by emitting light. Because they are energetically unfavorable, excited states are usually short-lived, on the order of 10^{-9}–10^{-6} s. Emission of light therefore occurs rapidly following excitation. One notable exception is the process of phosphorescence, described in Chapter 5; for this process the excited state lifetime can be as long as tens of seconds in some cases.

Absorption spectra are obtained when a molecule or atom absorbs radiant energy that satisfies the equation $\Delta E = h\nu = hc/\lambda$. The **absorption spectrum** for a substance shows us the energies (frequencies or wavelengths) of light absorbed as well as how much light is absorbed at each frequency or wavelength. The nature of the molecule or atom dictates the amount of light absorbed at a given energy. The complete spectrum of a substance consists of the set of energies absorbed and the corresponding intensity of light absorbed. A graph of the intensity of light amplitude change on the y-axis vs. the frequency or wavelength on the x-axis is constructed. It is this graph of intensity vs. energy that we call a spectrum. The IR absorption spectrum for polystyrene is shown in Fig. 4.1 in Chapter 4. The UV absorption spectrum of benzene is shown in Fig. 5.12 in Chapter 5.

Figure 2.4   Schematic energy level diagram for molecules. Each electronic energy level, $E_{0,1,2,...}$ has associated vibrational sublevels $V_{1,2,3,...}$ and rotational sublevels $J_{1,2,3,...}$.

     Emission spectra are obtained when an atom or molecule in an excited state returns to the ground state by emitting radiant energy. An emission spectrum can result from many different ways of forming an excited state. Atoms and molecules can be excited not only by absorption of electromagnetic radiation, but also by transfer of energy due to collisions between atoms and molecules, by addition of thermal energy, and by addition of energy from electrical discharges. Different excitation methods are used in several types of emission spectroscopy and will be discussed in detail in later chapters. A special term is used for the emission of electromagnetic radiation by either atoms or molecules following excitation by absorption of electromagnetic radiation. Such emission is called luminescence. In other words, if light is used as the source of excitation energy, the emission of light is called luminescence; if other excitation sources are used the emission of light is called simply emission.

2.2.   ATOMS AND ATOMIC SPECTROSCOPY

An atom consists of a nucleus surrounded by electrons. Every element has a unique number of electrons, equal to its atomic number for a neutral atom of that element. The electrons are located in atomic orbitals of various types and energies and the electronic energy states of atoms are quantized. The lowest energy, most stable electron configuration of an element is its ground state. The ground state is the normal electron configuration predicted from the "rules" for filling a many-electron atom, which you learned in general chemistry. These rules are based on the location of the atom in the periodic table, the *aufbau* principle, the Pauli exclusion principle, and Hund's rule. (It is important to keep in mind that the scientific rules or laws that we use were developed to explain observed experimental facts. The student should remember the observed experimental facts that gave rise to the rules and laws, not just the rules themselves. You may want to review your general chemistry text on the structure of the atom.) For example, the ground state electron configuration for sodium, atomic number 11, is $1s^2 2s^2 2p^6 3s^1$ based on its position in the third row, first group of the periodic table and the requirement to account for 11 electrons. The ground state electronic configuration for potassium is $1s^2 2s^2 2p^6 3s^2 3p^6 4s^1$, vanadium is $1s^2 2s^2 2p^6 3s^2 3p^6 4s^2 3d^3$, and so on. If energy of the right magnitude is provided to an atom, the energy may be absorbed and an outer (valence) electron promoted from the ground state orbital it is in to a higher energy orbital. The atom is now in a higher energy, less stable, excited state. Because the excited state is less stable than the ground state, the electron will return spontaneously to the ground state. In the process, the atom will emit energy; this energy will be equivalent in magnitude to the difference in energy levels between the ground and excited states (and equivalent to the energy absorbed initially). The process is shown schematically in Fig. 2.5. If the emitted energy is in the form of electromagnetic radiation, Eqs. (2.4) and (2.5) directly relate the wavelength of radiation absorbed or emitted to the electronic transition that has occurred:

$$\Delta E = E_{\text{final}} - E_{\text{initial}} = h\nu = \frac{hc}{\lambda} \tag{2.6}$$

Each element has a unique set of permitted electronic energy levels because of its unique electronic structure. The wavelengths of light absorbed or emitted by atoms of an element are characteristic of that element. The absorption of radiant energy by atoms forms the basis of AAS, discussed in Chapter 6. The absorption of energy and the subsequent emission of radiant energy by excited atoms form the basis of AES and atomic fluorescence spectroscopy, discussed in Chapter 7.

In practice, the actual energy level diagram for an atom is derived from the emission spectrum of the excited atom. Figure 2.6 shows an energy level diagram for mercury atoms. Notice that there are no rotational or vibrational sublevels in atoms! A free gas phase atom has no rotational or vibrational energy associated with it. When an electron is promoted to a higher atomic excited state, the change in energy is very well defined and the wavelength range absorbed (or emitted on **relaxation** to the ground state) is very narrow. The wavelengths of light involved in valence electronic transitions in atoms fall in the visible and UV regions of the spectrum. This region is often called the

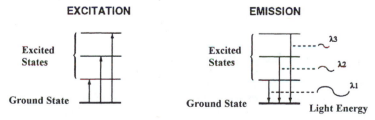

Figure 2.5 Energy transitions in atoms. Atoms may absorb energy and move a ground state valence electron to higher energy excited states. The excited atom may relax back to the ground state by emitting light of a wavelength equal to the difference in energy between the states. Three such emissions are shown. (From Beatty and Kerber, used with permission.)

Figure 2.6 Energy levels in mercury atoms (units of energy are electron volts, eV).

UV/VIS region for short. The energy level diagrams for all elements have been determined, and tables of wavelengths absorbed and emitted by atoms are available. Appendix 6.1 lists the absorption wavelengths used to measure elements by AAS, discussed in Chapter 6.

Knowing what wavelengths of light are absorbed or emitted by a sample permits qualitative identification of the elements present in the sample. Measuring the intensity of light absorbed or emitted at a given wavelength gives us information about how much of a given element is present (quantitative elemental analysis). All of the atomic spectroscopy methods—absorption, fluorescence, and emission—are extremely sensitive. As little as 10^{-12}–10^{-15} g of an element may be detected using atomic spectroscopy.

It is possible for atoms to absorb higher energy radiation, in the X-ray region; such absorption may result in the inner shell (core) electrons being promoted to an excited state, with the subsequent emission of X-ray radiation. This process forms the basis for qualitative and quantitative elemental analysis by XRF spectroscopy, as well as other X-ray techniques, discussed in Chapter 8.

2.3. MOLECULES AND MOLECULAR SPECTROSCOPY

The energy states associated with molecules, like those of atoms, are also quantized. There are very powerful spectroscopic methods for studying transitions between permitted states in molecules using radiation from the radiowave region to the UV region. These methods provide qualitative and quantitative information about molecules, including detailed information about molecular structure.

2.3.1. Rotational Transitions in Molecules

The ability of a molecule to rotate in space has associated rotational energy. Molecules may exist in only discrete (quantized) rotational energy states. Absorption of the appropriate energy causes transitions from lower energy rotational states to higher energy rotational states, in which the molecule rotates faster. This process gives rise to rotational absorption spectra. The rotational energy of a molecule depends on its angular velocity, which is variable. Rotational energy also depends on the molecule's shape and weight distribution, which change as bond angles change. While a change in shape is restricted in diatomic molecules such as O_2, molecules with more than two atoms, such as hexane, C_6H_{14}, have many possible shapes and therefore many possible rotational energy levels. Furthermore, the presence of more than one natural isotope of an atom in a molecule generates new sets of rotational energy levels. Such is the case with carbon, where a small percentage of the carbon atoms in a carbon-containing molecule are ^{13}C instead of ^{12}C. Consequently, even simple molecules have complex rotational absorption spectra. The energies involved in rotational changes are very small, on the order of 10^{-24} J per molecule. The radiation absorbed is therefore in the radiofrequency and microwave regions of the spectrum. Microwave spectroscopy has been largely unexploited in analytical chemistry because of the experimental difficulties involved and the complexity of the spectra produced. The technique is limited to the gas phase and has been used by radioastronomers to detect the chemical species in interstellar clouds.

2.3.2. Vibrational Transitions in Molecules

For the purposes of basic understanding of this branch of optical spectroscopy, molecules can be visualized as a set of weights (the atoms) joined together by springs (the chemical bonds). The atoms can vibrate toward and away from each other or they may bend at various angles to each other as shown in Fig. 2.7. Each such vibration has characteristic energy associated with it. The vibrational energy states associated with molecular vibration are quantized. Changes in the vibrational energy of a molecule are associated with absorption of radiant energy in the IR region of the spectrum. While absorption of IR radiation causes changes in the vibrations of the absorbing molecule, the increase in vibrational energy is also usually accompanied by increased molecular rotation. Remember, the rotational energy levels are sublevels of the vibrational energy levels, as we saw in Fig. 2.4. So in practice, absorption of IR radiation corresponds to a combination of changes in rotational and vibrational energies in the molecule. Because a molecule with more than two atoms has many possible vibrational states, IR absorption spectra are complex, consisting of multiple absorption bands. Absorption of IR radiation by molecules is one of the most important techniques in spectroscopy. Through IR absorption spectroscopy, the structure of molecules can be deduced, and both qualitative identification of molecules

Figure 2.7 Some possible vibrations of bonded atoms in a molecule.

and quantitative analysis of the molecular composition of samples can be performed. IR spectroscopy is discussed in Chapter 4.

2.3.3. Electronic Transitions in Molecules

A free gas phase atom has no rotational or vibrational energy associated with it, which results in the absorption or emission of very narrow wavelength ranges. When atoms combine to form molecules, the individual atomic orbitals combine to form a new set of molecular orbitals. Molecular orbitals with electron density in the plane of the bonded nuclei, that is, along the axis connecting the bonded nuclei, are called sigma (σ) orbitals. Those molecular orbitals with electron density above and below the plane of the bonded nuclei are called pi (π) orbitals. Sigma and pi orbitals may be of two types, bonding orbitals or antibonding orbitals. Bonding orbitals are lower in energy than the corresponding antibonding orbitals. When assigning electrons in molecules to orbitals, the lowest energy bonding orbitals are filled first. For a review of molecular orbital theory, see your general chemistry text, or the references by Umland and Bellama or Zumdahl and Zumdahl listed in the bibliography.

Under normal conditions of temperature and pressure, the electrons in the molecule are in the ground state configuration, filling the lowest energy molecular orbitals available. Absorption of the appropriate radiant energy may cause an outer electron to be promoted to a higher energy excited state. As was the case with atoms, the radiant energy required to cause electronic transitions in molecules lies in the visible and UV regions. And as with atoms, the excited state of a molecule is less stable than the ground state. The molecule will spontaneously revert (relax) to the ground state emitting UV or visible radiant energy. Unlike atoms, the energy states in molecules have rotational and vibrational sublevels, so when a molecule is excited electronically, there is often a simultaneous change in the vibrational and rotational energies. The total energy change is the sum of the electronic, rotational, and vibrational energy changes. Because molecules possess many possible rotational and vibrational states, absorption of UV or visible radiation by a large population of molecules, each in a slightly different state of rotation and vibration, results in absorption over a wide range of wavelengths, called an **absorption band**. The UV/VIS absorption spectra of molecules usually have a few broad absorption bands and are usually very simple in comparison with IR spectra. Molecular absorption and emission spectroscopy is used for qualitative identification of chemical species, especially for inorganic and organometallic molecules with metal atoms at their centers. This technique used to be one of the major methods for structural determination of organic molecules, but has been replaced by the more powerful and now commonly available techniques of NMR, IR spectroscopy, and MS. UV/VIS absorption spectroscopy is most often used for quantitative analysis of the molecular composition of samples. Molecular fluorescence spectroscopy is an extremely high sensitivity method, with the ability to detect single molecules! We will learn the laws governing absorption, which permit quantitative analysis by UV/VIS spectroscopy, in this chapter. The use of UV/VIS molecular spectroscopy will be discussed at greater length in Chapter 5.

2.4. ABSORPTION LAWS

The **radiant power** P of a beam of light is defined as the energy of the beam per second per unit area. A related quantity is the **intensity** I which is the power per unit solid angle. Both power and intensity are related to the square of the amplitude of the light wave, and the absorption laws can be written in terms of either power or intensity. We will use intensity I but you may see the same laws written with a P for power in other literature.

When light passes through an absorbing sample, the intensity of the light emerging from the sample is decreased. Assume the intensity of a beam of monochromatic (i.e., single wavelength) radiation is I_0. This beam is passed through a sample that can absorb radiation of this wavelength, as shown in Fig. 2.8. The emerging light beam has an intensity equal to I, where $I_0 \geq I$. If no radiation is absorbed by the sample, $I = I_0$. If any amount of radiation is absorbed, $I < I_0$. The **transmittance** T is defined as the ratio of I to I_0:

$$T = \frac{I}{I_0} \tag{2.7}$$

The transmittance is the fraction of the original light that passes through the sample. Therefore, the range of allowed values for T is from 0 to 1. The ratio I/I_0 remains relatively constant even if I_0 changes; hence, T is independent of the actual intensity I_0. To study the quantitative absorption of radiation by samples it is useful to define another quantity, the **absorbance** A where

$$A = \log\left(\frac{I_0}{I}\right) = \log\left(\frac{1}{T}\right) = -\log T \tag{2.8}$$

When no light is absorbed, $I = I_0$ and $A = 0$. Two related quantities are also used in spectroscopy, the **percent transmittance**, $\%T$, which equals $T \times 100$, and the **percent absorption**, $\%A$, which is equal to $100 - \%T$.

Suppose we have a sample of an aqueous solution of an absorbing substance in a rectangular glass sample holder with a length of 1.0 cm, as shown in Fig. 2.9. Such a sample holder is called a **sample cell**, or **cuvette**, and the length of the cell is called the **path length** b. The incident light has intensity, I_0, equal to 100 intensity units. If 50% of the light passing through the sample is absorbed, then 50% of the light is transmitted. The emerging light beam has an intensity denoted as $I_1 = 50$ intensity units. So the $\%T = 50$, and therefore:

$$T = \frac{\%T}{100} = \frac{I_1}{I_0}$$

$$T = \frac{50}{100} = 0.50$$

From T we calculate that absorbance equals

$$A = -\log T = -\log(0.50) = 0.30$$

Sample

Figure 2.8   Absorption of radiation by a sample.

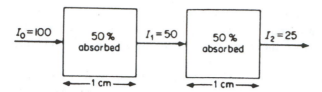

Figure 2.9 Absorption of radiation by two identical sample cells.

If a second identical cell with the same solution is placed in the path of beam I_1, 50% of the incident radiation, I_1, will be absorbed, and we have a new emerging beam, I_2. The intensity of I_1 is 50 intensity units and therefore I_2 must be 25 intensity units. So the transmittance for the second cell is

$$T = \frac{I_2}{I_1} = \frac{25}{50} = 0.50$$

The absorbance for just the second cell is $A = -\log 0.50 = 0.30$. The two cells are identical in their absorbance of light. Identical or "optically-matched" cells are required for accurate quantitative analysis using spectroscopy in many cases.

Now suppose we put the two cells back to back and consider them together. For our purpose, we will assume that these will behave as if there were no glass walls between the two cells; in other words, we have one "cell" that is 2.0 cm long. The path length for this experiment is now 2.0 cm. The incident light beam has intensity I_0, with $I_0 = 100$ intensity units. We know from passing light through the two cells (as described earlier) that the emerging light beam has intensity $I = 25$ intensity units. So we see that for a path length of 2.0 cm, T now equals 25/100 or 0.25 and $A = -\log 0.25 = 0.60$. If we put three cells in line (path length = 3.0 cm), the emerging beam has $I = 12.5$, $T = 0.125$, and $A = 0.90$. Four cells in line will give $I = 6.25$, $T = 0.063$, and $A = 1.20$. If we plot intensity I vs. the number of cells (i.e., the path length in cm) as shown in Fig. 2.10(a), it is clear that I decreases exponentially with increasing path length. If we plot absorbance A vs. path length, shown in Fig. 2.10(b), absorbance increases linearly with increasing path length. As analytical chemists, we find it better to work with linear equations rather than with exponential equations, because as you can see from Fig. 2.10, it is easier to interpolate and read data from a linear plot. A linear plot has a constant slope that greatly simplifies calculations. That is why the absorbance is such a useful quantity—it results in a linear relationship with quantities important to analytical chemists. This proportional relationship between sample thickness (the path length) and absorbance at constant concentration was discovered by P. Bouguer in 1729 and J. Lambert in 1760.

If we perform a similar experiment keeping the path length constant by using only one cell but change the **concentration** of the absorbing species, we find the same relationship between I, A, and concentration as we found for path length. A linear relationship exists between the absorbance A and the concentration c of the absorbing species in the sample, a very important quantity! Because A is linear with respect to path length b and concentration c we can write the following equation:

$$A = abc \tag{2.9}$$

The term "a" is a proportionality constant called the **absorptivity**. The absorptivity is a measure of the ability of the absorbing species in the sample to absorb light at the

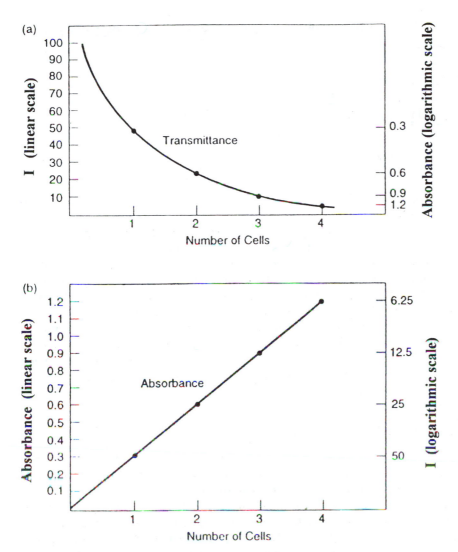

Figure 2.10 (a) Exponential relationship between intensity (or transmittance) and increasing number of cells (increasing path length). (b) Linear relationship between absorbance and increasing number of cells (increasing path length). $I_0 = 100$ for both plots.

particular wavelength used. Absorptivity is a constant for a given chemical species at a specific wavelength. If the concentration is expressed in molarity (mol/L or M), then the absorptivity is called the **molar absorptivity**, and is given the symbol ε. The usual unit for path length is centimeters, cm, so if the concentration is in molarity, M, the unit for molar absorptivity is $M^{-1}\,cm^{-1}$. The units for absorptivity a when concentration is expressed in any units other than molarity (for example, ppm, mg/100 mL, etc.) must be such that A, the absorbance, is dimensionless. A. Beer discovered the proportional relationship between concentration and absorbance at constant path length in 1852.

Equation (2.9), which summarizes the relationship between absorbance, concentration of the species measured, sample path length, and the absorptivity of the species is known as the **Beer–Lambert–Bouguer Law** or, more commonly, as **Beer's Law**.

Since $A = -\log T$, we have the following equivalent expressions for Beer's Law:

$$abc = -\log T \tag{2.10}$$

$$-\log\left(\frac{I}{I_0}\right) = abc \tag{2.11}$$

$$\log\left(\frac{I_0}{I}\right) = abc \tag{2.12}$$

$$\frac{I_0}{I} = 10^{abc} \tag{2.13}$$

$$\frac{I}{I_0} = 10^{-abc} \tag{2.14}$$

where a, the absorptivity, is equal to ε, the molar absorptivity, in all of the equations if c, the concentration of the absorbing species, is expressed as molarity.

Beer's Law shows mathematically, based on observed experimental facts, that there is a linear relationship between A and the concentration of an absorbing species if the path length and the wavelength of incident radiation are kept constant. This is an extremely important relationship in analytical spectroscopy. It forms the basis for the quantitative measurement of the concentration of an analyte in samples by quantitative measurement of the amount of absorbed radiation. The *quantitative* measurement of radiation intensity is called **spectrometry**. Beer's Law is used in all quantitative absorption spectrometry—IR absorption spectrometry, AAS, UV/VIS absorption spectrometry, and so on.

2.4.1. Deviations from Beer's Law

Beer's Law is usually followed at low concentrations of analyte for homogeneous samples. Absorbance is directly proportional to concentration for most absorbing substances when the concentration is less than about 0.01 M.

Deviations from linearity are common at high concentrations of analyte. There are several possible reasons for deviation from linearity at high concentrations. At low concentrations in a solution, the analyte would be considered the **solute**. As the solute concentration increases, the analyte molecules may begin to interact with each other, through intermolecular attractive forces such as hydrogen bonding and other van der Waals forces. Such interactions may change the absorptivity of the analyte, again resulting in a nonlinear response as concentration increases. At extremely high concentrations, the solute may actually become the **solvent**, changing the nature of the solution. If the analyte species is in chemical equilibrium with other species, as is the case with weak acids or weak bases in solution, changes in concentration of the analyte may shift the equilibrium (Le Chatelier's Principle). This may be reflected in apparent deviations from Beer's Law as the solution is diluted or concentrated.

Another source of deviation from Beer's Law may occur if the sample scatters the incident radiation. Solutions must be free of floating solid particles and are often filtered before measurement. The most common reason for nonlinearity at high analyte concentrations is that too little light is available to be absorbed. At low levels of analyte, doubling the concentration doubles the amount of light absorbed, say from 25% to 50%. If 99% of the light has already been absorbed, doubling the concentration still doubles the amount of remaining light absorbed, but the change is only from 99% to 99.5%. This results in the curve becoming flat at high absorbance.

It can be seen from Eq. (2.8) that $A = \log(I_0/I)$. If $I_0 = 100$ and $A = 1.0$, then $I = 10$. Only 10% of the initial radiation intensity is transmitted. The other 90% of the intensity is absorbed by the sample. If $A = 2.0$, $I = 1.0$, indicating that 99% of the incident light is absorbed by the sample. If $A = 3.0$, 99.9% of the incident light intensity is absorbed. As we shall see, the error in the measurement of A increases as A increases (or as I decreases). In practice, Beer's Law is obeyed for absorbance values less than or equal to 1.0.

2.5. METHODS OF CALIBRATION

Calibration is the process of establishing the relationship between the signal we measure (such as absorbance) and known concentrations of analyte. Once this relationship is established, it is possible to calculate the concentration of the analyte in an unknown sample by measuring its signal. The calibration methods discussed subsequently are applicable to most of the analytical instrumental methods discussed in this text, not just spectroscopic measurements. Historically, calibration data were collected manually by reading a meter or measuring a peak height with a ruler, and transcribing all the numbers into a lab notebook. The relationship between signal and concentration was plotted manually on graph paper. Modern instruments have software packages that fit the data points to the best-fit line or curve statistically, display the results on the computer screen and send them to the printer. Computerized data collection and processing greatly reduces transcription error, that is, copying the wrong figures into a notebook. The use of linear regression and other curve-fitting approaches provides greater accuracy and precision in the calibration equation and in sample results calculated from the equation.

2.5.1. Calibration with Standards

In order to use Beer's Law to determine the concentration of analyte in an unknown, it is necessary to establish the relationship between absorbance at a given wavelength and the concentration of the analyte. Solutions containing known concentrations of analyte are called **standard solutions** or more simply, **standards**. For some types of analyses, the standards may be in the form of solids or gases.

Standards must be prepared accurately from high purity materials so that the concentration of analyte is known as accurately as possible. A series of standards covering an appropriate concentration range is prepared. The standards should include one solution with no added analyte; the concentration of analyte in this standard is zero. This solution is called the reagent blank and accounts for absorbance due to impurities in the solvent and other reagents used to prepare the samples. It also accounts for the instrumental baseline. The absorbance of the reagent blank and each standard is measured. The absorbance of the reagent blank is subtracted from the absorbances of the other standards before any calculations are performed. The absorbances from which the blank absorbance has been subtracted are called "**corrected absorbances**". A plot is made of corrected absorbance on the y-axis vs. the known concentration of the standard on the x-axis. Such a plot used to be constructed manually on graph paper; now, plots are generated by computer software. A typical calibration curve of this type is shown in Fig. 2.11. This calibration curve shows the relationship between the absorbance of n-hexadecane, $CH_3(CH_2)_{14}CH_3$, a hydrocarbon found in petroleum, at 3.41 μm in the IR region and the concentration of solutions of n-hexadecane in tetrachloroethylene, C_2Cl_4. This measurement of the absorbance of solutions of n-hexadecane at 3.41 μm is a method used for determining petroleum

Figure 2.11 A typical external calibration curve for quantitative absorption spectrometry. The calibration standards follow Beer's Law since the relationship between concentration and absorbance is linear.

contamination in water, soil, and other environmental samples, because most hydrocarbons absorb at this wavelength. It is an official method developed by the US EPA (www.epa.gov) and relies on Beer's Law to permit the measurement of petroleum hydrocarbons in unknown samples. It is used to measure environmental contamination from oil spills, illegal dumping of oil, and leaking underground oil storage tanks.

Table 2.4 gives the values of the concentration and the measured and corrected absorbance for each standard. It is clear from Fig. 2.11 that the relationship between absorbance and concentration for this measurement follows Beer's Law since the plot results in a straight line ($y = mx + b$, where y is the absorbance and x is the concentration). Once the points have been plotted, the best straight line is fitted through the data points. As we learned in Chapter 1, the best-fit straight line is determined by linear regression (linear least squares) using a statistical program on your calculator or using a computer spreadsheet program such as Excel. It is of course possible to estimate the concentration that corresponds to a given absorbance visually from a calibration curve, but for accurate work the equation of the best-fit straight line must be determined.

Table 2.4 Calibration Data for Measurement of Petroleum Hydrocarbons by IR Absorption Spectrometry

Concentration of n-hexadecane (mg/100 mL solution)	Absorbance at 3.41 μm	Corrected absorbance
0.0	0.002	0.000
4.0	0.103	0.101
8.0	0.199	0.197
16.0	0.400	0.398
32.0	0.804	0.802

Performing a linear regression on the data in Table 2.4 provides us with the exact Beer's Law relationship for this method: $A = 0.0250x - 0.001$, where x is the concentration of n-hexadecane (in mg/100 mL). From the equation for the calibration curve, the concentration can be determined for any measured absorbance. For example, an unknown sample of contaminated soil is prepared according to Method 8440, and the absorbance of the sample solution is measured. The measured absorbance is 0.302, so the corrected absorbance would be $0.302 - 0.002 = 0.300$. From our calibration curve, we can see visually that this corresponds to a concentration of ~ 12.0 mg n-hexadecane/100 mL. The exact concentration can be calculated from the linear regression equation, and is found to be 11.96 mg/100 mL or 12.0 mg/100 mL rounded to three significant figures.

Suppose that in addition to the standards listed in Table 2.4, we also prepared n-hexadecane standards containing 60.0 mg n-hexadecane/100 mL solution and 100.0 mg n-hexadecane/100 mL solution. If we measure the absorbances for these standards, we obtain the data shown in Table 2.5. The additional points are plotted along with the original standards in Fig. 2.12 with the points shown as open circles. Clearly, we now see a deviation from Beer's Law at these high concentrations. The points no longer fit a straight line; the measured absorbances are lower than they should be if Beer's Law was followed at these concentrations. This shows why it is never a good idea to *extrapolate* (extend) a calibration curve beyond the range of the measured standards. If we had a sample with an absorbance of 1.30 (just like our 60.0 mg/100 mL standard) and had used our original calibration curve extrapolated to higher absorbances as shown by the black squares in Fig. 2.12, we would calculate a concentration of 51.9 mg/100 mL for the unknown, which would be erroneously low. What should we do if we have absorbances that are above the absorbance of our highest linear standard (in this case, above $A = 0.802$)? The best approach is to dilute the samples. Let us dilute our 60.0 mg/100 mL standard by a factor of 10, by taking 10.0 mL of the solution and diluting it with pure solvent to a total volume of 100 mL in a volumetric flask. If we measure the absorbance of the diluted solution, we find that the corrected absorbance = 0.153. The corresponding concentration is determined to be 6.07 mg/100 mL. Multiplied by 10 to account for the dilution, our original solution is calculated to contain 60.7 mg/100 mL, and we know that this is a reasonably accurate answer because we made up the standard to contain 60.0 mg/100 mL. So if we take our unknown sample solution with an absorbance of 1.30, dilute it by a factor of 10 and measure the absorbance of the diluted solution, we should get an accurate result for the sample as well.

Table 2.5 Calibration Data for Measurement of Petroleum Hydrocarbons by IR Absorption Spectrometry (Higher Concentration Standards Added)

Concentration of n-hexadecane (mg/100 mL solution)	Absorbance at 3.41 μm	Corrected absorbance
0.0	0.002	0.000
4.0	0.103	0.101
8.0	0.199	0.197
16.0	0.400	0.398
32.0	0.804	0.802
60.0	1.302	1.300
100.0	1.802	1.800

Figure 2.12 A calibration curve showing deviation from Beer's Law at high concentrations (high absorbance values). The open circles are the measured absorbance values for the standards and clearly deviate from linearity above $A = 0.8$. The black squares show the expected absorbance values if all standards followed Beer's Law. These points were obtained by extrapolation of the linear portion of the curve.

Preparing a calibration curve as just described, by making a series of standards of known concentrations of analyte, is called external calibration or calibration with external standards. No calculation of uncertainty has been included in this discussion. That subject is covered in Chapter 1. We never report analytical results without also reporting the uncertainty.

2.5.2. Method of Standard Additions

An alternate method of calibration is the Method of Standard Additions (MSA) calibration. This calibration method requires that known amounts of the analyte be added directly to the sample, which contains an unknown amount of analyte. The increase in signal due to the added analyte (e.g., absorbance, emission intensity) permits us to calculate the amount of analyte in the unknown. For this method of calibration to work, there must be a linear relationship between the concentration of analyte and the signal.

The MSA is often used if no suitable external calibration curve has been prepared. There may be no time to prepare calibration standards—for example, in an emergency situation in a hospital it may be necessary to measure sodium rapidly in a patient's serum. It may not be possible to prepare a valid set of calibration standards because of the complexity of the sample matrix or due to lack of sufficient information about the sample—for example, industries often require the analysis of "mystery" samples when something goes wrong in a process. MSA calibration is very useful when certain types of interferences are present in the sample matrix. MSA permits us to obtain accurate results without removing the interferences by performing the calibration in the presence of the interferences. It is often used when only one sample must be analyzed, and the preparation of external standards would be inefficient.

A typical example of the use of MSA is the determination of sodium by AES in an industrial plant stream of unknown composition. A representative sample of the plant stream is taken and split into four aliquots of 100 mL each. The first aliquot is left untreated; this is called the "no add" or "zero add" sample. To the second aliquot, 100 μg Na is added to the 100 mL sample in such a way as to not change the volume significantly. This can be done by adding a 10 μL volume of a 10,000 ppm Na solution to the sample. A 10,000 ppm Na solution contains 10,000 μg Na/mL, so a 10 μL portion contains 100 μg Na as shown:

$$\frac{10,000 \ \mu g \ Na}{mL \ solution} \times \frac{1 \ mL}{1000 \ \mu L} \times 10 \ \mu L = 100 \ \mu g \ Na$$

The second sample aliquot now contains an additional 1.0 ppm Na, since 100 μg Na/ 100 mL equals 1.0 ppm Na. To the third aliquot we add 0.020 mL of the 10,000 ppm Na solution; the third aliquot now contains an additional 2.0 ppm Na. To the fourth aliquot, an addition of 0.030 mL of the 10,000 ppm Na solution results in an additional 3.0 ppm Na in the sample aliquot. The maximum change in volume caused by the addition of Na solution is only 0.03%, an insignificant amount. It is important not to change the volume of the aliquots because a change in volume will cause a change in the concentration–signal relationship. All of the aliquots, untreated and the ones to which additions have been made, must have the same composition or MSA calibration will not produce accurate results. The concentrations of Na added and the sample aliquot numbers are listed in Table 2.6. The emission intensity for each of the four sample aliquots is measured at the 589.0 nm sodium emission line, using a flame AES or a flame photometer. The intensities measured are also listed in Table 2.6. In addition, the emission intensity from the flame is measured with no sample present. This measures "**background emission**"—a positive signal from the flame not due to the sample. The background emission signal must be subtracted from the sample intensities, just as a reagent blank is subtracted as we discussed earlier, to obtain the corrected intensities shown in Table 2.6. (we will learn more about background emission from flames in Chapters 6 and 7.)

The corrected emission intensity is plotted vs. added Na concentration (Fig. 2.13). It can be seen in Fig. 2.13 that the relationship between concentration added and signal is linear over the range examined. The quantity ($\Delta I_{emission}/\Delta$ ppm Na added) is the **slope** of the addition calibration line and is obtained by a linear regression calculation. In this case,

$$\frac{\Delta I_{emission}}{\Delta \ ppm \ Na \ added} = \frac{1.3}{1.0 \ ppm \ Na}$$

Table 2.6   MSA Calibration

Sample aliquot	Emission intensity (intensity units)[a]	Corrected intensity (intensity units)[a]	ppm Na added
1	2.9	2.4	0.0
2	4.2	3.7	1.0
3	5.5	5.0	2.0
4	6.8	6.3	3.0
Background[b]	0.5	0.0	

[a]Intensity units = 1000 counts/s.

[b]Flame only; no sample aspirated.

Figure 2.13 A typical MSA calibration curve. The *y*-axis shows corrected emission intensity. Intensity units = 1000 counts/s.

The emission intensity increases by 1.3 units for every 1.0 ppm Na present. Therefore, the concentration of Na in the untreated sample is calculated from the following equation:

$$\text{ppm Na}_{sample} = \frac{R}{\text{slope}} = R(\text{slope})^{-1}$$

where R = the corrected intensity measurement for sample aliquot 1 (with no added Na), and the slope = $(\Delta I_{emission}/\Delta$ ppm Na added) (i.e., the slope of the addition calibration curve). For our example, the Na concentration in the plant stream is:

$$\text{ppm Na}_{sample} = 2.4 \text{ intensity units} \times \frac{1.0 \text{ ppm Na}}{1.3 \text{ intensity units}}$$

$$\text{ppm Na}_{sample} = 1.85 \text{ ppm}$$

Alternatively, the addition calibration curve may be extrapolated to the intercept on the negative *x*-axis using the linear regression equation determined. The concentration of sodium in the sample is equal to the absolute value of the negative *x* intercept. The extrapolation is shown in Fig. 2.14 for the data in Table 2.6. The −x intercept occurs at −1.85, therefore the concentration of Na in the sample is |−1.85 ppm| = 1.85 ppm Na. The emission intensities plotted are the corrected intensities; the background must be subtracted first.

The MSA is a very powerful tool for obtaining accurate analytical results when it is not possible to prepare an external calibration curve. In some cases, it permits accurate results to be obtained even in the presence of interfering substances. MSA will not correct for background emission or background absorption, or other **spectral interferences**. These interferences and methods for correcting for background will be discussed in the appropriate later chapters.

If the amount of sample is limited, as it may very well be for a patient's serum, a one-point standard addition technique may be used. The technique is outlined for a sodium determination in serum by flame AES. The emission intensity from the sample is measured, and then a known concentration of Na is added, again without significantly

Figure 2.14 Determination of the concentration of analyte by extrapolation of the MSA curve to the negative x intercept. The y-axis shows corrected emission intensity. Intensity units $= 1000$ counts/s.

changing the volume of the sample. The emission intensity for the addition sample is measured, as is the flame background. Because the emission signal is directly proportional to the sodium concentration, we can write:

$$\frac{I_{unknown}}{I_{added}} = \frac{ppm\ Na_{unknown}}{ppm\ Na_{unknown} + ppm\ Na_{added}}$$

where $I_{unknown}$ is the corrected intensity for the original sample, I_{added} is the corrected intensity for the sample with added Na, and ppm Na_{added} is the concentration of the Na added to the original sample. For example, measurement of a serum sample gives a corrected intensity of 2.4 intensity units. Sufficient sodium is added to the serum sample to increase the concentration by 3.0 ppm, and the corrected intensity of this "addition" solution is 6.3. Setting $x =$ ppm $Na_{unknown}$,

$$\frac{3.7}{6.3} = \frac{x\ ppm}{(x + 3.0)\ ppm}$$

Solving for x,

$$x = 4.3\ ppm\ Na$$

While the use of a one-point addition may be required if sample is limited, it is generally good practice to make at least two additions whenever possible, to confirm that the additions are within the linear range of the analysis method. Again, results would need to be reported with the associated uncertainty.

2.5.3. Internal Standard Calibration

An **internal standard** is a known amount of a nonanalyte element or compound that is added to all samples, blanks, and standard solutions. Calibration with internal standardization is a technique that uses the signal from the internal standard element or compound to correct for interferences in an analysis. Calibration with internal standardization improves the accuracy and precision of an analysis by compensating for several sources of error.

For determination of analyte A, an internal standard S, which must not be present in the samples, is selected. The same concentration of S is added to all samples, standard solutions and blanks. The signals due to both A and S are measured. The ratio of the signal due to the analyte A to the signal due to the internal standard S, is calculated. The signal ratio, signal$_A$/signal$_S$, is plotted against the concentration ratio of A/S in the standards. The equation of the calibration curve, which should be linear for best results, is obtained by linear regression. The equation permits the calculation of the concentration ratio A/S in any unknown samples by measuring the signal of A and S in the sample and calculating the signal ratio for the sample.

The relationship between concentration and signal may be expressed as follows:

$$\frac{\text{Concentration ratio (A/S) in sample}}{\text{Concentration ratio (A/S) in standard}} = \frac{\text{signal ratio (A/S) in sample}}{\text{signal ratio (A/S) in standard}}$$

The method of internal standardization is widely used in spectroscopy, chromatography, MS, and other instrumental methods. The use of internal standards can correct for losses of analyte during sample preparation, for mechanical or electrical "drift" in the instrument during analysis, for volume change due to evaporation and other types of interferences. The internal standard must be chosen carefully, usually so that the chemical and physical behavior of the internal standard is similar to that of the analyte. The internal standard must not interact chemically or physically with the analyte. Whatever affects the signal from the analyte should affect the signal from the internal standard in the same way. The ratio of the two signals will stay constant, even if the absolute signals change; this provides more accuracy and precision than if no internal standard is used.

The determination of lead in drinking water by ICP-MS demonstrates how an internal standard can correct for a problem such as "instrumental drift", that is, a change in the signal over a period of time. The signal for the Pb-208 isotope (^{208}Pb) is monitored to determine lead. Two lead calibration standards are prepared, each containing 10.0 ppb Pb. One standard also contains 20.0 ppb bismuth as an internal standard. The Bi signal is measured at the Bi-209 isotope, along with the Pb-208 signal. Both standards are measured several times during the day and the resulting signals are listed in Table 2.7. As can be seen from Table 2.7, the signals for both Pb and Bi fluctuate during the day; such fluctuations could be due to changes in electrical line voltage to the instrument, for example. However, as the last column shows, the ratio of the analyte signal to the internal standard signal stays constant throughout the day. If the ICP-MS were calibrated without using an internal standard at 8 AM, samples run at 1 PM would give erroneously high Pb concentrations, because the signal for a given amount of lead has increased by a factor of 2.5. At 3 PM,

Table 2.7 Use of Internal Standard in Calibration

Time of measurement	Signal counts[a] 10 ppb Pb, $m/e = 208$	Signal counts 20 ppb Bi, $m/e = 209$	Ratio of counts Pb/Bi
8 AM	12,050	60,000	0.2008
10 AM	12,100	60,080	0.2013
1 PM	30,000	149,200	0.2010
3 PM	15,750	78,400	0.2009

[a]Both standards give the same signal for Pb, so only one column is shown. The counts for bismuth are from the second standard only.

samples would be approximately 30% higher than the true value. If 20.0 ppb Bi had been added to all the standards and samples, the ratio of the Pb signal/Bi signal would have remained constant and an accurate lead concentration would be determined. For example, the ICP-MS is calibrated at 8 AM with the 10.0 ppb Pb standard. The signal obtained for 10.0 ppb Pb is 12,050 counts. This is the calibration factor: 12,050 counts/ 10.0 ppb Pb. The second solution containing 10.0 ppb Pb and 20.0 ppb Bi is measured; the Pb counts are 12,050 (the same as the standard) and the counts for Bi are 60,000, as shown in Table 2.7. A drinking water sample to which 20.0 ppb Bi has been added as internal standard is measured throughout the day. The signals obtained are given in Table 2.8. At 8 AM, the Pb signal is equal to 6028 counts, so the concentration of Pb in the water may be calculated from our calibration factor for lead with no internal standard:

$$\text{ppb Pb in sample} = (6028 \text{ counts}) \frac{10.0 \text{ ppb Pb}}{12,050 \text{ counts}}$$

$$\text{ppb Pb in sample} = 5.00 \text{ ppb Pb}$$

This is the true value for lead in the water sample. The rest of the concentrations using the lead signal only are calculated the same way and the results are shown in the third column of Table 2.8. Due to the instrument "drift", the results obtained at 1 PM and 3 PM are clearly in error. For example, the 1 PM sample has a signal equal to 15,010 counts for Pb. The concentration is calculated using the lead calibration factor:

$$\text{ppb Pb in sample} = (15,010 \text{ counts}) \frac{10.0 \text{ ppb Pb}}{12,050 \text{ counts}}$$

$$\text{ppb Pb in sample} = 12.5 \text{ ppb Pb}$$

The percent error in this result due to instrumental drift is calculated:

$$\% \text{ Error} = \frac{\text{measured value} - \text{true value}}{\text{true value}} \times 100$$

$$\% \text{ Error} = \frac{12.5 - 5.00}{5.00} \times 100$$

$$\% \text{ Error} = +150$$

Table 2.8   Lead in Water With and Without Internal Standard

Time of measurement	Pb counts, $m/e = 208$	ppb Pb, no internal standard	Pb counts, $m/e = 208$	Bi counts, $m/e = 209$	Ratio of counts: Pb/Bi	ppb Pb, with internal standard
8 AM	6,028	5.00	6,028	60,010	0.1004	5.00
10 AM	6,063	5.03	6,063	60,075	0.1009	5.02
1 PM	15,010	12.5	15,010	149,206	0.1006	5.01
3 PM	7,789	6.46	7,789	78,398	0.0994	4.95
Mean		7.25				4.99
SD		3.57				0.03
% RSD		49.2				0.60
True value		5.00				5.00
% Error		45.0				−0.2

Note: SD = standard deviation, % RSD = % relative standard deviation = (SD/mean) × 100.

If the Bi internal standard signal is used and the ratio of the Pb signal to the Bi signal is calculated, the internal standard calibration factor is obtained:

$$10.0 \text{ ppb Pb} = \frac{12,050 \text{ counts Pb}}{60,000 \text{ counts Bi}} = 0.2008$$

The sample run at 8 AM also contained Bi as an internal standard and the counts for Pb and Bi are shown in Table 2.8. The ratio of the sample Pb counts to the sample Bi counts is 6028 counts Pb/60,010 counts Bi = 0.1004. The Pb concentration in the sample is obtained as follows:

$$\text{ppb Pb in sample} = (0.1004) \frac{10.0 \text{ ppb Pb}}{0.2008}$$

$$\text{ppb Pb in sample} = 5.00 \text{ ppb Pb}$$

This is the same result obtained without an internal standard. If the internal standard ratio is used for all of the samples run during the day, the instrument "drift" is taken into account and the correct results are obtained. For example, at 1 PM, the sample Pb counts are 15,010 and the sample Bi counts are 149,206. These are clearly much higher than the counts obtained at 8 AM, but the ratio of Pb counts to Bi counts is 15,010/149,206 = 0.1006, and the calculated Pb concentration in the 1 PM sample is:

$$\text{ppb Pb in sample} = (0.1006) \frac{10.0 \text{ ppb Pb}}{0.2008}$$

$$\text{ppb Pb in sample} = 5.01 \text{ ppb Pb}$$

The percent error in this result is

$$\% \text{ Error} = \frac{5.01 - 5.00}{5.00} \times 100$$

$$\% \text{ Error} = 0.2$$

An error of 0.2% is well within the expected accuracy for an instrumental method.

Compare this correct result to the 12.5 ppb Pb result at 1 PM calculated with no internal standard correction and the importance of using an internal standard when possible is clear.

2.5.4. Errors Associated with Beer's Law Relationships

All spectrometric measurements are subject to indeterminate (random) error, which will affect the accuracy and precision of the concentrations determined using spectrometric methods. A very common source of random error in spectrometric analysis is instrumental "noise". Noise can be due to instability in the light source of the instrument, instability in the detector, variation in placement of the sample in the light path, and is often a combination of all these sources of noise and more. Because these errors are random, they cannot be eliminated. Errors in measurement of radiation intensity lead directly to errors in measurement of concentration when using calibration curves and Beer's Law.

We can evaluate the impact of indeterminate error due to instrumental noise on the information obtained from transmittance measurements. The following discussion applies to UV/ VIS spectrometers operated in regions where the light source intensity is low or the detector sensitivity is low and to IR spectrometers where noise in the thermal detector is significant. From Beer's Law, it can be shown that:

$$\Delta c/c = \frac{0.434 \Delta T}{T \log T} \qquad (2.15)$$

where $\Delta c/c$ is the relative error in concentration and ΔT is the error in measurement of the transmittance. The value of ΔT can be estimated from a large number ($n > 20$) of replicate measurements of the same solution. If we assume that we have a constant error of 1% in the measurement of T, or $\Delta T = 0.01$, the relative error in concentration can be calculated using Eq. (2.15). Table 2.9 presents the relative error in concentration for a wide range of transmittance measurements when a constant error of $1\%T$ is assumed. It can be seen from Table 2.9 that the relative error in concentration is high when T is very low or very high; significant errors result when using Beer's Law at very low concentrations of analyte (high $\%T$) and at very high concentrations of analyte (low $\%T$).

We can plot the relative error data in Table 2.9 as a function of transmittance. The resulting plot is shown in Fig. 2.15. It can be seen from this plot that the minimum relative error occurs at $T = 0.37$ ($37\%T$), although satisfactory results can be obtained over the range of $15–65\%T$. This range corresponds to an absorbance range of $0.82–0.19$. For the greatest accuracy in quantitative absorption measurements, it is advisable to determine concentration from samples with absorbances between 0.82 and 0.19. Samples that are too concentrated ($A > 0.82$) should be diluted to bring their absorbance values below 0.8. Samples that are too dilute ($A < 0.19$) should be concentrated if possible, by evaporation or solvent extraction. If it is not possible to alter the sample solution, the analyst must be aware that the relative error will be large for very dilute or very concentrated samples when using an instrument with the limitations described.

Modern UV/VIS spectrometers are generally limited by "shot noise" in the photon detector as electrons cross a junction. In this case, the plot of relative uncertainty due to indeterminate instrument error looks very different from Fig. 2.15. For good quality, shot noise limited instruments, the relative error is high for very low values of A (high $\%T$), but absorbance values from 0.2 to above 2.0 have approximately the same low ($<1\%$) relative uncertainty. In other words, modern spectrometers are much more accurate and precise than older ones because of improvements in instrument components.

Another way of plotting spectrometric data is to use the Ringbom method in which the quantity ($100 - \%T$) is plotted against the logarithm of the concentration. The resulting S-shaped curve is called a Ringbom plot. A Ringbom plot for absorption by Mn as permanganate ion is shown in Fig. 2.16. The Ringbom plot shows the concentration range where the analysis error is minimal; this is the steep portion of the curve where the slope is nearly linear. The plot also permits the evaluation of the accuracy at any

Table 2.9　Relative Concentration Error from 1% Spectrometric Error

Transmittance (T)	Relative error in concentration $(\Delta c/c) \times 100$ (%)
0.02	12.8
0.08	4.9
0.15	3.5
0.30	2.8
0.37	2.7
0.45	2.8
0.65	3.6
0.80	5.6
0.97	33.8

Note: $\Delta T = 0.01$; $\Delta c/c = (0.434 \Delta T)/(T \log T)$.

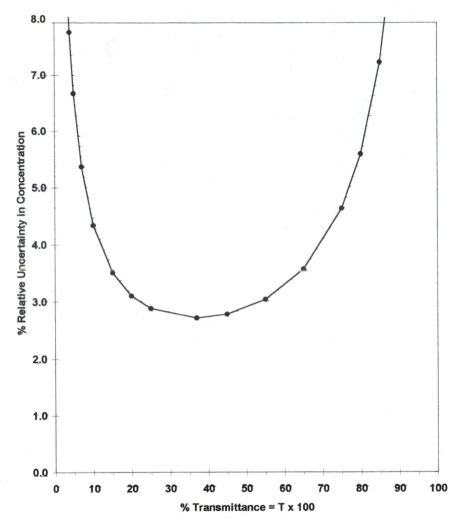

Figure 2.15 Relative uncertainty in measured concentration due to random error in spectrometric measurements due to some types of instrument noise. The data shown are for a constant 1% error in transmittance. The curve will have the same shape for other values of error in T, but the magnitude of the uncertainty percentage will change.

concentration level. From Beer's Law, it can be shown that the percent relative analysis error for a 1% transmittance error is given by:

$$\frac{\% \text{ relative analysis error}}{1\% \text{ transmittance error}} = \frac{100\Delta c/c}{1} = \frac{230}{1\Delta T/\Delta \log c} \qquad (2.16)$$

Because the quantity $(\Delta T/\Delta \log c)$ is the slope of the curve, the relative analysis error per 1% transmittance error at any point on the curve is equal to 230 divided by the slope at that point. The slope can be determined by constructing a tangent to the curve at the desired concentration. The difference in y for a 10-fold difference in x is calculated. This value, divided into 230, is the percent relative analysis error per 1% transmittance error. For example, the slope between the two points labeled A in Fig. 2.16 is determined by drawing a tangent which extends through the concentrations 2 and

Figure at left contains:

Evaluation of
range and accuracy

Range		%RE
A-A	4-20	2.8
B-B	2-25	5
C-C	1-35	10

Y-axis: 100 - %T

X-axis: Concentration, Mn, ppm

(log c)

Figure 2.16   Ringbom plot of permanganate solution measured at 526 nm in a 1.00 cm cell. %RE = percent relative error in concentration for a 1% error in transmittance. The magnitude of %RE is shown for three ranges of Mn concentration. Mn concentration is in units of ppm Mn in solution (1 ppm = 1 μg Mn/mL solution).

20 ppm (a 10-fold change) as shown. The values of y from the plot are 9% at 2 ppm and 90% at 20 ppm. The difference in y values is $90 - 9 = 81$; therefore $230/81 = 2.8$. The relative analysis error is 2.8% over this range. Other ranges and their respective errors are shown in Fig. 2.16 in the inset box at the top left. Of course, this calculation can be done more accurately using a computer than manually drawing tangent lines.

A practical application of the Ringbom plot is the determination of the concentration range over which the percent relative analysis error will not exceed a specified value. This sort of limit is often set for industrial analyses, where specifications are set on upper and lower limits of product composition based on spectrometric measurements.

Interpretation of the Ringbom plot leads to the same conclusions we deduced from Fig. 2.15. The error is lowest at approximately $100 - \%T = 63$, or $37\%T$. The relative analysis error per 1% transmittance is about 2.8%. The error is not significantly greater over the range $(100 - \%T)$ of 40–80%T, or between 20% and 60%T; this is the steep, nearly linear portion of the Ringbom plot. At very low and very high values of $100 - \%T$, the slope of the Ringbom plot approaches zero and therefore the % relative analysis error approaches infinity.

Table 2.10 provides a summary of the nomenclature, symbols, and definitions commonly used in spectroscopy. We will use these symbols throughout many of the later chapters.

2.6. OPTICAL SYSTEMS USED IN SPECTROSCOPY

In optical analytical spectroscopy the absorption or emission of radiation by a sample is measured. The instrumentation designed to measure absorption or emission of radiation must provide information about the wavelengths that are absorbed or emitted and the

Table 2.10 Nomenclature and Definitions for Spectroscopy

Term	Symbol	Definition
Transmittance	T, where $T = I/I_0$	Ratio of light intensity after passing through sample, I, to light intensity before passing through sample, I_0
Absorbance	A	$-\log T = abc$
Absorptivity	a	The proportionality constant a in Beer's Law, $A = abc$ where A is absorbance, c is concentration, and b is path length
Molar absorptivity	ε	The proportionality constant ε in Beer's Law, $A = \varepsilon bc$, where A is absorbance, b is path length, and c is concentration of the absorbing solution in molarity (M)
Path length	b	Optical path length through the sample
Sample concentration	c	Amount of sample (usually in terms of the absorbing species) per unit volume or mass. Typical units are g/mL, mol/L, ppm, %
Absorption maximum	λ_{max}	Wavelength at which greatest absorption occurs
Wavelength	λ	Distance between consecutive wave crests
Frequency	ν	Number of oscillations of a wave per second; the number of wave crests passing a given point per second
Wavenumber	$\bar{\nu}$	$1/\lambda$ or the number of waves per centimeter

intensity (I) or absorbance (A) at each wavelength. The instrumentation for spectroscopic studies from the UV through the infrared regions of the spectrum is very similar in its fundamental components. For the moment, the term **spectrometer** will be used to mean an instrument used for optical spectroscopy. More specific terms for instruments will be defined after the components are discussed.

Instruments for analytical spectroscopy require a radiation source, a wavelength selection device such as a monochromator, a sample holder transparent to the radiation range being studied, a detector to measure the intensity of the radiation and convert it to a signal, and some means of displaying and processing the signal from the detector. FT spectrometers, discussed subsequently, do not require a wavelength selection device. If emitted radiation is being measured, the sample, excited by some means, is the radiation source. If absorption, fluorescence, phosphorescence, or scattering of light is measured, an external radiation source is required. The specific arrangement of these components is referred to as the **optics** or **optical configuration** or **optical layout** of the instrument. The optical layout of a simple single-beam absorption spectrometer is shown schematically in Fig. 2.17. The placement of the sample holder and the wavelength

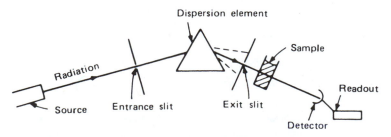

Figure 2.17 Schematic diagram of a single-beam absorption spectrometer.

selector may be inverted; in UV/VIS absorption spectrometry, the sample holder is usually placed after the wavelength selector, so that monochromatic light falls on the sample. For atomic absorption, IR, and fluorescence spectroscopy, the sample is usually placed in front of the wavelength selector.

2.6.1. Radiation Sources

An ideal radiation source for spectroscopy should have the following characteristics:

1. The source must emit radiation over the entire wavelength range to be studied.
2. The intensity of radiation over the entire wavelength range must be high enough so that extensive amplification of the signal from the detector can be avoided.
3. The intensity of the source should not vary significantly at different wavelengths.
4. The intensity of the source should not fluctuate over long time intervals.
5. The intensity of the source should not fluctuate over short time intervals. Short time fluctuation in source intensity is called "flicker".

The design of the radiation source varies with the wavelength range for which it is used (e.g., IR, UV, visible) and details of specific sources will be discussed in the appropriate instrumentation chapter. Most sources will have their intensities change exponentially with changes in voltage, so in all cases a reliable, steady power supply to the radiation source is required. Voltage regulators (also called line conditioners) are available to compensate for variations in incoming voltage. A double-beam optical configuration may also be used to compensate for variations in source stability, as described in Section 2.6.5.

There are two major types of radiation sources used in analytical spectroscopy, **continuum sources** and **line sources**. Continuum sources emit radiation over a wide range of wavelengths and the intensity of emission varies slowly as a function of wavelength. Typical continuum sources include the tungsten filament lamp which produces visible radiation (white light), the deuterium lamp for the UV region, high pressure mercury or xenon arc lamps for the UV region, and heated solid ceramics or heated wires for the IR region of the spectrum. Xenon arc lamps are also used for the visible region. Continuum sources are used for most molecular absorption and fluorescence spectrometric instruments. Line sources, in contrast, emit only a few discrete wavelengths of light, and the intensity is a strong function of the wavelength. Typical line sources include hollow cathode lamps and electrodeless discharge lamps, used in the UV and visible regions for AAS and atomic fluorescence spectrometry, sodium or mercury vapor lamps (similar to the lamps now used in street lamps) for lines in the UV and visible regions, and lasers. Lasers are high intensity **coherent** line sources; lasers are available with emission lines in the UV, visible, and IR regions. They are used as sources in Raman spectroscopy, molecular and atomic fluorescence spectroscopy.

2.6.2. Wavelength Selection Devices

2.6.2.1. Filters

The simplest and most inexpensive way to select certain portions of the electromagnetic spectrum is with a filter. There are two major types, absorption filters and interference filters. Absorption filters can be as simple as a piece of colored glass. In Section 2.1, we discussed how blue glass transmits blue wavelengths of the visible spectrum but absorbs red and yellow wavelengths. This is an example of an absorption filter for isolating the blue region of the visible spectrum. Colored glass absorption filters can be purchased that isolate various ranges of visible light. These filters are stable, simple, and cheap, so they are excellent for use in portable spectrometers designed to be carried into the field. The biggest limitation is that the range of wavelengths transmitted is broad compared with prisms and gratings which are also devices used to select a narrow wavelength range from a broad band polychromatic source. The transmission range may be 50–300 nm for typical absorption filters. Absorption filters are limited to the visible region of the spectrum and the X-ray region.

The second type of filter is the interference filter, constructed of multiple layers of different materials. The filter operates on the principle of constructive interference to transmit selected wavelength ranges. The wavelengths transmitted are controlled by the thickness and refractive index of the center layer of material. Interference filters can be constructed for transmission of light in the IR, visible, and UV regions of the spectrum. The wavelength ranges transmitted are much smaller than for absorption filters, generally 1–10 nm, and the amount of light transmitted is generally higher than for absorption filters.

2.6.2.2. Monochromator

A **monochromator** consists of a dispersion element, an entrance slit and an exit slit, plus lenses and mirrors for collimating and focusing the beam of radiation. The function of the dispersion element is to spread out in space, or *disperse*, the radiation falling on it according to wavelength. The two most common types of dispersion elements are prisms and gratings. You are probably already familiar with the ability of a prism to disperse white light into a rainbow of its component colors.

The entrance slit allows light from the source to fall on the dispersion element. The dispersed light falls on the exit slit of the monochromator. The function of the exit slit is to permit only a very narrow band of light to pass through to the sample and detector. One way to accomplish this is to rotate the dispersion element to allow dispersed light of different wavelengths to fall on the exit slit in sequence. For example, a white light source is dispersed into violet through red light by a prism or grating. The dispersion element is rotated slowly, allowing first violet light through the exit slit, then blue light, and so on all the way to red light. In this way, the monochromator sorts polychromatic radiation from a source into nearly monochromatic radiation leaving the exit slit.

Prisms.   Prisms are used to disperse IR, visible, and UV radiation. The most common prisms are constructed of quartz for the UV region, silicate glass for the visible and near-IR region, and NaCl or KBr for the IR region. Prisms are shaped like bars with triangular cross-sections. The $60°–60°–60°$ triangle, called a Cornu prism and shown in Fig. 2.17, is widely used, but other geometries are available. Polychromatic light passing through the entrance slit is focused on a face of the prism such that *refraction*, or bending, of the incident light occurs. Different wavelengths of light are refracted to different degrees, and the spatial separation of wavelengths is therefore possible. The refractive index of prism materials varies with wavelength. A quartz prism has a higher index of refraction for short wavelength radiation than for long wavelength radiation; therefore, short wavelength radiation is bent more than long wavelength radiation. In the visible region of the spectrum, red light would be bent less than blue light on passing through such a prism, as shown in Fig. 2.18. Prisms were historically the most used dispersion devices in monochromators, but they have been replaced by diffraction gratings or by FT systems.

Diffraction Gratings.   UV, visible, and IR radiation can be dispersed by a diffraction grating. A diffraction grating consists of a series of closely spaced parallel grooves cut

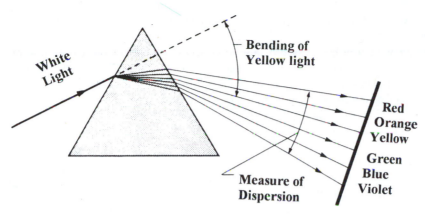

Figure 2.18   Dispersion of visible light by a prism.

(or *ruled*) into a hard glass, metallic, or ceramic surface (Fig. 2.19). The surface may be flat or concave, and is usually coated on the ruled surface with a reflective coating. A grating for use in the UV and visible regions will contain between 500 and 5000 grooves/mm, while a grating for the IR region will have between 50 and 200 grooves/mm. Traditionally, the grooves in a grating were cut mechanically with a diamond tipped tool, a time-consuming and expensive operation. Such a grating is called a *master*, and is used as a mold for casting *replica* gratings out of polymer resin. The replica gratings duplicate the grooves in the master and are coated with a reflective coating, such as aluminum, for use. Most instruments use replica gratings because they are much less expensive than a master grating. Most gratings are now produced by a holographic technique. The grating is made by coating the grating substrate with an optically flat photosensitive polymer film. The film is exposed to the interference pattern from laser beams and the interference pattern is "burned" into the film. The grooves from the interference pattern are then etched into the substrate to make the master grating, using chemical or ion etching to shape the grooves to the desired shape. The use of a laser interference pattern to form the grooves results in more perfect gratings at lower cost than mechanically ruled master gratings. These gratings, called holographic gratings, can be used in instruments directly or can serve as master gratings for the manufacture of replica holographic gratings. Holographic gratings can be made in many shapes other than the traditional plane or concave shape and

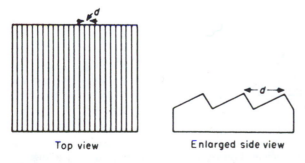

Top view   Enlarged side view

Figure 2.19   Highly magnified schematic view of a diffraction grating.

the grooves may be uniform or nonuniform in spacing, depending on the application. The size of a typical diffraction grating varies from about 25×25 to 110×110 mm.

Dispersion of light at the surface of a grating occurs by diffraction. Diffraction of light occurs because of constructive interference between reflected light waves. The path of one wave is shown in Fig. 2.20. Parallel waves can be envisioned on adjacent grooves. Constructive interference or diffraction of light occurs when

$$n\lambda = d(\sin i \pm \sin \theta) \tag{2.17}$$

where n is the order of diffraction (must be an integer: 1, 2, 3...); λ, the wavelength of the radiation; d, the distance between grooves; i, the angle of incidence of the beam of light; and θ, the angle of dispersion of light of wavelength λ.

The angle of incidence i and the angle of dispersion θ are both measured from the normal to the grating. For a given value of n, but different values of λ, the angle of dispersion θ is different. Separation of light occurs because light of different wavelengths is dispersed (diffracted) at different angles.

One problem with gratings is that light with several different wavelengths may leave the grating at the same angle of dispersion. For example, suppose that a beam of radiation falls on a grating at an angle i. The angle of dispersion of the radiation is described by Eq. (2.19). For a given angle of dispersion θ, the product $n\lambda$ is a constant. Any combination of n and λ that equals this constant will satisfy the equation. If $\lambda = 600$ nm and $n = 1$ gives an angle of dispersion $= \theta$; then, if $\lambda = 200$ nm and n is 3, the angle is also θ, and so on. In practice, radiation with each of these wavelengths is dispersed at an angle θ and travels down the same light path, as illustrated in Fig. 2.21. Wavelengths of light that are related in this way are said to be different *orders* of diffracted radiation. They are not separated by gratings, as is seen in Fig. 2.21. The wavelengths of radiation traveling the same path after dispersion are related by the number n, which may take the value of any whole number. On high quality spectrometers, different orders are separated by using a small prism or a filter system as an *order sorter* in conjunction with the grating (Figs. 2.22 and 2.23). It is common for IR instruments to use filters as order sorters. As the grating

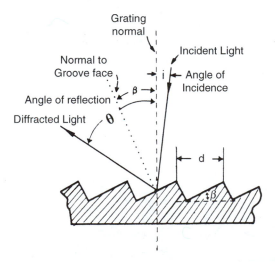

Figure 2.20   Cross-section diagram of a diffraction grating showing diffraction of a single beam of light. Symbols: i = angle of incidence, θ = angle of diffraction (or reflectance), β = blaze angle of the grating, d = grating spacing. (Modified from Dean and Rains, used with permission.)

Figure 2.21 The angle of diffraction of light from a grating depends not only on the wavelength but also on the *order of diffraction, n*. Wavelengths of 200, 300, and 600 nm are diffracted at different angles in first order ($n = 1$), but 200 nm light in third order ($n = 3$) and 300 nm light in second order ($n = 2$) diffract at the same angle as 600 nm light in first order. The three wavelengths overlap.

rotates to different wavelength ranges, the filters rotate to prevent order overlap, and only one wavelength reaches the detector. An excellent tutorial on diffraction gratings and the optics of spectroscopy is available on the Internet from the instrument company Jobin Yvon at www.jyhoriba.co.uk by typing "optics tutorial" into the search box.

2.6.2.3. *Resolution Required to Separate Two Lines of Different Wavelength*

Resolution of a Monochromator. The ability to disperse radiation is called *resolving power*. Alternative designations include *dispersive power* and *resolution*. For example, in order to observe an absorption band at 599.9 nm without interference from an absorption band at 600.1 nm, we must be able to resolve, or separate, the two bands.

The resolving power R of a monochromator is equal to $\lambda/\delta\lambda$, where λ is the average of the wavelengths of the two lines to be resolved and $\delta\lambda$ is the difference in wavelength between these lines. In the present example the required resolution is

$$R = \frac{\lambda}{\delta\lambda} \tag{2.18}$$

$$R = \frac{\text{average of 599.9 and 600.1}}{\text{absolute difference between 599.9 and 600.1}} = \frac{600}{0.2} = 3000$$

Figure 2.22 Prism used as an order sorter for a grating monochromator.

Figure 2.23   A filter is used as an order sorter to prevent higher order wavelengths from reaching the grating.

Resolution of a Prism.   The resolving power R of a prism is given by

$$R = t\frac{d\eta}{d\lambda} \tag{2.19}$$

where t is the thickness of the base of the prism and $d\eta/d\lambda$ is the rate of change of dispersive power (or *refractive index*) η of the material of the prism with wavelength. For the resolution of two beams at two wavelengths λ_1 and λ_2, it is necessary that the refractive index of the prism be different at these wavelengths. If it is constant, no resolution occurs. The resolving power of a prism increases with the thickness of the prism. Resolution can be maximized for a given wavelength region by choosing the prism material to maximize $d\eta/d\lambda$. For example, glass prisms disperse visible light better than quartz prisms. For maximum dispersion, a prism is most effective at wavelengths close to the wavelength at which it ceases to be transparent. At longer wavelengths, the resolving power decreases.

Resolution of a Grating.   The resolving power of a grating is given by

$$R = nN \tag{2.20}$$

where n is the order and N is the total number of grooves in the grating that are illuminated by light from the entrance slit. Therefore, longer gratings, smaller groove spacing, and the use of higher orders ($n > 1$) result in increased resolution. Suppose that we can obtain a grating with 500 lines/cm. How long a grating would be required to separate the sodium D lines at 589.5 and 589.0 nm in first order?

We know from Eq. (2.18) that the required resolution R is given by

$$R = \frac{589.25}{0.5} = 1178.5$$

The resolution of the grating must therefore be at least 1179 (to four significant figures). But R (for a grating) $= nN$; therefore, $1179 = nN$. Since we stipulated first order, $n = 1$; hence N, the total number of lines, is 1179. But the grating contains 500 lines/cm. It must be (1179/500) cm long, or 2.358 cm. This assumes that all of the grating surface is illuminated during use.

In a separate example, we may ask how many lines per centimeter must be cut on a grating 3.00 cm long to resolve the same sodium D lines, again assuming that the entire grating is illuminated. As before, the required resolution is 1179, and for first order, $nN = N = 1179$ total lines required. Therefore, the number of lines needed per cm is:

$$N/\text{cm} = 1179/3.00\,\text{cm} = 393\,\text{lines/cm}$$

It is not possible to cut a fraction of a line or to illuminate a fraction of a line; hence N must be a whole number in all calculations. This may require rounding off a calculated answer to the nearest whole number of lines.

Dispersion of a Grating Monochromator. The resolution of a monochromator measures its ability to separate adjacent wavelengths from each other. Resolution is related to a useful quantity called the reciprocal dispersion, or reciprocal linear dispersion, D^{-1}.

$$D^{-1} = \frac{d\lambda}{dy} \tag{2.21}$$

where the reciprocal linear dispersion equals the change in wavelength, $d\lambda$, divided by dy, the corresponding change in y, the distance separating the wavelengths along the dispersion axis. Units for D^{-1} are usually nm/mm. The reason D^{-1} is useful is that the **spectral bandpass** or **spectral bandwidth** of the light exiting a monochromator is directly related to D^{-1} and the slit width of the monochromator.

$$\text{Spectral bandwidth} = sD^{-1} \tag{2.22}$$

where s is the slit width of the monochromator. The spectral bandwidth represents the width of wavelength range of 75% of the light exiting the monochromator. For a monochromator that uses a grating as the dispersion device, the reciprocal linear dispersion for the grating is:

$$D^{-1} = \frac{d}{nF} \tag{2.23}$$

where d is the distance between two adjacent grooves on the grating, n is the diffraction order, and F is the focal length of the monochromator system. The reciprocal dispersion for a grating is therefore essentially constant with respect to wavelength.

The dispersion of a prism-based monochromator is a more complex calculation and will not be covered, due to the predominance of gratings in even inexpensive modern instruments.

Echelle Monochromator. From Figs. 2.19 and 2.20, you can see that the cuts shown on the surface of the gratings are not symmetrical v-shapes. Each cut has a short face and a long face. This type of grating is known as a blazed grating. The conventional blazed diffraction grating uses the long face of the groove, as seen in Fig. 2.20, and the angle of the cut, called the blaze angle β is generally optimized for first order diffraction. It is possible to rule a grating with a much higher blaze angle and to use the short side of the groove for diffraction; this type of grating is called an **echelle** grating. The angle of dispersion θ is much higher from an echelle grating than from a conventional grating. The echelle system improves dispersion by this increase in θ and by the use of higher orders (larger values of n). The result is a 10-fold improvement in resolution over a conventional grating monochromator of the same focal length. Because of the multiple high orders diffracted, it is necessary to use a second dispersing element to sort the overlapping orders. The second dispersing element, called a cross-disperser, is arranged to sort the light at right angles to the grating, so a two-dimensional (2D) spectrum results. An echelle optical layout for AES is shown in Fig. 2.24. An example of the 2D output, with wavelength plotted on the y-axis and diffraction order on the x-axis, is shown in Fig. 2.25. Commercial echelle spectrometers are used in AES and will be discussed in Chapter 7.

Figure 2.24 An Echelle spectrometer optical layout. The Echelle grating disperses the light to a second wavelength selector, called a cross-disperser. The cross-disperser may be a prism or a conventional grating. (From Boss and Fredeen, used with permission.)

Figure 2.25 Illustration of the 2D array of dispersed light produced by the Echelle spectrometer. (From Boss and Fredeen, used with permission.)

2.6.3.　Optical Slits

A system of slits (Fig. 2.17) is used to select radiation from the light beam both before and after it has been dispersed by the wavelength selector. The jaws of the slit are made of metal and are usually shaped like two knife edges. They can be moved relative to each other to change the mechanical width of the slit as desired. For the sake of simplicity, Fig. 2.17 does not show the system of lenses or mirrors used in a monochromator to focus and collimate the light as needed.

The *entrance slit* permits passage of a beam of light from the source. Radiation from the light source is focused on the entrance slit. Stray radiation is excluded. After being passed through the entrance slit, the radiation is collimated into a parallel beam of light, which falls onto and completely illuminates one side of the prism or the entire grating. The prism or grating disperses the light in different directions according to wavelength. At the setting selected for the dispersion device, one wavelength is refocused onto the exit slit. Light of other wavelengths is also focused, but not onto the exit slit. Ideally, the light is an image of the entrance slit. It is redirected and focused onto the detector for intensity measurement.

Lenses or front-faced mirrors are used for focusing and collimating the light. In the IR, front-faced mirrors are always more efficient than lenses and do not absorb the radiation. They are also easily scratched, since the reflecting surface is on the front and not protected by glass, as is the case with conventional mirrors. Back-faced mirrors are not used because the covering material (e.g., glass) may absorb the radiation. One type of monochromator system using mirrors for focusing and collimation and a grating for dispersion is presented in Fig. 2.26, with the entrance and exit slits shown.

The physical distance between the jaws of the slit is called the *mechanical slit width*. Instruments normally have a micrometer scale attached so that one can read off the

Figure 2.26　A grating monochromator showing the optical slits. The entrance slit is on the right and the exit slit on the left. (From Dean and Rains, used with permission.)

mechanical slit width directly; computer-controlled instruments set and read the slit width through the software that controls a stepper motor operating the slit mechanism. In UV absorption spectroscopy mechanical slit widths are of the order of $0.3-4\ \mu m$. In IR spectroscopy slit widths between 0.1 and 2.0 mm are common for dispersive instruments. There are no slits in FTIR spectrometers.

The wavelength range of the radiation that passes through the exit slit is called the *spectral bandpass* or *spectral bandwidth* or *spectral slit width*. This bandpass can be measured by passing an emission line of very narrow width through the slits to the detector. By rotating the dispersion element we can record the wavelength range over which response occurs. After correcting for the actual width of the emission line, we can calculate the spectral bandpass. For example, to measure the spectral bandpass for a monochromator system used as an AAS, we can use a cadmium hollow cathode lamp, which produces very narrow atomic emission lines from cadmium. One of those lines occurs at 228.8 nm. We move our dispersion device and monitor the signal at the detector. The emission line from cadmium gave a signal at all wavelengths from 228.2 to 229.4 nm. This means that the cadmium emission line reached the detector over a wavelength range that was 1.2 nm wide. Therefore, 1.2 nm is the spectral bandpass of this monochromator system. In this example, no correction was made for the actual width of the cadmium 228.8 nm line, which is about 0.001 nm and is negligible in this case. The signal that is measured in the above experiment has a Gaussian peak shape. The spectral bandwidth is usually defined as the width of the signal peak at one-half of the maximum peak height, called the full width at half maximum or fwhm. Spectral bandpasses are normally on the order of $0.3-4$ nm. Note that the spectral bandwidth is three orders of magnitude smaller than the physical slit width, nm vs. μm.

If the mechanical slit width were made wider, the spectral bandpass would simultaneously increase and vice versa. The spectral bandpass is one of the components of the spectrometer that affects resolution. For example, with the mechanical slit settings described, it would not be possible to resolve an emission line at 229.0 nm from the 228.8 nm Cd line, because both would pass through the slits. In practice, the slits are kept as narrow as possible to ensure optimum resolution; however, they must be wide enough to admit sufficient light to be measured by the detector. The final choice of slit width is determined by the analyst based on the particular sample at hand. A good rule of thumb is to keep the slits as narrow as possible without impairing the functioning of the detector or the ability to detect a specified amount of analyte.

By rotating the prism or grating (or by moving the exit slit across the light beam from the monochromator), the wavelength range passing through the exit slit can be changed. By continuously rotating the dispersion element from one extreme to another, the complete spectrum can be scanned.

2.6.4. Detectors

The detector is used to measure the intensity of the radiation that falls on it. Normally, it does this by converting the radiation energy into electrical energy. The amount of energy produced is usually low and must be amplified. The signal from the detector must be steady and representative of the intensity of radiation falling on it. If the signal is amplified too much, it becomes erratic and unsteady; it is said to be *noisy*. The degree of random variation in the signal is called the *noise level*. Amplifying the signal from the detector increases its response. In practice, the response can be increased until the noise level of the signal becomes too great; at this point the amplification is decreased until the noise level becomes acceptable.

There are a number of different types of photon detectors, including the photo-multiplier tube, the silicon photodiode, the photovoltaic cell, and a class of multichannel detectors called charge transfer devices. Charge transfer detectors include photodiode arrays, charge-coupled devices (CCDs), and charge-injection devices (CIDs). These detectors are used in the UV/VIS and IR regions for both atomic and molecular spectroscopy.

In addition to photon detectors, there are several important detectors that measure heat. These heat detectors or thermal detectors are particularly useful in the IR region, where the energy of photons is very low. The detectors will be discussed at length in the following chapters on specific techniques, for example, IR detectors in Chapter 4, photo-multiplier detectors and photodiodes in Chapter 5, and charge coupled devices and charge injection devices in Chapter 7.

2.6.5. Single-Beam and Double-Beam Optics

Single-beam optics, shown schematically in Fig. 2.17, are used for all spectroscopic *emission* methods. In emission procedures the sample is put where the source is located in Fig. 2.17. In spectroscopic *absorption* studies the intensity of radiation before and after passing through the sample must be measured. When single-beam optics are used, any variation in the intensity of the source while measurements are being made may lead to analytical errors. Slow variation in the average signal (not noise) with time is called *drift*, displayed in Fig. 2.27. Drift can cause a direct error in the results obtained. As shown in Fig. 2.27, a signal has been set to zero at Time 0 with no analyte present. As time increases toward Time 1, the signal with no analyte present (called the baseline signal) increases due to drift. At Time 1, a sample is measured and gives an increased signal due to analyte present (the peak shown above the baseline). The total signal, sample plus baseline, at Time 1 is 5 units. The baseline continues to drift upwards and at Time 2, the sample is measured again. As can be seen in the figure, the peak for the sample above the baseline is the same height as the peak at Time 1, but the total signal (peak plus baseline) is now 10 units. If the baseline drift were not accounted for, the analyst would conclude that the sample at Time 2 has twice as much analyte as the sample at Time 1—a direct error.

There are numerous sources of drift. The *radiation source* intensity may change because of line voltage changes, the source warming up after being recently turned on,

Figure 2.27   Error caused by baseline drift in a spectroscopic measurement.

or the source deteriorating with time. The *monochromator* may shift position as a result of vibration or heating and cooling causing expansion and contraction. The line voltage to the *detector* may change, or the detector may deteriorate with time and cause a change in response. Errors caused by drift lead to an error in the measurement of the emission signal or the absorption signal compared with the standards used in calibration. The problem can be reduced by constantly checking the light intensity or by using a standard solution measured at frequent intervals during the analysis. Single-beam optics are particularly subject to errors caused by drift. However, the problems associated with drift can be greatly decreased by using a double-beam system.

The double-beam system is used extensively for spectroscopic absorption studies. The individual components of the system have the same function as in the single-beam system, with one very important difference. The radiation from the source is *split* into two beams of approximately equal intensity using a *beam splitter*, shown in Fig. 2.28. One beam is termed the *reference beam*; the second beam, which passes through the sample, is called the *sample beam*. The two beams are then recombined and pass through the monochromator and slit systems to the detector. This is illustrated schematically in Fig. 2.28. In this schematic, there is a cell in the reference beam that would be identical to the cell used to hold the sample. The reference cell may be empty or it may contain the solvent used to dilute the sample, for example. This particular arrangement showing the monochromator after the sample is typical of a dispersive IR double-beam spectrophotometer. There are many commercial variations in the optical layout of double-beam systems.

As shown in Fig. 2.29(a), the beam splitter may be a simple mirror plate into which a number of holes are drilled. Light is reflected by the mirror plate and passes down the sample beam path. An equal portion of light passes through the holes in the plate and forms the reference beam. Another convenient beam splitter is a disk with opposite quadrants removed [Fig. 2.29(b)]. The disk rotates in front of the radiation beam and the mirrored surface reflects light into the sample path. The missing quadrants permit radiation to pass down the reference path. Each beam of light is intermittent and arrives at the detector in the form of an alternating signal. When no radiation is absorbed by the sample, the two beams are equal and recombine and form a steady beam of light. However, when radiation is absorbed by the sample the two beams are not equal, and an alternating signal arrives at the detector. This is illustrated in Fig. 2.30.

Using the double-beam system, we can measure the *ratio* of the reference beam intensity to the sample beam intensity. Because the ratio is used, any variation in the intensity of radiation from the source during measurement does not introduce analytical error.

Figure 2.28 Schematic diagram of a double-beam optical system.

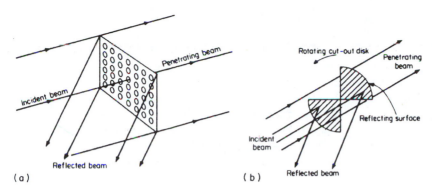

Figure 2.29 (a) Plate beam splitter. (b) Rotating disk beam splitter (or *chopper*).

This advantage revolutionized absorption spectroscopy. If there is a drift in the signal, it affects the sample and reference beams equally. The recombined beam will continue to give accurate signal information unless the drift is very great, in which case correction is not complete. Absorption measurements made using a double-beam system are virtually independent of drift and therefore more accurate.

2.6.6. Dispersive Optical Layouts

The configuration of the common components of dispersive spectroscopy systems is shown for the most used types of spectroscopy. In layouts 1 and 3, an external source of radiation is required, but for 3, the source is generally oriented at right angles to the sample. Emission, layout 2, does not require an external radiation source; the excited sample is the source. For absorption, fluorescence, phosphorescence, and scattering, the source radiation

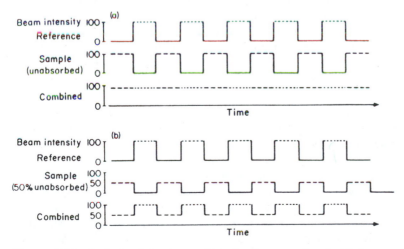

Figure 2.30 Radiation intensity reaching the detector using double-beam optics and a rotating disk beam splitter. (a) No absorption by the sample; (b) 50% absorption by the sample.

passes through the dispersive device, which selects the wavelength, and into the sample. The selected wavelength passes through the sample and reaches the detector, where the intensity of the signal is converted to an electrical signal. In emission, the radiation that emanates from the sample passes through the dispersive device, which selects one wavelength at a time to reach the detector. The detector signal in all types of spectroscopy is often processed (amplified, smoothed, derivatized, or otherwise transformed) and an output is obtained. The output may be graphical (e.g., a plot of intensity vs. wavelength or what we call a *spectrum*), tabular or both. In some absorption spectrometers, the position of the sample and the dispersive device may be reversed. In AAS for example, the sample is positioned between the source and the dispersive device for reasons discussed in Chapter 6.

1. Absorption spectroscopy:
   Source → Dispersive device → Sample → Detector → Data output
2. Emission spectroscopy:
   Sample → Dispersive device → Detector → Data output
3. Fluorescence, phosphorescence, and scattering spectroscopy:
   Sample → Dispersive device → Detector → Data output
   ↑
   Dispersive device
   ↑
   Source

2.6.7. Fourier Transform Spectrometers

The dispersive spectroscopy systems discussed above separate light into its component wavelengths and spread them into a spectrum. In these systems, the intensity can be measured at each point along a path where wavelength is proportional to position. The intensity over a narrow region around each point in the spectrum can be determined by slowly moving the grating so that each region of the dispersed spectrum passes by a single fixed detector or alternatively by simultaneously measuring all regions with a continuous array of detectors. The latter approach acquires more information in less time. It is achieved in the UV/VIS region by employing one-dimensional (1D) photodiode arrays or 2D CCDs, similar to those found in modern digital cameras. These will be discussed in Chapters 5–7.

Detectors for the less energetic IR wavelengths cannot be as easily miniaturized, so dispersive IR operates with the slow scanning approach. To obtain high wavelength resolution with scanning instruments requires restricting the wavelength region reaching the detector to a very narrow window. This in turn requires scanning the spectrum slowly to achieve a desired sensitivity.

Alternatively, one may measure the light at all wavelengths simultaneously in a manner that will permit reconstruction of the intensity vs. wavelength curve (i.e., the spectrum). If the wavelength information is encoded in a well-defined manner, such as by modulation of the light intensity using an interferometer, mathematical methods allow the information to be interpreted and presented as the same type of spectrum obtained from a dispersive instrument. An instrument that does this without a dispersive device is called a **multiplex** instrument. If all of the wavelengths of interest are collected at the same time without dispersion, the wavelengths and their corresponding intensities will overlap. The resulting overlapping information has to be sorted out in order to plot a spectrum. A common method of sorting or "deconvoluting" overlapping signals of varying frequency (or wavelength) is a mathematical procedure called Fourier analysis. The example presented here is to IR spectroscopy, its first application in instrumental analytical chemistry, but the principle is also employed with other techniques in which analytical data is displayed as a spectrum of response vs. frequency, for example, NMR and ion cyclotron resonance MS.

Fourier analysis permits any continuous curve, such as a complex spectrum of intensity peaks and valleys as a function of wavelength or frequency, to be expressed as a sum of sine or cosine waves varying with time. Conversely, if the data can be *acquired* as the equivalent sum of these sine and cosine waves, it can be *Fourier transformed* into the spectrum curve. This requires data acquisition in digital form, substantial computing power, and efficient software algorithms, all now readily available at the level of current generation personal computers. The computerized instruments employing this approach are called FT spectrometers—FTIR, FTNMR, and FTMS instruments, for example.

FT optical spectroscopy uses an **interferometer** similar in design to that of the Michelson interferometer shown schematically in Fig. 2.31. To simplify the discussion, we will initially consider a source that produces monochromatic radiation of wavelength λ. The source radiation strikes the beam splitter, which transmits half of the light to the fixed mirror and reflects the rest to a mobile mirror. The mobile mirror can be programmed to move at a precisely controlled constant velocity along the path of the beam. The beams are reflected from the mirrors back to the beam splitter. Half of each ray is directed through the sample holder to the detector. The other halves travel back in the direction of the source and do not need to be considered further. If the fixed and mobile mirrors are at exactly equal distances from the beam splitter, the two half beams will combine in phase. The combined wave will have twice the amplitude of each half and the detector signal will be at its maximum. If the mobile mirror then moves a distance equal to $\lambda/4$, the two half beams will combine 180° (i.e., $\lambda/2$) out of phase. The beams interfere destructively and the detector registers no signal. For all other values of path difference between the

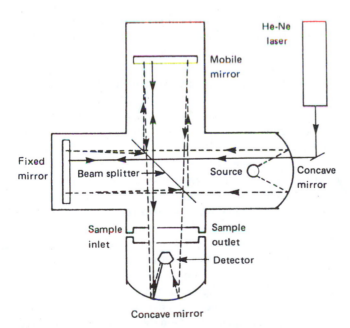

Figure 2.31   Schematic of a Michelson interferometer-based FTIR spectrometer.

mirrors, partial destructive interference occurs. As the mobile mirror moves at constant speed, the signal reaching the detector cycles regularly through this repetitive pattern of constructive and destructive interference. It maximizes when the path difference δ is an integral multiple of λ and goes to zero when δ is a half-integral multiple of λ. In FTIR, δ is called the *retardation*. For monochromatic light a plot of the signal (power, intensity) vs. δ is called an **interferogram** and has the form of a simple pure cosine curve:

$$P(\delta) = B(\bar{u}) \cos(2\pi\delta\bar{u}) \tag{2.24}$$

where $P(\delta)$ is the amplitude of the signal reaching the detector, $B(\bar{u})$ is a frequency dependent constant that accounts for instrumental variables such as detector response, the amount of light transmitted or reflected by the beam splitter, and the source intensity. The wavenumber $\bar{u}$ is equal to $1/\lambda$. The interferogram is the record of the interference signal reaching the detector. It is actually a "time-domain" spectrum; it records how the detector response changes with time. If the sample absorbs light at a specific frequency, the amplitude of that frequency changes. For a continuum source (a source with a continuously variable output of all wavelengths over the region of interest) the interferogram is a complex curve that can be represented as the sum of an infinite number of cosine waves and different amplitudes that reflect absorption of light by the sample. Although complex, if the interferogram is sampled at a sufficient number of points, modern computers using an FFT can process the interferogram and identify enough of the cosine waves to permit deconvolution of the data into the IR spectrum plot of intensity vs. wavelength. This will be discussed at greater length in Chapter 5 for FTIR, and in Chapter 3 for FTNMR.

2.6.7.1. Advantages of FT Systems

Compared with dispersive systems FT spectrometers produce better S/N ratios. This results from several factors. FT instruments have fewer optical elements and no slits, so the intensity of the radiation reaching the detector is much higher than in dispersive instruments. The increase in signal increases the S/N ratio. This is called the *throughput* advantage or Jacquinot advantage. All available wavelengths are measured simultaneously, so the time needed to collect all the data to form a spectrum is drastically reduced. An entire IR spectrum can be collected in less than 1 s. This makes practical the collection and signal averaging of hundreds of repetitions of the spectrum measurement. The theoretical improvement in S/N from signal averaging is proportional to the square root of the number of spectra averaged, $(n)^{1/2}$. This advantage is called the *multiplex* or Fellgett advantage.

FT spectrometers have high wavelength reproducibility. In an FTIR spectrometer, the position of the mobile mirror during its motion is continuously calibrated with extreme precision by parallel interferometry through the optical system using highly monochromatic light from a visible range laser, seen in Fig. 2.31. This accurate position measurement translates into accurate and reproducible analytical wavelength measurements after Fourier transformation of the interferogram. This accurate position measurement permits the precise addition of multiple spectra to achieve the multiplex advantage.

It should be noted that FT spectrometers are single-beam instruments. The background must be collected separately from the sample spectrum. The ratio of the two spectra results in a background-corrected spectra, similar to that obtained from a double-beam instrument. While the sample and background spectra are not collected at exactly the same time, because the spectra can be collected rapidly and processed rapidly, background spectra can be collected regularly to avoid the problems encountered with a single-beam dispersive instrument.

2.7. SPECTROSCOPIC TECHNIQUE AND INSTRUMENT NOMENCLATURE

Spectroscopy and spectroscopic instrumentation has evolved over many years. It is not surprising that the terminology used has also evolved and is in fact constantly evolving. Scientific terms are often defined by professional organizations and sometimes these organizations do not agree on the definitions, leading to the use of the same term to mean different things. Scientists (and students) have to keep up to date on the meaning of terms in order to communicate effectively but must also know the older usage of terms in order to understand the literature. The term spectroscopy means the study of the interaction of electromagnetic radiation and matter. The term spectrometry is used for quantitative measurement of the intensity of one or more wavelengths of electromagnetic radiation. **Spectrophotometry** is a term reserved for absorption measurements, where the ratio of two intensities (sample and reference) must be measured, either simultaneously in a double-beam system or sequentially in a single-beam system. The term is gradually being replaced by spectrometry; for example, *atomic absorption spectrometry* is now more common than *atomic absorption spectrophotometry.*

The terms used for instruments generally distinguish how wavelengths are selected or the type of detector used. An optical spectrometer is an instrument that consists of a prism or grating dispersion device, slits, and a photoelectric detector to measure transmittance or absorbance. However, the term spectrometer is now also applied to IR interferometer-based FT systems that are nondispersive and have no slits. Spectrophotometer used to mean a double-beam spectrometer; however, the term is now used for both single-beam and double-beam dispersive spectrometers used for absorption measurements. A **photometer** is a spectroscopic instrument that uses a filter to select the wavelength instead of a dispersive device. A **spectrograph** is an instrument with a dispersive device that has a large aperture instead of a tiny exit slit and uses either photographic film for detection (now almost obsolete) or a solid-state imaging detector.

BIBLIOGRAPHY

Ayres, G.H. *Quantitative Chemical Analysis*, 2nd ed.; Harper and Row: New York, 1968.

Beaty, R.D.; Kerber, J.D. *Concepts, Instrumentation and Techniques in Atomic Absorption Spectrophotometry*; PerkinElmer, Inc.: Norwalk, CT, 1993.

Boss, C.B.; Fredeen, K.J. *Concepts, Instrumentation and Techniques in Inductively Coupled Plasma Optical Emission Spectrometry*, 2nd ed.; PerkinElmer, Inc.: Norwalk, CT, 1997.

Dean, J.A.; Rains, T.C., Eds. *Flame Emission and Absorption Spectrometry*; Marcel Dekker, Inc.: New York, 1971; Vol. 2.

Harris, D.C. *Quantitative Chemical Analysis*, 5th ed.; W.H. Freeman and Company: New York, 1999.

Hollas, J.M. *Modern Spectroscopy*; John Wiley and Sons, Ltd.: Chichester, 1996.

Ingle, J.D.; Crouch, S.R. *Spectrochemical Analysis*; Prentice-Hall, Inc.: Englewood Cliffs, NJ, 1988.

Koenig, J.L. *Anal. Chem.* **1994**, *66* (9), 515A.

Meehan, E.J. *Optical Methods of Analysis. Treatise on Analytical Chemistry*, 2nd ed.; John Wiley and Sons, Ltd.: Chichester, 1981.

Settle, F.A., Ed. *Handbook of Instrumental Techniques for Analytical Chemistry*; Prentice-Hall PTR: Upper Saddle River, NJ, 1997.

Skoog, D.A.; Holler, F.J.; Nieman, T.A. *Principles of Instrumental Analysis*, 5th ed.; Harcourt, Brace and Company: Orlando, FL, 1998.

Umland, J.; Bellama, J. *General Chemistry*, 3rd ed.; Brooks/Cole Publishing Co.: Pacific Grove, CA, 1999.

Willard, H.H.; Merrit, L.L.; Dean, J.A.; Settle, F.A. *Instrumental Methods of Analysis*, 7th ed.; Van Nostrand: New York, 1988.

Zumdahl, S.S.; Zumdahl, S.A. *Chemistry*, 5th ed.; Houghton Mifflin: Boston, MA, 2000.

SUGGESTED EXPERIMENTS

2.1   You will need a UV/VIS spectrophotometer for this experiment and plastic or glass sample holders, either cuvettes or test tubes, depending on your instrument.

(a)   Prepare suitable standard solutions of (1) 0.1 g $KMnO_4$ per liter of water, (2) 1.0 g $K_2Cr_2O_7$ per liter of water, and (3) water-soluble red ink diluted 50% with water.

(b)   Measure the absorption spectrum from 700 to 350 nm and determine the wavelength of maximum absorption for each solution

(c)   Measure the transmittance I/I_0 at the wavelength of maximum absorption determined for each solution.

2.2   (a)   Choose one of the standard solutions prepared in Experiment 2.1(a) and measure the transmittance at the wavelength where maximum absorption occurs. Take 50 mL of this solution (solution A) and transfer it to a 100 mL volumetric flask; make up to volume with deionized water. This is solution B. Measure and record the transmittance of solution B. Dilute solution B by 50% to obtain solution C. Measure the transmittance of solution C. Repeat this process to produce solutions D, E, and F.

(b)   Prepare a graph correlating transmittance T and the concentrations of solutions A, B, C, D, E, and F. *Note*: You can use most commercial spreadsheet programs (e.g. Excel) or the software package on many spectrophotometers to do the curve-fitting and graph.

(c)   From the data obtained in step (a), calculate A, the absorbance of each solution. Prepare a graph correlating A, the absorbance, with the concentrations of solutions A, B, C, D, E, and F.

(d)   Is one graph preferred over the other for use in obtaining information about concentrations of unknown samples? Why or why not?

2.3   (a) Add 10.0 mL of the $K_2Cr_2O_7$ solution prepared in Experiment 2.1 to 10.0 mL of the $KMnO_4$ solution prepared in Experiment 2.1. Mix well and measure the absorption spectrum. (b) Add 10.0 mL of the $K_2Cr_2O_7$ solution prepared in Experiment 2.1 to 10.0 mL of deionized water. Mix well and measure the absorption spectrum. (c) Add 10.0 mL of the $KMnO_4$ solution prepared in Experiment 2.1 to 10.0 mL of deionized water. Mix well and measure the absorption spectrum. Using the wavelength of maximum absorption for each compound, answer the following questions. Is there a change in the absorbance at the wavelengths of maximum absorption for the solution containing both compounds compared with the solutions containing a single compound? Is the total amount of light absorbed by the single solutions equal to the amount absorbed by the mixture? Would this change in absorbance (if any) be a source of error? Is the error positive or negative? Can you think of

ways to correct for this error if you have to measure a mixture of potassium permanganate and potassium dichromate?

2.4   Measure the absorbance of a freshly prepared aqueous solution of $KMnO_4$ at its
       wavelength of maximum absorption. The concentration of the solution should be
       such that the absorbance is about 0.6–0.8. Make the solution in a volumetric
       flask and make your first measurement by pouring the solution into the
       sample holder directly from the flask. Now, pour the solution from the flask
       into a beaker or wide-mouth jar (you want to maximize the surface area).
       Leave the container open to the atmosphere for 5 min and then measure the
       absorbance of the solution again. Repeat measurements at 5 min intervals. (If
       no change is seen, cap the sample and shake it well, then uncap and allow it
       to sit in the air.) Plot the measured absorbance against the time exposed to the
       air. The change is caused by the chemical instability of the $KMnO_4$ (it reacts
       with the air). If it were being used as a standard solution for calibration, this
       change would be a source of error. Many standard solutions are subject to this
       sort of error to a greater or lesser extent, and precautions must be taken to
       prevent and avoid this source of trouble. Suggest two ways to avoid this
       problem with $KMnO_4$.

PROBLEMS

2.1   A molecule absorbs radiation of frequency 3.00×10^{14} Hz. What is the
       energy difference between the molecular energy states involved?

2.2   What frequency of radiation has a wavelength of 500.0 nm?

2.3   Describe the transition that occurs when an atom absorbs UV radiation.

2.4   Arrange the following types of radiation in order of increasing wavelength: IR,
       radiowaves, X-rays, UV, and visible light.

2.5   For a given transition, does the degree of absorption by a population of atoms
       or molecules depend on the number in the ground state or the excited state?
       Explain.

2.6   For a given transition, does the intensity of emission by a population of atoms
       or molecules depend on the number in the ground state or the excited state?
       Explain.

2.7   Briefly describe three types of transitions that occur in most molecules, includ-
       ing the type of radiation involved in the transition.

2.8   State the mathematical formulation of the Beer–Lambert–Bouguer Law and
       explain the meaning of each symbol in the equation.

2.9   (a)   Define transmittance and absorbance.
       (b)   What is the relationship between concentration and (1) transmittance,
              (2) absorbance?

2.10   Using Fig. 2.16, calculate the slope of the tangent drawn through the lower
       point marked B by extending the line to cover a 10-fold difference in concen-
       tration. Confirm that the range shown for B–B for 1% R.E. is correct by
       finding where on the upper portion of the curve you have a slope equal to
       the one you just calculated. Repeat the calculation for point C and confirm
       the C–C range.

2.11   The following data were obtained in a external standard calibration for the
       determination of iron by measuring the transmittance, at 510 nm and

1.00 cm optical path, of solutions of Fe^{2+} reacted with 1, 10-phenanthroline to give a red-colored complex.

Fe conc. (ppm)	%T	Fe conc. (ppm)	%T
0.20	90.0	3.00	26.3
0.40	82.5	4.00	17.0
0.60	76.0	5.00	10.9
0.80	69.5	6.00	7.0
2.00	41.0	7.00	4.5

(a) Calculate A, the absorbance, for each solution and plot A against concentration of iron. (You can do this using a spreadsheet program very easily.) Does the system conform to Beer's Law over the entire concentration range? (b) Calculate the average molar absorptivity of iron when it is determined by this method. (c) Plot $(100 - \%T)$ against log concentration (Ringbom method). (1) What are the optimum concentration range and the maximum accuracy (percent relative error per 1% transmittance error) in this range? (2) Over what concentration range will the relative analysis error per 1% transmittance error not exceed 5%?

2.12 The following data were obtained in a standard calibration for the determination of copper, as $Cu(NH_3)_4^{2+}$, by measuring the transmittance using a filter photometer.

Cu conc. (ppm)	%T	Cu conc. (ppm)	%T
0.020	96.0	0.800	27.8
0.050	90.6	1.00	23.2
0.080	84.7	1.40	17.2
0.100	81.4	2.00	12.9
0.200	66.7	3.00	9.7
0.400	47.3	4.00	8.1
0.600	35.8		

Calculate A, the absorbance, for each solution and plot A against concentration of copper. (You can do this using a spreadsheet program very easily.) Does the system, measured under these conditions, conform to Beer's Law over the entire concentration range? Is any deviation from the law of small or of large magnitude? Suggest a plausible cause for any deviation.

2.13 An amount of 0.200 g of copper is dissolved in nitric acid. Excess ammonia is added to form $Cu(NH_3)_4^{2+}$, and the solution is made up to 1 L. The following aliquots of the solution are taken and diluted to 10.0 mL: 10.0, 8.0, 5.0, 4.0, 3.0, 2.0, and 1.0 mL. The absorbances of the diluted solution were 0.500, 0.400, 0.250, 0.200, 0.150, 0.100, and 0.050, respectively. A series of samples was analyzed for copper concentration by forming the $Cu(NH_3)_4^{2+}$ complex and measuring the absorbance. The absorbances were (a) 0.450, (b) 0.300, and (c) 0.200. What were the respective concentrations in the three copper solutions? If these three samples were obtained by weighing out separately (a) 1.000 g, (b) 2.000 g, and (c) 3.000 g of sample, dissolving and diluting to 10.0 mL, what was the original concentration of copper in each sample?

2.14 Describe the standard addition method for measuring concentration of an unknown. What are the advantages of this method of calibration?

2.15 Describe the use of an internal standard for calibration. What characteristics must a species possess to serve as an internal standard? What are the advantages of the internal standard method?

2.16 Describe what you would do for samples whose absorbances fell above the absorbance of your highest calibration standard.

2.17 What range of % transmittance results in the smallest relative error for an instrument limited by (a) noise in the thermal detector of an IR spectrometer? (b) shot-noise?

2.18 What is A if the percentage of light absorbed is (a) 90%, (b) 99%, (c) 99.9%, and (d) 99.99%.

2.19 What is the purpose of having and measuring a reagent blank?

2.20 An optical cell containing a solution was placed in a beam of light. The original intensity of the light was 100 units. After being passed through the solution, its intensity was 80 units. A second similar cell containing more of the same solution was also placed in the light beam behind the first cell. Calculate the intensity of radiation emerging from the second cell.

2.21 The transmittance of a solution 1.00 cm thick of unknown concentration is 0.700. The transmittance of a standard solution of the same material is also 0.700. The concentration of the standard solution is 100.0 ppm; the cell length of the standard is 4.00 cm. What is the concentration of the unknown solution?

2.22 A solution contains 1.0 mg of $KMnO_4$/L. When measured in a 1.00 cm cell at 525 nm, the transmittance was 0.300. When measured under similar conditions at 500 nm, the transmittance was 0.350. (a) Calculate the absorbance A at each wavelength. (b) Calculate the molar absorptivity at each wavelength. (c) What would T be if the cell length were in each case 2.00 cm? (d) Calculate the absorptivity (if concentration is in mg/L) for the solution at each wavelength.

2.23 A series of standard ammoniacal copper solutions was prepared and the transmittance measured. The following data were obtained:

Cu concentration	Transmittance	Sample	Transmittance
0.20	0.900	1	0.840
0.40	0.825	2	0.470
0.60	0.760	3	0.710
0.80	0.695	4	0.130
1.00	0.635		
2.00	0.410		
3.00	0.263		
4.00	0.170		
5.00	0.109		
6.00	0.070		

Plot the concentration against absorbance (use your spreadsheet program). The transmittance of solutions of copper of unknown concentrations was also measured in the same way and the sample data in the above table were obtained. Calculate the concentration of each solution. What is missing from this experiment? List two things a good analytical chemist should have done to be certain that the results are accurate and precise.

2.24 List the components of a single-beam optical system for absorption spectroscopy. List the components of single-beam optical system for emission spectroscopy.

2.25 Describe the components in a grating monochromator. Briefly discuss the role of each component.

2.26 State the equation for the resolution of a grating.

2.27 (a) Define mechanical slit width.
(b) Define spectral bandpass or bandwidth.

2.28 What is the effect of mechanical slit width on resolution?

2.29 Write the expression for resolution of a grating ruled to be most efficient in second order. To resolve a given pair of wavelengths, will you need more or fewer lines if the grating were ruled in first order?

2.30 What resolution is required to separate two lines λ_1 and λ_2?

2.31 What resolution is required to resolve the Na D lines at 589.0 and 589.5 nm in first order?

2.32 How many lines must be illuminated on a grating to separate the Na D lines in second order?

2.34 A grating contains 1000 lines. Will it resolve two lines of λ 500.0 and 499.8 nm in first order if all 1000 lines are illuminated?

2.35 What are the components of a double-beam system? Describe two types of beam splitters.

2.36 How does a double-beam system correct for drift? Draw the alternating signal output from a double-beam system for a sample that absorbs 25% of the incident light.

2.37 Give an example of an absorption filter. Over what wavelength range do absorption filters function as wavelength selectors?

2.38 What are the advantages of absorption filters as wavelength selectors compared with gratings? What are the disadvantages?

2.39 Light of 300.0 nm is diffracted from a grating in first order at an angle of incidence normal to the grating (i.e., $i = 0°$). The grating contains 1180 grooves/mm. Calculate the angle of diffraction, θ, for this wavelength.

2.40 If an emission line for magnesium appears at 285.2 nm in first order, where will it appear in second order? Where will it appear in third order? If you needed to measure a first order iron emission line at 570 nm, will the presence of magnesium in the sample cause a problem? What can you do to solve the problem if one exists?

2.41 What are the major differences between an FT system and a dispersive system for spectroscopy?

2.42 Define the throughput advantage. How does it arise?

2.43 Define the multiplex advantage. How does it arise?

2.44 Draw two cosine waves of the same amplitude in phase. Draw the resulting wave if the two waves are combined. Draw two cosine waves 180° out of phase. Draw the resulting wave if these two waves are combined.

2.45 What is the difference between a spectrometer and a photometer? What is the difference between a spectrometer and a spectrograph?

3

Nuclear Magnetic Resonance Spectroscopy

3.1. INTRODUCTION

NMR spectroscopy is one of the most powerful techniques available for studying the structure of molecules. The NMR technique has developed very rapidly since the first commercial instrument, a Varian HR-30, was installed in 1952 at the Humble Oil Company in Baytown, Texas. These early instruments with small magnets were useful for studying protons (^{1}H) in organic compounds, but only in solution with high concentration of analyte or as neat liquids. That has now changed—much more powerful magnets are available. NMR instruments and experimental methods are now available that permit the determination of the 3D structure of proteins as large as 900,000 Da. "Magic angle" NMR instruments are commercially available for studying solids such as polymers, and ^{13}C, ^{19}F, ^{31}P, ^{29}Si, and other nuclei are measured routinely. NMR imaging techniques under the name *magnetic resonance imaging* (MRI) are in widespread use in noninvasive diagnosis of cancer and other medical problems. NMR instruments coupled to liquid chromatographs and mass spectrometers for separation and characterization of unknowns are commercially available. Resolution and sensitivity have both increased; detection and identification of ppm concentrations of substances with NMR is easily achieved in modern instruments and detection limits are approaching nanogram levels. NMR detection is being coupled with liquid chromatographic separation in HPLC-NMR instruments for identification of components of complex mixtures in the flowing eluant from the chromatograph, and NMR is now used as a nondestructive detector combined with mass spectrometry and chromatography in HPLC-NMR-MS instruments, an extremely powerful tool for organic compound separation and identification. In short, the field has broadened greatly in scope, especially since the 1970s, and gives every indication of continuing to advance for many years to come.

NMR involves the absorption of radiowaves by the nuclei of some combined atoms in a molecule that is located *in a magnetic field*. NMR can be considered a type of absorption spectroscopy, not unlike UV/VIS absorption spectroscopy. Radiowaves are low energy electromagnetic radiation. Their frequency is on the order of 10^7 Hz. The SI unit of frequency, 1 Hz, is equal to the older frequency unit, 1 cycle per second (cps) and has the dimension of s^{-1}. The energy of *radiofrequency* (RF) radiation can therefore be calculated from:

$$E = h\nu$$

where Planck's constant h is 6.626×10^{-34} J s and ν (the frequency) is between 4 and 1000 MHz (1 MHz = 10^6 Hz).

The quantity of energy involved in RF radiation is very small. It is too small to vibrate, rotate, or electronically excite an atom or molecule. It is great enough to affect the nuclear spin of atoms in a molecule. As a result, spinning nuclei of *some* atoms in a molecule *in a magnetic field* can absorb RF radiation and change the direction of the spinning axis. In principle, each chemically distinct atom in a molecule will have a different absorption frequency (or resonance) if its nucleus possesses a *magnetic moment*. The analytical field that uses absorption of RF radiation by such nuclei in a magnetic field to provide information about a sample is NMR spectroscopy.

In analytical chemistry, NMR is a technique that enables us to study the shape and structure of molecules. In particular, it reveals the different chemical environments of the NMR-active nuclei present in a molecule, from which we can ascertain the structure of the molecule. NMR provides information on the spatial orientation of atoms in a molecule. If we already know what types of compounds are present, NMR can provide a means of determining how much of each is in the mixture. It is thus a method for both qualitative and quantitative analyses, particularly of organic compounds. In addition, NMR is used to study chemical equilibria, reaction kinetics, the motion of molecules, and intermolecular interactions.

Three Nobel Prizes have been awarded in the field of NMR. The first was in 1952 to the two physicists, E. Purcell and F. Bloch, who demonstrated the NMR effect in 1946. In 1991, R. Ernst and W. Anderson were awarded the Nobel Prize for developing pulsed FTNMR and 2D NMR methods between 1960 and 1980. FTNMR and 2D experiments form the basis of most NMR experiments run today, even in undergraduate instrumental analysis laboratories. We will use the acronym NMR to mean FTNMR, since there are no other types of NMR instruments currently produced. The 2002 Nobel Prize in Chemistry was shared by three scientists for developing methods to use NMR and MS (MS is discussed in Chapters 9 and 10) in the analysis of large biologically important molecules such as proteins. K. Wüthrich, a Swiss professor of molecular biophysics, received the prize for his work in determining the 3D structure of proteins using NMR.

Since the 1970s, the technology associated with NMR has advanced dramatically. The theory, instrument design, and mathematics that make NMR so powerful are complex; a good understanding of quantum mechanics, physics, and electrical engineering is needed to understand modern NMR experiments. Fortunately, we do not need to completely understand the theory in order to make use of the technique. This chapter will address NMR in a simplified approach using a minimum of mathematics.

3.1.1. Properties of Nuclei

To understand the properties of certain nuclei in an NMR experiment, we must assume that nuclei rotate about an axis and therefore have a **nuclear spin**, represented as I, the spin quantum number. In addition, nuclei are charged. The spinning of a charged body produces a **magnetic moment** along the axis of rotation. For a nucleus to give a signal in an NMR experiment, it must have a nonzero spin quantum number and must have a magnetic dipole moment.

As a nucleus such as ^1H spins about its axis, it displays two forms of energy. Because the nucleus has a mass and because that mass is in motion (it is spinning), the nucleus has spin angular momentum and therefore mechanical energy. So the first form of energy is *mechanical energy*. The formula for the mechanical energy of the hydrogen nucleus is:

$$\text{spin angular momentum} = \frac{h}{2\pi}\sqrt{I(I+1)} \tag{3.1}$$

where I is the spin quantum number. For example, $I = 1/2$ for the proton ^1H.

The spin quantum number I is a physical property of the nucleus, which is made up of protons and neutrons. From observations of the nuclear spins of known nuclei, some empirical rules for predicting the spin quantum numbers can be tabulated. These rules are summarized in Table 3.1. For example, ^{12}C has atomic weight 12 and atomic number 6. Hence it has 6 protons (atomic number $= 6$) and 6 neutrons (atomic weight $-$ atomic number $=$ No. of neutrons, so $12 - 6 = 6$ neutrons). Since the mass and the number of protons are both even numbers, Table 3.1 predicts that the net spin quantum number of ^{12}C is zero, denoting no spin. Therefore the spin angular momentum [Eq. (3.1)] is zero and ^{12}C does not possess a magnetic moment. Nuclei with $I = 0$ do not absorb RF radiation when placed in a magnetic field and therefore do not give an NMR signal. NMR cannot measure ^{12}C, ^{16}O, or any other nucleus with both an even atomic mass and an even atomic number.

For ^{13}C, on the other hand, the atomic weight is 13 (i.e., $P + N = 13$), an odd number and the atomic number is 6, an even number. From Table 3.1, we predict that I for ^{13}C is therefore a half integer; like ^{1}H, ^{13}C has $I = 1/2$. So NMR can detect ^{13}C, and although ^{13}C represents only 1.1% of the total C present in an organic molecule, ^{13}C NMR spectra are very valuable in elucidating the structure of organic molecules.

The physical properties predict whether the spin number is equal to zero, a half integer, or a whole integer, but the actual spin number—for example, 1/2 or 3/2 or 1 or 2—must be determined experimentally. All elements in the first six rows of the periodic table have at least one stable isotope with a nonzero spin quantum number, except Ar, Tc, Ce, Pm, Bi, and Po. It can be seen from Table 3.1 and Appendix 10.1 that many of the most abundant isotopes of common elements in the periodic table cannot be measured by NMR, notably those of C, O, Si, and S, which are very important components of many organic molecules of interest in biology, the pharmaceutical industry, the polymer industry, and the chemical manufacturing industry. Some of the more important elements that can be determined by NMR and their spin quantum numbers are shown in Table 3.2. The two nuclei of most importance to organic chemists and biochemists, ^{13}C and ^{1}H, both have a spin quantum number $= 1/2$.

The second form of nuclear energy is *magnetic*. It is attributable to the electrical charge of the nucleus. Any electrical charge in motion sets up a magnetic field. The nuclear magnetic moment μ expresses the magnitude of the magnetic dipole. The ratio of the nuclear magnetic moment to the spin quantum number is called the **magnetogyric** (or gyromagnetic) **ratio** and is given the symbol γ. Therefore $\gamma = \mu/I$. This ratio has a different value for each type of nucleus. The magnetic field of a nucleus that possesses a nuclear magnetic moment can and does interact with other local magnetic fields. The basis of NMR is the study of the response of such magnetically active nuclei to an external applied magnetic field.

3.1.2. Quantization of ^{1}H Nuclei in a Magnetic Field

When a nucleus is placed in a very strong, uniform external magnetic field B_0, the nucleus tends to become lined up in definite directions relative to the direction of the magnetic

Table 3.1 Rules Predicting Spin Numbers of Nuclei

Mass ($P + N$) (atomic weight)	Charge (P) (atomic number)	Spin quantum number (I)
Odd	Odd or even	$1/2, 3/2, 5/2, \ldots$
Even	Even	0
Even	Odd	1, 2, 3

Table 3.2   NMR-Active Nuclei and Their Spin Quantum Numbers

Element isotope	I	Element isotope	I
^{13}C	1/2	^{35}Cl	3/2
^{17}O	5/2	^{37}Cl	3/2
^{1}H	1/2	^{79}Br	3/2
^{2}H (deuterium)	1	^{81}Br	3/2
^{3}H (tritium)	1/2	^{125}I	5/2
^{19}F	1/2	^{129}I	7/2
^{31}P	1/2	^{14}N	1
^{29}Si	$-1/2$	^{15}N	1/2
^{33}S	3/2	^{10}B	3
^{35}S	3/2	^{11}B	3/2

field. Each relative direction of alignment is associated with an energy level. Only certain well-defined energy levels are permitted; that is, the energy levels are **quantized**. Hence the nucleus can become aligned only in well-defined directions relative to the magnetic field B_0. (*Note*: The symbol B is the SI symbol for magnetic field; many texts still use the symbols H and H_0 for magnetic field.)

The *number of orientations* or number of magnetic quantum states is a function of the physical properties of the nuclei and is *numerically equal* to $2I + 1$:

$$\text{number of orientations} = 2I + 1 \tag{3.2}$$

In the case of ^{1}H, $I = 1/2$; hence the number of orientations is $2 \times (1/2) + 1 = 2$. The permitted values for the magnetic quantum states, symbolized by the magnetic quantum number, m, are $I, I - 1, I - 2, \ldots, -I$. Consequently, for ^{1}H only two energy levels are permitted, one with $m = 1/2$ and the other with $m = -1/2$. The splitting of these energy levels in a magnetic field is called nuclear Zeeman splitting. (This is analogous to the electronic Zeeman effect, the splitting of electronic energy levels in a magnetic field. The electronic Zeeman effect is discussed in Chapter 6.)

When a nucleus with $I = 1/2$, such as ^{1}H, is placed in an external magnetic field, its magnetic moment lines up in one of two directions, with the applied field or against the applied field. This results in two discrete energy levels, one of higher energy than the other, as shown in Fig. 3.1. The *lower* energy level is that where the magnetic moment is aligned *with* the field. The lower energy state is energetically more favored than the higher energy state, so the population of the nuclei in the lower energy state will be higher than the population of the higher energy state. The difference in energy between levels is proportional to the strength of the external magnetic field. The axis of rotation also rotates in a circular manner about the external magnetic field axis, like a spinning top, as shown in Fig. 3.2. This rotation is called **precession**. The direction of precession is either *with* the applied field B_0 or *against* the applied field.

So we have nuclei, in this case, protons, with two discrete energy levels. In a large sample of nuclei, more of the protons will be in the lower energy state. The basis of the NMR experiment is to cause a transition between these two states by absorption of radiation. It can be seen from Fig. 3.1 that a transition between these two energy states can be brought about by absorption of radiation with a frequency that is equal to ΔE according to the relationship $\Delta E = h\nu$.

The difference in energy between the two quantum levels of a nucleus with $I = 1/2$ depends on the applied magnetic field B_0 and the magnetic moment μ of the nucleus. The

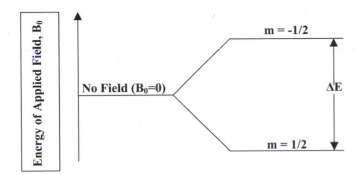

Figure 3.1 In the presence of an applied magnetic field, a nucleus with $I = 1/2$ can exist in one of two discrete energy levels. The levels are separated by ΔE. The lower energy level ($m = 1/2$) has the nuclear magnetic moment aligned with the field; in the higher energy state ($m = -1/2$), the nuclear magnetic moment is aligned against the field.

relationship between these energy levels and the frequency ν of absorbed radiation is calculated as follows. E is the expression for a given nuclear energy level in a magnetic field:

$$E = -m\left(\frac{\mu}{I}\frac{h}{2\pi}\right)B_0 = -m\left(\gamma\frac{h}{2\pi}\right)B_0 \tag{3.3}$$

where m is the magnetic quantum number; μ, the nuclear magnetic spin; B_0, the applied magnetic field; I, the spin angular momentum; γ, the magnetogyric ratio; and h, Planck's constant.

Equation (3.3) is the general equation for a given energy level for all nuclei that respond in NMR. However, if we confine our discussion to the ^{1}H nucleus, then $I = 1/2$. Therefore, there are only two levels.

For two energy levels with $m = +1/2$ and $-1/2$, respectively,

$$\Delta E = h\nu = -\left(+\frac{1}{2}\left(\frac{\mu}{I}\frac{h}{2\pi}\right)B_0\right) - \left(-\frac{1}{2}\left(\frac{\mu}{I}\frac{h}{2\pi}\right)B_0\right) \tag{3.4}$$

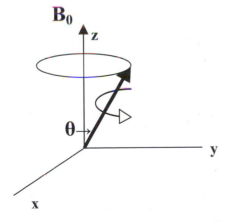

Figure 3.2 In the presence of an applied magnetic field, B_0, shown parallel to the $+z$-axis, a spinning nucleus precesses about the magnetic field axis in a circular manner. The nucleus is shown spinning counterclockwise (arrow with white head). The bold arrow with the black head represents the axis of rotation, which traces the circular path shown.

and

$$\Delta E = h v = \frac{\mu}{I} \frac{h}{2\pi} B_0 = \gamma \frac{h}{2\pi} B_0 \tag{3.5}$$

Therefore, the absorption frequency that can result in a transition of ΔE is:

$$v = \gamma \frac{B_0}{2\pi} \tag{3.6}$$

Equation (3.6) can be written as

$$\omega = \gamma B_0 \tag{3.7}$$

where ω is the frequency in units of rad/s.

Equation (3.7) is the **Larmor equation**, which is fundamental to NMR. It indicates that for a given nucleus there is a direct relationship between the frequency ω of RF radiation absorbed by that nucleus and the applied magnetic field B_0. This relationship is the basis of NMR.

The absorption process can be understood in terms of a classical approach to the behavior of a charged particle in a magnetic field. The spinning of the charged nucleus (Fig. 3.2) produces an angular acceleration, causing the axis of rotation to move in a circular path with respect to the applied field. As already noted, this motion is called precession. The frequency of precession can be calculated from classical mechanics to be $\omega = \gamma B_0$, the Larmor frequency. Both quantum mechanics and classical mechanics predict that the frequency of radiation that can be absorbed by a spinning charged nucleus in a magnetic field is the Larmor frequency.

As shown in Fig. 3.2, the axis of rotation of the precessing hydrogen nucleus is at an angle θ to the applied magnetic field. The energy of the precessing nucleus is equal to $E = \mu B_0 \cos \theta$. When energy in the form of RF radiation is absorbed by the nucleus, the angle θ must change. For a proton, absorption involves "flipping" the magnetic moment from being aligned with the field to being aligned against the applied field (Fig. 3.3). When the rate of precession equals the frequency of the RF radiation applied, absorption of RF radiation takes place and the nucleus becomes aligned *opposed* to the magnetic field and is in an excited state. To measure organic compounds containing protons by NMR, the sample is first put into a magnetic field and then irradiated with RF radiation. When the frequency of the radiation satisfies Eq. (3.7), the magnetic component of the radiant energy becomes absorbed. If the magnetic field B_0 is kept constant, we may plot the

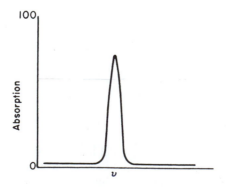

Figure 3.3 Absorption vs. the frequency of RF radiation.

absorption against the frequency v of the RF radiation. The resulting absorption curve should be similar to that shown in Fig. 3.3. The same experiment could be done by holding the RF frequency constant and varying B_0. Two actual NMR spectra of the compound toluene are shown in Fig. 3.4(a), the 300 MHz proton NMR spectrum at the bottom and the ^{13}C NMR spectrum at the top. Figure 3.4(b) is also the proton NMR spectrum of toluene, obtained at 60 MHz.

When a nucleus absorbs energy, it becomes excited and reaches an excited state. It then loses energy and returns to the unexcited state. Then it reabsorbs radiant energy and again enters an excited state. The nucleus alternately becomes excited and unexcited and is said to be in a state of *resonance*. This is where the term resonance comes from in nuclear magnetic resonance spectroscopy.

Magnetic field strengths are given in units of tesla (T) or gauss (G). The relationship between the two units is $1\ T = 10^4$ G. If the applied magnetic field is 14,092 G (or 1.41 T), the frequency of radiation (RF) absorbed by a proton is 60 MHz. The nomenclature 60 MHz NMR indicates the RF frequency for proton resonance and also defines the strength of the applied magnetic field if the nucleus being measured is specified. For example, the ^{13}C nucleus will also absorb 60 MHz RF radiation, but the magnetic field strength would need to be 56,000 G. Similarly, a 100 MHz proton NMR uses 100 MHz RF and a magnetic field of $14,092 \times 100/60 = 23,486$ G (2.35 T) for ^{1}H measurements. At that field strength, a ^{13}C nucleus would absorb at 25.1 MHz due to its very different magnetogyric ratio. If a frequency is specified for an NMR instrument without specifying the nucleus, the proton is assumed (e.g., a 500 MHz NMR would be assumed to refer to a proton absorbing at 500 MHz in order to calculate the magnetic field strength).

3.1.2.1. *Saturation and Magnetic Field Strength*

The energy difference ΔE between ground state and excited state nuclei is very small. The number of nuclei in the ground state is the number lined up with the magnetic field B_0. The ratio of excited nuclei to unexcited nuclei is defined by the Boltzmann distribution:

$$\frac{N^*}{N_0} = e^{-\Delta E/kT} = e^{-\gamma h B_0/2\pi kT} \tag{3.8}$$

where, N^* is the number of excited nuclei and N_0, the number of unexcited (ground state) nuclei. For a sample at 293 K in a 4.69 T magnetic field, the ratio $N^*/N_0 = 0.99997$. There are almost as many nuclei in the excited state as in the ground state because the difference between the two energy levels is very small. Typically, for every 100,000 nuclei in the excited state, there may be 100,003 in the ground state, as in this case. This is always the case in NMR; the Boltzmann ratio is always very close to 1.00. For this reason, NMR is inherently a low sensitivity technique.

A system of molecules in the ground state may absorb energy and enter an excited state. A system of molecules in an excited state may emit energy and return to the ground state. If the number of molecules in the ground state is equal to the number in the excited state, the net signal observed is zero and no absorption is noted. Consequently, a signal can be seen only if there is an excess of molecules in the ground state.

The excess of unexcited nuclei over excited nuclei is called the *Boltzmann excess*. When no radiation falls on the sample, the Boltzmann excess is maximum, N_x. However, when radiation falls on the sample, an increased number of ground-state nuclei become excited and a reduced number remain in the ground state. If the RF field is kept constant, a new equilibrium is reached and the Boltzmann excess decreases to N_s. When $N_s = N_x$, absorption is maximum. When $N_s = 0$, absorption is zero. The ratio N_s/N_x is called Z_0, the **saturation factor**.

Figure 3.4 (a) The 300 MHz proton NMR spectrum of toluene (bottom) and the ^{13}C NMR spectrum of toluene (top). Reprinted with permission of Aldrich Chemical Company, Inc. (b) The 60 MHz proton NMR spectrum of toluene. The structure of toluene is shown, with Me indicating a methyl group, —CH$_3$. Two absorption peaks are seen, one due to the protons of the methyl group, and the other to the aromatic ring protons. The spectrum is discussed in Section 3.4. (From Bhacca et al., courtesy of Varian Associates, Palo Alto, CA, www.varianinc.com.)

If the applied RF field is too intense, all the excess nuclei will be excited, $N_s \rightarrow 0$, and absorption $\rightarrow 0$. The sample is said to be **saturated**. The saturation factor Z_0 is:

$$Z_0 = (1 + \gamma^2 B_1^2 T_1 T_2)^{-1} \qquad (3.9)$$

where γ is the magnetogyric ratio, B_1 is the intensity of RF field, and T_1, T_2 are, respectively, the longitudinal and transverse relaxation times (discussed in Section 3.1.3.2).

As a consequence of this relationship, the RF field must not be very strong so as to avoid saturation. However, under certain experimental conditions, saturation of a particular nucleus can provide important structural information (Section 3.3).

From Eq. (3.8), we can derive the expression:

$$\frac{N^*}{N_0} = 1 - \frac{vhB_0}{2\pi kT} \qquad (3.10)$$

which shows that the relative number of excess nuclei in the ground state is related to B_0. As the field strength increases, the NMR signal intensity increases. This is the driving force behind the development of high field strength magnets for NMR.

3.1.3. Width of Absorption Lines

The resolution or separation of two absorption lines depends on how close they are to each other and on the absorption linewidth. The width of the absorption line (i.e., the frequency range over which absorption takes place) is affected by a number of factors, only some of which we can control. These factors are discussed below.

3.1.3.1. The Homogeneous Field

An important factor controlling the absorption linewidth is the applied magnetic field B_0. It is very important that this field be constant over all parts of the sample, which may be 1–2 in. long. If the field is not homogeneous, B_0 is different for different parts of the sample and therefore the frequency of the absorbed radiation will vary in different parts of the sample, as described by Eq. (3.7). This variation results in a wide absorption line. For qualitative analysis (i.e., structure determination), wide absorption lines are very undesirable, since we may get overlap between neighboring peaks and loss of structural information. The magnetic field must be constant within a few ppb over the entire sample and must be stable over the time required to collect the data. This time period is short for routine ^{1}H and ^{13}C measurements, on the order of 5–30 min, but may be hours or days for complex analyses. Most magnets used in NMR instruments do not possess this degree of stability. Several different experimental techniques are used to compensate for field inhomogeneity, such as spinning the sample holder in the magnetic field.

3.1.3.2. Relaxation Time

Another important feature that influences the absorption linewidth is the length of time that an excited nucleus stays in the excited state. The **Heisenberg uncertainty principle** tells us that:

$$\Delta E \Delta t = \text{constant} \qquad (3.11)$$

where ΔE is the uncertainty in the value of E and Δt is the length of time a nucleus spends in the excited state. Since $\Delta E \Delta t$ is a constant, when Δt is small, ΔE is large. But we know that $\Delta E = h\nu$ and that h is a constant. Therefore any variation in E will result in a variation in ν. If E is not an exact number but varies over the range $E + \Delta E$, then ν will not be exact but will vary over the corresponding range $\nu + \Delta \nu$. This can be restated as:

$$E + \Delta E = h(\nu + \Delta \nu) \tag{3.12}$$

We can summarize this relationship by saying that when Δt is small, ΔE is large and therefore $\Delta \nu$ is large. If $\Delta \nu$ is large, then the frequency range over which absorption takes place is wide and a wide absorption line results.

The length of time the nucleus spends in the excited state is Δt. This lifetime is controlled by the rate at which the excited nucleus loses its energy of excitation and returns to the unexcited state. The process of losing energy is called **relaxation**, and the time spent in the excited state is the **relaxation time**. There are two principal modes of relaxation: longitudinal and transverse. **Longitudinal relaxation** is also called spin–lattice relaxation; **transverse relaxation** is called spin–spin relaxation.

Longitudinal relaxation T_1. The entire sample in an NMR experiment, both absorbing and nonabsorbing nuclei, is called the **lattice**. An excited state nucleus (said to be in a *high spin* state) can lose energy to the lattice. When the nucleus drops to a lower energy (*low spin*) state, its energy is absorbed by the lattice in the form of increased vibrational and rotational motion. A very small increase in sample temperature results on spin–lattice (longitudinal) relaxation. This process is quite fast when the lattice molecules are able to move quickly. This is the state of affairs in most liquid samples. The excitation energy becomes dispersed throughout the whole system of molecules in which the sample finds itself. No radiant energy appears; no other nuclei become excited. Instead, as numerous nuclei lose their energy in this fashion, the temperature of the sample goes up very slightly. Longitudinal relaxation has a relaxation time, T_1, which depends on the magnetogyric ratio and the lattice mobility. In crystalline solids or viscous liquids T_1 is large because the lattice mobility is low.

Transverse relaxation T_2. An excited nucleus may transfer its energy to an unexcited nucleus nearby. In the process, a proton in the nearby unexcited molecule becomes excited and the previously excited proton becomes unexcited, for example. There is no net change in energy of the system, but the length of time that one nucleus stays excited is shortened because of the interaction. The average excited state lifetime decreases and line broadening results. This type of relaxation is called transverse relaxation or spin–spin relaxation, with a lifetime T_2.

It is found in practice that in liquid samples the net relaxation time is comparatively long and narrow absorption lines are observed. In solid samples, however, the transverse relaxation time T_2 is very short. Consequently ΔE and therefore $\Delta \nu$ are large. For this reason solid samples generally give wide absorption lines. As we will see, solid samples require a different set of experimental conditions than liquids to give useful analytical information from their NMR spectra. One approach is to make the solid behave more like a liquid. For example, solid polymer samples normally give broad NMR spectra. But if they are "solvated", narrower lines are obtained and the spectra are more easily interpreted. A sample is "solvated" by dissolving a small amount of solvent into the polymer. The polymer swells and becomes jelly-like but does not lose its chemical structure. The solvating process greatly slows down transverse relaxation and the *net* relaxation time is increased. The linewidth is decreased and resolution of the spectrum for structural information is better.

3.1.3.3. *Magic Angle Spinning*

A problem with the examination of solids is that the nuclei can be considered to be frozen in space and cannot freely line up in the magnetic field. The NMR signals generated are dependent, among other things, on the orientation of the nuclei to the magnetic field. Since the orientation of nuclei in solids is fixed, each nucleus (even chemically identical nuclei) "sees" a different applied magnetic field, resulting in broad NMR spectra. The effective magnetic field seen by a nucleus depends on the chemical environment in which the nucleus finds itself; the position at which a given nucleus resonates is called its **chemical shift**. As we will see subsequently, chemical shift due to different environments in a molecule is the key to structure determination by NMR. The phenomenon in solids of nuclei having different chemical shifts as a result of orientation in space is called **chemical shift anisotropy**.

The chemical shift due to magnetic anisotropy is directly related to the angle between the sample and the applied magnetic field. It has been shown theoretically and experimentally that by spinning the sample at an angle of 54.76°, the **magic angle**, to the magnetic field rather than the usual 90° for liquid sample analysis, the chemical shift anisotropy is eliminated and narrow line spectra are obtained. Figure 3.5 demonstrates the difference in ^{13}C NMR spectra of a solid with and without magic angle spinning. The spectrum in Fig. 3.5(a) was taken without spinning (i.e., static), on a crystalline powder sample of l-Dopa. The spectrum has broad, unresolved lines and does not provide much useful information. The same sample is then spun at 9.6 kHz at the magic angle, 54.76°. The spectrum obtained is shown in Fig. 3.5(b). The linewidths are dramatically decreased and seven carbon nuclei are resolved, providing significant structural information about the compound. The spinning is carried out at very high frequencies

Figure 3.5 (a) The ^{13}C NMR spectrum of solid powdered l-Dopa obtained without spinning the sample. (b) The ^{13}C spectrum of the same sample obtained with MAS at a frequency of 9.6 kHz. Note the dramatic decrease in linewidth and increase in resolution when using MAS to obtain the NMR spectrum of a solid sample. (Spectra provided courtesy of Dr. James Roberts, Department of Chemistry, Lehigh University, PA, and used with his permission.)

(5–15 kHz) for optimum performance. This spinning gives better resolution and improved measurement of chemical shift and spin–spin splitting. The functional groups and their positions relative to each other in the solid sample molecule can be determined, as will be discussed.

Special probes have been developed for solid-state NMR that automatically position the sample at the magic angle. Modern instruments with magic angle spinning (MAS) make the analysis of solid samples by NMR a routine analytical procedure. MAS, combined with two RF pulse techniques called cross-polarization and dipolar decoupling (discussed in Section 3.6.3), permits the use of the low abundance nuclei ^{13}C and ^{29}Si to analyze insoluble materials by NMR, including highly cross-linked polymers, glasses, ceramics, and minerals.

3.1.3.4. *Other Sources of Line Broadening*

Any process of *deactivating*, or relaxing, an excited molecule results in a decrease in the lifetime of the excited state. This in turn causes line broadening. Other causes of deactivation include: (1) the presence of ions—the large local charge deactivates the nucleus; (2) paramagnetic molecules such as dissolved O_2—the magnetic moment of electrons is about $10^3\times$ as great as nuclear magnetic moments and this high local field causes line broadening; and (3) nuclei with quadrupole moments. Nuclei in which $I > 1/2$ have quadrupole moments, which cause electronic interactions and line broadening; one important nucleus with a quadrupole field is ^{14}N, present in many organic compounds such as amines, amides, amino acids, and proteins.

3.2. THE FTNMR EXPERIMENT

The time required to record an NMR spectrum by scanning either frequency or magnetic field is Δ/R, where Δ is the spectral range scanned and R the resolution required. For ^1H NMR the time required is only a few minutes because the spectral range is small, but for ^{13}C NMR the chemical shifts are much greater; consequently the spectral range scanned is much greater and the time necessary to scan is very long. For example, if the range is 5 kHz and a resolution of 1 Hz is required, the time necessary would be $(5000 \text{ s})/1$ or 83 min, an unacceptably long time for routine analysis, and an impossible situation if rapid screening of thousands of compounds is needed, as it is in the development of pharmaceuticals.

This problem, and quite a few other problems with the NMR experiment were overcome with the development of FTNMR. The fundamentals of FT spectroscopy and the advantages gained through the use of FT spectroscopy were discussed in Chapter 2 for optical spectroscopy. In FTNMR, the RF frequency is applied to the sample as a short, strong "pulse" of radiation. The experiment is shown schematically in Fig. 3.6. Because there are slightly more nuclei lined up with the field, the excess nuclei in the ground state can be thought of as having a single magnetic moment lined up with the external field, B_0. This net magnetization is shown in Fig. 3.6(d) as M_z. The net magnetization behaves as a magnet. An electric current applied to a coil of wire surrounding a magnet will cause the magnet to rotate. An RF pulse through a coil of wire around the sample is used to generate a second magnetic field, B_1, at right angles to B_0; this provides the excitation step in the NMR experiment. In a modern NMR instrument, the field B_1 is applied as a **pulse** for a very short time, on the order of 10 μs, with a few seconds between pulses. The net magnetic moment of the sample nuclei is shifted out of alignment with B_0 by the pulse [Fig. 3.6(e)]. Most often, a 90° pulse is used; the net magnetization vector is shifted 90° (from the z-axis to the y-axis). Such a 90° change gives the largest signal

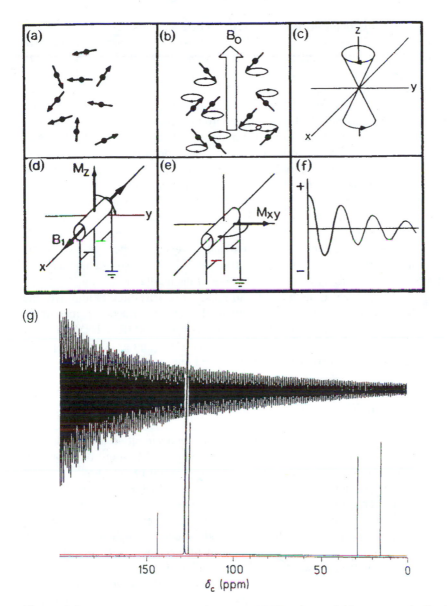

Figure 3.6 The pulsed NMR experiment. (a) NMR active nuclei have magnetic dipole moments. (b) In the presence of an external static field, B_0, the dipoles precess about the field axis, each with an average component either parallel or antiparallel to the field. (c) In this case there are two spin populations precessing in opposite directions represented by the two cones. At equilibrium, the top cone (the dipoles with components parallel to the field) has a slightly greater population. (d) The difference between the two populations is represented by the net magnetization, M_z. The NMR signal derives from M_z, which can be observed by rotating it into the plane of an RF coil using a resonant RF field, B_1. (e) B_1 is usually gated on just long enough to rotate M_z into the x–y plane. After B_1 is gated off, the net magnetization M_{xy} begins to rotate about the z-axis at a frequency equal to the difference in precession frequencies for the two populations. (f) This rotating M_{xy} induces an emf in the RF coil along the x-axis. The signal appears as a cosine wave decaying in time following the pulse, and is referred to as FID. (From Petersheim, used with permission.) (g) Free induction decay and resulting ^{13}C frequency spectrum after Fourier transformation for an organic compound containing multiple absorption frequencies due to chemically different carbon nuclei. (From Williams, used with permission.)

possible. The pulse is discontinued and the excited nuclei precess around the applied magnetic field at an angle to B_0. This "rotating magnet" induces a current in the wire coil. This induced current is the NMR signal and is picked up by the coil very quickly after the B_1 pulse ends. The signal undergoes free induction decay (FID); the current decreases with time as the freely precessing nuclei relax back to the ground state, as shown in Fig. 3.6(f). This FID signal is a time-domain signal and must be processed using the FT (or other mathematical transform) to produce the usual frequency-domain spectrum. Because all frequencies are excited simultaneously, the FID signal consists of one exponentially decaying sine wave for each frequency component in the spectrum. This type of pattern for a ^{13}C experiment is shown in Fig. 3.6(g), along with the Fourier transformation of the FID, resulting in the frequency-domain ^{13}C spectrum.

One advantage of the FTNMR experiment is that the entire spectrum is taken in a single pulse. This occurs because pulsing the RF field broadens the frequency distribution of the RF source. All resonances within several kHz of the source frequency are excited simultaneously. While the intensity of the FID signal is very low, the process may be repeated many times very rapidly, for example, 8192 times in 0.8 s. The signal increases linearly, but the noise increases only as the square root of the number of readings. The net effect is a significant improvement in the signal-to-noise ratio. This directly improves the sensitivity of the method. Signals can be obtained that are orders of magnitude more sensitive than those obtained with old "continuous wave" NMR. FTNMR is now the dominant form of NMR instrumentation; it is critical to obtaining ^{13}C NMR spectra and in performing 2D NMR experiments, both discussed later in the chapter.

3.3. CHEMICAL SHIFTS

From the Larmor equation it would seem that all protons in a sample in a given applied magnetic field would absorb at exactly the same frequency. NMR would therefore tell us only that protons were present in a sample. The same would be true for ^{13}C; NMR would tell us only that carbon was present. If NMR were suitable only for detecting and measuring the presence of hydrogen or carbon in organic compounds, it would be a technique with very limited usefulness. There are a number of fast, inexpensive methods for detecting and measuring hydrogen and carbon in organic compounds. Fortunately, this is not the case. In the NMR experiment, protons in different chemical environments within a molecule absorb at slightly different frequencies. This variation in absorption frequency is caused by a slight difference in the electronic environment of the proton as a result of different chemical bonds and adjacent atoms. The absorption frequency for a given proton depends on the chemical structure of the molecule. This variation in absorption frequency is called the chemical shift. The same type of chemical shift occurs for carbon in different chemical environments within a molecule. The physical basis for the chemical shift is as follows.

Suppose that we take a molecule with several different "types" of hydrogen atoms, such as the molecule ethanol, CH_3CH_2OH. This molecule has hydrogen atoms in three different chemical environments: the three hydrogen atoms in the terminal CH_3, the two hydrogen atoms in the CH_2 group, and the one in the OH group. Consider the nuclei of the different types of hydrogen. Each one is surrounded by orbiting electrons, but the orbitals may vary in shape and the bonds vary in electron density distribution. This changes the length of time the electrons spend near a given type of hydrogen nucleus. Let us suppose that we place this molecule in a strong magnetic field B_0. The electrons associated with the nuclei will be rotated by the applied magnetic field B_0.

This rotation, or *drift*, generates a small magnetic field σB_0, which opposes the much larger applied magnetic field B_0. The nuclei are *shielded* slightly from the applied magnetic field by the orbiting electrons. The extent of the shielding depends on the movement of the electrons caused by the magnetic field (not by the simple orbiting of the electrons). If the extent of this shielding is σB_0, then the nucleus is not exposed to the full field B_0, but to an *effective* magnetic field at the nucleus, B_{eff}:

$$B_{\text{eff}} = B_0 - \sigma B_0 \tag{3.13}$$

where B_{eff} is the effective magnetic field at the nucleus; B_0, the applied field, and; σB_0, the shielding by the drift of local electrons; σ is the *screening constant* or *diamagnetic shielding constant*.

In the case of ethanol, σB_0 is different for each type of hydrogen; therefore the effective field B_{eff} is different for each type of hydrogen. In order to get absorption at frequency v, we must compensate for this variation by varying B_0. In other words, resonance of the hydrogen atoms in different chemical environments takes place at slightly different *applied* magnetic fields. Another way of expressing this relationship is that if the applied field is kept constant, the nuclei in different chemical environments resonate at slightly different frequencies v. A shielded nucleus resonates or absorbs at a lower frequency than an unshielded nucleus. This change in frequency of absorption because of shielding is the chemical shift. Instead of one absorption signal for the protons in ethanol, we would predict three absorption signals at slightly different frequencies. Figure 3.7 is a low-resolution proton NMR spectrum of ethanol; indeed three absorption peaks are seen. Looking at the structure of ethanol, we can also predict what the ^{13}C NMR spectrum should look like. There are two chemically different C atoms in the molecule, the methyl carbon and the methylene carbon; two peaks should be seen in the ^{13}C NMR spectrum of ethanol. It is important to note that B_{eff} and the peak separations resulting from the chemical shift differences are directly proportional to the applied magnetic field strength. This is another reason for the development of high field strength magnets: better resolution and more structural information result.

The chemical shifts of nuclei are measured (and defined) relative to a standard nucleus. A popular standard for proton NMR is tetramethylsilane (TMS), which has the chemical formula $Si(CH_3)_4$. In this compound all 12 hydrogen nuclei are chemically equivalent; that is, they are all exposed to the same shielding and give a single absorption peak. The chemical shift for other hydrogen nuclei is represented as follows:

$$\text{Chemical shift} = \frac{v_S - v_R}{v_{\text{NMR}}} \tag{3.14}$$

Figure 3.7 Low resolution proton NMR absorption spectrum of ethanol, CH_3CH_2OH.

where v_S is the resonant frequency of a specific nucleus; v_R, the resonant frequency of the reference nucleus; and v_{NMR}, the spectrometer frequency.

The chemical shift is the difference in the observed shift between the sample and the reference divided by the spectrometer frequency. It is a dimensionless number. Typical values for the frequency difference between a nucleus and the reference are in the Hz or kHz range; the spectrometer frequency is in the MHz range. In order to make these numbers easier to handle, they are usually multiplied by 10^6 and then expressed in ppm. (This unit should not be confused with the concentration expression ppm used in trace analyses.) The symbol for the chemical shift is δ. It is expressed as:

$$\delta = \frac{v_S - v_R}{v_{NMR}} \times 10^6 \text{ ppm} \tag{3.15}$$

Looking again at Fig. 3.7, the x-axis has units of chemical shift in ppm. The TMS peak, not shown in this figure, would be a single peak located at 0.0 ppm, by definition. It is a convention that NMR spectra are presented with the magnetic field increasing from left to right along the x axis. Diamagnetic shielding therefore also increases from left to right. A nucleus that absorbs to the right hand side of the spectrum is said to be *more shielded* than a nucleus that absorbs to the left side of the spectrum. One reason TMS is used as the reference peak is that the protons in common organic compounds are generally deshielded with respect to TMS; the TMS peak appears at the far right of the spectrum. The student should be able to recognize that this means that δ decreases from left to right along the x-axis (look at the scale in Fig. 3.7). This is not the way one normally plots numbers; they are usually plotted increasing from left to right. An alternative scale was developed, where the chemical shift was plotted as τ; τ is defined as:

$$\tau = 10.00 - \delta \text{ (ppm)} \tag{3.16}$$

This tau scale is no longer used, but can be found in the literature through the 1970s, as can the now unused terms upfield and downfield to refer to resonance positions. The terminology that should be used is deshielded (higher δ) or shielded (lower δ). Shielding by the drifting electrons is modified by other nuclei in their vicinity. These in turn are affected by the chemistry, geometry, and electron density of the system. Consequently, some deshielding takes place; the end result is that chemically identical nuclei are shifted from chemically different nuclei. Deshielded nuclei would be moved to the left on the NMR plot. As a result, we are able to distinguish different functional groups in a molecule, even ones that contain the same atoms; methyl groups are separated from methylene groups, as seen in the ethanol spectrum. Two examples of how shielding and deshielding occur based on the geometry of the molecule are shown in Fig. 3.8. Figure 3.8(a) and (b) show the molecule acetylene. When the long axis of the molecule is aligned with the external magnetic field, the circulation of the π electrons in the triple bond, shown by the partially shaded curved arrow in Fig. 3.8(a), induces a magnetic field along the molecular axis that opposes the applied field. The induced magnetic field lines are shown in Fig. 3.8(b). The arrows show the direction of the induced magnetic field lines at the positions of the hydrogen atoms. Both of the protons in acetylene are shielded from the applied magnetic field by the induced magnetic field. The field these protons experience is $B_0 - B_{local}$, which is smaller than B_0. Therefore a higher frequency is needed to attain resonance. Figure 3.8(c) and (d) show the molecule benzene, a planar molecule with one proton on each carbon atom. The protons are located on the outside of the carbon ring in the plane of the ring. Only two of the six protons attached to the ring are shown here. The delocalized π electrons are depicted as the circle inside the carbon hexagon. When the benzene ring is perpendicular to the applied magnetic field, the delocalized electrons circulate as shown by the partially shaded curved arrow in Fig. 3.8(c)

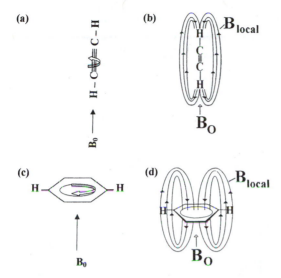

Figure 3.8 (a) and (b) Circulation of the π electrons in the C≡C bond in acetylene is shown by the partially shaded arrow in (a). The direction of the induced magnetic field, B_{local}, at the hydrogen atoms is shown by the arrows in (b). The direction of the applied field, B_0, is shown. The two protons are shielded from the applied field by the induced local field. (c) and (d) Circulation of the π electrons in the benzene ring, shown by the partially shaded arrow in (c), induces the magnetic field, B_{local}, shown in (d). Arrows show the induced field and the applied magnetic field directions. All six protons in the plane of the benzene ring are deshielded by the induced local magnetic field. (Fig. 3.8(b) and (d) are from McClure, used with permission.)

and generate an induced magnetic field as depicted in Fig. 3.8(d). In this case, the induced magnetic field shown reinforces the applied magnetic field outside the ring; all of the protons on the benzene ring feel a stronger field than B_0. They are deshielded by the induced field caused by the *ring current*. The field these protons experience is $B_0 + B_{local}$ so a lower frequency needs to be applied for resonance to occur. The term anisotropic effect or chemical shift anisotropy means that different chemical shifts occur in different directions in a molecule.

It is the chemical shift that allows us to identify which functional groups are present in a molecule or a mixture. Chemical shift enables us to ascertain what types of hydrogen nuclei are present in a molecule by measuring the shift involved and comparing it with known compounds run under the same conditions. For example, it is easy to distinguish between aromatic and aliphatic hydrogen atoms, alkenes, alkynes, or hydrogen bonded to oxygen or nitrogen, hydrogen on a carbon adjacent to ketones, esters, etc. Examples of typical chemical shifts for protons are given in Table 3.3. More extensive tables of chemical shifts can be found in the references by Silverstein and Webster, Pavia et al. and Lambert et al. listed in the bibliography.

Figure 3.7 shows the low-resolution spectrum of ethyl alcohol (or ethanol). This absorption spectrum, which is historically significant as one of the first NMR spectra to be recorded (Arnold et al., 1951), discloses that there are three types of hydrogen nuclei present in the ethanol molecule. Because NMR spectra provide valuable information about a molecule's structure, they are one of the most powerful tools available for characterizing unknown compounds or compounds for which we know the empirical formula but not the structure.

However, most proton NMR spectra show more peaks than one would predict based on chemical equivalency. With improved instrumentation resulting in higher resolution,

Table 3.3 Approximate Chemical Shifts of Protons in Common Organic Compounds

Type of hydrogen	Chemical shift (ppm)			
	a	b	c	d
Tetramethylsilane (TMS) CH_3 $\mid$ CH_3—Si—CH_3 $\mid$ CH_3		0		
$CH_3CH_2CH_2CH_3$ a b	1.0	1.2		
CH_3(c) $\mid$ CH_3CH_2—C—CH_2CH_3 a b $\mid$ b a CH_3 c	1.0	1.2	2.0	
CH_3CH_2Br a b	1.2	3.43		
CH_3CH_2I a b	1.8	3.2		
CH_3CH_2OH a b c	1.32	3.7	0.5–4.0	
CH_3C=O a $\mid$ OH b	2.1	11.4		
$CH_3CH_2CH_2NH_2$ a b c d	0.92	1.5	2.7	1.1
CH_3CH_2CH=CH_2 a b c d	.9	2.0	5.8	4.9
CH_3CH_2CH=$CHCH_2CH_3$ a b c c b a	1.0	1.3	5.4	
CH_3CH_2CH—CH_3 a b $\mid$c d NO_2	1.0	1.9	4.2	1.5
CH_3C—CH_2CH_3 a $\parallel$ b c O	2.1	2.5	1.1	
CH_3CH_2S—CH_3 a b c	1.3	2.53	2.1	
CH_3 CH_2—C≡CH a b c	1.0	2.2	2.4	
H(d) $\diagup$ $CH_3CH_2CH_2C$ a b c $\diagdown$ O	1.0	1.7	2.4	9.7

(continued)

Table 3.3   Continued

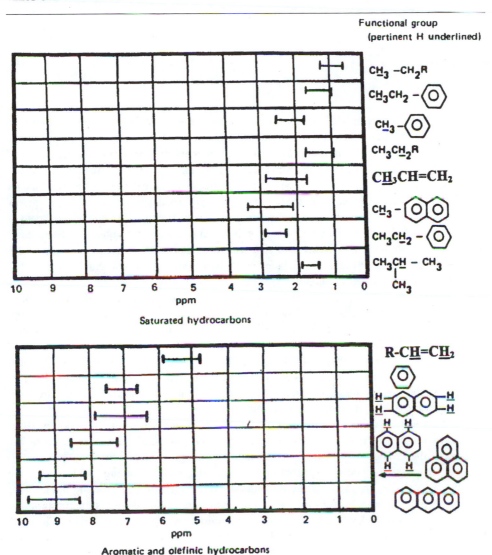

Saturated hydrocarbons

Aromatic and olefinic hydrocarbons

the absorption spectrum of ethyl alcohol has been found to be more complex than Fig. 3.7 indicates. When examined under *high* resolution, each peak in the spectrum can be seen to be composed of several peaks. This **fine structure** is brought about by **spin–spin splitting** or **spin–spin coupling**.

3.4.   SPIN–SPIN COUPLING

As we have already seen, the hydrogen nuclei of an organic molecule are spinning, and the axis of rotation may be with or against the applied magnetic field. Since the nucleus is magnetic, it exerts a slight magnetic field, which may be either with or against the applied magnetic field.

Suppose that we have a molecule such as the aldehyde shown, butanal, also called butyraldehyde:

$$
\begin{array}{cc}
& H \quad H \\
& | \quad | \\
CH_3-CH_2-\ & C-C=O \\
& | \\
& H
\end{array}
$$

The low resolution NMR spectrum for butanal is shown in Fig. 3.9. The protons on a given carbon atom are chemically equivalent to each other, but each group of protons is different from the rest. There are four chemically different groups of protons, labeled a–d in Fig. 3.9. The low resolution spectrum shows four peaks, as expected from the structure.

We will focus on the type c and type d protons first, and ignore the rest of the molecule for the moment. The type c protons are the two methylene protons adjacent to the aldehyde group; the type d proton is the single proton on the aldehyde carbon. The type d proton on the aldehyde group, represented as CHO, may be spinning either *with* or *against* the applied magnetic field. The spinning of the nucleus creates a small magnetic field, either with or against the applied magnetic field. This changes the effective field felt by the two protons of the adjacent CH_2 group. The CH_2 protons next to a CHO proton that is spinning *with* the field absorb at a slightly different frequency from that of the CH_2 protons next to a CHO proton spinning *against* the field. Statistically, a sample will contain as many protons spinning with the field as against it; so both groups will be equally represented. The single spinning proton of the CHO group splits the absorption peak (peak c in Fig. 3.9) for the adjacent protons in the CH_2 group into two peaks absorbing at slightly different frequencies but of equal intensity (1:1 peak area ratio). The spin–spin splitting is shown schematically in Fig. 3.10 (peak c). The two peaks are moved from the original frequency, one to a slightly higher frequency and one to a slightly lower frequency, but they are symmetrically located about the original frequency.

At the same time, the two protons of the CH_2 group are also spinning, and they affect the frequency at which the neighboring CHO proton absorbs. In this case, each of the protons of the CH_2 group can spin with or against the applied field. Several spin combinations are possible. These combinations may be depicted as in Fig. 3.11, where the arrows indicate the directions of the magnetic fields created by the spinning nuclei. In a typical NMR sample many billions of molecules are present, and each spin combination is represented equally by number. The three combinations (i), (ii), and (iii) in Fig. 3.11 modify the applied field felt by the CHO to three different degrees, and absorptions

Figure 3.9 Low resolution proton NMR spectrum of 1-butanal, showing four absorption peaks. The *x*-axis is chemical shift in ppm; the *y*-axis is absorption. The four groups of chemically nonequivalent protons are labeled a–d on the chemical structure. The peak corresponding to each type of proton is marked with the corresponding letter. Table 3.3 should be consulted to confirm the approximate chemical shifts for each type of proton.

Figure 3.10 A portion of the expected 1-butanal NMR spectrum showing spin–spin coupling between the aldehyde proton and the adjacent methylene group. The x-axis is chemical shift in ppm; the y-axis is absorption. Peaks (c) and (d) are shown, with peak (d) split into a triplet by the adjacent —CH_2— group and peak (c) split into a doublet by the aldehyde proton. (The rest of the molecule is ignored for the moment.)

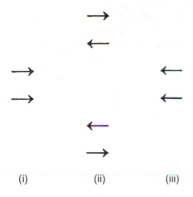

Figure 3.11 Possible alignments of the magnetic field of pairs of protons: (i) both aligned with the applied field; B_0 appears to be increased in magnitude; (ii) one aligned with the field and one against the field. There are two possible arrangements with equal probability. The net effect is that B_0 is unchanged; and (iii) both aligned against the field, B_0 appears to be reduced in magnitude.

occur at three frequencies for the CHO proton. It can be seen that there are two ways in which the combination of Fig. 3.11(ii) can exist, but only one way for the combination of Fig. 3.11(i) or (iii). There will therefore be twice as many nuclei with a magnetic field equal to the combination (ii) as of either (i) or (iii). As a result, the CH_2 group will cause the neighboring H of the CHO group to absorb at three slightly different frequencies with relative intensities in the ratio 1:2:1, as shown schematically in Fig. 3.10 (peak d).

The relative peak intensities in a multiplet can be worked out from the possible combinations of spin states, as was done in Fig. 3.11; the result is that the relative peak intensities in multiplets match the coefficients in the binomial expansion. Pascal's triangle can be used to rapidly assign peak intensities:

```
                              1
                        1           1
                  1           2           1
            1           3           3           1
      1           4           6           4           1
1           5           10          10          5           1
```

Figure 3.12 (a) Predicted proton NMR spectrum for butanal, showing spin–spin coupling. (b) Actual 60 MHz proton NMR spectrum of butanal. Note the peak for TMS and the presence of a small impurity peak at about 7 ppm; 60 MHz instruments required that chemical shifts greater than 8 ppm be plotted offset from the 0 to 8 ppm baseline; the aldehyde proton shift is plotted at the far left above the first baseline. (From Bhacca et al., courtesy of Varian Associates, Palo Alto, CA., www.varianinc.com.) (c) Actual 90 MHz proton NMR spectrum of 0.04 mL butanal dissolved in 0.5 mL CDCl₃. SDBS No. 1925HSP-04-071. (Courtesy of National Institute of Industrial Science and Technology, Japan, SDBSWeb:http://www.aist.go.jp/RIOBD/SDBS. Accessed 12/13/02.)

and so on. From the triangle relationship, it is seen that a quartet (fourth line from the top of the triangle) will have a 1:3:3:1 intensity ratio and a sextuplet (sixth line from the top) a 1:5:10:10:5:1 ratio.

The proton NMR spectrum of butanal, $CH_3CH_2CH_2CHO$, is shown in Fig. 3.12. Figure 3.12(a) is the predicted first order spectrum of butanal, Fig. 3.12(b) is an actual 60 MHz proton spectrum, and Fig. 3.12(c) is an actual 90 MHz proton spectrum. Looking at Fig. 3.12(a), the aldehydic proton, peak d, appears as the 1:2:1 triplet just predicted, but peak c is not the 1:1 doublet predicted from the splitting by the aldehyde proton alone. We of course ignored for the simple argument above that the CH_2 group adjacent to the CHO group is also adjacent to another CH_2 group on the opposite side.

Spin–spin splitting or coupling is the transmission of spin state information between nuclei through chemical bonds. Spin–spin splitting is quite strong between protons on *adjacent* carbons, but is generally negligible between protons farther removed than this. It is important to remember that spin–spin splitting by a given nucleus causes a change in the fine structure of peak for the *adjacent* protons in the molecule. For example, the CH_2 splits the H of the adjacent CHO group, and the H of the CHO group splits the adjacent CH_2, but the CH_2 *does not* split itself because that interaction is forbidden by quantum theory.

The number of peaks in the fine structure is termed the **multiplicity** of the peak. In Fig. 3.10, the multiplicity of the aldehyde proton peak is 3 and the multiplicity of the methylene peak is 2; these are also referred to as a triplet and a doublet, respectively. The multiplicity of proton peaks due to spin–spin splitting is

$$2nI + 1 \qquad (3.17)$$

where n is the number of *equivalent hydrogen atoms* on the carbon atoms adjacent to the one whose proton peak is being examined and I is the spin quantum number. For hydrogen, $I = 1/2$ and Eq. (3.17) simplifies to $n + 1$.

If two different adjacent groups cause splitting, the multiplicity is given by $(2nI + 1)(2n'I + 1)$, where n and n' are the numbers of equivalent protons in each group.

Using Eq. (3.17) and the structure of butanal we can work out the splitting patterns expected. The predicted spectrum is Fig. 3.12(a), which can be compared with the experimentally obtained spectra in Fig. 3.12(b) and (c). We predicted that the aldehyde proton should be split into a triplet by the adjacent CH_2 protons; the peak at 9.7 ppm is a triplet. We predicted that the aldehyde proton would split the adjacent CH_2 group into a doublet, but we did not take into account the protons on the other side of this CH_2 group. There are two methylene protons (the "b" protons) next to the "c" methylene protons. This is a case of having two different groups, b and d, adjacent to the "c" protons we are interested in. As stated earlier, the multiplicity is calculated from $(2nI + 1)(2n'I + 1)$, so we have $(2 + 1)(1 + 1) = 6$. The multiplicity of the peak for the "c" protons should be 6. We predict that peak c will consist of 6 peaks; in Fig. 3.12(b), the peak is found at 2.4 ppm and looks like a triplet with each triplet peak split into a doublet for a total of 6 peaks. The methyl protons (the "a" protons located at 1.0 ppm) are split into a triplet by the adjacent "b" methylene protons. The "b" methylene protons are split by both the methyl group (3 equivalent protons) and the "c" methylene protons (2 equivalent protons), so the multiplicity of the "b" proton peak should be $(3 + 1)(2 + 1) = 12$. While all 12 lines cannot be distinguished without expanding the scale around the peak at 1.7 ppm, it is clear, especially from the 90 MHz spectrum shown in Fig. 3.12(c), that there are more than 9 peaks present.

A few more examples of spin–spin coupling should help you to predict splitting patterns for simple compounds.

The structure of 1,1-dichloroethane is:

```
      H
      |
Cl — C — CH₃
      |
      Cl
```

The single proton on the carbon containing the two Cl atoms will be split into $(3 + 1)$ peaks by the three equivalent protons on the adjacent methyl group. The multiplicity of this peak is 4 and it is called a quartet. Each peak in the quartet will be separated by exactly the same frequency; they will be evenly spaced. We will get back to this in a bit. The relative peak intensities will be 1:3:3:1, which can be worked out as was done in Fig. 3.11. The peak due to the three methyl protons will be split into $(1 + 1)$ peaks, a doublet with a multiplicity of 2, by the adjacent single proton. The spacing between the two peaks in the doublet will be exactly the same as the peak spacing in the quartet and the ratio of peak intensities in the doublet will be 1:1.

If we replace one of the chlorine atoms with a hydrogen atom to give the compound chloroethane,

```
      H
      |
H — C — CH₃
      |
      Cl
```

the methyl group will split the peak due to the two protons on the chlorine-containing carbon into a quartet as in 1,1-dichloroethane, but the methyl group peak will appear as a triplet, since it is now split by two equivalent protons. We will look at the spectra for these two compounds in Section 3.6.2 and see if our predictions are correct.

The compound benzene, C_6H_6, has six chemically equivalent protons. We would predict that a single absorption peak should be seen located at a chemical shift of about 7 ppm (from Table 3.3). The 300 MHz proton spectrum of benzene (Fig. 3.13) confirms a single peak located at 7.3–7.4 ppm. We cannot tell anything about the number of protons giving rise to this peak; the relative area is obviously equal to 1 when only one peak is seen. Looking back at Fig. 3.4(a), the 300 MHz proton spectrum of toluene, the structure shows a methyl group substituting for one of the hydrogen atoms on a benzene ring. We should expect a singlet peak for the methyl group, since there are no protons adjacent to it. The chemical shift of the peak will be about 2 ppm (from Table 3.3). There is a singlet in that position in the spectrum. We also see a multiplet at about 7.3 ppm, where we would expect the aromatic ring protons to absorb. The five ortho, meta, and para protons on the ring are not chemically equivalent (as are the six protons on a benzene ring) and at 300 MHz we can tell this. In older spectra obtained at lower field strengths, such as the 60 MHz spectrum shown in Fig. 3.4(b), the electronic environments of the ortho, meta, and para protons are so similar as to cause the peaks to overlap, resulting in a single broadened peak. The overlap of peaks is often a problem in proton NMR, especially for alkanes. The expected area ratio should be $5/3$ and can be checked by the stepped line shown on the spectrum, as will be explained in Section 3.6.2. The aromatic compound naphthalene and its proton NMR spectrum are shown in Fig. 3.14(a); there are two types of protons, marked A and B on the structure. Each A proton is adjacent to a single B proton and vice versa. The spectrum shows the expected two doublets in the aromatic region of the spectrum. The relative area ratio of the doublets should be $1/1$, since there are equal numbers of A and B protons. Figure 3.14(b) is also the spectrum of naphthalene, with the relative peak areas

Figure 3.13 The 300 MHz proton NMR spectrum of benzene, C_6H_6, obtained in deuterated chloroform. The ^{13}C spectrum is also shown. (Reprinted with permission of Aldrich Chemical Co., Inc.)

shown by the stepped line. The two steps are equal in height, so there are the same numbers of A protons as B protons.

The spectra can be predicted for the alkanes butane and isobutane (or 2-methylpropane). The peaks should appear in the $1-1.5$ ppm chemical shift region according to Table 3.3. Butane, $CH_3CH_2CH_2CH_3$, has two types of protons as noted in Fig. 3.15(a). Isobutane also has two types of protons, shown in Fig. 3.15(b). Therefore both spectra should have two absorption peaks. In butane, the methyl protons should be split by the adjacent methylene protons into a triplet; the methylene protons would be split by the methyl protons into a quartet. We would predict that the proton NMR spectrum of butane would look like the schematic spectrum in Fig. 3.15(a), with the relative peak areas shown. Isobutane would show a very different splitting pattern. There are nine chemically equivalent protons (marked "b" on the structure) on the three methyl groups; the peak for these nine protons will be split into a doublet by the single "a" type proton on the middle carbon. The peak for the single proton will be split into $(9 + 1) = 10$ peak multiplet by the "b" type protons, with the relative peak areas as shown schematically in Fig. 3.15(b).

If we replace one of the methyl groups in isobutane with any other substituent, the resulting compound will contain an isopropyl group, $—CH(CH_3)_2$:

$$CH_3—\underset{\underset{H}{|}}{\overset{\overset{R}{|}}{C}}—CH_3$$

Figure 3.14 (a) The 400 MHz proton NMR spectrum of naphthalene obtained by dissolving 10.5 mg naphthalene in 0.5 mL deuterated chloroform. SDBS No. 1350HSP-40-209. (Courtesy of National Institute of Industrial Science and Technology, Japan, SDBSWeb:http://www.aist.go.jp/RIOBD/SDBS. Accessed 11/05/02.) (b) The 300 MHz proton spectrum and associated ^{13}C spectrum of naphthalene. The proton spectrum shows the peak integration. (Reprinted with permission of Aldrich Chemical Co., Inc.)

Figure 3.15 (a) The structure and predicted proton NMR spectrum for butane showing relative peak area and spin–spin coupling. The schematic spectrum is not to scale. (b) The structure and predicted proton NMR spectrum for isobutane showing relative peak area and spin–spin coupling. The schematic spectrum is not to scale. The chemical shift of the peaks can be predicted from Table 3.3.

where R represents a substituent. An example is the molecule cumene, which contains a benzene ring with one of the hydrogen atoms replaced by the isopropyl group. The structure of cumene and its proton NMR spectrum are presented in Fig. 3.16. The ring protons appear at about 7 ppm. The isopropyl group shows its very characteristic splitting pattern: a doublet equivalent to six protons located at about 1 ppm and a septuplet equivalent to one

Figure 3.16 The 90 MHz proton NMR spectrum of cumene, showing the characteristic isopropyl group splitting pattern. SDBS No. 1816HSP-00-081. (Courtesy of National Institute of Industrial Science and Technology, Japan, SDBSWeb:http://www.aist.go.jp/RIOBD/SDBS. Accessed 12/14/02.)

proton. The chemical shift of the septuplet depends on the nature of the R group, but the isopropyl splitting pattern is an easy one to spot in a spectrum.

If we consider the compound 1,1-dichloroethane again, and replace one of the Cl atoms with a methyl group, we have 2-chloropropane:

$$CH_3-\overset{\overset{\displaystyle H}{|}}{\underset{\underset{\displaystyle Cl}{|}}{C}}-CH_3$$

This is a compound containing a substituted isopropyl group, like cumene. The single proton on the carbon atom containing the chlorine will be split into $(6 + 1)$ peaks, a septuplet, by the six equivalent methyl protons. The two methyl groups are chemically equivalent and will absorb at the same frequency, so the single peak for all six protons will be split into a doublet by the single proton on the central carbon which is adjacent to both methyl groups. Figure 3.17 is a 300 MHz spectrum of 2-chloropropane. We see the characteristic isopropyl splitting pattern in both the cumene and the 2-chloropropane spectra, although the chemical shift position of the septuplet varies with R.

The NMR spectrum of very dry ethanol is presented in Fig. 3.18(a). The CH_3 protons are split by the CH_2 group into $(2 + 1) = 3$ peaks. The CH_2 is split by the OH proton and by the CH_3 protons, giving $(3 + 1)(1 + 1) = 8$ peaks. Finally, the OH proton is split into 3

Figure 3.17 The 300 MHz proton NMR spectrum of 2-chloropropane, showing the characteristic isopropyl splitting pattern with the 1:6 peak area ratio. Compare the position of the septuplet in this figure with Fig. 3.16. (Reprinted with permission of Aldrich Chemical Co., Inc.)

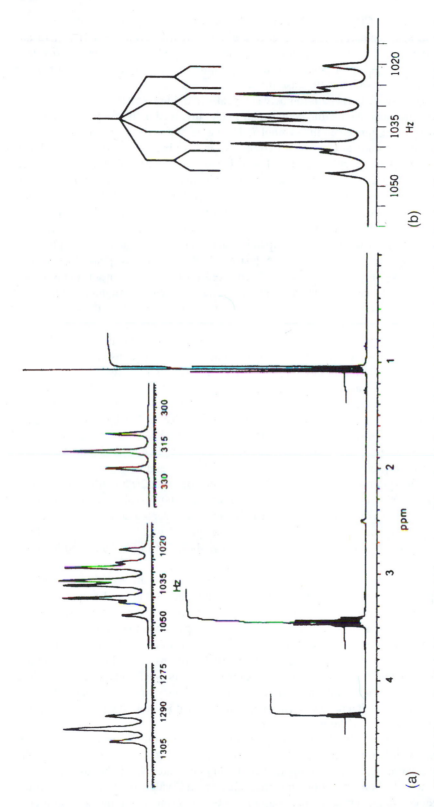

Figure 3.18 (a) The 300 MHz proton NMR spectrum of dry ethanol dissolved in deuterated dimethyl sulfoxide (DMSO). (From Silverstein and Webster, this material is used by permission of John Wiley and Sons, Inc.) (b) Detailed splitting pattern of the methylene protons. The J value for the methyl–methylene coupling is 7 Hz, which splits the methylene protons into a quartet. The J coupling constant for the hydroxyl–methylene coupling is 5 Hz, so each peak in the quartet is split into a doublet. (From Silverstein and Webster; this material is used by permission of John Wiley and Sons, Inc.)

peaks by the adjacent CH_2 group. Remember that n is the number of equivalent protons taking part in spin–spin splitting. For example, on a methyl group, $—CH_3$, there are three equivalent protons; on the methylene group $—CH_2—$ there are two equivalent protons. Also, the number of peaks due to spin–spin coupling equals $(n + 1)$ only for nuclei where $I = 1/2$, which is the case for ^1H.

The magnitude of the separation between peaks in the fine structure is a measure of strength of the magnetic interaction between the nuclei involved in the coupling. The magnitude of the separation is given the symbol J. The value of J is the **coupling constant** or **spin–spin coupling constant** between two nuclei, and is given the symbol J_{AB}, with A and B representing the coupled nuclei. The value of J is measured directly from the NMR spectrum by measuring the peak separation in the fine structure, and is usually expressed in Hz or cycles per second (cps), as shown in Fig. 3.18(a). Figure 3.18(a) is a proton spectrum collected at 300 MHz; the separation between the different peaks is improved over lower field instruments, but in order to see the splitting patterns the NMR spectroscopist must select and "expand" each peak. The expanded peak plots are shown either above the spectral peaks, or just off to the side of the peak. In very dry ethanol the hydroxyl proton can be considered fixed on the oxygen atom. (This is not the case in "normal" ethanol, which contains water. We will learn more about this later.) Therefore, the hydroxyl proton will be split by the adjacent methylene protons into a triplet, which appears at about 4.3 ppm. Looking at the expanded peak (inset above the hydroxyl peak), it can be seen that the triplet peaks occur at 1291, 1296, and 1301 Hz; that is, they are spaced 5 Hz apart. The J coupling constant between the methylene and hydroxyl protons, J_{AB}, is 5 Hz. The methyl group protons appear between 1.0 and 1.1 ppm; the peak is split into a triplet by the adjacent methylene protons. The J coupling constant between the methyl protons and the methylene protons is 7 Hz, as you can tell by measuring the distance in Hz between the peaks shown in the expanded inset just to the left of the methyl peak. Since we have used A for the hydroxyl proton, we would represent the methyl–methylene J as J_{BC}. We know that the methylene protons, at about 3.5 ppm, are split into eight peaks by the adjacent methyl and hydroxyl protons, since $(3 + 1)(1 + 1) = 8$. Because the J methyl–methylene constant, J_{BC}, is larger than the J methylene–hydroxyl constant, J_{AB}, the methylene peak will be split first into a quartet, with each peak in the quartet separated by 7 Hz, as shown schematically in Fig. 3.18(b). Then each peak in the quartet will be split into a doublet, with a peak spacing of 5 Hz. The resulting pattern is shown in Fig. 3.18(a) and in detail in Fig. 3.18(b).

In a compound such as chloroethane, CH_3CH_2Cl, there are only two groups of equivalent protons. The methyl protons split the methylene protons and vice versa. The coupling constant is given the symbol J_{AB}, where A represents the methyl protons and B the methylene protons. J_{AB} must equal J_{BA} and the spacing between the peaks in the quartet and the peaks in the triplet will be identical. Measurement of the coupling constants by measuring the peak spacing tells us which protons are splitting each other; this helps in deducing the structure of an unknown from its NMR spectrum. J coupling constants provide valuable information to physical chemists and to organic chemists interested in molecular interactions. The magnitude of J does not change if the applied magnetic field changes, unlike the chemical shift.

The magnitude of the J coupling constant between protons on adjacent single-bonded carbons is between 6 and 8 Hz. The magnitude of J decreases rapidly as the protons move farther apart in a compound containing saturated C—C bonds. The introduction of multiple bonds or aromatic rings changes the value of J and also permits longer-range coupling. There is no coupling between protons on the same carbon.

For example, in a compound such as butane,

$$CH_3\text{---}CH_2\text{---}CH_2\text{---}CH_3$$
$$\quad A \qquad B \qquad C \qquad D$$

the coupling between the protons on carbon A is zero; the coupling between the protons on carbon A and those on carbon B is about 6–8 Hz (this is written $J_{AB} = 6$–8). The coupling between the protons on carbon A and those on carbon C is never more than 1 Hz, that is, $J_{AC} \leq 1$, and J_{AD} is very small and usually negligible. If one of the C—C bonds is replaced by a C=C, the adjacent protons on the carbon atoms would have a coupling constant of 7–10 Hz if the protons are on the same side of the double bond (cis); the coupling constant for trans protons is even larger, 12–19 Hz. The magnitude of the coupling constant therefore provides structural information about the compound. Look back at Fig. 3.12(b), the spectrum of butanal. Peak a, the terminal methyl peak, shows the triplet expected from the adjacent methylene group, but each peak in the triplet is also split by a much smaller longer range coupling.

The ^{13}C nucleus is also an $I = 1/2$ nucleus; it would be expected that two adjacent ^{13}C nuclei would split the peak for each carbon into a doublet. Given that ^{13}C is only 1.1% of the naturally occurring carbon, the chance of finding more than one ^{13}C in a low molecular weight organic molecule is very small. The chance of finding two adjacent ^{13}C nuclei is even smaller. Consequently, C—C spin–spin coupling is not usually observed in carbon NMR spectra, unless special techniques or isotopically enriched molecules are studied. Values of J for carbon–carbon coupling have been measured using these special approaches and range from about 20 to 200 Hz.

Spin–spin interaction is also possible between ^{13}C and protons, and can be seen in both proton NMR and ^{13}C NMR spectra under the appropriate conditions. Owing to the low abundance of ^{13}C, the peaks due to ^{13}C–^1H interactions are very small. In the proton spectrum of a neat liquid, they can be observed as weak satellite peaks, one on either side of the central proton peak.

Interpretation of NMR spectra can be very difficult if we consider all the multiplicities that are possible from the interactions of all the nuclei. If we confine ourselves to spectra in which the chemical shift between interacting groups is large compared with the value of J, the splitting patterns and the spectra are easier to interpret. This is called utilizing *first order spectra*. The more complicated systems resulting in second order and higher order spectra will not be dealt with in this text. The term second order spectrum must not be confused with 2D NMR spectra, which will be discussed later in the chapter.

The capital letters of the alphabet are used to define spin systems that have strong or weak coupling constants (actually large or small values of $\Delta v/J$ where Δv is the difference in chemical shift between the nuclei, in Hz). For example, a system A_2B_3 indicates a system of two types of nuclei interacting strongly together of which there are two of type A and three of type B. For an AB system, the value of $\Delta v/J$ will be small; a value of 8 is an arbitrary limit. A break in the alphabetical lettering system indicates weak or no coupling, resulting in a large value for $\Delta v/J$. The system A_2X would indicate two protons of type A that weakly couple with one proton of type X. An AX spin system will give a first order splitting pattern. As the value of $\Delta v/J$ decreases, the splitting pattern becomes more complex.

The rules for interpreting first order proton spectra can be summarized as follows:

1. A proton spin-coupled to any equivalent protons will produce $n + 1$ lines separated by J Hz, where J is the coupling constant. The relative intensities of the lines are given by the binomial expansion $(r + 1)^n$, where n is the number of

equivalent protons. Splitting by one proton yields two lines of equal intensity, and by two protons yields three lines with a ratio of intensities 1:2:1. Three protons give four lines with a ratio of intensities 1:3:3:1. Four protons yield five lines with a ratio of intensities 1:4:6:4:1, and so on.

2. If a proton interacts with two different sets of equivalent protons, then the multiplicity will be the product of the two sets. For example, a proton split by both a methyl (CH_3) and a methylene (CH_2) group will be split into four lines by the methyl group. Each line will be split into three other lines by the methylene group. This will generate a total of 12 lines, some of which will probably overlap each other. The exact pattern (e.g., quartet of triplets vs. triplet of quartets) is determined by the magnitude of the coupling constants.

3. Equivalent protons do not split each other, the transition being forbidden. In practice, however, interactions do take place and can be seen in second order spectra.

Interpretation of simple NMR spectra will be discussed in Sections 3.6.2 and 3.6.3. It is not possible in this text to give complete instructions on the interpretation of NMR spectra, but it is hoped that by looking at some simple examples, it can be understood that NMR spectra give important and detailed structural information about molecules. For example, the chemical shift indicates the functional groups that are present, such as aromatics, halides, ketones, amines, alcohols, and so on. Spin–spin splitting indicates which groups are coupled to each other and therefore close to each other in the molecular structure. The multiplicity indicates how many equivalent protons are in the adjacent functional groups. The peak area gives us relative numbers of each type of nucleus. The 3D geometry of even large, complex molecules can be determined.

3.5. INSTRUMENTATION

For structural determination, a high resolution NMR is required and this type of instrument is discussed first. Low resolution instruments are discussed in Section 3.5.7. The most important parts of an FTNMR instrument are the magnet, the RF generator, and the sample chamber or probe, which not only houses the sample but also the RF transmission and detection coils. In addition, the instrument requires a pulse generator, an RF receiver, lots of electronics, and a computer for data processing. A block diagram of an FTNMR is shown in Fig. 3.19(a).

Older low field (e.g., 60 MHz) NMR instruments used a fixed RF and a permanent magnet or electromagnet with a set of Helmholtz coils in the pole faces of the magnet, shown schematically in Fig. 3.19(b). These coils could be adjusted to vary the applied magnet field slightly by passing a current through them, causing each chemically different nucleus to come into resonance sequentially. Such instruments were called *continuous wave* (CW) or *field sweep* instruments. The sample holder was placed in the magnetic field. Two RF coils surrounded the sample so as to be orthogonal to each other and to the applied magnetic field. One coil applied a constant RF frequency; the second coil detected the RF emission from the excited nuclei as they relaxed. These systems were simple and rugged, but limited in resolution and capability. There are no longer any manufacturers of CW NMR instrumentation. The increased sensitivity of FTNMR is so critical to the measurement of ^{13}C and other less abundant nuclei as well as to increased proton sensitivity

Figure 3.19 (a) Schematic block diagram of an FTNMR spectrometer. (b) Schematic diagram of a CW NMR spectrometer with a permanent magnet.

that all modern NMR spectrometers are FT instruments. The term NMR spectrometer therefore is used in the remainder of the chapter to mean a pulsed FT system.

A modern NMR magnet and probe are shown in Fig. 3.20(a). NMR spectrometers cost from about $200,000 for a 300 MHz instrument to $1,000,000 for a high field wide bore instrument for solids. Magnetic resonance imaging (MRI) and NMR research imaging instruments cost over $3 million dollars.

3.5.1. Sample Holder

The sample holder in NMR is normally tube-shaped and is therefore called the sample tube. The tube must be transparent to RF radiation, durable, and chemically inert. Glass or Pyrex tubes are commonly used. These are sturdy, practical, and cheap. They are usually about 6–7 in. long and ~1/8 in. in diameter, with a plastic cap to contain the

Figure 3.20 (a) Magnet and assembly for a modern NMR spectrometer. The superconducting coils (the primary solenoid and the shim coils, marked SC) are submerged in a liquid helium dewar (He) which is suspended in an evacuated chamber. A liquid nitrogen dewar (N_2) surrounds that to reduce the loss of the more expensive He. The levels of cryogenic fluids are measured with level sensors, marked LS. The room temperature shim coils (RS) and probe (P) are mounted in the bore of the magnet. A capped sample tube, shown inserted into the probe at the top, is introduced and removed pneumatically from the top of the bore. (b) A schematic probe assembly. A sample tube (uncapped, with a liquid sample as indicated by the meniscus) is shown inserted at the top of the probe. The RF coil for the observed nucleus (OC) is mounted on a glass insert closest to the sample volume. The RF coil for the lock signal (LC) is mounted on a larger glass insert. Variable capacitors (VC) are used to tune the appropriate circuit (lock, observe) to be in resonance with the appropriate RF. Only a small portion of the RF circuitry in the probe is shown. (From Petersheim, used with permission.)

sample. This type of tube is used for obtaining spectra of bulk samples and solutions. Sample tubes range in size from this "standard" size down to tubes designed to hold 40 μL of sample, such as the Nano Probe version from Varian Associates (www.varianinc. com). Flow-through cells are used for hyphenated techniques such as HPLC-NMR and on-line analysis.

3.5.2. Sample Probe

The sample chamber into which the sample holder is placed is called the **probe** in an NMR spectrometer. The probe holds the sample fixed in the magnetic field, contains an air turbine to spin the sample holder while the spectrum is collected and houses the coil(s) for transmitting and detecting NMR signals. A schematic of a probe is presented in Fig. 3.20(b).

The probe is the heart of the NMR system. The most essential component is the RF transmitting and receiving coil, which is arranged to surround the sample holder and is tuned to the precession frequency of the nucleus to be measured. Modern NMR probes use a single wire coil to both excite the sample and detect the signal. The coil transmits a strong RF pulse to the sample; the pulse is stopped and the same coil picks up the FID signal from the relaxing nuclei.

For maximum sensitivity, a fixed frequency probe is needed. This means that a separate probe is required for each nucleus to be studied: ^{1}H, ^{13}C, ^{19}F, and so on. A high-end probe costs on the order of $120,000. Much time can be lost in changing probes, which must be retuned before use, if one has to switch between nuclei frequently. Some variable frequency probe designs are available, but have decreased power, sensitivity, and spectral quality compared with fixed frequency probes. Probes for double resonance experiments require two concentric coils for the two RF sources. Triple resonance probes with many gradient options for liquids, solids, and flow experiments are available.

Probes usually have variable temperature control to run experiments at temperatures selected by the analyst. Cryogenically cooled probes can improve the resolution of a system, so that a 600 MHz spectrometer equipped with such a probe can provide resolution equivalent to a 700–800 MHz instrument. New probe designs with flow-through sample holders are commercially available, for use in coupled HPLC-NMR instruments and HPLC-NMR-MS instruments. These hyphenated instruments are discussed under applications later in the chapter.

The probe is installed in the spectrometer magnet so that the coils are centered in the magnet. The sample tube is inserted into the top of the probe and is moved by an air column through the magnet bore and centered among the probe coils. The tube exits the spectrometer at the top of the probe, again moved by an air column from the bore.

3.5.3. Magnet

The magnet in an NMR spectrometer must be strong, stable, and produce a homogeneous field. Homogeneous in this context means that the field does not vary in strength or direction from point to point over the space occupied by the sample. The field must be homogeneous to a few ppb within the sample area. It is common to express the magnetic field strength in terms of the equivalent proton frequency from the Larmor equation. A field strength of 1.4 T is equivalent to a proton frequency of 60 MHz. Commercial magnets range from the now obsolete 60 MHz (1.4 T) to 700 MHz (16.4 T) and higher. Varian Associates, the same company that introduced the first commercial NMR instrument in 1952, introduced a 900 MHz NMR instrument in 2001. The magnet is so large that this instrument comes with its own staircase so that the analyst can insert and remove the sample tube from the top of the probe. This instrument can be viewed at www.varianinc.com under the NMR link. This instrument costs approximately $4.5 million dollars. Other instrument companies have introduced similar high frequency instruments, with the trend now heading to GHz instruments.

Modern NMR spectrometers use superconducting solenoid magnets, as shown schematically in Fig. 3.20. The magnet consists of a main field coil made of superconducting Nb/Sn or Nb/Ti wire with a dozen or more superconducting **shim coils** wound around the main coil to improve field homogeneity. The superconducting coils must be submerged in liquid helium. The magnet and liquid helium reservoir are encased in a liquid nitrogen reservoir to decrease the evaporative loss of the more expensive liquid helium. As can be seen in Fig. 3.20, there is an open bore in the middle of the solenoid. The sample probe is mounted in the bore along with a set of room temperature shim coils. These coils are adjusted with every new sample placed in the probe to compensate for sample composition, volume, and temperature. Adjusting the field homogeneity, called "shimming", used to be a time-consuming task. Now computers with multivariate optimization procedures perform this task automatically. The magnetic field strength is held constant by a "frequency locking" circuit. The circuit is used to monitor a given nucleus, such as deuterium used in the solvent for liquid samples, and to adjust the magnetic field strength to keep this nucleus at a constant resonant frequency.

The bore also contains air conduits for pneumatic sample changing and spinning of the sample holder in the magnetic field. The size of the bore determines how large a sample can be introduced into the magnetic field. Conventional analytical NMRs generally have bore diameters of 5–10 cm, with the larger diameters used for NMR of solids by MAS; the wider bore is needed to accommodate the instrumentation required. Field homogeneity is better in narrow bore magnets. A very large bore size is that used in human whole-body MRI systems, where the bore is large enough to accommodate a table and the patient.

To see how an NMR magnet is constructed, the interested student should visit the JEOL, Inc. homepage at www.jeol.com, and follow the NMR links to "Magnet Destruction". A JEOL scientist, Dr. Michael Frey, cut open a 270 MHz magnet layer by layer. The pictures and commentary give a good appreciation for the complexity of the magnet construction. The file can be accessed directly at www.jeol.com.nmr/mag_view/magnet_destruction.html. In this particular magnet, over 12 miles of superconducting wire were used for the main coil alone.

The cryogenic fluids must be replenished on a regular basis, usually weekly for the liquid nitrogen and every 1–6 months for the liquid helium. The superconducting coils must be kept cold; if permitted to warm up they stop functioning. The magnet is said to have "quenched".

3.5.4. RF Generation and Detection

The RF radiation is generated by using an RF crystal oscillator. The output of the oscillator is amplified, mixed, and filtered to produce essentially monochromatic RF radiation for an NMR instrument. The RF radiation delivered to the sample is pulsed. A simple pulse might be a rectangular pulse of 500 MHz frequency for a 10 μs duration. The process of pulsing actually widens the RF band, providing a range of frequencies that is able to excite all nuclei whose resonances occur within the band of frequencies. All resonances within the band are excited simultaneously. An example of this is shown in Fig. 3.21. A 10 μs, 500 MHz rectangular pulse leads to a power distribution around the 500 MHz frequency as shown. A proton spectrum occurs over a chemical shift range of ~10 ppm, which corresponds to ± 2.5 kHz at 500 MHz. As seen in Fig. 3.21, all of the protons in the sample would see 98–100% of the power of the 500 MHz radiation delivered and all would be excited simultaneously. A pulse programmer is used to control the timing and shape of the RF pulses used to excite the sample. Square wave pulses are commonly used, but multipulse experiments and 2D NMR experiments with other pulse shapes are performed. There are hundreds of pulse sequences and 2D experiments that have been developed, with curious names like APT, DEPT, INEPT, INADEQUATE, COSY, and many more, some of which will be discussed later in the chapter. Each pulse sequence provides specific and unique NMR responses that enable the analyst to sort out the NMR spectrum and deduce the chemical structure of a molecule.

All signals are collected simultaneously. The RF pulse delivered is generally on the order of watts while the NMR signal collected is on the order of microwatts. The FID signal in the time domain must be converted to a frequency domain spectrum by application of a Fourier transformation or other mathematical transformation. Commercial instruments generally use quadrature phase-sensitive detection to avoid spectrum artifacts

Figure 3.21 The power distribution for a 10 μs rectangular pulse of 500 MHz RF radiation. (a) Distribution over a 400 kHz range. (b) The power level drops to 97% of the maximum at 5 kHz (10 ppm) on both sides of the center of the spectrum. This can affect the accuracy of integrals for resonances in different regions of the spectrum. (From Petersheim, used with permission.)

resulting from the mathematical transformation. The operational details of the electronics required to provide the strong pulse and detect the very weak signals are complex and beyond the scope of this text. The interested student should consult the text by Fukushima and Roeder or the references cited by Petersheim for more details.

3.5.5. Signal Integrator and Computer

All NMR spectrometers are equipped with a signal integrator to measure the area of peaks. The area often appears on the NMR spectrum as a step function superimposed on the spectrum, as seen in Fig. 3.18(a), but is also printed out in a data report as well.

All NMR spectrometers require a computer to process the data and much of the instrument is under computer control. Multitasking or networked computer workstations are normally used, so that data can be processed while long experiments are running on the instrument. A typical single NMR data file can be processed in less than 1 s. 2D and 3D NMR experiments generate megabytes to gigabytes of data that must be processed for each experiment. According to Petersheim, 2D NMR plot generation in the early 1980s required days of operator and computer time. The same 2D plots are now generated in minutes with today's improved software and faster computers.

Computer control of the room temperature shim currents that control the field homogeneity, of sample spinning rates, autosamplers, pulse sequences, and many other aspects of the instrument operation is a necessity.

3.5.6.   Wide-Line Benchtop NMR Spectrometers and Portable NMR Spectrometers

There is a class of low-resolution benchtop NMR spectrometers commercially available that are useful for dedicated quantitative analyses measurements. These are referred to as wide-line spectrometers. They generally have permanent magnets of low magnetic field strengths, for example, 7–20 MHz, but are pulsed, time domain instruments. These instruments are physically small and able to fit easily onto a laboratory bench. Examples of the use of these instruments are presented in Section 3.6.7. At the opposite end of the NMR instrument size range from the very large 900 MHz instruments are handheld NMR devices. Bruker Optics (www.minispec.com) has developed a palm-sized NMR device called the Minispec MOUSE that allows the analyst to bring the NMR to the sample instead of having the sample being inserted in the instrument. MOUSE stands for Mobile Surface Universal Explorer. The device has a small magnet and RF surface coil that directs the magnetic field and RF pulses into an object of any size. Measurements can be made approximately at the surface of the object or into the sample to a depth of several millimeters. Such a device can be used for medical applications, polymer, and rubber science and in production quality control.

3.6.   ANALYTICAL APPLICATIONS OF NMR

NMR spectroscopy is used for both qualitative and quantitative analyses. The applications of NMR are very diverse and only a very few examples can be given here. The student interested in applications of NMR is advised to look at journals such as Analytical Chemistry, published by the American Chemical Society, or Applied Spectroscopy, published by the Society for Applied Spectroscopy, for an overview of the wide range of uses for NMR. Particularly useful is the annual issue of Analytical Chemistry dedicated to Applications Reviews (odd-numbered years) or Fundamental Reviews (even-numbered years).

3.6.1.   Samples and Sample Preparation for NMR

Liquid samples are the simplest samples to analyze by NMR. Neat nonviscous liquids are run "as is" by placing about 0.5 mL of the liquid in a glass NMR tube. Liquids can be mixed in a suitable solvent and run as solutions; the analyte concentration is generally about 2–10%. For the examination of liquid samples, the sensitivity is sufficient to determine concentrations down to about 0.1%. NMR is not considered a "trace" analytical technique, but that is changing as instrumentation continues to improve. Microtubes with as little as 1 μg of sample in 100 μL of solvent are now in use. Samples in solution and neat liquids are degassed to remove oxygen and filtered to remove iron particles; both O_2 and iron are paramagnetic and cause undesired line broadening.

Soluble solid samples are dissolved in a suitable solvent for analyses. A typical sample size is 2–3 mg dissolved in 0.5 mL of solvent. Some solid polymer samples may be run under "liquid" conditions, that is, without MAS, by soaking the solid in solvent and allowing it to "swell". This gives enough fluidity to the sample that it behaves as a liquid with respect to the NMR experiment. Other solid samples must be run in an instrument equipped with MAS, as has been discussed.

NMR does not have sufficient sensitivity to analyze gas phase samples directly. Gases must be concentrated by being absorbed in a suitable solvent, condensed to the liquid phase, or adsorbed onto an appropriate solid phase.

A suitable solvent for NMR should meet the following requirements: (1) be chemically inert toward the sample and the sample holder, (2) have no NMR absorption spectrum itself or a very simple spectrum, and (3) be easily recovered, by distillation, for example, if the original sample is required for other testing. The best solvents for proton NMR contain no protons and therefore give no proton NMR signals. Carbon tetrachloride and carbon disulfide fall into this category. Replacing protons, ^1H, with deuterium, ^2H or D, will remove most of the proton signal for the solvent from the spectrum. Deuterated chloroform, $CDCl_3$, deuterated water, D_2O, and many other deuterated solvents are commercially available for use. Deuterated solvents have two drawbacks; they are expensive and they generally contain a small amount of ^1H, so some small signal from the solvent may be seen. A spectrum of the solvent (the blank spectrum) should be run regularly and whenever a new lot of solvent is used. Deuterated chloroform is the solvent used for most of the ^{13}C spectra shown in the chapter and the small signal from the naturally occurring ^{13}C in the solvent is visible in these spectra. An example will be pointed out when discussing these spectra.

3.6.2. Qualitative Analyses: Molecular Structure Determination

The primary use of NMR spectroscopy is for the determination of the molecular structure of compounds. These may be organic compounds synthesized or separated by organic chemists, pharmaceutical chemists, and polymer chemists; organic compounds isolated by biologists, biochemists, and medicinal chemists; and organometallic and inorganic compounds synthesized by chemists and materials scientists. The importance of NMR spectroscopy in deducing molecular structure cannot be overstated.

While some features of the proton NMR spectrum have been discussed earlier, we will go through some examples of how to use a proton NMR spectrum to work out the structure of some simple organic molecules. We need to understand a few more aspects of the NMR spectrum first.

3.6.2.1. Relationship Between the Area of a Peak and Molecular Structure

As we learned in Section 3.4, the multiplicity of a given peak tells us the number of adjacent equivalent protons. The multiplicity tells us nothing about the number of protons that give rise to the peak itself. That information comes from the peak area. The total area of an absorption peak is directly proportional to the number of protons that resonate at the frequency. By total area, we mean the area of all the peaks in the multiplet, if the peak is a multiplet, or the total area of the peak if the peak is a singlet.

In a sample of dry ethanol, Fig. 3.18(a), the methyl group, CH_3, is split by the methylene group to give three peaks. The area corresponding to the methyl group is the area enclosed by all three peaks of the triplet measured from a baseline. The two protons of the methylene group are split by the methyl group and by the proton of the OH into a total of eight peaks. The area contributed by the methylene protons is that enclosed by all eight peaks measured from a baseline on the spectrum. The total area of the methyl group, CH_3, will be $3/2\times$ the total area of the CH_2 group. The OH proton is split into a triplet by the methylene protons; the area corresponding to that proton is the entire area of its triplet measured from a baseline; the area corresponding to the hydroxyl proton should be $1/3$ that of the methyl group.

In practice, the peak areas are measured by integrating the signal area automatically. The computer prints out the relative area. Relative areas are often shown as step function traces on the spectrum printout, as has been done for the spectrum in Fig. 3.18(a); each peak has a line traced over it with a baseline at the bottom and at the top. The peak areas are **relative**; it is not possible to assign a molecular formula to an unknown from the NMR spectrum alone. For the ethanol example given earlier, the relative areas of the peaks should be 3:2:1 for the peaks as shown from right to left, that is, A:B:C. For ethanol, the relative area of the peaks is also the absolute area. We know this *only* because we know the sample is ethanol and we know the molecular formula for ethanol. The areas are shown as the stepped lines drawn across the spectrum. In the rare case that the integrator is not functioning, the relative areas can be obtained the old-fashioned way, by measuring the height of each step with a ruler. Take a ruler (marked in any units as long as they are small enough, such as $1/16$ in., cm, or mm divisions) and measure the height of the step for peak C, on the far left of the spectrum. It may help to extend the bottom and top plateaus using your ruler and a pencil. Measure the height of this step in mm and write down the value. Do the same for the other two steps for peaks B and A. Now it is necessary to divide each of the measurements by the smallest measurement to get the relative peak areas. For the methyl peak, divide the height of step A by the height of step C; the value should be equal to 3. Therefore, the area of peak A is 3 relative to peak C. As you can see by doing the other two ratios, the relative areas are 3:2:1 for peaks A, B, and C respectively. It is important to measure and read the step heights as accurately as you can for this method to work well.

From this relative area calculation, we can state that there are twice as many protons in the group giving rise to peak B as there are in the group giving rise to peak C and that there are $3\times$ as many protons in the peak A group as in the peak C group. We *cannot* say that peak C is due to one proton and peak A due to three protons without more information. The empirical formula can be determined if we have elemental analyses information and the molecular formula can be calculated if the molecular weight is known from another measurement, such as MS.

It is also important to keep in mind that the smallest peak may be equal to more than one proton. You can deduce this if your ratios look like 1.9:1.5:1, for example. Clearly, you cannot have a molecule with 1.5 protons on a carbon atom. Now, you need to multiply everything through by the same common factor, until you get whole number values. If you multiply 1.9, 1.5, and 1 by 2, you get 4, 3, and 2 protons, respectively, which is a reasonable set of whole numbers to work with in building a structure. A good practice example of this type is the spectrum of butylamine (Fig. 3.37).

3.6.2.2. Chemical Exchange

In a solution of methanol, CH_3OH, and water, or ethanol and water, the hydrogen of the alcohol OH group exchanges with hydrogen in the water, H_2O. Such a proton is said to be *labile*. If this physical exchange rate is greater than the change in resonance frequency for the nuclei involved, the nearby nuclei see only the average position of the nucleus, and spin–spin splitting due to the labile proton disappears. The exchange rate is affected by temperature, increasing with increased temperature. In addition, the proton on the OH group can participate in hydrogen bonding, which is also temperature dependent, concentration dependent, and very solvent dependent. The spectra of alcohols are therefore affected by any traces of water in the sample and by temperature. Figure 3.22 is the spectrum of normal reagent grade ethanol, which contains water. Compare this spectrum with that of dry ethanol in Fig. 3.18(a). There are three dramatic differences. First, the position of the hydroxyl proton has shifted to 2.6 ppm instead of its position at about 4.3 ppm in the very dry sample, so the peak on the far left of the normal ethanol spectrum is the methylene peak. The hydroxyl peak appears as a singlet, not a triplet, because the exchange rate is so fast that the hydroxyl proton is not split by the methylene protons. For the same reason, the methylene protons are not split by the hydroxyl proton, only by the methyl protons. Therefore, the methylene peak appears as a quartet in normal ethanol. The spacing between the peaks in the triplet and the peaks in the quartet should be equal since $J_{AB} = J_{BA}$.

For another example, the spectra of CH_3OH at 20°C and −40°C are shown schematically in Fig. 3.23, demonstrating the effect of temperature on the spectrum. The rate of exchange is so rapid at the higher temperature that no spin–spin splitting is seen. Changes in hydrogen bonding will dramatically affect the chemical shift position of the alcohol proton. The same is true for any proton capable of hydrogen bonding,

Figure 3.22   The proton NMR spectrum of normal reagent grade ethanol, which always contains water. Compare this spectrum with that in Fig. 3.18(a). Note that in the presence of water, the spin–spin coupling due to the OH proton disappears and the chemical shift of the hydroxyl proton changes. (From Silverstein and Webster; this material is used by permission of John Wiley and Sons, Inc.)

Figure 3.23   Schematic proton spectra of methanol obtained at 20°C, shown on the left side of the diagram, and at −40°C, on the right of the diagram. The fine structure (spin–spin splitting) is not observed at 20°C due to the rapid exchange rate of the hydroxyl proton. Cooling the sample to −40°C slows the exchange rate sufficiently that the hydroxyl proton can be considered fixed in place; spin–spin coupling is then observed.

which means that any compound with an NH or OH proton has a spectrum dependent on temperature, concentration, and solvent polarity.

   If the chemical exchange frequency is lower than the spectral frequency for the nuclei of interest, there will be a discrete set of resonances for each state. Processes with intermediate rates are studied by physical, organic, and inorganic chemists using NMR because the positions and shapes of the peaks can be used to estimate reaction kinetics and the lifetimes of reactants and products in a reaction.

   Chemical exchange is not limited to the exchange of hydrogen-bonded protons in solution. Chemical exchange includes conformational changes as well as actual bond-breaking and bond-forming changes that result in new chemical shifts for the nuclei involved. Examples include dimer formation and tautomerism. The important point is that chemical exchange can alter the appearance of an NMR spectrum, as is clear from the alcohol example. This must be remembered in interpreting NMR spectra.

3.6.2.3.   Double Resonance Experiments

When we first examine an NMR spectrum of an unknown sample, it is often not easy to tell which nuclei are coupled, especially if the splitting patterns are complex. Double resonance experiments employ two different RF sources and a variety of pulse sequences to sort out complex splitting patterns. It was discussed in Section 3.1.2 that a system of protons could become saturated by applying a strong RF field to the sample. When saturation occurs, the saturated protons do not give an NMR signal and they do not couple with and split the peaks for adjacent protons. The saturated proton is said to have been "decoupled". Advantage is taken of this phenomenon to simplify complicated NMR spectra. In a double resonance **spin decoupling** experiment for proton NMR, one RF source is scanned as usual. The second RF source selectively saturates one resonance frequency (one group or type of protons). This causes the collapse of all the splitting patterns of nuclei to which that group is coupled. Irradiating each resonance frequency in turn and observing which peaks have their fine structure disappear permits identification of all the coupled protons. For example, a peak that had been a triplet will collapse into a singlet if the methylene group that is causing the spin–spin splitting is saturated. A simple example of the use of spin decoupling by double resonance is shown in Fig. 3.24. The proton spectrum of ethylbenzene is shown in Fig. 3.24(a). If the methyl resonance is saturated, the

Figure 3.24 Homonuclear decoupling experiment. A 250 MHz ^{1}H NMR spectrum of ethylbenzene in deuterated chloroform obtained (a) without decoupling, (b) with irradiation of the methyl resonance, and (c) with irradiation of the methylene resonance. (From Bruch and Dybowski, used with permission.)

NMR spectrum obtained is that shown in Fig. 3.24(b). The signal for the methyl group at 1.2 ppm disappears and the methylene quartet at 2.6 ppm collapses into a singlet, because the methyl protons no longer split the methylene protons. If in a second experiment, the methylene resonance is saturated, the spectrum in Fig. 3.24(c) is obtained. The methylene signal at 2.6 ppm disappears and the methyl triplet collapses into a singlet. From these two experiments, it can be deduced that the methyl and methylene groups are coupled and therefore are adjacent to each other.

A more complex spectrum and decoupling experiments are shown for sucrose in Fig. 3.25. The structure of sucrose is shown in Fig. 3.25. Ignoring the hydroxyl protons, each methine proton and methylene group is chemically different, giving rise to 11

(a)

(b)

(c)

Figure 3.25 Homonuclear decoupling experiments of the 300 MHz proton NMR spectrum of sucrose dissolved in D$_2$O. The fully coupled spectrum is shown in (c). (a) Selective saturation of the triplet at 4.05 ppm collapses the doublet at 4.22 ppm, showing the coupling between the positions of protons a and b marked on the sucrose structure. (b) Saturation of the doublet at 5.41 ppm collapses the doublet of doublets at 3.55 ppm, leaving a doublet. The experiment shows the coupling between the protons marked c and d on the structure. (The spectra are from Petersheim, used with permission. The sucrose structure is that of D-(+)-sucrose, obtained from the SDBS database, courtesy of National Institute of Industrial Science and Technology, Japan, SDBSWeb:http://www. aist.go.jp/RIOBD/SDBS. Accessed 11/05/02.)

resonance peaks in the spectrum. The fully coupled spectrum for sucrose dissolved in deuterated water, D_2O, is shown in (c) and the exchange rate is fast enough that no coupling is seen from the hydroxyl protons. Look first at spectrum a, the top spectrum, and compare it with spectrum c. Saturation of the triplet at 4.05 ppm, which is due to the proton marked (a) in the structure, causes the triplet to disappear from spectrum a. It also collapses the doublet at 4.22 ppm, indicating that proton (a) is coupled to the proton causing the absorbance at 4.22 ppm. If we know that the peak at 4.05 is due to proton (a), it follows that the peak at 4.22 ppm must be from the proton marked (b). [Why? Why can it not be the proton on the carbon to the right of proton (a)? Hint: think about the splitting.] Now look at spectrum b. Saturation of the doublet at 5.41 ppm, due to proton (c), causes the doublet to disappear from spectrum b and also collapses the doublet of doublets at 3.55 ppm, leaving a doublet. This tells us that the resonance at 3.55 ppm is due to proton (d). [Why? Why does proton (d) appear as a doublet of doublets in the fully coupled spectrum?] The singlet peak at 4.8 ppm in each spectrum in Fig. 3.25 is due to residual HDO in the D_2O solvent.

By using double resonance experiments, one can greatly simplify the spectrum; coupling between different types of nuclei is confirmed by both the disappearance of the peak for the saturated nuclei and the collapse of the fine structure of the coupled nuclei. Nuclei of the same type can be decoupled, as in the proton–proton example given earlier; this is called **homonuclear decoupling**. It is of course possible to decouple unlike nuclei, such as ^1H–^{13}C decoupling; this is called **heteronuclear decoupling**. Both of these are examples of spin decoupling.

In modern instruments it is not uncommon to use double and triple resonance experiments to simplify the spectrum sufficiently for interpretation. One interesting result of the use of double resonance, the **nuclear Overhauser effect** (NOE), is very important in ^{13}C NMR and will be discussed in Section 3.6.4. Table 3.4 lists a few commonly used double resonance, multipulse, and 2D NMR experiments by name. Several of these experiments will be explained in more detail subsequently.

3.6.3. Interpretation of Proton Spectra

NMR absorption spectra are characterized by the chemical shift of peaks and spin–spin splitting of peaks. Recall that the chemical shift is caused by the drifting, not orbiting or spinning, of nearby electrons under the influence of the applied magnetic field. It is therefore a constant depending on the applied field (i.e., if the field is constant, the chemical shift is constant). The chemical shift therefore identifies the functional group, such as methyl, methylene, aldehydic H, aromatics, and so on (see Table 3.3). All proton spectra shown have TMS as the reference, with the TMS absorbance set at 0.0 ppm. The student should note that all of the 300 MHz proton NMR spectra provided by Aldrich Chemical Company, Inc. also include the 75 MHz ^{13}C spectrum at the top. ^{13}C NMR spectra are discussed in Section 3.6.4.

Spin–spin splitting is caused by adjacent nuclei and is transmitted through the bonds. It is independent of the applied field. The multiplicity is therefore a function of the number of equivalent ^1H nuclei in the adjacent functional groups. Numerically, it is equal to $(2nI + 1)$, where n is the number of equivalent H and I is the spin number (in this case, $I = 1/2$). For two adjacent groups the number is $(2nI + 1)(2n'I' + 1)$ where n and n' are the numbers of ^1H nuclei in each separate group and I and I' each equal $1/2$. It can readily be seen from the real spectra we have already looked at and will look at, that the intensity ratios in multiplets are often not the symmetrical intensities predicted from Pascal's triangle. Look, for example, at the peak (a) triplet in Fig. 3.12(b) and (c).

Table 3.4 Double Resonance, Multipulse, and 2D NMR Experiments

Acronym	Experiment description
APT: attached proton test	Multipulse sequence used to distinguish even and odd numbers of protons coupled to ^{13}C through one bond. Even numbers of bound protons give positive peaks; odd numbers give negative peaks
DEPT: distortionless enhancement by polarization transfer	Multipulse; conditions are chosen so that only ^{13}C nuclei with the same number of bound protons have enhanced resonances. Four experiments must be performed, but DEPT is more definitive than APT
INADEQUATE: incredible natural abundance double quantum transfer	Multipulse; allows observation of natural abundance ^{13}C–^{13}C coupling. Only 1 carbon atom in 10^4 carbon atoms is a ^{13}C bonded to another ^{13}C, hence the name "incredible"
COSY: correlated spectroscopy	Homonuclear 2D experiment; plot of chemical shift vs. chemical shift identifies spin-coupled resonances. Many variations permit measurements of J coupling constants, long-range connectivities, suppression, and enhancement of selected resonances
HETCOR: heteronuclear chemical shift correlated experiment	Heteronuclear 2D experiment; usually to connect ^1H resonances with ^{13}C resonances or ^1H–X, where X is another NMR-active nucleus. Plot is ^{13}C chemical shift (or X chemical shift) vs. ^1H chemical shift
NOESY: nuclear Overhauser effect spectroscopy	Identifies dipolar coupled nuclei within certain distances (e.g., within 0.4 nm for first order coupling) and identifies connectivities through cross-relaxation

Note: Table modified from Petersheim, used with permission.

It is certainly is not 1:2:1 or symmetrical. This deviation from theory actually provides us with more structural information, as we will see.

Another piece of information is obtained from the relative area of the absorption peaks in the spectrum, which tells us the relative number of protons in each group. So, a proton NMR spectrum should be examined for: (1) the number of proton resonances which tells you how many different types of protons are in the molecule; (2) the chemical shifts of the resonances which identifies the type of proton; (3) the multiplicity of the resonances which identifies the adjacent equivalent protons; and (4) the intensity (area) of the resonances which tells you the relative number of each type of proton. Some examples of common classes of organic compounds are discussed. Not every type of compound is covered in this brief overview. Students needing more detailed spectral interpretation should consult the references by Silverstein and Webster, Pavia et al., Lambert et al., or similar texts.

3.6.3.1. *Aliphatic Alkanes, Alkenes, Alkynes, and Alkyl Halides*

Alkane protons absorb in the region between 0.6 and 1.7 ppm. The spectrum of octane, $CH_3(CH_2)_6CH_3$, Fig. 3.26, is typical of straight chain alkanes. The methyl groups absorb at 0.88 ppm and the methylene groups at 1.26 ppm. In long chain alkanes, the methylene protons often overlap and spin–spin splitting cannot be resolved. The relative area ratio of the methylene peak/methyl peak is expected to be 2/1, which can be confirmed by measuring the relative height of the steps shown on the spectrum. The peak at 0.88 ppm due to the methyl groups is split into a triplet by the adjacent methylene protons as expected.

75 MHz ^{13}C NMR
In CDCl3

300 MHz ^{1}H NMR
In CDCl3

Figure 3.26 Proton and ^{13}C NMR spectra of octane, $CH_3(CH_2)_6CH_3$. (Reprinted with permission of Aldrich Chemical Co., Inc.)

Substitution of an electronegative halide atom, F, Cl, Br, or I, on an alkane will deshield any protons attached to the same carbon atom. The deshielding is a result of the electronegativity of the halogen atom. Fluorine shows the largest deshielding effect, iodine the smallest. Two halogen atoms on a single carbon have a greater deshielding effect than a single halogen atom. Since ^{19}F is an $I = 1/2$ nucleus, fluorine-substituted hydrocarbons will show H—F and F—F spin–spin coupling. The other halogens do not couple.

Figure 3.27(a) is the 60 MHz proton spectrum of bromoethane (or ethyl bromide). The structure and chemical shifts are shown. The methyl peak is split into a triplet by the methylene group and the methylene group is split into a quartet by the methyl group. The peak a/peak b area ratio is expected to be 3/2. The position of the methylene peak is shifted to a significantly higher resonance frequency by the deshielding due to Br. It occurs at about 3.4 ppm, compared with the normal position of 1.2–1.4 ppm for methylene protons in an alkane such as octane (Fig. 3.26). The electronegative Br also affects the resonance position of the adjacent methyl protons, moving them to a higher resonance frequency as well, 1.7 ppm vs. the normal alkane position of about 0.9 ppm. Now look at the triplet; you will note that it is not symmetrical. The intensity of the peak on the left side of the triplet is higher than that of the peak on the right. Look at the quartet; you should note that the two peaks on the right of the quartet are higher than the two peaks on the left side. The triplet appears to be "leaning" toward the quartet and the quartet is "leaning" toward the triplet. This is a clue that the two peaks are coupled together. The J coupling constant should be identical for the triplet and the quartet (i.e., $J_{AB} = J_{BA}$). You should be

(a)

(b)

Figure 3.27 (a) The 60 MHz proton spectrum of bromoethane. The methyl protons at 1.7 ppm are split into a triplet by the two methylene protons, since $2nI + 1 = (2)(2)(1/2) + 1 = 3$. The methylene protons are deshielded by the Br and moved to 3.4 ppm. They are split into a quartet by the adjacent methyl protons, $2nI + 1 = (2)(3)(1/2) + 1 = 4$. (b) The 300 MHz proton spectrum and ^{13}C spectrum for bromoethane. The 3/2 proton area ratio can be measured from this spectrum. (Reprinted with permission of Aldrich Chemical Co., Inc.)

able to measure it from this spectrum. Figure 3.27(b) is the 300 MHz proton spectrum of bromoethane and its ^{13}C spectrum. You can confirm the 3/2 area ratio from the height of the steps plotted on the 300 MHz spectrum.

The compound chloroethane (or ethyl chloride), CH_3CH_2Cl, would show the same splitting pattern as bromoethane; the position of the methylene peak would be slightly more deshielded (3.5 vs. 3.4 ppm) due to the greater electronegativity of Cl. In both of these compounds, the electronegative halogen has a deshielding effect on the adjacent methyl protons as well; they absorb at higher chemical shift (1.2–1.5) than they would in an alkane (0.7–1.0 ppm). Figure 3.28 shows the effect of two chlorine atoms on the same carbon; the proton on that carbon is moved to 5.9 ppm by the deshielding. Compare the position of this proton and the methyl protons to the chemical shifts given for chloroethane and shown for bromoethane (Fig. 3.27). You will see that the two chlorine atoms deshield the adjacent methyl protons as well and by a larger amount than a single chlorine atom. The methyl group is split into a doublet by the adjacent proton, while the single proton (the peak at 5.9 ppm) is split into a quartet by the methyl group. As a final example of this type of compound, Fig. 3.29 shows the 60 MHz spectrum of 1,3-dichloropropane. The "a" methylene protons are split into five peaks by the four equivalent "b" methylene protons. The "a" protons are also deshielded; they absorb at 2.2 ppm

Figure 3.28 The spectra of 1,1-dichloroethane, CH_3CHCl_2, showing the effect of two halogen atoms on the same carbon on the resonance frequencies of the protons. The CH_3 group absorbs at 2.06 ppm, while the proton on the chlorine-containing carbon is highly deshielded and absorbs at 5.89 ppm. The J_{AB} coupling constant is 6.0 Hz. (Reprinted with permission of Aldrich Chemical Co., Inc.)

Figure 3.29   The 60 MHz proton spectrum of 1,3-dichloropropane.

instead of the 1.2–1.4 ppm position in an alkane. The "b" methylene protons are split into a triplet by the two "a" protons and are more strongly deshielded since the Cl atom is on the same carbon atom as the "b" protons. The expected area ratio of peak a/peak b is 1/2. There is asymmetry in the multiplets; they are "leaning" toward one another as expected.

Alkenes have two characteristic types of protons. The vinyl protons are those attached to a double bond. Vinyl protons are deshielded by the double bond and generally appear at chemical shifts between 4.5 and 6 ppm. The spin–spin coupling of vinyl protons is complicated because they are generally not chemically equivalent due to the lack of free rotation about the double bond. Allylic protons are those located on carbon atoms adjacent to the double bond; these are slightly deshielded by the double bond and appear between 1.2 and 2.5 ppm. The presence of the double bond allows long-range coupling, so the allylic protons may couple to the protons on the far end of the double bond as well as to the adjacent protons. Figure 3.30 shows the spectrum of 1-octene, $CH_3(CH_2)_5CH{=}CH_2$ or C_8H_{16}. The structure of 1-octene is shown on the spectrum. Note the shift of the allylic protons (the "c" protons) to 2 ppm by deshielding from the double bond and the strong deshielding of the vinyl protons, "d, e, and f". All three vinyl protons are nonequivalent, resulting in the complex splitting pattern observed in the 4.9–5.8 ppm region. The terminal methyl group peak is a triplet, but the methylene groups in the middle of the chain overlap to give the broad peak from 1.0 to 1.5 ppm. The relative peak areas, from left to right, should be 1:2:2:8:3.

An alkyne with a triple bond at the end of a chain is called a terminal alkyne and the hydrogen atom at the end of the triple bond is referred to as an acetylenic hydrogen. This terminal proton is shielded by the anisotropy of the triple bond π electrons, as was shown in Fig. 3.8, and so absorbs at about 1.8 ppm. The protons on the carbon next to the triple bond are affected in the same way as allylic protons in alkenes and absorb in the same chemical shift range.

3.6.3.2.   Aromatic Compounds

The proton absorbances used to identify aromatic compounds are the protons on the aromatic ring itself, the *ring protons*, and the protons on the carbon atoms adjacent to the ring. The latter are called *benzylic protons*. Protons attached directly to an aromatic ring are highly deshielded by the π electrons, as shown in Fig. 3.8. Ring protons therefore absorb between 6.5 and 8.0 ppm. There are few other protons that absorb in this region, so

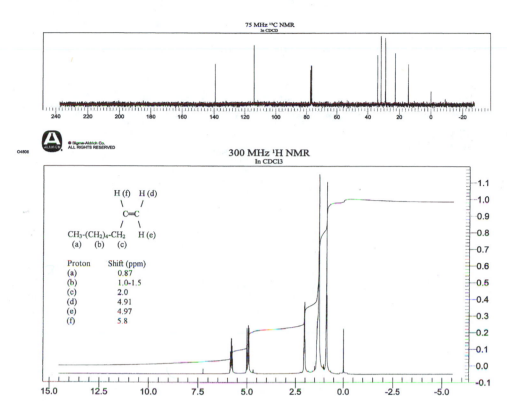

Figure 3.30 Spectra for 1-octene. The chemical shift of the terminal protons on the C=C bond is between 5 and 6 ppm, and the two protons are not equivalent because the double bond does not permit rotation of these protons. (Reprinted with permission of Aldrich Chemical Co., Inc. with the structure and chemical shift information added by the authors.)

an absorbance in this region is characteristic of aromatic compounds. The spectrum of toluene, Fig. 3.4(a) shows this characteristic absorbance; the ring protons give rise to the signal at 7.2 ppm. This peak shows a complex splitting pattern, because the ring protons are not equivalent; they are ortho, meta, and para to the substituent. The splitting pattern and J values can identify the number and position of substituents on the ring. A few examples will be shown but not discussed in detail. The patterns are second order and the details are beyond the scope of this text, but the interested student can consult the texts by Silverstein and Webster, Pavia et al., or Lambert et al. listed in the bibliography. One typical pattern is seen in Fig. 3.31, the proton spectrum of 1,2-dichlorobenzene, an ortho-disubstituted ring. The molecule contains a plane of symmetry, so that there are two sets of equivalent protons on the ring, as shown:

Figure 3.31 The 300 MHz proton spectrum of 1,2-dichlorobenzene. The only absorbance seen is due to the aromatic ring protons. The plane of symmetry in this molecule results in the characteristic "four line" splitting pattern in the aromatic ring signal. The ^{13}C spectrum is included. (Reprinted with permission of Aldrich Chemical Co., Inc.)

This results in a characteristic pattern of four lines in the aromatic region, as seen in Fig. 3.31. The same four-line splitting pattern would be expected for para-disubstituted aromatic compounds, such as 1-bromo-4-chlorobenzene, because such a molecule has a similar plane of symmetry as shown:

If both para substituents were the same, as in 1,4-dichlorobenzene, all four protons would be equivalent and a singlet peak would be seen.

The *benzylic protons* on carbon atoms adjacent to an aromatic ring are also deshielded by the ring but to a lesser extent than the ring protons. The characteristic absorbance values for benzylic protons are in the 2.2–2.8 ppm range. An example is ethylbenzene, C_6H_5—CH_2—CH_3, whose spectrum is shown in Fig. 3.32. The compound is a monosubstituted aromatic ring, so the ring protons at 7.2 ppm show the same complex splitting pattern we saw in toluene. The benzylic protons are the methylene protons, on

Figure 3.32 The spectra of ethylbenzene, showing the characteristic aromatic ring and benzylic proton absorbances typical of alkyl-substituted aromatic compounds. (Reprinted with permission of Aldrich Chemical Co., Inc.)

the carbon attached to the ring. The quartet appears at ∼2.6 ppm as expected. The methyl triplet is the peak at 1.2 ppm.

3.6.3.3. Oxygen-Containing Organic Compounds

Alcohols are organic compounds containing a hydroxyl group, —OH, attached to a carbon atom. As we have discussed and already seen in Figs. 3.18, 3.22, and 3.23, the chemical shift of the hydroxyl proton depends on variables such as temperature, solvent, and concentration. The range of chemical shifts covers 0.5–5.0 ppm for aliphatic alcohols. The hydroxyl proton on an —OH group attached to an aromatic ring is deshielded by the ring, as can be seen in the spectrum of phenol, C_6H_5OH (Fig. 3.33). Any other protons on the carbon to which the hydroxyl group is attached, —**CH**$_x$**OH**, shown in bold type, are shifted due to the electronegative oxygen atom. These protons absorb between 3.1 and 3.8 ppm. In Fig. 3.23, for example, we see these protons located at a chemical shift of ∼3.6 ppm. Exchange of the hydroxyl proton is usually rapid enough that no splitting is observed between the hydroxyl proton and the other protons on the same carbon.

Aldehydes have a terminal $-\overset{\overset{\text{O}}{\|}}{C}-H$ group. We have discussed the spectrum of butanal, shown in Fig. 3.12(b) and (c). The aldehydic proton occurs at ∼9.0–10 ppm. Hydrogens on the carbon adjacent to the C=O group are deshielded

Figure 3.33 The spectra of phenol, C_6H_5—OH, an aromatic alcohol. The hydroxyl proton is deshielded by the aromatic ring and appears at a chemical shift of 5.6 ppm. (Reprinted with permission of Aldrich Chemical Co., Inc.)

slightly, and occur at about 2–2.5 ppm. Ketones contain a nonterminal —C=O group, so a ketone is represented by $RCH_{(x)}$—C=O.
$$\overset{|}{R'}$$

The protons on the carbon adjacent to the C=O group, shown in bold, are in the same environment as those in an aldehyde. If you replace the R′ with H, you have the formula for an aldehyde. Therefore, the protons shown in bold type also absorb at about 2–2.5 ppm due to slight deshielding by the C=O. The spectrum of cyclohexanone, $C_6H_{10}O$, is presented in Fig. 3.34. The four equivalent protons on the two carbon atoms adjacent to the carbonyl group appear at 2.4 ppm.

Carboxylic acids contain the functional group —COH, also written —COOH. The acidic proton, the one on the —COOH group, is strongly deshielded and absorbs at 10–13 ppm. A peak in this position is a good indication of a carboxylic acid. As is the case for ketones and aldehydes, the protons on the carbon atom adjacent to the —COOH group are slightly deshielded and absorb in the 2.1–2.5 pm range. A simple example is acetic acid, CH_3COOH, whose spectrum is given in Fig. 3.35. The acid proton occurs at about 11.8 ppm, while the methyl protons, adjacent to the COOH group, appear at 2.1 ppm. Figure 3.36 is the 90 MHz spectrum of benzoic acid, showing the structure and assignments. The complex splitting pattern for a monosubstituted

Figure 3.34 The spectra of cyclohexanone. The only characteristic absorbance for ketones in the proton NMR spectrum is that of the protons adjacent to the carbonyl group, which are shifted to 2.4 ppm by slight deshielding. (Reprinted with permission of Aldrich Chemical Co., Inc.)

benzene ring is clearly seen in this spectrum. The ortho protons are more deshielded than either the meta or para protons; this is typical when a carbonyl group is bonded to the ring, and is seen in benzaldehyde, nitrobenzene, and similar compounds.

3.6.3.4. Nitrogen-Containing Organic Compounds

Proton NMR spectra of compounds such as amines, RNH_2, and amides, $RCONH_2$, which have protons bonded to the nitrogen atom, are complicated by several factors. The protons on nitrogen in these compounds, like the hydroxyl proton in alcohols, exhibit a very variable chemical shift. These protons can hydrogen-bond, so the chemical shift depends on temperature, solvent, and concentration. The ^{14}N nucleus is a spin = 1 nucleus, so in theory, a proton on a nitrogen atom should be split into $2(1) + 1 = 3$ peaks. Similarly, a proton on a carbon atom adjacent to a nitrogen atom should be split into 3 peaks. This is usually not seen for two reasons. Protons on electronegative nitrogen can undergo exchange (just like alcohols) and the rate of exchange will determine if the proton is coupled to the nitrogen. In aliphatic amines and amides, the exchange rate is fast enough that no splitting is seen. In addition, nitrogen has an electrical quadrupole moment. This quadrupole moment interacts in such a way as to broaden NH peaks, resulting in no observed splitting. The N—H peak in amines can vary in chemical shift from 0.5 to 4.00 ppm, that in amides from 5 to 9 ppm. The peaks can be sharp singlets or weak broadened signals. Coupling between *N—H, N—CH,* and *—HC—NH—* is usually not

Figure 3.35 The spectra of acetic acid. The acid proton absorbance at 11.8 ppm is characteristic of carboxylic acids. (Reprinted with permission of Aldrich Chemical Co., Inc.)

seen because of proton exchange and quadrupole interaction. The protons on the carbon atom adjacent to the nitrogen atom in amines are shifted to 2.3–3 ppm; the protons on the carbon atom adjacent to the amide group are shifted into the 2–2.5 ppm region, which is the same region where protons adjacent to a C=O absorb.

The chemical shift position of the proton on nitrogen in amines is not a good diagnostic tool because of its variability. In general, the peak due to the amine proton will be a sharp singlet in aliphatic amines because of fast proton exchange. A typical aliphatic amine spectrum is that of butylamine, $CH_3(CH_2)_2 CH_2NH_2$, Fig. 3.37, with the structure and peak assignments shown. The amine proton peak is peak d, the sharp singlet at 1.1 ppm. The relative peak areas from the step heights are, from left to right, 2:4:1:3. This is a good spectrum on which to practice the manual step height measurement technique, as was suggested earlier. The proton spectrum of propionamide, $CH_3CH_2CONH_2$, is shown in Fig. 3.45 and should be looked at. There are two very small broad peaks located at 6.3 and 6.6 ppm. These are due to the two protons on the nitrogen atom; the peaks are broad and therefore of low intensity due to the nitrogen quadrupole interaction. But the fact that there are two peaks means that these protons are not equivalent. But the formula for the molecule seems to show that the two protons on the nitrogen atom are chemically the same. We need to think back to general chemistry and Lewis structures. The two protons are not equivalent, because the nitrogen atom has an unshared pair of electrons on it. The unshared pair of electrons on nitrogen interacts with the unshared pairs of electrons on the oxygen. The interaction restricts rotation about the C—N bond, making the two protons on nitrogen nonequivalent; this results in two separate

Figure 3.36 The 90 MHz proton spectrum of benzoic acid. (Courtesy of National Institute of Industrial Science and Technology, Japan, SDBSWeb:http://www.aist.go.jp/RIOBD/SDBS. Accessed 11/5/02.)

peaks. The observation of these two peaks is typical of primary amides, that is, amides with an —NH_2 group.

Compounds containing a nitro group, —NO_2, have a characteristic shift for the protons on the carbon atom adjacent to the nitro group. These protons absorb between 4 and 4.5 ppm, as seen in the spectrum of nitropropane, $CH_3CH_2CH_2NO_2$, Fig. 3.38. The methylene group next to the nitro group gives the triplet peak at 4.3 ppm.

3.6.4. ^{13}C NMR

All nuclei with an odd mass number have a fractional (1/2, 3/2, etc.) spin number. Also, all nuclei with an even mass number and an odd atomic number have a unit (1, 2, 3, etc.) spin number. In fact, only those nuclei with an even mass and an even atomic weight have a zero spin number and therefore give no NMR signal. Unfortunately, this includes ^{12}C and ^{16}O—two important nuclei in organic chemistry. Carbon is the underlying "backbone" of organic molecules and knowledge of carbon atom locations in molecules is crucial to structural determination.

However, it can be readily seen that essentially all other elements have at least one isotope that can be examined by NMR. Examples were given in Table 3.1 and some additional isotopes are listed in Table 3.5, together with their sensitivity by NMR compared to the proton.

There has been great incentive to develop ^{13}C NMR because carbon is the central element in organic chemistry and biochemistry. However, useful applications were not forthcoming until the 1970s because of the difficulties in developing instrumentation. Two major problems were involved. First, the ^{13}C signal was very weak because the

Figure 3.37 The spectra of butylamine. The amine protons give rise to the sharp singlet at 1.1 ppm. The adjacent methylene protons are shifted to about 2.7 ppm by the electronegative nitrogen, while the other alkyl methylene and methyl protons appear at their usual chemical shift positions. (Reprinted with permission of Aldrich Chemical Co., Inc., modified by addition of the structure and peak identification.)

natural abundance of ^{13}C is low, only 1.1% of the total carbon present in a sample. Also, the ratio γ and the sensitivity are low compared to that for ^1H. The net result was a carbon NMR signal only 0.0002 as intense as a comparative ^1H signal. Second, the chemical shift range was up to 200 ppm using TMS as a standard. This increased range precluded a simple "add on" to the ^1H NMR instruments already commercially available. However, the incentive to develop such sensitive instruments was great. The introduction of FTNMR, which allowed the excitation of all ^{13}C nuclei simultaneously, made ^{13}C readily determinable by NMR. ^{13}C NMR is now a routine technique, providing important structural information about the carbon backbone of organic molecules.

There are several advantages to obtaining ^{13}C NMR spectra for structural identification of organic molecules. First, the wide range over which chemical shift occurs, 200 ppm for carbon compared with only 10 ppm for protons, greatly diminishes overlap between carbons in different chemical environments. The spectra are less crowded and a peak is usually seen for each unique carbon nucleus. Second, adjacent ^{12}C atoms do not induce spin–spin splitting, and the probability of two ^{13}C atoms being adjacent to each other is very low. Therefore spin–spin coupling between ^{13}C nuclei is not seen and the spectra are much simpler than proton spectra. ^{13}C NMR spectra are included with all of the Aldrich proton NMR spectra examples we have used. Some of these carbon NMR spectra will be discussed in detail.

Figure 3.38 The spectra of nitropropane. (Reprinted with permission of Aldrich Chemical Co., Inc.)

Coupling between ^{13}C and ^{1}H nuclei does occur, but there are techniques available to decouple these nuclei. Consequently, the ^{13}C NMR spectra are very simple, with a singlet seen for each chemically distinct carbon atom. This facilitates interpretation of the spectra. And, as we will see later, comparison of ^{13}C and ^{1}H spectra lead to data that can be interpreted with a high degree of confidence, thereby elucidating the structure of even

Table 3.5 Natural Abundance, Spin, and Sensitivity of Selected NMR-Active Nuclei

Nucleus	Natural abundance (%)	Spin	Sensitivity relative to ^{1}H
^{1}H	99.98	1/2	1.0
^{7}Li	92.6	3/2	0.3
^{13}C	1.1	1/2	0.0002
^{14}N	99.6	1	0.001
^{17}O	0.037	5/2	0.03
^{19}F	100	1/2	0.83
^{23}Na	15.9	3/2	0.09
^{25}Mg	10.1	5/2	0.003
^{27}Al	15.6	5/2	0.206
^{29}Si	4.7	1/2	0.008
^{31}P	24.3	1/2	0.07
^{33}S	0.8	3/2	0.002

very complicated molecules. Coupling also occurs between ^{13}C and D, which can be seen when deuterated solvents are used. Look at the ^{13}C spectra in Figs. 3.35 and 3.37, for example. Both spectra were acquired in $CDCl_3$. The triplet at 77 ppm is due to the natural ^{13}C in the solvent split by the D nucleus. In fact, the signal from the deuterated solvent, such as $CDCl_3$, is often used as the frequency-lock for the NMR. Since D is an $I = 1$ nucleus, the signal from the single C atom is split into a *triplet* by the single deuterium nucleus, not a doublet, because $2I + 1 = [(2 \times 1) + 1] = 3$.

The chemical shift of a ^{13}C nucleus is determined by its chemical environment (electronegativity and anisotropy) as for protons, but in a more complex manner. Chemical shifts for some ^{13}C functional groups are shown in Fig. 3.39. More detailed chemical shift tables can be found in the references on spectral interpretation listed in the bibliography.

3.6.4.1. Heteronuclear Decoupling

Coupling between ^{13}C and ^1H occurs and results in complex spectra with overlapping multiplets, but in practice, it is eliminated by broad band decoupling. The sample is irradiated with a wide RF frequency range to decouple all the protons at once. Each chemically different carbon atom should then give a single NMR resonance peak. Figure 3.40 shows the ^{13}C NMR spectrum of sucrose before proton decoupling (top) and after proton decoupling (bottom). As can be seen in the bottom spectrum, each of the 12 carbon atoms in sucrose gives rise to a single discrete absorption peak in the spectrum. Two of the resonances are fairly close (the ones at ~ 26.5 ppm) but they can be distinguished. The structure of sucrose is given in Fig. 3.24; you should confirm for yourself that each carbon atom is chemically nonequivalent.

3.6.4.2. The Nuclear Overhauser Effect

When broad band decoupling is used to simply the ^{13}C spectrum, it is noted that peak areas increase more than is expected from elimination of the peak splitting. This is the NOE. Direct dipolar coupling between a saturated nucleus and an unsaturated nucleus results

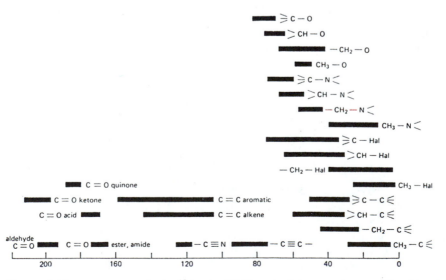

Figure 3.39 Chemical shifts for ^{13}C using TMS as the 0.0 ppm reference. The abbreviation Hal stands for halogen, (i.e., Cl, Br, or I.)

Figure 3.40 Heteronuclear decoupling of the 75 MHz ^{13}C spectrum of sucrose. The fully coupled spectrum is at the top. The bottom spectrum is the broadband ^{1}H decoupled spectrum. The structure of sucrose was given in Fig. 3.25. The molecule contains 12 nonequivalent carbon atoms; the decoupled spectrum clearly shows 12 single peaks, one for each nonequivalent C atom. (From Petersheim, used with permission.)

in a change in the ground state population of the unsaturated nucleus. This change in ground state population is a result of quantum transitions resulting in cross-relaxation between the nuclei; a transition in one nucleus induces a transition in the second nucleus. In ^{13}C NMR, the result of the NOE is that the signal for the low abundance ^{13}C nucleus is increased dramatically on proton decoupling. Under ideal conditions, the NOE can double the ^{13}C peak intensity; this decreases the time needed to collect a spectrum by a factor of 4. The maximum NOE between two isotopes can be expressed as:

$$\text{NOE} = \frac{\gamma_{\text{obs}}}{2\gamma_{\text{sat}}} + 1 \qquad (3.18)$$

where γ_{obs} is the magnetogyric ratio for the nucleus being measured and γ_{sat} is the magnetogyric ratio for the nucleus being saturated.

The major disadvantage of the NOE is that the relationship between peak area and number of carbon atoms giving rise to the peak is lost. An example of this is seen in Fig. 3.41, where the peaks for the protonated aromatic carbons (peaks 2 and 3) are more than twice the height of the peaks for the unprotonated aromatic carbons (peaks 1 and 4), even though each peak is due to a single carbon atom. The NOE can be eliminated experimentally, which must be done if quantitative analyses of the carbon spectral data is required. The method is discussed in Section 3.6.6. The NOE is also seen in homonuclear

Figure 3.41 (a) ^{1}H and (b) ^{13}C NMR spectra of poly(1,4-phenylene ether sulfone) in deuterated dimethylsulfoxide, DMSO-d_6. Spectrum (b) shows the NOE that occurs on proton decoupling of the ^{13}C spectrum. The peaks for the protonated carbon atoms 2 and 3 are enhanced by the NOE over the signals from the nonprotonated carbon atoms 1 and 4. The peak in the proton spectrum at 3.3 ppm is due to water in the solvent; the peak at 40 ppm in the ^{13}C spectrum is due to the natural ^{13}C in the solvent. (From Williams, used with permission.)

spin decoupling and can be used to determine the distances between nuclei, providing more structural information about a molecule.

3.6.4.3. ^{13}C NMR Spectra of Solids

Solid samples present a number of problems in ^{13}C NMR. Line broadening arises from chemical shift anisotropy, because of the many orientations the different carbon atoms have in a solid sample relative to the applied magnetic field. The chemical shift anisotropy can be eliminated by MAS at rotation frequencies $\sim$5–15 kHz around an axis forming an angle of 54.76° (the magic angle) with the applied magnetic field. This technique increases the resolution observed in the spectrum for a solid by averaging the chemical shift anisotropies to their isotropic values. Figure 3.5 shows how dramatically the use of MAS reduces line broadening in a solid sample spectrum. Line broadening also occurs as a result of interaction between ^{13}C and ^{1}H; decoupling of the dipolar interaction in a manner similar to spin decoupling also reduces the linewidth.

The spin–lattice relaxation time for ^{13}C in solids is very long (several minutes). Since the nuclei have to relax before another excitation pulse can be sent, this requires hours of instrument time in order to collect a spectrum of reasonable intensity. A pulse technique called **cross-polarization** can be used to reduce this effect by having the protons interact with the carbon nuclei, causing them to relax more rapidly. FTNMR systems for solid samples include the hardware and software to produce narrow line spectra from solid samples in a reasonable amount of time using high-power dipolar decoupling, MAS, and cross-polarization.

3.6.4.4. *Interpretation of ^{13}C Spectra*

A few examples of interpretation of ^{13}C spectra will be discussed. Figure 3.39 provides the chemical shift information for a variety of organic compounds and should be used to follow the discussion. Remember that you need to look at the carbon atoms in the structures, not the protons. From Fig. 3.39, we can see that alkane carbons are found between 0 and 75 ppm, aromatic and alkene carbons in the 100–160 ppm region, and the carbon of a carboxylic acid C=O group in a very narrow region between 170 and 180 ppm, for example.

Looking at the ^{13}C spectrum of acetic acid (at the top of Fig. 3.35), there are two single peaks, one at 21 ppm and one at 178 ppm. The small triplet at 77 ppm is due to the $CDCl_3$ solvent. The structure of acetic acid is CH_3COOH; it has two distinct carbon atoms, one alkyl carbon and one carboxylic acid carbon. From Fig. 3.39, the peak at 178 ppm is due to the acid carbon and the peak at 21 ppm is due to the CH_3 carbon. The peak for the alkyl carbon is much higher than the acid carbon peak. Proton decoupling has enhanced the protonated methyl carbon signal, while the intensity of the unprotonated acid carbon is not changed. The relative size of the peaks cannot be used to estimate the number of carbons because of the NOE.

The ^{13}C spectrum of benzene, C_6H_6, is given in Fig. 3.42. All six carbon atoms are chemically equivalent, so the spectrum consists of a single peak at 128 ppm, in the aromatic carbon region. Figure 3.43 shows the spectrum of cyclohexanol, $C_6H_{11}OH$, a cyclic aliphatic alcohol. There are six carbon atoms in the ring but they are not all equivalent. The structure and assignments are shown on the spectrum. The carbon to which the hydroxyl group is attached is unique, and gives the deshielded peak at 70 ppm. The carbon at the opposite end of the ring from the hydroxyl is also unique; it gives the peak at 25 ppm. There are two equivalent "b" carbons and two equivalent "d" carbons due to the symmetry of the molecule. The solvent triplet at 77 ppm is seen.

Monosubstituted benzene rings have the same pattern of symmetry as does cyclohexanol; the ortho carbons are equivalent, the meta carbons are equivalent, while the substituted carbon and the para carbon are each unique. Therefore, a monosubstituted benzene ring should show four carbon peaks in the aromatic region. Benzaldehyde, C_6H_5CHO, is an example of this pattern. The ^{13}C spectrum of benzaldehyde in Fig. 3.44 shows the expected four peaks in the aromatic region between 130 and 140 ppm. The smallest of the four peaks is the substituted carbon, the next largest is the para carbon, and the two tallest peaks are from the ortho and meta carbons. Although the NOE does not permit exact area/number of nuclei calculations, the height of a peak is still a function of the number of nuclei and the number of protons on the carbon. All else being equal, a single unprotonated carbon will give a smaller peak than a peak from multiple carbon

Figure 3.42 NMR spectra of benzene, C_6H_6. The six carbon atoms are equivalent, resulting in a single peak in the ^{13}C spectrum in the aromatic region at 128 ppm. The six protons are also equivalent, resulting in a single peak in the proton spectrum. (Reprinted with permission of Aldrich Chemical Co., Inc.)

nuclei bearing multiple protons. There is an additional signal in the spectrum from the carbon in the aldehyde group. From Fig. 3.39, we would expect to find the aldehydic carbon in the 190–210 ppm range; it appears in this spectrum at 192 ppm.

Figure 3.45 shows the NMR spectra for an amide, a class of compounds with —NH_2 substituted for the hydroxyl group of a carboxylic acid. The compound here is propionamide, $CH_3CH_2CONH_2$. There are three unique carbon atoms, the carbonyl carbon in the amide and two alkyl carbons. From Fig. 3.39, the carbonyl carbon in an amide is expected to absorb in the 165–175 ppm range; the peak occurs at 177 ppm in this spectrum. The methyl carbon is located at 10 ppm, while the methylene carbon appears at 29 ppm. The proton spectrum was discussed earlier, but it is worth noting again the two small broad peaks due to the nonequivalent amide protons at 6.3 and 6.6 ppm.

The student is encouraged to look at the ^{13}C spectra presented in the earlier figures in the chapter and try to work through the number and assignment of the peaks using Fig. 3.39 as a guide.

3.6.5. 2D NMR

High-resolution NMR spectra of organic compounds can be complex, with overlapping resonances and overlapping spin–spin couplings. The use of 2D NMR experiments and

Figure 3.43 NMR spectra of cyclohexanol, $C_6H_{11}OH$. (Reprinted with permission of Aldrich Chemical Co., Inc., modified by addition of the structure and peak identification by the authors.)

even 3D and 4D experiments extends the information obtained into a second (or third or fourth) frequency dimension. The spectrum becomes easier to interpret and much more structural information is usually provided. 2D and higher dimension experiments rely on the selective manipulation of specific nuclear spins, followed by interaction between nuclear spins. A series of such experiments can provide the entire molecular structure including the stereochemistry of the molecule.

A 2D experiment generally consists of the following: a pulse, followed by a time interval t_1, then a second pulse, followed by a time interval t_2. The first time interval t_1 is called the evolution period; t_2 is the acquisition period. It is during the evolution period that nuclear spins interact. A nucleus detected during the acquisition period has been frequency modulated by the nuclei it has interacted with during t_1. By varying the evolution period, t_1, in increments, and collecting the resulting FIDs, two frequencies are generated from a double Fourier transformation of the data. It is common to collect 1024 or more FIDs. Each one is Fourier transformed to give a frequency axis ν_2 obtained from t_2; a second FT is performed at right angles to the first one, resulting in a frequency axis ν_1 related to t_1. One frequency axis is the nucleus detected during the acquisition period; the other axis can be the same nucleus (a homonuclear experiment such as COSY) or a different nucleus (a heteronuclear experiment such as HETCOR). The data are plotted as frequency vs. frequency, usually presented as a contour plot.

For example, a homonuclear COSY experiment (see Table 3.4) is used to map the proton–proton J coupling in a molecule. Consider a simple system in which two protons are coupled to each other, CH—CH. The basic pulse sequence for a 2D

Figure 3.44 NMR spectra of benzaldehyde, showing the typical pattern of a monosubstituted benzene ring in the ^{13}C spectrum and the characteristic signal from an aldehydic carbon. (Reprinted with permission of Aldrich Chemical Co., Inc.)

COSY experiment consists of a relaxation or preparation period to establish spin equilibrium. A 90° pulse is applied and the spins precess at their characteristic frequencies during the evolution period, t_1. After the evolution period, a second 90° pulse, called the mixing pulse, is applied. This second pulse causes exchange of magnetization between J-coupled spins. The normal 1D proton spectrum is plotted on both the x-axis and the y-axis. For the simple system we are considering, the spectrum would show two doublets, one centered at a chemical shift of δ_A for proton A and one centered at a chemical shift of δ_x for proton X. The plot is shown schematically in Fig. 3.46. If the magnetization undergoes identical modulation during t_1 and t_2, the resulting frequencies will be the same. A plot of ν_1 vs. ν_2, where the two frequencies are identical, results in a point along the diagonal of the x–y graph. The contour peaks (points in the schematic diagram) that appear along the diagonal are the resonances in the "normal" spectrum and provide no additional information. The four points labeled 1–4 on the diagonal are just the frequencies of the four peaks (i.e., the two doublets) in the 1D NMR spectrum. Peaks 5–8 are called autocorrelation peaks and will not be discussed. What we are interested in are those results where the magnetization exhibits one frequency during t_1 and a different frequency during t_2, that is, peaks that have been frequency-modulated by interaction. In this case, peaks appear *off* the diagonal in the x–y plot; such peaks are called cross-correlation peaks, or cross-peaks. It is the cross-peaks that provide the additional information we are looking for. In the homonuclear COSY experiment, the cross-peaks tell us which

Figure 3.45 NMR spectra of propionamide, $CH_3CH_2CONH_2$. (Reprinted with permission of Aldrich Chemical Co., Inc.)

protons are coupled to each other. For this simple example, magnetization exchange results in eight peaks that appear as symmetric pairs off the diagonal. We can draw a connection between the two diagonal peaks from A and X and the symmetric pair of off-diagonal peaks, as shown in Fig. 3.47, proving that protons A and X are coupled. Any pair of diagonal peaks that can be connected through symmetric pairs of off-diagonal peaks are spin-coupled; in this way, the coupling throughout a complex spectrum can be traced. As shown on the figure, the fine structure or spacing between the peaks gives us J_{AX}. A COSY spectrum usually looks more like the simulated example presented in Fig. 3.48(a), where the peaks are represented as contour plots. In this simulated example, there are 6 protons shown along the diagonal. The connecting lines in Fig. 3.48(b) indicate that protons 1 and 2 are spin-coupled, protons 3 and 4 are spin-coupled, but protons 5 and 6 are not spin-coupled to any other protons. It is this type of information that helps to deduce the structure of an unknown; for this molecule, protons 5 and 6 cannot be adjacent to methyl, methylene, or methine protons, for example. The COSY spectrum of sucrose is shown in Fig. 3.49. The connecting lines in Fig. 3.49(b) confirm the couplings worked out earlier in the decoupling experiment (Fig. 3.25). COSY spectra for large molecules can be complex and difficult to interpret. Figure 3.50 shows just a small portion of the proton NMR spectrum (from 3.8 to 5.4 ppm) and the related COSY plot of a large glucopyranoside molecule.

Another popular 2D experiment is the HETCOR experiment, which identifies which protons are directly bonded to which ^{13}C nuclei. HETCOR stands for heteronuclear

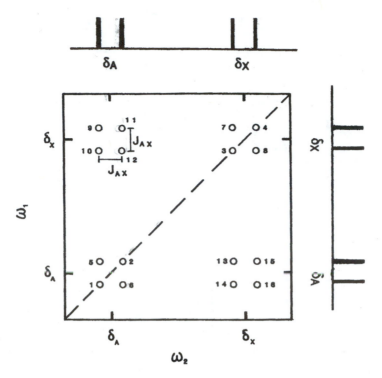

Figure 3.46 A simulated AX proton–proton COSY plot. (Modified from Bruch, used with permission.)

chemical shift correlation. The ^{13}C spectrum is plotted on one axis and the ^{1}H spectrum on the other axis. The HETCOR spectrum shows spots of intensity. If a straight line drawn from a carbon signal and a straight line drawn from a proton signal intersect at a spot on the HETCOR plot, the protons are attached to that carbon. The results of a 2D HETCOR experiment for sucrose is shown in Fig. 3.51. The glucose ring nuclei are marked with a G and the fructose ring nuclei with an F on both spectra; the structure of sucrose was given in Fig. 3.25. Look at the peak in the carbon spectrum marked G1 at 93 ppm. By drawing a vertical line from G1 in the carbon spectrum, we reach an intensity spot in the lower left hand portion of the HETCOR plot. A straight horizontal line from that point to the proton spectrum indicates that the proton(s) that absorb at 5.4 ppm are bonded to the G1 carbon. A vertical line from the F2 carbon does not intersect any spots; therefore the F2 carbon has no protons bonded to it. You should be able to identify the F2 carbon in the sucrose structure in Fig. 3.25 based on this knowledge.

The INADEQUATE experiment gives us the couplings between ^{13}C nuclei attached to each other and provides the carbon backbone of a molecule. As was mentioned in Table 3.4, the sensitivity of this experiment is low because of two factors: the low abundance of ^{13}C and the low probability that two ^{13}C nuclei are bonded to each other.

3.6.6. Qualitative Analyses: Other Applications

NMR spectra can be used to identify unknown compounds through spectral pattern matching. A number of companies, instrument manufacturers, government agencies, and other sources publish collections of reference spectra in electronic format and in hardcopy. These spectral databases may contain spectra of more than 200,000 compounds. The

Figure 3.47 The connection between the cross-peaks and the A and X peaks on the diagonal proves that A and X are spin-coupled and also provides a measurement of J_{AX}. (Modified from Bruch, used with permission.)

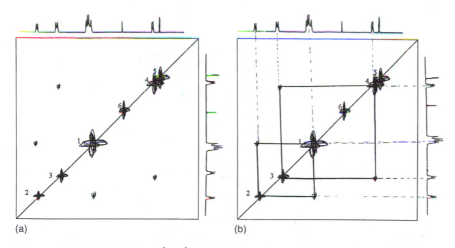

Figure 3.48 (a) Simulated ^1H–^1H COSY plot of an unknown compound. (b) Connecting lines indicate which protons are spin-coupled to each other. The COSY plot indicates that protons 5 and 6 are not coupled to any other protons. Any postulated structure for the unknown must be consistent with the coupling information from the COSY experiment.

unknown spectrum or some predetermined number of the strongest absorption bands from the unknown spectrum may be entered into a computerized search routine, which compares the unknown with stored spectra. It then retrieves all compounds from the database that may match the unknown spectrum, assigning a "goodness-of-fit" or probability to the suggested matches. The analyst then identifies the spectrum of the unknown based on spectral matching and chemical knowledge of the sample to rule out improbable compounds suggested by the search routine. A short list of reference spectra suppliers is located at the end of the bibliography. Most large spectral databases are expensive, but the amount of work required to compile these databases is considerable. Aldrich Chemical Company (www.sigma-aldrich.com) provides the Aldrich Spectral Library, which is the source of many of the spectra used in this chapter and Bio-Rad Laboratories

Figure 3.49   COSY profile and contour spectra of sucrose in D_2O. Note the strong contour connecting the diagonal peak for the proton at 5.41 ppm with the doublet of doublets at 3.55 ppm. In addition, the triplet at 4.05 ppm has strong off-diagonal contours indicating it is coupled to the doublet at 4.22 ppm. Compare with Fig. 3.25. (Modified from Petersheim, used with permission.)

Figure 3.50 A small portion of an actual 2D–COSY spectrum of a glucopyranoside. This is an expansion of the region from 3.6 to 5.6 ppm. (Modified from Bruch, used with permission.)

(www.bio-rad.com) provides the Sadtler spectra collection. More than 20 vendors market the NIST spectral database. A comprehensive spectral database, including ^{13}C and ^{1}H NMR spectra, from the Japanese National Institute of Advanced Industrial Science and Technology (www.aist.go.jp/RIODB/SDBS) can be searched for no charge as of this writing.

In addition to identification of unknowns, NMR can be used for conformational and stereochemical analyses. This includes the determination of **tacticity** in polymers, that is, whether the side chains are arranged regularly (isotactic and syndiotactic) or randomly (atatic) along the polymer backbone. Fundamental studies of bond distances from dipolar coupling and molecular motion from relaxation time measurements are used by physical chemists and physical organic chemists. In biology and biochemistry, the use

Figure 3.51 A 2D-HETCOR experiment. (a) 75 MHz ^{13}C spectrum of 1 M sucrose in D_2O. (b) 300 MHz ^{1}H spectrum of 1 M sucrose in D_2O. In both spectra the labels G and F refer to the glucose ring and the fructose ring, respectively. Structure of sucrose was given in Fig. 3.25. (c) The 2D-HETCOR spectrum of sucrose. Carbon F2 has no protons directly bonded to it because there is no spot of intensity in the HETCOR plot in line with the F2 chemical shift.

of isotope-labeled compounds and NMR can be used to study metabolism and understand metabolic pathways in living organisms. Chemical reaction rates can be measured and studied as a function of temperature directly in the NMR spectrometer.

Polymers of many chemical compositions are used to make materials and composites for use in appliances, electrical equipment, telecommunications equipment and fiber optics, computers, aircraft, aerospace, automobiles, food and beverage packaging, potable water delivery systems, and medical devices, to name a few applications. Adequate knowledge of the polymer structure and its relationship to the performance of these polymeric materials is crucial to their successful use. NMR is extremely useful in polymer characterization. ^{13}C and ^{1}H are the elements most commonly examined, followed by ^{29}Si, ^{19}F, ^{31}P, and ^{15}N. NMR can be used to determine the monomer sequence, branching, and end groups in polymers of many types. For example, Fig. 3.52 shows the ^{29}Si NMR spectrum of a polydimethylsiloxane polymer with trimethylsilyl end groups. Polydimethylsiloxanes (PDMS) are a major class of silicone polymers, used in many products from silicone caulk to shampoo. The NMR spectrum enables the analyst to identify the trimethylsilyl end groups (peak A) and the repeat unit of the chain, the dimethyl-substituted

Figure 3.52 The use of ^{29}Si NMR to characterize a silicone polymer. The polymer endgroups can be identified as trimethylsilyl groups and the measurement of peak areas permits the calculation of the average degree of polymerization. (From Williams, used with permission.)

SiO unit shown in parentheses, from the chemical shifts. The degree of polymerization can be determined from the areas of the peaks; the integrations are shown as the stepped lines on the spectrum, so quantitative information about the polymer is also obtained in this case.

Polymer composites or fiber-reinforced plastics (FRPs) have a wide range of applications in the aerospace and automotive industries. In these composite systems, fibers of carbon or silica are embedded in a polymer matrix to provide desirable physical properties, such as strength, while maintaining the low mass of a polymer. For example, polymer composite turbine blades are replacing heavier metal alloy blades in jet engines. Organic coupling agents are used to treat fiber surfaces to improve bonding between the fibers and the polymer matrix. High-resolution cross-polarization magic angle spinning (CP-MAS) NMR is used to observe structure, orientation, and interactions of coupling agents bound to surfaces.

Chemists and materials scientists are working to develop new materials with specific properties such as high strength and high modulus, resistance to temperature extremes, corrosion resistance, demanding optical or electrical properties, and the like. This requires detailed knowledge of the composition of the material, orientation of molecules, crystalline state, homogeneity, and other parameters. NMR is one tool that can provide much of the structural information necessary and often in a nondestructive manner. NMR is useful for studying both amorphous and crystalline materials, unlike X-ray diffraction spectroscopy, which requires a crystalline sample.

The chemical shift differences between reactants and products permit NMR to be used to follow the course of a reaction and to choose the optimum reaction conditions. In Fig. 3.53, the ^{29}Si NMR spectra show that NMR can follow the process of making β-SiC, a refractory ceramic, from polymethylvinylsilane and silicon metal. The NMR spectrum of the product silicon carbide (top spectrum) is clearly different from the spectrum of the starting mixture (bottom spectrum). In the bottom spectrum, the resonance at -18 ppm is due to the organosilane; the resonance at -82 ppm is the elemental silicon signal. All reactant and product signals are well separated in chemical shift, so any unreacted starting material can be measured in the product and the production process can be optimized.

Figure 3.53 (Bottom) The ^{29}Si NMR spectrum of polymethylvinylsilane and elemental silicon, the precursors to silicon carbide, SiC. (Top) Spectrum of the pyrolyzed product, β-SiC. (From Apple, used with permission.)

Solid-state NMR is proving to be a powerful technique for the study of reactions at surfaces. For example, NMR has been used in catalysis studies for determining the structure of chemisorbed molecules and for monitoring changes occurring in those structures as a function of temperature.

The need to determine the structures of large biological molecules like proteins is driving a new revolution in NMR. Extremely fast multidimensional NMR and new mathematical approaches, such as G-matrix FTNMR, are being developed to rapidly collect and process 4D and 5D NMR experiments on biological macromolecules. Articles on these developments can be found in Chemical and Engineering News, December 23, 2002, p.7 and January 27, 2003, p.15.

3.6.7. Quantitative Analyses

Both high and low resolution NMR are used for quantitative analyses of mixtures, quality control of both incoming raw materials and finished products, determination of percent purity of pharmaceuticals and chemicals, quantitative determination of fat and water *in vivo* in animals, and many other applications.

A significant advantage of NMR is that data can be obtained under experimental conditions where the area under each resonance is directly proportional to the number of nuclei contributing to the signal. No response factors are necessary to obtain quantitative results. A universal reference standard can be used for the analyses of most materials because the NMR response can be made the same for all components. This is a significant advantage over other methods of analyses. In cases where proton decoupling is used, such as in the analyses of polymers and organic compounds containing ^{13}C, ^{29}Si, ^{19}F, and other spin 1/2 nuclei, the NOE must be eliminated to re-establish the peak area to number of nuclei relationship. This is done by using gated decoupling (i.e., turning the decoupler off during the delay between RF pulses). This permits the return of normal equilibrium populations and results in the peak intensity (area) again relating to the number of nuclei giving rise to the resonance.

In polymers and ceramics, NMR can be used to determine quantitatively the amounts of amorphous and crystalline material in a sample. Molecules in amorphous regions "move" more than molecules in crystalline regions, so the relaxation times of molecules in these different environments is different. NMR can measure the difference in relaxation times and relate that to the percent crystallinity of the sample. Whether a sample is amorphous, crystalline, or a combination of both directly impacts the material's physical properties and behavior. It is an important piece of information for materials scientists, polymer chemists, and engineers to know.

One type of organic compound can be determined quantitatively in the presence of a different type, such as the percentage of alcohols in alkanes, amines in alcohols, aromatics and aliphatics in petroleum, olefins in hydrocarbon mixtures, organic halides and organometallic compounds in other organic compounds, or the number of side chains in a hydrocarbon. The method is not limited in the number of components that can be identified as long as there is at least one peak in the spectrum that is unique to each component.

NMR can be used to provide determination of chemical purity and quantitative measurements of impurities in materials. The accuracy and precision of quantitative NMR measurements are comparable to other instrumental analyses methods. Major components can be accurately determined with precisions better than 1% RSD while impurities in materials may be quantified at 0.1% or lower (Maniara et al.). The most common nuclei used for quantitative analyses are ^1H, ^{13}C, and ^{31}P. Quantitative ^{31}P NMR has been used to determine phospholipids, inorganic phosphorus, and organophosphorus compounds, such as phosphorus-based insecticides (Maniara et al.).

As an example of a simple quantitative analyses, suppose that we have a mixture of *n*-octane and 1-octene. The structure of *n*-octane, C_8H_{18}, which contains 18 protons, is composed of methyl and methylene groups. All atoms are joined by single bonds:

$$CH_3-CH_2-CH_2-CH_2-CH_2-CH_2-CH_2-CH_3$$

The structure of 1-octene, C_8H_{16}, which contains one terminal $C=C$ bond, is:

$$CH_3-CH_2-CH_2-CH_2-CH_2-CH_2-CH=CH_2$$

It can be seen that in 1-octene there are two protons on the *terminal* carbon that are olefinic in nature. These olefinic protons will absorb at about 5.0 ppm, according to Table 3.3. Quantitatively, they constitute 2 of 16 protons in 1-octene. If we measure the area of the peaks at 5.0 ppm (call it area A), then the total area in the whole spectrum due to the presence of 1-octene is equal to area A times 8:

$$\text{area due to octene} = \text{area } A \times 8$$

and

$$\text{area due to octane} = \text{total area} - \text{octene area}$$

Figure 3.54 The 75 MHz ^{13}C NMR spectrum of the aromatic region of a 57 mol% brominated poly(2,6-dimethyl-1,4-phenylene oxide) polymer. Assignments are shown for peaks used in determining the degree of bromination. (From Williams et al., used with permission.)

However, one molecule of octene contains 16 protons and one molecule of octane contains 18 protons. Hence, if these compounds were present in equimolar proportions, the ratio of the relative areas of the NMR absorption curves would be 16:18. A correction must be made for this difference in the final calculation. The mole ratio of octane to octene in the mixture would therefore be obtained as

$$\frac{\text{area due to octane}/18}{\text{area due to octene}/16}$$

that is, the mole ratio of octane/octene equals

$$\frac{[\text{total area} - (\text{area } A \times 8)]/18}{(\text{area } A \times 8)/16}$$

Sample calculation: In an actual experiment involving a mixture of octene and octane, it was found that the total area of all peaks = 52 units. The area of peaks at 5.0 ppm (area A) = 2 units. From the data and the relationships previously set out, we find that the area due to octene protons = $2 \times 8 = 16$ units. The area due to octane protons = $52 - 16 = 36$ units.

$$\text{Ratio of octane to octene} = \frac{36/18}{16/16} = \frac{2}{1}$$

Therefore the mole ratio of octane to octene in this sample is 2:1. Similar calculations can be made for many other combinations of compounds.

NMR can be used to determine the composition of mixtures, polymers, and other materials as long as there is at least one peak representing each different component. For example, the degree of bromination of a brominated polymer can be determined from the ^{13}C NMR spectrum. The degree of bromination can be calculated from a direct comparison of the integrated areas labeled g and h in Fig. 3.54 which represent the unbrominated and brominated aromatic rings or by using peak d to represent the unbrominated ring and the sum of peaks e and f to represent the brominated ring.

The end groups in a polymer are extremely important to the chemical and mechanical behavior of the polymer, but they are present at only low concentrations compared with the polymer chain repeat units. End groups may be less than 1 mol% of the total polymer. An example of end group identification in a silicone polymer was given earlier (Fig. 3.52). In cases where the polymer end groups are hydrocarbons, the less sensitive ^{13}C spectrum must often be used to identify the end groups because there is less likelihood of spectral overlap. The proton spectrum, with a chemical shift range of only 10 ppm, often has the proton signals overlapping and impossible to sort out for the low concentration of end groups in a large polymer. Under ideal conditions, ^{13}C NMR may be used to determine end group concentrations as low as 0.01 mol%.

NMR can be used to determine rates of reaction and reaction kinetics. In a chemical reaction such as

$$C_8H_{18} \xrightarrow[\text{catalyst}]{\text{Pd}} C_8H_{16} + H_2$$

one type of proton (paraffinic) is consumed and another type (olefinic) is formed. Samples of a reacting mixture can be taken at frequent intervals during the reaction, and the rate of disappearance or formation of the different types of protons can be measured. The results can be used to calculate the rate of the chemical reaction and the kinetics involved. With a modern FTNMR, many reactions can be monitored in the probe without the need for physically taking samples at intervals. Spectra are collected every few seconds, and the rate and order of the reaction are calculated after the data is processed.

The octane number of gasoline is a measure of the tendency of the gasoline to resist "knock". Octane number is determined using a standardized single-cylinder engine. The method requires the proper test engine, calibration with standard fuels, and then operation of the test engine with the sample gasoline. It has been shown that octane number can be determined directly on the liquid gasoline by NMR (Ichikawa et al.). In this procedure, various types of protons are measured quantitatively and the volume percentages of the components calculated. The octane number of the gasoline can be calculated using an empirical formula. A similar approach using IR spectroscopy (Chapter 4) has also been demonstrated.

In Section 3.5.6, benchtop wide-line NMRs were briefly described. One such instrument, the Bruker Optics Minispec NMR analyzer (www.minispec.com) has been optimized for the determination of whole body fat, lean tissue, and fluid in live mice. The pulse sequence and relaxation times distinguish protons in water from protons in fat. These measurements are critical to pharmaceutical and medical studies of obesity and the development of drugs to treat obesity. The mice are restrained but not anesthetized, and placed in the magnetic field. Although the magnet is low strength, the field penetrates the entire volume of the mouse. Total body fat, lean tissue, and fluid are measured in less than 2 min. The mice are not harmed during the NMR measurement, which makes this approach valuable for long-term clinical studies. The same mice can be studied for the entire duration of the clinical trial. The NMR method is more precise than methods currently in use for these measurements.

Dedicated benchtop NMR analyzers for a variety of applications are available. Bruker's Minispec mq series (www.minispec.com) includes an analyzer to determine fluoride in toothpaste quantitatively and another to determine water droplet size distribution in oil/water emulsions. Fluoride is often added to toothpaste as sodium fluoride or sodium monofluorophosphate to prevent tooth decay. The fluorine analyzer can determine fluorine and hydrogen at the level of a few hundred ppm. Toothpaste is squeezed into a glass sample tube and the quantitative determination of fluorine takes less than 1 min. The NMR method uses no solvents or reagents and is independent of the sample color and clarity, unlike the colorimetric methods and other instrumental methods such as ion chromatography that are used for this purpose. In the water droplet size distribution analyzer, droplets as small as 0.25 μm can be measured. The shelf life and palatability of products such as margarine, mayonnaise, salad dressings, and soft cheese depend on the size of water droplets in the water–oil emulsion. For example, products with multiple small droplets are less susceptible to bacterial growth than products with large droplet sizes. No sample preparation or dilution is required, the method is nondestructive and noninvasive and the NMR results agree with conventional methods such as laser light scattering.

Another example of the use of a wide-line NMR for quantitative analyses is in the measurement of surface coating applied to synthetic fibers and yarns (Rodgers, 1994). This surface coating is called "finish" and serves to lubricate and control static on the fibers. The amount of finish applied to fibers can be measured directly on a sample of nylon fiber by a low resolution (wide-line) pulsed benchtop NMR; in this study a 20 MHz proton QP20 from Oxford Instruments, Concord, MA, was used. Use of the NMR eliminated the need for costly and time consuming solvent extraction of the finish and subsequent determination by IR spectroscopy or gravimetry. The analyses required only 2–4 g of the solid fibers and the finish could be determined accurately over a range of 0.5–1.5%.

3.7. HYPHENATED NMR TECHNIQUES

Many samples of interest to researchers in biochemistry, pharmaceutical chemistry, medicinal chemistry, forensic chemistry, and industrial chemistry are not pure compounds. It is often necessary to go through complex extraction and separation procedures to isolate the compounds of interest before they can be studied. These separation procedures can be time consuming and labor intensive. Costly high purity solvents are used, which then must be disposed of (costing even more money). In the pharmaceutical industry, for example, **combinatorial chemistry** approaches are used to synthesize thousands of new compounds a day. These syntheses do not result in pure products; in fact many products and unreacted starting materials may be present in the sample to be analyzed. Various types of instrumental **chromatography** have been developed to expedite the separation and detection of compounds in complex mixtures. These techniques are discussed in detail in Chapters 11–13. One of the most important types of chromatography, especially for pharmaceutical, biochemical, and clinical chemists, is HPLC, covered in Chapter 13. HPLC is used for the separation of nonvolatile molecular compounds; there are methods available to separate compounds with low molecular weights, high molecular weights, low polarity, high polarity, and everything in between. Detectors for HPLC do not provide molecular structure or molecular weight, except mass spectrometers (Chp. 13).

We have seen that NMR can provide detailed molecular structure information. It is possible to join together or "couple" an HPLC instrument with an NMR spectrometer. The HPLC performs the separation of a complex mixture and the NMR spectrometer takes a spectrum of each separated component to identify its structure. We now have a "new" instrument, an HPLC-NMR instrument. We call a coupled instrument like this a **"hyphenated"** instrument. The coupling of two instruments to make a new technique with more capabilities than either instrument alone provides results in a **hyphenated technique** or hybrid technique. HPLC-NMR is made possible with a specially designed flow probe instead of the standard static probe. For example, Bruker Instruments (www.brukerbiospin.com) has a flow probe for proton and ^{13}C NMR with a cell volume of 120 μL. Complex mixtures of unknown alkaloids extracted from plants have been separated and their structures completely characterized by HPLC-NMR using a variety of 2D NMR experiments (Bringmann et al.). Only a simple aqueous extraction of the plant leaves, lyophilization, and dissolution in D_2O was required for injection into the HPLC-NMR, saving time and resources. HPLC-NMR has been used for the analyses of metabolites in body fluids; body fluid samples are very complex mixtures and are usually of limited volume (Lindon et al.).

HPLC-NMR and another hyphenated, more powerful instrument, HPLC-NMR-MS (the MS stands for mass spectrometry) are used in pharmaceutical research and development. These hyphenated techniques identify not only the structures of unknowns, but with the addition of MS, the molecular weight of unknown compounds. The HPLC-NMR-MS instrument separates the sample on the HPLC column, takes the NMR spectra as the separated components flow through the probe and then acquires the mass spectrum of each separated component to determine the molecular weight and additional structural information from the mass spectral fragmentation pattern. The MS must be placed after the NMR, since MS is a destructive technique. MS is covered in Chapters 9 and 10.

3.8. NMR IMAGING AND MRI

NMR is extensively used in imaging solid objects in a nondestructive, noninvasive manner. The most important application of this imaging is in medicine, where humans benefit tremendously from the power of NMR imaging. NMR imaging and **magnetic resonance imaging** (MRI) are the same technique. The term "nuclear" was dropped from instrumentation used for medical imaging so that patients would not mistakenly think this was a procedure that involved radioactive materials or gamma rays (X-rays). We will use the term MRI for NMR imaging applications in general.

MRI has found valuable applications in medical imaging of the human body. In this technique, a highly uniform magnetic field is used, but a linear magnetic field gradient is superimposed in three orthogonal directions using auxiliary coils. ^1H nuclei therefore respond at different frequencies at different physical locations, thus locating the physical position of the nuclei in three dimensions. By observing the NMR signal from numerous different angles simultaneously, the physical outline of the various body tissue components can be revealed. Abnormalities such as fractures or cancerous growths can then be located and measured in 3D at high resolution. The magnets used must have a large bore, large enough to accommodate a human patient lying on a table, or must use an open magnet design. MRI magnet designs are very different from NMR instrument magnet designs. Open magnet or short bore designs are used to avoid inducing claustrophobia in patients undergoing MRI. Field strengths of MRI medical imaging units range from 0.8 to 3 T, much lower than NMRs for chemical analyses. The cost of medical MRI units is extremely high compared with chemical analysis instrumentation; about $1.5 million is typical for a 1.5 T state-of the-art MRI unit. Figure 3.55(a) and (b) are photographs of two state of the art medical MRI systems from Toshiba Medical Systems, a high resolution long bore design and an open magnet (4 pole) design. The Toshiba Medical Systems website at http://medical.toshiba.com has numerous MRI images taken with its systems available for viewing.

X-rays have been used extensively for noninvasive medical imaging studies in the past, but the contrast between different forms of soft tissue is low and therefore abnormalities in soft tissues are hard to visualize using X-rays. Also, X-rays are ionizing radiation and can cause tissue damage at high exposure levels. In contrast, MRI, also a noninvasive procedure, has essentially no side effects and can more readily visualize very small differences in soft tissue. MRI can monitor *in vivo* concentrations of biologically important molecules like adenosine triphosphate (ATP) in a noninvasive manner; this permits

(a)

(b)

Figure 3.55 Two commercial medical MRI systems. (a) Female patient being placed in the bore of the Toshiba EXCELART™ MRI system. (b) The four pole open magnet Toshiba OPART™ MRI system. (Photos courtesy of Toshiba America Medical Systems; http://medical.toshiba.com.)

studies of the effect of drugs on metabolism, for example. MRI has proven to be a very valuable noninvasive medical tool for studying cancer, stroke, epilepsy, heart problems, arthritis, and many other conditions. Figure 3.56 illustrates the use of MRI to locate a brain tumor. The image in Fig. 3.56(a) shows the tumor, the gray oval object on the front left side of the brain behind the eye. The image in Fig. 3.56(b) is of the same tumor after a "contrast agent" containing gadolinium was given to the patient. The tumor absorbs more of the contrast agent because of its fast growth and high blood supply. The Gd ion is paramagnetic and changes the relaxation times of nuclei in its vicinity resulting in signal enhancement; the tumor now appears bright or "lit up" against the surrounding tissue. A brief discussion of gadolinium MRI contrast agents can be found in the reference by Skelly Frame and Uzgiris.

Another example of the use of MRI is seen in Fig. 3.57. This is the image of a fractured hip; the fracture is in the hip bone on the left side of the image and appears as the

(a)

(b)

Figure 3.56 (a) An unenhanced MRI image of a human brain tumor. The tumor is the gray circular object on the left side of the brain behind the eye. (b) MRI image of the same tumor after administration of a gadolinium-containing contrast agent. Note the brightness of the tumor in this image compared to (a). (Images courtesy of H.T. Alberry, D. Derico, and M. Farrell, Albany Advanced Imaging, Albany, NY.)

Figure 3.57 An MRI image of a hip fracture. The fracture in the hip on the left side of the image shows up as a bright area due to bleeding. The fracture was too small to be detected by conventional X-ray absorption. (Image courtesy of H.T. Alberry, D. Derico, and M. Farrell, Albany Advanced Imaging, Albany, NY.)

bright area in the region of the hip bone. Compare the right hip, which appears dark in this location. The brightness is due to blood in the fractured bone area; this fracture was too small to see by X-ray absorption. Two MRI views of a pituitary adenoma are shown in Fig. 3.58. Figure 3.58(a) is a side view of the head; the adenoma is the dark gray circle in the sinus area, behind the eyes. A view from the top of the head shows the large bright adenoma centered in front of the brain and behind the eyes. A final example of human soft tissue imaging is the image of a heart and its blood vessels (Fig. 3.59).

MRI instruments can be equipped to study polar ice and marine organisms in their salt-water environments, permitting the imaging of new marine species. The references by Bock and the website of the Alfred-Wegener-Institute for Polar and Marine Research offer amazing MRI pictures of ice microstructures and marine species *in vivo* (www.awi-bremerhaven.de). A few examples follow. Figures 3.60 and 3.61 are *in vivo* MRI images

(a)

(b)

Figure 3.58 (a) Side view of a pituitary gland adenoma. (b) Top view (slice) of the adenoma. (Images courtesy of H.T. Alberry, D. Derico, and M. Farrell, Albany Advanced Imaging, Albany, NY.)

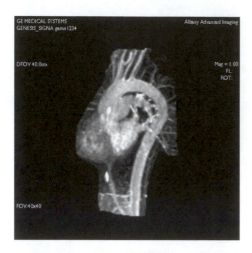

Figure 3.59 MRI image of a heart and associated blood vessels. (Image courtesy of H.T. Alberry, D. Derico, and M. Farrell, Albany Advanced Imaging, Albany, NY.)

and the NMR spectra of embryos in a pregnant fish. Figure 3.62 shows a stack plot of phosphorus-containing compounds in a living codfish that undergoes hypoxia (lack of oxygen). As hypoxia is induced, looking from the back of the plot to the front, the inorganic phosphate signal on the left increases. At the same time, the phosphocreatine signal decreases. As conditions return to normal, the inorganic phosphate disappears and the phosphocreatine returns to normal (as did the fish). These are all ^{31}P NMR spectra. The Bruker BioSpin® website at www.bruker-biospin.com offers a wide variety of MRI images, including a movie of a beating rat heart with a myocardial infarction imaged *in vivo*, an MRI of a living newly discovered saltwater fish and false color images showing temporal differences in brain activity during an epileptic seizure in a rat.

Figure 3.60 *In vivo* MR images of a North Sea fish, the eelpout. This fish is pregnant and embryonic fish are visible inside the uterus. Other tissues are also visible as marked. This fish was free-swimming in a salt-water filled flow-through chamber. (Courtesy of Dr. Christian Bock, Alfred-Wegener-Institute for Polar and Marine Research, Bremerhaven, Germany; www.awi-bremerhaven.de.)

Figure 3.61 *In vivo* localized proton NMR spectrum collected from the same fish embryos seen in Fig. 3.60. At location A on the image, signals identified in the embryo spectrum include ATP, glycine (Gly), and creatine (Cr) as well as large signals from lipids (Lip). (Courtesy of Dr. Christian Bock, Alfred-Wegener-Institute for Polar and Marine Research, Bremerhaven, Germany; www. awi-bremerhaven.de.)

Figure 3.62 Stack plot of *in vivo* ^{31}P NMR spectra from the muscle of a living codfish. Each spectrum was acquired over 5 min. When hypoxia (hypo) was induced in the fish by decreasing its oxygen supply, the inorganic phosphate levels (Pi) increased while phosphocreatine (PCr) decreased. As conditions return to normal, the Pi signal disappears, the PCr signal increases and the fish recovered its energy. (Courtesy of Dr. Christian Bock, Alfred-Wegener-Institute for Polar and Marine Research, Bremerhaven, Germany; www.awi-bremerhaven.de.)

MRI permits the noninvasive imaging of the interior of solid objects. This has been successfully utilized in the study of extruded polymers and foams and the study of spatial distributions of porosity in porous materials. The structure of ice in polar ice cores has been studied as noted above; unlike optical imaging, MRI imaging is nondestructive. Ice has sufficient mobile protons to be imaged with conventional MRI.

3.9. LIMITATIONS OF NMR

There are two major limitations to NMR: (1) it is limited to the measurement of nuclei with magnetic moments and (2) it may be less sensitive than other spectroscopic and chromatographic methods of analyses. As we have seen, although most elements have at least one nucleus that responds in NMR, that nucleus is often of low natural abundance and may have a small magnetogyric ratio, reducing sensitivity. The proton, ^{1}H, and fluorine, ^{19}F, are the two most sensitive elements.

Elements in the ionic state do not respond in NMR, but the presence of ions in a sample contributes to unacceptable line broadening. Paramagnetic contaminants such as iron and dissolved oxygen also broaden NMR lines. Nuclei with quadrupole moments, such as ^{81}Br, broaden the NMR signal. Line broadening in general reduces the NMR signal and hence the sensitivity.

BIBLIOGRAPHY

Ando, I.; Yamanobe, T.; Asakura, T. *Prog. NMR Spectrosc.* **1990**, *22*, 349.

Apple, T.M. NMR applied to materials analysis. *Appl. Spectrosc.* **1995**, *49*(6), 12A.

Arnold, J.T.; Dharmetti, S.S.; Packard, M.E. *J. Chem. Phys.* **1951**, *19*, 507.

Bell, A.T.; Pines, A., Eds. *NMR Techniques in Catalysis*; Marcel Dekker, Inc.: New York, 1994.

Bhacca, N.S.; Johnson, L.F.; Shoolery, J.N. *High Resolution NMR Spectra Catalog*; Varian Associates: Palo Alto, CA, 1962, Vol. 1. (This publication is the source of all the 60 MHz proton spectra used in this chapter, but is no longer available from Varian Associates).

Bock, C.; Frederich, M.; Wittig, R.M.; Pörtner, H.-O. *Magn. Reson. Imaging* **2001**, *19*, 1113–1124.

Bock, C.; Sartoris, F.-J.; Pörtner, H.-O. In vivo MR spectroscopy and MR imaging on nonanaesthetized marine fish: techniques and first results. *Magn. Reson. Imaging* **2002**, *20*, 165–172.

Bovey, F.A. *Nuclear Magnetic Resonance*, 2nd ed.; Academic Press: New York, 1988.

Bringmann, G.; Günther, C.; Schlauer, J.; Rückert, M. HPLC-NMR on-line coupling including the ROESY technique: direct characterization of naphthylisoquinoline alkaloids in crude plant extracts. *Anal. Chem.* **1998**, *70*, 2805.

Bruch, M.D.; Dybowski, C. Spectral editing methods for structure elucidation. In *NMR Spectroscopy Techniques*, 2nd ed.; Bruch, M.D., Ed.; Marcel Dekker, Inc.: New York, 1996.

Fukushima, E.; Roeder, S.B.W. *Experimental Pulse NMR: A Nuts and Bolts Approach*; Addison-Wesley: Reading, MA, 1981.

Fyfe, C.A. *Solid State NMR for Chemists*; CFC Press: Guelph, 1985.

Ichikawa, M.; Nonaka, N.; Amano, H.; Takada, I.; Ishimori, S.; Andoh, H.; Kumamoto, K. *Appl. Spectrosc.* **1992**, *46*, 1548.

Lambert, J.B.; Shurvell, H.F.; Lightner, D.; Cooks, R.G. *Introduction to Organic Spectroscopy*; Macmillan Publishing Company: New York, 1987.

Laupretre, F. *Prog. Polym. Sci.* **1990**, *15*, 425.

Lindon, J.C.; Nicholson, J.K.; Wilson, I.D. *Advances in Chromatography*; Marcel Dekker, Inc.: New York, 1996, Vol. 36; 315.

Maniara, G.; Rajamoothi, K.; Rajan, S.; Stockton, G.W. Method performance and validation for quantitative analysis by ^{1}H and ^{31}P NMR spectroscopy. Applications to analytical standards and agricultural chemicals. *Anal. Chem.* **1998**, *70*(23), 4921.

Mathias, L.J. *Solid State NMR of Polymers*; Plenum Press: New York, 1991.

McClure, C.K. In *NMR Spectroscopy Techniques*, 2nd ed.; Bruch, M.D., Ed.; Marcel Dekker, Inc.: New York, 1996.

Pavia, D.L.; Lampman, G.M.; Kriz, G.S. *Introduction to Spectroscopy*, 3rd ed.; Harcourt College Publishers: New York, 2001.

Petersheim, M. Nuclear magnetic resonance. In *Analytical Instrumentation Handbook*, 2nd ed.; Ewing, G.A., Ed.; Marcel Dekker Inc.: New York, 1997.

Robinson, J.W., Ed. *Handbook of Spectroscopy*; CRC Press: Boca Raton, FL, 1974, Vol. II.

Rodgers, J.E. Wide line nuclear magnetic resonance in measurement of finish-on-fiber of textile products. *Spectroscopy* **1994**, *9*(8), 40.

Shoolery, J.N. NMR spectroscopy in the beginning. *Anal. Chem.* **1993**, *65*(17), 731A.

Silverstein, R.M.; Webster, F.X. *Spectrometric Identification of Organic Compounds*, 6th ed.; John Wiley and Sons, Inc.: New York, 1998.

Skelly Frame, E.M.; Uzgiris, E.E. The determination of gadolinium in biological samples by ICP-AES and ICP-MS in evaluation of the action of MRI agents. *Analyst* **1998**, *123*, 675–679.

Skloss, T.W.; Kim, A.J.; Haw, J.F. High-resolution NMR process analyzer for oxygenates in gasoline. *Anal. Chem.* **1994**, *66*, 536.

Wendesch, D.A.W. *Appl. Spectrosc. Rev.* **1993**, *28*(3), 165.

Williams, E.A. Polymer molecular structure determination. In *Characterization of Materials*, Part I; Lifshin, E., Ed.; VCH Publishers, Inc.: New York, 1992.

Williams, E.A.; Skelly Frame, E.M.; Donahue, P.E.; Marotta, N.A.; Kambour, R.P. Determination of bromine levels in brominated polystyrenes and poly(2,6-dimethyl-1,4-phenylene oxides). *Appl. Spectrosc.* **1990**, *44*, 1107.

Yu, T.; Guo, M. *Prog. Polym. Sci.* **1991**, *15*, 825.

SPECTRAL DATABASES

This list is not complete. Many instrument manufacturers offer spectral databases packaged with their instruments. The publishers listed below offer their databases in both electronic and hardcopy formats, with CD versions and electronic versions becoming increasingly popular. The major drawback of the high-resolution databases for the beginner learning NMR spectral interpretation is the lack of peak expansion and area integrations on the spectra in many cases. The authors are deeply indebted to Aldrich Chemical Co., Varian Associates, and AIST for their permission to use their spectra in this chapter.

Aldrich Chemical Company, (www.sigma-aldrich.com), publishes 12,000 high resolution ^{13}C and proton spectra, in the volumes by Pouchert, C.J.; Behnke, J. *The Aldrich Library of ^{13}C and ^{1}H FT-NMR Spectra, 300 MHz*, Aldrich Chemical Company: Milwaukee, WI, 1993. *The Aldrich/ACD Library of FTNMR Spectra, Pro version*, is available on CD with 15,000 ^{13}C and ^{1}H 300 MHz spectra.

Bio-Rad Laboratories, Informatics Division, Philadelphia, PA, (www.bio-rad.com) publishes the Sadtler spectra collections of high resolution proton and ^{13}C NMR spectra.

National Institute of Advanced Industrial Science and Technology, Tsukuba, Ibaraki, Japan, publishes a free spectral database system for organic compounds. The spectra include IR, Raman, NMR, and MS for most compounds. The database may be accessed at www.aist.go.jp/RIODB/SDBS.

NIST, United States, publishes a comprehensive database of NMR, IR, and MS spectra. The NIST database is available for sale through 21 commercial distributors, such as Bio-Rad Laboratories.

The Varian Associates High Resolution NMR Spectra, Volumes 1 and 2, published in 1962 and 1963, which are the sources of the 60 MHz NMR spectra presented here are no longer in print.

SUGGESTED EXPERIMENTS

3.1   (For the instructor) Demonstrate to students in the laboratory the principal components of an NMR instrument. Indicate the steps taken in tuning the instrument.

3.2   Obtain the NMR spectrum of a straight-chain alkane, such as *n*-octane. Identify the methyl and methylene peaks. Note the chemical shift and the spin–spin splitting. Measure the total area of the methyl and methylene peaks and correlate this with the number of methyl and methylene protons in the molecule.

3.3   (a)   Obtain the NMR spectrum of ethyl alcohol (1) wet and (2) very dry.
      (b)   Identify the methyl, methylene, and alcohol protons. Note the chemical shift and spin–spin splitting.
      (c)   Measure *J*, the coupling constant, between the methyl and the methylene protons. Note the effect of water on the alcohol peak. Explain this phenomenon.

3.4   Integrate the peak areas obtained in Experiment 3.3. Measure the ratios of the areas and the numbers of hydrogen nuclei involved in the molecules. What is the relationship between the area and the number of hydrogen nuclei?

3.5   Obtain the NMR spectrum of a mixture of toluene and hexane. Integrate the peak areas of the different parts of the spectrum. From the ratio of the alcoholic hydrogen to the total hydrogen, calculate the percentage of toluene in the mixture.

3.6   Obtain the NMR spectra of organic halides, olefins, aromatics, and substituted aromatics. Correlate the peaks with the different types of protons present.

3.7   Identify an unknown compound from its NMR spectrum. The difficulty of the problem should match the level of competence of the student.

3.8   Obtain the NMR spectra of (a) hexane and (b) heptane. Integrate the areas under the peaks in each spectrum. Would it be possible to distinguish between these compounds based on their NMR spectra? Compare with IR spectroscopy when Chapter 4 is covered and with MS (Chapter 10).

3.9   Record the NMR spectrum of a sample of cooking oil. Measure the ratio of hydrogen in unsaturated carbon and that in saturated carbons. Compare the degrees of unsaturation of various commercial cooking oils.

3.10 Repeat Experiment 3.9 using margarine as the sample. First dissolve a known amount of margarine in CCl_4 and obtain the NMR spectrum. Compare the degrees of unsaturation of different brands of margarine.

3.11 Obtain the NMR spectra, proton and ^{13}C, for commonly available headache tablets and pain relievers. Most will dissolve in deuterated chloroform or acetone. Inert fillers may need to be filtered out of the solution. Read the labels—some are "pure" compounds such as aspirin, others contain more than one ingredient. Obtain the spectra of the pure components—acetylsalicylic acid, acetaminophen, and caffeine are common ingredients of these tablets. (The "pure" material can often also be purchased as a commercial tablet. Just be aware that you may see some impurity peaks if you are not using reagent grade materials.) Correlate the peaks with the structure of the compounds. In those tablets with more than one ingredient, note any spectral overlaps. Estimate the purity of different commercial brands of aspirin using your spectra.

3.12 Obtain the proton and ^{13}C spectra of geraniol. Perform the following 2D experiments: COSY, HETCOR, and INADEQUATE. Using the 2D data, work out the structure of geraniol.

3.13 Repeat experiment 3.12 with caffeine, acetaminophen, or butyl acetate.

PROBLEMS

3.1 Define the term chemical shift. Why does it occur? How is it measured for protons? For ^{13}C?

3.2 (a) Explain why spin–spin coupling occurs.
(b) A methylene group (CH_2) is adjacent to a CH group. Into how many peaks is the CH_2 peak split by the adjacent single hydrogen?

3.3 What does a J coupling constant tell you? Proton A is coupled to proton B with $J_{AB} = 9$. Proton A is also coupled to proton C with $J_{AC} = 2$. Draw the predicted splitting pattern for proton A.

3.4 Draw a schematic proton NMR spectrum for each of the following compounds. Indicate chemical shift and peak multiplicity.
(a) *n*-Butane   (b) Tetramethylmethane

3.5 Draw a schematic proton NMR spectrum for each of the following compounds. Indicate chemical shift and peak multiplicity.
(a) Benzene   (b) Benzaldehyde

3.6 The total area of the peaks in 1-butene shown

is 80 units. What will be the area of the peaks caused by (a) the CH_2 group on carbon b and (b) the CH group on carbon c?

3.7 Draw a schematic block diagram of an FTNMR instrument.

3.8 Define magnetogyric ratio.

3.9 What is the "magic angle"? Explain why MAS is used to acquire the NMR spectrum of a solid sample.

3.10 Explain why liquid samples are spun in the magnetic field while acquiring an NMR spectrum.

3.11 Aldehydic protons occur at about 9–10 ppm, meaning that they are highly deshielded compared with other protons. Using Fig. 3.8 as an example, show why the aldehyde proton is deshielded.

3.12 What are the rules for determining the spin number of an element?

3.13 Which of the following have a spin number $= 0$?
^{12}C, ^{13}C, ^{16}O, ^{17}O, ^{1}H, ^{2}H, ^{19}F, ^{28}Si, ^{35}Cl, ^{108}Ag, ^{96}Mo, ^{66}Zn, ^{65}Zn

3.14 What is the spin number for (a) ^{1}H and (b) ^{2}H? What is the number of orientations for (a) ^{1}H and (b) ^{2}H in a magnetic field (spin number $= 2$ $I + 1$)?

3.15 What information does the COSY experiment provide? Explain how you interpret a COSY plot.

3.16 What information does the HETCOR experiment provide? Explain how you interpret a HETCOR plot.

3.17 What information does the INADEQUATE experiment provide? What is plotted on the x- and y-axes in an INADEQUATE plot? What is the major limitation of this experiment?

3.18 What are the requirements for a standard reference material, such as TMS, in NMR?

3.19 Draw the proton NMR spectrum you would expect for the compound pro-pylamine, $CH_3CH_2CH_2NH_2$. Include the chemical shifts and spin–spin splitting. Explain your rationale.

3.20 What would be the spin–spin splitting caused by an NH group on an adjacent CH group? Why?

3.21 What would be the spin–spin splitting caused by an adjacent (i) CH_3, (ii) CH_2, (iii) CF_3, (iv) CF_2, (v) CH, (vi) CFH on a CH group?

3.22 What are (a) transverse relaxation and (b) longitudinal relaxation?

3.23 List the causes of relaxation in NMR.

3.24 How does the relaxation time affect linewidth?

3.25 What is the nuclear Overhauser effect? What advantages and disadvantages does the NOE result in for ^{13}C NMR?

3.26 What is the chemical exchange rate at the temperature at which multiplicity is lost?

3.27 What is the effect of viscosity on T_1?

3.28 A proton appears at a chemical shift of 7.0 ppm (vs. TMS) in an NMR spectrometer operated at 60 MHz. What is the resonance frequency of the proton in Hz?

3.29 What is meant by saturation of a nucleus?

3.30 Explain why the spectrum of naphthalene, Fig. 3.14, exhibits a "four line" splitting pattern for the aromatic protons. Do the same for the proton spectrum of the polymer in Fig. 3.41(a).

3.31 In the quantitative analysis example of a mixture of octane and 1-octene discussed in the text, the terminal olefinic protons that absorbed at 5.0 ppm were used to quantify the 1-octene. Could you have used any other resonance? (Look at Table 3.3.) If so, show how the calculation would change from the sample calculation in the text. Is there any advantage to one calculation over the other?

3.32 What are the requirements for solvents used for NMR studies of dissolved organic compounds?

3.33 If a disubstituted benzene ring has the same two substituents, for example, dibromobenzene, the number of unique carbon atoms in the ring can be 2, 3, or 4. The ^{13}C NMR spectrum depends on where the bromine atoms are located on the ring. (a) Draw the three possible dibromobenzene molecules. (b) Using the dichlorobenzene example in Section 3.6.3, draw a mirror plane in each of your dibromobenzene molecules. A "mirror plane" is a plane of symmetry that gives you identical halves of the molecule on each side of the plane, as was done for dichlorobenzene. Identify how many unique carbon atoms there are in each dibromobenzene molecule and predict how many peaks you will see in the ^{13}C NMR spectrum of each molecule.

Use the proton and ^{13}C spectra provided and the chemical shift tables in the chapter for Problems 3.34–3.48 and draw a reasonable structure for the compound, given the formula or molecular weight. All spectra below are reprinted with permission of the Aldrich Chemical Co. In some cases, the 300 MHz proton spectra have had peak "expansions" added to clarify the spin–spin splitting. (These "expansions" are actually 60 MHz signals from the Varian spectral library.) All proton spectra have a peak for TMS at 0.0 ppm and all ^{13}C spectra have a triplet from the CDCl$_3$ solvent at 77 ppm.

3.34 C$_{12}$H$_{10}$

3.35 C$_4$H$_9$I

3.36 C$_4$H$_8$O

3.37 Molecular weight = 120.97. The molecular formula contains a halogen atom.

3.38 Molecular weight = 156.23

3.39 C_3H_7O

75 MHz ^{13}C NMR
In CDCl3

300 MHz ^1H NMR
In CDCl3

3.40 $C_3H_5O_2Cl$

75 MHz ^{13}C NMR
In CDCl3

300 MHz ^1H NMR
In CDCl3

3.41 $C_3H_6Cl_2$

3.42 $C_3H_6Cl_2$

3.43 Molecular weight = 106.17

3.44 Molecular weight = 106.17

3.45 Molecular weight = 106.17

3.46 Molecular weight = 108.14

3.47 Molecular weight = 84.16. The compound is a hydrocarbon.

3.48 C$_4$H$_{11}$N

4

Infrared Spectroscopy

IR radiation was first discovered in 1800 by Sir William Herschel, who used a glass prism with blackened thermometers as detectors to measure the heating effect of sunlight within and beyond the boundaries of the visible spectrum. Coblentz laid the groundwork for IR spectroscopy with a systematic study of organic and inorganic absorption spectra. Experimental difficulties were immense; since each point in the spectrum had to be measured separately, it could take 4 h to record the full spectrum. But from this work came the realization that each compound had its own unique IR absorption pattern and that certain functional groups absorbed at about the same wavelength even in different molecules. The IR absorption spectrum provides a "fingerprint" of a molecule with covalent bonds. This can be used to identify the molecule. Qualitative identification of organic and inorganic compounds is a primary use of IR spectroscopy. In addition, the spectrum provides a quick way to check for the presence of a given functional group such as a carbonyl group in a molecule. IR spectroscopy and spectrometry as used by analytical and organic chemists is primarily *absorption* spectroscopy. IR absorption can also be used to provide quantitative measurements of compounds.

IR spectroscopy became widely used after the development of commercial spectrometers in the 1940s. Double-beam monochromator instruments were developed, better detectors were designed, and better dispersion elements, including gratings, were incorporated. These conventional spectrometer systems have been replaced in the last decade by FTIR instrumentation. This chapter will focus on FTIR instrumentation and applications of IR spectroscopy. In addition, the related techniques of near-IR (NIR) spectroscopy and Raman spectroscopy will be covered.

The wavelengths of IR radiation of interest to chemists trying to identify or study organic molecules fall between 2 and 20 μm. These wavelengths are longer than those in the red end of the visible region, which is generally considered to end at about 0.75 μm. IR radiation therefore is of lower energy than visible radiation, but of higher energy than radiowaves. The entire IR region can be divided into the *near-IR*, from 0.75 to 2.5 μm, the *mid-IR*, from about 2.5 to 20 μm, and the *far-IR*, from 20 to 200 μm. Visible radiation (red light) marks the upper energy end or minimum wavelength end of the IR region; the maximum wavelength end is defined somewhat arbitrarily; some texts consider the far-IR to extend to 1000 μm. The IR wavelength range tells us the IR frequency range from the equation (introduced in Chapter 2)

$$\nu = \frac{c}{\lambda} \tag{4.1}$$

where ν is the frequency, c is the speed of light, and λ is the wavelength. We also know that $\Delta E = h\nu$. When the frequency is high, λ is short and the energy of the radiation is high.

It is common to use *wavenumber*, symbolized by either $\tilde{\nu}$ or $\bar{\nu}$, with units of cm^{-1}, in describing IR spectra. The first symbol is called "nu tilde"; the second symbol is called "nu bar"; both symbols are used in the literature. The unit cm^{-1} is called a *reciprocal centimeter*. The wavenumber is the reciprocal of the wavelength. Wavenumber is the number of waves of radiation per centimeter, $1/\lambda$; frequency is the number of waves per second, c/λ. Wavelength and wavenumber are related by:

$$wavelength \ (\mu m) \times wavenumber \ (cm^{-1}) = 10{,}000 = 1 \times 10^4 \qquad (4.2)$$

Both wavenumbers and wavelengths will be used throughout the chapter, so it is important to be able to convert between these units. The older IR literature used the term *micron* and the symbol μ for wavelength in micrometers (μm).

4.1. ABSORPTION OF IR RADIATION BY MOLECULES

Molecules with covalent bonds may absorb IR radiation. This absorption is quantized, so only certain frequencies of IR radiation are absorbed. When radiation, (i.e., energy) is absorbed, the molecule moves to a higher energy state. The energy associated with IR radiation is sufficient to cause molecules to rotate (if possible) and to vibrate. If the IR wavelengths are longer than 100 μm, absorption will cause excitation to higher rotational states in the gas phase. If the wavelengths absorbed are between 1 and 100 μm, the molecule will be excited to higher vibrational states. Because the energy required to cause a change in rotational level is small compared to the energy required to cause a vibrational level change, each vibrational change has multiple rotational changes associated with it. Gas phase IR spectra therefore consist of a series of discrete lines. Free rotation does not occur in condensed phases. Instead of a narrow line spectrum of individual vibrational absorption lines, the IR absorption spectrum for a liquid or solid is composed of broad vibrational absorption **bands**. A typical IR spectrum for a condensed phase (liquid or solid) is shown in Fig. 4.1. This is the spectrum of a thin film of polystyrene; note that the absorption band at about 2950 cm^{-1} is more than 100 cm^{-1} wide at the top. The individual vibration–rotation lines can be seen in gas phase IR spectra. These narrow lines are clearly seen in Fig. 4.2, the gas phase spectrum of hydrogen chloride, HCl.

Molecules absorb radiation when a bond in the molecule vibrates at the same frequency as the incident radiant energy. After absorbing radiation, the molecules have more energy and vibrate at increased amplitude. The frequency absorbed depends on the masses of the atoms in the bond, the geometry of the molecule, the strength of the bond, and several other factors. Not all molecules can absorb IR radiation. The molecule must have a change in **dipole moment** during vibration in order to absorb IR radiation.

4.1.1. Dipole Moments in Molecules

When two atoms with different electronegativities form a bond, the electron density in the bond is not equally distributed. For example, in the molecule hydrogen fluoride, HF, the electron density in the bond shifts away from the H atom toward the more electronegative fluorine atom. This results in a partial negative charge on F and a partial positive charge on H. The bond is said to be **polar** when such charge separation exists. The charge separation can be shown as

$\delta^+ \quad \delta^-$
H——F

Figure 4.1   Fourier transform IR spectrum of a thin film of polystyrene. The *y* axis unit is %T, the *x* axis is in wavenumbers (cm^{-1}). Collected on a ThermoNicolet 6700 FTIR spectrometer with a DTGS detector. Courtesy of Thermo Electron Corp. (www.thermo.com).

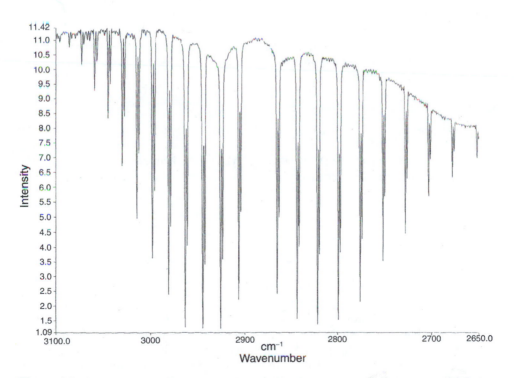

Figure 4.2   Vapor-phase FTIR spectrum of hydrogen chloride, HCl. Spectrum was collected in a 10 cm gas cell with NaCl windows on a Paragon 1000 FTIR spectrometer, PerkinElmer Instruments, Shelton, CT.

where δ indicates a partial charge. The dipole in the bond is indicated by a crossed arrow placed with the point on the more negative end of the bond as shown:

H—F

The HF molecule is a linear diatomic molecule with one polar bond; therefore, the molecule is polar and has a dipole moment. The dipole moment μ is the product of the charge Q and the distance between the charges, r:

$$\mu = Q \times r \tag{4.3}$$

The partial positive and negative charges in HF must be equal in magnitude but opposite in sign to maintain electrical neutrality. We can imagine that the HF molecule vibrates by having the bond stretch and compress over and over, just as a spring can be stretched and compressed by first pulling and then squeezing the spring. On vibration of the HF bond, the dipole moment changes because the distance between the charges changes. Because the dipole moment changes in a repetitive manner as the molecule vibrates, it is possible for HF to absorb IR radiation. It follows that diatomic molecules containing atoms of the same element such as H_2, N_2, O_2, and Cl_2 have no charge separation because the electronegativity of each atom in the bond is the same. Therefore they do not have dipole moments. Since a change in dipole moment with time is necessary for a molecule to absorb IR radiation, symmetrical diatomic molecules do not absorb IR radiation. Diatomic molecules made up of different atoms such as HCl and CO do have dipole moments. They would be expected to absorb IR radiation or to be "IR-active".

Molecules with more than two atoms may or may not have permanent dipole moments. It depends on the geometry of the molecule. Carbon dioxide has two equal C=O bond dipoles but because the molecule is linear the bond dipoles cancel and the molecule has no net dipole moment:

O=C=O

Water, H_2O, has two equal H—O bond dipoles. Water has a bent geometry and the vector sum of the bond dipoles does not cancel. The water molecule has a permanent net dipole moment.

Bond Dipoles in Water   Net Dipole in Water

Predicting the IR absorption behavior of molecules with more than two atoms is not as simple as looking at diatomic molecules. It is not the net dipole moment of the molecule that is important, but any change in the dipole moment on vibration. We need to understand how molecules vibrate. This is relatively simple for diatomic and triatomic molecules. Large molecules cannot be evaluated simply, because of their large number of vibrations and interactions between vibrating atoms. It can be said that most molecules do absorb IR radiation, which is the reason this technique is so useful.

4.1.2. Types of Vibrations in Molecules

The common molecular vibrations that are excited by IR radiation are **stretching** vibrations and **bending** vibrations. These are called **modes** of vibration. Stretching involves a change in bond lengths resulting in a change in interatomic distance. Bending involves a change in bond angle or a change in the position of a group of atoms with respect to the rest of the molecule. For a group of three or more atoms, at least two of which are the same type of atom, there are two stretching modes: symmetrical stretching and asymmetrical stretching. These two modes of stretching are shown, respectively, in Fig. 4.3 for a CH_2 group. In Fig. 4.3(a) the two H atoms both move away from the C atom—a symmetrical stretch. In Fig. 4.3(b) one H atom moves away from the C atom and one moves toward the C atom—an asymmetrical stretch. Bending modes are shown in Fig. 4.3(c)–(f). Scissoring and rocking are in-plane bending modes—the H atoms remain in the same plane as the C atom (i.e., in the plane of the page). Wagging and twisting are out-of-plane (oop) bending modes—the H atoms are moving out of the plane containing the C atom. The + sign in the circle indicates movement above the plane of the page toward the reader, while the − sign in the circle indicates movement below the plane of the page away from the reader. Bends are also called deformations and the term antisymmetric is used in place of asymmetric in various texts.

For the CO_2 molecule, we can draw a symmetric stretch, an asymmetric stretch, and a bending vibration, as shown:

$$\overset{\leftarrow \qquad \rightarrow}{O=C=O} \qquad\qquad \overset{\rightarrow \qquad \rightarrow}{\underset{\leftarrow}{O=C=O}} \qquad\qquad \overset{\uparrow \qquad \uparrow}{\underset{\downarrow}{O=C=O}}$$

The CO_2 molecule on the left is undergoing a symmetric stretch, the one in the middle an asymmetric stretch and the one on the right an in-plane bend. The symmetric

Figure 4.3 Principal modes of vibration between carbon and hydrogen in an alkane: (a) symmetrical stretching, (b) asymmetrical stretching and the bending vibrations, (c) scissoring, (d) rocking, (e) wagging, and (f) twisting.

stretching vibration does *not* change the dipole moment of the molecule. This vibrational mode does not absorb IR radiation—it is said to be *IR-inactive*. However, the other two modes of vibration do change the dipole moment—they are *IR-active*. The asymmetric stretching frequency occurs at $2350 \, \text{cm}^{-1}$ and the bending vibration occurs at $666 \, \text{cm}^{-1}$.

For diatomic and triatomic molecules, it is possible to work out the number and kind of vibrations that occur. To locate a point in three-dimensional space requires three coordinates. To locate a molecule containing N atoms in three dimensions, $3N$ coordinates are required. The molecule is said to have $3N$ degrees of freedom. To describe the motion of such a molecule, translational, rotational, and vibrational motions must be considered. Three coordinates or degrees of freedom are required to describe translational motion and three degrees of freedom are required to describe rotational motion about the molecule's center of gravity. This leaves $3N - 6$ degrees of freedom to describe vibrational motion. There are $3N - 6$ possible normal modes of vibration in a molecule of N atoms. For example, the water molecule contains 3 atoms, so it has $3 \times 3 = 9$ degrees of freedom and $(3 \times 3) - 6 = 3$ normal modes of vibration. For the water molecule, these normal modes of vibration are a symmetric stretch, an asymmetric stretch, and a scissoring (bending) mode. Linear molecules cannot rotate about the bond axis. As a result, only two degrees of freedom are needed to describe rotation, so linear molecules have $3N - 5$ normal modes of vibration. If we look at CO_2 above, three modes of vibration are shown, but $3N - 5 = 4$ normal modes of vibration, so one is missing. The fourth is a bending mode equivalent to that shown, but oop:

+   –   +
O═══C═══O

where $+$ indicates movement toward the reader and $-$ indicates movement away from the reader. The oop bending mode and the in-plane bending mode already shown both occur at $666 \, \text{cm}^{-1}$. They are said to be *degenerate*. Both are IR-active, but only one absorption band is seen since they both occur at the same frequency.

For simple molecules, this approach predicts the number of fundamental vibrations that exist. Use of the dipole moment rule indicates which vibrations are IR-active, but the IR spectrum of a molecule rarely shows the number of absorption bands calculated. Fewer peaks than expected are seen due to IR-inactive vibrations, degenerate vibrations, and very weak vibrations. More often, additional peaks are seen in the spectrum due to overtones and other bands.

The excitation from the ground state V_0 to the first excited state V_1 is called the *fundamental transition*. It is the most likely transition to occur. Fundamental absorption bands are strong compared with other bands that may appear in the spectrum due to overtone, combination, and difference bands. *Overtone* bands result from excitation from the ground state to higher energy states V_2, V_3, and so on. These absorptions occur at approximately integral multiples of the frequency of the fundamental absorption. If the fundamental absorption occurs at frequency ν, the overtones will appear at about 2ν, 3ν, and so on. Overtones are weak bands and may not be observed under real experimental conditions. Vibrating atoms may interact with each other. The interaction between vibrational modes is called *coupling*. Two vibrational frequencies may couple to produce a new frequency $\nu_3 = \nu_1 + \nu_2$. The band at ν_3 is called a *combination* band. If two frequencies couple such that $\nu_3 = \nu_1 - \nu_2$, the band is called a *difference* band. Not all possible combinations and differences occur; the rules for predicting coupling are beyond the scope of this text.

The requirements for the absorption of IR radiation by molecules can be summarized as follows:

1. The natural frequency of vibration of the molecule must equal the frequency of the incident radiation.
2. The frequency of the radiation must satisfy $\Delta E = h\nu$, where ΔE is the energy difference between the vibrational states involved.
3. The vibration must cause a change in the dipole moment of the molecule.
4. The amount of radiation absorbed is proportional to the square of the rate of change of the dipole during the vibration.
5. The energy difference between the vibrational energy levels is modified by coupling to rotational energy levels and coupling between vibrations.

4.1.3. Vibrational Motion

A molecule is made up of two or more atoms joined by chemical bonds. Such atoms vibrate about each other. A simple model of vibration in a diatomic molecule can be made by considering the bond to be a spring with a weight on each end of the spring (Fig. 4.4). The stretching of such a spring along its axis in a regular fashion results in simple harmonic motion. Hooke's Law states that two masses joined by a spring will vibrate such that:

$$\nu = \frac{1}{2\pi}\sqrt{\frac{f}{\mu}} \tag{4.4}$$

where ν is the frequency of vibration; f, the force constant of the spring (a measure of the stiffness of the spring); and μ, the reduced mass. The reduced mass is calculated from the masses of the two weights joined by the spring.

$$\mu = \frac{M_1 M_2}{M_1 + M_2} \tag{4.5}$$

where M_1 is the mass of one vibrating body and M_2 the mass of the other. From Eq. (4.4) it can be seen that the natural frequency of vibration of the harmonic oscillator depends on the force constant of the spring and the masses attached to it, but is independent of the

Figure 4.4 (a) The atoms and chemical bonds in methane, CH_4, presented as (b) a system of masses and springs.

amount of energy absorbed. Absorption of energy changes the amplitude of vibration, not the frequency. The frequency ν is given in hertz (Hz) or cycles per second (cps). During this time light travels a distance measured in cm/s (i.e., the speed of light). Therefore, if one divides ν by c, the result is the number of cycles per cm. This is $\bar{\nu}$, the wavenumber:

$$\bar{\nu} = \frac{\nu}{c} \tag{4.6}$$

Dividing both sides of Eq. (4.4) by c, we get:

$$\bar{\nu} = \frac{1}{2\pi c} \sqrt{\frac{f}{\mu}} \tag{4.7}$$

where $\bar{\nu}$ is the wavenumber of the absorption maximum in cm^{-1}; c, the speed of light in cm/s; f, the force constant of the bond in dyn/cm; and μ, the reduced mass in g. The term *frequency of vibration* is often used when the vibration is expressed in wavenumbers, but it must be remembered that wavenumber is directly proportional to the frequency ν, not identical to it. Equation (4.7) tells us the frequency of vibration of two atoms joined by a chemical bond. The bond vibrates according to the same equation, except that in this case the force constant f is a measure of the *strength* of the chemical bond and μ is the reduced mass of the vibrating atoms. The term f varies with bond strength; a simple but useful approximation is that a triple bond between two atoms is 3× as strong as a single bond between the same two atoms, so f would be 3× as large for the triple bond. A double bond is twice as strong as a single bond. The force constant f is directly proportional to the bond order, and depends on the electronegativity of the vibrating atoms and the mean distance between them. These are all physical constants and properties of the molecule. Since f and μ are constant for any given set of atoms and chemical bonds, the frequency of vibration ν is also constant. *The radiation absorbed by the system has the same frequency and is constant for a given set of atoms and chemical bonds*, that is, for a given molecule. The absorption spectrum is therefore a physical property of the molecule. Average values of the force constant for single, double, and triple bonds are given in Table 4.1. Using these values of f and the masses of given atoms, it is possible to estimate the wavenumber for fundamental stretching vibrations of given bonds as discussed. Table 4.2 presents frequencies for common vibrations.

From Eq. (4.7), we can deduce that a C—C bond vibrates at a lower wavenumber than a C=C bond, because the force constant for the C—C bond is smaller than that for C=C. For example, C—C vibrates at $1200 \, cm^{-1}$ and C=C vibrates at $1650 \, cm^{-1}$. In general, force constants for bending vibrations are lower than stretching vibrations. Resonance and hybridization in molecules also affect the force constant for a given bond.

Table 4.1   Average Values for Bond Force Constant

Bond order	Average force constant f (N/m)[a]
1 (single bond)	500
2 (double bond)	1000
3 (triple bond)	1500

[a]N/m is the SI unit; dyn/cm is the cgs unit. The single bond force constant in dyn/cm is about 5×10^5.

Table 4.2 Molecular Vibrations and Approximate Absorption Frequencies in Wavenumbers (cm^{-1})

Molecular group	Structure	Stretching $C \rightarrow H$	Bending $C \downarrow H$
C—H bonds			
Methylene (alkane)	R—C(H)(H)—R	2929, 2850	780–760
		1460	
Methyl (alkane)	R—C(H)(H)—H	2960, 2870	1375
		1450	
Alkene (terminal)	R—C(H)=C(H)(H)	3080, 2990	890
		1410	
Aromatic	(benzene ring)	3050	680–900
		1200	
Aldehyde	R—C(H)=O	2820, 2710	
		1390	

(continued)

Table 4.2 *Continued*

Molecular group	Structure	Stretching	Bending
C—H bonds		C → H	C ↓ H
Alkyne (terminal)	R—C≡CH	3300	615–680
Nitrogen bonds		N → H C → N	N ↓ H
Primary amine	R—N(H)(H)	3390, 3330 1065	1610
Secondary amine	R₁—NH—R₂	3330 1150, 850	1610
Tertiary amine	R₁—N(R₂)(R₃)	780	
Oxygen bonds		C ⇒ O O → H C → O	C—O ↓ H
Primary alcohol	H—C(H)—OH, R	3635 1050, 850	1300
Secondary alcohol	R—C(H)(R)—OH	3625 1100, 830	1300

Oxygen bonds		$O \rightarrow H$	$C \Rightarrow O$	$C \rightarrow O$	$C-O \downarrow H$
Tertiary alcohol	R_1, R_2-C-OH, R_3	3615		1150, 780	1300
Phenol	OH (on benzene ring)	3600		1220	1360
Carboxylic acid	O=C-OH (R)	2500–3000	1710		1420, 925
Ketone	R_1, R_2 C=O		1715		
Ether	H-C-R, H-C-H, R-C-O, H, H			1100, 860	
Aldehyde	H, R-C=O		1730		

(continued)

Table 4.2 *Continued*

Molecular group	Structure	Stretching			Bending
		C → C	C ⇒ C	C ≡ C	
C—C bonds					
Alkane	H H | | R—C—C—R | | H H	800–1200			
Alkene	R—C—C=CH₂ | H		1650		
Alkyne	R—C≡CH			2120	
Aromatic	(benzene ring)—CH₃		1590, 1450		
Chloroalkanes	Cl | R—C—R | H	C → Cl 550–850			

Note: A single arrow (→) denotes a single bond stretching vibration; a double arrow (⇒) denotes a double bond stretching vibration and so on. A vertical arrow (↓) denotes a bending vibration.

4.2. IR INSTRUMENTATION

Until the early 1980s, most IR spectrometer systems were double-beam dispersive grating spectrometers, similar in operation to the double-beam system for UV/VIS spectroscopy described in Chapter 2. These instruments have been replaced almost entirely by FTIR spectrometers because of the advantages in speed, signal-to-noise ratio, and precision in determining spectral frequency that can be obtained from a modern multiplex instrument. There are NIR instruments that are part of double-beam dispersive UV/VIS/NIR systems, but many NIR instruments are stand-alone grating instruments.

The first requirement for material used in an IR spectrometer is that the material must be transparent to IR radiation. This requirement eliminates common materials such as glass and quartz for use in mid-IR instruments because glass and quartz are not transparent to IR radiation at wavelengths longer than 3.5 μm. Second, the materials used must be strong enough to be shaped and polished for windows, samples cells, and the like. Common materials used are ionic salts, such as potassium bromide, calcium fluoride, sodium chloride (rock salt), and zinc selenide. The final choice among the compounds is determined by the wavelength range to be examined; for example, sodium chloride is transparent to radiation between 2.5 and 15 μm. This wavelength range was therefore termed the *rock salt region* when an ionic salt prism was used as the wavelength dispersion device in early instruments. Potassium bromide or cesium bromide can be used over the range of 2.1–26 μm, and calcium fluoride in the range of 2.4–7.7 μm. The wavelength ranges of some materials used for IR optics and sample holders are given in Table 4.3.

The major problem with the use of NaCl, KBr, and similar ionic salts is that they are very soluble in water. Any moisture, even atmospheric moisture, can dissolve the surface of a polished salt crystal, causing the material to become opaque and scatter light. Optics and sample containers made of salts must be kept desiccated. This limitation is one of the reasons salt prisms are no longer used in dispersive IR spectrometers.

4.2.1. Radiation Sources

A radiation source for IR spectroscopy should fulfill the requirements of an ideal radiation source, namely, that the intensity of radiation (1) be continuous over the wavelength range used, (2) cover a wide wavelength range, and (3) be constant over long periods of time. The most common sources of IR radiation for the mid-IR region are *Nernst glowers*, *Globars*, and *heated wires*. All of these heated sources emit continuous radiation, with a spectral output very similar to that of a blackbody radiation source. Spectral curves for blackbody radiators at several temperatures are shown in Fig. 4.5. The normal operating temperatures for IR sources are between 1100 and 1500 K. The range of light put out by mid-IR sources extends into both the NIR and far-IR regions, but intensity is at a maximum in the mid-IR region from 4000 to 400 cm^{-1}.

4.2.1.1. Mid-IR Sources

The two main types of sources for mid-IR radiation are electrically heated rigid ceramic rods and coiled wires.

The Nernst glower is a cylindrical bar composed of zirconium oxide, cerium oxide, and thorium oxide that is heated electrically to a temperature between 1500 and 2000 K. The source is generally about 20 mm long and 2 mm in diameter. The rare earth oxide ceramic is an electrical resistor; passing current through it causes it to heat and glow,

Table 4.3 Typical Materials Used in Mid-IR Optics

Material	Transmission range (μm)	Solubility (g/100 g water)	Refractive index	Comments
Sodium chloride (NaCl)	0.25–16	36	1.49	Most widely used; reasonable range and low cost
Potassium chloride (KCl)	0.30–20	35	1.46	Wider range than NaCl; used as a laser window
Potassium bromide (KBr)	0.25–26	65	1.52	Extensively used; wide spectral range
Barium fluoride (BaF$_2$)	0.2–11	0.1	1.39	Extremely brittle
Cesium iodide (CsI)	0.3–60	160	1.74	Transmits to 60 μm
Cesium bromide (CsBr)	0.3–45	125	1.66	
Thallium bromide/iodide eutectic (KRS-5)	0.6–40	<0.05	2.4	For internal reflection in the far IR when moisture is a problem
Strontium fluoride (SrF$_2$)	0.13–11	1.7×10^{-3}		Resistant to thermal shock
Silver chloride (AgCl)	0.4–25	1.5×10^{-4}	2.0	Darkens under UV light
Silver bromide (AgBr)	0.5–35	1.2×10^{-5}	2.2	Darkens under UV light
Germanium	2–11	Insoluble	4.00	
Fused silica	0.2–4.5	Insoluble	1.5 at 1 μm	
Magnesium fluoride (MgF$_2$)	0.5–9	Insoluble	1.34 at 5 μm	Most useful for NIR
Zinc sulfide (ZnS)	0.4–14.5	Insoluble	2.2	This and the next five materials are known commercially as Irtran® 1 through 6
Calcium fluoride (CaF$_2$)	0.4–11.5	Insoluble	1.3	
Zinc selenide (ZnSe)	0.5–22	Insoluble	2.4	
Magnesium oxide (MgO)	0.4–9.5	Insoluble	1.6 at 5 μm	
Cadmium telluride (CdTe)	0.9–31	Insoluble	2.7	

Figure 4.5 Radiant energy distribution curves for a blackbody source operated at various temperatures. (From Coates, used with permission.)

giving off continuous IR radiation. The Nernst glower requires an external preheater because of the negative coefficient of electrical resistance; it only conducts at elevated temperature. In addition, the Nernst glower can easily overheat and burn out because its resistance decreases as the temperature increases. The circuitry must be designed to control the current accurately. A related source, the Opperman, consists of a bar of rare earth ceramic material with a Pt or other wire running coaxially through the center of the ceramic. Electrical current through the wire heats the wire, and that heats the ceramic, providing a source similar to the Nernst glower without the preheating requirement. The Globar is a bar of sintered silicon carbide, which is heated electrically to emit continuous IR radiation. The Globar is a more intense source than the Nernst glower. These rigid cylinders were designed so that their shape matched the shape of the slit on a classical dispersive spectrometer. Modern FTIRs do not have slits, so the geometry of the source can now be made more compact. Commercial ceramic IR sources are available in a variety of sizes and shapes, as seen in Fig. 4.6. Typical spectral outputs from these commercial ceramic sources are compared with a blackbody radiator in Fig. 4.7.

Electrically heated wire coils, similar in shape to incandescent light bulb filaments, have also been used successfully as a light source. Nichrome wire is commonly used, although other metals such as rhodium are used as well. These wires are heated electrically in air to a temperature of ~1100°C. The main problem with these wire coils is "sagging" and embrittlement due to ageing, resulting in fracture of the filament, exactly the way a light bulb filament "burns out". Some coiled wire sources are wound around a ceramic rod for support; this results in a more uniform light output over time than that from an unsupported coil.

Modern sources for the mid-IR region are variants of the incandescent wire source or the Globar, but generally in a compact geometry. Commercial furnace ignitors and diesel engine heaters such as the silicon carbide tipped "glo-plug" have been adapted for use as IR sources because of their robustness, low operating voltage, and low current requirements.

Sources are often surrounded by a thermally insulated enclosure to reduce noise caused by refractive index gradients between the hot air near the source and cooler air in the light path. Short-term fluctuations in spectral output are usually due to voltage fluctuations and can be overcome by use of a stabilized power supply. Long-term changes occur as result of changes in the source material due to oxidation or other high temperature reactions. These

Figure 4.6 Commercial IR radiation sources. (Top) A variety of designs from ThermoOriel. Dimensions given are in inches (mm). Courtesy of Newport Corporation, Irvine, CA (www. newport.com). (Bottom) FTIR source element used in the PerkinElmer Spectrum One instrument. It is made of a proprietary ceramic/metallic composite and is designed to minimize hot spots to the end of the element. Only the last 5 mm on the end lights up. [Courtesy of PerkinElmer Instruments, Shelton, CT (www.perkinelmer.com).]

types of changes may be seen as hot or cold "spots" in the source, and usually require replacement of the source.

4.2.1.2. NIR Sources

As can be seen in Fig. 4.5, operating a mid-IR source at higher temperatures ($>$2000 K) increases the intensity of NIR light from the source. Operation at very high temperatures is usually not practical, due to the excessive heat generated in the instrument and premature burn-out of the source. For work in the NIR region, a quartz halogen lamp is used as the source. A quartz halogen lamp contains a tungsten wire filament and iodine vapor sealed in a quartz envelope or bulb. In a standard tungsten filament lamp, the tungsten evaporates from the filament and deposits on the lamp wall. This process reduces the light output as a result of the black deposit on the wall and the thinner filament. The halogen gas in a tungsten-halogen lamp removes the evaporated tungsten and redeposits it on the

Figure 4.7 Spectral output of a variety of commercial IR radiation sources, including a silicon carbide source (dashed line marked SiC) and an NIR quartz tungsten-halogen lamp (the dotted line marked QTH). A blackbody curve at 1273 K is included for comparison. [Courtesy of Newport Corporation, Irvine, CA (www.newport.com).]

filament, increasing the light output and source stability. The intensity of this source is very high compared to a standard tungsten filament incandescent lamp. The range of light put out by this source is from 25,000 to 2000 cm^{-1}. Figure 4.8 shows typical commercial quartz tungsten-halogen lamps and a plot of the spectral output of such a source.

4.2.1.3. Far-IR Sources

While some of the mid-IR sources emit light below 400 cm^{-1}, the intensity drops off. A more useful source for the far-IR region is the high pressure mercury discharge lamp. This lamp is constructed of a quartz bulb containing elemental Hg, a small amount of inert gas, and two electrodes. When current passes through the lamp, mercury is vaporized, excited,

Figure 4.8 (a) Commercial quartz tungsten-halogen lamps for use in the NIR region. The lamps are constructed of a doped tungsten coiled filament inside a quartz envelope. The envelope is filled with a rare gas and a small amount of halogen. (b) The spectral output of a model 6315 1000 W quartz tungsten-halogen lamp. The location and height of the peak depend on the model of lamp and the operating conditions. [Courtesy of Newport Corporation, Irvine, CA (www.newport.com).]

and ionized, forming a plasma discharge at high pressure (>1 atm). In the UV and visible regions, this lamp emits atomic Hg emission lines that are very narrow and discrete, but emits an intense continuum in the far-IR region.

4.2.1.4.   *IR Laser Sources*

A laser is a light source that emits very intense monochromatic radiation. Some lasers, called tunable lasers, emit more than one wavelength of light, but each wavelength emitted is monochromatic. The combination of high intensity and narrow linewidth makes lasers ideal light sources for some applications. Two types of IR lasers are available: gas phase and solid-state. The tunable carbon dioxide laser is an example of a gas phase laser. It emits discrete lines in the $1100-900 \text{ cm}^{-1}$ range. Some of these lines coincide with the narrow vibrational–rotational lines of gas phase analytes. This makes the laser an excellent source for measuring gases in the atmosphere or gases in a production process. Open path environmental measurements of atmospheric hydrogen sulfide, nitrogen dioxide, chlorinated hydrocarbons, and other pollutants can be made using a carbon dioxide laser.

Tunable gas phase lasers are expensive. Less expensive solid-state diode lasers with wavelengths in the NIR are available. Commercial instruments using multiple diode lasers are available for NIR analyses of food and fuels. Because of the narrow emission lines from a laser system, laser sources are often used in dedicated applications for specific analytes. They can be ideal for process analysis and product quality control, for example, but are not as flexible in their applications as a continuous source or a tunable laser.

4.2.2.   Monochromators and Interferometers

The radiation emitted by the source covers a wide frequency range. However, the sample absorbs only at certain characteristic frequencies. In practice, it is important to know what these frequencies are. To obtain this information we must be able to select radiation of any desired frequency from our source and eliminate that at other frequencies. This can be done by means of a monochromator, which consists of a dispersion element and a slit system, as discussed in Chapter 2. This type of system is called a *dispersive* spectrometer or spectrophotometer. Double-beam spectrophotometers were routinely used because both CO_2 and H_2O present in air absorb IR radiation. Changes in humidity and temperature would cause an apparent change in the source intensity if single beam optics were used, resulting in error in recording the spectrum. A double-beam system automatically subtracts the absorption by species in the air and also can subtract absorption by solvent if the sample is dissolved in a solvent. Double-beam systems for the mid-IR required that the optics be transparent to IR radiation. Lenses are rarely used because of the difficulty of grinding lenses from the ionic salts that are IR-transparent. Salt prisms and metal gratings are used as dispersion devices. Mirrors are generally made of metal and front surface polished. The IR spectrum is recorded by moving the prism or grating so that different frequencies of light pass through the exit slit to the detector. The spectrum is a plot of transmission intensity, usually as percent transmittance, vs. frequency of light. A dispersive system is said to record a spectrum in the *frequency domain*. It is estimated (Coates, 1997) that no more than 5% of the IR spectrometers in current use are dispersive instruments. Therefore, the discussion will focus on the FTIR based on a Michelson interferometer.

4.2.2.1.   *FT Spectrometers*

If two beams of light of the same wavelength are brought together in phase, the beams reinforce each other and continue down the light path. However, if the two beams are out of phase, destructive interference takes place. This interference is at a maximum when the two beams of light are 180° out of phase (Fig. 4.9). Advantage is taken of this fact in the FT instrument. The FT instrument is based on a Michelson interferometer; a schematic is shown in Fig. 4.10. The system consists of four optical arms, usually at right angles to each other, with a *beam splitter* at their point of intersection. Radiation passes down the first arm and is separated by a beam splitter into two perpendicular beams of approximately equal intensity. These beams pass down into other arms of the spectrometer. At the ends of these arms, the two beams are reflected by mirrors back to the beam splitter, where they recombine and are reflected together onto the detector. One of the mirrors is fixed in position; the other mirror can move toward or away from the beam splitter, changing the path length of that arm.

It is easiest to discuss what happens in the interferometer if we assume that the source is monochromatic, emitting only a single wavelength of light. If the side arm paths are equal in length there is no difference in path length. This position is shown in Fig. 4.10 as the zero path length difference (ZPD) point. For ZPD, when the two beams are recombined, they will be in phase, reinforcing each other. The maximum signal will be obtained by the detector. If the moving mirror is moved from ZPD by 1/8 of a wavelength, the total path difference on recombination is [2 × (1/8)λ] or (1/4)λ and partial interference will occur. If the moving mirror is moved from ZPD by 1/4 of a wavelength, then the beams will be one-half of a wavelength out of phase with each other; that is, they will destructively interfere with each other such that a minimum signal reaches the detector. Figure 4.11 shows the signal at the detector as a function of path length difference for monochromatic light. In practice, the mirror in one arm is kept stationary and that in the second arm is moved slowly. As the moving mirror moves, the net signal falling on the detector is a cosine wave with the usual maxima and minima when plotted against the travel of the mirror. The frequency of the cosine signal is equal to

$$f = \frac{2}{\lambda}(v) \tag{4.8}$$

where f is the frequency; v, the velocity of the moving mirror; and λ, the wavelength of radiation.

Wave Interactions (Interference)

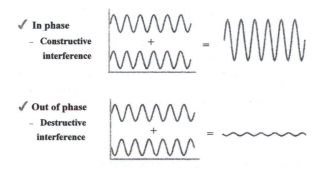

Figure 4.9 Wave interactions. (Top) Constructive interference occurs when both waves are in phase. (Bottom) Destructive interference occurs when both waves are out of phase. [Courtesy of ThermoNicolet, Madison, WI (www.thermonicolet.com).]

Figure 4.10 (Top) Schematic diagram of a Michelson interferometer. ZPD stands for zero path-length difference (i.e., the fixed mirror and moving mirror are equidistant from the beamsplitter). (From Coates, used with permission). (Bottom) A simple commercial FTIR spectrometer layout showing the He-Ne laser, optics, the source, as well as the source, interferometer, sample, and detector. [Courtesy of ThermoNicolet, Madison, WI (www.thermonicolet.com).]

The frequency of modulation is therefore proportional to the velocity of the mirror and inversely proportional to wavelength of the incident radiation. The frequency is therefore also proportional to the wavenumber of the incident radiation, as we know from the relationship between wavelength and wavenumber.

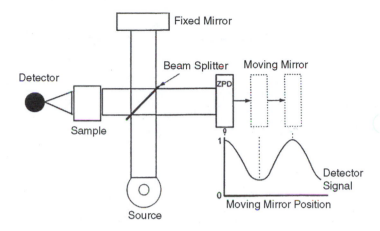

Signal at the Detector

Figure 4.11 (Top) A simplified schematic showing the generation of an interferogram from mono-chromatic light by displacement of the moving mirror. (Modified from Coates, used with permission). (Bottom) An enlarged view of the signal at the detector as a function of path difference between the moving and fixed mirrors for monochromatic light of wavelength λ. [Courtesy of ThermoNicolet, Madison, WI (www.thermonicolet.com).]

Real IR sources are polychromatic. Radiation of all wavelengths generated from the source travels down the arms of the interferometer. Each wavelength will generate a unique cosine wave; the signal at the detector is a result of the summation of all these cosine waves. An idealized interferogram from a polychromatic source is shown in Fig. 4.12. The "centerburst" is located in the center of the interferogram because modern FTIR systems scan the moving mirror symmetrically around ZPD. The interferogram holds the spectral information from the source (or sample) in a *time domain*, a record of intensity vs. time based on the speed of the moving mirror. The spectral information about the sample is obtained from the wings of the interferogram.

If the unique cosine waves can be extracted from the interferogram, the contribution from each wavelength can be obtained. These individual wavelength contributions can be reconstructed to give the spectrum in the frequency domain, that is, the usual spectrum obtained from a dispersive spectrometer. A Fourier transform is used to convert the

Figure 4.12   An idealized interferogram. (From Coates, used with permission.)

time-domain spectrum into a frequency-domain spectrum. Hence the term Fourier Transform infrared spectrometer for this type of system.

It is mechanically difficult to move the reflecting mirror at a controlled, known, steady velocity, and position variations due to temperature changes, vibrations, and other environmental effects must be corrected for. The position of the moving mirror must be known accurately. The position and the velocity are controlled by using a helium-neon (He-Ne) laser beam that is shone down the light path producing an interference pattern with itself. The cosine curve of the interference pattern of the laser is used to adjust the moving mirror in real time in many spectrometers. The He-Ne laser is also used to identify the points at which the interferogram is sampled for the Fourier transform, as shown schematically in Fig. 4.13.

Sampling the Interferogram

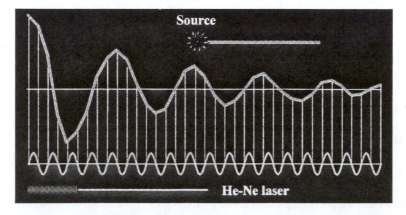

Figure 4.13   Schematic diagram showing how the interferogram is digitized by sampling it at discrete points based on the He-Ne laser signal, shown at the bottom. Each vertical line represents a sampling point. [Courtesy of ThermoNicolet, Madison, WI (www.thermonicolet.com).]

There are a number of advantages to the use of FTIR over dispersive IR. Because the sample is exposed to all source wavelengths at once, all wavelengths are measured simultaneously in less than 1 s. This is known as the *multiplex* or *Fellgett's advantage*, and it greatly increases the sensitivity and accuracy in measuring the wavelengths of absorption maxima. This multiplex advantage permits collection of many spectra from the same sample in a very short time. Many spectra can be averaged, resulting in improved signal-to-noise ratio. An FTIR is considerably more accurate and precise than a monochromator (*Connes' advantage*). Another advantage is that the intensity of the radiation falling on the detector is much greater because there are no slits; large beam apertures are used, resulting in higher energy throughput to the detector. This is called the *throughput* or *Jacquinot's advantage*. Resolution is dependent on the linear movement of the mirror. As the total distance traveled increases, the resolution improves. It is not difficult to obtain a resolution of 0.5 cm^{-1}. A comparison between FTIR and dispersive IR is given in Table 4.4.

The signal-to-noise improvement in FTIR comes about as a result of the multiplex and throughput advantages. These permit a rapid spectrum collection rate. A sample spectrum can be scanned and rescanned many times in a few seconds and the spectra added and averaged electronically. The IR signal (S) accumulated is additive, but the noise level (N) in the signal is random. The S/N ratio therefore increases with the square root of the number of scans (i.e., if 64 scans are accumulated, the S/N ratio increases $8\times$). FTIR has the potential to be many orders of magnitude more sensitive than dispersive IR. However, in practice the sheer number of scans necessary to continue to improve the

Table 4.4 Comparison of Dispersive IR and FTIR Instruments

Dispersive IR	FTIR
Many moving parts result in mechanical slippage	Only mirror moves during an experiment
Calibration against reference spectra required to measure frequency	Use of laser provides high frequency precision (to 0.01 cm^{-1}) (*Connes' advantage*)
Stray light within instrument causes spurious readings	Stray light does not affect detector, since all signals are modulated
In order to improve resolution, only small amount of IR beam is allowed to pass through the slits	Much larger beam aperture used; higher energy throughput (Throughput or *Jacquinot's advantage*)
Only narrow-frequency radiation falls on the detector at any one time	All frequencies of radiation fall on detector simultaneously; improved S/N ratio obtained quickly (Multiplex or *Fellgett's advantage*)
Slow scan speeds make dispersive instruments too slow for monitoring systems undergoing rapid change (e.g., GC effluents)	Rapid scan speeds permit monitoring samples undergoing rapid change
Sample subject to thermal effects from the source due to length of scan time	Short scan times, hence sample is not subject to thermal effects
Any emission of IR radiation by sample will fall on detector due to the conventional positioning of the sample before the monochromator	Any emission of IR radiation by sample will not be detected
Double-beam optics permit continuous real-time background subtraction	Single-beam optics; background spectrum collected separately in time from sample spectrum. Can result in error if background spectra not collected frequently

sensitivity limits the improvement. For example, 64 scans improve sensitivity $8\times$. It would require 4096 scans to increase the S/N 64-fold. A practical limit of one to two orders of magnitude sensitivity increase is therefore normal unless circumstances merit the additional time. Of course, the ability to process many spectra rapidly is a result of the improvement in computer hardware and software that has occurred over the past decade or so. Inexpensive powerful computers and commercially available user-friendly software allow this technology to be used in undergraduate laboratories as well as in industrial and academic research labs.

4.2.2.2. Interferometer Components

The schematic interferometer diagrams given do not show most of the optics, such as beam collimators and focusing mirrors. Mirrors in an FTIR are generally made of metal. The mirrors are polished on the front surface and may be gold-coated to improve corrosion resistance. Commercial FTIRs use a variety of flat and curved mirrors to move light within the spectrometer, to focus the source onto the beam splitter, and to focus light from the sample onto the detector

The beam splitter can be constructed of a material such as Si or Ge deposited in a thin coating onto an IR-transparent substrate. The germanium or silicon is coated onto the highly polished substrate by vapor deposition. A common beam splitter material for the mid-IR region is germanium and the most common substrate for this region is KBr. Both the substrate and the coating must be optically flat. KBr is an excellent substrate for the mid-IR region because of its transparency and its ability to be polished flat. Its major drawback is that it is hygroscopic; this limits the use of KBr as a substrate for field or process control instruments, where environmental conditions are not as well controlled as laboratory conditions. Germanium on KBr is also used for the long wavelength end of the NIR region, while Si coated on quartz can be used for the short wavelength end of the NIR region. A thin film of mylar is used as a beam splitter for the far-IR region. Other combinations of coatings and substrates are available, including complex multilayer materials, especially for applications where moisture may limit the use of KBr.

Ideally, the beam splitter should split all wavelengths equally, with 50% of the beam being transmitted and 50% reflected. This would result in equal intensity at both the fixed and moving mirrors. Real beam splitters deviate from ideality.

As noted earlier and discussed subsequently under background correction in IR spectroscopy, CO_2 and H_2O absorb IR radiation (Fig. 4.14). To reduce the spectral background from CO_2 and H_2O and increase the light intensity in the regions where these gases absorb, many spectrometers have sealed and desiccated optical systems. Only a small air path in the sample compartment remains. Alternately, some spectrometers allow the optics and the sample path to be purged with dry nitrogen or other dry gas, decreasing the H_2O and CO_2 in the light path.

IR spectrometers must be calibrated for wavelength accuracy. FTIRs are usually calibrated by the manufacturer and checked on installation. Wavelength calibration can be checked by the analyst by taking a spectrum of a thin film of polystyrene, which has well-defined absorption bands across the entire mid-IR region, as seen in Fig. 4.1. Polystyrene calibration standard films are generally supplied with an IR instrument or can be purchased from any instrument manufacturer. Recalibration of the spectrometer should be left to the instrument service engineer if required.

4.2.3. Detectors

Detectors for IR radiation fall into two classes: thermal detectors and photon-sensitive detectors. Thermal detectors include *thermocouples*, *bolometers*, *thermistors*, and

Figure 4.14 A background spectrum of air, showing the absorption bands due to water vapor and carbon dioxide. Collected on a Paragon 1000 FTIR spectrometer, PerkinElmer Instruments, Shelton, CT (www.perkinelmer.com).

pyroelectric devices. Thermal detectors tend to be slower in response than photon-sensitive semiconductors. The most common types of detectors used in dispersive IR spectroscopy were bolometers, thermocouples, and thermistors, but faster detectors are required for FTIR. FTIR relies on pyroelectric and photon-sensitive semiconducting detectors. Table 4.5 summarizes the wavenumber ranges covered by commonly used detectors.

4.2.3.1. Bolometer

A bolometer is a very sensitive electrical resistance thermometer that is used to detect and measure weak thermal radiation. Consequently, it is especially well suited as an IR detector. The bolometer used in older instruments consisted of a thin metal conductor, such as platinum wire. Incident IR radiation heats up this conductor, which causes its electrical resistance to change. The degree of change of the conductor's resistance is a measure of the amount of radiation that has fallen on the detector. In the case of platinum, the resistance change is 0.4% per °C. The change in temperature depends on the intensity of incident radiation and on the thermal capacity of the detector. It is important to use a small detector and to focus the radiation on it. The *rate* at which the detector heats up or cools down determines how fast the detector responds to a change in radiation intensity as experienced when an absorption band is recorded. This constitutes the *response time* of the detector. For these older types of bolometers, the response time is long, on the order of seconds. Consequently, a complete mid-IR scan using a dispersive instrument and bolometer could take 20 min.

Modern bolometers are micro-machined from silicon. This type of bolometer is only a few micrometers in diameter and is usually placed in one arm of a Wheatstone bridge for measurements. The modern micro-bolometer has a fast response time and is particularly useful for detecting far-IR radiation ($600-20$ cm^{-1}).

Table 4.5 Detectors for IR Spectroscopy

Near-IR (12,000–3800 cm^{-1}; 0.8–3 μm)	Mid-IR (4000–400 cm^{-1}; 2.5–25 μm)	Far-IR (400–20 cm^{-1}; 25–500 μm)
InGaAs (12,000–6,000 cm^{-1})	MCT (11,700–400 cm^{-1})	DTGS/PE (700–50 cm^{-1})
PbSe (11,000–2,000 cm^{-1})	DTGS/KBr (12,000–350 cm^{-1})	Si bolometer (600–20 cm^{-1})
InSb (11,500–1,850 cm^{-1})	Photoacoustic (10,000–400 cm^{-1})	
MCT (11,700–400 cm^{-1})	DTGS/CsI (6,400–200 cm^{-1})	
DTGS/KBr (12,000–350 cm^{-1})		

Note: KBr, CsI, and PE (polyethylene) are the window materials for the DTGS detectors. The MCT detector can vary in its spectral range depending on the stoichiometry of the material.
Source: Data courtesy of ThermoNicolet, Madison, WI.

Figure 4.15 Schematic diagram of a thermocouple.

4.2.3.2. *Thermocouples*

A thermocouple is made by welding together at each end two wires made from different metals (Fig. 4.15). If one welded joint (called the *hot junction*) becomes hotter than the other joint (the *cold junction*), a small electrical potential develops between the joints. In IR spectroscopy, the cold junction is carefully screened in a protective box and kept at a constant temperature. The hot junction is exposed to the IR radiation, which increases the temperature of the junction. The potential difference generated in the wires is a function of the temperature difference between the junctions and, therefore, of the intensity of IR radiation falling on the hot junction. The response time of the thermocouple detector is slow; thermocouples cannot be used as detectors for FTIR due to their slow response.

4.2.3.3. *Thermistors*

A thermistor is made of a fused mixture of metal oxides. As the temperature of the thermistor increases, its electrical resistance decreases (as opposed to the bolometer). This relationship between temperature and electrical resistance allows thermistors to be used as IR detectors in the same way as bolometers. The thermistor typically changes resistance by about 5% per °C. Its response time is also slow.

4.2.3.4. *Golay Detector*

The Golay detector was a pneumatic detector, a small hollow cell filled with a nonabsorbing gas such as xenon. In the center of the cell was a blackened film. Radiation was absorbed by the blackened film, causing an increase in temperature. In turn, the film heated the enclosed gas. Thermal expansion of the gas caused the internal pressure of the cell to increase. One wall of the cell was a thin convex mirror that was part of the optical system. As the pressure inside the cell increased, the mirror bulged. A change in radiation intensity falling on the Golay detector caused a change in the readout from the detector. An important advantage of this detector was that its useful wavelength range was very wide. The response was linear over the entire range from the UV through the visible and IR into the microwave range to wavelengths about as long as 7.0 mm. The Golay detector's response time was about 10^{-2} s, much faster than that of the bolometer, thermistor, or thermocouple. The detector was very fragile and subject to mechanical failure. While this detector is no longer in use as an IR detctor, a variation of the Golay detector, the photoacoustic detector, is used in photoacoustic spectroscopy, discussed later in this chapter.

4.2.3.5.   Pyroelectric Detectors

Pyroelectric materials change their electric polarization as a function of temperature. These materials may be insulators (dielectrics), ferroelectric materials, or semiconductors. A dielectric placed in an electrostatic field becomes polarized with the magnitude of the induced polarization depending on the dielectric constant. The induced polarization generally disappears when the field is removed. Pyroelectric materials, however, stay polarized and the polarization is temperature dependent.

A pyroelectric detector consists of a thin single crystal of pyroelectric material placed between two electrodes. It acts as a temperature-dependent capacitor. Upon exposure to IR radiation, the temperature and the polarization of the crystal change. The change in polarization is detected as a current in the circuit connecting the electrodes. The signal depends on the rate of change of polarization with temperature and the surface area of the crystal. These crystals are small; they vary in size from about 0.25 to 12.0 mm^2.

The most common pyroelectric material in use as an IR detector is deuterated triglycine sulfate (DTGS). The formula for triglycine sulfate is $(NH_2CH_2COOH)_3 \cdot H_2SO_4$; replacement of hydrogen with deuterium gives DTGS. DTGS with a cesium iodide window covers the $6400-200$ cm^{-1} range, which includes part of the NIR, all of the mid-IR, and some of the far-IR regions. With the use of a polyethylene window, a DTGS detector can be used as a far-IR detector ($700-50$ cm^{-1}). Other pyroelectric detectors for the mid-IR region include lithium tantalate, $LiTaO_3$, and strontium barium niobate.

Pyroelectric materials lose their polarization above a temperature called their Curie point. For DTGS, this temperature is low. DTGS detectors are cooled by thermoelectric cooling to between 20°C and 30°C to prevent loss of polarization. Lithium tantalate has a much higher Curie temperature and does not require cooling, but is less sensitive than DTGS by about an order of magnitude. Lithium tantalate has a high linear response range, unlike DTGS. DTGS does not respond linearly over the IR frequency range. Its response is inversely proportional to the modulation frequency of the source, resulting in lower sensitivity at the high frequency end of the spectral range than at the low frequency end.

4.2.3.6.   Photon Detectors

Semiconductors are materials that are insulators when no radiation falls on them but become conductors when radiation falls on them. Exposure to radiation causes a very rapid change in their electrical resistance and therefore a very rapid response to the IR signal. The response time of a semiconductor detector is the time required to change the semiconductor from an insulator to a conductor, which is frequently as short as 1 ns. The basic concept behind this system is that absorption of an IR photon raises an electron in the detector material from the valence band of the semiconductor across a band gap into the conduction band, changing its conductivity greatly. In order to do this, the photon must have sufficient energy to displace the electron. IR photons have less energy than UV or visible photons. The semiconductors chosen as IR detectors must have band gaps of the appropriate energy. The band gap of the detector material determines the longest wavelength (lowest wavenumber) that can be detected.

Materials such as lead selenide (PbSe), indium antimonide (InSb), indium gallium arsenide (InGaAs), and mercury cadmium telluride (HgCdTe, also called MCT) are intrinsic semiconductors commonly used as detectors in the NIR and mid-IR regions. Cooling of these detectors is required for operation. MCT requires operation at 77 K and must be cooled with liquid nitrogen; other detectors such as InGaAs can operate without cooling, but show improved S/N if cooled to -30°C or so with thermoelectric cooling. For the far-IR region, extrinsic semiconductors such as Si and Ge doped with low levels of copper, mercury, or other dopants are used. The dopants provide the electrons for conductivity and control the response range of the detector. These doped germanium or silicon detectors must be cooled in liquid helium, but are sensitive to radiation with wavelengths as long as 200 μm. The spectral response curves of some semiconductor detectors are shown in Fig. 4.16. The MCT material used is nonstoichiometric, and can be represented as $Hg_{(1-x)}Cd_xTe$. The actual spectral range of an MCT detector can be varied by varying the Hg/Cd ratio.

Semiconductor detectors are very sensitive and very fast. The fast response time has permitted rapid-scan IR to become a reality, as is needed in FT spectrometers and coupled techniques such as GC-IR that generate transient peaks. The sensitivity of these detectors has opened up the field of microsampling, IR microscopy and on-line IR systems for process control.

4.2.4. Detector Response Time

The length of time that a detector takes to reach a steady signal when radiation falls on it is called its *response time*. This varies greatly with the type of detector used and has a significant influence on the design of the IR instrument. Thermal detectors such as thermocouples, thermistors, and bolometers have very slow response times, on the order of seconds. Consequently, when a spectrum is being scanned, it takes several seconds for the detector to reach an equilibrium point and thus give a true reading of the radiation intensity falling on it. If the detector is not exposed to the light long enough, it will not reach equilibrium and an incorrect absorption trace will be obtained. It was normal for dispersive IR instruments with older-style thermal detectors to take on the order of 15 min to complete an IR scan. Attempts to decrease this time resulted in errors in the intensity of the absorption bands and recording of distorted shapes of the bands.

Figure 4.16 Spectral response of various semiconductor detectors. The operating temperature in kelvin is given next to the material.

The slow response time of thermal detectors is due to the fact that the detector temperature must change in order to generate the signal to be measured. When there is a change in radiation intensity, the temperature at first changes fairly rapidly, but as the system approaches equilibrium, the change in temperature becomes slower and slower and would take an infinitely long time to reach the true equilibrium temperature. It should also be remembered that when an absorption band is reached, the intensity falling on the detector decreases and the response depends on how fast the detector cools.

Semiconductors operate on a different principle. When radiation falls on them, they change from a nonconductor to a conductor. No temperature change is involved in the process; only the change in electrical resistance is important. This takes place over an extremely short period of time. Response times of the order of nanoseconds are common. This enables instruments to be designed with very short scanning times. It is possible to complete the scan in a few seconds using such detectors. These kinds of instruments are very valuable when put onto the end of a GC and used to obtain the IR spectra of the effluents. Such scans must be made in a few seconds and be completely recorded before the next component emerges from the GC column.

Response time is not the only detector characteristic that must be considered. Linearity is very important in the mid-IR region where wide variations in light intensity occur as a result of absorption by a sample. The ability of the detector to handle high light levels without saturating is also important. The MCT detectors saturate easily and should not be used for high intensity applications; DTGS, on the other hand, while not as sensitive as MCT, does not saturate as readily. DTGS can be used for higher intensity applications than MCT.

4.3. SAMPLING TECHNIQUES

IR spectroscopy is one of the few analytical techniques that can be used for the character-ization of solid, liquid, and gas samples. The choice of sampling technique depends upon the goal of the analysis, qualitative identification or quantitative measurement of specific ana-lytes, upon the sample size available, and upon sample composition. Water content of the sample is a major concern, since the most common IR-transparent materials are soluble in water. Samples in different phases must be treated differently. Sampling techniques are available for transmission (absorption) measurements and, since the advent of FTIR, for several types of reflectance (reflection) measurements. The common reflectance measurements are attenuated total reflectance (ATR), diffuse reflectance or diffuse reflectance infrared Fourier transform spectroscopy (DRIFTS), and specular reflectance. The term reflection may be used in place of reflectance and may be more accurate; specular reflection is actually what occurs in that measurement, for example. However, the term reflectance is widely used in the literature and will be used here.

4.3.1. Techniques for Transmission (Absorption) Measurements

These are the oldest and most basic sampling techniques for IR spectroscopy and apply to both FTIR and dispersive IR systems. Transmission analysis can handle a wide range of sample types and can provide both qualitative and quantitative measurements. Trans-mission analysis provides maximum sensitivity and high sample throughput at relatively low cost. There is in some cases substantial sample preparation required.

The sample or the material used to contain the sample must be transparent to IR radi-ation to obtain an absorption or transmission spectrum. This limits the selection of container materials to certain salts, such as NaCl or KBr, and some simple polymers. A final choice of the material used depends on the wavelength range to be examined. A list of commonly used materials is given in Table 4.3. If the sample itself is opaque to IR radiation, it may be possible to dissolve it or dilute it with an IR-transparent material to obtain a transmission spectrum. Other approaches are to obtain IR reflectance spectra or emission spectra from opaque materials.

4.3.1.1.   Solid Samples

Three traditional techniques are available for preparing solid samples for collection of transmission IR spectra: mulling, pelleting, and thin film deposition. First, the sample may be ground to a powder with particle diameters less than 2 μm. The small particle size is necessary to avoid scatter of radiation. A small amount of the powder, 2–4 mg, can be made into a thick slurry, or *mull*, by grinding it with a few drops of a greasy, viscous liquid, such as Nujol (a paraffin oil) or chlorofluorocarbon greases. The mull is then pressed between two salt plates to form a thin film. This method is good for qualitative studies, but not for quantitative analysis. To cover the complete mid-IR region it is often necessary to use two different mulling agents, since the mulling agents have absorption bands in different regions of the spectrum. The spectrum of the mulling agents alone should be obtained for comparison with the sample spectrum.

The second technique is the KBr pellet method, which involves mixing about 1 mg of a finely ground (<2 μm diameter) solid sample with about 100 mg powdered dry potassium bromide. The mixture is compressed under very high pressure (>50,000 psi) in a vacuum to form a small disk about 1 cm in diameter. An evacuable die is designed for use in a hydraulic press. A die consists of a body and two anvils that will compress the powdered mixture. The faces of the anvils are highly polished to give a pressed pellet with smooth surfaces. A schematic of an evacuable die is shown in Fig. 4.17. When done correctly, the KBr pellet looks like glass. The disk is transparent to IR radiation and may be analyzed directly by placing it in a standard pellet holder. There are small,

Figure 4.17   Schematic drawing of a typical IR pellet die showing the arrangement of the major components. (Reprinted from Aikens et al. by permission from Waveland Press, Inc. Long Grove, IL, 1984. All rights reserved.)

hand-operated presses available for making KBr pellets, but the quality of the pellet obtained may not be as good as that obtained with an evacuable die. The pellet often will contain more water, which absorbs in the IR region and may interfere with the sample spectrum. There are several types of handheld presses available. A common design consists of two bolts with polished ends that thread into a metal block or nut. The nut serves as the body of the die and also as the sample holder. One bolt is threaded into place. The KBr mix is added into the open hole in the nut so that the face of the inserted bolt is covered with powdered mix. The second bolt is inserted into the nut. Pressure is applied using two wrenches, one on each bolt. The bolts are then removed; the KBr pellet is left in the nut and the nut is placed in the light path of the spectrometer. The pellet should appear clear; if it is very cloudy, light scattering will result, giving a poor spectrum. The pellet is removed by washing it out of the nut with water. One disadvantage of this type of die is that the pellet usually cannot be removed from the nut intact; if pellets need to be saved for possible reanalysis, a standard die and hydraulic press should be used. Micropellet dies are available that produce KBr pellets on the order of 1 mm in diameter and permit spectra to be obtained on a few micrograms of sample. A beam condenser is used to reduce the size of the IR source beam at the sampling point.

It is critical that the KBr be dry; even then bands from water may appear in the spectrum because KBr is so hygroscopic. The KBr used should have its IR spectrum collected as a blank pellet; reagent grade KBr sometimes contains nitrate, which has IR absorption bands. IR-grade KBr should be used when possible. The quality of the spectrum depends on having small particle size and complete mixing. A mortar and pestle can be used for mixing, but better results are obtained with a vibrating ball mill such as the Wig-L-Bug®. It is also very important that the polished faces of the anvils not be scratched. The anvils should never have pressure applied to them unless powdered sample is present to avoid scratching the polished faces.

In the third method, the solid sample is deposited on the surface of a KBr or NaCl plate or onto a disposable "card" by evaporating a solution of the solid to dryness or allowing a thin film of molten sample to solidify. IR radiation is then passed through the thin layer deposited. It is difficult to carry out quantitative analysis with this method, but it is useful for rapid qualitative analysis. The thin film approach works well for polymers, for example. It is important to remove all traces of solvent before acquiring the spectrum. Disposable salt "cards" are available for acquiring the IR spectrum of a thin film of solid deposited by evaporation. These cards have an extremely thin KBr or NaCl window mounted in a cardboard holder, but are manufactured so that atmospheric moisture does not pose a storage problem (Real Crystal™ IR cards, International Crystal Laboratories, Garfield, NJ). Water can even be used as the solvent for casting films of polar organic molecules on these cards.

A new approach to collecting transmission spectra of solids is the use of a *diamond anvil cell*. Diamond is transparent through most of the mid-IR region, with the exception of a broad absorption around $2000 \ cm^{-1}$. A solid sample is pressed between two small parallel diamond "anvils" or windows to create a thin film of sample. A beam condenser is required because of the small cell size. Very high pressures can be used to compress solid samples because diamonds are very hard materials. As a result, the diamond anvil cell permits transmission IR spectra to be collected of thin films of very hard materials. Hard materials cannot be compressed between salt windows because the salt crystals are brittle and crack easily.

In general, spectra from solid samples are used for qualitative identification of the sample, not for quantitative analysis. The spectrum of a solid sample is generally collected when the sample is not soluble in a suitable IR-transparent solvent. There are some problems that can occur with spectra from solid samples. Many organic solids are crystalline materials. The mull and pellet approaches result in random orientation of the finely ground crystals; deposition of thin films by evaporation may result in a specific crystal orientation with respect to the light beam. Hence, thin film spectra may look different from the spectrum of the same material collected as a mull or a pellet. When possible, spectra of known materials obtained by the same sample preparation method should be compared when trying to identify an unknown. Use of a high-pressure hydraulic press for KBr pellets may cause crystal structure changes in some materials; again, standards and samples should have the same sample preparation method used if spectra are to be compared.

4.3.1.2. Liquid Samples

The easiest samples to handle are liquid samples. Many liquids may be analyzed "neat", that is, with no sample preparation. Neat liquids that are not volatile are analyzed by pressing a drop of the liquid between two flat salt plates to form a very thin film. The salt plates are held together by capillary action or may be clamped together. NaCl, KBr, AgCl, and similar salts are used as the plates. Volatile liquids may be analyzed neat by using a pair of salt plates with a thin spacer in a sealed cell. The path length of these cells depends on the spacer thickness. For neat liquids very small path lengths, less than 0.05 mm, must be used to avoid complete absorption of the source beam. Sample sizes used for the collection of neat spectra are on the order of a few milligrams of material.

The use of dilute solutions of material for IR analysis is the preferred choice for several reasons. Solutions give more reproducible spectra, and dilution in an IR-transparent solvent eliminates the problem of total absorption by the strong bands in a neat sample. Solvents commonly used for IR spectroscopy include carbon tetrachloride, carbon disulfide, methylene chloride, and some alkanes such as hexane. No one solvent is transparent over the entire mid-IR region, so the analyst must choose a solvent that is transparent in the region of interest. Figure 4.18 shows the IR-transparent regions for

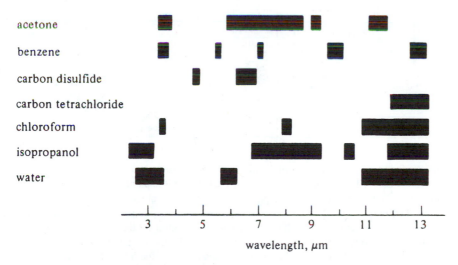

Figure 4.18 IR absorption characteristics of some common solvents. Regions of strong IR absorbance in 0.1 mm cells (except water, 0.01 mm cell) are shown as shaded areas. Longer cell paths will broaden the regions of absorption and in some cases introduce new regions where absorption is significant. (Reprinted from Aikens et al. by permission from Waveland Press, Inc. Long Grove, IL, 1984. All rights reserved.)

some common solvents. Liquid cells for solutions are sealed cells, generally with a path length of 0.1–1 mm and two salt windows. The path length is fixed by a spacer placed between the two salt windows. Some cells come with a single fixed path length; other cells can be purchased with a variety of spacers. These cells can be disassembled and the path length changed by inserting a different spacer [Fig. 4.19(a)]. The windows and spacer are clamped into a metal frame that has two ports: one inlet and one outlet port. The cell is filled by injecting sample solution with a syringe into one port and allowing it to flow until the solution exits the other port. Solution concentrations of 1–10% sample are used for most quantitative work. Solvent absorption peaks are compensated for in a double-beam dispersive IR by using matched cells. One cell is used to contain the sample solution, and the other cell to contain the solvent used to make the solution. Matched cells have the same window material, window thickness, and path length. In the single-beam FTIR, solvent absorption bands are corrected for by obtaining a blank spectrum of the solvent and subtracting the blank spectrum from the sample solution spectrum, just as the background is subtracted. In this case, the same cell can be used for both the blank spectrum and the sample spectrum.

Figure 4.19   (a) Standard demountable cell for liquid samples, shown in an "exploded" view. The spacer is of Teflon or metal. The width of the spacer used determines the pathlength of the assembled cell. The nuts screw onto the four threaded posts to seal the assembled cell. Once the cell is assembled, it is filled via syringe. The inlet port on the back plate is equipped with a fitting for a syringe (not shown); the outlet port is the hole in the back plate opposite the inlet hole. Sample is injected until the liquid appears at the top of the outlet port. Plugs are put into both inlet and outlet ports to seal the cell. Courtesy of PerkinElmer Instruments, Shelton, CT (www.perkinelmer.com). (b) and (c) show the absorbance spectra for two commercial disposable IR cards with polymer film windows. The choice of polymer depends on the region of the spectrum to be studied. PTFE (c) would be used if the C—H stretch region needs to be measured, while clearly polyethylene (b) is not suited for that use. [Courtesy of ThermoNicolet, Madison, WI (www.thermo.com).]

Most IR cells must be protected from water, because the salt plates are water soluble and hygroscopic. Organic liquid samples should be dried over an appropriate drying agent before being poured into the cells; otherwise, the cell surfaces become opaque and uneven. Such etching of the internal window surfaces is frequently a serious problem, particularly when quantitative analyses are to be performed. Light scattering will occur, the path length within the cell becomes uneven and erroneous quantitative results may be obtained.

It will be remembered that Beer's Law indicates that the absorbance $= abc$, where b is the path length through the sample, or in this case the width of the empty cell. In order for quantitative data to be reliable, b must be a constant, or at least measurable and correctable. A measurement of b may be performed by using a procedure based on interference fringes. An empty and dry cell is put into the light path, and the interferogram collected (or a suitable wavelength range is scanned if a dispersive instrument is used). Partial reflection of the light takes place at the inner surfaces, forming two beams. The first beam passes directly through the sample cell, and the second beam is reflected twice by the inner surfaces before proceeding to the detector. The second beam therefore travels an extra distance $2b$ compared with the first beam. If this distance is a whole number of wavelengths $(n\lambda)$, then the two emerging beams remain in phase and no interference is experienced. However, if $2b = (n + 1/2)\lambda$, interference is experienced and the signal reaching the detector is diminished. The interference signal generated is a sine wave, and each wave indicates an interference fringe. The path length of the sample holder can be measured by using the formula.

$$b \ (\mu m) = \frac{n}{2\eta} \left(\frac{\lambda_1 \lambda_2}{\lambda_2 - \lambda_1} \right) \tag{4.9}$$

where n is the number of fringes; η, the refractive index of the sample (or air, if empty lightpath); and λ_1 and λ_2, the wavelengths between which the number of fringes is measured.

If λ is measured in μm, b also has units of μm. For example, if $n = 14$, $\lambda_1 = 2 \ \mu m$, and $\lambda_2 = 20 \ \mu m$, b can be calculated as:

$$b = \frac{14}{2} \left(\frac{2 \times 20}{20 - 2} \right) = 15.5 \ \mu m \ \text{(assuming that } \eta = 1)$$

For quantitative analysis it is necessary to measure the path length in order to use calibration curves obtained with the same cell but at different times. If the cell becomes badly etched, the interference pattern becomes noisy and the cell windows have to be removed and repolished.

IR spectra of samples containing water can be accomplished using special cells with windows of barium fluoride, silver chloride or KRS-5. These materials are not very water-soluble (see Table 4.3). However, a more useful technique is to measure attenuated total reflection (Section 4.3.3.1).

Disposable IR cards with a thin polymer film window are available for the qualitative analysis of liquids. (These cards were originally manufactured by 3M®, but are now available from International Crystal Laboratories, Garfield, NJ, and other suppliers.) Two polymer substrates are available: polytetrafluoroethylene for the 4000–1300 wavenumber region and polyethylene for the lower wavenumber region. The absorption spectra for these two materials are displayed in Fig. 4.19(b) and (c). A thin film can be deposited onto the polymer window by evaporation from solution or by smearing the liquid onto the polymer. A major advantage of these cards is that the polymer films do not dissolve in water; therefore samples containing water can be analyzed. Absorption bands from the polymer substrate are subtracted from the sample spectra by running a blank card spectrum.

Microcells are available for the analysis of as little as 0.5 μL of liquid sample. These microcells also require a beam condenser as described for solid microsamples.

4.3.1.3.   *Gas Samples*

Gas sample cells have windows of KBr and cell bodies made of glass or metal, along with two ports with valves for evacuating the cell and filling the cell from an external gas source. Gases are much more dilute than liquids or solids; a gas has many fewer molecules per unit volume than does a condensed phase. To compensate for the small concentration of sample molecules in a gas (the c term in Beer's Law), the gas cells have longer path lengths (b is increased). The sample cavity of an IR spectrometer is generally about 10 cm long. There are gas cells with a single-pass 10 cm path length, but most gas cells make use of multiple reflections to increase the effective path length. Commercial gas cells with effective path lengths of 2, 10, 20, 40, and up to 120 m are available. The IR beam is reflected multiple times from internal mirrors in the cell. Such a cell is shown schematically in Fig. 4.20, where the multiple reflections make the effective path length 5× longer than the actual physical length of the cell. A single-pass 10 cm cell requires about 50 torr of sample pressure to obtain a good IR spectrum. However, multiple reflection cells with long path lengths permit the analysis of ppm concentrations of gases. Gas cells are also used to obtain the vapor-phase spectrum of highly volatile liquids. A drop or two of liquid is placed in the cell, the valves are closed and the sample is allowed to come to equilibrium. The vapor phase spectrum of HCl (Fig. 4.2) was collected by placing a few drops of concentrated hydrochloric acid in a 10 cm gas cell with a glass body and KBr windows. The gas sample must not react with the cell windows or surfaces. Temperature and pressure must also be controlled if quantitative measurements are being made.

4.3.2.   Background Correction in Transmission Measurements

The two main sources of background absorption (i.e., absorption from material other than the sample) are the solvent used for liquid solutions and the air in the optical light path. In a conventional double-beam dispersive system, comparing the sample beam to the reference beam and recording the difference spectrum in real time automatically eliminate absorption from air and solvent. If the sample is a liquid solution, a matching liquid cell with pure solvent is placed in the reference beam. The absorption from the solvent and from the air is measured simultaneously and subtracted from the sample beam signal.

1 m

Figure 4.20   Schematic gas absorption cell. Reflection of the light beam through the cell makes the effective path length longer than the cell length.

However, FTIR is a single-beam system and both air and solvent contribute to the signal, so corrections must be made in several steps.

4.3.2.1. Solvent Absorption

The solvent absorption spectrum is measured by putting pure solvent in the liquid sample cell and recording its spectrum. This spectrum is stored by the computer under an assigned file name (e.g., Spectrum A). The cell (or an identical cell) is then filled with the sample solution in that solvent, its spectrum taken, recorded, and stored under a file name (e.g., Spectrum B). Spectrum A is then subtracted from Spectrum B, giving the net spectrum of the sample. Of course, in this approach, any absorption by the air is also measured in both Spectrum A and Spectrum B, so the absorption by air is also corrected for.

4.3.2.2. Air Absorption

Gaseous CO_2 and H_2O vapor are both strong IR absorbers. Both occur in air in the optical light path and contribute to any IR absorption signal measured. This background signal may be corrected for in one of two ways. First, the air spectrum may be recorded by running a spectrum with no sample present. This constitutes the "blank" spectrum and is recorded and stored as a file (usually called BLANK). Samples of any type—mulls, pellets, or neat liquids—may be run and their total spectrum (air + sample) recorded and stored. The blank (air) spectrum is then subtracted by the computer, leaving the net sample spectrum. Any suspected changes in humidity or CO_2 content can be corrected by updating the blank spectrum at regular intervals. This is an easy and rapid measurement for an FTIR, and in routine analysis, the background spectrum should be collected and the file updated on a regular basis. The second method, *purging the optical path*, is more difficult. The optical system can be purged with dry N_2 or argon, removing CO_2 and H_2O in the process. This eliminates the necessity of correcting for the blank signal derived from impurities in the air if done effectively. However, the ease of collection and subtraction of the background with modern FTIR systems and the difficulty of purging the sample compartment completely makes the first option the more common approach.

A typical background spectrum of air taken by an FTIR spectrometer is shown in Fig. 4.14. The bands above $3000 \, cm^{-1}$ and between 1400 and $1700 \, cm^{-1}$ are due to water; the main CO_2 band is the band at about $2350 \, cm^{-1}$. The FTIR is a single-beam system; this background spectrum is collected and stored for subtraction from all subsequent sample spectra. However, the absorption of the source intensity by carbon dioxide and water reduces the energy available in the regions where they absorb. To reduce the spectral background from carbon dioxide and water and increase the light intensity in these regions, as already noted, many spectrometers have sealed and desiccated optical systems or a means of purging the optical path to remove the air.

4.3.3. Techniques for Reflectance and Emission Measurements

The sample techniques just described are designed for collection of transmission (absorption) spectra. This had been the most common type of IR spectroscopy, but it was limited in its applications. There are many types of samples that are not suited to the conventional sample cells and techniques just discussed. Thick, opaque solid samples, paints, coatings, fibers, polymers, aqueous solutions, samples that cannot be destroyed such as artwork or forensic evidence samples, hot gases from smokestacks—these materials posed problems for the analytical chemist who wanted to obtain an IR absorption spectrum. The use of reflectance techniques provides a nondestructive method for obtaining IR spectral information from materials that are opaque, insoluble, or cannot be placed into conventional sample cells. In addition, IR emission from heated samples can be used to characterize certain types of samples and even measure remote sources such as smokestacks. In reflectance and emission, the FTIR spectrometer system is the same as that for transmission. For reflectance, the sampling accessories are different and in some specialized cases contain an integral detector. The heated sample itself provides the light for emission measurements; therefore there is no need for an IR source. There may be a heated sample holder for laboratory emission measurements.

4.3.3.1. Attenuated Total Reflectance (ATR)

ATR or internal reflectance uses an optical element of high refractive index. This optical element is called the internal reflection element (IRE) or the ATR crystal. Light traveling in a high refractive index material is reflected when it encounters an interface with a material of a lower refractive index. The amount of light reflected depends upon the angle of incidence at the interface. When the angle of incidence is greater than the *critical angle*, which depends on the ratio of the two refractive indices, complete reflection of light occurs at the interface (i.e., total internal reflection). If a sample of material, such as a squishy polymer or rubber or a liquid is placed directly against the IRE, an interface is formed. The light beam traveling in the IRE will be completely reflected at the interface if the critical angle condition is met, but the beam of light actually penetrates a very short distance (generally less than 2 μm) into the lower refractive index material (in this case, the sample). This penetrating beam is called an **evanescent wave**. If the sample cannot absorb any of the light, all of the intensity is reflected. If the sample can absorb part of the light, the light beam is **attenuated**, that is, reduced in intensity, at the frequencies where the sample absorbs. This is the basis of the ATR sampling technique. A schematic representation of a multiple reflection ATR crystal is shown in Fig. 4.21. The interaction of the evanescent wave with the sample essentially provides an IR absorption spectrum of the sample.

Typical ATR crystal materials are listed in Table 4.6. Samples must be in actual intimate physical contact with the ATR crystal. The first ATR systems were designed to analyze solids that could be pressed against the surface of the crystal: polymers, films, moldable resins, textiles, canvas paintings, and the like. Little or no sample preparation is required. For example, the IR spectrum of a valuable painting could

Figure 4.21 Schematic ATR sampling accessory. The internal reflection crystal permits multiple reflections. At each reflection a small amount of IR energy penetrates the sample and absorption occurs at the vibrational frequencies for the sample. (Courtesy of Pattacini, Pattacini Associates, LLC, Danbury, CT.)

reflectance measurements in the NIR region. A commercial IR integrating sphere for NIR diffuse reflectance measurements is diagrammed in Fig. 4.23.

The diffuse reflectance experiment requires that the incident beam penetrate into the sample, but the path length is not well defined. The path length varies inversely with the sample absorptivity. The resulting spectrum is distorted from a fixed path absorbance spectrum and is not useful for quantitative analysis. Application of the Kubelka–Munk equation is a common way of making the spectral response linear with concentration.

The Kubelka–Munk relationship is:

$$f(R_\infty) = \frac{(1 - R_\infty)^2}{2R_\infty} = K'C \tag{4.10}$$

where R_∞ is the ratio of the sample reflectance spectrum at infinite sample depth to that of a nonabsorbing matrix such as KBr, K' is a proportionality constant, and C is the concentration of absorbing species. The Kubelka–Munk equation gives absorbance-like results for diffuse reflectance measurements, as can be seen by comparing it to Beer's Law, $A = abc = Kc$ for a fixed path length. In Beer's Law, K is a proportionality constant based on the absorption coefficient and the pathlength. K' is also a proportionality constant, but based on the ratio of absorption coefficient to scattering coefficient. The term $f(R_\infty)$ can be considered a "pseudoabsorbance".

4.3.3.4. IR Emission

Some samples are not amenable to transmission/absorption or reflectance spectroscopy. Samples can be characterized by their IR emission spectrum under certain conditions. If the sample molecules are heated, many will occupy excited vibrational states and will emit radiation upon returning to the ground state. The radiation emitted is characteristic of the vibrational levels of the molecule, that is, the IR spectrum, and can be used to identify the emitting sample. The IR emission from the sample is directed into the spectrometer instead of the usual IR light source. Very small samples can have their IR emission collected with an IR microscope, discussed later in this chapter. Large, physically remote samples can be imaged with a telescope arrangement and the emitted light directed into the spectrometer.

Figure 4.23 Schematic diagram of an NIR integrating sphere for DR. The sphere is placed in the sample compartment of a Hitachi model U-3410 dispersive UV/VIS/NIR spectrophotometer. The sphere design permits only diffuse reflectance to reach the detector; the specular component is reflected out through the same opening the light enters. [Courtesy of Hitachi High Technologies America, Inc., San Jose, CA (www.hitachi-hhta.com).]

IR emission can be used in the laboratory to study heated samples. Most modern research grade instruments offer an emission sampling port as an option. One significant advantage of IR emission spectrometry is that the sample can be remote—such as the emission from a flame or smokestack. Some typical applications of IR emission include analysis of gases, remote flames and smokestacks or other hot discharges, process measurements, photochemical studies, and the analysis of thin films and coatings. IR emission measurements are nondestructive and do not suffer from atmospheric background problems, since the room temperature water and carbon dioxide in air do not emit radiation. The major limitation is that thick samples cannot be measured due to reabsorption of the emitted radiation by cool parts of the sample.

4.4. FTIR MICROSCOPY

FTIR instruments with sensitive MCT detectors have permitted the development of the IR microscope, which extends IR spectroscopy to the examination of very small samples with detection limits up to two orders of magnitude better than can be achieved with dispersive instruments. An IR microscope uses two light beams, one visible and the other IR, that travel through the microscope optics to the sample following identical paths, as shown in Fig. 4.24. The sample is viewed optically and the exact region to be studied is centered and focused using the microscope controls. In some microscope designs, the visible beam is then moved out of the light path and the IR beam is moved in. Microscopes designed with dichroic optics allow both beams to reach the sample so that the analyst can view the sample while the IR spectrum is collected. It is possible to collect an IR spectrum, in either transmission or reflectance mode, from an area as small as 10 μm in diameter. The IR signal from the sample passes to a dedicated MCT detector designed for small samples.

Figure 4.24 IR microscope schematic with the detector integrated into the microscope. The microscope is usually coupled to a light port on the side of the FTIR spectrometer. The FTIR spectrometer supplies a modulated, collimated beam of light to the microscope. Courtesy of PerkinElmer Instruments, Shelton, CT (www.perkinelmer.com). (From Coates, used with permission.)

To obtain a transmission spectrum, the sample must be prepared. A microtome is used for cutting a very thin slice of the sample through which radiation can penetrate. Sample thickness must be in the range of 15 μm and the sample must be flat. The quality of the spectrum depends on the sample preparation. All of the reflectance modes are available for microscopy, including ATR and grazing angle analysis. These generally require little or no sample preparation. The sensitivity obtainable is subnanogram quantities.

Modern FTIR microscopes are available with computer-controlled microscope stages and video imaging systems that permit a 2D picture of the sample to be displayed, and areas containing a specified functional group to be highlighted using "false color" to show differences in composition with respect to position in the sample. Microscopes are available that allow the use of polarized light for imaging and that can obtain fluorescence images. These are useful to improve the contrast in samples that lack features under normal illumination.

A prime example of the use of FTIR microscopy is in the examination of polymers, a very important class of engineering materials. The physical properties of polymers are very dependent on their molecular structure. The presence of impurities, residual monomers, degree of crystallinity, size, and orientation of crystalline regions (the microstructure of a polymer) greatly affects their mechanical behavior. FTIR microscopy can identify polymers, additives, and determine the presence of impurities.

Food-packaging materials may be made up of several layers of different polymers, called a laminate, to provide a single plastic sheet with the desired properties. Typical layers are between 10 and 200 μm thick. Using an automated FTIR microscope it is possible to obtain acceptable spectra from each layer and identify the polymers involved. As an example, a cross-section of a polymer laminate, compressed between NaCl plates, is shown is shown in Fig. 4.25. Three layers were seen under magnification. The sample was moved in a straight line, as shown, and IR spectra were collected every 2 μm across the sample. The spectra collected from the laminate can be displayed in a variety of formats, such as the "waterfall display" presented in Fig. 4.26. This display gives the analyst a very clear picture of the differences in the three layers. If we look at the band on the left, between 3200 and 3400 cm^{-1}, we see that it is high in the layer plotted at the front of the display, as we move back (along the sample), we reach the thin middle

Figure 4.25 Micrograph of a polymer laminate, showing two broad layers and a narrow middle layer. The sample was mounted in an NaCl compression cell and spectra were collected automatically in 2 μm steps along the white line indicated. A Centaurμs Analytical FTIR Microscope System from ThermoNicolet was used for the automatic data collection and results presented in this figure and in Figs. 4.26–4.28. [Courtesy of ThermoNicolet, Madison, WI (www.thermonicolet.com).]

Figure 4.26 The spectra collected from the polymer laminate are displayed as a function of position along the sample, in a "waterfall display". The chemical differences in the layers are clearly seen. For example, the top layer has a large broad peak at about 3400 cm^{-1} (the peak on the far left); that peak disappears as the middle and bottom layers are scanned. [Courtesy of ThermoNicolet, Madison, WI (www.thermonicolet.com).]

layer. Note the band is still there, but much less intense. Then, moving back into the third layer, the band disappears. The same thing happens to the intense band at about 1700 cm^{-1}. Three distinct IR spectra were obtained, one from each layer (Fig. 4.27). The front layer was identified as a polyamide polymer by matching its spectrum to a known spectrum. The back layer is identified as polyethylene—note that this spectrum does not show the bands seen in the polyamide spectrum at 3400 and 1700 cm^{-1}. We will come back to these spectra later in the chapter. The middle layer was not immediately identified. A search of a computerized IR spectral library matched the spectrum of the middle layer to a urethane alkyd, as shown in Fig. 4.28. Figure 4.28 shows what a spectral search routine does-it picks a series of possible "fits" to the unknown from its database and

Figure 4.27 One spectrum from each layer is displayed. The spectrum from the top layer matches that of a polyamide; that on the bottom is the spectrum of polyethylene. The middle spectrum has not yet been identified. [Courtesy of ThermoNicolet, Madison, WI (www.thermonicolet.com).]

Index	Match	Compound Name	Library Name
1 175	77.00	Urethane alkyd	HR Hummel Polymer and Additives
2 1767	76.48	PENTAERYTHRITOL MONORICINOLEATE #2	HR Polymer Additives and Plasticizers
3 1768	76.16	PENTAERYTHRITOL TETRARICINOLEATE	HR Polymer Additives and Plasticizers
4 1597	75.35	PROPYLENE GLYCOL MONOOLEATE	HR Polymer Additives and Plasticizers
5 1769	74.91	PROPYLENE GLYCOL MONORICINOLEATE	HR Polymer Additives and Plasticizers

Figure 4.28 The results of a library search of the spectrum from the middle layer of the laminate. The top spectrum is that collected from the sample; the bottom spectrum, urethane alkyd, is the best match found in the search of a polymer database. Other possible compounds are suggested by the search routine and listed in the box below the spectra. Note the match number—the higher the number, the better the agreement between the sample spectrum and the library spectrum. [Courtesy of ThermoNicolet, Madison, WI (www.thermonicolet.com).]

assigns a goodness-of-fit or match number. In this case, the urethane alkyd spectrum has the highest match number, 77, of the spectra in this database.

In forensic science, FTIR microscopy has been used to examine paint chips from automobile accidents. An example of a paint chip spectrum is shown in Fig. 4.29. Hit-and-run drivers frequently leave traces of paint on cars with which they collide. Identification of the paint can help to identify the car. Other uses of an IR microscope in forensic analysis include the examination of fibers, drugs, and traces of explosives.

Figure 4.29 Transmission spectrum of a blue paint chip from an American car measured using a miniature diamond anvil cell. [Courtesy of PerkinElmer Instruments, Shelton, CT (www.perkinelmer.com).]

IR microscopy is used in the characterization of pharmaceuticals, catalysts, minerals, gemstones, adhesives, composites, processed metal surfaces, semiconductor materials, fossils, and artwork. Biological samples such as plant leaves and stems, animal tissue, cells, and similar samples can be imaged. Frequently, such information cannot be obtained by any other means. A microscope that combines both IR and Raman measurements will be discussed in the section on Raman spectroscopy.

4.5. NONDISPERSIVE IR SYSTEMS

In industry it is often necessary to monitor the quality of a product on a continuous basis to make certain the product meets its specifications. This on-line, real-time approach to analysis is called **process analysis**. IR spectroscopy is often the method of choice for process monitoring of organic chemical, polymer, and gas production. It is usually not feasible to use laboratory IR instruments under production conditions because they are too delicate, too big, and too expensive. Nondispersive systems have therefore been developed that are much sturdier and can be left running continuously. Many nondispersive systems have been designed for the NIR region. These will be discussed in Section 4.7. The mid-IR region is used mainly for monitoring gas streams.

Nondispersive IR spectrometers may use filters for analysis of gaseous substances. Each filter is designed to measure a specific compound. Figure 4.30 presents a commercial filter photometer for the mid-IR region with a filter wheel containing multiple narrow bandpass filters. The compound measured is selected by turning the wheel to put the proper filter in the light path. Other photometers have been designed for dedicated measurement of a single gaseous species. A schematic diagram of such a dedicated nondispersive IR instrument is shown in Fig. 4.31. The system consists essentially of the radiation source and two mirrors that reflect two beams of light, which pass through the sample and reference cells, respectively, to two detectors. These detectors are transducers similar in design to the Golay detector; each contains the gas phase of the compound being determined. The detector is therefore selective for the compound. For example, imagine that the two detectors are filled with gas phase carbon tetrachloride. If there is no carbon tetrachloride vapor in the sample call, detectors A and B will absorb the IR radiation equally and consequently their temperatures will increase. The temperature difference between the two detectors is measured and is at a minimum when there is no sample in the light path. When carbon tetrachloride vapor is introduced into the sample cell, it absorbs radiation. The light falling on detector A is therefore decreased in intensity and the temperature of the detector decreases. The temperature

Figure 4.30   A schematic of a 2-filter wheel, multiwavelength filter photometer. (From Coates, used with permission.)

Figure 4.31 Schematic of a positive-filter nondispersive IR system.

difference $T_2 - T_1$ between the two detectors increases. As the concentration of sample increases, the temperature of detector A decreases and $T_2 - T_1$ increases. The relationship between $T_2 - T_1$ and sample concentration is positive in slope. The reference cell may be a sealed cell containing N_2 gas, which does not absorb in the IR, or a flow-through cell that is purged with a nonabsorbing gas. Clearly, the system can be made specific for any IR-absorbing gas by putting that vapor in the detectors. The response of the system is usually better if the light beams are chopped, and it is common to chop the light as it leaves the source so that the sample and reference beams are chopped equally. This provides an ac signal that helps correct for changes in room temperature during operation, instrument drift, and other types of noise in the system.

A problem may be encountered if an interfering material is present in the sample that has absorption bands overlapping those of the sample. This will result in a direct interference in the measurement. The problem can be overcome by placing a cell containing a pure sample of the interfering material in the sample arm. In this fashion all the absorbable radiation at a common wavelength is absorbed and this eliminates any variation in absorption due to the impurity in the sample because all the light at this wavelength has been removed.

Nondispersive IR systems are good for measuring concentrations of specific compounds under industrial and other similar circumstances. As can be readily understood, they are not generally used as research instruments and do not have scanning capabilities. However, they are robust and enduring and can be used for the continuous monitoring of specific compounds. Nondispersive systems are very common in NIR applications, as discussed in Section 4.7.

4.6. ANALYTICAL APPLICATIONS OF IR SPECTROSCOPY

The two most important analytical applications of IR spectroscopy are the qualitative and quantitative analyses of organic compounds and mixtures. We pointed out at the beginning of this chapter that the frequencies of radiation absorbed by a given molecule are characteristic of the molecule. Since different molecules have different IR spectra that depend on the structure and mass of the component atoms, it is possible, by matching the absorption spectra of unknown samples with the IR spectra of known compounds, to identify the unknown molecule. Moreover, functional groups, such as $-CH_3$, $-C=O$, $-NH_2$, and

—OH, act almost as separate groups and have characteristic absorption frequencies relatively independent of the rest of the molecule they are part of. This enables us to identify many of the functional groups that are important in organic chemistry, and provides the basis for **qualitative structural identification** by IR spectroscopy. By examining the absorption spectra of an unknown sample and comparing the bands seen with the characteristic absorption frequencies of known functional groups, it is possible to classify the sample as, say, a ketone or a carboxylic acid very quickly, even if it is not possible to identify the compound exactly. In some cases, the structure of an unknown can be deduced from its IR spectrum, but this requires much practice and is not always possible, even for an expert. In addition to identifying a molecule, or its functional groups, we can acquire information about structural and geometrical isomers from IR spectroscopy. IR spectroscopy has been coupled to chromatography to identify separated compounds and to thermogravimetric analyzers to identify compounds or degradation products that volatilize as a sample is heated.

We can measure the extent of absorption at a specific frequency for an analyte of known concentration. If now we were to measure the extent of absorption at the same frequency by a sample solution of unknown concentration, the results could be compared. We could determine the sample's concentration using Beer's Law. Thus, as a *quantitative tool*, IR spectroscopy enables us to measure the concentration of analytes in samples.

The introduction and widespread use of FTIR has resulted in considerable extension of the uses of IR in analytical chemistry. With regard to wavelength assignment, speed of analysis, and sensitivity, FTIR has opened new fields of endeavor. Some of these uses are described.

Typical analyses include the detection and determination of paraffins, aromatics, olefins, acetylenes, aldehydes, ketones, carboxylic acids, phenols, esters, ethers, amines, sulfur compounds, halides, and so on. From the IR spectrum it is possible to distinguish one polymer from another, or determine the composition of mixed polymers, or identify the solvents in paints. Atmospheric pollutants can be identified while still in the atmosphere. Another interesting application is the examination of old paintings and artifacts. It is possible to identify the varnish used on the painting and the textile comprising the canvas, as well as the pigments in the paint. From this information fake "masterpieces" can be detected. Modern paints and textiles use materials that were not available when many masterpieces were painted. The presence of modern paints or modern synthetic fabrics confirms that the painting must have been done recently. In a similar manner, real antiques can be distinguished from modern imitations.

As already discussed, paints and varnishes are measured by *reflectance analysis*, a process wherein the sample is irradiated with IR light and the reflected light is introduced into an IR instrument. The paint, or other reflecting surface, absorbs radiation in the same manner as a traversed solution. This technique can be used to identify the paint on appliances or automobiles without destroying the surface. Scraps of paint from automobiles involved in wrecks can be examined. From the data obtained, the make and year of the car may be able to be determined.

In industry, IR spectroscopy has important uses. It is used to determine impurities in raw materials. This is necessary to ensure good products. It can be used for quality control by checking the composition of the product, either in batch mode or continuously (on-line or process analysis). On-line IR analyzers can be used to control the process in real time, a very cost-effective way of producing good products. IR spectroscopy is used in the identification of new materials made in industrial research laboratories and in the analysis of materials made or used by competitors (a process called "reverse engineering").

4.6.1.   Qualitative Analyses and Structural Determination by Mid-IR Absorption Spectroscopy

Qualitative analysis of unknown samples is a major part of the work of an analytical chemist. Since it is better to give no answer than an incorrect answer, most analytical chemists perform qualitative analysis using an array of techniques that overlap and confirm each other, providing in the sum more information than could be obtained with the separate individual methods. For qualitative analysis of an unknown organic compound the most commonly used spectroscopic methods are: IR spectroscopy to tell which functional groups are present; NMR to indicate the relative positions of atoms in the molecule and the number of these atoms; and MS to provide the molecular weight of the unknown and additional structural information. UV spectroscopy was used in the past to study unsaturated or substituted compounds; it has been almost entirely replaced for qualitative structural information by NMR and MS, which are commonly available in undergraduate chemistry labs. Each technique provides an abundance of valuable information on molecular structure, but a combination of methods is used to ensure more reliable identification. In addition to spectroscopy, real samples may be submitted to chromatography to determine if the unknown is a pure substance or a mixture, to determine the number of compounds present, and to separate and purify the compound of interest.

The value to qualitative analysis of prior knowledge about the sample cannot be overemphasized. Before trying to interpret an IR spectrum, it is important to find out as much as possible about the sample, For example, to identify the products of an organic reaction, it is very valuable to have information about the materials that were present before the reaction started, the compounds the reaction was expected to produce, the possible degradation products that may come about after the reaction, and so on. Armed with as much of this information as possible, we may be able to identify the molecules in the sample from their IR spectra.

The general technique for qualitative analysis is based on the characteristics of molecular structure and behavior mentioned at the beginning of the chapter. That is, the frequency of vibration of different parts of a molecule depends on the weight of the vibrating atoms (or groups) and the bond strength. Many groups can be treated as isolated harmonic oscillators and their vibrational frequencies calculated. More commonly, vibrational frequencies for functional groups are identified by the collection of spectra from hundreds of different compounds containing the desired functional group. These characteristic group vibrational frequencies are tabulated in correlation tables or correlation charts. Table 4.2 is a short list of functional groups and their relevant vibrational frequencies. Since the absorption frequency is the same as the vibration frequency, the presence of absorption at a given frequency is an indication that the functional group may be present. More tables are found later and very detailed tables are found in the references listed in the bibliography by Silverstein and Webster, Pavia et al., Lambert et al., Colthup et al., Dean, Robinson, and in the CRC Handbook of Chemistry and Physics.

Qualitative analysis is carried out by matching the wavelengths of the absorption bands in the spectrum of the sample with the wavelengths of functional groups listed in a correlation table. Before a positive identification can be made, all the absorption bands typical of the functional group must be observed. More importantly, the lack of an absorption band where one should be can be used to rule out certain functional groups. For example, as we will see later, if there is no strong absorption at about $1700 \, \text{cm}^{-1}$ due to the C=O stretch in a pure unknown compound, we can state with certainty that the compound does not contain a C=O group and therefore is not a ketone, aldehyde, amide, ester, or carboxylic acid.

Because the IR spectrum of each compound is unique, matching the IR spectrum of an unknown peak for peak to a reference spectrum of a known material is a very good way to identify the unknown. This is often done with the aid of computerized spectral libraries and search routines, as we saw for the polymer laminate (Fig. 4.28). A number of companies, instrument manufacturers, government agencies, and other sources publish collections of reference spectra in electronic format and in hardcopy. These spectral databases may contain spectra of more than 200,000 compounds, with subsets of the database available for specific fields of endeavor, such as environmental chemistry, pharmaceuticals, polymers, and forensic science. The unknown spectrum or some predetermined number of the strongest absorption bands from the unknown spectrum may be entered into a computerized search routine, which compares the unknown with stored spectra. It then retrieves all compounds from the database that may match the unknown spectrum, assigning a goodness-of-fit or probability to the suggested matches. The analyst then identifies the spectrum of the unknown based on spectral matching and chemical knowledge of the sample to rule out improbable compounds suggested by the search routine. A short list of reference spectra suppliers is located at the end of the bibliography. Most large spectral databases are expensive to buy. Many small companies or individuals can now access these by a "pay for what you use" approach. The KnowItAll^{TM} system from the Informatics Division, Bio-Rad Laboratories (www.bio-rad.com) and the FTIRsearch.com system from ThermoGalactic and ThermoNicolet (www.thermo.com or www.ftirsearch.com) are two examples of this very new and cost-effective approach to spectral matching. In addition, there are some free databases that allow the user to view spectra of known compounds. These sources include FTIR and FTNMR spectra from Sigma-Aldrich (www.sigma-aldrich.com), gas-phase IR spectra from NIST in the US (www.nist.gov), and a comprehensive spectral database, including IR spectra, from the Japanese National Institute of Advanced Industrial Science and Technology (www.aist.go.jp/RIODB/SDBS/menu-e.html). If a database or spectral library is not available or the spectra do not match exactly, the analyst must try to identify the compound from its spectrum. Even when electronic databases are available, it is useful for the analyst to understand how to look at and interpret an IR spectrum. This process is described subsequently for common classes of organic compounds. Only a limited number of examples of the most common types of compounds are discussed here. For detailed basic IR spectral interpretation, the texts by Colthup et al., Pavia et al., Lambert et al., or Silverstein and Webster should be consulted. The ability to interpret an IR spectrum and deduce molecular structure requires a great deal of practice and experience as well as detailed correlation tables and knowledge of organic chemistry and molecular geometry. But even experienced analysts never try to assign every peak in an IR spectrum!

The first region to look at is the $4000-1300 \text{ cm}^{-1}$ region, called the **functional group region**. This is the region where strong absorptions due to stretching from the hydroxyl, amine, carbonyl, and CH_x groups occur. The region also has areas of weak absorptions that are nonetheless very informative. The $2000-1660 \text{ cm}^{-1}$ region will show a set of weak overtone/combination bands if an aromatic ring is present. The intensity pattern of these weak bands can identify how the ring is substituted (i.e., ortho, meta, or para). Weak absorptions from triple bonds occur in this region, identifying alkynes (C≡C), cyano groups (C≡N), and diazonium salts (N≡N); also occurring in the region are absorptions by single bonded heteroatom groups such as S—H, Si—H, and P—H.

The region $1300-910 \text{ cm}^{-1}$ is called the **fingerprint region** because the complex absorption patterns are really what make the IR spectrum unique—a molecular "fingerprint". These absorptions are not easily interpreted because they arise from interactions between vibrations. There are some very important bands in this region, especially the

C—C—O band of alcohols and the C—O—C band of esters. These should be confirmed in conjunction with the appropriate bands (OH or C=O) in the functional group region of the spectrum.

The low frequency end of the spectrum, 910–650 cm^{-1}, is sometimes called the aromatic region. If there are no strong absorptions in this region, the structure is probably not aromatic. Strong absorptions in this region are due to oop bending of aromatic ring C—H bonds. Broad absorption bands in this region are usually due to nonaromatic amines and amides or due to carboxylic acid dimers. Below 800 cm^{-1}, absorption due to C—Cl and C—Br bonds occurs.

Before trying to interpret an IR spectrum, there are some things the analyst should note. The method of collecting the spectrum should be stated—mull, thin film, KBr pellet, solution, and the solvent—because the appearance of the spectrum may change as has been discussed earlier. The analyst should compare reference spectra collected under the same conditions when possible. Older spectra were printed with the spectrum displayed linear in wavelength (on the x-axis); modern grating and FTIR spectra are generally plotted linear in wavenumber. The two plots look very different for the same spectrum; for example, bands will appear to have expanded or contracted depending on their position in the spectrum. It is important that the analyst pay attention to the scale when comparing spectra from the older literature. It should also be noted that in grating IR spectra, the scaling changes at 5 μm. The spectrum should be of a pure compound (at least 95% pure when possible), should have adequate intensity and resolution, and should have been collected on a wavelength-calibrated instrument in order for the interpretation to be useful. The y-axis units should also be noted. Until recently, the y-axis for IR absorption spectra was "Transmittance" or "% Transmittance", with 100%T at the top of the spectrum. Transmittance is the ratio of radiant power transmitted by the sample to the radiant power incident on the sample, P/P_0 or I/I_0. 100%T is the transmittance multiplied by 100. Transmittance ranges from 0 to 1.0; %T ranges from 0 to 100. The absorption peaks therefore are pointing toward the bottom of the spectrum as printed. The y-axis could be given in A, where A is defined as $= -\log T$. If 0.00 absorbance is at the top of the y-axis, the spectrum is similar to the standard %T plot, but the contrast between strong and weak bands is not as good because A ranges from infinity to 0. However, it is becoming more common to see IR spectra plotted with A on the y-axis and with 0.00 A at the bottom of the y-axis, resulting in the peaks pointing up to the top of the plot. This is the inverse of the traditional %T spectrum. One reason for this is that Raman spectra are plotted with peak intensity increasing from bottom to top of the plot; having the complementary IR spectrum in the same format may make comparisons easier.

4.6.1.1. Hydrocarbons

Alkanes, alkenes, and alkynes. Figure 4.32 is the IR spectrum of n-hexane, a typical straight-chain hydrocarbon with only single C—C and C—H bonds. A hydrocarbon with all single C—C bonds is called an alkane or paraffin. All of the absorption bands in the spectrum of hexane or any other alkane must be due to the stretching or bending of C—H and C—C bonds, since these are the only types of bonds in the molecule. The C—C stretching vibrations are weak and are widely distributed across the fingerprint region; they are not useful for identification. The C—C bending vibrations generally occur below 500 cm^{-1}; they also are not useful for identification. Therefore in the IR spectrum of an alkane, all of the bands observed are due to stretching and bending of C—H bonds. The approximate frequencies of some common C—H vibrational modes are given in Table 4.7. The

Figure 4.32 Hexane, C_6H_{14}, a normal alkane. (Copyright Bio-Rad Laboratories, Informatics Division, Sadtler Software and Databases, 1999. All rights reserved.)

notation sym for symmetrical and asym for asymmetrical will be used in the charts. A horizontal arrow between atoms indicates a stretch, with C → H symbolizing a C—H stretch; a vertical arrow between two atoms indicates a bending mode, with C ↓ H used for a C—H bend. (The arrows do not specify the type of bend or the symmetry of the vibrational mode.) The relative strengths of the bands are not the same for all vibrations and the intensity of the absorptions provides valuable information about structure. The abbreviations for intensity used in the tables are: s, strong; m, medium; w, weak. Peaks not marked are of variable intensity depending on the structure of the molecule.

We can assign the peaks in the *n*-hexane spectrum by looking at the table. The band between 2840 and 3000 cm^{-1} contains peaks from the symmetric and asymmetric stretching of the methyl C—H and methylene C—H bonds. The band at about 1460 cm^{-1} is due to the overlapped asymmetric methyl bend and the methylene bending mode called scissoring. The symmetric methyl bend is located at 1375 cm^{-1}, while the peak at 725 cm^{-1} is due to the methylene "rocking" bending mode. This band only appears if there are four or more adjacent methylene groups in an acyclic chain. These peaks are seen in the spectrum of any straight-chain alkane, regardless of the number of C—C and C—H bonds. Mineral oil, also known as Nujol®, is a paraffin oil often used

Table 4.7 Alkanes

Functional group	Vibrational mode and strength	Wavenumber (cm^{-1})
Methyl (—CH$_3$)	C—H stretching, s	2870 (sym); 2960 (asym)
Methylene (—CH$_2$—)	C—H stretching, s	2860 (sym); 2930 (asym)
Methyl (—CH$_3$)	Bending, m	1375 (sym); 1450 (asym)
Methylene (—CH$_2$—)	Bending, m (scissoring, rocking)	1465; 720 (if four or more CH$_2$ groups in an open chain); weak bands in the 1100–1350 region
gem-Dimethyl[a]	Bending, m	Doublet (two peaks); 1380–1390 and 1365–1370
Aldehyde (—CHO)[b]	Aldehydic C—H stretching, m	2690–2830 (two peaks often seen)
Aldehyde (—CHO)[b]	Aldehydic C—H bending, m	1390

[a]The term *gem*-dimethyl refers to two methyl groups attached to the same carbon atom, such as in an isopropyl group.
[b]An aldehyde group has a hydrogen bonded to a carbon atom that also has double bond to the oxygen atom.

Figure 4.33 Cyclohexane, C_6H_{12}, a cyclic alkane. The structure shown on the spectrum is the typical shorthand notation—each point of the hexagon is a CH_2 group and the sides of the hexagon are the single covalent bonds between the six CH_2 groups. (Copyright Bio-Rad Laboratories, Informatics Division, Sadtler Software and Databases, 1999. All rights reserved.)

for preparing mulls of solids for IR analysis. The peaks seen in the IR spectrum of mineral oil are virtually identical to those in hexane. The other regions of the IR spectrum are "clear", so that other functional groups can be observed without interference.

Figure 4.33 shows the spectrum of cyclohexane, a cyclic alkane that has no methyl groups, only methylene groups. Cyclic alkanes show similar absorption peaks to the straight-chain (acyclic) alkanes. Note the differences in this spectrum compared with that of hexane. The C—H stretching band appears narrower (no broadening due to overlap of CH_3 stretches with the methylene stretches), the two peaks clearly match the methylene peaks listed in the table (lower wavenumber than the corresponding methyl peaks), there is no peak at 1375 cm^{-1} which confirms the absence of methyl groups, and there are two peaks at about 900 and 860 cm^{-1} due to ring deformation. The band at 725 cm^{-1} is also missing, because there is not an open long chain of methylene groups.

Aliphatic hydrocarbons containing one or more C=C double bonds are called alkenes or olefins. The IR spectra of alkenes contain many more peaks than those of alkanes. The major peaks of interest are due to the stretching and bending of the C—H and C=C bonds shown in Table 4.8, but the position and intensity of these vibrations is affected by the substituents on the C=C carbons. This includes the geometry of substitution, cis vs. trans, for example. In addition to the vibrations themselves, many of the strong vibrations give rise to overtones, adding to the complexity of the spectrum. A strong bending mode at 900 cm^{-1} may give rise to an overtone at about 1800 cm^{-1}.

Table 4.8 Alkenes and Alkynes

Functional group	Vibrational mode and strength	Wavenumber (cm^{-1})
C=C → H	Stretching, w	3000–3040
C=C ↓ H	Bending, s (oop)	650–1000
C=CH$_2$	Vinyl stretching, m	2990 (sym), 3085 (asym)
C=CH$_2$	Vinyl bending, s	910, 990
C≡C → H	Stretching, s	3285–3320
C≡C ↓ H	Bending, s	610–680
C⇒C	C=C stretching, s[a]	1600–1680
C⇒C	C≡C stretching, m-w[a]	2120 (terminal); 2200 (nonterminal)

[a]The C=C and C≡C stretch is weak or absent in symmetric molecules.

The C═C stretch in alkenes occurs in the region noted in the table. If the molecule is symmetrically substituted about the C═C bond, no change in dipole moment occurs and the band will not appear in the IR spectrum. Molecules that are close to symmetrical, such as those with symmetrically disubstituted C═C bonds, may show only a weak absorption for the C═C stretch. The position of the C═C stretch shifts if the bond is in a ring or exocyclic to a ring, with the shift depending on the size of the ring. Clearly, the spectrum of an alkene can provide many clues to the structure of the molecule. The oop bending mode for the ═C—H bond is often the strongest band in the spectrum of an alkene.

The spectrum shown in Fig. 4.34 is that of an olefin, 1-decene. Looking at the C—H stretch region, it should be noted that the weak but sharp absorption above 3000 cm^{-1} is due to the olefinic ═C—H stretch. The olefinic C—H bends occur at 910 and 990 cm^{-1} and the C═C stretch at about 1640 cm^{-1} is typical of a monosubstituted olefin.

Figure 4.35 is the spectrum of an alkyne, 1-hexyne. Alkynes contain at least one carbon–carbon triple bond, denoted as C≡C. The carbon triple bond peak appears in the region 2100–2200 cm^{-1}, depending on its position in the molecule. Very few other functional groups absorb in this region; nitrile, C≡N, is probably the most common. Other less common groups that can absorb in the 2000–2500 cm^{-1} range are isocyanates and other nitrogen-containing conjugated double bond systems, and molecules with P—H or Si—H bonds. As is the case with the C═C stretch, symmetrical or nearly symmetrical substitution on the C≡C carbons will result in a weak or missing band. The other important peak in the spectrum of a terminal alkyne is the ≡C—H stretch, occurring near 3300 cm^{-1}.

Aromatic hydrocarbons. Aromatic compounds are those in which some valence electrons are delocalized over the molecule, as opposed to aliphatic compounds with localized electron pairs in covalent bonds. A typical example of an aromatic compound is benzene, a six-membered carbon ring. All the carbon atoms have sp^2 hybridization, giving a planar ring. The six remaining p orbitals overlap above and below the plane of the ring to give a "ring" of delocalized electrons. This can be contrasted with cyclohexane, also a six-membered carbon ring where all the carbons are sp^3 hybridized. The ring is not planar, and the electrons can be considered to be localized in pairs in the covalent bonds between the atoms. Cyclohexane is aliphatic while benzene is aromatic. Cyclohexane is often shown as a plain hexagon (Fig. 4.33); aromatic rings such as benzene may be shown either with a circle in the middle of the hexagon to denote the ring of delocalized electrons or in the older "resonance" format. Benzene can be drawn, as it is in Fig. 4.36, with three double bonds to account for the six delocalized electrons. The three bonds can be shifted by one position each clockwise in the ring, giving an equivalent structure.

Figure 4.34 1-Decene, C$_{10}$H$_{20}$, a linear alkene. (Copyright Bio-Rad Laboratories, Informatics Division, Sadtler Software and Databases, 1999. All rights reserved.)

Figure 4.35 1-Hexyne, C_6H_{10}, an alkyne. (Copyright Bio-Rad Laboratories, Informatics Division, Sadtler Software and Databases, 1999. All rights reserved.)

The two equivalent structures, which differ only in the location of the double bonds, are called resonance structures. The student should remember that there are no real double bonds in an aromatic ring; the drawing is just a convention.

Aromatic hydrocarbons have many IR active vibrations, resulting in complex spectra. The C → H aromatic absorption occurs above 3000 cm^{-1}, but in the same region as that for alkenes. The aromatic bands are often weak or appear as shoulders on the aliphatic bands. More useful for structural identification are the aromatic C=C ring stretching absorptions in the 1400–1600 cm^{-1} region which often appear as doublets and the aromatic C ↓ H oop bands in the 690–900 cm^{-1} region. These oop bending peaks are very strong and result in weak overtone and combination bands between 1660 and 2000 cm^{-1}. The exact frequencies and intensities of these peaks are very dependent on substitution of the ring. Table 4.9 lists these ring deformation modes used to identify the positions substituted on a single benzene ring. The IR spectrum of benzene itself is shown in Fig. 4.36.

Figure 4.37 depicts the overtone peaks in the 1660–2000 cm^{-1} region for substituted benzene rings. The number of peaks and peak intensities are very characteristic for the substitution on the ring and are used in conjunction with the oop bending information in a spectrum to confirm the location of substituents. Table 4.9 and Fig. 4.37 may not hold for aromatic rings with very polar groups such as acids and nitro groups attached directly to the ring.

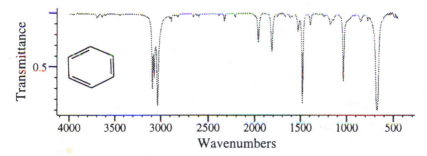

Figure 4.36 Benzene, C_6H_6, an aromatic hydrocarbon. The structure shown on the spectrum is one way of representing the delocalized electrons; they can be shown as three double bonds in one of two resonance structures. A more accurate way of representing them is to draw a circle inside the hexagon, with the circle representing the six delocalized electrons in the molecular orbital. (Copyright Bio-Rad Laboratories, Informatics Division, Sadtler Software and Databases, 1999. All rights reserved.)

Table 4.9 Aromatic C—H Ring Deformation Modes and Benzene
Ring Substitution

Substitution positions	Wavenumber (cm^{-1})	Band strength
Monosubstitution	730–770	s
	680–720	s
1,2 Disubstitution (ortho)	740–770	s
1,3 Disubstitution (meta)	860–890	m
	760–810	s
	680–710	s
1,4 Disubstitution (para)	800–855	s
1,2,3 Trisubstitution	740–790	s
	680–720	m
1,2,4 Trisubstitution	860–900	m
	790–830	s
1,3,5 Trisubstitution	820–910	s
	670–700	m

Spectra for several aromatic compounds are shown in Figs. 4.38–4.41. Note the aromatic C—H stretch in these spectra—a sharp peak just above 3000 cm^{-1}. Just to the right, in the 2900 cm^{-1} region, are the methyl C—H stretching bands. Toluene clearly shows the pattern of four peaks in the 1650–2000 cm^{-1} region and the pair of strong peaks for the oop C—H bend at about 690 and 740 cm^{-1} expected for a monosubstituted benzene ring. o-Xylene shows only a single strong oop C—H bend at about 745 cm^{-1}, as expected from the table, and the pattern in the 1650–2000 cm^{-1} region matches that in Fig. 4.37 for an ortho substitution pattern. The student should compare the m-xylene

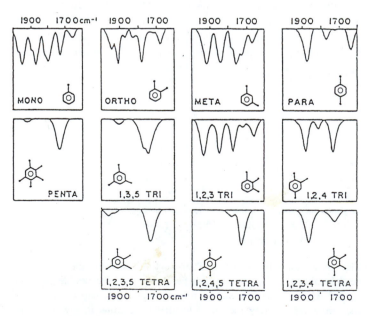

Figure 4.37 Characteristic absorption bands in the 1700–2000 cm^{-1} region for various substituted benzene rings.

Figure 4.38 Toluene. (Copyright Bio-Rad Laboratories, Informatics Division, Sadtler Software and Databases, 1999. All rights reserved.)

and *p*-xylene spectra with Fig. 4.37 and Table 4.9 as has been done for toluene and *o*-xylene. Can you distinguish all four compounds based on their IR spectra?

Figure 4.1, the first spectrum in this chapter, is that of polystyrene. Polystyrene is a polymer containing a monosubstituted benzene ring in each repeating unit. There is a pair of strong peaks between 700 and 800 cm^{-1}, characteristic of the monosubstituted oop bending pattern. The overtone pattern of four peaks in the 2000–1650 cm^{-1} region confirms that the ring is monosubstituted. The sharp peaks on the far left of the CH stretch area occur at >3000 cm^{-1}, indicative of a bond between hydrogen and the sp^2 hybridized carbon found in aromatic and alkene compounds. The peaks in the 1400–1500 cm^{-1} region are the aromatic C=C stretching bands.

4.6.1.2. Organic Compounds with C—O Bonds

There are many classes of organic compounds that contain carbon–oxygen bonds, including alcohols, carboxylic acids, ethers, peroxides, aldehydes, ketones, esters, and acid anhydrides. Not all classes of compounds are included in the following discussion, so the interested student is advised to look at the more detailed references already mentioned for additional information.

Alcohols and carboxylic acids. Alcohols are organic compounds that contain an OH group bonded to carbon. Aromatic alcohols, with an OH group substituted on a benzene ring, are called phenols. Carboxylic acids contain an OH group bonded to a carbon that also has a double-bond to a second oxygen atom; the functional group

Figure 4.39 *o*-Xylene. (Copyright Bio-Rad Laboratories, Informatics Division, Sadtler Software and Databases, 1999. All rights reserved.)

Figure 4.40 *m*-Xylene. (Copyright Bio-Rad Laboratories, Informatics Division, Sadtler Software and Databases, 1999. All rights reserved.)

is written as COOH, but it must be remembered that the OH group is attached to the carbon atom, not to the other oxygen atom: —C—O—H.

$$\underset{\displaystyle O}{\overset{\displaystyle \|}{}}$$

Carboxylic acids also show a very distinctive absorption band due to the carbonyl C=O group. Typical absorption frequencies for the OH group and related C—O vibrations in alcohols and carboxylic acids are shown in Table 4.10. Carbonyl frequencies are given in Table 4.11.

The appearance and frequency of the OH stretching band is very dependent on hydrogen bonding. Hydrogen bonding is an attractive force between a hydrogen atom bonded to either oxygen or nitrogen and a nearby electronegative oxygen or nitrogen atom. Hydrogen bonding may occur between molecules (intermolecular) or within the same molecule if the geometry permits (intramolecular). The force is weaker than a covalent bond, but is the strongest of the intermolecular attractive forces (van der Waals forces). A hydrogen bond is depicted as O—H···N or O—H···O, where the dotted line indicates the hydrogen bond and the solid dash a covalent bond. Alcohols and carboxylic acids are capable of intermolecular hydrogen bonding unless prevented by steric hindrance. The OH band in neat (pure) aliphatic alcohols appears broad and centered at about 3300 cm^{-1} due to intermolecular hydrogen bonding. In very dilute solutions of these compounds in a nonpolar solvent or in the gas phase, there is minimal hydrogen bonding; the OH band then appears "free" as a sharp peak at about 3600 cm^{-1}. The free OH peak shifts to slightly lower frequency, by about 10 cm^{-1}, in the order 1° > 2° > 3° for aliphatic alcohols. The free OH band in phenols appears slightly below that for a 3°

Figure 4.41 *p*-Xylene. (Copyright Bio-Rad Laboratories, Informatics Division, Sadtler Software and Databases, 1999. All rights reserved.)

Table 4.10 OH and Related C—O Group Frequencies for Alcohols and Carboxylic Acids

Functional group	Vibrational mode and strength	Wavenumber (cm^{-1})
Aliphatic alcohols		
CO → H	OH stretch (free), s	3600–3640
	OH stretch (hydrogen-bonded), s	3200–3400
C—O ↓ H	C—O—H bend, w	1200–1400
C—C → O	C—C—O stretch (out-of-phase, asym), s	1000–1260
Phenols		
CO → H	OH stretch (free), s	3600
	OH stretch (hydrogen-bonded), s	3100–3300
C—O ↓ H	C—O—H bend, s	1300–1400
C—C → O	C—C—O stretch (oop, asym), s	1160–1300
Carboxylic acids		
CO → H	OH stretch (free)	Not usually observed
	OH stretch (hydrogen-bonded), s	2500–3200
C—O ↓ H	C—O—H bend (in-plane), m	1400–1440
	C—O—H bend (dimer, oop), m	900–950
C—C → O	C—C—O stretch (dimer), m	1200–1320

aliphatic alcohol, again by about 10 cm^{-1}. Carboxylic acids generally exist as dimers in all but extremely dilute solutions; most acids show only the broad hydrogen-bonded band listed in Table 4.10. The H-bonded OH band in phenols is about 100 cm^{-1} lower than that for aliphatic alcohols and even lower for carboxylic acids, as seen in Table 4.10. The C—O—H bend is strong and broad in phenols, as is the C—O stretch.

The C—O stretch is also very useful for structural diagnosis of alcohols. The band is due to the C—O stretch coupled to the C—C stretch, and is called an oop or asymmetric stretch in some references. The band shifts position in aliphatic alcohols depending on the substitution on the OH-bearing carbon. The band increases in frequency by about 50 cm^{-1}

Table 4.11 Carbonyl Stretching Frequencies for Selected Compound Classes

Functional group	Vibrational mode and strength	Wavenumber (cm^{-1}) (± 10 cm^{-1})
Aldehydes		
Aliphatic C ⇒ O	C=O stretch, s	1730
Aromatic C ⇒ O	C=O stretch, s	1690
Amides		
C ⇒ O	C=O stretch, s	1690
Carboxylic acids		
Aliphatic C ⇒ O	C=O stretch (dimer), s	1710
Aromatic C ⇒ O	C=O stretch, s	1690
Esters		
Aliphatic C ⇒ O	C=O stretch, s	1745
Aromatic C ⇒ O	C=O stretch, s	1725
Ketones		
Aliphatic C ⇒ O	C=O stretch, s	1710
Aromatic C ⇒ O	C=O stretch, s	1690

in the order $1° < 2° < 3°$. As noted, the band is strong and broad in phenols and is shifted about $50 \, \text{cm}^{-1}$ higher than a $3°$ aliphatic alcohol. This C—O band is shifted to a lower frequency by about $100–200 \, \text{cm}^{-1}$ in aromatic carboxylic acids from the position given in Table 4.10. Note that the characteristic bands for a carboxylic acid arise from its existence as a dimer due to strong intermolecular hydrogen bonding.

The spectrum of neat ethanol is shown in Fig. 4.42(a). This spectrum was collected from a thin liquid film of ethanol between salt plates. The broad hydrogen-bonded OH stretch covers the region from 3100 to $3600 \, \text{cm}^{-1}$, centered at about $3350 \, \text{cm}^{-1}$. The strong C—C—O stretch at $1048 \, \text{cm}^{-1}$ is characteristic of a primary alcohol. Coupling of the weak OH bend to the methylene bending mode and overlap of the methyl

Figure 4.42 (a) IR spectrum of ethanol, collected as a liquid film between salt plates. [Courtesy of http://www.aist.go.jp/RIODB/SDBS.] (b) IR spectrum of ethanol vapor. [Courtesy of NIST (http://webbook.nist.gov/chemistry).] Note the difference in the position and appearance of the OH band in the hydrogen-bonded liquid vs. the non-hydrogen-bonded gas phase.

Figure 4.43 Phenol. (Copyright Bio-Rad Laboratories, Informatics Division, Sadtler Software and Databases, 1999. All rights reserved.)

bending vibrations gives the broad, weak peaks between 1200 and $1500 \, cm^{-1}$. Figure 4.42(b) is the spectrum of ethanol in the gas phase. Note the dramatic change in the OH stretch peak. It is shifted to $3650 \, cm^{-1}$ and is now very sharp, because there is no significant hydrogen bonding in the gas phase.

Figure 4.43 shows the absorption spectrum of phenol. Phenol has one OH group substituted on a benzene ring, so you need to consider not only the peaks in Table 4.10, but also Table 4.9 and the earlier discussion of aromatic hydrocarbons. Note that the CO $\rightarrow$ H is very broad due to hydrogen bonding and the C $\rightarrow$ H at about $3050 \, cm^{-1}$, on the right side of the OH band, is a short sharp peak typical of aromatics. The C—C $\rightarrow$ O band appears at about $1225 \, cm^{-1}$, shifted up from the position in ethanol, a primary alcohol, as expected. The broad, moderately strong band at about $1350 \, cm^{-1}$ is the C—O—H bend. The sharp peaks between 1450 and $1600 \, cm^{-1}$ are from stretching of the aromatic ring carbon bonds; note that there is a doublet, as previously discussed. Characteristic bands at 745 and $895 \, cm^{-1}$ and the pattern of four peaks between 1650 and $2000 \, cm^{-1}$ typify a monosubstituted benzene.

Figure 4.44 is the spectrum of heptanoic acid, an aliphatic carboxylic acid. As expected from the table, the broad OH band appears below that of phenol, centered at about $2900 \, cm^{-1}$. The C—H stretching bands from the methyl and methylene groups occur in the same region; they are the peaks sticking out at the bottom of the broad OH band. The appearance of this region is very characteristic of an aliphatic carboxylic acid. The strong band at $1710 \, cm^{-1}$ is due to the C═O stretch, discussed in the next section. The peak at $1410 \, cm^{-1}$ is the in-plane C—O—H bend and that about

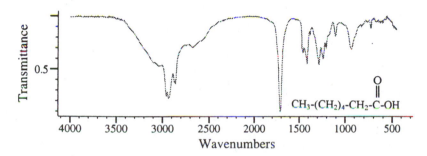

Figure 4.44 Heptanoic acid. (Copyright Bio-Rad Laboratories, Informatics Division, Sadtler Software and Databases, 1999. All rights reserved.)

930 cm^{-1} is the oop C—O—H bend. The C—C—O stretch from the dimer is at 1280 cm^{-1}.

Carboxylic acids, esters, ketones, and aldehydes. These compounds, along with acid anhydrides, acid halides, amides, and others, all possess a double bond between carbon and oxygen, C=O, called a carbonyl group. The IR spectra of these compounds are easily recognized from the very strong C ⇒ O stretching band between 1650 and 1850 cm^{-1}. The position of this band is very characteristic of the class of compound.

The carbonyl stretch position depends on its environment; electronegative substituents, the position of substituents relative to the C=O bond, position in a ring and the size of the ring, resonance, conjugation, and both intermolecular and intermolecular hydrogen bonding affect the vibrational frequency. The band is usually extremely strong. The use of related bands can narrow the possible chemical class of an unknown; for example, an ester will also show a characteristic C—O—C stretch, an acid halide will show the C—X band, a carboxylic acid will show an OH band, amides will show the NH stretch bands, and so on. Only a few examples of the huge variety of carbonyl-containing compounds will be discussed.

The spectrum for acetone is shown in Fig. 4.45. Acetone is a ketone. The most intense peak in the spectrum is the C=O stretching band at 1710 cm^{-1} (Table 4.11). The peak at 1200 cm^{-1} is due to bending of the carbonyl group. The peaks between 1350 and 1450 cm^{-1} and the peaks around 2900 cm^{-1} are the C—H bend and stretch modes, respectively. The C—H peaks tell us something more. The low intensity of the CH stretch band and the lack of a "long chain" band tell us we have no long aliphatic chains and few methyl groups in the molecule. What is the small, sharp peak at 3400 cm^{-1} due to? Remember that it is possible to see overtones (generally approximately whole number multiples of a given frequency) and combinations of strong bands in a spectrum. The peak at 3400 cm^{-1} is an overtone of the intense C=O band at 1710 cm^{-1}. The analyst needs to keep these possibilities in mind when interpreting a spectrum.

The spectrum for 3-phenylpropionaldehyde (Fig. 4.46) shows a variety of features and is quite complex. This compound is an aldehyde (hence its name), and contains an aromatic ring and a short aliphatic chain. The aldehyde functional group, —CHO, has the structure —C—H, as shown on the spectrum.

$$\overset{\parallel}{\underset{O}{}}$$

The most intense band is the C=O stretch, at 1730 cm^{-1}, indicating that the aldehyde carbonyl is part of the aliphatic portion of the molecule. Looking at the C—H stretching region, the sharp aldehydic C—H bands appear at 2725 and 2825 cm^{-1}. The doublet

Figure 4.45   Acetone. (Copyright Bio-Rad Laboratories, Informatics Division, Sadtler Software and Databases, 1999. All rights reserved.)

Figure 4.46 3-Phenylpropionaldehyde (or hydrocinnimaldehyde). (Copyright Bio-Rad Laboratories, Informatics Division, Sadtler Software and Databases, 1999. All rights reserved.)

appears due to resonance with an overtone of the aldehydic C—H bend at 1390 cm^{-1}. This is called Fermi resonance and is always seen in aldehydes. The sharp peaks above 3000 cm^{-1} are the aromatic C—H stretching peaks and the peaks between the aromatic and aldehydic peaks are due to the aliphatic C—H stretches. The sharp peaks in the $1400–1600 \text{ cm}^{-1}$ region are due to carbon—carbon aromatic ring stretches. The strong peak near 700 cm^{-1} is attributed to ring bending, and is seen in other substituted benzenes, such as in Fig. 4.51, nitrobenzene.

The spectrum for heptanoic acid (Fig. 4.44) was already examined, but the student should confirm that the position of the C=O stretch is consistent with that for an aliphatic carboxylic acid.

4.6.1.3. *Nitrogen-Containing Organic Compounds*

There are many classes of nitrogen-containing organic compounds, including amines, nitriles, pyridines, azides, and others. In addition, there are a variety of classes of compounds containing both nitrogen and oxygen, such as amides, oximes, nitrates, nitrites, and others. Absorptions from NH, CN, NN, and NO vibrational modes result in many IR spectral bands. The peaks listed in Table 4.12 are just a small sample of some of the bands in nitrogen-containing compounds.

Primary amines have two modes of "bending" of the NH_2 group, a scissoring mode and the low frequency wagging mode. Secondary amines have only one H on the nitrogen; they cannot scissor, but display the NH "wagging" band also at low frequency. Tertiary amines have no NH bands, and are characterized by C—N stretching modes in the regions $1000–1200 \text{ cm}^{-1}$ and $700–900 \text{ cm}^{-1}$. Primary, secondary, and tertiary amides are similar to their amine counterparts with the addition, of course, of the C=O stretching band. In primary and secondary amides, the C=O stretch is often called the Amide I band, while the adjacent NH combination band is called the Amide II band.

Figure 4.47 shows the spectrum of a primary aliphatic amine, represented generally as RNH_2. This is the spectrum of propylamine, $C_3H_7NH_2$. The N—H stretch doublet characteristic of a primary amine is seen at 3370 and 3291 cm^{-1}; the primary NH bends occur at 1610 and about 800 cm^{-1}. The C—N stretching band around 1100 cm^{-1} is weak and not very useful. The rest of the bands are from C—H stretching and bending, as expected. Figure 4.48 is the spectrum of a related compound, dipropylamine, a secondary amine, $(C_3H_7)_2NH$ or R_2NH. Note the single NH stretch characteristic of a secondary amine at 3293 cm^{-1} and the expected shift to lower frequency of the broad NH bend, from about 800 cm^{-1} in propylamine to 730 cm^{-1}. The spectrum of

Table 4.12 Frequencies of Selected Nitrogen-Containing Functional Groups

Functional group	Vibrational mode and strength	Wavenumber (cm^{-1})
Aliphatic amines		
Primary R—NH_2	N → H stretch (sym and asym), m	3300–3500 Doublet
Secondary R_2NH	N → H stretch, w	3300 Singlet
Tertiary R_3N	No NH	
Primary R—NH_2	N ↓ H bend, s	1550–1650; 800
Secondary R_2NH	N ↓ H bend, s	700
Tertiary R_3N	No NH	
Aromatic amines		
Primary Ar—NH_2	N → H stretch, m	3500
Secondary Ar_2NH	N → H stretch, m	3400
Tertiary Ar_3N	No NH	
Primary Ar—NH_2	N ↓ H bend, s-m	1620, 650
Nitriles, isonitriles		
C—C≡N	C≡N stretch, w-m	2220–2260
C—N≡C	N≡C stretch, w-m	2110–2180
Nitro compounds		
Aliphatic RNO_2	NO_2 stretch, asym, s	1530–1580
	NO_2 stretch, sym, s	1350–1390
Aromatic $ArNO_2$	NO_2 stretch, asym, s	1500–1550
	NO_2 stretch, sym, s	1285–1350
Nitrates $RONO_2$	NO_2 stretch, asym, s	1620–1650
	NO_2 stretch, sym, s	1280
Azides	N≡N stretch, m	2140

tripropylamine, a tertiary amine with the formula $(C_3H_7)_3N$, is presented in Fig. 4.49. Note the absence of the NH stretch and bend bands since there is no N—H group in a tertiary amine, R_3N. The peaks observed are due to C—H stretches and bends, both methyl and methylene, and a C—N stretch at about 1100 cm^{-1}.

The spectrum in Fig. 4.50 is that of butyronitrile, $CH_3CH_2CH_2C≡N$. Note the characteristic sharp C≡N stretch at 2260 cm^{-1}. Figure 4.51 is the spectrum of nitrobenzene, a monosubstituted benzene ring. The nitro group is very polar and attached directly to the ring; the normal "rules" for substitution are not useful, as was mentioned earlier. The peaks due to the nitro group are located at 1350 and 1510 cm^{-1}, and the position of the sharp aromatic C—H stretching peaks above 3100 cm^{-1} is typical of an aromatic ring,

Figure 4.47 Propylamine. (Copyright Bio-Rad Laboratories, Informatics Division, Sadtler Software and Databases, 1999. All rights reserved.)

Figure 4.48 Dipropylamine. (Copyright Bio-Rad Laboratories, Informatics Division, Sadtler Software and Databases, 1999. All rights reserved.)

but the overtone bands in the $1700-2000$ cm^{-1} region do not show the expected monosubstitution pattern.

Amino acids. These are the building blocks of proteins and as such are very important in biochemistry and natural products chemistry. As their name implies, amino acids contain both an amine group and a carboxylic acid group, and are represented as RCH(NH$_2$)COOH. Amino acids exist as *zwitterions*, that is, the H ion from the acid protonates the basic amine group, resulting in the formation within a single molecule of a positively charged ammonium cation and a negatively charged carboxylate anion. The result is that the IR absorption spectrum of an amino acid has an appearance related to salts of acids and salts of amines. Important absorption bands are as follows (Table 4.13).

The bands are broad and overlap with each other and with the CH bands. A broad band at 2100 cm^{-1} often appears in these spectra. The spectrum of valine, Fig. 4.52, shows the broad NH stretch overlapping the CH and OH bands ($2600-3200$ cm^{-1}); trying to sort out the methyl and CH bands from the OH and NH bands is not practical. The carbonyl stretch is typical of the carboxylate ion, shifted to a lower wavenumber than a typical carbonyl stretch, broader and overlapped by the NH bend. The symmetric NH bend and symmetric carboxylate ion stretch can be seen at 1500 and 1400 cm^{-1}, respectively.

4.6.1.4. Functional Groups Containing Heteroatoms

Halogenated compounds. The halogens F, Cl, Br, and I are usually represented in organic formulas by the letter X, as in CHX$_3$ for chloroform, CHCl$_3$ and bromoform,

Figure 4.49 Tripropylamine. (Copyright Bio-Rad Laboratories, Informatics Division, Sadtler Software and Databases, 1999. All rights reserved.)

Figure 4.50 Butyronitrile. (Copyright Bio-Rad Laboratories, Informatics Division, Sadtler Software and Databases, 1999. All rights reserved.)

$CHBr_3$. The C → X stretch absorption bands occur in the fingerprint and aromatic regions and are quite strong (Table 4.14). Several halogen atoms on the same C atom cause an increase in absorption intensity, more absorption peaks and a shift to higher wavenumbers (higher frequency or shorter wavelength) for the C—X stretch.

Compounds containing the —CH_2X group, where X = Cl, Br, or I have a C—H bending mode that occurs in the 1150–1280 cm^{-1} region. Absorption due to this bend can be quite strong. As is to be expected, the overtones of the very strong stretch and bend modes may often be observed, at twice the wavenumber of the fundamental band. The spectrum shown in Fig. 4.53 is of chloroform, $CHCl_3$. The bands for C → H at 3000 cm^{-1}, C ↓ H (from the —CH_2X group) at 1200 cm^{-1} and the very strong, broad C → Cl at 760 cm^{-1} can be seen. Overtones (weak) at about 1510 and 2400 cm^{-1} are also visible.

Sulfur, phosphorus, and silicon compounds. Some absorption bands of the heteroatoms S, P, and Si commonly seen in organic compounds are listed in Tables 4.15–4.17. There are many classes of compounds containing S, P, and Si atoms; detailed discussions of these compounds may be found in the texts already recommended.

Little information is obtained on sulfides or disulfides in the IR region. The compound carbon disulfide, S=C=S, is a thiocarbonyl, but atypical. It shows an intense band at about 1500 cm^{-1}, as seen in Fig. 4.54, but the rest of the IR region is clear, making CS_2 a useful IR solvent. Figure 4.55 is the spectrum of benzylmercaptan and shows the SH stretch between 2500 and 2600 cm^{-1}, along with the expected aromatic ring peaks and the characteristic monosubstituted benzene pattern between 1600 and 2000 cm^{-1}. While the SH stretch is weak in the IR spectrum, it is strong in the Raman

Figure 4.51 Nitrobenzene. (Copyright Bio-Rad Laboratories, Informatics Division, Sadtler Software and Databases, 1999. All rights reserved.)

Table 4.13 Amino Acid Absorptions

Functional group	Vibrational mode and strength	Wavenumber (cm^{-1})
Amine cation, RNH_3^+	$N \rightarrow H$, s	2700–3100
Amine cation, RNH_3^+	$N \downarrow H$, s	1600 (asym); 1500 (sym)
Carboxylate ion	CO_2^- ion stretch, s	1600 (asym); 1400 (sym)

spectrum, indicating the complementary nature of the two techniques, as discussed later in the chapter.

One reason for concern about the IR absorption of organosilicon compounds is that silicone polymers are widely used in many laboratories and in many consumer products. Silicone lubricants, greases, caulks, plastic tubing, and o-rings may dissolve in the solvents used for IR absorption or may contaminate samples extracted in an apparatus that uses silicone polymer o-rings or lubricants. Hand cream, moisturizers, hair care products, and lubricants in protective gloves and similar products often contain silicones, so the analyst may contaminate the sample. Many common silicones are poly(dimethyl-siloxanes), polymers with long Si—O—Si—O chains and two methyl groups on each Si atom. The methyl CH stretch is usually very sharp, the major peak is a strong broad band (doublet) between 1000 and 1100 cm^{-1} due to the Si—O—Si stretch and the Si—C stretch appears at about 800 cm^{-1}.

Now that you have a basic understanding of the information that is available in an IR absorption spectrum, it is possible, with some additional information, to work out the likely structure of an unknown for small molecules (MW < 300). For unknowns where the molecular weight has been determined by mass spectrometry the analyst generally takes the following approach.

Using the IR spectrum,

1. Identify the major functional groups present from the strong absorption peaks in the spectrum, for example, C=O, C—OH, NH_2, C—O, NO_2, C≡N, etc. Note their formula weight (FW).
2. Identify if the compound is aromatic (look for the out of plane bends in the region below 900 cm^{-1} and the sharp but weak CH stretch above 3000 cm^{-1}). Remember that the "aromatic region" also may have bands due to the halogens Cl and Br—these are important to identify.

Figure 4.52 Valine. (Copyright Bio-Rad Laboratories, Informatics Division, Sadtler Software and Databases, 1999.)

Table 4.14 Halogen-Containing Functional Groups

Functional group	Vibrational mode and strength	Wavenumber (cm^{-1})
Aliphatic halide	$C \rightarrow F$, s	720–1400
	$C \rightarrow Cl$, s	530–800
	$C \rightarrow Br$, s	510–690
	$C \rightarrow I$, s	>600
Aromatic halide[a]	$C \rightarrow F$, s	1100–1250
	$C \rightarrow Cl$, s	1000–1100
	$C \rightarrow Br$, s	1000–1090

[a]The halide atom is substituted directly on a carbon atom in an aromatic ring.

3. Subtract FW of all functional groups identified from the given molecular weight of the compound.
4. Look for the other unique CH bands, such as aldehyde and vinyl. Look for the $C{\equiv}C$ and $C{=}C$ stretching peaks.
5. Accommodate the difference between step 3 and the molecular weight as aliphatic or aromatic hydrocarbon components, depending on your answers from steps 2 and 4.

For example, an unknown has been determined to have a molecular weight of 94. The IR spectrum shows absorptions due to OH and a monosubstituted aromatic ring pattern. The OH group has a formula weight of 17. Subtracting 17 from the MW of 94 leaves 77 mass units to be accounted for. The FW of a benzene ring minus a hydrogen atom is 77, which is consistent with the IR spectrum. The unknown structure is a benzene ring with one hydrogen replaced by an OH group—therefore it is probably phenol. (As an analyst, what other tests can you do to prove that the unknown is phenol?)

Another IR spectrum of a compound with MW = 60 shows $C \Rightarrow O$ and OH peaks consistent with a carboxylic acid group, COOH. The FW of COOH = 45. Then, $MW - FW = 60 - 45 = 15$. This does not give us too many possibilities—the 15 mass units are due most likely to a CH_3 group. Hence, CH_3 and COOH give us acetic acid, CH_3COOH.

The third sample has MW = 147. The IR spectrum shows a very strong peak due to an aromatic C—Cl stretch, and the rest of the IR spectrum is consistent with an aromatic hydrocarbon and nothing else. A benzene ring minus one hydrogen has FW = 77. One Cl atom FW = 35.5. That adds up to 112.5, leaving a difference of 34.5. This is an example

Figure 4.53 Chloroform, $CHCl_3$. (Copyright Bio-Rad Laboratories, Informatics Division, Sadtler Software and Databases, 1999. All rights reserved.)

Table 4.15 Frequencies in Select Sulfur Compounds

Functional group	Vibrational mode and strength	Wavenumber (cm^{-1})
Mercaptan, RSH	S → H, w	2575
Sulfide, R_2S	C → S, w	>700; not useful
Disulfide, RSSR	S → S	>500; not observed
Thiocarbonyl, R_2C=S	C ⇒ S, w	1000–1250
Sulfoxide, R_2S=O	S ⇒ O, s	1050
Sulfone, R_2SO_2[a]	O=S=O, s (note that there are two other single bonds, not shown, from the sulfur atom to the two R groups)	1325 (asym); 1140 (sym)

[a]Many classes of compounds contain the O=S=O group found in sulfones. These include sulfonyl chlorides, sulfonates, sulfonamides, and sulfonic acids and their salts. The exact positions of the symmetric and asymmetric stretches vary for each class.

that demonstrates a limitation of IR spectroscopy. You know there is Cl present from the IR spectrum; you do not know how many Cl atoms there are. What if there are two Cl atoms substituted on the benzene ring? We need to subtract another hydrogen, giving us an FW = 76 for the benzene ring minus 2 hydrogen atoms. Two Cl atoms = 2 × 35. 5 = 71. 71 + 76 = 147, the MW of the unknown. So the unknown is probably a dichlorobenzene. We need to look at the 1600–2000 cm^{-1} region to determine if it is ortho, meta, or para substituted.

These are very simple examples, but provide a plan of attack for the problems at the end of the chapter. It is highly unlikely that an analyst can identify a complete unknown by its IR spectrum alone (especially without the help of a spectral library database and computerized search). For most molecules, not only the molecular weight, but also the elemental composition (empirical formula) from combustion analysis and other classical analysis methods, the mass spectrum, proton and ^{13}C NMR spectra, possibly heteroatom NMR spectra (P, Si, and F), the UV spectrum, and other pieces of information may be required for identification. From this data and calculations such as the unsaturation index, likely possible structures can be worked out.

4.6.2. Quantitative Analyses by IR Spectrometry

The quantitative determination of various compounds by mid-IR absorption is based on the measurement of the concentration of one of the functional groups of the analyte compound. For example, if we have a mixture of hexane and hexanol, the hexanol

Table 4.16 Frequencies in Select Phosphorus Compounds

Functional group	Vibrational mode and strength	Wavenumber (cm^{-1})
Phosphines, RPH_2	P → H, m	2275–2350
	P ↓ H, m	890–990
	↓PH_2, m	1080; 900
Phosphine oxides	P ⇒ O, s	1150–1200
Phosphate esters	P ⇒ O, s	1250–1290
	P → O, m	720–850

Table 4.17 Frequencies in Select Silicon Compounds

Functional group	Vibrational mode and strength	Wavenumber (cm^{-1})
Si—X	Si → F	800–1000
	Si → Cl	>650
Si—H	Si → H, m	2100–2200
	Si ↓ H	850–950
Si—C	Si → C, m	700–820
Si—O	Si → OR, s	1000–1100
	Si → O → Si, s	1000–1100
	Si → OH, s	820–920

may be determined by measuring the intensity of absorption that takes place near $3300 \ cm^{-1}$ by the OH band. The spectrum of pure hexanol would be used to determine the exact wavenumber corresponding to the absorption maximum. Alternatively, the intensity of absorption due to the C → O at about $1100 \ cm^{-1}$ could be used. From this the concentration of alcohol can be calculated, once the intensity from a set of hexanol standards of known concentrations has been measured at the same wavenumber. Whenever possible, an absorption band unique to the sample molecule should be used for measuring purposes. This reduces the problem of overlapping bands, although there are mathematical approaches for deconvoluting overlapping peaks. The quantitative calculation is based on Beer's Law

$$A = abc \tag{4.11}$$

where A is the absorbance; a, the absorptivity of the sample; b, the internal path length of the sample cell; and c, the concentration of the solution.

If the same sample cells are used throughout, b is a constant. Also, the absorptivity a is a property of the molecular species being determined and can be taken as a constant. Therefore A is proportional to c. The intensity of the peak being measured in each sample must be measured from the same baseline. Usually, a straight baseline is drawn from one side of the peak to the other, or across multiple peaks or a given wavenumber range, to correct for sloping background. Quantitative measurements are generally made on samples in solution, because scattered radiation results in deviations from Beer's Law. KBr pellets have been used for quantitative work, but the linear working range may be small due to light scatter if the pellets are not well-made. A hydraulic press is recommended for preparation of pellets for quantitative analysis.

Figure 4.54 Carbon disulfide, CS_2. (Copyright Bio-Rad Laboratories, Informatics Division, Sadtler Software and Databases, 1999. All rights reserved.)

Figure 4.55 Benzylmercaptan. (Copyright Bio-Rad Laboratories, Informatics Division, Sadtler Software and Databases, 1999. All rights reserved.)

To find the proportionality constant between A and c, calibration curves are constructed from measurements of intensity of absorption by solution standards at different known concentrations of the analyte. From these data the relationship between absorption and concentration is obtained. To carry out quantitative analysis, the absorbance A of a sample of unknown concentration is measured. In comparison with the calibration curve, the sample composition can be determined. Quantitative analysis using mid-IR absorption is not as accurate as that using UV or visible absorption. The sample cells for mid-IR absorption have much smaller path lengths than those for UV and must be made with material transparent to IR radiation, such as NaCl or KBr. These materials are very soft and easily etched by traces of water or distorted by clamping the cell together; as a result, the sample path length may vary significantly from one sample to the next. In quantitative analysis, $A = abc$, where b is the sample path length; hence any change in b between samples or between samples and standards causes an error in the calculation of c, the concentration of the analyte. In addition, nonlinear response is more of a problem in the IR when dispersive spectrometers are used than in the UV/VIS, due to stray radiation, external IR radiation, slit width (really the spectral bandpass) in regions of low source intensity, or low detector response, among other causes. Nonlinear response means that Beer's Law is not obeyed, especially at high absorbance. Samples must often be diluted to keep the absorbance below 0.8 for quantitative measurements with a dispersive instrument. FTIR instruments do not suffer from stray or external radiation and they do not have slits, so it might be expected that accurate absorbance measurements can be made on solutions with absorbance values greater than 1.0. However, the detector and the apodization function limit accuracy in FTIR instruments. It is still advisable to stay below 1.0 absorbance units for FTIR quantitative work. An external calibration curve covering the range of concentrations expected in the samples should be used for the best accuracy, instead of relying on a calculation of the proportionality factor from one standard. An example of quantitative analysis by ATR is the determination of water in a polyglycol. The wavelength measured, peak intensities, and Beer's Law plot are displayed in Fig. 4.56.

Hydrocarbon contamination from oil, grease, and other sources in drinking water and wastewater is determined quantitatively by extraction of the hydrocarbons and measurement of the CH stretching band. The water is extracted with trichlorotrifluoroethane or other suitable solvent, the IR spectrum of the extracted solution is obtained and the absorbance is measured at $2930\ \text{cm}^{-1}$ after a baseline is drawn as described earlier. A series of standards of known oil diluted in the same solvent is prepared as the external calibration curve. The method can measure as little as 0.2 g oil/L of water, with a precision of about 10% RSD. Precisions of 5–10% RSD are typical for

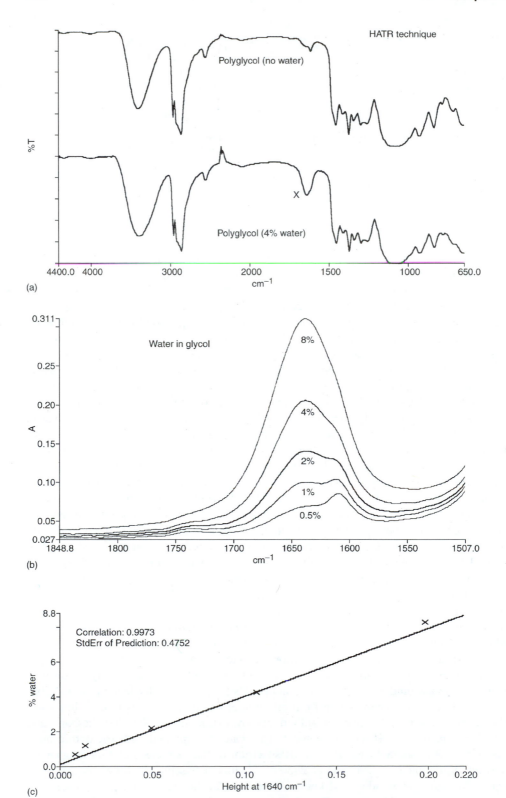

Figure 4.56

Figure 4.56 (p. 284) (a) The IR spectra of a polyglycol with no water present and with 4% water. The peak marked with an X is the O—H bend at 1640 cm^{-1} that will be used for quantitation. (b) Standards of the polyglycol containing 0.5–8% water were scanned and the spectra saved. The spectra of the standard solutions were collected using a horizontal attenuated total reflectance (HATR) liquid sample cell using a ZnSe crystal. All spectra were collected on a PerkinElmer Instruments Paragon 1000 FTIR spectrometer at 8 cm^{-1} resolution. (c) The data from (b) were input into the Spectrum® Beer's Law quantitative analysis software. The calibration curve is shown. (Spectrum® is a registered trademark of PerkinElmer Instruments, Shelton, CT). (Data and spectra courtesy of Pattacini, Pattacini Associates, LLC, Danbury, CT.)

IR measurements. The entire method can be found in Standard Methods for the Examination of Water and Wastewater, listed in the bibliography.

Quantitative analysis of multiple components in a mixture can be done by assuming that Beer's Law is additive for a mixture at a given frequency. For a mixture with two components, the total absorbance of the mixture at a given frequency is the sum of the absorbance of the two components, compound F and compound G, at that frequency.

$$A_{\text{total}} = A_F + A_G = a_F b c_F + a_G b c_G \tag{4.12}$$

The absorptivity of F, a_F, and the absorptivity of G, a_G, are determined at two different wavenumbers, $\bar{\nu}_1$ and $\bar{\nu}_2$ from a series of mixtures with known amounts of F and G. This results in four absorptivity values, a_{F1}, a_{F2}, a_{G1}, and a_{G2}; the symbol for wavenumber is eliminated for simplicity. The concentrations of F and G in an unknown mixture can be calculated from two absorbance measurements at $\bar{\nu}_1$ and $\bar{\nu}_2$. Two simultaneous linear equations with two unknowns are constructed and solved for c_F and c_G:

$$A \text{ at } \bar{\nu}_1 = a_{F1} b c_F + a_{G1} b c_G$$
$$A \text{ at } \bar{\nu}_2 = a_{F2} b c_F + a_{G2} b c_G$$

This approach can be used for multicomponent mixtures by applying matrix algebra. This is generally done with a software program and even nonlinear calibrations can be handled with statistical regression methods.

In addition to the quantitative analysis of mixtures, measurement of the intensity of an IR band can be used to determine reaction rates of slow to moderate reactions. The reactant or product has to have a "clean" absorption band and the absorbance concentration relationship must be determined from calibration standards. The reaction cannot be extremely fast because the band has to be scanned and even FTIR spectrometers are not instantaneous. FTIR spectrometers have permitted the determination of the kinetics of reactions much more rapid than could be handled by dispersive instruments.

4.7. NEAR IR SPECTROSCOPY

The NIR region covers the range from 0.75 μm (750 nm or 13,000 cm^{-1}) to about 2.5 μm (2500 nm or 4000 cm^{-1}). This is the range from the long-wavelength end of visible light (red) to the short-wavelength side of the mid-IR region. The bands that occur in this region are generally due to OH, NH, and CH bonds. The bands in this region are primarily overtone and combination bands and are less intense than the fundamental bands in the mid-IR region. While weaker absorptions and limited functional group information might seem to limit the usefulness of the NIR region, there are some inherent advantages to working in the range. High-intensity sources such as tungsten-halogen lamps give strong, steady radiation over the entire range. Very sensitive detectors, such as lead sulfide photodetectors can be used. These detectors need not be operated in liquid N_2. The third important advantage is that quartz and fused silica can be used both in optical systems and as sample containers. Long path length cells can be used to compensate for the weaker absorptions and can be used to advantage for process analysis applications. The NIR region is used primarily for quantitative analysis of solid and liquid samples for compounds containing OH, NH, and CH bonds, such as water and proteins. This is in contrast to the mid-IR region, which is used primarily for qualitative analysis.

4.7.1. Instrumentation

NIR instrumentation is very similar to UV/VIS instrumentation, discussed in Chapters 2 and 5. Quartz and fused silica optics and tungsten-halogen lamps identical to those used in UV/VIS systems may be used, as discussed earlier in the chapter. Commercial research grade double-beam dispersive grating spectrometers are available that cover the optical range from the UV (180 or 200 nm) through the visible up to the NIR long wavelength limit of 2500 or 3000 nm. In addition, many customized portable filter-based NIR instruments, FT-NIR instruments, and other designs for dedicated applications are commercially available. Dispersive, nonscanning spectrographs are also used in NIR spectroscopy, with a silicon photodiode array detector. Figure 4.57 shows a schematic diagram of such a system. These are used for dedicated applications, such as moisture analyzers. Absorption, transmission, and reflection measurements are made in NIR spectrometry.

Quartz optical fibers are transparent to NIR radiation and are often used to interface the sample and spectrometer or spectrograph. The low OH-content quartz fiber used for many NIR applications is single filament, between 100 and 600 μm in diameter. Fibers can be bundled together when it is necessary to collect light over a large area. Fiber optic probes can be used for remote sampling and continuous monitoring of bulk flowing streams of commercial products as well as for simple "dip" probes. This greatly simplifies sample preparation and in many cases, eliminates it completely. Dilute liquid solutions can be analyzed in the standard 1 cm quartz cuvets used for UV/VIS spectrometry.

Figure 4.57 Idealized layout for a simple diode array-based spectrograph. (From Coates, used with permission.)

4.7.2. NIR Vibrational Bands and Spectral Interpretation

The primary absorption bands seen in the NIR are:

C—H bands between 2100 and 2450 nm and between 1600 and 1800 nm
N—H bands between 1450 and 1550 nm and between 2800 and 3000 nm
O—H bands between 1390 and 1450 nm and between 2700 and 2900 nm

Bands for S—H, P—H, C≡N, and C=O also appear in the NIR region. Water has several distinct absorption peaks at 1400, 1890, 2700, and 2750 nm. These bands enable the determination of hydrocarbons, amines, polymers, fatty acids, proteins, water, and other compounds in a wide variety of materials.

There is some correlation between molecular structure and band position for certain bands, but because these are often overtone and combination bands, their positions are not as structure-dependent as the fundamental bands in the mid-IR. For example, primary amines, both aliphatic and aromatic, have two absorption bands, one at about 1500 nm and the second at about 1990 nm. Secondary amines have only one band at about 1500 nm. As expected, a tertiary amine has no NH band. Amides with an —NH_2 group can be distinguished from R—NH—R' amides by the number and position of the N—H bands. The reference by Goddu has a detailed table of NIR structure–wavelength correlations.

The molecular absorption coefficients (molar absorptivities) for NIR bands are up to three orders of magnitude lower than the fundamental bands in the mid-IR. This results in reduced sensitivity. Greater sample thickness can be used to compensate for this, giving more representative results with less interference from trace contaminants. Sample pathlengths of 0.1 mm–10 cm are common.

4.7.3. Sampling Techniques for NIR Spectroscopy

NIR spectroscopy has a real advantage over many spectroscopic techniques in that many plastic and glass materials are transparent to NIR radiation at the common thicknesses encountered for films, packaging materials, and coatings. It is practical to take an NIR spectrum of a sample without even opening the sample container in many cases; the spectrum is collected through the plastic or glass bottle or through the plastic film on food or the plastic bubble packs used for pharmaceutical tablets. In many cases, no sample preparation is required at all. When diffuse reflectance measurements are made, not only is no sample preparation required, but the method is also nondestructive.

4.7.3.1. Liquids and Solutions

With a fiber optic dip probe, many liquids and solutions can be analyzed with no sample preparation. The use of a dip probe for transmission measurements requires that the liquid or solution be free from small particles or turbidity. Suspended particles scatter light and reduce the sensitivity of the measurement. The already low absorptivity of NIR bands makes transmission measurements of limited use for liquid samples that are not clear.

There are a number of solvents that can be used to prepare solutions for NIR measurements. Carbon tetrachloride and carbon disulfide are transparent over the entire NIR range. Many other organic solvents are transparent up to 2200 nm, with only a short region between 1700 and 1800 nm obscured by the solvent. Solvents as varied as acetonitrile, hexane, dimethyl sulfoxide (DMSO), and dibutyl ether fit into this category. Methylene chloride and chloroform can be used up to 2600 nm, with short "gaps" at about 1700 and 2300 nm.

Liquids, gels, and solutions can be poured into cuvets for transmission measurements in either a scanning NIR instrument or an FT-NIR instrument. Modern FT-NIR transmission instruments permit the analysis of liquid samples in the cylindrical sample vials used for chromatography. For example, the 7 mm disposable glass vials used for LC can be used for transmission NIR measurements. Pharmaceutical laboratories often need to collect the IR spectrum as well as performing HPLC or HPLC-MS on samples. The ability to fill one type of sample vial and use it for multiple measurements is very valuable. It increases sample throughput by eliminating transfer of samples to multiple sample holders and by eliminating washing of sample cells or cuvets. It also conserves sample, decreases waste, and saves money.

4.7.3.2. Solids

For reflectance measurements, most solids require no sample preparation. Powders, tablets, textiles, solid "chunks" of material, food, and many other solids are analyzed "as is". Samples can be analyzed through plastic bags, glass vials, or in sample cups. Examples will be discussed in the applications section. NIR reflectance spectra of solids are often plotted as the second derivative of the spectrum. This format shows small differences between samples more clearly and eliminates scatter and slope in the baseline.

Polymer films can be measured in either reflectance or transmission mode. For transmission, the film may be taped across an IR transmission card or cardboard slide mount.

4.7.3.3. Gases

Gas samples are handled by filling the same type of gas cell used for mid-IR gas analysis, except that the cell windows are of quartz instead of salt.

4.7.4. Applications of NIR Spectroscopy

The most important bands are overtones or combinations of the stretching modes of C—H, O—H, and N—H. These bands enable the quantitative characterization of polymers, chemicals, foods, and agricultural products for analytes such as water, fatty acids, proteins, and the like. In many cases, the use of NIR reflectance spectroscopy has been able to replace time consuming, classical "wet" chemical analyses, such as the Kjeldahl method for protein nitrogen and the Karl Fischer titration for water content. The NIR region has been used for qualitative studies of hydrogen bonding, complexation in organometallic compounds, and solute–solvent interactions because the NIR absorptions are sensitive to intermolecular forces.

Polymer characterization is an important use of NIR spectrometry. Polymers can be made either from a single monomer, as is polyethylene, or from mixtures of monomers, as are styrene–butadiene rubber from styrene and butadiene and nylon 6-6, made from hexamethylenediamine and adipic acid. An important parameter of such copolymers is the relative amount of each present. This can be determined by NIR for polymers with the appropriate functional groups. Styrene content in a styrene–butadiene copolymer can be measured using the aromatic and aliphatic C—H bands. Nylon can be characterized by the NH band from the amine monomer and the C=O band from the carboxylic acid monomer. Nitrogen-containing polymers such as nylons, polyurethanes, and urea formaldehyde resins can be measured by using the NH bands. Block copolymers, which are typically made of a "soft block" of polyester and a "hard block" containing aromatics, for example, polystyrene, have been analyzed by NIR. These analyses have utilized the

C=O, aromatic and aliphatic C—H bands. NIR is used to measured hydroxyl content in polymers with alcohol functional groups (polyols), both in final product and online for process control.

Fibers and textiles are well suited to NIR reflectance analysis. Analyses include identifying the type of fiber, the uptake of dyes, the presence of processing oil in polyester yarns, and the presence of fabric "sizing" agents.

Proteins in foods can be measured by NIR reflectance spectrometry with no sample preparation. This has replaced the standard Kjeldahl protein nitrogen determination, which required extensive sample preparation to convert protein nitrogen to ammonia, distillation of the released ammonia, and subsequent titration of the ammonia. The replacement of the Kjeldahl method for routine analysis by NIR has permitted online measurement of protein in food and beverage products. The Kjeldahl method is required for assaying the materials used to calibrate the NIR and for method validation.

The determination of ppm amounts of water in many chemicals is critical. Organic solvents used for organic synthesis may have to be very dry so that traces of water do not interfere with the desired reaction, for example. The classic method for measuring water at low levels in organic solvents and other chemicals is the Karl Fischer titration, a time-consuming procedure requiring expensive reagents. Using the O—H bands characteristic of water, NIR has been used to measure water quantitatively in materials from organic solvents to concentrated mineral acids.

The use of NIR for process analysis and real-time analysis of complex samples is impressive. Grains such as wheat and corn can be measured with no sample preparation for protein content, water content, starch, lipids, and other components. An NIR spectrometer is now commercially available on grain harvesting machines to measure protein, moisture, and oil as the grain is being harvested. The analysis is performed in real-time, as the

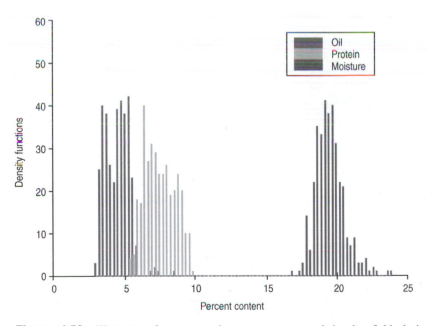

Figure 4.58 Histogram for a corn harvest as measured in the field during harvest by NIR spectroscopy. The oil content is centered about 4%, the protein is the lighter group of lines between 5% and 10% and the moisture is centered at about 19%. (Reprinted from von Rosenberg et al., with permission from Advanstar Communications, Inc.)

farmer is driving around the field harvesting the grain. The data, displayed in Fig. 4.58, is coupled with a global positioning unit on the harvester, resulting in a map of the farmer's field, showing the quality of the grain harvested from different points in the field (Fig. 4.59). This gives the farmer important information about where more fertilizer or water is needed. The real-time field analyzer gives excellent accuracy and precision, as shown by comparison to standard laboratory analyses (Fig. 4.60).

Pharmaceutical tablets packaged in plastic "blister packs" can be analyzed non-destructively by NIR spectrometry right through the package. In the quality control laboratory, this permits rapid verification of product quality without loss of product. In forensic analysis, unknown white powders in plastic bags seized as evidence in a drug raid can be identified by obtaining the NIR spectrum nondestructively right through the bag, eliminating the need to open the bag. This avoids possible contamination or spillage of the evidence.

The cost savings provide by the use of NIR instead of titrations for water and protein are enormous, not just in labor cost savings but in the cost of buying and then properly disposing of expensive reagents. NIR permits increased efficiency and increased product quality by online and at-line rapid analysis in the agricultural, pharmaceutical, polymer, specialty chemicals, and textile industries, to name a few.

4.8. RAMAN SPECTROSCOPY

Raman spectroscopy is a technique for studying molecular vibrations by light scattering. Raman spectroscopy complements IR absorption spectroscopy because some vibrations, as we have seen, do not result in absorptions in the IR region. A vibration is only seen in the IR spectrum if there is a change in the dipole moment during the vibration. For a vibration to be seen in the Raman spectrum, only a change in polarizability is necessary.

Figure 4.59   This figure is a false-color map of the field during a corn harvest showing the differ-ences in protein content for corn from different parts of the field. The bar under the map runs from blue on the left to red on the right. The top rectangle of the map is primarily blue (dark) and green (light), indicating 6–7% protein. The irregular spot in the middle left section is yellow (about 8% protein), while the rest of the lower portion is red and orange (8.5–9% protein). (Reprinted from von Rosenberg et al., with permission from Advanstar Communications, Inc.)

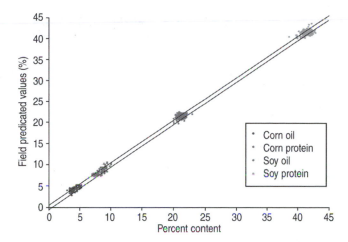

Figure 4.60 Correlation plot of the oil and protein values measured by the NIR field grain analyzer and the laboratory reference method (combustion analysis). (Reprinted from von Rosenberg et al., with permission from Advanstar Communications, Inc.)

That is, only a distortion of the electron cloud around the vibrating atoms is required. Distortion becomes easier as a bond lengthens and harder as a bond shortens, so the polarizability changes as the bound atoms vibrate. As we learned earlier in the chapter, homonuclear diatomic molecules such as Cl_2 do not absorb IR radiation, because they have no dipole moment. The Cl—Cl stretching vibration is said to be *IR-inactive*. Homonuclear diatomic molecules do change polarizability when the bond stretches, so the Cl—Cl stretch is seen in Raman spectroscopy. The Cl—Cl stretching vibration is said to be *Raman-active*. Some molecular vibrations are active in IR and not in Raman, and vice versa; many modes in most molecules are active in both IR and Raman. Looking at CO_2 again, shown below, the mode on the left is the IR-inactive symmetric stretch, while the other two asymmetric vibrations are both IR-active. The symmetric stretch is Raman-active.

4.8.1. Principles of Raman Scattering

When radiation from a source is passed through a sample, some of the radiation is scattered by the molecules present. For simplicity, it is best to use radiation of only one frequency and the sample should not absorb that frequency. The beam of radiation is merely dispersed in space. Three types of scattering occur. They are called *Rayleigh scattering*, *Stokes scattering*, and *anti-Stokes scattering*. Most of the scattered radiation has the same frequency as the source radiation. This is Rayleigh scattering, named after Lord Rayleigh, who spent many years studying light scattering. Rayleigh scattering occurs as a result of elastic collisions between the photons and the molecules in the sample; no energy is lost on collision. However, if the scattered radiation is studied closely, it can be observed that slight interaction of the incident beam with the molecules occurs. Some of the photons are scattered with less energy after their interaction with molecules and some photons are scattered with more energy. These spectral lines are called *Raman lines*, after Sir C.V. Raman, who first observed them in 1928. Only about 1 photon in a million will scatter with a shift in wavelength. The Raman–Stokes lines are from those photons scattered with less energy than the incident radiation; the Raman–anti-Stokes lines are from the photons scattered with more energy. The slight shifts in energy and therefore, slight shifts in the frequencies of these scattered photons are caused by inelastic collisions with molecules. The differences in the energies of the scattered photons from the incident photons have been found to correspond to vibrational transitions. Therefore the molecules can be considered to have been excited to higher vibrational states, as in IR spectroscopy, but by a very different mechanism. Figure 4.61 shows a schematic diagram of the Rayleigh and Raman scattering processes and of the IR absorption process.

The energy of the source photons is given by the familiar expression $E = h\nu$. If a photon collides with a molecule, the molecule increases in energy by the amount $h\nu$. This process is not quantized, unlike absorption of a photon. The molecule can be thought of as existing in an imaginary state, called a virtual state, with an energy between the ground state and the first excited electronic state. The energies of two of these virtual states are shown as dotted lines in Fig. 4.61. The two leftmost arrows depict increases in energy through collision for a molecule in the ground state and a molecule in the 1st excited vibrational state, respectively. The arrows are of the same length, indicating that the interacting photons have the same energy. If the molecule releases the absorbed energy, the scattered photons have the same energy as the source photons. These are the Rayleigh scattered photons, shown by the two middle arrows. The molecules have returned to the same states they started from, one to the ground vibrational state and the other to the first excited vibrational state. The arrows are the same length; therefore the scattered photons are of the same energy.

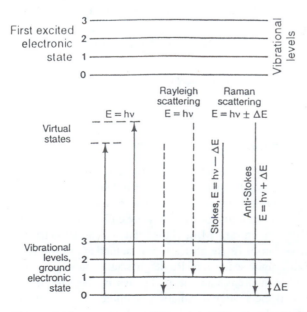

Figure 4.61   The process of Rayleigh and Raman scattering. Two virtual states are shown, one of higher energy. Rayleigh and Raman scattering are shown from each state. Normal IR absorption is shown by the small arrow on the far right marked ΔE, indicating a transition from the ground state vibrational level to the first excited vibrational level within the ground electronic state.

If the molecule begins to vibrate with more energy after interaction with the photon, that energy must come from the photon. Therefore, the scattered photon must decrease in energy by the amount equal to the vibrational energy gained by the molecule. That process is shown by the second arrow from the right. Instead of returning to the ground vibrational state, the molecule is now in the first excited vibrational state. The energy of the scattered photon is $E - \Delta E$, where ΔE is the difference in energy between the ground and first excited vibrational states. This is Raman scattering, and the lower energy scattered photon gives rise to one of the Stokes lines. Note that ΔE is equal to the frequency of an IR vibration; if this vibration were IR-active, there would be a peak in the IR spectrum at a frequency equal to ΔE. In general, the Raman–Stokes lines have energies equal to $E - \Delta E$, where ΔE represents the various possible vibrational energy changes of the molecule. This relationship can be expressed as:

$$E - \Delta E = h(v - v_1) \tag{4.13}$$

where v is the frequency of the incident photon and v_1 is the *shift* in frequency due to an energy change ΔE. Several excited vibrational levels may be reached, resulting in several lines of energy $h(v - v_1)$, $h(v - v_2)$, $h(v - v_3)$, and so on. These lines are all shifted in frequency from the Rayleigh frequency. The Stokes lines, named after Sir George Gabriel Stokes, who observed a similar phenomenon in fluorescence, are shifted to lower frequencies than the Rayleigh frequency. The Raman shifts are completely independent of the wavelength of the excitation source. Sources with UV, visible, and NIR wavelengths are used, and the same Raman spectrum is normally obtained for a given molecule. There are exceptions due to instrumental variations and also if a resonance or near-resonance condition applies at certain wavelengths (Section 4.8.4).

Less commonly, the molecule *decreases* in vibrational energy after interacting with a photon. This might occur if the molecule is in an excited vibrational state to begin with and relaxes to the ground vibrational state. In this case, the molecule has given energy to the scattered photon. The photon is shifted to a frequency higher than the incident radiation. These higher frequency lines, the Raman–anti-Stokes lines, are less important to analytical chemists than the Stokes lines because of their lower intensity. One exception to this is for samples that fluoresce strongly. Fluorescence interferes with the Stokes lines but to a much lesser extent with the anti-Stokes lines.

It is convenient to plot the Raman spectrum as intensity vs. shift in wavenumbers in cm^{-1}, because these can be related directly to IR spectra. The Raman shift in cm^{-1} is identical to the IR absorption peak in cm^{-1} for a given vibration, because both processes are exciting the same vibration. The Raman spectrum for benzene is shown in Fig. 4.62, along with the related IR transmission spectrum.

4.8.2. Raman Instrumentation

A Raman spectrometer requires a light source, a sample holder or cell, a wavelength selector (or interferometer), and a detector, along with the usual signal processing and display equipment. Since Raman spectroscopy measures scattered radiation, the light source and sample cell are usually placed at $90°$ to the wavelength selector, as shown schematically in Fig. 4.63. The radiation being measured in Raman spectroscopy is either visible or NIR; therefore spectrometer optics, windows, sample cells, and so on can be made of glass or quartz. It is critical in Raman spectroscopy to completely exclude fluorescent room lights from the spectrometer optics. Fluorescent lights give rise to numerous spurious signals.

Figure 4.62   (a) The IR spectrum and Raman spectrum of benzene. (b) The IR spectrum of ethanol. (c) The Raman spectrum of ethanol. (b) and (c) are not plotted on the same scale. (The ethanol spectra are courtesy of http://www.aist.go.jp/RIODB/SDBS.)

4.8.2.1.   Light Sources

Monochromatic light sources are required for Raman spectroscopy. The light sources used originally were simple UV light sources, such as Hg arc lamps; however, these were weak sources and only weak Raman signals were observed. The Raman signal is directly

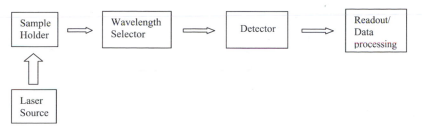

Figure 4.63   Idealized layout of a Raman spectrometer.

proportional to the power of the light source, which makes the laser, which is both mono-chromatic and very intense, a desirable light source. It was the development in the 1960s of lasers that made Raman spectroscopy a viable and useful analytical technique. Modern Raman instruments use a laser as the light source. The use of these intense light sources has greatly expanded the applications of Raman spectroscopy, because of the dramatically increased intensity of the signal and a simultaneous improvement of the signal-to-noise ratio. Lasers and excitation wavelengths commonly used for Raman instruments include visible wavelength helium/neon lasers and ion lasers such as the argon ion laser (488 nm) and the krypton ion laser (531 nm). The intensity of Raman scattering is pro-portional to the fourth power of the excitation frequency or to $1/\lambda^4$, so the shorter wave-length blue and green ion lasers have an advantage over the red helium/neon laser line at 633 nm. The disadvantage of the shorter wavelength lasers is that they can cause the sample to decompose on irradiation (photodecomposition) or fluoresce, an interference discussed subsequently. NIR lasers, such as neodymium/yttrium aluminum garnet, Nd/YAG, with an excitation line at 1064 nm, are used to advantage with some samples because they do not cause fluorescence or photodecomposition.

4.8.2.2.   Dispersive Spectrometers Systems

Traditional Raman spectrometers used a monochromator with two or even three gratings to eliminate the intense Rayleigh scattering. The optical layout is similar to that for the UV/VIS single grating monochromators discussed in Chapters 2 and 5. Holographic inter-ference filters, called *super notch filters*, have been developed that dramatically reduce the amount of Rayleigh scattering reaching the detector. These filters can eliminate the need for a multiple grating instrument unless spectra must be collected within $150\ \text{cm}^{-1}$ of the source frequency. Dispersive systems generally use a visible laser as the source.

The traditional detector for these systems was a photomultiplier tube. Multichannel instru-ments with photodiode array (PDA), CID, or CCD detectors are commonly used today. All three detectors require cryogenic cooling. The PDA has the advantage of having the fastest response but requires more complicated optics than the other detectors. The CID has the advantage over both the PDA and CCD of not "blooming", that is, not having charge spill over onto adjacent pixels in the array, which would be read in error as a signal at a frequency where no signal exists. CCDs are the slowest of the three array detectors because they have to be read out by trans-ferring the stored charge row by row, but they are also the least expensive of the detector arrays. Sensitivity is improved in newer CCD designs as well. A dispersive Raman spectrometer with a CCD detector is shown schematically in Fig. 4.64.

4.8.2.3.   FT Raman Spectrometers

FT-Raman systems generally use an NIR laser source, such as the Nd/YAG laser, and a Michelson interferometer. A schematic FT-Raman spectrometer is shown in Fig. 4.65.

Figure 4.64 Schematic of a dispersive Raman spectrometer. (Reprinted from Weesner and Longmire, with permission from Advanstar Communications, Inc.)

The NIR laser source line is at 1064 nm, so the Raman–Stokes lines occur at longer wavelengths. This is beyond the detection range of the materials used in array detectors. The detector for an NIR FT-Raman system is a liquid nitrogen-cooled photoconductive detector such as Ge or InGaAs.

FT-Raman has many of the advantages of FTIR. There is high light throughput, simultaneous measurement of all wavelengths (the multiplex advantage), increased signal-to-noise ratio by signal averaging, and high precision in wavelength. A major advantage is in the use of the NIR laser, which dramatically reduces fluorescence in samples. Fluorescence occurs when the virtual states populated by excitation overlap excited electronic states in the molecule. Then, the molecule can undergo a radiationless transition to the lowest ground state of the excited electronic state before emitting a fluorescence photon on relaxation to the ground state. The fluorescence photon is of lower energy than the exciting radiation, and so fluorescence occurs at longer wavelengths, interfering with the Stokes scattering lines. The NIR laser is of low energy and does not populate virtual states that overlap the excited electronic states, as higher energy visible lasers

Figure 4.65 Schematic of an FT-Raman spectrometer. (Reprinted from Weesner and Longmire, with permission from Advanstar Communications, Inc.)

can. As an example, the Raman spectrum of cocaine is shown in Fig. 4.66. The spectrum in Fig. 4.66(a) was collected with an FT-Raman spectrometer using an NIR laser, while that in Fig. 4.66(b) was collected with a dispersive Raman system and a visible laser. Figure 4.66(b) shows a large fluorescence band that obscures most of the Raman spectrum below 2000 cm^{-1}. With appropriate mathematical "smoothing" algorithms and multipoint baseline correction, it is possible to extract a useable Raman spectrum from samples that exhibit strong fluorescence, as shown in Fig. 4.67. One consideration in FT-Raman is that the laser line at 1064 nm is very close to a water absorption band. While this does not prevent aqueous solutions from being studied by FT-Raman, aqueous solutions cannot be studied as easily as they can with dispersive Raman.

4.8.2.4. *Samples and Sample Holders for Raman Spectroscopy*

Because the laser light source can be focused to a small spot, very small samples can be analyzed by Raman spectroscopy. Samples of a few microliters in volume or a few milligrams are sufficient in most cases. Liquid samples can be held in beakers, test tubes, glass capillary tubes, or NMR tubes. Aqueous solutions can be analyzed since water is a very weak Raman scatterer. This is a significant advantage for Raman spectroscopy over IR. Other solvents that can be used for Raman studies include chloroform, carbon tetrachloride, acetonitrile, and carbon disulfide. Solid powders can be packed into glass capillary tubes, NMR tubes, or glass vials for analysis. The spectra are obtained through the glass. Solid samples can also be mounted at the focal point of the laser beam and their spectra obtained "as is" or pressed into pellets. Gas samples do not scatter radiation efficiently, but can be analyzed by being placed into a multipath gas

Figure 4.66 Analyses of crack cocaine using (a) FT-Raman and (b) 785 nm dispersive Raman. Note the lack of fluorescence in the FT-Raman spectrum and the rich spectral information between 2500 and 3300 cm^{-1}. This information is obscured by the fluorescence band in the dispersive spectrum. (Reprinted from Weesner and Longmire, with permission from Advanstar Communications, Inc.)

Figure 4.67 It is possible to extract data from a Raman spectrum that exhibits fluorescence interference by the application of mathematical data treatments as shown here. The data treatments were applied sequentially to give the final spectrum shown in (d). (a) is the raw spectrum of a weak scatterer with a fluorescence background, (b) is the spectrum after Savitzky–Golay smoothing, (c) after a multipoint baseline correction performed by the analyst, and (d) after a Fourier smoothing. (Reprinted from Kawai and Janni, with permission from Advanstar Communications, Inc.)

cell, with reflecting mirrors at each end. The body of the gas cell must be of glass to allow collection of the scattered light at 90°.

The sample must be placed at the focal point of an intense laser beam, and some samples may be subject to thermal decomposition or photodecomposition. Accessories that spin the sample tube or cup are available, to distribute the laser beam over the sample and reduce heating of the sample. Spinning or rotating the sample minimizes thermal decomposition, but does not stop photodecomposition. Sample spinning is required for *resonance Raman spectroscopy*, discussed later.

Raman spectroscopy does not suffer interference from atmospheric water vapor or carbon dioxide, as does IR. Gases do not scatter well, so even though Raman-active bands occur for these gases, the contribution to the Raman signal from air in the optical path is insignificant. Materials in the optical path outside of the laser focus also have negligible scattering.

4.8.3. Applications of Raman Spectroscopy

Quantitative and qualitative analyses of inorganic and organic compounds can be performed by Raman spectroscopy. Raman spectroscopy is used for bulk material characterization, online process analysis, remote sensing, microscopic analysis, and chemical imaging of inorganic, organic and organometallic compounds, polymers, biological systems, art objects, and much more. Raman spectra have fewer lines and much sharper lines than the corresponding IR spectra. This makes quantitative analysis, especially of mixtures, much simpler by Raman spectroscopy than by IR spectroscopy.

Quantitative analysis had not been as common as in IR spectroscopy until recently, due to the high cost of Raman instruments. With prices for Raman systems dropping below $40,000, and even as low as $10,000, the use of Raman spectroscopy for quantitative analysis is increasing. Quantitative analysis requires measurement of the intensity of the Raman peaks and the use of a calibration curve to establish the concentration–intensity relationship. The intensity of a Raman peak is directly proportional to the concentration:

$$I = KJv^4c \tag{4.14}$$

where I is the intensity of Raman signal at frequency v; K, the proportionality constant including instrument parameters such as laser power; J, the scattering constant for the given Raman peak; v, the scattering frequency of the Raman peak; and c, concentration of analyte.

K, J, and v are all constants for a given sample measurement. The frequency and intensity of the desired Raman lines are measured, and the intensity of an unknown compared to the calibration curve. It is common to use an internal standard for Raman analysis, because of the dependence of the signal on the laser power (in the K term). Without an internal standard, the laser power, sample alignment, and other experimental parameters must be carefully controlled. If an internal standard is used, the intensity of the internal standard peak is also measured, and the ratio of intensities plotted vs. concentration. Upon division, this reduces Eq. (4.14) to:

$$\frac{I_{unk}}{I_{std}} = K'c \tag{4.15}$$

The use of the internal standard minimizes the effect of changes in instrumental parameters and can result in better accuracy and precision. Quantitative analyses that can be done by Raman spectroscopy include organic pollutants in water, inorganic oxyanions and organometallic compounds in solution, aromatic/aliphatic hydrocarbon ratios in fuels, antifreeze concentration in fuel, and concentration of the active pharmaceutical ingredient in the presence of excipients such as microcrystalline cellulose. Mixtures of compounds in pharmaceutical tablets can be determined quantitatively, without dissolving the tablets. Raman sensitivity varies greatly, depending on the sample and the equipment. In general, analyte concentrations of at least 0.1–0.5 M are needed to obtain good signals.

Another use of Raman spectroscopy for quantitative analysis is the determination of percent crystallinity in polymers. Both the frequency and intensity of peaks can shift on going from the amorphous to the semicrystalline state for polymers. The percent crystallinity can be calculated with the help of chemometrics software.

Qualitative analysis by Raman spectroscopy is very complementary to IR spectroscopy and in some cases has an advantage over IR spectroscopy. The Raman spectrum is more sensitive to the organic framework or backbone of a molecule than to the functional groups, in contrast to the IR spectrum. IR correlation tables are useful for Raman spectra, because the Raman shift in wavenumbers is equal to the IR absorption in wavenumbers for the same vibration. Raman spectral libraries are available from commercial and government sources, as noted in the bibliography. These are not as extensive as those available for IR, but are growing rapidly.

The same rules for number of bands in a spectrum apply to Raman spectra as well as IR spectra: $3N-6$ for nonlinear molecules and $3N-5$ for linear molecules. There may be fewer bands than theoretically predicted due to degeneracy and nonactive modes. Raman spectra do not usually show overtone or combination bands; they are much weaker than in IR. A "rule of thumb" that is often true is that a band that is strong in IR is weak in Raman and vice versa. A molecule with a center of symmetry, such as CO_2, obeys another rule: if a band is present in the IR spectrum, it will not be present in the Raman spectrum. The reverse is also true. The detailed explanation for this is outside the scope of this text, but the rule "explains" why the symmetric stretch in carbon dioxide is seen in the Raman spectrum, but not in the IR spectrum, while the asymmetric stretch appears in the IR spectrum but not in the Raman spectrum.

Structural identification of an unknown is usually done with both IR and Raman spectra, or by matching Raman spectra to spectral libraries of known compounds. Subtraction of known spectra from the spectrum of an unknown mixture to identify the components of the mixture works better for Raman spectra than for IR spectra, because there are fewer Raman peaks, the peaks are sharp and their position and shape are not affected by hydrogen-bonding. For example, it is possible to identify the components of a commercial pain relief tablet by spectral subtraction from the Raman spectrum of the intact tablet, as shown in Fig. 4.68. Another example of spectral subtraction is the identification of cocaine in a mixture of cocaine and lactose, seen in Fig. 4.69.

Raman spectroscopy is particularly useful for studying inorganic and organometallic species. Most inorganic, oxyanionic, and organometallic species have vibrations in the far-IR region of the spectrum, which is not easily accessible with commercial IR equipment. These metal–ligand and metal–oxygen bonds are Raman-active and are easily studied in aqueous solutions.

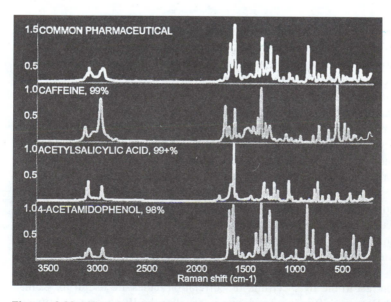

Figure 4.68 The analysis of a common multicomponent pharmaceutical tablet by Raman spectroscopy with spectral subtraction. The top spectrum is the entire tablet. The spectra for acetylsalicylic acid and caffeine were subtracted, resulting in the spectrum of the third component, identified as 4-acetamidophenol. [Courtesy of ThermoNicolet, Madison, WI (www.thermonicolet.com).]

Figure 4.69   Raman spectral subtraction can be used to identify the "cutting agent", lactose, used to dilute cocaine. The reference spectrum for lactose is shown as well as the spectrum of the mixture. Subtraction of the lactose spectrum from the mixture results in a spectrum that matches pure cocaine. [Courtesy of ThermoNicolet, Madison, WI (www.thermonicolet.com).]

As noted earlier, fused silica optical fiber is used for remote NIR measurements. The same type of fiber optic probe can be used for Raman spectroscopy, and enables remote measurement of samples and online process measurements. *In situ* reaction monitoring by Raman spectroscopy has been used to study catalytic hydrogenation, emulsion polymerization, and reaction mechanisms. Remote sensing of molecules in the atmosphere can be performed by Raman scattering measurements using pulsed lasers.

4.8.4.   The Resonance Raman Effect

When monochromatic light of a frequency that cannot be absorbed by the sample is used, the resulting Raman spectrum is the **normal** Raman spectrum. Normal Raman spectroscopy, as has been noted, is an inefficient process resulting in low sensitivity and it suffers from interfering fluorescence in many samples. If a laser excitation wavelength is used that falls within an excited electronic state of the molecule, the intensity of some Raman lines increases by as much as $10^3 - 10^6$ over the normal Raman intensity. This is known as the *resonance Raman effect*, and the technique is called resonance Raman spectroscopy. The molecule must possess a *chromophore* that can absorb visible or UV radiation (discussed in Chapter 5). The process that occurs is shown schematically in Fig. 4.70. The resulting spectra are even simpler than normal Raman spectra, because only totally symmetric vibrations associated with the chromophore are enhanced in intensity. This makes resonance Raman a very selective probe for specific chromophores. Resonance Raman spectroscopy has been used to study low concentrations of biologically important molecules such as hemoglobin.

Lasers that have wavelengths in the UV and visible regions of the spectrum are used for resonance Raman spectroscopy. Tunable dye lasers are often used; these lasers can

First excited
electronic state
with 4 vibrational
levels shown

Ground electronic
state with 4
vibrational levels
shown

Figure 4.70   The resonance Raman process.

be set to a selected wavelength over the UV/VIS range of 200–800 nm. This permits
maximum flexibility in the choice of excitation wavelength.

4.8.5.   Surface-Enhanced Raman Spectroscopy (SERS)

Surface-enhanced Raman spectroscopy (SERS) is another technique for obtaining strong
Raman signals from surfaces and interfaces, including species adsorbed onto surfaces.
Fleischmann and coworkers developed SERS in 1974. The SERS technique requires adsorption of the species to be studied onto a "rough" metal or metal colloid surface, where the
roughness is at the atomic level. Inorganic and organic species adsorbed onto such surfaces
show enhancement of Raman signals by up to 10^6 over normal Raman signals. The reasons
for the enhancement are not yet well understood. Metals used as surfaces include gold,
silver, and copper. Metal electrodes, metal films, and metal colloids have been used. The
adsorbed or deposited analyte molecule must be less than 50 Å from the surface for the
enhancement to be observed. Samples have been deposited electrolytically onto electrode
surfaces, or mixed with colloidal metal suspensions. Samples separated on thin layer
chromatography (TLC) plates have been sprayed with metal colloid solution.

   The enhancement leads to the ability to detect extremely small amounts of
material, making SERS an effective tool for a variety of problems, from corrosion studies
to detection of chemical warfare agents. Detection limits for SERS are in the nanogram range.

4.8.6.   Raman Microscopy

Raman spectroscopy coupled to a microscope permits the analysis of very small samples
nondestructively. The use of Raman microscopy allows the characterization of specific

domains or inclusions in heterogeneous samples with very high spatial resolution. With dispersive Raman microscopy, the spatial resolution is often better than 1 μm. FT-Raman microscopy is limited to spot sizes of about 2–10 μm, but with no interference from fluorescence. The use of a confocal microscope (Fig. 4.71) allows only the light at the sample focus to pass into the detector; all other light is blocked. This permits non-destructive depth profiling of samples without the need for cross-sectioning of the sample. For example, confocal Raman microscopy can identify the polymers in complex layered structures, such as multilayer films used for food packaging. Characterization of heterogeneous materials includes inclusions in minerals, the pigments, and other components in cosmetics, and the study of pigments, resins, and dyes in art and archeological objects. Using Raman microscopy, it is possible to identify if a red pigment in a painting is expensive cinnabar (HgS), cheaper hematite (Fe_2O_3), a mixture of the two, or an organic dye. The article by Edwards provides a detailed table of Raman bands from common minerals used in art and an overview of the use of Raman microscopy for the study of art objects. Raman microscopy is particularly useful for art and archeological objects, because there is no sample preparation required and the natural water present in paintings, manuscripts, and ancient textiles does not interfere, as it does in IR spectroscopy.

FTIR microscopy was discussed earlier in the chapter, and given the complementary nature of IR and Raman, it is reasonable that laboratories performing IR microscopy might well need Raman microscopy and vice versa. Two microscope systems were required and the sample had to be moved from one system to the other. The difficulty of relocating the exact spot to be sampled can be imagined. A new combination dispersive Raman and FTIR microscopy system was introduced in 2002. The system, called the LabRam-IR (JY Horiba, Edison, NJ), allows both Raman and IR spectra to be collected at exactly the same location on the sample. The resolution depends on the

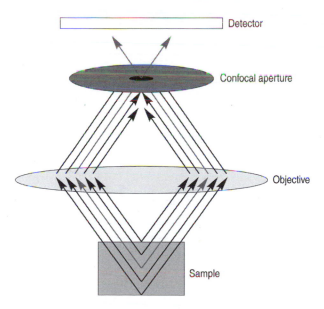

Figure 4.71 A simplified illustration of confocal microscopy. A spectrum is collected only from the area of the sample denoted by the middle arrow; only that light exits the confocal aperture. Information from all other sample areas is blocked. (Reprinted from Weesner and Longmire, with permission from Advanstar Communications, Inc.)

Figure 4.72 The Raman and FTIR spectra of a 10 μm fiber of a nylon 6-polyethylene glycol block copolymer. Spectra were collected at exactly the same spot on the fiber with the JYHoriba LabRam-IR microscope with Same Spot™ technology. [Courtesy of Jobin Yvon, Horiba Group, Edison, NJ (www.jyhoriba.com).]

Figure 4.73 Raman and FTIR spectra of a gallstone. The spectra were recorded with the JYHoriba LabRam-IR microscope using a 532 nm laser. The FTIR spectrum is more sensitive to the OH bands, while the Raman spectrum starts below 600 cm⁻¹ and shows details of the cholesteric species and the C=C bands. [Courtesy of Jobin Yvon, Horiba Group, Edison, NJ (www.jyhoriba.com).]

Figure 4.74 A schematic representation of a 3D spectral data cube. Sample spatial positions are represented by the *x* and *y* coordinates; the spectral wavelength is represented by the λ axis. (Reprinted from McLain et al., with permission from Advanstar Communications, Inc.)

Figure 4.75 Use of FTIR imaging. (a) The visible light image of an inclusion in polypropylene film. Sample size is approximately 450×450 μm. (b) The IR image of the same inclusion showing the distribution of a carbonyl contaminant through the inclusion. The image contains more than 5000 spectra at 6.25 μm pixel size. The total data collection time was 90 s. Collected on a Spectrum Spotlight 300 FT-IR Imaging System. [Courtesy of PerkinElmer Instruments, Shelton, CT (www.perkinelmer.com).]

wavelength observed, because resolution is limited by diffraction, but is <1 μm for the Raman spectrum and 10–20 μm for the IR spectrum. Examples of the type of data that can be obtained with this combination microscopy system are shown in Figs. 4.72 and 4.73.

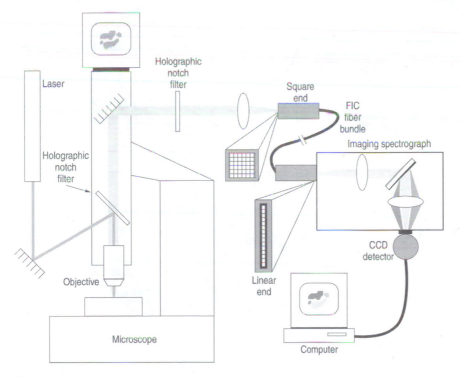

Figure 4.76 Schematic of the NIRIM. (Reprinted from McLain et al., with permission from Advanstar Communications, Inc.)

4.9. CHEMICAL IMAGING USING NIR, IR, AND RAMAN SPECTROSCOPY

The most recent breakthrough in the use of vibrational spectroscopy for chemical analysis is in the area of *chemical imaging*. Chemical imaging is the use of 2D or 3D detectors to collect spectral data from a large number of locations within a sample and then using the variations in the spectral data to map chemical differences within the sample. The chemical differences are often displayed as a false-color image of the sample. The use of chemical imaging technology in NMR has been described in Chapter 3; it is the technique more commonly called MRI.

FTIR imaging has been commercially available since 1996. The usual "detector" is an MCT array detector, called a focal plane array (FPA) detector, used in conjunction with an FTIR microscope. A 64 × 64 FPA detector has 4096 detector elements and allows 4096 interferograms to be collected simultaneously. Because each pixel in the detector array generates a spectrum, there are three dimensions in the data set. These data sets are often referred to as data cubes or image cubes. The x and y coordinates of the cube are the spatial positions while the z coordinate represents the wavelength, as shown in Fig. 4.74. The data can be handled in many ways, including the use of library searching, principal component analysis, and more, making this a powerful technique. As a simple example, the intensity of the carbonyl-stretching band in each pixel can be "reassembled" into the visual image seen through the microscope, giving a distribution of carbonyl-containing material in the sample. Such an image is shown in Fig. 4.75, collected with the Spectrum Spotlight 300 FTIR Imaging System (PerkinElmer Instruments,

Figure 4.77 Identification of a small molybdenum sulfide inclusion in a larger boric acid crystal using the NIRIM system. (Reprinted from McLain et al., with permission from Advanstar Communications, Inc.)

Shelton, CT, www.perkinelmer.com). Samples can be measured in reflectance mode or transmission mode. Solid samples such as biological tissues may need to be sliced into thin sections for transmission analysis. FTIR imaging can be performed on a wide variety of sample matrices, including polymers, pharmaceutical tablets, fibers, and coatings.

Commercial Raman and NIR imaging microscope systems are also available. Raman imaging of polymer blend surfaces, Raman and NIR imaging of silicon integrated circuits, and NIR imaging of whole pharmaceutical tablets are a few applications. The FALCON™ Raman Chemical Imaging Microscope (ChemIcon, Pittsburgh, PA, www.chemimage.com) can be equipped with fiber optic probes for remote monitoring and for high-temperature remote monitoring, such as in a heated process stream. The ChemIcon CONDOR™ Macro Chemical Imaging System can do NIR absorption imaging as well as fluorescence and visible emission imaging. A fast near-IR Raman Imaging microscope system (NIRIM) is described by McLain and coauthors that uses a fiber optic bundle and CCD detector to collect a complete 3D Raman data cube from a sample in 1 s or less. A schematic of the NIRIM is shown in Fig. 4.76. The system has been used to image inorganic inclusions in crystals, mixtures of metal oxides, amino acid mixtures and to perform surface-enhanced Raman imaging of catalyst and nanoparticle surfaces, among other studies. Two examples are shown in Figs. 4.77 and 4.78.

Chemical imaging provides a nondestructive and noninvasive way to map the chemical composition of a wide variety of samples very rapidly and at the microscopic

Figure 4.78 Imaging and identification of three different amino acid crystals in a mixture using the NIRIM system. (Reprinted from McLain et al., with permission from Advanstar Communications, Inc.)

level. Applications range from real-time process monitoring and biomonitoring to materials characterization, engineering materials fabrication, and failure analysis. Chemical imaging systems are now available for about $150,000, making this technology feasible for even routine analysis.

BIBLIOGRAPHY

Aikens, D.A.; Bailey, R.A.; Moore, J.A.; Giachino, G.G.; Tomkins, R.P.T. *Principles and Techniques for an Integrated Chemistry Laboratory*; Waveland Press, Inc.: Prospect Heights, IL, 1978. (Reissued 1984).

Asher, S.A. UV resonance Raman spectroscopy for analytical, physical and biophysical chemistry, Pts.1 and 2. Anal. Chem. **1993**, *65* (2), 59A–66A and 201A–210A.

Burns, D.R.; Ciurczak, E.W., Eds., *Handbook of Near Infrared Analysis*; Marcel Dekker, Inc.: New York, 1993.

Coates, J. Vibrational spectroscopy. In *Analytical Instrumentation Handbook*, 2nd Ed.; Ewing, G.W., Ed.; Marcel Dekker, Inc.: New York, 1997.

Colthup, N.B.; Daly, L.H.; Wiberley, S.E., *Introduction to Infrared and Raman Spectroscopy*, 3rd Ed.; Academic Press: New York, 1990.

Cooke, P.M. Chemical microscopy. Anal. Chem. **1996**, *68* (12), 339R.

De Blase, F.J.; Compton, S. IR emission spectroscopy: a theoretical and experimental review. Appl. Spectrosc. **1991**, *45* (4), 611.

Edwards, H.G.M. Raman microscopy in art and archeology. Spectroscopy **2002**, *17* (2), 16.

Ferraro, J.R.; Nakamoto, K. *Introductory Raman Spectroscopy*; Academic Press: New York, 1994.

Fleischmann, M.; Hendra, P.J.; McQuillan, A.J. Chem. Phys. Lett. **1974**, *26*, 163.

Garrell, R.L. Surface-enhanced Raman spectroscopy. Anal. Chem. **1989**, *61* (6), 401A.

Goddu, R.F. Near infrared spectrophotometry. In *Advances in Analytical Chemistry and Instrumentation*; Reilly, C.N., Ed.; John Wiley and Sons, Inc.: New York, 1960; Vol. 1.

Grasselli, J.G.; Bulkin, B.J., Eds. *Analytical Raman Spectroscopy*; John Wiley and Sons, Inc.: New York, 1991.

Grasselli, J.G.; Snavely, M.K.; Bulkin, B.J. *Chemical Applications of Raman Spectroscopy*; John Wiley and Sons, Inc.: New York, 1981.

Greenberg, A.E.; Clesceri, L.S.; Eaton, A.D., Eds. *Standard Methods for the Examination of Water and Wastewater*, 18th Ed.; American Public Health Association: Washington, DC, 1992.

Griffiths, P.R.; de Haseth, J.A. *Fourier Transform Infrared Spectrometry. Chemical Analysis*; Wiley-Interscience: New York, 1986; Vol. 83.

Humecki, H.J., Ed. Practical guide to infrared microspectroscopy. *Practical Spectroscopy Series*; Marcel Dekker, Inc.: New York, 1995; Vol. 19.

Hsu, C.-P.S. Infrared spectroscopy. In *Handbook of Instrumental Techniques for Analytical Chemistry*, Settle, F.A., Ed.; Prentice Hall PTR: Upper Saddle River, NJ, 1997.

Ingle, J.D., Jr.; Crouch, S.R. *Spectrochemical Analysis*; Prentice-Hall, Inc.: Englewood Cliffs, NJ, 1988.

Kawai, N.; Janni, J.A. Chemical identification with a portable raman analyzer and forensic spectral database. Spectroscopy, **2000**, *15* (10), 32.

Lambert, J.B.; Shurvell, H.F.; Lightner, D.; Cooks, R.G. *Introduction to Organic Spectroscopy*; Macmillan Publishing Company: New York, 1987.

Marbach, R.; Kosehinsky, T.; Gries, F.A.; Heise, H.M. Noninvasive blood glucose assay by NIR diffuse reflectance spectroscopy of the human inner lip. Appl. Spectrosc. **1983**, *47* (7), 875.

McLain, B.L.; Hedderich, H.G.; Gift, A.D.; Zhang, D.; Jallad, K.; Haber, K.S.; Ma, J.; Ben-Amotz, D. Fast chemical imaging. Spectroscopy **2000**, *15* (9), 28.

Metzel, D.L.; LeVine, S.M. In-situ FTIR microscopy and mapping of normal brain tissue. Spectroscopy **1993**, *8* (4), 40.

Mirabella, F.M., Ed. Internal reflection spectroscopy: theory and applications. *Practical Spectroscopy Series*; Marcel Dekker, Inc.: New York, 1992; Vol. 15.

Morris, M.D., Ed. *Microscopic and Spectroscopic Imaging of the Chemical State*; Marcel Dekker: New York, 1993.

Nakamoto, K. *Infrared and Raman Spectra of Inorganic and Coordination Compounds*, 5th Ed.; John Wiley and Sons, Inc.: New York, 1996.

Pattacini, S. *Solving Analytical Problems Using Infrared Spectroscopy Internal Reflectance Sampling Techniques*; Pattacini Associates, LLC: Danbury, CT.

Pavia, D.L.; Lampman, G.M.; Kriz, G.S. *Introduction to Spectroscopy*, 3rd Ed.; Harcourt College Publishers: New York, 2001.

Raman, C.V. Indian J. Phys. **1928**, *2*, 387.

Robinson, J.W., Ed., *Handbook of Spectroscopy*; CRC Press: Boca Raton, FL, 1974; Vol. II.

Robinson, J.W., *Practical Handbook of Spectroscopy*; CRC Press: Boca Raton, FL, 1991.

Schultz, C.P. Precision infrared spectroscopic imaging. Spectroscopy **2001**, *16* (10), 24.

Silverstein, R.M.; Webster, F.X., *Spectrometric Identification of Organic Compounds*, 6th Ed.; John Wiley and Sons, Inc.: New York, 1998.

Smith, A.L. *Infrared Spectroscopy, Treatise on Analytical Chemistry*, Part 1; John Wiley and Sons, Inc.: New York, 1981; Vol. 7.

Strommen, D.P. Raman spectroscopy. In *Handbook of Instrumental Techniques for Analytical Chemistry*; Settle, F.A., Ed.; Prentice Hall PTR: Upper Saddle River, NJ, 1997.

von Rosenberg, Jr., C.W.; Abbate, A.; Drake, J.; Mayes, D.M. A rugged near-infrared spectrometer for the real-time measurement of grains during harvest. Spectroscopy **2000**, *15* (6), 35.

Weesner, F.; Longmire, M. Dispersive and fourier transform raman. Spectroscopy **2001**, *16* (2), 68.

Note: The articles cited from *Spectroscopy* magazine are available online at www. spectroscopyonline.com, where the color pictures give a much better idea of what imaging can do than the grayscale copies used in the text.

SPECTRAL DATABASES

This list is by no means complete. Many instrument manufacturers offer spectral databases packaged with their IR and Raman instruments. Many government agencies, including the US NIST (www.nist.gov) and the US EPA (www.usepa.gov), and the Georgia State Crime Lab publish specialized spectral databases. The Canadian government publishes a forensic spectral database.

Bio-Rad Laboratories, Informatics Division, Philadelphia, PA (www.bio-rad.com) publishes the Sadtler Infrared and Raman spectra collections of over 150,000 spectra. They are available in electronic format, hardcopy and in a variety of specialized subsets.

Sigma-Aldrich Chemical Company (www.sigma-aldrich.com) publishes over 18,000 FTIR spectra in both electronic and hardcopy formats.

National Institute of Advanced Industrial Science and Technology, Tsukuba, Ibaraki, Japan, publishes a free spectral database system of organic compounds. The spectra include IR, Raman, NMR, and MS for most compounds. The database may be found at www.aist.go.jp/RIODB/SDBS.

SUGGESTED EXPERIMENTS

4.1   Record the IR absorption spectrum of hexane. Identify the absorption bands caused by the C—H stretching frequency, the C—H bending frequency, and the C—C stretching frequency.

4.2   Record the IR spectrum of heptane. Note the similarity with the spectrum obtained in Experiment 4.1. Would it be possible to distinguish between these compounds based on their IR spectra?

4.3   Record the IR spectrum of *n*-butanol. Note the O → H stretching peak and the C → OH peak. Repeat with *iso*-butanol and *tert*-butanol. Note any changes in the positions of these peaks. Is there a peak that can be used to distinguish among primary, secondary, and tertiary alcohols?

4.4   Record the IR spectrum of *n*-butylamine. Note the N → H and C → N stretching peaks. Repeat with *sec*-butylamine and *tert*-butylamine. Compare the spectra to what you expect from your reading.

4.5   Place a few drops of a volatile organic liquid (toluene, xylene, carbon tetrachloride, etc.) in a 10 cm gas cell and close the valves. Allow the cell to sit for several minutes. Record the gas phase spectrum of the compound. Compare the spectrum to the liquid phase spectrum and explain your observations.

4.6   Make up several different solutions of known concentrations of ethanol in carbon tetrachloride over the concentration range of 1–30%. Measure the intensity of the C—OH absorption band. Plot the relationship between absorbance and the concentration of ethanol. (Alternatively, run the quantitative "oil and grease" method from Standard Methods for the examination of Water and Wastewater, or the similar EPA method 413.2 or 418.1.)

4.7   Repeat Experiment 4.6, using acetone as the sample. Note the sharp C ⇒ O stretching band and use the absorbance of this band for quantitative studies as suggested in Experiment 4.6.

4.8   Take a sample of unsaturated cooking oil and record the IR absorption spectrum. Note the C ⇒ C stretching frequency. Repeat with several brands of cooking oil. Based on the IR absorption spectrum, which brand was the most unsaturated?

PROBLEMS

4.1   What types of vibrations are encountered in organic molecules?

4.2   What materials are used for making the cell windows used in IR spectroscopy? Why must special precautions be taken to keep these materials dry?

4.3   Why do organic functional groups resonate at characteristic frequencies?

4.4   How can primary, secondary, and tertiary alcohols be distinguished by their IR absorption spectra?

4.5   Indicate which C—H and C—C stretching and bending vibrations can be detected in an IR absorption trace of range of 2.5–16 μm. What would be the frequency of each vibration?

4.6   In discussing quantitative analysis using the mid-IR region, it was suggested that either the OH stretching band or the C—O stretching band could be used to measure hexanol in a mixture of hexanol and hexane. Which band would you choose to give more accurate results? Why?

4.7   A solution is known to contain acetone and ethyl alcohol. Draw the expected IR absorption curve for each compound separately. Which absorption band could be used to identify the presence of acetone in the mixture?

4.8   What requirements must be met before a molecule will absorb IR radiation?

4.9   In preparing a calibration curve for the determination of methyl ethyl ketone (MEK), solutions with different concentrations of MEK were prepared in chloroform. The absorbance at the C=O stretching frequency was measured. The measured absorbance A for each solution is given. A blank (the pure solvent) had an absorbance = 0.00.

MEK concentration (%)	Absorbance
2	0.18
4	0.36
6	0.52
8	0.64
10	0.74

(a) Does the relationship between A and sample concentration deviate from Beer's Law? (b) Several unknown samples containing MEK were measured at the same wavelength as that used for the calibration curve. The results were as follows:

Sample	Absorbance
A	0.27
B	0.44
C	0.58
D	0.69

What were the concentrations of MEK in the solutions A–D?

4.10   Explain what is meant by the Fellgett advantage.

4.11   What are (a) fundamental and (b) overtone vibrational bands?

4.12   Is FTIR a single-beam or double-beam technique? How is background correction achieved?

4.13   How does a semiconductor detector such as an MCT detector compare to a thermocouple detector for use in IR spectroscopy?

4.14   Describe the components of an FTIR spectrometer. Which detectors are used for FTIR?

4.15   List the advantages of FTIR over dispersive IR spectroscopy.

4.16   Describe how attenuated total reflectance works to give an IR absorption spectrum.

4.17   Give two examples of the use of FTIR microscopy for chemical analysis.

4.18   What wavelength range is covered by NIR? What bands occur here?

4.19   What is the advantage of using NIR compared with mid-IR? What are the disadvantages?

4.20   What is meant by the term "virtual state"?

4.21   Diagram the processes that give rise to Rayleigh, Stokes, and anti-Stokes scattering.

4.22   What are the requirements for a molecule to be Raman-active?

4.23   Explain why fluorescence is a problem in normal Raman spectroscopy. Give two examples of how the fluorescence interference in Raman spectroscopy can be minimized or eliminated.

4.24   Describe the resonance Raman process and its advantages.

For Problems 4.25–4.41, deduce a reasonable structure for a compound consistent with the IR spectrum and the other information provided.

4.25 The molecular weight (MW) of the unknown = 86.

4.26 The molecular weight (MW) of the unknown = 128.

4.27 The molecular weight (MW) of the unknown = 46.

4.28 The molecular weight (MW) of the unknown = 154.

4.29 The molecular weight (MW) of the unknown = 106.

4.30 The molecular weight (MW) of the unknown = 93. (Hint: note that the MW is an odd number.)

4.31 The molecular weight (MW) of the unknown = 284.

4.32 The molecular weight (MW) of the unknown = 107.

4.33 The molecular weight (MW) of the unknown = 104. Molecular formula from elemental analysis is $C_3H_6NO_3$.

4.34 The molecular weight (MW) of the unknown = 153. Molecular formula from elemental analysis is $C_8H_{11}NO_2$.

4.35 The molecular weight (MW) of the unknown = 60.

4.36 The molecular weight (MW) of the unknown = 124. (Hint: the molecule contains a heteroatom.)

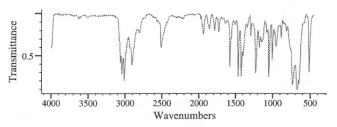

4.37 The molecular weight (MW) of the unknown = 87.

4.38 The molecular weight (MW) of the unknown = 126.

4.39 The molecular weight (MW) of the unknown = 135.5 There are no C atoms in the molecule.

4.40 The molecular weight (MW) of the unknown = 160. (Hint: there are more C atoms than H atoms.)

4.41 You have three compounds, A, B, and C, giving rise to three spectra, a, b, and c, respectively. Note that spectrum a and spectrum b look very similar. (But they are not identical; an easy way to see what the differences are is to make a copy of each spectrum, overlay them and hold the pages up to a light. Align the baselines so that they match, and spectral differences will be more apparent.) The MW of Compound A is 102. The MW of Compound

B is 114. Compound C also has MW = 114 and has the exact same molecular formula as Compound B. However, spectrum b and spectrum c are very different. Draw three plausible structures for Compounds A, B, and C.

5

Visible and Ultraviolet Molecular Spectroscopy

5.1. INTRODUCTION

Probably the first physical method used in analytical chemistry was based on the quality of the color in colored solutions. The first things we observe regarding colored solutions are their *hue*, or color, and the color's *depth*, or *intensity*. These observations led to the technique historically called *colorimetry*; the color of a solution could identify species (qualitative analysis) while the intensity of the color could identify the concentration of the species present (quantitative analysis). This technique was the first use of what we now understand to be absorption spectroscopy for chemical analysis. When white light passes through a solution and emerges as red light, we say that the solution is red. What has actually happened is that the solution has allowed the red component of white light to pass through, whereas it has absorbed the complementary colors, yellow and blue. The more concentrated the sample solution, the more yellow and blue light is absorbed and the more intensely red the solution appears to the eye. For a long time, experimental work made use of the human eye as the detector to measure the hue and intensity of colors in solutions. However, even the best analyst can have difficulty comparing the intensity of two colors with slightly different hues, and there are of course people who are color-blind and cannot see certain colors. Instruments have been developed to perform these measurements more accurately and reliably than the human eye. While the human eye can only detect visible light, this chapter will focus on both the ultraviolet (UV) and the visible (VIS) portions of the spectrum.

The wavelength range of UV radiation starts at the blue end of visible light (about 400 nm) and ends at approximately 200 nm for spectrometers operated in air. The radiation has sufficient energy to excite valence electrons in many atoms and molecules; consequently, UV radiation is involved with electronic excitation. Visible light, considered to be light with wavelengths from 800 to 400 nm, acts in the same way as UV light. It is also considered part of the electronic excitation region. For this reason we find commercial spectroscopic instrumentation often operates with wavelengths between 800 and 200 nm. Spectrometers of this type are called UV/Visible (or UV/VIS) spectrometers. The vacuum UV region of the spectrum extends below 200 nm to the X-ray region of the spectrum, at $\sim$100 Å. It is called the vacuum UV region because oxygen, water vapor, and other molecules in air absorb UV radiation below 200 nm, so the spectrometer light path must be free of air to observe wavelengths <200 nm. The instrument must be evacuated (kept under vacuum) or purged with an appropriate non-UV absorbing gas such as helium for this region to be used. Vacuum UV radiation is also involved in electronic excitation but the spectrometers are specialized and not commonly found in undergraduate or routine analytical laboratories. For our purposes the term UV will mean radiation between 200 and 400 nm, unless stated otherwise. The major types of analytical spectroscopy operating within this wavelength range are listed in Table 5.1. This chapter will focus on molecular spectroscopy—the absorption and emission of UV and visible radiation by molecules and polyatomic species. We will also look at the use of scattering of visible light to provide information about macromolecules and particles. AAS is covered in Chapter 6 and atomic emission spectroscopy in Chapter 7.

The interaction of UV and visible radiation with matter can provide qualitative identification of molecules and polyatomic species, including ions and complexes. Structural information about molecules and polyatomic species, especially organic molecules, can be acquired. This qualitative information is usually obtained by observing the UV/VIS **spectrum**, the absorption of UV and visible radiation as a function of wavelength by molecules. A typical UV absorption spectrum is shown in Fig. 5.1. The spectrum may be plotted as wavelength vs. absorbance, transmittance, or molar absorptivity, ε. The molar absorptivity is defined subsequently. In Fig. 5.1, the absorption spectrum of pyridine dissolved in ethanol is plotted as log ε vs. wavelength in ångstroms (Å).

Quantitative information can also be obtained by studying the absorption or emission of UV and visible radiation by molecules or polyatomic species. As a very simple example, we can look at the absorption spectrum of a red solution such as red ink in water (Fig. 5.2). It can be seen that with a colorless sample of pure water, shown as the dotted line, all wavelengths of white light, including all of the red wavelengths, are transmitted through the sample. If we add one drop of red ink to water, to make a solution that appears pale red, the spectrum shows that some of the blue and some of the yellow light

Table 5.1 Spectroscopy Using UV and Visible Light

Function	Analytical field	Analytical application
Atomic Spectroscopy		
Absorption of UV/VIS radiation	Atomic absorption spectrometry	Quantitative elemental analysis
Emission of UV/VIS radiation	Flame photometry, atomic emission spectrometry	Qualitative and quantitative multielemental analysis
Emission of UV/VIS radiation	Atomic fluorescence spectrometry	Quantitative elemental analysis of ultratrace concentrations (sub-ppb)
Molecular Spectroscopy		
Absorption of UV/VIS radiation	UV/VIS Molecular absorption spectroscopy, spectrophotometry	Qualitative and quantitative determinations of aromatic and unsaturated organic compounds, including natural products; direct and indirect quantitative determination of inorganic ions, organic molecules, and biochemicals
Emission of UV/VIS radiation	Molecular fluorescence, Molecular phosphorescence	Detection of small quantities (<ng) of certain aromatic compounds and natural products; analysis of gels and glasses; determination of organic and inorganic species by "tagging"

Figure 5.1 A typical UV/VIS absorption spectrum for an organic molecule in solution. The spectrum is that of pyridine dissolved in 95% ethanol. The absorption wavelength is plotted on the *x*-axis and the logarithm of the molar absorptivity is plotted on the *y*-axis. (From Jaffé and Orchin. Reprinted with the permission of Professor M. Orchin.)

have been absorbed, but all of the red light has been transmitted. If we add more red ink to the water to make a dark red solution, most of the blue and yellow light has been absorbed, but again all of the red light has been transmitted. The amount of red light falling on the eye or the detector is the same in each case; the amount of ink in the solution is related to the blue and yellow light absorbed, not to the color transmitted. We could construct a series of known amounts of red ink in water and quantitatively measure other ink solutions by measuring the amount of light absorbed at, for example, 450 nm. Concentrations of species in samples, especially solutions, are often measured using UV/VIS absorption spectrometry or fluorescence spectrometry. The measurement of concentrations or changes in concentrations can be used to calculate equilibrium constants, reaction kinetics, and stoichiometry for chemical systems. Quantitative measurements by UV/VIS spectrometry are important in environmental monitoring, industrial process control, pharmaceutical quality control, and clinical chemistry, to name a just a few areas. The emission of radiation by molecules may occur in several ways following excitation of the molecule; two processes are fluorescence and phosphorescence. These processes will be discussed in Sections 5.8–5.10.

5.1.1. Electronic Excitation in Molecules

Molecules are composed of atoms that are held together by sharing electrons to form chemical bonds. Electrons in molecules move in molecular orbitals at discrete energy

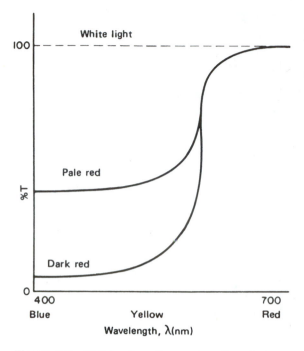

Figure 5.2   Visible absorption spectra of a colorless sample of pure water (dotted line), a pale red solution of red ink in water, and a dark red solution of red ink in water. Note that none of the samples absorbs red light.

levels as defined by quantum theory. When the energy of the electrons is at a minimum, the molecules are in the lowest energy state, or ground state. The molecules can absorb radiation and move to a higher energy state, or *excited state.* When the molecule becomes excited, an outer shell (valence) electron moves to an orbital of higher energy. The process of moving electrons to higher energy states is called *electronic excitation.* For radiation to cause electronic excitation, it must be in the visible or UV region of the electromagnetic spectrum.

The frequency absorbed or emitted by a molecule and the energy of radiation are related by $\Delta E = h\nu$. The actual amount of energy required depends on the difference in energy between the ground state E_0 and the excited state E_1 of the electrons. The relationship is described by

$$\Delta E = E_1 - E_0 = h\nu \tag{5.1}$$

where E_1 is the energy of the excited state and E_0 is the energy of the ground state.

You may want to review the topics of bonding, molecular orbitals, Lewis structures, and organic chemistry in your general chemistry textbook or in the texts by Chang or Zumdahl listed in the bibliography to help you understand the material discussed subsequently. The discussion will focus on organic molecules, as the bonding is relatively easy to understand. Inorganic molecules also undergo absorption and emission of UV and visible radiation, as do complexes of organic molecules with metal ions, but the bonding in inorganic molecules and complexes of the transition metals and heavier elements is complicated due to electrons in the d and f orbitals.

   Three distinct types of electrons are involved in valence electron transitions in molecules. First are the electrons involved in single bonds, such as those between carbon and hydrogen in alkanes. These bonds are called *sigma* (σ) *bonds*. The amount of energy required to excite electrons in σ bonds is usually more than UV photons of wavelengths >200 nm possess. For this reason, alkanes and other saturated compounds (compounds with only single bonds) do not absorb UV radiation and are therefore frequently very useful as transparent solvents for the study of other molecules. An example of such a nonabsorbing compound is the alkane hexane, C_6H_{14}.

   Next we have the electrons involved in double and triple (unsaturated) bonds. These bonds involve a *pi* (π) *bond*. Typical examples of compounds with π bonds are alkenes, alkynes, conjugated olefins, and aromatic compounds (Fig. 5.3). Electrons in π bonds are excited relatively easily; these compounds commonly absorb in the UV or visible region.

   Electrons that are not involved in bonding between atoms are the third type of electrons in molecules. These are called n electrons, for nonbonding electrons. In saturated hydrocarbons the outer shell electrons of carbon and hydrogen are all involved in bonding; hence these compounds do not have any n electrons. Organic compounds containing nitrogen, oxygen, sulfur, or halogens, however, frequently contain electrons that are nonbonding (Fig. 5.4). Because n electrons are usually excited by UV or visible radiation, many compounds that contain n electrons absorb UV/VIS radiation.

Figure 5.3 Examples of organic molecules containing π bonds. Note that benzene rings can be drawn showing three π bonds (the Kekulé structure) or with a circle inside the ring, as has been done for ethylbenzene, to more accurately depict the delocalized nature of the π electrons in aromatic compounds.

2-bromopropene

aminobenzene

Figure 5.4 Examples of organic molecules with nonbonding electrons. The n electrons are represented as pairs of dots around the atom on which they are located. For the carbonyl compound, if R = H, the compound is an aldehyde; if R = an organic group, the compound is a ketone.

A schematic energy diagram of two s electrons in atomic orbitals on adjacent atoms combining to form a σ bond is shown in Fig. 5.5. Orbitals are conserved, therefore two molecular orbitals are formed, a sigma bonding orbital and a higher energy sigma anti-bonding orbital. The antibonding orbital is denoted σ^*. The energy difference between σ and σ^* is equal to ΔE, shown by the large arrow. Remember that each atom has three 2p atomic orbitals. One of those p orbitals can overlap with a p orbital on an adjacent atom to form a second set of sigma orbitals. Sideways overlap of the other two p orbitals is possible, resulting in pi bonding and antibonding orbitals. The schematic energy diagram for the formation of one set of π orbitals is shown in Fig. 5.6, and the energy difference between the π orbital and the antibonding π^* orbital is shown by the large arrow. If a p orbital is filled with a pair of electrons, it will have no tendency to form a bond. Figure 5.7 shows that a filled atomic p orbital (in the atom on the right) may form a nonbonding n orbital that is unshifted in energy from the atomic orbital, while

Figure 5.5 Schematic energy diagram of two s orbitals on adjacent atoms forming a σ bonding orbital and a σ^* antibonding orbital.

Figure 5.6 Schematic energy diagram of two p orbitals on adjacent atoms forming a π bonding orbital and a π^* antibonding orbital.

the partially filled p orbitals on each atom overlap to form a pair of pi bonding and antibonding orbitals.

A relative energy diagram of σ, π, and n electrons is shown in Fig. 5.8, although there are exceptions to this general order. It can be seen that the energy required to excite an electron from a σ to a σ^* orbital is considerably greater than that required to excite an electron from a π to a π^* orbital or an n electron to either a σ^* or a π^* orbital. As a consequence, the energy necessary to excite σ electrons to σ^* orbitals is greater than that available in the UV region, but usually UV radiation is sufficient to excite electrons in π orbitals to π^* antibonding orbitals or n electrons to π^* or σ^* antibonding orbitals.

5.1.2. Absorption by Molecules

Quantum mechanics provides a theoretical basis for understanding the relative energy levels of molecular orbitals and how they vary with structure. Quantum mechanics also generates a set of "selection rules" to predict what transitions occur in molecules. The transitions that occur in molecules are governed by quantum mechanical selection rules. Some transitions are "allowed" by the selection rules, while others are "forbidden". The selection rules are beyond the scope of this text, but may be found in most physical chemistry texts or in the

Figure 5.7 The relative energy levels of the π, π^*, and nonbonding (n) orbitals formed from p orbitals on adjacent atoms.

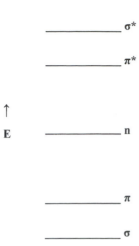

Figure 5.8 The relative energy levels of a set of σ, π, and n orbitals and the associated antibonding orbitals.

text by Ingle and Crouch listed in the bibliography. As is often the case with rules, there are exceptions, and many forbidden transitions do occur and can be seen in UV/VIS spectra.

When molecules are electronically excited, an electron moves from the highest occupied molecular orbital to the lowest unoccupied orbital, which is usually an antibonding orbital. Electrons in π bonds are excited to antibonding π^* orbitals, and n electrons are excited to either σ^* or π^* orbitals.

Both organic and inorganic molecules may exhibit absorption and emission of UV/VIS radiation. Molecular groups that absorb visible or UV light are called *chromophores*, from the Greek word *chroma*, color. For example, for a $\pi \rightarrow \pi^*$ transition to occur, a molecule must possess a chromophore with an unsaturated bond, such as C=C, C=O, C=N, and so on. Compounds with these types of chromophores include alkenes, amides, ketones, carboxylic acids, and oximes, among others. The other transition that commonly occurs in the UV/VIS region is the $n \rightarrow \pi^*$ transition, so organic molecules that contain atoms with nonbonded electrons should be able to absorb UV/VIS radiation. Such atoms include nitrogen, oxygen, sulfur, and the halogen atoms, especially Br and I. Table 5.2 presents some typical organic functional groups that serve as chromophores. Table 5.3 lists types of organic compounds and the wavelengths of their absorption maximum,

Table 5.2 Organic Functional Groups that can Absorb UV/VIS Radiation

Functional group	Chemical structure	Electronic transitions
Acetylenic	—C≡C—	$\pi \rightarrow \pi^*$
Amide	—CONH$_2$	$\pi \rightarrow \pi^*$, $n \rightarrow \pi^*$
Carbonyl	>C=O	$\pi \rightarrow \pi^*$, $n \rightarrow \pi^*$
Carboxylic acid	—COOH	$\pi \rightarrow \pi^*$, $n \rightarrow \pi^*$
Ester	—COOR	$\pi \rightarrow \pi^*$, $n \rightarrow \pi^*$
Nitro	—NO$_2$	$\pi \rightarrow \pi^*$, $n \rightarrow \pi^*$
Olefin	>C=C<	$\pi \rightarrow \pi^*$
Organoiodide	R—I	$n \rightarrow \sigma^*$
Thiol	R—SH	$n \rightarrow \sigma^*$

Note: R = any organic group (e.g., CH$_3$, C$_2$H$_5$, C$_6$H$_5$, etc.).

Table 5.3 Absorption Wavelengths of Typical Organic Functional Groups

Chromophore	System	Wavelength of absorption maximum, λ_{max} (nm)
Amine	$-NH_2$	195
Bromide	$-Br$	208
Iodide	$-I$	260
Thioketone	$>C=S$	460
Thiol	$-SH$	220
Ester	$-COOR$	205
Aldehyde	$-CHO$	210
Carboxylic acid	$-COOH$	200–210
Nitro	$-NO_2$	210
Azo	$-N=N-$	285–400
Conjugated olefins	$(-HC=CH-)_2$	210–230
	$(-HC=CH-)_3$	260
	$(-HC=CH-)_5$	330
	$(-HC=CH-)_{10}$	460
Benzene		198 255
Naphthalene		210 220 275

that is, the wavelength at which the most light is absorbed. Some compounds have more than one absorption peak, so several "maxima" are listed. Compounds such as alkanes (also called paraffins) contain only σ bonds, which do not absorb radiation in the visible or UV region.

Transition metal compounds are often colored, indicating that they absorb light in the visible portion of the spectrum. This is due to the presence of unfilled d orbitals. The exact wavelength of the absorption band maximum depends on the number of d electrons, the geometry of the compound, and the atoms coordinated to the transition metal.

5.1.3. Molar Absorptivity

Beer's Law, which relates absorbance of a sample to the path length and concentration of absorbing species, was covered in Chapter 2. The proportionality constant, a, in Beer's Law is the absorptivity of the absorbing species.

The *absorptivity*, a, of a molecule defines how much radiation will be absorbed by that molecule at a given concentration and at a given wavelength. If the concentration is expressed in molarity (mol/L, M), the absorptivity is defined as the *molar absorptivity*, ε. The absorptivity can be calculated directly from the measured absorbance using Beer's Law:

$$A = abc = \varepsilon bc \tag{5.2}$$

where A is the absorbance, b, the path length; and c, the concentration of the absorbing species.

If b is in units of cm and c has units of molarity, then the proportionality constant is the molar absorptivity and is given the symbol ε, with units of $L\,mol^{-1}cm^{-1}$. Commonly

$\varepsilon \approx 10^4 - 10^5 \, \text{L mol}^{-1} \, \text{cm}^{-1}$ for an allowed transition and is on the order of $10-100$ for a forbidden transition. The magnitude of the absorptivity is an indication of the probability of the electronic transition. High values of ε give rise to strong absorption of light at the specified wavelength; low values of ε result in weak absorption of light. Both a and ε are constants for a given wavelength and are physical properties of the molecule. The molar absorptivity may be specified for any wavelength, but is usually tabulated for the wavelength at which maximum absorption of light occurs for a molecule. The wavelength of maximum absorption is symbolized by λ_{max} and the associated ε is symbolized as ε_{max}. Table 5.4 presents typical values for λ_{max} and ε_{max} for some common organic molecules.

The absorptivity is not a direct measure of the probability that a given electronic transition will occur. This is because absorbance is measured over a wavelength range that is much smaller than the width of the absorption band. The absorptivity will differ at different wavelengths over the band profile. There are several fundamental quantities that are directly related to the transition probability; these include the transition probability, R^2, Einstein coefficients, and the oscillator strength, f. The text by Ingle and Crouch presents the derivation of these quantities for the interested student.

5.1.4. The Shape of UV Absorption Curves

Figure 5.1 shows a "typical" UV absorption spectrum. The spectrum appears to be very simple, with a broad absorption "band" over a wide wavelength range instead of the numerous, narrower absorptions seen in IR spectra (Chapter 4). The absorption bands are broad because each electronic energy level has multiple vibrational and rotational energy levels associated with it. Excitation from the ground electronic state can occur to more than one vibrational level and to more than one rotational level. A schematic representation of an electronic transition with vibrational and rotational sublevels is shown in Fig. 5.9. In the ground state, only the lowest vibrational level is shown, with four rotational sublevels. At room temperature, most molecules are in the ground state in the lowest vibrational state. In the excited state, four vibrational sublevels are shown, slightly separated, with four rotational sublevels in each. Only four of the many possible transitions are shown; each arrow represents an absorption wavelength. The electronic transition consists of a large number of wavelengths that overlap to give the "continuous" absorption band observed. Even though each separate transition is quantized, the close energy spacing

Table 5.4 Typical Absorption Maxima and Molar Absorptivities for Common Chromophores

Chromophore	Compound	λ_{max} (nm)	ε_{max} (L mol^{-1} cm^{-1})
>C=C—C=C<	$H_2C=C—C=CH_2$	210	2.5×10^4
—NO_2	CH_3NO_2	210	1.0×10^4
		280	10
—N=N—	$CH_3N=NCH_3$	<250	$>1.0 \times 10^5$
—Br	CH_3Br	205	1.8×10^3
—SH	C_2H_5SH	230	160
Aromatic ring	Benzene (C_6H_6)	198	8.0×10^3
		255	200
	Naphthalene	221	1.0×10^5
		285	9.0×10^3
		300	290
>C=S	$(CH_3)_2C=S$	460	<10

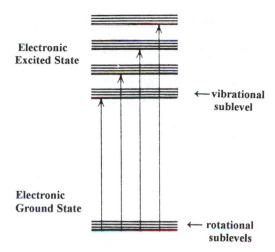

Figure 5.9 An electronic transition occurs over a band of energy due to the multiple vibrational and rotational sublevels associated with each electronic state. This schematic depicts four of the many possible transitions that occur. The length of the arrow is proportional to the energy required for the transition, so a molecular electronic transition consists of many closely spaced transitions, resulting in a band of energy absorbed rather than a discrete line absorption.

of the vibrational levels and the even more closely spaced rotational sublevels cause the electronic transition to appear as a broad band. This is shown schematically in Fig. 5.10.

An absorption band is characterized by its shape, that is, by its width and intensity. The shape of the band is determined primarily by the vibrational energy level spacing and the intensity of each vibrational transition. The intensity distribution is related to the probability of the transition to a given vibrational sublevel. The transition probabilities can be determined using the Franck–Condon principle. The texts by Hollas and Lambert et al. may be consulted for details. Suffice it to say that if we have a million molecules, even if they are mostly in the ground vibrational state before excitation, they may be in various vibrational states after excitation. The radiant energy required to cause electronic excitation to each vibrational energy level is slightly different and is further modified by rotational energy changes. For this reason, when UV radiation falls on the million molecules, it is absorbed at numerous wavelengths. The total range of the absorption wavelengths may stretch over 100 nm. The effect of the rotational energy of the molecule is to add even more

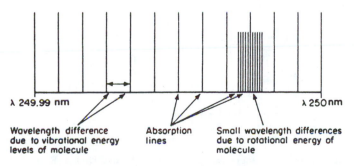

Figure 5.10 Illustration of a UV absorption band greatly expanded.

absorption lines to the single band. The increased number of absorption lines makes the lines even closer together, but it does not appreciably increase the total range of the band. This is because the energy involved in rotation is very small compared to vibrational energy and extremely small compared to electronic excitation energy. UV radiation is therefore absorbed in *absorption bands* rather than at discrete wavelengths.

In some cases, the UV/VIS spectra will show the different energies associated with the vibrational sublevels. For example, simple molecules in the gas phase often show the vibrational levels superimposed on the electronic transitions, as seen in Fig. 5.11, the gas phase spectrum for benzene. The sharp peaks on top of the broad bands are called vibrational "fine structure". This fine structure is usually lost at high temperatures in the gas phase due to population of higher vibrational energy levels in the ground state electronic level, with the result that many more lines are seen. Molecules in solution (such as the spectrum shown in Fig. 5.1) usually do not exhibit vibrational structure due to interactions between the solvent and the solute molecules. Compare the gas phase spectrum of benzene (Fig. 5.11) to the solution spectrum for benzene (Fig. 5.12) and note the loss of much of the fine structure in solution. Fine structure due to rotational sublevels is never observed in routine UV/VIS spectra; the resolution of commercial instrumentation is not high enough to separate these lines.

5.1.5. Solvents for UV/VIS Spectroscopy

Many spectra are collected with the absorbing molecule dissolved in a solvent. The solute must be soluble in the solvent and the solvent must be transparent over the wavelength range of interest. A molecule will dissolve in a solvent if the formation of a solution leads to a lower energy system. The intermolecular attractive forces between the solute and solvent must be greater than solute–solute and solvent–solvent attractive forces. The forces involved in solution formation are dipole–dipole attraction, hydrogen bonding, and van der Waals forces. Polarity plays a major role, and gives rise to the "like dissolves like" rule. Polar substances dissolve more readily in polar solvents than

Figure 5.11 The gas phase absorption spectrum of benzene. (From Jaffé and Orchin. Reprinted with the permission of Professor M. Orchin.)

Figure 5.12 The absorption spectrum of benzene in solution. The solvent is cyclohexane. Note the loss of fine structure when compared to Fig. 5.11. (From Jaffé and M. Orchin. Reprinted with the permission of Professor M. Orchin.)

in nonpolar solvents. It is important for the solute to be dissolved completely; undissolved particles can scatter light from the light source. This can result in serious errors in qualitative and quantitative analyses.

The solvent may affect the appearance of the spectrum, sometimes dramatically. Polar solvents generally wipe out the vibrational fine structure in a spectrum. Solvents may also shift the position of the absorption band, as will be discussed. For the visible region of the spectrum any colorless solvent can be used in which the sample is soluble.

The common solvents used in UV/VIS spectroscopy are listed in Table 5.5, along with their low wavelength cutoff. At wavelengths shorter than the cutoff wavelength, the solvent absorbs too strongly to be used in a standard 1 cm sample cell. The cutoff is affected by the purity of the solvent. For spectroscopy, the solvents should be of spectral or spectrochemical grade, conforming to purity requirements set by the American Chemical Society.

5.2. INSTRUMENTATION

5.2.1. Optical System

Spectrometers are instruments that provide information about the intensity of light absorbed or transmitted as a function of wavelength. Both single-beam and the double-beam optical systems (see the schematics in Chapter 2) are used in molecular absorption spectroscopy. Single-beam systems and their disadvantages were discussed in Chapter 2. Most commercial instruments for absorption spectrometry are double-beam systems, so these will be reviewed.

Table 5.5 Common UV Solvents and their
Lower Wavelength Cutoffs

Solvent	Lower wavelength cutoff (nm)
Acetone	330
Pyridine	306
Toluene	285
Carbon tetrachloride	265
Chloroform	245
Diethyl ether	215
Methanol	205
Cyclohexane	205
95% Ethanol	204
Hexane	195
Isooctane	195
Water	195
Acetonitrile	190

In the double-beam system, the source radiation is split into two beams of equal intensity. The two beams traverse two light paths identical in length; a *reference* cell is put in one path and the *sample* cell in the other. The intensities of the two beams after passing through the cells are then compared. Variation in radiation intensity due to power fluctuations, radiation lost to the optical system (e.g., cell surfaces, mirrors, etc.), radiation absorbed by the solvent, and so on should be equal for both beams, correcting for these sources of error. A dispersive spectrometer used for absorption spectroscopy that has one or more exit slits and photoelectric detectors that ratio the intensity of two light beams as a function of wavelength is called a spectrophotometer.

Commercial UV/VIS spectrometers are designed to operate with air in the light path over the range of 200–800 nm. Purging the spectrometer with dry nitrogen may permit wavelengths as low as 175 nm to be observed. For lower wavelengths, as mentioned, the spectrometer must be put under vacuum or purged with a nonabsorbing gas. Analytically, the vacuum UV region has been of minor importance for routine analysis because of the difficulties and expense inherent in instrumentation requiring a vacuum.

Simple optical systems using filters for wavelength selection and a photoelectric detector are called photometers. Photometers are used for both the visible and the UV region. For example, UV photometers were commonly used as detectors in HPLC but have been superceded by PDAs. HPLC detectors will be discussed in greater detail in Chapter 13.

All spectrometers for absorption measurements require a light source, a wavelength selection device, a sample holder, and a detector.

5.2.2. Radiation Sources

Radiation sources for molecular absorption measurements must produce light over a continuum of wavelengths. Ideally, the intensity of the source would be constant over all wavelengths emitted. Traditionally, the two most common radiation sources for UV/VIS spectroscopy were the tungsten lamp and the deuterium discharge lamp. The *tungsten lamp* is similar in functioning to an ordinary electric light bulb. It contains a tungsten filament heated electrically to white heat, and generates a continuum spectrum.

It has two shortcomings: the intensity of radiation at short wavelengths (<350 nm) is low; furthermore, to maintain a constant intensity, the electrical current to the lamp must be carefully controlled. However, the lamps are generally stable, robust, and easy to use. Typically, the emission intensity varies with wavelength as shown in Fig. 5.13. The shape of these curves is typical of the continuum output of a solid heated to incandescence. An incandescent solid that produces a curve of this type is called a blackbody radiator. The continuum emission is due to thermally excited transitions in the solid, in this case, the tungsten filament. The intensity vs. wavelength plot for a blackbody radiator is dependent on the temperature of the emitting material, not on its chemical composition. The tungsten lamp is most useful over the visible range and is therefore commonly used in spectrophotometry, discussed subsequently. Because it is used only in the visible region, the bulb (i.e., the lamp envelope) can be made of glass instead of quartz. Quartz is required for the transmission of UV light. The tungsten-halogen lamp, similar to the lamp in modern auto headlights, has replaced the older tungsten lamp in modern instruments. The tungsten-halogen lamp has a quartz bulb, primarily to withstand the high operating temperatures of the lamp. This lamp is much more efficient than a W lamp and has a significantly longer lifetime. The wavelength/intensity output of a tungsten halogen lamp is presented in Fig. 5.14.

The *deuterium arc lamp* consists of deuterium gas (D_2) in a quartz bulb through which there is an electrical discharge. The molecules are excited electrically and the excited deuterium molecule dissociates, emitting UV radiation. The dissociation of the deuterium molecule into atoms results in UV photon emission over a continuous range of energies from zero up to the energy of excitation of the molecule. This causes the lamp to emit a continuum (broadband) UV spectrum over the range of 160–400 nm rather than a narrow line atomic emission spectrum. The lamps are stable, robust, and widely used. The use of deuterium (D_2) instead of hydrogen gas results in an increase in the emission intensity

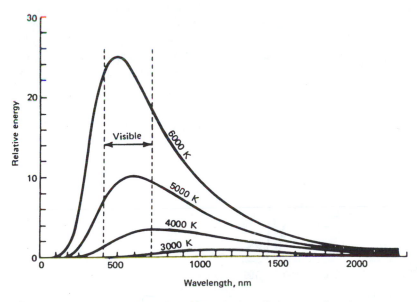

Figure 5.13 Emission intensity of blackbody radiation at various temperatures as a function of wavelength: 3000 K is equivalent to a tungsten filament lamp (an incandescent lamp); 6000 K is equivalent to a xenon arc lamp.

Figure 5.14 Emission spectrum of a commercial tungsten-halogen lamp. [Courtesy of Agilent Technologies (www.agilent.com).]

by as much as a factor of three at the short-wavelength end of the UV range. Deuterium is more expensive than hydrogen, but is used to achieve the high intensity required of the source. Figure 5.15 presents the emission spectrum of a deuterium arc lamp.

Xenon arc lamps operate in a manner similar to deuterium lamps. A passage of current through xenon gas produces intense radiation over the 200–1000 nm range. They provide very high radiation intensity and are widely used in the visible region and long-wavelength end of the UV range. This lamp is used in fluorescence spectrometry and the lamp schematic and spectrum are shown in Section 5.9.2.

5.2.3. Monochromators

The purpose of the monochromator is to disperse the radiation according to wavelength and allow selected wavelengths to illuminate the sample. Diffraction gratings are used to disperse light in modern instruments, as discussed in Chapter 2. The monochromators in modern systems, such as the Cary spectrophotometers from Varian, Inc. (www.varianinc. com) can scan at rates up to 2000–3000 nm/min, with slew rates (the time to move between wavelengths without taking measurements) as high as 16,000 nm/min to accommodate the high throughput measurements needed in pharmaceutical and biotechnology laboratories.

Figure 5.15 Emission spectrum of a commercial deuterium arc lamp. [Courtesy of Agilent Technologies (www.agilent.com).]

5.2.4. Detectors

The earliest detector used for visible light spectroscopy was the human eye. There are still *spectroscopes* and *color comparators* designed for visual observation of color and intensity.

Most modern instruments rely on **photoelectric transducers**, detection devices that convert photons into an electrical signal. Photoelectric transducers have a surface that can absorb radiant energy. The absorbed energy either causes the emission of electrons, resulting in a photocurrent or moves electrons into the conduction band of a solid semi-conductor, resulting in an increase in conductivity. There are several common forms of these detectors including barrier layer cells, photomultiplier tubes, and semiconductor detectors.

5.2.4.1. Barrier Layer Cell

In a barrier layer cell, also called a photovoltaic cell, a current is generated at the interface of a metal and a semiconductor when radiation is absorbed. For example, silver is coated onto a semiconductor such as selenium (see Fig. 5.16) that is joined to a strong metal base, such as iron. To manufacture these cells, the selenium is placed in a container and the air pressure reduced to a vacuum. Silver is heated electrically, and its surface becomes so hot that it melts and vaporizes. The silver vapor coats the selenium surface, forming a very thin but evenly distributed layer of silver atoms. Any radiation falling on the surface generates electrons and holes at the selenium–silver interface. A barrier seems to exist between the selenium and the iron that prevents electrons from flowing into the iron; the electrons flow to the silver layer and the holes to the iron. The electrons are collected by the silver. These collected electrons migrate through an external circuit toward the holes. The photocurrent generated in this manner is proportional to the number of photons striking the cell.

Barrier layer cells are used as light meters in cameras and in low cost, portable instruments. The response range of these cells is 350–750 nm. These detectors have two main disadvantages: they are not sensitive at low light levels and they show *fatigue*, that is, the current drops gradually under constant exposure to light. On the plus side, they require no external electrical power and they are very rugged.

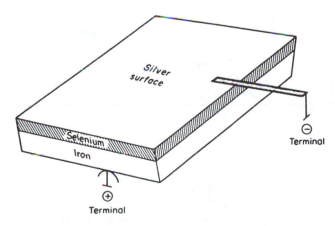

Figure 5.16 Barrier layer cell.

Figure 5.17   A vacuum phototube.

5.2.4.2.   Photomultiplier Tube

The most common detector is the photomultiplier tube (PMT). A PMT is a sealed, evacuated transparent envelope (quartz or glass) containing a *photoemissive cathode*, an anode, and several additional electrodes called *dynodes*. The photoemissive cathode is a metal coated with an alkali metal or a mixture of elements (e.g., Na/K/Cs/Sb or Ga/As) that emits electrons when struck by photons. The PMT is a more sophisticated version of a vacuum phototube (Fig. 5.17), which contained only a photoemissive cathode and an anode; the photocurrent was limited to the electrons ejected from the cathode. In the PMT (Fig. 5.18), the additional dynodes "multiply" the available electrons. The ejected electrons are attracted to a dynode that is maintained at a positive

Figure 5.18   Schematic of a PMT, looking down through the tube. Impinging photons pass through the quartz envelope and liberate electrons from the light-sensitive cathode. The electrons are accelerated to the first dynode, where each electron liberates several electrons on impact. The process is repeated at the other dynodes, resulting in a cascade of electrons for every photon hitting the PMT.

voltage with respect to the cathode. Upon arrival at the dynode, each electron strikes the dynode's surface and causes several more electrons to be emitted from the surface. These emitted electrons are in turn attracted to a second dynode, where similar electron emission and more multiplication occurs. The process is repeated several times until a shower of electrons arrives at the anode, which is the collector. The number of electrons falling on the collector is a measure of the intensity of light falling on the detector. In the process, a single photon may generate many electrons and give a high signal. The dynodes are therefore operated at an optimum voltage that gives a steady signal. A commercial photomultiplier tube may have nine or more dynodes. The gain may be as high as 10^9 electrons per photon. The noise level of the detector system ultimately limits the gain. For example, increasing the voltage between dynodes increases the signal, but if the voltage is made too high, the signal from the detector becomes erratic or *noisy*. In practice, lower gains and lower noise levels may be preferable for accuracy.

PMTs are extremely sensitive to UV and visible radiation. In fact, they are so sensitive that care must be taken not to expose PMTs to bright light, to avoid damage. There are a wide variety of photoemissive surfaces available, which respond to different wavelength ranges. A plot of detector signal vs. wavelength is called a *response curve*. Figure 5.19 displays some response curves for commercial PMTs. The PMT detector should be chosen so that it has maximum response to the wavelength range of interest. For example, the IP28 is not useful at 800 nm, but the R136 and Ga/As PMT detectors respond in this range. PMTs have very fast response times, but they are limited in sensitivity by their *dark current*. Dark current is a small, constant signal from the detector when

Figure 5.19 Response curves for various commercial PMTs. Note the variable response among different models and the sharp drop-off in response outside the useable range.

no radiation is falling on it. Dark current can be minimized or eliminated by cooling the detector housing. Cooling devices are available commercially for this purpose.

5.2.4.3. *Semiconductor Detectors—Diodes and Diode Array Systems*

Solid semiconducting materials are extremely important in electronics and instrumentation, including their use as radiation detectors. To understand the behavior of a semiconductor, it is necessary to briefly describe the bonding in these materials.

When a large number of atoms bond to form a solid, such as solid silicon, the discrete energy levels that existed in the individual atoms spread into *energy bands* in the solid. The valence electrons are no longer localized in space at a given atom. The width of the energy bands increases as the interatomic spacing in the solid decreases. The highest band that is at least partially occupied by electrons is called the *valence band*; the energy band immediately above the valence band is called the *conduction band*. The valence and conduction bands are separated by a forbidden energy range (forbidden by quantum mechanics); the magnitude of this separation is called the *band gap, E_g*. A set of energy bands and the band gap are shown schematically in Fig. 5.20. If the valence band of a solid is completely filled at a temperature of 0 K, the material is a **semiconductor** or an **insulator**. The difference between a semiconductor and an insulator is defined by the size of the band gap. If $E_g \leq 2.5$ eV, the material is a semiconductor; if $E_g > 2.5$ eV, the material is an insulator. The third type of material, a **conductor**, has a partially filled valence band at 0 K.

The two elements most used for semiconductor devices are silicon and germanium; both are covalently bonded in the solid state and both belong to group 4A of the periodic table. [This group is also called group 14 in the new International Union of Pure and Applied Chemistry (IUPAC) nomenclature and group IVA in some texts.] Other semiconductors include GaAs, CdTe, InP, and other inorganic and organic compounds. Most semiconductors are covalently bonded solids. Band gap energies for semiconductors are tabulated in the CRC Handbook of Chemistry and Physics.

Silicon has the valence electronic structure $3s^2 3p^2$. The partially filled p orbitals might lead one to suppose that silicon has a partially filled valence band and would therefore be an electrical conductor. Because silicon is covalently bonded, the two 3s electrons and the two 3p electrons occupy sp^3 hybrid orbitals. This results in a solid with two electron energy bands, each with four closely spaced sublevels, one for each electron in the valence shell of Si. The four electrons occupy and fill the valence band at 0 K and are therefore nonconducting. However, at temperatures above 0 K, a few electrons can

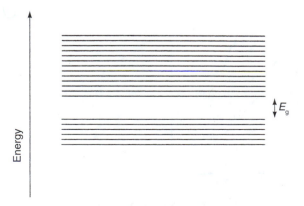

Figure 5.20   Schematic of energy bands in a solid material separated by a band gap of energy E_g.

be thermally promoted from the valence band into the conduction band; there they become conductors of electricity. When an electron leaves the valence band, it leaves behind a positive hole that is also mobile, thus producing an electron–hole pair. Both the electron and the hole are charge carriers in a semiconductor. Semiconductors such as Si and Ge are called intrinsic semiconductors; their behavior is a result of the band gap and band structure of the pure material.

The conductivity can be increased by doping either one of these elements with a group 5A element, such as arsenic or antimony, or a group 3A element, such as indium or gallium. Doping means to add another species to the *host* material; the added species is referred to as the *dopant*. The electrons associated with the dopant atom do not have the same energy levels as the host and may lie at energies forbidden to the host. Conductivity caused by addition of a dopant is called *extrinsic conduction*. A group 5A element has an extra electron (or extra negative charge). This electron is not held as tightly as the covalently bonded electrons of the host and requires less energy to move it into the conduction band. This is an n-type semiconductor. Similarly, adding a group 3A element leads to "missing" electrons, which can be considered to be the generation of extra positive holes. These positive holes from the dopant atom can accept electrons from the valence band. The energy needed to move an electron into an acceptor hole is less than the energy needed to move an electron into the conduction band. This is a p-type semiconductor. In an n-type semiconductor the electron is mobile, and in the p-type the positive hole is mobile. In an intrinsic semiconductor, two charge carriers are formed for every excitation event. In extrinsic semiconductors, either n or p type, only one charge carrier is formed per excitation event.

Semiconductors can be used as detectors for electromagnetic radiation. A photon of light with $E > E_g$ is sufficient to create additional charge carriers in a semiconductor. Additional charge carriers increase the conductivity of the semiconductor. By measuring the conductivity, the intensity of the light can be calculated. Selection of a material with the appropriate band gap can produce light detectors in the UV, visible, and IR regions of the spectrum.

5.2.4.4. Diodes

A diode or rectifier is an electronic device that permits current to flow in only one direction. If we put together a p-type semiconductor and an n-type semiconductor, the junction between the two types is a pn junction, as shown in Fig. 5.21. It is formed from a single piece of semiconductor by doping one side to be a p-type and the other side to be an n-type semiconductor. The junction is formed where the two types meet. Before any potential is applied to the device, holes will be the major charge carriers on the p side and electrons will be the major charge carriers on the n side. If we apply a positive potential to the p-type side and a negative potential to the n-type side, as shown in Fig. 5.22, positive charges (holes) flow from the p region to the junction and negative charges flow from the n region to the junction. At or near the junction, the holes and

Figure 5.21 A p–n junction with no applied electrical potential.

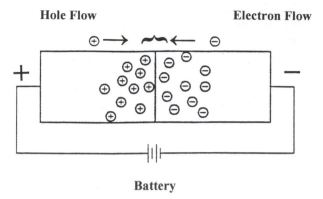

Figure 5.22 A p–n junction under forward bias. The positive battery terminal is connected to the p side.

electrons recombine and are annihilated. This is called a *forward bias*, and under these conditions current flows easily across the semiconductor. The annihilation of the electron–hole pair produces energy.

However, if the applied voltage were in the reverse direction, the flow of carriers would be in the opposite direction, as shown in Fig. 5.23(a). These are the conditions of *reverse bias*. The junction region is depleted of mobile charge carriers, recombination cannot occur, and no significant flow of current occurs. There is always a small flow of current due to the intrinsic conductivity. In short, the pn junction acts as a rectifier and permits significant current flow only under forward bias.

If a diode is held under reverse bias, and photons of energy greater than the band gap energy fall on the diode junction as shown in Fig. 5.23(b), electron–hole pairs are formed in the depleted region. These carriers will move through the diode, producing a current that is proportional to the intensity of the light falling on the diode. These detectors cover spectral ranges from the UV (about 190 nm) into the NIR (about 1000 nm), but are not as sensitive as PMTs. They have limited dynamic range compared to PMTs and when they deviate from linearity, they do so precipitously.

5.2.4.5.   *Diode Arrays*

In UV/VIS spectroscopy a complete absorption spectrum can be obtained by scanning through the entire wavelength range and recording the spectrum with a PMT, one wavelength at a time. This takes time using a conventional scanning monochromator system, although modern instruments are much faster than old instruments. The absorption at one wavelength is measured at a different time from that at another wavelength.

There are two conditions under which scanning optical systems do not work very well. The first is when there is a rapid chemical reaction taking place and conventional scanning is too slow to follow the reaction. The second is when the sample is available only for a limited time and complete scanning is not possible. Examples of the latter include the eluent from a liquid chromatographic separation, the flowing stream in a flow injection system, or a process stream in a chemical or pharmaceutical production plant. In cases like this, many wavelengths need to be monitored simultaneously. Ideally, intensity over the entire spectral range of interest should be measured at the same instant.

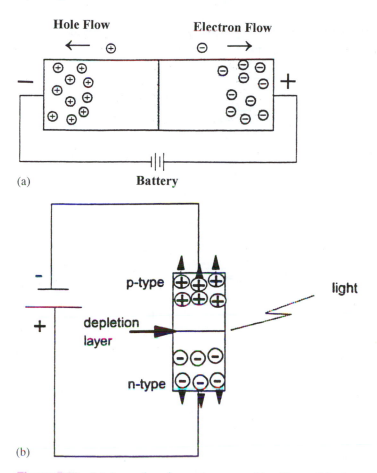

Figure 5.23 (a) A p–n junction under reverse bias. The positive battery terminal is connected to the n side. A depletion layer forms along the junction. (b) Diagram of a photodiode showing light incident upon the depletion layer. (From Brown, used with permission.)

The linear photodiode array (LPDA) is a transducer developed to enable simultaneous measurement of light intensity at many wavelengths. The diode array consists of a number of semiconductors embedded in a single crystal in a one-dimensional linear array. A common procedure is to use a single crystal of doped silicon that is an n-type semiconductor. A small excess of a group 3A element, such as arsenic, is embedded into the surface at regular intervals. This creates local p-type semiconductors. The semiconductor device ideally has a cross-section such as that shown in Fig. 5.24. The surface contains a linear series or array of pn junctions, each of which is a photodiode. The individual diodes are called *elements, channels,* or *pixels.*

The PDA is arranged as part of a circuit. A reverse bias is created across the pn junction by charging a capacitor in the circuit. Radiation falling on the array creates charge carriers in both p and n regions. The electrons will then flow to the nearest p-type semiconductor and the holes are collected in the p-type region. The current flow partially discharges the capacitor. At the end of the measurement cycle the capacitor is recharged; the charging current results in a voltage that is directly related to the light intensity. The number of charge carriers formed in a particular element depends on the light intensity falling on the array in the immediate vicinity of that particular element. By

TOP VIEW

SIDE VIEW

Figure 5.24   A PDA. The top view shows the face that the light would fall on. The side view shows that the p-type elements are embedded in a continuous layer of n-type semiconductor. (From Brown, used with permission.)

measuring the charges on each individual element, it is possible to get an instantaneous measurement of light intensity vs. wavelength of the whole spectral range, but in discrete elements. This is tantamount to a digital UV absorption spectrum.

The optical layout of a commercial multichannel instrument is shown schematically in Fig. 5.25. In this system radiation from a source, which may be a deuterium lamp or other UV/VIS light source, passes through the sample to a holographic grating, where the radiation is separated by wavelengths and directed to the diode array detector. No exit slit is used. The entire spectral region is measured in much less than 1 s. In practice, the spectrum is usually acquired for more than 1 s, and stored by the computer. This practice improves the signal-to-noise ratio for the measurement. By acquiring multiple measurements, the signal can be accumulated and sensitivity considerably increased.

PDAs are available to cover the wavelength range between 190 and 1100 nm. Simultaneous use of the entire wavelength range provides the multiplex advantage and improves the resolution of the system. The resolution of the system is limited by the

Figure 5.25   Optical diagram of the Agilent 8453 diode array spectrophotometer, showing both UV and visible light sources and a 1024 element PDA detector. [Courtesy of Agilent Technologies (www.agilent.com).]

number of diode elements involved. Typical diode spacing is 0.025 mm. Each one can be thought of as covering a finite spectral range. Detectors have been developed with as many as 4096 elements in the array, although 1024 is probably the most common number.

The most important applications for diode array systems are in molecular spectroscopy, since in general they do not have the resolution necessary for atomic spectroscopy. In molecular spectroscopy the most useful areas of application are for (1) scanning fast reactions to determine kinetics, (2) applications involving low light levels because spectra can be stored and added to each other, increasing the intensity, and (3) detectors for HPLC and capillary electrophoresis (CE). HPLC and CE are discussed in Chapter 13.

5.2.5. Sample Holders

Samples for UV/VIS spectroscopy can be solids, liquids, or gases. Different types of holders have been designed for these sample types.

5.2.5.1. Liquid and Gas Cells

The cells or *cuvettes* (also spelled *cuvets*) used in UV absorption or emission spectroscopy must be transparent to UV radiation. The most common materials used are quartz and fused silica. Quartz and fused silica are also chemically inert to most solvents, which make them sturdy and dependable in use. (*Note*: Solutions containing hydrofluoric acid or very strong bases, such as concentrated NaOH should never be used in these cells. Such solutions will etch the cell surfaces, making them useless for quantitative work.) Quartz and fused silica cells are also transparent in the visible and into the NIR region, so these could be used for all work in the UV and visible regions. These are also the most expensive cells, so if only the visible portion of the spectrum is to be used, there are cheaper cell materials available.

Some typical cell types are shown in Fig. 5.26. Cells are available is many sizes. The standard size for spectrophotometry is the 1 cm path length rectangular cell, which holds about 3.5 mL of solution, shown in the upper left of Fig. 5.26. There are microvolume cells (second from the left in the top row of Fig. 5.26) with volumes as small as 40 μL, flow-through cells for process streams or the routine analysis of large numbers of samples, microflow cells for chromatographic systems, and larger path length/volume cells for gases and highly dilute solutions. Two flow-through cells are shown in the middle of the bottom row in Fig. 5.26. In general, gas cells are long path cells, such as the one shown on the upper right of Fig. 5.26 and must be able to be closed once filled with a gas sample.

For spectrophotometric analysis in the visible region of the spectrum, glass or disposable plastic cells may be used. These are less expensive than quartz or fused silica but cannot be used at UV wavelengths. Plastic cells cannot be used with any organic solvent in which the plastic is soluble. Disposable plastic cells are not suitable for accurate quantitative work. Price differences are significant between the materials. For example, a high-quality 1 cm quartz cuvet for use in UV costs about $70, the same size glass cuvet for use in the visible region costs about $30 and a 1 cm plastic disposable cuvet costs about 10 cents. Microvolume cells, flow cells, and other specialty cells are expensive, with costs of $200–500 per cell. Some spectrometers are designed to use ordinary glass test tubes as the "cells". These test tube "cells" should not be used for accurate quantitative work. Transparencies of some typical cell materials are presented in Fig. 5.27.

Figure 5.26 Cells for liquid samples, showing just a few of the wide variety of types and sizes available. From left to right, top row: standard 1 cm spectrophotometer cuvet with two optical faces and two frosted faces; semimicro 0.7 mL cuvet; 10 μL submicro cell; constant temperature cell with a jacket for circulating a temperature-controlling fluid. From left to right, bottom row: 5 mm fluorometer cuvet (all four faces are optically clear); in-line continuous flow cell for process monitoring (sample flow is from bottom to top); 10 mm flow cell; cylindrical cell. [Courtesy of Starna Cells, Inc., Atascadero, CA (www.starna.com).]

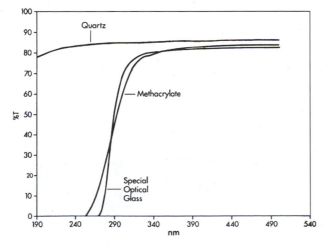

Figure 5.27 Transparencies of materials used to make sample cells for UV/VIS spectroscopy.

It is important that cells be treated correctly in order to achieve best results and to prolong their lifetime. To that end, the analyst should (1) always choose the correct cell for the analysis; (2) keep the cell clean, check for stains, etch marks, or scratches that change the transparency of the cell; (3) hold cells on the nontransparent surfaces if provided; (4) clean cells thoroughly before use and wash out the cell with a small amount of the sample solution before filling and taking a measurement; (5) not put strongly basic solutions or HF solutions into glass, quartz, or fused silica cells; (6) check for solvent compatibility with disposable plastic cells before putting them into the spectrometer; (7) for nondisposable cells, always dry carefully and return to their proper storage case; and (8) never wipe the optical surfaces with paper products, only lens cleaning paper or cloth material recommended by the manufacturer. At all times when not in use, cells should be kept clean and dry, and stored so that the optical surfaces will not become scratched.

5.2.5.2. Matched Cells

When double-beam instrumentation is used, two cells are needed: one for the reference and one for the sample. It is normal for absorption by these cells to differ slightly. This causes a small error in the measurement of the sample absorption and can lead to analytical error. For most accurate quantitative work, *optically matched cells* are used. These are cells in which the absorption of each one is equal to or very nearly equal to the absorption of the other. Large numbers of cells are manufactured at one time and their respective absorptivities measured. Those with very similar absorptivities are designated as optically matched cells. These are sold in pairs or sets of four. Matched cells are usually etched near the top with an identification mark and must be kept together. It is important for the analyst to understand that even closely matched cells will show small differences in absorption. The proper use of matched cells is to fill both the sample and the reference cells with the solvent and run a baseline spectrum, which is stored by the instrument computer system. The sample cell is then cleaned and sample solution put into it, while the reference cell and its solvent are left in place. After measuring the sample spectrum, the baseline is subtracted from the sample spectrum by the computer. This approach will correct for small differences in the cells. It is also important that the sample cell be reinserted facing in the same direction it was facing when the background was obtained. The etch mark on the top of the cell helps to facilitate this.

Modern cell manufacturing practices have improved greatly, and high-quality cell manufacturers are now producing cells with tolerances for window flatness, parallelity of windows, polish, and path length precision better than older "matched" cells. These modern cells are in effect optically matched by the nature of the manufacturing process. Tolerances for modern cells can be found on the Starna website (www.starnacells.com) as one example. It is still necessary for the analyst to check the cells on a routine basis by measuring a dilute solution of an absorbing material in all cells. This will identify any possible problems with microscopic scratches, residual film or deposits on the windows, and so on. Matched cells are not needed for qualitative analysis, such as obtaining the spectrum of a compound to help identify its structure.

5.2.5.3. Flow-Through Samplers

For routine analysis of large numbers of samples, the filling, cleaning, and emptying of large numbers of cells is time consuming. A flow cell and a peristaltic pump can be used to measure sample solutions directly from their original containers. This eliminates the need for sample handling and can minimize errors from sample handling, as well as eliminating the need to wash many cuvets. For example, Varian, Inc. makes a flow-through sampler for routine analysis for their Cary UV/VIS spectrometers that can send samples to a thermostatted flow cell as small as 80 μL sample volume. Dedicated flow-injection systems and segmented flow systems are available for specific routine analyses, such as nitrate, sulfate, and fluoride in drinking water. These systems are automated to take the sample, add and mix the reagents, and send the absorbing solution through a fixed wavelength spectrometer for completely unattended quantitative analysis.

5.2.5.4. Solid Sample Holders

The absorption spectrum of thin films of transparent solid materials, such as polymers, can be obtained by using a film holder. The simplest type of holder can be a paper slide mount with the sample taped to the slide mount. However, producer of films, gels, and other sheet materials are often interested in the homogeneity of the film or sheet. The film holder accessory for the Cary line of spectrometers (Varian, Inc.) allows samples up to 160 mm in length to be mounted. The absorption spectrum can be collected and then the sample moved automatically to produce a plot of absorption vs. position along the sample length.

Gel electrophoresis is an important technique for separating high molecular weight biological molecules, such as DNA, lipoproteins, immunoglobulins, and enzyme complexes. The classic method for visualizing the separate molecules is to stain them with colored dyes. Slices of electrophoresis gels can be mounted in a holder (Fig. 5.28) and the absorption spectrum collected as a function of distance along the gel using the same device used to move film samples. The holder can be moved in increments of 0.25 mm and gels up to 100 mm long can be analyzed in this fashion.

5.2.5.5. Fiber Optic Probes

In all the cells described earlier, the sample had to be brought to the spectrometer and placed in the light path (or pumped into the light path). Modern fiber optics have enabled the spectrometer to be brought to the sample. Using a fiber optic probe such as the one shown in Fig. 5.29, the absorption spectrum can be collected from a very small sample volume in a microcentrifuge tube. Fiber optic probes can be used to collect a spectrum from inside almost any container—an open beverage can, a 55 gallon drum of material, a tanker truck, or railroad car full of liquid. Probes are made in various path lengths, just as cells are, but eliminate the need to collect a sample and put it into a cell for measurement. This is especially useful for unknown and possibly hazardous samples.

Gel Boat Holder

Figure 5.28 Gel boat holder for UV/VIS absorption spectroscopy of electrophoresis gels. [Courtesy of Varian, Inc., Walnut Creek, CA. (www.varianinc.com).]

Figure 5.29 The photo on the left shows a fiber optics microprobe taking a spectrum from a 120 μL sample in a 500 μL tube. The photo on the right shows the fiber optics probe system for a Cary 50 spectrometer. [Courtesy of Varian, Inc., Walnut Creek, CA (www.varianinc.com).]

5.3. UV ABSORPTION SPECTRA OF MOLECULES

As has been discussed, the shape and intensity of UV/VIS absorption bands are related to the electronic structure of the absorbing species. This discussion will focus of the relationship of the absorption to the structure of simple organic molecules. Tables 5.3 and 5.4 list the approximate absorption maxima of common organic chromophores, functional groups that absorb UV and/or visible light. Of course, the chromophore is not an isolated unit; it is part of a molecule. The molecule is often dissolved in a solvent to acquire the spectrum. We will look at how we can use the absorption maximum of a chromophore and a set of guidelines to predict the position of the absorption maximum in a specific molecule. We will also consider how the solvent affects the spectrum of some molecules. As a reminder, the transitions that give rise to UV/VIS absorption by organic molecules are the $n \rightarrow \sigma^*$, $\pi \rightarrow \pi^*$, and $n \rightarrow \pi^*$ transitions.

Some terms need to be defined. A **chromophore** is a group of atoms (part of a molecule) that gives rise to an electronic absorption. An **auxochrome** is a substituent that contains unshared (nonbonding) electron pairs, such as OH, NH, and halogens. An auxochrome attached to a chromophore with π electrons shifts the absorption maximum to longer wavelengths. A shift to *longer* wavelengths is called a **bathochromic shift** or **red shift**. A shift to *shorter* wavelengths is called a **hypsochromic shift** or **blue shift**. An increase in the *intensity* of an absorption band (that is, an increase in $\varepsilon_{\max}$) is called **hyperchromism**; a decrease in intensity is called **hypochromism**. These shifts in wavelength and intensity come about as a result of the structure of the entire molecule or as a result of interaction between the solute molecules and the solvent molecules.

5.3.1. Solvent Effects on UV Spectra

5.3.1.1. *Bathochromic or Red Shift*

There is a general observation that many molecules that absorb radiation due to a $\pi \rightarrow \pi^*$ transition exhibit a shift in the absorption maximum to a longer wavelength when the molecule is dissolved in a polar solvent compared to a nonpolar solvent. The shift to a longer wavelength is called a bathochromic or red shift. This does not mean that the solution turns red or that the absorption occurs in the red portion of the visible spectrum, merely that the wavelength is shifted toward the red or longer wavelength end of the spectrum. This observation can be used to confirm the presence of $\pi \rightarrow \pi^*$ transitions in a molecule. The confirmation is carried out by dissolving a sample in two different solvents. For example, one solution can be made in a nonpolar solvent such as hexane and the second in a polar solvent such as alcohol. The spectra of the two solutions are recorded. If the absorption maximum in the alcohol solution occurs at a longer wavelength than the absorption maximum in the hexane solution, the compound exhibits a red shift. We would conclude that a $\pi \rightarrow \pi^*$ transition is present in the molecule and that the molecule must therefore have unsaturated bond(s). One major exception to this observation is the $\pi \rightarrow \pi^*$ transition in dienes and other hydrocarbon polyene molecules; the spectra of these molecules are not shifted by solvent polarity.

The reason for the shift in wavelength is related to the energy level of the excited state. The excited π^* state is more affected by attractive dipole–dipole interactions and hydrogen-bonding than the unexcited π state. Therefore, if a molecule is dissolved in a polar solvent, the energy level of the π^* antibonding orbital will decrease more than the energy level of the π bonding orbital. This is illustrated in Fig. 5.30. The energy difference in the polar solvent is less than the energy difference when the molecule is in a nonpolar solvent. As a consequence, the absorption maximum is changed to a longer wavelength in a polar solvent.

If the π^* energy level is decreased by attractive forces in polar solvents, it should be expected that the $n \rightarrow \pi^*$ transition will also show a red shift in polar solvents. It does, but there is a much more important interaction that overcomes the red shift for the $n \rightarrow \pi^*$ transition.

5.3.1.2. Hypsochromic or Blue Shift

The intermolecular attractive force known as hydrogen-bonding can occur when a molecule contains hydrogen covalently bound to oxygen or nitrogen. The O—H or N—H bond is highly polarized; the electrons in the covalent bond are pulled toward the electronegative atom, while the hydrogen can be thought of as having a partial positive charge. This hydrogen is able to form a "hydrogen bond" with an atom containing a pair of nonbonded electrons in an adjacent molecule. The hydrogen bond is the strongest of the intermolecular attractive forces. Examples of solvent molecules capable of hydrogen-bonding

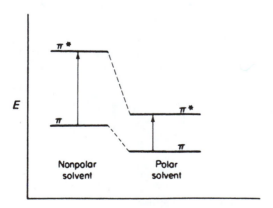

Figure 5.30   The energy difference between the π and π^* levels is decreased in the polar solvent. The absorption wavelength therefore increases. This is a bathochromic or red shift.

include water, alcohols such as ethanol and methanol, and molecules such as amines that contain an N—H bond.

The n electrons in a molecule are highly affected by hydrogen bond formation. The energy levels of n electrons decrease significantly in a solvent that has the ability to form hydrogen bonds. The result is an increase in the energy difference between the n orbital and the π^* orbital. This causes a shift in the absorption maximum of an n $\rightarrow \pi^*$ transition to shorter wavelengths, a blue shift, by as much as 25–50 nm. The decrease in the energy of the n electrons is almost equal to the energy of the hydrogen bond formed. The n $\rightarrow \pi^*$ transition also shows a hypsochromic shift as solvent polarity increases even in non-hydrogen-bonding solvents. This is thought to be due to the increased solvation of the n electrons; the energy level of the solvated electrons is lowered. The energy involved in the transition n $\rightarrow \pi^*$ when the solvent is nonpolar is less than the energy involved in the same transition when the molecule is in a polar or hydrogen-bonding solvent.

The hypsochromic shift effect on n electrons is much larger than the n $\rightarrow \pi^*$ bathochromic shift due to lowering of the π^* orbital described above. As a consequence, the absorption maximum for the n $\rightarrow \pi^*$ transition in a molecule which contains a lone pair of electrons will move to a shorter wavelength when dissolved in ethanol vs. hexane. Again, blue shift does not mean the absorption becomes blue, merely that the absorption wavelength becomes shorter. This information can be used to confirm the presence of n electrons in a molecule. The sample is dissolved in a non-hydrogen-bonding solvent such as hexane and also dissolved in a hydrogen-bonding solvent such as ethanol. If the absorption spectrum of the ethanol solution exhibits a blue shift (absorption maximum at a shorter wavelength) compared to the spectrum in hexane, n electrons are present in the sample molecule. Both n $\rightarrow \pi^*$ and n $\rightarrow \sigma^*$ transitions are blue-shifted by hydrogen bonding solvents. However, there are few n $\rightarrow \sigma^*$ transitions above 200 nm to begin with; a hypsochromic shift puts these absorptions further into the vacuum UV, so they are rarely observed in routine analysis.

A hypothetical example of how we can use this solvent effect information follows. A compound with molecules that contain both π and n electrons may exhibit two absorption maxima and may show both a red shift and a blue shift with a change in solvent polarity. In general, $\pi \rightarrow \pi^*$ transitions absorb approximately 10 times more strongly than n $\rightarrow \pi^*$ transitions. A molecule that contains both π and n electrons and is dissolved in a nonpolar solvent such as hexane might have an absorption spectrum similar to the spectrum in Fig. 5.31. By comparing the relative intensities of the two peaks, we would suspect that the absorption at 250 nm was due to a $\pi \rightarrow \pi^*$ transition and the absorption at 350 nm was due

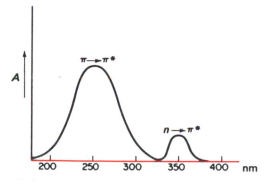

Figure 5.31 Expected absorption spectrum of a molecule in a nonpolar solvent exhibiting $\pi \rightarrow \pi^*$ and n $\rightarrow \pi^*$ transitions. Note the difference in absorptivity of the two transitions.

to an n → π^* transition. This is because the absorption coefficient of the $\pi \rightarrow \pi^*$ transition is considerably greater than that of the n → π^* transition, thus generating a higher degree of absorption. Also, in general, n → π^* transitions occur at longer wavelengths than $\pi \rightarrow \pi^*$ transitions because the energy difference for the n → π^* transition is lower.

If we now needed to confirm these assignments, we would put the compound in a solvent such as ethanol, which is both polar and capable of hydrogen-bonding. The polar nature of the solvent would induce a red shift in the $\pi \rightarrow \pi^*$ transition and hydrogen-bonding would induce a blue shift in the n → π^* transition. If our assignments were correct, then the absorption spectrum of the compound in ethanol might be as shown in Fig. 5.32. The combined evidence of the relative intensity (degree of absorption) and the blue and red shifts occurring in ethanol strongly supports the idea that the molecule contains both π bonds and n electrons.

5.4. UV SPECTRA AND THE STRUCTURE OF ORGANIC MOLECULES

Empirical rules based on thousands of laboratory observations have been developed over the years relating the wavelengths of the UV absorption maxima to the structures of molecules. R.B. Woodward in 1941 and then L. Fieser and M. Fieser developed rules for predicting the absorption maxima of dienes, polyenes, and conjugated ketones by studying terpenes, steroids, and other natural products. The rules are known as Woodward's Rules or the Woodward–Fieser Rules.

There are essentially four organic molecular systems of interest. The principal parent chromophore systems are (1) conjugated dienes, (2) monosubstituted benzene rings, (3) disubstituted benzene rings, and (4) conjugated carbonyl systems. The method of calculation is to identify a parent system and assign an absorption maximum. The parent system is then modified by the presence of other systems within the molecule. From these modifications, the absorption maximum of a specific molecular structure can be calculated.

5.4.1. Conjugated Diene Systems

The parent diene of conjugated diene systems is C=C—C=C. This system in a hexane solvent absorbs at 217 nm. If the conjugated system is increased, the wavelength of the

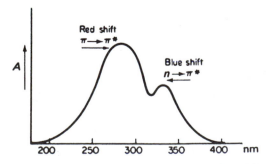

Figure 5.32 Expected absorption spectrum for the molecule in Fig. 5.31 dissolved in a polar solvent such as ethanol. There is a bathochromic (red) shift in the $\pi \rightarrow \pi^*$ transition and a hypsochromic (blue) shift in the n → π^* transition.

Table 5.6 Empirical Rules for Calculating the Absorption

	nm
Maxima of conjugated dienes	
Absorption of parent diene system C=C—C=C	217
Shift to longer λ	
Double bond extension to diene system	30
Diene system within a ring	36
Exocyclic nature of double bond in conjugated system	5
Each alkyl substituent or ring residue	5
Auxochrome is	
O-acyl	0
O-alkyl	6
S-alkyl	30
N-alkyl$_2$	60
Cl, Br	5

Source: With permission from A.I. Scott.

absorption maximum is increased by 30 nm for every double bond extension. Similarly, an alkyl group attached to the conjugated system increases the absorption maximum by 5 nm. Other substitutions, such as O-alkyl, cause an increase in the absorption maximum, as does inclusion within a ring or the exocyclic character of a double bond. These shifts are listed in Table 5.6. It should be emphasized that these results are empirical and not theoretical. They result from experimental observations.

    Some applications of these rules are demonstrated. The first compound is shown in Example 5.1. We have a diene (the two double bonds) in a ring with an alkyl group attached to one of the diene carbon atoms. The predicted absorption maximum is calculated to be 268 nm. How was this done? Since we have a diene, we use Table 5.6. In Example 5.1, the value assigned to the parent diene is 217 nm. This diene is within a ring (a diene within a ring is called a "homoannular" diene); therefore we add 36 nm to the absorption maximum. It is not so clear that there are *three* alkyl substituents on the diene. One is obviously the ethyl group, —C$_2$H$_5$, but the "groups" in positions A and B are also alkyl substituents. In each case carbons at A and B are attached to the diene system and therefore contribute to the electron density and to the shift to a longer wavelength. The fact that carbons A and B are attached to each other does not change their effect on the shift of the absorption maximum. Each alkyl group adds 5 nm to the absorption maximum, so the peak should appear at $217 + 36 + 3(5) = 268$ nm.

Example 5.1

Diene	217 nm
Within a ring	36
Alkyl substituent 3x5	15
Predicted λ_{max}	268 nm

Example 5.2 is a little more complicated. The parent diene is at 217 nm. The diene system is within a ring, as in Example 5.1, so we add 36 nm. There is one double bond extension to the system. As you can see, the conjugated system consists of three double bonds rather than two double bonds. This adds

Example 5.2

Diene	217 nm
Within a ring	36
Double bond extension	30
Exocyclic double bond	5
Alkyl substituents (5)	25
λ_{max}	313 nm

30 nm to the absorption maximum, from Table 5.6. There is one exocyclic double bond between carbons C and D, which adds 5 nm to the absorption maximum. This double bond is *touching* the adjacent ring and is within the conjugated system. The double bond between carbons E and F is also within a ring, but neither carbon *touches* the other ring, so this double bond is *not* exocyclic to a ring. The same thing is true for the double bond between carbons A and B; since neither carbon touches the adjacent ring, this bond is not exocyclic to a ring. There are five alkyl substitutes as follows: two on carbon A, one on carbon B, one on carbon D, and one on carbon F. The alkyl substituents on carbons D and F are designated by asterisks. The five alkyl substituents add another 25 nm to the absorption maximum. So we predict the absorption maximum to be $(217 + 36 + 30 + 5 + 25) = 313$ nm.

Examples 5.3 and 5.4 are molecules with very similar formulas, but different structures. The double bonds are in different positions. The wavelength of the absorption maximum for Example 5.3 is 309 nm, which is 54 nm longer than in Example 5.4. The major difference between the two compounds is that the molecule in Example 5.3 has a two double bonds within one ring and a conjugated system of three double bonds, that is, the double bonds are each separated by one single bond. That results in the addition of 36 nm for the diene in a ring and 30 nm for the double bond extension. In Example 5.4, there are three double bonds, but the system is not conjugated. The double bond farthest to the right (on the upper part of the ring containing the two double bonds) is separated by two single bonds from the parent diene and is not part of a conjugated system.

Example 5.3

Parent diene	217 nm
Within a ring (1)	36
Double bond extension (1)	30
Exocyclic double bond (1)	5
Alkyl substituents 3x5	15
-OCH$_3$	6
λ_{max}	309 nm

The only contributions to the absorption maximum in Example 5.4 are the exocyclic double bond (the diene bond on the right at the bottom of the ring containing the two double bonds) and the alkyl substituents. These two compounds can readily be distinguished by a simple UV absorption spectrum because of the large difference in λ_{max}.

Example 5.4

Parent diene	217 nm
Within a ring (0)	—
Double bond extension (0)	—
Exocyclic double bond (1)	5
Alkyl substituents 3x5	15
-OCH$_3$	6
λ_{max}	243 nm

With the rules in Table 5.6, the absorption maximum in Example 5.5 would be calculated as follows:

Example 5.5

Parent diene	217 nm
One exocyclic double bond (A)	5
Three alkyl substituents (C, D, E)	15
Calculated λ_{max}	237 nm

The experimentally observed value is 235 nm. It should be noted that there is no homo-annular diene system because the complete diene system is not contained in a single ring. Also, double bond B is not exocyclic to the ring, because it is not attached to a carbon that is part of another ring.

5.4.2. Conjugated Ketone Systems

The parent system is

$$C=C-C=C-C=O$$
$$\delta \quad \gamma \quad \beta \quad \alpha$$

The absorption maximum assigned to this parent system is 215 nm. In a manner similar to that for conjugated dienes, the wavelengths of the absorption maxima for conjugated ketones are modified by extension of the double bond substitution and position relative to rings and relative to the carbonyl group. The carbons are labeled α, β, γ, and δ, and substitutions in these positions change the shift of the absorption maximum. The empirical values used for calculating the absorption maxima of different compounds are shown in Table 5.7.

An example of the calculation of the wavelength of the absorption maximum using Table 5.7 is shown in Example 5.6. In this example the calculations are similar to

Example 5.6

Parent ketone	215 nm
Alkyl substituent γ(1)	18
δ(2)	36
Within a ring (1)	39
Double bond extension	30
Calculated λ_{max}	338 nm

those used in the conjugated diene system. Note that in order to affect the absorption maximum the substituents must be attached to the parent or conjugated system. The β carbon has two alkyl substituents and therefore increases the absorption maximum by 36 nm. The double bond between the γ and δ carbons is not an exocyclic double bond: it is within two different rings, but is not exocyclic to either of them. The OH group is *not* attached to the conjugated system and therefore does *not* contribute to its spectrum.

An isomer of the molecule used in Example 5.6 is shown in Example 5.7. The calculated absorption maximum is 286 nm. This spectrum is different in several ways from Example 5.6. For example, the carbon at the δ position has only one alkyl substitution.

Table 5.7   Rules for α,β-Unsaturated Ketone and Aldehyde Absorptions

$$\overset{\delta}{C}=\overset{\gamma}{C}-\overset{\beta}{C}=\overset{\alpha}{C}-C=O$$

Value assigned to parent α,β-unsaturated six-ring or acyclic ketone: 215
Value assigned to parent α,β-unsaturated five-ring ketone: 202
Value assigned to parent α,β-unsaturated aldehyde: 207

Increments added for	Shift to longer λ (nm)
Each exocyclic double bond	5
Diene within a ring	39
Double bond extending the conjugation	30
Each alkyl substituent	
$\quad \alpha$	10
$\quad \beta$	12
$\quad \gamma$ and higher	18
Each	
$\quad$ OH	
$\quad\quad \alpha$	35
$\quad\quad \beta$	30
$\quad\quad \delta$	50
$\quad$ O—Acyl	
$\quad\quad \alpha, \beta, \delta$	6
$\quad$ O—Me	
$\quad\quad \alpha$	35
$\quad\quad \beta$	30
$\quad\quad \gamma$	17
$\quad\quad \delta$	31
$\quad$ S-alkyl	
$\quad\quad \beta$	85
$\quad$ Cl	
$\quad\quad \alpha$	15
$\quad\quad \beta$	12
$\quad$ Br	
$\quad\quad \alpha$	25
$\quad\quad \beta$	30
$\quad$ NR$_2$	
$\quad\quad \beta$	95

Source: With permission from A.I. Scott.

The double bond between the γ and δ positions is exocyclic to a ring in this case and therefore increases the wavelength of the absorption maximum. The conjugated system is not within a ring, and therefore contributions from ring currents are not observed. These two isomers could be distinguished from each other by their UV spectra.

Example 5.7

Parent ketone	215 nm
Alkyl substituents γ(1)	18
δ(1)	18
Within a ring	0
Double bond extension	30
Exocyclic to a ring	5
Calculated λ_{max}	286 nm

The calculation of the absorption wavelength maximum for Example 5.8 is shown to be 286 nm. The calculation follows the steps used in previous examples. It should again be emphasized that the parts of the molecule that do not touch the conjugated or parent system do not shift the absorption maximum. This is true even for complex molecules, as some of these examples demonstrate. This makes prediction of the absorption maximum easier but the absorption maximum gives no information as to the structure of the entire molecule.

Example 5.8

Parent ketone	215 nm
Alkyl substituents γ(1)	18
δ(1)	18
Exocyclic to a ring	5
Double bond extension	30
Calculated λ_{max}	286 nm

The idea behind being able to predict absorption maxima based on structure is the hope that the absorption spectrum will tell us about the structure of an unknown. Examples 5.7 and 5.8 demonstrate the shortcoming of using the UV absorption maximum to deduce molecular structure. These are two very different compounds, easily distinguished by MS or NMR, but each gives the same UV absorption maximum. In fact, one could draw many different organic molecules that would give the same predicted absorption maximum. The UV absorption spectrum is only one of many tools needed to elucidate the structure of an organic molecule.

5.4.3. Substitution of Benzene Rings

Benzene is a strong absorber of UV radiation and particularly in the gas phase shows considerable fine structure in its spectrum (Fig. 5.11). Substitution on the benzene ring causes a shift in the absorption wavelengths. It is not uncommon for at least two bands to be observed, and frequently several more. The observed wavelengths of the absorption maxima of some substituted benzene rings are given in Table 5.8. These are experimental data and may be insufficient to completely identify unknown compounds. If the benzene ring is disubstituted, then calculations are necessary to predict the absorption maximum,

Table 5.8 Absorption Maxima of Monosubstituted Benzene Rings Ph—R

R	λ_{max} (nm) (solvent H_2O or MeOH)	
	Band 1	Band 2
—H	203	254
—NH_3^+	203	254
—Me	206	261
—I	207	257
—Cl	209	263
—Br	210	261
—OH	210	270
—OMe	217	269
—SO_2NH_2	217	264
—CO_2	224	268
—CO_2H	230	273
—NH_2	230	280
—O^-	235	287
—NHAc	238	
—COMe	245	
—CH=CH_2	248	
—CHO	249	
—Ph	251	
—OPh	255	
—NO_2	268	
—CH'=$CHCO_2H$	273	
—CH'=CHPh	295	

Note: Me = methyl, Ph = phenyl.
Source: From H.H. Jaffé and M. Orchin. Used with permission of Professor Milton Orchin.

because a list containing all the possible combinations would be very long and unwieldy and would need further experimental supporting evidence. There are some rules that help us understand disubstitution of benzene rings. These are as follows.

1. An electron-accepting group, such as NO_2, and an electron-donating group, such as OH, situated ortho or para to each other tend to cancel each other out and provide a spectrum not very different from the monosubstituted benzene ring spectrum (Table 5.9).
2. Two electron-accepting groups or two electron-donating groups para to each other produce a spectrum little different from the spectrum of the monosubstituted compound.
3. An electron-accepting group and an electron-donating group para to each other cause a shift to longer wavelengths. A para-disubstituted benzene ring illustrating this is shown in Example 5.9.

Example 5.9

Parent system
λ_{max} 230 nm

Parent system		230 nm
Para	NH_2	58
Calculated λ_{max}		288 nm

To become an expert in the interpretation of UV spectra requires reading of more detailed texts, as well as practice. It can be seen that the absorption maxima can be calculated with reasonable accuracy if the structure is known, usually within 5 nm or so. A complete study of this subject is beyond the scope of this book. More detailed treatments can be found in the texts by Creswell and Runquist, Pavia, Lampman and Kriz, or Silverstein, Bassler and Morrill (5th edition or earlier) listed in the bibliography.

5.5. ANALYTICAL APPLICATIONS

5.5.1. Qualitative Structural Analysis

As described at the beginning of the chapter, the types of compounds that absorb UV radiation are those with nonbonded electrons (n electrons) and conjugated double bond systems (π electrons) such as aromatic compounds and conjugated olefins. Unfortunately, such compounds absorb over similar wavelength ranges, and the absorption spectra overlap considerably. As a first step in qualitative analysis, it is necessary to purify the sample to eliminate absorption bands due to impurities. Even when pure, however, the spectra are often broad and frequently without fine structure. For these reasons, UV absorption is much less useful for the qualitative identification of functional groups or particular molecules than analytical methods such as MS, IR, and NMR. UV absorption is rarely used for organic structural elucidation today in modern laboratories because of the ease of use and power of NMR (Chapter 3), IR (Chapter 4) and MS (Chapters 9 and 10).

When UV spectra are used for qualitative identification of a compound, the identification is carried out by comparing the unknown compound's absorption spectrum

Table 5.9 Absorption Maxima of Disubstituted Benzene Derivatives

Substituent	Orientation	Shift for each substituent, γ, in EtOH (nm)
X = alkyl or ring residue		246
X = H		250
X = OH or O-alkyl		230
R = alkyl or ring residue	o-, m-	3
	p-	10
R = OH, O—Me, O-alkyl	o-, m-	7
	p-	25
R = O	o-	11
	m-	20
	p-	78
R = Cl	o-, m-	0
	p-	10
R = Br	o-, m-	2
	p-	15
R = NH$_2$	o-, m-	13
	p-	58
R = NH—Ac	o-, m-	20
	p-	45
R = NH—Me	p-	73
R = NMe$_2$	o-, m-	20
	p-	85

Note: o — ortho, m = meta, p = para; Me = methyl.
Source: From H.H. Jaffé and M. Orchin. Used with permission of Professor Milton Orchin.

with the spectra of known compounds. Compilations of UV absorption spectra in electronic and printed formats can be found from commercial sources such as the Informatics Division, Bio-Rad Laboratories (www.bio-rad.com) or the American Petroleum Institute (API) indices. Computer searching and pattern matching are the ways spectra are compared and unknowns identified in modern laboratories. Many academic libraries still maintain the printed spectra collections, which must be searched manually.

One area where qualitative UV/VIS spectroscopy is still very useful is in the rapid screening of samples for UV absorbing compounds. In a high throughput environmental lab, UV absorption can be used qualitatively to screen for samples that may have high levels of organic compounds, to avoid contaminating a sensitive instrument, or to evaluate dilution factors needed for quantitative analysis. In today's very high throughput pharmaceutical laboratories, UV absorption spectra can provide a quick survey of synthesized compounds, to screen out those that do not have the expected absorbance and therefore probably do not have the desired structure.

5.5.2. Quantitative Analysis

UV and visible absorption spectrometry is a powerful tool for quantitative analysis. It is used in chemical research, biochemistry, chemical analysis, and industrial processing. Quantitative analysis is based on the relationship between the degree of absorption and the concentration of the absorbing material. Mathematically, it is described for many chemical systems by Beer's Law, $A = abc$, as discussed in Chapter 2. The term applied to quantitative absorption spectrometry by measuring intensity ratios is **spectrophotometry**. The use of spectrophotometry in the visible region of the spectrum used to be referred to as *colorimetry*. (The term colorimetry appears in much of the older literature, but the term is also used for an unrelated measurement, the specification of color in samples using the tri-stimulus system.) To avoid confusion, the term spectrophotometry should be used for both UV and visible regions when quantitative determination of an analyte species is meant. While qualitative analysis is most useful for organic molecules and some transition and rare earth compounds, quantitative UV/VIS spectrophotometry is useful for determination of organic molecules, inorganic molecules, metal and nonmetal ions, and organometallic complexes.

UV/VIS spectrophotometry is a widely used spectroscopic technique. It has found use everywhere in the world for research, clinical analysis, industrial analysis, environmental analysis, and many other applications. Some typical applications of UV absorption spectroscopy include the determination of (1) the concentrations of phenol, nonionic surfactants, sulfate, sulfide, phosphates, fluoride, nitrate, a variety of metal ions, and other chemicals in drinking water in environmental testing; (2) natural products, such as steroids or chlorophyll; (3) dyestuff materials; and (4) vitamins, proteins, DNA, and enzymes in biochemistry.

Quantitative UV/VIS spectrophotometry has been used for the determination of impurities in organic samples, such as in industrial plant streams using flow-through cells. For example, it can be used to determine traces of conjugated olefins in simple olefins or aromatic impurities in pure hexane or similar paraffins. It has also been used in the detection of possible carcinogenic materials in foods, beverages, cigarette smoke, and air. In the field of agriculture, UV/VIS spectrophotometry is used for the determination of nitrogen- and phosphorus-containing fertilizers. In the medical field, it is used for the determination of enzymes, vitamins, hormones, steroids, alkaloids, and barbiturates. These measurements are used in the diagnosis of diabetes, kidney damage, and myocardial infarction, among other ailments. In the pharmaceutical industry, it can be used to measure the purity of drugs during manufacture and the purity of the final product. For example, aspirin, ibuprofen, and caffeine, common ingredients in pain relief tablets, all absorb in the UV and can be determined easily by spectrophotometry.

Spectrophotometry is used routinely to determine the concentrations of metal and nonmetal ions in a wide variety of samples. Spectrophotometry in the UV region of the spectrum is used for the direct measurement of many organic compounds, especially those with aromatic rings and conjugated multiple bonds. There are also colorless inorganic species that absorb in the UV. A good example is the nitrate ion, NO_3^-. A rapid screening method for nitrate in drinking water is performed by measuring the absorbance of the water at 220 and at 275 nm. Nitrate ion absorbs at 220 but not at 275 nm; the measurement at 275 nm is to check for interfering organic compounds that may be present. Spectrophotometric analysis in the visible region can be used whenever the sample is colored. Many materials are inherently colored without chemical reaction (e.g., inorganic ions such as dichromate, permanganate, cupric ion, and ferric ion) and need no further chemical reaction to form colored compounds. Colored organic compounds, such as dyestuffs, are also naturally colored. Solutions of such materials can be analyzed directly. The majority of metal and nonmetal ions, however, are colorless. The presence of these ions in a sample solution can be determined by first reacting the ion with an organic reagent to form a strongly absorbing species. If the product of the reaction is colored, absorbance can be measured in the visible region; alternatively, the product formed may be colorless but absorb in the UV. The majority of spectrophotometric determinations result in an increase in absorbance (darker color if visible) as the concentration of the analyte increases. However, there are analyses that cause a bleaching of color (decrease in absorbance) with increasing concentration of analyte.

As was mentioned in the introduction, the color we observe in a colored solution is the color that is transmitted by the solution. The color absorbed is the complementary color. The relationship between the color of light absorbed and the color observed is given in Table 5.10.

There are thousands of possible compounds and complexes that can be formed by reacting analyte species with organic reagents. Ideally, the reagent chosen should be selective; that is, it should react with only one ion or molecule under the conditions present. Second, the reagent should cause an abrupt color change or absorbance change when mixed with the analyte. This imparts high sensitivity to the method. Third, this intensity of color or UV absorbance should be related to the concentration of ions in the sample. Spectrophotometric reagents have been developed for almost all metal and nonmetal ions and for many molecules or classes of molecule (i.e., for functional groups). Many of these reactions are both sensitive and selective. Several examples of these reagents and their uses are given in Table 5.11. The books by Boltz and by Sandell and Onishi listed in the bibliography are classic reference sources, but may be hard to locate. The analytical literature contains thousands of direct and indirect methods for quantitative analysis of metals and nonmetals. A good summary of methods with literature references for most metal and nonmetal ions may be found in the handbook by Dean listed in the bibliography.

Quantitative analysis by absorption spectrophotometry requires that the samples be free from particulates, that is, free from **turbidity**. The reason for this is that particles can scatter light. If light is scattered by the sample away from the detector, it is interpreted as an absorbance. The absorbance will be erroneously high if the sample is turbid. We can make use of the scattering of light to characterize samples as discussed in Section 5.7, but particulates must be avoided for accurate absorbance measurements.

Quantitative analysis by spectrophotometry generally requires the preparation of a calibration curve, using the same conditions of pH, reagents added, and so on for all of the standards, samples, and blanks. It is critical to have a reagent blank that contains everything that has been added to the samples (except the analyte). The absorbance is measured for all blanks, standards, and samples. The absorbance of the blank is subtracted from all other absorbances and a calibration curve is constructed from the standards. The concentrations of analyte in the samples are determined from the calibration curve. The highest accuracy results from working in the linear region of the calibration curve. These quantitative methods can be quite complicated in the chemistry involved, the number of steps required (extraction, back-extraction, pH-adjustment, precipitation, masking, and many other types of operations may be involved in a method), and the analyst must pay attention

Table 5.10 Absorbed and Observed Colors

Color absorbed	Color observed
Red	Green
Orange	Blue-green
Yellow	Violet
Blue-green	Red
Blue	Orange
Violet	Yellow

Table 5.11   Typical Reagents Used in Spectrophotometry

Aluminon (also called ammonium aurintricarboxylate)

This compound reacts with aluminum in a slightly acid solution (pH 4–5) to form an intense red color in solution. It detects 0.04–0.4 μg/mL (ppm) of Al. Other elements, such as Be, Cr, Fe, Zr, and Ti, also react with aluminon. These elements must be removed if a sample is being analyzed for Al. The absorbance of the red solution is measured at 525 nm. The red color is the result of formation of a metal–dye complex called a "lake".

4-Aminophenazone (also called 4-aminoantipyrine)

This compound reacts with a variety of phenols to give intensively colored compounds and will detect 0.02–6.4 ppm of phenol in water. Drinking water is steam-distilled to separate the volatile phenols from interfering compounds. The distillate is treated with the reagent, and the colored complex is extracted into $CHCl_3$. The absorbance of the chloroform solution is measured at 460 nm. The reagent does not react with some para-substituted phenols, such as paracresol. This reaction is an example of the determination of organic compounds by spectrophotometric analysis following reaction with a color-producing reagent.

Thiourea
H_2NCSNH_2

Thiourea will react with osmium, a very toxic element, in sulfuric acid solution to form a colored product. The absorbance is measured at 460 nm with a detection range of 8–40 ppm Os. The only interferences are Pd and Ru. Compared with the other reagents in this table, the sensitivity of this reagent is low. Interestingly, under different analytical conditions (different acid, pH) thiourea reacts with Bi. The absorbance of this product is measured at 322 nm, and detects 0.06–0.6 ppm of Bi. A number of elements such as Ag, Cu, Hg, and Pb interfere. This is an example of a reagent that works under different chemical conditions to produce a low-sensitivity determination for one element and a high-sensitivity determination for another.

Chloranilic acid

Chloranilic acid forms solutions that are intensely red. The addition of calcium to the solution precipitates the chloranilic acid and the intensity of the red diminishes. The change (loss) in color is a measure of the quantity of calcium added. Numerous other elements interfere with the procedure. This is an example of spectrophotometric analysis by loss of color after addition of the sample.

(continued)

Table 5.11 Continued

Quinalizarin

This reagent gives intensely colored solutions in aqueous solutions. In 93% w/w H_2SO_4/H_2O, the color is red. The presence of borate causes the color to become blue. Numerous other ions, such as Mg^{2+}, Al^{3+}, and Be^{3+}, also react with quinalizarin. This is an example of a change of color of the reagent after reaction with the sample.

Curcumin

Curcumin is a sensitive reagent for boron, detecting 0.01–0.1 ppm of B by absorbance at 555 nm. Fluoride, nitrate and nitrite interfere, but can be eliminated by separating the boron from the sample by distillation of B as a methyl borate ester.

SPADNS [also known as 4,5-dihydroxy-3-(2-hydroxy-5-sulfophenylazo)-2, 7-naphthalenedisulfonic acid]

SPADNS (pronounced "spa-dens") is used to determine fluoride ion in drinking water. The SPADNS dye reacts with zirconium to form a dark red Zr–dye "lake". The F^- ion reacts to dissociate the Zr–dye complex and form (ZrF_6^{2-}), which is colorless. The color of the solution decreases with increasing fluoride ion concentration. The absorbance is measured at 570 nm, with a range of 0.2–1.40 ppm F^-. There are both positive and negative interferences from chlorine, chloride, phosphate, sulfate, and other species in drinking water. This is another example of a reaction where the color is "bleached" with increasing concentration of analyte.

Source: Examples were extracted from Standard Methods for the Examination of Water and Wastewater, and the references by Dean and Dulski.

to all the details to achieve accurate and precise results. There is both science and art involved in performing many of these analyses. Many standard or regulatory methods (e.g, from Standard Methods for the Examination of Water and Wastewater, ASTM, EPA, etc.) have published precision and accuracy data in the methods. These are the precisions and accuracies that can be achieved by an experienced analyst.

5.5.3. Multicomponent Determinations

It has been seen that UV/VIS absorption peaks are generally broad, so if there are two compounds, X and Y, in solution, it is likely that they will not be completely resolved from each other. That is, both X and Y contribute to the absorbance at most wavelengths. It is possible to calculate the concentrations of X and Y from a series of measurements. Measurements must be made at a number of wavelengths equal to the number of components in the mixture. In this case, there are two components, so two wavelengths are needed.

Four calibration curves need to be prepared: X at λ_1, X at λ_2, Y at λ_1, and Y at λ_2. All calibration curves should be blank corrected to pass through the origin. The absorbance of the sample mixture is measured at λ_1 and at λ_2.

Two equations can be written:

$$A_1 = C_X S_{X1} + C_Y S_{Y1}$$
$$A_2 = C_X S_{X2} + C_Y S_{Y2}$$

(5.3)

where A_1 is the absorbance of the unknown at λ_1; A_2, the absorbance of the unknown at λ_2; C_X, the concentration of X in the unknown; C_Y, the concentration of Y in the unknown; S_{X1}, the slope of the calibration curve for X at λ_1; S_{X2}, the slope of the calibration curve for X at λ_2; S_{Y1}, the slope of the calibration curve for Y at λ_1; and S_{Y2}, the slope of the calibration curve for Y at λ_2.

The absorbances and slopes are known; this leaves us with two equations and two unknowns, C_X and C_Y. The equations can be solved for the concentrations of X and Y in the unknown mixture. Dulski (see bibliography) gives an example of this approach with a method for the simultaneous determination of niobium and titanium by reaction with hydroquinone and measurement at 400 and 500 nm.

This same approach can be used for a mixture of three components. More complex mixtures can be unraveled through computer software that uses an iterative process at multiple wavelengths to calculate the concentrations. Mathematical approaches used include partial least squares, multiple least squares, principle component regression, and other statistical methods. Multicomponent analysis using UV absorption has been used to determine how many and what type of aromatic amino acids are present in a protein and to quantify five different hemoglobins in blood.

5.5.4. Other Applications

5.5.4.1. *Reaction Kinetics*

In common with other spectroscopic techniques, UV spectroscopy can be used to measure the kinetics of chemical reactions, including biochemical reactions catalyzed by enzymes. For example, suppose that two compounds A and B react to form a third compound C. If the third compound absorbs UV radiation, its concentration can be measured continuously. The original concentrations of A and B can be measured at the start of the experiment. By measuring the concentration of C at different time intervals, the kinetics of the reaction $A + B \rightarrow C$ can be calculated. Enzyme reactions are important biochemically and also analytically; an enzyme is very selective, even specific, for a given compound. The compound with which the enzyme reacts is called the substrate. If the enzyme assay is correctly designed, any change in absorbance of the sample will result only from reaction of the substrate with the enzyme. The rate of an enzyme reaction depends on temperature, pH, enzyme concentration and activity, and substrate concentration. If conditions are selected such that all of the substrate is converted to product in a short period of time, the amount of substrate can be calculated from the difference between the initial absorbance of the solution and the final absorbance. Alternatively, the other experimental variables can be controlled so that the rate of the enzyme reaction is directly proportional to substrate concentration.

5.5.4.2. *Spectrophotometric Titrations*

Many titration procedures in volumetric analysis use an indicator that changes color to signal the endpoint of the titration. For example, acid–base titrations are often performed

Figure 5.33   The structure of phenolphthalein in acidic solution and basic solution. The change in structure results in a change in the light absorbed by the molecule.

with indicators such as phenolphthalein. Figure 5.33 shows the structure of phenolphthalein in an acid solution and in a basic solution. As can be seen, the loss of protons results in a change in the structure of the molecule. As we know, this should result in a change in the energy levels in the molecule. In phenolphthalein, the energy level difference gives rise to the absorption of visible radiation when it is in an alkaline solution, but not in an acid solution. Phenolphthalein appears red in basic solution but colorless in acidic solution. Such structure changes and energy level changes are the basis of many acid–base indicators. Use of the human eye to detect the color change at the end of a titration is subject to the problems described at the beginning of the chapter. Each analyst may "see" the endpoint slightly differently from other analysts, leading to poor precision and possible errors. The use of a spectrophotometer to detect the color change is more accurate and reproducible. Use of the spectrophotometer also permits any change in absorbance in the UV or visible region by the titrant, analyte, or product to be used to determine the endpoint of the titration, so the method is not limited to reactions that use a colored indicator.

Spectrophotometric titrations have been used for redox titrations, acid–base titrations, and complexation titrations. The spectrophotometer can be used in a light scattering mode to measure the endpoint for a precipitation titration by turbidimetry. Spectrophotometric titrations can be easily automated.

5.5.4.3.   *Spectroelectrochemistry*

Oxidation–reduction reactions of inorganic and organic compounds can be studied by using a combination of electrochemistry (Chapter 15) and spectroscopy. Diode array systems are usefully employed when transparent thin electrodes are used to study these reaction mechanisms. By taking the absorption spectra in rapid succession and accumulating the data, it is possible to detect and measure intermediates formed in complex reactions. This is much more reliable than using absorption at a single wavelength to measure the reactions, since the choice of the single wavelength is often made with the assumption that the intermediates and end products are well known and suitable absorption wavelengths are therefore easily chosen. This is often not the case. Using the diode array system, the complete UV absorption spectra can be obtained, and much more information on the identity and concentration of species is therefore available.

5.6. ACCURACY AND PRECISION IN UV/VIS ABSORPTION SPECTROMETRY

There are three major factors that affect the accuracy and precision of quantitative absorption measurements: the instrument, the skill of the analyst, and the method variables. Instruments vary in the quality of their optical, mechanical and electrical systems and also in their data processing. Each instrument has fixed limitations; these must be understood by the analyst and optimized when possible. Wavelength calibration must be checked routinely using recognized wavelength standards. Holmium oxide standards are commonly used for this purpose. Stray light, transmittance, resolution, and other instrument parameters should be checked regularly. The analyst must optimize slit widths if the instrument is equipped with variable slits; too narrow a slit width may result in errors due to a low signal-to-noise ratio, while too wide a slit width causes both loss of resolution and negative deviations from Beer's Law. Sample cells are very often the cause of error; they must be cleaned and handled properly for the best accuracy. Method variables include the quality of the reagents used, pH, temperature control, color stability, reaction kinetics, and stoichiometry. It may be necessary to remove interferences, to buffer the sample, to control exposure to air and light, and perform other chemical manipulations to achieve accurate results. The analyst must be trained to operate the instrument and to perform all the chemical manipulations required. Attention to detail, accurate recordkeeping, routine use of replicates, spiked samples, or reference materials, and the preparation and measurement of appropriate blanks and standards are the analyst's responsibility. The accuracy of a method will depend on the analyst, the method specificity, the removal or control of interferences, and finally on the spectrophotometer itself.

Spectrophotometric analyses are capable of being performed with relative standard deviations as low as 0.5%. Detection limits depend on the molar absorptivity of the transition being measured, but are often 0.05 ppm or lower for many analytes. The linear working range for spectrophotometry is generally only one to two orders of magnitude. This is a short linear range compared to fluorescence, as will be seen.

5.7. NEPHELOMETRY AND TURBIDIMETRY

Much of the theory and equipment used in spectrophotometry applies with little modification to **nephelometry** and **turbidimetry**. These fields involve the **scattering** of light by nontransparent particles suspended in a liquid; examples of such particles include fine precipitates and colloidal suspensions. In *nephelometry* we measure the amount of radiation *scattered* by the particles; in *turbidimetry* we measure the amount of light *not scattered* by the particles. These processes are illustrated in Figs. 5.34 and 5.35. The applications of nephelometry include the estimation of the clarity of drinking water, beverages, liquid pharmaceuticals, and other products where the transparency is important and in the determination of species that can be precipitated, such as calcium or barium by precipitation as the phosphate or sulfate insoluble salt. The quantity of calcium or barium present is measured by the amount of radiation scattered by the precipitated compound. From the intensity of scattered radiation, the original concentration of calcium or barium can be determined. Conversely, sulfate and phosphate can be determined by precipitation as the barium compound. Process analyzers using nephelometry or turbidimetry can be used to monitor the clarity of a plant stream or water treatment facility stream on a continuous basis.

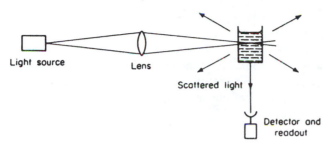

Figure 5.34   Schematic optical system for nephelometry.

When using nephelometry or turbidimetry for quantitative analysis, standard suspensions or standard turbid solutions are required for calibration. The precipitate or suspension standards must be prepared under rigidly controlled conditions. This is essential because the scattering of light depends on the *size, shape,* and *refractive index* of the *particles* involved, as well as on the concentration of particles. Some particles also absorb light, which will cause an error in the turbidity measurement. It is necessary for a given solution to produce the same number of particles of the same size and other properties listed for the degree of light scattering to be meaningful. Interferences include dirty sample cells, and any colored or absorbing species that will remove light from the light path. Any absorbance of light will result in an erroneously high turbidity, just as turbidity results in an erroneously high absorbance.

The wavelength of the light scattered most efficiently depends on the physical size of the scattering particles. From this, it can be reasoned that the size of the scattering particle may be determined if the wavelength of scattered light is accurately known. This type of light scattering forms the basis for the measurement of polymer molecular weights from the size of polymer molecules.

For water analysis, the formulation of turbid standards is very difficult, so most water laboratories use a synthetic polymer suspension as a standard. The formazin polymer suspension is easy to make and more stable and reproducible than adding clay or other particles to water to prepare standards. Alternatively, suspensions of polymer beads of the appropriate size can be used as scattering standards. (See Standard Methods for the Examination of Water and Wastewater for details.)

In the determination of a given species by a precipitation reaction, it is critical to control the experimental conditions. Two identical samples of equal concentration of analyte will scatter light equally only if they form the same number and size distribution of particles when they are precipitated. This depends on many experimental conditions, including the sample temperature, the rate at which the precipitant and the sample are

Figure 5.35   Schematic optical system for turbidimetry.

mixed, the degree of agitation or stirring, and the length of time the precipitates are allowed to stand before measurement. Procedures usually call for the use of a stopwatch to make all measurements at same point in time, such as 60 s after the reagent was added. Interferences include other particles, and colored or absorbing species. Sulfate in drinking water can be determined turbidimetrically by precipitation as barium sulfate over the range of 1–40 mg/L sulfate, with precision of about 2% RSD and accuracy, estimated by recovery of spiked samples, of about 90%.

5.8. MOLECULAR EMISSION SPECTROMETRY

5.8.1. Fluorescence and Phosphorescence

If a "black light" (UV light) is shone onto certain paints or certain minerals in the dark, they give off visible light. These paints and minerals are said to *fluoresce*. An energy diagram of this phenomenon is shown in Fig. 5.36. For fluorescence to occur, a molecule must absorb a photon and be promoted from its ground state to an excited vibrational state in a higher electronic state. There are actually *two* possible electronic transitions. Electrons possess the property of spin; we can think of this simplistically as the electron rotating either clockwise or counterclockwise. For two electrons to occupy the same orbital, their spins must be opposite to each other; we say that the spins are paired. If one electron is raised to the excited level without changing its spin, the electron in the excited level is still opposite in spin to the electron left behind in the valence level. This excited state of the molecule in which electron spins are paired is called a *singlet* state. If the electron spins are parallel (both spinning in the same direction as a result of the excited electron reversing its spin), the excited state is called a *triplet* state. Each "excited state" has both a singlet and corresponding triplet state. Singlet state energy levels are higher than the corresponding triplet state energies. Singlet states are designated S_1, S_2, S_3, and so on; triplet states are designated T_1, T_2, T_3, and so on. The ground state is a singlet state, S_0. Figure 5.36 shows a ground state with the first excited singlet and triplet states. Some vibrational sublevels of the excited states are also shown.

 The molecule absorbs energy and an electron is promoted to one of the higher vibrational levels in the singlet state; this is a vibrationally excited electronic state. The vibrationally excited molecule will rapidly "relax" to the lowest *vibrational* level of the

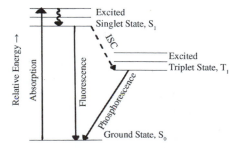

Figure 5.36 Schematic diagram of the ground state, excited singlet state, and excited triplet state, of a molecule. The wavy line denotes a radiationless transition from a higher vibrational level in the excited singlet state to the lowest vibrational level in the excited singlet state. The dotted arrow marked ISC shows the radiationless intersystem crossing from the excited singlet state to the excited triplet state. The length of the solid arrows denotes the relative energy of the transitions: absorption > fluorescence > phosphorescence. This results in $\lambda_{phosphorescence} > \lambda_{fluorescence} > \lambda_{absorption}$.

electronic excited state S_1. This relaxation or loss of energy is a radiationless process, shown by the wavy arrow. Energy decreases but no light is emitted. Now the molecule can return to the ground state by emitting a photon equal to the energy difference between the two levels. This is the process of *fluorescence*: excitation by photon absorption to a vibrationally excited state, followed by a rapid transition between two levels with the same spin state (singlet to singlet, in this case) that results in the emission of a photon. The emitted photon is of lower energy (longer wavelength) than the absorbed photon. The wavelength difference is due to the radiationless loss of vibrational energy, depicted by the wavy line in Fig. 5.36. This type of fluorescence, emission of a longer wavelength than was absorbed, is what is usually seen in solutions; it is called *Stokes* fluorescence. The lifetime of the excited state is very short, on the order of $1-20$ ns, so fluorescence is a virtually instantaneous emission of light following excitation. However, the lifetime of the fluorescent state is long enough that time-resolved spectra can be obtained with modern instrumentation. A molecule that exhibits fluorescence is called a *fluorophore*.

The transition from the singlet ground state to a triplet state is a forbidden transition. However, an excited singlet state can undergo a radiationless transition to the triplet state by reversing the spin of the excited electron. This is an energetically favorable process since the triplet state is at a lower energy level than the singlet state. This radiationless transition, shown schematically in Fig. 5.36 and 5.37, is called *intersystem crossing* (*ISC*). The molecule can relax to the ground state from the triplet state by emission of a photon. This is the process of *phosphorescence*: excitation by absorption of light to an excited singlet state, then an ISC to the triplet state, followed by emission of a photon due to a triplet–singlet transition. The photon associated with phosphorescence is even lower energy (longer wavelength) than the fluorescence photon, as seen from the relative energy levels in Figs. 5.36 and 5.37. Because the triplet–singlet transition is forbidden, the lifetime of the triplet excited state is long, up to 10 s in some cases. The sample will "glow" for some time after the excitation light source is removed. "Glow in the dark" paint is an example of phosphorescent material.

Fluorescence and phosphorescence are both types of luminescence. They are specifically types of *photoluminescence*, meaning that the excitation is achieved by absorption of light. There are other types of luminescence. If the excitation of a molecule and emission of light occurs as a result of chemical energy from a chemical reaction, the luminescence is called *chemiluminescence*. The light emitted by a firefly is an example of *bioluminescence*.

As shown in Fig. 5.37, there are other ways for molecules to return to the ground state. Excited molecules may collide with other molecules; it is possible for energy to

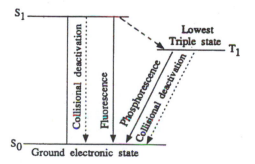

Figure 5.37 Processes by which an excited molecule can relax to the ground state. (Adapted from Guilbault, used with permission.)

be transferred during this collision. The molecule returns to the ground state but does not emit radiation. This is called collisional deactivation or *quenching*. Quenching occurs in solution by collision of the excited analyte molecule with the more numerous solvent molecules. Quenching in fluorescence is often a serious problem, but with care it can be minimized. Quenching by collision with the solvent molecules can be reduced by decreasing the temperature, thus reducing the number of collisions per unit time. The same result can be achieved by increasing the viscosity—for example, by adding glycerine. Dissolved oxygen is a strong quenching agent. It can be removed by bubbling nitrogen through the sample. Phosphorescence is very susceptible to quenching; the molecule in a triplet state has an extended lifetime in the excited state, so it is quite likely that it will collide with some other molecule and lose its energy of excitation without emitting a photon. Phosphorescence is almost never seen in solution at room temperature because of collisional deactivation. Low temperatures must be used and the analyte must be constrained from collision. This can be done for fluorescence and phosphorescence by converting the sample into a gel (highly viscous state), glass, or by adsorption of the analyte onto a solid substrate. "Organized" solvents such as surfactant micelles have been used successfully to observe room temperature phosphorescence and to greatly enhance fluorescence by reducing or eliminating collisional deactivation. Even with the appropriate experimental care, only a small fraction of available analyte molecules will actually fluoresce or phosphoresce, since radiationless transitions are very probable.

5.8.2. Relationship Between Fluorescence Intensity and Concentration

The intensity of fluorescence F is proportional to the amount of light absorbed by the analyte molecule. We know from Beer's Law that

$$\frac{I_1}{I_0} = e^{-abc} \tag{5.4}$$

so, subtracting each side of the equation from 1 gives:

$$1 - \frac{I_1}{I_0} = 1 - e^{-abc} \tag{5.5}$$

We multiply each side by I_0:

$$I_0 - I_1 = I_0(1 - e^{-abc}) \tag{5.6}$$

Since

$$I_0 - I_1 = \text{amount of light absorbed}$$

the fluorescence intensity, F, may be defined as

$$F = (I_0 - I_1)\Phi \tag{5.7}$$

where Φ is the quantum efficiency or quantum yield. The quantum yield, Φ, is the fraction of excited molecules that relax to the ground state by fluorescence. The higher the value of Φ, the higher the fluorescence intensity observed from a molecule. A nonfluorescent molecule has $\Phi = 0$.

Therefore, fluorescence intensity is equal to:

$$F = I_0(1 - e^{-abc})\Phi \tag{5.8}$$

From Eq. (5.8), it can be seen that fluorescence intensity is related to the concentration of the analyte, the quantum efficiency, the intensity of the incident (source) radiation, and the absorptivity of the analyte. Φ is a property of the molecule, as is the absorptivity, a. A table of typical values of Φ for fluorescent molecules is given in Table 5.12. The absorptivity of the compound is related to the fluorescence intensity [Eq. (5.8)]. Molecules like saturated hydrocarbons that do not absorb in the UV/VIS region do not fluoresce.

The fluorescence intensity is directly proportional to the intensity of the source radiation, I_0. In theory, the fluorescence intensity will increase as the light source intensity increases, so very intense light sources such as lasers, mercury arc lamps, or xenon arc lamps are frequently used. There is a practical limit to the intensity of the source because some organic molecules are susceptible to photodecomposition.

When the term abc is <0.05, which can be achieved at low concentrations of analyte, the fluorescence intensity can be expressed as:

$$F = I_0 abc\Phi \tag{5.9}$$

That is, F, total fluorescence, $= kI_0c$, where k is a proportionality constant. At low concentrations, a plot of F vs. concentration should be linear. But only a portion of the total fluorescence is monitored or measured; therefore,

$$F' = Fk' \tag{5.10}$$

where F' is the measured fluorescence
and

$$F' = k'I_0c \tag{5.11}$$

where k' is another proportionality constant.

A plot of F vs. c is shown in Fig. 5.38. It is linear at low concentrations. The linear working range for fluorescence is about five orders of magnitude, from 10^{-9} to 10^{-4} M.

At higher concentrations the relationship between F and c deviates from linearity. The plot of F vs. c rolls over as seen in Fig. 5.39. It can be seen that at higher concentrations the fluorescence intensity actually decreases because the molecules in the outer part of the sample absorb the fluorescence generated by those in the inner part of the

Table 5.12 Fluorescence Quantum Yields, Φ

Compound	Solvent	Φ
9-Aminoacridine	Ethanol	0.99
Anthracene	Hexane	0.33
9,10-Dichloroanthracene	Hexane	0.54
Fluorene	Ethanol	0.53
Fluorescein	0.1 N NaOH	0.92
Naphthalene	Hexane	0.10
1-Dimethylaminonaphthalene-4-sulfonate	Water	0.48
Phenol	Water	0.22
Rhodamine B	Ethanol	0.97
Sodium salicylate	Water	0.28
Sodium sulfanilate	Water	0.07
Uranyl acetate	Water	0.04

Note: Solutions are 10^{-3} M, temperatures 21–25°C.
Source: Guilbault, used with permission.

Figure 5.38 Dependence of fluorescence on the concentration of the fluorescing molecule. (From Guilbault, used with permission.)

sample. This is called the "inner cell" effect or self-quenching. In practice, it is necessary to recognize and correct for this effect. It is impossible to tell directly if the fluorescence measured corresponds to concentration A or concentration B as shown in Fig. 5.39. Both concentrations would give the same fluorescence intensity. Diluting the sample slightly can solve the dilemma. If the original concentration were A, then the fluorescence intensity would sharply decrease on dilution. On the other hand, if the concentration were B, then the fluorescence should increase on slight dilution of the sample.

5.9. INSTRUMENTATION FOR LUMINESCENCE MEASUREMENTS

A schematic diagram of a spectrofluorometer is shown in Fig. 5.40.

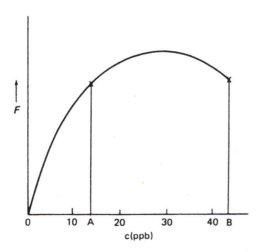

Figure 5.39 Fluorescence intensity at high concentrations of analyte. Note the reversal of fluorescence at high concentration. Concentrations A and B give the same fluorescence intensity and could not be distinguished from a single measurement.

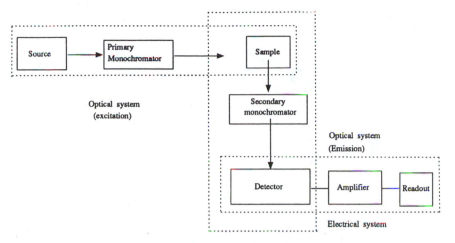

Figure 5.40 Block diagram of the optical components of a typical fluorometer. (From Guilbault, used with permission.)

5.9.1. Wavelength Selection Devices

Two monochromators are used, the primary or excitation monochromator, and the secondary or fluorescence monochromator. These are generally grating monochromators, although filters can be used for specific analyses. The excitation monochromator selects the desired narrow band of wavelengths that can be absorbed by the sample. The sample emits light in all directions. The second monochromator is placed at 90° to the incident light beam. The second monochromator is set to pass the fluorescence wavelength to the detector. The 90° orientation of the second monochromator is required to avoid the detector "seeing" the intense incident light, thus eliminating the background caused by the light source. Unlike absorption spectrophotometry, the measurement is not of the small difference between two signals, but of a signal with essentially no background. This is one reason for the high sensitivity and high linearity of fluorescence. Most fluorescence instruments are single beam instruments. This means that changes in the source intensity will result in changes in the fluorescence intensity. To compensate for changes in the source intensity, some instruments split off part of the source output, attenuate it, and send it to a second detector. The signals from the two detectors are used to correct for drift or fluctuations in the source.

The 90° geometry is the most common orientation for measuring fluorescence and works very well for solution samples that do not absorb strongly. Other angles are used in specific applications. For strongly absorbing solutions or for solid samples such as thin layer chromatography plates, fluorescence is measured from the same face of the sample illuminated by the source. This is called front-surface geometry. It is shown schematically for a solid sample in Fig. 5.41.

5.9.2. Radiation Sources

The fluorescence intensity is directly proportional to the intensity of the light source. Therefore intense sources are preferred. Excitation wavelengths are in the UV and visible regions of the spectrum, so some of the same sources used in UV/VIS absorption spectrometry are used for fluorescence. The optical materials will of course be the same—quartz for the UV, glass for the visible region.

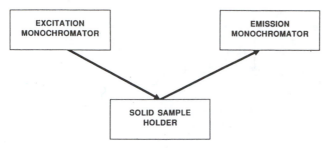

Figure 5.41 Front surface fluorescence geometry for a solid sample. (From Froelich and Guilbault, used with permission.)

Mercury or xenon arc lamps are used. A schematic of a xenon arc lamp is given in Fig. 5.42. The quartz envelope is filled with xenon gas, and an electrical discharge through the gas causes excitation and emission of light. This lamp emits a continuum from 200 nm into the IR. The emission spectrum of a xenon arc lamp is shown in Fig. 5.43. Mercury lamps under high pressure can be used to provide a continuum, but low-pressure Hg lamps, which emit a line spectrum, are often used with filter fluorometers. The spectrum of a low-pressure Hg lamp is presented in Fig. 5.44.

Because of their high intensity, laser light sources are ideal sources for fluorescence. The laser must exhibit a wide range of emission wavelengths, so tunable dye lasers have

Figure 5.42 Compact xenon arc lamp used in fluorometers. The quartz envelope is filled with xenon gas. The lamp is ignited by a 10–20 kV pulse across the electrodes. (From Froelich and Guilbault, used with permission.)

Figure 5.43 Spectral output of a compact xenon arc lamp. (From Froelich and Guilbault, used with permission.)

Figure 5.44 Spectral output of a mercury arc lamp, used as a source in fluorometers. (From Froelich and Guilbault, used with permission.)

been the only choice until recently. The dye lasers are generally pumped by an Nd:YAG laser. Nd stands for neodymium and YAG is yttrium aluminum garnet. These pumped dye laser systems are very expensive and complicated to operate. They have much greater intensity output than lamps and so enable lower detection limits to be achieved. Recent advances in solid-state lasers have made small, less expensive visible wavelength lasers available. Solid-state UV lasers are available, but do not have the required intensity for use as a fluorescence source.

5.9.3. Detectors

The most common detector in use is the PMT. The operation of the PMT was described earlier in this chapter. Because the signal is small due to the low concentrations of analyte used, the PMT is often cooled to subambient temperature to reduce noise. The limitation of the PMT is that it is a single wavelength detector. This requires that the spectrum be scanned. As we have discussed, scanning takes time and is not suitable for transient signals such as those from a chromatographic column. Diode array detectors are now used to collect the entire spectrum at once instead of scanning. The CCD, a 2D array detector, is another alternative to scanning in fluorescence spectrometry. Both PDA and CCD detectors can be used with LC or CE systems for separation and detection of fluorescent compounds or "tagged" compounds in mixtures. LC and CE are discussed in Chapter 13.

5.9.4. Sample Cells

The most common cell for solutions is a 1 cm rectangular quartz or glass cuvet with **four** optical windows. For extremely small volumes, fiber optic probes, microvolume cells, and flow cells are available. Gas cells, and special sample compartments for solid samples are commercially available.

5.10. ANALYTICAL APPLICATIONS OF LUMINESCENCE

Fluorescence occurs in molecules that have low energy $\pi > \pi^*$ transitions; such molecules are primarily aromatic hydrocarbons and polycyclic aromatic compounds. Examples include those in Table 5.12, as well as compounds like indole and quinoline. Molecules with rigid structures exhibit fluorescence; the rigidity evidently decreases the probability of a radiationless deactivation. Some organic molecules increase their fluorescence intensity on complexation with a metal ion. The resulting complex structure is more rigid than the isolated organic molecule in solution. Molecules that fluoresce can be measured directly; the number of such molecules is estimated to be between 2000 and 3000 from the published literature. There are several compounds that exhibit strong fluorescence; these can be used to derivatize, complex or "tag" nonfluorescent species, thereby extending the range of fluorescence measurements considerably. Other analytes are very efficient at quenching the fluorescence of a fluorophore; there are quantitative methods based on fluorescence quenching.

A fluorometric analysis results in the collection of two spectra, the excitation spectrum and the emission spectrum. The excitation spectrum should be the same as the absorption spectrum obtained spectrophotometrically. Differences may be seen due to instrumental factors, but these are normally small, as seen in Fig. 5.45, which shows the absorption and excitation spectra for Alizarin garnet R, a fluorometric reagent for aluminum ion and fluoride ion. The longest wavelength absorption maximum in the excitation spectrum is chosen as the excitation wavelength; this is where the first monochromator is set to excite the sample. It would seem reasonable to choose the wavelength that

Figure 5.45 Absorption and fluorescence spectra of the aluminum complex with acid Alizarin garnet R (0.008%): Curve A, the absorption spectrum; Curve B, the fluorescence excitation spectrum; Curve C, the fluorescence emission spectrum. (From Guilbault, used with permission.)

provides the most intense fluorescence as the excitation wavelength, but often short wavelengths from the high intensity sources used can cause a compound to decompose. The emission spectrum is collected by the second monochromator. The emission or fluorescence spectrum for Alizarin garnet R is shown in Fig. 5.45. Similar excitation and emission spectra are shown in Fig. 5.46 for quinine and anthracene. Note, especially for anthracene, that the fluorescence (emission) spectrum is almost a mirror image of the excitation spectrum. The shape of the emission spectrum and wavelength of the fluorescence maximum do not depend on the excitation wavelength. The same fluorescence spectrum is obtained for any wavelength the compound can absorb. However, the intensity of the fluorescence is a function of the excitation wavelength.

Fluorometry is used in the analysis of clinical samples, pharmaceuticals, natural products, and environmental samples. There are fluorescence methods for steroids, lipids, proteins, amino acids, enzymes, drugs, inorganic electrolytes, chlorophylls, natural and synthetic pigments, vitamins, and many other types of analytes. The detection limits in fluorometry are very low. Detection limits of 10^{-9} M and lower can be obtained. Single molecule detection has been demonstrated under *extremely* well controlled conditions. This makes fluorometry one of the most sensitive analytical methods available. Therefore, the technique is widely used in quantitative trace analysis. For example, Table 5.10 indicated that Al^{3+} could be detected spectrophotometrically using Aluminon at about 0.04 ppm in solution and fluoride could be detected at 0.2 ppm with SPADNS. Using fluorometry and Alizarin garnet R, whose structure is shown in Fig. 5.47, Al^{3+} can be determined at 0.007 ppm and F^- at 0.001 ppm. The strongly fluorescent compounds like fluorescein can be detected at part per trillion levels (ng/mL in solution), so use of

Figure 5.46 Absorption and fluorescence spectra of anthracene and quinine: Curve A, anthracene absorption; Curve B, quinine absorption; Curve C, anthracene fluorescence; Curve D, quinine fluorescence. (From Guilbault, used with permission.)

Figure 5.47 Structure of Alizarin garnet R.

such a compound as a "tag" can result in a very sensitive analytical method for many analytes.

5.10.1. Advantages of Fluorescence and Phosphorescence

The advantages of fluorescence and phosphorescence for analyses of molecules include extremely high sensitivity, high specificity, and a large linear dynamic range. The sensitivity is a result of the direct measurement of the fluorescence or phosphorescence signal against a zero background signal, as described. Specificity is a result of two factors: first, not all molecules fluoresce; therefore, many molecules are eliminated from consideration; and second, two wavelengths, excitation and emission, are used in fluorometry instead of one in spectrophotometry. It is not likely that two different compounds will emit at the same wavelength, even if they absorb the same wavelength and vice versa. If the fluorescing compounds have more than one excitation or fluorescent wavelength, the difference in either the emission spectrum or the excitation spectrum can be used to measure mixtures of compounds in the same solution. In Fig. 5.46, for example, the excitation spectra of quinine and anthracene overlap, but they do not emit at the same wavelengths, so the two compounds could be measured in a mixture. The linear dynamic range in fluorometry is six to seven orders of magnitude compared to one to two orders of magnitude that can be achieved in spectrophotometry.

5.10.2. Disadvantages of Fluorescence and Phosphorescence

Other compounds that fluoresce may need to be removed from the system if the spectra overlap. This can be done, for example, by column chromatography. Peaks may appear in the fluorescence spectrum that are due to other emission and scattering processes; Rayleigh, Tyndall, and Raman scattering may be seen because of the high intensity of the light source used. Peaks due to fluorescent impurities may occur.

Reversal of fluorescence intensity or self-quenching at high concentrations is a problem in quantitative analysis but can be eliminated by successive dilutions. Quenching by impurities can also occur and can cause significant problems in analysis. Changes in pH can frequently change structure, as we saw with phenolphthalein in Fig. 5.33, and thereby change fluorescence intensity; pH must therefore be controlled. Temperature and viscosity need to be controlled as well for reproducible results.

Photochemical decomposition or photochemical reaction may be induced by the intense light sources used. In general, the approach of using the longest excitation wavelength possible and the shortest measurement time possible will minimize this problem.

BIBLIOGRAPHY

Agilent Technologies, *Fundamentals of Modern UV–Visible Spectroscopy*, Publication Number 5980-1397E and the companion *Fundamentals of Modern UV–Visible Spectroscopy Workbook*, Publication 5980-1398E, Agilent Technologies, 2000 (Both publications may be accessed as pdf files at www.chem.agilent.com).

Boltz, D.F., Ed. *Colorimetric Determination of Nonmetals*; Interscience Publishers: New York, 1958.

Brown, C. *Analytical Instrumentation Handbook*, 2nd Ed.; Ewing, G.W., Ed.; Marcel Dekker, Inc.: New York, 1997.

Burgess, C.; Knowles, A., Eds. *Techniques in Visible and Ultraviolet Spectrometry*; Chapman and Hall: London, 1981.

Callister, W.D., Jr. *Materials Science and Engineering: An Introduction*, 5th Ed.; John Wiley and Sons: New York, 2000.

Chang, R. *Essential Chemistry*, 2nd Ed.; McGraw-Hill Companies, Inc: New York, 2000.

Creswell, C.J.; Runquist, O. *Spectral Analysis of Organic Compounds*; Burgess: Minneapolis, MN, 1970.

Dean, J.A. *Analytical Chemistry Handbook*; McGraw-Hill, Inc.: New York, 1995.

Dulski, T.R. *A Manual for the Chemical Analysis of Metals*; American Society for Testing and Materials: West Conshohocken, PA, 1996.

Ewing, G.W., Ed. *Analytical Instrumentation Handbook*, 2nd Ed.; Marcel Dekker, Inc.: New York, 1997.

Froelich, P.M.; Guilbault, G.G. In *Practical Fluorescence*, 2nd Ed.; Guilbault, G.G., Ed.; Marcel Dekker, Inc.: New York, 1990.

Guilbault, G.G., Ed. *Practical Fluorescence*, 2nd Ed.; Marcel Dekker, Inc.: New York, 1990.

Handbook of Chemistry and Physics, 61st Ed.; CRC Press: Boca Raton, FL, 1980.

Hollas, J.M. *Modern Spectroscopy*, 3rd Ed.; John Wiley and Sons, Ltd.: England, 1996.

Huber, L.; George, S.A., Eds. *Diode Array Detection in HPLC*; Marcel Dekker, Inc.: New York, 1993.

Ingle, J.D., Jr.; Crouch, S.R. *Spectrochemical Analysis*; Prentice-Hall, Inc.: Englewood Cliffs, NJ, 1988.

Jaffé, H.H.; Orchin, M. *Theory and Applications of Ultraviolet Spectroscopy*; John Wiley and Sons: New York, 1962.

Lambert, J.B.; Shurvell, H.F.; Lightner, D.; Cooks, R.G. *Introduction to Organic Spectroscopy*; Macmillan Publishing Company: New York, 1987.

Meehan, E.J. Optical methods of analysis. In *Treatise on Analytical Chemistry*; Elving, P.J., Meehan, E., Kolthoff, I.M., Eds.; John Wiley and Sons: New York, 1981, Vol. 7.

Pavia, D.L.; Lampman, G.M.; Kriz, G.S. *Introduction to Spectroscopy: A Guide for Students of Organic Chemistry*, 3rd Ed.; Harcourt College Publishers: Fort Worth, 2001.

Pisez, M.; Bartos, J. *Colorimetric and Fluorometric Analysis of Organic Compounds and Drugs*; Marcel Dekker, Inc.: New York, 1974.

Sandell, E.B.; Onishi, H. *Colorimetric Determination of Traces of Metals*, 4th Ed.; Interscience: New York, 1978.

Settle, F.A., Ed. *Handbook of Instrumental Techniques for Analytical Chemistry*; Prentice Hall, Inc.: Upper Saddle River, NJ, 1997.

Schaffer, J.P.; Saxena, A.; Antolovich, S.D.; Sanders, T.H., Jr.; Warner, S.B. *The Science and Design of Engineering Materials*, 2nd Ed.; WCB/McGraw-Hill: Boston, MA, 1999.

Scott, A.I. *Interpretation of the Ultraviolet Spectra of Natural Products*; Pergamon: Oxford, 1964.

Shackelford, J.F. *Introduction to Materials Science for Engineers*, 4th Ed.; Prentice Hall, Inc.: Upper Saddle River, NJ, 1996.

Silverstein, R.M.; Bassler, G.C.; Morrill, T.C. *Spectrometric Identification of Organic Compounds*, 5th Ed.; John Wiley and Sons: New York, 1991. [*Note*: The most recent edition of this text (Silverstein, R.M.; Webster, F.X., 6th ed., John Wiley and Sons: New York, 1998) has eliminated entirely the topic of UV spectroscopy.]

Standard Methods for the Examination of Water and Wastewater, 18th Ed.; American Public Health Association: Washington, DC, 1992.

Zumdahl, S.S.; Zumdahl, S.A. *Chemistry*, 5th Ed.; Houghton Mifflin Co.: Boston, MA, 2000.

SUGGESTED EXPERIMENTS

5.1 Add a drop of toluene to a UV absorption cell and cap or seal the cell. Record the absorption spectrum of toluene vapor over the UV range (220–280 nm) several times, varying the slit widths but keeping the scan speed constant. For example, slit widths of 0.1, 0.5, 1, and 5 nm can be used. Explain what happens to the spectral resolution as the slit width is changed.

5.2 Record the absorption spectrum of a solution of pure octane. Record the absorption spectrum of a 0.02% v/v toluene in octane solution from 220 to 280 nm. For a double-beam spectrometer, pure octane can be put into the reference cell. Compare with Experiment 5.1. Explain your observations. Change the slit width as in Experiment 5.1 and observe what happens to the resolution.

5.3 Record the absorption spectrum between 400 and 200 nm of (a) l-octene, (b) 1,3-butadiene, and (c) a nonconjugated diolefin (e.g., 1,4-pentadiene). What is the effect of a conjugated system on the absorption spectrum? What does the spectrum tell you about the relative energy of the molecular orbitals in each compound?

5.4 Record the absorption spectrum of a polynuclear aromatic compound such as anthracene and of a quinonoid such as benzoquinone. How does the structure of the compound affect the spectrum?

5.5 From Experiment 5.2, choose a suitable absorbance wavelength (or wavelengths) for toluene. Based on the maximum absorbance for your 0.02% solution, prepare a series of toluene in octane solutions of higher and lower concentrations. (For example, 0.1%, 0.05%, 0.01% v/v toluene in octane might be suitable.) Measure the absorbance of each solution (using octane as the reference) at your chosen wavelength.

5.6 Using the absorbance data obtained from Experiment 5.5, plot the relationship between the absorbance A and concentration c of toluene at each wavelength chosen. Indicate the useful analytical range for each wavelength.

5.7 Prepare a standard solution of quinine by dissolving a suitable quantity of quinine in water. Record the UV absorption spectrum of the solution between 500 and 200 nm. Record the absorption spectra of several commercial brands of quinine water (after allowing the bubbling to subside). Which brand contained the most quinine? (Tonic water contains quinine.)

5.8 Prepare ammonium acetate buffer by dissolving 250 g of ammonium acetate in 150 mL of deionized water and then adding 700 mL of glacial acetic acid. Prepare a 1,10-phenanthroline solution by dissolving 100 mg 1,10-phenanthroline monohydrate in 100 mL deionized water to which two drops of conc. HCl have been added. 1 mL of this reagent will react with no more than 100 μg of ferrous ion, Fe^{2+}. Prepare a stock ferrous iron solution containing 1 g/L of ferrous sulfate. By taking aliquots of the stock solution and diluting, prepare four standard solutions and a blank in 100 mL volumetric flasks as follows: Pipet 0, 100, 200, 300, and 400 μg of Fe^{2+} into 100 mL flasks, then add 2 mL conc. HCl to each flask. Add 10 mL of ammonium acetate buffer solution and 4 mL of 1,10-phenanthroline solution. Dilute to the mark with deionized water. Mix completely and allow to stand for 15 min for color development. Measure the absorbance at 510 nm. Correlate the absorbance with the concentration of iron in the solutions (remember to subtract the blank) and prepare a calibration curve. *Note*: all reagents used should be low in iron or trace metal grade. The use of this method for determining total iron, ferric iron, and ferrous iron in water may be found in Standard Methods for the Examination of Water and Wastewater. Sample preparation is required for real water samples, as the reagent only reacts with ferrous ion.

PROBLEMS

5.1   What types of molecules are excited by UV radiation? Why?

5.2   Indicate which of the following molecules absorb UV radiation and explain why: (a) heptane, (b) benzene, (c) 1,3-butadiene, (d) water, (e) 1-heptene, (f) 1-chlorohexane, (g) ethanol, (h) ammonia, and (i) *n*-butylamine.

5.3   Draw a schematic diagram of a double-beam spectrophotometer. Briefly explain the function of each major component.

5.4   List the principal light sources used in UV/VIS spectrometry.

5.5   Radiation with a wavelength of 640 nm is dispersed by a simple grating monochromator at an angle of 20°. What are the other wavelengths of radiation that are dispersed at the same angle by this grating (lowest wavelength 200 nm)?

5.6   Explain the operating principle of the photomultiplier tube.

5.7   What are the limitations of UV absorption spectroscopy as a tool for qualitative analysis?

5.8   (a)   Plot a calibration curve for the determination of monochlorobenzene from the data listed below.

      (b)   Three samples of monochlorobenzene were brought in for analysis. The samples transmitted (1) 90%, (2) 85%, and (3) 80% of the light under the conditions of the calibration curve just prepared. What was the concentration of monochlorobenzene in each sample?

Concentration (ppm)	Absorbance
1.2	0.24
2.5	0.50
3.7	0.71
5.1	0.97
7.2	1.38
9.8	1.82

5.9   Several samples of monochlorobenzene were brought to the laboratory for analysis using the calibration curve in Problem 5.8. The absorbance of each sample is listed below.

Sample	Absorbance
A	0.400
B	0.685
C	0.120
D	0.160
E	3.0

(a) What are the respective concentrations of monochlorobenzene in samples A–D? (b) What is the problem with Sample E? How could the analysis of sample E be obtained?

5.10   Which of the following absorb in the UV region? (a) N_2, (b) O_2, (c) O_3, (d) CO_2, (e) CH_4, (f) C_2H_4, (g) I_2, (h) Cl_2, (i) Cyclohexane, and (j) C_3H_6.

5.11   Why does phenolphthalein change color when going from an acid to a basic solution?

5.12   Why do UV absorption spectra appear as broad bands?

5.13   What causes the blue shift and the red shift in spectra?

5.14   Why do D_2 lamps emit a continuum and not line spectra?

5.15   How does a pn diode work?

5.16   Describe a diode array.

5.17   Describe the processes of UV molecular fluorescence and phosphorescence.

5.18   What is the relationship between fluorescence and excitation light intensity I_0?

5.19   Explain the reversal of fluorescence intensity with increase in analyte concentration. How is this source of error corrected?

5.20   Draw a schematic diagram of the instrumentation used for measuring UV fluorescence intensity.

5.21   (a)   What interferences are encountered in UV fluorescence?
         (b)   Why is phosphorescence not used as extensively as fluorescence for analytical measurements?

5.22   From the emission spectra of quinine and anthracene (Fig. 5.46), pick a wavelength that will permit you to determine quinine in a mixture of quinine and anthracene. Do the same for anthracene. Can you use the excitation spectra to distinguish between the two compounds? Explain.

Calculate the wavelength of the absorption maximum of the following compounds:

5.23   (a) CH_3 (b) CH_3

5.24

5.25

5.26

5.27

5.28

5.29

5.30

5.31

5.32

5.33

5.34

5.35

5.36

5.37

5.38

5.39

5.40

5.41

5.42 If a solution appears blue when a white light is passed through it, what colors has the solution absorbed?

5.43 State Beer's Law. What conditions must be met for Beer's Law to apply? Complete the following table:

Solution	Absorption (%)	T	A	Concentration (ppm)
1	1			1
2	13			6
3	30			15
4	55			34
5	80			69

5.44 Assuming the data obtained in Problem 5.43 were for a calibration curve and the same cell was used for all measurements, complete the following table:

Solution	Absorption (%)	T	Concentration (ppm)
A	30		
B	3		
C	10		
D	50		
E	70		

5.45 What is the relationship between the absorption cell length b and the absorbance A? Complete the following table (the concentration c was equal in all cases):

Sample	Path length b (cm)	A
1	0.1	0.01
2	0.5	
3	1.0	
4	2.0	
5	5.0	

5.46 Name three reagents used for quantitative UV/VIS spectrometric analysis and the elements they are used to determine.

5.47 Name two fluorometric reagents. What are the structural characteristics that make a molecule fluoresce?

5.48 Below are the absorption spectra of naphthalene and anthracene (from Jaffé and Orchin, with permission). Their structures are shown on the spectra.

These molecules are polycyclic aromatic hydrocarbons, formed by fusing together benzene rings.

(a) Tabulate the wavelengths for the absorption maxima for these two compounds and in the spectrum of benzene (Fig. 5.12 in the text). What trend do you observe? (b) What transition is causing the peaks observed in these compounds? (c) Explain the trend you observe in the absorption maxima. (d) The next larger molecule in this family is naphthacene, a four-ring compound, with the structure shown here:

Predict where the absorption maxima will occur for naphthacene. Explain your prediction.

5.49   The UV/VIS absorption spectrum shown here is the spectrum of holmium oxide (from Starna Cells, Inc., www.starna.com, with permission). It is a rare earth oxide and is available in high purity.

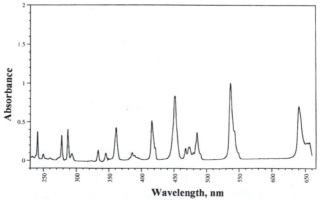

(a) Qualitatively, what differences do you see between this spectrum and the spectrum of an organic molecule such as pyridine (Fig. 5.1)? Why do you think they are different in appearance? (b) Consider the spectrum. Think of how you might use holmium oxide to check on the operation of your UV/VIS spectrometer. What could you check?

6

Atomic Absorption Spectrometry

The basis of atomic absorption spectrometry (AAS) is the absorption of discrete wavelengths of light by ground state, gas phase free atoms. Free atoms in the gas phase are formed from the sample by an "atomizer" at high temperature. AAS was developed in the 1950s by Alan Walsh and rapidly became a widely used analytical tool. AAS is an **elemental analysis** technique capable of providing quantitative information on $\sim$70 elements in almost any type of sample. As an elemental analysis technique, it has the significant advantage in many cases (but not all) of being practically independent of the chemical form of the element in the sample. A determination of cadmium in a water sample is a determination of the total cadmium concentration. It does not matter whether the cadmium exists as the chloride, sulfate, or nitrate, or even if it exists as a complex or an organometallic compound, if the proper analysis conditions are used. Concentrations as low as ppt levels of some elements in solution can be determined, and AAS is used routinely to determine ppb and ppm concentrations of most metal elements. Another principal advantage is that a given element can be determined in the presence of other elements, which do not interfere by absorption of the analyte wavelength. Therefore it is not necessary to separate the analyte from the rest of the sample (the matrix). This results in rapid analysis times and eliminates some sources of error. This is not to say that AAS measurements are completely free from interferences; both chemical and spectral interferences do occur and must be compensated for, as will be discussed. The major disadvantages of AAS are that no information is obtained on the chemical form of the analyte (no "speciation") and that often only one element can be determined at a time. This last disadvantage makes AAS of very limited use for qualitative analysis. AAS is used almost exclusively for quantitative analysis of elements, hence the use of the term "spectrometry" in the name of the technique instead of "spectroscopy".

6.1. ABSORPTION OF RADIANT ENERGY BY ATOMS

AAS is based on the absorption of radiant energy by free gas phase atoms. In the process of absorption, an atom changes from a low-energy state to a higher energy state as discussed in Chapter 2. Gas phase atoms do not vibrate in the same sense that molecules do. Also, they have virtually no rotational energy. Hence no vibrational or rotational energy is involved in the electronic excitation of atoms. As a result, atomic absorption spectra consist of a few very narrow absorption lines, in contrast to the wide bands of energy absorbed by molecules in solution.

Each element has a specific number of electrons "located" in an orbital structure that is unique to each element. The lowest energy electronic configuration of an atom is called the ground state. The ground state is the most stable electronic state. If energy ΔE of exactly the right magnitude is applied to a free gas phase atom, the energy will be absorbed. An outer electron will be promoted to a higher energy, less stable excited state. The frequencies and wavelengths of radiant energy capable of being absorbed by an atom are predicted from $\Delta E = h\nu = hc/\lambda$. The energy absorbed, ΔE, is the difference between the energy of the higher energy state and the lower energy state. As shown schematically in Fig. 6.1, this atom has four electronic energy levels. E_0 is the ground state, and the other levels are higher energy excited states. If the exact energies of each level are known, the three wavelengths capable of being absorbed can be calculated as follows:

$$\Delta E' = hc/\lambda_1 = E_1 - E_0$$

$$\Delta E'' = hc/\lambda_2 = E_2 - E_0$$

$$\Delta E''' = hc/\lambda_3 = E_3 - E_0$$

The calculated wavelengths λ_1, λ_2, and λ_3 all arise from transitions from the ground state to excited states. Absorption lines due to transitions from the ground state are called **resonance lines**. It is possible for an electron in an excited state to absorb radiant energy and move to an even higher excited state; in that case, we use the ΔE values for the appropriate energy levels involved. As we will see, in AAS most absorptions do arise from the ground state.

Quantum theory defines the electronic orbitals in an atom and predicts the lowest energy configuration (from the order of filling the orbitals). For example, the 11 electrons in sodium have the configuration $1s^2 2s^2 2p^6 3s^1$ in the ground state. The inner shells (principal quantum number, $n = 1$ and 2) are filled and there is one electron in the $n = 3$ shell. It is this outer shell electron that is involved in atomic absorption transitions for sodium. UV and visible wavelengths are the range of radiant energies absorbed in AAS. UV/VIS radiation does not have sufficient energy to excite the inner shell electrons, only the electrons in the outermost (valence) shell are excited. This is true of all elements: only the outermost electrons (valence electrons) are excited in AAS. While atomic spectroscopy considers the energy state of the atom and considers quantized leaps from one state to another, a simplified picture can be developed for the electronic transitions that are of interest in atomic absorption. Details of the quantum mechanics, spectroscopic selection rules, and designation of electronic states are topics that are covered in Physical Chemistry courses and are beyond the scope of this text.

The number of energy levels in an atom can be predicted from quantum theory. The actual energy differences of these levels have been deduced from studies of atomic spectra. These levels have been graphed in *Grotrian diagrams*, which are plots for a given atom showing energy on the *y*-axis and the possible atomic energy levels as horizontal lines. A partial Grotrian diagram for sodium is shown in Fig. 6.2. The energy levels are split

Figure 6.1   Schematic electronic energy levels in a free atom.

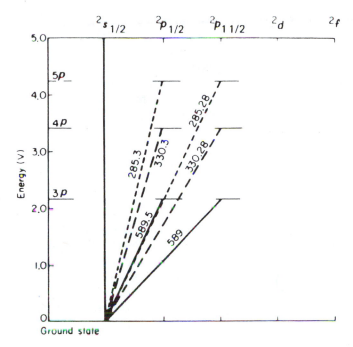

Figure 6.2 Partial Grotrian diagram for sodium.

because the electron itself may spin one way or another, resulting in two similar energy levels and therefore two possible absorption lines rather than a single line (a singlet). For the transition from the ground state to the first excited state of sodium, the electron moves from the 3s orbital to the empty 3p orbital. The latter is split into two levels, designated $^2P_{1/2}$ and $^2P_{3/2}$, by the electron spin, so two transitions are possible. The levels differ very slightly in energy because of the interaction of the electron spin and the orbital motion of the electron. The wavelengths that are associated with these transitions are 589.5 and 589.0 nm, respectively, the well-known sodium D lines.

Under the temperatures encountered in the atomizers used in commercial AAS systems, a large majority of the atoms exist in their lowest possible energy state, the ground state. Very few atoms are normally in the higher energy states. The ratio of atoms in an upper excited state to a lower energy state can be calculated from the Maxwell–Boltzmann equation (also called the Boltzmann distribution):

$$\frac{N_1}{N_0} = \frac{g_1}{g_0} e^{-\Delta E/kT} \tag{6.1}$$

where N_1 is the number of atoms in the upper state; N_0, the number of atoms in the lower state; g_1, g_0, the number of states having equal energy at each level 0, 1, etc. (g is called the degeneracy of the level); ΔE, the energy difference between the upper and lower states (in joules); k, the Boltzmann constant $= 1.381 \times 10^{-23}$ J/K; and T, the absolute temperature (in kelvin).

For example, it can be calculated from the Boltzmann distribution that if zinc vapor (Zn^0 gas) with resonance absorption at 213.9 nm is heated to 3000 K, there will be only one atom in the first excited state for every 10^{10} atoms in the ground state. Zinc atoms need a considerable amount of energy to become excited. On the other hand, sodium atoms are excited more easily than the atoms of most other elements. Nevertheless, at 3000 K only 1 sodium atom is excited for every 1000 atoms in the ground state. In a normal atom population there are very few atoms in states E_1, E_2, E_3, and higher. The total amount of radiation absorbed depends, among other things, on how many atoms are available in the lower-energy state to absorb radiation and become excited. Consequently, the total amount of radiation absorbed is greatest for absorptions from the ground state. Excited to excited state transitions are very rare, because there are so few excited atoms; only the ground state resonance lines are useful analytically in AAS.

For practical purposes, all absorption in AAS is by atoms in the ground state. This greatly restricts the number of absorption lines that can be observed and used for measurement in atomic absorption. Quite frequently only three or four useful lines are available in the UV/VIS spectral region for each element, and in some cases fewer than that. The wavelengths of these absorption lines can be deduced from the Grotrian diagram of the element being determined, but are more readily located in AAS instrument methods manuals (called "AAS cookbooks") available from the major instrument manufacturers. A list of the most intense absorption wavelengths for flame AAS determination of elements is given in Appendix 6.1.

AAS is useful for the analysis of approximately 70 elements, almost all of them metal or metalloid elements. Grotrian diagrams correctly predict that the energy required to reach even the first excited state of *nonmetals* is so great that they cannot be excited by normal UV radiation (> 190 nm). The resonance lines of nonmetals lie in the vacuum UV region. Commercial AAS systems generally have air in the optical path, and the most common atomizer, the flame, must operate in air. Consequently, using flame atomizers, atomic absorption cannot be used for the direct determination of nonmetals. However, nonmetals have been determined by indirect methods, as will be discussed in the applications section.

6.1.1. Spectral Linewidth

According to the Bohr model of the atom, atomic absorption and emission linewidths should be infinitely narrow, because there is only one discrete value for the energy of a given transition. However, there are several factors that contribute to line broadening. The natural width of a spectral line is determined by the Heisenberg uncertainty principle and the lifetime of the excited state. Most excited states have lifetimes of 10^{-8}–10^{-10} s, so the uncertainty in the energy of the electron slightly broadens the spectral line. This is called the *natural* linewidth, and is on the order of 10^{-4} Å. (1.0 Å $= 1.0 \times 10^{-10}$ m)

Collisions with other atoms in the atomizer lead to *pressure (Lorentz) broadening*, on the order of 0.05 Å. *Doppler broadening*, due to random kinetic motion toward and away from the detector, results in broadening of the spectral line on the order of 0.01–0.05 Å. Doppler and collisional broadening are temperature-dependent. In an atomization source with high concentrations of ions and electrons (such as in a plasma), *Stark broadening* occurs as a result of atoms encountering strong local electrical fields. In the presence of a magnetic field, *Zeeman splitting* of the electronic energy levels also occurs. Localized magnetic fields within atomizers from moving ions and electrons are negligibly small and their effects are generally not seen. However, as we will see later, by adding an external magnetic field we can use Zeeman splitting to assist in the correction of background absorption. The width of atomic absorption lines is on the order of 0.002 nm. These are very narrow lines, but not infinitely narrow.

6.1.2. Degree of Radiant Energy Absorption

The fraction of incident light absorbed by atoms at a particular wavelength is proportional to the number of atoms, N, that can absorb the wavelength and to a quantity called the oscillator strength f. The oscillator strength f is a dimensionless quantity whose magnitude expresses the transition probability for a specific transition. The **oscillator strength** is a constant for a particular transition; it is an indicator of the probability of absorbing the photon that will cause the transition. N is the number of ground state atoms in the light path, since most atoms are in the ground state at normal atomizer temperatures.

As discussed in Chapter 2, the fraction of incident light absorbed by a species can be expressed as the absorbance, A. The relationship between absorbance and the amount of analyte, in this case, atoms, in the light path is given by Beer's Law:

$$A = abc = (\text{constant} \times f)(b)\ (N/\text{cm}^3) \tag{6.2}$$

The proportionality constant a is called the absorptivity and includes the oscillator strength f. The term b is the length of the light path, and c is the concentration of ground state atoms in the light path (i.e., atoms/cm^3).

The amount of radiation absorbed is only slightly dependent on temperature as shown by Eq. (6.1). This is an advantage for AAS over atomic emission spectrometry and flame photometry (Chapter 7) where the signal intensity is highly temperature-dependent. Although the temperature does not affect the *process* of absorption by atoms, it does affect the efficiency with which free atoms are produced from a sample and therefore indirectly affects the atomic absorption signal. This effect can sometimes be significant. Some atoms, particularly those of the alkali metals, easily ionize. Low ionization energies and high temperatures result in the formation of ions rather than atoms. Ions do not absorb at atomic absorption wavelengths. Atoms that become ionized are effectively removed from the absorbing population, resulting in a loss of signal.

6.2. INSTRUMENTATION

A schematic block diagram of the instrumentation used for AAS is shown in Fig. 6.3. The components are similar to those used in other spectroscopic absorption methods as discussed in Chapters 2 and 5. Light from a suitable source is directed through the **atomizer**, which serves as the sample cell, into a wavelength selector and then to a detector. The detector measures how much light is absorbed by the sample. The sample, usually in solution

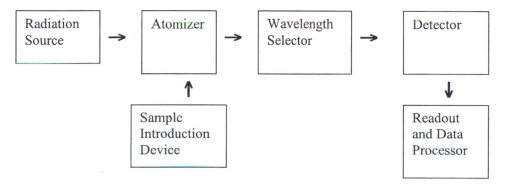

Figure 6.3 Block diagram of the components of an AAS.

form, is introduced into the atomizer by some type of introduction device. The atomizer converts the sample to gas phase ground state atoms that can absorb the incident radiation.

6.2.1. Radiation Sources

Two radiation sources are commonly used in commercial AAS instruments, the hollow cathode lamp (HCL) and the electrodeless discharge lamp (EDL). Both types of lamps are operated to provide as much intensity as possible while avoiding line-broadening problems caused by the collision processes described earlier.

6.2.1.1. Hollow Cathode Lamp (HCL)

As we have already mentioned, atomic absorption lines are very narrow (about 0.002 nm). They are so narrow that if we were to use a continuous source of radiation, such as a hydrogen or deuterium lamp, it would be very difficult to detect any absorption of the incident radiation at all. Absorption of a narrow band from a continuum is illustrated in Fig. 6.4, which shows the absorption of energy from a deuterium lamp by zinc atoms absorbing at 213.9 nm. The width of the zinc absorption line is exaggerated for illustration purposes. The wavelength scale for the deuterium lamp in Fig. 6.4 is 50 nm wide, and is controlled by the monochromator bandpass. If the absorption line of Zn were 0.002 nm wide, its width would be $0.002 \times 1/50 = 1/25,000$ of the scale shown. Such a narrow line would be detectable only under extremely high resolution (i.e., very narrow bandpass), which is not encountered in commercial AAS equipment.

   With the use of slits and a good monochromator, the bandwidth falling on the detector can be reduced to about 0.2 nm (much less than the earlier 50 nm example). If the light source is continuous, the entire 0.2 nm bandwidth contributes to the signal falling on the detector. If an absorption line 0.002 nm wide were absorbed from this light source, the signal reaching the detector would be reduced by only 1% from the original signal. Since this is about the absorption linewidth of atoms, even with complete absorption of the radiation at 213.9 nm by Zn atoms, the total signal from a continuous light source would change by only 1%. This would result in an insensitive analytical procedure of little practical use.

   What is needed for a light source in AAS is a source that produces very narrow emission lines at the exact wavelengths capable of being absorbed by analyte atoms. The problem was solved by the development of the HCL, shown schematically in Fig. 6.5. The cathode is often formed by hollowing out a cylinder of pure metal or

Figure 6.4 Width of an atomic absorption line (Zn 213.9 nm line), greatly exaggerated, compared with the emission bandwidth from a continuum source such as a deuterium lamp.

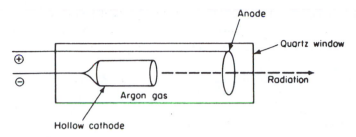

Figure 6.5 Schematic diagram of an HCL. The anode and cathode are sealed in a glass cylinder filled with argon or neon gas at low pressure. The window must be transparent to the emitted radiation.

making an open cylinder from pure metal foil. The metal used for the cathode is the metal whose spectrum will be emitted by the lamp. If we want to determine Cu in our AAS experiment, the lamp cathode must be a copper cylinder; if we want to determine gold, the cathode must be a gold cylinder, and so on. The cathode and an inert anode are sealed in a glass cylinder filled with Ar or Ne at low pressure (the "filler gas"). A window of quartz or glass is sealed onto the end of the lamp; a quartz window is used if UV wavelengths must be transmitted. Most HCLs have quartz windows, because most elements have emission and absorption lines in the UV. Glass can be used for some elements, such as sodium, where all the strong absorption lines are in the visible region of the spectrum.

The HCL emits narrow, intense lines from the element that forms the cathode. Applying a high voltage across the anode and cathode creates this emission spectrum. Atoms of the filler gas become ionized at the anode and are attracted and accelerated toward the cathode. The fast-moving ions strike the surface of the cathode and physically dislodge some of the surface metal atoms (a process called "sputtering"). The displaced atoms are excited by collision with electrons and emit the characteristic atomic emission spectrum of the metal used to make the cathode. The process is shown in Fig. 6.6. The emitted atomic lines are extremely narrow. Unlike continuum radiation, the narrow emission lines from the HCL can be absorbed almost completely by unexcited atoms. Using this light source, atomic absorption is easily detected and measured. Narrow line sources such as the HCL provide not only high sensitivity, but also specificity. If only Cu atomic emission lines are produced by the Cu HCL, there are few species other than

Figure 6.6 The HCL process, where Ar^+ is a positively charged argon ion, M^0 is a sputtered ground state metal atom, M^* is an excited state metal atom, and λ is emitted radiation at a wavelength characteristic for the sputtered metal. [From Beaty and Kerber, used with permission of PerkinElmer, Inc., (www.perkinelmer.com).]

Cu atoms that can absorb these lines. Therefore, there are few spectral interferences in AAS. The emitted spectrum consists of all the emission lines of the metal cathode, including many lines that are not resonance absorption lines, but these other lines do not interfere in the analysis.

Each hollow cathode emits the spectrum of metal used in the cathode. For this reason, a different HCL must be used for each different element to be determined. This is an inconvenience in practice and is the primary factor that makes AAS a technique for determining only one element at a time. The handicap is more than offset, however, by the advantage of the narrowness of the spectral lines and the specificity that results from these narrow lines.

It is possible to construct a cathode from more than one element. These are called "multielement" HCLs, and can be used for the determination of all of the elements in the cathode. This can be done sequentially but without having to change the lamp, which saves some time. In general, multielement cathodes do not perform as well for all of the elements in the cathode as do single HCLs for each element. The multielement cathode may have reduced intensity for one or more of the elements. All of the elements present will emit their atomic emission spectrum, resulting in a more complex emission than from a single element lamp. This may require that a less-sensitive absorption line be chosen to avoid a spectral interference. The obvious reason to use a multielement lamp is in the hope that more than one element can be measured simultaneously, making AAS a multielement technique. In fact, there are a few commercial simultaneous multielement AAS systems available for measuring up to eight elements or so. Most use a bank of single element lamps all focused on the atomizer rather than multielement cathodes. The disadvantage with this approach is that only one set of conditions in the atomizer can be used, and this set of atomization conditions may not be optimum for each element.

HCLs have a limited lifetime, usually due to loss of filler gas atoms through several processes. Adsorption of filler gas atoms onto the lamp surfaces causes decreased sputtering and decreased intensity of emission; eventually the number of filler gas atoms becomes so low that the lamp will not "light". The sputtering process causes atoms to be removed from the cathode; these metal atoms often recondense elsewhere inside the lamp, trapping filler gas atoms in the process and decreasing lamp life. This is particularly a problem for HCLs of volatile metals like Cd and As. HCLs operated at currents higher than recommended will have shorter lifetimes than those operated according to the manufacturer's recommendation. Operating at higher currents results in more intensity in the lamp output, but also may increase noise, which impacts both precision and limit of detection. Since we are measuring the *ratio* of light absorbed to incident light, there is little to be gained by increasing the lamp current.

Single element HCLs cost between $200 and 400 per lamp, while multielement lamps cost between $300 and 400 each.

6.2.1.2.   *Electrodeless Discharge Lamp (EDL)*

It is difficult to make stable hollow cathodes from certain elements, particularly those that are volatile, such as arsenic, germanium, or selenium. The HCLs of these elements have short lifetimes and low intensities. An alternative light source has been developed in the EDL. A commercial EDL design is shown in Fig. 6.7. A small amount of metal or a salt of the element whose spectrum is desired is sealed into a quartz bulb with a low pressure of Ar gas. The bulb is shown centered inside the coils in Fig. 6.7. The coils are part of a self-contained RF generator. When power is applied to the coils, the RF field generated will "couple" with the metal or salt in the quartz bulb. The coupled energy will vaporize

Figure 6.7 Electrodeless discharge lamp. [Courtesy of PerkinElmer Inc., Shelton, CT (www. perkinelmer.com).]

and excite the metal atoms in the bulb. The characteristic emission spectrum of the metal will be produced.

EDLs are very intense, stable emission sources. They provide better detection limits than HCLs for those elements that are intensity-limited either because they are volatile or because their primary resonance lines are in the low-UV region. Some elements like As, Se, and Cd suffer from both problems. For these types of elements, the use of an EDL can result in a limit of detection that is two to three times lower than that obtained with an HCL. EDLs are available for many elements, including antimony, arsenic, bismuth, cadmium, germanium, lead, mercury, phosphorus, selenium, thallium, tin, and zinc. Older EDLs required a separate power supply to operate the lamp. Modern systems are self-contained. EDL lamps cost slightly more than the comparable HCL.

6.2.2. Atomizers

The atomizer is the sample cell of the AAS system. The atomizer must produce the ground state free gas phase atoms necessary for the AAS process to occur. The analyte atoms are generally present in the sample as salts, molecular compounds, or complexes. The atomizer must convert these species to the reduced, free gas phase atomic state. The atomizer generally does this via thermal energy and some chemistry. The two most common atomizers are flame atomizers and electrothermal (furnace) atomizers.

6.2.2.1. Flame Atomizers

To create a flame, we need to mix an oxidant gas and a fuel gas, and light the mixture. In modern commercial flame AAS, two types of flames are used. The first is the air–acetylene flame, where air is the oxidant and acetylene is the fuel. The second type of flame is the nitrous oxide–acetylene flame, where nitrous oxide is the oxidant and acetylene is again the fuel. The fuel and oxidant gases are mixed in a burner system, called a premix burner. An exploded view of one type of commercial flame atomic absorption burner is shown in Fig. 6.8(a). In this design, the fuel gas is introduced into the mixing chamber through one inlet (not shown) while the oxidant gas is introduced through the sidearm on the *nebulizer*. The premix burner design generates laminar gas flow and this results in a very steady flame. The steady flame generates less noise due to "flame flicker"; this improves precision. Mixing the gases in the mixing chamber eliminates

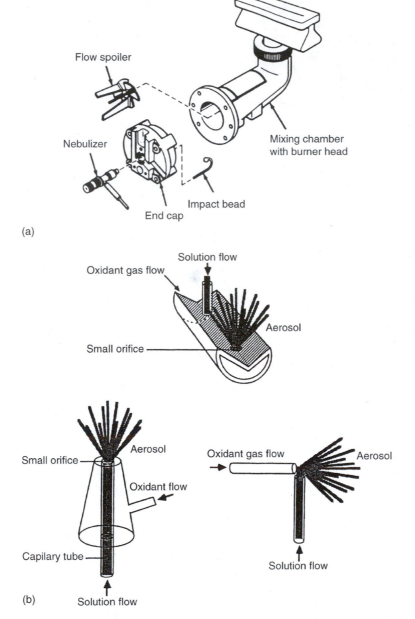

Figure 6.8 (a) Premix burner system. [Courtesy of PerkinElmer Inc., Shelton, CT (www.perkinelmer.com).] (b) Schematic nebulizer designs. (Top): modified Babington type; (left) concentric (the most common in FAAS); (right) cross-flow type. [From Parsons, used with permission.]

the safety hazard of having a combustible gas mixture piped through the laboratory. The flame burns just above the burner head, along the slot shown in the figure.

The sample is introduced into the burner in the form of a solution. The solution is aspirated into the nebulizer, which is basically a capillary tube. The nebulizer sprays the solution into the mixing chamber in the form of a fine aerosol. Three nebulizer designs are shown schematically in Fig. 6.8(b). The term "to nebulize" means to convert to a fine mist, like a cloud. The solution exits the capillary tube at high velocity and breaks into tiny droplets as a result of the pressure drop created. Kinetic energy transfer from the nebulizer gas overcomes the surface tension and cohesive forces holding the liquid together. The fuel and oxidant gases carry the sample aerosol to the base of the flame. In the flame the sample aerosol is desolvated, vaporized, and atomized to form free gas phase atoms of the analyte. The process will be discussed in detail later.

When the sample solution passes through the nebulizer, an aerosol is formed, but the droplets are of different sizes. As a droplet enters the flame, the solvent (water or organic solvent) must be vaporized, the residue must vaporize, and the sample molecules must dissociate into atoms. The larger the droplet, the more inefficient this process is. The two devices shown in Fig. 6.8(a) are used to overcome this problem. The impact bead is made of glass, quartz, or ceramic, and is placed directly in front of the nebulizer spray inside the mixing chamber. The impact bead improves nebulization efficiency by breaking larger droplets into smaller ones through collision of the spray with the bead. The flow spoiler is a piece of polymer or other corrosion-resistant material machined into three or more vanes. The flow spoiler is placed in the mixing chamber, about midway between the end cap and the burner. It physically removes larger droplets from the aerosol through collision while smaller droplets pass through the openings between the vanes. The larger droplets drain from the mixing chamber through a drain opening (not shown). The aim of this system is to produce an aerosol with droplets 4 μm in diameter. The two devices may be used alone or in combination. The drain opening is connected to a liquid waste container with a length of polymer tubing. It is extremely important that there be a trap between the drain opening and the waste container to prevent the free flow of flame gases out of the burner assembly. The presence of the trap helps to eliminate "flashback", discussed subsequently. The trap is often a simple water-filled loop in the drain tubing itself.

The nebulizer described is a self-aspirating, pneumatic nebulizer and is the one shown schematically on the lower left of Fig. 6.8(b). The nebulizer capillary is usually made of stainless steel but other materials such as Pt, Ta, and polymers may be used for corrosive solvents when contamination from the elements in steel must be avoided. A variety of other nebulizer designs have been developed for specific applications but are not commonly used in AAS. These other nebulizers are often used in atomic emission spectrometry and will be described in Chapter 7.

The burner head is constructed either of stainless steel or titanium. Different sizes and geometries of burner heads are used for various flames. The single slot burner head shown in Fig. 6.8 with a 10 cm long slot is used for air–acteylene flames. The 10 cm long flame is the sample path length for the AA spectrometer in this case. A smaller 5 cm long single slot burner head with a narrower slot is used for nitrous oxide–acetylene flames. Usually, the slot is oriented parallel to the light beam from the radiation source, so the path length is as long as possible to achieve the highest sensitivity (remember Beer's Law), as shown in the AAS layout in Fig. 6.9. The modern burner head design coupled with the use of a liquid-filled trap between the mixing chamber drain and the waste container prevents the possibility of a flashback. The gas mixture is ignited above the burner head and the flame, a highly energetic chemical reaction between the fuel and oxidant, propagates rapidly. The flame is supposed to propagate up from the burner head. It will do so if the linear gas flow rate through the burner slot is higher than the **burning velocity** of the flame. Burning velocity is a characteristic of the flame type; both nitrous oxide–acetylene and air–acetylene flames have low burning velocities. If the premixed gas flow rate is

Figure 6.9 Basic AAS system. [From Beaty and Kerber, used with permission of PerkinElmer, Inc. (www.perkinelmer.com).]

less than the burning velocity, a flashback can occur when the gas mixture is ignited. Flashback is the very undesirable and extremely hazardous propagation of the flame below the burner head slot and back into the mixing chamber. A flashback results in an explosion in the mixing chamber; it can destroy the burner assembly, create flying debris, rupture the fuel and oxidant lines thereby releasing combustible gases into the laboratory and cause injury to the analyst or other people in the laboratory. Early burner designs and high burning velocity flames such as those using pure oxygen as oxidant were often prone to flashback. It is imperative that AAS systems be operated according to the manufacturer's directions, that only the correct gases, properly regulated, and the correct burner head be used, and that the trap and all safety interlocks be in place and functioning.

One reason for the long path length used in flame AAS (compared with the typical 1 cm path length in UV/VIS or IR absorption spectrometry) is that the premix burner and nebulizer system used is very inefficient and wasteful of sample. Sample solution is aspirated into the nebulizer at $\sim$5 mL/min but only a small amount ($<$5%) of that solution reaches the burner. The path length has to be as long as possible to compensate for the loss of sample in the mixing chamber. Flame AAS is very popular because it is fast, has high element selectivity, and the instruments are easy to operate, but the inefficiency of the sample introduction system combined with noise inherent in the system restricts detection limits in flames to the ppm range for most elements.

6.2.2.2. Electrothermal Atomizers

In order to measure ppb concentrations of metals, a different type of atomizer is needed. A furnace or electrothermal atomizer (ETA) was developed less than 10 years after the technique of AAS was developed. In 1961 B.V. L'vov built a heated carbon tube atomizer, illustrated in Fig. 6.10. The system used a carbon tube heated by electrical resistance (hence, an *electrothermal* atomizer). The tube was lined with Ta foil and purged with argon gas. After the tube or furnace reached an elevated temperature, the sample, on a carbon electrode, was inserted at the bottom, as shown in the figure. Atomization took place and the analyte atoms absorbed the light beam passing through the carbon tube. His system was orders of magnitude more sensitive than flame atomizers, but was difficult to control. Other workers in the field, particularly West, Massman, and Robinson, refined carbon rod and furnace atomizers. Most commercial instruments use some version of a carbon tube atomizer for electrothermal atomization. The most common atomizer of this type uses a tube of graphite coated with pyrolytic graphite and heated by electrical resistance; therefore, this commercial ETA is called a **graphite furnace atomizer**. The acronym GFAAS tells the reader that the atomizer used is a graphite furnace, as opposed to a flame.

A commercial graphite furnace atomizer is illustrated in Fig. 6.11. This atomizer consists of a graphite tube, approximately 6 mm in diameter and 25–30 mm long, the

Figure 6.10 High-temperature furnace atomizer designed and used by L'vov.

Figure 6.11 Graphite furnace atomizer. This is a longitudinally heated furnace design. The graphite tube is shown in greater detail in Fig. 6.20. [Courtesy of PerkinElmer Inc., Shelton, CT (www.perkinelmer.com).]

electrical contacts required to heat the tube, a system for water-cooling the electrical contacts at each end of the tube, and inert purge gas controls to remove air from the furnace. An inert gas is used to prevent the graphite from being oxidized by air during the heating process. Quartz windows at each end of the furnace assembly permit the light from the HCL or EDL to pass through the furnace and out to the spectrometer. A small amount of sample solution, between 5 and 50 μL, is dispensed into the graphite tube through a small hole in the top of the tube. The furnace is heated in a series of programmed temperature steps to evaporate the solvent, decompose (ash, char) the sample residue, and finally to atomize the sample into the light path. Details of the graphite furnace atomization process and temperature program will be discussed later. The diagram in Fig. 6.11 shows a longitudinally heated graphite furnace. The electrical contacts are at each end of the tube and they must be water-cooled. This longitudinal heating results in a temperature gradient in the heated furnace—the ends are cooler than the center. This may result in recondensation of vaporized species at the ends of the tube. This can be a problem for the next sample analysis if material from the previous sample is still in the furnace. The problem is called "carry-over" or a "memory effect", and can result in poor accuracy and precision. To overcome this problem, new graphite furnaces that are heated transversely have been developed. A transverse graphite tube is shown in Fig. 6.12. The electrical contacts are transverse to the light path, and the tube is heated across the circumference. This results in even heating over the length of the furnace and reduces the carry-over problem significantly.

Modern graphite furnace atomizers have a separate power supply and programmer that control the electrical power, the temperature program, the gas flow, and some spectrometer functions. For example, the spectrometer can be programmed to "read", that is, collect absorbance data, only when the furnace reaches the atomization temperature. This saves data storage space and data processing time.

Researchers have developed other types of ETAs over the years, including filaments, rods, and ribbons of carbon, tantalum, tungsten, and other materials, but the only commercial ETA available is the graphite furnace atomizer.

6.2.2.3. Other Atomizers

Two additional commercially available atomizers (really analysis techniques with unique atomizers) must be discussed, because they are extensively used in environmental and clinical analysis. They are the **cold vapor-AAS technique** (CVAAS) for determination of the element mercury, Hg, and the **hydride generation technique** (HGAAS) for several elements that form volatile hydrides, including As, Se, and Sb. These elements are toxic; federal and state laws regulate their concentrations in drinking water, wastewater, and air, so their measurement at ppb concentrations is very important. Because

Figure 6.12   A graphite tube for a transversely heated furnace. [Courtesy of PerkinElmer Inc., Shelton, CT (www.perkinelmer.com).]

the CVAAS and HGAAS "atomizers" are analysis techniques they will be discussed under applications of AAS.

An atomizer based on glow discharge techniques is commercially available for the analysis of solid metal samples by AAS. It will be discussed in the applications sections under solid sample analysis.

6.2.3. Spectrometer Optics

6.2.3.1. Monochromator

A monochromator is required to separate the absorption line of interest from other spectral lines emitted from the HCL and from other elements in the atomizer that are also emitting their spectra. Because the radiation source produces such narrow lines, spectral interference is not common. Therefore the monochromator does not need high resolution.

A typical monochromator is shown in Fig. 6.13. The most common dispersion element used in AAS is a diffraction grating. The grating disperses different wavelengths of light at different angles, as discussed in Chapter 2. The grating can be rotated to select the wavelength that will pass through the exit slit to the detector. All other wavelengths are blocked from reaching the detector. The *angle of dispersion* at the grating is a function of the density of lines ruled on the grating. The more lines/mm on the grating, the higher is the dispersion. Higher dispersion means greater separation between adjacent lines. A high dispersion grating permits the use of wider entrance and exit slits on the monochromator to achieve the same resolution. Wider slits allow more light to pass through the system to the detector. Large high-quality gratings with high dispersion are expensive, but they offer better energy throughput than cheaper low-dispersion gratings. In addition to the number of lines ruled on the grating, the blaze angle affects the intensity of diffracted wavelengths. Gratings can be blazed for any wavelength; the farther away a given wavelength is from the wavelength for which the grating is blazed, the more light intensity will be lost in diffraction. The analytically useful wavelength range for AAS is from 190 to 850 nm. If a grating is blazed for the middle of this range, there will be loss in intensity at both extremes. That is particularly bad for elements such as Se, As, P, Cd, and Zn, with resonance lines at the low UV end, and for the alkali metals, with resonance lines at the high end of the visible region. To overcome this problem, the monochromator can be equipped with two gratings, one blazed for the UV and one for the visible. Modern spectrometers automatically control the movement and alignment of the grating or gratings, so that the correct one is used for the wavelength being measured.

Figure 6.13 Typical grating monochromator layout. [From Beaty and Kerber, used with permission of PerkinElmer, Inc. (www.perkinelmer.com).]

Most commercial AAS systems have the monochromator, optics, and detector designed for the measurement of one wavelength at a time; they are single-element instruments. There are a few systems available that do perform multielement determinations simultaneously, using an Echelle spectrometer (discussed in Chapter 2) and a bank of HCLs all focused on the atomizer. The limitation to this approach is not the sources or the spectrometer or the detector, but the atomizer. The atomizer can only be at one set of conditions, and those conditions will not necessarily be optimum for all of the elements being measured. There will be a tradeoff in detection limits for some of the elements.

6.2.3.2.   Optics and Spectrometer Configuration

The spectrometer system for AAS can be configured as a single-beam system, as shown in Fig. 6.8, as a double-beam system, shown in Fig. 6.14, or as a pseudo-double-beam system, which will not be discussed. (See the reference by Beaty and Kerber for a description of this system.) Note that in AAS the sample cell is placed in front of the monochromator, unlike UV/VIS spectrometers for molecular absorption or spectrophotometry, where the sample is placed after the monochromator.

A single-beam system is cheaper and less expensive than a double-beam system, but cannot compensate for instrumental variations during analysis. In a double-beam system, part of the light from the radiation source is diverted around the sample cell (flame or furnace atomizer) to create a *reference beam*, as shown in Fig. 6.14. The reference beam monitors the intensity of the radiation source and electronic variations (noise, drift) in the source. The signal monitored by the detector is the ratio of the sample and reference beams. This makes it possible to correct for any variations that affect both beams, such as short-term changes in lamp intensity due to voltage fluctuations in the power lines feeding into the instrument. Compensation for these variations is performed automatically in modern double-beam spectrometers. Even though part of the source radiation is directed to the reference beam, modern double-beam instruments have the same signal-to-noise ratio as single-beam instruments with the advantage of more accurate and precise absorbance measurements.

Lenses or mirrors are used to gather and focus the radiation at different parts of the optical system. This avoids losing too much signal as a result of the light beams being non-parallel and focuses the light beam along the flame so that as much light as possible passes through the sample. Any light that does not pass through the sample cannot be absorbed and results in a loss in sensitivity. Quartz lenses have been used for this purpose, but most instruments today use front-surfaced concave mirrors, which reflect and focus light from

Figure 6.14   Double-beam AAS configuration. [From Beaty and Kerber, used with permission of PerkinElmer, Inc. (www.perkinelmer.com).]

their faces and do not lose much radiation in the process. The monochromator has two slits, an entrance slit and an exit slit. The entrance slit is used to prevent stray radiation from entering the monochromator. Light passes from the entrance slit to the grating. The entrance slit should be as wide as possible to permit as much light as possible to fall on the grating, but must be narrow enough to isolate the wavelength of interest. After dispersion by the grating, the radiation is directed toward the exit slit. At this point the desired absorption line is permitted to pass, but the other lines emitted from the source and atomizer are blocked from reaching the detector by the monochromator exit slit. The system of slits and grating enables the analyst to choose the wavelength of radiation that reaches the detector. There is rapid loss in sensitivity as the mechanical slit width, and hence the spectral bandpass (spectral slit width), is increased.

Commercial AAS instrumentation may be purchased with fixed slits or with variable slits. Fixed mechanical slit widths are available so that the resolution and sensitivity are acceptable for most analytical purposes at lower cost than instruments with variable slit widths. Variable slit widths are desirable for maximum flexibility, especially if samples are varied and complex. Instruments that have both flame and graphite furnace atomizers often have separate sets of slits of different heights for each atomizer. The furnace slits are usually shorter to avoid having emission from the small diameter incandescent furnace reach the detector. In general, the analyst should use the widest slit widths that minimize the stray light that reaches the detector while spectrally isolating a single resonance line for the analyte from the HCL.

6.2.4. Detectors

The common detector for AAS is the PMT. The construction and operation of a PMT has been described in Chapter 5. While PMTs are the most common detectors, solid-state single and multichannel detectors such as PDAs (discussed in Chapter 5) and CCDs (discussed in Chapter 7) are increasingly being used in AAS spectrometers. Many small systems, particularly those dedicated to one element such as a dedicated CVAAS mercury analyzer, use solid-state detectors instead of PMTs. Multielement simultaneous AAS systems also use multichannel solid-state detectors to measure more than one wavelength at a time.

6.2.5. Modulation

Many metals, when atomized in a flame or furnace, emit strongly at the same wavelength at which they absorb. The emission signal can cause a serious error when the true absorption is to be measured. This problem is illustrated in Fig. 6.15. Emission by the metal in the atomizer is at precisely the same wavelength as the absorption wavelength of the metal because the same electronic transition is involved. Better resolution cannot improve the situation. Furthermore, since we are trying to measure I_1, the interference by the emission from the flame will result in a direct error. Unless a correction is made, the signal recorded will be $(I_1 + E)$, where E is the emission intensity.

This problem is overcome most simply by modulation of the radiation source. **Modulation** means that the source radiation is switched on and off very rapidly. This can be done by using a rotating mechanical *chopper* placed directly in front of the source. A chopper is shown in Figs. 6.8 and 6.14. The mechanical chopper is a circle of metal sheet with opposite quadrants cut out. Every quarter turn of the chopper alternately blocks and passes the source radiation. Another way to modulate the source is by pulsing the power to the lamp at a given frequency. When modulated, the signal from the source is

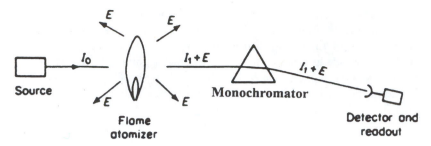

Figure 6.15 Emission by excited atoms in the atomizer can occur at the resonance wavelength, resulting in a direct error in the AAS measurement.

now an alternating current (AC) signal. Therefore the signal for both I_0 and I_1 is AC but the sample emission E is not modulated. It is a direct current (DC) signal. The detector electronics can be "locked in" to the frequency of the AC signal. The modulated absorption signal will be measured and the DC emission signal will be ignored. This eliminates emission from the atomizer as a source of error. Modulation of the source is required for accurate results in AAS. All modern commercial instruments have some means of source modulation in the instrument.

6.3. THE ATOMIZATION PROCESS

6.3.1. Flame Atomization

Most samples we want to examine by AAS are solid or liquid materials. Examples of solid samples are soil, rock, biological tissues, food, metal alloys, ceramics, glasses, and polymers. Examples of liquid samples are water, wastewater, urine, blood, beverages, oil, petroleum products, and organic solvents. For flame AAS (FAAS) and most GFAAS determinations, the sample must be in the form of a solution. This requires that most samples be prepared by acid digestion, fusion, ashing, or other forms of sample preparation (Chapter 1) to give us an aqueous, acidic solution, or a solution in a combustible organic solvent. We will look at aqueous acidic solutions, since they are the most common form of sample for FAAS. Metals are present in aqueous acidic solutions as dissolved ions; examples include Cu^{2+}, Fe^{3+}, Na^+, and Hg^{2+}.

   To measure an atomic absorption signal, the analyte must be converted from dissolved ions in aqueous solution to reduced gas phase free atoms. The overall process is outlined in Fig. 6.16. As described earlier, the sample solution, containing the analyte as dissolved ions, is aspirated through the nebulizer. The solution is converted into a fine mist or aerosol, with the analyte still dissolved as ions. When the aerosol droplets enter the flame, the solvent (water, in this case) is evaporated. We say that the sample is "desolvated". The sample is now in the form of tiny solid particles. The heat of the flame can melt (liquefy) the particles and then vaporize the particles. Finally the heat from the flame (and the combustion chemistry in the flame) must break the bonds between the analyte metal and its anion, and produce free M^0 atoms. This entire process must occur very rapidly, before the analyte is carried out of the observation zone of the flame. After free atoms are formed, several things can happen. The free atoms can absorb the incident radiation; this is the process we want. The free atoms can

1) Nebulization $M^+ + A^-$ (Solution)

2) Desolvation $M^+ + A^-$ (Aerosol)

3) Liquefaction   MA   (Solid)

4) Vaporization   MA   (Liquid)

5) Atomization   MA   (Gas)

6) Excitation $M^o + A^o$ (Gas)

7) Ionization M^* (Gas)

    $M^+ + e^-$ (Gas)

Figure 6.16 The processes occurring in a flame atomizer. M^+ is a metal cation; A^- is the associated anion. M^o and A^o are the ground state free atoms of the respective elements. [From Beaty and Kerber, used with permission of PerkinElmer, Inc. (www.perkinelmer.com).]

be rapidly oxidized in the hostile chemical environment of the hot flame, making them unable to absorb the resonance lines from the lamp. They can be excited (thermally or by collision) or ionized, making them unable to absorb the resonance lines from the lamp. The analyst must control the flame conditions, flow rates, and chemistry to maximize production of free atoms and minimize oxide formation, ionization, and other unwanted reactions.

The flame is responsible for production of free atoms. Flame temperature and the fuel/oxidant ratio are very important in the production of free atoms from compounds. Many flame fuels and oxidants have been studied over the years, and temperature ranges for some flames are presented in Table 6.1. In modern commercial instruments, only air–acetylene and nitrous oxide–acetylene flames are used.

Lower temperature flames are subject to interferences from incomplete atomization of the analyte. The air–acetylene flame is useful for many elements, but the hotter and more chemically reducing nitrous oxide–acetylene flame is needed for refractory elements and elements that are easily oxidized, such as Al and Si. Appendix 6.1 lists the usual flame chosen for each element. This appendix also gives the fuel/oxidant ratio conditions used for maximum sensitivity for each element. Flames are classified as oxidizing (excess oxidant) or reducing (excess fuel). Air–acetylene flames can be used in either an oxidizing mode or a reducing mode; nitrous oxide–acetylene flames are usually run in a reducing mode. In general, excess oxidant helps to destroy organic material in samples. However, excess oxidant can react with elements that exist as stable oxides to form oxide molecules. These oxide molecules cannot undergo atomic absorption. Elements that form stable oxides, such as aluminum, boron, molybdenum, and chromium, are therefore determined using reducing flame conditions, usually with the high-temperature nitrous oxide–acetylene flame, to prevent the formation of oxide molecules. The chemistry that occurs depends on what part of the flame is observed; the base of the flame differs from the inner core of the flame and both differ from the outer mantle of the flame.

If we measure the absorbance signal of an atom with respect to the height of the signal above the burner, we arrive at a relationship called a flame profile. The burner assembly can be moved up and down with respect to the light source, usually by turning a knob manually or through the software that controls the burner position. The beam of light from the source is fixed in position. The flame profile is determined by aspirating a solution of an element into a flame with the burner height set so that the light beam from the lamp is at the base of the flame, just above the burner head. Then the burner head is slowly lowered, so that the light beam passes through higher and higher regions of the flame. The absorbance is plotted vs. the height above the burner. Flame profiles for Cr, Mg, and Ag in an oxidizing air–acetylene flame are shown in Fig. 6.17. Let us look at each of these profiles. The signal for magnesium, Mg, starts off low at the base of the flame, because free atoms have not yet formed from the sample. The absorbance increases with increasing height in the flame (above the burner head), to a maximum. Moving farther up in the flame causes the absorbance to decrease. This curve is brought about by the complicated reactions that take place in a flame, including atomization, oxidation, and ionization. All of these processes compete in the flame. The maximum sensitivity for Mg will be obtained when the light beam is at the position of maximum absorbance, above the burner head, but not very high up in the flame. The decrease in absorbance as we go higher up in the flame is due to the formation of oxide molecules. Many elements have flame profiles similar to Mg, but not all, as can be seen in Fig. 6.17. Chromium forms very stable oxides that are hard to atomize. As can be seen from its profile, the absorbance by Cr atoms is actually highest at the base of the flame. Cr atoms combine with oxygen in air almost as soon as they are formed, and the absorbance decreases rapidly with height above the burner head (i.e., with increasing exposure to air). Chromium is an element that would be better determined in a fuel-rich (reducing) flame to minimize the formation of chromium oxide. Silver, on the other hand, starts with low absorbance at the base of the flame just like Mg. Then, the absorbance increases with increasing height in the flame, indicating that silver atoms are stable once formed; there is no decrease in absorbance due to oxide formation or ionization.

In practice, the optimum position for taking absorbance measurements is determined in exactly this way. A solution of the element to be determined is aspirated, the absorbance is monitored continually and the burner head is moved up and down (and forward and back) with respect to the light beam until the position of maximum absorbance is located. Most instruments have a ruler mounted along the burner compartment to indicate where the burner head is with respect to the light beam to assist in adjusting the height of the burner. Figure 6.17(b) shows the 2D distribution of atoms in an air–acetylene flame for several elements. These are side views of the flames, as would be seen by the detector. The contour lines differ by 0.1 in absorbance, with the maximum absorbance at the center of the profile. Fuel-rich (reducing) and fuel-lean (oxidizing) conditions are

Table 6.1 Temperatures Obtained in Various Premixed Flames (°C)

Fuel	Oxidant	
	Air	N_2O
H_2	2000–2100	
Acetylene	2100–2400	2600–3000
Propane (natural gas)	1700–1900	

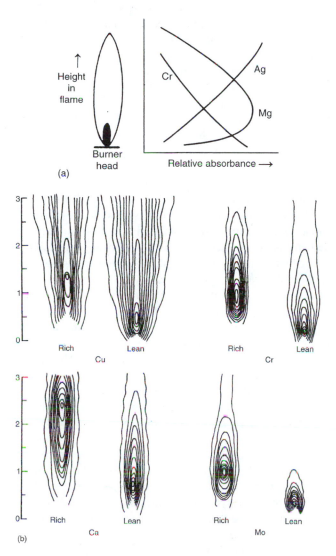

Figure 6.17 (a) Flame profiles for Cr, Mg, and Ag demonstrating differences in free atom formation as a function of height above the burner in an oxidizing air–acetylene flame. (b) Two-dimensional atom distributions in a 10 cm air–acetylene flame. Fuel-rich (reducing) and fuel-lean (oxidizing) results are shown. The scale on the left is the height above the burner head (in cm). Maximum absorbance is shown as the smallest circle in the plot with the contour lines separated by 0.1 in absorbance. For Cr, the fuel-lean plot shows maximum absorbance close to the burner, in accordance with the results shown in (a). The fuel-rich flame shows that the maximum absorbance for Cr moves up above the burner head by approximately 1 cm. Molybdenum is even more dramatically affected by the fuel/oxidant ratio. The diagrams in (b) are adapted and reprinted with permission from Rann and Hambly, (Copyright 1965, American Chemical Society.)

displayed. As we discussed, Cr shows the highest absorbance in a fuel-lean flame just above the burner head; in a fuel-rich flame, the maximum absorption position is moved up in the flame. Molybdenum shows the same response, but even more strongly. Calcium can be determined in either a fuel-rich or fuel-lean flame, but the position of maximum absorbance is very different for each flame. In the fuel-rich flame, the optimum position is more than 2 cm above the burner, while in an oxidizing flame the position is between 0.5 and 1 cm above the burner. The need to optimize the horizontal position of the burner with respect to the light source can be seen from the width of the intervals of the atom distributions.

The flame profile results from the complex physical and chemical processes occurring in the flame. The sample, introduced into the base of the flame, goes through the process of evaporation, forming solid particles, which are then decomposed by the hot flame to liberate free atoms. The latter can then oxidize, forming metal oxides in the upper and outer parts of the flame. The formation of free atoms depends on the flame temperature and on the chemical form of the sample. Chemical species with small dissociation energies at high temperatures will dissociate to form free atoms. For a given flame temperature, free atom formation depends on the chemical species and its dissociation constant. If the metal exists in the sample in a stable chemical form (small dissociation constant), it may be difficult to decompose, and **atomization efficiency** is low. On the other hand, if the metal is in a chemical form that is easily decomposed, the number of atoms formed is high and the atomization efficiency is high. The atomization efficiency is a measure of how many free atoms are formed from all the possible species containing the atom. It can be expressed as the fraction of the total element in the gaseous state that is present as free atoms at the observation height in the atomizer or as the fraction of the total element present as both free atoms and ions. By total element we mean all atomic, molecule, and ionic species containing the atom of interest, M (e.g., M, MX, MO, M^+). The atomization efficiency can vary from 0 to 1, and the value will depend on where in the atomizer the observation of the fraction M/(total M-containing species) is made. The interested student should consult the more detailed discussions in the Handbook of Spectroscopy, Vol. 1, edited by Robinson, or the references by Dean and Rains (Vol. 1), or Ingle and Crouch listed in the bibliography.

The loss of free atoms in the atomizer is also a function of the chemistry of the sample. If the oxide of the analyte element is readily formed, the free atoms will form oxides in the flame and the population of free atoms will simultaneously decrease. This is the case with elements such as chromium, molybdenum, tungsten, and vanadium. On the other hand, some metal atoms are stable in the flame and the free atoms exist for a prolonged period. This is particularly so with the noble metals platinum, gold, and silver. Adjusting the fuel/oxidant ratio can change the flame chemistry and atom distribution in the flame as shown in Fig. 6.17(b). Atoms with small ionization energies will ionize readily at high temperatures (and even at moderate temperatures). In an air−acetylene flame, it can be shown that moderate concentrations of potassium are about 50% ionized, for example. Ions do not absorb atomic lines.

The maximum absorbance signal depends on the number of free atoms in the light path. These free atoms are in dynamic equilibrium with species in the flame; they are produced continuously by the flame and lost continuously in the flame. The number produced depends on the original concentration of the sample and the atomization efficiency (Table 6.2). The number lost depends on the formation of oxides, ions, or other nonatomic species. The variation of atomization efficiency with the chemical form of the sample is called *chemical interference*. It is the most serious interference encountered in AAS and must always be taken into account. Interferences are discussed in detail in Section 6.4. Table 6.2 can be used as a guide for choosing which flame to use for AAS determination of an element. For example, Al will definitely give better sensitivity in a nitrous oxide−acetylene flame than in air−acetylene, where hardly any free atoms are formed; the same is true of Ba. But for potassium, clearly an air−acetylene flame is a

Table 6.2   Efficiencies of Atomization[a] of
Metals in Flames

Metal	Flame	
	Air–C_2H_2	N_2O–C_2H_2
Ag	1.0	0.6
Al	<0.00005	0.2
Au	1.0	0.5
B	<0.0005	0.004
Ba	0.001	0.2*
Be	0.00005	0.1
Bi	0.2	0.4
Ca	0.07	0.5*
Cd	0.5	0.6
Co	0.3	0.3
Cu	1.0	0.7
Cr	0.07	0.6
Cs*	0.7	—
Fe	0.4	0.8
Ga	0.2	0.7
In	0.6	0.9
K*	0.4	0.1
Li*	0.2	0.4
Mg	0.6	1.0
Mn	0.6	0.8
Na*	1.0	1.0
Pb	0.7	0.8
Rb*	1.0	—
Si	—	0.06
Sn	0.04	0.8
Sr	0.08	0.03
Ti	—	0.2
Tl	0.5	0.56
V	0.01	0.3
Zn	0.7	0.9

Source: Modified from Robinson, *Handbook of Spectroscopy*.
[a]The efficiency of atomization in the flames has been measured as the fraction of total element in the gaseous state present as free neutral atoms or ionized atoms at the observation height in the atomizer. That is, efficiency of atomization = (neutral atoms + ions)/total element. Entries marked * obtained using ionization suppression, discussed in Section 6.4.1. The data were obtained under a variety of conditions by multiple researchers and may not be directly comparable between elements. The reference should be consulted for details.

better choice in terms of sensitivity than the hotter nitrous oxide–acetylene flame. This is because K is very easily ionized at high temperatures and ions do not give atomic absorption signals. Based on ionization energy trends, you should expect that Cs and Rb would also be more sensitive in the cooler air–acetylene flame. As you can see, there are no entries in the table for Cs or Rb in a nitrous oxide–acetylene flame; they ionize to such an extent that the nitrous oxide–acetylene flame cannot be used for determining Cs or Rb by FAAS.

6.3.2. Graphite Furnace Atomization

There are many differences in the atomization process in a flame and in a graphite furnace. One very important difference to keep in mind is that in FAAS, the sample solution is aspirated into the flame continuously for as long as it takes to make the absorbance measurement. This is usually not long—about 30 s once the flame has stabilized after introducing the sample solution, but it is a continuous process. GFAAS is not a continuous process, as will be seen; the atomization step produces a transient signal that must be measured in less than 1 s. We will again consider an aqueous acidic solution of our sample.

A small volume of solution, between 5 and 50 μL, is injected into the graphite tube via a micropipette or an autosampler. The analyte is once again in the form of dissolved ions in solution, and the same process outlined in Fig. 6.16 must occur for atomic absorption to take place. The graphite furnace tube is subjected to a multistep temperature program. The program controls the temperature ramp rate, the final temperature at each step, the length of time the final temperature is held at each step, and the nature and flow rate of the purge gas through the furnace at each step. A typical graphite furnace program consists of six steps: (1) dry, (2) pyrolyze (ash, char), (3) cool, (4) atomize, (5) clean out, and (6) cool down.

A generic temperature program for GFAAS might look like that in Fig. 6.18. The process of atomization is extremely fast and must be rigidly controlled. The temperature program is therefore very carefully controlled, both with respect to the times used for each section of the heating program and the temperature range involved in each step. It is vital to avoid loss of sample during the first two steps but it is also extremely important to eliminate as much organic and other volatile matrix material as possible.

Figure 6.18 Typical temperature program for a graphite furnace atomizer. [From Beaty and Kerber, used with permission of PerkinElmer, Inc. (www.perkinelmer.com).]

The "dry" step is used to remove the solvent. The solvent must be removed without splattering the sample by rapid boiling, which would result in poor precision and accuracy. A slow temperature ramp from room temperature to about 110°C is used for aqueous solutions. The upper temperature is held for about a minute. The purge gas during this step is the normal inert gas (nitrogen or argon) at its maximum flow of about 300 mL/min to remove the solvent vapor from the furnace. The "pyrolyze" step is also called the ashing step or the "char" step. Its purpose is to remove as much of the matrix as possible without volatilizing the analyte. The "matrix" is everything except the analyte; it may consist of organic compounds, inorganic compounds, or a mixture of both. The sample is again subjected to a temperature ramp. The upper temperature is chosen to be as high as possible without losing the analyte and held for a short time. The gas flow is normally 300 mL/min and is usually the inert purge gas. With some organic sample matrices, switching to air is done in this step to help oxidize the organic materials.

A cool down step before the atomization step is used for longitudinally heated furnaces. This has been shown to improve sensitivity and reduce peak "tailing" for some refractory elements. This improves the accuracy of the measurement for these elements. The cool down step is not used in transversely heated furnaces.

The atomization step must produce gas phase free analyte atoms. The temperature must be high enough to break molecular bonds. In general, the temperature is raised very rapidly for this step, as shown in Fig. 6.18, and the purge gas flow is stopped to permit the atoms to remain in the light path. Stopping the purge gas flow increases the sensitivity of the analysis. The atomization occurs very rapidly and the signal generated is a transient signal; as the atoms form, the absorbance increases. As the atoms diffuse out of the furnace, the absorbance decreases, resulting in a nearly Gaussian peak-shaped signal. An idealized signal is presented in Fig. 6.19(a). Atomization signals for molybdenum from two different types of graphite furnace tubes are shown in Fig. 6.19(b). The spectrometer system is programmed to begin to acquire the absorbance data as soon as the atomization step begins. The integrated area under the absorbance peak is used for quantitative measurements.

Finally, the furnace is taken to a temperature higher than the atomization temperature to burn out as much remaining residue as possible; this is the "clean out" step. The furnace is allowed to cool before the next sample is injected. The entire program for one replicate of one sample is usually about 2 min long.

Early GFAAS instruments operated by having the sample pipetted onto the bottom wall of the tube. The tube was heated in a programmed fashion as has been described and atomization occurred "off the wall" of the tube. Poor precision and a variety of interferences were serious problems with this approach. Modern graphite furnace atomizers make use of a pyrolytic graphite platform, first introduced by L'vov, inserted into the graphite tube or fabricated as an integral part of the tube. The sample is pipetted onto the platform and is atomized from the platform, not from the wall of the tube. The reasons for use of the L'vov platform will be discussed in Section 6.4. In addition, modern GFAAS methods add one or more chemicals to the sample in the furnace. These chemicals are called "modifiers" and are used to control interferences.

6.4. INTERFERENCES IN AAS

Interferences are physical or chemical processes that cause the signal from the analyte in the sample to be higher or lower than the signal from an equivalent standard. Interferences can therefore cause positive or negative errors in quantitative analysis. There are two major classes of interferences in AAS, spectral interferences and nonspectral

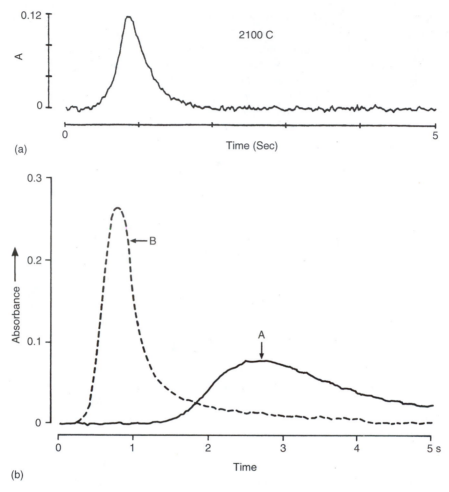

Figure 6.19 (a) An ideal Gaussian-shaped atomization signal from a graphite furnace. (b) Atomization signals for molybdenum in a graphite furnace atomizer. A is the signal resulting from an uncoated (normal) graphite tube atomizer. B is the signal from a graphite tube coated with pyrolytic graphite. The absorbance signal is transient in GFAAS. [Courtesy of PerkinElmer, Inc., Shelton, CT (www.perkinelmer.com).]

interferences. Nonspectral interferences are those that affect the formation of analyte free atoms. Nonspectral interferences include chemical interference, ionization interference, and solvent effects (or matrix interference). Spectral interferences cause the amount of light absorbed to be erroneously high due to absorption by a species other than the analyte atom. While all techniques suffer from interferences to some extent, AAS is much less prone to spectral interferences and even nonspectral interferences than atomic emission spectrometry and XRF, the other major atomic spectroscopic techniques.

6.4.1. Nonspectral Interferences

6.4.1.1. Chemical Interference

A serious source of interference is *chemical interference.* Chemical interference occurs when some species in the sample affects the atomization efficiency of the sample compared with the standard solution. The result is either an enhancement or a depression in the analyte signal from the sample compared with that from the standard. This effect is associated most commonly with the predominant anions present in the sample. The anion affects the stability of the metal compound in which the analyte is bound, and this, in turn, affects the efficiency with which the atomizer produces metal atoms. For example, a solution of calcium chloride, when atomized in an air–acetylene flame, decomposes to calcium atoms more readily than a solution of calcium phosphate. Calcium phosphate is more thermally stable than calcium chloride. A solution of calcium chloride containing 10 ppm Ca will give a higher absorbance than a solution of calcium phosphate containing 10 ppm Ca. If phosphate ion is added to a solution of calcium chloride, the absorbance due to Ca will decrease as the concentration of phosphate increases. This is a chemical interference. It occurs in the atomization process. Chemical interference is a result of having insufficient energy in the flame or furnace to break the chemical bonds in molecules and form free atoms.

There are three ways of compensating for chemical interference. The first approach is to match the matrix of the standards and samples; that is, to have the same anion(s) present in the same concentrations in the working standards as in the samples being analyzed. This supposes that the samples have been thoroughly characterized and that their composition is known and constant. This may be the case in industrial production of a material or chemical, but often the sample matrix is not well characterized or constant.

A second approach is to add another metal ion that forms an even more stable compound with the interfering anion than does the analyte ion. Such an ion is called a "releasing agent" because it frees the analyte from forming a compound with the anion and permits it to atomize. For example, lanthanum forms a very thermally stable phosphate, more stable than calcium phosphate. To determine Ca in solutions that contain an unknown or variable amount of phosphate, such as those from biological samples, the analyst can add a large excess of lanthanum (as the chloride or nitrate salt) to all standards and samples. The lanthanum "ties up" the phosphate by forming lanthanum phosphate. If all of the phosphate is now present as lanthanum phosphate, this eliminates the dependence of the formation of Ca atoms on the phosphate concentration. The exact concentration of phosphate does not have to be measured; it is only necessary to add enough La to completely react with the phosphate in the solution to be analyzed. Usually 500–2000 ppm La is sufficient for most types of samples. The same amount of La must be added to all the solutions, including the blank.

The third approach is to eliminate the chemical interference by switching to a higher-temperature flame, if possible. For example, when a nitrous oxide–acetylene flame is used, there is no chemical interference on Ca from phosphate, because the flame has sufficient energy to decompose the calcium phosphate molecules. Therefore, no lanthanum addition is required.

A fourth possible approach is the use of the MSA, discussed in Chapter 2. This approach can correct for some chemical interferences but not all. For example, in the graphite furnace, if the analyte is present as a more volatile compound than the added analyte compound, MSA may not work. If the analyte form in the sample is lost prior to atomization as a result of volatilization, while the added analyte compound remains in the furnace until atomization, the standards additions method will not give accurate analytical results.

6.4.1.2. *Matrix Interference*

Other potential sources of interference are the sample **matrix** and the solvent used for making the sample solution. The sample matrix is anything in the sample other than the analyte. In some samples, the matrix is quite complex. Milk, for example, has a matrix that consists of an aqueous phase with suspended fat droplets and suspended micelles of milk protein, minerals, and other components of milk. The determination of calcium in milk presents matrix effects that are not found when determining calcium in drinking water. Sample solutions with high concentrations of salts other than the analyte may physically trap the analyte in particles that are slow to decompose, interfering in the vaporization step and causing interference. Differences in viscosity or surface tension between the standard solutions and the samples, or between different samples, will result in interference. Interference due to viscosity or surface tension occurs in the nebulization process for FAAS because different volumes of solution will be aspirated in a given period of time and nebulization efficiency will change as a result of the solvent characteristics. Metals in aqueous solutions invariably give lower absorbance readings than the same concentration of such metals in an organic solvent. This is due in part to enhanced nebulization efficiency of the organic solvent, resulting in higher absorbance. In aqueous acidic solutions, higher acid concentrations often lead to higher solution viscosity and a decrease in absorbance due to decreased sample uptake. Matrix and solvent effects are often seen in GFAAS as well as FAAS, and may be more severe in furnaces than in flames. The presence of matrix interference can be determined by comparing the slope of an external calibration curve with the slope of an MSA curve (discussed in Chapter 2). If the slopes of the two calibrations are the same (parallel to each other) there is no matrix interference; if the slopes are different (not parallel), interference exists and must be corrected for.

The solvent may interfere in the atomization process. If the solvent is an organic solvent, such as a ketone, alcohol, ether, or a hydrocarbon, the solvent not only evaporates rapidly, but may also burn, thus increasing the flame temperature. The atomization process is more efficient in a hotter flame. More free atoms are produced and a higher absorbance signal is registered from solutions in organic solvents than from aqueous solutions, even though the metal concentration in the two solutions is equal.

Matching the solutions in the working standards to the sample solutions can compensate for matrix or solvent interferences. Type of solvent (water, toluene, methyl isobutyl ketone, etc.), amount and type of acid (1% nitric, 5% HCl, 20% sodium chloride, etc.), and any added reagents such as releasing agents must be the same in calibration standards and samples for accurate results.

Alternatively, the MSA may be used to compensate for matrix interferences. This calibration method uses the sample to calibrate in the presence of the interference. Properly used, MSA will correct for the solvent interference but care must be taken. The assumption made in using MSA is that whatever affects the rate of formation of free atoms in the sample will affect the rate of formation of free atoms from the added analyte spike in the same way. This may not be the case. MSA *will not* correct for spectral interference or for ionization interference. MSA can correct for interferences that affect the slope of the curve but not for interferences that affect the intercept of the curve.

6.4.1.3. Ionization Interference

In AAS the desired process in the atomizer should stop with the production of ground state atoms. For some elements, the process continues as shown in Fig. 6.16 to produce excited state atoms and ions. If the flame is hot enough for significant excitation and ionization to occur, the absorbance signal is decreased because the population of ground state atoms has decreased as a result of ionization and excitation. This is called ionization interference. Ionization interferences are commonly found for the easily ionized alkali metal and alkaline earth elements, even in cool flames. Ionization interferences for other elements may occur in the hotter nitrous oxide–acetylene flame, but not in air–acetylene flames.

Adding an excess of a more easily ionized element to all standards and samples eliminates ionization interference. This addition creates a large number of free electrons in the flame. The free electrons are "captured" by the analyte ions, converting them back to atoms. The result is to "suppress" the ionization of the analyte. Elements often added as ionization suppressants are potassium, rubidium, and cesium. For example, in the AAS determination of sodium, it is common to add a large excess of potassium to all samples and standards. Potassium is more easily ionized than sodium. The potassium ionizes preferentially and the free electrons from the ionization of potassium suppress the ionization of sodium. The detection limit of the sodium determination thereby decreases. The ionization suppression agent, also called an *ionization buffer*, must be added to all samples, standards, and blanks at the same concentration for accurate results. An example of the use of ionization suppression is shown in Fig. 6.20. Absorbance at a barium resonance line (atomic absorption) and absorbance at a barium ion line (by barium ions in the flame) are plotted as a function of potassium added to the solution. As the potassium concentration increases, barium ionization is suppressed; the barium stays as barium atoms. This results in increased atomic absorption at the resonance line and a corresponding decrease in absorbance at the ion line. The trends in absorbance at the atom and ion lines very clearly show that barium ion formation is suppressed by the addition of 1000 ppm of the more easily ionized potassium.

6.4.1.4. *Nonspectral Interferences in GFAAS*

Graphite furnace atomizers experience significant nonspectral interference problems, some of which are unique to the furnace. Compensation or elimination of these interferences is different than what is done in flame atomizers.

Some analytes react with the graphite surface of the atomizer at elevated temperatures to form thermally stable carbides. These carbides do not atomize, thereby decreasing the sensitivity of the analysis. Carbide formation also results in poor precision and poor accuracy, because the amount of carbide formation depends on the condition of the graphite atomizer surface. Elements that tend to form carbides include W, Ta, B, Nb, and Mo, among others. The use of a dense pyrolytic graphite coating on the tube wall is

Figure 6.20   The suppression of barium ionization in a flame atomizer by addition of the more easily ionized element potassium. [From Beaty and Kerber, used with permission of PerkinElmer, Inc. (www.perkinelmer.com).]

highly recommended to minimize carbide formation. Normal graphite has a very porous structure that permits solution to soak into the graphite tube, thereby increasing the graphite/solution contact area and allowing more carbide formation at high temperatures. Pyrolytic graphite forms an impervious surface that prevents the solution from entering the graphite structure, decreasing carbide formation. Comparing the atomization profiles for Mo in Fig. 6.19(b), it is apparent that the atomization from an uncoated tube (peak A) appears at a much later time than that from a pyrolytic graphite-coated tube. This time delay is due to the need for Mo to diffuse out of the porous uncoated graphite. The result is not only a time delay, but also a very broad absorbance signal that is difficult to integrate accurately. As you can see, the peak has not returned to the baseline, even after 5 min at the atomization temperature. The signal from the pyrolytic graphite-coated tube rises rapidly and returns to baseline rapidly, evidence that no significant diffusion of Mo into the graphite has occurred. This results in a higher absorbance signal and a peak that can be integrated accurately. It also permits a much shorter atomization step and increases tube life; both contribute to higher sample throughput.

When an analyte is atomized directly from the wall of the graphite tube, chemical and matrix interferences can be severe. The analyte, either as atoms or gas phase molecules, is released from the hot wall at the atomization temperature into a cooler inert gas atmosphere inside the tube. The atmosphere is heated by conduction and convection from the tube walls, so the temperature of the atmosphere lags behind that of the walls. The atoms or molecules enter this cooler atmosphere where a number of processes may occur. For example, the atoms may recombine with matrix components into molecules or the temperature may be low enough that vaporized analyte-containing molecules fail to atomize. In either case, nonspectral interference occurs. A solution to this problem is the use of a platform insert in the tube, shown schematically in Fig. 6.21. The platform surface is pyrolytic graphite and is deliberately designed to be in poor contact with the tube. The platform is heated by radiation from the tube walls. When an analyte is placed on the platform, it does not atomize when the wall reaches the right temperature, but later, when the platform and the atmosphere are both at the same, higher temperature. The difference in temperature of the gas phase inside the atomizer for wall and platform atomization is shown in Fig. 6.22. This higher temperature improves the atomization of molecules and prevents recombination of free atoms. The result is a significant reduction in interferences, and a significant improvement in precision.

6.4.1.5. Chemical Modification

Chemical modification, also commonly called **matrix modification**, is the addition of one or more reagents to the graphite tube along with the sample. The use of these chemical modifiers is to control nonspectral interferences by altering the chemistry occurring inside the furnace. The reagents are chosen to enhance the volatility of the matrix or to decrease the volatility of the analyte or to modify the surface of the atomizer. The use of a large amount of chemical modifier may, for example, convert all of the analyte into a single compound with well-defined properties. The end result of the use of chemical modifiers is to improve the accuracy and precision of the analysis by permitting the use of the highest possible pyrolysis temperature. This permits removal of the matrix with no loss of analyte.

Cadmium is one of the heavy metal elements whose concentrations in the environment are regulated by law due to the toxicity of Cd. Drinking water, blood, urine, and environmental samples are analyzed routinely for Cd at low ppb concentrations by GFAAS. Many common cadmium compounds, such as cadmium chloride, are relatively

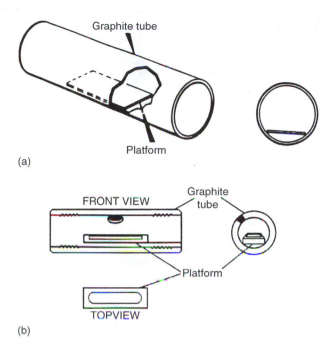

(a)

(b)

Figure 6.21   (a) Schematic of an L'vov platform inserted into a standard longitudinal graphite tube. The left diagram is a cutaway view of the platform inside the tube. The right diagram is an end-on view of the tube and platform looking along the light path. (b) Commercial platforms have a shallow depression into which the sample is pipetted through the opening in the top of the tube. The opening is shown as a dark area on the front and end-on views. [Courtesy of PerkinElmer Inc., Shelton, CT (www.perkinelmer.com).]

volatile. With no modifiers, Cd starts to volatilize out of the furnace at pyrolysis temperatures as low as 400°C, as seen in Fig. 6.23. By adding either single compounds or mixtures of compounds, the pyrolysis temperature can be increased significantly, to over 1000°C. The addition of the modifier(s) converts cadmium in the sample to a much less volatile form. In general, this improves both the precision and accuracy of the analysis. If, for example, cadmium had been present in the sample in two different

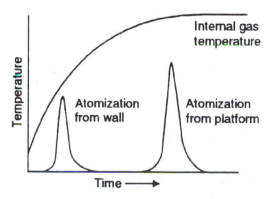

Figure 6.22   Tube wall and platform temperature profiles. [Courtesy of PerkinElmer Inc., Shelton, CT (www.perkinelmer.com).]

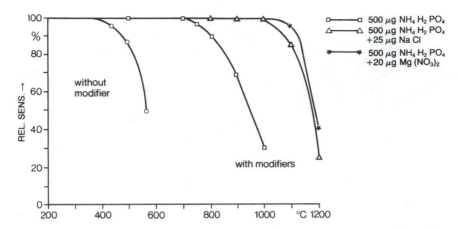

Figure 6.23 Effect of matrix modification on pyrolysis temperatures for cadmium. Without modification, Cd begins to volatilize out of the furnace at temperatures below 500°C. By adding modifiers or combinations of modifiers, Cd is retained in the furnace at much higher temperatures. [Courtesy of PerkinElmer, Inc., Shelton, CT (www.perkinelmer.com).]

chemical forms, such as an organometallic Cd compound and an inorganic Cd compound, these two compounds might atomize at different rates and at different temperatures. The result would be multiple atomization signals, as shown in Fig. 6.24. This can cause a serious error in the measurement of the signal. Multiple analyte peaks can usually be eliminated through the use of an appropriate matrix modifier that converts the analyte to one common species.

In the case of a modifier used to stabilize the analyte is the use of nickel nitrate in the determination of selenium in biological tissues. Addition of nickel nitrate as a modifier retains Se in the furnace as nickel selenide while allowing the pyrolysis and removal of the organic matrix. Without the addition of nickel nitrate, the pyrolysis temperature must be 350°C or less to prevent loss of Se. With the use of nickel nitrate, temperatures up to 1100°C can be used.

In some cases, the matrix itself can be made more volatile through chemical modification. For example, in the determination of traces of elements in seawater the matrix in the graphite tube after drying would be primarily NaCl. NaCl is a high melting ionic compound and requires relatively high temperatures to volatilize it. The NaCl molecule stays intact on volatilization, increasing the background absorbance (a spectral interference). If all the NaCl could be volatilized out of the furnace, very volatile analyte elements like Hg,

Figure 6.24 Multiple atomization signals caused by different chemical forms of analyte in the sample. [Courtesy of PerkinElmer, Inc., Shelton, CT (www.perkinelmer.com).]

As, Se, Pb, and many more would have been lost in the process. By adding an excess of ammonium nitrate as a *matrix modifier* to the sample in the graphite tube, the NaCl matrix can be converted to the much more volatile compounds ammonium chloride and sodium nitrate. Most of the matrix can be removed at a temperature below 500°C, which substantially reduces background absorption without loss of volatile analytes.

A generic "mixed modifier" of palladium and magnesium nitrate improves the GFAAS determination of many elements. The AAS cookbook of methods provided with commercial instruments should contain the recommendations for matrix modification for standard GFAAS determination of each element. Of course, it is imperative that very high-purity reagents be used for matrix modification and blank determinations that include the modifiers must be run.

6.4.2. Spectral Interferences

Spectral interferences occur when absorption of the hollow cathode resonance line occurs by species other than the element being determined. For example, the Pb 217.0 nm line may be absorbed by the components of a flame even though no lead is present in the sample or in the flame. This is a spectral interference.

6.4.2.1. Atomic Spectral Interference

The resonance absorption lines of the various elements are very narrow, on the order of 0.002 nm, and at discrete wavelengths. Direct overlap between absorption lines of different elements is rare and can usually be ignored as a source of error. Absorption by the wings of the absorption lines of interfering elements present in high concentrations has been observed, but this is also a rare occurrence. A table of reported atomic spectral overlaps can be found in the handbook by Dean cited in the bibliography. The only cures for direct atomic spectral interference are (1) to choose an alternate analytical wavelength or (2) to remove the interfering element from the sample. Extracting the interfering element away from the analyte or extracting the analyte away from the interfering element can accomplish the last option. There are many successful methods in the literature for the separation of interferences and analytes, but care must be taken not to lose analyte or contaminate the sample in the process. The extraction approach also permits the analyst to concentrate the analyte during the extraction, improving the accuracy and the sensitivity of the analysis when performed correctly.

6.4.2.2. Background Absorption and its Correction

A common occurrence that results in spectral interference is absorption of the HCL radiation by molecules or polyatomic species in the atomizer. This is called "background absorption"; it occurs because not all of the molecules from the sample matrix and the flame gases are completely atomized. This type of interference is more commonplace at short wavelengths (<250 nm) where many compounds absorb, as discussed in Chapter 5. Incomplete combustion of organic molecules in the atomizer can cause serious background interference problems. If a flame atomizer is used, incomplete combustion of the solvent may take place, particularly if the flame is too reducing (fuel-rich). The extent of the interference depends on flame conditions (reducing or oxidizing), the flame temperature used, the solvent used, and the sample aspiration rate. Background interference is much more severe when graphite furnace atomizers are used because the pyrolysis step is limited to a maximum temperature that does not volatilize the analyte. Consequently, many matrix molecules are not thermally decomposed. They then volatilize into the light path as the higher atomization temperature is reached and absorb significant amounts of the source radiation.

We have seen that atoms absorb over a very narrow wavelength range and that overlapping absorption by other atoms of the resonance lines of the analyte is extremely unlikely. However, molecular absorption occurs over broad bands of wavelengths and is observed in flame and graphite furnace atomizers. The molecular absorption may come from hydroxyl radicals generated from water in the flame, incomplete combustion of organic solvent, stable molecular residues, metal oxides, and so on. Solid particles in flame atomizers may scatter light over a wide band of wavelengths; this is not absorption, but results in less light reaching the detector. Scattering of light therefore causes a direct error in the absorption measurement. If these broad absorption (and scattering) bands overlap the atomic absorption lines, they will absorb the resonance line from the hollow cathode and cause an increase in absorbance, a spectral interference. There are several ways of measuring this background absorption and correcting for it.

6.4.2.3. Two-Line Background Correction

An early manual method of measuring background and applying a correction for it used the absorption of a nearby nonresonance line. The emission spectrum from a hollow cathode is quite rich in emission lines and contains the resonance lines of the element of interest plus many other nonresonance emission lines from the element and the filler gas. The analyte atoms do not absorb these nonresonance lines; however, the broad molecular background absorbs them. Two measurements are made. The absorbance at a nonresonance line close to the analyte resonance line but far enough away that atomic absorption does not occur is measured. Then the absorbance at the resonance line of the analyte is measured. The absorbance at the nonresonance line is due *only* to background absorption. The absorbance at the resonance line is due to *both* atomic and background absorption. The difference in absorbance between the resonance and nonresonance lines is the net atomic absorption. The two-line method of background correction was convenient when all AAS instruments were manual instruments, since no change in light source or complicated instrumentation was necessary. However, this method is not always accurate. Unless the nonresonance line is very close to the atomic absorption line, inaccurate correction of the background will result. Modern computerized instrumentation has made this technique obsolete.

6.4.2.4. Continuum Source Background Correction

Let us remind ourselves why we do not use a broadband, continuum lamp as a light source in AAS. The normal monochromator and slit system results in a spectral slit width about 0.2 nm wide. If we have a continuum source, the light from the source will fill the entire spectral window, as shown in the lower left-hand side drawing in Fig. 6.25. When the continuum emission passes through the flame, the atoms can only absorb that portion of it exactly equal to the resonance wavelength. The atomic absorption line has a total width of about 0.002 nm. Consequently, if the atoms absorb all of the radiation over that linewidth, they will absorb only 1% of the radiation from the continuum lamp falling on the detector, as shown in the lower center drawing of Fig. 6.25. All light within the 0.20 nm bandwidth, but not within the 0.002 nm absorption line, will reach the detector and not be absorbed by the sample. In other words, 99% or more of the emitted light reaches the detector. Consequently, the effect of atomic absorption on the continuum lamp is negligible and we can say that atoms do not measurably absorb the continuum

Figure 6.25 (Top row) Atomic and background absorption with a line source (HCL, EDL). (Bottom row) Atomic and background absorption with a continuum source (deuterium lamp, tungsten-filament lamp). The width (x-axis) of each diagram is the spectral slit width. [From Beaty and Kerber, used with permission of PerkinElmer, Inc. (www.perkinelmer.com).]

lamp emission. It is this observation that permits the use of a continuum light source to measure and correct for broadband molecular background absorption.

In the continuum source background correction method, when an HCL source is used, the absorption measured is the total of the atomic and background absorptions. When a continuum lamp source is used, only the background absorption is measured. The continuum lamp, which emits light over a range or band of wavelengths, is placed into the spectrometer system as shown in Fig. 6.26. This setup allows radiation from both the HCL and the continuum lamp to follow the same path to reach the detector. The detector observes each source alternately in time, either through the use of mirrored choppers or through pulsing of the lamp currents. When the HCL lamp is in position as the source, the emission line from the HCL is only about 0.002 nm wide; in other words it fills only about 1% of the spectral window as shown in the upper left-hand side drawing in Fig. 6.22. When the HCL radiation passes through the flame, both the free atoms at the

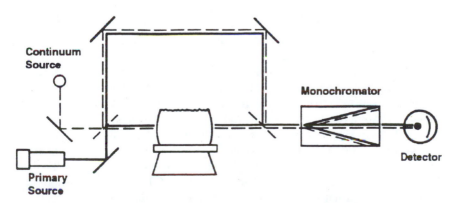

Figure 6.26 Continuum background corrector in a double-beam AAS system. [From Beaty and Kerber, used with permission of PerkinElmer, Inc. (www.perkinelmer.com).]

resonance line and the broadband absorbing molecules will absorb the line, resulting in significant attenuation of the light reaching the detector, as shown in the upper center and right-hand side drawing in Fig. 6.25. When the continuum source is in place, the continuum lamp emission fills the entire spectral window as shown in the lower left-hand side drawing in Fig. 6.25. Any absorption of the radiation from the continuum lamp observed is broadband background absorption, since it will be absorbed over the entire 0.2 nm as seen in the lower right-hand side drawing in Fig. 6.25. We showed above that any *atomic* absorption of the continuum lamp is negligible. The absorbance of the continuum source is therefore an accurate measure of background absorption. An advantage of this method is that the background is measured at the same nominal wavelength as the resonance line, resulting in more accurate correction.

This correction method is totally automated in modern AAS systems. The HCL or EDL and the continuum lamp are monitored alternately in time; instrument electronics compare the signals, and calculate the net atomic absorption.

The lamps used as continuum sources are a deuterium lamp for the UV region and a tungsten filament lamp for the visible region. Commercial instruments using continuum background correction normally have both of these continuum sources installed to cover the complete wavelength range; the continuum lamp used is computer-selected based on the analyte wavelength.

Continuum background correction is used for FAAS and can be used for furnace AAS. There are limitations to the use of continuum background correction. If another element (not the analyte) in the sample has an absorption line within the spectral bandpass used, especially if this element is present is large excess, it can absorb radiation and result in inaccurate background correction. The background correction is made using two different sources. The correction can be inaccurate if the sources do not have similar intensities, and if they are not aligned properly to pass through exactly the same region of the atomizer. This is especially true when the background levels are high. Continuum sources can generally correct for background absorbance up to $A = 0.8$. A problem can arise when the background absorption spectrum is not uniform over the spectral bandpass; such background absorption is said to have "fine structure". Absorption with fine structure may not be corrected properly using a continuum lamp corrector; the correction may be too large or too small. These limitations are particularly problematic in GFAAS which is prone to high background levels.

6.4.2.5.   *Zeeman Background Correction*

Atomic absorption lines occur at discrete wavelengths because the transition that gives rise to the absorption is between two discrete energy levels. However, when a vapor phase atom is placed in a strong magnetic field the electronic energy levels split. This gives rise to several absorption lines for a given transition in place of the single absorption line in the absence of a magnetic field. This occurs in all atomic spectra and is called **Zeeman splitting** or the Zeeman effect. In the simplest case, the Zeeman effect splits an absorption line into two components. The first component is the π component, at the same wavelength as before (unshifted); the second is the σ component, which undergoes both a positive and negative shift in wavelength, resulting in two equally spaced lines on either side of the original line. This splitting pattern is presented in Fig. 6.27. The splitting results in lines that are separated by approximately 0.01 nm or less depending on the field strength. The strength of the magnetic field used is between 7 and 15 kgauss. Background absorption and scatter are usually not affected by a magnetic field.

The π and σ components respond differently to polarized light. The π component absorbs light polarized in the direction parallel to the magnetic field. The σ components

Figure 6.27 Zeeman effect causing shifts in electronic energy levels. (a) Hg resonance line in the absence of a magnetic field occurs as a single line at 253.7 nm. (b) Hg resonance line in the presence of a magnetic field. The π component is at the original, unshifted wavelength (253.7 nm); the $\pm\sigma$ components are shifted equally away from the original line to higher and lower wavelengths.

absorb only radiation polarized 90° to the applied field. The combination of splitting and polarization differences can be used to measured total absorbance (atomic plus background) and background only, permitting the net atomic absorption to be determined.

A permanent magnet can be placed around the furnace to split the energy levels. A rotating polarizer is used in front of the HCL or EDL. During that portion of the cycle when the light is polarized parallel to the magnetic field, both atomic and background absorptions occur. No atomic absorption occurs when the light is polarized perpendicular to the field, but background absorption still occurs. The difference between the two is the net atomic absorption. Such a system is a DC Zeeman correction system.

Alternately, a fixed polarizer can be placed in front of the light source and an electromagnet can be placed around the furnace. By making absorption measurements with the magnetic field off (atomic plus background) and with the magnetic field on (background only), the net atomic absorption signal can be determined. This is a transverse AC Zeeman correction system. The AC electromagnet can also be oriented so that the field is along the light path (a longitudinal AC Zeeman system) rather than across the light path. No polarizer is required in a longitudinal AC Zeeman system.

The use of a polarizer in either DC or AC Zeeman systems cuts the light throughput significantly, adversely affecting sensitivity and precision. DC Zeeman systems require less power, but have poorer linear working range and sensitivity than AC systems. AC Zeeman systems are more expensive to operate but have higher sensitivity and larger linear working ranges than DC systems.

Zeeman correction can also be achieved by having an alternating magnetic field surround the hollow cathode, causing the emission line to be split and then not split as the field is turned on and off. By tuning the amplifier to this frequency it is possible to discriminate between the split and unsplit radiation. A major difficulty with the technique is that the magnetic field used to generate Zeeman splitting also interacts with the ions in the hollow cathode. This causes the emission from the hollow cathode to be erratic, which in turn introduces imprecision into the measurement. Most commercial instruments with Zeeman background correctors put the magnet around the atomizer. Zeeman correction systems can be used with flame atomizers and are commercially available; one disadvantage to using this approach is that the magnet blocks the analyst's view of the flame. This makes it impossible to visually check that the flame conditions are correct, since the analyst cannot see the color of the flame. The analyst cannot see if the burner slot is partially blocked or if charred material is building up on the burner when organic solvents are run. These unobserved problems might result in errors in the analysis. The significant advantage of Zeeman background correction is to compensate for the high background absorption in graphite furnace atomizers.

The advantages of Zeeman background correction are numerous. Only one light source is required. Since only one light source is used, there is no need to match intensity or align multiple sources. The physical paths of the split and unsplit light are identical, as opposed to the use of a continuum lamp, where the radiation may not pass along the identical path as the HCL radiation. The background correction is made very close to the resonance line and at the same absorption bandwidth as the atomic absorption signal. Therefore the correction is generally very accurate, even for background with fine structure and for high background levels, up to absorbance = 2.0. The Zeeman correction system can be used for all elements at all wavelengths. One limitation to the use of Zeeman background correction is that it shortens the analytical working range. The relationship between absorbance and concentration becomes nonlinear at lower absorbances than in a non-Zeeman corrected system, and then the absorbance–concentration relationship "rolls over". That is, as concentration increases, absorbance first increases linearly, then nonlinearly until a maximum absorbance is reached. After the maximum, absorbance actually *decreases* with increasing concentration. It is possible for two completely different concentrations, one on either side of the absorbance maximum, to exhibit the same absorbance value. The roll-over point must be established by carefully calibrating the instrument and unknown samples may need to be run at more than one dilution factor to ensure that an unknown is not on the "wrong side" of the absorbance maximum. If this well-understood potential problem is kept in mind, the use of Zeeman background correction with modern GFAAS instrumentation and methodology provides accurate and precise GFAAS results.

6.4.2.6.   *The Smith–Hieftje Background Corrector*

It will be remembered that the HCL functions by the creation of excited atoms that radiate at the desired resonance wavelengths. After radiating, the atoms form a cloud of neutral atoms that, unless removed, will absorb resonance radiation from other emitting atoms.

If the HCL is run at a high current, an abundance of free atoms form. These free atoms absorb at precisely the resonance lines the hollow cathode is intended to emit; an example is shown in Fig. 6.28, with the free atoms absorbing exactly at λ over an extremely narrow bandwidth [Fig. 6.28(b)]. The result is that the line emitted from the HCL is as depicted in Fig. 6.28(c) instead of the desired emission line depicted in Fig. 6.28(a). The emitted lines are broadened by the mechanisms discussed in Section 6.1.1. The phenomenon of absorption of the central portion of the emission line by free atoms in the lamp is called "self-reversal". Such absorption is not easily detectable, because it is at the very center of the emitted resonance line and very difficult to resolve. It is, of course, exactly the radiation that is most easily absorbed by the atoms of the sample. In practice, if the HCL is operated at too high a current, the self-reversal decreases the sensitivity of the analysis by removing absorbable light. It also shortens the life of the lamp significantly.

The Smith–Hieftje background corrector has taken advantage of this self-reversal phenomenon by pulsing the lamp, alternating between high current and low current. At low current, a normal resonance line is emitted and the sample undergoes normal atomic absorption. When the HCL is pulsed to a high current, the center of the emission

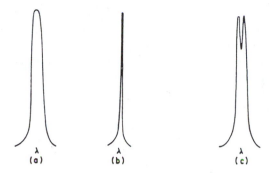

Figure 6.28 Distortion of spectral line shape in an HCL due to self-reversal. (a) Shape of the spectral line emitted by the HCL. (b) Shape of the spectral energy band absorbed by cool atoms inside the HCL. (c) Shape of the net signal emerging from an HCL, showing self-absorption of the center of the emission band.

line is self-absorbed, leaving only the wings of the emitted resonance line; the emitted line would look similar to Fig. 6.28(c). The atoms of the sample do not significantly absorb such a line. The broad background, however, absorbs the wings of the line. Consequently, the absorption of the wings is a direct measurement of the background absorption at the wavelength of the atomic resonance absorption. The Smith–Hieftje technique can be used to automatically correct for background. In practice, the low current is run for a fairly short period of time and the absorbance measured. The high current is run for a very short, sharp burst, liberating intense emission and free atoms inside the hollow cathode and the absorbance is again measured. The difference between the measurements is the net atomic absorption. There is then a delay time to disperse the free atoms within the lamp before the cycle is started again.

The advantages of the Smith–Hieftje technique are that it can be used in single-beam optics and it is not necessary to align the beam measuring background absorption and atomic absorption since it is the same beam. In addition, the electronics are much simpler than that used in a Zeeman background correction system. The major disadvantage of the technique is that it shortens the life of the HCL. The Smith–Hieftje background correction system is no longer in use in commercial AAS instrumentation as of this writing. A more detailed comparison of the methods used for background correction is given in the reference by Carnrick and Slavin listed in the bibliography.

6.4.2.7. Spectral Interferences in GFAAS

Graphite furnace atomizers have their own particular spectral interference problems. Any emission from the light source or atomizer that cannot be absorbed by the analyte should be prevented from reaching the detector. Usually, the emission from flames is low. At atomization temperatures in excess of 2200°C, the graphite furnace is a white-hot emission source. If this intense visible radiation reaches the PMT, noise increases, which decreases precision. The emission from the hot graphite is essentially blackbody radiation, extending from 350 to 800 nm and varying in intensity with wavelength. The extent of emission interference caused by this radiation falling on the detector varies with analyte wavelength. Elements with resonance lines in the 400–600 nm region like Ca and Ba can be seriously affected. The control of emission interference lies in spectrometer design, which is not controlled by the analyst, and in the correct installation, alignment, and maintenance of the furnace and optical windows, which are very definitely controlled by the analyst.

Background absorption can be severe in GFAAS. Decreasing the sample volume and choosing alternate wavelengths can be used to reduce background levels in some cases. Control of the chemistry in the furnace is key to reducing background. The pyrolysis step must be designed to volatilize as much of the matrix as possible without loss of analyte. Ideally, the matrix would be much more volatile than the analyte, so that 100% matrix removal could occur in the pyrolysis step with 100% retention of analyte in the furnace. This situation rarely occurs. It is possible to control the volatility of the matrix and analyte to some extent through matrix modification as discussed earlier. The use of matrix modifiers seems to control background (a spectral interference) by controlling nonspectral interferences.

A potential problem with any automatic background corrector is that the analyst may be unaware of the extent to which the background is present. It should be remembered that when absorbance $A = 2$, the transmittance $T = 1/100$, and the absorbed radiation is 99% of the available radiation. Therefore the total signal falling on the detector is only 1% of I_0. If quantitative analysis is carried out using this very small signal, major errors may result. A 1% error in measuring the background may result in a 100% error in measuring the atomic absorption by the sample. Instrument manufacturers claim quantitative atomic absorption measurements while correcting background absorption in excess of 99% using Zeeman background correction. Analysts should be aware of the magnitude of the background signal, because a small error in measuring the background becomes a major error in the net atomic absorption measurement. Steps such as the use of L'vov platforms, matrix modification, alternate wavelengths, reduction in sample volume, and other techniques to minimize background should be taken to keep background absorbance as low as possible to obtain accurate results.

6.5. ANALYTICAL APPLICATIONS OF AAS

AAS is a mature analytical technique. There are thousands of published methods for determining practically any element in almost any type of sample. There are books and journals devoted to analytical methods by AAS and other atomic spectrometry techniques. The bibliography provides a list of some texts on AAS. Journals such as Analytical Chemistry, Applied Spectroscopy, Journal of Analytical Atomic Spectroscopy, The Analyst, Spectroscopy Letters, and others are sources of peer-reviewed articles, but many applications articles can be found in specialized journals on environmental chemistry, food analysis, geology, and so on. The applications discussion here is necessarily limited, but the available literature is vast.

AAS is used for the determination of all metal and metalloid elements. Nonmetals cannot be determined directly because their most sensitive resonance lines are located in the vacuum UV region of the spectrum. Neither flame nor furnace commercial atomizers can be operated in a vacuum. It is possible to determine some nonmetals indirectly by taking advantage of the insolubility of some compounds. For example, chloride ion can be precipitated as insoluble silver chloride by adding a known excess of silver ion in solution (as silver nitrate). The silver ion remaining in solution can be determined by AAS and the chloride ion concentration calculated from the change in the silver ion concentration. Similar indirect approaches for other nonmetals or even polyatomic ions like sulfate can be devised.

6.5.1. Qualitative Analysis

The radiation source used in AAS is an HCL or an EDL, and a different lamp is needed for each element to be determined. Because it is essentially a single-element technique, AAS is not well suited for qualitative analysis of unknowns. To look for more than one element requires a significant amount of sample and is a time-consuming process. For a sample of unknown composition, multielement techniques such as XRF, ICP-MS, inductively coupled plasma-optical emission spectrometry, and other atomic emission techniques are much more useful and efficient.

6.5.2. Quantitative Analysis

Quantitative measurement is one of the ultimate objectives of analytical chemistry. AAS is an excellent quantitative method. It is deceptively easy to use, particularly when flame atomizers are utilized.

6.5.2.1. Quantitative Analytical Range

The relationship between absorbance and concentration of the analyte being determined in AAS follows Beer's Law (discussed in Chapter 2) over some concentration range. There is an optimum linear analytical range for each element at each of its absorption lines. The *minimum* of the range is a function of the detection limit of the element under the operating conditions used. The ultimate limiting factor controlling the detection limit is the noise level of the instrument being used. The LOD in AAS is defined by the IUPAC and by many regulatory agencies as the concentration giving a signal equal to $3\times$ the standard deviation of a blank solution measured under the same operating conditions being used for the analysis. However, the LOD is never used as the minimum for reporting quantitative data. The LOQ is defined as that concentration giving a signal equal to $10\times$ the standard deviation of a blank solution measured under the operating conditions for the analysis. This is the lowest quantitative concentration that should be reported. Concentrations between the LOD and the LOQ should be reported as "detected but not quantifiable" and concentrations below the LOD should be reported as "not detected". The *maximum* of the analytical range is determined by the sample. The linear working range for AAS is small; generally only one to two orders of magnitude at a given wavelength. The calibration curve deviates from linearity, exhibiting a flattening of the slope at high absorbance values. With a flat slope, changes in concentration of the sample produce virtually no changes in absorption. Hence, it is impossible to measure concentrations accurately at extremely high absorbance values. Many calibration curves deviate from linearity below $A = 0.8$, and Zeeman background correction shortens the analytical working range even more. With modern computerized instruments, nonlinear curve fitting can be used to some advantage to measure concentrations beyond the linear range. However, it is advisable to keep the absorbance values of samples and standards below 0.8, as was discussed in Chapter 2. When a concentration range is quoted for an analytical method, it is important to know how the lower and upper limits were determined. Approximate upper limits for the linear range of elements determined by FAAS are listed in Appendix 6.1. Detection limits for elements by FAAS, GFAAS, cold vapor Hg and hydride generation AAS, and other atomic spectroscopic techniques discussed in later chapters are given in Appendix 6.2. For AAS, it had been customary to report *method sensitivity*, where sensitivity was defined as the concentration of analyte that gives an absorbance of 0.0044 (equal to 1% absorption), so this term may be encountered in the literature. Sample sensitivities for FAAS and GFAAS are tabulated in Appendix 6.2. Both the LOD and the sensitivity are highly dependent on the sample matrix, operating conditions, the particular instrument used, and the way in which the data are processed.

Modern AAS instruments have computerized data collection and data processing systems. These computer-based systems have the ability to fit different types of calibration curves to the data using a variety of equations, including linear, quadratic, and higher-order polynomial functions. Such systems are capable of calculating concentration results from nonlinear calibration curves. The accuracy of such calculated results depends on the equations used for calibration and on the number and concentrations of the standards used to provide the calibration data. It is not advisable to use a nonlinear calibration curve without verification that the results are accurate. When in doubt, dilute the samples into the linear region, especially if using Zeeman background correction.

6.5.2.2. Calibration

Calibration of AAS methods can be performed by the use of an external calibration curve or by MSA; both calibration methods were presented in Chapter 2. Internal standards are not used in AAS, because it is usually a single-element technique; we cannot measure an internal standard element at the same time we measure the analyte.

External calibration curves are prepared from solutions of known concentrations of the sample element. High-purity metals dissolved in high-purity acids are used to make the stock standard solution. For AAS, stock standard concentrations are either 1000 or 10,000 ppm as the element. Working standards are diluted from the stock standard. For example, if we wanted to make a calibration curve for copper determination in the range of 2–20 ppm, we would make a stock standard solution of 1000 ppm Cu by dissolving Cu metal in nitric acid. Then we would dilute this solution to make a series of working standards. The concentrations of the working standards must cover the complete range of 0–20 ppm. A calibration blank (the 0.0 ppm standard) is of course required to correct for any analyte present in the water or reagents used to prepare the standard solutions. In this example, the calibration blank would be prepared from deionized water and the same high purity nitric acid used to make the standard solutions. Three to five working standards are typical, so we might make Cu solutions of 0, 2, 5, 10, 15, and 20 ppm. These standard solutions would then be aspirated into the flame and the absorbance of each standard would be measured. The absorbance would be plotted vs. concentration. A calibration curve like that in Fig. 6.29 results. It should be noted that the

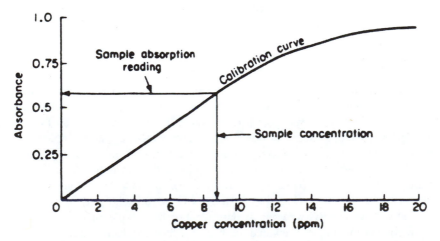

Figure 6.29 AAS calibration curve for the copper 324.8 nm line showing linear and nonlinear regions of the curve.

relationship between absorbance and concentration is linear over the range of 0–10 ppm Cu, but at higher concentrations the relationship deviates from linearity. Provided that the slope of the curve is fairly steep, it may possible to distinguish between concentrations outside the linear range, such as between 12 and 14 ppm. Above 16 ppm the curve becomes so flat that accurate concentration results are difficult or impossible to obtain.

When a quantitative analysis is to be performed, the sample is atomized and the absorbance measured under exactly the same conditions as those used when the calibration standards were measured. The calibration standards are usually measured first, and the sample solutions should be measured immediately following calibration. If the sample has been prepared using reagents such as acids, it is necessary to prepare a method blank that is carried through the same sample preparation steps as the sample. The method blank is used to correct for any contamination that may have occurred during sample preparation. The absorbance of the method blank is subtracted from the sample absorbance to give the corrected or net sample absorbance. The concentration of the unknown copper solution is determined from the calibration curve. For example, suppose that the corrected absorbance reading of the sample solution was 0.58. Using the calibration curve shown in Fig. 6.29, it can be deduced that the concentration of the solution is about 8.7 ppm.

Stock standard solutions of single elements at 1000 and 10,000 ppm can be purchased from a number of companies making these solutions. Purchasing the stock solutions saves a great deal of time and effort, and is often more accurate and less expensive than making these solutions "in-house". The US NIST also sells Standard Reference Material (SRM) solutions of many elements, certified for their concentration. NIST SRM solutions are expensive, but are often used as verification standards to confirm that the analyst has made the working standard solutions correctly as part of a method quality control process. Stock standards of most elements have a shelf life of at least 1 year; commercial solutions will have the expiration date marked on the bottle. Working standards should be prepared fresh daily or as often as needed as determined by a stability test. Low concentrations of metal ions in solution are not stable for long periods of time; they tend to adsorb onto the walls of the container. In addition, evaporation of the solvent over time will change the concentration of the solution, making it no longer a "standard".

It is most important in preparing external calibration curves that the solution matrix for samples and standards be as similar as possible. To obtain reliable quantitative data the following should be the same for the samples and standards:

1. The same solvent (e.g., water, 5% nitric acid, alcohol, MEK); same matrix modifier if used.
2. The same predominant anion (e.g., sulfate, chloride) at the same concentration.
3. The same type of flame (air–acetylene or nitrous oxide–acetylene) or the same graphite tube/platform.
4. Stable pressure in the flame gases during the analysis.
5. Absorbance measured at the same position in the flame/furnace.
6. Background correction carried out on each sample, blank, and standard using the same correction technique.

Sample solutions should be measured immediately following calibration. If the instrument is shut down for any reason (the gas tank runs out, the power fails in a thunderstorm, the lamp burns out, the nebulizer clogs up and needs to be cleaned, the graphite tube cracks, the analyst goes to lunch, etc.) the calibration must be repeated when the instrument is turned back on to be sure that items 3–6 are exactly the same for samples and standards. For GFAAS, the platform and tube must be the same for the calibration curve and the samples. If a tube cracks during a run, a new tube and platform are inserted, conditioned per the manufacturer's directions and the calibration standards rerun before samples are analyzed.

For extremely complex sample matrices, it may not be possible to make external standards with a similar matrix. In this case, the MSA should be used for quantitative analysis. The use of standard additions can correct for some types of interference but not for spectral interference.

The signal in FAAS is a continuous, steady state signal; as long as sample is aspirated into the flame, the absorbance stays at a constant value. Measurement of the absorbance in FAAS is fairly simple. In GFAAS, the signal is transient and the shape and height of the peak depend on the rate of atom formation. That rate can be affected by interferences from the sample matrix. The *integrated peak area* is independent of the rate of atom formation. For quantitative analysis by GFAAS, the concentration should be plotted against the peak area, not the peak height, for the most accurate results. Modern instruments are equipped with integrators to measure absorbance over the atomization period.

The accuracy can be excellent for both FAAS and GFAAS. The precision of FAAS is usually less than 1% RSD. The precision of GFAAS can be in the 1–2% RSD range, but is often 5–10% for complex matrices.

6.5.3. Analysis of Samples

6.5.3.1. *Liquid Samples*

Liquid solutions are the preferred form for sample introduction into flame and furnace atomizers. Frequently liquid samples can be analyzed directly or with minimal sample preparation, such as filtration to remove solid particles. Typical samples that have been analyzed directly include urine, electroplating solutions, petroleum products, wine, fruit juice and, of course, water and wastewater. If the samples are too concentrated, they may be diluted prior to analysis. If they are too dilute, the solvent may be evaporated or the analyte concentrated by solvent extraction or other methods.

Milk may be analyzed for trace metals by FAAS by adding trichloroacetic acid to precipitate the milk proteins. The precipitate is removed by filtration or centrifugation. The standards should also contain trichloroacetic acid to match the samples. Trace metals can be determined in seawater, urine, and other high salt liquid matrices by complexing the metals with a chelating agent and extracting the metal complexes into organic solvent. For example, ammonium pyrrolidine dithiocarbamate (APDC) will complex Cu, Fe, Pb, and other metals. Complexation allows the metals to be extracted into MIBK. The standards are made in MIBK and aspirated directly into the flame. This permits not just extraction, but preconcentration; for example, 1 L of seawater can be extracted into 50 mL of MIBK, resulting in a 20-fold concentration factor. Seawater can be analyzed for the major cations by preparing an artificial seawater solution for making calibration standards in a matrix similar to the samples. Artificial seawater contains calcium carbonate, magnesium oxide, potassium carbonate, and sodium chloride at levels that reflect the cation and anion levels found in real seawater. Oils may be diluted with a suitable organic solvent such as xylene and the soluble metals determined directly. Organometallic standards soluble in organic solvent are commercially available for preparing calibration curves for these types of analyses.

There are liquid sample matrices, such as blood, serum, very viscous oils, and the like that require sample preparation by acid digestion or ashing prior to FAAS determination. In many cases, these matrices can be analyzed without digestion or prior ashing by GFAAS.

6.5.3.2. Solid Samples

Solid samples are usually analyzed by dissolving the sample to form a liquid solution that can be introduced into the flame or furnace. Dissolution can be accomplished by mineral acid digestion ("wet ashing"), fusion of solids with molten salts and dissolution of the fusion bead, dry ashing of organic solids with acid dissolution of the residue, combustion in oxygen bombs, and other procedures too numerous to mention. Dissolution is time consuming, even with the fast microwave digestion and ashing systems, automated fusion fluxers, and other automated sample preparation devices now available. Dissolution entails the possibility of introducing impurities or losing analyte, causing error. This is particularly problematic at ppm and ppb levels of analyte. However, hundreds of types of solid samples are dissolved for AAS analysis on a daily basis; it is still the most common approach to AAS analysis of solids. Types of samples that have been analyzed after dissolution include metals, alloys, soils, animal tissue, plant material, fertilizer, ores, cements, polymers, cosmetics, pharmaceuticals, coal, ash, paint chips, and many solid industrial chemicals. Solid particulates in air, gas and fluid streams can be collected by filtration and analyzed by digesting the filter and the collected particulates.

For example, grains, plant tissues, and many other organic materials can be digested on a hot plate in a mixture of nitric and other mineral acids to destroy the organic matrix. The cooled, clear solution is diluted to volume and analyzed against aqueous acidic calibration standards. Dry ashing of food, plant, and biological tissue is performed by placing 10–50 g of the sample in a suitable crucible and heating in a muffle furnace for several hours to burn off the organic material. The ash is dissolved in mineral acid (nitric, HCl, or acid mixtures) and diluted to volume with deionized water. An aqueous acidic calibration curve would be used. Inorganic materials, ceramics, and geological materials often require fusion in molten salts at high temperatures to convert them to soluble forms. For example, bauxite, an aluminum ore, can be fused in a Pt crucible with a mixture of sodium carbonate and sodium borate. The mixture is heated over a Bunsen or Meker burner to form a clear melt. The fusion salts convert the aluminum, silicon, and titanium compounds in bauxite to salts that will dissolve in aqueous HCl. The calibration standards must contain HCl, sodium chloride, and sodium borate at levels that match those in the diluted fused samples. It is critical that a blank be prepared in the same manner as samples are prepared, to correct for any traces of analytes that might be present in the acids, salts, and other reagents used and for contamination from "the environment". Dust in the air is often a major source of contamination of samples by "the environment", especially for analytes such as Al and Si.

The analysis of solids is time consuming and is prone to many sources of error. Analytical chemists have been trying for many years to analyze solid samples directly, without having to dissolve them. For some types of samples, this can be done by AAS. Solids can be analyzed using *glow discharge atomizers*, by inserting small pieces or particles of sample directly into the flame or furnace, or by the use of *laser ablation*. Some of these approaches are also used in atomic emission spectrometry and in ICP-MS and are discussed in greater detail in Chapters 7, 9, and 10, and in the references by Sneddon, Robinson, and Skelly Frame and Keliher listed in the bibliography.

One approach to the direct analysis of solids is to form a *slurry* of the sample in a suitable solvent and introduce the slurry directly into a graphite furnace atomizer. A slurry is a suspension of fine solid particles in a liquid. Slurry preparation requires that the sample either is in the form of a fine powder or can be ground to a fine powder without contamination from the grinding process. This can be done successfully for many types of samples, such as foods, grains, pharmaceuticals, and sediments. A key development in achieving reproducible results from slurry samples introduced into graphite furnace atomizers was to keep the slurry "stirred" with an ultrasonic agitator during sampling, as discussed in the reference by Miller-Ihli listed in the bibliography. A commercial ultrasonic mixer for graphite furnace autosamplers is available that permits unattended analysis of slurries.

A commercial glow discharge source, the AtomSource® (Leeman Labs, Inc., Hudson, NH, www.leemanlabs.com) is available for the analysis of solid metals and metal alloys. A glow discharge is more commonly used as an atomic *emission* source, but this source acts as a unique atomizer for conductive samples. The source is shown in Fig. 6.30(a). The atomizer cell is a vacuum chamber with the light path along the axis. The conductive sample is positioned as shown and is bombarded with six streams of ionized argon gas. This process sputters the sample, releasing free atoms into the light path, as shown in Fig. 6.30(b). The method does not rely on a high temperature for atomization, so refractory metals such as boron, tungsten, zirconium, niobium, and uranium can be atomized easily. In addition to ground state atoms, the source also produces excited atoms. The emission from the excited atoms can be used to measure elements not able to be measured by AAS. Carbon in steels can be determined by emission, for example, while the other steel components are determined by absorption. The AtomSource atomizer is shown incorporated into the entire spectrometer system in Fig. 6.31.

(a)

(b)

Figure 6.30   (a) The AtomSource glow discharge atomizer for conductive solid samples. (b) Schematic of the sputtered atom cloud produced by bombardment of the sample surface by ionized argon. [Courtesy of Leeman Labs, Inc., Hudson, NH (www.leemanlabs.com).]

Figure 6.31 The AtomSource glow discharge atomizer incorporated into an analyzer for solid metals and metal alloys (The Pulsar system). [Courtesy of Leeman Labs, Inc., Hudson, NH (www.leemanlabs.com).]

6.5.3.3. Gas Samples

There are some metal-containing compounds that are gaseous at room temperature and in principle the gas sample can be introduced directly into a flame atomizer. More commonly, the gas is bubbled through an appropriate absorbing solution and the solution is analyzed in the usual manner.

The introduction of a gas phase sample into an atomizer has significant advantages over the introduction of solids or solutions. The transport efficiency may be close to 100%, compared to the 5–15% efficiency of a solution nebulizer. In addition, the gas phase sample is homogeneous, unlike many solids. There are two commercial analysis systems with unique atomizers that introduce gas phase sample into the atomizer. They are the cold vapor technique for mercury and the hydride generation technique. Both are used extensively in environmental and clinical chemistry laboratories.

6.5.3.4. Cold Vapor Mercury Technique

Mercury is unusual in that it exists as gas phase free atoms at room temperature. Elemental mercury is a liquid at room temperature, but a liquid with a very high vapor pressure. So it is not necessary to apply heat from a flame or furnace to measure mercury vapor by AAS.

The cold vapor (CV or CVAAS) technique requires the *chemical reduction* of mercury in a sample solution to elemental Hg. This is usually done with a strong reducing agent such as sodium borohydride or stannous chloride in a closed reaction system external to the AA spectrometer. The Hg atoms are then purged out of solution (sparged) by bubbling an inert gas through the solution or pumping the post-reduction solution through a gas/liquid separator. The gas stream containing the free Hg atoms is passed into an unheated quartz tube cell sitting in the AAS light path. The cell is usually clamped into the position normally occupied by the burner head. Mercury atoms will absorb the HCL or EDL Hg wavelength and the measurement is made. The cell is sometimes heated to prevent water condensation in the cell, but no heat is required to atomize the Hg. The mercury atomization process is a chemical reduction reaction. CVAAS can be performed manually or can be automated using a technique called flow injection (see subsequently). Because Hg determinations are required in many environmental samples including drinking water, instrument companies have built small AAS cold vapor systems just for Hg measurements. Figure 6.32 shows a schematic of an automated dedicated mercury analyzer that uses a gas–liquid separator, a mercury lamp, a long optical path quartz cell, and a solid-state detector. Samples are pumped from the autosampler tray and mixed automatically with the stannous chloride reductant. Stannous chloride will reduce mercuric ion, Hg^{2+}, to Hg^0, but will not reduce organomercury compounds or mercurous ion. Samples are normally predigested with a strong oxidizing agent to ensure that all Hg in the sample is in the form of mercuric ion before adding the stannous chloride.

The sensitivity of the CVAAS technique for Hg is about 0.02 ppb, much better than that obtained from FAAS or GFAAS. The reasons are simple; the sampling is 100% efficient and a large sample volume is used. All of the reduced Hg atoms are sent into the light path of the spectrometer, unlike the 5% or so efficiency of a flame nebulizer system. Unlike GFAAS, the cold vapor technique uses large sample volumes (10–100 mL of sample) compared with the microliter sample volume put into a graphite furnace. Incorporating a gold amalgamation accessory can increase the sensitivity of the CVAAS technique even further. The mercury vapor can be trapped and concentrated in the gold as an amalgam at room temperature, and then released into the optical path as a concentrated "plug" of vapor by rapidly heating the gold.

6.5.3.5. Hydride Generation Technique

The hydride generation (HGAAS) technique is similar in many ways to CVAAS. The atomizer is a quartz tube cell sitting in the light path of the AA spectrometer. In the hydride generation technique, the cell must be heated. Having the cell clamped above the burner head and lighting an air–acetylene flame accomplishes this. The flame surrounds the cell and heats it. Alternatively, some systems have electrically heated cells.

Sample solutions are reacted with sodium borohydride in a closed reaction system external to the AA spectrometer, as in CVAAS. Some elements will react with the sodium borohydride to form volatile hydride compounds. Arsenic, for example, forms

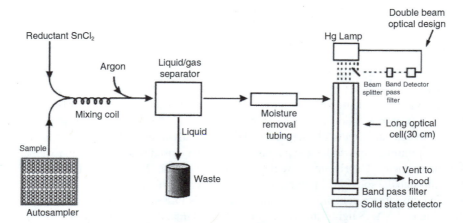

Figure 6.32 Automated cold vapor mercury analyzer. Mercury in the sample is reduced to Hg. The Hg vapor is separated from the solution by a gas–liquid separator and carried to the optical cell. The Hg vapor absorbs the 253.7 nm emission line from the Hg lamp and the amount of atomic absorption is measured using a solid-state detector. [Courtesy of Leeman Labs, Inc., Hudson, NH (www.leemanlabs.com).]

arsine gas, $AsH_3(g)$. The volatile hydrides are sparged from solution by an inert gas and sent to the heated quartz cell. Arsine is a molecule, not an atom, so to measure atomic absorption by arsenic, the hydride species must be decomposed (by breaking bonds) and reduced to free atoms. That is done in the heated quartz cell, and the absorption signal measured as usual. The elements that form volatile hydrides include As, Bi, Ge, Pb, Sb, Se, Sn, and Te. The HGAAS technique can also be operated in manual mode or automated through the use of flow injection techniques. The sensitivity for As, Se, and other elements is excellent because of the efficient transport of the hydride to the atomizer and the larger sample volumes used. Detection limits for CVAAS and HGAAS are given in Appendix 6.2.

6.5.3.6. *Flow Injection Analysis*

The term **flow injection analysis** (FIA) describes an automated continuous analytical method in which an aliquot of sample is injected into a continuously flowing carrier stream. The carrier stream usually contains one or more reagents that react with the sample. A transient signal is monitored as the analyte or its reaction product flows past a detector. We have just described CVAAS, where a reduction reaction occurs and the analyte must be swept in a carrier stream into the optical path. This results in a transient signal. All of the components for an FIA are present—sample, reagent, carrier stream, and transient signal measurement. The same is true of HGAAS. The implementation of FIA for CVAAS and HGAAS permits these determinations to be carried out completely automatically, from sample uptake to printout of concentrations measured.

An FIA system consists of four basic parts: a pump or pumps for regulation of flow, an injection valve to insert sample volumes accurately and reproducibly into the carrier stream, a manifold, and a flow-through detector. A manifold is the term used for the tubing, fittings, mixing coils, and other apparatus used to carry out the desired reactions. The flow-through detector in AAS is the atomizer/detector combination in the spectrometer.

FIA techniques for AAS are described in detail in the reference by Tyson. In addition to automated CVAAS and HGAAS, FIA has been used to automate online dilution for the preparation of calibration curves, online matrix matching of solutions, online preconcentration and extraction for GFAAS, automated digestion of samples, and much more. Commercial FIA systems are available for most AAS instruments.

BIBLIOGRAPHY

Analytical Methods for Atomic Absorption Spectrometry; PerkinElmer, Inc.: Shelton, CT, 1994.

Beaty, R.D.; Kerber, J.D. *Concepts, Instrumentation and Techniques in Atomic Absorption Spectrophotometry*; PerkinElmer, Inc.: Shelton, CT, 1993.

Burguera, J.L., Ed. *Flow Injection Atomic Spectroscopy*; Marcel Dekker, Inc.: New York, 1989.

Carnrick, G.R.; Slavin, W. *Appl. Spectrosc.* **1983**, *37* (1), 1.

Dean, J.A. *Analytical Chemistry Handbook*; McGraw-Hill, Inc.: New York, 1995.

Dean, J.A.; Rains, T.C., Eds. *Flame Emission and Atomic Absorption Spectrometry*; Marcel Dekker, Inc.: New York, 1969; Vol. 1 (1971; Vol. 2; 1975; Vol. 3).

Ingle, J.D., Jr.; Crouch, S.R. *Spectrochemical Analysis*; Prentice Hall: Englewood Cliffs, NJ, 1988.

Jackson, K.W., Ed. *Electrothermal Atomization for Analytical Atomic Spectrometry*; John Wiley and Sons, Inc.: New York, 1999.

Kirkbright, G.F.; Sargent, M. *Atomic Absorption and Fluorescence Spectroscopy*; Academic Press: New York, 1974.

Lajunen, L.H.J. *Spectrochemical Analysis by Atomic Absorption and Emission*; Royal Society of Chemistry: Cambridge, UK, 1992.

L'vov, B.V. *Atomic Absorption Spectrochemical Analysis*; Elsevier: New York, 1970.

Miller-Ihli, N. *J. Anal. Atomic. Spectrosc.* **1988**, *3*, 73.

Parsons, M.L. Atomic absorption and flame emission. In *Analytical Instrumentation Handbook*, 2nd Ed.; Ewing, G.A., Ed.; Marcel Dekker, Inc.: New York, 1997.

Rann, C.S.; Hambly, A.N. *Anal. Chem.* **1965**, *37*, 879.

Robinson, J.W., Ed. *Handbook of Spectroscopy*; CRC Press: Boca Raton, FL, 1975; Vol. 1.

Robinson, J.W., Ed. *Atomic Spectroscopy*; Marcel Dekker, Inc.: New York, 1990.

Schlemmer, G.; Radziuk, B. *A Laboratory Guide to Graphite Furnace Atomic Absorption Spectroscopy*; Birkhauser Verlag: Basel, 1999.

Skelly Frame, E.M.; Keliher, P.N. Atomic spectrometry. In *Materials Science and Technology*; Cahn, R.W.; Haasen, P.; Kramer, E.J., Eds.; VCH Publishers, Inc.: New York, 1992; Vol 2A.

Slavin, W. *Graphite Furnace Source Book*; PerkinElmer, Inc.: Shelton, CT, 1984.

Smith, S.B., Jr.; Hieftje, G.M. *Appl. Spectrosc.* **1983**, *37*, 419.

Smith, S.B., Jr.; Hieftje, G.M. Science **1983**, *220*, 183.

Sneddon, J., Ed. *Sample Introduction in Atomic Spectroscopy*; Elsevier: Amsterdam, 1990.

Sneddon, J. Ed. *Advances in Atomic Spectroscopy*; JAI Press: Greenwich, CT, 1992, Vol. I (1994; Vol. II).

Tyson, J.F. Flow injection techniques for atomic spectrometry. In *Advances in Atomic Spectroscopy*; Sneddon, J., Ed.; JAI Press: Greenwich, CT, 1992; Vol. I.

Walsh, A. Spectrochim. Acta **1955**, *7*, 108.

Welz, B.; Sperling, M. *Atomic Absorption Spectrometry*, 3rd Ed.; John Wiley and Sons: New York, 2002.

SUGGESTED EXPERIMENTS

For the following experiments, prepare at least 25 mL–100 mL of each solution in acid-cleaned volumetric flasks. If the solutions must be stored before analysis, be certain the solutions are acidified and that an acid blank is prepared, stored, and analyzed with the samples.

6.1   Prepare solutions of zinc chloride in deionized water containing 1% (v/v) HNO_3. The solutions should contain known concentrations of zinc over the range of 0.5–50.0 ppm, and an acid blank of 1% (v/v) nitric acid containing no added zinc should be prepared. Aspirate each solution in turn into the burner. Measure the absorbance of the solutions at 213.9 nm, the zinc resonance line. Plot the relationship between the absorbance and the zinc concentrations. Note the deviation from Beer's Law. Indicate the useful analytical range of the calibration curve. If your instrument permits it, try different curve-fitting algorithms and evaluate how they affect the working range of the analysis.

6.2   (a)   Look up the wavelength, preferred flame conditions, and linear range for calcium in Appendix 6.1 or your instrument "cookbook" and set up the AAS to measure Ca. Prepare 1 L of a stock standard solution of 1000 ppm calcium using calcium phosphate. Prepare 1 L of a 10,000 ppm La standard using lanthanum chloride. Make two sets of calcium working standards. The two sets should have the same concentrations of Ca, but to one set of standards (including the blank) add enough of the La stock solution to make each of these solutions 2000 ppm in La. The La must be added before the Ca standards are made to volume. For example: to make a 10 ppm Ca standard, take 1 mL of the 1000 ppm Ca stock and dilute it to 100 mL in a volumetric flack. To make the 10 ppm Ca standard with added La, take 1 mL of the Ca stock solution and add it to an empty 100 mL volumetric flask. Add 20 mL of the stock La solution to the same flask, and then add deionized water to bring the solution to volume. Optimize the burner position for maximum absorbance. Run both sets of solutions and plot absorbance vs. concentration of Ca. Explain your observations.

(b)   Using one of the Ca working standards with La and the same concentration solution without La, aspirate each solution in turn while moving the burner position. Start with the HCL beam just above the burner head (at the base of the flame) and then lower the burner in increments, making note of the absorbance. Construct a flame profile for Ca with and without La by plotting absorbance vs. height above the burner. Are the two profiles different? Give a reasonable explanation for the observed flame profiles, based on your reading of the chapter.

6.3   Prepare a series of standard solutions containing sodium at concentrations of 1, 3, 5, 7, 9, and 10 ppm. Measure the absorbance of each solution and plot the absorbance against the sodium concentration. Take water samples from various sources, such as tap water, bottled drinking water, distilled water, river water, distilled water stored in a polyethylene bottle, and distilled water stored in a glass bottle. Determine the sodium concentration in each sample. (If you choose any carbonated bottled waters as samples, allow the sample to "go flat" or loosen the cap and shake it slightly to release the dissolved gases to avoid poor precision in your measurements. If you want to make the comparison, analyze a carbonated water sample that is not flat. The reproducibility of the absorbance measurements should be better in the "flat" sample.) All samples should be at room temperature when measured.

6.4   Prepare a solution containing 20 ppm of Pb as $Pb(NO_3)_2$. Prepare similar solutions containing 20 ppm of Pb as (a) $PbCl_2$, (b) lead oxalate, and (c) lead acetate. Measure the absorbance by the solutions of the Pb resonance line at 283.3 nm. Note the change in absorbance as the compound changes. This is chemical interference. Find a literature method for the elimination of chemical interference using excess EDTA. Following the literature method, see if the use of EDTA does eliminate the chemical interference you observed.

6.5   Prepare an aqueous 1% HNO_3 solution containing 5 ppm of NaCl; also prepare five separate aqueous 1% HNO_3 solutions (100 mL each) containing 500 ppm of (a) Ca, (b) Mg, (c) Fe, (d) Mn, and (e) K. Take an aliquot of each of these solutions (10–25 mL, but use the same volume for all of the solutions). Spike each aliquot with 5 ppm Na. Be sure to prepare an acid blank and use the same acid for all solution preparation. Measure the absorbance at the Na resonance line at 589.0 nm by all solutions. Compare the absorbance of the Na spiked solutions (a) through (e) with the aqueous acidic Na solution (i.e., compare all the 5 ppm Na solutions in all matrices). Are any of the signals enhanced compared to the aqueous acidic solution? Which ones? Explain. Are any suppressed? Again, explain. For the solutions (a), (b), (c), (d), and (e) with and without Na, plot two-point standard addition calibration curves, and calculate the amount of Na present in the unspiked solutions. Any sodium present was caused by a Na impurity in the original (500 ppm) solutions. Look at the slopes of all five plots. Are they the same? If not, why not?

6.6   For laboratories with a graphite furnace atomizer, the following experiments can be run. The instrument should be setup for the element according to the instrument "cookbook". (a) Dilute the lead solutions from Exp. 6.4 to give 10 ppb lead solutions (or some other concentration within the linear range of your furnace). Using the "cookbook" furnace program and background correction but *no matrix modifiers*, run triplicate injections of each solution (20 μL is a usual injection volume) and observe the peak shape, appearance time of the peak maximum, the peak height and area. Do the peak maxima appear at the same time? Do all peaks have the same shape? If not, why not? Repeat using a lower pyrolysis temperature. If the background absorbance is printed out, compare the background absorbance values at both sets of conditions. How is the background affected by pyrolysis temperature? Does it depend on the compound? (b) Using background correction *and the matrix modifiers* and furnace program recommended by the "cookbook" for Pb, rerun the 10 ppb solutions of lead compounds. Compare the peak appearance time, shape and so on with and without matrix modification. Comment on the approach you would use to obtain accurate results, based on your observations. (c) If your system has the ability to run atomization from the tube wall and from a L'vov platform, run the lead nitrate solution both ways (wall and platform). Perform 20 replicate injections at each condition and compare the precision of wall vs. platform measurements. Explain your results.

PROBLEMS

6.1   Why are atomic absorption lines very narrow?

6.2   Why must HDLs or EDLs be used as the radiation source for AAS? Illustrate schematically an HDL for lead (Pb). How would you make the cathode?

6.3   Why is "modulation" of the source necessary for accurate results? How is modulation achieved?

6.4   If the radiation source is not modulated, will emission from the analyte in the atomizer result in a positive or negative error? Show your calculation to support your answer.

6.5   Why is atomic absorption not generally used for qualitative analysis?

6.6   What causes chemical interference in AAS? Give three examples.

6.7   How are solid samples analyzed by AAS?

6.8   Several standard solutions of copper were prepared. These were aspirated into a flame and the absorption measured with the following results. Prepare a calibration curve from the data.

Sample concentration (ppm)	Absorbance
Blank (0.0)	0.002
0.5	0.045
1.0	0.090
1.5	0.135
2.0	0.180
2.5	0.225
3.0	0.270

6.9 Samples of polymer were brought to the lab for a determination of their copper content. They were ashed, the residue was dissolved in mineral acid and diluted to a known volume. The absorbance of each solution and a digestion blank was measured. Using the calibration curve from Problem 6.8, calculate the copper concentration in each solution and the blank and fill in the table below.

Sample	Absorbance	Concentration (ppm)
Blank	0.006	
A	0.080	
B	0.105	
C	0.220	
D	0.250	

Exactly 2.00 g of each sample were digested and the final solution volume for each sample and the blank was 100.0 mL. Calculate the concentration of copper in the original polymer for each sample.

6.10 What is shown in a Grotrian diagram?

6.11 Why can nonmetal elements not be determined directly by AAS?

6.12 What is the relationship between the amount of light absorbed and the oscillator strength of the transition involved?

6.13 How can the population distribution of atoms in various energy levels be calculated?

6.14 What is the basis for concluding that at temperatures up to 3000 K the great majority of an atom population is in the ground state?

6.15 Describe the two major light sources used in AAS—the HCL and the EDL.

6.16 Why are EDLs used? List the elements that benefit from use of an EDL instead of an HCL.

6.17 What is meant by a flame profile?

6.18 Describe the atomization process that takes place in a flame.

6.19 How does the rapid formation of a stable oxide of the analyte affect its flame profile?

6.20 Define chemical interference. Give an example of how this interference is corrected.

6.21 What is the source of background absorption?

6.22 How is the background corrected?

6.23 Describe the operation of a Zeeman background corrector. Discuss its advantages and disadvantages compared with use of a deuterium lamp for background correction.

6.24 Describe the operation of a deuterium (D_2) lamp background corrector.

6.25 Draw a schematic of a typical graphite furnace atomization tube.

6.26 Describe the L'vov platform.

6.27 What is the advantage of the L'vov platform?

6.28 What are the advantages of graphite furnace atomizers over flames?

6.29 What are the disadvantages of graphite furnace atomizers vs. flames?

6.30 Define ionization interference and give an example. How is this interference corrected?

6.31 You need to determine potassium in serum samples by FAAS. What will you add to correct for ionization interference in the determination?

6.32 Distinguish between spectral and nonspectral interference in AAS.

6.33 The indirect determination of chloride ion by precipitation as silver chloride was described. What metal element could you use to determine sulfate ion in water indirectly by AAS? Describe how you would do this experiment and give an example calculation for a sulfate solution containing 200 mg SO_4^{2-}/L. (You may need to consult some external references on this question.)

APPENDIX 6.1 ABSORPTION WAVELENGTHS, PREFERRED FLAMES, AND CHARACTERISTIC CONCENTRATIONS FOR ELEMENTS BY FAAS

Abbreviations Used

A.A.: Air–acetylene flame

N.A.: Nitrous oxide–acetylene flame

ox.: oxidizing flame (excess oxidant, either air or nitrous oxide). An oxidizing A.A. flame is clear and blue in color. Oxidizing conditions are rarely used for N.A. flames.

red.: Reducing flame (excess fuel, acetylene.) A reducing A.A. flame is yellow in color; a reducing N.A. flame is red in color.

EDL: Electrodeless discharge lamp

Char. conc.: Characteristic concentration in ppm (mg/L) in aqueous solution equivalent to an absorbance = 0.2.

Linear range: The upper concentration limit in ppm of the linear region of the concentration–absorbance plot.

Absorption wavelength (nm)	Preferred flame	Char. conc. (ppm)	Linear range (ppm)
Aluminum			
309.3	N.A. (red.)	1.1	100.0
396.2	N.A. (red.)	1.1	150.0
308.2	N.A. (red.)	1.5	150.0
394.4	N.A. (red.)	2.2	—
237.3	N.A. (red.)	3.3	—
236.7	N.A. (red.)	4.8	—
257.5	N.A. (red.)	6.7	—
Antimony			
217.6	A.A. (ox.)	0.55	30.0
206.8	A.A. (ox.)	0.85	50.0
231.2	A.A. (ox.)	1.3	50.0
Arsenic[a]			
189.0	A.A. (ox.)	0.78	180.0
193.7	A.A. (ox.)	1.0	100.0
197.2	A.A. (ox.)	2.0	250.0
Barium			
553.5	N.A. (red.)	0.46	20.0
350.1	N.A. (red.)	5.6	—
Beryllium			
234.8	N.A. (red.)	0.025	1.0

Absorption wavelength (nm)	Preferred flame	Char. conc. (ppm)	Linear range (ppm)
Bismuth			
223.1	A.A. (ox.)	0.45	20.0
306.8	A.A. (ox.)	1.3	100.0
206.2	A.A. (ox.)	3.7	—
227.7	A.A. (ox.)	6.1	—
Boron			
249.7	N.A. (red.)	13.0	400.0
208.9	N.A. (red.)	27.0	—
Cadmium			
228.8	A.A. (ox.)	0.028	2.0
326.1	A.A. (ox.)	11.0	—
Calcium			
422.7	A.A. (ox.)	0.092	5.0
239.9	A.A. (ox.)	13.0	800.0
Cerium			
520.0	N.A.	30	—
Cesium			
852.1	A.A. (ox.)	0.21	15.0
894.4	A.A. (ox.)	0.40	15.0
455.5	A.A. (ox.)	25.0	—
459.3	A.A. (ox.)	94.0	—
Chromium			
357.9	A.A. (red.)	0.078	5.0
359.3	A.A. (red.)	0.10	7.0
360.5	A.A. (red.)	0.14	7.0
425.4	A.A. (red.)	0.20	7.0
Cobalt			
240.7	A.A. (ox.)	0.12	3.5
242.5	A.A. (ox.)	0.15	2.0
241.2	A.A. (ox.)	0.22	3.0
252.1	A.A. (ox.)	0.28	7.0
352.7	A.A. (ox.)	3.2	—
Copper			
324.8	A.A. (ox.)	0.077	5.0
327.4	A.A. (ox.)	0.17	5.0
216.5	A.A. (ox.)	0.17	20.0
222.6	A.A. (ox.)	1.1	50.0
249.2	A.A. (ox.)	5.8	100.0
224.4	A.A. (ox.)	14.0	—
244.2	A.A. (ox.)	24.0	—
Dysprosium			
421.2	N.A. (red.)	0.70	20.0
404.6	N.A. (red.)	0.97	50.0
418.7	N.A. (red.)	1.0	—
419.5	N.A. (red.)	1.3	—
Erbium			
400.8	N.A. (red.)	0.068	40.0
415.1	N.A. (red.)	1.2	150.0
389.3	N.A. (red.)	2.3	150.0

Absorption wavelength (nm)	Preferred flame	Char. conc. (ppm)	Linear range (ppm)
Europium			
459.4	N.A. (red.)	0.67	200.0
462.7	N.A. (red.)	0.79	300.0
466.2	N.A. (red.)	0.98	—
Gadolinium			
407.9	N.A. (red.)	19.0	1000
368.4	N.A. (red.)	19.0	1000
405.8	N.A. (red.)	22.0	—
Gallium			
287.4	N.A. (red.)	1.3	100.0
294.4	N.A. (red.)	1.1	50.0
417.2	N.A. (red.)	1.5	100.0
403.3	N.A. (red.)	2.8	400.0
Germanium			
265.1	N.A. (red.)	2.2	200.0
259.3	N.A. (red.)	5.2	300.0
271.0	N.A. (red.)	5.0	400.0
275.5	N.A. (red.)	6.1	—
Gold			
242.8	A.A. (ox.)	0.33	50.0
267.6	A.A. (ox.)	0.57	60.0
274.8	A.A. (ox.)	210.0	—
312.8	A.A. (ox.)	210.0	—
Hafnium			
286.6	N.A. (red.)	11.0	500.0
294.1	N.A. (red.)	14.0	—
307.3	N.A. (red.)	16.0	—
Holmium			
410.4	N.A. (red.)	0.87	50.0
405.4	N.A. (red.)	1.0	50.0
416.3	N.A. (red.)	1.4	70.0
Indium			
304.9	A.A. (ox.)	0.76	80.0
325.6	A.A. (ox.)	0.80	60.0
410.5	A.A. (ox.)	2.5	200.0
Iridium			
264.0	A.A. (red.)	12.0	1000.0
266.5	A.A. (red.)	13.0	—
285.0	A.A. (red.)	15.0	—
Iron			
248.3	A.A. (ox.)	0.11	6.0
252.3	A.A. (ox.)	0.18	10.0
248.8	A.A. (ox.)	0.19	10.0
302.1	A.A. (ox.)	0.40	10.0
296.7	A.A. (ox.)	0.81	20.0
Lanthanum			
550.1	N.A. (red.)	48.0	1000.0
418.7	N.A. (red.)	63.0	2000.0
495.0	N.A. (red.)	72.0	3000.0

Absorption wavelength (nm)	Preferred flame	Char. conc. (ppm)	Linear range (ppm)
Lead			
217.0	A.A. (ox.)	0.19	20.0
283.3	A.A. (ox.)	0.45	20.0
205.3	A.A. (ox.)	5.4	—
202.2	A.A. (ox.)	7.1	—
Lithium			
670.8	A.A. (ox.)	0.035	3.0
323.3	A.A. (ox.)	10.0	—
610.4	A.A. (ox.)	150.0	—
Lutetium			
336.0	N.A. (red.)	6.0	500.0
331.2	N.A. (red.)	11.0	—
337.7	N.A. (red.)	12.0	—
451.9	N.A. (red.)	66.0	—
Magnesium			
285.2	A.A. (ox.)	0.0078	0.50
202.6	A.A. (ox.)	0.19	10.0
Manganese			
279.5	A.A. (ox.)	0.052	2.0
279.8	A.A. (ox.)	0.067	5.0
280.1	A.A. (ox.)	0.11	5.0
Mercury[a]			
253.7	A.A. (ox.)	4.2	300.0
Molybdenum			
313.3	N.A. (red.)	0.67	40.0
317.0	N.A. (red.)	1.1	30.0
319.4	N.A. (red.)	1.4	60.0
390.3	N.A. (red.)	2.9	—
Neodymium			
492.4	N.A. (red.)	7.3	200.0
463.4	N.A. (red.)	11.0	200.0
471.9	N.A. (red.)	19.0	—
Nickel			
232.0	A.A. (ox.)	0.14	2.0
231.1	A.A. (ox.)	0.20	5.0
352.5	A.A. (ox.)	0.39	20.0
341.5	A.A. (ox.)	0.40	—
Niobium			
334.4	N.A. (red.)	15.0	600.0
334.9	N.A. (red.)	15.0	600.0
408.0	N.A. (red.)	20.0	—
412.4	N.A. (red.)	26.0	—
353.5	N.A. (red.)	42.0	—
374.0	N.A. (red.)	47.0	—
Osmium			
290.9	N.A. (red.)	1.0	10–200
305.9	N.A. (red.)	1.6	—
263.7	N.A. (red.)	1.8	—
301.8	N.A. (red.)	3.0	—

Absorption wavelength (nm)	Preferred flame	Char. conc. (ppm)	Linear range (ppm)
Palladium			
244.8	A.A. (ox.)	0.22	10.0
247.6	A.A. (ox.)	0.25	15.0
276.3	A.A. (ox.)	0.74	—
340.5	A.A. (ox.)	0.72	—
Phosphorus[a]			
178.3	N.A. (red.)	5.0	2000.0
213.6	N.A. (red.)	290.0	10000.0
214.9	N.A. (red.)	460.0	20000.0
Platinum			
265.9	A.A. (ox.)	2.2	60.0
306.5	A.A. (ox.)	3.2	—
262.8	A.A. (ox.)	4.2	—
283.0	A.A. (ox.)	5.4	—
293.0	A.A. (ox.)	6.1	—
Potassium			
766.5	A.A. (ox.)	0.043	2.0
769.9	A.A. (ox.)	0.083	20.0
404.4	A.A. (ox.)	7.8	600.0
Praseodymium			
495.1	N.A. (red.)	39.0	5000.0
513.3	N.A. (red.)	61.0	—
492.5	N.A. (red.)	79.0	—
Rhenium			
346.0	N.A. (red.)	14.0	500.0
346.5	N.A. (red.)	24.0	500.0
345.2	N.A. (red.)	36.0	1000.0
Rhodium			
343.5	A.A. (ox.)	0.20	15.0
369.2	A.A. (ox.)	0.35	20.0
339.7	A.A. (ox.)	0.45	15.0
350.3	A.A. (ox.)	0.65	20.0
Rubidium			
780.0	A.A. (ox.)	0.11	5.0
794.8	A.A. (ox.)	0.19	5.0
420.2	A.A. (ox.)	8.7	—
421.6	A.A. (ox.)	19.0	—
Ruthenium			
349.9	A.A. (ox.)	0.66	20.0
372.8	A.A. (ox.)	0.86	20.0
379.9	A.A. (ox.)	1.6	—
392.6	A.A. (ox.)	7.5	—
Samarium			
429.7	N.A. (red.)	6.7	400.0
476.0	N.A. (red.)	12.0	—
511.7	N.A. (red.)	14.0	—

Absorption wavelength (nm)	Preferred flame	Char. conc. (ppm)	Linear range (ppm)
Scandium			
391.2	N.A. (red.)	0.30	25.0
390.8	N.A. (red.)	0.40	25.0
402.4	N.A. (red.)	0.41	25.0
408.2	N.A. (red.)	2.1	25.0
Selenium[a]			
196.0	A.A. (ox.)	0.5	50.0
204.0	A.A. (ox.)	1.5	—
206.3	A.A. (ox.)	6.0	—
207.5	A.A. (ox.)	20.0	—
Silicon			
251.6	N.A. (red.)	2.1	150.0
251.9	N.A. (red.)	3.0	200.0
250.7	N.A. (red.)	5.9	—
252.9	N.A. (red.)	6.1	—
252.4	N.A. (red.)	7.0	—
Silver			
328.1	A.A. (ox.)	0.054	4.0
338.3	A.A. (ox.)	0.11	10.0
Sodium			
589.0	A.A. (ox.)	0.012	1.0
330.2	A.A. (ox.)	1.7	—
Strontium			
460.7	N.A. (red.)	0.11	5.0
407.8	N.A. (red.)	2.0	20.0
Sulfur			
180.7	A.A. (ox.)	9.0	—
Tantalum			
271.5	N.A. (red.)	11.0	1200.0
260.9	N.A. (red.)	23.0	—
277.6	N.A. (red.)	24.0	—
Technetium			
261.4	A.A. (red.)	3.0	70.0
260.9	A.A. (red.)	12.0	1000.0
429.7	A.A. (red.)	20.0	1000.0
Tellurium			
214.3	A.A. (ox.)	0.43	20.0
225.9	A.A. (ox.)	4.4	—
238.6	A.A. (ox.)	18.0	—
Terbium			
432.6	N.A. (red.)	5.9	200.0
431.9	N.A. (red.)	6.8	200.0
390.1	N.A. (red.)	9.5	400.0
406.2	N.A. (red.)	11.0	200.0
Thallium			
276.8	A.A. (ox.)	0.67	40.0
377.6	A.A. (ox.)	1.6	100.0
238.0	A.A. (ox.)	3.7	200.0
258.0	A.A. (ox.)	13.0	—

Absorption wavelength (nm)	Preferred flame	Char. conc. (ppm)	Linear range (ppm)
Thorium			
324.6	N.A. (red.)	500	—
Thulium			
371.8	N.A. (red.)	0.45	60.0
410.6	N.A. (red.)	0.66	50.0
374.4	N.A. (red.)	0.74	60.0
409.4	N.A. (red.)	0.80	100.0
Tin[a]			
224.6	N.A. (red.)	1.7	300.0
235.5	N.A. (red.)	2.2	—
286.3	N.A. (red.)	3.2	400.0
Titanium			
364.3	N.A. (red.)	1.8	100.0
365.4	N.A. (red.)	1.9	100.0
320.0	N.A. (red.)	2.0	200.0
363.5	N.A. (red.)	2.4	200.0
Tungsten			
255.1	N.A. (red.)	9.6	500.0
294.4	N.A. (red.)	13.0	1500.0
268.1	N.A. (red.)	12.0	1000.0
272.4	N.A. (red.)	13.0	500.0
Uranium			
351.5	N.A. (red.)	110.0	5000.0
358.5	N.A. (red.)	47.0	—
356.7	N.A. (red.)	76.0	—
Vanadium			
318.4	N.A. (red.)	1.9	200.0
306.6	N.A. (red.)	4.6	200.0
306.0	N.A. (red.)	4.7	400.0
305.6	N.A. (red.)	6.2	500.0
320.2	N.A. (red.)	13.0	—
390.2	N.A. (red.)	13.0	—
Ytterbium			
398.8	N.A. (red.)	0.12	15.0
346.4	N.A. (red.)	0.45	15.0
246.4	N.A. (red.)	1.0	—
267.2	N.A. (red.)	5.0	—
Yttrium			
410.2	N.A. (red.)	1.6	50.0
407.7	N.A. (red.)	1.9	50.0
412.8	N.A. (red.)	1.9	50.0
362.1	N.A. (red.)	2.9	100.0
Zinc			
213.9	A.A. (ox.)	0.018	1.0
307.6	A.A. (ox.)	79.0	—

Absorption wavelength (nm)	Preferred flame	Char. conc. (ppm)	Linear range (ppm)
Zirconium			
360.1	N.A. (red.)	7.0	600.0
303.0	N.A. (red.)	11.0	600.0
301.2	N.A. (red.)	11.0	600.0
298.5	N.A. (red.)	13.0	—
362.4	N.A. (red.)	17.0	—

[a]The Char. conc. obtained by using EDL.

Source: The data are from *Analytical Methods for Atomic Absorption Spectrometry*, PerkinElmer, Inc., Shelton, CT, 1994, courtesy of PerkinElmer, Inc., except as noted below. At least the three most intense lines (where available) are listed for each element. Some elements (e.g., Hg, S, P,) have their most intense lines in the vacuum UV region ($<$190 nm); these lines are not accessible on most commercial AAS systems. Data not in the above reference are courtesy of the late Dr. Fred Brech, of Thermo Jarrell Ash (now ThermoElemental, Franklin, MA).

APPENDIX 6.2

Table A1 Representative Atomic Spectroscopy Detection Limits (μg/L)

Element	Flame AA	Hg/ hydride	Graphite furnace	ICP emission	ICP-MS
Ag	1.5		0.02	0.9	0.003
Al	45		0.1	3	0.006
As	150	0.03	0.2	50	0.006
Au	9		0.15	8	0.001
B	1000		20	0.8	0.09
Ba	15		0.35	0.09	0.002
Be	1.5		0.008	0.08	$<$0.015
Bi	30	0.03	0.25	30	0.0005
Br					0.2
C				75	150
Ca	1.5		0.01	0.02	0.5[a]
Cd	0.8		0.008	1	0.003
Ce				5	0.0004
Cl					10
Co	9		0.15	1	0.0009
Cr	3		0.03	2	0.02
Cs	15			0.05	0.0005
Cu	1.5		0.1	0.4	0.003
Dy	50			2	0.001
Er	60			1	0.0008
Eu	30			0.2	0.0007
F					10,000
Fe	5		0.1	2	0.005[a]
Ga	75			4	0.001
Gd	1800			0.9	0.002
Ge	300			20	0.003

(continued)

Table A1 Continued

Element	Flame AA	Hg/ hydride	Graphite furnace	ICP emission	ICP-MS
Hf	300			4	0.0006
Hg	300	0.009	0.6	1	0.004
Ho	60			0.4	<0.0005
I					0.008
In	30			9	<0.0005
Ir	900		3.0	5	0.0006
K	3		0.008	20	0.015[a]
La	3000			1	0.0005
Li	0.8		0.06	0.3	0.0001[a]
Lu	1000			0.2	<0.0005
Mg	0.15		0.004	0.07	0.007
Mn	1.5		0.035	0.4	0.002
Mo	45		0.08	3	0.003
Na	0.3		0.02	3	0.003[a]
Nb	1500			10	0.0009
Nd	1500			2	0.002
Ni	6		0.3	5	0.005
Os	120			6	
P	75,000		130	30	0.3
Pb	15		0.06	10	0.001
Pd	30		0.8	3	0.003
Pr	7500			2	<0.0005
Pt	60		2.0	10	0.002
Rb	3		0.03	30	0.003
Re	750			5	0.0006
Rh	6			5	0.0008
Ru	100		1.0	6	0.002
S				30	70
Sb	45	0.15	0.15	10	0.001
Sc	30			0.2	0.02
Se	100	0.03	0.3	50	0.06
Si	90		1.0	3	0.7
Sm	3000			2	0.001
Sn	150		0.2	60	0.002
Sr	3		0.025	0.03	0.0008
Ta	1500			10	0.0006
Tb	900			2	<0.0005
Te	30	0.03	0.4	10	0.01
Th					<0.0005
Ti	75		0.35	0.4	0.006
Tl	15		0.15	30	0.0005
Tm	15			0.6	<0.0005
U	15,000			15	<0.0005
V	60		0.1	0.5	0.002
W	1500			8	0.001
Y	75			0.3	0.0009
Yb	8			0.3	0.001

(*continued*)

Table A1 Continued

Element	Flame AA	Hg/ hydride	Graphite furnace	ICP emission	ICP-MS
Zn	1.5		0.1	1	0.003
Zr	450			0.7	0.004

[a]Denotes detection limits obtained under cold plasma conditions. [Data courtesy of PerkinElmer Inc., Shelton, CT (www.perkinelmer.com).]

Note: All detection limits were determined using elemental standards in dilute aqueous solution. All detection limits are based on a 98% confidence level (3 SD). Atomic absorption (Model 5100) detection limits were determined using instrumental parameters optimized for the individual element and EDL where available. ICP emission (Optima 3000) detection limits were obtained under simultaneous multielement conditions with a radial plasma. Detection limits obtained with an axial plasma are typically 5–10 times lower. Cold vapor mercury AA detection limits were determined with a FIAS[TM]-400 flow injection system with an amalgamation accessory. Hydride detection limits were determined with an MHS-10 Hydride system. Furnace AA (Model 5100/ZL Zeeman furnace) detection limits were determined using a L'vov platform and 50 µl sample volumes. ICP-MS (Elan 6100) detection limits were determined using a 3 s integration.

Table A2 Sensitivity (1% Absorption) in FAAS

Element	λ (nm)	Sensitivity (ppm)	Element	λ (nm)	Sensitivity (ppm)
Al	309.2	1.0	Mo	313.3	0.1
Sb	217.6	0.1	Nd	463.4	10.0
As	193.7	1.0	Ni	232.0	0.1
Ba	553.5	0.2	Nb	334.9	20.0
Be	234.9	0.1	Pd	247.6	0.5
Bi	223.1	0.1	Pt	265.9	1.0
B	249.7	30.0	K	766.5	0.1
Ca	442.7	0.05	Pr	495.1	10.0
Cs	852.1	0.1	Re	346.0	15.0
Cr	357.9	0.1	Rh	343.5	0.1
Co	240.7	0.1	Rb	780.0	0.1
Cu	324.7	0.1	Ru	349.9	0.8
Dy	421.2	1.0	Sm	429.7	10.0
Er	400.8	1.0	Sc	391.2	1.0
Eu	459.4	2.0	Se	196.1	1.0
Gd	368.4	20.0	Si	251.6	0.8
Ga	287.4	1.0	Ag	328.1	2.5
Ge	265.2	2.0	Na	589.0	2.5
Au	242.8	1.0	Sr	460.7	0.1
Hf	307.2	10.0	Ta	471.4	10.0
Ho	410.4	2.0	Te	214.3	0.5
In	304.0	0.1	Th	377.6	0.4
Fe	248.3	0.1	Sn	235.4	0.5
La	392.8	75.	Ti	364.3	1.0
Pb	217.0	0.05	W	400.9	1.0
Li	670.7	0.03	U	351.5	100.
Mg	285.2	0.001	V	318.4	1.0
Mn	279.5	0.05	Y	398.8	2.0
Hg	253.7	1.0	Zn	218.9	0.01
			Zr	360.1	50.0

Note: Data includes air–acetylene and nitrous oxide–acetylene flames as appropriate for the element (see Appendix 6.1 for standard flame).

Table A3 Sensitivity (1% Absorption) in the Graphite Tube Furnace (HGA-70)

Element	Absolute sensitivity $(g \times 10^{-12})$	20 µL solution (µg/mL)
Al	150	0.007
As	160	0.008
Be	3.4	0.0002
Bi	280	0.014
Ca	3.1	0.05
Cd	0.8	0.00004
Co	120	0.006
Cr	18	0.01
Cs	71	0.004
Cu	45	0.02
Ga	1,200	0.06
Hg	15,000	1.5
Mn	7	0.01
Ni	330	0.10
Pb	23	0.001
Pd	250	0.013
Pt	740	0.02
Rb	41	0.002
Sb	510	0.15
Si	24	0.10
Sn	5500	0.2
Sr	31	0.0015
Ti	280	0.5
Tl	90	0.1
V	320	0.2
Zn	2.1	0.0001

Note: Data obtained using a PerkinElmer HGA-70 graphite furnace. Courtesy of PerkinElmer, Inc., Shelton, CT.

7
Atomic Emission Spectroscopy

Atomic emission spectroscopy is the study of the radiation emitted by excited atoms and monatomic ions. Excited atoms and ions relax to the ground state, as discussed in Chapter 2 and shown schematically in Fig. 7.1. Relaxation often results in the emission of light, producing line spectra in the visible and UV regions of the spectrum, including the vacuum UV region. These emitted atom and ion lines can be used for the qualitative identification of elements present in a sample and for the quantitative analysis of such elements at concentrations ranging from low parts per billion (ppb) to percent. Atomic emission spectroscopy has relied in the past on flames and electrical discharges as excitation sources, but these sources have been overtaken by plasma sources, such as the inductively coupled plasma (ICP) source. Atomic emission spectroscopy is a multielement technique with the ability to determine metals, metalloids, and some nonmetal elements simultaneously. The major difference between the various types of atomic emission spectroscopy techniques lies in the source of excitation and the amount of energy imparted to the atoms or ions (i.e., the excitation efficiency of the source). For many years, the acronym AES was used for atomic emission spectroscopy; however, the same acronym is used for Auger electron spectrometry, discussed in Chapter 14. To minimize confusion, the term *optical emission spectroscopy* (OES) is now used for atomic emission spectroscopy.

Flame sources impart relatively low quantities of energy to the atoms produced in the flame. Electrons are promoted to only a few low-energy excited states in flames; this results in simple emission spectra. In practice, flame emission spectroscopy is most useful for the easily excited alkali metals and alkaline earth metals. Electrical discharges such as arcs and sparks provide significantly more energy than flames. Most elements can be atomized and ionized in these discharges; the higher energy input causes electrons to populate many higher-energy levels. This results in spectra with many emission lines. Plasma sources, such as the ICP and the direct current plasma (DCP), are high-energy sources that permit the excitation of most elements, both metals and nonmetals, and result in very line-rich spectra. This is one of the major advantages of emission spectroscopy; many elements, as atoms and ions, emit multiple wavelengths simultaneously. The analyst has a choice of several wavelengths for each element and the ability to measure multiple elements concurrently. The disadvantage is that as the number of emission wavelengths increases, so does the *spectral interference* from overlapping lines. This mandates the use of high-resolution spectrometers, which are more expensive than the spectrometers needed for flame emission systems or for atomic absorption (Chapter 6). Atomic emission spectroscopy provides elemental analysis and can measure most metals, metalloids, and nonmetals in a wide variety of liquid, solid, and gaseous samples.

449

Figure 7.1 Schematic diagram of the excitation and emission process. The energies a and b represent atomic excitation, c represents ionization and d represents ionization and excitation. Four possible emission energies and their respective wavelengths are shown; e is ionic emission and f, g, and h are atomic emission. The emission wavelength and energy are related by $\Delta E = hc/\lambda$. [From Boss and Fredeen, courtesy of PerkinElmer Inc. (www.perkinelmer.com).]

The related subject of atomic fluorescence spectrometry (AFS), the emission of photons by excited gas phase atoms following excitation by absorption of photons, is also covered in this chapter.

7.1. FLAME ATOMIC EMISSION SPECTROSCOPY

Flame atomic emission spectroscopy, also called flame photometry, is based on the measurement of the emission spectrum produced when a solution containing metals or some nonmetals such as halides, sulfur, or phosphorus is introduced into a flame. In early experiments, the detector used was the analyst's eye. Those elements that emitted visible light could be identified qualitatively, and these "flame tests" were used to confirm the presence of certain elements in the sample, particularly alkali metals and alkaline-earth metals. A list of visible colors emitted by elements in a flame is given in Table 7.1.

Table 7.1 Flame Colors from Atomic Emission

Flame color	Flame color through blue glass[a]	Element
Carmine red	Purple	Lithium
Dull red	Olive green	Calcium
Crimson	Purple	Strontium
Golden yellow	(Absorbed)	Sodium
Greenish yellow	Bluish green	Barium, molybdenum
Green		Copper, P as phosphate, B as borate
Blue		Copper
Violet	Violet red	Potassium

[a]The blue glass is used to absorb the intense yellow emission from sodium, which is a common constituent of most samples.

The human eye is a useful detector for qualitative analysis but not for quantitative analysis. Replacing the human eye with a spectrometer and photon detector such as a PMT or CCD permits more accurate identification of the elements present because the exact wavelengths emitted by the sample can be determined. In addition, the use of a photon detector permits quantitative analysis of the sample. The wavelength of the radiation indicates what element is present, and the radiation intensity indicates how much of the element is present. Flame atomic emission spectrometry is particularly useful for the determination of the elements in the first two groups of the periodic table, including sodium, potassium, lithium, calcium, magnesium, strontium, and barium. The determination of these elements is often called for in medicine, agriculture, and animal science. Remember that the term *spectrometry* is used for quantitative analysis by the measurement of radiation intensity.

7.1.1. Instrumentation for Flame OES

Flame OES can be performed using most modern atomic absorption spectrometers (discussed in Chapter 6). No external lamp is needed since the flame serves as both the atomization source and the excitation source. A schematic diagram of a flame emission spectrometer based on a single-beam atomic absorption spectrometer is shown in Fig. 7.2. For measurement of the alkali metals in clinical samples such as serum or urine, only a low-resolution filter photometer is needed because of the simplicity of the spectra. The filter photometer is discussed in Section 7.1.1.2. Both instruments require a burner assembly, a flame, a wavelength selection device, and a detector.

7.1.1.1. Burner Assembly

The central component of a flame emission spectrometer is the burner assembly. The assembly has a device to nebulize the sample and then introduces the sample aerosol into the flame. In the flame, free atoms are formed and then excited, which causes them to emit radiant energy. For analytical purposes, it is essential that the emission intensity be steady over reasonable periods of time (1–2 min). The Lundegardh or premix burner is the most commonly used and is depicted in Fig. 6.8(a).

In the premix burner, the sample, in solution form, is first aspirated into a nebulizer where it forms an aerosol or spray. An impact bead or flow spoiler is used to break the droplets from the nebulizer into even smaller droplets. Larger droplets coalesce on the sides of the spray chamber and drain away. Smaller droplets and vapor are swept into the base of the flame in the form of a cloud. An important feature of this burner is that only a small portion (about 5%) of the aspirated sample reaches the flame. The droplets that reach the flame are, however, very small and easily decomposed. This results in an efficient atomization of the sample in the flame. The high atomization efficiency leads

Figure 7.2 Schematic atomic emission spectrometer. [From Boss and Fredeen, courtesy of PerkinElmer Inc. (www.perkinelmer.com).]

Table 7.2   Process Generating Atomic Emission in Flames

to increased emission intensity and increased analytical sensitivity compared with other burner designs. The process that occurs in the burner assembly and flame is outlined in Table 7.2. This process is identical to the atomization process for AAS, but now we want the atoms to progress beyond ground state free atoms to the excited state.

The common nebulizers used in flame emission spectroscopy are the pneumatic nebulizer and the cross-flow nebulizer, just as in AAS.

7.1.1.2.   Wavelength Selection Devices

Two wavelength selectors are used in flame OES, monochromators and filters.

Monochromators.   These consist of slits and dispersion elements and are described in Chapters 2 and 6. The common dispersion element in modern flame atomic absorption and emission spectrometers is a diffraction grating.

Filters.   The alkali metals in a low-temperature flame emit only a few lines and therefore have a simple emission spectrum. In this case, wider wavelength ranges may be allowed to fall on the detector without causing errors, so an optical filter may replace the more expensive diffraction grating. Filters are built with materials that are transparent over a narrow spectral range. The transparent spectral range is designed to be the one in which emission from a given element occurs. When a filter is placed between the flame and the detector, radiation of the desired wavelength from the sample is allowed to reach the detector and be measured. Other radiation is absorbed by the filter and is not measured. Therefore, a separate filter is required for each element to be measured. For example, a filter transparent to radiation with a wavelength of 766 nm is used to determine potassium, while a filter transparent to radiation of 589 nm permits the determination of sodium. The potassium filter absorbs sodium emission and the sodium filter absorbs potassium emission, so the two elements can be measured in the same sample without interference from the other element. Instruments that use filters as wavelength selectors are very convenient for simple repetitive analysis but are limited with regard to the number of elements for which they can be used unless a large number of filters are employed. Most flame photometers are single channel instruments; one element at a time is determined, then the filter is changed and a new element can be determined. Multichannel instruments have been designed where the emission from the flame falls on two or more filters. Each filter transmits the radiation for which it has been designed, and the transmitted radiation falls on a PMT behind the filter. The intensity of the radiation is measured for calibration standards, blanks, and samples. A calibration curve of concentration vs. emission intensity is made for the standards and the concentration of the sample is determined by comparison to the calibration curve. This multichannel approach permits the use of an internal standard to improve precision. For example, a multichannel instrument with filters for lithium and sodium may be used for the determination of lithium in serum. Lithium compounds are used in drugs to treat depression. Patients taking such a drug need to be monitored to be certain that the prescribed dose is effective, so the Li concentration in serum samples from patients is measured using flame photometry. Alternatively, the Li channel could be used as an internal standard channel; lithium is added to all samples and standards to be analyzed for Na. The lithium internal standard corrects for instrument drift, changes in sample uptake due to viscosity differences, and other interferences as described in Chapter 2. Filter photometers designed for use in hospital or veterinary clinical laboratories often have autosamplers and autodilutors attached, permitting the unattended analysis of many samples per hour.

7.1.1.3. Detectors

The detectors in common use for these systems are the PMT or solid-state detectors such as CCDs and CIDs. PMTs and PDAs are discussed in Chapter 5. More detailed discussion of solid-state detectors is covered in Section 7.2.

7.1.1.4. Flame Excitation Source

A flame is the result of the exothermic chemical reaction between two gases, one of which serves as the fuel and the other as the oxidant. The reaction is an oxidation–reduction reaction, with the oxidant oxidizing the fuel. In the process, the reaction generates a great deal of heat. Common oxidants for modern AAS/OES systems are air and nitrous oxide. The only fuel used in modern AAS/OES systems is acetylene, although commercial filter photometer systems can use propane, natural gas, or butane as fuel. When a liquid sample is introduced into a flame, a complex process to produce excited state atoms occurs.

Spectral emission lines are generated by the excited atoms formed during the process of combustion in a flame. Emission lines are characterized by wavelength and intensity. The wavelengths emitted depend on the atoms present. Each element has a different set of quantized energy levels (such as those shown schematically in Fig. 7.1) and will emit different, characteristic wavelengths of light. The intensity of the emission depends on several factors including (1) the concentrations of the elements in the sample and (2) the rate at which excited atoms are formed in the flame. The latter depends on (3) the rate at which the sample is introduced into the flame, (4) the composition of the flame, and (5) the temperature of the flame. The intensity–concentration relationship is the basis for quantitative analysis by flame OES.

Flame temperature is probably the most important single variable in flame photometry. The type of fuel and oxidant used controls the temperature. In general, an increase in flame temperature causes an increase in emission intensity for most elements. This does not happen with elements that ionize easily, such as sodium, potassium, and lithium. If these elements are heated at too high a temperature, they become ionized; the outer electrons move to higher and higher energy states until they leave the atom completely, forming an ion. If the atoms ionize, the valence electrons are lost and therefore cannot return to the ground state and emit atomic radiation in the process. A loss of atomic emission intensity results. These elements must be determined in low-temperature flames to minimize ionization.

The ratio of the number of atoms in an upper excited state to the number of atoms in a lower-energy state can be calculated from the Maxwell–Boltzmann equation (also called the Boltzmann distribution):

$$\frac{N_1}{N_0} = \frac{g_1}{g_0} e^{-\Delta E/kT} \tag{7.1}$$

where N_1 is the number of atoms in the upper state; N_0, the number of atoms in the lower state; g_1, g_0, the number of states having equal energy at each level 0, 1, etc. (g is called the **degeneracy** of the level); ΔE, the energy difference between the upper and lower states (in joules); k, the Boltzmann constant, $= 1.381 \times 10^{-23}$ J/K; and T, the absolute temperature (in kelvin).

The Boltzmann distribution assumes the system is in thermal equilibrium. The emission intensity is related to the number of atoms in the higher excited state, N_1, since we are looking at emission as the atom relaxes from a higher state to a lower state.

Using the Boltzmann equation, we can calculate the ratio of the number of excited state atoms at two different temperatures. For potassium atoms, the major atomic emission line occurs at 766 nm. The energy of this transition in joules is:

$$\Delta E = \frac{hc}{\lambda}$$
$$= (6.626 \times 10^{-34}\,\mathrm{J\,s})(2.998 \times 10^8\,\mathrm{m/s})/(766\,\mathrm{nm})(1 \times 10^{-9}\,\mathrm{m/nm})$$
$$= 2.59 \times 10^{-19}\ \mathrm{J}$$

The temperature in a typical air–acetylene flame is about 2200°C or 2473 K. The ratio of excited state potassium atoms at 2498 vs. 2473 K is calculated by dividing the Boltzmann equation at 2498 K by that at 2473 K. N_0 and the degeneracy terms cancel and we are left with:

$$\frac{N_{2225}}{N_{2200}} = \frac{\exp[-2.59 \times 10^{-19}(\mathrm{J})/1.38 \times 10^{-23}(\mathrm{J/K})(2498\,\mathrm{K})]}{\exp[-2.59 \times 10^{-19}(\mathrm{J})/1.38 \times 10^{-23}(\mathrm{J/K})(2473\,\mathrm{K})]}$$
$$= \frac{5.458 \times 10^{-4}}{5.059 \times 10^{-4}} = 1.08$$

This tells us that a 25 K increase in flame temperature results in an 8% increase in the excited state population of potassium atoms that give rise to this emission line. The intensity of the emission line is directly proportional to the excited state population, even for systems not in thermal equilibrium. The relationship between emission intensity, S, and excited state population can be expressed as

$$S = kN \tag{7.2}$$

where S is the intensity, k is a proportionality constant that includes a number of factors such as the transition probability and energy of the emitted photon, and N is the excited state atom population. S is related directly to the number of atoms in the excited state. As this number increases, the intensity of radiation increases. As the absolute temperature increases, the number of atoms in the excited state increases, therefore the emission intensity increases. Even a small change in temperature results in a significant change in excited state population, as just shown. Atomic emission spectrometry is very sensitive to changes in temperature. The temperature in the atomizer must be carefully controlled for quantitative measurement of emission intensity. This is not the case in atomic absorption spectrometry, where transitions from the ground state are measured. The ground state population is relatively unaffected by small changes in temperature. Table 7.3 lists representative maximum temperatures for some common flames.

As the energy required to cause excitation increases, it is more difficult to excite the atoms, and the number of atoms in the excited state decreases. As a consequence of the relationship $\Delta E = h\nu = hc/\lambda$, a decrease in the wavelength of the emission line indicates that more energy is required to excite the atom. The process becomes more difficult, fewer

Table 7.3 Maximum Flame Temperatures (°C)

Fuel	Oxidant		
	Air	O_2	N_2O
H_2	2100	2900	
Acetylene	2200	3100	3200
Propane	1900	2800	
Coal gas	1800		

atoms are excited and the intensity of radiation decreases. Consequently, elements with emission lines in the short-wavelength part of the spectrum give weak emission signals in low temperature flames. For these elements, the high-temperature nitrous oxide–acetylene flame is favored, or the high-energy electrical or plasma excitation sources discussed later in the chapter should be used.

Another factor that influences emission intensity is the ratio of fuel to oxidant in the flame. The highest flame temperature is obtained when a stoichiometric mixture of the two is used. In a stoichiometric flame, the number of moles of fuel and oxidant present react completely and there is no excess of either after combustion. Any excess of oxidant or fuel results in a decrease in the temperature of the flame. However, some atoms are unstable in certain kinds of flames. Aluminum atoms oxidize very quickly to aluminum oxide in a stoichiometric flame or a flame with an excess of oxidant. Aluminum oxide emits molecular radiation that is not at the same wavelength as the line emission associated with aluminum atoms. This results in a direct loss of atomic emission intensity. To prevent the formation of aluminum oxide, the flame is usually run in a "reducing state", that is, with an excess of fuel. The excess fuel "mops up" free oxidant and minimizes oxidation of the aluminum in the flame. Some elements emit more strongly in oxidizing flames. Consequently, an oxidizing flame is recommended for these elements. In addition, the excess oxygen helps decompose other materials present in the sample and reduces the molecular background. Manufacturers of flame emission instruments provide a list of recommended flame compositions for elements measured by OES. Table A1 in Appendix 7.1 provides flame OES detection limits for many elements and identifies the flame used.

7.1.2. Interferences

The radiation intensity measured may not represent the concentration of analyte in the sample accurately because of the presence of interferences. Interferences fall into two categories: spectral and nonspectral. Three principal sources of interference are encountered in flame OES; they are the same interferences that occur in FAAS.

7.1.2.1. Chemical Interference

If the analyte element is present in the sample with anions with which it combines strongly, it will not decompose to free atoms easily. If the anions present in solution combine only weakly with the analyte element, decomposition and formation of free atoms will be easier. For example, if calcium is to be determined, a given concentration of calcium in the presence of phosphate ion will give a lower signal than the same amount of calcium in a chloride solution. It is easier to decompose calcium chloride into free calcium atoms than it is to decompose calcium phosphate in the same flame. This effect is called *chemical interference.* Chemical interference is a nonspectral interference. It can be reduced or eliminated in a number of ways. The analyte ion may be extracted away from the sample matrix by using a chelating agent or the interfering anion may be removed by ion exchange (discussed in Chapter 13). The standards can be made from the same salt (phosphate, chloride, etc.) as is present in the sample, compensating for effect of the interference. A *releasing agent* may be added to the sample solution. Releasing agents reduce or minimize chemical interferences, often by forming a more stable salt with the anion than is formed by the analyte metal. For example, adding a large excess of lanthanum releases calcium from the effect of phosphate ion by preferentially forming lanthanum phosphate.

7.1.2.2.   *Excitation and Ionization Interferences*

Excitation and ionization interferences are nonspectral interferences. When a sample is aspirated into a flame, the elements in the sample may form neutral atoms, excited atoms, and ions. These species exist in a state of dynamic equilibrium which gives rise to a steady emission signal. If the samples contain different amounts of elements, the position of equilibrium may be shifted for each sample. This may affect the intensity of atomic emission. For example, if sodium is being determined in a sample that contains a large amount of potassium, the potassium atoms may collide with unexcited sodium atoms in the flame, transferring energy in the collision and exciting the sodium atoms in the process. This results in an increased number of excited sodium atoms and increased atomic emission signal compared to a solution of sodium atoms at the same concentration with no potassium atoms in the solution. This is *excitation interference*, and is generally restricted to the alkali metals. It can be overcome by matrix-matching the samples and standards, that is, the standards must be made up so that the standard solutions contain concentrations of the nonanalyte elements similar to those present in the sample. Matrix matching is often done in routine industrial analysis where the sample composition is known. It cannot be done for samples of unknown or highly variable composition. In cases where matrix matching is not practical, the method of standard additions can be used. An example of the MSA is given in Section 7.1.3.

A related problem is *ionization interference.* If the analyte atoms are ionized in the flame, they cannot emit atomic emission wavelengths, and a reduction in atomic emission intensity will occur. For example, if potassium is ionized in the flame, it cannot emit at its atomic emission line at 766.5 nm and the sensitivity of the analysis will decrease. If a large amount of a more easily ionized element, such as cesium, is added to the solution, the cesium will ionize preferentially and suppress the ionization of potassium. The potassium ions will capture the electrons released by the cesium, reverting to neutral potassium atoms. The intensity of emission at 766.5 nm will increase for a given amount of potassium in the presence of an excess of cesium. The added cesium is called an *ionization suppressant.* Ionization interference is a problem with the easily ionized elements of groups 1 and 2. The use of ionization suppressants is recommended for the best sensitivity and accuracy when determining these elements. Of course, as ionization increases, *ion emission* line intensity increases. It may be possible to use an ionic emission line instead of an atomic emission line for measurements.

7.1.2.3.   *Spectral Interferences*

There are two types of spectral interferences in flame OES: (1) background radiation and (2) overlapping line emission from different elements. Excited molecules and radicals in the flame emit over broad regions of the spectrum; it is this broad *band emission* that we call background radiation. The molecules and radicals may be combustion products from the flame gases, oxides, and hydroxides formed from elements in the sample and similar species. For example, species such as BaOH, CaOH, SrOH, MnOH, CaO, CN, and the rare earth oxides all emit band spectra when introduced into a flame. These band spectra can in fact be used to identify these species in a gas phase form of molecular UV/VIS emission spectroscopy, but interfere when atomic emission is to be measured. An example of background emission from an air–acetylene flame is illustrated in Figure 7.3. Atomic emission lines of magnesium, nickel, sodium, and potassium are superimposed on the broad molecular background. In order to accurately determine the intensity emitted by the atom of interest, the intensity due to the background emission must be measured and subtracted from the total intensity. A blank solution (no analyte present) is aspirated and the intensity at the analyte wavelength is measured. This intensity is due to the background emission, since no analyte is present. Then the sample and standard solutions are aspirated and their intensity measured. The sample or standard intensity is due to the sum of the background emission plus the analyte emission. Subtraction of the blank intensity from the sample or standard intensity results in the intensity due to analyte. A sample calculation is given in Section 7.1.3.

All samples and standards must be corrected for background emission. A second method for measuring background intensity is to measure the background intensity of the actual sample at a wavelength very close to the analyte emission line, either slightly higher in wavelength or slightly lower in wavelength than the emission line. This intensity is then subtracted from the intensity of the sample measured at the analyte emission wavelength. This second approach can be less accurate than the first method if the background spectrum is not "flat" in the region of interest. For example, look at Fig. 7.3 in the region of the Ni emission line. The background slopes sharply in the region of the Ni line; it is much higher on the high wavelength side than on the low wavelength side. The use of an "off-line" background correction for Ni could result in a high degree of error because of the rapid change in background intensity with wavelength in this region. On the other hand, the background region around the potassium emission line is relatively flat; the method

Figure 7.3 Emission spectrum of a flame. The broad background emission bands from polyatomic species such as OH and CH are seen. Superimposed on the broad background emission are the very narrow atomic emission lines from the elements Mg, Ni, Na, and K.

of "off-line" background correction would result in accurate background correction for potassium.

The other type of spectral interference is the emission by another element of the same wavelength as the analyte element. This is a direct source of error. The analyst may choose another analyte wavelength, extract the interfering element, or apply a correction factor if the concentration of the interfering element is known. This is not much of a problem in flame OES, due to the low temperature of flames and the relatively few excited states that are populated, but it is a major source of interference in high temperature excitation sources such as electric discharges and plasmas. By "same wavelength" we mean that the instrument we are using cannot resolve the interfering line from the analyte line, even if the actual wavelengths are slightly different. The spectral bandpass is such that both wavelengths pass through the exit slit of the system to the detector. For this reason, high-resolution spectrometers are needed for non-flame OES.

7.1.3. Analytical Applications of Flame OES

7.1.3.1.   Qualitative Analysis

Unlike AAS, atomic emission spectroscopy is an excellent qualitative method for determining multiple elements in samples. Flames are not the optimum emission source for most elements because of the limited temperatures available but they can be used. The presence of elements in a sample is determined qualitatively by observing emission at the wavelength characteristic of the element. Table A1 in Appendix 7.1 lists common characteristic emission lines for flame OES, and a more comprehensive list can be found in the Handbook of Spectroscopy (Robinson) cited in the bibliography. Emission at 766.5 nm indicates that potassium is present in a sample, while the bright yellow emission of sodium at 598 nm indicates its presence. While some elements can be detected visually as we noted in Table 7.1, it is much more accurate to use a spectrometer to identify the wavelengths emitted. The use of a flame source is not as reliable as the use of higher energy excitation sources where multiple wavelengths can be observed for each element, thereby providing more proof of the presence of a given element. For this reason, plasma sources have become the preferred atomic emission source for most elements. Flame OES is a fast, simple method for the qualitative identification of the group 1 and 2 elements and can be used for any element that emits radiation in a flame provided care is taken to discriminate the emission lines from any spectral interference.

7.1.3.2.   Quantitative Analysis

Flame OES can be used to determine the concentrations of elements in samples. The sample usually must be in solution form. Generally, one element is determined at a time if using an AAS system in emission mode. Multichannel instruments are available for the simultaneous determination of two or more elements. Detection limits can be very low as seen in Appendix 7.1, Table A1. Detection limits for the alkali metals are in the ppt concentration range when ionization suppression is used. One part per trillion in an aqueous solution is 1 pg of analyte per mL of solution or 1×10^{-12} g/mL. Most elements have detection limits in the high ppb to low ppm range.

An important factor in quantitative analysis by flame OES is the solvent used for the samples and standards. When water is the solvent, the process of atomization is endothermic and relatively slow. If the solvent is organic, the reactions in the flame are exothermic and atomization is rapid. Other things being equal, more free atoms are liberated and emission intensity increased in organic solvent than in aqueous solution. For accuracy it is therefore critical that the samples and standards be prepared in the same solvent.

For a quantitative measurement, the sample is aspirated into the flame and the intensity of radiation is measured at the emission wavelength of the analyte element to be determined. The concentration of the element in the sample is calculated from comparison with an external calibration curve or by the standard addition method. Both of these methods have been discussed in Chapter 2. A typical external calibration curve for lithium is shown in Fig. 7.4. The calibration curve was constructed by aspirating Li standard solutions of 5 and 10 ppm Li and the blank solution (0.0 ppm Li). Each intensity was measured and a plot of intensity vs. concentration made. Modern systems use a statistical curve-fitting program to construct the calibration curve and to calculate the concentrations of unknowns. Note that an emission signal was detected in the standard that contained no lithium (the zero concentration standard or calibration blank). This is called the *background signal* or *blank emission signal.* The source of the background signal was discussed in Section 7.1.2.2. In this case, the background emission was not subtracted from each standard, but was plotted as the 0.0 ppm Li standard. To use this curve, an unknown sample is aspirated, its emission intensity is measured and the concentration read (or calculated by the computer) from the calibration curve. The assumption in this case is that the samples and standards all have the same background emission as the blank. If the blank is subtracted from all of the standards and the calibration curve then plotted, the line will go through the origin. To use that type of calibration curve, the blank signal must be subtracted from the unknown sample signals. In one way or another, the background emission signal must be corrected for in the final calculation of analyte concentration in the sample. The relationship between emission intensity and lithium concentration is linear in this example. The calibration curve follows Beer's Law at low concentrations of analyte. Deviations from linearity at high concentrations are commonly observed in flame emission calibration curves. As an example, assume that we have the calibration curve in Fig. 7.4. The data points for this curve

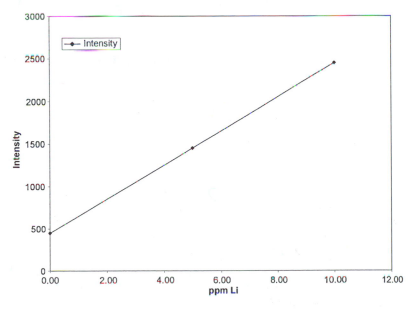

Figure 7.4 Calibration curve for lithium by flame emission using external calibration standards. Note the nonzero intercept for the blank due to background emission from the flame.

Table 7.4   Calibration Curve for Li by Flame OES

Li concentration (ppm)	Intensity
0.00	450
5.00	1450
10.00	2450

Linear regression equation: $y = mx + b = 200x + 450$

and the equation of the calibration curve determined by linear regression are given in Table 7.4.

If we aspirate an unknown sample solution and measure an intensity of 1626 counts, we can use the equation to calculate the Li concentration in the unknown. Inserting 1626 as y in the equation and solving for x, the concentration x is calculated to be 5.88 ppm Li in the unknown solution.

If we subtracted the blank first, the calculation would be as shown in Table 7.5. Now in order to use the blank-corrected calibration curve, we have to subtract the background intensity from our sample intensity. The corrected intensity is shown in Table 7.5. Inserting $y = 1176$ into the equation shown and solving for x gives us 5.88 ppm Li. As expected, the results are the same since the background was accounted for in both approaches (and assumed to be the same in all samples and standards).

Suppose we need to determine Li in human serum to monitor an Li-based anti-depressant medication. We would not expect the background from a serum sample to be the same as the background from an aqueous standard solution. We might make our calibration standards in lithium-free serum, but if we only have one sample to run, that is not efficient in terms of cost or time. This is a good example of when to use the MSA. In MSA, the calibration curve is made up in the sample itself.

The sample is split into three aliquots of equal volume. Two aliquots have known amounts of Li added, while the third is just the sample aliquot itself, or the "zero addition" sample. As we discussed in Chapter 2, the Li is added in such a way that the total volume of the three aliquots does not change significantly. The intensity of emission for all three solutions is measured, as presented in Table 7.6 and the intensity plotted against added Li, as shown in Fig. 7.5. The concentration of Li in the original sample can be calculated using the calibration curve and the background-corrected sample intensity. The equation for the line obtained is included in Table 7.6. The sample concentration is the absolute value of y at $x = 0.0$ (no added Li). The calculated value for x at $y = 0.0$ is -9.1, so the sample Li concentration is 9.1 ppm Li.

The sample concentration can also be read from the intercept of the line with the negative x-axis from a graphical plot of the calibration curve. In Fig. 7.5, the sample

Table 7.5   Blank Corrected Calibration Curve for Li by Flame OES

Li concentration (ppm)	Intensity	Blank-corrected intensity
0.00	450	0
5.00	1450	1000
10.00	2450	2000
Unknown	1626	1176

Linear regression equation: $y = mx + b = 200x + 0 = 200x$

Table 7.6 Data for Li in Serum by MSA

Standard addition, ppm Li added	Intensity	Blank corrected intensity
0	1432	1247
5.0	2117	1932
10.0	2802	2617
Blank	185	0

Linear regression equation: $y = mx + b = 137x + 1247$
Correlation coefficient = 1.00

can be seen to contain 9.1 ppb Li, because the intercept occurs at -9.1 ppm. Taking the absolute value results in 9.1 ppm Li in agreement with the calculated value. It is critical to measure the background emission from the flame and to subtract the background emission signal from all the aliquot signals before plotting the standard addition curve. If this is not done, a positive error in the reported concentration will result.

In order to obtain a steady emission signal of constant intensity, each step in the process shown in Table 7.2 must be controlled. Modern equipment is able to do this with little attention from the analyst. The skillful analyst, however, is always aware of the chemical and physical processes involved and is thus able to recognize and correct any problems that may arise. Variations in emission intensity can be caused by several factors, including (1) blockage in the nebulizer or burner, preventing the flow of sample; (2) viscosity differences in samples or between samples and standards; (3) change of solvent in the sample; (4) change in the fuel or oxidant flow rate to the burner; and (5) change of the position of the burner in the instrument, causing the radiation to be displaced from the light path. The operating conditions must be optimized for the analysis, including the selection of the flame to use, the wavelength to monitor, the

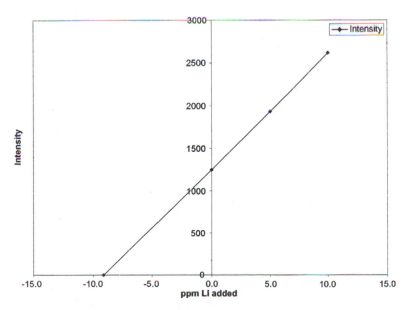

Figure 7.5 Determination of lithium in a sample by the MSA. The background emission from a blank reading was subtracted from all sample aliquots before plotting the data.

Table 7.7 Detection Limits for Flame Photometry in Commercial Instruments

Element	LOD (ppm) (Model PFP-7)	Range[a] (mmol/L) or LOD[b] (ppm) (Model PFP-7C)	LOD (ppt) (Model 300)
Na	≤0.2	120–190[a]	8
K	≤0.2	0–10.0[a]	15
Li	≤0.25	≤0.25[b]	5
Ca	≤15	≤15[b]	
Ba	≤30	≤30[b]	

Note: ppt = part per trillion, or ng/L. The Model PFP-7C is designed specifically for clinical chemistry and displays the Na and K concentrations in mmol/L. The Model 300 is designed for high sensitivity analysis of power plant feedwater.
Source: Data courtesy of Buck Scientific, Inc., East Norwalk, CT. (www. bucksci.com).

fuel/oxidant ratio, the sample uptake rate, the slit widths, and the position of the burner in the light path.

The major use of flame emission spectroscopy is in the determination of Na, K, Li, Ca, Mg, and Ba in biological tissues and fluids by medical laboratories. The same elements are often measured in plant tissue, animal feed and forage, and food. While flame OES is a simple, rapid method for the determination of the group 1 and 2 elements, it has been replaced for the determination of most other elements in liquid samples by ICP-OES. A typical commercial flame photometer for clinical determination of Na, K, Li, and other elements is capable of detection limits as shown in Table 7.7. Buck Scientific, Inc. (www.bucksci.com) makes a specialized flame photometer, their Model 300, for the measurement of ions in power plant feedwater, with detection limits of 5–15 ppt for Na, K, and Li, with a linear working range of from 0.1 to 20 ppb. The high sensitivity comes from an optimized low temperature hydrogen flame. For quantitative analysis, the precision, as expressed by the percent relative standard deviation, is about 1%. There are many flame emission applications published in the literature before plasma emission systems became dominant in the early 1980s.

7.2. ATOMIC OPTICAL EMISSION SPECTROSCOPY

Radiation from excited atoms and ions may also be emitted by a sample when it is intro-duced into an electrical discharge, a glow discharge, or a plasma. Because these excitation sources are of higher energy than a flame source, all metallic and semimetallic (metalloid) elements can be detected in low concentration, including refractory elements such as boron, tungsten, tantalum, and niobium, and some nonmetals can be measured, including C, N, H, Cl, Br, and I. Liquids and solids can be analyzed quite easily using a plasma source; electrical and glow discharge sources are used for the analysis of solids only. Pure gas phase elements, such as hydrogen, helium, neon, and mercury vapor, can be identified by the emission spectra generated by an electrical discharge through the gas confined in a sealed quartz tube. This same process generates the light in "neon" signs and mercury vapor streetlights.

Because the temperatures of electrical discharges and plasmas are much higher than temperatures that can be achieved in flames, the emission spectra from nonflame excitation sources are very complex. Two types of line spectra are generated from these systems. First are the *atomic emission spectra* from neutral atoms. These lines are designated in tables of emission lines by a roman numeral I following the wavelength. Under spark and plasma conditions ions are often formed. A second electron in the ion may then become excited and enter one of the higher energy states. From these states, the ion relaxes and emits a photon in the process. The energy levels of the ions are not the same as the energy levels of the atoms. They therefore emit different emission lines, called *ion lines*. Lines from singly charged ions are designated by a roman numeral II and lines from doubly charged ions by a roman numeral III following the wavelength. Several reference books that list tables of emission lines are listed in the bibliography.

Ion lines are less sensitive than atomic lines but are not subject to reversal due to self-absorption, discussed below. They are used for quantitative analyses when a sufficient concentration of the analyte is available. Ion lines are seldom used for qualitative analysis because of their lack of sensitivity.

7.2.1. Instrumentation for Emission Spectroscopy

In simple terms, an emission spectrometer with an electrical discharge source (Fig. 7.6) functions as follows. An electrical source produces an electrical discharge in a spatial "gap" between two electrodes, the *sample electrode* and the *counterelectrode*. A solid machined piece of the metal to be analyzed serves as the sample electrode; the counterelectrode is an "inert" electrode of tungsten or graphite. Material from the sample electrode is introduced into the discharge, where it becomes vaporized and excited. The excited atoms emit radiation, which is detected and measured by the detector readout system. The wavelengths of the emitted lines identify the elements present and the intensity of the emission at each wavelength can be used to determine the amount of each element present. The individual components of the instrumentation are described in the following sections.

7.2.1.1. Electrical Excitation Sources

Historically, several types of electrical excitation sources have been used since the early part of the 1900s, among them the DC arc, the AC arc, and the AC spark. Commercial instruments using electrical excitation sources became available about 1940 with PMT detectors; prior to this, emission instruments used a photographic plate or film as the detector. Modern instruments are either DC arc emission instruments or high voltage spark

Figure 7.6 Schematic diagram of an emission spectrometer with an electrical discharge excitation source. The prism is meant only to illustrate a dispersive device; a diffraction grating is used in all modern spectrometers with a single dispersive device. Echelle spectrometers use two dispersive devices, either a prism and a grating or two gratings.

emission instruments capable of operating under "spark-like" or "arc-like" conditions as necessary. A significant advantage of these electrical excitation sources is that conductive solids can be analyzed directly, avoiding the need for tedious sample dissolution and the degradation of detection limits by having to dissolve and therefore dilute the sample. Nonconductive solids can often be mixed with conductive graphite, which does dilute the sample to some extent, but not as much as dissolution.

DC arc.   In this source, a DC voltage is applied across a pair of graphite or metal electrodes separated by a small gap. The typical voltage for a DC source is 200 V with currents of 3–20 A. The arc must be initiated by the formation of ions, usually by a low current spark. Ions are formed in the gap, electrical conduction occurs, and the motion of electrons and ions formed by thermal ionization sustains the arc. A *plasma* forms into which the sample is vaporized by electrical and thermal energies. The term *burning* is used to describe the process, although no combustion is involved. The sample, in the form of solid powder, chips, or filings, is generally packed into the cupped end of a graphite electrode [Fig. 7.7(a) and (b)]. This electrode would be at the bottom of the gap, as seen in Fig. 7.6. The top electrode, sometimes referred to as a *counterelectrode*, is a solid graphite electrode. A photograph of the electrode assembly in the AtomComp 2000 DC Arc emission spectrometer is shown in Fig. 7.8(a). If the sample is a solid piece of conductive metal, it may be machined into the proper shape and serve as the bottom electrode. In DC arc emission, the entire sample packed into the electrode cup is usually vaporized or "burned" over a period of several minutes. The arc is a *continuous* discharge, and the

Figure 7.7   (a) Diagram of a direct current arc source. The solid sample is packed into the cupped end of the lower graphite electrode. The graphite counterelectrode is also shown. (From Hareland, used with permission.) (b) A schematic of the lower electrode used to hold the sample.

(a)   (b)

Figure 7.8   (a) Photograph of the electrode assembly in the AtomComp 2000 DC Arc emission spectrometer. (b) Photograph of the DC Arc in operation. The discharge is established in the gap between the electrodes. [Courtesy of ThermoElemental, Franklin, MA (www.thermo.com).]

emission intensity is collected and integrated over the burn period. An operating DC arc is shown in Fig. 7.8(b).

The temperature of the arc depends upon the composition of the plasma and varies with the nature of the sample. If the sample is made of material with low ionization energy, the temperature of the plasma will be low; if the ionization energy of the material is high, the temperature will be high. In addition, the temperature is not uniform in either the axial or radial directions. This results in matrix effects and self-absorption. Arc temperatures are on the order of 4500 K with a range of 3000–8000 K. Emission spectra from arc sources contain primarily atomic lines with few ion lines. The DC arc can excite more than 70 elements.

A significant problem with DC arc sources is the poor stability of the emission signals. Generally, the plasma will link to a point or projection on the surface of the solid sample and will continuously erode the sample, vaporizing it in the process. A pit is formed in the sample by the erosion, and finally the pit becomes so large that the discharge can no longer be maintained at that point. The discharge then "wanders" to another nearby point on the sample surface. In practice, the wandering of the discharge is quite fast and causes the signals to fluctuate. Numerous such discharge points are created at the same time. Local hot spots are formed, which cause the sample in the immediate vicinity to evaporate rapidly and become excited easily; however, these hot spots are not continuous and soon die away, and the local emission intensity is reduced. As a consequence of this behavior, the total emission intensity is somewhat erratic and the signals produced lack precision. On the other hand, the signals are very intense, and the analytical sensitivity is better than when other types of electrical discharges are used.

Another problem with the DC arc is that volatile elements will selectively vaporize and enter the plasma before the less volatile elements in a sample. The electrode temperature increases slowly from the initiation of the arc. If the sample is the anode, which heats more rapidly and to a higher temperature than the cathode, volatile elements will rapidly enter the arc. For example, tungsten powder for light bulb wire contains added potassium and silicon. In a DC arc analysis, the low melting potassium would volatize first, followed later (possibly several minutes later) by the high melting silicon and tungsten. This can cause serious problems with quantitative analysis if standards are used that do not match the volatilization behavior of the sample matrix. The problem may be overcome by mixing the samples and standards with an excess of a "spectrochemical buffer". This is a salt or mixture of salts such as NaF and Na_2CO_3, often mixed with graphite. The "buffer" increases conductivity and provides a uniform matrix that keeps the arc temperature constant. Mixing with pure graphite is also used to give a common matrix. The purity of the graphite or other buffer materials used is very important, given the ppb detection limits that can be reached.

There is a positive aspect to the selective volatility of low melting elements, in that spectral interferences are likely to be less at the beginning of a burn (low temperature) than at high temperatures when line-rich elements such as Fe start to vaporize. In modern DC arc sources, this problem of selective volatility is addressed by having the polarity of the electrodes reversed for the first minute or so of the burn. With the sample as the cathode, the temperature is lower and the rate of heating lower. This slows the rate of evolution of volatile species and improves sensitivity and precision for the volatile elements. Sensitivity for the nonvolatile elements is somewhat reduced when the sample is the cathode. A common sequence for analysis would consist of a burn of 60 s with the sample as cathode (reverse polarity) at 8–10 A, a 60 s burn of normal polarity (sample as anode) at 8–10 A, and then 60 s at 15 A to volatilize the very refractory elements.

The DC arc is used primarily for semiquantitative and qualitative analysis of solids, including powdered solids, chips, and filings. Most elements can be determined. The spectrometer operates from about 190–800 nm which covers the UV and visible range in which most atomic emission occurs. The precision attainable in a modern DC arc spectrometer is 5–10%, with detection limits in the 10–100 ppb range in solid samples. Detection limits for trace elements in high purity copper by DC Arc emission using a CID detector are given in Table 7.8. These detection limits are comparable to the instrumental detection limits of GFAAS and ICP-OES, both of which generally require dissolution of the solid and dilution to form a solution. The effective detection limits in the solid sample therefore would be poorer by GFAAS or ICP-OES than the DC arc detection limits. The

Table 7.8 Detection Limits for Elements in High Purity Copper Using DC Arc Emission Spectrometry

Element	DC Arc with PMT detection (ppm)	DC Arc with CID detection (ppm)
Ag	0.2	0.15
As	1.0	0.20
Bi	0.1	0.012
Fe	1.0	0.15
Mn	0.1	0.06
Ni	0.1	0.02
Pb	0.1	0.008
Sb	1.0	0.25
Sn	0.5	0.02
Te	1.0	0.15
Zn	1.0	0.02

Source: Data courtesy of Thermo Elemental, Franklin, MA (www.thermo.com). The DC Arc/CID data were collected using the Thermo Elemental AtomComp 2000.

limitation to DC arc is the requirement that the solid sample be electrically conductive. The linear working range of a DC arc source is about three orders of magnitude, limited by the relatively low energy of the source.

Spark source. A schematic diagram of a high-energy spark source is shown in Fig. 7.9. When the switch is closed, the capacitor is charged. When the switch is opened, the capacitor discharges through the inductor and the resistor and produces an electrical discharge in the spark gap. The voltage in the circuit shown in a modern instrument is between 400 and 1000 V. This voltage is not high enough to cross the 3−4 mm gap between the sample and the counterelectrode. A separate ignitor circuit, similar to those used in automobiles, is used to create a very high voltage (10 kV). This high voltage initiates electrical excitation across the gap. Once the electrical path is initiated, the operating voltage sustains the discharge. The spark source is an interrupted discharge, unlike the continuous DC arc discharge.

In a modern spark source, the discharge is only in one direction, from the counterelectrode to the sample. This is done by including a diode across the capacitor so that current flows in only one direction. The switch is opened and closed electronically many times per second. This is the source frequency and is equal to the number of discharges per second. Spark source frequencies are generally in the 200−600 Hz range. A modern spark source can be used in either spark-like or arc-like mode. Spark-like refers to a high-current discharge that comes in short bursts. Arc-like refers to a lower current, almost continuous discharge. A diagram of current in the gap vs. time for the two types of discharges shows the difference between the two modes (Fig. 7.10). The spark frequency is hundreds of sparks per second. The sparks strike the sample surface randomly, resulting in an averaging of the composition and improved precision. The spark source is flushed with argon gas to eliminate oxidation of samples and to permit measurement of wavelengths below 200 nm. The flow of argon through the spark stand is as high as 250 L/h when intensity measurements are taken.

The temperature in a spark-like discharge is much higher than the temperature in a DC arc discharge. Temperatures are not stable in either time or space in the spark, but may be as high as 40,000 K. This permits excitation of most elements, including oxygen, nitrogen, the halogens, and the group 18 gases (Ne, Ar, etc.). The higher temperature also results in many ion emission lines being produced. Spark spectra contain ion lines for singly ionized elements and for doubly ionized elements. Spark spectra therefore are more complex than DC arc spectra. The spark-like discharge is more reproducible than the DC arc; this makes the spark a better choice for quantitative analysis. However, signal intensity is lost

Figure 7.9 Schematic spark excitation source. S is a switch, C is a capacitor, L is an inductor, R is a resistor, and V is a voltage source. (From Thomsen, used with permission from ASM International®.)

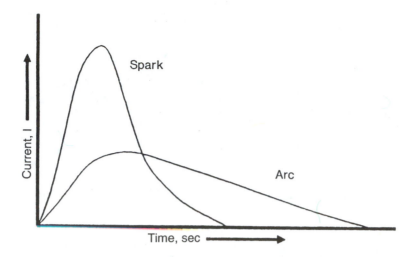

Figure 7.10 Current–time relationship for a spark discharge and an arc discharge. (From Thomsen, used with permission from ASM International®.)

with this technique, with a consequent loss of analytical sensitivity. Spark-like conditions are used for the best precision; arc-like conditions for the best sensitivity.

While the primary function of the electrical excitation source is to generate a discharge that will vaporize and excite sample atoms, the source serves a second function. It provides a high-energy prespark (HEPS), a high current spark discharge that homogenizes the sample surface by essentially remelting it. The emission during the HEPS period (formerly called a preburn) is not integrated. After the HEPS, the sample is more homogeneous and the integration of the signals from the sparking period results in higher precision. A typical spark cycle for a metal sample would include a 10–20 s HEPS period, a 5 s measurement under spark-like conditions for major and minor constituents, and then a 5 s measurement under arc-like conditions for trace elements. Typical excitation source parameters are given in Table 7.9.

7.2.1.2. Sample Holders

The function of the sample holder is to introduce the sample into the electrical discharge. There are two types of sample holders, those for solid samples and those for liquid

Table 7.9   Typical Spark Excitation Source Parameters

Parameter	HEPS	Spark	Arc
R (ohms)	1	1	15
L (μhenry)	30	130	30
C (μfarad)	12	2	12
V (volt)	400	400	400
f (Hz)	300	300	300
I (amp)	130	60	20
t (μsec)	100	90	600

Source: From Thomsen, used by permission of ASM International®.

samples. More than 50 types and sizes of electrodes for holding samples are commercially available.

Solid sample holders. Solid samples, especially nonmetallic materials, may be analyzed as powders. High-purity carbon electrodes are available with a small well drilled in one end to hold powdered samples, such as the electrode shown in Fig. 7.7(b). Powdered samples may be loaded into the carbon electrode and packed or tamped down. In some cases, the sample powder is mixed with a large amount of a high-purity matrix powder, such as alumina or silica mixed with graphite. This ensures a constant matrix and therefore a constant plasma temperature, which improves the precision of the analysis. Another approach is to mix the sample powder with a large excess of a conducting powder, such as high-purity graphite or copper, and then to press a pellet of the mixture to serve as the electrode. The conducting powder also serves as a constant matrix, improving the precision of the measurements.

Metallic or electrically conductive samples, such as sheets, wires, or rods of alloys or pure metals, may be used directly as one or both of the electrodes. Some preparation may be required, such as milling the metal into the desired shape or melting the metal and casting it into a mold. The end of the electrode is either polished flat or tapered at one end. The formed electrode must be cleaned to remove any surface contamination before analysis. The most common approach used in modern spark instruments is to have a polished flat metal sample serve as the cathode and a tungsten electrode as the anode, shown schematically in Fig. 7.11.

Liquid sample holders. Oil samples may be analyzed directly by spark emission spectrometry using a rotating disk electrode. The rotating disk electrode, or rotrode (Fig. 7.12) is a graphite disk that rotates through the liquid sample, which wets the surface of the disk. The rotating wet surface carries the sample into the discharge at a steady rate. Calibration data must be obtained under identical conditions and an identical number of revolutions. This approach had been used routinely for the determination of metals in lubricating and fuel oils. The analysis of oil for metal content indicates wear of the engine and other lubricated parts and is an important quality control measure in aircraft and commercial and military vehicle maintentance. The rotrode was used with spark sources, not with DC arc systems. Many modern oil analysis laboratories have switched

Figure 7.11 A schematic spark source, showing a flat metal sample as the cathode and a tungsten counterelectrode as anode. (From Hareland, used with permission.)

Figure 7.12 A schematic rotating disk electrode (or rotrode) used for the introduction of liquid samples into an electrical discharge. The system is used primarily for the determination of metals in oils.

to DCP or ICP instruments for these determinations, because of the improved precision and lower detection limits that can be obtained.

7.2.1.3. *Spectrometers*

Simultaneous multichannel spectrometers are required for most arc and spark emission work because the instability of the source or the interrupted nature of the discharge requires that emission signals be integrated for 10 s or longer. A scanning spectrometer is therefore not practical. There are three types of spectrometers in use, spectrographs, polychromators, and array-based systems, also called electronic spectrographs. The basic components of spectrometers were discussed in Chapter 2 and more detail on atomic spectrometers was covered in Chapter 6. A significant difference between systems for AAS and atomic emission spectrometry is in the spectral bandwidth or band-pass used. The bandpass in AAS is generally in the range of 0.2–0.5 nm, while for atomic emission, it is usually less than 0.1 nm. The reason for this difference is left as a problem for the student at the end of the chapter.

Spectrographs.   Spectrographs are spectrometers that use photographic film or photographic plates to detect and record emitted radiation. Spectrographs were introduced in the 1930s and served as the fundamental elemental analysis instrument in industries for many years, especially in the production of steel, other metals, and alloys. While there are still a few spectrographs in operation as of this writing, no new spectrographs of this type have been built for years and spare parts are becoming very hard to find. The photographic plates are expensive because they are now a nonroutine product and must be ordered a year or more in advance. Because of this and because of the hours of time and the skill required to record spectra, develop the plates, and analyze the film images, most industries have turned to array-based systems or PMT-based polychromator systems.

Polychromators.   Polychromators are multichannel spectrometers with PMTs as detectors. Instruments with up to 64 PMTs are commercially available. These instruments generally use a concave diffraction grating as the dispersion device, as shown schematically in Fig. 7.13. As can be seen, a concave grating focuses light of different wavelengths, λ_1, λ_2, λ_3, and so on, at different points on the circumference of a circle.

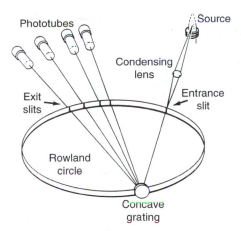

Figure 7.13 A Rowland circle polychromator. This configuration is called a Paschen–Runge mount. The source shown is an inductively coupled plasma, but any of the electrical excitation sources can be substituted. Multiple photomultiplier detectors are placed at the appropriate positions on the circumference of the circle to measure dispersed wavelengths. [From Boss and Fredeen, courtesy of PerkinElmer Inc. (www.perkinelmer.com).]

For a given radius of curvature of the grating, the entrance slit, the surface of the grating, and the focal points of the dispersed wavelengths lie on a circle called the *Rowland circle*. The *diameter* of the Rowland circle is equal to the radius of curvature of the surface of the concave grating. Photomultiplier detectors are put behind fixed slits at the focal points of up to 64 different wavelengths as illustrated in Fig. 7.13, which shows only four slits and detectors. The limitation on the number of channels (wavelengths monitored) is set by the physical size of the PMTs in a conventional Rowland circle system. The instrument buyer chooses the wavelengths, so each instrument is custom-built. The advantage of this system is that routine multielement analysis for many elements can be performed rapidly. Once chosen, the wavelengths and therefore the elements detected, are fixed, so the instrument lacks flexibility if nonroutine samples need to be analyzed. The precision of these instruments with a spark source is excellent, often better than 1% RSD, and quantitative analysis can be performed rapidly. If a DC arc source is used, rapid semiquantitative analysis with ppb detection limits can be performed.

In order to measure 70 or 80 elements simultaneously, it is necessary to cover the entire range of wavelengths from the vacuum UV through the visible region (120–800 nm). Using a single Rowland circle, the number of PMTs is limited by the size of the detectors. The spark spectrometer shown schematically in Fig. 7.14 uses fiber optics to carry the light from the excited sample to up to four separate polychromators, each optimized for a specific wavelength region. This greatly increases the number of wavelengths that can be measured; up to 128 separate PMTs can be arranged in this system. The UV polychromator has a sealed optical system filled with nitrogen, permitting wavelengths as short as 120 nm to be measured without the need for vacuum pumping. A steel sample can be completely analyzed quantitatively, including determination of C and N in the steel, in less than 30 s using this system. A sealed purged optical system, continuous inert gas purging or a vacuum spectrometer is required to measure wavelengths in the 120–190 nm range.

Echelle monochromators. In the past 10 years, Echelle monochromators have become increasingly common in multielement emission systems. These are often

Figure 7.14 Schematic diagram of the SPECTROLAB spark emission spectrometer system. [Courtesy of Spectro Analytical Instruments, Inc., Fitchburg, MA (www.spectro-ai.com).]

coupled with array-based detectors. The Echelle design uses two dispersion devices in tandem, a prism and an Echelle grating, or two gratings. The design of the Echelle grating is discussed in Chapter 2. The two dispersion devices are placed so that the prism disperses the light in one plane. This light falls upon the grating, which disperses in a plane at right angles to the plane of dispersion of the prism. This is illustrated in Fig. 7.15(a). The final dispersion takes place over a 2D plane rather than along the single line of the Rowland circle optics. The radiation intensity can be measured using either PMT detectors or array detectors. As shown in Fig. 7.15(b), wavelengths can be isolated by the use of an aperture grid, which is a 2D array of multiple individual exit slits. Numerous photomultipliers can be located behind the grid. More commonly, the entire 2D "echellogram" is imaged using a 2D solid-state array detector, such as the CID or CCD discussed subsequently. Figure 7.16 shows a commercial Echelle spectrometer with a CID detector. A schematic representation of the resulting echellogram, wavelength vs. diffraction order, is shown in Fig. 7.17. With new array detectors, up to 5000 different wavelengths can be monitored simultaneously, many more than could possibly be detected using PMTs. An additional advantage of an Echelle system with a solid-state detector is the ability to correct for background emission in real time (i.e., at exactly the same time as the analyte intensity is measured). These array-based Echelle systems are electronic versions of the old photographic imaging systems, and while they are often called spectrometers, the term electronic spectrograph can be applied as well.

7.2.1.4. Detectors

PMTs and linear photodiode array detectors are discussed in detail in Chapter 5. This section will cover the 2D array detectors used in arc/spark and plasma emission spectrometers. In order to take advantage of the 2D dispersion of wavelengths from an Echelle spectrometer, a 2D detector is required. The detector should consist of multiple

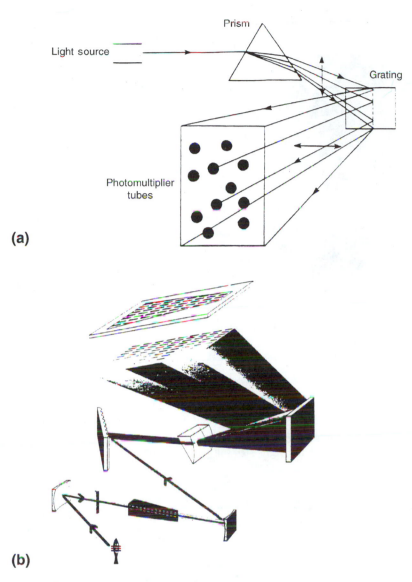

Figure 7.15 (a) A schematic diagram of an Echelle spectrometer. Two dispersive devices are used. The prism disperses light in the vertical plane; the grating disperses the vertically dispersed light in the horizontal plane. The physical arrangement of the grating and prism can be reversed as shown in (b) or two gratings may be used. (b) Schematic of a commercial Echelle spectrometer system, the Profile ICP spectrometer, which uses an aperture grid to isolate the 2D dispersed wavelengths. This system uses PMT detection (not shown). [Diagram (b) courtesy of Leeman Labs, Inc., Hudson, NH (www.LeemanLabs.com).]

individual detectors positioned (arrayed) so that different wavelengths fall on each individual detector. Such an array detector is called a multichannel detector, whether the detector units are arranged linearly or in two dimensions. The charge-injection device (CID) and the charge coupled device (CCD) are both silicon-based sensors that respond to light. They belong to a broader class of devices called charge transfer

Figure 7.16   Photograph of the interior of a commercial Echelle system with a CID detector. This spectrometer is used in the ThermoElemental DC arc system and the IRIS ICP system. [Courtesy of ThermoElemental, Franklin, MA (www.thermo.com).]

devices (CTD). The two detectors discussed can be considered to be 2D arrays of silicon photodiode detectors. CTDs store the charge created when electromagnetic radiation strikes them and permit the integrated charge to be read. The amount of charge measured is proportional to the intensity of the light striking the detector units.

Figure 7.17   Exit plane representing the 2D array of wavelengths produced by the Echelle configuration. [From Boss and Fredeen, courtesy of PerkinElmer Inc. (www.perkinelmer.com).]

An insulating layer of SiO_2 is grown onto a crystalline silicon substrate and thin conducting electrodes are placed on top of the silica layer to create metal oxide-semiconductor (MOS) capacitors. The crystalline silicon has Si—Si bonds arranged in a 3D lattice. A bond can be broken by a photon of sufficient energy, releasing an electron into the silicon lattice and creating a "hole" simultaneously. This is exactly analogous to the formation of an electron–hole pair in a photodiode, described in Chapter 5. In the presence of an applied electric field across the device, the electrons will move in one direction and the holes will move in the opposite direction, leaving an area depleted of positive charge (a depletion region or potential well). The electrons accumulate and are stored in the depletion region. The more photons absorbed by the silicon, the more electrons are accumulated in the potential well. Wells can hold as many as 10^6 charges before "leaking" or "blooming" into an adjacent storage unit. The CTD storage units, or pixels, are arranged in a 2D array on a silicon wafer. The number of pixels varies from 512×512 to 4096×4096. The difference between the CID and the CCD is the manner in which the charge is obtained, accessed, and stored. Both are multichannel detectors, permitting multielement analysis using several hundred wavelengths. A significant advantage to these multichannel detectors is that the background is measured simultaneously with the analyte emission, providing more accurate background correction and faster analysis times.

CID. Each pixel in a CID can be accessed randomly to determine the amount of charge accumulated during a given period of time. The time period is called the integration time. Positive charges from the photon production of an electron–hole pair are stored in a potential well beneath a negatively biased capacitor. Reading the amount of charge does not change the amount of charge; the CID readout is nondestructive. The charge can be accessed even while the pixel is exposed to light, that is, during the integration time. This helps to eliminate blooming, the spilling over of charge from one pixel to another. The random access and nondestructive readout are advantages for the CID. Another advantage is that in its operation, the CID provides complete wavelength coverage, just like a photographic plate. It is possible to measure more than 250,000 lines simultaneously. Background correction and spectral interference correction are performed by processing the data after it is collected. The CID has a high noise level and must be cooled to liquid nitrogen temperature for operation. CIDs are available to cover the UV and visible regions of the spectrum (190–800 nm). The dynamic range for the CID used in the spectrometer shown in Fig. 7.16 is eight orders of magnitude, significantly more than the one to two orders of magnitude offered by photographic film. A picture of this CID and a real echellogram image collected by the CID are shown in Fig. 7.18.

CCD. The CCD operates in a manner identical to a linear photodiode array detector and similar to the CID, except for the way the individual detectors are accessed and controlled. The CCD depicted in Fig. 7.19 shows the depletion region (potential well) and stored negative charge from photoelectrons generated when light hits a pixel. In a CCD, the accumulated charge must be read sequentially. In the process of reading it, the charge is destroyed. The charge is transferred from one pixel to the next in a row in a "bucket brigade" manner. The readout "buckets" are at the edge of the CCD chip. As each row is read, the charge from the next row is shifted to the corresponding pixels in the empty row, a process called parallel charge transfer. The readout electronics know the location in the original array of the pixel being dumped into the end bucket and match the measured charge with that position. This allows the CCD to provide intensity–wavelength correlation even though the charge has been moved. The readout is very fast and has low noise associated with it. CCDs can be cooled by liquid nitrogen, but also by more convenient thermoelectric cooling devices and still have

(a)   (b)

Figure 7.18 (a) Photograph of a CID. This solid-state array detector captures an electronic image of the complete emission spectrum. (b) The echellogram of a sample of zinc oxide collected by a CID detector from a DC arc source. [Courtesy of ThermoElemental, Franklin, MA (www.thermo.com).]

good signal-to-noise ratios. They are more susceptible to blooming than CIDs at high levels of illumination; this limits their dynamic range, but makes them good detectors for weak sources.

While PMT detectors are more sensitive and operate with lower signal-to-noise than CID or CCD detectors, they cannot match the multichannel capacity of the CTD array detectors. The references by Harnly and Fields, Sweedler et al., Epperson et al. and Pilon et al. can provide more detail on CID and CCD detectors.

7.2.2. Interferences in Arc and Spark Emission Spectroscopy

In most modern laboratories, arc and spark emission is limited to the analysis of solid samples, because liquids are much more easily analyzed by ICP or DCP emission spectroscopy.

Arc and spark emission spectroscopy is widely used for the qualitative, semiquantitative, and quantitative determination of elements in geological samples, metals, alloys, ceramics, glasses, and other solid samples. Quantitative analysis of more than 70 elements

Figure 7.19 Schematic of a CCD. In this CCD, three electrodes define a pixel. [Courtesy of Jobin Yvon, Inc, Horiba Group, Edison, NJ (www.jyhoriba.com).]

at concentration levels as low as 10–100 ppb can be achieved with no sample dissolution. The determination of critical nonmetal elements in metal alloys, such as oxygen, hydrogen, nitrogen, and carbon can be performed simultaneously with the metal elements in the alloys.

7.2.2.1. Matrix Effects and Sample Preparation

The matrix, or main substance, of the sample greatly affects the emission intensity of the trace elements present. For example, the emission intensity from 5 ppm of iron in a sample of pure aluminum may be very different from the intensity of the same 5 ppm of iron in a sample of aluminum–copper alloy. This is because the matrix (i.e., the aluminum or the mix of aluminum and copper) has different properties such as thermal conductivity, crystal structure, boiling point (or heat of vaporization). The matrix affects the temperature of the plasma and the heat build-up in the electrode, which affect the rate of vaporization of the iron into the electrical discharge. The rate of vaporization into the discharge affects the number of iron atoms excited per unit time, and therefore the emission intensity from the iron. If the iron volatilizes from the sample matrix faster than it does from the matrix of the iron standards used for calibration, a positive error will result. If the iron volatilizes more slowly from the sample than from the iron standards, a negative error will result. There are other causes for this matrix or **interelement effect**, including the physical state of the analyte element. Some elements exist in solid solution in metals, others occur as precipitated elements or compounds in the sample. Some alloys consist of two or more solid phases, and the proportion of these phases will change with how a metal alloy has been treated or worked. The physical state of the analyte in the sample will often affect the emission intensity. The text by Slickers discusses this in detail, especially for metals. Interelement effects result in changes in the slope of the calibration curve. This differs from the effect of spectral interference from an overlapping line. In the case of line overlap, the entire calibration curve is shifted parallel to a calibration curve with no spectral interference; no change in slope occurs.

To minimize this problem, it has been common practice to convert all samples to a common matrix, especially for arc emission analysis. This may be done by converting the sample to an oxide powder through either wet ashing with oxidizing acid or dry ashing. The sample oxide powder is then mixed with a large excess of a common inert matrix, such as graphite powder. Organic compounds, such as animal tissue, body fluids, plant materials, plastics, paper, and textiles are analyzed in this manner. Metal powders are often oxidized before analysis. Mixing with a matrix powder allows the sample to enter the electrical discharge at a steady rate. It avoids sudden "flares" of emission from the sample, which give rise to erratic results. The common matrix permits the accurate and precise determination of the elements present.

Bulk solid samples must be machined or polished into the correct shape, and this must be done without contaminating the surface from the grinding material or altering the surface composition by polishing out a soft phase, for example. The surface must be cleaned of any lubricant or oil used in the machining process before analysis.

Advances in understanding of the spark and arc sources and improvement in electronics have led to redesign of the sources and spectrometers, so that in modern instruments, structure-induced interelement effects can be minimized by proper choice of excitation and measurement conditions. This has reduced the need for a common matrix and has permitted, in some cases, the use of one "global" calibration curve for multiple materials. However, if an interelement effect does exist, correction factors can be calculated to compensate for the effect. The text by Thomsen has a detailed explanation of how this is done.

7.2.2.2.   Spectral Interference

As discussed in flame emission spectrometry, there are two types of spectral interference: background interference and line overlap. Background in arc/spark emission can be due to thermal radiation (light emitted by a heated object), emission by molecular or polyatomic species such as CN from atmospheric nitrogen or the edges of adjacent emission lines reaching the detector. Background from the edges of adjacent emission lines will vary depending on the elements present. In general, the background for atomic lines from an arc source is lower than the background for the same lines in a spark source. Background must be measured as close to the analyte emission line as possible and the analyte intensity corrected for accurate results.

Spectral line overlap is a serious problem with high-temperature excitation sources. Thousands of wavelengths are emitted from metal alloys such as steels and from heavy elements such as molybdenum and tungsten. Overlapping lines are unavoidable. The resolution of the spectrometer needs to be as high as possible to separate analytical lines of interest. Even with high resolution, spectral overlap will occur. The analyst needs to use published tables of emission lines, such as the MIT wavelength tables (see the bibliography) to choose an analytical line that is not directly overlapped by another element in the sample. For example, Mn has an intense emission line at 293.306 nm; the MIT tables indicate that uranium emits at exactly the same wavelength, so clearly this line cannot be used to measure Mn in uranium or U in manganese. There are five emission lines within ± 0.008 nm of the Mn 293.306 nm line. None of the lines is from Cu or Al, so this Mn line could be used to measure Mn in copper alloys or Al alloys, but not in U, Th, W, or Ir.

If there is no suitable analytical line free of spectral overlap, corrections can be made for the interfering element, but only if the overlapping element's intensity is less than 10% of the analyte intensity. Less than 1% is even better, if possible. Corrections cannot be made if the overlapping line is one of the major elements in the sample (e.g., overlap from an iron line on Cr in a steel sample cannot be corrected mathematically).

7.2.2.3.   Internal Standard Calibration

Apart from matrix effects, which must be controlled to obtain accurate analyses, such factors as physical packing and the size of particles cause a variation in the quantity of sample vaporized into the discharge from one sample to the next. Instrument drift, slight changes in the source energy, and other system problems can cause variations in intensity from measurement to measurement. Because arc/spark emission spectroscopy is a multielement technique, these problems may be compensated for by using *internal standard calibration*. In this calibration method, described in Chapter 2, the ratio of analyte intensity at its chosen wavelength to the intensity of an internal standard element at its emission wavelength is measured. The intensity ratio is plotted vs. the concentration ratio of analyte to internal standard. The use of the ratio should compensate for some variations if the internal standard element and its wavelength are chosen correctly. If something causes the analyte intensity to increase, the internal standard intensity should also increase and the ratio will remain constant.

For bulk solid samples such as metals and alloys, the easiest way to do this is to choose an emission line of the major matrix element as the internal standard line. For example, an iron emission line would be used in steels, a Cu line in copper alloys, and so on. If the sample is a powder, either the major element can be used or a separate element added to all standards and samples.

The following guidelines are used when choosing an element and wavelength to serve as an internal standard:

1. The element used as an internal standard should be similar in properties such as enthalpy of vaporization and ionization energy to the element to be determined.
2. The wavelength of the emission line from the internal standard should be *homologous* with the wavelength of the analyte. Homologous means that the lines should behave similarly with respect to excitation. Atomic internal standard lines should be used for atomic analyte lines; ion internal standard lines for ion analyte lines. (The roman numerals after the wavelengths chosen should be the same. Examples of homologous line pairs for elements in steel are shown in Table 7.10.)
3. The concentration of the internal standard added should be similar to that of the analyte measured.
4. For spectrometers with more than one polychromator, the internal standard line and analytical line must be measured on the same polychromator.

The guidelines should be adhered to as far as possible to obtain the best accuracy. It is often not possible to meet all of the guidelines with one internal standard element. For example, if the analyst is required to measure 23 trace elements in a given sample, including transition metals, rare earth elements, nonmetals such as B, S, P, and refractory metals such as Nb, W, and Zr, it is not likely that one element chosen as an internal standard will meet all of the guidelines. More than one internal standard element may be added or more than one matrix wavelength chosen, but the possibility of spectral interference must be kept in mind. The more elements present, the more emission lines there will be and the greater the possibility of spectral overlap.

7.2.3. Applications of Arc and Spark Emission Spectroscopy

7.2.3.1. Qualitative Analysis

Qualitative analysis is performed by recording the emission spectrum of the sample. The sample elements are then identified by comparing their emission spectra with previously recorded spectra from known elements. Generally, for trace amounts, the *raies ultimes*, or *RU lines*, must be present and identified in the emission spectra as discussed below. For higher concentrations, the RU lines are present together with many other emission

Table 7.10 Some Homologous Line Pairs in Steel

Analyte line (nm)	Fe II 273.0 nm internal std. line	Fe I 281.3 nm internal std. line
Mn II 293.3	×	
Cr II 298.9	×	
Ni II 376.9	×	
Mo II 281.6	×	
Si I 251.6		×
Ni I 352.4		×

Note: The roman numerals denote the type of line (atomic, ionic) not the valence state.
Source: Adapted from Thomsen, with permission of ASM International[®].

lines. The emission spectra encountered in arc and spark spectroscopy are rich and intense for most elements in the periodic table, in contrast to those observed in flame photometry. As a result, these techniques are widely used in elemental qualitative analysis of solid samples. Most elements can be detected at low concentrations (ppm or ppb) with a high degree of confidence. In practice, emission spectra are now "matched" to identify the elements present using a computer database of emission lines contained in the instrument software.

7.2.3.2. Raies Ultimes

When an element is excited in an electrical discharge, it emits a complex spectrum that consists of many lines, some strong, some weak. When the concentration of the element is decreased, the weaker lines disappear. Upon continuing dilution of the element, more and more lines disappear until only a few are visible. The final lines that remain are called the raies ultimes or RU lines, although this term is not used in many modern publications. The RU lines are the most intense emission lines and invariably involve an electronic transition from an excited state to the ground state. These lines can be detected even when small concentrations of the element are present. Conversely, if the element is present in a sample, whatever other spectral lines are emitted from the sample, the RU lines should always be detectable. Some elements and their RU lines are listed in Table 7.11.

Table 7.11 RU Lines of Some Common Elements

Element	Wavelength of RU lines (nm)	Element	Wavelength of RU lines (nm)	Element	Wavelength of RU lines (nm)
Ag	328.07	Co	344.36	Mo	379.63
	520.91		240.73		386.41
Al	396.15	Cr	399.86	Na	589.59
	394.40		425.43		568.8
As	193.76	Cs	852.11	Ni	341.48
	189.04		894.35		352.45
	286.04	Cu	324.75	P	253.57
Au	242.80		327.40		255.33
	267.60	Fe	385.99	Pb	405.78
B	249.77		371.99		368.35
	249.68	Ge	265.12	Pt	265.95
Ba	265.05		303.90		292.98
	553.55	Hg	185.0	Se	196.03
Be	234.86		253.65		206.28
	313.11		435.84		203.99
Bi	306.77	K	766.49	Si	288.16
	289.80		769.90		251.61
C	193.09	La	624.99	Sn	286.33
	247.86		579.13		284.00
Ca	422.67	Li	670.78	Sr	460.73
	445.48		610.36		483.21
Cd	228.80	Mg	285.21	Ti	498.18
	326.11		383.83		399.86
Ce	569.92	Mn	403.08	V	318.40
	404.0		403.31		437.92

Generally, for confirmation of the presence of a trace amount of a given element, two or three raies ultimes should be detected. Confirmation based on a single RU line is unreliable because of the possibility of spectral interference from other elements.

Because the plasma is not uniform in temperature throughout the discharge, there will be unexcited atoms of the analyte element in the cooler regions of discharge. These free atoms absorb radiation at the same wavelength as the emitted RU lines from other excited atoms of the same element since they involve the same transition. The greater the concentration of a given element, the greater the number of free atoms in the discharge. As a result, such emission lines are reabsorbed by neighboring unexcited atoms in a process called self-absorption or self-reversal. (This is the same process of self-reversal that occurs in HCLs and serves as the basis for the Smith–Hieftje background correction technique discussed in Chapter 6.) The result is seen as curvature in the emission calibration curve for the element at high concentrations. If the concentration is high enough, the self-absorption is so effective that the net emission intensity actually decreases as shown in Fig. 7.20 for the beryllium 313 nm line, an RU line. A plot of emission intensity against concentration will go through a maximum and then decrease. The intensity therefore "reverses" at higher concentrations. Such lines are said to be *reversible lines*. They cannot be used for quantitative analysis at higher concentrations, but they are useful for measuring very small concentrations, where reversal is not a problem. In fact these lines are usually the most sensitive; the Be 313 nm line can be used for determining ppb levels of Be, a highly toxic element, in biological samples. Ion lines do not suffer from self-reversal, because the ion energy states are not the same as the atomic energy states.

7.2.3.3. *Quantitative Analysis*

Quantitative analysis is carried out by measuring the intensity of one emission line of the spectrum of each element to be determined. The choice of the line to be measured depends on the expected concentration of the element to be determined. For extremely low concentrations, the RU lines must be used. These are the most intense lines and must be used when high sensitivity is required. These lines are reversible, and their intensity does not increase linearly with concentration. Therefore, for high concentrations a nonreversible, less intense line must be used in order to stay within the linear calibration range.

7.2.3.4. *Preburn Time*

When a sample is introduced into an electrical discharge, there is a lag in time before the emission signal becomes steady. In the first few seconds of discharge, the intensity of emission is erratic and difficult to control. In order to obtain reproducible quantitative results, it is advisable to ignore these erratic signals. This is done by not recording the emission until the signal is judged stable. By plotting emission intensity from a standard against time, this period can be measured. For example, for tin this period may be 15 s. When a sample is analyzed for tin, the first 15 s of the discharge are not recorded. This period is called the *preburn time*, that is, the time during which the emission intensity is not stable and is not recorded. The term used for modern spark instruments is HEPS and is usually set for 10–20 s. In modern DC arc, as already mentioned, the preburn is performed under conditions of reverse polarity for about 60 s. A preburn time is always used if accurate results are required. In modern spectrometers, the preburn time is set in the software and performed automatically.

Figure 7.20 A plot of emission intensity vs. %Be at the 313.0 nm Be line. This is a reversible line for Be. Note that the emission intensity reaches a maximum and then decreases as the Be concentration increases. Such lines cannot be used for high concentrations of analyte but are excellent for low concentrations. (Modified from Thomsen, used with permission from ASM International®.)

The relative accuracy and precision obtained by arc and spark emission spectroscopy is commonly about 5%, but may be as poor as 20–30%. Arc emission is much more prone to matrix effects than spark emission due to the lower temperature of the discharge. Both arc and spark excitation may require matrix matching of sample and standards for accurate analyses, and usually require the use of an internal standard.

In practice, the sample is excited in the electrical discharge. Intensities of emission lines are measured using the detector(s) in the spectrometer and concentrations of elements are determined by comparison to the intensities obtained from standard calibration curves. External calibration (discussed in Chapter 2) can be used for many materials, with the internal standard being the major matrix element as has been described. Certified metal alloy standards in flat polished forms and bulk solid pieces are available from government organizations such as the US NIST and from private standards companies. It is possible to purchase sets of solid steel standards, brass standards, nickel alloy standards, and so on for use as calibration standards and quality control check samples. High-purity metal powders are available from commercial suppliers as well. The accuracy of the analysis will be only as good as the accuracy of the standards used for calibration of the instrument. Using commercially available automatic readout systems, the simultaneous determination of the elements present in a sample can be carried out in a minute or less.

Emission spectroscopy with arc and spark excitation has been used since the 1930s for many industrial analyses. In metallurgy, for example, the presence in iron and steel of the elements nickel, chromium, silicon, manganese, molybdenum, copper, aluminum, arsenic, tin, cobalt, vanadium, lead, titanium, phosphorus, and bismuth have been determined on a routine basis. Modern instruments can also measure oxygen, nitrogen, and carbon in metals, which used to require separate measurements with dedicated high-temperature analyzers. Frequently the analysis can be carried out with very little sample preparation, apart from physically shaping the sample to fit the source of the discharge unit. Alloys of aluminum, magnesium, zinc, and copper (including brass and bronze), lead and tin alloys (including solders), titanium alloys, high-purity metals such as gold, platinum, palladium, and many others are routinely analyzed during production by arc and spark spectrometry. Alloy identification for metal scrap sorting and recycling has become so important that portable, handheld spark emission spectrometers have been developed. These units are battery operated, and contain a built-in computer database of alloy compositions, so that a metal recycling facility can immediately separate aluminum alloys from nickel alloys, for example, even when the metal pieces are out in a field or as they are unloaded from a truck.

Production and quality control of uranium fuel rods used in nuclear power plants are monitored by DC arc emission spectroscopy. Trace elements in high-purity metal powders are measured for quality control purposes. Tungsten powder used to make light bulb filament can be analyzed for trace elements by arc/spark emission spectroscopy without the need to dissolve the tungsten; this eliminates the use of expensive and hazardous hydrofluoric acid.

In the oil industry, lubricating oils can be analyzed for iron, nickel, chromium, manganese, iron, silicon, copper, aluminum, and other elements. These metals get into the lubricant when bearings and pistons wear. Such analyses give the engine designer valuable information on the parts of the system that are likely to fail or need protection from wear and corrosion. The information is used in aircraft and railway engine maintenance to warn of bearing wear in the engine. The bearings may be replaced based on lube oil analysis rather than taking the engine apart for physical inspection, which is a long and expensive procedure. Oil feed stocks to petroleum refinery catalytic cracking units are analyzed for iron, nickel, and vanadium, which can poison the catalyst and interfere in the production of the branched hydrocarbons desired in fuel. Many of the oil analysis applications are now performed by ICP or DCP emission spectrometers, since these systems handle liquids more easily.

Arc and spark excitation sources are still widely used in the analysis of solid materials by emission spectroscopy, especially if the solids are difficult to dissolve or rapid analysis for quality control and production is required. They are still the mainstay of analysis in foundries, where samples are easily cast into electrodes and the range of compositions analyzed is well known.

7.3. PLASMA EMISSION SPECTROSCOPY

A *plasma* is a form of matter that contains a significant percentage ($>1\%$) of electrons and ions in addition to neutral species and radicals. Plasmas are electrically conductive and are affected by a magnetic field. The plasmas used in emission spectroscopy are highly energetic, ionized inert gases. The most common plasma in commercial use is the argon ICP. Two other commercial plasma sources are the DCP, also usually supported in argon and the helium microwave induced plasma (MIP). The temperature of a plasma excitation source is very high, from 6500 to 10,000 K, so almost all elements are atomized or ionized and excited to multiple levels. The resulting emission spectra are very line-rich, which necessitates the use of high-resolution spectrometers to avoid spectral overlap.

7.3.1. Instrumentation for Plasma Emission Spectrometry

7.3.1.1. Excitation Sources

RF ICP Source. In ICP-OES, the sample is usually introduced to the instrument as a stream of liquid. The sample solution is nebulized and the aerosol transported to the plasma. In the plasma the sample undergoes the same process outlined in Table 7.2. The argon plasma serves to atomize, ionize, and excite the elements in the sample. The emitted radiation is sorted by wavelength in a spectrometer and the intensity is measured at each wavelength. A schematic of a typical ICP-OES system is presented in Fig. 7.21.

In Fig. 7.21, the argon plasma is the flame-like object at the top of the ICP torch, above the coils (the dark lines) from the RF generator. Figure 7.22 depicts a cross-section of a typical ICP torch. This torch contains three concentric tubes for argon flow and sample aerosol introduction. The two outer tubes are normally made of quartz. The inner tube, called the injector tube, may be made of quartz, alumina, sapphire, or other ceramic. Surrounding the torch is a water-cooled copper load coil or induction coil, which acts as an antenna to transmit power to the plasma gas from the RF generator. The power required to generate and sustain an argon plasma ranges from 700 to 1500 W. RF generators on older commercial instruments operated at 27.12 MHz; most modern instruments operate at 40.68 MHz, which results in better coupling efficiency and lower background emission intensity. The plasma initiation sequence is depicted in Fig. 7.23. When RF power is applied to the load coil, an alternating current oscillates within the coil at the frequency of the generator. The oscillating electric field induces an oscillating magnetic field around the coil. Argon is swirled through the torch and slightly ionized using a Tesla coil. The few ions and electrons formed are immediately affected by the magnetic field. Their translational motion changes rapidly from one direction to the other, oscillating at the same frequency as the RF generator. The rapid movement in alternating directions is *induced* energy (adding energy in this manner is called *inductive coupling*). The high-energy electrons and ions collide with other argon atoms and cause more ionization. This continues in a rapid chain reaction to convert the argon gas into a plasma of argon ions, free electrons, and neutral argon atoms.

The ICP torch is designed with narrow spacing between the two outermost tubes, so that the gas emerges at a high velocity. The outer tube is designed so that the argon flow in

Figure 7.21 Major components and layout of a typical ICP-OES instrument. [From Boss and Fredeen, courtesy of PerkinElmer Inc. (www.perkinelmer.com).]

Plasma

Emission region

Induction coil

Magnetic field

Quartz tubes

Argon tangential flow

Auxiliary gas flow

Auxiliary gas flow

Sample or Nebulizer gas flow

Figure 7.22 Cross-section of an ICP torch. The emission is viewed in the region above the coils when the torch is in an axial position.

this tube, called the plasma flow, follows a tangential path as shown in Fig. 7.22. This flow keeps the quartz tube walls cool and centers the plasma. A typical flow rate for the plasma flow is 7–15 L argon/min. The argon flow in the middle channel is called the auxiliary flow and can be 0–3 L argon/min. The auxiliary gas flow serves several purposes, including that of reducing carbon deposits at the injector tip when organic solvents are being analyzed. The gas flow that carries the sample aerosol into the plasma goes through the center or injector tube. It is called the nebulizer flow or sample flow and is typically about 1 L/min. The tangential or radial flow spins the argon to create a toroidal or doughnut-shaped region at the base of the plasma through which the sample aerosol passes. The temperatures for various regions of the plasma are shown in Fig. 7.24(a). Immediately

Figure 7.23 Cross-section of an ICP torch and the load coil depicting an ignition sequence. (a) Argon gas is swirled through the torch. (b) RF power is applied to the load coil. (c) A spark produces some free electrons in the argon. (d) The free electrons are accelerated by the RF field causing more ionization and formation of a plasma. (e) The nebulizer flow carries sample aerosol into the plasma. [From Boss and Fredeen, courtesy of PerkinElmer Inc. (www.perkinelmer.com).]

above the load coil, the background emission is extremely high. The background signal drops with distance from the load coil, and emission is usually measured slightly above the load coil, where the optimum signal to background ratio is achieved. This area is called the "normal analytical zone", shown in Fig. 7.24(b).

The advantage of the argon ICP as an excitation source lies in its high temperature and its stability. The gas temperature in the center of the plasma is about 6800 K, which permits the efficient atomization, ionization, and excitation of most elements in a wide range of samples. In addition, the high temperature reduces or eliminates many of the chemical interferences found in lower temperature electrical sources and flames, making the ICP relatively free from matrix effects. Another important advantage of the ICP is that the sample aerosol is introduced through the center of the plasma and is exposed to the high temperature of the plasma for several milliseconds, longer than in other excitation sources; this contributes to the elimination of matrix effects. Elements are atomized and excited simultaneously. The stability of the ICP discharge is much better than arc or spark discharges, and precision of less than 1% RSD is easily achieved. The dynamic range of an ICP source is approximately four to six orders of magnitude. It is often possible to measure major, minor, and trace elements in a single dilution with an ICP source.

DCP. The design of the DCP source is shown in Fig. 7.25. Two jets of argon issue from graphite anodes. These join and form an electrical bridge with the cathode, which is made of tungsten. In operation, the three electrodes are brought into contact, voltage is

Figure 7.24 (a) Temperature regions in a typical argon ICP discharge. (b) Zones of the ICP discharge. IR, induction region; PHZ, preheating zone; IRZ, initial radiation zone; NAZ, normal analytical zone. [From Boss and Fredeen, courtesy of PerkinElmer Inc. (www.perkinelmer.com).]

Figure 7.25 The inverted Y configuration of a direct current plasma jet. [Courtesy of Thermo-Elemental, Franklin, MA (www.thermo.com).]

applied, and the electrodes are then separated. Ionization of the argon occurs and a plasma is generated, forming a steady discharge shaped like an inverted letter Y.

As shown in Fig. 7.25, the region that is monitored for analytical measurements is at the junction of the anode argon streams. This small region gives good emission intensity from analyte atoms and has a background lower than the regions in the immediate vicinity. The sample is injected into this region with a stream of argon from a separate injection system. In the excitation region, the effective electronic temperature is about 5000 K. This results in spectra that are simpler than ICP emission spectra, with fewer ions lines. Sensitivities are within an order of magnitude of those reported for ICP sources for most elements and for the alkali metals and alkaline earth elements are as good as ICP. The power supply is less expensive and the quantity of argon required for DCP operation is much less than what is necessary to operate an ICP. A DCP can be operated for several days on a single cylinder of argon gas; an ICP would use the cylinder up in less than 8 h. Because of the way sample is introduced into the plasma, the DCP can analyze solutions with a high dissolved solids content better than some ICP sample introduction systems. There are several disadvantages to the DCP source. The graphite electrodes must be replaced frequently and the wearing away of the electrodes during analysis contributes to long-term drift in the signal. The dynamic range of a DCP is about three orders of magnitude, less than that of an ICP. The residence time of sample in the plasma is short because the plasma is small compared to an ICP source, making it difficult to atomize and excite highly refractory elements.

MIP. The MIP operates at lower power than the ICP and at microwave frequencies instead of the radiofrequencies used for ICP. Because of the low power, an MIP cannot desolvate and atomize liquid samples. Therefore, MIPs have been limited to the analysis of gaseous samples or very fine (1–20 μm diameter) particles. Helium is the usual plasma gas for an MIP source. A helium MIP has been used as an element-specific detector for GC. This detector is shown in Fig. 7.26 and the use of it for GC will be covered in greater detail in Chapter 12. The effluent from the GC column consists of carrier gas and separated gas-phase chemical compounds. The separated compounds flow through the plasma contained in the discharge tube shown. A compound in the plasma is decomposed, atomized, excited, and emits the wavelengths characteristic of the elements present. The light from the plasma is sent to a grating monochromator with a PDA detector, as shown. Another unique commercial instrument that uses a helium MIP as the excitation source is a particle analysis system designed to both count and identify the chemical composition

Figure 7.26   Atomic emission detector for GC using a helium MIP as the excitation source. This is Model G2350A, Agilent Technologies. It uses a scanning diffraction grating spectrometer and PDA detector to measure atomic emission from samples excited in the MIP. The wavelength range covered is 171–837 nm. [Courtesy of Agilent Technologies, Palo Alto, CA (www.agilent.com).]

of particles. Particles that have been collected on a filter are "vacuumed" into the He MIP source, where the particles are atomized, excited, and emit the characteristic radiation from the elements present. The instrument is discussed in Section 7.5. Graphite furnaces, hydride generation instruments, and other devices have been used to generate gas-phase samples or desolvated particles for introduction into MIPs.

Electronic excitation temperatures in a helium MIP are on the order of 4000 K, permitting the excitation of the halogens, C, N, H, and other elements that cannot be excited in a flame atomizer. The lower temperature results in less spectral interference from direct line overlap than in ICP or high-energy sources, but also causes more chemical interference.

7.3.1.2. Spectrometer Systems for ICP and DCP

Spectrometer systems for ICP and DCP include all of the dispersive devices and designs discussed in Chapter 2 and earlier in this chapter. Scanning monochromators of all optical designs are used in *sequential* spectrometer systems. In a sequential instrument, one wavelength at a time is measured, and the grating must be moved if more than one element is to be determined. Scanning rates can be very rapid in commercial instruments, but sequential instruments are slower than *simultaneous* spectrometers. Simultaneous systems contain either a polychromator or an Echelle spectrometer, and measure multiple wavelengths at the same time as has been described for arc/spark instruments. *Combination instruments* are available, with both polychromators and a monochromator, to take advantage of the speed of the simultaneous system for routine work, but to have the flexibility of the sequential system if a nonroutine wavelength needs to be accessed. An example of the optical layout of a combination instrument is shown in Fig. 7.27.

Figure 7.27 A combination sequential–simultaneous ICP emission spectrometer. Such a combination permits rapid multielement analysis using the polychromator and preselected wavelengths. The monochromator adds the flexibility to monitor additional elements or alternate wavelengths in case of spectral interferences. [Courtesy of Jobin Yvon, Inc., Horiba Group, Edison, NJ (www.jyhoriba.com).]

The detectors used in ICP and DCP systems include PMTs, CCDs, and CIDs. One variation of the CCD used for ICP is the segmented array CCD or SCD. The SCD has individual small subarrays positioned on a silicon substrate in a pattern that conforms to the echellogram pattern (Fig. 7.28). More than 200 subarrays are used; they cover approximately 236 of the most important wavelengths for the 70 elements routinely measured by ICP emission spectrometry. The SCD differs from a standard CCD in that the individual subarrays can be rapidly interrogated in random order (much like a CID). The SCD detector responds from 160 to 782 nm.

The DCP source is oriented in only one configuration within the spectrometer. The ICP source had traditionally been viewed from the side of the plasma, as shown in Fig. 7.21. This configuration is called "radial viewing". However, looking along the plasma axis instead of across the diameter provides a longer optical path length and that should result in lower detection limits. Unfortunately, the longer optical path includes more temperature zones, so more chemical interference results. Radial viewing was chosen to avoid the problems of increased spectral interference, matrix interferences, and self-absorption in the cooler tip of the plasma that would result from axial viewing. Axial viewing of the plasma does, however, provide improved detection limits for many elements. Axial or end-on viewing of the plasma was made practical by use of a flow of argon, called the shear gas, to cut off the cool tip of the plasma, as shown in Fig. 7.29. This reduces self-absorption and matrix or chemical interferences. There are commercial ICP systems available with only radial viewing, with only axial viewing, with simultaneous radial and axial viewing (dual view) and with the ability to flip the torch into either radial or axial position. The advantage of the axial view mode, improved detection limits, is especially important for environmental laboratories, which may be able to run analyses by ICP alone, instead of having to use both ICP and GFAAS instruments to meet regulatory detection limits for elements such as Se and As. A comparison of axial and radial detection limits for some elements, as well as a comparison of various nebulizers is presented in Table 7.12.

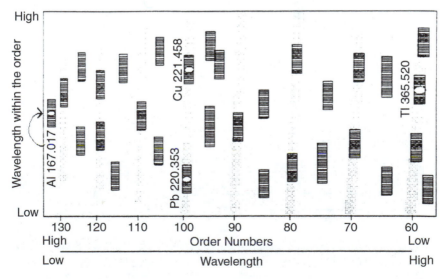

Figure 7.28   SCD detector. [From Boss and Fredeen, courtesy of PerkinElmer Inc. (www.perkinelmer. com).]

Figure 7.29 Axial (end-on) ICP torch position showing how the shear gas cuts off the cool plasma tail plume. Light emitted from the plasma passes through the ceramic interface into the spectrometer on the right side of the diagram (not shown). [From Boss and Fredeen, courtesy of PerkinElmer Inc. (www.perkinelmer.com).]

7.3.1.3. Sample Introduction Systems

Liquids are the most common form of sample to be analyzed by plasma emission. These are usually introduced with a nebulizer and spray chamber combination, similar to that used for AAS. An aerosol is formed and introduced into the plasma by the nebulizer gas stream through the injector tube. Nebulizers and spray chambers come in a variety of designs to handle aqueous solutions, high salt (high total dissolved solids) solutions, HF-containing solutions, and organic solvents.

There are sample introduction systems that can handle slurries of particles suspended in liquids. Powders can be injected directly into the plasma for analysis. Lasers, sparks, and graphite furnaces (exactly the same as AAS graphite furnaces) are used to generate gaseous samples from solids for introduction into the plasma. Hydride generation for As and Se and cold-vapor Hg introduction are used for ICP as for AAS; these two techniques were discussed in Chapter 6.

Liquid sample introduction. Solutions are introduced into the plasma through a nebulizer. The nebulizer is usually connected to a spray chamber. While some nebulizers discussed below can draw in the solution by self-aspiration, like the nebulizers used in AAS, it is common to pump the sample solution at a fixed rate into all nebulizers for ICP and DCP measurements. A pump, nebulizer and spray chamber arrangement are shown schematically in Fig. 7.21. Use of a pump eliminates sample uptake changes due to viscosity differences and permits rapid washing out of the nebulizer and spray chamber using a fixed fast flow rate in between samples. The type of pump used in most ICP systems is a peristaltic pump. This pump uses a series of rollers that push the sample solution through tubing. Only the tubing comes into contact with the sample. Peristaltic pump tubing must be compatible with the sample being pumped and most instrument manufacturers provide different types of tubing to handle highly acidic solutions, various organic solvents, highly basic solutions, and so on. The analyst should check the chemical compatibility of a given polymer tubing with the samples to be run through it and should also check the tubing for leachable elements by running blanks before analyzing samples. Some polymers have traces of metal catalysts that might cause erroneous results to be obtained if the catalyst metal is one of the analytes. Silicone rubber tubing should not be used if determination of silicon in samples is required.

Table 7.12 Detection Limit Comparisons of Radial and Axial ICP with Pneumatic and Ultrasonic Nebulization

Element	Wavelength (nm)	Radial[a] Pneumatic nebulizer	Radial[a] Ultrasonic nebulizer	Axial[b] Pneumatic nebulizer	Axial[b] Ultrasonic nebulizer
Ag	328.068	2	0.3	1	0.03
Al	396.153	9	0.5	2	0.06
As	193.69	18	2		
As	188.979			3	0.7
Ba	233.527			0.5	0.01
Be	313.107	0.3	0.02	0.1	0.009
Bi	233.061			2	0.2
Ca	317.933			2	0.03
Cd	214.43	2	0.2		
Cd	228.802			0.1	0.02
Co	228.616	5	0.3	0.2	0.02
Cr	267.716	4	0.2	0.2	0.01
Cu	324.754	2	0.3	0.6	0.02
Fe	238.204			0.1	0.02
Mg	285.213			0.5	0.06
Mn	257.610			0.1	0.03
Mo	202.031	8	0.3	0.6	0.3
Ni	231.604	10	0.4	0.4	0.06
Pb	220.353	27	1	2	0.2
Sb	206.836	12	3	2	0.3
Se	196.026	20	1.3	3	0.5
Sn	189.927			1	0.4
Ti	334.940			0.2	0.006
Tl	190.801	22	5	2	0.5
V	292.40	2	0.2		
V	290.880			2	0.02
Zn	213.857	2	0.2	0.2	0.03

Note: The ultrasonic nebulizer used in both cases was the CETAC U-5000AT+. The pneumatic cross-flow nebulizer used for radial ICP was the Thermo Jarrell Ash ICAP 61. The pneumatic cross-flow nebulizer used for axial ICP was the GemCone™ PerkinElmer 4300DV.
Source: Data courtesy of CETAC Technologies, Omaha, NE (www.cetac.com).
[a]Radial detection limits based on 3σ, 10 s integration time.
[b]Axial detection limits based on 3σ, 20 s integration time.

Nebulizers. The nebulizer converts a liquid into an aerosol of small droplets that can be transported to the plasma. Two approaches for aerosol formation are used commercially for ICPs, pneumatic nebulization, and ultrasonic nebulization. Pneumatic nebulizers use high-speed gas flows to create the aerosol. There are three commonly used pneumatic nebulizers: the concentric nebulizer, the cross-flow nebulizer, and the Babington nebulizer. Pneumatic nebulizers produce a range of aerosol droplet sizes. Large droplets cannot be efficiently desolvated, so the aerosol is passed through a *spray chamber* to remove large droplets. As is the case with AAS, less than 5% of the original sample liquid actually reaches the plasma.

The *concentric nebulizer* works by having the liquid solution introduced through a capillary tube into a region of low pressure created by a concentric gas flow past the tip of the capillary at high speed. Concentric means that the flow of liquid and flow of gas are parallel to each other. The liquid breaks up into a fine mist as it exits the capillary tip. The typical nebulizer gas flow rate is about 1 L/min. The concentric nebulizers have very small orifices, which result in highly efficient aerosol formation; this results in excellent sensitivity and stability. The most common glass concentric nebulizer is the Meinhard® nebulizer, shown in Fig. 7.30. The glass Meinhard design is easily clogged by solutions containing as little as 0.1% dissolved solids; the solids dry on the capillary tip and block it. The glass nebulizer cannot be used with HF-containing solutions, since hydrofluoric acid will dissolve the glass. High-solids concentric nebulizers are available that can run solutions with up to 20% dissolved solids without clogging. Nebulizers made of polymer are also available for HF solutions. The sample uptake rate of a standard concentric nebulizer is about 1–3 mL solution/min.

The *cross-flow nebulizer* uses a high-speed gas flow perpendicular to the sample flow capillary tip, as shown in Fig. 7.31(a), to form the aerosol. Droplets formed by cross-flow nebulizers are larger than those from concentric nebulizers, so the cross-flow nebulizer is not as efficient. However, the sample capillary is a larger diameter and this minimizes clogging of the nebulizer by salt deposits. The cross-flow nebulizer has an inert polymer body and sapphire and ruby tips on the capillaries; it is more rugged and chemically resistant than a glass nebulizer.

The *Babington nebulizer* was originally developed to aspirate fuel oil. A variation of the Babington nebulizer, called a *V-groove nebulizer*, is commonly used for ICP sample introduction. A V-groove nebulizer is shown in Fig. 7.31(b). The liquid sample flows down a smooth-surfaced groove in which there is a small hole. The high flow of gas from the hole shears the film of liquid into small drops. This nebulizer is the least prone to clogging and is used for viscous solutions, high salt solutions, and can be made of polymer for use with HF solutions.

Microconcentric nebulizers have been developed that put solution directly in the ICP torch sample capillary, thereby eliminating the need for a spray chamber. These nebulizers, also called direct insertion nebulizers (DIN), can be used for very small (microliter) sample volumes and for direct coupling of liquid chromatographs to the ICP as an element-specific detector.

Ultrasonic nebulizers (USN) operate by having the liquid sample pumped onto an oscillating quartz plate. The frequency of oscillation of the quartz plate for a commercial USN is 1.4 MHz, for example. The rapid oscillations of the quartz plate break the liquid film into an aerosol. This aerosol is of very fine, uniform droplets, unlike the aerosol from a pneumatic nebulizer. The fine aerosol contains a significant amount of solvent that must be

Figure 7.30 Meinhard glass concentric nebulizer, showing a close up of the tip.

Figure 7.31 (a) A cross-flow nebulizer, with a close up of the nebulizer tips. (b) A V-groove nebulizer. [From Boss and Fredeen, courtesy of PerkinElmer Inc. (www.perkinelmer.com).]

removed before the aerosol reaches the plasma. The aerosol flows through a desolvation unit consisting of a heated section to vaporize the solvent and then a chiller to condense the solvent so that a dry aerosol reaches the plasma. The efficiency of an ultrasonic nebulizer is 5–10 times greater than that of a pneumatic nebulizer. This results in improved sensitivities and better detection limits because more sample reaches the plasma. USNs have good stability, but are prone to carry-over or memory effects from one sample to the next. Carry-over is caused by one sample not being washed out of the nebulizer completely before the next sample is introduced. It is often related to the ease of adsorption of certain elements on nebulizer surfaces and is a problem if the concentrations of analyte in samples vary from trace to major. USNs cannot handle high total dissolved solids solutions and are not HF-resistant. They are very good at removing solvent from the sample before it reaches the plasma. Organic solvents can present problems in analysis by ICP because they tend to extinguish the plasma. Also, molecular species and radicals emit broad-band spectra, which give rise to high background radiation. The USN removes the solvent by cryogenic trapping in the chiller. A commercial USN is shown schematically in Fig. 7.32. Comparative detection limits for some elements determined using this nebulizer are presented in Table 7.12.

Figure 7.32 Schematic of a commercial USN, the CETAC U-5000AT + and desolvation system. [Courtesy of CETAC Technologies, Omaha, NE (www.cetac.com).]

Spray chambers. Pneumatic nebulizers generally require a spray chamber placed in between the nebulizer and the plasma. The spray chamber is designed to remove large droplets from the aerosol, because large droplets are not efficiently desolvated, atomized, and excited. In addition, the spray chamber smoothes out pulsations in the aerosol flow that arise from the pumping action of the peristaltic pump. Droplets with diameters of 10 μm or less generally pass through the spray chamber to the plasma. Only about 5% of the sample solution makes it into the plasma; the rest is drained to waste. The spray chamber shown in Fig. 7.21 is a Scott double pass spray chamber. New more efficient cyclonic spray chambers are now available. The spray chamber can be made of glass or inert polymer depending on the application. Water-cooled spray chambers may be used for volatile organic solvents, to decrease the amount of solvent vapor entering the plasma. Figure 7.33(a) shows the individual parts needed for aqueous liquid sample introduction. A three part demountable torch of quartz is shown, with a Tracey cyclonic spray chamber and a glass concentric nebulizer. Figure 7.33(b) shows the components assembled for use. The fittings are for argon gas connection to provide the auxiliary and plasma gas flows. The demountable quartz torch parts are held in a polymer torch body.

Solid sampling systems. The direct analysis of solid materials is the goal of every analyst who must deal with solid samples. Dissolution of solid samples is time consuming and error prone. There have been devices designed by researchers to place solid samples directly into the argon plasma of an ICP for analysis but these have not been commercialized with the exception of the slurry nebulizer. A slurry is a suspension of fine particles in liquid. Samples that are powders or can be finely ground into powder can be suspended in water or an organic solvent and aspirated directly into the plasma. Slurry nebulizers are available in both concentric and V-groove types. Two of these available from Glass Expansion Pty. Ltd. (www.geicp.com) are the Slurry® concentric nebulizer for analysis of samples with particles up to 150 μm in diameter and the VeeSpray® nebulizer, a V-groove type in either quartz or HF-resistant ceramic that can handle particulates up

(a)   (b)

Figure 7.33   (a) The components of a liquid sample introduction system for ICP: glass concentric nebulizer, demountable three piece quartz torch, torch body, and glass cyclonic spray chamber. (b) The components assembled for use. This torch design is typical of that used in Jobin Yvon ICP instruments. [Courtesy of Glass Expansion Pty, Ltd., Australia (www.GEICP.com).]

to 300 μm in diameter. Slurry introduction is used for analysis of suspensions of cement powder and milk powder, for example.

The other commercial solid sampling systems all vaporize a portion of the solid sample first; the vapor or fine particulate is transported to the plasma where it is atomized and excited. The three techniques commercially available for solid sampling are electrothermal vaporization (ETV), spark ablation, and laser ablation. The advantage of separating the sampling step (using the ETV, spark or laser) from the atomization and excitation step (in the ICP) is that each process can be optimized and controlled separately. The disadvantage of direct solid sampling is calibration of the system to achieve quantitative results. It is often necessary to purchase a series of well-characterized solids similar to the sample matrix as standards; stainless steel alloys for stainless steel samples, nickel alloys for nickel alloy samples, certified glass standards for glass samples, and so on. There are limits as to the kinds of certified solid reference materials available and they are generally very costly.

ETV.   ETV techniques adapted from graphite furnace AAS have been successfully applied as a means of introducing the vapor from a solid sample into plasma. The technique is particularly useful for organic matrices such as polymers and biological tissues since it eliminates the need to digest the sample. The electrothermal vaporizer follows the same type of temperature program used in AAS, drying, ashing, and "atomizing" the sample in sequence, although in this case, vaporization is really all that needs to be accomplished since the ICP also serves as an atomizer. The vapor formed is swept into the ICP by a flow of gas, where it is atomized and excited. The method can be used with small volume liquid samples and solid samples. About 10–50 μL of liquid or a few milligrams of solid sample or slurry can be analyzed in this fashion. Detection limits for electrothermal vaporization-ICP-OES are given in Table A2, Appendix 7.1. The same table compares GFAAS detection limits and the use of ETV with ICP-MS. The ICP-MS method will be discussed in Chapters 9 and 10.

Spark ablation. Spark ablation solid sampling uses the same type of spark source already described. The function of the spark in this case is to vaporize the solid sample; the ICP plasma can atomize any nonatomic vapor reaching it. Spark ablation is limited to the analysis of solids that conduct electricity. It is very useful for metals and alloys because it eliminates time-consuming sample dissolution and costly high-purity acids.

Laser ablation. Another technique for direct sampling of solids is laser ablation. A high-energy laser beam is directed at the solid sample. The photon energy is converted to thermal energy on interaction with the sample. This thermal energy vaporizes surface material and the vaporized material is swept into the plasma. Lasers can vaporize most types of solid surfaces, not just conductive materials, so laser ablation can be used to sample metals, glass, ceramics, polymers, minerals, and other materials. The sample does not have to be of any particular shape or size, as long as it fits into the sample chamber and a flat polished surface is not required. Laser beams can be focused to small spot sizes, so microscopic inclusions in a surface can be analyzed. The laser beam can be moved across the surface (a process called "rastering") for bulk analysis of a material. The laser beam can be focused on one position on the sample and the emission signals measured as material is vaporized from deeper and deeper below the surface, creating a *depth profile* of the sample composition. A depth profile is useful for the analysis of coatings on surfaces and multilayered composite materials, for example. Gas-filled excimer lasers such as XeCl and ArF, which emit in the UV region, can be used for laser ablation, but excimer lasers are costly and complex. Recent advances in solid-state laser design have resulted in a commercial Nd:YAG solid-state laser. This solid-state laser can be operated at its primary wavelength, 1064 nm, which is in the IR region, or can be frequency doubled or quadrupled to make lasers that operate at 532 or 266 nm. The UV solid-state laser shows efficient coupling of energy with most types of materials, high sensitivity, good precision, and controlled ablation characteristics. With ICP-OES, the laser ablation technique is used primarily for bulk analysis of solid samples, although depth profiling of materials is possible. The analysis of inclusions, multilayered materials, and microscopic phases in solids is often done by laser ablation and the more sensitive technique of ICP-MS (Chapters 9 and 10). A schematic of a commercial UV Nd:YAG laser system for laser ablation is shown in Fig. 7.34.

7.3.2. Interferences and Calibration in Plasma Emission Spectrometry

Calibration is achieved by the same techniques used for AAS: external calibration with standard solutions or MSA. Plasma emission does not suffer from significant chemical interference because of the high temperatures achieved in the plasma, but it is good practice to matrix-match the standard solutions and calibration blank to the samples to be analyzed. Matrix-matching can correct for differences in viscosity and suppression or enhancement of analyte signal due to interelement effects. Matrix-matching enables accurate background corrections to be made. The same solvent should be used in standards and samples, the same concentration of acid, and when possible, the same major chemical species in solution. For example, in the analysis of tungsten metal for trace elements, if the sample solutions contain 100 mg of tungsten, the standards and calibration blank should contain 100 mg of high purity tungsten. Standards for the analysis of steels should contain the appropriate amount of high purity iron, and so on.

Figure 7.34 A commercial laser ablation system for solid samples, the CETAC Technologies LSX-500. [Courtesy of CETAC Technologies, Omaha, NE (www.cetac.com).]

Most samples for analysis by plasma emission spectrometry are dissolved in acids, because metal elements are more soluble in acids. The acidic nature of the solution prevents elements from adsorbing onto the surface of the glassware. Some bases, such as tetramethyl ammonium hydroxide (TMAH), are used to prepare solutions but care needs to be taken to avoid the formation of insoluble metal hydroxides. Organic solvents may be used to dissolve organometallic compounds directly or to extract chelated metals from aqueous solution into the solvent as part of separation and preconcentration steps in sample preparation.

Calibration standards are made by dissolving very high-purity elements or compounds accurately in aqueous acid solution. In general, a concentrated stock standard is prepared, and then serially diluted to prepare the working calibration standards. Many vendors now supply 1000 or 10,000 ppm single element standards in aqueous acidic solution, with certified concentration values. It is easier, more accurate, and more efficient to purchase these stock solutions from a high-quality supplier than to make them from scratch. Because emission is a multielement technique, the analyst combines single element standards to make the necessary working standings that contain all of the analytes of interest. If the same set of multielement standards is used routinely, these can be ordered from the standards vendors. Multielement solutions for environmental analyses are stocked by vendors, for example. Custom mixtures can be ordered for routine analyses in industry.

For most analyses, standard solutions are aqueous solutions containing several percent of a mineral acid such as HCl or nitric acid. When mixing different elements in acids, it is important to remember basic chemistry and solubility rules for inorganic compounds. The elements must be compatible with each other and soluble in the acid used so that no precipitation reactions occur. Such reactions would change the solution concentration of the elements involved in the reaction and make the standard useless. Combinations to be avoided are silver and HCl, barium and sulfuric acid, and similar combinations that result in insoluble precipitates. The use of certain reagents, such as HF or strong bases, can etch glass and contaminate solutions prepared in glass with Si, B, and Na. HF is required to keep certain elements such as tungsten and silicon in solution, so the analyst needs to be aware that purchased stock solutions of these types of elements contain small amounts of HF that can etch glassware and nebulizers.

Because emission is inherently a multielement technique, accuracy and precision can be improved through the use of an *internal standard.* The internal standard is an element that is not present is the samples, is soluble in the sample and standard solutions, and that does not emit at a wavelength that interferes with the analyte emission lines. Ideally, the internal standard element should behave similarly to the analyte in the plasma; matching the ionization energies of the analyte and internal standard is a good approach. The use of internal standard ion lines to match analyte ion lines and internal standard atom lines to match analyte atom lines is also a good practice. In a multi-element analysis, it may be necessary to use more than one internal standard element to achieve this "ideal" behavior. The internal standard element is added in the same concentration to all blanks, standards, and samples. The emission intensities of the analyte elements and the internal standard element are measured. The calibration curve is plotted as the ratio of the analyte intensity to the emission intensity vs. the ratio of analyte concentration to internal standard concentration. Use of the ratio of intensities corrects for instrument drift, sample uptake rate changes, and other instrumental sources of error. Use of an internal standard does not correct for background interferences or spectral interferences.

7.3.2.1. *Chemical and Ionization Interference*

Chemical interferences are rare in plasma emission spectroscopy because of the efficiency of atomization in the high-temperature plasma. For example, in flame AAS, there is a severe suppression of the calcium signal in samples containing high amounts of Al, as seen in Fig. 7.35. The same figure shows the lack of chemical interference in the ICP emission signal for Ca in solutions containing Al up to very high Al/Ca ratios. In the few cases where chemical interference is found to occur, increasing the RF power to the plasma and optimizing the Ar flow rates usually eliminates the problem.

The alkali metals Li, Na, and K are easily ionized in flames and plasmas. Concentrations of an easily ionized element (EIE) greater than 1000 ppm in solution can result in suppression or enhancement of the signals from other analytes. The EIE effect is wavelength dependent; ion lines are affected differently than atom lines. The EIE effect can be minimized by optimizing the RF power and plasma conditions, by matrix-matching, by choosing a different wavelength that is not subject to the EIE effect, or by application of mathematical correction factors.

7.3.2.2. *Spectral Interference and Correction*

Spectral interference is much more common in plasmas than in flames due to the great efficiency of excitation in plasmas. Elements such as Fe, Mn, Ta, Mo, W, and U emit thousands of lines in a plasma source. Ideally, the analyte wavelength chosen should have no interference from other emission lines, but this is often not possible.

The most common type of spectral interference is a shift in the background emission intensity between samples and standards. Figure 7.36 shows a background shift that is a constant broad emission, from a solution containing only 1000 ppm Al, measured at the tungsten 207.911 nm emission line. The tungsten emission spectrum shown is from an

Figure 7.35 A comparison of the chemical interference effect of Al on the determination of Ca by flame AAS and ICP-OES. The flame AAS signal for Ca is severely depressed by small increases in the Al/Ca ratio, while the emission signal from the high temperature plasma is not significantly affected. [From Boss and Fredeen, courtesy of PerkinElmer Inc. (www.perkinelmer.com).]

aqueous acid solution containing no Al. The signal intensity increases from bottom to top in the Figure. The spectral window being examined is 0.25 nm wide, centered on the tungsten emission line. This type of background emission difference can result in two positive errors. If we were trying to measure W in an Al alloy and used aqueous W standards (no Al), the continuum emission from the Al solution would result in a positive error in a determination of tungsten if not corrected. We would be measuring both the background and the

Figure 7.36 A simple background shift caused by 1000 ppm Al at the tungsten 207.911 nm line. A 1000 ppm Al solution would give an erroneous positive reading for tungsten due to the increased background from the aluminum solution. (The baseline of the tungsten solution on either side of the W emission line can be assumed to be zero intensity.) [From Boss and Fredeen, courtesy of PerkinElmer Inc. (www.perkinelmer.com).]

tungsten signal, and assuming it was all due to tungsten. Another error could occur if we were determining tungsten in mineral samples and did not know that some samples had a high Al content. Even if these Al-containing samples had no tungsten in them, we would erroneously report that tungsten was present, because we were measuring the background intensity inadvertently. Because the background is shifted uniformly in this example, the background signal can be corrected for by picking a wavelength near but not on the tungsten emission line as shown in Fig. 7.37. The background correction point could be on either side of the tungsten peak in this case. The intensity at this correction point is measured and subtracted from the total intensity measured at the tungsten peak. Computer software is used to set the background point and automatically make the correction.

A more complicated background is seen in Fig. 7.38. The Cd 214.438 nm emission line is overlapped by a nearby broad Al emission line, so the background intensity is not the same on both sides of the Cd peak. If Cd were to be measured in an aluminum alloy, this asymmetrical background would result in a positive error. When the background emission intensity is asymmetrical, two background correction points are needed, one on either side of the emission peak, as shown in Fig. 7.39. The computer software uses the two chosen points to draw a new baseline for the peak and automatically corrects for the background emission.

Figure 7.40 is an example of a very complex background pattern at the Au 267.595 nm emission line caused by closely spaced emission lines from a line-rich matrix, tungsten. The two-point approach will not work here, because of the severe overlap by the intense tungsten line on the left side of the Au line. Modern instrument software is available with mathematical programs that will model or "fit" the background and then deconvolute complex backgrounds. If such spectral deconvolution software is not available, the first choice of the analyst should be to find an alternate gold emission line with less interference. If there are no lines without significant interference, it may be necessary to separate the analyte from the matrix or vice versa using classical "wet" chemistry, chelation and extraction techniques, ion exchange methods, or electrochemical

Figure 7.37 An example of single-point background correction for the simple background shift caused by Al at the tungsten emission line. A point on the left-side of the tungsten emission is chosen and the intensity at this point subtracted from the intensity at the tungsten peak. This will have the effect of subtracting the increased background due to the Al matrix. The new baseline, shown as the dotted line, will return to the same baseline intensity (shown as the lower line) given by the tungsten solution in Fig. 7.36. Because the baseline is symmetric with respect to the emission line, a background point on the right side of the tungsten peak could have been chosen instead; the result would be the same. [From Boss and Fredeen, courtesy of PerkinElmer Inc. (www.perkinelmer.com).]

Figure 7.38 An aluminum background spectrum representing a sloping background shift at the Cd 241.438 nm line. This asymmetric background cannot be corrected using the one-point correction shown in Fig. 7.37, because the background intensity is different of each side of the emission line. [From Boss and Fredeen, courtesy of PerkinElmer Inc. (www.perkinelmer.com).]

methods. The chemical literature should be consulted for advice, since there are thousands of published papers on analysis by ICP-OES.

Figure 7.41 shows a case of direct spectral overlap between two emission lines, one from Pt and one from Cr. A high-resolution spectrometer will limit the number of direct spectral overlaps encountered, but not eliminate them entirely, as in this case. If the analyst has a choice of wavelengths, the first thing to do is to look for alternate emission lines for Pt and Cr. If no other analyte λ is available, the technique of *interelement correction* (IEC) may be used. The emission intensity of the interfering element at another wavelength is measured and a calculated correction factor is applied to the intensity at the wavelength that has the interference. The correction factor has to be determined through a series of prior experimental measurements. The use of the technique requires one interference-free emission line for each interfering element in the sample. This can be quite complicated and time consuming to set up, but is useful for the repetitive analysis of the same sample type when an interference-free analyte line cannot be found. Alternatively, the analyte or the interfering element may be extracted from the sample.

Figure 7.39 The sloping background shown in Fig. 7.38 requires the use of a two-point background correction, with one correction point on each side of the Cd emission line. The peak is corrected using a straight line fit between the background correction points as shown by the dotted line (the new baseline). [From Boss and Fredeen, courtesy of PerkinElmer Inc. (www.perkinelmer.com).]

Figure 7.40 The complex, line-rich emission spectrum of tungsten severely overlaps the Au emission line at 267.595 nm. The overlap is very severe on the left of the Au peak, so a two-point approach will not work for accurate determination of trace amounts of Au in tungsten. Finding a gold emission line with less interference or extracting the gold from the tungsten matrix is a possible solution to this problem. [From Boss and Fredeen, courtesy of PerkinElmer Inc. (www.perkinelmer.com).]

7.3.3. Applications of ICP and DCP Atomic Emission Spectroscopy

The applications of plasma emission spectrometry are very broad. The technique is used for clinical chemistry, biochemistry, environmental chemistry, geology, specialty and bulk chemical production, materials characterization of metal alloys, glasses, ceramics, polymers and composite materials, atmospheric science, forensic science, conservation and restoration of artworks by museums, agricultural science, food and nutrition science, industrial hygiene, and many other areas. The versatility of plasma emission spectrometry comes from its ability to determine a large number of elements rapidly in a wide variety of sample matrices.

The strength of plasma emission spectrometry is rapid, multielement qualitative and quantitative analyses. ICP emission spectrometry has a linear dynamic range of four to six orders of magnitude, meaning that elements from trace level to major constituents can

Figure 7.41 Direct spectral overlap of Pt and Cr emission lines. No background correction technique can solve this problem. A line with no interference must be found, an interelement correction factor must be applied or the elements must be separated chemically. [From Boss and Fredeen, courtesy of PerkinElmer Inc. (www.perkinelmer.com).]

usually be analyzed in one run. This eliminates the need for dilution of samples in many cases. By contrast, the linear range of AAS is usually one to two orders of magnitude. Precision of ICP measurements is usually better than 2% (relative) and with a minor amount of optimization, can often be better than 1% (relative). Aqueous solutions, organic solvent solutions and high salt solutions can be analyzed with virtually the same precision and accuracy.

Plasma emission is used in agricultural science to study elements in soils, plant tissue, animal tissue, fertilizers, and feed. Samples are usually acid-digested in a heating block or by using a microwave digestion system, or ashed in a muffle furnace and the ash dissolved in acid. The samples may be extracted instead of being totally dissolved; extraction results in the measurement of "available" elements. Such extracted available elements are usually considered to be biologically available to plants and animals, while digestion or ashing gives total element content. Applications include the determination of metals in beer, wine, infant formula, and fruit juice. It is possible to identify the country of origin of some food products based on their trace elemental "fingerprint".

Plasma emission is used to study elements in environmental and geological samples. The analysis of soil, water, and air for industrial pollutants is a common application. Contaminated soil and water can be analyzed as well as soil and water that have been treated to remove heavy metals, for example. This will verify if the treatment worked and provides the data engineers need to optimize and improve their removal processes. The soil and water content of naturally occurring potentially toxic metals such as As and Se can be studied as well as the uptake of As and Se by plants and animals grazing on the plants or living in the water. Particulates in air can be trapped on filters and the entire filter digested in acid or ashed prior to dissolution of the residue. This approach is used to study coal fly ash and Pb in house dust from lead-based paint. Geologists use ICP to measure major, minor, and trace components of minerals and rocks, to identify sources of metal ores, to study marine geochemistry, and to identify the origins of rocks. A picture of a mineral specimen, garnet, that has been sampled by laser ablation is shown in Fig. 7.42. Laser ablation permits the spatial distribution of elements in minerals to be studied. The surface of the garnet was studied by rastering the laser across the face of the sample. Two lines of 50 μm diameter laser spots can be seen.

Biological and clinical chemistry applications of plasma emission spectrometry include determinations of those metals required for proper functioning of living systems,

Screen capture of garnet after two spot traverses.
50 μm spots, 4 Hz, 200 shots. Crossed polarized light.

Figure 7.42 A garnet sample analyzed by laser ablation to determine spatial distribution of elements in the mineral. Two traverses of the laser beam can be seen. Each ablated spot is 50 μm in diameter. [Courtesy of CETAC Technologies, Omaha, NE (www.cetac.com).]

such as Fe, Cu, K, Na, P, S, and Se, in urine, blood, serum, bone, muscle, and brain tissue. Aluminum exposure was suspected of playing a role in Alzheimer's disease and Al concentrations in blood and tissue can be determined by emission spectrometry. No link between exposure to aluminum and Alzheimer's disease has been found. In addition to essential elements, plasma emission spectrometry can monitor metal-based therapeutic drugs in patients. Gadolinium compounds are used in patients undergoing some forms of MRI (discussed in Chapter 3) to enhance the contrast between normal tissue and tumors. The optimal dose of the Gd "contrast agent" can be determined from studies where the tissues are analyzed for Gd by ICP-OES (Skelly Frame and Uzgiris).

Plasma emission spectrometry is widely used to study metals, alloys, ceramics, glass, polymers, and engineered composite materials. The major difficulty with metals and alloys is the possibility of spectral interference from the rich emission spectra of the major metal elements, but high-resolution instruments permit the routine determination of trace elements in steels, nickel alloys, copper alloys, tungsten, and other refractory metals. Most metals and alloys can be dissolved in one or more acids for analysis. Elements like tungsten and silicon require HF for dissolution; it is therefore critical to use a polymeric sample introduction system for analysis, not a quartz nebulizer and spray chamber. Glass and ceramic samples are often fused in molten salt for dissolution; the resulting bead of molten salt is dissolved in acid and water for analysis. This fusion is done at high temperatures over a burner in a suitable crucible, often a Pt or Pt/Au crucible. There are commercial automatic fusion fluxers that will fuse up to six samples at once.

Polymers, oils, and other organic materials can often be analyzed by dissolution in an organic solvent. Lubricating oil and petroleum feedstock analysis was discussed in Section 7.2.4. Most oils and fuels are now analyzed by ICP or DCP instead of arc or spark emission instruments, due to the better precision that can be obtained. Lead in gasoline is determined by dilution of the gasoline in organic solvent and analysis of the solvent solution. Metals in cooking oils can be determined by dilution with organic solvent and analysis of the organic solution. Organophosphates and organochlorine compounds from environmental and agricultural samples can be determined after extraction into organic solvents. The silicon in silicone polymers can be measured by dissolution of the silicone in organic solvent and analysis of the solvent solution. Such a method can be used to measure the amounts of silicone used on the two components of self-stick labels; both the adhesive on the label and the release coating on the paper backing are organosilicon polymers. In general, the ICP is very tolerant of organic solvents. Toluene, xylene, hexane, tetrahydrofuran, acetonitrile, methylene chloride, and many other solvents work perfectly well in this type of analysis. In some cases, it may be necessary to increase the RF power to the plasma or to add a small amount of oxygen to the plasma to help with combustion of the organic matrix.

7.3.4. Chemical Speciation with Hyphenated Instruments

Plasma emission spectroscopy is an elemental analysis technique; that is, it provides no information on the chemical form or oxidation state of the element being determined. The identification of the chemical state of an element in a sample is called **speciation**. For example, in environmental samples, mercury may exist in a variety of species: mercuric ion, mercurous ion, methymercury compounds, and the extremely toxic compound dimethylmercury. Determination of mercury by ICP-OES results in total mercury concentration; chemical speciation would tell us how much mercury is present in each of the different forms. Arsenic is another element of environmental and health interest because of its toxicity. Arsenic, like mercury, exists in multiple organoarsenic compounds and multiple oxidation states as inorganic arsenic ions. Why is speciation so important? The most common arsenic compound found in shellfish is not metabolized by humans; therefore it is not toxic. Estimating toxicity from a determination of total arsenic concentration would vastly overestimate the danger of eating shellfish, for example. In order to perform chemical speciation, we need to separate the different species in the sample and measure each of them. Chromatography is a separation method; it permits different chemical compounds in a sample to be separated from one another by the processes and instruments described in Chapters 11–13. The use of chromatography to separate chemical species and then introduction of the separated species into an ICP sequentially can identify different chemical forms of the same element. An instrument combination formed by the connection of two separate types of instrumentation is called a hyphenated instrument. A very common type of hyphenated instrument is one that couples a separation technique such as chromatography with a spectroscopic technique such as NMR, FTIR, atomic emission spectrometry or MS. GC, HPLC, ion chromatography, and molecular exclusion chromatography instruments have all been interfaced to plasma emission spectrometers to identify chemical species in a sample.

For example, an HPLC coupled to a sequential ICP set to measure a silicon emission line has been used to separate and detect the small organosilicon molecules that result from the hydrolysis of high molecular weight silicone polymers (Dorn and Skelly Frame). The chromatograph separates the compounds in solution by the selective interaction of the compounds with the chromatographic **column stationary phase** and the **liquid mobile phase** that carries the sample through the column. The mobile phase with the now separated compounds flows from the column into the ICP. The ICP serves as the detector for this system; it detects and measures any silicon-containing compound that enters the plasma. This permits the study of silicone polymer degradation in a variety of environments, including degradation inside the human body. Leaking silicone polymer implants in humans have been suspected of causing serious health problems. The coupling of HPLC and ICP is one of the very few ways to separate and measure trace quantities of organosilicon compounds. This is only one of many applications of coupled chromatography-atomic emission spectrometry.

The use of a helium MIP as a detector for compounds separated by GC has been mentioned and will be discussed in Chapter 12 in greater detail.

7.4. GLOW DISCHARGE EMISSION SPECTROMETRY

The glow discharge (GD) is a reduced-pressure gas discharge generated between two electrodes in a tube filled with an inert gas such as argon. The sputtered atom cloud in a glow discharge source consists of excited atoms, neutral atoms, and ions. The emission spectrum can be used for emission spectrometry in the technique of GD-OES, but the GD source can also be used for AAS, atomic fluorescence spectrometry, and MS. The source can be used with any of the types of spectrometers discussed for plasma emission: sequential monochromator, Rowland circle polychromator, Echelle spectrometer, or combination sequential–simultaneous designs. The detectors used are the same as described for plasma emission spectrometry: PMTs, CCDs, or CIDs.

7.4.1. DC and RF GD Sources

A schematic DC GD source is shown in Fig. 7.43. The gas is present at a pressure of a few torr. The DC GD source can be operated with a DC potential of 800–1200 V applied between the electrodes. The sample is in electrical contact with and serves as the cathode as seen in Fig. 7.43. The applied potential causes spontaneous ionization of the

Figure 7.43 A DC GD source. A flat conducting sample serves as part of the cathode. [Courtesy of Jobin Yvon, Inc., Horiba Group, Edison, NJ (www.jyhoriba.com).]

Ar gas to Ar^+ ions. The Ar^+ ions are accelerated to the cathode and a discharge is produced by argon ions colliding with argon atoms. The resulting plasma is called a **glow discharge**. The electric field accelerates some Ar^+ ions to the sample surface, where impact causes neutral sample atoms to be *sputtered* or uniformly removed from the sample surface. Therefore, the GD source is both the sample introduction system and the atomizer. Dissociation, atomization, and excitation of the sputtered cathode material in the GD occurs by collision. The emission spectrum of the cathode, that is, of the sample, is produced. The sample must be machined to have a flat surface and must be electrically conductive. The Grimm GD source (also called a GD lamp) was designed so that a flat conductive sample served as the cathode. The cathode could be easily changed for analysis of different samples. The Grimm design results in lower self-absorption and low material redeposition on the sample surface.

A DC GD source can only be used to sputter conductive samples. Some applications to nonconductive materials were made by mixing the nonconductive sample powder with graphite or pure copper powder and pressing flat pellets. In the 1990s, a new GD source based on RF excitation was developed (Marcus and Winchester et al.). This RF GD source permits the sputtering and excitation of nonconductive materials, such as glass and ceramics. This source still has low Ar pressure and two electrodes, a planar sample as the cathode and a tubular copper anode facing the sample. The sample is subjected to an RF frequency of 13.56 MHz. The sample acts as a capacitor with respect to the alternating RF voltage. This results an alternating charge on the sample surface through each cycle. When the sample surface is negatively charged, the positive Ar^+ ions are accelerated to the negative surface and the sample material is sputtered into the glow discharge. The charge alternates very rapidly, so the sputtering process is very efficient. Excitation and emission occur as described for the DC source.

7.4.2. Applications of GD Atomic Emission Spectrometry

7.4.2.1. Bulk Analysis

The first application of GD-OES using a DC source was for direct multielement analysis of solid metals and alloys, much like a spark source. The bulk composition of the sample was determined. The DC GD source for analysis of conductive solids has several advantages over spark emission spectrometry. The spectra produced by the GD exhibit lower background and narrower emission lines than spark sources due to less Doppler broadening. Emission lines that appear in the spectrum are almost exclusively atom lines. Linear calibration curves are achieved over a wide concentration range, which decreases the number of standards required for calibration. The GD source exhibits much lower levels of interelement and matrix interferences than arc or spark sources, since the sputtering step and the excitation step are separate events. This allows the use of one set of calibration standards for different families of materials, because there is less need to matrix-match for accurate results. Detection limits are on the order of 0.1–10 ppm for most elements, with a precision of 0.5–2% RSD.

Applications now include the analysis of nonconductive materials such as polymers, ceramics, and glasses using the RF Marcus-type source. The advantages are similar to those just discussed, with a major improvement in detection limits and a decrease in sources of error. With a DC GD source, a nonconductive material had to be diluted with a conductive powder; this decreased the amount of analyte that could be detected in the sample. Use of an excess of conductive powder and the process of blending and pressing always introduced the possibility of contamination of the sample. This source of error has been eliminated by direct analysis of nonconductive materials.

7.4.2.2. Depth Profile Analysis

Because the GD sputtering process removes uniform layers of materials from the sample surface, it can be used to examine coatings and multilayered materials. The analysis of composition of a sample as a function of depth from the surface is called *depth profiling*. Uniform removal of the sample can be seen in Fig. 7.44, which shows a GD sputtered spot in a steel sample. Layers only nanometers thick can be analyzed quantitatively. Applications of depth profiling are many. Paint layers on automobile bodies, the thickness of the Zn layer on galvanized steel, carbon and nitrogen concentrations as a function of depth in nitrocarburized steel, the structure of multilayer semiconductor materials, nonconductive coatings of nitrides or carbides on alloys, thermal barrier coatings on alloys and many other complex samples can be easily characterized by depth profiling with GD-OES. Several examples are shown in Figs. 7.45–7.47. Many other examples and application notes can be found on the Jobin Yvon, Inc. website at www.jyhoriba.com and the LECO website at www.leco.com.

Figure 7.44 A GD "sputter spot" in a steel sample showing the uniform removal of the surface. It is this uniform sputtering process that permits accurate depth profiling of layered materials by glow discharge OES. [Courtesy of LECO Corporation, St. Joseph, MI (www.leco.com).]

Figure 7.45 A GD depth profile of an Al/Zn/Fe coating on steel. The *y*-axis is concentration in %, the *x*-axis is depth in micrometers from the surface. [Courtesy of Jobin Yvon, Inc., Horiba Group, Edison, NJ (www.jyhoriba.com).]

7.5. PARTICLE CHARACTERIZATION USING A HELIUM MIP SYSTEM

A helium microwave cavity plasma or MIP serves as the heart of a unique instrument dedicated to the analysis of particles with diameters between 1 and 20 μm. The analysis of small particles is becoming increasingly important in fields as diverse as the semiconductor industry where extremely low levels of contaminants can destroy semiconductor chips, to indoor and outdoor air quality and the health impact of respirable particles. This instrument is basically an atomic emission spectrometer, but designed to count particles, calculate particle size, and determine the chemical composition of each particle analyzed, one by one.

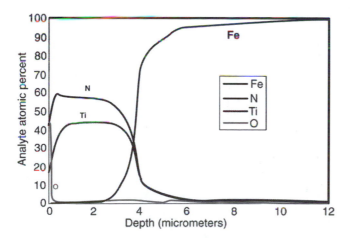

Figure 7.46 A GD depth profile of a titanium nitride coating on steel. TiN is a hard, brittle material often used to modify the surface of steel. The quantitative depth profile software on this system can verify the stoichiometry of the coating layer. [Courtesy of LECO Corporation, St. Joseph, MI (www.leco.com).]

Figure 7.47 A quantitative GD depth profile of a painted, galvanized steel sheet. This profile shows an initial layer containing Ti, Si, C, H, and O. The paint is white, so titania and silica are probable ingredients and latex paints would certainly contain hydrocarbons. An intermediate layer containing Sr and Cr is seen, then the third layer (with the Zn peak) shows the galvanization and the increasing Fe signal shows where the steel base begins. [Courtesy of Jobin Yvon, Inc., Horiba Group, Edison, NJ (www.jyhoriba.com).]

An atmospheric pressure helium MIP is generated using a 2.45 GHz microwave generator and an electromagnetic cavity resonator, called a Beenakker cavity. The helium gas is passed through a discharge tube placed in the cavity, as seen in Fig. 7.48. The plasma is initiated by a spark from a Tesla coil. The electrons produced by the spark oscillate in the

Figure 7.48 A helium MIP. [Courtesy of Horiba, Ltd. (www.horiba.com) and Yokogawa Electric Company.]

microwave field and ionize the helium gas by collision, producing a plasma. The microwave energy is coupled to the gas stream in the discharge tube by the external cavity. The plasma is centered in the discharge tube, and is represented in Fig. 7.48 by the shaded spheroid shape. The helium MIP has much higher excitation energy than an argon ICP. Theoretically, every element can be excited in this discharge, and the He MIP permits measurement of emission lines from the halogens, carbon, nitrogen, oxygen, and other nonmetal elements. Particulates are collected on polycarbonate filters using standard air sampling pumps. The filter is placed in a helium atmosphere and the particles are "vacuumed up" by an aspirator nozzle that travels over the filter. The particles are entrained in a stream of He and carried into the plasma, one particle at a time. The particles decompose, are atomized, excited, and emit their characteristic radiation. One filter in position, the aspirator nozzle, and the plasma are shown in Fig. 7.49. The emitted light is collected by fiber optics and sent to four monochromators, each set to measure a different wavelength using a PMT detector. A block diagram of the instrument, the Horiba DP 1000, is shown in Fig. 7.50.

The emission intensity should be proportional to the mass of the material in the plasma. If the particles of the material are spherical, the emission intensity is proportional to the diameter of the particle raised to the third power or:

$$D = kI^{1/3} \tag{7.3}$$

where D is the equivalent particle size (diameter of a spherical particle); k, the proportionality constant; and I, the emission intensity.

From the emission intensity, an equivalent spherical particle size can be calculated for sample particulates. One material which is available as pure, spherical particles of well-controlled particle size is silica, used for chromatography column packing. A micrograph of 3 μm diameter SiO_2 beads collected on a polycarbonate filter is shown in Fig. 7.51(a). While relatively uniform in diameter (monodisperse), you can see that not all the beads are perfect single spheres. The silica particles are aspirated into the He

Figure 7.49 Particles collected on a filter are aspirated one at a time in to the helium stream and carried to the plasma, where they are atomized, excited, and emit characteristic radiation. [Courtesy of Horiba, Ltd. (www.horiba.com) and Yokogawa Electric Company.]

Figure 7.50 Block diagram of the Horiba DP 1000. [Courtesy of Horiba, Ltd. (www.horiba.com).]

MIP and the emission intensity at a silicon emission line is measured. The correlation between emission intensity and true particle size measured from the electron micrograph is excellent [Fig. 7.51(b)]. Only 397 individual particles were measured to obtain the size distribution of the material; the number of particles is determined from the number of Si photons registered by the PMT detector.

If a particle in the plasma is composed of two elements, the signals for the emission intensity of both elements will occur at the same time. That is, the signals from a compound will be correlated in time. If the sample is composed of a mixture of two compounds, the signals for the elements from each compound will not occur at the same time, since particles enter the plasma individually. For example, the important semi-conductor material gallium arsenide, GaAs, contains both gallium and arsenic in every particle. A plot of the time correlation between Ga and As in gallium arsenide powder shows that the signals are perfectly correlated (Fig. 7.52). If a mixture of gallium oxide powder and arsenic oxide powder were analyzed, the resulting correlation plot is shown in Fig. 7.53; there is no correlation at all between the Ga and As, since they do not occur in the same particles. In this way, the instrument is able to differentiate between compounds and mixtures.

The slope of a linear correlation plot is related to the stoichiometry of the compound. This allows rapid comparison of different materials. Figure 7.54 shows the correlation plots for sodium and boron in three glass powders. It is clear that one glass (the open tri-angles) differs in stoichiometry from the other two glass materials. This type of analysis would be difficult by XRF due to the low sensitivity of boron by XRF, and would require time-consuming dissolution or fusion of the glass for ICP and similar methods. The DP 1000 required small amounts of the glass powder placed on filters using an impact pump (of the type used for collecting air particulates). The instrument is very useful for the rapid qualitative comparison of complex industrial powders. Figure 7.55 shows the comparison of four fluorescent lamp phosphors. The phosphor is the white pow-dered material that coats the inside of fluorescent lamp tubes and is responsible for the "white light" spectrum emitted by the lamp; it usually contains a mixture of rare earths and other elements. These materials are not easily dissolved and methods such as XRF are not sensitive for B and Li. Figure 7.55 shows clear differences in the relative amounts of several of the elements in these four phosphors with no sample preparation required. As a final example, the instrument easily counted the particles and analyzed

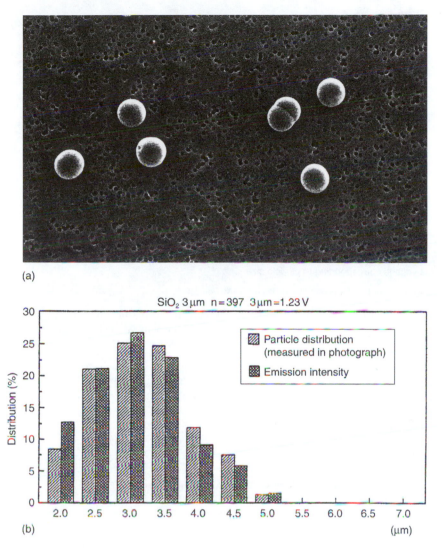

(a)

(b)

Figure 7.51 (a) Silica microspheres of 3 μm diameter supported on a polycarbonate filter. The small holes in the background are the submicron pores of the filter. (b) Agreement between the actual particle sizes of the silica measured from the electron micrograph and the calculated particle size from the Si emission intensity determined by the DP 1000. [Courtesy of Horiba, Ltd. (www. horiba.com).]

the composition of the particles found in a semiconductor clean room atmosphere after cigarette smokers and nonsmokers entered the clean room. Figure 7.56(a) shows the number of particles (y-axis) and elements detected after nonsmokers entered the clean room. The correlation plot is not shown, but the most highly correlated elements were Na and Si, from normal outdoor "dust". Figure 7.56(b) shows the number of particles and the number of elements detected increased significantly when smokers enter the clean room immediately after smoking.

Many industrial materials today are composed of a coating of one material on a second material. Toner particles for copy machines are one example; coated time-release drug formulations are another example. The DP 1000 easily distinguishes coated particles

Correlation between Ga and As Emission Intensities for GaAs Powder

Figure 7.52 Correlation in time between Ga and As emission intensities in gallium arsenide powder. The total correlation indicates that all particles contain both Ga and As. [Courtesy of Horiba, Ltd. (www.horiba.com) and Yokogawa Electric Company.]

from compounds, which give linear correlation plots. The coating intensity is related to the surface area of the particle, while the particle base material is related to the total volume of the particle. For a coated spherical particle, the correlation plot is not linear; it is a curve corresponding to the base particle radius raised to the power of 2/3.

Figure 7.53 Simulated plot of the correlation between a mixture of gallium oxide and arsenic oxide. There are no particles that contain both Ga and As. All of the As intensity is found on the y-axis, where Ga = 0; similarly, the Ga intensity is located on the x-axis, where As = 0. This lack of correlation indicates that the elements are present as a mixture of separate As-containing and Ga-containing materials, not as a compound of As and Ga.

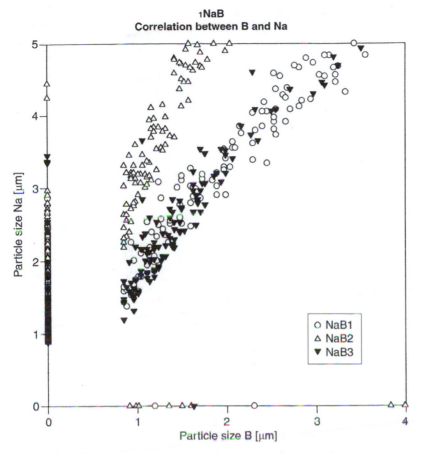

Figure 7.54 Correlation plots for commercial glass powders containing sodium and boron. The glass corresponding to the open triangles has a different Na:B stoichiometry than the other two glasses, based on the difference in slope.

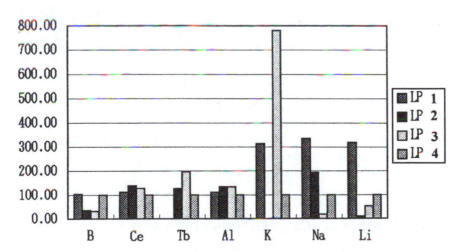

Figure 7.55 Comparison of commercial fluorescent lamp phosphors, showing clear differences in the chemical composition of the materials. The sample powders required no sample preparation prior to analysis.

Figure 7.56 (a) Numbers of particles (*y*-axis) and elements detected when nonsmokers enter a semiconductor clean room vs. (b) the numbers of particles and elements detected when cigarette smokers entered the clean room immediately after smoking. [Courtesy of Horiba, Ltd. (www. horiba.com) and Yokogawa Electric Company.]

This particle analyzer system has been used to measure contaminants in clean rooms for the semiconductor industry, evaluate air quality, study gas–solid reaction mechanisms to optimize particulate removal from industrial smokestacks, to characterize environmental samples including soils, and to characterize industrial products that are in the form of fine particles, such as toner for copy machines, pigments and inorganic fillers for polymers, inert fillers for pharmaceutical tablets, and many other applications. The instrument has advantages over conventional atomic emission in that no sample preparation is required for fine powders, and the design of the instrument to aspirate single particles into the plasma allows compounds to be distinguished from mixtures, coatings to be distinguished from compounds, total numbers of particles to be counted, and equivalent particle size to be calculated. Applications can be found in the article by Skelly Frame, Suzuki, and Takamatsu cited in the bibliography or on the Horiba Ltd. website at www.horiba.com.

7.6. ATOMIC FLUORESCENCE SPECTROMETRY (AFS)

The process of atomic fluorescence involves the emission of a photon from a gas phase atom that has been excited by the absorption of a photon, as opposed to excitation by thermal or electrical means. Figure 7.57 shows a few of the various types of fluorescence

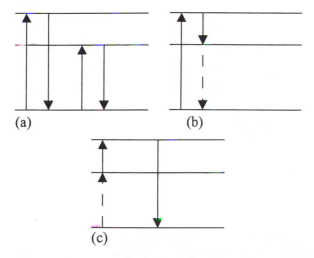

Figure 7.57 Example of atomic fluorescence transitions: (a) resonance fluorescence, (b) Stokes direct-line fluorescence, and (c) anti-Stokes direct-line fluorescence.

transitions that can occur. The solid arrows are photon absorptions or emissions; the dotted arrows are transitions that do not involve a photon. The lowest energy level represents the ground state. Two **resonance fluorescence** transitions are shown in Fig. 7.57(a), where the initial and final energy levels are involved in the absorption and emission steps. The absorption and emission wavelengths are the same in resonance fluorescence. Figure 7.57(b) depicts **Stokes direct-line fluorescence**, while Fig. 7.57(c) shows an **anti-Stokes direct-line fluorescence** transition. There are other transitions that can occur, including stepwise transitions and multiphoton processes.

We know from Chapter 6 and the Boltzmann distribution that most atoms are in the ground state even at the temperatures found in graphite furnaces. Therefore, most analytical AFS uses ground-state resonance fluorescence transitions because these have the greatest transition probability and thus result in the highest fluorescent yield. Even so, the fluorescent yield is only a fraction of the total excitation source power, because not all the excitation source power is absorbed, not all absorbed photons result in fluorescence, and only a small amount of the total fluorescence is measured. A major disadvantage of the use of resonance fluorescence lines is that scattered light from the source has exactly the same wavelength as the fluorescence emission and is a direct interference.

The calculation of fluorescence yields for AFS are similar to those for molecular fluorescence (Chapter 5). Ingle and Crouch present an extensive discussion of theory of atomic fluorescence. Given the limited commercial applications of AFS, the theory will not be covered here. It is sufficient to understand that for a resonance transition and low analyte concentration, the fluorescence signal is proportional to the analyte concentration and to the intensity of the source. This assumption is valid for sources that do not alter the population of the analyte states. Intense laser sources can deplete the population of lower-energy states, including the state from which excitation occurs. This condition is called **saturation** and is discussed under applications of AFS in Section 7.6.3.

7.6.1. Instrumentation for AFS

A block diagram for an AFS spectrometer is shown in Fig. 7.58. The fluorescence signal is generally measured at an angle of 90° with respect to the excitation source to minimize

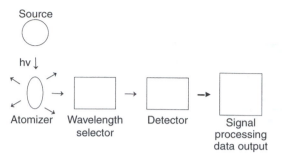

Figure 7.58 Schematic AFS spectrometer system showing the typical 90° orientation between the source and the wavelength selector and detector to minimize scattered radiation.

scattered radiation from the source entering the wavelength selector, as was done in the measurement of molecular fluorescence (Chapter 5). While the fluorescence radiation is emitted in all directions from the atomizer, only a small fraction of it is collected and sent to the detector. This combined with the low fluorescent yield results in a very small signal in most cases.

The atomizer used can be a flame, a plasma, a graphite furnace, or a quartz cell atomizer used for hydride generation and CVAAS systems. The wavelength selector can be any of the devices discussed in earlier chapters, from grating monochromators to nondispersive filter systems. The detector commonly used is the PMT because of its high sensitivity. The fluorescence signal is directly proportional to the intensity of the excitation source at low concentration of analyte, so intense stable light sources are needed for AFS. The most common sources are line sources, and include EDLs and HCLs (described in Chapter 6); the HCLs are generally modified to be high intensity lamps. Laser excitation is ideal for AFS since lasers provide extremely intense monochromatic radiation and permit the use of less sensitive nonresonance fluorescence transitions. The primary disadvantage to lasers is the cost and complexity of operation of some types of lasers. In all cases, the source is pulsed or mechanically chopped (discussed in Chapter 6) to eliminate the interference from emission by the analyte caused by thermal or other means of excitation. Resonance emission occurs at the same wavelength as resonance fluorescence; however, the fluorescence signal will be modulated at the source frequency while the emission signal (from thermal or other excitation) will be a steady signal. A lock-in amplifier set to the source modulation frequency will measure only the fluorescence signal.

Commercial instruments are available for the determination of mercury by cold vapor AFS and for the determination of As, Se, and other hydride-forming elements by hydride generation AFS. The chemistry of these methods is the same as for their AAS counterparts (discussed in Chapter 6). The instruments use either a cold quartz cell for Hg or a heated quartz cell for As, Se, Sb, Bi, Te, and other hydride-forming elements. PMT detectors with high sensitivity to the UV region are used, since the most sensitive fluorescence lines for Hg and the hydride-forming elements are in the 180–260 nm region. These commercial instruments are equipped with autosamplers and automated sample/reagent mixing systems. Flow injection systems exactly as described for AAS in Chapter 6 are used for some systems. The Hg systems are often equipped with an amalgamation trap to preconcentrate the mercury vapor on gold. The concentrated Hg vapor is released by heating the gold amalgamation trap. A major research laboratory tool (not commercially available) is the use of a graphite furnace atomizer with a laser excitation source; the technique is called graphite furnace laser-excited atomic fluorescence spectrometry (GF-LEAFS).

7.6.2. Interferences in AFS

AFS would be expected to suffer from chemical interferences and spectral interferences, as do the other atomic emission techniques we have discussed. The focus will be on the commercial instruments available and on the graphite furnace-laser system.

7.6.2.1. Chemical Interference

Chemical interference in the cold vapor Hg AFS method is not a significant problem. The chemistry of the reaction is well understood and the interferences are both few and rare. Large amounts of gold, and several other metals interfere in the release of Hg atomic vapor, but the interferences are not encountered commonly in the biological and environmental samples routinely analyzed for Hg. Chemical interference in hydride generation is also well understood. The elements must be converted to the appropriate oxidation state, for example. For routine analysis of biological and environmental samples using standard analytical protocols, AFS determination of As and Se does not suffer from significant chemical interference. GF-LEAFS using modern furnace techniques (Chapter 6) has not been reported to suffer from serious chemical interference.

7.6.2.2. Spectral Interference

Spectral interferences can be significant in AFS and some of them are unique to AFS. The total fluorescence signal at the detector can include light scattered from the source, fluorescence from nonanalyte atoms and molecules, background emission, analyte emission, and analyte fluorescence. To measure only analyte fluorescence, the other spectral interferences must be eliminated or corrected for. As noted, if the source is modulated, analyte emission is not measured. Fluorescence from nonanalyte atoms and molecules, or background fluorescence is rare. The multistep nature of the fluorescence process makes it unlikely that nonanalyte species will absorb and fluoresce at the same wavelengths as the analyte. Any background fluorescence that occurs can be compensated for by having a suitable blank and subtracting the blank signal from the measured total signal. Background emission and scattered light are the major sources of spectral interference in AFS. Scattered light and background emission can both be compensated for by subtraction of a suitable blank. Scattered light is more significant in resonance fluorescence, where the excitation and fluorescence wavelengths are the same, than in nonresonance fluorescence. Wavelength selection can distinguish between the source scattering and the analyte signal in nonresonance fluorescence because the fluorescence wavelength is not the same as the excitation wavelength. The advantage to the use of a laser as the excitation source is that less sensitive nonresonance transitions can be used, and the scattered light eliminated as a source of interference by wavelength selection.

Spectral interferences are minimal for cold vapor Hg AFS and for hydride generation AFS, due to the chemical separation step that is used prior to excitation. Background signals in GF-LEAFS are smaller than those from GFAAS, but Zeeman background correction, described in Chapter 6, has been used for GF-LEAFS to provide accurate background correction.

7.6.3. Applications of AFS

The AFS phenomenon has been studied extensively since the late 1950s. An overview of the research into AFS can be found in the book by Ingle and Crouch, the chapter by Michel, and in the handbook by Settle (see the bibliography). AFS is primarily used for quantitative analysis. The accuracy and precision of AFS depends on the atomizer used but is generally comparable to cold vapor AAS for Hg or hydride generation for As and Se, on the order of 1–2% RSD. Use of a graphite furnace as the atomizer results in precisions typical of GFAAS, on the order of 5% RSD for solutions to 10–30% RSD for solid samples. Linear ranges vary from three to eight orders of magnitude.

7.6.3.1. Graphite Furnace Laser-excited Atomic Fluorescence Spectrometry (GF-LEAFS)

The use of a graphite furnace atomizer and a laser excitation source combine to give the technique called GF-LEAFS. While this is a research technique limited at present to a few laboratories, the technique has dramatic advantages over other atomic spectroscopic methods. Small solid samples, such as polymer pieces, can be analyzed directly as described by Lonardo et al., with detection limits that are two to four orders of magnitude better than GFAAS and one to two orders of magnitude better than ICP-MS. There are definite advantages to direct solid sampling, such as the elimination of costly high purity acids for dissolution, reduction in contamination of the sample and minimization of loss of volatile elements during preparation, and faster sample analysis. Direct solid sampling has been used to determine a number of elements, including Pb, Tl, and Te in alloys, while many other elements have been determined on dissolved samples. References to these applications, from the research group of Professor Robert Michel, can be found at http://chemweb.chem.uconn.edu.

The use of an intense laser source can change the population of states in the analyte. In particular, the lower state, from which excitation occurs, can be depleted and an excited state populated using high-intensity sources such as pulsed dye lasers. For a simple two-level system (ground state and first excited state), complete saturation is defined as having one half of the atoms in the excited state. As we know from the Boltzmann distribution, this is not the usual distribution of atoms. If we graph the logarithm of the relative fluorescence signal vs. the logarithm of the source intensity, we obtain the **saturation curve**, shown schematically in Fig. 7.59. In the region of low source intensity, which would occur with low laser power or with conventional EDL or HCL sources, the signal is linearly related to the source intensity. This region demonstrates the need for high-intensity conventional sources. The higher the intensity, the higher the signal in the linear region of the curve for conventional sources. With very high intensity sources, the signal becomes independent of the source intensity because the population of ground state and excited state atoms has been altered. The intense laser source has altered the population of atoms and caused the rates of absorption and fluorescence to become equal. This condition is called saturation or optical saturation. Saturated fluorescence gives several advantages. The fluorescence signal is the maximum obtainable signal, so the detection limits are the lowest obtainable. The signal is independent of fluctuations in the source intensity. Saturation eliminates quenching of the fluorescence signal by collisional deactivation and improves linearity by decreasing self-absorption of the fluorescence signal. In practice, sufficient laser power is used to just saturate a given transition. This provides the advantages just listed while minimizing scattered radiation.

Figure 7.59 Schematic saturation curve for AFS. At low intensity, the signal is linear with source intensity. At high source intensity, the signal is independent of the source intensity. This is the saturation region.

7.6.3.2. Mercury Determination and Speciation by AFS

Commercial CV mercury AFS systems are available because of the dramatic improvement in detection limits for Hg using AFS. Detection limits for Hg on one commercial instrument using amalgamation to preconcentrate the mercury are reported to be 0.1 pg Hg or less than 0.1 ppt. That is at least 10 times more sensitive than CVAAS. The linear working range is up to 10 ppm Hg. These instruments are used to determine low levels of Hg in biological fluids and tissues, environmental samples, air, and gases. The commercial instruments can be coupled to an LC or GC for speciation of mercury compounds in samples. Application notes and system diagrams are available from PS Analytical Ltd., UK on their website at www.psanalytical.com.

7.6.3.3. Hydride Generation and Speciation by AFS

The hydride-forming elements, including As, Se, Sb, Bi, Te, and others, can be determined routinely using commercial hydride-generation AFS instruments. A number of these elements, like Hg, are important analytes because of their toxicity. Detection limits for most of these elements are in the low ppt range, with linear ranges up to 10 ppm. The detection limits are in general 5–10 times better than the equivalent hydride generation AAS detection limits. Speciation for these elements can be accomplished by coupling an LC with the AFS system. Application notes and system diagrams are available from PS Analytical Ltd., UK, on their website at www.psanalytical.com.

7.7. COMMERCIAL ATOMIC EMISSION SYSTEMS

7.7.1. Arc and Spark Systems

Commercial DC arc emission spectrometers and commercial spark spectrometers are available as stand-alone laboratory instruments. This list is not comprehensive, but examples include a DC arc CID-based Echelle system from Thermo Elemental (www.thermoelemental.com), a multiple polychromator spark system with PMT detectors from Spectro Analytical Instruments (www.spectro-ai.com), and a spark spectrometer polychromator system with PMT detectors from Jobin Yvon, Inc. (www.jyhoriba.com). Many instrument companies make portable spark instruments, some small enough to be handheld, for use in sorting metal alloys at scrap yards, confirming the identity of alloys received at loading docks, and for at-line quality control in production. Many instrument manufacturers also offer an interchangeable spark ablation stand for analysis of solid samples using an ICP spectrometer (which is generally used for liquid samples as described in Section 7.3).

7.7.2. ICP and DCP Systems

Sequential and simultaneous ICP systems of many types are available from numerous instrument manufacturers. Most are available in radial or axial view, or both. Most spectrometers can be purged to reach 160 nm. Examples of some of the companies who make ICPs and DCPs are given. Jobin Yvon, Inc. makes a line of PMT-based high-resolution sequential systems, simultaneous systems, and combination sequential–simultaneous systems. PerkinElmer Inc. makes a scanning ICP with CCD detection, and a line of simultaneous CCD-based Echelle systems. Leeman Laboratories, Inc. makes an Echelle ICP spectrometer with an aperture grid available as a sequential system with rapid scanning PMT detection, a simultaneous system with a PMT array, and a combination system. Varian, Inc. makes a sequential ICP with PMT detection and a simultaneous system with CCD detection. ThermoElemental offers ICP Echelle spectrometers with CID detection and a sequential diffraction grating instrument with PMT detection.

Spectrometer systems are now available with nitrogen purged, sealed optics for routine analysis from 120 to 800 nm. ICP systems that have this capability include instruments from Jobin Yvon, Inc. and from Spectro Analytical Instruments. There is one commercial DCP available from ThermoElemental, an Echelle spectrometer with CID detection. The ability to use lines in the "vacuum UV" permits the determination of the halogens and other elements whose most intense lines occur at <160 nm.

Most systems, especially for use in industry, include an autosampler that allows unattended, overnight operation of the instrument. Computer software in commercial instruments controls the autosampler and the instrument, collects the data, performs the calculations, prints out the results, and shuts down the instrument when the analysis is completed. Most instruments have computer-controlled safety interlocks that shut down the plasma or the instrument and autosampler in the event a problem is detected.

This list is not inclusive or comprehensive, but is meant to give the student a feeling for the types of instruments commercially available at the present time. Typical detection limits for elements determined by ICP emission spectrometry can be found in Appendix 6.2, Table A1, and in Table 7.12.

7.7.3. GD Systems

Commercial GD spectrometers with RF sources are available from Jobin Yvon, Inc. (www.jyhoriba.com), and from LECO (www.leco.com); DC sources are available from LECO and from Spectro Analytical Instruments (www.spectro-ai.com), among others.

7.7.4. AFS Systems

PS Analytical Ltd. (www.psanalytical.com) makes commercial AFS systems for Hg and hydride generation, as well as systems for speciation of Hg, As, Se, and related elements using coupled chromatography-AFS.

7.8. ATOMIC EMISSION LITERATURE AND RESOURCES

There are many journals that publish articles on atomic emission spectroscopy. Analytical Chemistry, Applied Spectroscopy, Spectrochimica Acta Part B, and The Analyst publish articles on atomic emission spectroscopy as well as other analytical methods. The Journal of Analytical Atomic Spectrometry is a more focused journal, as the name implies. Applications articles that use atomic emission spectroscopy for analysis of specific materials may be found in journals related to the field of application, such as geology, agriculture, food science, pharmaceutical science, polymer science, and the like.

Methods and applications can be found on line from government, academic, and instrument manufacturer sites. The US EPA methods that use atomic emission spectrometry for analysis of environmental samples can be found at www.usepa.gov, for example. Most of the instrument companies whose websites are given in the chapter have applications notes and methods on their websites. Many also have tutorials on the various techniques.

Tables of atomic spectral lines may be found in print in the books and handbooks by Boumans, Harrison, Robinson and Zaidel listed in the bibliography. Atomic spectral wavelengths can be found at the US NIST website, www.nist.gov.

7.9. COMPARISON OF ATOMIC SPECTROSCOPIC AND ICP-MS TECHNIQUES

The atomic spectroscopic techniques discussed in Chapters 6 and 7 and ICP-MS, discussed in Chapters 9 and 10, each have advantages and disadvantages in the determination of elements in real samples. A summary of the capabilities of each technique and a comparison of the techniques is given in Appendix 7.2.

BIBLIOGRAPHY

American Society for Testing and Materials *Annual Book of ASTM Standards*; ASTM: West Conshohocken, PA, 2000–2001; Vols. 3.05 and 3.06.

Beaty, R.D.; Kerber, J.D. *Concepts, Instrumentation and Techniques in Atomic Absorption Spectrophotometry*; PerkinElmer Instruments: Shelton, CT, 1993.

Boss, C.B.; Fredeen, K.J. *Concepts, Instrumentation and Techniques in Inductively Coupled Plasma Optical Emission Spectrometry*; PerkinElmer Instruments: Shelton, CT, 1999.

Boumans, P.W.J.M. *Line Coincidence Tables for Inductively Coupled Plasma Emission Spectrometry*; Pergamon Press: New York, 1984; Vols. I and II.

Boumans, P.W.J.M. *Inductively Coupled Plasma Emission Spectroscopy*; Wiley: New York, 1987; Parts 1 and 2.

Broekaert, J.A.C. Glow discharge atomic spectroscopy. *Appl. Spectrosc.* **1995**, *49* (7), 12A.

Busch, K.W.; Busch, M.A. *Multielement Detection Systems for Spectrochemical Analysis*; Wiley: New York, 1990.

Chandler, C. *Atomic Spectra*, 2nd Ed.; D. Van Nostrand Co. Inc.: Princeton, NJ, 1964.

Dean, J.A. *Flame Photometry*; McGraw-Hill: New York, 1960.

Dean, J.A.; Rains, T.E. *Flame Emission and Atomic Absorption Spectroscopy*; Marcel Dekker, Inc.: New York, 1965–1971; Vols. 1–3.

Dorn, S.B.; Skelly Frame, E.M. Development of a high-performance liquid chromatographic-inductively coupled plasma method for speciation and quantification of silicones: from silanols to polysiloxanes. *Analyst* **1994**, *119*, 1687–1694.

Dulski, T.R. *A Manual for the Chemical Analysis of Metals*; ASTM: West Conshohocken, PA, 1996.

Epperson, P.M.; Sweedler, J.V.; Billhorn, R.B.; Sims, G.R.; Denton, M.B. *Anal. Chem.* **1988**, *60*, 327A.

Hareland, W. Atomic emission spectroscopy. In *Analytical Instrumentation Handbook*, 2nd Ed.; Euring, G.W., Ed.; Marcel Dekker, Inc.: New York, 1997.

Harrison, G.R. *MIT Wavelength Tables*; The MIT Press: Cambridge, MA, 1969.

Harnly, J.M.; Fields, R.E. Solid-state array detectors for analytical spectrometry. *Appl. Spectrosc.* **1997**, *51* (9), 334A.

Ingle, J.D., Jr.; Crouch, S.R. *Spectrochemical Analysis*; Prentice Hall: Englewood Cliffs, NJ, 1988.

Kirkbright, G.F.; Sargent, M. *Atomic Absorption and Fluorescence Spectroscopy*; Academic Press: London, 1974.

Lajunen, L.H.J. *Spectrochemical Analysis by Atomic Absorption and Emission*; The Royal Society of Chemistry: Cambridge, UK, 1992.

Lonardo, R.F.; Yuzefovsky, A.; Yang, K.X.; Michel, R.G. *JAAS* **1996**, *4*, 279.

Marcus, R.K., Ed. *Glow Discharge Spectroscopies*; Plenum Press: New York, 1993.

Michel, R.G. Atomic fluorescence spectrometry. In *Metallobiochemistry, Part A*; Riordan, J.F., Vallee, B.L., Eds.; Academic Press: San Diego, CA, 1988; Vol. 158.

Montaser, A.; Golightly, D.W., Eds. *Inductively Coupled Plasmas in Analytical Atomic Spectrometry*, 2nd Ed.; VCH: New York, 1992.

Parsons, M.L.; Foster, A. *An Atlas of Spectral Interferences in ICP Spectroscopy*; Plenum Press: New York, 1980.

Pilon, M.J.; Denton, M.B.; Schleicher, R.G.; Moran, P.M.; Smith, S.B. *Appl. Spectrosc.* **1990**, *44*, 1613.

Robinson, J.W., Ed. *Handbook of Spectroscopy*; CRC Press: Boca Raton, FL, 1974; Vol. 1.

Robinson, J.W. *Atomic Spectroscopy*; Marcel Dekker: New York, 1990.

Sacks, R.D. Emission spectroscopy. In *Treatise on Analytical Chemistry*; Elving, P.J., Meehan, E., Kolthoff, I., Eds.; Wiley: New York, 1981; Vol. 7.

Settle, F., Ed. *Handbook of Instrumental Techniques for Analytical Chemistry*; Prentice Hall PTR: Upper Saddle River, NJ, 1997.

Skelly Frame, E.M. *Application of the Yokogawa PT1000 Particle Analysis System to Materials Characterization*, GE Report 96CRD126, Class 1, 1996, GE Corporate Research and Development: Schenectady, NY. (Note: The Yokogawa PT 1000 is now the Horiba DP 1000. The report can be accessed as a pdf file at www.research.ge.com/cooltechnologies.)

Skelly Frame, E.M.; Suzuki, T.; Takamatsu, Y. Particle characterization by helium microwave induced plasma spectrometry. *Spectroscopy* **1996**, *11*, 1.

Skelly Frame, E.M.; Uzgiris, E.E. The determination of gadolinium in biological samples by ICP-AES and ICP-MS in evaluation of the action of MRI agents. *Analyst* **1998**, *123*, 675–679.

Slickers, K.A. *Automatic Emission Spectroscopy*, 2nd Ed.; Bruehlische University Press, Giessen, 1993.

Sneddon, J., Ed. *Sample Introduction in Atomic Spectroscopy*; Elsevier: Amsterdam, 1990.

Sneddon, J., Ed. *Advances in Atomic Spectroscopy*; JAI Press: Greenwich, CT, 1992, 1994; Vols. I and II.

Sweedler, J.V.; Jalkian, R.D.; Denton, M.B. *Appl. Spectrosc.* **1989**, *43*, 953.

Syty, A. Flame photometry. In *Treatise on Analytical Chemistry*; Elving, P.J., Meehan, E., Kolthoff, I., Eds.; Wiley: New York, 1981; Vol. 7.

Thomsen, V.B.E. *Modern Spectrochemical Analysis of Metals: An Introduction for Users of Arc/Spark Instrumentation*; ASM International: Materials Park, OH, 1996.

Winchester, M.R.; Lazik, C.; Marcus, R.K. *Spectrochim. Acta* **1991**, *46B*, 483.

Zaidel, A.N.; Prokofev, V.K.; Raiskii, S.M.; Slavnyi, V.A.; Shreider, E.Y. *Table of Spectral Lines*, 3rd Ed.; IFI/Plenum Press: New York, 1970.

SUGGESTED EXPERIMENTS

Flame Emission

7.1   Warm up the flame photometer or flame atomic absorption spectrometer in emission mode, following manufacturer's directions. Using an air–acetylene burner, set the flow rates of air:acetylene to (a) oxidizing flame, (b) stoichiometric flame, and (c) reducing flame, following manufacturer's directions.

Note the change in flame color, shape, and size. With each flame measure the emission at 589.0 and 589.5 nm. This is from sodium contamination in the flame gases and from dust in the atmosphere. Aspirate into the flame (a) deionized water from a plastic container, (b) deionized water from a soda-lime glass container (your typical "glass" jar or bottle), and (c) deionized water from a borosilicate glass container (e.g., Pyrex® or Kimax®). (Allow water to sit in each container overnight). Compare with freshly drawn deionized water. Compare the relative sodium contamination of the water by the containers (emission intensity at 589.0 nm).

7.2 Prepare aqueous solutions containing amounts of potassium varying from 1.0 to 100.0 ppm of K^+. Aspirate each sample into the flame. Measure the intensity of emission at 766.0 and 404.4 nm. Plot the relationship between the emission intensity of each line and the concentration of the solution aspirated into the burner.

7.3 Repeat Experiment 7.2, but add 500 ppm of Cs to each solution. Note the change in the emission intensity. Explain.

7.4 Prepare solutions containing 20 ppm of Mg, Ca, Zn, Na, Ni, and Cu. Measure the emission intensity of Mg at 285.2 nm, Ca at 422.7 nm, Zn at 213.9 nm, Na at 589.0, Ni at 341.4 nm, and Cu at 324.7 nm. Note the wide divergence in emission intensity, even though the metal concentration was constant. Compare the sensitivity of your results with that of the results of Experiments 7.1 and 7.2. (This experiment can be done with an ICP-OES spectrometer as well as a flame emission spectrometer.) If both a flame emission system and an ICP system are available, do Experiments 7.1–7.4 on both instruments and compare the results. Explain your observations in terms of the excitation source temperature, type of nebulizer, and any other instrument variables that apply.

7.5 Take a sample containing an unknown quantity (about 3 ppm) of sodium. Split it into four 10 mL aliquots. To the first aliquot add 10 mL of standard sodium solution containing 2 ppm Na. To other aliquots add solutions containing 4 and 6 ppm of Na (10 mL of each, one per aliquot). To the remaining aliquot add 10 mL of deionized water so that all solutions now have a volume of 20 mL. Measure the intensity of Na emission at 589.0 and 589.5 nm. Using the MSA, calculate the Na concentration in the original sample.

7.6 Determine the sodium content of the drinking water in your chemistry department or city.

Arc/Spark or ICP Emission

For those of you who still own working emission spectrographs with photographic plate detectors, a series of experiments is suggested. Some of these experiments can be modified to suit other emission sources for solid samples—GD, spark, DC arc with modern PMT or CCD/CID detectors. Experiments 7.11–7.14 can be made suitable for ICP-OES by dissolution of the solid samples in acid. For example, the point of Experiment 7.10 for an emission spectrograph is that the matrix can be a problem. Does that problem exist for ICP-OES if the samples are all in solution? Matrix effects (Experiment 7.10) can be studied in ICP-OES by running one of the copper salts in 5%, 10%, and 20% nitric acid; then 5% nitric acid plus varying amounts of NaCl (e.g., 5–20% NaCl), or by running Experiment 7.12 using external calibration standards in acid alone and then using the MSA for brass, bronze, and so on to see if there is a difference in slope which would indicate a matrix effect. Note that if you want to dissolve SiO_2 (Experiment 7.10), you need to use hydrofluoric acid and have an HF-resistant sample introduction system in your ICP. HF is EXTREMELY toxic and must only be used with the permission and under the supervision of your instructor. Alternatively, go through a molten salt fusion or substitute another material.

7.7   Take several 1 g portions of dry graphite powder to be used as the matrix. Add powdered anhydrous copper sulfate to make a powder containing 0.01% Cu. Load this into the graphite electrode of an emission spectrograph and fire under spark conditions for 3 min. Record the spectrum on the photographic plate.

7.8   Repeat Experiment 7.7, but record the spectrum after exposures of 1, 3, 5, 7, 10, 15, and 20 min. Measure the intensity of the emission lines at 327.4 and 324.7 nm. Plot the intensity of each line and a background intensity at a suitable position against the exposure time. Note that after a certain period of time, the intensity of the line does not increase, but the background does.

7.9   Using graphite powder and anhydrous copper sulfate powder, prepare powders that contain 0.01, 0.05, 0.1, 0.2, 0.5, 0.7, and 1.0% Cu. Insert each powder separately into the emission spectrograph and expose for 3 min. Measure the intensity of the lines at 324.7 and 327.4 nm. Plot the relationship between intensity and the concentration of the copper. Note that the intensity of the lines goes through a maximum. These are reversible lines.

7.10 Repeat Experiment 7.9, but use (a) MgO, (b) SiO_2, and (c) powdered Al metal as the matrix powder. Note that the changes in the matrix compound cause a variation in the emission from the copper, even though other conditions are constant. This is the matrix effect.

7.11 Repeat Experiment 7.9, but use (a) $CuCl_2$, (b) $Cu(NO_3)_2$, (c) Cu acetate, (d) CuO, and (e) CuS as the source of copper. Note the variation in the intensity of the copper emission when different salts are used, even though other conditions are constant. What causes this effect?

7.12 Determine the percentage of copper in (a) brass, (b) bronze, (c) a quarter (or other suitable coin), and (d) a nickel (or other suitable coin). Record the emission spectra of these samples.

7.13 Record the emission spectra of (a) Zn, (b) Sn, (c) Ni, (d) Fe, (e) Co, and (f) Cd.

7.14 Based on the emission spectra obtained in Experiments 7.12 and 7.13, what elements are present in (a) brass, (b) bronze, (c) a quarter, and (c) a nickel?

ICP Emission

7.15 Determine the elements present in powdered infant formula by weighing a known amount of formula, dissolving the formula in deionized water and aspirating it into an ICP. Aqueous multielement standards can be used for calibration. Compare your results to the label. Compare your results from this dissolution procedure to a sample of formula you prepare by wet-ashing or dry-ashing.

7.16 Determine the elements present in a multivitamin tablet prepared by digestion in concentrated nitric acid and dilution to a suitable volume in deionized water. You may need to filter out a white insoluble powder (or allow it to settle) after digestion. Pharmaceutical tablets often contain silica or titania, which will not dissolve in nitric acid. Aqueous multielement standards in the same concentration of nitric acid should be used for calibration.

PROBLEMS

7.1 What is the difference between a monochromator and a polychromator?

7.2 When a sample is introduced into a flame, what processes occur that lead to the emission of radiant energy?

7.3 Does a spectral interference in the standards for atomic emission cause a positive or negative intercept in the calibration curve? Briefly discuss the possibilities.

7.4 List three elements that are commonly analyzed by flame photometry. What is the major use of flame photometry?

7.5 In preparing a calibration curve for the determination of potassium by atomic emission spectrometry, the following data were obtained:

Concentration of K (ppm)	Emission intensity
0.0	10
0.1	410
0.2	800
0.4	1620
0.6	2450
0.8	3180
1.0	3600
1.2	4850

(a) Plot the results using a spreadsheet. Determine the equation of the calibration curve using linear regression. (b) Which of the results appear(s) to be in error? (c) What should the emission intensity of these point(s) have been assuming the relationship between concentration and intensity is linear? (d) What are you going to do with the data point(s) that appear to be in error? Explain your reasoning (you may want to reread Chapter 1).

7.6 The emission intensities from five samples were, respectively, 1020, 2230, 2990, 4019, and 3605 units. The blank prepared with the samples has an intensity of 398. Using the final calibration curve prepared in Problem 7.5 after you have decided what to do with any suspect data, calculate the potassium concentration of the five samples.

7.7 The following results were obtained using the standard addition method:

	Emission intensity from calcium
Sample alone	5.1
Sample + 0.1 ppm Ca	6.2
Sample + 0.2 ppm Ca	7.3
Sample + 0.5 ppm Ca	10.6
Flame alone	0.7

What was the concentration of calcium in the original sample?

7.8 Can a given nonanalyte element present in a sample cause both spectral and nonspectral interferences in atomic emission spectrometry? Explain, with diagrams as necessary.

7.9 In preparing a calibration curve for the determination of lithium, the following data were obtained:

Standard concentration (ppm Li)	Radiation intensity
0.1	360
0.2	611
0.4	1113
0.6	1602
0.8	2091
1.0	2570

Plot the curve. What was the radiation intensity of the flame with no lithium present (i.e., what is your y intercept)? The radiation intensity from a sample containing lithium was measured three times; the intensity readings were 1847, 1839, and 1851. What was the lithium concentration in the sample? Express the result including the uncertainty.

7.10 Spectral emission lines in flames are the same width as atomic absorption lines in flames, on the order of 0.01 nm. Why is the spectral bandpass in an atomic emission spectrometer much smaller than that in an AAS?

7.11 What causes background emission in atomic emission spectrometry? Consider all of the excitation sources discussed.

7.12 What is meant by chemical interference? Compare flame photometry, arc/spark, ICP, and GD sources in the amount of chemical interference exhibited by each technique.

7.13 How does emission spectrometry differ from flame photometry? How does this difference affect the number of elements that are detectable by each method? Explain.

7.14 Describe and illustrate a graphite electrode used for analysis of powdered samples by DC arc emission spectrometry.

7.15 What effect does the matrix have on the intensity of the emission signal in arc/spark emission, ICP emission, and GD emission?

7.16 Define the preburn time. Why is it used? Would you expect that (a) gallium and (b) tungsten could be satisfactorily determined in steel using the same preburn time? Explain your answer.

7.17 Sketch the components of a Rowland circle monochromator. What is the advantage of this design over a scanning monochromator?

7.18 Describe the components and operation of an Echelle monochromator. What are its advantages over a Rowland circle design?

7.19 What is the advantage to using a CCD or CID detector for an Echelle spectrometer instead of a PMT or multiple PMTs?

7.20 What is meant by matrix effect in atomic emission spectrometry? Give at least two examples.

7.21 How is an ICP formed? What gas is used to form the plasma? What other kinds of plasmas are there?

7.22 Draw a cross-flow nebulizer and explain its operation. What are its advantages and disadvantages when compared with an USN?

7.23 Describe how laser ablation is used for sample introduction. What kinds of samples would you run by laser ablation? Explain your answer.

7.24 What is the dynamic analytical range of a typical ICP? How does this compare to AAS? Is there any advantage to using ICP based on its dynamic range?

7.25 What are the advantages of using a GD source over a spark source for analysis of solids?

7.26 Explain what is meant by a depth profile and how it can be used in analysis of materials.

7.27 Compare atomic absorption and atomic emission for qualitative analysis. What are the advantages and disadvantages of each approach?

7.28 Compare atomic absorption and atomic emission for quantitative analysis. What are the advantages and disadvantages of each approach?

7.29 Draw the processes that occur for (a) resonance fluorescence, (b) Stokes direct-line fluorescence, and (c) anti-Stokes direct-line fluorescence.

7.30 What are the advantages of using resonance fluorescence? What is the major disadvantage to its use?

7.31 Why are high intensity line sources preferred for AFS over less intense or broad band sources?

7.32 What are the advantages of using a laser as the source for AFS? What are the disadvantages?

7.33 Briefly describe the spectral interferences in AFS and how they are eliminated or accounted for.

7.34 What is meant by saturated fluorescence? Why is saturated fluorescence used for analytical AFS?

7.35 Compare atomic fluorescence with atomic absorption and atomic emission. What are the advantages and disadvantages of each technique?

APPENDIX 7.1

Table A1 Flame Atomic Emission Detection Limits

Element	Wavelength (nm)	Type of flame	Flame conditions	Detection limits (ppm)
Aluminum	396.2	NA	R	0.005
Antimony	259.8	NA	R	1.0
Arsenic	235.0	AA	Ox.	2.2
Barium	553.6	NA	R	0.001
Beryllium	234.9	NA	R	
Bismuth	223.1	NA	R	6.4
Boron	249.7	NA	R	7
Cadmium	326.1	NA	R	0.5
Calcium	422.7	NA	R	0.0001
Cerium	569.9	NA	R	30
Cesium	852.1	AA	Ox.	0.5
Chromium	425.4	NA	R	0.005
Cobalt	345.4	NA	R	0.5
Copper	327.4	NA	R	0.01
Dysprosium	404.6	NA	R	0.07

(continued)

Table A1 Continued

Element	Wavelength (nm)	Type of flame	Flame conditions	Detection limits (ppm)
Erbium	400.8	NA	R	0.04
Europium	459.4	NA	R	0.0006
Gadolinium	440.2	NA	R	1.0
Gallium	403.3	NA	R	0.01
Germanium	265.1	NA	R	0.5
Gold	267.6	NA	R	0.5
Hafnium	368.2	NA	R	
Holmium	405.4	NA	R	0.02
Indium	451.1	NA	R	0.002
Iridium	380.0	NA	R	30
Iron	371.9	NA	R	0.05
Lanthanum	579.1	NA	R	0.1
Lead	405.8	NA	R	0.2
Lithium	670.8	AA	Ox.	5×10^{-6a}
Lutetium	451.9	NA	R	1.0
Magnesium	285.2	NA	R	0.005
Manganese	403.1	NA	R	0.005
Mercury	253.7	NA	R	2.5
Molybdenum	390.3	NA	R	0.2
Neodymium	492.4	NA	R	0.2
Nickel	341.5	NA	R	0.1
Niobium	405.9	NA	R	1.0
Osmium	442.0	NA	R	
Palladium	363.5	NA	R	0.1
Phosphorus	253.0	NA	R	1.0
Platinum	265.9	NA	R	2.0
Potassium	766.5	AA	Ox.	15×10^{-6a}
Praseodymium	495.1	NA	R	1.0
Rhenium	346.0	NA	R	0.3
Rhodium	369.2	NA	R	0.1
Rubidium	780.0	AA	Ox.	0.001
Ruthenium	372.8	NA	R	0.02
Samarium	476.0	NA	R	
Scandium	402.0	NA	R	0.012
Silicon	251.6	NA	R	4.0
Silver	328.1	NA	R	0.02
Sodium	589.0	AA	Ox.	8×10^{-6a}
Strontium	460.7	NA	R	0.001
Tantalum	481.3	NA	R	5
Tellurium	238.3	NA	R	2.0
Terbium	431.9	NA	R	0.4
Thallium	535.0	NA	R	0.6
Thorium	570.7	NA	R	0.02
Thulium	371.8	NA	R	0.02
Tin	284.0	NA	R	0.5
Titanium	399.9	NA	R	0.2
Tungsten	400.9	NA	R	0.5
Uranium	591.5	NA	R	

(continued)

Table A1   Continued

Element	Wavelength (nm)	Type of flame	Flame conditions	Detection limits (ppm)
Vanadium	437.9	NA	R	0.01
Ytterbium	398.8	NA	R	0.002
Yttrium	362.1	NA	R	0.3
Zinc	213.9	NA	R	77
Zirconium	360.1	NA	R	3

Note: AA, air–acetylene flame; NA, nitrous oxide–acetylene flame; R, reducing flame conditions; Ox, oxidizing flame conditions. Wavelengths and flame conditions from Analytical Methods for Atomic Absorption Spectrometry, PerkinElmer Inc., Shelton, CT, 1994, used with permission and Robinson, used with permission.
[a]Data from Model 300 flame photometer, courtesy of Buck Scientific Company.

Table A2   Approximate Detection Limits for ETA Techniques

Elements	ETA-ICP-MS ng/mL (2 μL)	pg	GFAAS ng/mL (10 μL)	pg	ETA-ICP-OES ng/mL (50 μL)	pg
^{107}Ag	0.08	0.16	0.01	0.1	0.1–30	1–300
^{27}Al	0.03	0.05	0.2	2	1.5–13,000	7.5–26,000
^{75}As	0.05	0.1	1	10	20–5000	200–10,000
^{44}Ca	0.7	1.4	0.5	5	0.002–70,000	0.02–139,000
^{114}Cd	0.15	0.3	0.01	0.1	0.2–1640	1–3280
^{52}Cr	0.1	0.2	1	10	0.3–790	1.5–1580
^{39}K	1.5	3	0.1	1	110–83,400	550–167,000
^{23}Na	0.2	0.4	1	10	80–38,000	400–76,000
^{58}Ni	0.47	0.93	1	10	0.9–1050	4.5–2100
^{208}Pb	0.1	0.3	0.2	2	2–1400	20–2800
^{78}Se	5.7	11.4	1	10	6–600	600–3000
^{28}Si	2.7	5.4	5	50	10–500	100–2500
^{64}Zn	0.2	0.4	0.05	0.5	0.05–1700	0.25–3540

Note: This table presents a comparison of graphite furnace AAS with the use of a graphite furnace (electrothermal atomizer) as the sample vaporization step in conjunction with ICP-MS and ICP-OES. The solution detection limit, the sample volume, and the absolute detection limit in picograms are given for each technique. The isotope measured is specified for the ICP-MS technique; the isotope number has no meaning for the AAS or ICP-OES results.

APPENDIX 7.2

A Comparison of Atomic Spectroscopic Analytical Techniques

Atomic spectroscopy is the most widely used technique for determining elements in almost every conceivable matrix. The techniques commercially available and in wide use are FAAS, GFAAS, and ICP-OES. Commercial AFS is limited to the determination of Hg, As, and Se using the cold vapor and hydride chemistry discussed in Chapter 6. Elemental mass spectrometry using an argon ICP as the ionization source (ICP-MS) is now also widely used for ultratrace determination of elements and is discussed in Chapters 9 and 10. Although ICP-MS complements atomic spectroscopy, it is not a spectroscopic method; mass/charge ratio is determined, not radiant energy.

Metal and metal alloy producers and materials research laboratories routinely use spark and arc emission spectrometry and GD spectrometry has been gaining in popularity because it can analyze nonconducting solids directly. It is difficult to make direct sensitivity and accuracy comparisons between AAS, ICP, ICP-MS, and the arc/spark/glow discharge methods, since the latter techniques have sensitivities based on solid samples (LODs in μg/g) and the former are solution techniques (LODs in μg/mL). While the solution detection limits of graphite furnace or ICP may appear to be "better" than arc/spark or GD, the actual dilution factor for a real solid sample needs to be considered to compare the techniques for analysis of solids.

Atomic absorption, atomic fluorescence, and atomic emission spectrometry have the same goal—the qualitative or quantitative determination of elements, particularly metals and semimetals. These methods have a number of things in common and a number of differences. The choice of methods for a particular analysis depends on the type of sample to be analyzed and the elements to be determined. While some generalizations can be made, it is important for the student to understand that the generalized statements below about sensitivity and other parameters may not hold true for all elements in all matrices and may not hold true for all instrument manufacturers of a given type of instrument. Detection limits based on signal-to-noise ratio for AAS, ICP, and ICP-MS instruments, primarily from one manufacturer, are listed in Appendix 6.1. It can be seen that ICP-MS holds a distinct advantage over the other techniques as far as sensitivity is concerned for most elements in solution. LODs for other atomic spectroscopic techniques are given in Table 7.12, and Tables A1 and A2 of Chapter 7. The literature should be searched for specific applications when needed. The precision of these techniques is similar when operating in the linear region and above the limit of quantitation, that is, above $10\times$ the 3σ detection limit. The relative precision is 1–2% for flame AAS, 1–2% for ICP-AES, 1–10% for furnace AAS, and 1–2% for ICP-MS. It is possible to achieve precisions less than 1% for most of these techniques with simple matrices and proper optimization of the operating parameters.

Table A3 lists several of the parameters that are important in the selection of the instrumental method to be used for an analysis.

Flame Atomic Absorption

Flame AAS is an excellent choice for a small laboratory. The instrumentation is relatively inexpensive and easy to operate. Flame AAS is a mature technique with many published

Table A3   Analytical Parameters

Analytical accuracy
Analytical precision
Analytical concentration range
Amount of sample available
Total analytical time per sample
Total analytical cost per sample
Bulk analysis or region specific analysis
Elemental information or speciation
Instrument cost (initial and operating costs)
Degree of automation
Skills required to develop methods and interpret data
Analytical interferences
Difficulties with contamination

---

textbooks, applications, and comprehensive guides for analysis and method development (the "cookbooks" from instrument manufacturers). Spectral interferences are few and well documented. "Flame" AAS instrumentation is also a platform for cold vapor Hg and hydride generation determination of As, Se, and Tl by AAS, so the instrumentation is a staple of environmental laboratories. On the negative side, the refractory elements like boron, tungsten, zirconium, and tantalum are difficult to determine because the flame is not hot enough to atomize these analytes. The low temperature of the flame results in chemical interference in many samples. Elimination of chemical interference may require the addition of reagents or the use of the slower method of standard additions calibration or even extraction of the analyte from the interfering substances. If more than two or three analytes must be determined in each sample, even automated AAS is a much slower process than ICP-OES. Several milliliters of solution are required per analyte, so multiple analytes per sample may need 25 mL or more of solution. The operating cost of a flame AAS is low; the gases used are relatively inexpensive. The cost of HCLs and EDLs may be significant, depending on what elements are required, but most lamps last for years if operated correctly.

GFAAS

The analytical advantages of furnace AAS are that its detection limits are about two orders of magnitude better than either flame AAS or ICP-OES and that very small sample volumes can be analyzed. The entire analytical process is slow, but is completely automated in modern instruments. Environmental labs rely on automated unattended GFAAS analysis of the priority pollutant metals in drinking water and wastewater. Like flame AAS, it usually determines a single element at a time. Furnace AAS can suffer from severe interferences, including high background signals, possible loss of volatile analytes, and formation of refractory carbides by analytes such as B, W, Ta, and Nb. The use of matrix modifiers, L'vov platforms, and transverse heating designs for the atomizer has alleviated some of these problems. Slurries can be analyzed directly and automatically by GFAAS, eliminating sample preparation time for some classes of materials. A GFAAS instrument costs more than an FAAS instrument; a modern instrument with both flame and graphite furnace atomizers costs about the same as an Echelle ICP-OES instrument, approximately $60–70K. Operating costs for furnace AAS are significant due to the high cost of the graphite parts, especially the tubes and platforms, which must be replaced frequently.

ICP-OES

The emission signal can be recorded for all elements of interest either simultaneously or in rapid sequence, thus making ICP-OES much faster than AAS techniques if many elements must be determined in each sample and requiring less sample volume in the process. The linear working range for ICP-OES is much greater than that of AAS, so trace to major components of a sample can often be determined at the same time. However, the measurement of analyte emission at high temperature requires a monochromator of better resolution than required for AAS; thus ICP-OES instrumentation is more expensive than FAAS. The high-temperature plasma is very efficient at atomizing and exciting all of the metal and semimetal elements, as well as some nonmetal elements. The high temperature of the plasma virtually eliminates chemical interferences of the type found in AAS. The sensitivity and signal-to-noise ratio is better than that for FAAS. Because of the high-temperature source, ICP-OES may suffer from spectral interference due to the large number of emission lines associated with each element. This is particularly troublesome when materials to be analyzed have very complex spectra, like iron, nickel, uranium, tungsten, molybdenum, and the rare earths. With a fixed polychromator system, it may be necessary to measure and apply correction factors if an interference-free line cannot be found for an analyte. In most sequential systems and in the simultaneous Echelle systems with solid-state detectors, it is usually possible to find a suitable analyte line; if not, the interference must be determined and corrected for. Operating costs are high; the instruments require a large amount of argon, and may require a high-purity purge gas, the torches and quartz nebulizers are expensive and fragile, pump tubing must be replaced at least daily.

ICP-MS

ICP-MS is discussed in Chapters 9 and 10, but it is a powerful trace elemental analysis method and many analytical texts cover it with atomic spectroscopy. Its advantages and disadvantages are compared here for completeness.

For most elements, the ICP ionizes a significant amount of the elemental vapor. This serves as the ionization source for the MS. Coupled together, ICP-MS measures the analyte concentration by the mass spectrum, that is, by the ratio of mass/charge for separated ions rather than by the emission of light. The instrumentation is more expensive than AAS, and slightly more expensive than high-end ICP-OES instrumentation. ICP-OES instruments generally cost $65–130K; ICP-MS instruments cost $160–210K as of this writing. Much of the cost is due to the requirement for a high vacuum system. ICP-MS possesses the fast, multielement capabilities of ICP-OES and better detection limits than the graphite furnace. The linear working range is very broad, and detection limits are in the ppt range for most elements. The extreme sensitivity of ICP-MS means that "environmental" contamination of samples is a serious problem; sample preparation areas and the instrument itself may need to be housed in a "clean lab" with filtered air. The ability of the mass spectrometer to measure the elements' individual isotopes with great accuracy adds some very special applications for ICP-MS, such as calibration by isotope dilution and the use of stable isotopes as tracers instead of more hazardous radioactive isotopes. Specific applications are discussed in Chapter 10. ICP-MS is not suitable for the analysis of solutions high in dissolved solids, unlike ICP-OES. Sensitivities for some elements are not as good as those by ICP-OES; S, Si, C, and Ca are examples. Calcium suffers from a direct isobaric interference from Ar at its most abundant isotope. Isobaric polyatomic Ar-containing interferences can limit LODs for As, Se, and some other elements, depending on the design of the ICP-MS. Operating costs are similar to those for ICP-OES; additional very costly expense items are the skimmer and sampler cones, especially if Pt cones are required. The installation cost and upkeep of a metal-free clean lab are very high.

8

X-Ray Spectroscopy*

8.1. ORIGIN OF X-RAY SPECTRA

X-rays were discovered in 1895 by Wilhelm Conrad Röntgen who received the first Nobel Prize in Physics, awarded in 1901, for his discovery. X-ray absorption, emission, and fluorescence spectra are used in the qualitative and quantitative determination of elements in solid and liquid samples. X-ray absorption is used in the nondestructive evaluation of flaws in objects, including voids or internal cracks in metals, cavities in teeth, and broken bones in humans, a technique called radiography or X-ray fluoroscopy. This same technique is used to perform security screening of baggage at airports. A computerized version of radiography, computed tomography (CT) scanning or computed axial tomography (CAT) scanning, provides a powerful, high-resolution medical diagnostic tool by giving a 3D cross-sectional image of body tissues. Diffraction of X-rays by crystalline materials, a technique called X-ray crystallography, provides crystal structure identification, orientation of atomic planes in materials, and other physical information about samples. X-ray astronomy uses cosmic X-rays to study the universe and X-ray spectrometers have been sent to the moon and Mars to study the surface rocks *in situ*. This chapter will focus on X-ray fluorescence spectrometry (XRF) and X-ray diffractometry (XRD), the techniques of most use to analytical chemists.

X-rays consist of electromagnetic radiation with a wavelength range from 0.005 to 10 nm (0.05–100 Å). X-rays have shorter wavelengths and higher energy than UV radiation. X-rays are generated in several ways, such as when a high-speed electron is stopped by a solid object or by electronic transitions of inner core electrons.

8.1.1. Energy Levels in Atoms

An atom is composed of a nucleus and electrons. The electrons are arranged in shells around the nucleus with the valence electrons in the outer shell. The different shells correspond to the different principal quantum numbers of the possible quantum states. The principal quantum number, n, can have integral values beginning with 1. The shells are named starting with the shell closest to the nucleus, which is called the K shell. The K shell is the lowest in energy and corresponds to the quantum level with $n = 1$. The shells moving out from the nucleus are named the L shell, M shell, and so on alphabetically. The letters used for the two lowest shells are historical; K is from the German word

---

*Dedicated to the memory of Dr. Ron Jenkins, an outstanding X-ray spectroscopist, teacher, and mentor.

kurz, meaning short, L is from the German word *lang*, meaning long. An atom is shown schematically in Fig. 8.1(a), with Φ_K, Φ_L, and Φ_M representing the energy of the K, L, and M shells, respectively. A partial list of elements and their electron configurations is given in Table 8.1. For example, a sodium atom contains filled K and L shells and one electron in the M shell.

When an X-ray or a fast-moving electron collides with an atom, its energy may be absorbed by the atom. If the X-ray or electron has sufficient energy, it knocks an electron out of one of the atom's inner shells (e.g., the K shell) and the atom becomes ionized as shown in Fig. 8.1(b). An electron from a higher-energy shell (e.g., the L-shell) then falls into the position vacated by the dislodged inner electron and an X-ray photon is emitted as the electron drops from one energy level to the other [Fig. 8.1(c)]. The wavelength of this emitted X-ray is characteristic of the element being bombarded.

A fourth process can also occur, as shown in Fig. 8.1(d). Instead of emitting an X-ray photon, the energy released knocks an electron out of the M shell. This electron is called an Auger electron. This Auger process is the basis for a sensitive surface analysis technique. Auger electron spectroscopy and the related method of X-ray photoelectron spectroscopy, based on the measurement of the emitted electron shown in Fig. 8.1(b), are discussed in Chapter 14.

If we plot the energy levels of the K, L, and M shells for a given element, we get a diagram similar to Fig. 8.2. Note that the K shell has only one energy level, while the

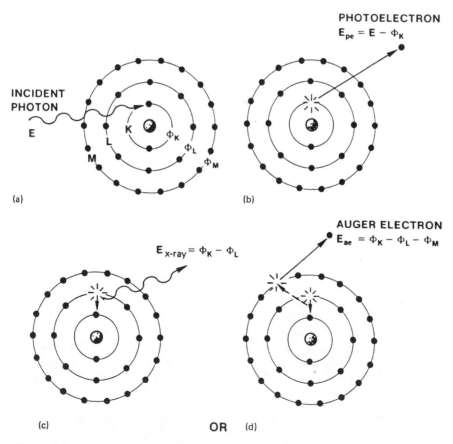

Figure 8.1 A schematic atom showing the steps leading to the emission of an X-ray photon (c) or an Auger electron (d). (From Jenkins et al., 1981, used with permission.)

Table 8.1 Electron Configurations of Various Elements

| Element | Z | K | L | | M | | | N | |
		1s	2s	2p	3s	3p	3d	4s	4p
H	1	1							
He	2	2							
Li	3	2	1						
Be	4	2	2						
B	5	2	2	1					
C	6	2	2	2					
N	7	2	2	3					
O	8	2	2	4					
F	9	2	2	5					
Ne	10	2	2	6					
Na	11	Neon core (10)			1				
Mg	12				2				
Al	13				2	1			
Si	14				2	2			
P	15				2	3			
S	16				2	4			
Cl	17				2	5			
Ar	18				2	6			
K	19	Argon core (18)						1	
Ca	20							2	
Sc	21						1	2	
Ti	22						2	2	
V	23						3	2	
Cr	24						5	1	
Mn	25						5	2	
Fe	26						6	2	
Co	27						7	2	
Ni	28						8	2	
Cu	29						10	1	
Zn	30	Cu^+ core (28)						2	
Ga	31							2	1
Ge	32							2	2
As	33							2	3
Se	34							2	4
Br	35							2	5
Kr	36							2	6

higher shells have sublevels within each shell. If an electron is dislodged from the K shell, an electron from an L or an M shell may replace it. The resulting ion emits radiation with energy E equal to the energy difference between the electronic energy levels, such as:

$$E_{\text{X-ray}} = \Phi_L - \Phi_K \tag{8.1}$$

where Φ_L is the energy of the electron in a specific electronic state within the L shell that "drops" to the K shell. Similar equations may be written for other transitions, such as between an M shell sublevel and an L shell sublevel, using the appropriate energy of the electron in the M shell sublevel that drops into the L shell and so on. As we know

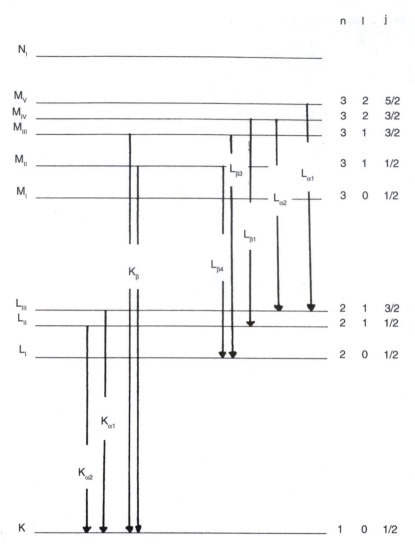

Figure 8.2 Atomic energy levels and symbols used for some common X-ray transitions. (Modified from Parsons, 1997, used with permission.)

from Chapter 2,

$$E = h\nu = hc/\lambda$$

Therefore, for the X-ray photon released when an L electron in a specific sublevel drops down to fill a vacancy in the K shell,

$$h\nu = hc/\lambda = \Phi_L - \Phi_K$$

Hence the frequency of the emitted X-ray is

$$\nu = \frac{\Phi_L - \Phi_K}{h} \tag{8.2}$$

The frequency or wavelength for transitions between other sublevels and shells is calculated in the same manner. Transitions are not possible between all available energy levels.

As in all forms of spectroscopy, transitions are governed by quantum mechanical *selection rules*. Some transitions are allowed by these rules while others are forbidden. For a brief discussion of the selection rules, the interested student should consult the texts by Jenkins or Bertin listed in the bibliography.

X-ray emission lines from electron transitions **terminating** in the K shell are called K lines, lines from transitions terminating in the L shell are called L lines, and so on. There are three L levels differing by a small amount of energy and five M levels. These sublevels are different quantum states, as shown in Fig. 8.2; the quantum numbers and states will not be discussed in detail. An electron that drops from an L shell sublevel to the K shell emits a photon with the energy difference between these quantum states. This transition results in a K_α line. There are two possible K_α lines for atoms with atomic number >9: $K_{\alpha 1}$ and $K_{\alpha 2}$, which originate in different sublevels of the L shell. The K_α lines are often not resolved, and only one peak is seen. These lines are illustrated in Fig. 8.2. The use of a Greek letter and numerical subscript to identify an X-ray emission line is called the *Siegbahn* notation. For the purposes of this text, the notation is just a "name" for the peak. An electron that drops from an M shell sublevel to the K shell generates a K_β X-ray. There is more than one K_β line, but the energy differences are so small between $K_{\beta 1}$ and $K_{\beta 3}$ that only a single K_β line is seen unless a high-resolution spectrometer is used. If an electron is ejected from an L shell, an electron from an M shell may fall into its place and emit an X-ray of characteristic wavelength with energy equivalent to the difference between the L and M shell sublevels. These are designated as L lines. A number of L lines are possible, as indicated by Fig. 8.2. Table 8.2 indicates the actual transition that gives rise to selected X-ray emission lines. Electrons originating in an N or O shell and falling into the L shell also generate L lines. The energy levels of the K, L, M, and higher shells are characteristic of the element being examined and the sharp emission lines resulting from electronic transitions are called *characteristic lines* or characteristic radiation. A schematic X-ray emission spectrum obtained under certain conditions by bombarding a solid metal, such as rhodium or lead or tungsten, with high-energy electrons is shown in Fig. 8.3. The characteristic lines are shown as sharp peaks on a broad continuous background. The characteristic K X-ray emission lines from some elements are given in Table 8.3. A more comprehensive table of K and L lines for the elements is found in Appendix 8.1 and in handbooks such as the CRC Handbook of Chemistry and Physics and the CRC Handbook

Table 8.2 Electron Transitions for Selected X-Ray Emission lines

Siegbahn line designation	Electron transition	Siegbahn line designation	Electron transition
K_{α_1}	$L_{III} \rightarrow K$	L_{β_1}	$M_{IV} \rightarrow L_{II}$
K_{α_2}	$L_{II} \rightarrow K$	L_{β_3}	$M_{III} \rightarrow L_I$
K_{β_1}	$M_{III} \rightarrow K$	L_{β_4}	$M_{II} \rightarrow L_I$
K_{β_3}	$M_{II} \rightarrow K$	L_η	$M_I \rightarrow L_{II}$
K_{β_5}	$M_{IV, V} \rightarrow K$	$L_{\gamma_{2,3}}$	$N_{II, III} \rightarrow L_I$
K_{β_2}	$N_{II, III} \rightarrow K$	L_{β_6}	$N_I \rightarrow L_{III}$
K_{β_4}	$N_{IV, V} \rightarrow K$	M_{ζ_1}	$N_{III} \rightarrow M_V$
L_{α_1}	$M_V \rightarrow L_{III}$	M_{ζ_2}	$N_{II} \rightarrow M_{IV}$
L_{α_2}	$M_{IV} \rightarrow L_{III}$		

Note: Not all lines are seen for all elements, and many of the lines are not resolved with standard X-ray spectrometers. Many $M \rightarrow M$, $N \rightarrow M$, $O \rightarrow L$, and $O \rightarrow M$ transitions have no Siegbahn notation associated with them.

Figure 8.3 X-ray emission spectrum obtained by bombarding rhodium metal (Rh) with electrons. Both a broad continuum (bremsstrahlung) and sharp emission lines from rhodium are seen. [Courtesy of ThermoNoran (www.thermo.com).]

of Spectroscopy, Vol. 1. Not all X-ray lines have a Siegbahn designation, so the IUPAC established a new identification system for X-ray lines in 1991. Appendix 8.1 contains a list of the Siegbahn notation for lines and the IUPAC designation for these lines.

The wavelengths of the characteristic lines depend only on the element, because the inner electrons do not take part in bonding. Therefore, the lines are independent of oxidation state, bonding, and physical state, making the use of the characteristic lines an *elemental* analysis technique. No molecular information is obtained from these lines.

The broad continuous "background" emission of X-radiation seen in Fig. 8.3 is due to a second process that occurs when high-energy electrons strike a solid metal. The continuous radiation results from the collision of electrons with the atoms of the solid. At each

Table 8.3 Wavelengths of Absorption Edges and Characteristic Emission Lines of Various Elements

Element	K absorption edge ($\mathring{A}$)	Emission ($\mathring{A}$)	
		$K_{\beta 1,3}$[a]	$K_{\alpha 1,2}$[b]
Mg	9.512	9.559	9.890
Ti	2.497	2.514	2.748, 2.752
Cr	2.070	2.085	2.290, 2.294
Mn	1.896	1.910	2.102, 2.106
Ni	1.488	1.500	1.658, 1.662
Ag	0.4858	0.4971, 0.4977	0.5594, 0.5638
Pt	0.1582	0.1637, 0.1645	0.1855, 0.1904
Hg	0.1492	0.1545, 0.1553	0.1751, 0.1799

[a]When more than one number is listed, $K_{\beta 1}$ is listed first.
[b]When more than one number is listed, $K_{\alpha 1}$ is listed first.

collision, the electron loses some energy, and decelerates with the production of an X-ray photon. The energy of the photon is equal to the kinetic energy difference of the electron as a result of the collision. Each electron generally undergoes a series of collisions with each collision resulting in a photon of slightly different energy. The result of these many collisions is emission of a continuum of X-rays over a wide wavelength range. This continuous radiation is called *bremsstrahlung* or *white radiation*.

When all the energy of the impinging electrons is turned into X-rays, as would occur if the electrons transferred all their energy in one collision, the wavelength of the emitted photons is the shortest attainable. This is termed the minimum λ or λ_{min}. The radiation with the highest energy (and therefore the shortest wavelength) is deduced as follows. When all the energy of the electrons is converted to radiant energy, then the energy of the electrons equals the energy of the radiation. The energy of the radiation is given by $E = h\nu$, whereas the energy of the electrons is given by $E = eV$. When they are equal, $h\nu = eV$, where e is the charge of the electron; V, the applied voltage; and ν, the frequency of the radiation. But:

$$\nu = \frac{c}{\lambda}$$

where c is the speed of light and λ is the wavelength of radiation. Therefore,

$$h\nu = \frac{hc}{\lambda} = eV \tag{8.3}$$

Rearranging, we get

$$\lambda = \frac{hc}{eV} \tag{8.4}$$

When all the energy of the electron is converted to x-radiation, the wavelength of the radiation is a minimum and we achieve minimum λ conditions:

$$\lambda_{min} = \frac{hc}{eV} \tag{8.5}$$

Inserting the values for h, c, and e, which are constants, we have the **Duane–Hunt Law**,

$$\lambda_{min} = \frac{(6.626 \times 10^{-34}\,\text{J s})(3.00 \times 10^8\,\text{m/s})(10^{10}\,\text{Å/m})}{(1.60 \times 10^{-19}\,\text{C}) \times V} = \frac{12,400}{V} \tag{8.6}$$

where h is Planck's constant; c, the speed of light; e, the charge of an electron; V, the applied voltage (in volts); and λ_{min}, the shortest wavelength of X-rays radiated (in angstroms). The continuum radiation spectrum from a solid metal therefore has a well-defined short wavelength limit. This limit is a function of the accelerating voltage, but not of the solid metal. The same λ_{min} would be obtained by bombardment of lead or tungsten or rhodium at the same accelerating voltage.

The Duane–Hunt Law gives the conversion factor between energy and wavelength. Most X-ray systems express wavelength in angstroms and energy in keV. To convert between these units, Eq. (8.6) gives:

$$\text{Energy (keV)} = \frac{12.4}{\lambda\,(\text{Å})} \tag{8.7}$$

An X-ray emission spectrum is similar for all elements, in that K_α, K_β, L_α, and L_β lines may be seen, if the element possesses enough electrons to populate the appropriate levels. However, the actual wavelengths of these lines vary from one element to another, depending on the *atomic number* of the particular element. A mathematical relationship was discovered between the wavelengths of the K series and the atomic number of the element, and similar relationships were found for the L lines, and others.

8.1.2. Moseley's Law

Henry Moseley, a young graduate student working at Cambridge, UK, in 1913, discovered the relationship between wavelength for characteristic X-ray lines and atomic number. After recording the X-ray spectra from numerous elements in the periodic table, he deduced the mathematical relationship between the atomic number of the element and the wavelength of the K_α line. A similar relationship was found between the atomic number and the K_β line, the L_α line, and so on. The relationships were formulated in **Moseley's Law**, which states that

$$v = c/\lambda = a(Z - \sigma)^2 \tag{8.8}$$

where c is the speed of light; λ, the wavelength of the X-ray; a, a constant for a particular series of lines (e.g., K_α or L_α lines); Z, the atomic number of the element; and σ, a screening constant that accounts for the repulsion of other electrons in the atom. A partial Moseley's Law plot for the K_α, K_β, L_α, and L_β emission lines is shown in Fig. 8.4. Shortly after this monumental discovery, Moseley was killed in action in World War I. The impact of Moseley's Law on chemistry was substantial, in that it provided a method of unequivocally assigning an atomic number to newly discovered elements, of which there were several at that time. In addition, it clarified disputes concerning the positions of all known elements in the periodic table, some of which were still in doubt in the early part of the 20th century.

8.1.3. X-Ray Methods

There are several distinct fields of X-ray analysis used in analytical chemistry and materials characterization; namely, X-ray absorption, X-ray diffraction, X-ray fluorescence, and X-ray emission. X-ray emission is generally used for microanalysis, with either an electron

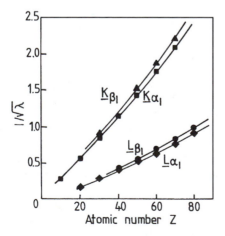

Figure 8.4 Partial Moseley's Law plots for selected K and L lines, showing the relationship between the X-ray emission wavelength and atomic number of the element. Using this relationship it was possible to predict undiscovered elements and to correctly assign atomic numbers to known elements. (From Helsen and Kuczumow, 2nd ed., used with permission.)

microprobe (Chapter 14) or a scanning electron microscope. The basic principles of each are described below.

8.1.3.1. The X-ray Absorption Process

The absorption spectrum obtained when a beam of X-rays is passed through a thin sample of a pure metal is depicted in Fig. 8.5. As is the case with other forms of radiation, some of the intensity of the incident beam may be absorbed by the sample while the remainder is transmitted. We can write a Beer's Law expression for the absorption of X-rays by a thin sample:

$$I(\lambda) = I_0(\lambda)\, e^{-\mu_m \rho x} \tag{8.9}$$

where $I(\lambda)$ is the transmitted intensity at wavelength λ; $I_0(\lambda)$, the incident intensity at the same wavelength; μ_m, the **mass absorption coefficient** (in cm^2/g); ρ, the density of the sample (in g/cm^3); and x, the sample thickness (in cm). The mass absorption coefficient is a constant for a given element at a given wavelength and is independent of both the chemical and physical state of the element. Tables of mass absorption coefficients can be found in the text by Bertin or in handbooks listed in the bibliography.

Of course, most samples do not consist of a single pure element. The total mass absorption coefficient for a sample can be calculated by adding the product of the individual mass absorption coefficients for each element times the weight fraction of the element present in the sample. That is, for a metal alloy like steel,

$$\mu_{total} = w_{Fe}\mu_{Fe} + w_{Cr}\mu_{Cr} + w_{Ni}\mu_{Ni} + \cdots \tag{8.10}$$

where w_{Fe} is the weight fraction of iron and μ_{Fe} is the mass absorption coefficient for pure iron, w_{Cr} is the weight fraction of chromium, and so on for all the elements in the

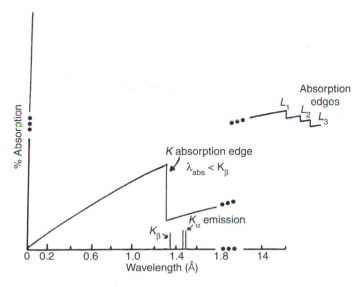

Figure 8.5 The X-ray absorption spectrum of a pure metal. Longer wavelengths are more readily absorbed than shorter wavelengths. The absorption spectrum is characterized by absorption edges, which are abrupt increases in absorption at energies sufficient to eject an electron from one of the atomic shells. The K absorption edge occurs at an energy sufficient to eject an electron from the K shell of the given metal.

alloy. For accurate quantitative work, the *mass attenuation coefficient* is used in place of the mass absorption coefficient. The mass attenuation coefficient takes into account both absorption and scattering of X-rays by the sample.

The amount of light absorbed increases as the wavelength increases. This is reasonable since longer wavelengths have less energy and a less energetic photon has less "penetrating power" and is more likely to be absorbed. Only a few absorption peaks are seen in an X-ray absorption spectrum, but there is a very distinct feature in these spectra. An abrupt change in absorptivity (or the mass absorption coefficient) occurs at the wavelength of the X-ray necessary to eject an electron from an atom. These abrupt changes in X-ray absorptivity are termed *absorption edges*. Looking at Fig. 8.5, it can be seen that radiation with a wavelength of 1.8 Å has a certain percent absorption value. As the wavelength of the X-ray *decreases*, its energy increases, its penetrating power increases, and the percent absorption decreases. This can be seen by the downward slope of the absorption trace, moving to the left along the *x*-axis from a wavelength of 1.8 Å. As the wavelength decreases further, the X-ray eventually has sufficient energy to displace electrons from the K shell. This results in an abrupt *increase* in absorption. This is manifested by the **K absorption edge**. After the absorption edge, the penetrating power continues to increase as the wavelength decreases further until finally the degree of absorption is extremely small at very small wavelengths. At wavelengths less than 0.2 Å, penetrating power is extremely great and we are approaching the properties of interstellar radiation such as cosmic rays, which have extremely high penetrating power. Wavelengths shorter than the K absorption edge have sufficient energy to eject K electrons; the bombarded sample will exhibit both continuum radiation and the characteristic K lines for the sample. This process is called XRF and will be discussed in detail. Wavelengths just slightly longer than the K absorption edge do not have enough energy to displace K electrons. The absorption spectrum is unique for each element; portions of the absorption spectrum showing the position of the K absorption edge for several pure elements are shown in Fig. 8.6.

Another way of looking at X-ray absorption is to plot the mass absorption coefficient as a function of wavelength or energy. For a thin sample of pure metal and a constant incident intensity, Eq. (8.9) indicates that if the percent absorption changes as a function of wavelength, it must be that μ_m changes. A plot of μ_m vs. X-ray energy for the element lead is shown in Fig. 8.7. The K, L, and M absorption edges are seen.

The wavelengths of the absorption edges and of the corresponding emission lines do not quite coincide. This is because the energy required to dislodge an electron from an atom (the absorption edge energy) is not quite the same as the energy released when the dislodged electron is replaced by an electron from an outer shell (emitted X-ray energy). The amount of energy required to displace the electron must dislodge it from its orbital and remove it completely from the atom. This is more than the energy released by an electron in an atom falling from one energy level to another. A few absorption edge values are given in Table 8.3. Figure 8.5 shows that the energy of the K absorption edge is greater than the energy of the K emission lines. As opposed to emission spectra, only one K absorption edge is seen per element, since there is only one energy level in the K shell. Three absorption edges of different energies are observed for the L levels, five for the M levels, and so on; these can be seen in Fig. 8.7. A comprehensive table of absorption edge wavelengths is located in Appendix 8.2.

8.1.3.2.   *The X-ray Fluorescence (XRF) Process*

X-rays can be emitted from a sample by bombarding it with electrons or with other X-rays. When electrons are used as the excitation source, the process is called X-ray emission.

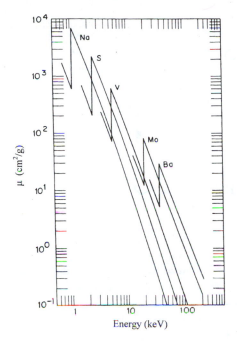

Figure 8.6 Energies of the K absorption edge for several pure elements. [Courtesy of ORTEC (Ametek) (www.ortec-online.com). From Jenkins et al., 1981, used with permission.]

This is the basis of X-ray microanalysis using an electron microprobe (Chapter 14) or a scanning electron microscope. When the excitation source is a beam of X-rays, the process of X-ray emission is called fluorescence. This is analogous to molecular fluorescence discussed in Chapter 5 and atomic fluorescence discussed in Chapter 7,

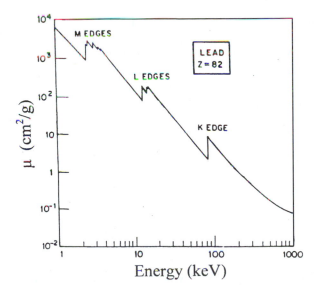

Figure 8.7 The mass absorption coefficient for Pb as a function of energy. The K, L, and M absorption edges are seen. [Courtesy of ORTEC (Ametek) (www.ortec-online.com). From Jenkins et al., 1981, used with permission.]

because the wavelength of excitation is shorter than the emitted wavelengths. The beam of exciting X-rays is called the *primary* beam; the X-rays emitted from the sample are called *secondary* X-rays. The use of an X-ray source to produce secondary X-rays from a sample is the basis of XRF spectroscopy. The primary X-ray beam must have a λ_{min} that is shorter than the absorption edge of the element to be excited.

8.1.3.3.   *The X-ray Diffraction (XRD) Process*

Crystals consist of atoms, ions or molecules arranged in a regular, repeating 3D pattern, called a crystal lattice. This knowledge came from the pioneering work of German physicist Max von Laue and the British physicists, W.H. Bragg and W.L. Bragg. Max von Laue demonstrated in 1912 that a crystal would diffract X-rays, just as a ruled grating will diffract light of a wavelength close to the distance between the ruled lines on the grating. The fact that diffraction occurs indicates that the atoms are arranged in an ordered pattern, with the spacing between the planes of atoms on the order of short wavelength electromagnetic radiation in the X-ray region. The diffraction pattern could be used to measure the atomic spacing in crystals, allowing the determination of the exact arrangement in the crystal, the *crystal structure*. The Braggs used von Laue's discovery to determine the arrangement of atoms in several crystals and to develop a simple 2D model to explain XRD.

If the spacing between the planes of atoms is about the same as the wavelength of the radiation, an impinging beam of X-rays is reflected at each layer in the crystal, as shown in Fig. 8.8. Assume that the X-rays falling on the crystal are parallel waves that strike the crystal surface at angle θ. That is, the waves O and O' are in phase with each other and reinforce each other. In order for the waves to emerge as a reflected beam after scattering from atoms at points D and B, they must still be in phase with each other, that is, waves Y and X must still be parallel and coherent. If the waves are completely out of phase, their amplitudes cancel each other, they are said to interfere destructively, and no beam emerges. In order to get reinforcement, it is necessary that the two waves stay in phase with each other after diffraction at the crystal planes.

It can be seen in Fig. 8.8 that the lower wave travels an extra distance AB + BC compared with the upper wave. If AB + BC is a whole number of wavelengths, the emerging beams Y and X will remain in phase and reinforcement will take place. From this deduction, we can calculate the relationship between the wavelengths of X-radiation, the distance *d* between the lattice planes, and the angle at which a diffracted beam can emerge. We employ the following derivation.

Figure 8.8   Diffraction of X-rays by crystal planes.

X-ray waves O and O′ are parallel. The extra distance traveled by wave O′ in traveling through the crystal is AB + BC. For diffraction to occur it is necessary that this distance be a whole number of wavelengths, n; that is,

$$\text{distance AB} + \text{BC} = n\lambda \tag{8.11}$$

but

$$\text{AB} + \text{BC} = 2\text{AB} \tag{8.12}$$

and

$$\text{AB} = \text{DB} \sin \theta \tag{8.13}$$

where θ is the angle of incidence of the X-ray beam with the crystal; therefore

$$\text{AB} = d \sin \theta \tag{8.14}$$

where d is the distance between the crystal planes, called the interplanar distance. (Note from Fig. 8.8 that $d = \text{DB}$).

Therefore

$$\text{AB} + \text{BC} = n\lambda = 2\text{AB} = 2d \sin \theta$$

or

$$n\lambda = 2d \sin \theta \tag{8.15}$$

The equation $n\lambda = 2d \sin \theta$ is known as the **Bragg equation**. The important result of this equation is that at any particular angle of incidence θ, only X-rays of a particular wavelength fulfill the requirement of staying in phase and being reinforced, and are therefore diffracted by the crystal. Diffraction of X-rays by crystals forms the basis of XRD for crystal structure determination and is also the reason XRF spectrometry is possible, as will be seen.

8.2. INSTRUMENTATION

Instrumentation for X-ray spectrometry requires a source, a wavelength (or energy) selector, a detector, collimators, and filters. The component parts of the instrument are similar for XRF, XRD, and the other fields, but the optical system varies for each one. For example, in XRF spectrometry, either the energies or wavelengths of emitted X-rays are measured to characterize the elements emitting them. In the *wavelength-dispersive mode* of analysis (WDXRF), a dispersing device separates X-rays of differing wavelength by deflecting them at different angles proportional to their wavelength. In the *energy-dispersive mode* (EDXRF), there is no dispersing device, and a detector measures and records the energies of each individual detected X-ray photon.

The low energy X-rays emitted by elements with atomic numbers less than sodium ($Z < 11$) are easily absorbed by air. Therefore most X-ray systems operate either under vacuum or purged with helium. The entire spectrometer, including the source, sample, optics, and most detectors are within the vacuum/purge chamber. Liquid samples cannot be analyzed under vacuum, so most systems permit the analyst to switch from a vacuum to a helium purge as needed, usually in less than 2 min.

Commercial X-ray spectrometers may be equipped with automatic sample changers for unattended analysis of multiple samples. Computer-controlled spectrometers permit the identification of the position of liquid samples in the sample changer, and automatically switch to a helium purge to avoid exposing the liquid samples to vacuum.

8.2.1. X-Ray Source

Three common methods of generating X-rays for analytical use in the laboratory are:

1. Use of a beam of high-energy electrons to produce a broad band *continuum* of X-radiation resulting from their deceleration upon impact with a metal target as well as element-specific X-ray *line* radiation by ejecting inner core electrons from the target metal atoms. This is the basis of the X-ray tube, the most common source used in XRD and XRF.
2. Use of an X-ray beam of sufficient energy (the primary beam) to eject inner core electrons from a sample to produce a secondary X-ray beam (XRF).
3. Use of a radioactive isotope which emits very high energy X-rays (also called gamma radiation) in its decay process.

A fourth method of producing X-rays employs a massive, high-energy particle accelerator called a synchrotron. These are available at only a few locations around the world, such as the Brookhaven National Laboratory or the Stanford Accelerator Center in the US, and are shared facilities servicing a large number of clients. X-rays may be generated when alpha particles or other heavy particles bombard a sample; this technique is called particle-induced X-ray emission (PIXE) and requires a suitable accelerator facility. The use of an electron beam to generate X-rays from a microscopic sample as well as an image of the sample surface is the basis of X-ray microanalysis using an electron microprobe or scanning electron microscopy.

These different X-ray sources may produce either broad band (continuum) emission or narrow line emission, or both simultaneously, depending on how the source is operated. Figure 8.3 displays the simultaneous generation of both a broad continuum of X-ray energies with element-specific lines superposed upon it, obtained by bombarding rhodium metal with electrons in an **X-ray tube**.

8.2.1.1. The X-ray Tube

A schematic X-ray tube is depicted in Fig. 8.9. The X-ray tube consists of an evacuated glass envelope containing a wire filament cathode and a pure metal anode. A thin beryllium window sealed in the glass envelope allows X-rays to exit the tube. The glass envelope is encased in lead shielding and a heavy steel jacket with an opening over the window, to protect the tube. When a cathode (a negatively charged electrode) in the form of a metal wire is electrically heated by the passage of current, it gives off electrons, a process called thermionic emission. If a positively charged metallic electrode (called an anode) is placed

Figure 8.9   Schematic diagram of an X-ray tube.

near the cathode *in a vacuum*, the negatively charged electrons will be accelerated toward
the anode. Upon striking it, the electrons give up their energy at the metallic surface of the
anode. If the electrons have been accelerated to a high enough velocity by a sufficiently
high voltage between the cathode and anode, energy is released as radiation of very
short wavelength (0.1–100 Å), called X-radiation or X-rays. X-ray tubes are generally
operated at voltage differentials of 4–50 kV between the wire filament cathode and the
anode.

The cathode is normally a tungsten wire filament. The anode is called the target. The
X-ray tube is named for the anode; a copper X-ray tube has a copper anode and a tungsten
wire cathode, a rhodium tube has a rhodium anode and a tungsten wire cathode, a tungsten
tube has a tungsten anode and a tungsten wire cathode. Numerous metals have been used
as target materials, but common target elements are copper, chromium, molybdenum,
rhodium, gold, silver, palladium, and tungsten. The wavelengths of the X-ray line radiation emitted by the target depend on the metal used. The voltage between the anode
and cathode determines how much energy the electrons in the beam acquire, and this in
turn determines the overall intensity of the wide range of X-ray intensities in the continuum distribution and the maximum X-ray energy (shortest wavelength). Figure 8.10
shows how the intensity of the continuum radiation from a tungsten tube and the short
wavelength cutoff vary as the applied voltage to the tube varies. Figure 8.11 displays
the variation in intensity of one of the characteristic tungsten lines from a tungsten
X-ray tube as the applied voltage is changed.

In choosing the element to be used for the target, it should be remembered that it is
necessary for the energy of the X-rays emitted by the source to be greater than that
required to excite the element being irradiated in an XRF analysis. As a simple rule of
thumb, the target element of the source should have a greater atomic number than the
elements being examined in the sample. This ensures that the energy of radiation is more
than sufficient to cause the sample element to fluoresce. This is not a requirement in X-ray

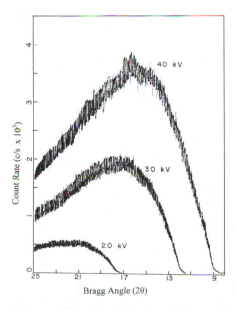

Figure 8.10 The intensity of the continuum radiation from an X-ray tube and the short
wavelength cut-off vary as the applied voltage varies. This plot is of a tungsten X-ray tube.

Figure 8.11 The characteristic radiation from an X-ray tube also varies as the applied voltage varies. Below 20 kV, the intensity of the tungsten L_{α_1} line is very low. The intensity of this characteristic line increases as the applied voltage increases.

absorption or XRD, where excitation of the analyte atoms is not necessary. In many X-ray tube designs, the anode, or target, gets very hot in use, because it is exposed to a constant stream of high-energy electrons, with most of the energy being dissipated as heat on collision. This problem is overcome by water-cooling the anode. Modern X-ray tubes have been designed to operate at lower voltages and do not require water-cooling of the anode.

The exit window of the X-ray tube is usually made of beryllium, which is essentially transparent to X-rays. The Be window is thin, generally 0.3–0.5 mm thick, and is very fragile. The window may be on the side of the tube, as shown in Fig. 8.9, or in the end of the tube. Side window tubes are common, but end-window tubes permit the use of a thinner beryllium window. This makes end-window tubes good for low energy X-ray excitation by improving the low-energy output of the tube. A commercial end-window tube is depicted in Fig. 8.12(a). Figure 8.12(b) shows a close up of the anode and cathode designs in this tube.

X-ray tubes must provide adequate intensity over a relatively wide spectral range for XRF in order to excite a reasonable number of elements. In some applications, monochromatic or nearly monochromatic X-rays are desired; that is accomplished by using filters or a monochromator as described below or by using a secondary fluorescent source, described subsequently. The tube output must be stable over short time periods for the highest precision and over long time periods for accuracy without frequent recalibration. The X-ray emission lines from the anode element must not interfere with the sample spectrum. Tube lines can be scattered into the detector and be mistaken for an element present in the sample.

8.2.1.2. *Secondary XRF Sources*

If it is necessary to prevent the continuum emission from an X-ray tube from falling on a sample, a standard tube can be used to excite another pure metal target. The resulting XRF from the secondary target is used as the source of X-ray excitation for the sample. Such an example is shown in Fig. 8.13. A standard tungsten X-ray tube is used to produce the emission spectrum on the left, with the tungsten characteristic lines superimposed on the continuum radiation. The radiation from this tube is used to strike a secondary pure copper target. The resulting emission from the copper is the copper XRF spectrum on the right.

Figure 8.12 (a) Schematic of the 4.0 kW ceramic end window X-ray tube, called the Super Sharp Tube™. (b) Close up of the window end of the Super Sharp Tube™, showing the circular cathode design of this tube. [Courtesy of PANalytical, Inc., The Netherlands (www.panalytical.com).]

This source emits very little or no continuum radiation but does emit quite strongly at the copper K and L lines. Of course, the metal used in the target of the first source must have a higher atomic number than copper to generate fluorescence. The Cu lines then can be used as an excitation source, although the intensity of the secondary source is much less than that of a Cu X-ray tube. However, when monochromatic or nearly monochromatic radiation is required, the loss of intensity is more than offset by the low background from the secondary source.

8.2.1.3. Radioisotope sources

X-radiation is a product of radioactive decay of certain isotopes. The term gamma ray is often used for an X-ray resulting from such a decay process. Alpha and beta decay and electron capture processes can result in the release of gamma rays. Table 8.4 lists some common radioisotopes used as XRF sources.

The advantages of radioisotope sources are that they are small, rugged, portable, and do not require a power supply. They are ideal for obtaining XRF spectra from bulky samples that do not fit into conventional spectrometers (and cannot have pieces cut from them), such as aircraft engines, ship hulls, art objects, and the like. The disadvantage is that the intensity of these sources is weak compared with that of an X-ray tube, the source cannot be optimized by changing voltage as can be done with an X-ray tube,

Figure 8.13 (a) The primary X-ray emission from a tungsten target. (b) The secondary emission from a copper target. Note the removal of the continuum radiation and the nearly monochromatic output from the secondary target. (From Parsons, used with permission.)

and the intensity of the source drops off with time, depending on the half-life of the isotope. In addition, the source cannot be turned off. This requires care on the part of the analyst to avoid exposure to the ever-present ionizing radiation.

8.2.2. Collimators

The X-rays emitted by the anode are radially directed. As a result, they form a hemisphere with the target at the center. In WDXRF spectroscopy or XRD structural determination, the monochromator's analyzing crystal or the crystalline substance undergoing structure determination requires a nearly parallel beam of radiation to function properly. A narrow, nearly parallel beam of X-rays can be made by using two sets of closely packed metal plates separated by a small gap. This arrangement absorbs all the radiation except the narrow beam that passes between the gap. Decreasing the distance between the plates or increasing the total length of the gap decreases the divergence of the beam of X-rays (i.e., it collimates, or renders them parallel). The use of a collimator increases the wavelength resolution of a monochromator's analyzing crystal, cuts down on stray X-ray emission, and reduces background. Commercial instruments use multiple tube or multiple slit collimator arrangements, often both before the analyzing crystal (the primary collimator) and before the detector (the secondary collimator). The collimator positions in a sequential WDXRF spectrometer are shown schematically in Fig. 8.14. In many wavelength dispersive instruments, two detectors are used in tandem, and a third auxiliary collimator may be required. Such an arrangement is shown in Fig. 8.15.

Table 8.4 Characteristics of Radioisotope Sources for XRF Spectrometry

Isotope	Primary decay mode	Half-life (years)	Useful photon energies emitted	% Theoretical yield (photons per 100 decay transformations)	Typical activity (mCi)
^{55}Fe	Electron capture	2.7	5.9, 6.4 keV Mn K X-rays	28.5	5–100
^{109}Cd	Electron capture	1.3	22.2–25.5 keV Ag K X-rays	102	0.5–100
			88.2 keV γ-ray	4	
^{241}Am	Alpha	458	14–21 keV Np L X-rays	37	1–50
			59.6 keV γ-ray	36	
^{57}Co	Electron capture	0.74	6.4, 7.1 keV Fe K X-rays	51	1
			14.4 keV γ-ray	8.2	
			122 keV γ-ray	88.9	
			136 keV γ-ray	8.8	
$^{3}H^{a}$	Beta	12.3	Bremsstrahlung source, endpoint at 18.6 keV		3000–5000
^{147}Pm	Beta	2.6	Bremsstrahlung source, endpoint at 225 keV		500

Source: Table from Jenkins et al., used with permission.
[a]Radioactive tritium gas is adsorbed on nonradioactive metal foil, such as titanium foil.

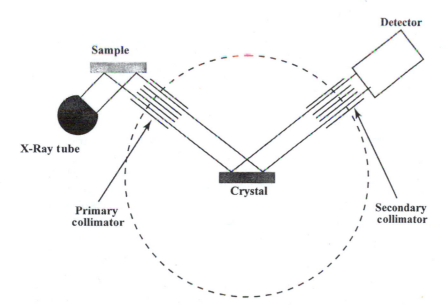

Figure 8.14 Schematic of the optical path in a wavelength-dispersive sequential spectrometer, showing the positions of the collimators. [Courtesy of PANalytical, Inc., The Netherlands (www. panalytical.com).]

Figure 8.15 A sequential spectrometer with two tandem dectors, showing the placement of the collimators in the optical path. (From Jenkins et al., 1981, used with permission.)

Collimators are not needed for curved crystal spectrometers where slits or pinholes are used instead nor are they needed for energy dispersive spectrometers.

8.2.3. Filters

One of the problems of using the X-ray tube illustrated in Fig. 8.9 is that both continuum and characteristic line radiation is generated at certain operating voltages, as seen in Fig. 8.3. For many analytical uses, only one type of radiation is desired. Filters of various materials can be used to absorb unwanted radiation but permit radiation of the desired wavelength to pass by placing the filter between the X-ray source and the sample.

A simple example of how a filter is used is shown in Fig. 8.16. The solid line spectrum is the output of a Rh tube operated at 20 kV with no filter between the tube and the detector. The Rh L_α line at 2.69 keV is seen, along with a broad continuum of X-rays from 4 to 19 keV. If the Rh L_α line gets scattered into the detector, as it can from a crystalline sample, it can be mistaken for an element in the sample or may overlap another line, causing spectral interference. Placing a cellulose filter over the tube window causes the low energy Rh characteristic line to be absorbed; only the continuum radiation reaches the detector, as shown by the dotted line spectrum.

Alternatively, when monochromatic radiation is desired, a filter is chosen with its absorption edge between the K_α and the K_β emission lines of the target element. The filter then absorbs the K_β line and all shorter wavelengths, including much of the continuum; the light reaching the sample is essentially the K_α line of the target. Filters are commonly thin metal foils, usually pure elements, but some alloys such as brass and materials like cellulose are used. Varying the foil thickness of a filter is used to optimize peak-to-background ratios. Commonly used filters for various targets are listed in Table 8.5. Figure 8.17 shows a commercial sequential X-ray spectrometer with a series of selectable beam filters located between the X-ray tube and the sample. The filters are computer-controlled and are changed automatically according to the analytical program set up in the instrument software.

Tube voltage = 20 kV

Figure 8.16 The use of a filter to remove unwanted radiation from entering the spectrometer is demonstrated. A cellulose filter placed between a Rh X-ray tube and the sample removes the Rh L_α line at 2.69 keV and allows only the continuum radiation to excite the sample. [Courtesy of ThermoNoran (www.thermo.com).]

8.2.4. WDXRF Spectrometers

Schematics of sequential WDXRF spectrometers are shown in Figs. 8.14, 8.15, and 8.17. In the configurations shown, the source is placed under the sample; the sample is presented surface-down to the X-ray beam. Some instruments have the tube above the sample, with the sample surface facing up. There are advantages and disadvantages to both designs, as we shall see. The sample fluoresces as a result of excitation by the source. The sample fluorescence is directed through the primary collimator to the *analyzing crystal*. Diffraction

Table 8.5 Filters for Commonly Available X-Ray Tubes

Target	Target K_α (Å)	Target K_β (Å)	Filter element	Filter K-edge (Å)	Foil thickness (μm)	% K_β absorbed
Cr	2.289	2.085	V	2.269	15.3	99.0
Fe	1.936	1.757	Mn	1.896	12.1	98.7
Co	1.789	1.621	Fe	1.743	14.7	98.9
Ni	1.658	1.500	Co	1.608	14.3	98.4
Cu	1.541	1.393	Ni	1.488	15.8	97.9
Mo	0.709	0.632	Zr	0.689	63.0	96.3
Ag	0.559	0.497	Pd	0.509	41.3	94.6
W	0.209	0.185				

Note: No suitable filter is available for tungsten.
Source: From Parsons, used with permission.

Figure 8.17 The schematic layout of a commercial sequential X-ray spectrometer, the MagiX, showing a series of selectable beam filters located between the X-ray tube and the sample. [Courtesy of PANalytical, Inc., The Netherlands (www.panalytical.com).]

occurs at the crystal planes according to Bragg's Law and X-rays of different wavelengths are diffracted at different angles. The diffracted X-rays are passed through another collimator to one or more detectors. Figure 8.15 depicts two detectors in tandem, one behind the other. The analyzing crystal is mounted on a turntable that can be rotated through θ degrees (see the arrow marked θ on the lower left side of the diagram). The detector(s) are connected to the crystal turntable so that when the analyzing crystal rotates by θ degrees, the detector rotates through 2θ degrees, as shown by the marked arrow. Therefore the detector is always in the correct position (at the Bragg angle) to detect the dispersed and diffracted fluorescence. This crystal positioning system is called a **goniometer**. Figure 8.18 shows the turntable and the concentric circles made by the crystal and the detector. In most systems, the maximum diffraction angle attainable is $75°$ θ (or $150°$ 2θ).

In some systems the rotation of the crystal and the detector is mechanically coupled with gears. Other systems have no mechanical coupling but use computer-controlled stepper motors for the crystal and the detector. The newest systems use optical position control by optical sensors or optical encoding devices. Optical position control permits very high angular precision and accuracy and very fast scanning speeds.

8.2.4.1. The Analyzing Crystal

As we have discussed, a crystal is made up of layers of ions, atoms, or molecules arranged in a well-ordered system, or lattice. If the spacing between the layers of atoms is about the same as the wavelength of the radiation, an impinging beam of X-rays is reflected at each layer in the crystal (Fig. 8.8). Bragg's Law [Eq. (8.14)] indicates that at any particular angle of incidence θ, only X-rays of a particular wavelength fulfill the requirement of staying in phase and being reinforced, and are therefore diffracted by the crystal. If an X-ray beam consisting of a range of wavelengths falls on the crystal, the diffracted beams of different wavelengths emerge at different angles. The incident beam is thus split up by the crystal into its component X-ray wavelengths, just as a prism or grating splits up white light into a spectrum of its component colors. The principle is illustrated

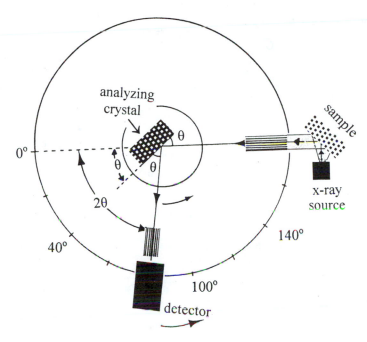

Figure 8.18   Goniometer layout for a sequential XRF spectrometer.

in Fig. 8.19. Figure 8.19 shows schematically that two detectors placed at the proper locations could detect the two wavelengths simultaneously. Alternatively, the detector or analyzing crystal could move, allowing each wavelength to be detected sequentially. Both types of spectrometers are commercially available. Some crystals in common use for dispersion of X-rays in commercial XRF spectrometers are listed in Table 8.6.

The analyzing crystal is an X-ray monochromator. The crystal separates X-rays of different wavelengths by diffracting them at different angles. Bragg's Law fixes the spectral range of a given crystal. Since the maximum value of sin θ is 1.00, the upper spectral limit $\lambda_{max} = 2d$. The diffraction efficiency and the resolution depend on the purity and perfection of the crystal lattice. The crystal should be as perfect as possible, so that the d spacing for a given plane will be constant in all parts of the crystal. As is clear from Table 8.6, different crystals are needed to measure different elements. Commercial sequential XRF systems have a computer-controlled multiple crystal holder inside the spectrometer, with positions for as many as 8–10 crystals in some instruments. The crystals are interchanged but the same goniometer is used to select the diffraction angle, meaning that only one wavelength can be measured at a time with a sequential system.

Figure 8.19   The analyzing crystal as a monochromator.

Table 8.6 Crystals Used for X-Ray Monochromators

Crystal	Orientation (Miller indices)	$2d$ (Å)	Useful element range (atomic number)	Remarks
Lithium fluoride	420	1.80	Ni (28)–U (92)	High resolution
	220	2.85	V (23)–U (92)	High resolution, good intensity
	200	4.02	K (19)–U (92)	General purpose, highest intensity
Silicon	111	6.26	K (19), S (16), Cl (17)	
Germanium	111	6.53	K (19), S (16), Cl (17)	Good for low and intermediate Z elements
Pyrolytic graphite	002	6.71	K (19), S (16), Cl (17)	Good intensity, poor resolution
Indium antimonide	111	7.84	Si (14)	
Pentaerythritol	002	8.72	Al (13)–Cl (17)	Good intensity, soft
Thallium hydrogen phosphate (TlAP)	100	5.75	O (8)–Mg (12)	Poisonous
Synthetic Multilayers[a]				
PX-1		51	O (8)–Mg (12)	Low resolution, good intensity
PX-3		195	B (5)	Low resolution, good intensity
PX-4		122	C (6), N (7), O (8)	Low resolution, good intensity
OVO 55		55	Mg (12), Na (11), F (9)	Low resolution, good intensity
OVO H300		300	Be (4), B (5)	One of the largest spacings available

Source: Data from Helsen and Kuczumow, used with permission.
[a]The designations listed for these synthetic multilayers are commercial tradenames from different instrument manufacturers. A more generic approach is to classify them according to their d spacing.

A serious limitation in XRF was the lack of natural crystals with d spacings large enough to diffract the low energy X-rays from low atomic number elements. That limitation has been overcome by the synthesis of multilayer "pseudocrystals". These are made from alternating layers of materials with high and low optical densities deposited on a silicon or quartz flat. The PX3 multilayer is made from B_4C alternating with Mo, for example. These engineered multilayers are stable, commercially available, and permit the routine determination of elements as light as Be in samples.

The analyzing crystal shown schematically in Fig. 8.14 has a flat surface. Flat crystals are used in scanning (sequential) spectrometers. Curved crystals, both natural and synthetic multilayers, are used in simultaneous spectrometers, electron microprobes, and for synchrotron X-ray spectrometry. The advantage to a curved crystal is that the X-rays are focused and the collimators replaced by slits, resulting in much higher intensities than with flat crystal geometry. This makes curved crystals excellent for analysis of very small samples. The use of a curved crystal and slits in a simultaneous spectrometer is illustrated schematically in Fig. 8.20. The curved crystal spectrometer geometry should remind you of the Rowland circle geometry for optical emission spectrometers discussed in Chapter 7.

8.2.4.2. Detectors

X-ray detectors transform photon energy into electrical pulses. The pulses (and therefore, the photons) are counted over a period of time. The *count rate*, usually expressed as counts per second, is a measure of the intensity of the X-ray beam. Operating the detector as a photon counter is particularly useful with low-intensity sources, as is often the case with X-radiation.

There are three major classes of X-ray detectors in commercial use: gas-filled detectors, scintillation detectors, and semiconductor detectors. Semiconductor detectors will be discussed with EDXRF equipment. Both WDXRF and EDXRF detection makes use of a signal processor called a pulse height analyzer or selector in conjunction with the detector, and discussed subsequently.

8.2.4.3. Gas-Filled Detectors

Suppose we take a metal cylinder, fit it with X-ray transparent windows, place in its center a positively charged wire, fill it with inert filler gas, such as helium, argon, or xenon, and seal it. If an X-ray photon enters the cylinder, it will collide with and ionize a molecule of the filler gas by ejecting an *outer shell electron*, creating a *primary ion pair*. With helium as a filler gas, the ion pair would be He^+ and a photoelectron e^-. A sealed gas-filled detector of this type is illustrated in Fig. 8.21. The interaction

$$h\nu + He \longrightarrow He^+ + e^- + h\nu$$

takes place inside the tube. The electron is attracted to the center wire by the applied potential on the wire. The positive charge causes the wire to act as the anode, while the positive ion, He^+ in this case, migrates to the metal body (the cathode). The ejected photoelectron has a very high kinetic energy. It loses energy by colliding with and ionizing many

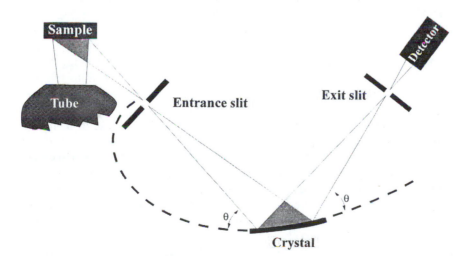

Figure 8.20 Schematic of the optical path in a curved crystal spectrometer. [Courtesy of PANalytical, Inc., The Netherlands (www.panalytical.com).]

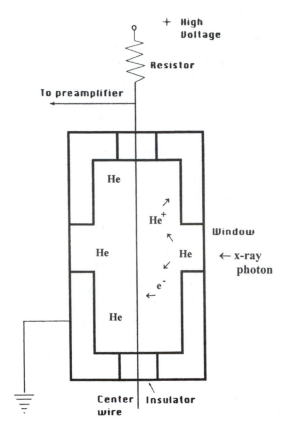

Figure 8.21 Schematic diagram of a gas-filled X-ray detector tube. He filler gas is ionized by X-ray photons to produce He^+ ions and electrons, e^-. The electrons move to the positively charged center wire and are detected. (Modified from Parsons, used with permission.)

additional gas molecules as it moves to the center wire. A plot of the number of electrons reaching the wire vs. the applied potential is given in Fig. 8.22.

With no voltage applied, the electron and the positive ion (He^+) recombine and no current flows. As the voltage is slowly increased, an increasing number of electrons reach the anode, but not all of them; recombination still occurs. This is the sloping region marked A in Fig. 8.22. At the plateau marked B in Fig. 8.22, all the electrons released by a single photon reach the anode and the current is independent of small changes in the voltage. A detector operating under these voltage conditions is known as an *ionization counter*. Ionization counters are not used in X-ray spectrometers because of their lack of sensitivity.

As the voltage increases further, the electrons moving toward the center wire are increasingly accelerated. More and more electrons reach the detector as a result of an avalanche of secondary ion pairs being formed and the signal is greatly amplified. In the region marked C in Fig. 8.22, the current pulse is proportional to the energy of the incoming X-ray photon. This is the basis of a *proportional counter*. In X-ray spectrometry, gas-filled detectors are used exclusively in this range, that is, as proportional counters. The amplification factor is a complex function that depends on the ionization potential of the filler gas, the anode potential, the mean free path of the photoelectrons, and other factors. It is critical that the applied potential, filler gas pressure, and other

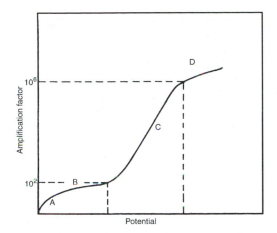

Figure 8.22 Gas-filled detector response vs. potential. A detector operating at the plateau marked B is an ionization counter. A proportional counter operates in the sloping region marked C where the response is proportional to the energy of the incoming photon. The plateau marked D represents the response of a Geiger counter. (Modified from Helsen and Kuczumow, used with permission.)

factors be kept constant to produce accurate pulse amplitude measurements. There are two main types of proportional counter: flow proportional counters and sealed proportional counters.

As shown in Fig. 8.22, if the voltage is further increased, electrons formed in primary and secondary ion pairs are accelerated sufficiently to cause the formation of more ion pairs. This results in huge amplification in electrons reaching the center wire from each X-ray photon falling on the detector. The signal becomes independent of the energy of the photons and results in another plateau, marked D. This is called the Geiger-Müller plateau; a detector operated in this potential range is the basis of the *Geiger counter* or *Geiger-Müller* tube. It should be noted that a Geiger counter gives the highest signal for an X-ray beam without regard to the photon energy. However, it suffers from a long *dead time*. The dead time is the amount of time the detector does not respond to incoming X-rays. It occurs because the positive ions move more slowly than the electrons in the ionized gas, creating a *space charge*; this stops the flow of electrons until the positive ions have migrated to the tube walls. The dead time in a Geiger counter is on the order of 100 μs, about 100 times longer than the dead time in a proportional counter. Due to the long dead time compared with other detectors, Geiger counters are not used much for quantitative X-ray spectrometry. They are, however, very important portable detectors for indicating the presence or absence of X-rays. Portable radiation detectors equipped with Geiger counters are used to monitor the operation of equipment that creates or uses ionizing radiation to check for leaks in the shielding.

8.2.4.4. *Escape Peaks*

Ionization of the filler gas by an X-ray photon usually results in the ejection of an outer shell electron. However, it is possible for ionization to occur by ejection of an inner shell electron. When this happens, the incoming X-ray photon is absorbed and the filler gas emits its characteristic K or L lines. This will result in peaks appearing at an energy E' equal to:

$$E' = E \text{ (incoming X-ray)} - E \text{ (filler gas characteristic X-ray)} \qquad (8.16)$$

As an example, if the detector filler gas is Ar, the Ar K line has energy of about 3 keV (or a wavelength of 3.87 Å). If an incoming X-ray has a wavelength shorter than 3.87 Å, it can eject an argon K electron. Assuming that the incoming X-ray is the Fe K_α line at 6.3 keV, a peak will appear at (6.4 − 3) keV or about 3.4 keV. This peak at 3.4 keV is called an **escape peak** (and can be called either the Fe Kα escape peak or the argon escape peak). An escape peak appears at constant distance from the parent fluorescence X-ray (in this case, Fe K_α) on the low-energy side. Escape peaks can often be very intense and can be useful in identifying elements. Escape peaks form in proportional and scintillation counters and in semiconductor detectors.

8.2.4.5. Pulse Height Distribution and Selection

In the detector, a photon generates a number of ion pairs, that is, a current pulse with a certain magnitude or pulse height. The pulse height in a proportional detector depends on the energy of the photon. Unfortunately, the height of the current pulse that results is not exactly the same for photons of the same energy. Formation of ion pairs and secondary ion pairs is a statistically random process, so a Gaussian distribution of pulse heights centered on the most probable value results. A series of pulses and their heights is shown in Fig. 8.23. On the left side, this figure shows a series of current pulses from photons of two different energies counted over a period of time. If the pulses are plotted by height (amplitude), the result is a Gaussian **pulse height distribution**, shown on the right side of the figure. Two Gaussian pulse height distributions are seen since we had two photons of different energies reaching the detector. The width of the distribution is measured at half of the maximum height; this is called the full-width at half-maximum (FWHM). The FWHM is a measure of the energy resolution of a detector. Energy resolution is best in semiconductor detectors and worst in scintillation detectors, with gas-filled proportional counters in the middle.

In a wavelength dispersive instrument, the analyzing crystal separates the wavelengths falling on it, but as Bragg's Law tells us, it is possible for higher order ($n > 1$) lines of other elements to reach the detector. A higher-order line from a different element will have a very different energy and will result in a second pulse height distribution centered at a different energy reaching the detector. This would result in an error, since the signal would be misinterpreted as coming from just one element. This problem is eliminated by the use of a **pulse height selector**. A pulse height selector is an electronic circuit that throws away signals that are below and above a preselected

Figure 8.23 Amplitude or pulse height and time record of signals from the detector is on the left. Transformation of the data into a pulse height distribution is shown on the right. The FWHM measurement is shown for the higher peak. (From Helsen and Kuczumow, used with permission.)

energy window. It is very similar in its end result to the use of filters in UV/VIS spectrometers. The use of a pulse height selector in WDXRF eliminates noise as well as higher-order lines from the sample.

8.2.4.6. *Proportional Counters*

Flow Proportional Counter. The flow proportional counter covers a wide wavelength range and is generally used for wavelengths longer than 2 Å (elements with $Z < 27$). This detector is illustrated in Fig. 8.24(a). The windows are thin (<6 μm) polymer film, coated on the inside surface with aluminum to permit a homogeneous electric field to be established within the detector. The thin windows allow the filler gas to leak out; therefore a supply of filler gas is constantly provided to the detector through the inlet as shown in Fig. 8.24(a). The filler gas for a flow proportional counter is often 10% CH_4, 90% Ar, a mixture called P10 gas. The pressure, flow, and temperature of the gas must be precisely controlled for accurate detector response.

The operating voltage range for a flow proportional counter is $1-3$ kV. As seen in Fig. 8.22, the amplification factor, which is the number of ion pairs discharged at the electrodes divided by the number of primary ion pairs formed, is 10^2-10^6. The current pulse is converted to a voltage pulse, is processed through a pulse height selector or discriminator and counted. Maximum count rate for a flow proportional counter is on the order of 2×10^6 cps.

Sealed Proportional Counter. A sealed proportional counter is shown schematically in Fig. 8.24(b). The windows are thicker, so they do not leak. Window materials include polymers, mica, aluminum, and beryllium. The filler gas used in a sealed proportional counter may be Ne, Kr, or Xe. Window and gas combinations are optimized for the wavelength of radiation to be detected; Al and Ne would be best for light elements, for example.

Multiple proportional counters are used in simultaneous X-ray spectrometers, while one proportional counter is often used in tandem with a scintillation counter in a sequential

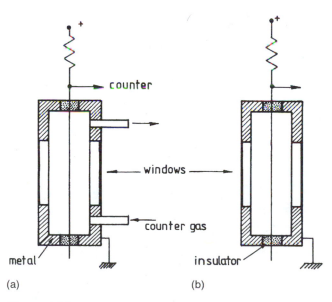

(a) (b)

Figure 8.24 Schematics of (a) a flow proportional counter and (b) a sealed proportional counter. (From Helsen and Kuczumow, used with permission.)

system. It is for this reason that the detector has two windows as shown in Fig. 8.24. X-ray photons pass through the proportional counter to the scintillation counter located behind it, as illustrated in Fig. 8.15 and signals are obtained from both detectors. It should be noted that this tandem arrangement does not permit independent optimization of both detectors. There are sequential spectrometer systems available with independent proportional and scintillation detectors.

Scintillation Counter.   Photomultiplier detectors, discussed in Chapter 5, are very sensitive to visible and UV light, but not to X-rays, to which they are transparent. In a *scintillation detector* the X-radiation falls on a compound that absorbs X-rays and emits visible light as a result. This phenomenon is called *scintillation*. A PMT can detect the visible light scintillations. The scintillating compound or phosphor can be an inorganic crystal, an organic crystal or an organic compound dissolved in solvent.

The most commonly used commercial scintillation detector has a thallium-doped sodium iodide crystal, NaI(Tl), as the scintillating material. A single crystal of NaI containing a small amount of Tl in the crystal lattice is coupled to a PMT, shown in Fig. 8.25. When an X-ray photon enters the crystal, it causes the interaction

$$I^- \longrightarrow I^0 + e^-$$

and the ejection of photoelectrons, as in the gas-filled detector. The ejected photoelectrons cause excited electronic states to form in the crystal by promotion of valence band electrons. When these excited electrons drop back to the ground state, flashes of visible light (scintillations) are emitted. The excited state lies about 3 eV above the ground state, so the emitted light has a wavelength of 410 nm. The intensity of the emitted light pulse from the crystal is proportional to the number of electrons excited by the X-ray photon. The number of electrons excited is proportional to the energy of the X-ray photon; therefore the scintillation intensity is proportional to the energy of the X-ray.

The scintillations (visible light photons) from the crystal fall on the cathode of the PMT, which is made of a photoemissive material such as indium antimonide. Photoemissive

(a)

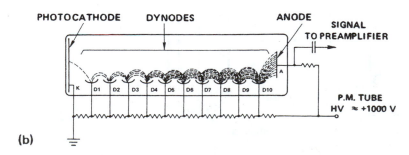

(b)

Figure 8.25   The NaI(Tl) scintillation detector. (a) The assembled detector; (b) a schematic representation of the photomultiplier and its circuitry. [Courtesy of ORTEC (Ametek) (www.ortec-online.com). From Jenkins et al., 1981, used with permission.]

materials release electrons when struck by photons. Electrons ejected from the cathode are accelerated to the first dynode, generating a larger number of electrons. The electron multiplication process occurs at each successive dynode, resulting in approximately 10^6 electrons reaching the anode for every electron that strikes the cathode. The amplitude of the current pulse from the photomultiplier is proportional to the energy of the X-ray photon causing the ionization in the crystal.

To summarize, the scintillation detector works by (1) formation of a photoelectron in the NaI(Tl) crystal after an X-ray photon hits the crystal, (2) emission of visible light photons from an excited state in the crystal, (3) production of photoelectrons from the cathode in the photomultiplier, and (4) electron multiplication.

The NaI(Tl) scintillation detector is most useful for short-wavelength X-rays, <2 Å ($Z > 27$), so it complements the proportional counter. It also has the potential for escape peaks caused by the iodine K line (about 30 keV or 0.374 Å). Incoming X-rays with wavelengths less than 0.374 Å will result in escape peaks about 30 keV lower in energy than the true energy. The major disadvantage of the NaI(Tl) detector is that its resolution is much worse than that of the proportional counter. This is due to the wider pulse height distribution that results in the output pulse because of the multiple steps involved in the operation of this detector.

8.2.5. Sample Holders

XRF is used for the analysis of solid and liquid samples. For quantitative analysis the surface of the sample must be as flat as possible, as will be discussed in the applications section. There are two classes of sample holders, cassettes for bulk solid samples and cells for loose powders, small drillings, and liquids. A typical cassette for a flat bulk solid such as a polished metal disk, a pressed powder disk, a glass or polymer flat is shown in Fig. 8.26(a). The cassette is a metal cylinder, with a screw top and a circular opening or aperture, where the sample will be exposed to the X-ray beam. The maximum size for a bulk sample is shown. The sample is placed in the cassette. For a system where the sample is analyzed face down, the cassette is placed with the opening down and the

Figure 8.26 (a) Sample cassette for bulk solid samples. (b) Position of a bulk solid in the face-down configuration used in many spectrometers. [Courtesy of Thermo ARL (www.thermoARL.com).]

bulk sample sits in the holder held in position by gravity, as shown in Fig. 8.26(b). If the system requires the sample face up, the body of the cassette must be filled with an inert support (often a block of wood) to press the sample surface against the opening. These cassettes are available with a variety of apertures, usually from 8 to 38 mm in diameter, to accommodate samples of different diameters. Other types of solid samples, such coatings on a solid substrate can be placed directly in this type of cassette.

The analysis of liquids, loose powders, or small pieces requires a different holder. The cells for these types of samples are multipart plastic holders, shown in Fig. 8.27(a) and require squares or circles of thin polymer film to hold the sample in the cell. The body of the cell is a cylinder open on both ends. One end of the cylinder is covered with the plastic film (or even clear plastic adhesive tape) and the film or tape is clamped into place by a plastic ring. The cell is placed with the film down and the sample of

(a)   (b)

Figure 8.27 Cells for liquid samples, loose powders and chips. (a) Two types of disposable polyethylene sample cups for liquids and loose samples, consisting of a cup and snap-ring to hold the polymer film cover. The cells with polymer film in place are shown at the top of the photograph. The disassembled cup and ring pieces are shown at the bottom of the photograph. [Courtesy of SPEX Certiprep, Inc. Metuchen, NJ (www.spexcsp.com).] (b) Liquid and other loose samples in cells such as those shown in the photo, and then inserted into a sample cassette of the type shown in Fig. 8.26(a) for a face-down configuration spectrometer. As shown, dust sampled on impact filters or thin layer samples may be inserted directly into the sample cassette. [Courtesy of Thermo ARL (www.thermoARL.com).]

liquid, powder, or filings is added. The film surface should be completely covered, as uniformly as possible. A plastic disk that just fits into the cell is inserted and pressed against the sample to obtain as flat a surface as possible and a top cap is screwed or pressed on. For liquid samples, a vented top is used to avoid pressure build-up from heating of the sample by the X-ray beam. This assembled cell may be used "as is" or may be inserted into a standard cassette, as shown in Fig. 8.27(b) in a face-down configuration. As you can imagine, if the thin polymer film breaks, samples of loose powder, chips, or liquid will spill into the interior of the spectrometer, contaminating the analyzing crystal and the rest of the system and possibly breaking the Be window of the X-ray tube, if the tube is below the sample. It is for this reason that liquid samples cells are vented and that a vacuum is not used. This is the main disadvantage of the face-down configuration; for anything other than bulk samples, there is a risk of contaminating the instrument if the film covering the sample ruptures. Figure 8.27(b) shows that in the face-down configuration, a liquid naturally assumes a flat surface. Imagine what the liquid sample would look like face-up. An air bubble will form at the film surface if a sealed cell is used and not filled completely. A bubble may form at the surface by heating of the sample in the X-ray beam if the cell is filled completely. If this occurs, the intensity of X-ray fluorescence from the sample will drop dramatically and the possibility of film rupture as the pressure in the cell builds increases dramatically. So, if liquid samples must be analyzed, the face-down configuration gives better quantitative results, even at the risk of contaminating the spectrometer.

The sample cassette is moved into position, either manually or with the automatic sample changer. In position, the sample is spun, generally at about 30 rpm, to homogenize the surface presented to the X-ray beam.

Polymer films used to cover the cell opening must be low in trace element impurities, strong enough to hold the sample without breaking, thermally stable, and chemically inert. They certainly must not be soluble in any liquid samples to be analyzed. Films of polyester (Mylar®), polyimide (Kapton®), polycarbonate, polypropylene, and fluoropolymer (Teflon®) are commonly used, with film thickness ranging about 3–8 μm. Films of different composition and thickness transmit X-rays to varying degrees (Fig. 8.28), and the film chosen must transmit the wavelengths for the elements to be measured in the sample.

Figure 8.28 The X-ray transmission characteristics of some thin polymer films used as sample holder covers for liquid, loose powder, and similar samples. [Courtesy of SPEX Certiprep, Inc. Metuchen, NJ (www.spexcsp.com).]

8.2.6. Simultaneous WDXRF Spectrometers

A simultaneous WDXRF system uses multiple channels, with each channel having its own crystal/detector combination optimized for a specific element or background measurement. Instruments with as many as 40 fixed crystal/detector channels or as few as two are available. These systems are designed for specific applications, such as the analysis of steel in a production facility where hundreds of samples must be analyzed for the same suite of elements every day. They have the advantage of being very fast compared with a sequential system, but are not flexible.

Most simultaneous systems have the X-ray tube above the sample, with the sample facing up. As discussed earlier, this makes the analysis of liquids difficult or impossible. Several instrument manufacturers offer combination systems with a simultaneous set of channels as well as a sequential monochromator. These systems offer the speed needed for routine analysis and the flexibility needed for nonroutine analysis, but are expensive.

8.2.7. EDXRF Spectrometers

In EDXRF spectrometry, there is no physical separation of the fluorescence radiation from the sample. There is no dispersing device prior to the detector, as seen in Fig. 8.29. All of the photons of all energies arrive at the detector simultaneously. The semiconductor detector used in EDXRF is a proportional detector with very high intrinsic energy resolution. In this system, the detector resolves the spectrum. The signal pulses are collected, integrated and displayed by a multichannel analyzer (MCA). As Fig. 8.29 shows, a primary beam filter is often used to improve the signal-to-noise ratio for given energy regions. As with WDXRF systems, most EDXRF systems have a series of selectable filters. Typical filters used in EDXRF and their ranges of use are listed in Table 8.7.

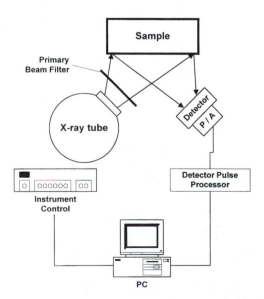

Figure 8.29 A schematic energy dispersive XRF system with an X-ray tube source. There is no dispersion device between the sample and the detector. Photons of all energies are collected simultaneously. (From Ellis, used with permission.)

Table 8.7 Typical Primary Beam Filters and Range of Use in EDXRF Systems

Filter	Thickness (μm)	X-ray tube range (kV)	Elements	Comments
None		4–50	Na–Ca	Optimum for light elements, 4–8 kV excitation[a]
Cellulose		5–10	Si–Ti	Suppresses tube L lines (see Fig. 8.16.)[a]
Al, thin	25–75	8–12	S–V	Removes tube L lines[a]
Al, thick	75–200	10–20	Ca–Cu	Used for transition elements
Anode element, thin	25–75	25–40	Ca–Mo	Good for trace analysis[b]
Anode element, thick	100–150	40–50	Cu–Mo	Trace analysis with heavy element L lines[b]
Cu	200–500	50	>Fe	Suppresses tube K lines

Source: Table modified from Ellis, used with permission.
[a]He purge or vacuum path needed to avoid attenuation of low energy lines.
[b]Serves as a secondary fluorescence source, also called a monochromatizing filter; preferentially transmits tube K lines.

8.2.7.1. *Semiconductor Detectors*

When an X-ray falls on a semiconductor, it generates an electron ($-$e) and a hole ($+$e) in a fashion analogous to the formation of a primary ion pair in a proportional counter. Based on this phenomenon, semiconductor detectors have been developed and are now of prime importance in EDXRF and scanning electron microscopy. The principle is similar to that of the gas ionization detector as used in a proportional counter, except that the materials used are in the solid-state. The total ionization caused by an X-ray photon striking the detector is proportional to the energy of the incident photon.

The most common semiconductor detector for laboratory EDXRF systems is the *lithium-drifted silicon diode*, represented as Si(Li). (It is called a "silly" detector for short). A schematic diagram of a silicon lithium-drifted detector is shown in Fig. 8.30. A cylindrical piece of pure, single crystal silicon is used. The size of this piece is 4–19 mm in diameter and 3–5 mm thick. The density of free electrons in the silicon is very low, constituting a p-type semiconductor. If the density of free electrons is high in a semiconductor, then we have an n-type semiconductor. Semiconductor diode detectors always operate with a combination of these two types.

The diode is made by plating lithium onto one end of the silicon. The lithium is drifted into, that is diffused into, the silicon crystal by an applied voltage. The high concentration of Li at the one end creates an n-type region. In the diffusion process, all electron acceptors are neutralized in the bulk of the crystal, which becomes highly nonconducting. This is the *"intrinsic"* material. The lithium drifting is stopped before reaching the other end of the silicon crystal, leaving a region of pure Si (p-type), as shown in Fig. 8.30. Submicron gold layers are applied at each end as electrical contacts. The detector is reverse-biased, removing any free charge carriers from the intrinsic region. Under this condition no current should flow since there are no charge carriers in the intrinsic region. However, the band-gap between the valence band and the conduction band is small, only 1.1 eV for Si(Li). At room temperature, thermally generated charge carriers cross this barrier easily and become conductive even with no X-ray photons striking the detector. This causes a high noise level. To decrease this noise

(a)

(b)

Figure 8.30 The Si(Li) semiconductor detector. (a) Schematic shows the n-type Si region on one end of the Si crystal, a central charge depleted intrinsic region and p-type Si on the other end. (b) The actual detector has 200 Å layers of gold as electric contacts on each end of the crystal. An X-ray photon striking the intrinsic region generates electron–hole pairs within the diode. [Fig. 8.30(b) courtesy of Thermo ARL (www.thermoARL.com).]

and increase the sensitivity of the detector, the temperature of the system must be decreased significantly. This is accomplished by cooling the detector to 77 K with liquid nitrogen, which must be replenished regularly. In exactly the same fashion, germanium, also in group IV of the periodic table, can be used instead of silicon, making a Ge(Li) drifted detector (you might guess this is called a "jelly" detector). The Ge(Li) detector also requires liquid nitrogen cooling, since its band gap is only 0.66 eV.

An X-ray photon striking the detector produces multiple electron–hole pairs in the intrinsic region [Fig. 8.30(b)]. The number of electron–hole pairs produced is proportional to the photon energy. The energy required to make an electron–hole pair is 3.86 eV in Si(Li), so the number of electron–hole pairs formed is approximately

$$n = \frac{E}{\varepsilon} = \frac{E}{3.65\,\text{eV}} \tag{8.17}$$

where n is the number of electron–hole pairs; E, the energy of the incident X-ray photon (in eV); and ε, the energy to form an electron–hole pair in eV.

For a similar Ge lithium-drifted detector, the energy required for ionization is 2.96 eV. This is much less than the energy required for ionization in a proportional counter or a NaI(Tl) scintillation detector.

Under the influence of an applied voltage, the electrons move toward the positive end and the holes toward the negative end of the detector. The total charge collected at the positive contact is:

$$Q = nq_e \tag{8.18}$$

where Q is the total charge in coulombs (C); n, the number of electron–hole pairs = E/ε; and q_e, the charge on one electron = 1.69×10^{-19} C/electron.

The collection of charge results in a voltage pulse. Since the total charge is proportional to the energy of the incident photon, the amplitude of the voltage pulse produced is also directly proportional to the energy of the incident photon. The voltage pulses are amplified and "shaped" electronically and sent to a *multichannel pulse height analyzer* to be sorted by pulse height and counted.

The operation of a multichannel analyzer can be modeled in a simple fashion. Assume that we have a pulse height analyzer of a given total voltage range with the ability to change the voltage in small increments. As an example, the total voltage range is 10 V and the interval of change is 0.1 V. X-rays of short wavelengths (high energies) must be separated from X-rays of long wavelengths (low energies). That is what a pulse height analyzer does; it rejects energy signals that are higher or lower than a selected energy window. If the analyzer window can be changed in small energy increments, only photons with that energy will pass through. Those photons are counted and stored in that energy window location in the analyzer memory. Each energy window location is called a *channel*. Then the energy window (voltage) is changed by 0.1 V and only photons corresponding to the new energy window will pass through and be counted and stored in a second channel. Sweeping the voltage range in steps of 0.1 V permits us to distinguish between X-rays of various energies. If the X-ray photons are counted by energy, we can obtain I, the X-ray intensity at given energy. This permits us to plot I vs. wavelength (energy), which gives us an energy spectrum of the XRF from the sample. An EDXRF spectrum is in the form of a histogram, usually plotted as "counts" on the y-axis, where counts means the number of photons counted in a given channel, vs. energy on the x-axis. In practice, an EDXRF is equipped with a pulse height analyzer with many channels and complicated signal processing circuitry. A typical multichannel pulse height analyzer may have 1024 channels, each corresponding to a different energy interval.

Resolution in a semiconductor detector EDXRF system is a function of both the detector characteristics and the electronic pulse processing. The energy resolution of semiconductor detectors is much better than either proportional counters or scintillation counters. Their excellent resolution is what makes it possible to eliminate the physical dispersion of the X-ray beam; without the energy resolution of semiconductor detectors, EDXRF would not be possible.

While Si(Li) is still the most common semiconductor detector material for laboratory-based EDXRF systems, Ge(Li), high-purity Ge, mercuric iodide (HgI_2), and compound materials such as GaAs and CdTe have been used. The lithium-drifted detectors must be maintained at liquid nitrogen temperatures at all times, to minimize noise and to prevent Li migration in the crystal. High-purity Ge and the other compound materials do not required constant cooling. Signal-to-noise is often improved by operating these detectors at subambient temperatures, but mercuric iodide detectors work well at ambient temperatures. This makes mercuric iodide an excellent detector for portable, handheld EDXRF

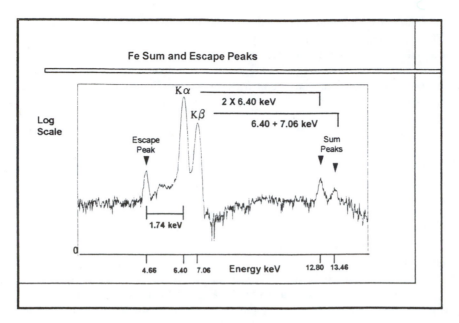

Figure 8.31 Artifacts in an EDXRF spectrum. The spectrum of pure iron, measured with a Si(Li) detector, shows a peak lower in energy than the Fe K_α peak by an amount exactly equal to the energy of the Si K_α line. Some of the Fe photon energy is transferred to the Si detector atoms; the amount of energy absorbed by an Si atom has *escaped* from the Fe photon. This type of peak is called an escape peak. Sum peaks also appear in EDXRF spectra when two intense photons arrive at the detector simultaneously. A sum peak from two K_α photons is shown along with a sum peak from one K_α and one K_β photon. [Courtesy of Thermo ARL (www.thermo ARL.com).]

analyzers where carrying around a large dewar (a fancy thermos bottle) of liquid nitrogen is not practical and for the X-ray spectrometers sent on space missions, where refilling the liquid nitrogen dewar every week is not possible.

8.2.7.2. Escape Peaks and Sum Peaks

Spectrum artifacts may appear in the energy dispersive spectrum. These are peaks that are not from elements in the sample. The Si escape peak, from the Si K_α line, results in an artifact peak 1.74 eV lower than the parent peak when a Si(Li) detector is used. Such an escape peak is shown in the EDXRF spectrum of an iron sample in Fig. 8.31. Similar escape peaks at different energies appear for Ge if a Ge detector is used.

Sum peaks in the EDXRF spectrum occur when two high-intensity peaks arrive so close in time that the signal processing electronics cannot separate them. A single peak is registered at an energy that is the sum of the two peaks. Figure 8.31 displays this type of artifact. The major elements in the sample (e.g., iron in steel) are usually the source of the sum peaks.

Most EDXRF systems come with software that automatically corrects for escape and sum peaks.

8.3. ANALYTICAL APPLICATIONS OF X-RAYS

There are three distinct fields of X-ray analysis: X-ray absorption, which varies with atomic weight; XRD, which depends on the crystal properties of solids; and XRF, which is characteristic of the elements present and their concentrations in the sample. Their analytical uses are described in this section.

8.3.1. X-Ray Absorption

If the wavelength of an X-ray beam is short enough (high energy), it will excite an atom that is in its path. In other words, the atom absorbs X-rays that have enough energy to cause it to become excited. As a rule of thumb, the X-rays emitted from a particular element will be absorbed by elements with a lower atomic number. The ability of each element to absorb increases with atomic number.

Beer's Law indicates that

$$\log\left(\frac{P_0}{P_x}\right) = \mu_x \tag{8.19}$$

where μ_x is the linear absorption coefficient; x, the path length through the absorbing material; P_0, the X-ray power before entering sample; and P_x the X-ray power leaving sample.

However,

$$\left(\frac{\mu_x}{\rho}\right) = \mu_m \tag{8.20}$$

where μ_m is the mass absorption coefficient and ρ is the density.

But

$$\mu_m = \left(\frac{CN_0 Z^4 \lambda^3}{A}\right) \tag{8.21}$$

where C is a constant; N_0 the Avogadro's number; Z, the atomic number; A, the atomic weight; and λ, the wavelength of the radiation.

It can be seen that at a given wavelength, μ_m is proportional to Z^4 divided by the atomic weight. This relationship is shown in Fig. 8.32. The set-up for X-ray absorption is slightly different than that for XRF, as seen in Fig. 8.33. The sample is placed directly in line with the X-ray tube, in a configuration very similar to UV/VIS absorption spectrometry. The parameter measured is the decrease in intensity of the incident

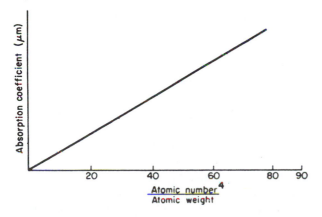

Figure 8.32 Relationship between atomic number and X-ray absorption coefficient.

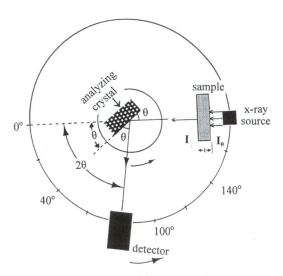

Figure 8.33 X-ray absorption. The sample is placed between the X-ray tube and the detector. The intensity of the source is I_0. The intensity of light reaching the detector after passing through a sample of thickness t is I. I will be less than I_0 if absorption occurs.

beam after passing through the sample. The same detectors described for XRF may be used for X-ray absorption spectrometry. Older systems and current medical systems use photographic film for detection (radiography).

The most familiar example (and the oldest use) of X-ray absorption does not provide chemical information, but rather physical information. That is the use of X-ray absorption in medical radiography, but it is based on the relationship between absorption coefficient and atomic number. For example, the human arm consists of flesh, blood, and bone. The flesh or muscle is made up primarily of carbon, nitrogen, oxygen, and hydrogen. These are all low atomic number elements, and their absorptive power is very low. Similarly, blood, which is primarily water, consists of hydrogen, oxygen, plus small quantities of sodium chloride and trace materials. Again, the absorptive power of blood is quite low. In contrast, bone contains large quantities of calcium and phosphorus, primarily as calcium phosphate. The atomic numbers of these elements are considerably higher than those mentioned before, and so the absorptive power is considerably higher. When an X-ray picture is taken of an arm, the x-radiation penetrates the muscle tissue and blood quite readily, but is absorbed significantly by the bone. A photograph of this absorption indicates the location of the bone in the arm. The procedure is routinely used in medicine to detect broken bones.

Another application of X-ray absorption in medicine is to define the shapes of arteries and capillaries. Normally the blood absorbs only poorly; however, it is possible to inject a solution of strongly absorbing cesium iodide into the veins. The material is then swept along with the blood and follows the contours of the arteries. An X-ray video is recorded as the highly absorbing cesium iodide flows through the arteries, showing the contours of the arteries. This can be used to identify breaks in the veins or arteries that could cause internal bleeding. Such internal bleeding can be the cause of a stroke. The technique may also be used to indicate a buildup of coating on the inside of the veins. This is particularly dangerous in the heart, where deposits of cholesterol restrict the flow of blood through the heart. If this is left unchecked, a heart attack will result.

X-ray absorption can be used to diagnose this problem and to locate exactly the position of deposits. Surgery is made much easier by this technique.

In the field of metallurgy, applications of X-ray absorption include the detection of voids or the segregation of impurities, such as oxides, in welds and other joints. Figure 8.34 shows an idealized X-ray absorption photograph of a mechanical weld that contains voids or internal holes. Such holes indicate that the weld is mechanically weak and might break in use. If the weld is weak, it must be strengthened to form a sufficiently strong joint. This type of nondestructive testing is used to check the manufacturing quality of ships, aircraft, bridges, and buildings. It also used to check these structures during routine maintenance. X-ray absorption is routinely used for measuring the thickness of thin metal films.

The technique of X-ray absorption can also be used to determine the levels of liquids in enclosed vessels or pipes without opening or breaking them. The same process can be used to detect metal supports or metal fillings inside constructed objects as diverse as buildings and small works of art. A major advantage is that X-ray techniques are usually nondestructive. Sometimes artists paint over old paintings, using the canvas for their own work and covering unrecognized masterpieces in the process. Using X-ray absorption, it is possible to reveal the covered painting without removing the top painting. When used to examine a metal horse sold for several million dollars as an ancient Greek art piece, X-ray absorption showed that the horse contained internal metal supports and was therefore a fake. This was done without destroying the art piece, in case it had been authentic.

For elemental analysis, X-ray absorption is not particularly useful. As we saw in Eq. (8.10), the mass absorption coefficient needed for the Beer's Law calculation [Eq. (8.19)] must be calculated from the weight fractions of elements present in the sample. The weight fractions are usually unknown. Quantitative analysis by X-ray absorption is usually only used for the determination of a high atomic number element in a matrix of lower atomic number elements. Examples include the determination of lead or sulfur in hydrocarbon fuels, and the determination of Pt catalyst in polymers, where the difference in mass absorption coefficients between analyte and matrix is large. One approach to quantitative analysis using X-ray absorption is based on the measurement of the intensities of two or more monochromatic X-rays passed through the sample. This is called X-ray preferential absorption analysis or dual-energy transmission analysis. The analysis depends on the selective absorption of the transmitted X-rays by the analyte compared with absorption by the rest of the sample (the matrix). The sensitivity of the analysis also depends on the difference in mass absorption coefficients of the analyte and sample matrix for the transmitted X-rays; a big difference results in a more sensitive analysis. The analyte

Figure 8.34 X-ray absorption photograph of a mechanical weld.

concentration calculation in any absorption method requires that the thickness and density of the sample be known and requires a homogenous matrix for accurate quantitative results.

8.3.1.1. EXAFS

A recent development in the use of X-ray absorption is a technique called EXAFS, extended X-ray absorption fine structure spectroscopy. A sample is placed in a beam of X-rays and the incident and transmitted intensities are measured as the energy of the X-ray beam is varied. A plot of the absorption vs. energy gives us the position (energy) and exact shape of the absorption edge for the element being measured. The exact energy of the absorption edge and its shape, the "fine structure", does change slightly depending on the oxidation state of the element and the number and type of nearest neighbor atoms. This change in position of the absorption edge is analogous to the chemical shift seen in NMR. EXAFS does provide oxidation state information and molecular structure information, unlike normal X-ray absorption or X-ray fluorescence, which are strictly elemental analysis techniques. Details of the spectral interpretation of EXAFS are beyond the scope of this book. The interested student can consult the text by Teo and Joy listed in the bibliography.

8.3.2. X-Ray Diffraction

X-ray diffraction or X-ray diffractometry (XRD) is a technique that is useful for the analysis of solid crystalline or semicrystalline materials. Most organic and inorganic compounds, minerals, metals, and alloys, and many types of polymers form crystals and can be analyzed by XRD. XRD can provide the exact crystal structure of a pure single crystal material. In addition, XRD can provide the qualitative and quantitative identification of the molecules present in pure crystalline powders or mixtures of crystalline powders.

The ions or molecules that make up a crystal are arranged in well-defined positions, called a crystal lattice. Figure 8.35 is an electron micrograph of the (110) plane of crystalline silicon. Three coordinates, called Miller indices, identify the plane in space; the Miller indices for this plane are 1,1, and 0. The light spots are individual Si atoms. As can be seen, they are arranged in a very regular pattern in the 2D plane. The dark area is the empty space or interstitial space between the atoms in the lattice. A crystal is a 3D well-ordered array of atoms. An illustration of a typical crystal structure, greatly magnified, is shown in Fig. 8.36(a). As we examine the structure of the crystal, we see that the ions or atoms or molecules form planes in three dimensions. You can imagine stacking identical planes of Si atoms on top of each other to create a 3D crystal, for example.

Figure 8.35 An electron micrograph of the (110) plane in crystalline silicon.

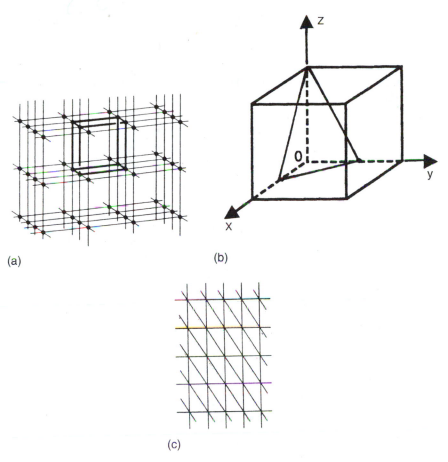

(a)

(b)

(c)

Figure 8.36 (a) A portion of a 3D crystal lattice. The unit cell, or basic repeating unit, of the lattice is shown in heavy outline. The black dots represent the atoms or ions or molecules that make up the crystal. (b) A cubic unit cell, with the corners of the cell located at 1 unit from the origin (o). The triangular plane drawn within the unit cell intersects the x-axis at $1/2$, the y-axis at $1/2$, and the z-axis at 1. This plane has Miller indices of (221). (c) A family of planes shown in a 2D lattice.

The **unit cell**, shown in heavy outline in the lattice, can be moved in three dimensions to recreate the entire crystal lattice. The unit cell is the smallest volume that can be used to describe the entire lattice. A Cartesian coordinate system is used to locate points, directions, and planes in a crystal lattice. A unit cell has its origin at the intersection of the three axes, and is designated by its edge lengths in the x, y, and z directions and by three angles. An atom (molecule or ion) in the crystal lattice is a point, identified by its x, y, and z coordinates. A plane is identified by its Miller indices, the reciprocals of the intersection points of the plane with the x-, y-, and z-axes. For example, suppose the unit cell is a cube, with edges equal to 1 unit of length on each axis as shown in Fig. 8.36(b). A triangular plane is shown within the unit cell. The plane intersects the x-axis at $1/2$, the y-axis at $1/2$, and the z-axis at 1; it has intercepts of $1/2$, $1/2$, 1. The reciprocals are 2,2, and 1, so the Miller indices for this plane are (221). A plane that is parallel to a given axis has an intercept of infinity; the reciprocal of infinity is 0. [What axis is the (110) plane in Si parallel to? Draw the (110) plane in a cubic unit cell such as the one shown in Fig. 8.36(b)]. A crystal lattice will have many parallel planes, each uniformly spaced from each other. Such groups of planes are called families of planes and will have related Miller indices [e.g., the (110), (220), (330), (440) planes are a family of planes]. These planes in a given family are all parallel, as shown in Fig. 8.36(c), just at different distances from the origin specified for the coordinate system. The (110) plane is the farthest from the origin and the (440) plane is the closest to the origin of the set of planes (110), (220), (330), and (440).

If a monochromatic X-ray beam falls on such a crystal, each atomic plane reflects the beam. Each separate reflected beam interacts with other reflected beams. If the beams are not in phase, they destroy each other and no beam emerges. Other beams reinforce each other and emerge from the crystal. The net result is a **diffraction pattern** of reinforced beams from many planes. It is the atomic planes that are important in X-ray diffraction. It is of course possible to draw an infinite number of planes in three dimensions, but only those planes with electron density on them reflect X-rays. For example, you could draw a plane extending into the page along one of the dark diagonals in Fig. 8.35, but that plane contains no atoms and therefore will not diffract X-rays.

In Fig. 8.37 radiation from the source falls on the crystal, some on the top atomic plane, some on the second plane. Since the two beams are part of the same original beam, they are in phase on reaching the crystal. However, when they leave the crystal, the part leaving the second plane has traveled an extra distance ABC. If ABC is a whole number of wavelengths, the two beams leaving the crystal will be in phase and the light is coherent. If ABC is not a whole number of wavelengths, the two beams come together out of phase and by destructive interference the light is destroyed.

As we derived in Section 8.1.3, $n\lambda = 2d \sin \theta$. This is the Bragg equation, which states that coherence occurs when $n\lambda = 2d \sin \theta$. It can be used to measure d, the distance between planes of electron density in crystals, and is the basis of **X-ray crystallography**, the determination of the crystal structure of solid crystalline materials. Liquids, gases, and solids such as glasses and amorphous polymers have no well-ordered structure; therefore they do not exhibit diffraction of X-rays.

For any given crystal, d is constant for a given family of planes; hence for any given angle θ and a given family of planes, $n\lambda$ is constant. Therefore, if n varies, there must be a corresponding change in λ to satisfy the Bragg equation. For a given diffraction angle, a number of diffracted lines are possible from a given family of planes; n is known as the **order** of diffraction. As an example, if $2d \sin \theta$ equals 0.60, each of the conditions of Table 8.8 will satisfy the Bragg equation. Radiation of wavelength 0.60, 0.30, 0.20, or

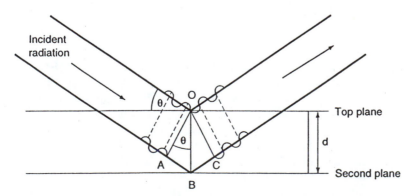

Figure 8.37 Reinforcement of light diffracted from two crystal planes.

Table 8.8 Order of Diffraction for Values of n, λ

n	λ (Å)	$n\lambda$	Order
1	0.60	0.60	First
2	0.30	0.60	Second
3	0.20	0.60	Third
4	0.15	0.60	Fourth

0.15 Å will diffract at the same angle θ in first, second, third, or fourth order, respectively, as seen in Fig. 8.38. This is called order overlap and can create difficulty in interpretation of crystal diffraction data.

It should be noted that radiation of 0.30 Å would also be diffracted at a different angle in first order from the same family of planes (same d value), as shown in Fig. 8.38. Wavelengths corresponding to low orders such as first and second order give observable diffraction lines. Consequently, a single plane will generate several diffraction lines for each wavelength. Each of the planes in the three dimensions of the crystal will give diffraction lines. The sum total of these diffraction lines generates a diffraction pattern. From the diffraction pattern it is possible to deduce the different distances between the planes as well as the angles between these planes in each of the three dimensions. Based on the diffraction pattern, the physical dimensions and arrangement of the atomic planes in the crystal can be identified.

8.3.2.1. X-Ray Diffractometer

The schematic layout of a single crystal diffractometer is given in Fig. 8.39. This system uses an X-ray tube, a sample specimen, and a detector that rotates in an arc described by a Rowland circle. Note that the single crystal sample takes the place of the analyzing crystal

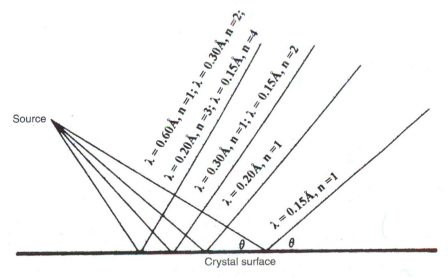

Figure 8.38 Diffraction of radiation of different wavelengths. Overlap can occur when different orders are diffracted at the same angle.

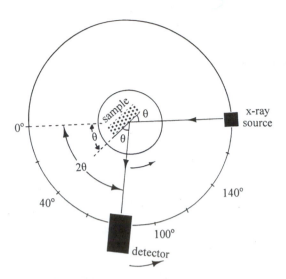

Figure 8.39 Schematic layout of a single-crystal XRD.

in a WDXRF analyzer. The goniometer mounting for a single crystal diffractometer is very complex, because the crystal must be moved in three dimensions to collect data from many planes.

In order to determine the structure of a single crystal, such a crystal must be grown from the material to be studied. The growth of single crystals of materials often is not easy. Simple inorganic salts and small organic molecules can be crystallized as single crystals by very slow evaporation of a supersaturated solution of the salt or compound. Once one tiny single crystal forms, it will grow in preference to the formation of more small crystals. Proteins and other biomolecules are more difficult to grow as single crystals because they are complex. One method that often works is to suspend a drop of protein solution over a reservoir of buffer solution. Water diffusion from the drop often results in single crystal growth. Different techniques are required to form metal crystals. The interested student can find many references and resources on the Internet, by searching the term "X-ray crystallography".

For any given experiment, λ is the known wavelength of the monochromatic X-ray beam, θ is controlled and varied by the goniometer. From this information d can be calculated. By rotating the goniometer and examining various sides of the crystal, hundreds (or thousands) of diffracted X-rays are collected. This data is processed to identify the positions of the planes and atoms in the crystal in three dimensions. Modern single crystal diffractometers use computers to control the goniometer and to process the data; even with a computer, the data processing can take days. The diffraction data is usually converted to a 2D electron density map by Fourier transformation. The electron density map shows the location of atoms. A 2D electron density map is produced for each angle. The computer program uses the 2D maps plus the rotation angle data to generate the 3D coordinates for atoms (molecules, ions) in the crystal. The mathematical treatment of the experimental data to produce a crystal structure from an unknown single crystal diffraction pattern is complicated and beyond the scope of this text.

The diffraction pattern of a single crystal of an inorganic salt is shown in Fig. 8.40. This inorganic salt always gives the same diffraction pattern, and from this pattern we can determine the spacing between planes and the arrangement of planes in the salt crystal. Also, qualitative identification can be obtained by matching this pattern to previously

Figure 8.40   Diffraction pattern in two dimensions of a single crystal of an inorganic salt.

identified patterns. This type of 2D pattern was collected using X-ray film as the detector, as shown schematically in Fig. 8.41. Modern instruments are equipped with 2D imaging detectors such as CCDs.

Powdered crystalline samples can also be studied by XRD. The sample is loaded onto the specimen holder, which is placed in the X-ray beam in a setup similar to that used for single crystal XRD. The sample must be powdered by hand or by mechanical grinding and is pressed into a sample holder to form a flat surface or packed into a thin glass or polymer capillary tube. After mounting, the specimen is rotated relative to the X-ray source at a rate of (degrees θ)/min. Diffracted radiation comes from the sample according to the Bragg equation. The detector is simultaneously rotated at 2θ/min.

A powdered crystalline material contains many thousands of tiny crystals. These crystals are oriented in all possible directions relative to the beam of X-rays. Hence, instead of the sample generating only single diffraction spots, it generates cones of diffracted X-rays, with the point of all of the cones at the sample (Fig. 8.42). Each family of planes will have a different circular diameter, so the result is a series of concentric

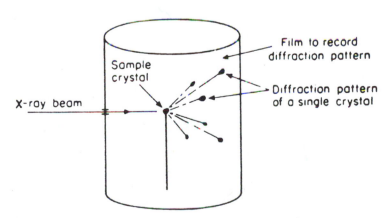

Figure 8.41   Schematic of a single crystal diffractometer using X-ray film to record the diffraction pattern. The film is curved along the circumference of the Rowland circle, replacing the movable detector shown in Fig. 8.39.

Figure 8.42   Schematic of diffraction from a powdered crystalline sample. The powdered sample generates the concentric cones of diffracted X-rays because of the random orientation of crystallites in the sample. The X-ray tube exciting the sample is not shown in this diagram. The cones of diffracted light intersect X-ray film curved to fit the diameter of the Rowland circle. The result is a series of curved lines on the X-ray film.

cones radiating from the sample. Imagine that inside the Rowland circle opposite the sample, we have a strip of X-ray film as shown in Fig. 8.41. The circular cones of X-rays will hit the film, resulting in a series of curved lines on the film (Fig. 8.42). A typical diffraction pattern from a powdered sample collected on film is shown in Fig. 8.43. These are called Laue photographs, after Max von Laue, the German scientist who developed the technique. Film has been replaced by automated scanning with a standard X-ray detector as discussed for XRF or by the use of imaging detectors such as CCDs or image plates to give a 2D image similar to the images obtained from film. A cylindrical image plate detector used by Rigaku MSC has an active area of 454 mm × 256 mm, a pixel size of 100 μm × 100 μm, extremely rapid readout, and a sensitivity of 1 X-ray photon per pixel for Cu K_α radiation. One X-ray photon per pixel is a quantum efficiency of 100%; the quantum efficiency of film is much less. (Details of the Rigaku MSC detector may be found at www.RigakuMSC.com) The major advantage of these imaging detectors is that the images can be stored and manipulated electronically, and without the need for a photographic film-developing lab.

8.3.2.2.   Applications of XRD

The analytical applications of XRD are numerous. The method is nondestructive and gives detailed information on the structure of crystalline samples. Comparing powder diffraction

Spot pattern from a single crystal becomes a line when a powder is used

Figure 8.43   A typical diffraction photograph, called a Laue photograph, from a powdered crystalline sample.

patterns from crystals of unknown composition with patterns from crystals of known compounds permits the identification of unknown crystalline compounds. The number of peaks or lines, intensities of peaks or lines, and the angular positions of peaks or lines (in terms of 2θ) are used to identify the material.

Diffraction patterns are unique for each compound and serve as a fingerprint for a crystalline material. For example, as shown in Fig. 8.44(a) and (b), pure crystals of compound A and pure crystals of compound B give different diffraction patterns. A *mixture* containing both A and B will show diffraction peaks from both pure compounds [Fig. 8.44(c)]. If we had a mixture of 15% KCl and 85% NaCl, the diffraction pattern would show strong NaCl peaks with a weak pattern of KCl intermixed. A mixture containing 15 percent NaCl and 85% KCl would show the diffraction pattern of KCl with a weak pattern of NaCl. Such a mixture is a *multiphase material*, and interpretation of multiphase patterns is more difficult than for single-phase (pure) materials. If, on the other hand, the crystal were a *mixed crystal* of sodium potassium chloride, in which the sodium and potassium ions are in the *same crystal lattice*, there would be changes in the crystal's lattice size from that of pure NaCl or pure KCl. However, the mixed crystal is a single-phase material, resulting in a unique diffraction pattern as shown in Fig. 8.44(d). X-ray powder diffraction therefore can be used to distinguish between a *mixture of crystals*, which would show both diffraction patterns, and a *mixed crystal*, which would give a separate unique diffraction pattern. The exact crystallographic lattice constants can be measured using XRD. Powder diffraction pattern matching to identify unknowns is now

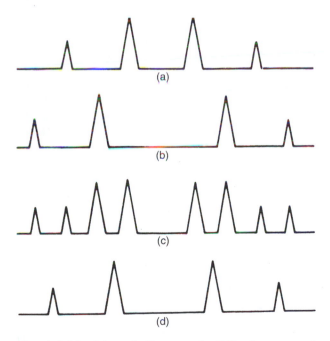

Figure 8.44 Schematic X-ray powder diffraction patterns for simple inorganic salts. (a) Pure salt A. (b) Pure salt B. (c) A physical mixture of salt A and salt B. Note that the peaks for both pure salts can be seen in the mixture; every peak matches a peak in either (a) or (b). (d) A diffraction pattern for a mixed crystal containing the same elements present in both A and B, but chemically combined in the same crystal lattice. The diffraction pattern for the mixed crystal (a unique structure) is unique; it does not match either pure A or pure B.

done with a computer, software, and a powder diffraction pattern spectral library or database that can be searched by the computer. An example is shown in Fig. 8.45. The International Centre for Diffraction Data (ICDD), located in Newtown Square, PA, USA maintains a database of more than 50,000 single-phase powder XRD patterns (www.icdd.com).

In polymer characterization, it is possible to determine the degree of crystallinity of semicrystalline polymers. The noncrystalline (amorphous) portion simply scatters the X-ray beam to give a continuous background, whereas the crystalline portion gives diffraction lines. A typical schematic diffraction spectrum of a semicrystalline polymer is shown in Fig. 8.46. The ratio of the area of diffraction peaks to scattered radiation is proportional to the ratio of crystalline to noncrystalline material in the polymer. The ultimate quantitative analysis must be confirmed using standard polymers with known percent crystallinity and basing the calculation on the known ratio of crystalline diffraction to amorphous scattering.

Using an XRD pattern and looking at the intensity of peaks, it is possible to determine if the crystals of a polymer or a metal are oriented in any particular direction. "Preferred orientation" can occur after the material has been rolled out into a sheet, for example. This is sometimes a very undesirable property, since the material may be very weak in one direction and strong in another, with the result that it tears easily in one direction. Sometimes, however, this is a desirable property, as, for example, in packaging material that we may wish to tear easily in one direction to open the package.

XRD at different temperatures can be used to study phase transitions between different crystallographic forms of a material (e.g., tetragonal vs. monoclinic forms of yttria-stabilized zirconia). This approach can be used to measure thermal expansion coefficients and to study crystalline-to-amorphous transitions in materials.

A property of metals that can be determined by XRD is the state of anneal. Well-annealed metals are in an ordered crystal form and give sharp diffraction lines. If the metal is subjected to drilling, hammering, or bending, it becomes "worked" or "fatigued"; the crystal structure and the diffraction pattern change. Working a metal initially increases its strength, but continued deformation (fatigue) weakens the metal and can result in the metal breaking. (Try bending a paper clip slowly—you should note that it becomes harder to bend after one or two workings, but then if you continue to bend it at the same point, it will eventually become fatigued and break.) Disorder in metals and

Figure 8.45 The X-ray powder diffraction pattern of an unknown material is shown in the upper spectrum. A search of a computerized database identified the unknown as silicon, based on the match to the stored spectrum for silicon (lower spectrum).

Pattern (peaks) from crystalline material

Scatter from
the amorphous material

Scatter from the instrument

Figure 8.46 Schematic diffraction pattern from a semicrystalline polymer, showing how both crystalline and amorphous phases may be detected. The amorphous portion results in broad scattering while the crystalline portion shows a typical diffraction peak pattern. A totally amorphous polymer would show no diffraction peaks. (There are no 100% crystalline polymers.) The units on the x-axis are degrees θ.

alloys can also result from rapid cooling from the molten state. Disorder results in lower density and a more brittle material. Electrical conductivity of metals is also affected by the order in the metal crystal. XRD can be used to distinguish between ordered and disordered materials and provide valuable information on the suitability of a material for a given use.

A very important use of XRD is in the determination of the structure of single crystals, that is, identifying the exact position in 3D space of every atom (molecules, ion) in the crystal. Single crystal XRD was a major tool in elucidating the structure of ribonucleic acid (RNA) and deoxyribonucleic acid (DNA), insulin, vitamins, and proteins. Single crystal diffractometry is used for structural determination of biomolecules, natural products, pharmaceuticals, inorganic coordination complexes, and organometallic compounds.

8.3.2.3. Analytical Limitations of XRD

Amorphous materials cannot be identified by XRD, so many polymeric and glassy materials cannot be studied. XRD is not a trace analysis technique. A component should be present at 3–5% by weight at a minimum in order for diffraction peaks to be detected. The unique pattern from a mixed crystal, where one atom has replaced another in the lattice, was discussed earlier. A mixed crystal is analogous to a solid solution of a contaminant in a pure material. The contaminated material will appear to be single phase, since it is a solid solution, but the lattice of the pure material will be expanded or contracted as the result of contaminant atoms of the "wrong" size in the lattice. There will be a unique diffraction pattern, but such a pattern may be hard to match. Mixtures may be difficult to identify because of overlapping peaks.

8.3.3. X-Ray Fluorescence (XRF)

When a sample is placed in a beam of X-rays, some of the X-rays are absorbed. The absorbing atoms become excited and emit X-rays of characteristic wavelength. This process is called *X-ray fluorescence*. Since the wavelength (energy) of the fluorescence is characteristic of the element being excited, measurement of this *wavelength* enables us to *identify* the fluorescing element. Tables of X-ray lines are given in the appendices of this chapter. The *intensity* of the fluorescence depends on *how much* of that element is in the sample. For most laboratory XRF equipment, the energy of the emitted X-ray is independent of the chemical state of the element; therefore, XRF is generally considered to be an elemental analysis method.

Modern XRF instruments permit the determination of all elements from fluorine ($Z = 9$) to uranium ($Z = 92$). Some systems allow measurement of elements from Be to U. Elements with atomic numbers between 12 and 92 can be analyzed in air. Elements with atomic numbers 3 (beryllium) through 11 (sodium) fluoresce at long wavelengths (low energy), and air absorbs the fluorescence. Analysis of this group must be carried out in a vacuum or in a helium atmosphere. The detection limits are in the ppm range for most elements with a wide linear working range. XRF thus permits multielement analysis of alloys and other materials for major, minor, and trace elements. Sensitivity is poorest for the low Z elements and best for the high Z elements.

Both solid and liquid samples can be analyzed by XRF as described earlier in the chapter. Very flat surfaces are required for quantitative analysis, as discussed subsequently. Liquids flow into flat surfaces, but cannot be run under vacuum. The best solvents are H_2O, HNO_3, hydrocarbons, and oxygenated carbon compounds, because these compounds contain only low atomic number elements. Solvents such as HCl, H_2SO_4, CS_2, and CCl_4 are undesirable because they contain elements with higher atomic numbers; they may reabsorb the fluorescence from lower-Z elements and will also give characteristic lines for Cl or S. This will preclude identification of these elements in the sample. Organic solvents must not dissolve or react with the film used to cover the sample.

Solid samples that can be cut and polished to give a flat surface can be analyzed after polishing. Care must be taken not to contaminate the sample with the cutting tool or polishing compound. For example, cutting a flat piece of polymer with a steel razor blade can result in iron being detected in the polymer sample. Other solids should be ground to a powder, preferably using a ball mill or similar device to pulverize the sample. Again, the grinding tools must not contaminate the sample, so boron carbide is often used to contain and grind samples. The sample powder may be pressed "as is" or mixed with lithium borate salt, borax, wax, or other suitable "binder" and formed into a briquette. This procedure provides a sample that can be easily handled and has the advantage that the borate salts or hydrocarbon binders provides a standard matrix for the sample. Furthermore, the matrix is composed of low atomic weight elements, which interact only slightly with the X-ray beam. The ultimate preparation is to mix the sample and lithium metaborate or lithium tetraborate, and heat the mixture in a Pt crucible over high heat. The process is called fusion and is shown in the photographs in Fig. 8.47(a)–(c). The heat melts the sample and salt, and the molten mixture is poured into a Pt/Au mold with a flat bottom. When the melt is allowed to cool, a glassy flat "fused bead" is formed, suitable for quantitative XRF analysis with the advantage that the matrix is now the same for all samples. Fusion is usually required for analysis of geological samples to eliminate mineralogical effects, for example. Calibration standards are prepared by fusing known amounts of the analyte with the borate salt and casting standard beads. Automated fusion devices, called fluxers, are available, as are a wide variety of devices for grinding and powdering samples. A handbook of XRF sample preparation methods and equipment, with pictures of the various fluxers, crucibles, molds, and grinders, is available on the Web from SpexCertiprep at www.spexcsp.com.

8.3.3.1. Qualitative Analysis by XRF

Each element fluoresces at its own characteristic wavelengths (energies); the fluorescing element can be identified from a table of wavelengths (energies) such as those in

Figure 8.47 Preparing a fusion bead for analysis. (a) The sample and a salt such as lithium tetraborate are heated until liquid in a Pt crucible over a burner. The crucible tongs are tipped with Pt to avoid contamination of the crucible. (b) The molten mixture is poured into a mold, in this case a ring placed on a smooth marble slab. (c) Cooled beads removed from the molds. The flat bottom surface, shown turned up, is the side used for analysis. (From Jenkins et al., 1981, used with permission.)

Appendix 8.1. An example of qualitative analysis is shown in Fig. 8.48. The EDXRF spectrum is that of a zinc-coated iron washer. The spectra are much simpler than those from atomic emission spectrometry. Fe gives hundreds of strong emission lines in an ICP or DC arc emission experiment; here only two lines are seen for Fe in this XRF spectrum. Even with the relatively simple spectra, spectral interference does occur in XRF. The major sources of spectral interference are: scattered radiation from the tube, higher order lines diffracted in a wavelength dispersive system, and L lines of higher atomic number elements overlapping K lines of lower atomic number elements. Two examples are given in Fig. 8.49. The peaks in each spectrum are labeled, but you should use the tables in Appendix 8.1 to confirm that they have been labeled correctly and to give yourself some practice in figuring out what elements are present from the energy positions of the peaks. The top spectrum shows the overlap of the spectral lines from pure Fe and pure Mn; the spectral interference occurs between the Fe K_α and Mn K_β peaks. The peak maxima are not at exactly the same energy, so in a mixture or alloy of the two elements, the Fe K_α line can be identified, and if the Mn content of a mixture or alloy were high, the Mn K_β peak might appear as a shoulder on the Fe K_α peak. The overlap is not critical in a two-component system; as you can see, the unobstructed Mn K_α, the strong Fe K_α and the unobstructed Fe K_β peaks can easily identify the presence of both elements. The bottom spectrum is that of a more complex alloy containing Cr, Mn, Fe, Co, Ni, and W. Qualitatively, it is easy to see that Cr, Fe, W, and Co are present. Cr and W have unobstructed lines that tell us they are in the alloy. Fe is clearly present, because if there is any Mn, it must be in low concentration compared to the Cr and Fe. The Mn K_α peak overlaps the Cr K_β peak; we know there is Cr present from its K_α peak, so some (at least) of the intensity of the small peak at 5.8 keV is due to Cr. If Mn is present but in low

F.S. = 16K

This washer actually has much more iron
than zinc, but the zinc is present as a
surface coating. If the sample was completely
homogenous iron would be present at about
66%. Remember, EDXRF is a surface sensitive
technique.

0 1 2 3 4 5 6 7 8 9 10 11 12 13 14 15 16 17 18 19 20
KEV

(a)

F.S. = 16K

This is a spectrum of the same washer
but it has been sanded to remove some of
the zinc surface coating.

0 1 2 3 4 5 6 7 8 9 10 11 12 13 14 15 16 17 18 19 20
KEV

(b)

Figure 8.48 Qualitative EDXRF spectrum of a zinc-coated iron washer. (a) The coated washer. The peaks present are, from left to right, Fe K_α, Fe K_β, Zn K_α, and Zn K_β. (Confirm these peaks by looking in Appendix 8.1, Table 2.) (b) The same washer, but after sanding to remove some of the surface. Note the decrease in the intensity of the Zn peaks and the increase in the Fe peaks. XRF is a surface sensitive technique because the depth of penetration depends on the exciting radiation and the sample composition (atomic number). These spectra were collected for 50 s with an Rh tube at 30 kV and 0.01 mA, under vacuum, using a filter to remove the X-ray tube lines. [Courtesy of ThermoNoran (www.thermo.com).]

Figure 8.49 Spectral overlap in EDXRF spectra. (a) A simple overlap demonstrated by overlaying the spectra of pure Fe and pure Mn. There is overlap of the Fe K_α line with the Mn K_β line. (b) The spectrum of a multielement alloy showing multiple spectral interferences due to line overlaps. Despite the overlaps, knowledge of the relative line intensities permits qualitative identification of most of the elements present.

concentration, then most of the intensity at about 6.4 keV is from Fe, since the Mn K_β peak is much smaller than its K_α peak, as seen in the top spectrum. Looking at the peak to the right of the Fe K_α peak, we see a much bigger peak than would be expected from Fe K_β alone. (Compare the Fe peak heights in the top spectrum.) Therefore, most of the peak at about 6.9 keV must be from Co K_α. That indicates that Co is present in the alloy. From the Mn and Fe peak ratios in the top spectrum, it is reasonable to assume that the Co K_β peak is less intense than the Co K_α peak. But the peak at approximately 7.5 keV, about where we would expect the Co K_β peak, is actually more intense than the Co K_α peak (and the "Co K_α" peak intensity also includes some interfering Fe K_β intensity). Therefore, the high intensity of the 7.5 keV peak tells us we have another element, Ni, in the alloy. The one element that is not clearly present is Mn; the intensity ratio for the two Cr lines needs to be ascertained. If the "Cr K_β" peak is too high relative to the K_α peak, then Mn is probably in the alloy as well. So even though only two elements, Cr and W, have lines with no spectral interference, knowledge of the relative intensities of lines allows us to state with certainly that Fe, Co, and Ni are present, and that Mn may be present in low concentration. We will come back to this spectrum in the discussion of quantitative analysis by XRF.

The XRF method is nondestructive, an important feature when the sample is available in limited amounts or when it is valuable or even irreplaceable, as in the case with works of art, antiques, or rocks from the moon. The nondestructive nature of XRF coupled with the fact that sample preparation may not be required means that direct multi-element analysis can be performed *in situ*. Portable handheld XRF analyzers are used to sort metal scrap at landfills and recycling centers and to check incoming lots of material before unloading a ship or a truck. Antiques, jewelry, gems, and art objects can be characterized. Original art and copies of masterpieces can be distinguished from each other based on elemental composition of pigments. Natural gems can be distinguished from synthetic gems in many cases, because the laboratory synthesis often adds specific elements not usually found in the natural gems or may even produce gems with fewer trace elements than real gems. Plated jewelry can be distinguished from solid gold or solid silver jewelry. Museums rely heavily on this method for examining works of art. The sample is usually unaffected physically or chemically by the analytical process. It is possible for some sensitive materials (plastics, paper) to turn yellow or brown upon prolonged exposure to X-rays, so some care must be taken to prevent this.

8.3.3.2. Quantitative Analysis by XRF

Quantitative analysis can be carried out by measuring the intensity of fluorescence at the wavelength characteristic of the element being determined. The method has wide application to most of the elements in the periodic table, both metals and nonmetals and many types of sample matrices. It is comparable in precision and accuracy to most atomic spectroscopic instrumental techniques. The sensitivity limits are of the order of 1–10 ppm, although sub-ppm detection limits can be obtained for the heavier-Z elements under appropriate conditions. Accuracy depends on a thorough understanding of the sample heterogeneity, the sample matrix, and the absorption and fluorescence processes.

X-rays only penetrate a certain distance into a sample and the fluorescing X-rays can only escape from a relatively shallow depth (otherwise they would be reabsorbed by the sample). While XRF is considered a "bulk" analysis technique, the X-rays measured are usually from no deeper than 1000 Å from the surface. True surface analysis techniques such as Auger only measure a few angstroms into the sample, so in that respect XRF does measure the bulk sample, but only if the surface is homogeneous and representative of the entire sample composition.

The relationship between the measured X-ray intensity for a given peak and the concentration of the element can be written:

$$C = K(I)(M)(S) \tag{8.22}$$

where C is the concentration of the element. The factor K is a function of the spectrometer and the operating conditions. I, the intensity of the measured signal, is the net intensity after subtraction of the background. M represents interelement effects such as absorption and enhancement effects. Interelement effects must be corrected for or minimized to obtain accurate concentration values. S is the specimen homogeneity, and the value of S depends on the average particle size of the sample and the penetration depth of the X-rays. The only way to control S is to control the sample particle size by using suitable sample preparation techniques. The classical approach to quantitative analysis is to use an external calibration curve. A set of standards at various concentrations is prepared in a manner similar to the sample preparation, intensity is carefully measured for all standards and samples using the same instrumental conditions, and a plot of C vs. I is made. M and S are considered to be constant (a good assumption only if the standards reflect the matrix and particle size distribution of the samples). Precision depends only on careful measurement of intensity; accuracy depends on the elimination of M, the interelement effects.

Peak and background intensities are measured by counting photons at the appropriate wavelength. When the peak to background ratio exceeds 10:1, the net counting error is small and almost constant, so in wavelength dispersive systems, background is ignored if the ratio of peak to background is greater than 10:1. In energy dispersive systems, background subtraction is usually performed. The background is very matrix-dependent in XRF, so background correction is more difficult than in other types of spectroscopy. The correction is usually an estimate at best.

Calibration methods include the use of external standards, and may include the use of internal standards to improve precision. The internal standard may be an added element or elements not present in the sample, but this is a very time-consuming endeavor, both in making the standards and samples and in measuring all the lines. Frequently, the scattered radiation from the X-ray tube is used as an internal standard; this is a reasonable approach because the scattered radiation (background) is very dependent on the sample matrix. The ratio of analyte line to scattered radiation results in better precision than the measurement of the analyte line alone.

A variety of mathematical approaches to the correction of absorption-enhancement effects and calibration are now in use, due to the availability of inexpensive, powerful computers, and powerful software programs. One approach is the **fundamental parameters method**. The method requires the measurement of pure element samples for every element that needs to be measured (most of the periodic table). Tables of values for three *fundamental parameters*, the primary spectral distribution, the absorption coefficient, and the fluorescent yield, are used by a complex software program to calculate concentrations from measured analyte line intensity. If done correctly, no external matrix-matched calibration standards are needed, so the fundamental parameters method is also called "standardless analysis". The sample must have a planar surface, must be homogeneous, and must be infinitely thick (no transmission of X-rays through the sample). Often, especially for failure analysis or forensic science, the sample may be very small, not planar, inhomogenous, and so on. Most commercial XRF systems have a fundamentals parameters software program as part of their data processing package. Early versions of these mathematical approaches with no calibration standards were really "semiquantitative" methods; improvements have been and are still being made. The analyst should, when possible, verify the accuracy of any mathematical approach with known materials that reflect the samples being analyzed. More information on these approaches can be found in several of the introductory XRF texts (e.g., those by Bertin, Jenkins et al. and Herglotz and Birks listed in the bibliography). The advantage to the fundamental parameters (FP) approach is that many industrial materials do not have readily available matrix-matched calibration standards commercially available. Preparation of good, stable calibration standards takes a lot of time and money even when it is possible to make such standards. Even when well-defined standards are available commercially, as they are for steels, brasses, bronzes, nickel-based superalloys, and many other types of alloys, some glasses, and ceramics, the standards are expensive and must be handled carefully to avoid scratching or contaminating their surfaces. Sources for prepared XRF standards include government standards agencies such as NIST in the US, as well as commercial firms.

Modern XRF systems can measure the entire periodic table in about 20 min and calculate the composition of a complete unknown using an FP program. One commercial instrument uses a fast scan approach that permits the entire periodic table to be measured in about 2 min, due to improvements in counting electronics.

When quantitative analysis is needed on complex samples with peak overlaps, a method of extracting the individual intensities of each element is needed. These mathematical corrections are now performed in instrument data processing software, and may include overlap corrections, interelement corrections, matrix corrections, corrections for non-Gaussian peak shapes, and different statistical fitting routines. A combination of these corrections was used to determine the concentration of the elements in the alloy shown in Fig. 8.49.

8.3.3.3. *Applications of Quantitative XRF*

Quantitative XRF is used in virtually every industry for almost any type of liquid or solid sample. XRF is used daily to analyze minerals, metals, paper, textiles, ceramics, cement, polymers, wood, environmental samples, food, forensic samples, cosmetics and personal care products, and more. Only a few examples will be given here.

The petroleum industry uses XRF to measure sulfur in fuels, the elemental composition of catalysts used to "crack" petroleum hydrocarbons, and lubricating oil additives. For example, cracking catalysts usually contain more than 12 elements, including transition metals, rare earth elements, alkali and alkaline earth elements at concentrations varying from 0.005 to 35 wt%. The composition of the catalyst can be determined accurately after fusion into borate beads; the total analysis takes about 1 h. Sulfur in petroleum products and gasoline can be quantified using EDXRF by procedures described in ASTM Methods D4294 and D6445 (in the Annual Book of ASTM Standards). Method D4294 has a working range of 0.015–5 wt% S, while Method D6445 for gasoline can determine 5–1000 ppm S. The amount of sulfur in fuel must be known not only for processing the fuel but also to meet air pollution emission standards for oxides of sulfur.

Engine oil used to be monitored routinely for metal content, since the metal content indicated how parts of the engine were wearing away in use. The wear metals analysis indicated what parts needed to be replaced during maintenance. This analysis could be performed by XRF or by atomic emission spectrometry (e.g., ICP or DCP emission). However, filters in modern engines are now so efficient at trapping wear metal particles that the analysis of oil is no longer a good indicator of engine wear. XRF can be used to measure the metals on the filters themselves to predict engine failure and part replacement.

In metallurgy, alloy composition can be rapidly determined and unknown samples identified rapidly. XRF has an advantage over wet chemistry in that all of the components can be measured due to the wide dynamic range of XRF. For example, in the analysis of nickel alloys, a wet chemical approach would measure all the other elements and calculate the Ni content as the balance. With XRF, the major element, nickel, as well as the minor and trace components can be measured accurately.

XRF is used to measure the amounts of phosphorus-based flame retardants in textiles and the amount of antimony-based flame retardant in plastics. For textiles, a piece of fabric is cut into a square piece and stretched across a standard sample cup and held in place by the sample container ring. This results in a flat specimen for analysis. Phosphorus levels in the range of 0.3–3% P can be measured in a sample in as little as 60 s. Details of the method are available from SPECTRO Analytical Instruments (www.spectro-ai.com).

The ceramics industry routinely measures 6–12 elements quantitatively in both pressed pellet and fusion bead form for quality control. The elements (reported as oxides) vary in concentration from 0.01 to 70 wt%. Using a modern XRF spectrometer, aluminum (reported as Al_2O_3) can be determined at the 1 wt% level in a magnesium oxide based ceramic with a standard deviation of 0.006 and a detection limit of about 8 ppm.

In medicine, researchers can measure lead levels in bone *in vivo* by taking the ratio of the Pb lines to the elastic scatter peak from the bone (as an internal standard). Levels of lead in the 3–30 ppm range have been measured. Trace elements in soil and sediment can be measured to collect geological, agricultural, and environmental data both in the lab and in the field, using portable XRF analyzers.

Online XRF analyzers are available for monitoring process streams for metals in plating baths, metals in hydrocarbons, silicon in adhesive coatings on paper, acid leaching solutions, effluent discharge monitoring, and similar applications. The thickness of coatings can be measured on a wide variety of materials, such as paint on steel for the automotive industry and silicone release coating on paper (the shiny nonstick backing paper you peel adhesive labels from is coated with a silicone polymer called a "release coating").

8.3.4. Electron Probe Microanalysis

A beam of electrons striking a target results in the emission of characteristic X-rays. This is the basis of the X-ray tube, as was discussed earlier in the chapter. A beam of electrons striking a sample will also generate characteristic X-rays from the sample. The use of a small diameter electron beam, on the order of 0.1–1.0 μm, to excite a sample is the basis of **electron probe microanalysis**. An electron probe microanalyzer is an X-ray emission spectrometer. The small diameter electron beam excites an area of the surface of the sample that is about 1 μm in diameter. Elemental composition and variation of composition on a microscopic scale can be obtained.

Two different instruments are available for microanalysis. The electron microprobe analyzer (EMA) uses high electron beam currents to provide elemental analysis of samples, with moderate spatial resolution and low magnification of the sample. The intensity of emitted X-rays from the sample is high. The scanning electron microscope (SEM) is designed to provide high-resolution, high-magnification images of a sample. The SEM operates at low electron beam currents; the characteristic X-rays are of lower intensity than in the EMA but still provide elemental composition. In addition to X-ray spectra, other information may be obtained as a result of interactions of the electron beam with the sample. Some samples will exhibit UV or visible fluorescence (called *cathodoluminescence*); secondary electrons may be ejected, such as Auger electrons and backscattered electrons are used to provide information about the sample topography and composition.

Both systems operate on similar principles. A tungsten filament emits electrons, which are focused by an electron optical system. The electron beam can scan the sample surface and can provide composition at a point, along a line or over a rectangular area, by *rastering* the beam across the surface in a series of parallel lines. The sample is mounted on a stage that can be accurately moved in the x and y directions and in the z direction, normal to the plane of the sample. The system has an optical microscope, to permit alignment of the sample and selection of the area or feature of the sample to be analyzed.

The X-ray analysis system for the EMA is a wavelength dispersive spectrometer with gas proportional counter detectors. In the SEM, an energy dispersive X-ray spectrometer with a Si(Li) detector is used. The entire electron and X-ray optical systems are operated under a vacuum of about 10^{-5} torr. Modern systems are completely automated with computer control of the instrument parameters, specimen stage movement, data collection and data processing.

Samples must be solid and may be in almost any form. Thin films, bulk solids, particles, powders, machined pieces, and small objects (including biological specimens) can be analyzed. All elements from beryllium ($Z = 4$) to U can be determined at concentrations of about 100 ppm or greater. For qualitative analysis, the surface finish of the sample is not important. For quantitative analysis the surface of the specimen must be flat. A common method for achieving a flat surface for an SEM sample is to embed the sample in epoxy and then carefully polish the hardened epoxy to expose a flat surface of the sample. Calibration standards should have flat surfaces as well, and the composition of the standards should be similar to that of the samples. Alternatively, a fundamental parameters approach using pure element standards can be used.

Applications of electron probe microanalysis are numerous in analytical chemistry, metallurgy, geology, medicine, biology and materials science. In materials science, not only can the composition of the material be determined, but the location of elements in the solid structure. Segregation of a particular element in grain boundaries between crystals can be seen; such segregation may have a dramatic effect on the properties of a material. Pure tungsten wire is too brittle to make a useful light bulb. The wire fractures easily as the light bulb is turned on (wire heats up) and off (wire cools down) because the crystals separate along the grain boundaries. By adding certain elements to tungsten, an impure tungsten wire that is much more ductile results; this ductile wire can be heated and cooled many times without fracturing. Analysis of the impure tungsten using SEM shows that the added elements are located in the boundaries between the crystals. The elements in the grain boundaries permit the tungsten wire to expand and contract on heating without fracturing. Thin films and multilayer materials can be analyzed and film thickness measured. Small particles filtered from air and water can be analyzed, and both particle count and particle size determined. The number, size, and type of particles in air is a measure of air quality and is very important in the manufacturing of semiconductor devices.

The distribution of elements in biological samples, mineral samples, soils, and other heterogeneous materials can be determined. Many other applications can be found in the literature. This is one of the few techniques that can provide spatial variation of composition at the micrometer scale. Other related techniques are discussed in Chapter 14.

BIBLIOGRAPHY

American Society for Testing and Materials. *Annual Book of ASTM Standards*; ASTM: West Conshohocken, PA, 2002.

Application Reviews. *Anal. Chem.* **1994**, *June*.

Bertin, E.P. *Principles and Practice of X-Ray Spectrometric Analysis*, 2nd Ed.; Plenum Press: New York, 1975.

Bertin, E.P. *Introduction to X-Ray Spectrometric Analysis*; Plenum Press: New York, 1978.

Burr, A. In *Handbook of Spectroscopy*; Robinson, J.W., Ed.; CRC Press: Boca Raton, FL, 1994; Vol. 1.

Clark, G.L., Ed. *Encyclopedia of X-rays and Gamma Rays*; Reinhold Publishing: New York, 1963.

Criss, J.W.; Birks, L.S. *Anal. Chem.* **1968**, *40*, 1080–1086.

Ellis, A.T. *Handbook of X-ray Spectrometry*, 2nd Ed.; Van Griekin, R.E., Markowicz, A.A., Eds.; Marcel Dekker, Inc.: New York, 2002.

Formica, J. X-ray diffraction. In *Handbook of Instrumental Techniques for Analytical Chemistry*; Settle, F., Ed.; Prentice-Hall, Inc.: Upper Saddle River, NJ, 1997.

Havrilla, G.J. X-ray fluorescence spectrometry. In *Handbook of Instrumental Techniques for Analytical Chemistry*, Settle, F., Ed.; Prentice-Hall, Inc.: Upper Saddle River, NJ, 1997.

Helson, L.A.; Kuczumov, A. In *Handbook of X-Ray Spectrometry*, 2nd Ed.; Van Griekin, R.E.; Markowicz, A.A., Eds.; Marcel Dekker, Inc.: New York, 2002.

Herglotz, H.K.; Birks, L.S. *X-Ray Spectrometry*; Marcel Dekker, Inc.: New York, 1978.

Jenkins, R. X-ray fluorescence. *Anal. Chem.* **1984**, *36*, 1009A.

Jenkins, R. *X-ray Fluorescence Spectrometry*, 2nd Ed.; John Wiley and Sons, Inc.: New York, 1999.

Jenkins, R.; Snyder, R.L. *Introduction to X-ray Powder Diffractometry*; John Wiley and Sons, Inc.: New York, 1996.

Jenkins, R.; Gould, R.W.; Gedcke, D. *Quantitative X-ray Spectrometry*; Marcel Dekker, Inc.: New York, 1981.

Jenkins, R.; Manne, R.; Robin, J.; Senemaud, C. *Pure Appl. Chem.* **1991**, *63* (5), 785.

Lubhofsky, H.A.; Schweikert, E.A.; Myers, E.A. *Treatise on Analytical Chemistry*, 2nd Ed.; Wiley Interscience: New York, 1986.

Markowicz, A.A. In *Handbook of X-Ray Spectrometry*, 2nd Ed.; Van Griekin, R.E., Markowicz, A.A., Eds.; Marcel Dekker, Inc.: New York, 2002.

Parsons, M.L. X-ray methods. In *Analytical Instrumentation Handbook*, 2nd Ed.; Ewing, G.W., Ed.; Marcel Dekker, Inc.: New York, 1997.

Robinson, J.W. *Handbook of Spectroscopy*; CRC Press: Boca Raton, FL, 1974; Vol. 1.

Robinson, J.W. *Practical Handbook of Spectroscopy*; CRC Press: Boca Raton, FL, 1991.

Stout, G.H.; Jensen, L.H. *X-ray Structure Determination*; Macmillan Publishing Co., Inc.: New York, 1968.

Teo, B.K.; Joy, D.C., Eds. *EXAFS Spectroscopy: Techniques and Applications*; Plenum Press: New York, 1981.

Van Griekin, R.E.; Markowicz, A.A., Eds. *Handbook of X-ray Spectrometry*, 2nd Ed.; Marcel Dekker, Inc.: New York, 2002.

SUGGESTED EXPERIMENTS

SAFETY NOTE: The student is reminded that X-radiation poses significant health hazards. Safety is a very important issue with X-ray methods. X-radiation is ionizing radiation, a significant health hazard because it can damage or kill living cells and produce genetic mutations. Some instrumentation uses radioactive isotopes as sources; these sources are always producing ionizing radiation and cannot be "turned off". Modern commercial instrumentation is equipped with the proper shielding and safety interlocks that will shut off the ionizing radiation automatically or protect the analyst from exposure when operated properly. Operators of X-ray instrumentation usually must undergo specialized safety training and wear radiation dosimeters to ensure that they are not being exposed to damaging X-rays. Only commercial equipment equipped with safety interlocks should be used and only after appropriate training and under the guidance of your instructor or radiation safety official at your facility. Under no circumstances should safety interlocks be tampered with. Experiments should not be performed with unconfined or unshielded X-ray sources.

X-Ray Absorption

8.1 Take several metal disks made of different pure elements, each about 1 in.^{2} in area and 1/8 inch in thickness. It is important that the thickness of each disk be constant. The disks should be made from (a) Mg, (b) Al, (c) Fe, (d) Ni, (e) Cu, (f) Sn, and (g) Pb. Expose them to an X-ray beam simultaneously on a single sheet of X-ray photographic film. Develop this film. Note that the absorption of X-rays by each disk is proportional to the atomic weight of the particular element.

8.2 Take a piece of glass known to contain an entrapped bubble of air. Place it in front of X-ray film and expose it to X-rays. Note that the bubble can be detected on the film as a dark patch. Similar holes in metal castings (or other metal objects) can be detected by X-ray absorption, even though the holes are not visible to the naked eye.

X-Ray Fluorescence

8.3 Using the metal disks used in Experiment 8.1, record the fluorescence spectra of the different metals. Identify the K and L lines for each element. Plot the relationship between the wavelengths of these lines and the atomic numbers of the metals using a spreadsheet program such as Excel. Explain your results.

8.4 Record the fluorescence spectrum of metal disks of unknown composition. By measuring the wavelength of the K_α line, identify the major components. Use disks of brass, bronze, stainless steel, and aluminum and coins of various denominations, for example.

8.5 Using silver powder and copper powder, prepare mixtures with various ratios of these metals. Record the XRF spectrum of each mixture. Relate the silver/copper ratio to the ratio of the Ag K_α and Cu K_α lines. Measure the Ag/Cu ratio of various US silver coins minted (a) before 1964 and (b) after 1964.

8.6 Record the fluorescence spectrum of sulfur. Locate the S K_β line. Record the fluorescence spectra of various grades of motor oil. Compare the sulfur contents of the oils by the intensities of the S K_β lines. The oil may be held in a polymer "zipper lock" bag during analysis if no liquid sample cups are available. Run a blank spectrum of the polymer bag as well to be certain it contains no detectable sulfur.

8.7 Analyze small chips of various rocks, or samples of sand or soil from different geographic locations. Identify the elements present in each sample. Can you distinguish between sands or soils from different locations by their XRF spectra? Look for any trace or minor elements that may differentiate the samples. (Remember that peak intensities will vary due to surface roughness, so particle size differences between samples may account for peak height differences in the spectra.)

8.8 Take the spectrum of various pieces of jewelry, supported on cellophane tape or a suitable heavy polymer film so that they do not fall into the spectrometer. Identify the elements present. Is your gold ring solid gold or gold-plated?

X-Ray Diffraction

8.9 Use a tungsten or copper X-ray source and a powder diffractometer. Load crushed NaCl powder into the sample holder. Measure the diffraction pattern of the NaCl. Repeat with powdered KCl. Note the difference in spectra. Make mixtures by varying the amounts of NaCl and KCl. Can you distinguish the mixtures and the relative amounts of each component from the powder patterns?

8.10 For the student interested in crystal growth, potassium aluminum sulfate, also called alum, is a salt available in most pharmacies. It will form single crystals relatively easily from a supersaturated solution. Alum crystals are octahedral in shape and colorless. The related compound, chromium aluminum sulfate dodecahydrate, called chrome alum, forms dark purple octahedral crystals. Mixing various proportions of supersaturated alum and chrome alum solutions allows the growth of mixed crystals that resemble amethysts in color. (The more alum in the mix, the lighter the purple color of the crystal.) Take the powder patterns of crushed pure alum and crushed pure chrome alum, and a mixture of the two compounds. After you have grown some mixed crystals, crush some and take their powder pattern. Is it different from the two pure materials and from the mixture? Explain your observations.

PROBLEMS

Note: Problems 8.23–8.30 are based on EDXRF spectra obtained using a Rh tube and the conditions listed for each spectrum. To identify the element(s) present, the student needs to use the X-ray line tables in energy units (keV) in Appendix 8.1. Looking at the conditions under which the spectrum was obtained will help. Note if vacuum or air was used, for example. These spectra are reproduced courtesy of ThermoNoran (www.thermo.com).

8.1 Draw a schematic diagram of an X-ray tube, label the major parts and describe its operation.

8.2 What is the origin of the K_α X-ray lines? The K_β lines? The L lines?

8.3 State Moseley's Law. What was the importance of this law to chemistry in general?

8.4 What are the three major analytical fields of X-ray spectroscopy? State three analytical applications of each field.

8.5 The intensity of XRF is weak compared to some other spectroscopic techniques. What instrumental changes are made to maximize the fluorescence signal?

8.6 What elements can be detected by XRF? What elements cannot be detected by XRF in air? Explain.

8.7 Can X-rays from a tungsten target be used to excite copper atoms? Can X-rays from a copper target be used to excite tungsten atoms? Explain.

8.8 Look at Figs. 8.3 and 8.16. The spectrum in Fig. 8.3 was obtained at a tube voltage of 30 kV; that in Fig. 8.16 with *no* filter was obtained at 20 kV. Explain the difference in the spectra.

8.9 What is the relationship between the wavelengths of the absorption edges and the related emission lines? Explain.

8.10 Why can XRF not distinguish between Cu^{2+} and Cu^{1+} in a sample?

8.11 Explain the difference between the two spectra in Fig. 8.16. Why is a filter useful?

8.12 Derive the Bragg equation using reflection from two parallel planes as the model.

8.13 (a) Plot the signal-to-voltage relationship in an ionization counter and proportional counter.
(b) Which is most sensitive? Why?

8.14 Diagram a scintillation detector and describe its principle of operation.

8.15 Describe how a Si(Li)-drifted detector operates.

8.16 (a) What is meant by the mass absorption coefficient?

(b) What is the relationship between the mass absorption coefficient, atomic number, and atomic weight at the same wavelength?

8.17 Describe the equipment used to collect XRD patterns from a single crystal.

8.18 Why are X-ray powder diffraction patterns useful in analytical chemistry? What advantage does a powdered sample have over a single crystal in terms of identifying an unknown sample?

8.19 Describe the basic components of a goniometer.

8.20 Look up the Ge line energies in Appendix 8.1 and predict where you would find a Ge escape peak in an EDXRF spectrum of pure Cu using a Ge(Li) detector. Where would the escape peak occur if a Si(Li) detector were used for the same sample?

8.21 Predict what sum peaks you might see as artifacts in the EDXRF spectrum of pure Cu.

8.22 Looking at the cell in Fig. 8.36(b) with the origin in the back lower left corner (marked O), explain why the (440) plane is closer to the origin than the (110) plane. Remember that the Miller indices are the reciprocals of the intercepts. Draw the planes. (You may find drawing a 2D picture is easier than a 3D one).

8.23 Identify the binary inorganic salt that gives the following spectrum.

TUBE VOLTAGE : 14 KV   FILTER USED : ALUMINUM
TUBE CURRENT : 0.30 mA   LIVETIME   : 50 SEC
ATMOSPHERE   : VACUUM

F.S. = 8K

0 1 2 3 4 5 6 7 8 9 10 11 12 13 14 15 16 17 18 19 20
KEV

8.24 Identify the elements present in this spectrum.

```
TUBE VOLTAGE : 12 KV          FILTER USED : NO FILTER
TUBE CURRENT : 0.05 mA        LIVETIME    : 50 SEC
ATMOSPHERE   : VACUUM
                                            F.S. = 2K
```

8.25 Identify the elements present in this spectrum.

```
TUBE VOLTAGE : 26 KV          FILTER USED : FIVE
TUBE CURRENT : 0.35 mA        LIVETIME    : 50 SEC
ATMOSPHERE   : VACUUM
                                            F.S. = 2K
```

8.26 Identify the elements present and label the lines.

8.27 Identify the elements present-there are at least seven. (Hint: many are high Z elements).

8.28 More high Z elements to identify. Label the lines.

```
TUBE VOLTAGE : 50 KV       FILTER USED : SIX (0.6mm Cu)
TUBE CURRENT : 0.35 mA     LIVETIME    : 50 SEC
ATMOSPHERE   : VACUUM
                                            F.S. = 2K
```

8.29 This is the spectrum of the X-ray tube. The line spectrum was obtained with NO filter in the X-ray tube path. The dotted spectrum was obtained using a 0.63 mm Cu filter. (1) Identify the X-ray tube anode. (2) What do we call the radiation emitted by the filtered X-ray tube? (3) What elements will be excited using the filtered tube radiation? (4) Why do you want the tube voltage so high? (50 kV is high voltage). (5) Why not use the unfiltered X-ray tube spectrum to excite your sample? Can you list two elements that could not be determined using the unfiltered X-ray tube, other than the anode element itself?

```
TUBE VOLTAGE : 50 KV       FILTER USED : NO FILTER
TUBE CURRENT : 0.00 mA     LIVETIME    : 295 SEC
ATMOSPHERE   : VACUUM
                                            F.S. = 8K
```

No filter

Cu filter

8.30 A sample with many elements. Identify as many as you can.

```
TUBE VOLTAGE : 30 KV          FILTER USED : FIVE
TUBE CURRENT : 0.35 mA        LIVETIME    : 200 SEC
ATMOSPHERE   : VACUUM
```

APPENDIX 8.1. CHARACTERISTIC X-RAY WAVELENGTHS (Å) AND ENERGIES (KEV)*

Table A1 K Series Lines (Å)

Line		$\alpha_{1,2}$	α_1	α_2	β_1		β_3
Approximate intensity		150	100	50		15	
Li	3	230					
B	4	113					
B	5	67					
C	6	44					
N	7	31.603					
O	8	23.707					
F	9	18.307					
Ne	10	14.615			14.460		
Na	11	11.909			11.574		11.726
Mg	12	9.889			9.559		9.667
Al	13	8.339	8.338	8.341	7.960		8.059

(*continued*)

*Appendices 8.1 and 8.2 are modified from Markowicz, used with permission. The data in these tables were obtained from the following sources: Clark, 1963, Burr, 1974, Jenkins et al., 1991.

Table A1 Continued

Line Approximate intensity		$\alpha_{1,2}$ 150	α_1 100	α_2 50	β_1	15	β_3
Si	14	7.126	7.125	7.127		6.778	
P	15	6.155	6.154	6.157	5.804		
S	16	5.373	5.372	5.375	5.032		
Cl	17	4.729	4.728	4.731	4.403		
Ar	18	4.192	4.191	4.194	3.886		
K	19	3.744	3.742	3.745	3.454		
Ca	20	3.360	3.359	3.362	3.089		
Sc	21	3.032	3.031	3.034	2.780		
Ti	22	2.750	2.749	2.753		2.514	
V	23	2.505	2.503	2.507		2.285	
Cr	24	2.291	2.290	2.294		2.085	
Mn	25	2.103	2.102	2.105		1.910	
Fe	26	1.937	1.936	1.940		1.757	
Co	27	1.791	1.789	1.793		1.621	
Ni	28	1.659	1.658	1.661		1.500	
Cu	29	1.542	1.540	1.544	1.392		1.393
Zn	30	1.437	1.435	1.439	1.296		
Ga	31	1.341	1.340	1.344	1.207		1.208
Ge	32	1.256	1.255	1.258	1.129		1.129
As	33	1.177	1.175	1.179	1.057		1.058
Se	34	1.106	1.105	1.109	0.992		0.993
Br	35	1.041	1.040	1.044	0.933		0.933
Kr	36	0.981	0.980	0.984	0.879		0.879
Rb	37	0.927	0.926	0.930	0.829		0.830
Sr	38	0.877	0.875	0.880	0.783		0.784
Y	39	0.831	0.829	0.833	0.740		0.741
Zr	40	0.788	0.786	0.791	0.701		0.702
Nb	41	0.748	0.747	0.751	0.665		0.666
Mo	42	0.710	0.709	0.713	0.632		0.633
Tc	43	0.676	0.675	0.679		0.601	
Ru	44	0.644	0.643	0.647	0.572		0.573
Rh	45	0.614	0.613	0.617	0.546		0.546
Pd	46	0.587	0.585	0.590	0.521		0.521
Ag	47	0.561	0.559	0.564	0.497		0.498
Cs	48	0.536	0.535	0.539	0.475		0.476
In	49	0.514	0.512	0.517	0.455		0.455
Sn	50	0.492	0.491	0.495	0.435		0.436
Sb	51	0.472	0.470	0.475	0.417		0.418
Te	52	0.453	0.451	0.456	0.400		0.401
I	53	0.435	0.433	0.438	0.384		0.385
Xe	54	0.418	0.416	0.421	0.369		0.360
Cs	55	0.402	0.401	0.405	0.355		0.355
Ba	56	0.387	0.385	0.390	0.341		0.342
La	57	0.373	0.371	0.376	0.328		0.329
Ce	58	0.359	0.357	0.362	0.316		0.317
Pr	59	0.346	0.344	0.349	0.305		0.305
Nd	60	0.334	0.332	0.337	0.294		0.294

(continued)

Table A1   Continued

Line		$\alpha_{1,2}$	α_1	α_2	β_1		β_3
Approximate intensity		150	100	50		15	
Pm	61	0.322	0.321	0.325		0.283	
Sm	62	0.311	0.309	0.314	0.274		0.274
Eu	63	0.301	0.299	0.304	0.264		0.265
Gd	64	0.291	0.289	0.294	0.255		0.256
Tb	65	0.281	0.279	0.284	0.246		0.246
Dy	66	0.272	0.270	0.275	0.237		0.238
Ho	67	0.263	0.261	0.266	0.230		0.231
Er	68	0.255	0.253	0.258	0.222		0.223
Tm	69	0.246	0.244	0.250	0.215		0.216
Yb	70	0.238	0.236	0.241	0.208		0.208
Lu	71	0.231	0.229	0.234	0.202		0.203
Hf	72	0.224	0.222	0.227	0.195		0.196
Ta	73	0.217	0.215	0.220	0.190		0.191
W	74	0.211	0.209	0.213	0.184		0.185
Re	75	0.204	0.202	0.207	0.179		0.179
Os	76	0.198	0.196	0.201	0.173		0.174
Ir	77	0.193	0.191	0.196	0.168		0.169
Pt	78	0.187	0.158	0.190	0.163		0.164
Au	79	0.182	0.180	0.185	0.159		0.160
Hg	80	0.177	0.175	0.180	0.154		0.155
Tl	81	0.172	0.170	0.175	0.150		0.151
Pb	82	0.167	0.165	0.170	0.146		0.147
Bi	83	0.162	0.161	0.165	0.142		0.143
Po	84	0.185	0.156	0.161	0.138		
At	85		0.152	0.157	0.134		0.135
Rn	86		0.148	0.153	0.131		0.132
Fr	87		0.144	0.149	0.127		0.128
Ra	88		0.144	0.149	0.127		0.128
Ac	89		0.140	0.145	0.124		0.125
Th	90	0.135	0.133	0.138	0.117		0.118
Pa	91		0.131	0.136	0.115		0.116
U	92	0.128	0.126	0.131	0.111		0.112

Table A2   Energies of Principal K and L X-ray Emission Lines (keV)

Atomic number	Element	$K\beta_2$	$K\beta_1$	$K\alpha_1$	$K\alpha_2$	$L\gamma_1$	$L\beta_2$	$L\beta_1$	$L\alpha_1$	$L\alpha_2$
3	Li			0.052						
4	Be			0.110						
5	B			0.185						
6	C			0.282						
7	N			0.392						
8	C			0.523						
9	F			0.677						
10	Ne			0.851						

(*continued*)

Table A2 Continued

Atomic number	Element	$K\beta_2$	$K\beta_1$	$K\alpha_1$	$K\alpha_2$	$L\gamma_1$	$L\beta_2$	$L\beta_1$	$L\alpha_1$	$L\alpha_2$
11	Na		1.067	1.041						
12	Mg		1.297	1.254						
13	Al		1.553	1.487	1.486					
14	Si		1.832	1.40	1.739					
15	P		2.136	2.015	2.014					
16	S		2.464	2.309	2.306					
17	Cl		2.815	2.622	2.621					
18	Ar		3.192	2.957	2.955					
19	K		3.589	3.313	3.310					
20	Ca		4.012	3.691	3.688			0.344	0.341	
21	Sc		4.460	4.090	4.085			0.399	0.395	
22	Ti		4.931	4.510	4.504			0.458	0.492	
23	V		5.427	4.952	4.944			0.519	0.510	
24	Cr		5.946	5.414	5.405			0.581	0.571	
25	Mn		6.490	5.898	5.887			0.647	0.636	
26	Fe		7.057	6.403	6.390			0.717	0.704	
27	Co		7.647	6.930	6.915			0.790	0.775	
28	Ni	8.328	8.264	7.477	7.460			0.866	0.849	
29	Cu	8.976	8.904	8.047	8.027			0.943	0.928	
30	Zn	9.657	9.571	8.638	8.615			1.032	1.009	
31	Ga	10.365	10.263	9.251	9.234			1.122	1.096	
32	Ge	11.100	10.981	9.885	9.854			1.216	1.166	
33	As	11.863	11.725	10.543	10.507			1.517	1.282	
34	Se	12.651	12.495	11.221	11.181			1.419	1.379	
35	Br	13.465	13.290	11.923	11.877			1.526	1.480	
36	Kr	14.313	14.112	12.648	12.597			1.638	1.587	
37	Rb	15.184	14.960	13.394	13.335			1.752	1.694	1.692
38	Sr	16.083	15.834	14.164	14.097			1.872	1.806	1.805
39	Y	17.011	16.736	14.957	14.882			1.996	1.922	1.920
40	Zr	17.969	17.666	15.774	15.690	2.302	2.219	2.124	2.042	2.040
41	Nb	18.951	18.261	16.614	16.520	2.462	2.367	2.257	2.166	2.163
42	Mo	19.964	19.607	17.478	17.373	2.623	2.518	2.395	2.293	2.290
43	Tc	21.012	20.585	18.410	18.328	2.792	2.674	2.538	2.424	2.420
44	Ru	22.072	21.655	19.278	19.149	2.964	2.836	2.683	2.558	2.554
45	Rh	23.169	22.721	20.214	20.072	3.144	3.001	2.834	2.696	2.692
46	Pd	24.297	23.816	21.175	21.018	3.328	3.172	2.990	2.838	2.833
47	Ag	25.454	24.942	22.162	21.988	3.519	3.348	3.151	2.994	2.978
48	Cd	26.641	26.093	23.172	22.982	3.716	3.528	3.316	3.133	3.127
49	In	27.859	27.274	24.207	24.000	3.920	3.713	3.487	3.287	3.279
50	Sn	29.106	28.483	25.270	25.042	4.131	3.904	3.662	3.444	3.436
51	Sb	30.387	29.723	26.357	26.109	4.347	4.100	3.543	3.605	3.595
52	Te	31.698	30.993	27.471	27.200	4.570	4.301	4.029	3.769	3.758
53	I	33.016	32.292	28.610	28.315	4.800	4.507	4.220	3.937	3.926
54	Xe	34.446	33.644	29.802	29.485	5.036	4.720	4.422	4.111	4.098
55	Cs	35.819	34.984	30.970	30.623	5.280	4.936	4.620	4.286	4.272
56	Ba	37.255	35.376	32.191	31.815	5.531	5.156	4.828	4.467	4.451
57	La	38.728	37.799	33.440	33.033	5.789	5.384	5.043	4.651	4.635
58	Ce	40.231	39.255	34.717	34.276	6.052	5.613	5.262	4.840	4.823
59	Pr	41.772	40.746	36.023	35.548	6.322	5.850	5.489	5.034	5.014
60	Nd	43.298	42.269	37.359	36.845	6.602	6.090	5.722	5.230	5.208
61	Pm	44.955	43.945	38.649	38.160	6.891	6.336	5.956	5.431	5.408
62	Sm	46.553	45.400	40.124	39.523	7.180	6.587	6.206	5.636	5.609
63	Eu	48.241	47.027	41.529	40.877	7.478	6.842	6.456	5.846	5.816

(continued)

Table A2 Continued

Atomic number	Element	$K\beta_2$	$K\beta_1$	$K\alpha_1$	$K\alpha_2$	$L\gamma_1$	$L\beta_2$	$L\beta_1$	$L\alpha_1$	$L\alpha_2$
64	Gd	49.961	48.718	42.983	42.280	7.788	7.102	6.714	6.039	6.027
65	Tb	51.737	50.391	44.470	43.737	8.104	7.368	6.979	6.275	6.241
66	Dy	53.491	52.178	45.985	45.193	8.418	7.638	7.249	6.495	6.457
67	Ho	55.292	53.934	47.528	46.686	8.748	7.912	7.528	6.720	6.680
68	Er	57.088	55.690	49.099	48.205	9.089	8.188	7.810	6.948	6.904
69	Tm	58.969	57.576	50.730	49.762	9.424	8.472	8.103	7.181	7.135
70	Yb	60.959	59.352	52.360	51.326	9.779	8.758	8.401	7.414	7.367
71	Lu	62.946	61.282	54.063	52.959	10.142	9.048	8.709	7.654	7.604
72	Hf	64.936	63.209	55.757	54.579	10.514	9.346	9.021	7.898	7.843
73	Ta	66.999	65.210	57.524	56.270	10.892	9.649	9.341	8.145	8.087
74	W	69.090	67.233	59.310	57.973	11.283	9.959	9.670	8.396	8.333
75	Re	71.220	69.298	61.131	59.707	11.684	10.273	10.008	8.651	8.584
76	Os	73.393	71.404	62.991	61.477	12.094	10.596	10.354	8.910	8.840
77	Ir	75.605	73.549	64.886	63.278	12.509	10.918	10.706	9.173	9.098
78	Pt	77.866	75.736	66.820	65.111	12.939	11.249	11.069	9.441	9.360
79	Au	80.165	77.968	68.794	66.980	13.379	11.582	11.439	9.711	9.625
80	Hg	82.526	80.258	70.821	68.894	13.828	11.923	11.823	9.987	9.896
81	Tl	84.904	82.558	72.860	70.320	14.288	12.268	12.210	10.266	10.170
82	Pb	87.343	94.922	74.957	72.794	14.762	12.620	12.611	10.549	10.448
83	Bi	89.833	87.335	77.097	74.805	15.244	12.977	13.021	10.836	10.729
84	Po	92.386	89.809	79.296	76.868	15.740	13.338	13.441	11.128	11.014
85	At	94.976	92.319	81.525	78.956	16.248	13.705	13.873	11.424	11.304
86	Rn	97.616	94.877	83.800	81.080	16.768	14.077	14.316	11.724	11.597
87	Fr	100.305	97.483	86.119	83.243	17.301	14.459	14.770	12.029	11.894
88	Ra	103.048	100.136	88.485	85.446	17.845	14.839	15.233	12.338	12.194
89	Ac	105.838	102.846	90.894	87.681	18.405	15.227	15.712	12.650	12.499
90	Th	108.671	105.592	93.334	89.942	18.977	15.620	16.200	12.966	12.808
91	Pa	111.575	108.408	95.851	92.271	19.559	16.022	16.700	13.291	13.120
92	U	114.549	111.289	98.428	94.648	20.163	16.425	17.218	13.613	12.438
93	Np	117.533	114.181	101.005	97.023	20.774	16.837	17.740	13.945	12.758
94	Pu	120.592	117.146	103.653	99.457	21.401	17.254	18.278	14.279	14.082
95	Am	123.706	120.163	106.351	101.932	22.042	17.667	18.829	14.618	14.411
96	Cm	126.875	123.235	109.098	104.448	22.699	18.106	19.393	14.961	14.743
97	Bk	130.101	126.362	111.896	107.023	23.370	18.540	19.971	15.309	15.079
98	Cf	133.383	129.544	114.745	109.603	24.056	18.980	20.562	15.661	15.420
99	Es	136.724	132.781	118.646	112.244	24.758	19.426	21.166	16.018	15.764
100	Fm	140.122	136.075	120.598	114.926	27.475	19.879	21.785	16.379	16.113

Note: The conversion equation between energy in keV and wavelength in Å is: $E(\text{Key}) = 12.4/\lambda$ (Å).

Table A3 Correspondence Between Old Siegbahn and New IUPAC Notation X-ray Diagram Lines

Siegbahn	IUPAC	Siegbahn	IUPAC	Siegbahn	IUPAC
K_{α_1}	$K-L_3$	L_{α_1}	L_3-M_5	L_{γ_1}	L_2-N_4
K_{α_2}	$K-L_2$	L_{α_2}	L_3-M_4	L_{γ_2}	L_1-N_2
K_{β_1}	$K-M_3$	L_{β_1}	L_2-M_4	L_{γ_3}	L_1-N_3
$K_{\beta_2^I}$	$K-N_3$	L_{β_2}	L_3-N_5	L_{γ_4}	L_1-O_4
$K_{\beta_2^{II}}$	$K-N_2$	L_{β_3}	L_1-M_3	$L_{\gamma_4'}$	L_1-O_2
K_{β_3}	$K-M_2$	L_{β_4}	L_1-M_2	L_{γ_5}	L_2-N_1
$K_{\beta_4^I}$	$K-N_5$	L_{β_5}	$L_3-O_{4,5}$	L_{γ_6}	L_2-O_4

(continued)

Table A3 Continued

Siegbahn	IUPAC	Siegbahn	IUPAC	Siegbahn	IUPAC
$K_{\beta_4^{II}}$	$K-N_4$	L_{β_6}	L_3-N_1	L_{γ_8}	L_2-O_1
$K_{\beta_{4x}}$	$K-N_4$	L_{β_7}	L_3-O_1	$L_{\gamma_8'}$	$L_2-N_{6(7)}$
$K_{\beta_5^{I}}$	$K-M_5$	$L_{\beta_7'}$	$L_3-N_{6,7}$	L_η	L_2-M_1
$K_{\beta_5^{II}}$	$K-M_4$	L_{β_9}	L_1-M_5	L_l	L_3-M_1
		$L_{\beta_{10}}$	L_1-M_4	L_s	L_3-M_3
		$L_{\beta_{15}}$	L_3-N_4	L_t	L_3-M_2
		$L_{\beta_{17}}$	L_2-M_3	L_u	$L_3-N_{6,7}$
				L_v	$L_2-N_{6(7)}$
				M_{α_1}	M_5-N_7
				M_{α_2}	M_5-N_6
				M_β	M_4-N_6
				M_γ	M_3-N_5
				M_ζ	$M_{4,5}-N_{2,3}$

APPENDIX 8.2. ABSORPTION EDGE WAVELENGTHS AND ENERGIES

See Table A4 on next page.

Table A4 Critical Absorption Wavelengths and Critical Absorption Energies

Atomic number	Element	K edge		L$_1$ edge		L$_2$ edge		L$_3$ edge		M$_4$ edge		M$_5$ edge	
		Å	keV	Å	keV	Å	keV	Å	keV	Å	keV	Å	keV
1	H	918	0.014										
2	He	504	0.025										
3	Li	226.953	0.055										
4	Be	106.9	0.116										
5	B	64.6	0.192										
6	C	43.767	0.283										
7	N	31.052	0.399										
8	O	23.367	0.531										
9	F	18.05	0.687										
10	Ne	14.19	0.874	258	0.048	564	0.022	564	0.022				
11	Na	11.48	1.08	225	0.055	365	0.034	365	0.034				
12	Mg	9.512	1.303	197	0.063	248	0.050	253	0.049				
13	Al	7.951	1.559	143	0.087	170	0.073	172	0.072				
14	Si	6.745	1.837	105	0.118	125	0.099	127	0.098				
15	P	5.787	2.142	81.0	0.153	96.1	0.129	96.9	0.128				
16	S	5.018	2.470	64.2	0.193	75.6	0.164	79.1	0.163				
17	Cl	4.397	2.819	52.1	0.238	61.1	0.203	61.4	0.202				
18	Ar	3.871	3.202	43.2	0.287	50.2	0.247	50.6	0.245				
19	K	3.437	3.606	36.4	0.341	41.8	0.297	42.2	0.294				
20	Ca	3.070	4.037	30.7	0.399	35.2	0.352	35.5	0.349				
21	Sc	2.757	4.495	26.8	0.462	30.2	0.411	30.8	0.402				
22	Ti	2.497	4.963	23.4	0.530	27.0	0.460	27.3	0.454				
23	V	2.269	5.462	20.5	0.604	23.9	0.519	24.2	0.512				
24	Cr	2.070	5.987	18.3	0.679	21.3	0.583	21.6	0.574				
25	Mn	1.896	6.535	16.3	0.762	19.1	0.650	19.4	0.639				
26	Fe	1.743	7.109	14.6	0.849	17.2	0.721	17.5	0.708				
27	Co	1.608	7.707	13.3	0.929	15.6	0.794	15.9	0.779				

28	Ni	1.488	8.329	12.22	1.015	14.2	0.871	14.5	0.853		
29	Cu	1.380	8.978	11.27	1.100	13.0	0.953	13.3	0.933		
30	Zn	1.283	9.657	10.33	1.200	11.87	1.045	12.13	1.022		
31	Ga	1.196	10.365	9.54	1.30	10.93	1.134	11.10	1.117		
32	Ge	1.117	11.100	8.73	1.42	9.94	1.248	10.19	1.217		
33	As	1.045	11.860	8.107	1.529	9.124	1.358	9.39	1.32		
34	Se	0.980	12.649	7.506	1.651	8.416	1.473	8.67	1.43		
35	Br	0.920	13.471	6.97	1.78	7.80	1.59	8.00	1.55		
36	Kr	0.866	14.319	6.46	1.92	7.21	1.72	7.43	1.67		
37	Rb	0.816	15.197	5.998	2.066	6.643	1.865	6.89	1.80		
38	Sr	0.770	16.101	5.583	2.220	6.172	2.008	6.387	1.940		
39	Y	0.728	17.032	5.232	2.369	5.755	2.153	5.962	2.079		
40	Zr	0.689	17.993	4.867	2.546	5.378	2.304	5.583	2.220		
41	Nb	0.653	18.981	4.581	2.705	5.026	2.467	5.223	2.373		
42	Mo	0.620	19.996	4.298	2.883	4.718	2.627	4.913	2.523		
43	Tc	0.589	21.045	4.060	3.054	4.436	2.795	4.632	2.677		
44	Ru	0.561	22.112	3.83	3.24	4.180	2.965	4.369	2.837		
45	Rh	0.534	23.217	3.626	3.413	3.942	3.144	4.130	3.001		
46	Pd	0.509	24.341	3.428	3.616	3.724	3.328	3.908	3.171		
47	Ag	0.486	25.509	3.254	3.809	3.514	3.527	3.698	3.351		
48	Cd	0.464	26.704	3.085	4.018	3.326	3.726	3.504	3.537		
49	In	0.444	27.920	2.926	4.236	3.147	3.938	3.324	3.728		
50	Sn	0.425	29.182	2.777	4.463	2.982	4.156	3.156	3.927		
51	Sb	0.407	30.477	2.639	4.695	2.830	4.380	3.000	4.131		
52	Te	0.390	31.800	2.511	4.937	2.687	4.611	2.855	4.340		
53	I	0.374	33.155	2.389	5.188	2.553	4.855	2.719	4.557		
54	Xe	0.359	34.570	2.274	5.451	2.429	5.102	2.592	4.780		
55	Cs	0.345	35.949	2.167	5.719	2.314	5.356	2.474	5.010		
56	Ba	0.331	37.399	2.068	5.994	2.204	5.622	2.363	5.245	15.56	0.7967
57	La	0.318	38.920	1.973	6.282	2.103	5.893	2.258	5.488	15.89	0.7801

(continued)

Table A4 Continued

Atomic number	Element	K edge Å	K edge keV	L$_1$ edge Å	L$_1$ edge keV	L$_2$ edge Å	L$_2$ edge keV	L$_3$ edge Å	L$_3$ edge keV	M$_4$ edge Å	M$_4$ edge keV	M$_5$ edge Å	M$_5$ edge keV
58	Ce	0.307	40.438	1.889	6.559	2.011	6.163	2.164	5.727				
59	Pr	0.295	41.986	1.811	6.844	1.924	6.441	2.077	5.967	13.122	0.9448	13.394	0.9257
60	Nd	0.285	43.559	1.735	7.142	1.843	6.725	1.995	6.213	12.459	0.9951	23.737	0.9734
61	Pm	0.274	45.207	1.665	7.448	1.767	7.018	1.918	6.456				
62	Sm	0.265	46.833	1.599	7.752	1.703	7.279	1.845	6.719	11.288	1.0983	11.552	1.0732
63	Eu	0.256	48.501	1.536	8.066	1.626	7.621	1.775	6.981	10.711	1.1575	11.013	1.1258
64	Gd	0.247	50.215	1.477	8.391	1.561	7.938	1.710	7.250				
65	Tb	0.238	51.984	1.421	8.722	1.501	8.256	1.649	7.517				
66	Dy	0.231	53.773	1.365	9.081	1.438	8.619	1.579	7.848				
67	Ho	0.223	55.599	1.317	9.408	1.390	8.918	1.535	8.072				
68	Er	0.216	57.465	1.268	9.773	1.338	9.260	1.482	8.361	8.601	1.4415	8.847	1.4013
69	Tm	0.209	59.319	1.222	10.141	1.288	9.626	1.433	8.650			8.487	1.4609
70	Yb	0.202	61.282	1.182	10.487	1.243	9.972	1.386	8.941				
71	Lu	0.196	63.281	1.140	10.870	1.199	10.341	1.341	8.239				
72	Hf	0.190	65.292	1.100	11.271	1.155	10.732	1.297	9.554				
73	Ta	0.184	67.379	1.061	11.681	1.114	11.128	1.255	9.374	6.87	1.804	7.11	1.743
74	W	0.178	69.479	1.025	12.097	1.075	11.533	1.216	10.196	6.59	1.880	6.83	1.814
75	Re	0.173	71.590	0.990	12.524	1.037	11.953	1.177	10.529	6.33	1.958	6.560	1.890
76	Os	0.168	73.856	0.956	12.968	1.001	12.380	1.140	10.867	6.073	2.042	6.30	1.967
77	Ir	0.163	76.096	0.923	13.427	0.967	12.817	1.106	11.209	5.83	2.126	6.05	2.048

Z													
78	Pt	0.158	78.352	0.893	13.875	0.934	13.266	1.072	11.556	5.59	2.217	5.81	2.133
79	Au	0.153	80.768	0.863	14.354	0.903	13.731	1.040	11.917	5.374	2.307	5.584	2.220
80	Hg	0.149	83.046	0.835	14.837	0.872	14.210	1.008	12.3	5.157	2.404	5.36	2.313
81	Tl	0.415	85.646	0.808	15.338	0.843	14.695	0.979	12.655	4.952	2.504	5.153	2.406
82	Pb	0.141	88.037	0.782	15.858	0.815	15.205	0.950	13.041	4.757	2.606	4.955	2.502
83	Bi	0.137	90.420	0.757	16.376	0.789	15.713	0.923	13.422	4.572	2.711	4.764	2.603
84	Po	0.133	93.112	0.732	16.935	0.763	16.244	0.897	13.817				
85	At	0.130	95.740	0.709	17.490	0.739	16.784	0.872	14.215				
86	Rn	0.126	98.418	0.687	18.058	0.715	17.337	0.848	14.618				
87	Fr	0.123	101.147	0.665	18.638	0.693	17.904	0.825	15.028				
88	Ra			0.645	19.229	0.671	18.478	0.803	15.439				
89	Ac	0.116	106.759	0.625	19.842	0.650	19.078	0.782	15.865				
90	Th	0.113	109.741	0.606	20.458	0.630	19.677	0.761	16.293	3.557	3.485	3.729	3.325
91	Pa	0.110	112.581	0.588	21.102	0.611	20.311	0.741	16.731	3.436	3.608	3.618	3.436
92	U	0.108	115.610	0.56	21.764	0.592	20.938	0.722	17.160	3.333	3.720	3.497	3.545
93	Np	0.105	118.619	0.553	22.417	0.574	21.596	0.704	17.614				
94	Pu	0.102	121.720	0.537	23.097	0.557	22.262	0.686	18.066				
95	Am	0.099	124.816	0.521	23.793	0.540	22.944	0.669	18.525				
96	Cm	0.097	128.088	0.506	24.503	0.525	23.640	0.653	18.990				
97	Bk	0.094	131.357	0.491	25.230	0.509	24.352	0.637	19.461				
98	Cf	0.092	134.683	0.477	25.971	0.494	25.080	0.622	19.938				
99	Es	0.090	138.067	0.464	26.729	0.480	25.824	0.607	20.422				
100	Fm	0.088	141.510	0.451	27.503	0.466	26.584	0.593	20.912				

9

Mass Spectrometry I: Principles and Instrumentation

Mass spectrometry (MS) is a technique for creating gas phase ions from the molecules or atoms in a sample, separating the ions according to their mass-to-charge ratio, m/z, and measuring the abundance of the ions formed. MS is currently one of the most rapidly advancing fields of instrumental analysis. It has developed from an inorganic method used to prove that most elements exist as isotopes of differing masses to one of the cornerstone techniques used to elucidate the structure of biomolecules, especially high molecular weight proteins. MS provides the analyst with information as diverse as the structure of complex organic and biomolecules to the quantitative determination of ppb concentrations of elements and molecules in samples.

MS is an analytical technique that provides qualitative and quantitative information, including the mass of molecules and atoms in samples as well as the molecular structure of organic and inorganic compounds. MS can be used as a qualitative analytical tool to identify and characterize different materials of interest to the chemist or biochemist. MS is used routinely for the quantitative analysis of mixtures of gases, liquids, and solids. MS is used in many fields in addition to identifying and quantifying organic and biological molecules. These include atomic physics, physical chemistry, and geology.

This chapter will focus on MS principles and instrumentation for both organic and inorganic analysis. Chapter 10 will cover applications of MS for organic and inorganic analyses as well as interpretation of simple mass spectra for structural identification of organic compounds.

9.1. PRINCIPLES OF MS

The mass spectrometer is an instrument that separates gas phase ionized atoms, molecules, and fragments of molecules by the difference in their mass-to-charge ratios. The mass-to-charge ratio is symbolized by m/z, where the mass m is expressed in **unified atomic mass units** and z is the number of charges on the ion. The statement found in many texts that MS separates ions based on *mass* is only true when the charge on the ion is $+1$; in this case m/z is numerically equal to the mass of the ion. While some MS methods do generate mostly $+1$ charged ions, many new techniques generate ions with multiple charges.

The mass of an ion is given in **unified atomic mass units**, **u**. One unified atomic mass unit is equal to $1/12$ of the mass of the most abundant, stable, naturally occurring isotope of carbon, ^{12}C. The mass of ^{12}C is defined as exactly 12 u. The abbreviation **amu**, for atomic mass unit, is now considered obsolete but may still be encountered in the literature.

A synonym for the unified atomic mass unit is the dalton (Da); 1 u = 1 Da. In the SI unit of mass, 1 u = 1.665402×10^{-27} kg. Table 9.1 presents the exact masses for some common isotopes encountered in organic compounds.

The term z symbolizes the *number of charges* on the ion; this number may be positive or negative, such as $+1$, -1, $+2$, $+10$, and so on. The number of charges is not the same as the total charge of the ion in coulombs. The total charge $q = ze$, where e is the magnitude of the charge on the electron, 1.6×10^{-19} C.

For simplicity, the following discussion will focus on what happens to an organic molecule in one type of MS experiment. One method of forming ions from sample molecules or atoms is to bombard the sample with electrons. This method is called **electron ionization** (EI):

$$M + e^- \longrightarrow M^{\bullet+} + 2e^- \tag{9.1}$$

where M, is the analyte molecule; e^-, the electron; and $M^{\bullet+}$, the ionized analyte molecule; this species is called the **molecular ion**.

In many cases, only ions with a single positive charge are formed; that is, the number of charges on the ion, z, equals $+1$. The MS separates ions based on the mass-to-charge ratio, m/z; for ions with a single positive charge, m/z equals the mass of the ion in unified atomic mass units. For a molecular ion, m/z is related to the molecular weight (also called the relative molecular mass) of the compound. The symbol $M^{\bullet+}$ indicates that a molecular ion from an organic compound is a radical cation formed by the loss of one electron. In most cases, molecular ions have sufficient energy as a result of the ionization process to undergo *fragmentation* to form other ions of lower m/z. All these

Table 9.1 Comparison of Atomic Weights and Measured Accurate Isotope Masses for Some Common Elements in Organic Compounds

Element	Atomic weight	Isotope	Mass[a]	% Abundance
Hydrogen	1.00794(7)	^{1}H	1.007825	99.99
		^{2}H (or D)	2.01410	0.01
Carbon	12.0107(8)	^{12}C	12.000000 (defined)	98.91
		^{13}C	13.00336	1.1
Nitrogen	14.0067(2)	^{14}N	14.0031	99.6
		^{15}N	15.0001	0.4
Oxygen	15.9994(3)	^{16}O	15.9949	99.76
		^{17}O	16.9991	0.04
		^{18}O	17.9992	0.20
Fluorine	18.99840	F	18.99840	100
Phosphorus	30.97376	P	30.97376	100
Sulfur	32.065(5)	^{32}S	31.9721	95.02
		^{33}S	32.9715	0.76
		^{34}S	33.9679	4.22
Chlorine	35.453(2)	^{35}Cl	34.9689	75.77
		^{37}Cl	36.9659	24.23
Bromine	79.904(1)	^{79}Br	78.9183	50.5
		^{81}Br	80.9163	49.5

[a]Many of the isotope masses have been determined by MS accurately to seven or more decimal places. Numbers in parentheses in the atomic weights represent the first uncertain figure. Atomic weights based on the values in *Pure Appl.Chem.* **2001**, *73*, 667–683 as tabulated at http://www.chem.qmul.ac.uk/iupac/AtWt/. Used with permission.

ions are separated in the mass spectrometer according to their m/z values and the abundance of each is measured.

A **magnetic sector** mass spectrometer is shown schematically in Fig. 9.1. This instrument is a *single-focus* mass spectrometer. The gas phase molecules from the sample inlet are ionized by a beam of high energy (i.e., high velocity) electrons passing closely among them. The rapidly changing electric field can both eject electrons from the atom or molecule (ionization) and transfer sufficient energy to the molecule to cause its bonds to rupture (fragmentation). The ions are then accelerated in an electric field at a voltage V. If each ion has a single positive charge, the magnitude of the charge is equal to $+1$. The energy of each ion is equal to the charge z times the accelerating voltage, zV. The energy acquired by an ion on acceleration is kinetic energy. It is important to note that the kinetic energy of an ion accelerated through a voltage V depends only on the charge of the ion and the voltage, not on the mass of the ion. The translational component of the kinetic energy is equal to $1/2mv^2$. The kinetic energy of all singly charged ions is the same for a given accelerating voltage; therefore those ions with small masses must travel at higher velocity than ions with larger masses. That is, for

Figure 9.1 (a) Schematic magnetic sector mass spectrometer. (b) Schematic magnetic sector mass spectrometer showing a gas sample reservoir with fill and inlet valves and an EM detector. (c) Forces and energies associated with ions traveling through a magnetic sector mass spectrometer. For an ion to reach the detector, Bzv must equal mv^2/r.

ions with a single positive charge:

$$1/2mv^2 = zV \tag{9.2}$$

hence

$$v = \left(\frac{2zV}{m}\right)^{1/2} \tag{9.3}$$

where m is the mass of the ion; v, the velocity of the ion; z, the charge of the ion and V, the accelerating voltage.

As m varies, the velocity v changes such that $1/2mv^2$ remains a constant. This relation can be expressed for two different ions as follows:

$$zV = 1/2m_1v_1^2 \text{ (ion 1)}$$

$$zV = 1/2m_2v_2^2 \text{ (ion 2)} \tag{9.4}$$

$$1/2m_1v_1^2 = 1/2m_2v_2^2$$

where m_1 is the mass of ion 1; v_1, the velocity of ion 1; m_2, the mass of ion 2 and v_2, the velocity of ion 2.

The velocity of an ion depends on its mass; the velocity is inversely proportional to the square root of the mass of the ion.

After an applied voltage has accelerated the charged ions, they enter a curved section of a magnet of homogeneous magnetic field B and a fixed radius of curvature. This magnetic field acts on the ions, making them move in a circle. The attractive force of the magnet equals Bzv. The centrifugal force on the ion is equal to mv^2/r, where r is the radius of the circular path traveled by the ion. If the ion path is to follow the radius of curvature of the magnet, these two forces must be equal:

$$\frac{mv^2}{r} = Bzv \tag{9.5}$$

or

$$\frac{1}{r} = \frac{Bzv}{mv^2}$$

and

$$r = \frac{mv}{zB} \tag{9.6}$$

Substituting for v [Eq. (9.3)], we get

$$r = \frac{m}{zB}\left(\frac{2zV}{m}\right)^{1/2}$$

Squaring both sides, we have

$$r^2 = \frac{m^2}{z^2B^2}\frac{2zV}{m} = \frac{m2V}{zB^2} \tag{9.7}$$

which can be rearranged to give:

$$\frac{m}{z} = \frac{B^2 r^2}{2V} \tag{9.8}$$

That is, the radius of the circular path of an ion depends on the accelerating voltage V, the magnetic field B, and the ratio m/z. When V and B are kept constant, the radius of the circular path depends on the m/z value of the ionized molecule. Ions of different m/z travel in circles with different radii; this is the basis of the separation by m/z. Ions with different paths are shown as the dotted lines in Fig. 9.1(a); only one particular m/z has the right r to pass through the mass spectrometer under a given V and B. The others, as shown, follow paths that cause them to hit the sides of the instrument and be lost. By varying V or B, we can select which m/z will pass through the system. Voltage scanning is cheaper and was used in early instruments with permanent magnets. Scanning the magnetic field strength is more sensitive and is easily done with modern electromagnets. (In reality, a curved magnet with a path of significant width could accommodate a range of different m/z values, and it is the width and position of the exit slit that determine precisely which ions finally get through to the detector in a real scanning magnetic sector mass spectrometer.)

In all modern magnetic sector mass spectrometers, the applied magnetic field B of the electromagnet is varied while the accelerating voltage V is held constant. The radius of curvature of the magnetic sector of a given instrument is a constant, so only ions with a trajectory of radius r pass through. For a particular magnetic field strength, then, only ions with an m/z value that satisfies Eq. (9.8) will exit the chamber. Consequently, for different values of B, ions with different mass-to-charge ratios will pass through the instrument to the detector and by varying B, we can scan the range of mass-to-charge ratios of the sample ions and measure their abundance with a detector which either counts and sums individual ions or measures a current produced by their impact on the detector. A plot of m/z vs. abundance is called a **mass spectrum**. The mass spectral data can be shown as a plot or as a table. A digitized mass spectrum of benzene, C_6H_6, is shown in Fig. 9.2(a); the plot is shown as a histogram. The most abundant peak in the spectrum, called the **base peak**, is scaled to 100, so the y-axis represents the **relative abundance** of the ions of each m/z value. The nominal mass of the benzene molecular ion is equal to the sum of the nominal masses of the C and H isotopes, so $6(12) + 6(1) = 78$; the most abundant peak in this benzene mass spectrum is the molecular ion peak at m/z 78. A number of fragment ions at lower values of m/z are also seen in this spectrum. Table 9.2 presents the same mass spectral data for benzene as shown in the plot. Figure 9.2(b) is a mass spectrum of cocaine, which exhibits many fragment ions. This example shows that the molecular ion is not always the most abundant ion. The base peak in the cocaine mass spectrum is the fragment ion at m/z 82. The m/z values and the fragmentation pattern enable the analyst to determine the molecular weight and structure of organic compounds by MS. Detailed interpretation of simple organic mass spectra will be covered in Chapter 10.

There is a difference between the mass of an atom (or ion) and what chemists are used to thinking of as the atomic weight or molecular weight of a species. Atomic weight is the average weight of all of the isotopes of an element and molecular weight is calculated from these average atomic weights. This is the value we would need to use if we wished to weigh out an exact number of moles of a substance whose component atoms have a distribution of different isotopes, which is usually the case. Each stable isotope (nuclide) of an element has an exact mass (more properly called a measured

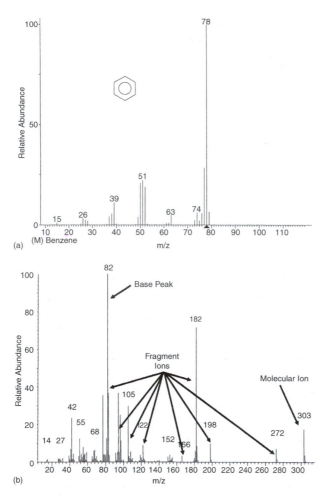

Figure 9.2 (a) A mass spectrum of benzene, C_6H_6, (MW $= 78$), plotted as m/z vs. relative abundance, with the most abundant peak set to a relative abundance of 100. In this spectrum, the most abundant peak (or base peak) is the molecular ion peak at m/z 78, marked by the dark triangle on the x-axis. Some fragment ion m/z values are also marked on the spectrum. (b) A mass spectrum of cocaine. In this spectrum the molecular ion peak is not the base peak; the fragment ion at m/z 82 is the most abundant ion. (The cocaine spectrum is courtesy of Dr. Robert Kobelski, CDC, Atlanta, GA.)

accurate mass), which has been determined accurately by MS. MS separates compounds containing ^{12}C atoms from compounds with ^{13}C atoms precisely because of the difference in mass between the isotopes. Table 9.1 gives a few examples of the atomic weights of some of the elements found in organic compounds as well as the measured accurate mass and abundance of the isotopes of the element. In a mass spectrum, a given ion is monoisotopic, not a weighted average. For example, the molecular weight of acetone, C_3H_6O, is calculated from the atomic weights of the elements to be $(3 \times 12.011) + (6 \times 1.00797) + (1 \times 15.9994) = 58.0795$ g/mol. The molecular ion of acetone as measured by MS has a mass that consists of only contributions from the one most abundant isotope of each element, that is, H, ^{12}C, and ^{16}O; its mass can be calculated from the formula $^{12}C_3^1H_6^{16}O$ to be $(3 \times 12.000) + (6 \times 1.00783) + (1 \times 15.9949) = 58.0419$ u or 58.0419 g/mol if we had a mole of the monoisotopic compound.

Table 9.2 Mass Spectral Data for Benzene

m/z	Relative abundance	m/z	Relative abundance
37	4	53	0.80
37.5	1.2	63	2.9
38	5.4	64	0.17
38.5	0.35	73	1.5
39	13	74	4.3
39.5	0.19	75	1.7
40	0.37	76	6.0
48	0.29	77	14
49	2.7	78	100
50	16	79	6.4
51	18	80	0.18
52	19		

The term m/z is the correct term to use for the mass-to-charge ratio of an ion. Older literature used the term m/e; however, the symbol e is used for the charge on the electron in coulombs and is *not* what goes into the divisor when the mass is given in unified atomic mass units, u. Two terms used in the older literature that are no longer acceptable for use in MS are "parent ion" for molecular ion and "daughter ion" for fragment ions. Ions do not have gender. The terms molecular ion and fragment ion should be used; "precursor ion" and "product ion" are used for tandem MS-MS experiments described later in the chapter.

9.1.1. Resolving Power and Resolution of a Mass Spectrometer

The **resolving power** of a mass spectrometer is defined as its ability to separate ions of two different m/z values. Numerically, the resolution is equal to the mass of one singly charged ion, M, divided by the difference in mass between M and the next m/z value that can be separated. For example, to separate two singly charged ions of 999 and 1001 Da requires a resolving power of $999/(1001 - 999) = 500$. That is:

$$\text{resolving power} = \frac{M}{\Delta M} \tag{9.9}$$

In practice, it is found that if we wish to distinguish between ions of 600 and 599 Da, the resolving power required is 600, or 1 Da in 600 Da. As a rule of thumb, if we wish to distinguish between ions differing by 1 Da in the 600 mass range, we need a resolving power of 600. If we need to distinguish between ions differing by 1 Da in the 1200 Da range, we need a resolving power of at least 1200. This is not a very high resolution. Some isotopes of *different* elements or molecular fragment ions composed of different combinations of atomic isotopes adding up to close to the same mass may produce ions differing by *much less* than 1 Da. To distinguish between these types of species, high-resolution mass spectrometry, with resolution in the 20,000–100,000 or higher range, is required.

The resolving power is determined by actual measurement of the mass spectral peaks obtained. The method for calculating ΔM must also be specified. Two methods are commonly used to indicate the separation between peaks and these are shown in Fig. 9.3. One definition is that the overlap between the peaks is 10% or less for two peaks of equal height; that is, the height of the overlap should not be more than 10% of

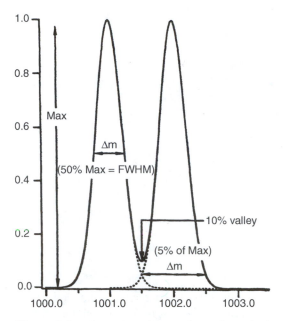

Figure 9.3 Illustration of the peaks used to calculate the resolving power of a mass spectrometer showing the location of the FWHM, the 10% valley, and the 5% valley. (From Sparkman, used with permission of the publisher.)

the peak height. A second method is to use the full width at half maximum (FWHM) as a measure of ΔM. The FWHM method results in a resolving power twice that of the 10% overlap method, so it is important to state how the calculation was performed.

Resolving power for commercial mass spectrometers depends on the instrument design, and can range from 500 to more than 1×10^6. In general, the higher the resolving power, the greater the complexity and cost of the MS instrument.

Resolution is the value of ΔM at a given M and is often expressed in ppm. For the spectrometer with a resolving power of 600 earlier described, the resolution would be 1 part in 600 parts. To distinguish between $^{14}N_2^+$, with an exact mass of 28.0061 Da and $^{12}C^{16}O^+$, with an exact mass of 27.9949 Da, we would need a resolution of $(28.0061 - 27.9949) = 0.0112$ Da in 28 Da. To convert this resolution to ppm, divide 0.0112 by 28 and multiply by 1×10^6; a resolution of 400 ppm is required. The resolving power needed would be $27.9949/0.0112 = 2500$.

9.2. INSTRUMENTATION

All mass spectrometers require a sample input system, an ionization source, a mass analyzer, and a detector. All of the components with the exception of some sample input systems or ion source volumes are under vacuum (10^{-6}–10^{-8} torr for that portion where ions are separated by mass, i.e., the **analyzer**, or 10^{-4}–10^{-5} torr in some **ion sources**, where the ions are initially formed), so vacuum pumps of various types are required. Modern mass spectrometers have all of the components under computer control, with a computer-based data acquisition and processing system. A block diagram of a typical mass spectrometer is shown in Fig. 9.4.

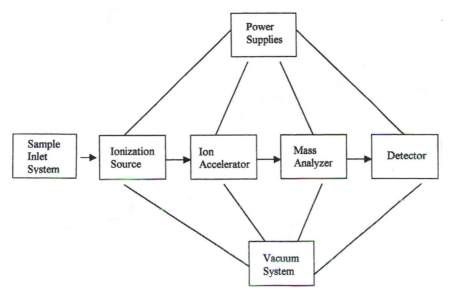

Figure 9.4   Block diagram of a mass spectrometer.

9.2.1. Sample Input Systems

9.2.1.1. Gas Expansion

This method of sample introduction is useful for gases and for liquids with sufficiently high vapor pressures. The gas or vapor is allowed to expand into an evacuated, heated vessel. The sample is then "leaked" into the ionization source through pin holes in a gold foil seal. This is sometimes termed a **"molecular leak" inlet**. Vacuum pumps control the system so that the pressure in the ionization source is at the required 10^{-6}–10^{-8} torr.

9.2.1.2. Direct Insertion and Direct Exposure Probes

A **direct insertion probe** is used for introduction of liquids with high boiling points and solids with sufficiently high vapor pressure. The sample is put into a glass capillary that fits into the tip of the probe shown in Fig. 9.5. The probe is inserted into the ionization source of the mass spectrometer and is heated electrically, vaporizing sample into the electron beam where ionization occurs. A problem with this type of sample introduction is that the mass spectrometer can be contaminated because of the volume of sample ionized.

A **direct exposure probe** usually has a rounded glass tip. The sample is dissolved in solvent, a drop of the solution is placed on the end of the probe and the solvent is allowed to evaporate. A thin film of sample is left on the glass tip. The tip is inserted into the ion source and heated in the same manner as the direct insertion probe. Much less sample is introduced into the ion source and the spectrometer is less likely to be contaminated as a result.

9.2.1.3. Chromatography and Electrophoresis Systems

The appropriate chromatographic instrument can separate mixtures of gases and liquids and the separated components are then introduced sequentially into a mass spectrometer for detection. Mass spectrometrists consider the chromatograph to be a "sample inlet" for their spectrometers while chromatographers consider the mass spectrometer to be "a detector" for their chromatographs. The truth is that these hyphenated (or coupled)

Figure 9.5 Schematic diagram of a direct insertion probe for solids and high boiling liquids. The sample is placed into the cavity in the tip of the probe. When not in use, the cutoff valve is closed and gas samples can be introduced by opening the stopcock shown. The black dot at the center of the square shows the point at which the electron beam impacts the sample. (From Ewing, used with permission.)

chromatography-mass spectrometry systems are extremely powerful analytical techniques, much more powerful than either instrument alone. The major problem to be overcome in coupling these two types of instruments is that the chromatography systems run at atmospheric pressure with large amounts of "carrier gas" or solvent. The mass spectrometer operates at very low pressures as noted above. The carrier gas or solvent must be removed without losing the analyte before the analyte can be introduced into the evacuated ionization source or analyzer regions. Alternatively one may use an ionization source which can ionize the target analytes under the conditions of higher pressure produced by the accompanying carrier fluid, and then selectively extract the ions into a lower-pressure region while diverting and discarding the vast majority of that vaporized fluid. Interfaces for these systems have been developed and are described in the chapters on the individual chromatographic techniques.

GC coupled with MS is known as GC-MS. It is a well-established method for separating gases or volatile compounds; the mass spectrum of each component in a mixture can be obtained and the components measured quantitatively. The interfacing, operation and applications of GC-MS are discussed in Chapter 12.

Several types of LC and one nonchromatographic separation system for liquids have been interfaced with MS. HPLC is widely used to separate nonvolatile organic compounds of all polarities and molecular weights. Coupled to a mass spectrometer, the technique is called LC-MS. Supercritical fluid chromatography (SFC) and the nonchromatographic separation technique of capillary electrophoresis (CE) are also used with mass spectrometric detection. The interfacing, ionization sources, operation, and applications of these hyphenated methods are covered in Chapter 13.

9.2.2. Ionization Sources

9.2.2.1.   Electron Ionization (EI)

The EI source is a commonly used source for organic MS. Electrons are emitted from a heated tungsten or other metal filament. The electrons are accelerated by a potential of 50–100 V toward the anode (Fig. 9.6). As shown in this figure, the paths of the electrons and sample molecules meet at right angles. Ionization of the sample molecules and fragmentation into smaller ions occurs as a result of interaction with the high-energy electrons.

Figure 9.6 Cross-section of an EI source. The filament and anode define the electron beam. The ions are formed in the space above the two repellers (the solid color blocks). A positive charge on the repellers together with a negative potential on the focus electrodes, cause positive ions to be accelerated upward in the diagram, into the mass analyzer. (Modified from Ewing, used with permission.)

Ions are accelerated out of the center of the source into the mass analyzer by an accelerating voltage of about 10^4 V.

The EI source forms both positive and negative ions, so it can be used as a source for negative ion MS. Negative ions form from molecules containing acid groups or electronegative atoms. The high energy imparted to the ions by the EI source causes significant fragmentation of organic molecules. This type of high-energy ionization source is referred to as a *hard* ionization source. The fragmentation of the molecular ion into smaller ions is very useful in deducing the structure of a molecule. However, fragmentation can be so significant for some types of molecules that no molecular ions remain. This means that the molecular weight of the compound cannot be determined although deduction of the structure of an unknown compound is greatly facilitated by knowing the molecular weight.

Collisions between ions and molecules in the source can result in the formation of ions with higher m/z values than the molecular ion. A common ion–molecule reaction is that between a proton, H^+, and an analyte molecule, M, to give a protonated molecule, MH^+ or $(M + H)^+$. Such a species has a $+1$ charge and a mass that is 1 u greater than that of the molecule and is called an $(M + 1)$ peak. One reason for keeping the sample pressure low in the EI ionization source is to prevent reactions between ions and molecules that would complicate interpretation of the mass spectrum.

(The electron ionization source used to be called the electron impact source and the term EI meant electron impact ionization. The use of the term *impact* is now considered archaic. Ionization and fragmentation are more often induced by the close passage of the energetic electron and the consequent large fluctuation in the electric field as opposed to a physical "impact", and an energetic electron may well perform this function multiple times on different molecules. As the student will note throughout the chapter, MS terminology has changed in recent years as a result of agreements by professional scientific organizations to standardize definitions and use of terms to avoid confusion but not all organizations have agreed to the same terms and definitions. The recommendations from Sparkman (see the bibliography) have been followed but even current literature will be found that uses "archaic" terminology. The old or alternate terms will be provided when necessary.)

9.2.2.2. Chemical Ionization (CI)

A chemical ionization (CI) source is considered a *soft* ionization source; it results in less fragmentation of analyte molecules and a simpler mass spectrum than that resulting from EI. Most importantly, the molecular ion is much more abundant using CI, allowing the determination of the molecular weight. If the CI process is "soft" enough the spectrum may consist almost entirely of only the molecular ion. Such a lack of fragmentation provides less structural information than an EI spectrum. Concentration of the charge on mostly this one ion in CI-MS improves the sensitivity of detection when the method of selected ion monitoring (SIM) is employed for quantitative measurements in GC-MS (Chapter 12, Section 12.8.1). If a fragmented EI spectrum is absent a molecular ion, then combining data from a CI spectrum containing a strong molecular ion will greatly assist interpretation of an unknown compound's spectra. The two modes complement one another for identification and quantitation of unknowns.

In CI, a large excess of **reagent gas** such as methane, isobutane, or ammonia is introduced into the ionization region. The pressure in the ion source is typically several orders of magnitude higher than in EI ion sources. A CI source design will be more enclosed with smaller orifices for the source vacuum pump to remove the reagent gas, allowing the higher source pressure to be maintained. The mixture of reagent gas and sample is subjected to electron bombardment. The reagent gas is generally present at a level $1000-10,000\times$ higher than the sample; therefore the reagent gas is most likely to be ionized by interaction with the electrons. Ionization of the sample molecules occurs indirectly by collision with ionized reagent gas molecules and proton or hydride transfer. A series of reactions occurs. Methane, for example, forms $CH_4^{\bullet+}$ and CH_3^+ on interaction with the electron beam. These ions then react with additional methane molecules to form ions as shown:

$$CH_4^{\bullet+} + CH_4 \longrightarrow CH_5^+ + CH_3^{\bullet}$$
$$CH_3^+ + CH_4 \longrightarrow C_2H_5^+ + H_2$$

Collisions between the ionic species CH_5^+ or $C_2H_5^+$ and a sample molecule M causes ionization of the sample molecule by proton transfer from the ionized reagent gas species to form MH^+ or by hydride transfer from the sample molecule to form $(M - H)^+$, also written as $(M - 1)^+$:

$$M + CH_5^+ \longrightarrow MH^+ + CH_4$$
$$M + C_2H_5^+ \longrightarrow MH^+ + C_2H_4$$
$$M + C_2H_5^+ \longrightarrow (M - H)^+ + C_2H_6$$
$$M + CH_5^+ \longrightarrow (M - H)^+ + CH_4 + H_2$$

Hydride transfer from M occurs mainly when the analyte molecule is a saturated hydrocarbon. In addition, the ionized reagent gas can react with M to form, for example, an $(M + C_2H_5)^+$ ion with $m/z = (M + 29)$. The presence of such an *adduct* ion of mass 29 Da above a candidate molecular ion in a methane CI mass spectrum is a good confirmation of the identity of the molecular ion.

Many commercial sources are designed to switch from EI to CI rapidly to take advantage of the complementary information obtained from each technique. The main advantage of CI is that fragmentation of the sample molecule is greatly reduced and significant peaks at $m/z = (M + 1)$ or $(M - 1)$ are seen, permitting the identification of the molecular weight of the analyte.

It is possible to introduce a sample directly into the chemical ionization source on a tungsten or rhenium wire. A drop of sample in solution is applied to the wire, the solvent is allowed to evaporate, and the sample inserted into the CI source. The sample molecules are desorbed by passing a current through the wire, causing it to heat. The analyte molecules then ionize by interaction with the reagent gas ions as has been described. This technique is called **desorption chemical ionization** and is used for nonvolatile compounds.

9.2.2.3.   *Atmospheric Pressure Ionization (API) Sources*

There are two major types of ionization sources that operate at atmospheric pressure, electrospray ionization (ESI) and atmospheric pressure chemical ionization (APCI). A modified version of the ESI source is the ion spray source. These sources are described in detail in Chapter 13, Section 13.1.6.1, because they are used to interface LC with MS for the separation and mass spectrometric analysis of mixtures of nonvolatile high molecular weight compounds, especially in the fields of pharmaceutical chemistry, biochemistry, and clinical biomonitoring. ESI will be described briefly so that its use may be demonstrated but more detail will be found in Chapter 13.

When a strong electric field is applied to a liquid passing through a metal capillary, the liquid becomes dispersed into a fine spray of positively or negatively charged droplets, an electrospray. The electric field is created by applying a potential difference of 3–8 kV between the metal capillary and a counter electrode. The highly charged droplets shrink as the solvent evaporates until the droplets undergo a series of "explosions" due to coulombic interactions. Each "explosion" forms smaller and smaller droplets. When the droplets become small enough, the analyte ions desorb from the droplets and enter the mass analyzer. A schematic *ESI source* is shown in Fig. 9.7. The ESI source is at atmospheric pressure. The droplets and finally the analyte ions pass through a series of orifices and skimmers. These serve to divert and exclude unevaporated droplets and excess vaporized solvent from the higher vacuum regions where analyte ions are accelerated and analyzed

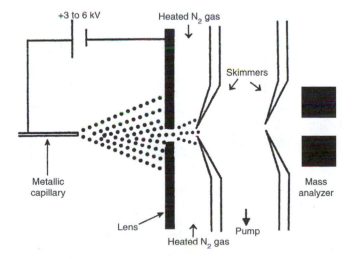

Figure 9.7   Schematic diagram of an ESI source. This source is shown with a lens system and skimmers to focus the ions and heated nitrogen gas to desolvate the ions.

by m/z. A flow of gas such as nitrogen or argon serves to desolvate the droplets and to break up ion clusters. The skimmers act as velocity filters. Heavier ions have less velocity from random thermal motions transverse to the direction of voltage acceleration through the orifices than lighter ions and continue in a straight path to the mass analyzer while the lighter ions (and solvent vapors and gases) are pumped away, permitting the pressure to be reduced without affecting the ion input to the mass analyzer. Liquid flow through the metal capillary is in the range of $1-10$ μL/min for the standard ESI design. For the increasingly important HPLC-MS instrumentation used in analysis of biomolecules, orthogonal spray ESI interfaces operate at 1 mL/min and can handle flow of up to $4-8$ mL/min from monolithic HPLC columns.

The advantage of ESI lies in the fact that large molecules, especially biomolecules like proteins, end up as a series of multiply charged ions, M^{n+} or $(M + nH)^{n+}$ with little or no fragmentation. For example, a given analyte M might form ions of M^{9+}, M^{10+}, M^{11+}, and so on. If the mass of the analyte M is 14,300 Da, then peaks would appear in the mass spectrum of this analyte at m/z values of $(14,300/9) = 1588.9$, $(14,300/10) = 1430.0$, and $(14,300/11) = 1300.0$. These ions are at much lower m/z values than would be the case if we had a singly charged M^+ ion at m/z 14300. One advantage to having multiply charged ions with low m/z values is that less-expensive mass analyzers with limited mass range can be used to separate them. Another is that high m/z ions such as high MW biomolecules with a only single charge leave a CI source with low velocities; these low velocities result in poor resolution due to dispersion and other processes in the mass spectrometer. Ions with low m/z values due to high charge are easily resolved.

Examples of mass spectra of biological molecules obtained with ESI are shown in Figs. 9.8 and 9.9. In reality, the analyst does not know the numerical charge on the peaks in the mass spectrum, but the successive peaks often vary by 1 charge unit. Computer-based algorithms have been developed for deconvoluting the sequence of m/z values of the multiply charged ions into the equivalent mass of single charged ions; such a deconvolution has been done in Fig. 9.8. This permits identification of the molecular weight of the analyte. Applications of LC-ESI-MS are described in the Chapter 13 Section 13.1.6.1. Dr. John Fenn, one of the inventors of ESI, received the Nobel Prize in Chemistry in 2002.

9.2.2.4.   *Desorption Ionization*

Large molecules, such as proteins and polymers, do not have the thermal stability to vaporize without decomposing. Desorption ionization sources permit the direct ionization of solids, facilitating the analysis of large molecules. There are several types of desorption sources in which solid samples are adsorbed or placed on a support and then ionized by bombardment with ions or photons. Desorption CI, one form of desorption ionization, has already been described. The important technique of **secondary ion mass spectrometry (SIMS)** is used for surface analysis as well as characterization of large molecules; SIMS is covered in detail in Chapter 14, Section 14.4. Several other important desorption sources are described subsequently.

Laser Desorption and Matrix-Assisted Laser Desorption Ionization (MALDI).   The use of a pulsed laser focused on a solid sample surface is an efficient method of ablating material from the surface and ionizing the material simultaneously. A variety of lasers have been used; examples include IR lasers such as the CO_2 laser ($\lambda = 10.6$ μm) and UV lasers such as Nd:YAG ($\lambda = 266$ nm, 355 nm). (YAG stands for yttrium aluminum garnet). Selective ionization is possible by choosing the appropriate laser wavelength. The laser can be focused to a small spot, from submicron to several

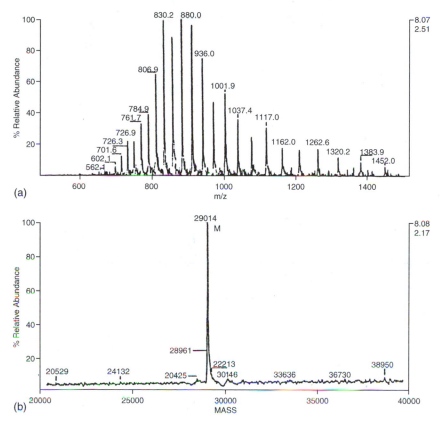

(a)

(b)

Figure 9.8 ESI mass spectrum (a) and deconvoluted mass spectrum (b) of carbonic anhydrase. (Reprinted with permission from Finnigan.)

microns in diameter. This permits the investigation of inclusions and multiple phases in solids as well as bulk analysis. The laser pulses generate transient signals, so a simultaneous detection system or a time-of-flight mass analyzer or a FT mass spectrometer is required. The laser provides large amounts of energy to the sample. This energy must be quickly dispersed within the molecule without fragmenting the molecule. Until the development of matrix-assisted laser desorption, the use of a laser resulted in fragmentation of biological molecules with molecular masses above about 1000 Da.

By mixing large analyte molecules with a "matrix" of small organic molecules, a laser can be used to desorb and ionize analyte molecules with molecular weights well over 100,000 Da with little fragmentation. The function of the matrix is to disperse the large amounts of energy absorbed from the laser, thereby minimizing fragmentation of the analyte molecule. This technique of "matrix-assisted" laser desorption ionization (MALDI) has revolutionized the mass spectrometric study of polymers and large biological molecules such as peptides, proteins, and oligosaccharides. The actual process by which ions are formed using the MALDI approach is still not completely understood.

Typical matrices and optimum laser wavelengths are shown in Table 9.3. A matrix is chosen that absorbs the laser radiation but at a wavelength at which the analyte absorbs moderately or not at all. This diminishes the likelihood of fragmenting the analyte molecule.

Figure 9.9 ESI mass spectra of two proteins: (a) hen egg-white lysozyme and (b) equine myoglobin.

Table 9.3 MALDI Experimental Conditions

Matrix		Wavelength	
Nicotinic acid	266 nm	2.94 μm	10.6 μm
2,5-Dihydroxy	266 nm	337 nm	355 nm
benzoic acid	2.8 μm	10.6 μm	
Succinic acid	2.94 μm	10.6 μm	
Glycerol (liquid)	2.79 μm	2.94 μm	10.6 μm
Urea (solid)	2.79 μm	2.94 μm	10.6 μm

A typical sample is prepared for MALDI by mixing $1-2$ μL of sample solution with $5-10$ μL of matrix solution. A drop (<2 μL) of the mixture is placed on the MALDI probe and allowed to dry at room temperature. The solvent evaporates and the now crystallized solid is placed into the mass spectrometer. The analyte molecules are completely separated from each other by the matrix, as shown in Fig. 9.10.

The matrix material must absorb strongly at the wavelength of the laser used for irradiation. In addition, the matrix must be stable in a vacuum, must not react chemically and must have a low enough mass to sublime. Matrix compounds used for MALDI include 2,5-dihydroxybenzoic acid, 3-hydroxypicolinic acid, and 5-chlorosalicylic acid for the UV region of the spectrum and carboxylic acids, alcohols, and urea for the IR region of the spectrum. Intense pulses of laser radiation are aimed at the solid on the probe. The laser radiation is absorbed by the matrix molecules and causes rapid heating of the matrix. This heating causes desorption of entire analyte molecules along with the matrix molecules. Desolvation and ionization of the analyte occurs; several processes have been suggested for the ionization, such as ion–molecule reactions but the MALDI ionization process is not completely understood. A useful if simplistic analogy is to think of the matrix as a mattress and the analyte molecules as china plates sitting on the mattress. The laser pulses are like an energetic person jumping up and down on the mattress. Eventually, the oscillations of the mattress will cause the china to bounce up into the air without breaking. The plates (i.e., molecules) are then whisked into the mass analyzer intact.

MALDI acts as a soft ionization source and generally produces singly charged molecular ions from even very large polymers and biomolecules, although a few multiple-charge ions and some fragment ions may occur (Fig. 9.11).

Fast Atom Bombardment. Fast atom bombardment (FAB) uses a beam of fast-moving neutral inert gas atoms to ionize large molecules. In this technique, the sample is dissolved in an inert, nonvolatile solvent such as glycerol and spread in a thin layer on a metal probe. The probe is inserted into the mass spectrometer through a vacuum interlock.

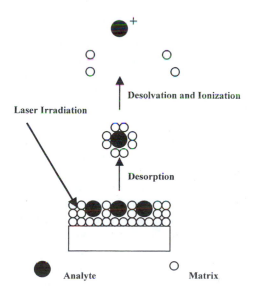

Figure 9.10 The MALDI process. Isolated analyte molecules are desorbed from a bed of matrix molecules by laser irradiation of the matrix. Subsequent desolvation and ionization of the analyte molecule occur by processes that are not completely understood.

Figure 9.11 The MALDI mass spectrum of a mouse monoclonal antibody. The matrix used was nicotinic acid; the laser radiation used was 266 nm. (Reprinted from Karas and Bahr, with permission from Elsevier.)

The beam of fast atoms is directed at the probe surface (the target). Argon is commonly used as the bombarding atom, although xenon is more effective (and more expensive).

Argon ions are produced in a heated filament ion source or gun, just as in SIMS, a surface analysis technique discussed in Chapter 14. The ions are accelerated through a cloud of argon atoms under an electrostatic field of 3–8 kV toward the target. The fast-moving ions exchange their charge with slow-moving argon atoms but lose no kinetic energy in the process. This results in a beam of fast-moving argon atoms and slow-moving argon ions. The latter are repelled and excluded from the system using a negatively charged deflection plate. The fast-moving atoms now strike the target, liberating molecular ions of the sample from the solvent matrix. The process may again be visualized using the analogy of china dishes atop a mattress. Instead of responding to repetitive laser pulses the matrix (mattress) absorbs, moderates, and transfers impact energy from the heavy fast atoms to the analyte molecules (dishes). If the analytes have surfactant character they will preferentially concentrate at the liquid matrix surface, in a location optimal for being lofted into the vapor state. Positively charged M^+ or negatively charged M^- may be produced, so positive ion or negative ion mass spectra may be collected. The process is shown schematically in Fig. 9.12.

There are several advantages to the FAB technique. The instrumentation is simple and the sensitivity is high. Analytes such as surfactants have been measured quantitatively at concentrations as low as 0.1 ppb. It is difficult to get very large molecules into the gas phase because of their low volatility, and it is difficult to ionize large molecules and retain the molecular ion in many ionization sources. The FAB process works at room temperature; volatilization is not required, so large molecules and thermally unstable molecules can be studied. The duration of the signal from the sample is continuous and very stable over a long period.

Sample fragmentation is greatly reduced in the FAB process, resulting in a large molecular ion, even with somewhat unstable molecules. This provides information on the molecular weight of the molecule, which is particularly important in biological samples such as proteins. Spectra from molecules with molecular weights greater than 10,000 have been obtained. Although a strong molecular ion is obtained with

Figure 9.12 Schematic FAB ionization source. The sample, dissolved in solvent, is spread in a thin film on the end of a metal probe and bombarded by fast-moving argon atoms. Both positive and negative ions are produced.

FAB, fragmentation patterns are also obtained, providing structural information on biologically important molecules such as proteins.

An important advantage of FAB is that only those analyte molecules sputtered from the glycerol are lost; the remainder can be recovered for other analyses. Consequently, samples as small as 1 μg can be placed on the probe, and after the mass spectrum is obtained, a significant amount of the sample can be recovered.

A modification of the FAB technique is **continuous flow FAB (CFFAB)**. In this approach, the sample in solution is introduced into the mass spectrometer through a fused silica capillary. The tip of the capillary is the target. The solution is bombarded by fast atoms produced as described earlier. Solvent is flowing continuously and the liquid sample is introduced by continuous flow injection (Fig. 9.13). The mass spectrum produced has the same characteristics as that from conventional FAB, but with low background. Typically, the solvent used is 95% water and 5% glycerol. The ability to inject aqueous samples is an enormous advantage in biological and environmental studies.

Figure 9.13 Schematic of continuous flow FAB MS operated in the flow-injection mode.

Very frequently, these types of samples are aqueous in nature, such as blood, urine, and other body fluids, water, and wastewater.

The sensitivity at the lower molecular weight range (1500 Da) is increased by two orders of magnitude over conventional FAB. Further, the background is reduced because of the reduced amount of glycerol present. In addition, when the solvent alone is injected, a background signal can be recorded. This can be subtracted from the signal due to sample plus solvent, and the net signal of the sample is obtained. This is especially valuable for trace analysis; concentrations as low as 10^{-12} g have been detected using CFFAB.

The CFFAB system can be incorporated into LC-MS systems. The mobile phase is the solvent used. The effluent from the LC is transported directly into the mass spectrometer and the MS obtained by CFFAB. This provides a mass spectrum of each separated peak in mixtures. (See Chapter 13 for a detailed discussion of LC/MS.)

In summary, FAB and CFFAB have greatly increased the potential of mass spectrometry by increasing the molecular weight range of molecules whose molecular ion can be determined. The system can be directly attached to LC, permitting identification of the components of a solution. Also, trace analysis is possible. The method can be applied to the important research areas of the health sciences, biology, and environmental science, as well as to chemistry.

9.2.2.5. *Ionization Sources for Inorganic MS*

The following ionization sources are used mainly in inorganic (atomic) MS, where the elemental composition of the sample is desired. The glow discharge (GD) and spark sources are used for solid samples, while the inductively coupled plasma (ICP) is used for solutions. All three sources are also used as atomic emission spectroscopy sources; they are described in more detail with diagrams in Chapter 7.

GD Sources and Spark Sources. The GD source and spark source are both used for sputtering and ionizing species from the surface of solid samples and have been discussed in Chapter 7 for use as atomic emission sources. As MS ionization sources, they are used primarily for atomic MS to determine the elements present in metals and other solid samples. The DC GD source has a cathode and anode in 0.1–10 torr of argon gas. The sample serves as the cathode. When a potential of several hundred volts is imposed across the electrodes, the argon gas ionizes forming a plasma. Electrons and positive argon ions are accelerated toward the oppositely charged electrodes. The argon ions impact the cathode surface, sputtering off atoms of the cathode material. The sample atoms are then ionized in the plasma by electrons or by collision with excited argon (Penning ionization). The sample ions are extracted from the plasma into the mass analyzer by a negatively charged electrode with a small aperture. The DC GD source is used for the analysis of conductive samples including metals, alloys, and semiconductors. The sample must be conductive to serve as the cathode. RF GD sources have been developed that enable the sputtering of electrically nonconductive samples such as ceramics and other insulators. Spark sources also can be used for sputtering of solids, but the GD source produces a more stable ion beam with better signal-to-noise ratio. The GD source sputters more material from a sample and gives more representative and more quantitative results of the elemental bulk composition than the spark source.

ICP Source. The argon ICP source has also been described in Chapter 7 for use with atomic emission spectrometers. The source produces ions from the elements introduced into the plasma as well as radiation; these ions can be extracted into a mass analyzer. The ICP torch is usually mounted horizontally with the tip of the plasma at the entrance to the mass analyzer as shown in Fig. 9.14. Most of the plasma gas is deflected by a metal

Figure 9.14 Argon ICP torch used as an ionization source for ICPMS. (From Ewing, used with permission.)

cone with a small orifice in its center, called the sampling cone. The gas that enters through the orifice expands into an evacuated region. The central portion passes through another metal cone, the skimmer cone, into the evacuated mass analyzer. Singly charged positive ions are formed from most elements, metallic and nonmetallic. The ICP has a high ionization efficiency, which approaches 100% for most of the elements in the periodic table. The mass spectra are very simple and elements are easily identified from the m/z values and the isotope ratios observed. Background ions from the solvent and from the argon gas used to form the plasma are usually observed. Such ions include Ar^+, ArH^+, ArO^+, and polyatomic ions from water and the mineral acids used to dissolve most samples.

9.2.3. Mass Analyzers

The mass analyzer is at the core of the mass spectrometer. Its function is to differentiate among ions according to their mass-to-charge ratio. There are a variety of mass analyzer designs. *Magnetic sector* mass analyzers and *quadrupole* mass analyzers are scanning instruments; only ions of a given mass-to-charge ratio pass through the analyzer at a given time. The m/z range is scanned over time. Other mass analyzers allow simultaneous transmission of all ions; these include *time-of-flight (TOF)*, *ion trap*, and *ion cyclotron resonance* mass analyzers as well as *dispersive magnetic* mass analyzers. Tandem mass spectrometers are instruments with several mass analyzers in sequence; these allow the selection of one ion in the first analyzer (the precursor ion) and the analysis of fragmentation or decomposition of that ion into product ions in the second analyzer.

9.2.3.1. Magnetic and Electric Sector Instruments

The principle of operation of a simple single-focusing magnetic sector mass analyzer was described briefly in Section 9.1. An ion moving through a magnetic field B will follow a circular path with radius r [Eq. (9.6)]. Changing B as a function of time allows ions of different m/z values to pass through the fixed radius flight tube sequentially. This scanning magnetic sector sorts ions according to their masses, assuming that all ions have a $+1$ charge and the same kinetic energy. A schematic of a 90° sector instrument is shown in Fig. 9.15. A variety of other magnetic mass spectrometers are shown in Fig. 9.16; some of these will be discussed later. The sector can have any apex angle, but 60° and 90° are common. It can be demonstrated that a divergent beam of ions of a given m/z will be brought to a focus by passing through a sector shaped magnetic field, as shown by the three ion paths in Fig. 9.15.

Figure 9.15 A 90° magnetic sector mass spectrometer. (From Ewing, used with permission.)

A dispersive magnetic sector mass analyzer does not use a flight tube with a fixed radius. Since all ions with the same kinetic energy but different values of m/z will follow paths with different radii, advantage can be taken of this. The ions will emerge from the magnetic field at different positions and can be detected with a position-sensitive detector such as a photoplate or an array detector. Examples of dispersive magnetic sector systems are shown in Fig. 9.16(c) and (d).

Figure 9.16 Early mass spectrometer designs. (a) Aston, 1919; (b) Dempster, 1918; (c) Mattauch–Herzog, 1935; (d) Bainbridge, 1933. In each case, B signifies the magnetic field. Spectrometers (c) and (d) are dispersive mass spectrometers; (c) the Mattauch–Herzog design is also a double sector instrument, using an electric sector before the magnetic field.

A single-focusing instrument such as the system shown in Fig. 9.16(b) has the disadvantage that ions emerging from the ion source do not all have exactly the same velocity. This is due to several factors. The ions are formed from molecules that have a Boltzmann distribution of energies to begin with. The ion source has small variations in its electric field gradient, causing ions formed in different regions of the source to experience different acceleration. Also, when fragmentation occurs, kinetic energy is released. This results in a distribution of velocities and adversely affects the resolution of the instrument by broadening the signal at the detector.

However, ions in a radial electrostatic field also follow a circular trajectory. The electrostatic field is an *electric sector* and separates ions by kinetic energy, not by mass (Fig. 9.17). The ion beam from the source can be made much more homogeneous with respect to velocities of the ions if the beam is passed through an electric sector before being sent to the mass analyzer. The electric sector acts as an energy filter; only ions with a very narrow kinetic energy distribution will pass through.

Most magnetic sector instruments today combine both an electric sector and a magnetic sector. Such instruments are called double-focusing mass spectrometers. One common commercial double-focusing design is the Nier–Johnson design (Fig. 9.18), introduced in 1953; a second common design using two sectors is the Mattauch–Herzog dispersive design, shown in Fig. 9.16(c).

Mass ranges for magnetic sector instruments are in the m/z 1–1400 range for single-focusing instruments and m/z 5000–10,000 for double-focusing instruments. Very high mass resolution, up to 100,000, is possible using double-focusing instruments.

9.2.3.2. Time of Flight (TOF) Analyzer

A TOF analyzer does not use an external force to separate ions of different m/z values. Instead, pulses of ions are accelerated into an evacuated field free region called a drift tube. If all ions have the same kinetic energy, then the velocity of an ion depends on its mass-to-charge ratio, or on its mass, if all ions have the same charge. Lighter ions will travel faster along the drift tube than heavier ions and are detected first. The process is shown schematically in Fig. 9.19.

A schematic TOF mass spectrometer is shown in Fig. 9.20. The drift tube in a TOF system is approximately 1–2 m in length. Pulses of ions are produced from the sample using pulses of electrons, secondary ions, or laser pulses (e.g., MALDI). Ion pulses are produced with frequencies of 10–50 kHz. The ions are accelerated into the drift tube by a pulsed electric field, called the ion-extraction field, because it extracts (or draws out) ions into the field-free region. Accelerating voltages up to 30 kV and extraction pulse frequencies of 5–20 kHz are used.

Source

+

−

Exit
Slit

Figure 9.17 A cylindrical electrostatic-sector energy filter. (From Ewing, used with permission.)

Figure 9.18 A Nier–Johnson double focus mass spectrometer. (From Ewing, used with permission.)

Ions are separated in the drift tube according to their velocities. The velocity of an ion, v, can be expressed as:

$$v = \sqrt{\frac{2zV}{m}} \tag{9.10}$$

where V is the accelerating voltage. If L is the length of the field-free drift tube and t is the time from acceleration to detection of the ion (i.e., the flight time of the ion in the tube),

$$v = \frac{L}{t} \tag{9.11}$$

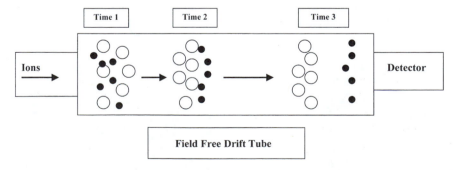

Figure 9.19 A pulse of ions of two different m/z values enters the field free drift tube of a TOF mass spectrometer at time 1. The large white circles have $m/z >$ than the small dark circles. As they travel down the tube, the lighter ions move faster, and by time 3, have been separated from the heavier ions.

Figure 9.20　Schematic TOF mass spectrometer.

and the equation that describes ion separation is:

$$\frac{m}{z} = \frac{2Vt^2}{L^2} \tag{9.12}$$

The flight time, t, of an ion is:

$$t = L\sqrt{\frac{m}{2zV}} \tag{9.13}$$

Eq. (9.13) can be used to calculate the difference in flight time between ions of two different masses. Actual time separations of adjacent masses can be as short as a few nanoseconds, with typical flight times in microseconds.

　　TOF instruments were first developed in the 1950s, but fell out of use because of the inherent low resolution of the straight drift tube design (as in Fig. 9.20). The drift tube length and flight time are fixed, so resolution depends on the accelerating pulse. Ion pulses must be kept short to avoid overlap of one pulse with the next, which would cause mass overlap and decrease resolution. Interest in TOF instruments resurfaced in the 1990s with the introduction of MALDI and rapid data acquisition methods. The simultaneous transmission of all ions and the rapid flight time means that the detector can capture the entire mass spectral range almost instantaneously.

　　The resolution of a TOF analyzer can be enhanced by the use of an ion mirror, called a **reflectron**. The reflectron is used to reverse the direction in which the ions are traveling and to energy-focus the ions to improve resolution. The reflectron's electrostatic field allows faster ions to penetrate more deeply than slower ions of the same m/z value. The faster ions follow a longer path before they are turned around, so that ions with the same m/z value but differing velocities end up traveling exactly the same distance and arrive at the detector together. The use of a curved field reflectron permits the focusing of ions over a broad mass range to collect an entire mass spectrum from a single laser shot. In a reflectron TOF, the ion source and the detector are at the same end of the spectrometer; the reflectron is at the opposite end from the ion source. The ions traverse the drift tube twice, moving from the ion source to the reflectron and then back to the detector. A schematic of a commercial reflectron TOF mass analyzer is shown in Fig. 9.21.

　　The mass range of commercial TOF instruments is up to 10,000 Da. Resolution depends on the type of TOF and ranges from 1000 for instruments designed as dedicated detectors for GC (GC-TOFMS) to 20,000 for reflectron instruments. One limitation to the use of a conventional reflectron instrument is a loss in sensitivity; about 10% of the ions are lost with a conventional wire grid reflectron.

Push Pulse Plate
Electron Focusing Optics
Ion Source Chamber
Electron Beam
Vial with Calibration Compound
Heated Transfer line
Turbomolecular Pump
Ion Focusing Optics
GC Effluent
Signal Out
Detector
Steering and Deflection Plates
Barrier Grid
Einzel Lens
Steering Plates
Flight Tube
Turbomolecular Pump
Ion Mirror

Figure 9.21 A commercial reflectron TOF mass analyzer, the Pegasus III from LECO. The analyzer is shown with sample introduction from a GC. [Diagram courtesy of LECO Corporation (www.leco.com).]

The rapid collection of the entire mass spectrum made possible by the TOF makes it ideal for interfacing with a chromatograph. It is especially useful when combined with fast GC, which requires the rapid collection of hundreds of mass spectra. For example, the LECO Pegasus 4D GC-TOFMS collects the entire mass range from 1 to 1000 Da in 170 μs and collects up to 500 mass spectra/s. (A detailed description of this instrument can be found at www.leco.com.)

9.2.3.3. *Quadrupole Mass Analyzer*

The quadrupole mass analyzer does not use a magnetic field to separate ions. The quadrupole separates ions in an electric field (the quadrupole field) that is varied with time. This field is created using an oscillating radio frequency (RF) voltage and a constant direct current (DC) voltage applied to a set of four precisely machined parallel metal rods (Fig. 9.22). This results in an AC potential superimposed on the DC potential. The ion beam is directed axially between the four rods.

The opposite pairs of rods A and B, and C and D, are each connected to the opposite ends of a DC source, such that when C and D are positive, A and B are negative. The pairs of electrodes are then connected to an electrical source oscillating at RFs. They are connected in such a way that the potentials of the pairs are continuously 180° out of phase with each other. The magnitude of the oscillating voltage is greater than that of the DC source, resulting in a rapidly oscillating field. The RF voltage can be up to 1200 V while the DC voltage is up to 200 V. The rods would ideally be hyperbolic instead of circular in cross-section to provide a more uniform field. Under these conditions, the potential at any point between the four poles is a function of the DC voltage and the amplitude and frequency of the RF voltage. The shape of the rods varies with different manufacturers; cheaper circular cylindrical rods are often used instead of hyperbolic rods.

(a)   (b)

Figure 9.22 (a) Transmission quadrupole mass spectrometer. Rods A and B are tied together electrically, as are Rods C and D. The two pairs of rods, AB and CD, are connected both to a source of direct potential and a variable RF excitation such that the RF voltages are 180° out of phase. (b) The geometry of the rods.

An ion introduced into the space between the rods is subjected to a complicated lateral motion due to the DC and RF fields. Assume that the x direction is the line through the midpoint of the cross-sections of rods A and B; the y direction is the line through the midpoint of the cross-sections of rods C and D, as shown in Fig. 9.22(b). The forward motion of the ion in the z direction (along the axis between the rods) is not affected by the field. The following equations describe the lateral motion of the ion:

$$\frac{d^2x}{dt^2} + \frac{2}{r^2(m/z)}(V_{DC} + V_{RF}\cos 2\pi ft)y = 0 \tag{9.14}$$

$$\frac{d^2y}{dt^2} + \frac{2}{r^2(m/z)}(V_{DC} + V_{RF}\cos 2\pi ft)x = 0 \tag{9.15}$$

where V_{DC} is the voltage of the DC signal; V_{RF}, the amplitude of the voltage of the RF field; f, the frequency of oscillation of the RF field (rad/s); r, the half the distance between the inner edges of opposing poles such as A and B as shown in Fig. 9.22(b); and t, the time.

The motion is complex because the velocity in the x direction is a function of the position along y and vice versa. In order for an ion to pass through the space between the four rods, every time a positive ion is attracted to a negatively charged rod, the AC electric field must be present to push it away; otherwise, it will collide with the rod and be lost. The coordination between the oscillating (AC) field and the time of the ion's arrival at a rod surface over the fixed distance between the rods is critical to an ion's movement through the quadrupole. As a result of being alternately attracted and repelled by the rods, the ions follow an oscillating or "corkscrew" path through the quadrupole to the detector. For a given amplitude of a fixed ratio of DC to RF at a fixed frequency, only ions of a given m/z value will pass through the quadrupole. If the mass-to-charge ratio of the ion and the frequency of oscillation fit Eqs. (9.14) and (9.15), the ion will oscillate toward the detector and eventually reach it. If the m/z value and the frequency do not meet the conditions required by Eqs. (9.14) and (9.15), these ions will oscillate with an increasingly wide path until they collide with the rods or are pulled out by the vacuum system. In any case, the ions will not progress to the detector. Only a single m/z value can pass through the quadrupole at a given set of conditions. In this respect, the quadrupole acts like a filter, and is often called a mass filter.

The separation of ions of different m/z can be achieved by several methods. The frequency of oscillation of the RF field can be held constant while varying the potentials of the DC and RF fields in such a manner that their ratio is kept constant. It can be shown mathematically that the best resolution is obtained when the ratio V_{DC}/V_{RF} is equal to 0.168. If the ratio is greater than this number, a stable path cannot be achieved for any mass number; if the ratio is lower than this number, resolution is progressively lost.

The resolution of the system is dependent on the number of oscillations an ion undergoes in the drift chamber. Increasing the rod lengths, therefore, increases resolution and extends the use of the system to higher molecular weight compounds. Increasing the frequency of the RF field can bring about this same improvement. The rod diameter is also important. If the diameter is increased, the sensitivity is greatly increased, but then the mass range of the system is decreased. The manufacturer must come to a compromise with these factors when designing an instrument for analytical use. The resolution achievable with the quadrupole mass spectrometer is approximately 1000; the m/z range for a quadrupole mass analyzer is 1–1000 Da. As with other mass spectrometers, the sample must be available in the gas phase and must be ionized.

Quadrupole mass analyzers are found in most commercial ICP-MS instruments, in most GC-MS instruments (Chapter 12) and in many LC-MS instruments (Chapter 13). Quadrupoles are also used in MS-MS systems as mass analyzers and ion lenses. This use will be described in Section 9.2.3.4.

Although the quadrupole mass analyzer does not have the range or resolution of magnetic sector instruments, it is very fast. It can provide a complete mass spectrum in less than 100 ms. This property and its wide angle of acceptance make it suitable for coupling to transient signal sources such as those from chromatography or laser ablation. In addition the quadrupole mass analyzer is inexpensive, compact, and rugged. Most GC-MS and LC-MS instruments with quadrupoles are small enough to fit on a benchtop. Quadrupoles are the most common mass analyzer in commercial use. The term **transmission quadrupole** mass spectrometer is sometimes used for this mass analyzer to avoid confusion with the **quadrupole ion trap** mass spectrometer discussed in Section 9.2.3.5.

9.2.3.4. MS-MS and MS^n Instruments

Many analytical questions require the mass spectrometrist to obtain more information about the structure of fragment ions or about ion–molecule reactions than can be obtained from the initial ionization of an analyte. In such cases, the technique of MS-MS, also called tandem MS may be useful. MS-MS is a mass spectral technique that uses two (or more) stages of mass analysis combined with a process that causes a change in mass of the ion of interest, such as dissociation into lighter fragment ions by collision with an inert gas or conversion into a heavier ion by reaction with a neutral molecule.

The stages of mass analysis may be performed by two physically separate mass analyzers, such as two quadrupoles coupled in series; this type of arrangement for MS-MS is called "tandem in space". Figure 9.23 shows a quadrupole MS-MS instrument with three quadrupoles for "tandem in space" analysis. Alternatively, ion traps, discussed in Sections 9.2.3.5 and 9.2.3.6, may be used to perform MS-MS experiments within the same mass analyzer; this type of MS-MS experiment is called "tandem in time".

Using Fig. 9.23, we will look at a simple MS-MS experiment. For example, an analyte may be ionized as usual by the ion source. One ion of a particular m/z value is

Figure 9.23 A commercial quadrupole tandem MS-MS instrument. [Courtesy of Thermo Electron Corporation (www.thermo.com).]

of interest. This ion is called the **precursor** ion. The precursor ion is selected by the first quadrupole, which is operating as a mass analyzer. The precursor ion enters the second quadrupole. This second quadrupole is the reaction region and acts as a collision cell and ion lens, not as a mass analyzer. An inert gas may be added in this region to cause collision-induced fragmentation of the precursor ion into lighter product ions or a reactive reagent gas may be introduced to form heavier product ions through ion–molecule reactions. The second quadrupole also serves to focus the product ions; that is:

$$\text{M}^+_{\text{precursor}} \longrightarrow \text{M}^+_{\text{product}}$$

where the precursor and product ions have different m/z values. The product ions then undergo mass analysis as usual in the third quadrupole. This type of design, where the first and third quadrupoles are used for mass analysis and the center quadrupole is used for collision and focusing, is often abbreviated as a QqQ design, to indicate that there are only two stages of mass analysis symbolized by the capital Q.

If we had an instrument with three mass analyzers, the fragmentation process could be repeated before final analysis. A precursor ion is selected, fragmented, a given product ion is selected and fragmented again before mass analysis of its product ions; that is:

$$\text{M}^+_{\text{precursor}} \longrightarrow \text{M}^+_{\text{product 1}} \longrightarrow \text{M}^+_{\text{product 2}}$$

where all three ions have different m/z values. This is an example of MS-MS-MS or MS^3; the number of steps can be increased to give an MS^n experiment. It is not practical to build "tandem in space" instruments with large numbers of mass analyzers; three or four is the upper limit. Commercial MS-MS instruments are limited to two mass analysis stages. Ion trap instruments are used for higher order experiments. In general, $n = 7$ or 8 is a practical upper limit in ion trap instruments.

Tandem mass spectrometers have been built with three quadrupoles as shown in Fig. 9.23, and with other combinations of sector and TOF mass analyzers. Electric and magnetic sector analyzers have been combined with quadrupoles and with TOF analyzers.

9.2.3.5. Quadrupole Ion Trap

An **ion trap** is a device where gaseous ions can be formed and/or stored for periods of time, confined by electric and/or magnetic fields. There are two commercial types of ion traps in use in MS, the *quadrupole ion trap* (QIT) and the *ion cyclotron resonance trap* (ICR).

The QIT mass spectrometer is also called a **Paul ion trap** or more commonly, just an ion trap. This analyzer uses a quadrupole field to separate ions, so "quadrupole" is used in the name to distinguish this system from the ICR trap discussed in the next section. The QIT is shown schematically in Fig. 9.24. A ring-shaped electrode and two end cap electrodes, one above and one below the ring-shaped electrode, are used to form a 3D field. A fixed frequency RF voltage is applied to the ring electrode while the end caps are either grounded or under RF or DC voltages. Ions are stored in the trap by causing them to move in stable trajectories between the electrodes under the application of the field. This is done by varying the potentials, so that just before an ion collides with an electrode the potential changes sign and repels the ion. Ions with a very broad range of m/z values can be stored simultaneously in the ion trap.

Ionization of the sample can take place outside of the ion storage area of the ion trap; such external ionization is required for LC-MS using an ion trap and may be used for GC-MS. Alternatively, ionization can take place inside the ion storage area; this internal ionization approach can be used for GC-MS. Inert gas may be introduced into the trap after initial ionization for MS-MS experiments using collision-induced dissociation.

Ions are extracted from the trap by changing the amplitude of the ring electrode RF. As the amplitude increases, the trajectory of ions of increasing m/z becomes unstable. These ions move toward the end caps, one of which has openings leading to the detector. Ions of a given m/z value pass through the end cap sequentially and are detected.

The use of various RF and DC waveforms on the end caps allows the ion trap to selectively store precursor ions for MS-MS experiments or to selectively store analyte ions while eliminating ions from the matrix. This can result in improved detection limits in analysis. The ion trap has limitations. Because the stored ions can interact with each other (a space–charge effect), thereby upsetting stability of trajectories, the concentration

Figure 9.24 Cross-section of a quadrupole ion trap mass spectrometer. This schematic shows a gas phase sample introduced from a GC and ionized inside the trap by electrons from the filament. (From Niessen and van der Greef, used with permission.)

of ions that can be stored is low. This results in a low dynamic range for ion trap mass spectrometers. Trace level signals from a target analyte ion at one mass can be destabilized by the presence of great excesses of contaminant ions, even if these are of sufficiently different mass to be well resolved from the ion of interest. Ion trap MS instruments are less forgiving of "dirty samples" than are quadrupoles, which "throw away" such unwanted ions as they are measuring the target ion. The stored ion interaction also limits the accuracy of the mass-to-charge ratio measurement. Resolution of commercial QIT mass spectrometers is on the order of $0.1-1$, with an m/z range of $10-1000$.

9.2.3.6. *Fourier Transform Ion-Cyclotron Resonance (FTICR)*

The ICR instrument, also called a **Penning ion trap**, uses a magnetic field to trap and store ions. As shown in Fig. 9.25, six conducting plates arranged as a cube serve as the ion trap. The cubic cell is about 100 mm on a side, is under high vacuum ($<10^{-8}$ torr) and is located inside a strong magnetic field produced by a superconducting magnet. Sample is introduced into the cell and ionized by an external ion source such as an electron beam passing through the trap. Ions in the presence of a magnetic field move in circular orbits perpendicular to the applied field, at a frequency called the cyclotron frequency:

$$\omega_c = \frac{z}{m}(eB) = \frac{v}{r} \tag{9.16}$$

where ω_c is the frequency of rotation of ions (radians/s); e, the charge on electron (coulombs); B, the magnetic field (tesla); z, the charge on the ion; m, the mass of the ion; v, the velocity of the ion; and r, the radius of orbit.

The frequency of motion of an ion depends on the inverse of its m/z in a fixed magnetic field. Mass analysis is performed by applying an RF pulse of a few milliseconds duration to the transmitter plates. The RF pulse provides energy to the ions, causing them to move in larger circular orbits at the same frequency. For a given m/z value, a pulse at a frequency of ω_c causes all ions of that m/z value to absorb energy and increase their orbit of rotation. When the RF pulse is off, the motion of the ions is detected by current induction in the receiver plates. As a group of positive ions approaches the receiver

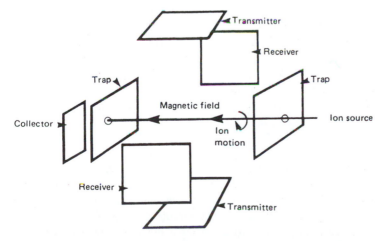

Figure 9.25 "Exploded" view of an ICR ion trap. The ICR has been the primary mass analyzer used in FTMS, both alone and in newer "hybrid" FTMS instruments.

plate, its charge attracts electrons to the inside surface of the plate. As the group recedes, the electrons are released. This induced current, called an "image current" is a sinusoidal signal with frequency ω_c. The larger the orbit, the larger is the induced current. The frequency provides the m/z information about the ion and the current amplitude depends on the number of ions of that m/z value, providing information about the concentration of ions.

It would be possible to scan the RF and measure the magnitude of the image current at each m/z value to obtain the mass spectral information but the process would be very slow. Instead, an RF pulse is used that contains a range of frequencies. The range of frequencies is chosen to excite the desired m/z range. When the pulse is off, all of the exited ions induce image currents in the receiver plates as they rotate. The output current, which contains all of the frequency and magnitude information from all of the ions present can be converted mathematically to a mass spectrum by application of the Fourier transform (FT). The use of an ICR ion trap and Fourier transformation is called Fourier transform ion-cyclotron resonance mass spectrometry (FTICRMS) or just FTMS. As of early 2003, this was the only type of FTMS instrument commercially available.

There are several advantages to the ICR. One is that the ion detection is non-destructive. Therefore, signals can be accumulated by averaging many cycles, resulting in greatly improved S/N and signal-to-background as well as very low detection limits. Detection of attomoles of analyte is possible. Frequency can be measured very accurately, so the mass accuracy of these FTMS systems can be very high, on the order of 1 ppb for a mass of 100 Da. In order to acquire sufficient information to achieve such high resolutions by the FT process, the data must be acquired over a longer period. In order that collisions with residual gas atoms in the ICR trap not remove the ions during this period, it must be operated at very high vacuum (e.g., $<10^{-8}$ torr), if such high resolution is to be attained. The ICR can also be used for MS-MS and MS^n experiments, by storing precursor ions and fragmenting them in the trap using a collision gas, lasers, or ion beams. An advantage of the FTMS system is that it is nondestructive, so ions at all stages of an MS^n experiment can be measured. A QIT instrument expels ions to be analyzed, so only ions in the final step can be measured.

The major disadvantages of the ICR are a limited dynamic range due to the same space–charge effect described for the quadrupole ion trap, a more complex design, and high instrument cost.

Despite the high cost of the FTICR instrument, new "hybrid" FTMS instruments costing significantly more than 1 million US dollars were introduced commercially in 2003 because of their ability to determine the structure of proteins. Protein structure determination is critical to fundamental biology, genomics, proteomics, and the understanding of drug–biomolecule interactions for development of pharmaceuticals. "Hybrid" FTMS instruments combining either an ion trap or quadrupole(s) on the front end with the FTICR on the back end exhibit both high sensitivity and high resolution.

9.2.4. Detectors

Most mass spectrometers measure one m/z value at a time. A single channel ion detector is used for these instruments, either an **electron multiplier** or a **Faraday cup**. TOF, ion trap, and FTICR mass spectrometers have the ability to extract ions with many m/z values simultaneously, so simultaneous detection of these ions is desirable. One approach to multiple ion detection has been to use multiple detectors. Multiple detectors are also used for high-resolution magnetic sector MS instruments designed for very precise isotope ratio determination and for quantitative analysis using isotope dilution. Instruments with multiple detectors are called "multicollectors". New detector developments in array detectors hold the promise of simultaneous m/z measurement over a wide mass range.

9.2.4.1. Electron Multiplier (EM)

The most common detector used for ions in mass spectrometers is the electron multiplier (EM). The EM is very similar in concept to the photomultiplier tube for optical detection. It is very sensitive and has fast response. The EM is based on the dynode, which has a surface that emits electrons when struck by fast-moving electrons, positive ions, negative ions, or neutrals. A **discrete-dynode EM** uses a series of 12–24 dynodes, each biased more positively than the preceding dynode. A collision releases several electrons from the dynode surface. These electrons are then accelerated to a second such surface, which, in turn, generates several electrons for each electron that bombards it. This process is continued until a cascade of electrons (an amplified current) arrives at the collector. The process is shown schematically in Fig. 9.26. Typically, one ion can produce 10^5 electrons or more; this ratio of electrons measured per ion is referred to as the gain. The gain of the detector can be adjusted, with operating gains of 10^4–10^8 used, depending on the application. Figure 9.26(b) shows a commercial discrete-dynode electron multiplier. A **continuous-dynode EM**, also called a channel electron multiplier (CEM) uses a continuous glass tube, either lead-doped or coated on the inside with a conductive surface of high electrical resistance, such as those shown in Fig. 9.27. A potential difference is applied across the tube ends so that the potential varies in a linear manner along the tube. Each incident ion releases electrons that are accelerated and strike the tube again, resulting in the same cascade effect seen in the discrete-dynode EM. The curved or coiled form is designed to reduce electrical noise by preventing positive ions from returning upstream.

A disadvantage to dynode-based detectors is that the number of secondary electrons released depends on the type of incident primary particle, its angle and energy. The dependence of the number of secondary electrons emitted on incident energy is shown for electron impact in Fig. 9.26(c); the same plot for ion impact would be similar. Therefore, they can exhibit *mass discrimination* due to differences in ion velocity. Heavy ions from quadrupole mass analyzers and from QIT mass analyzers impact the dynode surface at lower velocities than light ions. EM detectors for these instruments must be designed to overcome the difference in velocities, often by accelerating the ions prior to them striking the first electron-emitting dynode. An excellent source of information on how discrete dynode electron multipliers work is the SGE website at www.sge.com, which describes their ETP electron multipliers. Similarly, the Burle Technologies website at www.burle.com provides technical information on their Channeltron® continuous-dynode electron multiplier.

9.2.4.2. Faraday Cup

The least-expensive ion detector is the Faraday cup, a metal or carbon cup that serves to capture ions and store the charge. The resulting current of a few microamperes is measured and amplified. The cup shape decreases the loss of electrons from the metal due to ion impact. The Faraday cup is an absolute detector and can be used to calibrate other detectors. The current is directly proportional to the number of ions and to the number of charges per ion collected by the detector. Unlike dynode-based detectors, the Faraday cup does not exhibit mass discrimination. The detector does have a long response time, which limits its utility. The Faraday cup detector is used for making very accurate measurements in isotope-ratio MS, where the ion currents do not change rapidly. The

Figure 9.26 (a) A schematic discrete-dynode electron multiplier, showing the electron gain at each successive dynode after impact of an ion on the first dynode surface. The electron cascading process results in gains of up to 10^8 being achieved with approximately 21 dynodes. (b) An ETP electron multiplier schematic showing the position of the dynodes in the detector. (c) Dependence of the number of secondary electrons emitted on impact energy. [Images courtesy of SGE, Inc. (Austin, TX) and ETP Electron Multipliers Pty Ltd, a division of SGE (Sydney, Australia). (www.etpsci.com and www.sge.com).]

Faraday cup detector has no gain associated with it, unlike dynode-based detectors. This limits the sensitivity of the measurement.

High-precision isotope ratio mass spectrometers are designed with combinations of multiple Faraday cup detectors and multiple miniature electron multipliers (used as ion counters) for simultaneous isotope measurement. For example, the TRITON and NEPTUNE multicollector mass spectrometers from Thermo Electron Corporation can be configured with up to nine Faraday cups and eight ion counters to detect 17 ion beams simultaneously. Details of these instruments can be found at www.thermo.com. The use of multicollector instruments improves precision by two to three orders of

(a) Secondary Electron cascade — Electrons

(b) Electrons (c)

Figure 9.27 (a) A schematic channel electron multiplier (CEM), consisting of a glass or interior-coated ceramic tube that emits secondary electrons upon ion impact. (b) A schematic curved CEM. The curved shape minimizes ion feedback noise. (c) Photo of the Channeltron® electron multiplier, showing the curved glass tube without the associated electronics. [Courtesy of BURLE Electro-Optics (www.burle.com).]

magnitude over a single collector magnetic sector instrument, and this high precision is needed for isotope ratio measurements.

9.2.4.3. Array Detectors

The microchannel plate is a spatially resolved array detector formed of 10^5–10^7 continuous-dynode electron multipliers, each only 10–100 μm in diameter. This detector is used in focal plane mass spectrometers as a replacement for photograph plate detectors and is used in some TOFMS instruments.

The *focal plane camera* (FPC), still in initial development, consists of an array of 31 Faraday cups, each 145 μm wide. Up to 15 m/z values can be measured simultaneously. This detector shows improved precision compared with single channel detectors and has the ability to measure fast transient signals such as those from laser ablation. The detector design is described in the references by Barnes et al. and Knight et al. cited in the bibliography.

BIBLIOGRAPHY

Alexander, M.L.; Hemberger, P.H.; Cesper, M.E.; Nogar, N.S. Laser desorption in a quadrupole ion trap: mixture analysis using positive and negative ions. *Anal. Chem.* **1993**, *65*, 1600.

Application Reviews. *Anal. Chem.* **1993**, *June*.

Barker, J. *Mass Spectrometry: Analytical Chemistry by Open Learning*, 2nd ed.; Wiley: Chichester, UK, 1999.

Barnes, J.H., IV; Hieftje, G.M.; Denton, M.B.; Sperline, R.; Koppenaal, D.W.; Baringa, C. A mass spectrometry detector array that provides truly simultaneous detection. *Am. Lab* **2003**, *October*, 15.

Buchanan, M.V.; Hettich, R.L. Fourier transform mass spectrometry of high-mass biomolecules. *Anal. Chem.* **1993**, *65* (5), 245.

Caprioli, R.M. Continuous flow FAB MS. *Anal. Chem.* **1990**, *62* (8), 177A.

Cody, R.N.; Tamura, L.; Musselman, B.D. Electrospray ionization/magnetic sector MS calibration and accuracy. *Anal. Chem.* **1992**, *64*, 1561.

Duncan, W.P. *Res. Dev.* **1991**, *April*, 57.

Ewing, G.W. Mass spectrometry. In *Analytical Instrumentation Handbook*, 2nd ed.; Ewing, G.W., Ed.; Marcel Dekker, Inc.: New York, 1997.

Fenn, J.B.; Mann, M.; Meng, C.K.; Wong, S.E.; Whitehouse, C. *Science* **1989**, *64*, 246.

Finnigan, M. *Anal. News*, **1990**, *February*.

Goldner, H.J. Electrospray excites scientists with its amazing potential. *R&D Mag.* **1993**, *10*, 43.

Karas, M.; Bahr, U. *Trends Anal. Chem.* **1990**, *9*, 323.

Karas, F.H.; Beavis, R.C.; Chait, B.T. Matrix-assisted laser desorption ionization mass spectrometry of biopolymers. *Anal. Chem.* **1991**, *63* (24), 1193A.

Knight, A.K.; Sperline, R.P.; Hieftje, G.M., et al. *Int. J. Mass Spectrom.* **2002**, *215*, 131.

Lambert, J.B.; Shurvell, H.F.; Lightner, D.; Cooks, R.G. *Introduction to Organic Spectroscopy*; MacMillan Publishing Company: New York, 1987.

McLafferty, F.W.; Tureček, F. *Interpretation of Mass Spectra*, 4th ed.; University Science: Mill Valley, CA, 1993.

Mahoney, J.; Perel, J.; Taylor, S. Primary ion source for FABMS. *Am. Lab.* **1984**, *March*, 92.

Montaser, A., Ed. *Inductively Coupled Plasma Mass Spectrometry*; VCH: Berlin, 1998.

Niessen, W.M.A.; van der Greef, J. *Liquid Chromatography–Mass Spectrometry*; Marcel Dekker, Inc.: New York, 1992.

Silverstein, R.M.; Webster, F.X. *Spectrometric Identification of Organic Compounds*, 6th ed.; Wiley: New York, 1981.

Smith, R.D.; Wahl, J.H.; Goodlett, D.R.; Hofstadler, S.A. *Anal. Chem.* **1993**, *65* (13). 574 A.

Smith, R.M.; Busch, K.L. *Understanding Mass Spectra–A Basic Approach*; Wiley: New York, 1999.

Sparkman, O.D. *Mass Spectrometry Desk Reference*; Global View Publishing: Pittsburgh, PA, 2000.

Voreos, L. Electrospray mass spectrometry. *Anal. Chem.* **1993**, *66* (8), 481A.

Watson, J.T. *Introduction to Mass Spectrometry*, 3rd ed.; Lippincott-Raven: Philadelphia, PA, 1997.

PROBLEMS

9.1   What is meant by "molecular ion"? What is the importance of identifying it to a chemist?

9.2   How are resolving power and resolution defined for a mass spectrometer?

9.3   What is the difference between hard ionization and soft ionization?

9.4   Describe how an EI source forms ions from analyte molecules. Is this a hard or soft ionization source? What are the advantages and disadvantages of this source?

9.5 Describe how a CI source forms ions from analyte molecules. Is this a hard or soft ionization source? What are the advantages and disadvantages of this source?

9.6 Draw a block diagram of a typical mass spectrometer.

9.7 Consider an ion with $z = 1$. Calculate the kinetic energy added to the ion by acceleration through a potential of 1000 V in an electron ionization source. Does the kinetic energy acquired depend upon the mass of the ion? Explain.

9.8 In an electron ionization source, does the mass of the ion affect its velocity? Explain.

9.9 Describe two methods for ionizing liquid samples for mass spectral analysis.

9.10 Diagram a transmission quadrupole mass analyzer. What are the advantages and disadvantages of this mass analyzer compared with a double focus mass spectrometer?

9.11 What resolution is necessary to differentiate between
    (a) an ion with $m/z = 84.0073$ and one with $m/z = 84.0085$
    (b) $ArCl^+$ and As^+
    (c) $C_4H_6O_2^+$ ($m/z = 86.0367$) and $C_5H_{10}O^+$ ($m/z = 86.0731$)

9.12 How would you expect the EI and CI mass spectra of stearyl amine (MW = 269) to differ? If the CI mass spectrum is collected using methane as a reagent gas, would you expect to see any peaks in the spectrum at $m/z > 269$? Explain.

9.13 Diagram the process that occurs in MALDI. What are the advantages of this technique?

9.14 Diagram a simple TOF mass spectrometer and explain how mass separation is achieved in a TOF mass analyzer. What is the advantage to a reflectron TOF over a TOF with a straight drift tube?

9.15 Explain the steps in an MS-MS experiment. Why are MS^n experiments valuable?

9.16 Diagram an ESI source and describe its uses.

9.17 Diagram a discrete-dynode electron multiplier and describe its operation.

9.18 Diagram a CEM and describe its operation.

9.19 What is meant by "gain" in an EM?

9.20 Compare the advantages and disadvantages of the Faraday cup detector and electron multipliers.

9.21 What is meant by FTMS? What is the most common type of FTMS instrument?

9.22 Compare the principles of operation, advantages, and disadvantages of the QIT and the ICR. How are they used in MS^n experiments?

10

Mass Spectrometry II: Spectral Interpretation and Applications

J.J. Thompson in 1913 first used MS to demonstrate that neon gas consisted of a mixture of nonradioactive isotopes, ^{20}Ne and ^{22}Ne. The atomic weight of neon listed in a modern periodic table is 20.18. Thompson obtained two peaks in the mass spectrum of neon, at masses of 20 and 22 with a roughly 10:1 intensity ratio, but no peak at mass 20.18. This work was revolutionary because it demonstrated that elements existed as isotopes with different atomic weights and simultaneously explained why the apparent atomic weight of an element based on chemical reactions was not a whole number. Neon in fact has three natural isotopes, but ^{21}Ne is present in much smaller amounts than the other two isotopes. In 1923, Francis W. Aston used a higher resolution instrument he designed to determine the atomic weights of the elements and the isotope ratios of each particular element. This was extremely useful to inorganic chemists and helped solve many of the problems concerning the position of elements in the periodic table at that time. During World War II, Nier at the University of Minnesota developed the high-resolution double-focusing instrument that permitted the analysis and separation of ^{235}U from ^{238}U, aiding in the development of the atomic bomb.

In the 1940s, the first commercial mass spectrometers were developed for petroleum analysis. Subsequent instrument developments, many of them only in the past decade, have led to the widespread use of MS in many branches of science. It has been estimated (Busch) that a billion mass spectra are recorded daily.

MS is a powerful analytical tool with vast applications in organic, inorganic, environmental, polymer and physical chemistry, physics, geology, climatology, paleontology, archaeology, materials science, biology, and medicine. Advances in MS instrumentation have made possible major advances in our understanding of the human genome, protein structure, and drug metabolism. For example, intact viruses of millions of daltons have been analyzed by MS using ESI with retention of viral activity and structure (Fuerstenau et al.). Commercial hyphenated GC-MS^n and LC-MS^n systems permit rapid, sensitive biomonitoring of humans for exposure to chemicals, including chemicals used by terrorists. It is impossible to cover all applications of MS in one chapter, but examples of important uses of both molecular and atomic mass spectrometry will be presented. In addition, this chapter introduces the interpretation of simple mass spectra for the identification of molecules.

10.1. INTERPRETATION OF MASS SPECTRA: STRUCTURAL DETERMINATION OF SIMPLE MOLECULES

The major reasons for learning to interpret mass spectra of molecules are so that the structure of an unknown compound can be deduced and an unknown molecule can be completely and unambiguously identified. For even fairly small organic molecules, a thorough knowledge of physical organic chemistry and lots of practice are required to do this. Fortunately, simple mass spectra often can be interpreted using basic arithmetic, and some topics you learned in general chemistry: chemical bonding, valency, isotopes, atomic weights, and molecular weight calculations. For completely unknown molecules, more than one mass spectral technique may need to be combined with other analytical techniques to positively identify a molecule. As will be seen, there are some types of molecules that cannot be identified using MS alone.

A mass spectrum is a plot or table of the mass-to-charge ratio, m/z, of detected ions vs. their relative abundance (relative concentration). A typical mass spectral plot for a small organic molecule, benzene, is presented in Fig. 10.1. The m/z values are plotted on the x-axis; relative abundance is plotted on the y-axis. The *most abundant peak* in the spectrum is called the **base peak**. The base peak is assigned an abundance of 100% and the other peak heights are plotted as percentages of that base peak. A tabular form of benzene mass spectral data is given in Table 10.1. The tabular data have the advantage that very low abundance ions can be listed, such as the ions at $m/z = 64$ and 80 which are too small to be seen on the normalized plot.

The mass spectra we will study have been obtained by electron ionization (EI) since other techniques enhance the molecular ion, but greatly reduce fragmentation, making structural analysis difficult. Consequently, the following discussion is concerned mostly with EI spectra of pure compounds. The ions that appear in these spectra are all positively charged ions. These are also low-resolution mass spectra; the m/z values are measured

Figure 10.1   A mass spectrum of benzene, C_6H_6. [From NIST Mass Spec Data Center, S.E. Stein, director (http://webbook.nist.gov), used with permission.]

Table 10.1 Mass Spectral Data for Benzene

m/z	Relative abundance	m/z	Relative abundance
37	4.0	53	0.80
37.5	1.2	63	2.9
38	5.4	64	0.17
38.5	0.35	73	1.5
39	13	74	4.3
39.5	0.19	75	1.7
40	0.37	76	6.0
48	0.29	77	14
49	2.7	78	100
50	16	79	6.4
51	18	80	0.18
52	19		

only to the units place. (We will explain the half-integral values in Table 10.1 later in the chapter.) Exact molecular weights, which are not integer values, are obtained by high-resolution MS, also discussed later in the chapter.

There are two ways to interpret such spectra. The first is to compare the spectrum you have with those in a searchable computerized mass spectral database. The second is to evaluate the spectrum using the interpretation procedure described subsequently. In either case, once an unknown compound has been "identified" from its mass spectrum, the pure compound should be obtained and analyzed under the same conditions as the sample for confirmation. Over 10 million chemical compounds have been identified. No mass spectral database contains spectra for every possible compound, although mass spectral databases of over 400,000 spectra are available. The mass spectral database from the US National Institute for Standards and Technology (NIST) may be purchased from a number of licensed vendors. Limited mass spectra from NIST are available online in the NIST Chemistry WebBook (http://webbook.nist.gov). Commercial vendors and publishers offer specialized mass spectral libraries of compounds, such as environmental compounds, pharmaceuticals, natural products, oil industry compounds, and the like.

In practice, the analyte spectrum is entered into the computer, which compares it to the spectra in the stored database using a search algorithm. There are a number of algorithms currently available, including Probability Based Matching, designed by Professor F.W. McLafferty and coworkers at Cornell University, the INCOS dot-product algorithm, and the NIST library search algorithm. These algorithms use pattern matching, peak intensities, weighting factors, and other information from the spectrum to compare the candidate spectrum to spectra in the library database. The search will result in a list of possible candidate compounds with a probability attached to the "match". The analyst should visually compare the candidate spectra to that of the analyte. Using knowledge, judgment, and experience, the analyst then chooses which of the candidate compounds is the unknown. This spectral matching method in theory requires little training on the part of the analyst to identify the compound, but requires pure compounds, a good mass spectrum of the sample, and a comprehensive mass spectral database. The technique of identifying a pure compound by comparing its spectrum with known spectra works well if we already know something about the compound (e.g., odor, color, melting point, functional groups present from an IR spectrum, elements present from combustion analysis) and if the compound's mass spectrum is in our database. This enables us to make an informed deduction as to the identity of the compound by direct comparison with the library spectrum. Unfortunately, this is not always the case; more frequently, the nature of the sample is not known. Furthermore, the sample may not be pure, and therefore direct comparison of spectra will not constitute a valid confirmation of the compound's identity. It is possible that spectra in a database are not identified correctly. It is not prudent to rely completely on a library match, especially for a complete unknown, since there are many cases of multiple compounds with very similar mass spectra. In these circumstances it is necessary to take the spectrum and deduce the information from it. Therefore, an analyst should understand how to identify the compound from the information in the mass spectrum; this ability will also enable the analyst to evaluate the validity of the spectral database search results.

The procedure for interpreting a mass spectrum consists of the following steps:

Identify the molecular ion if present
Apply the "nitrogen rule"
Evaluate for "A + 2" elements
Calculate "A + 1" and "A" elements
Look for reasonable loss peaks from the molecular ion
Look for characteristic low mass fragments
Postulate a possible formula
Calculate "rings + double bonds"
Postulate a reasonable structure

We will define the terms in quotation marks as we go through the process. Interpretation of mass spectra is much like solving a jigsaw puzzle: all or most of the pieces are there, but putting them together involves a lot of educated guesswork followed by confirmation.

10.1.1. The Molecular Ion and Fragmentation Patterns

When a molecule is ionized by EI, a **molecular ion** forms by loss of an electron from the molecule. This radical cation, symbolized $M^{\bullet+}$, has the same mass as the neutral molecule, because the loss in mass of one electron is too small to measure. The molecular ion generally has absorbed excess energy in the ionization process. The excess energy causes the molecular ion to break apart. The fragments formed may be ions, neutral molecules, radicals, and the like. Fragments may undergo more fragmentation into even smaller pieces. The ions that appear in the mass spectrum are called fragment ions. If the ionization conditions are kept constant, a given molecule will always produce the same fragments. The mass and abundance of these fragments are called the molecule's **fragmentation pattern**. Each molecule has its own characteristic fragmentation pattern under particular ionization conditions.

The molecular ion, if present, will provide the molecular weight of the compound to the units place, a very valuable piece of information. Identification of the molecular ion is important because from the molecular weight we can immediately conclude a great deal about the compound. For example, for an organic compound with a molecular weight $= 54$, we would be able to conclude that the compound can only have a limited number of empirical formulas, such as C_2NO, $C_2H_2N_2$, C_3H_2O, C_3H_4N, or C_4H_6. Some of these possibilities can be eliminated by inspection, because it is not possible to have a neutral molecule with certain empirical formulas. For example, C_2NO cannot exist as a neutral molecule. Thus, if we know the molecular weight, we have considerable information about the possible empirical formula of the molecule. Mass spectrometry is one of the most rapid and reliable methods of determining the molecular weight of compounds. Unfortunately, in EI spectra, the molecular ion is often of low abundance and is not always observed. In short, the m/z value of the molecular ion indicates the molecular weight of the molecule; the fragmentation pattern is used to deduce the structure of the molecule. The molecular ion is not necessarily the base peak in the spectrum; particularly in EI spectra, the base peak is usually not the molecular ion.

In this text, we shall emphasize compounds with molecular ions that can be identified or deduced with reasonable certainty. If the molecular ion is present, it must have the highest m/z in the spectrum, excluding the effects of isotopes. Examples are shown for methane (Fig. 10.2, Table 10.2), methanol (Fig. 10.3, Table 10.3), and benzene (Fig. 10.4, Table 10.1). In each case, the molecular ion was very abundant and not difficult to identify. This is not always the case, as will be seen in later examples. The student should note that in most of the mass spectra used as examples, the molecular ion m/z value is marked by a black triangle on the x-axis. The x-axis is in units of m/z while the y-axis is relative abundance, even though these units are not marked on the spectra. The most intense peak is set to 100% and the rest of the peaks normalized to that peak. The structure of the compound is also shown on the spectrum, using a shorthand method that does not show the hydrogen atoms or the carbon atoms. The methanol spectrum (Fig. 10.3) demonstrates clearly that the molecular ion is not the base peak in the spectrum; the fragment ion at $m/z = 31$ is the most abundant ion.

When a molecule is ionized by electron impact, it undergoes the reaction:

$$M + e^- \longrightarrow M^{\bullet+} + e^- + e^-$$

Thus, a molecular ion is always a radical cation, usually with a single positive charge. It is evident that if an organic molecule loses an electron, it must be left with an unpaired electron (i.e., it is a radical ion), as shown by the dot representing the unpaired electron. This radical ion with an unpaired electron is termed an *odd electron ion. The molecular ion is always an odd electron ion.* For example, in the reaction $C_2H_6 + e^- \rightarrow C_2H_6^{\bullet+} + e^- + e^-$, the $C_2H_6^{\bullet+}$ is the molecular ion.

However, the molecule may fragment in such a way as to leave a pair of electrons behind on one ion. This is an *even electron ion.* Such an ion arises by *fragmentation* and *cannot* be the molecular ion. Hence, the molecular ion is *never* an even electron ion. As an example, CH_3OH may ionize and fragment as follows:

$$CH_3OH + e^- \longrightarrow CH_3OH^{\bullet+} \ (m/z = 32) + e^- + e^-$$
$$CH_3OH^{\bullet+} \longrightarrow CH_2OH^+ \ (m/z = 31) + H^{\bullet}$$

The CH_2OH^+ ion is an even electron ion and cannot be the molecular ion.

Figure 10.2 A mass spectrum of methane, CH_4.

Table 10.2   Mass Spectral Data
for Methane

m/z	Relative abundance
1	3.1
2	0.17
12	1.0
13	3.9
14	9.2
15	85
16	100
17	1.11

While a detailed discussion of fragmentation patterns and how to identify a molecular ion peak is beyond the scope of this text, there are certain classes of organic compounds that tend to give stable molecular ions and classes that do not. Aromatic compounds and conjugated hydrocarbons give more intense molecular ion peaks than alkanes; aliphatic alcohols, nitrates, and highly branched compounds tend not to give peaks for the molecular ion. Texts listed in the bibliography such as those by Silverstein and Webster, McLafferty and Tureček, Pavia et al., or Lambert et al. should be consulted for more detail.

10.1.2. The Nitrogen Rule

The "nitrogen rule" can be used to help decide if a peak is molecular ion peak. Because of the atomic weights and valences of elements commonly present in organic molecules, it transpires that the m/z value of the molecular ion *is always an even number if the molecular ion contains either no nitrogen atoms or an even number of nitrogen atoms.* If the molecular ion contains an odd number of nitrogen atoms, the m/z value (and the molecular weight) must be an odd number. This is a very important rule to remember and it is valid for organic compounds containing C, H, N, O, S, the halogens, P, Si and many other elements. If the highest m/z value in a mass spectrum is an odd number, the ion must contain an odd number of nitrogen atoms if it is the molecular ion. To use this rule, molecular weights are calculated using the most abundant isotope and unit

Figure 10.3   A mass spectrum of methanol, CH_3OH.

Table 10.3 Mass Spectral Data
for Methanol

m/z	Relative abundance
12	0.3
13	0.7
14	2.4
15	13.0
16	0.2
17	1.0
28	6.3
29	64.0
30	3.8
31	100
32	66
33	1.0
34	0.14

atomic masses. This calculation is discussed in the next section. Some examples of the rule
follow:

Compound	Number of nitrogen atoms	MW	
CH_4	0	16	Even
HCN	1	27	Odd
H_2NNH_2	2	32	Even
C_2H_5OH	0	46	Even
C_6H_6	0	78	Even
$C_2H_5NH_2$	1	45	Odd
$C_6H_5NH_2$	1	93	Odd
$C_6H_4(NH_2)_2$	2	108	Even
C_9H_7N	1	129	Odd

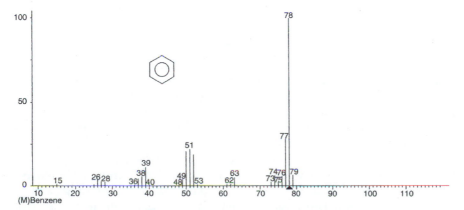

Figure 10.4 Another mass spectrum of benzene, C_6H_6. Compare with Fig. 10.1.

10.1.3. Molecular Formulae and Isotopic Abundances

If you look at the mass spectrum for methane (Fig. 10.2) and the corresponding Table 10.2, you note from the marked peak on the spectrum that the molecular ion is at $m/z = 16$. The molecular weight to the units place for methane is, of course, equal to the atomic weight of carbon plus four times the atomic weight of hydrogen, or $(1)(12) + (4)(1) = 16$. The table shows the presence of an additional small peak at $m/z = 17$. The peak at 17 exists because the element carbon has more than one isotope. The two stable isotopes of carbon are ^{12}C and ^{13}C. ^{12}C constitutes 98.90% of all naturally occurring carbon atoms; ^{13}C is 1.10% of all naturally occurring carbon. (A third carbon isotope, ^{14}C, is radioactive and does not need to be accounted for in organic structure determination, but is very important for other applications, such as radiocarbon dating.) Therefore, in a compound containing one carbon atom, 98.90% of the molecules will have a ^{12}C atom while 1.10% will have a ^{13}C atom.

In every organic compound (natural or synthesized from natural sources), the ^{13}C will result in a peak that is one mass number greater than the mass of the molecular ion. This ^{13}C-containing peak is generally designated as M + 1. For example, with methane, CH_4, the molecular weight is 16, assuming that carbon has an atomic weight of 12 (i.e., $^{12}C^1H_4 = 12 + 4 = 16$). Let the abundance of this peak be 100. But the abundance of ^{13}C is 1.1% of ^{12}C, therefore $^{13}C^1H_4 = 13 + 4 = 17$, and its abundance is 1.1% of $^{12}C^1H_4$, as shown:

Peak designation	Isotopic formula	MW	Rel. abundance
M	$^{12}C^1H_4$	16	100
M + 1	$^{13}C^1H_4$	17	1.1

Similarly, for ethane, C_2H_6, which has two carbon atoms, we can designate the MS pattern as:

Peak designation	Isotopic formula	MW	Rel. abundance
M	$^{12}CH_3{}^{12}CH_3$	30	100
M + 1	$^{12}CH_3{}^{13}CH_3$	31	1.1
M + 1	$^{13}CH_3{}^{12}CH_3$	31	1.1
Total M + 1		31	2.2
M + 2	$^{13}CH_3{}^{13}CH_3$	32	0.012

With C_2H_6, we have $^{12}C_A{}^{12}C_B{}^1H_6$ for a molecular weight of 30 and a corresponding $m/z = 30$ for the molecular ion. But we also will have ions of $^{13}C_A{}^{12}C_B{}^1H_6$ ($m/z = 31$) and $^{12}C_A{}^{13}C_B{}^1H_6$ ($m/z = 31$). Both combinations are equally possible. If the abundance at $m/z = 30$ is 100, then $^{13}C^{12}C^1H_6$ is 1.1% of that and $^{12}C^{13}C^1H_6{}31$ is also 1.1% of that, so the total abundance at $m/z = 31$ is 2.2% of $m/z = 30$. The small peak at two mass units above the molecular ion peak occurs when both carbon atoms are ^{13}C atoms.

If our molecule contains only carbon and hydrogen, since there is 1.1% relative natural abundance of ^{13}C compared to 100% ^{12}C, and our mass spectrum shows a 1.1% abundance of M + 1 to M, only one carbon atom can be present in the molecule. That is, if we are dealing with a hydrocarbon and only one carbon atom is present, then the ratio $(M + 1)/M = 1.1\%$. If two carbons are present in the molecule, as there are in ethane, the probability of ^{13}C being present is twice as great and $(M + 1)/M = 2.2\%$.

10.1.3.1. Counting Carbon Atoms

There is a definite relationship between the peaks at $M + 1$ and M that is directly related to the number of carbons present in hydrocarbon molecules:

$$\frac{M + 1}{M} = 1.1\% \times \text{number of C atoms in the molecule} \tag{10.1}$$

This relationship is very valuable in characterizing an unknown compound. The calculation just presented is for hydrocarbons (compounds containing only carbon and hydrogen) and ignores the contribution from the ^{2}H isotope of hydrogen, called deuterium. In rigorous calculations, the contribution from deuterium would be taken into account.

For benzene, C_6H_6, (Table 10.1), the molecular ion $m/z = 78$, and the $(M + 1)$ $m/z = 79$. The relative abundance of the $(M + 1)$ peak is 6.6%; dividing 6.6% by 1.1% confirms that there are 6 carbons in the molecule.

Using a similar process for C_3H_8, $MW = 44$, the abundance of an M+1 peak at $m/z = 45$ will be predicted to be $3 \times 1.1\%$ of the abundance at $m/z = 44$. The same calculation can be made for any hydrocarbon.

A list of common isotopes found in organic compounds and their relative abundance in nature is given in Table 10.4. A complete table of the relative abundance of natural isotopes for all elements is located in Appendix 10.1. As can be seen in Table 10.4, the natural abundance of deuterium is only 0.016% of the ^{1}H abundance, so it can usually be ignored in calculations.

However, there are a number of common isotopes that we cannot ignore for many organic molecules, such as those of oxygen, nitrogen, sulfur, and the halogens. This gives rise to the question of what atomic weights to use when calculating molecular weights for MS so that we have the correct mass for the molecular ion.

There are three expressions for mass that are important to know. The **integer mass** of the most abundant naturally occurring stable isotope is the **nominal mass** of the element. The **exact mass** of an isotope is determined by high-resolution MS; the exact mass of ^{35}Cl is 34.9689 Da and that for ^{37}Cl is 36.9659 Da. The **atomic weight** is the

Table 10.4   Relative Natural Isotope Abundances of Common Elements in Organic Compounds

Isotope	Mass	Relative abundance	Mass	Relative abundance	Mass	Relative abundance
H	1	100	2	0.016		
C	12	100	13	1.11		
N	14	100	15	0.37		
O	16	100	17	0.04	18	0.20
F	19	100				
Si	28	100	29	5.07	30	3.36
P	31	100				
S	32	100	33	0.78	34	4.39
Cl	35	100			37	32.7
Br	79	100			81	97.5
I	127	100				

average mass of the element. For example, if we have a compound containing chlorine, most periodic tables list the atomic weight of chlorine as 35.45 or 35.5. The molecular weight of CH_2Cl_2 is therefore equal to $12 + 2(1) + 2(35.5) = 85$. But a mass spectrum of dichloromethane does not have an intense peak at $m/z = 85$. The "atomic weight" of chlorine in the periodic table is a weighted average of the exact masses of ^{35}Cl and ^{37}Cl, and ^{35}Cl is about three times as abundant as ^{37}Cl. If chlorine were analyzed by MS, there would be a peak at $m/z = 35$ and a peak about $1/3$ as intense at $m/z = 37$, but no peak at $m/z = 35.5$. For MS, molecular weights are calculated using the most abundant isotope of the element with its mass rounded to the nearest integer value; that is, the nominal mass is used. For most organic compounds the most abundant isotope is the lowest mass isotope. Therefore, the MW of dichloromethane would be calculated as $12 + 2(1) + 2(35) = 84$. There is a significant peak at $m/z = 84$ in the mass spectrum of dichloromethane. At what higher values of m/z would you expect to see peaks in this spectrum based on the isotopes present?

10.1.3.2. Counting Carbon, Nitrogen, and Sulfur Atoms

From Table 10.4, it is clear that not only carbon, but also other elements contribute to the $M + 1$ peak intensity, notably nitrogen and sulfur when they are present. Silicon obviously will impact the $M + 1$ peak in organosilicon compounds such as silicone polymers. A more general equation for the ratio of the $M + 1$ to M peak is:

$$\frac{M + 1}{M} = 1.1(\#C \text{ atoms}) + 0.016(\#H \text{ atoms}) + 0.37(\#N \text{ atoms})$$

$$+ 0.78(\#S \text{ atoms}) + \cdots \qquad (10.2)$$

where #C atoms is the number of carbon atoms in the molecule, and so on. The general expression for compounds containing only C, H, N, and O, and F, P, and I since these have only one isotope, would use only the terms for carbon and nitrogen. It can be seen that the formula is of most use in confirming molecular formulas, since we need to know what elements are present to use the equation effectively. We may also have previous information about what elements are present from combustion analysis; that type of analysis would indicate oxygen, nitrogen, or sulfur in the molecule, for example. The contribution from hydrogen is small enough that it is usually ignored in calculations.

10.1.3.3. Counting Oxygen Atoms

Similarly, the $M + 2$ peak contains intensity contributions from oxygen, Cl and Br, and multiple heavy isotopes of carbon, hydrogen, and so on. A general formula for compounds that contain only C, H, N, O, F, P, and I can be used to ascertain the number of oxygen atoms, in theory. Examination of Table 10.4 shows that oxygen has two important isotopes, ^{16}O and ^{18}O, separated by 2 Da, with a relative abundance $^{18}O/^{16}O$ of 0.2%. The number of oxygen atoms in a molecular ion can be calculated from:

$$\frac{M + 2}{M} = 0.20(\#O \text{ atoms}) + \frac{(1.1(\#C \text{ atoms}))^2}{200} \qquad (10.3)$$

This equation again ignores the contribution from hydrogen. Compounds containing the halogens Cl and Br will be discussed later in the chapter. Iodine does not affect the equation, since iodine is monoisotopic.

For example, the mass spectrum of methanol, CH_3OH, is shown in Fig. 10.3 and the mass spectral data for methanol is shown in Table 10.3. Figure 10.3 notes with the triangle

Figure 10.5   A mass spectrum of water, H_2O.

that the molecular ion is at $m/z = 32$. (This can be deduced with practice; aliphatic alcohols often fragment by loss of a proton or water. Is there an M − 18 peak in the methanol spectrum corresponding to loss of water?) The ratio of $(M + 1)/M$, the $m/z = 33$ peak to that at $m/z = 32$, is $1.0/66$ or 1.5%, indicating not more than one carbon atom. Also, the $(M + 2)/M$ ratio, $m/z\ 34/32$, is 0.21%; inserting 1 carbon atom and 0.21 for the ratio into Eq. (10.3) shows us that only one oxygen atom is present in the molecular ion.

There are some other interesting points in the methanol spectrum. For instance, the most abundant peak (the base peak) is at $m/z = 31$, indicating that methanol very easily loses one hydrogen atom and forms CH_3O. Another abundant peak is at $m/z = 29$, indicating that methanol loses three hydrogen atoms to form COH. The next most abundant peak is at 15, which is equivalent to a CH_3 ion, indicating breaking of the OH bond to the methyl group. The fragment that would correspond to the hydroxyl ion would be at $m/z = 17$; it is not abundant, indicating that if OH ions are formed, they do not reach the detector. Since we are detecting positively charged ions in EI MS, is this surprising?

The mass spectrum of water is shown in Fig. 10.5 and Table 10.5. To calculate the number of carbon and oxygen atoms in a molecule, we assume that only the elements C, H, N, O, F, P, and I are present, and that there are no interferences from other ions. The molecular ion $^1H_2^{16}O$ is at $m/z = 18$, but there is also a peak listed in the table at $m/z = 20$, the M+2 peak caused by $^1H_2^{18}O$. Its abundance is approximately 0.2% of the abundance of 18 ($^1H_2^{16}O$) confirming that only one oxygen atom is present in the molecular ion. The M + 1 peak is only 0.08% of the M peak, which does not allow for a carbon atom in the molecular ion.

Table 10.5   Mass Spectral Data for H_2O

m/z	Relative abundance
1	<0.1
16	1.0
17	21
18	100
19	0.08
20	0.22

While Eq. (10.3) can be used, in theory, to calculate the number of oxygen atoms in a compound, in reality the ^{18}O isotope is of low abundance and oxygen-containing compounds often fragment so that the molecular ion is of low intensity or not detected. In practice, the observation of isotope information for oxygen is often difficult.

10.1.4. Compounds with Heteroatoms

The elements shown in Table 10.4 can be classified into three categories, following the recommended nomenclature of McLafferty. "A" elements are those that are monoisotopic, such as F, P, and I. Hydrogen is also classified as an A element, because the natural abundance of deuterium is so low. "A + 1" elements are those with two isotopes, one heavier than the most abundant isotope by 1 Da. Carbon and nitrogen are A + 1 elements. "A + 2" elements have an isotope that is 2 Da heavier than the most abundant isotope. Cl, Br, and O are the three most common of these; S and Si are also A + 2 elements that must be considered if their presence is detected by elemental combustion analysis or other analytical methods.

The A + 2 elements are the easiest to recognize, so they should be looked for first when presented with an unknown mass spectrum. Oxygen has already been discussed, so look at the isotope ratios of Cl and Br from Table 10.4. Cl has two isotopes, ^{35}Cl and ^{37}Cl, in a ratio of 100:33. A compound that contains 1 Cl atom would have an M + 2 peak $\sim 1/3$ as intense as the M peak. Br has two isotopes, ^{79}Br and ^{81}Br, in a ratio of $\sim 1:1$. A compound with 1 Br atom would have approximately equal M and M + 2 peaks. These are very distinctive patterns.

Figure 10.6 and Table 10.6 show two strong peaks at $m/z = 38$ (M + 2) and 36 (M) in an abundance ratio of $\sim 1/3$. (We know that the peak at $m/z = 36$ is the M peak because it is marked by the triangle.) This is typical of the pattern seen when 1 Cl atom is present. If a chlorine atom is present, then $m/z = 36$ arises from ^{35}Cl plus one ^{1}H atom (only ^{1}H could increase the mass from 35 to 36). Then $m/z = 38$ would be due to $^{37}Cl^{1}H$. This is the mass spectrum of hydrochloric acid, HCl. The peaks at $m/z = 35$ and 37 are of course due to the ionized chorine isotopes which have lost the hydrogen through fragmentation. The ratio of the m/z 37/35 peaks is also 1/3, supporting the identification of a single Cl atom.

Figure 10.6 A mass spectrum of hydrogen chloride, HCl, showing the characteristic 3:1 ratio of the M/(M + 2) peaks for a compound containing a single Cl atom.

Table 10.6 Mass Spectral Data for HCl

m/z	Relative abundance
35	12
36	100
37	4.1
38	33
39	0.01

For organochlorine compounds with one Cl atom, such as CH_3Cl, we will expect to see a peak at $M + 2$ because there will be ^{37}Cl present. The ratio $(M + 2)/M$ will be the same as the natural ratio of ^{37}Cl to ^{35}Cl, that is, 32.7%, resulting in the same $1/3$ pattern seen in HCl.

Looking at Fig. 10.7 and Table 10.7, the almost equal intensity of the two peaks at $m/z = 94$ and 96 is characteristic of the presence of 1 bromine atom. The presence of 1 Br atom is confirmed by the peaks at $m/z = 79$ and 81 in approximately equal abundance. These two peaks can reasonably be attributed to ^{79}Br and ^{81}Br ions. If the molecular weight is 94 as indicated by the identification of $m/z = 94$ as the molecular ion (note the triangle), a mass of ^{79}Br subtracted from 94 shows that atoms equivalent to a group of mass 15 must be present. This is most likely a methyl group, CH_3-, confirmed by an abundant peak at $m/z = 15$. Based on the m/z 16/15 ratio, which is 0.62/ $59 = 1.05\%$, the ion of $m/z = 15$ contains 1 C and is probably CH_3. Consecutive peaks at masses of 13, 14, 13, 12 confirm 3 hydrogen atoms, ending in a C ion with no H attached. Hence, the molecule is most likely composed of a methyl group and one Br atom: $Br + CH_3$ or CH_3Br. This is bromomethane (also called methyl bromide).

10.1.5. Halogen Isotopic Clusters

If more than one chlorine atom or bromine atom is present in a molecule, distinctive "isotope cluster patterns" are seen in the mass spectrum. Figure 10.8 gives a graphical

Figure 10.7 A mass spectrum of bromomethane (or methyl bromide), CH_3Br, showing the characteristic 1:1 ratio of the $M/(M + 2)$ peaks for a compound containing a single Br atom.

Table 10.7 Mass Spectral Data for CH_3Br

m/z	Relative abundance	m/z	Relative abundance
12	1.2	48	0.95
13	1.4	79	10
14	3.8	81	10
15	59	91	4.2
16	0.62	92	2.4
39.5	0.19	93	6.8
40.5	0.20	94	100
46	1.3	95	0.9
46.5	0.30	96	96
47	2.3	97	1.1
47.5	0.28		

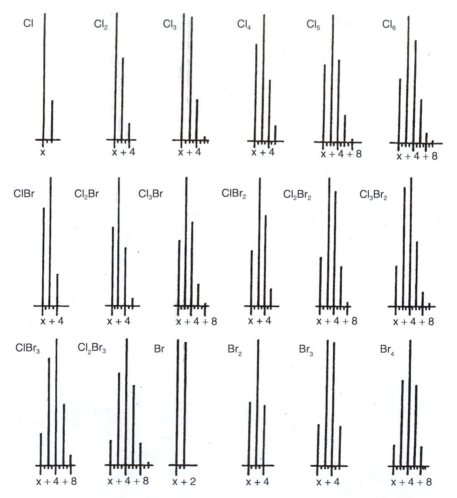

Figure 10.8 Graphical representation of relative isotope peak intensities for any given ion containing the indicated number of chlorine and/or bromine atoms. The numeric values are given in Table 10.8. (From Sparkman, used with permission.)

Table 10.8 Chlorine and Bromine Isotope Abundance Ratios

Atoms of Cl and Br	X	X + 2	X + 4	X + 6	X + 8	X + 10
Cl	100	32.5				
Cl_2	100	65.0	10.6			
Cl_3	100	97.5	31.7	3.4		
Cl_4	76.9	100	48.7	0.5	0.9	
Cl_5	61.5	100	65.0	21.1	3.4	0.2
Cl_6	51.2	100	81.2	35.2	8.5	1.1
ClBr	76.6	100	24.4			
Cl_2Br	61.4	100	45.6	6.6		
Cl_3Br	51.2	100	65.0	17.6	1.7	
$ClBr_2$	43.8	100	69.9	13.7		
Cl_2Br_2	38.3	100	89.7	31.9	3.9	
Cl_3Br_2	31.3	92.0	100	49.9	11.6	1.0
$ClBr_3$	26.1	85.1	100	48.9	8.0	
Cl_2Br_3	20.4	73.3	100	63.8	18.7	2.0
Br	100	98				
Br_2	51.0	100	49.0			
Br_3	34.0	100	98.0	32.0		
Br_4	17.4	68.0	100	65.3	16.0	

Source: From Sparkman, used with permission.

representation of the isotope peak intensity patterns for Cl and Br. The numerical values for the peak ratios are given in Table 10.8. The patterns arise as follows.

If we have two Cl atoms present in an ion, the distribution of masses in each ion may be $^{35}Cl^{35}Cl$, $^{37}Cl^{35}Cl$, $^{35}Cl^{37}Cl$, or $^{37}Cl^{37}Cl$. The $^{35}Cl:^{37}Cl$ abundance is 100:33, so the probability of these isotope distributions occurring is approximately 100:66:11.

Consequently, any molecule RCl_2 containing two Cl atoms will exhibit masses of R + 70, R + 72, and R + 74 with a relative abundance due to Cl isotopes of 100:66:11, due to the presence of $R^{35}Cl^{35}Cl$, $R^{37}Cl^{35}Cl$, $R^{35}Cl^{37}Cl$, and $R^{37}Cl^{37}Cl$. This pattern in the mass spectrum is a Cl_2 *isotope cluster*. The same pattern will be seen not only for the molecular ion, but also for any fragment ion that contains both Cl atoms. Similarly, if three chlorine atoms are present (RCl_3), a cluster will occur with masses R + 105, R + 107, R + 109, and R + 111 with approximate abundances of 100:98:32:3.

Similar sets of clusters can be determined for bromine compounds and for compounds containing both Br and Cl. Bromine has two isotopes, ^{79}Br and ^{81}Br, and they are approximately equal in abundance. A compound RBr containing 1 Br atom will have masses of R + 79 and R + 81, and they will be about equal in abundance. If 2 Br atoms are present, as in RBr_2, there will be masses of R + 158, R + 160, and R + 162 from $R^{79}Br^{79}Br$, $R^{79}Br^{81}Br$, $R^{81}Br^{79}Br$, $R^{81}Br^{81}Br$, and their relative abundances will be 51:100:49, based purely on probability.

The mass spectrum in Fig. 10.9, shows a molecular ion at $m/z = 130$, with a pattern of peaks at m/z values of 130, 132, 134, and 136. Looking at the patterns shown in Fig. 10.8, this group of four peaks most closely resembles the pattern from 3 Cl atoms. Using a ruler to measure the peak heights, and dividing all four peak heights by the largest (the peak at 130) gives a 100:95:31:3 ratio, very close to the values predicted in Table 10.8. There is another cluster of peaks at $m/z = 95$, 97, and 99. Using Fig. 10.8, it is apparent that this cluster is due to a fragment with only 2 Cl atoms; and indeed the

Figure 10.9   A mass spectrum of trichloroethylene, C_2HCl_3, showing the typical isotope cluster pattern for a compound containing 3 Cl atoms.

difference between 130 and 95 is equal to the mass of one ^{35}Cl atom. Similarly, the pair of peaks at m/z 60 and 62 is due to a fragment containing only one Cl atom, as shown by the 3:1 peak height ratio. Three ^{35}Cl atoms have a mass of 105, leaving a mass of $130 - 105 = 25$ to be accounted for in the molecule. The MW is an even number, so there are no nitrogen atoms in the molecule (two nitrogen atoms would have a mass of 28, more than is possible based on the molecular ion mass). The peak ratio m/z 131/ 130 is about 2.3%, indicating two carbon atoms at most. Two carbon atoms (total mass 24) would leave room for only one H atom. Putting together 2 C atoms, one H atom and three Cl atoms gives a molecular formula of C_2HCl_3. A reasonable structure is $HClC{=}CCl_2$, the structure is shown in Fig. 10.9 (the hydrogen atom is not shown, but would be bonded to the carbon atom on the left). This compound is trichloroethylene, a commercial solvent.

10.1.6.   Rings Plus Double Bonds

The formula for *n*-hexane is C_6H_{14}. The formula for cyclohexane is C_6H_{12}, the same as the formula for hexene. The presence of a ring or double bond (db) resulted in a change in the ratio of carbon to hydrogen. This change is a general property and based on it the following formula is derived.

The number of rings plus double bonds in a molecule of formula $C_xH_yN_zO_m$ is

$$x - \frac{1}{2}y + \frac{1}{2}z + 1 \tag{10.4}$$

For example, for *n*-hexane, $x = 6$, $y = 14$, so the number of rings + db is $6 - (1/2 \times 14) + 0 + 1 = 0$. For cyclohexane, $x = 6$, $y = 12$, therefore rings + db = $6 - (1/2 \times 12) + 0 + 1 = 1$. For benzene, C_6H_6, rings + db = $6 - (1/2 \times 6) + 0 + 1 = 4$, i.e., 3 double bonds plus 1 ring. Acetylene, C_2H_2, contains a triple bond between the two carbon atoms. Equation (10.4) applied to acetylene yields $2 - (1/2 \times 2) + 0 + 1 = 2$. A triple bond is equivalent to two double bonds. The equation does not tell us whether we are dealing with a ring or a double bond or 2 double bonds vs. a triple bond, or, as is the case with benzene, a delocalized molecular orbital that can be represented for bonding purposes as three double bonds. We need to use the equation in conjunction with the molecular formula (and any other information we have from IR, NMR, etc.) to postulate a reasonable structure.

When atoms other than CHNO are present, the heteroatoms are matched according to their lowest absolute valence. Si takes 4 bonds, like C; P is equivalent to N, while the halides, with a lowest absolute valence of 1, act like H. Hence the general equation is:

$$\text{Rings and double bonds} = x - \frac{1}{2}y + \frac{1}{2}z + 1 \tag{10.5}$$

where $x = $ C, Si; $y = $ H, F, Cl, Br, I; and $z = $ N, P.

10.1.7. Common Mass Losses on Fragmentation

When a molecule ionizes it follows the reaction $M + e^- \rightarrow M^{\bullet +} + e^- + e$, as described earlier. When an odd electron molecular ion fragments it can do so in variety of ways, and resulting odd electron fragment ions can also fragment in similar ways. Simple cleavage results in the expulsion of a neutral fragment that is an odd electron radical ($OE^\bullet$) and formation of an even electron ion (EE^+) as shown:

$$M^{\bullet +} \longrightarrow EE^+ + OE^\bullet$$

Multicentered fragmentation results in the expulsion of an even electron neutral species (EE) and formation of an odd electron ion ($OE^{\bullet +}$):

$$M^{\bullet +} \longrightarrow OE^{\bullet +} + EE$$

In the first case we would detect the EE^+ ion and in the second case the $OE^{\bullet +}$ ion. The neutral species lost on fragmentation can be identified by the difference between the mass of $M^{\bullet +}$ and the fragment ion. Common losses from the molecular ion are given in Table 10.9. Tables 10.10 and 10.11 list more common neutral fragments expelled by simple cleavage and by multicentered fragmentations.

10.2. MASS SPECTRAL INTERPRETATION: SOME EXAMPLES

The general appearance of the mass spectrum depends on the type of compound analyzed. Seeing the patterns that distinguish, say, a normal alkane from an aromatic hydrocarbon,

Table 10.9 Common Neutral Losses from the Molecular Ion

M − 1	Loss of hydrogen radical	$M - {}^\bullet H$
M − 15	Loss of methyl radical	$M - {}^\bullet CH_3$
M − 29	Loss of ethyl radical	$M - {}^\bullet CH_2CH_3$
M − 31	Loss of methoxyl radical	$M - {}^\bullet OCH_3$
M − 43	Loss of propyl radical	$M - {}^\bullet CH_2CH_2CH_3$
M − 45	Loss of ethoxyl radical	$M - {}^\bullet OCH_2CH_3$
M − 57	Loss of butyl radical	$M - {}^\bullet CH_2CH_2CH_2CH_3$
M − 2	Loss of hydrogen	$M - H_2$
M − 18	Loss of water	$M - H_2O$
M − 28	Loss of CO or ethylene	$M - CO; M - C_2H_4$
M − 32	Loss of methanol	$M - CH_3OH$
M − 44	Loss of CO_2	$M - CO_2$
M − 60	Loss of acetic acid	$M - CH_3CO_2H$

Table 10.10 Some Neutral Fragments Expelled by Simple Cleavage

Mass	Fragment (OE$^{\bullet}$)	Mass	Fragment (OE$^{\bullet}$)
1	$^{\bullet}H$	46	$^{\bullet}NO_2$
15	$^{\bullet}CH_3$	47	$^{\bullet}CH_2SH$
16	$^{\bullet}NH_2$	57	$^{\bullet}C_4H_9, ^{\bullet}C_3H_5O$
17	$^{\bullet}OH, ^{\bullet}NH_3$	58	$^{\bullet}C_3H_8N$
19	$^{\bullet}F$	61	$^{\bullet}C_2H_5S$
26	$^{\bullet}CN$	69	$^{\bullet}CF_3$
29	$^{\bullet}C_2H_5, ^{\bullet}CHO$	71	$^{\bullet}C_4H_7O$
30	$^{\bullet}CH_3NH, ^{\bullet}CH_2NH_2$	79	$^{\bullet}Br, ^{\bullet}C_6H_7$
31	$^{\bullet}CH_3O, ^{\bullet}CH_2OH$	81	$^{\bullet}C_6H_9, ^{\bullet}C_5H_5O$
35	$^{\bullet}Cl$	91	$^{\bullet}C_7H_7$
45	$^{\bullet}C_2H_5O, ^{\bullet}COOH$	127	$^{\bullet}I$

Source: From *Mass Spectral Interpretation Quick Reference Guide*, Varian, Inc., courtesy of Varian, Inc. (www.varianinc.com).

requires practice and lots of it. The following examples include simple gas molecules and simple compounds representative of a variety of organic chemicals.

Looking at the spectrum in Fig. 10.10 (data in Table 10.12), assume that $m/z = 44$ is the molecular ion, since it has the highest m/z (excepting isotope peaks) in the spectrum. Since the ratio of $m/z\ 45/44 = 1.2\%$, it appears that not more than 1 carbon atom is present. $M - 12 = 32$ Da to be accounted for. The peak ratio $m/z\ 46/44 = 0.42\%$, which is consistent with 2 oxygen atoms, for an additional $2(16) = 32$ Da. $12 + 32$ equals the assumed MW, therefore, a reasonable suggestion is that the compound is CO_2. Looking at the rest of the spectrum, the peak at $m/z = 28$ could be due to formation of a CO ion by loss of one oxygen atom from the molecular ion, while the peak at $m/z = 16$ can be attributed to an ionized oxygen atom. Using the rings + double bonds equation, the compound is predicted to have 2 double bonds, as it does.

Table 10.11 Some Neutral Fragments Expelled by Multicentered Fragmentations

Mass	Fragment (EE)	Mass	Fragment (EE)
2	H_2	45	C_2H_7N
17	NH_3	46	$C_2H_6O, H_2O + C_2H_4$
18	H_2O	48	CH_4S
20	HF	54	C_4H_6
27	HCN	56	C_4H_8, C_3H_4O
28	CO, C_2H_4	58	C_3H_6O
30	CH_2O	59	C_3H_9N
31	CH_5N	60	$C_3H_8O, C_2H_4O_2$
32	CH_4O	62	C_2H_6S
34	H_2S	74	$C_3H_6O_2$
36	HCl	76	C_6H_4
42	C_3H_6, C_2H_2O	78	C_6H_6
44	CO_2	80	HBr

Source: From *Mass Spectral Interpretation Quick Reference Guide*, Varian, Inc., courtesy of Varian, Inc. (www.varianinc.com).

Figure 10.10 A mass spectrum of carbon dioxide, CO_2.

Now consider the mass spectrum in Fig. 10.11 and compare it to the carbon dioxide spectrum. The spectrum in Fig. 10.11 is one that every mass spectrometrist should know. It is the spectrum of air, a mixture of components. (Why should a mass spectrometrist recognize this spectrum?) The peak at $m/z = 44$ is from carbon dioxide, while the other peaks are due to water (18), nitrogen (28), and oxygen (32). The peak at m/z 28 in the CO_2 spectrum is due to a CO ion, while in the air spectrum, it is due to nitrogen plus any carbon monoxide that is in the sample. At this resolution, it is not possible to distinguish between N_2 and CO. It is important to remember that not all mass spectra are of a single pure component. In the real world, mixtures and impure samples are much more commonly encountered.

Consider the mass spectrum in Fig. 10.12 (data in Table 10.13), and assume the molecular ion is the peak at $m/z = 27$. If the MW is 27, there must be an odd number of nitrogen atoms present in the molecule. There cannot be three, because the mass would be too high. Therefore, there must be 1 nitrogen atom in the molecule. The ratio of the $(M + 1)/M$ peaks at $28/27 = 1.5$, indicating 1 carbon atom. The mass accounted for is 1 C $(12) + 1$ N $(14) = 26$ Da. Since the MW is 27 Da, the only possibility is the presence of one hydrogen atom, giving a molecular formula of HCN, hydrogen cyanide. The mass and the M + 2 peak are too small to allow for any oxygen atoms. The rings + db calculation results in $1 - 1/2 + 1/2 + 1 = 2$. Hydrogen cyanide has a triple bond between the C and N, as shown in Fig. 10.12. A triple bond is equivalent to two double bonds, as we learned.

Table 10.12 Mass Spectral
Data for CO_2

m/z	Relative abundance
12	8.7
16	9.6
22	1.9
28	9.8
29	0.13
30	0.02
44	100
45	1.2
46	0.42

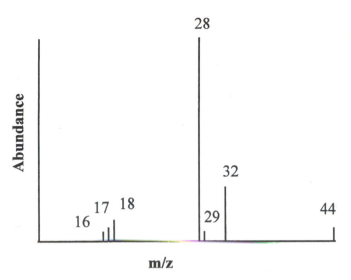

Figure 10.11 A mass spectrum of air, a mixture of gases. (Courtesy of Dr. Robert Kobelski, CDC, Atlanta, GA.)

10.2.1. Mass Spectra of Hydrocarbons

The general appearance of the mass spectrum depends on the type of compound analyzed. We will look first at the normal alkanes (saturated straight-chain hydrocarbons of the formula C_nH_{2n+2}, also called paraffins). Figure 10.13 shows the spectra of n-C_{16} and n-C_{30} alkanes. It can be seen that these spectra are very similar and would be hard to distinguish even though one molecule is twice as big as the other. Small n-alkanes generally show an intense molecular ion peak. The molecular ion may be very small for large n-alkanes using electron ionization. (To obtain the molecular ion, chemical ionization could be used.)

It can also be noted that with each compound, there are groups of fragments differing by 14 Da from each other. This is caused by the loss of successive numbers of CH_2 (methylene) groups as the molecules fragment by breaking the carbon—carbon bonds.

Figure 10.12 A mass spectrum of hydrogen cyanide, HCN.

Table 10.13 Mass Spectral Data for HCN

m/z	Relative abundance
12	4.2
18	1.7
13.5	0.88
14	1.6
15	0.12
26	17
27	100
28	1.6
29	0.06

For the homologous series of alkanes, peaks at $m/z = 15, 29, 43, 57$, etc. will be noted starting at 15 (CH_3). For other homologous series such as straight-chain alcohols or amines, the same general appearance is observed in the spectrum but the m/z values will be displaced by the difference in mass of the functional group. Table 10.14 shows the terminal group and the first several m/z peaks of such series.

Branched alkanes fragment more readily than straight chain alkanes, so branched alkanes are less likely to show a molecular ion peak than n-alkanes. Cycloalkanes show a strong molecular ion peak and also show the characteristic peaks separated by 14 Da. Figure 10.14 is the mass spectrum of hexane, C_6H_{14}, MW = 86, and Fig. 10.15 is the mass spectrum of cyclohexane, C_6H_{12}, MW = 84. Cyclohexane has a stronger molecular ion

Figure 10.13 (a) Mass spectrum of n-hexadecane, $C_{16}H_{34}$ and (b) n-tridecane, $C_{30}H_{62}$. The MW of n-hexadecane is 226 Da; the molecular ion is visible in spectrum (a) but small. The MW of n-tridecane is 422 Da; the spectrum does not extend far enough to see the molecular ion, but it would be very small as well. Both n-alkanes show a similar pattern of peaks separated by 14 Da, due to the loss of CH_2 groups and both show very similar patterns at the low m/z end of the spectrum.

Table 10.14 Homologous Series Separated by CH_2

Functional group	End group	m/z Series
Amine	$—CH_2—NH_2$	30, 44, 58, ...
Alcohol	$—CH_2—OH$	31, 45, 59, ...
Aldehyde	$—C{\overset{\displaystyle O}{\underset{\displaystyle H}{}}}$	29, 43, 57, 71, ...
Ketone	$—\overset{O}{\underset{\|}{C}}—CH_3$	43, 57, 71, ...

peak than hexane, although both are clearly present. Cycloalkanes preferentially fragment by loss of ethene, C_2H_4. This preferred fragmentation gives rise to the base peak for cyclohexane at $m/z = 56$. Look at the mass spectrum of 2,2-dimethylpropane, a branched alkane as shown by the structure in Fig. 10.16. The molecular ion at $m/z = 72$ is too weak to be observed; the strongest peak in the spectrum is due to the loss of a methyl group.

Alkenes and alkynes show strong molecular ion peaks because the double and triple bonds are able to absorb energy. Alkenes with more than 4 carbon atoms often show a strong peak at $m/z = 41$ due to formation of an allyl ion. Alkynes show strong (M − 1) peaks. The mass spectra of alkene isomers are very similar, especially the mass spectra of cis- and trans-isomers, so it is not usually possible to locate the position of the double bonds in these compounds through mass spectrometry. Figures 10.17 and 10.18 show the mass spectra of 1-hexene and acetylene, respectively.

The mass spectral data for acetylene is given in the following table. The molecular formula can be deduced as follows. If the molecular ion is at $m/z = 26$, then the ratio of M + 1 to M is 2.2%, indicating 2 carbon atoms. Their total mass is 24 Da. The missing mass of 2 Da can only be two hydrogen atoms. Hence, the formula is C_2H_2. Note that the M + 2 peak is too small to indicate the presence of oxygen, and the even numbered m/z value

Figure 10.14 A mass spectrum of a normal alkane, hexane, C_6H_{14}.

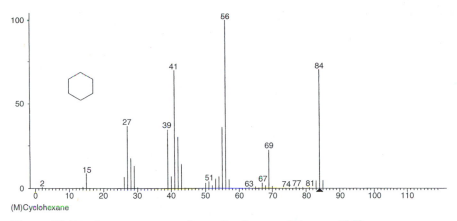

(M)Cyclohexane

Figure 10.15 A mass spectrum of a cyclic alkane, cyclohexane, C_6H_{12}.

for the molecular ion indicates zero (or two) nitrogen atoms. Two nitrogen atoms have a mass of 28 Da, so there cannot be two nitrogen atoms in this molecule if the MW is 26 Da. The rings + db calculation results in a value of two, which can only be a triple bond between the two carbon atoms. Also, the strong (M − 1) peak at $m/z = 25$, due to the loss of one hydrogen, is typical of alkynes.

m/z	Relative abundance
12	0.91
13	3.6
14	0.10
24	6.1
25	23
26	100
27	2.2
28	0.02

(M)Propane, 2,2-dimethyl-

Figure 10.16 A mass spectrum of a branched alkane, 2,2-dimethylpropane, $(CH_3)_4C$. The structure is shown schematically in the figure. The hydrogen atoms are not shown.

Figure 10.17 A mass spectrum of an aliphatic alkene, 1-hexene, C_6H_{12}.

Aromatic compounds present quite different spectra from alkanes, as can be seen in Fig. 10.4 (benzene) and Fig. 10.19 (naphthalene). Aromatic molecules are very stable and do not fragment easily. The molecular ion peak for aromatic compounds is very intense, as seen in both Figs. 10.4 and 10.19.

Using naphthalene (Fig. 10.19 and Table 10.15) as an example, we see an intense molecular ion peak at $m/z = 128$ and relatively little fragmentation. This should suggest an aromatic compound just from the appearance of the spectrum. The ratio $(M + 1)/M$ is $11/100$, indicating 10 carbon atoms, although a formula with 9 carbon atoms cannot be completely ruled out based on isotopic abundances alone. If we assume 10 carbon atoms, their mass is $10 \times 12 = 120$ Da. The molecule, therefore, has $128 - 120 = 8$ Da to be accounted for; 8 hydrogen atoms are the most reasonable suggestion. The formula would be $C_{10}H_8$ and is probably naphthalene, whose structure is shown in the figure. Can you work out the rings + db calculation and explain it?

It will also be noted in Table 10.1 for benzene and in Table 10.15 for naphthalene, peaks at noninteger m/z values are observed. For naphthalene, there are peaks at m/z values of 64.5, 63.4, and 62.5. These are not fragments with fractional masses. Rather, they are doubly charged ions. It will be remembered (Chapter 9) that

$$m/z = \frac{B^2 r^2}{2V} \tag{10.6}$$

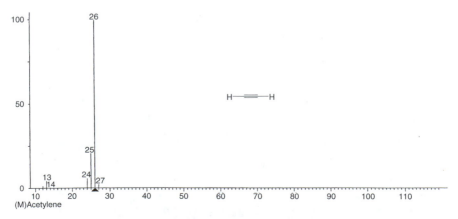

Figure 10.18 A mass spectrum of an alkyne, acetylene, C_2H_2.

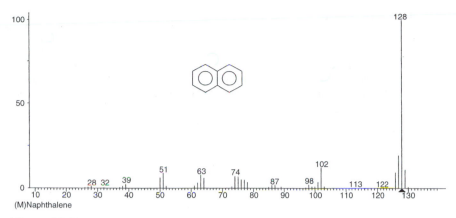

(M)Naphthalene

Figure 10.19 A mass spectrum of an aromatic hydrocarbon, naphthalene, $C_{10}H_8$.

and it is normally assumed for EI spectra that $z = +1$. But if the ion is doubly charged to M^{++}, it appears to have half the mass to satisfy Eq. (10.6). Therefore, these ions really have masses of $64.5 \times 2 = 129$ Da, $63.5 \times 2 = 127$ Da, and $62.5 \times 2 = 125$ Da. Large M^{++} peaks are also typical of aromatic compounds. The apparent mass at $m/z = 42.67$

Table 10.15 Mass Spectral Data for Naphthalene

m/z	Relative abundance	m/z	Relative abundance
38	1.8	64.5	1.1
39	3.9	65	0.24
40	0.19	66	0.02
41	0.01	74	4.7
42	0.06	75	4.9
42.67	0.10	76	3.3
43	0.16	77	4.1
43.5	0.02	78	2.7
49.5	0.24	79	0.19
50	6.4	87	1.4
50.5	0.09	88	0.22
51	12	89	0.73
51.5	0.37	90	0.06
52	1.6	101	2.7
53	0.27	102	7.1
54	0.04	103	0.64
55	0.14	104	0.15
55.5	0.03	110	0.06
56	0.12	111	0.09
56.5	0.06	112	0.02
61	1.4	113	0.18
61.5	0.08	125	0.85
62	2.7	126	6.1
62.5	0.20	127	9.8
63	7.4	128	100
63.5	0.95	129	11
64	10	130	0.52

is due to a triply charged ion (i.e., M^{+++}), and its true mass is $52.67 \times 3 = 128$ Da. This should drive home the message that the x-axis in a mass spectrum is not mass, but m/z. In order to confirm that this is the mass spectrum of naphthalene, a known sample should be run.

There is another phenomenon that gives rise to peaks with noninteger masses. Ions with lower internal energy than those that produce the "normal" product ions can fragment after they leave the ion source but before entering the mass analyzer. These **metastable ions** have the velocity of the precursor ion, m_1, but the mass of the product ion, m_2. In a magnetic sector instrument, a **metastable peak** will appear at m* where

$$m^* = \frac{m_2{}^2}{m_1} \tag{10.7}$$

The value of m* is generally not an integer. The spectral interpretation texts cited in the bibliography should be consulted for more details.

The lack of significant fragmentation and the intense molecular ion illustrate how stable aromatic molecules are. As a general rule, increased unsaturation increases stability. It is harder to break double bonds than single bonds, and harder to break apart the very stable delocalized molecular orbitals in aromatic compounds than to break bonds in aliphatic compounds. Also, rings are more stable than straight chains.

Aromatic rings with aliphatic side chains will show characteristics of both types of compounds. In particular, benzene rings with an alkyl group as a substituent, such as a methyl group, often show an intense peak at $m/z = 91$, due to the formation of a seven-membered carbon ring, $C_7H_7^+$. This ion is called the **tropylium ion** and results from a rearrangement of the benzyl cation, $C_6H_6CH_2^+$. This and many other well-known rearrangement reactions are detailed in the specialized texts listed in the bibliography on interpretation of mass spectra.

10.2.2. Mass Spectra of Other Organic Compound Classes

Functional groups are of two main types: those that have a single bond to a heteroatom and those that have a multiple bond to a heteroatom. Alcohols, halides, ethers, and amines are examples of the first type, while aldehydes, ketones, and carboxylic acids are examples of the second type.

10.2.2.1. Alcohols

The mass spectra of three aliphatic alcohols, butanol, 2-butanol, and 2-methyl-2-propanol are given in Figs. 10.20–10.22, respectively. These three alcohols are isomers; they have the same chemical formula, $C_4H_{10}O$ also written as C_4H_9OH to emphasis the alcohol OH functional group, but different structures, as shown:

$$CH_3\text{-}CH_2\text{-}CH_2\text{-}CH_2\text{-}OH$$

Butanol
(*n*-butanol)

$$\begin{array}{c} CH_3 \\ | \\ CH_3CH_2-CH-OH \end{array}$$

2-butanol
(*sec*-butanol)

$$\begin{array}{c} CH_3 \\ | \\ CH_3-C-OH \\ | \\ CH_3 \end{array}$$

2-methyl-2-propanol
(*tert*-butanol)

The names in parentheses are old names for these alcohols. The *n* stands for normal (meaning a primary alcohol; the OH group is attached to a carbon atom attached to only

(M)1-Butanol

Figure 10.20   A mass spectrum of a normal (straight chain) alcohol, butanol, $C_4H_{10}O$.

(M)2-Butanol

Figure 10.21   A mass spectrum of a secondary alcohol, 2-butanol, $C_4H_{10}O$.

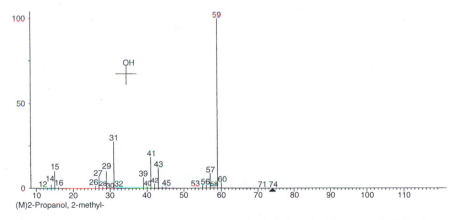

(M)2-Propanol, 2-methyl-

Figure 10.22   A mass spectrum of a tertiary alcohol, 2-methyl-2-propanol, $C_4H_{10}O$.

one other carbon atom), *sec* for secondary (note that the OH group is attached to a carbon atom attached to two other carbon atoms), and *tert* for tertiary (the OH group is on a carbon atom attached to three carbon atoms). The terms primary, secondary, and tertiary to describe the position of the functional group are still used. Table 10.16 presents relative

Table 10.16 Mass Spectral Distribution of the Fragments of Isomers of Butanol[a] Using Electron Ionization at 70 eV

m/z	Butanol	2-Butanol	2-Methyl-2-propanol
15	8.4	6.80	13.3
27	50.9	15.9	9.9
28	16.2	3.0	1.7
29	29.9	13.9	12.7
31	*100	20.31	35.5
33	8.5	0	0
39	15.6	3.4	7.7
41	61.6	10.1	20.8
42	32.4	1.7	3.3
43	61.4	9.8	14.5
45	6.6	*100	0.6
55	12.3	2.0	1.6
56	99.9	1.0	1.5
57	6.7	2.7	9.0
59	0.3	17.7	*100
60	0	0.7	3.2
74	0.8	0.3	0

[a]Molecular weight for all isomers = 74 Da. Asterisks in table body denote the most abundant fragment (the base peak). The abundance is normalized to 100 for the base peak in each pattern, but the actual abundances of these peaks probably differ from each other.

abundances for these compounds (again remembering that the spectra and the tabular data are from different sources and may not be exactly comparable).

Molecular ions for primary and secondary aliphatic alcohols are usually weak or absent; tertiary aliphatic alcohols typically show no molecular ion peak.

In the butanol spectrum the most abundant fragment is at $m/z = 31$, corresponding to CH_2OH^+, with a structure as shown: $H_2C=OH^+$. The fragment lost was $74 - 31 = 43$, probably as a propyl radical, $C_3H_7^{\cdot}$. This indicates that fragmentation between the first and second carbon atoms is favored as shown below.

$$CH_3CH_2CH_2\!-\!CH_2OH \quad \rightarrow \quad CH_3CH_2CH_2^{\cdot} + H_2C=OH^+$$

The reasons for this favored fragmentation, a process called **alpha cleavage**, meaning cleavage at the bond adjacent to the carbon to which the functional group is attached, are discussed in detail in the text by McLafferty and Tureček. The fragmentation can be thought of as loss of an alkyl group, with loss of the largest alkyl group most likely. In butanol, the propyl group is the largest possible alkyl fragment. Alpha cleavage occurs for many functional groups in addition to alcohols, as shown schematically:

bond cleavage here
↓
$$(R\!-\!CH_2\!-\!OR')^{\cdot +} \quad \rightarrow \quad R^{\cdot} + (CH_2\!=\!OR')^+$$
↑
α carbon

As shown, cleavage occurs at the first C—C bond counting from the functional group.

However, the spectrum of butanol also shows an almost equally abundant peak at $m/z = 56$, due to loss of water ($74 - 18 = 56$) from the molecular ion. Notice how much more likely the loss of water is than the loss of just a neutral OH fragment, which gives rise to the smaller peak at $m/z = 57$.

The butanol spectrum shows the primary decomposition paths for aliphatic alcohols: alpha cleavage and loss of water. Alcohols containing more than four carbons often lose both water and ethylene simultaneously (total loss of $18 + 28$ Da). Detailed interpretation of the peaks requires extensive knowledge of the reaction mechanisms, which requires a good understanding of physical organic chemistry. For example, the peak in the butanol spectrum at $m/z = 43$ can be due to the formation of a C_3H_7 ion by loss of the CH_2OH fragment as a neutral species. Part of the peak intensity may also be due to a secondary decomposition of $C_nH_{2n+1}O^+$ ions, in this case the $(M - H)^+$ ion. The observation of a peak at $m/z = 18$ (water) should be interpreted with caution, because water is a common contaminant in samples or may have come from air getting into the instrument.

With 2-butanol, the most abundant peak is at $m/z = 45$, indicating loss of a neutral fragment, $74 - 45 = 29$, probably C_2H_5. Favored fragmentation is again by alpha cleavage, with the largest alkyl group (ethyl) being lost and formation of a $CH_3CH{=}OH^+$ ion.

$$\begin{array}{c} CH_3 \\ | \\ (\,CH_3CH_2{-}CH{-}OH\,)^{\cdot +} \quad \rightarrow \quad CH_3CH_2^{\cdot} \; + \; CH_3CH{=}OH^+ \\ \uparrow \\ \text{bond cleavage here} \end{array}$$

The loss of CH_3 from the secondary alcohol is a common fragmentation, since the peak at $m/z = 59$ is more abundant than with butanol. (Is this also alpha cleavage?) With 2-methyl-2-propanol, the most abundant peak is 59, showing a loss of $74 - 59 = 15$ Da, a methyl group. Looking at the structure below, it is easy to understand, since any one of the CH_3 groups may break off. All three methyl groups are equivalent, and each CH_3 group is attached to the alpha carbon so this is again an example of alpha cleavage. Tertiary alcohols also tend to lose OH rather than water, accounting for the $M - 17$ peak at $m/z = 57$ in the spectrum.

$$\begin{array}{c} CH_3 \qquad\qquad\qquad CH_3 \\ | \qquad\qquad\qquad\quad | \\ (\,CH_3{-}C{-}OH\,)^{\cdot +} \quad \rightarrow \quad CH_3^{\cdot} \; + \; CH_3C{=}OH^+ \\ | \\ CH_3 \end{array}$$

Aromatic alcohols such as phenols usually show strong molecular ion peaks and the characteristic loss of CO and CHO to give $(M - 28)$ and $(M - 29)$ peaks. A substituted phenol mass spectrum is given in Fig. 10.23. Two features of this spectrum should stand out. There is a strong molecular ion peak (the base peak) and not much fragmentation, suggesting an aromatic compound; the almost equal heights of the peaks at $m/z = 172$ and 174 tell us there is a bromine atom present in the compound. The ratio of the $M + 1$ to M peaks is about 7%, allowing for 6 carbon atoms. If we assume that we have a substituted benzene ring with one Br atom as a substituent, $172 - 79 = 93$ Da to account for. A disubstituted benzene ring would have six carbon atoms and four H atoms, for a mass of 76 Da. The difference between 93 and 76 is 17 Da, which could be an OH group. This would suggest one of the bromophenol compounds. The rings + double bonds calculation gives an answer of four, consistent with a disubstituted benzene ring containing a hydroxyl group. (Remember that Br is counted along

Figure 10.23   A mass spectrum of a halogen-substituted aromatic alcohol, 4-bromophenol, C_6H_5OBr.

with H for this calculation and of course the H atom in the OH group is also counted.) The spectrum should be matched to those for the possible bromophenol compounds; it is in fact 4-bromophenol.

10.2.2.2.   Ethers

Ethers are compounds of the formula R—O—R'. Like alcohols, they undergo alpha cleavage as well as cleavage of the C—O bond to form a hydrocarbon ion and an oxygen-containing radical. The molecular ion is usually weak but present.

10.2.2.3.   Ketones and Aldehydes

Ketones and aromatic aldehydes show strong molecular ion peaks, while aliphatic aldehydes give a weak but measurable molecular ion peak. Both ketones and aldehydes fragment by alpha cleavage. Aldehydes also fragment by beta cleavage, at the bond shown:

bond cleavage here
  ↓
$(R—CH_2—CH_2—OR')^{\cdot+} \rightarrow R^+ + CH_2\!=\!CH—O^{\cdot}$
  ↑
β carbon

A very important fragmentation with rearrangment occurs with aldehydes, ketones, and other classes of compounds. This is the **McLafferty rearrangement**. The ion must be able to form a six-membered cyclic transition state. A transfer of hydrogen to the oxygen atom occurs and an alkene molecule splits off. The McLafferty rearrangement is shown for a ketone:

The rearrangement product detected is the ionic product (an enol) shown; a neutral alkene is also produced. At least one of the alkyl chains on the ketone and the alkyl chain on the

(M)3-Heptanone

Figure 10.24 A mass spectrum of an aliphatic ketone, 3-heptanone, $C_7H_{14}O$.

aldehyde must have three or more carbon atoms for the McLafferty rearrangement to occur. The mass spectrum of 3-heptanone is shown in Fig. 10.24. The peak at $m/z = 72$ is due to the McLafferty rearrangement; propene is the neutral alkene that splits off. The peak at $m/z = 57$ is due to alpha cleavage on the left side of the carbonyl group as the structure is shown in Fig. 10.24 to form $(C_4H_9)^+$ or $(C_3H_5O)^+$; cleaving the bond gives two possible ions with the same m/z value. Alpha cleavage can also occur on the right side, forming $(CH_3CH_2)^+$ with $m/z = 29$ or $(C_5H_9O)^+$ with $m/z = 85$. Both ions are seen in the spectrum.

Let us look at the spectrum of an aromatic ketone in Fig. 10.25 (data in Table 10.17). The general appearance of the spectrum indicates an aromatic compound because there are few peaks indicating little fragmentation. The molecular ion is at $m/z = 182$ and is strong, as expected for an aromatic ketone. The value of $(M + 1)/M$ is 12.8, indicating 12 or 13 carbon atoms. The $(M + 2)/M$ value is 1.18. Using Eq. (10.3) and an assumed value of 12 C atoms based on the $M + 1$ peak, there is no more than one oxygen atom present. The same conclusion is drawn if 13 C atoms are used in Eq. (10.3). (Put in the numbers and demonstrate to yourself that this is the case.) This highlights the fact that the $M + 1$ and $M + 2$ peaks do not always give us a definite answer for numbers of atoms.

The most abundant peak is $m/z = 105$, showing the loss of a neutral fragment of $182 - 105 = 77$ Da, probably a phenyl group, C_6H_5, based on the value of the

(M)Benzophenone

Figure 10.25 A mass spectrum of an aromatic ketone, benzophenone, $C_{13}H_{10}O$.

Table 10.17 Mass Spectral Data for Benzophenone

m/z	Relative abundance	m/z	Relative abundance	m/z	Relative abundance	m/z	Relative abundance
27	1.3	74	2.0	105	100	153	1.8
28	1.0	75	1.7	106	7.8	154	1.4
38	0.37	75.5	0.21	107	0.51	155	0.16
39	1.1	76	4.3	119	0.02	164	0.06
50	6.2	76.5	0.33	126	0.63	165	0.44
51	19	77	62	127	0.40	166	0.06
52	1.4	78	4.2	128	0.13	181	8.2
53	0.29	79	0.16	139	0.17	182	60
63	1.3	91	0.08	151	1.1	183	8.5
64	0.61	104	0.48	152	3.4	184	0.71
65	0.09						

$m/z = 78$ peak. The $m/z = 78$ peak can be thought of as the M + 1 peak for the fragment at $m/z = 77$. The idea of the M + 1 and M + 2 peak ratios can be used to count carbon atoms and oxygen atoms in fragments, but with caution. These peaks may arise from some other fragmentation pathway, in which case the answer will be wrong and misleading. The fragment at $m/z = 105$ probably contains 7 carbon atoms, based on the value of the $m/z = 106$ peak. Using this information, calculating the ratio of peaks $107/105$ and Eq. (10.3) leads to the conclusion that this fragment (the 105 fragment) contains 7 C atoms and one oxygen atom.

The peak at $m/z = 77$ is very strong. It is *not* a good assumption in general that lower m/z peaks result from sequential fragmentation of higher m/z fragments. However, in this case it is notable that the difference between 105 and 77 Da is 28, the mass of a C=O group. CO is a commonly lost neutral species. The compound is aromatic, we were told it was a ketone and the compound has at least one phenyl ring. What aromatic ketone can we make from 13 carbon atoms and an oxygen atom? We could postulate two phenyl rings and a carbonyl group. One possibility that would account for the spectrum is benzophenone, $C_{13}H_{10}O$. Alpha cleavage of one phenyl group would leave an ion consisting of the other phenyl ring substituted with CO, mass = 105 Da. Loss of CO from this ion would result in a second phenyl ring fragment. To confirm our hypothesis, we would either run a spectral library search to match the spectrum or run a sample of benzophenone on our mass spectrometer and match the spectrum.

10.2.2.4. Carboxylic Acids and Esters

Carboxylic acids are compounds with the general formula:

$$\overset{\displaystyle O}{\overset{\|}{RC}}\!-\!OH$$

while esters are compounds with the acidic hydrogen of the acid replaced by another alkyl or aromatic group to give the formula:

$$\overset{\displaystyle O}{\overset{\|}{RC}}\!-\!OR'$$

Figure 10.26 A mass spectrum of a carboxylic acid, 2-propenoic acid, $C_3H_4O_2$.

Aliphatic carboxylic acids and small aliphatic esters (those with four to five carbon atoms) have a weak but measurable molecular ion in their mass spectra. Larger esters may show no molecular ion peak. Aromatic carboxylic acids give strong molecular ions.

The acids typically lose OH and COOH through alpha cleavage, giving fragment ions of M − 17 and M − 45. The $COOH^+$ ion at $m/z = 45$ also forms and is a very characteristic peak in the spectra of these acids. The mass spectrum of 2-propenoic acid, $H_2C=CH-COOH$, shown in Fig. 10.26, displays the characteristic peaks of a carboxylic acid. The molecular ion occurs at $m/z = 72$ and is actually quite strong. Loss of the OH group from the molecular ion results in the M − 17 ion at $m/z = 55$, while loss of the COOH group from the molecular ion produces the M − 45 ion at $m/z = 27$. The $COOH^+$ ion arising from loss of the rest of the molecule as a radical gives the peak at $m/z = 45$. A table of mass spectral data (again from a different source than the spectrum shown in Fig. 10.26) for 2-propenoic acid is presented in Table 10.18. From the M + 1 peak, three carbon atoms are likely; from the M + 2 peak and Eq. (10.3), we would deduce the presence of two oxygen atoms. Three carbon atoms plus two oxygen atoms have a mass of 68 Da, leaving 4 Da to be accounted for as hydrogen atoms. Therefore,

Table 10.18 Mass Spectral Data for 2-Propenoic Acid

m/z	Relative abundance	m/z	Relative abundance
25	5.2	45	32
26	38	46	2.5
26.5	0.15	47	0.14
27	74	52	1.4
27.5	0.26	53	6.0
28	12	54	2.3
29	4.3	55	74
30	0.13	56	2.6
31	0.48	57	0.19
41	1.2	71	4.3
42	1.3	72	100
43	5.8	73	3.5
44	14	74	0.48

the formula is most likely $C_3H_4O_2$. Rings + db: $3 - 1/2(4) + 1 = 2$. One of the double bonds is the C=O bond in the acid; the other must be a C=C bond.

Esters undergo alpha cleavage to form an RCO^+ ion. Those esters with four or more carbon atoms (butyrates and higher) undergo the McLafferty rearrangement. For methyl esters, where the R' is CH_3, the ion formed as a result of the McLafferty rearrangement is:

$$(CH_3O{-}\overset{\overset{\textstyle OH}{|}}{C}{=}CH_2)^+$$

which has an m/z value of 74. This is a characteristic peak in the spectra of methyl esters.

10.2.3. Compounds Containing Heteroatoms

Compounds containing the halogen atoms Cl and Br have been discussed in Sections 10.1.4 and 10.1.5. The two other important halogen atoms, fluorine and iodine, are monoisotopic and have no isotope cluster patterns. It is important to remember that they do need to be counted in the rings + double bonds equation. In general, bromine and iodine compounds tend to fragment by loss of the halogen, giving weak or no molecular ion. Chlorine compounds tend to lose HCl. Fluorine compounds undergo some unique reactions with rearrangements that are beyond the scope of this text, although we will look at a fluorinated inorganic compound in the next section. One very characteristic indication of the presence of fluorine, mass = 19 Da, in a molecule is the loss of fluorine, resulting in an M − 19 peak.

While nitrogen, sulfur, silicon, and phosphorus are important heteroatoms, we will look briefly only at nitrogen and sulfur compounds. In general, when the heteroatom is incorporated into an aromatic ring, strong molecular ions are seen. Table 10.4 should be reviewed to remind yourself of the isotope ratios for these heteroatoms to help in interpreting spectra.

10.2.3.1. Nitrogen-Containing Compounds

There are a number of classes of organic compounds that contain nitrogen, including amines, amides, nitro compounds, and nitriles. All of these compounds follow the Nitrogen Rule, which tells us that a compound with an odd number of nitrogen atoms has a molecular weight that is an odd number. This rule is very useful in determining if a compound contains nitrogen atoms. Unfortunately, many nitrogen-containing compounds give no detectable molecular ion. Aliphatic amines, nitriles, and nitro compounds rarely give detectable molecular ions. Amides, cyclic aliphatic amines, and aromatic amines, nitriles and nitro groups do give measurable molecular ions.

The primary fragmentation reaction for aliphatic amines is alpha cleavage. A primary amine, RCH_2NH_2, gives the $CH_2NH_2^+$ ion, $m/z = 30$, as a result of alpha cleavage and this serves as a diagnostic peak. The spectrum (Fig. 10.27) of a secondary amine, 2-butanamine (MW = 73), shows the peaks expected from alpha cleavage on both sides of the amine group, an M − 15 peak due to loss of the methyl radical at $m/z = 58$ and the intense peak at $m/z = 44$ due to cleavage and loss of the ethyl radical (M − 29). Remember that the longest alkyl group is preferentially lost as a neutral species, which explains why the peak at $m/z = 44$ is larger than at $m/z = 58$. However, the molecular ion peak ($m/z = 73$) is very small for this compound, as is typical of aliphatic amines. Its odd number value is consistent with the nitrogen rule and gives a great deal of information for so small a peak.

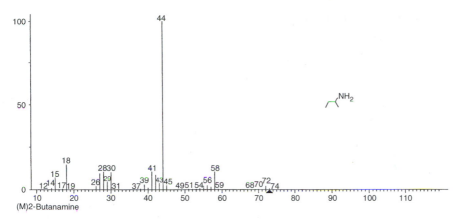

Figure 10.27 A mass spectrum of a secondary aliphatic amine, 2-butanamine, $C_4H_9NH_2$.

Amides have fragmentation patterns similar to their corresponding carboxylic acids. Nitro compounds often have the ions NO^+ ($m/z = 30$) and NO_2^+ ($m/z = 46$) in their spectra. Aromatic nitro compounds have characteristic peaks at M − 30 and M − 46, due to loss of the radicals $NO^\bullet$ and $NO_2^\bullet$. Heteroaromatic nitrogen compounds like pyrrole often fragment to lose a neutral HCN molecule; this loss of HCN is also seen with aromatic amines such as aniline.

Inorganic molecules can also be determined by MS. The compound nitrogen fluoride, NF_3, exhibits characteristics of both a nitrogen-containing compound and a fluorine-containing compound in its mass spectrum (Fig. 10.28). The molecular ion occurs at $m/z = 71$, an odd number, indicating that the molecule contains an odd number of nitrogen atoms. The base peak shows a loss of 19 Da from the M^+ peak; this is characteristic of the loss of fluorine. The peak at $m/z = 19$ confirms the presence of fluorine. The peak at $m/z = 33$ results from a loss of two fluorine atoms from the molecular ion or another fluorine atom from the $m/z = 55$ fragment. Two fluorine atoms and one nitrogen atom account for a mass of 52 Da. The difference between 71 and 52 is 19, leading to the conclusion that there are three fluorine atoms and one nitrogen atom in the compound. The formula is NF_3.

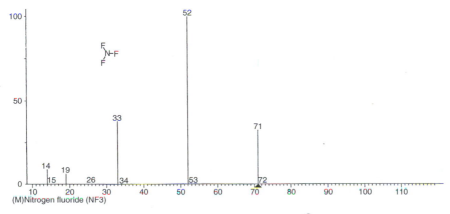

Figure 10.28 A mass spectrum of the inorganic compound, nitrogen fluoride, NF_3.

10.2.3.2.   Sulfur-Containing Compounds

Thiols, also called aliphatic mercaptans, with the formula RSH, are the sulfur analogs of alcohols. Thiols generally show stronger molecular ion peaks than the equivalent alcohols. Looking back at Table 10.4, sulfur has a significant ^{34}S isotope. This gives rise to an enhanced M + 2 peak in the mass spectrum of sulfur-containing compounds.

Primary thiols lose H_2S just as alcohols lose H_2O on fragmentation. This results in a characteristic peak at M − 34. The fragmentation patterns for thiols are similar to those for alcohols, with alpha cleavage and loss of the larger fragment being common. Hetero-aromatic compounds such as thiophene, with the sulfur atom in the ring, give strong molecular ion peaks. Other sulfur-containing organic compounds show similar fragmentation patterns to their oxygen-containing counterparts, but often with stronger molecular ion peaks.

Inorganic sulfur compounds can be determined by MS, as illustrated by the spectrum of SO_2, sulfur dioxide. The spectrum is shown in Fig. 10.29. The molecular ion is shown at $m/z = 64$ and is the base peak in the spectrum. Using a ruler, the M + 1 and M + 2 peaks can be measured, but not very accurately. Doing so gives an $(M + 1)/M$ value that is about 0.6, much less than 1.1 and an $(M + 2)/M$ value of 4.8. Consulting Table 10.4, there cannot be any carbon present due to the small M + 1 peak, but both the M + 1 and M + 2 peaks are consistent with the presence of one sulfur atom. The M + 2 peak is actually larger than the 4.4 due to a sulfur atom, so oxygen may also be present. The peak in the spectrum at $m/z = 48$ is an M − 16 peak, due to the loss of an oxygen atom. The peak at $m/z = 32$ is sulfur and is also 16 Da lower than the $m/z = 48$ peak, indicating loss of another oxygen atom from that fragment. Oxygen is confirmed by the peak at $m/z = 16$. Two oxygen atoms and one sulfur atom account for the MW of 64 needed. The formula is therefore SO_2.

We have only touched on a few examples of select organic compound classes and a few inorganic compounds. The student is advised to consult the texts on mass spectral interpretation cited in the bibliography for more detail. Interpreting mass spectra requires a great deal of practice coupled with knowledge of the major fragmentation reactions and rearrangement reactions, which we cannot cover in detail in this text. It is rare that a practicing mass spectrometrist actually tries to assign an unknown structure by evaluating every peak in the mass spectrum and the isotope ratios, as we have outlined here. Matching to a spectral library plays a predominant role in modern MS with confirmation performed by analysis of a known sample. However, with much practice, knowledge of the molecular

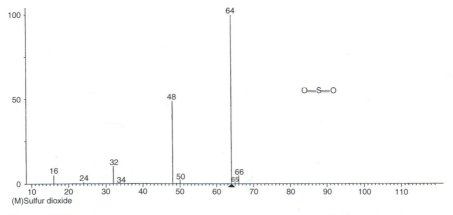

Figure 10.29   A mass spectrum of the inorganic compound, sulfur dioxide, SO_2.

ion m/z value, fragmentation patterns, isotope ratios, and the like, it may be possible to come to well-founded conclusions about the empirical formula and the molecular weight of an unknown compound. MS alone often cannot achieve confirmation of the identity of an unknown, even when a pure sample of the candidate compound is analyzed by MS. Corroborative evidence is often necessary from IR, which identifies the presence of many functional groups, and NMR, which confirms functional groups and, by spin–spin splitting patterns, the placement of these groups. Elemental analysis to determine the C, H, N, O, and heteroatom content is usually performed on pure compounds to assist in the assignment of an empirical formula. Optical activity measurements may be needed for chiral compounds. When used in conjunction with other analytical methods, such as elemental analysis, IR and NMR, MS makes it possible to identify unknown compounds. Combined with a separation method like chromatography, as in GC-MS or LC-MS, even impure samples and mixtures can be analyzed and components identified. GC-MS and LC-MS are described in Chapters 12 and 13, respectively.

10.3. APPLICATIONS OF MOLECULAR MS

The applications for molecular MS have grown enormously in recent years. MS (molecular and atomic) is used by clinical chemists, pharmaceutical chemists, synthetic organic chemists, petroleum chemists, geologists, climatologists, molecular biologists, marine biologists, environmental scientists, forensic chemists, food scientists, nuclear scientists, materials scientists, inorganic chemists, organometallic chemists, analytical chemists, and scientists in many other fields. MS is not just used in laboratories, but online MS analyzers and field-portable MS systems are available commercially for process control and on-site detection of volatile pollutants, explosives, and gases. Applications are as diverse as determination of molecular weight, molecular structure, reaction kinetics, dating of minerals, fossils and artifacts, fundamental studies of ion–molecule reactions, and quantitative analysis of elements and compounds at sub-ppb concentrations. Inorganic and organic solids, liquids and gases can be analyzed. Subcellular structures such as whole ribosomes can be studied. Some applications examples will be discussed but the field is vast and cannot be adequately covered in this text. In addition to the scientific literature, many applications of MS can be found on the websites of the instrument manufacturers. Some of the websites are given subsequently.

10.3.1. High-Resolution Mass Spectrometry

Double-focusing magnetic sector mass spectrometers and Fourier transform (FT) mass spectrometers (described in Chapter 9) are capable of mass measurements with high resolution. The most important use of high-resolution MS is the direct determination of molecular formulas by exact mass measurements.

The atomic masses of the individual isotopes (nuclides) of the elements are nominally whole numbers, H = 1 and 2, C = 12 and 13, N = 14 and 15, O = 16, 17, and 18, and so on. If these masses are measured with sufficient accuracy, we find that actually they are only close to being whole numbers. This is due to the "mass defect", discovered by the early MS pioneer Aston in the 1920s. The mass defect is characteristic of a given isotope. If we use ^{12}C as the standard, the atomic masses of some common isotopes are as shown in Table 10.19. As a consequence, two molecular formulas may have the same nominal unit molecular weight but actually differ slightly but significantly. For example, $C_{20}H_{26}N_2O$ and $C_{16}H_{28}N_3O_3$ have the same unit molecular weight, 310 Da,

Table 10.19   Exact Mass of Common Isotopes

Element	Atomic weight	Isotope (nuclide)	Exact mass
Hydrogen	1.00794	^{1}H	1.007825
		^{2}H (or D)	2.014102
Carbon	12.01115	^{12}C	12.000000
		^{13}C	13.003355
Nitrogen	14.0067	^{14}N	14.003074
		^{15}N	15.0001
Oxygen	15.9994	^{16}O	15.994915
		^{17}O	16.9991
		^{18}O	17.999159
Fluorine	18.998405	^{19}F	18.998405

but if we take into account the exact atomic weights of the nuclei involved, the molecular weights are actually 310.204513 and 310.213067, respectively. Numerous other formulas with this nominal mass can be written; they also have slightly different actual molecular weights. With sufficient resolution it is possible to distinguish between all reasonable formulas. More importantly, knowing the exact molecular weight allows us to identify the molecular formula of the sample. Not only can the formula of the molecular ion be determined, but also the formulas of fragments of the molecule. This provides valuable information on large fragments of very large molecules and on fragments when no molecular ion is observed. Much of the pioneer work in this field was carried out by F.W. McLafferty at Cornell University.

Resolution was defined in Chapter 9 and a graphical example of the effect of resolution on the measured mass is shown in Fig. 10.30. The compound $C_{101}H_{145}N_{34}O_{44}$ has a nominal mass of 2537 Da, calculated from the unit masses of the most abundant stable isotopes, ^{12}C, ^{1}H, ^{14}N, and ^{16}O. A spectrum with resolution of 200 (low resolution) would give the single broad peak shown. The mass measured from this peak at the peak maximum would be 2539.5 Da (or 2540 Da for unit resolution). The measured average mass is slightly higher than the nominal mass because the large number of atoms in the molecule makes it probable that there will be higher isotopes present, most likely ^{13}C. So the measured average mass at low resolution could be misleading in trying to assign the elemental composition. With a resolution of 2500, the monoisotopic peaks that contribute to the average mass are separated, allowing much more exact mass measurement as shown. If the resolution were infinite, each monoisotopic peak would be a line, as shown.

Loss of resolution (peak broadening) is largely caused by the narrow beam of ions spreading out while traveling through the sector. Similar masses may spread out enough to actually overlap each other and therefore be unresolved. Double focus instruments refocus the ion beam to prevent overlap of peaks. Exact mass measurements are made by comparing the position of the unknown peak with the position of an internal standard of known composition and known exact mass. The ions of the known internal standard are used to calibrate the mass scale. High-resolution MS can be used to identify the formulas of fragments of large molecules such as proteins. It can also be used to assign empirical formulas to newly synthesized organic compounds or complex molecules isolated from natural products.

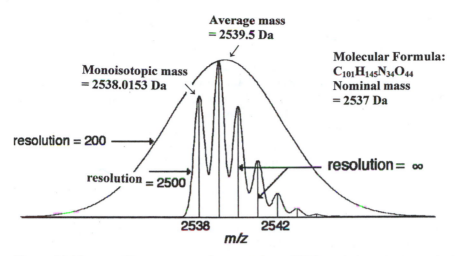

Figure 10.30 The difference between low-resolution and high-resolution MS. The nominal mass of the molecule $C_{101}H_{145}N_{34}O_{44}$ = 2537 Da. At the low resolution of 200, a single peak is measured with an average mass at the peak maximum = 2539.5 Da. A high-resolution instrument (resolution = 2500) separates the monoisotopic peaks and permits measurement of exact masses. [Courtesy of Professor Gary Siuzdak, Scripps Research Institute Center for Mass Spectrometry (www.masspec.scripps.edu). The website has a number of tutorials on MS.]

10.3.2. Quantitative Analysis of Compounds and Mixtures

Quantitative analysis of compounds can be performed by measuring the relative intensities of a spectral peak or peaks unique to the compound and comparing the intensity of the sample to a series of known standards of the compound. The use of **isotope dilution** for quantitative measurements in MS is common. Isotope dilution is a special case of calibration using an internal standard; this calibration approach was covered in Chapter 2. For molecular analysis, the internal standard is the analyte molecule, but with some of the atoms replaced by heavier isotopes. Such a compound has been **isotopically labeled**. Labeling with deuterium (^{2}H) and ^{13}C is common, but isotopes of heteroatoms can also be used. For example, orotic acid is a compound with two nitrogen atoms (MW = 372 Da); it is a metabolite present in only trace amounts in human urine, but an increase in orotic acid in urine can signify diseases that disrupt the urea cycle. If both ^{14}N atoms in natural orotic acid are replaced with ^{15}N atoms, the isotopically labeled orotic acid has MW = 374 Da, but behaves in all respects like natural orotic acid during analysis. An isotopically labeled compound added in a known amount to known amounts of the natural compound is the perfect calibration standard in this respect. A quantitative mass spectral determination of orotic acid in urine can be performed by measuring ions with m/z = 357 for natural orotic acid and m/z = 359 for the labeled acid. The ratio of m/z 357/359 is plotted vs. the concentration of orotic acid in a set of calibration standards. (Details of this analysis can be found in Briand et al., Varian GC-MS Application Note 72, at www.varianinc.com.)

Quantitative analysis can also be performed by external calibration or by the use of an internal standard that is not a labeled analyte molecule but a compound that is not present in the sample.

In some cases, components in a mixture can be determined quantitatively without prior separation if the mass spectrum of each component is sufficiently different from the others. Suppose that a sample is known to contain only the butanol isomers listed in Table 10.16. It can be seen from Table 10.16 that the peak at $m/z = 33$ is derived from butanol, but not from the other two isomers. A measurement of the $m/z = 33$ peak intensity compared to butanol standards of known concentration would therefore provide a basis for measuring the butanol content of the mixture. Also, we can see that the abundances of the peaks at $m/z = 45, 56$, and 59 vary greatly among the isomers. Three simultaneous equations with three unknowns can be obtained by measuring the actual abundances of these three peaks in the sample and applying the ratio of the abundances from pure compounds. The three unknown values are the percentages of butanol, 2-butanol, and 2-methyl-2-propanol in the mixture. The three equations can be solved and the composition of the sample determined. Computer programs can be written to process the data from multicomponent systems, make all necessary corrections, and calculate the results.

Complex mixtures of molecules such as biological fluids, natural products, foods, and beverages will result in mass spectra that are quite complicated, even if soft ionization is used to minimize fragmentation. Mixtures are more commonly analyzed by using GC-MS^n or LC-MS^n to separate the components of the mixture and to obtain mass spectral information on the separated components. Standards are run, often using isotope dilution and internal standard calibration and the peak intensities or intensity ratios of appropriately selected peaks are used to make a calibration curve from which unknown concentrations in samples can be determined.

GC-MS and LC-MS instrumentation and applications are described in more detail in Chapters 12 and 13, but a few examples will be given here. A JEOL GCmate II^{TM} high resolution GC-MS with a magnetic sector mass analyzer was used to analyze Scotch whiskey and tequila samples. Over 50 compounds were identified in the whiskey using exact mass measurements (see Section 10.3.1) from EI high-resolution spectra. Figure 10.31 shows the gas chromatogram for a sample of whiskey; the peaks are the separated compounds, identified by their retention time. Figure 10.32 presents the high-resolution exact mass measurements and the elemental compositions of the identified compounds. Such an analysis would permit distillers, food scientists, and regulators to compare brands, compare different batches for quality, and monitor contaminants. One of the compounds identified in the whiskey was a plasticizer that may have leached from packaging material. The GC-high-resolution MS system was also able to separate

Figure 10.31 Gas chromatogram of compounds in a sample of whiskey analyzed on a GCmate II^{TM} high-resolution magnetic sector GC-MS instrument. [From JEOL Applications Note MS-1126200-A. Courtesy of JEOL USA, Inc., Peabody, MA (www.jeol.com).]

Compound	Meas. m/z	Composition	Ion	Calc. m/z	Error (mmu)
Ethyl acetate	88.05077	C4 H8 O2	M+.	88.05243	-1.7
1-Propanol, 2-methyl	74.07131	C4 H10 O	M+.	74.07317	-1.9
Diethoxyethane	103.0752	C5 H11 O2	[M-CH3]+	103.0759	-0.7
Furfural	96.02313	C5 H4 O2	M+.	96.02113	2
Furfural	95.01513	C5 H3 O2	[M-H]+	95.01331	1.8
1-Butanol, 3-methyl acetate	87.04336	C4 H7 O2	Fragment	70.07825	-0.2
1-Butanol, 3-methyl acetate	70.07805	C5 H10	Fragment	87.04461	-1.2
Hexanoic acid, ethyl ester	88.05026	C4H8O2	Fragment	88.05243	-2.2
Hexanoic acid, ethyl ester	99.07999	C6 H11 O	Fragment	99.081	-1
Hexanoic acid, ethyl ester	101.0588	C5 H9 O2	Fragment	101.0603	-1.5
Hexanoic acid, ethyl ester	115.0717	C6 H11 O2	Fragment	115.0759	-4.2
Hexanoic acid, ethyl ester	117.0923	C6 H13 O2	Fragment	117.0916	0.7
Hexanoic acid, ethyl ester	144.1134	C8 H16 O2	M+.	114.115	-1.6
Phenylethyl alcohol	122.0713	C8 H10 O	M+.	122.0732	-1.9
Phenylethyl alcohol	92.06038	C7 H8	Fragment	92.0626	-1
Phenylethyl alcohol	91.05377	C7 H7	Fragment	91.05478	-1
Decanoic acid	172.1481	C10 H20 O2	M+.	172.1463	1.8
Decanoic acid, ethyl ester	200.1776	C12 H24 O2	M+.	200.1776	0
Dodecanoic acid	200.1764	C12 H24 O2	M+.	200.1776	-1.2
Dodecanoic acid, ethyl ester	228.2078	C14 H28 O2	M+.	228.2089	-1.1

Figure 10.32 The exact masses, elemental compositions, and identification of the compounds in a whiskey sample (Fig. 10.31) analyzed by high-resolution GC-MS. [From JEOL Applications Note MS-1126200-A. Courtesy of JEOL USA, Inc., Peabody, MA (www.jeol.com).]

and measure trace amounts of polychlorinated biphenyls (PCBs) in crude oil, quite a complicated mixture. (These and the following JEOL applications, along with many others, can be found on the JEOL website at www.jeol.com.)

While we have focused on positive ion MS, there are classes of compounds that give much better mass spectral detection limits as negative ions. The use of a highly selective ionization source, a tunable-energy electron beam (the TEEM™ from JEOL), allows negative ions to be formed directly from analyte molecules. The electron beam energy can be tuned in the range of 0–25 eV; this permits selective ionization of the analyte, not matrix molecules. The combination of the tunable-energy electron ionization source with the JEOL GC-MS system into an instrument called the TEEMmate™ has permitted the determination of less than 10 ppb of explosives such as TNT, nitroglycerin, pentaerythritol tetranitrate, and RDX in complex matrices like soils by negative ion mass spectrometry. An example is shown in Fig. 10.33. The instrument has also been applied to the detection of chemical warfare agents, bacterial spores, and environmental contaminants.

10.3.3. Protein-Sequencing Analysis (Proteomics)

Proteins make up some of the most important components of the human body and indeed all animal life. They are major components in all living cells and are therefore of great importance in all chemical studies of life sciences. It is estimated that in human blood plasma alone there are about 100,000 proteins ranging in concentration from millimolar levels for proteins such as albumin to femtomolar levels for proteins such as tumor necrosis factor.

Proteins are made up of amino acids, which, as the name implies, include both an amino group and a carboxylic acid. There are 20 essential amino acids and the many hundreds of thousands of proteins are made up of these building blocks. What

Figure 10.33 Determination of less than 10 ppb of a mixture of explosives spiked into a "dirty" matrix by negative ion MS using a JEOL TEEMmate™ with a highly selective tunable energy electron ionization source. [Courtesy of JEOL USA, Inc., Peabody, MA (www.jeol.com).]

distinguishes one protein from another is the sequence in which these amino acids are arranged in the molecule. Sequencing has become a cornerstone of the research into proteomics, the study of protein structure and function. There are several approaches that can be taken to identify the sequence of amino acids in a protein by MS. The reference by de Hoffman and Stroobant contains a large section on analysis of biomolecules by MS. 2D gel electrophoresis separation followed by MALDI-TOF MS and digestion of the protein into peptide fragments followed by HPLC-MS-MS are two ways in which protein sequences and structure can be obtained. A detailed description of the HPLC-MS^n approach to protein sequencing is given in Chapter 13 (Section 13.1.7). The analysis generally requires digestion of the protein with trypsin to form short peptide fragments. Such a tryptic digest of a pure 14 kDa protein was analyzed on a TOF-based LC-MS system, the JEOL Accu-TOF™ using nanoelectrospray ionization. As expected for ESI, multiply charged peptides were seen in the mass spectrum. A database search of a protein library (ProFound, available on the Internet at http://prowl.rockefeller.edu/cgi-bin/ProFound) resulted in 10 possible proteins matches, one of which had a 100% probability.

MS is used to identify proteins, protein complexes with DNA, and even intact ribosomes (a multiple protein–RNA structure that produces proteins), but the analysis is slow. Due to the large number of possible proteins and the wide concentration range found in samples, current MS methods can identify only about 1000 proteins in up to 10 samples a month.

10.3.4. Gas Analysis

The direct measurement of gases is an important use of MS.

One direct application of MS in medicine is in blood gas analysis. The speed of the analysis allows surgeons to monitor the blood of patients during operations. The concentration in the blood of compounds such as CO_2, CO, O_2, N_2, and anesthetics such as N_2O can be controlled in this way.

Mass spectrometric analysis of gases has many industrial applications. In petroleum chemistry applications, MS is used to identify chemicals in hydrocarbon fractions, to detect and identify impurities such as ethane in polymer feedstocks such as ethylene, and to determine CO, CO_2, H_2, and N_2 in feed streams used for making alcohols. MS is also used for analysis of jet and automobile exhausts. Online quadrupole MS gas analyzers, also called residual gas analyzers (RGA), are used to monitor noble gases, solvent vapors, corrosive acid vapors, inorganic gases, and other species in semiconductor and optoelectronics manufacturing processes. The composition of feed, exhaust and stack gases, evolved gas, and headspace gas in many industries can be determined using these types of mass spectrometers. Systems such as the MKS Instruments, Inc. (Andover, MA) Mini-Lab™ online analyzers can track gas concentrations over a wide dynamic range (ppb to percent levels) in seconds. These systems are used to monitor trace contamination in bulk gases, to study catalysis, fuel cells, and semiconductor wafer processing. For example, in the deposition of epitaxial silicon layers on a silicon wafer, a Mini-Lab™ quadrupole mass analyzer was used to measure oxygen and water vapor, which can interfere with deposition on the wafer surface. The unit was also used to study the deposition of Si. The deposition of Si occurs by hydrogen reduction of $SiHCl_3$, a gas, to form solid Si. The $SiHCl_3$, gaseous intermediates such as $SiCl_2$ and gaseous products such as HCl can all be measured using the mass spectrometer. (This and other applications as well as a useful tutorial with pictures of the quadrupole and other parts of the mass spectrometer can be found on the MKS Instruments, Inc. website at www.mksinst.com). Even smaller quadrupole mass analyzer systems are available as field portable gas analyzers; quadrupoles in these field portable systems are as small as 0.5 inches in length.

10.3.5. Environmental Applications

Molecular MS, especially GC-MS, is widely used in the identification and quantitative measurement of compounds that are of concern in the environment. PCBs, dioxins, dibenzofurans, and pesticides can be measured in the tissues of shellfish, fish, birds, and other animals, as well as in water, soil, sediments, and materials using MS, generally coupled to GC.

Brominated flame-retardants are under increasing scrutiny for their environmental impact. Common brominated flame-retardants are the polybrominated diphenyl ethers (PBDEs), of which there are 209 individual compounds (congeners) just as there are for polychlorinated biphenyls. GC-high-resolution MS is the method of choice for determining PBDEs in the environment. PCBs or PBDE congeners with the same number of halogen substituents (from 1 to 10 halogens are possible) are isomers with identical molecular weights and very similar mass spectra. The retention time information from the GC portion of the GC-MS will be needed to distinguish among such isomers. The isomers cannot be identified based only on the mass spectra.

GC-MS is the basis for most US Environmental Protection Agency (EPA) official analytical methods for organic pollutants in water and wastewater. Detection limits in the ppb to ppt ranges are common. Accuracy of $\pm 20\%$ relative standard deviation is usual.

10.3.6. Other Applications of Molecular MS

Biomolecules of all types can be analyzed by MS. The molecular weight and sequence of bases in nucleotides can be determined by MALDI-TOF MS or by MS-MS. This allows the comparison of normal and mutated genes, for example. Analysis of nucleotides is important in medicine, agriculture, environmental science, and molecular biology. Oligosaccharides, long chains of sugar units, are extremely complex in structure, more so than proteins or nucleotides. These compounds can be linear or branched, and the sugar units have isomers and specific configurations at the bonds between units. All of this structural information can be obtained from mass spectral analysis of the oligosaccharides. FT-ICR and MS^n instruments are valuable in determining the structure of these complex molecules. Lipids such as fatty acids and steroids can be determined by MS. Drugs, toxins, and other compounds often undergo metabolic changes in the human body; MS is used to measure metabolites of compounds as well as the compounds themselves.

In organic and inorganic chemistry, MS can be used to identify reaction products and byproducts. Impurities at concentrations as low as parts per trillion (ppt) can be detected; MS is widely used for this purpose. In inorganic chemistry, with special inlet techniques, the elemental composition of materials as diverse as crystals and semiconductors can be determined. Reaction kinetics and ion–molecule reactions can be studied using MS.

Polymers are routinely characterized by MS, with MALDI now being the most common form of sampling and ionization of polymeric materials, using MALDI-TOF MS or MALDI-FTMS. Synthetic polymers actually have a distribution of chain lengths and so have a distribution of molecular weights. Polymer molecular weights are reported as the average MW, that is the center of the MW distribution. The average mass of polymers can be determined more accurately by MS than by the more commonly employed method of gel permeation chromatography (Chapter 13, Section 13.4).

10.3.7. Limitations of Molecular MS

The major limitation to the use of MS is that compounds must be volatile or must be able to be put into the gas phase without decomposing. A variety of ionization sources have been developed to handle materials that are volatile, semivolatile, nonvolatile, or thermally labile, as described in Chapters 9 and 13. For compounds that are not volatile, chemical derivatization to a more volatile form can be performed to make them suitable for MS or GC-MS, as described in Chapter 12. Carboxylic acids can be converted to the corresponding volatile methyl esters, for example; trimethylsilane is another common derivatizing agent used to make volatile ether derivatives.

Certain isomers cannot be distinguished by MS alone. Some isomers, such as the PCBs and PBDE isomers discussed in Section 10.3.5, may be able to be separated by chromatography prior to analysis by MS.

10.4. ATOMIC MS

The original use of MS was for the detection and determination of elements. The elements have different masses, so MS provided a method of determining atomic weights. The various elements as well as their isotopes could be separated from each other with this technique. This made it possible to obtain the isotope distribution of pure elements (Appendix 10.1). The application of MS to the determination of atomic weights and isotope distribution was crucial in the development of atomic chemistry and physics. While there are several types of ionization sources for atomic MS, such as the glow discharge (GD) and spark source (described in Chapters 7 and 9), used for atomic mass spectrometric analysis of solids, it is the development of the inductively coupled plasma

(ICP) and the quadrupole mass analyzer that has caused a recent enormous increase in the use of atomic MS as an analytical tool.

10.4.1. Inductively Coupled Plasma Mass Spectrometry (ICP-MS)

ICP-MS permits determination of most of the elements in the periodic table at very low concentration levels. More than 60 different elements from lithium to uranium can be determined in a few seconds using an ICPMS with quadrupole mass analyzer. This technique has great advantages over other techniques for elemental analysis in that ICP-MS can be used to determine most elements at high sensitivity and over a wide range of concentrations. In addition, isotope ratios can be obtained, providing geochemical and geochronological information, for example, that cannot be obtained by other elemental analysis techniques.

There are several properties of the mass spectrometer as an analyzer and the argon ICP as an ionization source that make ICP-MS an attractive combination. The ICP has a high ionization efficiency, which approaches 100% for most of the elements in the periodic table, and it produces mainly singly charged positive ions. The mass spectra therefore are very simple and elements are easily identified from the masses and the isotope ratios measured. ICP-MS in this respect overcomes the major problem of ICP optical emission spectrometry (i.e., the line-rich complicated optical emission spectrum with high and variable background). In addition the mass spectrometer is very sensitive, with detection limits up to three orders of magnitude better than ICP-OES (see Appendix 7.1) and is linear over a wide dynamic range of up to five orders of magnitude. Table 10.20 presents some of the aspects of ICP-OES and MS that complement each other. The ability to rapidly analyze solutions of dissolved inorganic materials (metals, rocks, bones, ceramics, ash, and the

Table 10.20 Complementary Aspects of MS and ICP-OES: Why ICP-MS Evolved

ICP-OES	MS
Efficient but mild ionization source (produces mainly singly charged ions)	Ion source required
Sample introduction for solutions of inorganic salts is rapid and convenient	Sample introduction can be difficult for inorganic samples (generally not volatile). Thermal ionization, spark source, glow discharge and lasers restricted to solid samples and are time consuming
Sample introduction is at atmospheric pressure. Efficiency is poor (5% of sample into plasma; rest to waste using conventional nebulizer)	Requires reduced-pressure sample introduction
Matrix or solvent interelement effects are observed but relatively large amounts of dissolved solids can be tolerated	Limited to small quantities of sample and low amounts ($<1\%$) of dissolved solids
Complicated spectra with frequent spectral overlaps	Relatively simple spectra
Detectability is limited by relatively high background continuum over much of the useful wavelength range	Very low background level throughout a large section of the mass range
Moderate sensitivity (ppm to ppb range)	Excellent sensitivity (ppb to ppt range)
Isotope ratios cannot be determined	Isotope ratio determinations are possible

Figure 10.34 Schematic diagram of an ICP-MS system with a quadrupole mass analyzer.

like) combined with the simple mass spectra, low background and excellent sensitivity led to the rapid spread in the use of ICP-MS in many fields.

The RF ICP source was described in Chapter 7, Section 7.3.1, and the quadrupole mass analyzer was described in Chapter 9. The interfacing between the ICP source and the quadrupole mass analyzer will be looked at with a little more detail to explain how ICP-MS provides the information it does.

Simplistically the instrument is an ICP interfaced with a quadrupole MS as shown in Fig. 10.34. The ICP source is horizontal, with the argon plasma concentric to the mass spectrometer inlet. The ICP operates at atmospheric pressure and at a temperature of about 10,000 K. On the other hand, the mass spectrometer requires a high vacuum $(10^{-4}-10^{-6}$ torr) and operates at room temperature. Interfacing of the two systems is therefore the critical problem to be overcome. Most modern ICP-MS systems have an interface similar to the one shown in Fig. 10.35. The argon ICP plasma is on the right side of the diagram. Ions from the plasma enter into the mass spectrometer through a two-stage interface. The plasma is centered on the *sampler cone*, and ions and plasma

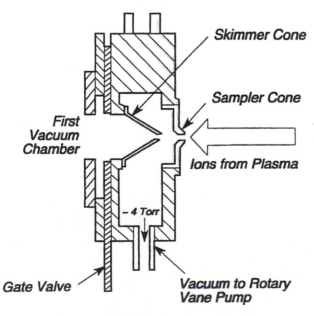

Figure 10.35 Close-up of the ICP-MS interface, showing the sampler and skimmer cones. [Courtesy of PerkinElmer Instruments, Shelton, CT (http://las.perkinelmer.com).]

gas pass through the orifice in the cone into a vacuum-pumped region. Most of the argon gas is pumped away in this region. The remaining ions pass through the *skimmer cone* into the mass spectrometer. The skimmer cone is a sharper-angled cone with an orifice of about 0.9 mm in diameter. The design of this cone restricts the flow of ions into the mass spectrometer to the central part of the flow initially coming from the plasma. The region behind the skimmer cone is evacuated to a pressure of about 10^{-4} torr by a turbomolecular pump. This region can be isolated from the higher pressure of the interface region by a gate valve. This permits the sampler and skimmer cones to be removed for cleaning without breaking the vacuum in the mass spectrometer. Both sampler and skimmer cones are made of either nickel or platinum, and are water cooled by contact with water flowing within the interface chamber.

A series of ion-focusing elements (ion lenses) similar to those developed for double-focusing mass spectrometers have been utilized to introduce the ions into the quadrupole. Also, photon blockers have eliminated interference effects due to the presence of large numbers of photons reaching the detector. Therefore, background signals have been largely eliminated in modern ICP-MS instruments.

Other mass analyzers are used with the ICP ionization source, including high-resolution magnetic sectors and TOFs. ICP-TOF mass analyzer systems as well as other designs have the advantage of simultaneous measurement of ions. Simultaneous measurement is critical for accurate determination of isotope abundances as well as for accurate quantitative work using isotope-dilution. The article by Blades contains an excellent review of atomic MS instruments and an extensive bibliography. However, the instrumentation is constantly improving. In 2003, Varian, Inc. (www.varianinc.com) introduced a new quadrupole-based ICP-MS design with an ion mirror that reflects the analyte ions through 90°, while the neutrals and photons pass through. This results in exceptionally high S/N and extremely low detection limits compared with older quadrupole instrument designs. The instrument uses an all-digital extended range detector that gives nine orders of linear dynamic range, meaning that concentrations from ppt to hundreds of ppm can be measured reliably.

10.4.2. Applications of Atomic MS

The high-temperature plasma decomposes the sample into its elements. A high percentage of these elements are ionized in the plasma and therefore do not need to pass through an additional ionization source. ICP-MS is particularly useful for rapid multielement analysis of metals and nonmetals at concentrations of ppb and ppt.

Only unit mass (low) resolution is required to discriminate between different elements, because isotopes of different elements differ by 1 u as can be seen in Appendix 10.1. There are only a few isotopic overlaps between elements, so one can usually find an isotope to measure for any given element. In fact, there is only one element that cannot be definitely identified by ICP-MS, the element indium. (One of the problems at the end of the chapter asks you to explain why.) The abundance of each isotope is a quantitative measure of that element's concentration in the original sample. The isotope patterns for the elements are shown in Fig. 10.36.

The applications discussed are from many forms of atomic MS, including ICP-MS, glow discharge MS (GDMS), and coupled chromatography-ICP-MS. The websites of the major instrument manufacturers (Agilent, LECO, PerkinElmer, Thermo Electron, and Varian, to name a few for ICP-MS) are an excellent source for applications of the techniques.

Some of the many uses for ICP-MS include analysis of environmental samples for ppb levels of trace metals and nonmetals, the analysis of body fluids for elemental toxins such as lead and arsenic, determination of trace elements in geological samples, metals and alloys,

Figure 10.36 Element isotope patterns. [Modified from graphic by PE Sciex, courtesy of PerkinElmer Life and Analytical Sciences, Shelton, CT (http://las.perkinelmer.com).]

determination of isotope ratios, analysis of ceramics and semiconductors, analysis of pharmaceutical and cosmetic samples, determination of platinum group catalysts in polymers, forensic analysis, elemental analysis in the petroleum and chemical industries, and metals determinations in clinical chemistry and food chemistry.

The most common samples analyzed by ICP-MS are aqueous solutions. The sample is dissolved in acid, digested or fused in molten salt (all described in Chapter 1), and then diluted to volume with water. All acids, bases, reagents, and water must be of extremely high purity, given the sensitivity of the ICP-MS technique. "Ultratrace metals" grade acids, solvents, and deionized water systems are all commercially available. The aqueous solution is introduced into the plasma using a peristaltic pump, nebulizer, and spray chamber system identical to those used for ICP-OES (Chapter 7, Section 7.3.1.3).

Solid samples can be analyzed by laser ablation ICP-MS or by coupling an Electrothermal Vaporizer (ETV) (also called a graphite furnace) to the ICP-MS. Both sample introduction systems are described in Chapter 7, Section 7.3.1.3, and operate the same way they do for ICP-OES. The direct analysis of solid samples is certainly desirable as it eliminates sample preparation and the errors that go along with sample preparation. Table 10.21 compares the detection limits of ETV-ICP-MS with those of GFAAS. Solids can also be analyzed by GDMS and spark source MS. Laser ablation-ICP-MS is used to measure elements in fluid inclusions in rocks, microscopic features in heterogeneous materials, and individual crystals in samples such as granite. It can be used for analysis of artworks and jewelry, since the small laser spot size (<10 μm) available with modern lasers results in minimal damage to the object.

Quantitative analysis by ICP-MS is usually done with external calibration standards and the addition of internal standards to all standards and samples. When a large number of

Table 10.21 Approximate Detection Limits for Electrothermal Atomizer-ICP-MS

Elements	ETA-ICP-MS ng/mL (2 μL)	pg	GFAAS ng/ml (10 μL)	pg
^{107}Ag	0.08	0.16	0.01	0.1
^{27}Al	0.03	0.05	0.2	2
^{75}As	0.05	0.1	1	10
^{44}Ca	0.7	1.4	0.5	5
^{114}Cd	0.15	0.3	0.01	0.1
^{52}Cr	0.1	0.2	1	10
^{39}K	1.5	3	0.1	1
^{23}Na	0.2	0.4	1	10
^{58}Ni	0.47	0.93	1	10
^{208}Pb	0.1	0.3	0.2	2
^{78}Se	5.7	11.4	1	10
^{28}Si	2.7	5.4	5	50
^{64}Zn	0.2	0.4	0.05	0.5

Note: This table presents a comparison of GFAAS with the use of a graphite furnace (electrothermal atomizer) as the sample vaporization step in conjunction with ICP-MS. The solution detection limit, the sample volume, and the absolute detection limit in picograms are given for each technique. The isotope measured is specified for the ICP-MS technique; the isotope number has no meaning for the AAS results.

elements across the periodic table are to be determined, it is usual to add Li, Y, In, Tb, and Bi, and measure the ions ^{6}Li, ^{89}Y, ^{115}In, ^{159}Tb, and ^{209}Bi as internal standards (unless you need to measure one of these elements as an analyte). Not all of the internal standard elements are used to quantitate every analyte. The internal standard that is most closely matched in first ionization potential to the analyte is generally used, since this will compensate for ionization interferences in matrices containing easily ionized elements such as Na. Results obtained using this approach are generally very accurate and precise. Fig. 10.37 presents typical spike recovery and precision data for ICP-MS determination of 25 elements in a certified reference material (CRM) "Trace Elements in Drinking Water" from High Purity Standards, Charleston, SC.

10.4.2.1. Geological and Materials Characterization Applications

Atomic MS has the ability to measure the isotope ratios of the elements. It is of interest to note that this was the original application of mass spectrometry both by Aston and by Thompson. Both quantitative measurement of trace elements in minerals and accurate isotope ratio determinations are extremely important in geology, climatology, and earth science, among other fields. Most elements have fixed ratios because their isotopes are stable, but some elements, such as uranium, carbon, lead, and strontium have isotopes that vary in their abundance because of natural radioactive decay processes. The isotope ratios of elements such as these can be used for estimating the age of a sample or for identifying the source of the element. The application of atomic MS to geological and fossil samples for accurate isotope ratios permits the dating of samples (geochronology), as well as applications in paleothermometry, marine science, and climatology.

The incorporation rates of the elements uranium, magnesium, and strontium into corals appear to be a function of water temperature. Accurate measurements of the elements and their ratios in fossil corals can provide information to climatologists about historical sea temperatures (paleothermometry). An ICP-MS method for determining these ratios accurately using internal standard isotope dilution is described by Le Cornec and Correge and can be read online (see the bibliography). As these authors note, the precision of the ICP-MS instrument must be very high, because the rate of change of the Sr/Ca ratio is only 0.9% per °C and yearly variations in the sea surface temperature are only 5–6°C.

A long-standing method for thermometry has been to measure the stable oxygen isotope ratios in carbonate rocks using MS. Uranium/lead isotope ratios, determined by MS, are used to estimate the age of the earth.

Accurate isotope ratios often require high-precision mass spectrometers. Instruments capable of such high precision are either ICP-TOF-MS or high-resolution magnetic sector ICP-MS instruments of either the single detector or multicollector type. An excellent fundamental discussion of high precision isotope ratio measurements using the LECO Renaissance^{TM} TOF-based instrument is the cited technical brief by Allen and Georgitis. Despite the high cost of the magnetic sector instruments (approximately two to six times the cost of a quadrupole ICP-MS), it is estimated that in 2003 there were more than 350 magnetic sector single detector ICP-MS instruments and more than 100 multicollector magnetic sector ICP-MS units in operation around the world, testifying to the need for high-precision, high mass resolution measurements. The high mass resolution permits analysis of complex minerals and other matrices. The high sensitivity of these instruments makes them especially suitable for laser ablation analysis of geological specimens. (The instrument numbers and cost estimates are courtesy of Chuck Douthitt of Thermo Electron, a manufacturer of magnetic sector based high-resolution instruments.)

Precision and Recovery Data for High Purity "Trace Metals in Drinking Water" (CRM)

Analyte	Isotope	Average Measured Conc (µg/L)	Std. Dev.	Rel. % Dif.	High Purity Certified Value	Recovery of Certified Value (%)	Spike Level (µg/L)	Average Spike Recovery (%)	Std. Dev. of Spike Rec.	Rel. % Dif.
Be	9	17.17	0.13	0.75	20	85.8	50	92.1	1.0	1.1
Na	23	5711.45	113.71	1.99	6000	95.2	—	—	—	—
Mg	24	8544.22	11.45	0.13	9000	94.9	—	—	—	—
Al	27	113.60	0.37	0.33	115	98.8	50	106.7	3.8	3.5
K	39	2421.85	22.01	0.91	2500	96.9	—	—	—	—
Ca	44	33871.41	611.84	1.81	35000	96.8	—	—	—	—
V	51	28.92	0.08	0.28	30	96.4	50	101.0	1.3	1.2
Cr	52	18.99	0.04	0.22	20	94.9	50	103.2	5.6	5.4
Mn	55	33.58	0.73	2.18	35	95.9	50	99.0	2.6	2.6
Co	59	23.71	0.81	3.41	25	94.8	50	100.9	2.2	2.2
Ni	60	60.46	0.26	0.43	60	100.8	50	102.6	3.9	3.8
Cu	63	19.32	0.51	2.64	20	96.6	50	101.1	4.6	4.5
Zn	66	68.98	1.29	1.87	70	98.5	50	101.2	3.2	3.2
As	75	78.67	0.54	0.68	80	98.3	50	108.4	0.0	0.0
Se	82	9.21	0.15	1.61	10	92.1	50	91.7	0.6	0.7
Mo	98	94.93	3.23	3.40	100	94.9	50	105.4	1.1	1.0
Ag	107	2.34	0.04	1.78	2.5	93.6	50	99.4	1.0	1.0
Cd	111	11.35	0.01	0.12	12	94.6	50	96.3	1.1	1.2
Sb	121	9.55	0.03	0.31	10	95.5	50	99.8	1.0	1.0
Ba	135	48.49	0.73	1.51	50	97.0	50	117.5	0.3	0.2
Hg	202	0.11	0.03	27.27	Not Available	—	1.5	100.7	0.04	0.04
Tl	205	10.21	0.14	1.34	10	102.1	50	106.3	2.1	2.0
Pb	208	34.85	0.73	2.10	35	99.6	50	109.7	4.8	4.4
Th	232	0.01	0.01	79.75	Not Available	—	50	107.6	0.8	0.8
U	238	10.16	0.20	1.97	10	101.6	50	106.7	2.7	2.5

Figure 10.37 Precision and recovery data for "Trace Metals in Drinking Water" Certified Reference Material (CRM) by ICP-MS. The ICP-MS used was a PE SCIEX Elan® 6000 and the CRM is from High Purity Standards, Charleston, SC. (From Wolf et al. 1995, used with permission.)

Trace levels of all of the rare earth elements (REEs) can be determined in rock samples and their distribution provides critical information to geologists. Some of the REE concentrations in basalt samples are in the sub-ppm range, but can be measured accurately in basalt samples that have been fused with lithium metaborate and diluted 5000-fold. The high dilution is necessary to keep the total dissolved solids <2000 mg/L, as the ICP-MS instrument does not tolerate high salt solutions well. The application is described by Ridd (see the bibliography). Use of ICP-MS for the REEs gives spectra that are simpler than those obtained by ICP-OES and all of the REEs can be measured, which is not the case with techniques such as neutron activation analysis.

An interesting overlap of geology and synthetic materials characterization is the result of improvements in recent years in the ability to synthesize gem quality diamonds in large quantities and in rare and therefore valuable colors. In addition, it is now possible to chemically diffuse trace amounts of metals like Ti and Be into low quality sapphires and rubies, dramatically improving their color and value. Rubies and sapphires are both Al_2O_3 minerals, and as few as five Be atoms in the Al_2O_3 crystal lattice can result in dramatic color change in the gem. In most cases, these synthetic diamonds and chemically treated gemstones cannot be detected by traditional gemological testing. Laser ablation ICP-MS is now being used to quantify Be and other trace elements in gems to distinguish natural gemstones from treated or synthetic ones. (Fig. 7.42 in Chapter 7 shows two laser ablation spots in a garnet crystal, for example.) The cited article by Yarnell and the website of the Gemological Institute of America (www.gia.org) provide details of this application and other spectroscopic techniques used to examine gems.

GDMS is used for the characterization of solid materials, especially coated and layered materials. The GD ablates the sample from the surface inward, providing the ability to depth profile materials. Bulk analysis of solid metals with detection limits in the low ppb range is possible. Coatings and layered materials a few nanometers thick can be characterized by GDMS. Examples are given by Broekaert. The major disadvantage to GDMS is its relatively poor precision and the severe matrix effects encountered in analysis. Matrix effects are described below in Section 10.4.3.1. GDMS measurements generally have 20–30% RSD compared with 5% RSD or even less for solutions analyzed by ICP-MS. The poor precision and severe matrix effects make GDMS suitable only for qualitative and semi-quantitative work unless matrix-matched standards are available. In the experience of one of the textbook authors, GDMS determination of an easily ionized trace element in a heavy metal matrix gave results that were $100\times$ higher than results obtained by ICP-OES and AAS after dissolution of the sample. The GDMS results were in error because of a matrix effect; the analyte volatized and ionized from the metal sample much differently than it volatilized and ionized from the glass material being used as the calibration standard.

10.4.2.2.   Speciation by Coupled Chromatography-ICP-MS

The coupling of chromatography or capillary electrophoresis (CE) to ICP-MS allows the separation of complex mixtures and speciation by compound or oxidation state of the elements present. GC is described in Chapter 12; liquid chromatography (LC) including HPLC, ion chromatography (IC), and the nonchromatographic separation method of CE are described in Chapter 13. All of these separation instruments have been interfaced to ICP-MS systems. Coupling a separation technique to a quadrupole ICP-MS may require the use of additional software that can handle transient signals; such software is commercially available.

Chong and Houk used an argon ICP-mass spectrometer as a GC detector for chemical compounds containing nitrogen, oxygen, phosphorus, sulfur, carbon, chlorine, bromine, boron, and iodine.

An application of ion chromatographic separation of ionic species coupled to ICP-MS is the determination of halogen oxyanions such as IO_4^-, IO_3^-, BrO_3^-, and ClO_3^-. IC generally uses a conductivity detector (described in Chapter 13), which is not a specific detector. Identification of ionic species must be made based on retention time and comparison to known standards. There is always some risk that two ionic species will have the same retention time, which could lead to misidentification of the species. Coupling the IC to an ICP-MS allows the anions to be separated, their retention times measured and the specific halogens confirmed by ICP-MS. The mass spectrometer identified the halogen in the species; for example, for the iodine oxyanions, the MS monitored the ^{127}I isotope, confirming that the ions giving rise to these peaks in the ion chromatogram were iodine-containing ions. The ions can be measured quantitatively using either the IC conductivity detector or the much more sensitive detector on the mass spectrometer, depending on the concentrations.

Organic solvents and organic petroleum fractions often need to be analyzed by ICP-MS for trace element content. HPLC separations are often carried out in organic solvents ranging from nonpolar solvents like xylene to polar alcohol–water mixtures. Organic solvents can be analyzed by ICP-MS with the same limitations that organic solvents present in ICP-OES. The RF power, the temperature of the ICP spray chamber, the nebulizer gas flow rate and sample uptake rate may need to be adjusted depending on the volatility of the solvent to avoid extinguishing the plasma or causing carbon deposition on the sampler cone. Running organic solvents can cause specific polyatomic interferences, discussed in Section 10.4.3. The reference by Elliot at the bottom of Table 10.22 discusses analysis of organic solvents for trace metals and can be found online.

Speciation of compounds of environmental concern is another important application of coupled GC-ICP-MS and LC-ICP-MS. Arsenic compounds in shellfish are one example of the importance of being able to speciate the arsenic containing compounds, that is, to determine the exact chemical forms of arsenic present. Most of the arsenic in shellfish (80–99%) in is the form of organoarsenic compounds, including arsenobetaine, monomethylated, and dimethylated arsenic acids. Arsenobetaine is not toxic because humans do not metabolize it, while the other organoarsenic compounds are much less toxic than inorganic arsenic compounds. To assess the risk of eating shellfish based on a determination of the total amount of arsenic present would greatly overestimate the danger of eating shellfish. Inorganic arsenic is present as well, in two oxidation states, As(III) and As(V). As(III) is more toxic than As(V). The ability to separate all of these arsenic species and detect them at extremely low levels by LC-ICP-MS permits a much better assessment of exposure and hazards from consumption of water and food than a determination of total As by elemental techniques such as ICP-OES, AAS, AFS, or ICP-MS. The cited articles by Milstein et al. describe arsenic speciation using IC-ICP-MS.

10.4.2.3.   *Applications in Food Chemistry, Environmental Chemistry, Biochemistry, Clinical Chemistry, and Medicine*

Foods and beverages of many types have been analyzed by ICP-MS. Solid or semisolid samples are generally digested with mineral acid, as are some beverages. Peanut butter, commercial breakfast cereal, dried milk, fish and shellfish, wine, beer, and the like have been analyzed for trace elements such as Cu, Fe, Se, and Zn for nutritional purposes as well as for toxic metals like As and Pb. Al has been determined in many foods because dietary Al was being studied for a possible link to Alzheimer's disease.

Whole blood and serum have been analyzed by ICP-MS for Al, Cu, Se, Zn, As, Cd, Mn, and Pb among other elements. Blood lead (Pb) measurements are done to determine exposure to lead-based paint in infants and small children. Pb, like many of the heavy metals, is a central nervous system toxin, and ingestion of lead-based paint chips or exposure to lead-containing dust has been implicated in learning disabilities in small children. ICP-MS is sensitive enough to permit multielement quantitative analysis of blood using sample volumes of 100 μL or less. In addition, the isotopic ratio pattern of the four Pb isotopes can be used like a fingerprint to match the source of lead (home, environment, industrial exposure) to the lead in the patient's blood (Allen and Georgitis).

The inherent magnetic resonance imaging (MRI) of tumors cannot reliably distinguish tumors from normal tissue. However, gadolinium compounds targeted at tumors can enhance the MRI signal from tumors as seen in Fig. 10.38. Gadolinium affects the T1 relaxation time in MRI, making the tissue that has absorbed the Gd compound "stand out" from adjacent tissue. The subject in Fig. 10.38 is a rat with a mammary tumor. The tumor is at the top of both images just to the left of center. Before injection with the Gd-containing contrast agent, the tumor cannot be distinguished from the surrounding tissue very well. Following injection with the contrast agent the tumor is clearly enhanced (seen as a much brighter area) because of preferential uptake of the contrast agent by the tumor. The other bright area in the lower right of the post-injection image is the rat's liver/kidneys. The use of ICP-MS to accurately determine the amount of Gd in rat tissues, blood, plasma and serum is discussed in the article by Skelly Frame and Uzgiris. Accurate concentrations were needed so that the dose of contrast agent could be optimized for use in humans to obtain the best image with the lowest possible dose. Blood, serum, and plasma samples of 30–100 μL and tissue samples of 0.1–1.0 g wet weight were microwave digested in high purity nitric acid and analyzed after dilution. The Gd isotopes 156 and 158 were measured; the detection limit was ∼0.005 ppb Gd.

Environmental applications of ICP-MS are numerous, and include analysis of water, wastewater, soil, sediment, air particulates, and so on. A typical environmental analysis is to determine the leachable metals from soil or sediment; the solid is not dissolved but leached or extracted to determine labile elements. These labile or leachable elements are the ones that might be mobilized from a landfill into a drinking water supply, for example. Figure 10.39 gives an example of determining leachable metals from an NIST Standard Reference Material (SRM) soil sample by ICP-MS.

As is the case with isotopic labeling of molecules, enriched levels of stable isotopes of elements can be used as tracers. Isotopes of elements can be used as nutritional supplements for plants or animals to trace absorption, assimilation, and metabolism of elements (Allen and Georgitis). Processes such as biomethylation of elements like mercury and arsenic in the environment can be studied using isotopically enriched elements. In some cases, methylated metals are more toxic than the inorganic species, and generally accumulate up the food chain.

10.4.3. Interferences in Atomic MS

Mass spectral interferences are observed in atomic MS. These interferences are of two types, direct m/z overlap between two ions, and matrix interferences, which cause enhancement or suppression of analyte signals.

10.4.3.1.   Matrix Effects

The introduction of the sample into the plasma suffers from the same problems in ICP-MS as in ICP emission spectrometry. These include the complicated process of nebulization and atomization. These processes occur before introduction of the sample into the plasma-mass spectrometer system.

Figure 10.38 MRI of a tumor in a rat before injection with a Gd-containing contrast agent (a) and after injection (b). The tumor is at the top of each image slightly to the left of center. The contrast agent is preferentially taken up by the tumor and appears enhanced (brighter) compared to the surrounding muscle tissue. Images courtesy of Dr. E.E. Uzgiris.

The interface between the two systems includes the 1 torr region where deposition of products can occur. Severe suppression of signal has been observed when high concentrations of dissolved solids are present. This may be caused by suppression of ionization by other more easily ionized elements present in the sample. The exact cause of this interference is not clear, but the fact remains that interference does take place. The problem can be overcome to some extent by limiting the concentration in the samples to less than 0.2% total solids. This can be a serious limitation, particularly when body fluids or fused minerals are being examined.

The solvent and other elements present in the sample cause matrix effects. These affect atomization efficiency, ionization efficiency, and therefore the strength of the MS signal. This directly impacts quantitative results. Signals may be suppressed or enhanced by matrix effects. Aqueous solutions act very differently from organic solvents, which in turn act differently from each other. The problem can be overcome for the most part by matrix matching (i.e., the standards used for calibration are matched for acid concentration, major elements, viscosity, etc. to the matrix of the samples being analyzed). This is similar to atomic absorption and atomic emission spectrometry where the same requirement in matching solvent and predominant matrix components is required for accurate quantitative analysis. The use of internal standards will also compensate for some matrix effects and will improve the accuracy and precision of ICP-MS measurements.

		Measured Conc (mg/kg)	STD DEV	RSD	RPD	NIST Leach Value (mg/kg)	Range Low	High	Spike Amount (ppb)	Spike Recovery (%)
Analyte	Mass									
Be	9	1.1	0.03	2.54	3.59				100.0	104
Al	27	20 066.5	621.82	3.10	4.38	18 000.0	12 000.0	23 000.0	100.0	–
V	51	48.2	1.63	3.39	4.79	42.0	34.0	50.0	100.0	98
Cr	52	23.7	0.23	0.95	1.35	20.0	15.0	25.0	100.0	97
Mn	55	493.0	14.82	3.01	4.25	490.0	400.0	620.0	100.0	110
Co	59	8.1	0.12	1.44	2.03	8.2	7.0	12.0	100.0	98
Ni	60	17.1	0.01	0.08	0.11	16.0	14.0	20.0	100.0	96
Cu	63	104.1	3.62	3.48	4.92	100.0	91.0	110.0	100.0	99
Zn	66	315.8	8.79	2.78	3.94	310.0	290.0	340.0	100.0	111
As	75	94.0	1.93	2.06	2.91	90.0	88.0	110.0	100.0	103
Se	82	2.2	0.11	4.92	6.96	NR			100.0	109
Mo	98	1.2	0.02	1.48	2.09	<2			100.0	105
Ag	107	4.3	0.03	0.67	0.94	4.0	2.5	5.5	100.0	102
Cd	111	40.0	0.91	2.27	3.20	40.0	32.0	46.0	100.0	102
Sb	123	3.9	0.14	3.71	5.25	<10			100.0	98
Ba	135	192.2	6.64	3.45	4.88	200.0	170.0	260.0	100.0	98
Tl	205	1.8	0.06	3.58	5.07				100.0	105
Pb	208	1087.3	35.25	3.24	4.58	1100.0	930.0	1500.0	100.0	132
Na	23	320.3	5.82	1.82	2.57	260.0	200.0	290.0		–
Mg	24	7726.8	201.46	2.61	3.69	0100.0	7200.0	8900.0		–
K	39	5064.4	128.65	2.54	3.59	3800.0	2600.0	5300.0		–
Ca	44	20 742.0	166.68	0.80	1.14	21 000.0	20 000.0	25 000.0		–
Fe	54	21 662.2	526.25	2.43	3.44	22 000.0	17 000.0	26 000.0		–

Results for NIST SRM 2711 – Moderately Contaminated Montana Soil

Figure 10.39 ICP-MS results from leaching of NIST SRM 2711 Moderately Contaminated Montana Soil. The ICP-MS used was a PE SCIEX Elan® 6000. (From Wolf et al. 1996, used with permission.)

10.4.3.2. *Spectral (Isobaric) Interferences*

In ICP-MS interferences are seen from argon ions and argon-containing polyatomic species, since these are present in high abundance in the argon plasma. Argon is composed of three isotopes, but ^{40}Ar is 99.600% of the total, so $^{40}Ar^+$ is the ion of most concern. $^{40}Ar^+$ and $^{40}Ca^+$ have the same nominal mass, and we can say that they overlap. A quadrupole MS cannot distinguish between the two. Such mass overlaps are called isobaric interferences; they are spectral interferences. Unfortunately for us, $^{40}Ca^+$ is the most abundant Ca isotope, constituting 96.941% of Ca (see Appendix 10.1). To determine Ca using a standard quadrupole ICP-MS instrument, another less-abundant Ca isotope must be chosen for measurement. Polyatomic ions such as ArH^+, ArN^+, $ArCl^+$, $ArOH^+$, and Ar_2^+ are commonly formed in the plasma and interfere with a number of possible analytes, as shown in Tables 10.22 and 10.23. These polyatomic spectral interferences are very problematic when the element affected is monoisotopic, as is the case with ^{75}As and ^{55}Mn or when the isotope affected comprises more than 90% abundance of the element, as happens with ^{56}Fe, ^{39}K, ^{40}Ca, and ^{51}V.

Polyatomic ionic species also form from air, solvents, and matrix components, such as oxides of metals present in large amounts in the sample as well as from the acids and other reagents used to digest samples. For example, titanium has five isotopes, ^{46}Ti, ^{47}Ti, ^{48}Ti, ^{49}Ti, and ^{50}Ti, which can form $Ti^{16}O$ oxide ions that interfere with ^{62}Ni, ^{63}Cu, ^{64}Zn, ^{65}Cu, and ^{66}Zn. Examples in addition to the argon ions mentioned earlier include CO_2^+, CO^+, SH^+, SO^{2+}, NOH^+, ClO^+, ArS^+, and so on, which are derived from the plasma gas, and from reactions of Ar with water, carbon, and other elements present in the solvent or the sample. Based on studies of these interfering ions, nitric acid is considered to be the most attractive acid to be used in preparing sample solutions. Unfortunately, not all elements are soluble in nitric acid alone. No ions occur, however, with a mass greater than 82 from any of the common acids. Therefore, the higher part of the mass range is unaffected by these interfering ions.

The analysis of organic solvents presents some unique polyatomic interferences. Tables 10.22 and 10.23 list potential polyatomic interfering species and the affected

Table 10.22　Potential Interferents in a Carbon Rich Plasma with Oxygen Addition

Potential interfering species	m/z	Affected element
$^{12}C_2$	24	Mg
$^{12}C^{13}C$	25	Mg
$^{40}Ar^{16}O$	56	Fe
$^{12}C^{16}O$	28	Si
$^{12}C^{16}O_2$	44	Ca
$^{40}Ar^{12}C$	52	Cr
$^{40}Ar^{12}C^{16}O$	68	Zn

Source: Modified from S. Elliott, *Analysis of organic solvents, including naphtha*, by ICP-MS, Varian Applications Note ICP-MS-19, Oct. 1998 (www.varianinc.com). All species would be in the form of ions as measured, used with permission.

element. Some analysts add oxygen to the plasma when running organic solvents to minimize carbon (soot) formation on the cones. The additional oxygen can not only create the polyatomic species listed in Table 10.22 but can also react with some elements to form refractory oxides, as happens in aqueous solution. Examples include $^{46}Ti^{16}O$, which interferes with ^{63}Cu and the rare earth elements, such as $^{143}Nd^{16}O$, which interferes with ^{159}Tb.

One approach to minimizing the formation of polyatomic species is to adjust plasma and operating conditions. A "cool plasma" or "cold plasma" can be formed by decreasing the power to the plasma. Using a cool plasma eliminates formation of or reduces many of the argon polyatomic ions. The drawbacks to changing the operating conditions to a cool

Table 10.23　Common Polyatomic Interferences in ICP-MS

Polyatomic ion interference	m/z	Affected element
$^{14}N_2$	28	^{28}Si
$^{12}C^{16}O$	28	^{28}Si
$^{14}N^{16}O^1H$	31	^{31}P
$^{16}O^{16}O$	32	^{32}S
$^{38}Ar^1H$	39	^{39}K
^{40}Ar	40	^{40}Ca
$^{35}Cl^{16}O$	51	^{51}V
$^{36}Ar^{16}O$	52	^{52}Cr
$^{40}Ar^{12}C$	52	^{52}Cr
$^{40}Ar^{14}NH$	55	^{55}Mn
$^{38}Ar^{16}O^1H$	55	^{55}Mn
$^{40}Ar^{16}O$	56	^{56}Fe
$^{23}Na^{40}Ar$	63	^{63}Cu
$^{40}Ar^{35}Cl$	75	^{75}As
$^{40}Ar^{40}Ar$	80	^{80}Se

plasma are that detection limits for some elements may be poorer than in a normal "hot plasma" and matrix effects may increase. It may be necessary to run samples under two sets of operating conditions to optimize detection limits for all elements.

Appendix 10.1 indicates that there are also direct spectral interferences of one element on another. ^{58}Ni overlaps with ^{58}Fe; even though ^{58}Fe is the least abundant Fe isotope at 0.28%, trying to measure trace ^{58}Ni (the most abundant Ni isotope at 68%) in an iron sample would be difficult. It would be better to measure ^{60}Ni, which is not interfered with by Fe. Other overlaps can be seen in this table; Zr and Mo have three isotopes that directly interfere with each other, Lu, Hf, and Yb all have a 176 isotope, while Ta, W, and Hf all have a 180 isotope. Although most ions formed in the plasma have a single positive charge, some doubly charged ions form with the same apparent mass as a singly charged ion of another element. Remember that we are still measuring m/z in atomic mass spectrometry, not mass. When the doubly charged ion is the most abundant isotope and is present in large quantities in the sample, this can cause significant problems. For example, ^{88}Sr^{2+} interferes with ^{44}Ca^{+} and ^{86}Sr^{2+} interferes with ^{43}Ca^{+}, ^{112}Cd^{2+} interferes with ^{56}Fe^{+}, and ^{138}Ba^{2+} interferes with ^{69}Ga^{+}. An analyst needs to consider the relative abundances and nature of the sample matrix and acids or other reagents used when assessing potential isobaric interferences in ICP-MS.

10.4.4. Instrumental Approaches to Eliminating Interferences

There are three primary instrumental approaches to eliminating or reducing interferences in atomic mass spectrometry: (1) use of high mass resolution ICP-MS (HR-ICP-MS); (2) use of a collision cell to break apart polyatomic interferences; (3) use of gas phase chemical reactions in a "reaction cell" to eliminate polyatomic interferences. The approach of changing the instrument operating conditions to form a cool or cold plasma has been discussed.

10.4.4.1.   High-Resolution ICP-MS (HR-ICP-MS)

As is the case with high-resolution molecular mass spectrometers, these magnetic sector or TOF based instruments permit accurate mass determination. Quadrupole systems are limited to unit mass resolution. Therefore, the use of HR-ICP-MS eliminates many of the spectral interferences seen with quadrupoles because the "overlaps" do not occur; the resolution is sufficient to separate the exact masses of the polyatomic species and analyte elements. HR-ICP-MS is the only way to measure accurately low levels of elements that suffer from significant interferences, including S, P, Si, F, and Cl, using MS. HR-ICP-MS allows interferences to be characterized using exact mass measurements. The high S/N of these instruments results in extremely low detection limits (parts per quintillion limits are claimed by one manufacturer). This allows direct analysis of low concentration samples without preconcentration, allows dilution of samples, and allows the use of low flow rates, minimizing matrix effects.

10.4.4.2.   Collision and Reaction Cells

A collision cell is often a small hexapole or quadrupole inserted between the ion optics and the mass analyzer of the instrument. As the ions, including polyatomic ions, pass through the cell, they are bombarded with gas molecules, usually helium or hydrogen. The collisions break apart the polyatomic ions, greatly reducing their numbers. A simple hexapole collision cell can reduce the interference on ^{56}Fe^{+} by the polyatomic ArO^{+} ion by 10^{2}.

The Dynamic Reaction Cell^{TM} (DRC) from PerkinElmer uses specific gas phase chemical reactions to greatly reduce polyatomic interferences. The DRC is a quadrupole

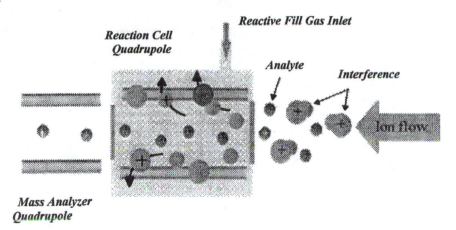

Reactive Fill Gas Inlet

Reaction Cell Quadrupole

Analyte

Interference

Ion flow

Mass Analyzer Quadrupole

Figure 10.40 The Dynamic Reaction Cell^{TM} eliminates polyatomic interferences using ion–molecule reactions. [Courtesy of PerkinElmer Life and Analytical Sciences, Shelton, CT (http://las.perkinelmer.com).]

placed inside an enclosed reaction chamber. The cell is positioned between the ion optics and the mass analyzer, as shown schematically in Fig. 10.40. The gas inlet allows a small volume of reactive gas into the cell. Some of the gases used in the DRC are ammonia, methane, hydrogen, and helium. The gas is selected based upon its ability to undergo a gas phase chemical reaction with the interfering species. Examples are given in Table 10.24. The gas phase ion–molecule reaction converts the interfering ions into non-interfering products. These reaction products are ejected from the analyte ion path either by the DRC quadrupole or by the analyzer quadrupole. Using this approach, the ArO$^+$ interference on ^{56}Fe$^+$ is reduced by 10^6.

10.4.5. Limitations of Atomic MS

The major limitations to atomic MS at this point are the inefficient introduction system to the ICP (a common problem with the ICP), matrix effects, and isobaric interference. In

Table 10.24 Examples of Spectral Interferences Addressed Using the Elan$^{®}$ DRC^{TM} ICP-MS

Isotope	Abundance (%)	Interfering species	Typical reaction gas
^{39}K$^+$	93.3	^{38}Ar^{1}H$^+$	NH$_3$
^{40}Ca$^+$	96.9	^{40}Ar$^+$	NH$_3$
^{51}V$^+$	99.8	^{35}Cl^{16}O$^+$	NH$_3$
^{52}Cr$^+$	83.8	^{36}Ar^{16}O$^+$	
		^{40}Ar^{12}C$^+$	
		^{35}Cl^{16}O^{1}H$^+$	
^{55}Mn$^+$	100	^{38}Ar^{16}O^{1}H$^+$	NH$_3$
^{56}Fe$^+$	91.7	^{40}Ar^{16}O$^+$	NH$_3$
^{63}Cu$^+$	69.2	^{23}Na^{40}Ar	NH$_3$
^{75}As$^+$	100	^{40}Ar^{35}Cl$^+$	H$_2$/Ar mix
^{80}Se$^+$	49.6	^{40}Ar^{40}Ar$^+$	CH$_4$

Source: Courtesy of PerkinElmer, Inc., Shelton, CT (www.perkinelmer.com).

addition, quadrupoles are limited in sensitivity for trace amounts of light elements such as S, Cl, F, P, and Si due to significant interferences from C, N, and O based ions.

The flow rate, gas mixture, and solvent used affect formation of polyatomic species and therefore the degree of interference from polyatomic species including refractory oxides. For reproducible results, the operating conditions must be rigorously controlled. Plasma conditions can be chosen that minimize the formation of interferences but may cause loss of sensitivity.

The isobaric interference problem can be assessed and perhaps avoided by measuring a different isotope of the analyte. That may require measuring a less-abundant isotope, reducing the sensitivity of the analysis. The use of high-resolution MS can eliminate the mass overlap problem, but the cost of such instruments is very high compared with quadrupole instruments. The use of collision cells and gas phase reaction cells in a quadrupole instrument can be used to eliminate or reduce polyatomic interferences.

BIBLIOGRAPHY

Allen, L.A.; Georgitis, S.J. *High Precision Isotope Ratio Measurements by the LECO Renaissance*[TM] *TOF-ICP-MS*, LECO Technical Brief, available at www.leco.com.

Application reviews. *Anal. Chem.* American Chemical Society, published biennially in odd years.

Barker, J. *Mass Spectrometry: Analytical Chemistry by Open Learning*, 2nd Ed.; Wiley: Chichester, UK, 1999.

Blades, M.W. Atomic mass spectrometry. *Appl. Spectrosc.* **1994**, *48* (11), 12A. (Reprinted in *Focus on Analytical Spectrometry*; Holcombe, J.A., Hieftje, G.M., Majidi, V., Eds.; Society for Applied Spectroscopy: Frederick, MD, 1998 (www.s-a-s.org).)

Borman, S.; Russell, H.; Siuzdak, G. A mass spec timeline. *Today's Chemist at Work*, September 2003, 47 (www.tcawonline.org).

Broekaert, J.A.C. Glow discharge atomic spectroscopy. *Appl. Spectrosc.* **1995**, *49* (7), 12A. (Reprinted in *Focus on Analytical Spectrometry*; Holcombe, J.A., Hieftje, G.M., Majidi, V., Eds.; Society for Applied Spectroscopy: Frederick, MD, 1998 (www.s-a-s.org).)

Buchanan, M.V.; Hettich, R.L. Fourier transform mass spectrometry of high-mass biomolecules. *Anal. Chem.*, **1993**, *65* (5), 245.

Busch, K.L. Units in mass spectrometry. *Current Trends in Mass Spectrometry (Spectroscopy Magazine)*, 18(5S), May 2003, S32 (www.spectroscopyonline.com)

Chong, N.S.; Houk, R.S. *Appl. Spectrosc.* **1987**, *41*, 66.

Ewing, G.W. Mass spectrometry. In *Analytical Instrumentation Handbook*, 2nd Ed.; Ewing, G.W., Ed.; Marcel Dekker, Inc.: New York, 1997.

Fenn, J.B.; Mann, M.; Meng, C.K.; Wong, S.E.; Whitehouse, C. *Science* **1989**, *64*, 246.

Fuerstenau, S.D., et al. *Angew. Chem. Intl. Ed.* **2001**, *40*, 541.

de Hoffmann, E.; Stroobant, V. *Mass Spectrometry: Principles and Applications*, 2nd Ed.; John Wiley and Sons LTD: Chichester, 1999.

Jarvis, K.E.; Gray, A.L.; Houk, R.S. *Handbook of Inductively Coupled Plasma Mass Spectrometry*; Blackie: London, 1992.

Karas, F.H.; Beavis, R.C.; Chait, B.T. Matrix-assisted laser desorption ionization mass spectrometry of biopolymers. *Anal. Chem.* **1991**, *63* (24), 1193A.

Lambert, J.B.; Shurvell, H.F.; Lightner, D.; Cooks, R.G. *Introduction to Organic Spectroscopy*; MacMillan Publishing Company: New York, 1987.

Le Cornec, F.; Correge, T. *A new internal standard isotope dilution method for the determination of U/Ca and Sr/Ca ratios in fossil corals by ICP-MS.* Varian ICP-MS Applications Note ICP-MS-18, September 1998 (www.varianinc.com).

McLafferty, F.W.; Tureček, F. *Interpretation of Mass Spectra*, 4th Ed.; University Science: Mill Valley, CA, 1993.

Milstein, L.S.; Essader, A.; Fernando, R.; Akinbo, O. *Environ. Int.* **2002**, *28* (4), 277–283.

Milstein, L.S.; Essader, A.; Murrell, C.; Fernando, R.; Akinbo, O. *JOAC* **2003a**, *51* (15), 4180–4184.

Milstein, L.S.; Essader, A.; Fernando, R.; Levine, K.; Akinbo, O. *Envr. Health Persp.* **2003b**, *111* (3), 293–296.

Montaser, A., Ed. *Inductively Coupled Plasma Mass Spectrometry*; VCH: Berlin, 1998.

Niessen, W.M.A.; van der Greef, J. *Liquid Chromatography–Mass Spectrometry*; Marcel Dekker, Inc.: New York, 1992.

NIST Mass Spec Data Center. Mass spectra. In *NIST Chemistry WebBook, NIST Standard Reference Database Number 69*; Lindstorm, P.J., Mallard, W.D., Eds.; NIST: Gaithersburg, MD, March 2003, 20899.

Pavia, D.L.; Lampman, G.M.; Kriz, G.S. *Introduction to Spectroscopy: A Guide for Students of Organic Chemistry*, 3rd Ed.; Harcourt College Publishers: Fort Worth, TX, 2001.

Ridd, M. *Determination of trace levels of rare earth elements in basalt by ICP-MS*. Varian ICP-MS Applications Note ICP-MS-2, April 1994 (www.varianinc.com).

Silverstein, R.M.; Webster, F.X. *Spectrometric Identification of Organic Compounds*, 6th Ed.; Wiley: New York, 1981.

Skelly Frame, E.M.; Uzgiris, E.E. The determination of gadolinium in biological samples by ICP-AES and ICP-MS in evaluation of the action of MRI agents. *Analyst* **1998**, *123*, 675–679.

Smith, R.M.; Busch, K.L. *Understanding Mass Spectra—A Basic Approach*; Wiley: New York, 1999.

Sparkman, O.D. *Mass Spectrometry Desk Reference*; Global View Publishing: Pittsburgh, PA, 2000.

Vorcos, L. Electrospray mass spectrometry. *Anal. Chem.* **1994**, *66* (8), 481A.

Watson, J.T. *Introduction to Mass Spectrometry*, 3rd Ed.; Lippincott-Raven: Philadelphia, PA, 1997.

Wolf, R.; Denoyer, E.; Grosser, Z. *EPA Method 2008 for the Analysis of Drinking Waters, Application Note ENVA-300*; PerkinElmer Life and Analytical Sciences, Shelton, CT, July 1995.

Wolf, R.; Denoyer, E.; Sodowski, C.; Grosser, Z. *RCRA SW-846 Method 6020 for the ICP-MS Analysis of Soils and Sediments, Application Note ENVA-301*; PerkinElmer Life and Analytical Sciences, Shelton, CT, January 1996.

Yarnell, A. The many facets of man-made diamonds. *Chem. Eng. News* **2004**, *82* (5), 26.

PROBLEMS

10.1   Why can the element Indium not be identified definitively by atomic MS?

10.2   Identify the compound and the isotopic composition of each fragment given the following mass spectral data:

m/z	1	16	17	18	19	20
Relative abundance	<0.1	1.0	21	100	0.15	0.22

10.3   Identify the compounds and the isotopic composition of each fragment given the following mass spectral data:

(a)

m/z	12	16	28	29	30	44	45	46
Relative abundance	8.7	9.6	9.8	0.13	0.02	100	1.2	0.42

(b)

m/z	35	36	37	38	39
Relative abundance	12	100	4.1	33	0.01

Identify the following compounds from their mass spectra and the data provided:

10.4

m/z	28	29	30	31	32	33	34
Relative abundance	6.3	64	3.8	100	66	0.98	0.14

10.5

m/z	14	15	16	19	31	32	33	34	35
Relative abundance	17	100	1.0	2.0	10	9.3	89	95	1.1

10.6

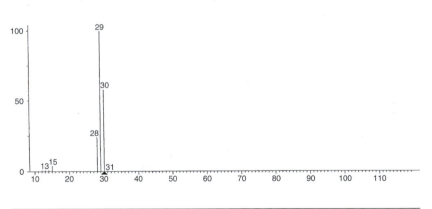

m/z	12	13	14	15	16	28	29	30	31	32
Relative abundance	3.3	4.3	4.4	0.07	1.7	31	100	89	1.3	0.21

10.7

m/z	16	24	25	32	34	48	50	64	65	66	67
Relative abundance	5.2	0.8	0.04	11	0.42	49	2.3	100	0.88	4.9	0.04

10.8

m/z	14	19	33	52	53	71	72
Relative abundance	5.2	8.4	42	100	0.39	30	0.11

10.9

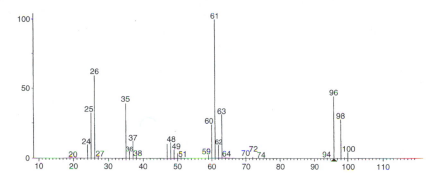

m/z	12	13	14	15	24	25	26	27	30	31	35	36	37	38	47	47.5	48	48.5
Relative abundance	2.7	3.0	0.63	0.05	4.0	15	34	1.2	0.06	0.32	7.0	1.9	2.3	0.68	6.5	0.22	5.9	0.19

m/z	49	50	51	59	60	61	62	63	64	74	95	96	97	98	99	100	101
Relative abundance	4.2	1.5	0.31	2.6	24	100	9.9	32	0.67	0.02	3.0	67	3.3	43	1.2	7.0	0.40

10.10

m/z	15	16	26	27	28	29	38	39	40	41	42	43
Relative abundance	6.3	0.20	1.4	15	2.4	38	1.4	13	1.4	41	2.3	1.6

m/z	51	52	53	54	55	56	57	58	59	71	72
Relative abundance	1.0	0.27	1.2	0.20	2.8	4.3	100	4.4	0.08	0.04	0.01

10.11

m/z	15	16	26	27	28	29	30	31	39	40	41
Relative abundance	2.8	0.03	1.4	10	5.2	6.0	0.33	4.5	5.8	1.0	7.2

m/z	42	43	44	45	46	47	57	58	59	60	61
Relative abundance	4.2	19	3.9	100	2.5	0.19	0.41	0.20	4.3	0.51	0.03

10.12

m/z	12	14	19	24	26	27	31	32
Relative abundance	13	2.1	2.0	2.7	11	0.11	22	0.28

m/z	38	50	69	70	76	77	95	96
Relative abundance	6.2	25	100	1.08	46	1.0	2.4	0.06

In the spectra below, remember that the molecular ion peak is indicated by a dark triangle. A *white triangle* indicates where the molecular ion should appear in those spectra that do not show a peak for the molecular ion.

10.13

10.14

10.15

10.16 Hint: This compound contains one halogen atom.

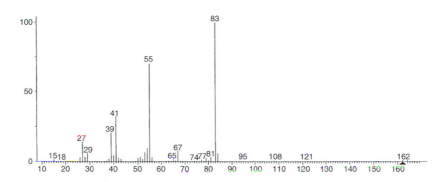

10.17 Hint: This compound contains three halogen atoms.

10.18

10.19 Hint: This compound is an ether.

10.20 Hint: This compound is a carboxylic acid.

These are a little more challenging. Draw the structure and name the compound.

10.21 The empirical formula is $C_4H_{11}N$.

10.22 The empirical formula is $C_7H_6N_2O_4$.

10.23 The empirical formula is $C_5H_{10}O$.

10.24 Hint: This compound has 6 "rings plus double bonds".

10.25

10.26 Hint: This compound is a ketone and contains three halogen atoms.

10.27 What are the advantages of high-resolution MS for molecular MS? What are the disadvantages?

10.28 What are the advantages and disadvantages of high-resolution atomic MS?

10.29 Why is it important to understand the difference between atomic weight and exact mass in interpreting mass spectra?

10.30 Explain how quantitative analysis is performed using isotope dilution. What is the advantage of an isotope-labeled compound as a calibration standard?

10.31 Diagram an ICP-MS interface. What is the purpose of the sampler and skimmer cones?

10.32 Ni and Pt are the two common materials for making ICP-MS cones. What do you think the advantages and disadvantages of the two materials are for this application?

10.33 Which is easier to interface to a mass spectrometer, a gas chromatograph or a liquid chromatograph? Why?

10.34 What mass analyzers are used for ICP-MS instruments?

10.35 What advantages are there to a TOF-based ICP-MS? (You may need to review the TOF mass analyzer, Chapter 9.)

10.36 Explain the two major types of interference in atomic MS.

10.37 Describe two instrumental approaches for eliminating or reducing poly-atomic interferences in ICP-MS.

APPENDIX 10.1 RELATIVE ABUNDANCE OF THE NATURAL ISOTOPES

Isotope	Element	%	Element	%	Element	%
1	H	99.985				
2	H	0.015				
3	He	0.000137				
4	He	99.999863				
5						
6	Li	7.5				
7	Li	92.5				
8						
9	Be	100				
10	B	19.9				
11	B	80.1				
12	C	98.90				
13	C	1.10				
14	N	99.643				
15	N	0.366				
16	O	99.762				
17	O	0.038				
18	O	0.200				
19	F	100				
20	Ne	90.48				
21	Ne	0.27				
22	Ne	9.25				
23	Na	100				
24	Mg	78.99				
25	Mg	10.00				
26	Mg	11.01				
27	Al	100				
28	Si	92.23				
29	Si	4.67				
30	Si	3.10				
31	P	100				
32	S	95.02				
33	S	0.75				
34	S	4.21				
35	Cl	75.77				
36	S	0.02	Ar	0.337		
37	Cl	24.23				
38	Ar	0.063				
39	K	93.2581				
40	Ar	99.600	K	0.0117	Ca	96.941
41	K	6.7302				
42	Ca	0.647				
43	Ca	0.135				
44	Ca	2.086				
45	Sc	100				
46	Ca	0.004	Ti	8.0		
47	Ti	7.3				
48	Ca	0.187	Ti	73.8		
49	Ti	5.5				
50	Ti	5.4	V	0.250	Cr	4.345
51	V	99.750				
52	Cr	83.789				
53	Cr	9.501				
54	Cr	2.365	Fe	5.8		
55	Mn	100				
56	Fe	91.72				
57	Fe	2.2				
58	Fe	0.28	Ni	68.077		
59	Co	100				
60	Ni	26.223				
61	Ni	1.140				
62	Ni	3.634				
63	Cu	69.17				
64	Ni	0.926	Zn	48.6		
65	Cu	30.83				
66	Zn	27.9				
67	Zn	4.1				
68	Zn	18.8				
69	Ga	60.108				
70	Zn	0.6	Ge	21.23		
71	Ga	39.892				
72	Ge	27.66				
73	Ge	7.73				
74	Ge	35.94	Se	0.89		
75	As	100				
76	Ge	7.44	Se	9.36		
77	Se	7.63				
78	Kr	0.35	Se	23.78		
79	Br	50.69				
80	Kr	2.25	Se	49.61		
81	Br	49.31				
82	Kr	11.6	Se	8.73		
83	Kr	11.5				
84	Kr	57.0	Sr	0.56		
85	Rb	72.165				
86	Kr	17.3	Sr	9.86		
87	Rb	27.835	Sr	7.00		
88	Sr	82.58				
89	Y	100				
90	Zr	51.45				
91	Zr	11.22				
92	Zr	17.15	Mo	14.84		
93	Nb	100				
94	Zr	17.38	Mo	9.25		
95	Mo	15.92				
96	Zr	2.80	Mo	16.68	Ru	5.52
97	Mo	9.55				
98	Mo	24.13	Ru	1.88		
99	Ru	12.7				
100	Mo	9.63	Ru	12.6		
101	Ru	17.0				
102	Pd	1.02	Ru	31.6		
103	Rh	100				
104	Pd	11.14	Ru	18.7		
105	Pd	22.33				
106	Pd	27.33	Cd	1.25		
107	Ag	51.839				
108	Pd	26.46	Cd	0.89		
109	Ag	48.161				
110	Pd	11.72	Cd	12.49		
111	Cd	12.80				
112	Cd	24.13	Sn	0.97		
113	Cd	12.22	In	4.3		
114	Cd	28.73	Sn	0.65		
115	In	95.7	Sn	0.34		
116	Cd	7.49	Sn	14.53		
117	Sn	7.68				
118	Sn	24.23				
119	Sn	8.59				
120	Sn	32.59	Te	0.096		
121	Sb	57.36				
122	Sn	4.63	Te	2.603		
123	Sb	42.64	Te	0.908		
124	Sn	5.79	Te	4.816	Xe	0.10
125	Te	7.139				
126	Te	18.95	Xe	0.09		
127	I	100				
128	Te	31.69	Xe	1.91		
129	Xe	26.4				
130	Te	33.80	Ba	0.106	Xe	4.1
131	Xe	21.2				
132	Ba	0.101	Xe	26.9		
133	Cs	100				
134	Ba	2.417	Xe	10.4		
135	Ba	6.592				
136	Ba	7.854	Ce	0.19	Xe	8.9
137	Ba	11.23				
138	Ba	71.70	La	0.0902	Ce	0.25
139	La	99.9098				
140	Ce	88.48				
141	Pr	100				
142	Ce	11.08	Nd	27.13		
143	Nd	12.18				
144	Nd	23.80	Sm	3.1		
145	Nd	8.30				
146	Nd	17.19				
147	Sm	15.0				
148	Nd	5.76	Sm	11.3		

(continued)

Appendix 10.1 Continued

Isotope	%	%	%	Isotope	%	%	%	Isotope	%	%	%	Isotope	%	%	%
149		Sm 13.8		186	Os 1.58	W 28.6		223							
150	Nd 5.64	Sm 7.4		187	Os 1.6		Re 62.60	224							
151			Eu 47.8	188	Os 13.3			225							
152	Gd 0.20	Sm 26.7		189	Os 16.1			226							
153			Eu 52.2	190	Os 26.4		Pt 0.01	227							
154	Gd 2.18	Sm 22.7		191		Ir 37.3		228							
155	Gd 14.80			192	Os 41.0		Pt 0.79	229							
156	Gd 20.47	Dy 0.06		193		Ir 62.7		230							
157	Gd 15.65			194			Pt 32.9	231	Pa 100						
158	Gd 24.84	Dy 0.10		195			Pt 33.8	232	Th 100						
159			Tb 100	196	Hg 0.15		Pt 25.3	233							
160	Gd 21.86	Dy 2.34		197		Au 100		234	U 0.0055						
161		Dy 18.9		198	Hg 9.97		Pt 7.2	235	U 0.7200						
162	Er 0.14	Dy 25.5		199	Hg 16.87			236							
163		Dy 24.9		200	Hg 23.10			237							
164	Er 1.61	Dy 28.2		201	Hg 13.18			238	U 99.2745						
165			Ho 100	202	Hg 29.86										
166	Er 33.6			203			Tl 29.524								
167	Er 22.95			204	Hg 6.87	Pb 1.4									
168	Er 26.8	Yb 0.13		205			Tl 70.476								
169			Tm 100	206		Pb 24.1									
170	Er 14.9	Yb 3.05		207		Pb 22.1									
171		Yb 14.3		208		Pb 52.4									
172		Yb 21.9		209	Bi 100										
173		Yb 16.12		210											
174		Yb 31.8	Hf 0.162	211											
175	Lu 97.41			212											
176	Lu 2.59	Yb 12.7	Hf 5.206	213											
177			Hf 18.606	214											
178			Hf 27.297	215											
179			Hf 13.629	216											
180	Ta 0.012	W 0.13	Hf 35.100	217											
181	Ta 99.988			218											
182		W 26.3		219											
183		W 14.3		220											
184	Os 0.02	W 30.67		221											
185			Re 37.40	222											

Source: Isotopic composition of the elements 1989. *Pure Appl. Chem.,* 1991, 63(7), 991–1002. © 1991 IUPAC

11

Principles of Chromatography

11.1. INTRODUCTION TO CHROMATOGRAPHY

Many of the techniques of spectroscopy we have discussed in preceding chapters are selective for certain atoms or functional groups and structural elements of molecules. Often this selectivity is insufficient to distinguish compounds of closely related structure from each other. It may be insufficient for measurement of low levels of a compound in a mixture of others which produce an interfering spectral signal. Conversely, when we wish to measure the presence and amounts of a large number of different compounds in a mixture, even the availability of unique spectral signatures for each analyte may require laborious repeated measurements with different spectral techniques to characterize the mixture fully. Living systems have evolved large protein molecules (e.g., receptors, enzymes, immune system antibodies) with unique 3D structures which can bind strongly only to very specific organic compounds. Analytical procedures such as immunoassay, enzyme mediated assays, and competitive binding assays, employ the extreme selectivity of such protein macromolecules to measure particular biomolecules in very complex mixtures. Microarrays of thousands of these, each with unique selectivity, can rapidly screen complex mixtures for a list of expected components. However, precise quantitation of each detected component is difficult to achieve this way.

Analysis of complex mixtures often requires separation and isolation of components, or classes of components. Examples in noninstrumental analysis include extraction, precipitation, and distillation. These procedures partition components between two phases based on differences in the components' physical properties. In liquid–liquid extraction components are distributed between two immiscible liquids based on their similarity in polarity to the two liquids (i.e., "like dissolves like"). In precipitation, the separation between solid and liquid phases depends on relative solubility in the liquid phase. In distillation the partition between the mixture liquid phase and its vapor (prior to recondensation of the separated vapor) is primarily governed by the relative vapor pressures of the components at different temperatures (i.e., differences in boiling points). When the relevant physical properties of the two components are very similar, their distribution between the phases at equilibrium will result in slight enrichment of each in one of the phases, rather than complete separation. To attain nearly complete separation the partition process must be repeated multiple times, and the partially separated fractions recombined and repartitioned multiple times in a carefully organized fashion. This is achieved in the laborious *batch processes* of *countercurrent liquid–liquid extraction*, *fractional crystallization*, and *fractional distillation*. The latter appears to operate continuously, as the vapors from a single equilibration chamber are drawn off and recondensed, but the equilibration in each of the chambers or "plates" of a fractional distillation tower represents a discrete equilibration at a characteristic temperature.

A procedure called chromatography automatically and simply applies the principles of these "fractional" separation procedures. Chromatography can separate very complex mixtures composed of many very similar components. The various types of chromatographic instrumentation which have been developed and made commercially available over the past half century now comprise a majority of the analytical instruments for measuring a wide variety of analytes in mixtures. They are indispensable for detecting and quantitating trace contaminants in complex matrices. A single chromatographic analysis can isolate, identify, and quantitate dozens or even hundreds of components of mixtures. Without their ability to efficiently characterize complex mixtures of organic biochemicals at trace levels from microscale samples, research in molecular biology as we know it today would be impossible. We will be discussing another powerful separation technology, **electrophoresis**, together with chromatography, because much of the way electrophoresis operates, and the appearance of its separations, is analogous to chromatography, even though the underlying separation principle is different. Chromatographic-type separation is at the heart of the instrumentation used to separate fragmented DNA molecules and to sequence the order of the "bases" which form the genomes of all living creatures on Earth. The heroic sequencing of around 3 billion DNA bases in the human genome, completed in the year 2001, could never have been envisioned, let alone completed, without chromatographic-type instrumentation. This forms the basis of the new field of *genomics*. Even more extensive analytical separation, identification, and quantitation problems arise in the next new field of molecular biological research, coming in the so-called "post-genomic" era; namely, *proteomics*. This requires separation, identification, and quantitation of the thousands of proteins that may be produced ("expressed") by the genome of each type of living cell. The contents and proportions of this mix will depend on what the cell is doing, and whether it is in a healthy or diseased state. The amounts of analytes may be at the nanomolar scale or less, and unlike DNA analyses, there is no handy polymerase chain reaction (PCR) procedure to amplify the amounts of analytes. If genomics provides a complete parts list for an organism, then proteomics will be necessary to understand the assembly and operating manual. Billions of dollars of chromatography-based instrumentation, and hundreds of thousands of jobs using it, will be required for research in these fields. Environmental monitoring and regulation has the potential of becoming another multimillion dollar market which can benefit from chromatography's ability to isolate, identify, and measure trace pollutants in complex environmental mixtures.

11.2. WHAT IS THE CHROMATOGRAPHIC PROCESS?

The Russian botanist Mikhail Tswett invented the technique and coined the name chromatography. Early in the 20th century he placed extracts containing a mixture of plant pigments on the top of glass columns filled with particles of calcium carbonate. Continued washing (**elution**) of the mixtures through the columns with additional solvent (called the **eluent**) resulted in the separation of the pigments, which appeared as colored bands adsorbed on the solid particles, moving along the column at different rates, thus appearing at different distances down the column. He named this technique "chromatography", from the Greek for "color writing". By collecting separated components as they exit the bottom of the column, one could isolate pure material. If a detector in the exiting fluid stream (called the **effluent**) can respond to some property of a separated component other than color, then neither a transparent column nor colored analytes are necessary to separate and measure materials by "chromatography".

When there is a great difference in the retention of different components on the material filling the column, a short column can be used to separate and isolate a rapidly moving component from highly retained material. This is the basis of a useful sample preparation technique called **solid phase extraction (SPE)**.

Real instrumental chromatography employs highly engineered materials for the **stationary phase** past which the **mobile phase** fluid carrying the mixture of analytes passes, with continuous partitioning of the analytes between the two phases. The column needs to be sufficiently long and the eluent flow sufficiently slow for the process to approach equilibrium conditions. Then partitioning will be repeated a large number of times. Even very slight differences among mixture components in the ratio of the amounts which would exist in each phase at equilibrium will result in their separation into bands along the column. These bands can be separately detected or collected as they elute off the column.

The critical defining properties of a chromatographic process are:

1. immiscible stationary and mobile phases;
2. an arrangement whereby a mixture is deposited at one end of the stationary phase;
3. flow of the mobile phase toward the other end of the stationary phase;
4. different rates or ratios of partitioning for each component of the mixture, and many cycles of this process during elution;
5. a means of either visualizing **bands** of separated components on or adjacent to the stationary phase, or of detecting the eluting bands as **peaks** in the mobile phase effluent.

    (Note that in some applications the bands are also referred to as **zones**.)

The first major implementation of instrumental chromatography was based on partition between a liquid, nonvolatile, stationary phase supported on inert solid particles packed in metal columns, and an inert gaseous mobile phase flowing through the column from a pressurized tank. The various partition ratios between the phases, which affected separation, were primarily related to the components' different volatilities (i.e., boiling points). The initial application of this procedure, **gas chromatography (GC)**, to petroleum hydrocarbon samples, led to facile separation of many more components than could be achieved by the fractional distillation process used in petroleum refining. Instead of just petroleum "fractions", individual specific hydrocarbons could be resolved and isolated, if the GC column was long enough. The analogy to the theoretically well-studied multiple plate fractional distillation columns used in refining processes led to the initial plate theory of the chromatographic process. This assumed that the separation of the bands as they migrated down the column proceeded as a series of sequential, discrete equilibrium partitionings, instead of the actual situation, which is a continuous, not-quite-in-equilibrium process. Nevertheless, this **plate theory** proved fairly accurate for most cases. It contained useful terms [e.g., $N =$ *the number of plates* in a column as a descriptor of its separation efficiency, and the **"height equivalent to a theoretical plate" (HETP)**]. These remain central to the simplified theoretical characterization of the chromatographic process, despite the fact that continuous, non-equilibrium kinetic (or rate) theory is more accurate, and there exist no actual plates in a chromatographic column. GC was not suitable for separation and measurement of more polar, and less volatile or less thermally stable large biomolecules. Optimized chromatographic separations of these molecules at ambient temperatures were achieved later using more polar, water miscible liquid mobile phases. The instruments and stationary phases which enabled this improvement on the traditional vertical **gravity column LC** apparatus, form the basis of the even more widely used **HPLC**.

11.3. CHROMATOGRAPHY IN MORE THAN ONE DIMENSION

In GC the gaseous mobile phase must be confined in a column, so that a pressure gradient can cause it to flow past the stationary phase and eventually elute the separated bands out the effluent end of the column. This is inherently a 1D separation, along the column, from one end to the other. This dimensionality applies even should the column be coiled to fit in a GC oven rather than vertically straight, like Tswett's gravity flow liquid mobile phase column. However, unlike gases, liquids as mobile phases do not always require confinement to move in a desired direction or retain their volume. If they are in contact with porous beds of small particles or fiber mats, surface forces (capillary attraction) can often induce them to flow. Thus it is possible to carry out the LC process on a stationary phase arrayed as a thin surface layer, usually a planar, 2D surface. Examples include the matted cellulose fibers of a sheet of paper, or a thin layer of silica gel or alumina particles on a planar support (e.g., a pane of glass).

An example of paper chromatography, beloved of grade-school science fairs for its simplicity, is to place a drop of colored ink in the center of a sheet of paper. After drying, a series of drops of organic solvent are slowly added to the spot. Capillary attraction causes a radial eluent flow away from the spot in all directions. The different colored dyes in the ink adhere more or less strongly to the multiple hydroxyl groups on the cellulose molecule chains through polar or hydrogen-bonding interaction. Therefore the capillary eluent flow carries some further than others, resulting in a pattern of colored circles of different radii. This recreates Tswett's original 1D column chromatography experiment on a 2D surface using his original definition. Being on a piece of paper, it seems even more like "color writing".

The radial separation in the earlier example is not the most efficient way to perform surface chromatographic separations. A square planar **thin layer chromatography (TLC)** plate (not to be confused with the "theoretical plates" discussed previously!) may have a line of spots containing sample mixtures and reference standard materials deposited just above one edge. Submerge that edge in solvent to a level just below the line of spots, and capillary attraction will produce a unidirectional, ascending, eluent flow which will move and separate each spot's components vertically up the plate in separate **lanes**. If they are not already colored they may be sprayed with a reagent which will react to "develop" a color, just as a photographic plate may be developed chemically to reveal a latent image. Alternatively the separated component spots may be visualized, or even quantified, by observing their fluorescence emission induced by exposing them to UV light, or by their quenching of the natural fluorescence of a silica gel TLC layer. Note that an advantage to this form of surface chromatography is that multiple 1D (i.e., in one direction) chromatographic separations of samples or standard mixtures are carried out simultaneously in parallel (literally).

If a mixture has enough components, and some are sufficiently similar in structure and physical properties, it is likely that chromatography in one dimension, either on a column or a surface, will not separate all of them from each other. Many column bands or surface spots will contain multiple unseparated components. These are said to **coelute** in the chromatographic system. If we repeat the previous TLC plate separation, but this time place just one spot containing a mixture to be separated near one corner of the plate, we will again achieve a line of eluted spots in a lane along a vertical edge. Some of these spots may contain still unseparated, coeluting components. Now let us rotate the plate ninety degrees and place this line of separated spots back in the tank just above *a new eluting solvent* with polarity or pH or some property very different from the first solvent. This will elute each of the partially separated spots from the first chromatographic separation vertically up the plate perpendicular (or orthogonal) to the initial separation dimension.

If the nature of the solvent used for the second, orthogonal elution produces different stationary/mobile phase partition ratios for components that coeluted in spots from the first separation, they will now separate in the dimension perpendicular to that of the first chromatographic separation. This is an example of *true 2D chromatography*. It is most easily achieved in the planar surface mode described here. A 2D *column* chromatography instrument, necessary for 2D GC separations, requires more complex hardware. The effluent from one column may be sampled one time, or in continuous segments, by a device which collects, reconcentrates, and redirects it to a second column. The second column must have separation selectivity different from the first one. If the coeluting components in just a *single* peak isolated from the first column are separated on the second one, the procedure is **heart-cutting, 2D chromatography**. If segments of the effluent from the first column are *continuously* and very rapidly separated by the second column at a pace matching that at which they are sampled, the resulting detector signal may be transformed by computer to yield **comprehensive 2D chromatograms**, displayed as a planar array of spots like those observed directly on the TLC plate. These examples of multidimensional chromatography should not be confused with the use of a spectroscopic detector (e.g., a mass spectrometer at the end of a GC column, or a UV/VIS diode array spectrometer monitoring the effluent of a liquid chromatograph). Such **hyphenated systems**, so called from their abbreviations such as GC-MS or LC-diode array, are said to add *a second, spectral, dimension* to the initial chromatographic separation. If components coeluting in a chromatographic peak have sufficiently different spectral characteristics, they may be measured separately in the peak despite their coelution.

11.4. VISUALIZATION OF THE CHROMATOGRAPHIC PROCESS AT THE MOLECULAR LEVEL: ANALOGY TO "PEOPLE ON A MOVING BELT SLIDEWAY"

Molecules undergoing chromatographic separation may be likened to people standing on a moving slideway, such as those which transport people between terminals in an airport. Imagine a very broad moving belt, capable of accepting a large number of passengers across its width, with no side rails to prevent them from stepping on and off along its length. If several ranks of people step on together at one end and rigidly hold their positions, the group will proceed to the end in the formation in which it boarded. This is like a crystalline particle suspended in a mobile phase flowing down a column. The time, (t_m), taken to transit the belt or column is the product of their speed and length. If the people are free to mill about while on the belt, like molecules dissolved in a flowing mobile phase, they will tend to spread out along the length of the belt as they transit. The narrow, compact band of *people/molecules* which was put on at the head of the *belt/column* becomes increasingly wider, and the degree of spread is proportional to the time elapsed before the exit is reached. In the case of a narrow band of molecules, this spreading occurs by a process of **random diffusion**, resulting in a **Gaussian distribution** about the mean, producing a peak with the "**bell-shaped curve**" familiar from statistical description. The center, and maximum, concentration of people/molecules still travels at the velocity of the belt/mobile phase. Now imagine that during their progress down the *slideway/column*, the people/molecules can *step-on-and-off/partition* to the *nonmoving floor/stationary phase*. This happens continuously, rapidly, and repeatedly during the trip. During their visits to the floor/stationary phase, they are delayed from moving toward the exit, so they come off the slideway/column later than if they had spent all their time on it. We can characterize the extent of the effect of the interaction between the people/molecules and the floor/stationary phase by the **adjusted retention time**: $(t_r')=t_r - t_m$, where t_m is the time it takes people/molecules *not* getting-on-and-off/partitioning to go from one end to the other, and t_r is the time required when these processes take place.

Imagine we have two different categories of people (e.g., call them men and women) in our initial compact group at the start, and men are slightly more inclined than women to linger on the floor when they step off the moving slideway. On average, the women will reach the end of the system sooner, and their t_r and t_r' times will be shorter. If the tendency of men to linger is enough greater than that of women, the two sexes will separately exit the slideway. However, do not forget that the groups will both mill about and spread out, mainly while on the slideway. If the distance that they spread out is close to the average difference by which they separate while moving down the system as a whole, they will still partially overlap at the end. Several factors affect how much separation can be achieved:

1.   How rapidly does the *milling-about/molecular diffusion* spread the bands along the length of the system compared with how rapidly the bands separate on it? Suppressing this improves separation (i.e., the ability of the system to *resolve* a pair of components).

2.   How much time do these two processes have to compete with each other (i.e., how much time do a closely separating pair of bands spend on the system). This depends directly on its length, and inversely on the speed of the slideway/ mobile phase.

3.   How different is the relative attraction of the slideway to the floor, and therefore the relative times spent in each, between the men and the women. The greater the difference, the more selective and complete the separation. Imagining a politically correct rationale for why the men prefer to linger for more or less longer periods in various stationary floor environments is an exercise left to the student.

4.   How far along the slideway does the crowd to be separated extend when it gets on at the start? Even without any further band spreading during movement to the exit, the differential separation process must cause the bands to separate at least by this much before they exit the system to obtain complete separation.

Note that while in the slideway/mobile phase, all people/molecules move at its rate and therefore spend the same time in it. What varies is the amount of time different members of the mixture stay off it in the nonmoving environment. In the chromatography side of this analogy, the variations in the chemical interactions of the mixture components with differing stationary phases vary these times and achieve different separations. Different gas mobile phases in GC have much the same affinity for given compounds. The major variable affecting distribution between the phases is temperature, which controls vapor pressure. This largely controls the proportions in the vapor mobile phase and the "liquid" stationary phase. When the mobile phase is a liquid, a variety of chemical interactions equivalent to or even greater than those of the stationary phases can be chosen to optimize separations.

Let us stretch our moving belt slideway analogy to chromatography a bit further before it snaps (no pun intended). We assume a very wide belt, capable of holding many people across its width (after all, molecules are very small compared to dimensions in a chromatography column). Another process which could affect band broadening is that the people randomly milling about will sometimes drift toward or away from the edge of the belt where they could step on and off. Sufficiently slow, pressure-driven fluid flow across small dimensions is **laminar** (in smooth straight parallel lines, not swirling with turbulence). In unobstructed narrow columns its rate is variable across the width of the tube. It is stationary at the tube wall, and increases to a maximum at the center. This is actually the design for multibelt moving slideways envisioned by science fiction writers for very high traffic loads in some distant future. One could step on and off slow moving belts at the outer edge, and step over with small increases in velocity by moving to successively faster belts toward the center. In our analogy, people randomly milling about not only along the direction of travel but across it would undergo additional spreading out from traveling at a range of different forward speeds. If we use a column packed with small particles to support the stationary phase, we are less likely to achieve laminar flow, and additional band broadening factors come into play. Imagine the people on our moving slideway confronting thick obstructing columns hanging down from the ceiling. Some will dodge one way and some another, and the different paths taken during these dodges will cause some people to travel further and to be held up longer than others. In dodging they may bump one another more than in random milling about, and form swirls and eddys of humanity. This spreads the bands even more, and works against maintaining the separation of populations achieved by different rates of stepping on and off the belt to the floor.

In our chromatographic analogy, separation of the people by categories is affected only by the relative rates of stepping on and off the belt, working in opposition to processes which spread out or broaden the initial band. On actual slideways many people often walk purposefully in one direction or another (usually forward). If different categories of people walked forward or backward at specific rates, while some simply stood still, the system could achieve separation without the need for the people to step on and off to the side at different rates. If it were instead molecules in an open tubular column with flowing fluid that could be induced to behave this way, we could achieve column separations without the actual chromatographic process. That process by definition requires continuous partition of analytes between a mobile and a stationary phase. If some of our analytes are positively or negatively charged species together with uncharged species in solution, the charged species can in fact be separated from the uncharged species and from each other in just this way!

In a procedure called **capillary zone electrophoresis (CZE)** the *anions* will appear to "walk backwards" at different rates, the *cations* to "walk forward", and uncharged molecules, or those with a balance of positive and negative charged sites (**zwitterions**), will flow along with the solvent for the ride. Many of the same processes of band broadening that apply in chromatography occur here, although not all. For example, the mobile fluid flow is not pressure driven, and its speed is constant across the width of the column, removing one of chromatography's sources of band broadening. Although its mechanism is distinctly different from that of chromatography, electrophoresis as a separation method shares many of its features. Therefore it will be discussed in Chapter 13, along with other forms of chromatography which employ a liquid mobile phase.

In a perfectly operating chromatography system or its people-on-a-moving belt analogy, the moving people/molecules form gradually spreading, symmetrical bands advancing through the system. One way this can fail is for there to be insufficient room on the floor to accommodate all the people ready to step off, or in the stationary phase for molecules to partition into. When this happens, a portion of them will be forced to remain on the moving portion of the system, and they will advance further until they reach less crowded areas where they can resume normal partitioning. This will distort or skew the shape of the bands, with more people/molecules moving in advance of the highest concentration in the band than behind it. When the band elutes off the column the resulting peak is said to display a "**leading**" shape. The greater the "**overload**" of the stationary phase, the larger the percentage of people/molecules which cannot be accommodated, and the more pronounced is this leading asymmetry.

Imagine that on the floor by the moving belt there are limited numbers of comfortable chairs. Many people may step back on to the belt without approaching one, but those who get close enough to sit down remain for a much longer time than the rest. They will lag well behind people who do not sit down. Depending on the ratio of people to available chairs, the band advancing along the system will lose its symmetry in the other direction; that is, a portion of its members will lag behind. It will be said to display a "tailing" shape. In chromatography, the analogy to our chairs is a limited number of highly retentive sites on the solid that supports the liquid stationary phase where normal equilibrium partitioning of molecules takes place. Both leading due to overloading and tailing due to "active sites" cause additional band broadening, and degrade resolution of closely eluting peaks.

In some chromatography, such as Tswett's color bands on the transparent column or the spots on a TLC plate, we measure separated components directly on the separating system. More generally, in column chromatography, a detector which responds to separated components in the mobile phase effluent as they elute off the column gives a signal which is plotted as a "**peak**". In our moving belt analogy we must imagine the "**detector cell**" as being a continuation of the belt without any more floor for people to step off to. This is different from the end of actual moving walkways, where people step off the end of the belt and walk away. In the case of chromatography of molecules there are two types of detector cells.

1. The "**mass detector**", actually an "amount detector", gives a signal proportional to the total amount of molecules exiting the column per unit time. In our analogy we might imagine a trap door that opens continuously at regular intervals in the detector cell and drops people in a bin to be counted. If the belt moves more rapidly, more people will be collected and the peak will be higher and narrower. Peak area is independent of mobile phase velocity. If the belt stops, people will not continue to be counted. The signal falls to zero.
2. The "**concentration detector**" measures the total number of people in a given volume (say that of the detector cell) at a given time. If the belt moves faster the peak will be narrower, but its height will remain the same. Peak area depends on mobile phase velocity. If the belt stops, the people are still in the measurement volume, and the signal stays constant.

Note: The observed peak width also depends on the rate at which the band corresponding to it moves through the chromatographic system (i.e., the mobile phase/ stationary phase combination). This is proportional to the rate at which it comes off the column. More retained bands of people/molecules exit the system over a longer period, giving rise to peaks broader in time. *This is the major cause of peak broadening*, and it is separate from the process of *band broadening* due to mobile phase diffusion, non-equilibrium partitioning, multipath travel, turbulence, and so on. These latter processes affect whether or how much two closely eluting bands can be separated as resolved peaks as they elute from the system into the detector cell. Contrary to common misconception they are not the primary reason that later eluting peaks appear broader. This confusion arises from a tendency to speak of peaks and bands interchangeably, and to forget that a band has a dimension along the column of *length*, while a peak has a corresponding dimension in the detector signal of time.

11.5. A DIGRESSION ON THE CENTRAL ROLE OF SILICON–OXYGEN COMPOUNDS IN CHROMATOGRAPHY

Elemental silicon is central to the vast industry of solid-state electronics. Appropriately doped with other elements it forms a variety of semiconductors that constitute most transistors and integrated circuits. Other elements and compounds such as germanium or gallium arsenide have also found a niche as semiconductors in electronics, but silicon occupies the prime position. How fortunate that it is the second most abundant element in the Earth's crust. Its compounds with the most abundant crustal element, oxygen, are equally central in many different aspects of chromatography. Silica, silica gel, "glass", quartz, fused silica, silicones—all have a remarkable variety of key roles to play in chromatography. Let us familiarize ourselves with some of their relevant properties.

Crystalline silicon dioxide, SiO_2, *quartz*, is a transparent, very high melting, 3D network of silicon–oxygen tetrahedra. When melted and cooled it forms a noncrystalline, high-melting, glass known as *fused silica*. This no longer has the regular crystalline lattice of quartz. Being a glass it can easily be drawn while molten into long fibers. If other elements, such as calcium, sodium, or boron are mixed in, the lower melting glasses can be more easily shaped, blown, molded, and so on. Their transparency, moldability, and ability to be ground and polished lead to many applications in optics, which are well represented in optical spectroscopic instrumentation. Their resistance to heat and most chemical reagents make them the primary material for the vessels in which chemists perform and observe reactions. In chromatography, we shall see that the lower melting glasses with their guest atoms in the silica structure are generally less desirable than pure silica itself.

Silica can also be precipitated from aqueous solutions of silicic acid, H_4SiO_4. Quartz itself forms in nature from very slow hydrothermal crystallization from hot solutions under great pressure deep underground. A more porous open structure called *silica gel* can be precipitated quickly in the laboratory. Particles of this can be coated as the stationary phase on a glass substrate to prepare one of the most useful types of TLC plates. A vast variety of very pure, monolithic, or porous silica particles, as carefully controlled mono-disperse (all about the same dimension, 3–50 μm) spheres can be synthesized for use as supports for the liquid-like stationary phase of packed columns in GC or LC. At the surface of silica, those oxygen atoms which are attached to one silicon by only one bond generally have an H atom on the other. The resulting —Si—OH on the surface of silica interacts with water both by polar and hydrogen-bonding interactions to bind it tightly. Thus, unless vigorously heated and kept in a very dry environment, silica has a bound surface layer of adsorbed water. This may serve as the very polar stationary phase in a chromatographic system. Many early simple gravity flow liquid–solid column chromatography systems used silica or similar particles such as *alumina* (Al_2O_3), eluting with and partitioning between a less polar, water-immiscible mobile phase. Thus this was called "**normal phase**" chromatography.

The great majority of molecules we want to separate by LC are more-polar organic biomolecules, which are more suitably eluted in an aqueous mobile phase against a less-polar stationary phase. We can still use the array of specially tailored silica particles. We simply replace adsorbed water by chemically attaching a less-polar organic substituent, such as a long hydrocarbon chain, to the dangling —OH groups on the silica surface. Using this method we have replaced very polar water at the surface with nonpolar hydrocarbon. Since such columns reverse the relative polarity of the stationary and mobile phases from what used to be considered "normal" for column chromatography, this is called "**reversed phase**" LC. It becomes increasingly difficult to force more of these large chains to attach to every last —OH group on the silica surface, so some are left uncovered. Analytes penetrating the less-polar stationary phase layer during the chromatographic partitioning process may burrow down to bind strongly to these "active" sites. They are the "chairs" of our slideway analogy. They can be deactivated and blocked by reacting with a small molecule reagent that attaches as a methyl group, a process called "end capping". Glass or silica columns, or particular supports with which a stationary phase is coated but not chemically bonded, are similarly pretreated to deactivate all such sites which could cause tailing peaks. As the pH of an aqueous mobile phase in LC increases (becomes more basic), the higher $[OH^-]$ combines with the H on the surface of $-Si-OH$, and at high enough pH slowly dissolves and destroys the silica surface and its attached stationary phase. To operate at high pH, a more resistant substrate, such as **zirconia, (ZrO_2)**, or the use of polymer resins, is required.

Chromatographers learned that coating a stationary phase on a long narrow tube of *capillary dimension* (i.e., internal diameter about 0.1–1.0 mm) yielded exceptionally efficient separations, especially for GC. Metal capillaries were replaced by glass, which was more inert. Glass still had some activity due to its dissolved metal ions, and was brittle and broke easily when being installed into instruments. The billion dollar fiber optic communication industry learned to draw fibers from rods of extremely pure silica, which absorbed light much less than glass, and was much more flexible when incorporated in fiber optic cable bundles. If an open tube of such pure fused silica were drawn by similar machinery, an inert, flexible capillary perfectly suited for supporting a chromatographic stationary phase was formed. A deliberate scratch across the capillary allows it to be broken cleanly by hand to a desired length. Protection against inadvertant scratches which might induce unwanted breaks is provided by fusing a high-temperature resistant *polyimide coating* to the outside of the capillary. As a result, instead of appearing clear, fused silica capillary tubing for chromatography has a brown to golden hue.

The separation technique of CZE has no stationary phase deposited or bonded to the interior of the capillary. As discussed in Section 11.4 it is not a chromatographic process, but it yields a separation into bands or peaks in a fashion that looks much like chromatography. The $-Si-OH$ on the surface actually begins to lose some H^+ at pH above 2, so the silica surface becomes more and more negatively charged with $-Si-O^-$ units as the pH rises. As we will see, in CE a high electrical voltage between the ends of the capillary causes charged analytes to move in different directions at different speeds through the aqueous solvent. Because of the charge on the silica surface, the solvent as a whole moves from one end of the capillary to the other, carrying the separating charged analytes past a detector and out of the column, even those which move in a direction opposite to the solvent flow!

When charges flow through a CE column great frictional heat is released. This could result in excessive band broadening and loss of resolution. Fused silica has much higher heat conductivity than glass, and the small diameters and thin walls of the capillaries enable this heat to be efficiently dissipated to a surrounding bath. Conversely, when used for GC, a fused silica chromatographic column rapidly equilibrates with temperature changes produced by the column oven.

Finally, chains of alternating silicon and oxygen form the backbone of a class of polymers called *silicones*, whose temperature stability and variable polarity make them the workhorse of GC stationary phases. Their general structure is:

$$
\begin{array}{ccccccccc}
 & R_1 & & R_1 & & R_1 & & R_1 & & R_1 \\
 & | & & | & & | & & | & & | \\
R_3- & Si & -O- & Si & -O- & Si & -O- & Si & -O- & Si & -R_3 \\
 & | & & | & & | & & | & & | \\
 & R_2 & & R_2 & & R_2 & & R_2 & & R_2
\end{array}
$$

where R_1 and R_2 are organic groups.

If R_1 and R_2 are both methyl groups, this is **polydimethylsiloxane**, the prototype and most widely used silicone polymer. If the chains are much longer than the 5 unit one illustrated here, the material is very viscous and nonvolatile. It can be coated on silica particles or on the wall of a fused silica capillary. It will dissolve small molecules and partition them to liquid or gaseous mobile phases. Changing the R-groups will affect the polarity of the polymer and thus the relative separability of pairs of molecules dissolving in it. The silicon–oxygen backbone is more resistant to higher temperatures than many other polymers used as stationary phases. At high-enough temperatures the ends of the chains can cyclize with themselves to successively shorten the chain by losing 4 or 5-membered rings of —Si(R$_x$)$_2$—O. Substituting a bulky aromatic hydrocarbon group for occasional O atoms in the chain can interrupt this decomposition and make a phase with a higher temperature limit. Reactive functional groups on some of the R-groups can cross-link the polymer chains and bond them to —Si—OH on the silica support to further increase phase stability. The remaining —Si—OH on the silica surface can then be reacted with reagents like trimethylchlorosilane to deactivate them to —Si—O—Si—(CH$_3$)$_3$.

11.6. THE BASIC EQUATIONS DESCRIBING CHROMATOGRAPHIC SEPARATIONS

We will state or derive a bare minimum of the equations most useful for understanding and describing how the most easily measured or controlled variables affect separations in chromatography. Many of the factors in these equations can be derived or calculated from more fundamental parameters, such as diffusion coefficients of analytes in the two chromatographic phases, column dimensions, or variables defined in the statistical theory of random variation. Such details are covered in more advanced texts.

The chromatographic process for a component X is governed by its **equilibrium partition ratio, K_x,** (also called the *distribution ratio*). This is the ratio of its concentrations at distribution equilibrium between the two immiscible phases [A in Eq. (11.1)]. In chromatography these are the stationary (s) and mobile (m) phases. In the chromatographic separation process, we are not interested so much in knowing the ratio of concentrations in each phase as in the ratio of the amounts. These are related to volumes of each phase at each point along the separation path as derived in (B) and (C) [Eq. (11.1)]. We define the amount ratios and volume ratios in expression (C) by the new terms in (D); namely k' and β. The ratio of the amounts distributing at equilibrium, k', is called the **capacity factor** or **retention factor**. It tells us the relative capacity of each phase in the particular chromatographic system, under the particular operating conditions (e.g., temperature in GC), for each particular compound. Instead of measuring these directly in the phases, we most often infer them from measurements of retention times, thus the use of "retention factor". The β factor is the (volume) phase ratio. Note that the terms for (s) and (m) are in opposite order in these two ratios.

$$\quad\;\text{(A)}\qquad\quad\text{(B)}\qquad\qquad\text{(C)}\qquad\qquad\text{(D)}$$

$$K_x = \frac{[X]_s}{[X]_m} = \frac{\text{Amt X}_s/V_s}{\text{Amt X}_m/V_m} = \frac{\text{Amt X}_s}{\text{Amt X}_m} \times \frac{V_m}{V_s} = k' \times \beta \tag{11.1}$$

Manufacturers of open tubular GC columns often specify **column radius r_c** and stationary phase **film thickness (d_f)**, which is usually much smaller. Simple geometry enables us to calculate the phase ratio:

$$\beta = \frac{r_c}{2d_f} \qquad \text{(the limit when } d_f \text{ is much smaller than } r_c) \tag{11.2}$$

For packed GC columns one might have weighed the stationary phase applied to the particulate support, calculated its volume knowing its density, and measured the gas flow rate through the column, and combined it with the time for an unretained component to transit the column to get the mobile phase volume. In the case of LC columns where a molecular scale layer of stationary phase is bonded to an unknown surface area of support, stationary phase volume is very difficult to estimate.

Figure 11.1 illustrates a very simple chromatogram with a single retained component peak. The y-axis represents detector signal response, while the x-axis has either mobile phase volume eluted, or time, increasing from left to right. In chromatography the more accessible measurement is the retention time of a component, t_r, which is measured from its introduction at one end, until elution of the peak maximum at the center of the eluting band. In the figure point O is the start. An **unretained component** (i.e., one that does not partition to the stationary phase, and so travels through the system at the speed of the mobile phase) gives a small peak at point D at time (t_m). This could be air, methane, or dichloromethane in GC depending on the detector and conditions, while in LC it might be a large ionized dye molecule that does not interact with the stationary phase. The retention time of the component peak is t_r, at point B. Tangents at maximum slope to the peak intersect the **baseline** at points A and C. Since the peak merges imperceptibly at each end with the baseline, time segment AC can be used to define the **peak width at base (w_b)**. If AB = BC, the peak has perfect Gaussian shape. If AB > BC, it displays some degree of **leading** or **fronting**, often associated with stationary phase overload. If AB < BC, it displays a **tailing** character, associated with some adsorption on active sites.

How much a component is retained by the stationary phase is better described by $(t_r - t_m) = t_r'$, which is defined as the "**adjusted retention time**" (i.e., we count time from the introduction on the column, and then adjust it by subtracting the time it would take to flow through the column if no stationary phase interaction occurred). The capacity or retention factor described in Eq. (11.1) can be calculated from these time measurements using Eq. (11.3)

$$k' = \frac{(t_r')}{(t_m)} = \frac{(t_r - t_m)}{t_m} \tag{11.3}$$

If a component is unretained, $k'=0$. If $k' < 2$, all peaks are packed together too closely at the beginning of the chromatogram. If k' is between 2 and about 10, good separations are observed. If $k' > 10$, the peaks are separated, but they take too long to elute, have become wider and lower, and if close to baseline noise in magnitude, they may become difficult to distinguish from it. These wide peaks with high k' have spent the same amount of time in the mobile phase, where most band spreading takes place, as have the narrower ones eluting

Figure 11.1 Chromatogram with a single eluted peak; peak position and shape terms.

earlier. Their greater width is due to slower elution of their band off the column to form the peak.

The amount by which we can separate two components as peaks in time, or bands in distance, on a chromatographic system is determined by the selectivity of the stationary phase. It is defined for specific conditions; namely, temperature in GC, or mobile phase composition in LC. Selectivity, α, is related to the ratio of partition constants of components A and B, or their more easily measured adjusted retention times, as illustrated in Fig. 11.2 and calculated from Eq. (11.4).

$$\alpha = \frac{K_B}{K_A} = \frac{(V'_R)_B}{(V'_R)_A} = \frac{k'_B}{k'_A} = \frac{(t'_r)_B}{(t'_r)_A} \tag{11.4}$$

When the selectivity is close to 1, the peaks elute closely together, as illustrated in Fig. 11.2. They may even overlap one another, in which case we say that they are incompletely resolved. This concept of **peak resolution R_S** is put on a quantitative basis by ratioing twice the **separation d** between two closely eluting peaks (determined by the column's selectivity), to the sum of their two widths at the base, w_b, (Fig. 11.2).

$$R_S = \frac{2d}{(w_b)_A + (w_b)_B} \quad \text{or} \quad \frac{d}{w_b} \quad \text{for adjacent peaks of equal area and width} \tag{11.5}$$

Resolution is a measure of the efficiency of the column. Another parameter which defines the efficiency is the **number of theoretical plates**, N. Review the discussion of plates in chromatography at the end of Section 11.2. The determination of N from measurements of retention time and peak width in time units will not be derived, but will simply be given here in Eq. (11.6). An advantage of N as a measure of efficiency is that it may be calculated from measurements on a single peak. Unlike R_s, it does not require a pair of peaks and is independent of their relative selectivity α. It may be calculated using either the peak width at base, w_b, or peak width at half the peak height, w_h. The latter is sometimes better, as w_b can often be subtly broadened by overloading (leading) or the effects of adsorption (tailing).

$$N = 16\left(\frac{t_r}{w_b}\right)^2 = 5.54\left(\frac{t_r}{w_h}\right)^2 \tag{11.6}$$

The plates may be imagined as distances along the column where a complete equilibration of the sample between the two phases takes place. The greater the number of plates for a given column length, the shorter the **height equivalent to a theoretical plate, H**, and the more *efficiently* the column operates. This distance along the column is described as a height, since early chromatographic columns (and fractional distillation

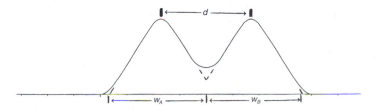

Figure 11.2 Closely resolved peaks of equal size; parameters defining resolution (R_S).

towers) were oriented vertically.

$$H = \frac{L}{N} \tag{11.7}$$

where L is the length of the column.

Since d in Eq. (11.5) equals $(t_r)_B - (t_r)_A$, we can solve for w_b in Eq. (11.6), insert it into the second expression of Eq. (11.5) and relate resolution R_S to plate number N as follows:

$$R_S = \frac{(t_r)_B - (t_r)_A}{w_b} = \frac{(t_r)_B - (t_r)_A}{t_r} \times \frac{\sqrt{N}}{4} \tag{11.8}$$

Students adept at algebraic transformations can substitute appropriate versions of Eqs. (11.3) and (11.4) into (11.8) and rearrange to get Eq. (11.9)

$$R_S = \frac{\sqrt{N}}{4} \times \left(\frac{\alpha - 1}{\alpha}\right) \times \left(\frac{k'_B}{1 + k'_B}\right) \text{ or approx.}$$

$$R_S = \frac{\sqrt{N}}{4} (\alpha - 1) \times \left(\frac{k'}{1 + k'}\right) \tag{11.9}$$

where k' = the average of k'_A and k'_B

Equation (11.9) enables us to estimate the degree of resolution of two closely eluting peaks if we have information on the *column's* **selectivity**, [α, Eq. (11.3)] and **efficiency**, [N, Eq. (11.6)], and the *peaks'* **average capacity factor** [k', Eqs. (11.1) and (11.3)] under the conditions of the separation.

11.7. HOW DO COLUMN VARIABLES AFFECT EFFICIENCY (PLATE HEIGHT)?

At the microscopic, molecular level, very complex theoretical equations are required to describe the chromatographic process. These include expressions for laminar or turbulent fluid flow; random walk, diffusional broadening of analyte bands in both the mobile and stationary phases; and the kinetics of near-equilibrium mass transfer between the phases. Such discussions are beyond the scope of this text.

Fortunately Dutch scientists in the 1950s related the performance of columns in terms of H, the height equivalent to a theoretical plate, to a single variable, the **linear mobile phase velocity**, u. This could be calculated from column dimensions and volume flow rates, or more simply measured directly using the retention time of an unretained analyte, t_m, and the measured column length, L. Three constants, the *ABC*s of chromatographic column efficiency, combine in the **Van Deempter Equation** [Eq. (11.10)], to describe how H varies with u for a particular geometry and construction. Three terms sum together, one independent of u, one inversely proportional to it, and one directly proportional to it.

$$H = A + \frac{B}{u} + Cu \tag{11.10}$$

A generalized plot of the Van Deempter Equation is illustrated in Fig. 11.3. Each of the terms is plotted individually ($A = 0$ here, as this figure illustrates a plot for an open tubular column) and their sum is shown as the dashed line with a distorted U-shape. The important feature of the plot is that each chromatographic system will have a *minimum* value for H as a

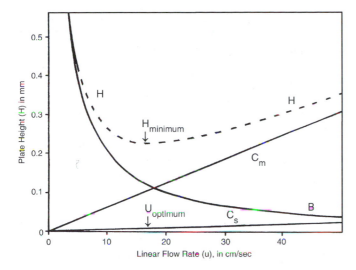

Figure 11.3 Van Deempter plot for open tubular column as sum of three terms: *B* and *C* [as Sum of C_m (mobile phase) and C_s (stationary phase)] with *A* term = 0.

function of *u*. This value defines the flow conditions which will produce the highest possible resolution. When there is high selectivity of a column for all pairs of analytes, a desire to decrease analysis time may encourage operation at flows above the Van Deempter minimum. The coefficients of the three terms are governed by the following processes:

1. *The multipath term* (*A*): This term applies to columns packed with support particles. It becomes zero for open tubular columns when the mobile phase velocity is slow enough for the flow to be laminar (i.e., without turbulent eddies). In a packed column, the paths of individual analyte molecules will differ as they take different routes through the spaces between the particles. Thus they will travel varying distances before they exit the column, and the difference between these distances contributes to band broadening. The relative magnitude of the multipath term depends on the particle and column dimensions. If Fig. 11.3 depicted a packed column, *A* would be a constant value for all values of *u*, and would appear as a horizontal line raising the curve by a constant amount. The multipath process is illustrated in Fig. 11.4.

2. *The longitudinal diffusion term* (*B*): This term accounts for the spreading of molecules in both directions from the band center along the length of the column as a result of random-walk diffusion. This occurs primarily in the mobile phase in GC, but significantly in both phases in LC (as the analytes

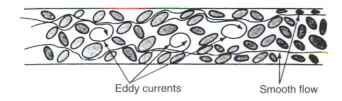

Eddy currents   Smooth flow

Figure 11.4 Mechanism of origin of multipath term *A* of Van Deempter equation.

dissolved in the stationary phase behave much as if they were in a liquid). The faster the linear mobile phase velocity u, the less time the analytes have to spread in the mobile phase. Thus band spreading, and H, are inversely proportional to u. The B term is much more significant in GC than in LC, since diffusion rates in gases are much higher than in liquids.

3. *The mass transfer term (C)*: Diffusion in both the mobile and stationary phases also occurs perpendicular to the direction of flow. If the distance molecules traveled into and back out of each phase were infinitely small, and the mobile phase flow were infinitely slow, there would be time at every point along the column for the molecules to achieve perfect equilibration between the phases. However, the faster the mobile phase moves, the more some analytes move ahead in it due to insufficient time for diffusion to allow them to come to equilibrium, and conversely the further some analytes are left behind in the stationary phase as they penetrate more deeply into it. Thus this mechanism for band spreading is directly proportional to u. Additionally, under laminar flow conditions in small channels, the flow rate will actually vary from zero at the boundary between the two phases to a maximum at the channel center. Molecules diffusing further away will be carried ahead even faster. In Fig. 11.3 we can see how B and C work against each other as u varies to produce a minimum H value. This figure displays a Van Deempter Curve for open tubular column GC. The C term displayed there is actually divided into contributions due to mass transfer in the mobile phase, C_m, and in the stationary phase, C_s, where $C = C_m + C_s$. Because diffusion is so much greater in the gas mobile phase than in the "liquid" stationary phase, C_m dominates. In HPLC, where both phases are liquid, C_m and C_s are similar and C is smaller, so the Van Deempter minimum is less sharply defined and the curves are more "L-shaped".

11.8. PRACTICAL OPTIMIZATION OF CHROMATOGRAPHIC SEPARATIONS

As analysts, how do we employ the principles summarized in the equations listed in Sections 11.7 and 11.8? We want to arrive at the best tradeoff between optimum resolution and separation of mixture components, and the time to complete the separation. Even if time were of no concern, recall from the discussion following Eq. (11.3), that components with very high retention time and k' elute off the column more slowly. This spreads out the peak in the detector, making the signal lower and more difficult to measure above the baseline, thus decreasing sensitivity of detection. On the other hand, if we operate with conditions that make k' and retention times too small, many of the earliest eluting peaks will be crowded together, and we will lose resolution. Note what happens to the value of R_S in Eq. (11.9) as k' becomes much less than 2 in the last factor. The first step is to select a stationary phase (and in LC, a mobile phase solvent mixture) which optimizes selectivity for the analytes of interest. The principles governing this choice will be treated in Chapters 12 and 13 for the various type of chromatography. Catalogs and application manuals from suppliers of chromatography columns publish extensive descriptions of optimized separations and the most suitable stationary phases for all classes of analytes. Consult these, or compilations on CD-ROM, or their online web sites, or their application help phone lines for suggested chromatographic separations of commonly encountered analytes or ones structurally similar to yours.

From Eq. (11.7), we see that greater efficiency of the column is indicated by lower values of H. The number of plates is directly proportional to the length L but, from

Eq. (11.8), the resolution is proportional only to the square roots of either N or L. The retention time increases directly with L. Thus if all other conditions are unchanged, doubling the column length doubles the analysis time but only increases the resolution by the square root of two (i.e., 1.4 times). We could change the temperature in GC or the polarity of the mobile phase in LC to keep the analytes longer in the mobile phase and thus get them off the column faster, but as k' gets near 2 or below, the last factor in Eq. (11.9) rapidly becomes much smaller than its maximum value of 1, and resolution decreases. With many analytes spread out in retention time, bringing later eluting pairs into the region of k' near 2 will force even earlier eluting pairs into the region of severely degraded resolution. Speeding up the mobile phase flow rate causes us to climb up the right-hand side branch of the Van Deempter Curve, which also results in decreased resolution by decreasing the number of theoretical plates. The ability to do this is also limited by the instrument's ability to supply a column head pressure necessary to increase the flow rate. The more efficient columns with narrower channels for passage of the mobile phase require more pressure per unit length to maintain a given linear flow rate, and the required pressure also increases linearly with increasing column length. For these reasons, going to a longer column should be a last resort to improve resolution. A set of mobile and stationary phases with improved selectivity should be sought first, followed by optimization of k' and u (i.e., operating close to the minimum of the system's Van Deempter Curve).

Increasing column temperature often improves efficiency in LC systems, because the resulting decrease of the C term is greater than the increase of the B term (i.e., improvement in mass transfer outweighs the effect of diffusive band broadening in liquids). The decrease in liquid viscosity with increasing temperature means less pressure is required to achieve optimum values of u if longer columns are used. To properly employ a liquid mobile phase at elevated temperature, one must be careful to assure that it is preheated to the desired operating temperature before entering the column. If not, a temperature and flow rate gradient will be set up across the width of the column which will act to broaden bands and reverse the desired resolution improvement. Contrary to our intuitive understanding that increasing temperature lowers the viscosity of liquids, the effect is the opposite with gases. Additionally, temperature is the primary variable that affects retention of analytes in GC, so it cannot be independently changed to affect resolution only through the terms in the Van Deempter Equation which affect plate height and efficiency. It is important to recall that in GC, increasing the temperature increases the mobile phase viscosity and thus decreases flow rate. Even though analytes will come off more quickly because k' becomes much smaller, u will have decreased. If we were originally operating with optimum efficiency at the Van Deempter minimum, the smaller u will shift operation to the steeply rising left hand branch, resulting in a significant loss of resolution. In GC we often employ "temperature programming" (see Section 12.6), raising the column temperature at a constant rate during the analysis to more rapidly elute mixtures of analytes with a wide range of boiling points. In these circumstances it is advisable to start with a u value on the less steeply rising right-hand side branch of the Van Deempter Curve, so that raising the temperature during the run causes u to shift back toward, but not much past, the position of the curve's minimum.

11.9. EXTRA-COLUMN BAND BROADENING EFFECTS

In Sections 11.7–11.9 we have seen how interacting parameters of column efficiency and linear flow velocity (the Van Deempter Equation), selectivity, and capacity factor (related to the distribution of analytes between phases under operating conditions) affect our ability to resolve mixture components. These descriptions apply to columns generating ideal symmetrical Gaussian peaks. We know that processes of overloading or adsorption on active sites can lead to peak fronting or tailing, which degrade this ideal resolution. We have not treated these processes occurring in the column quantitatively. There are other sources of band or peak broadening which occur "outside" the column that we need to take care to avoid. Working backwards from the chromatographic instrument's detector to its injector we may lose resolution in the following ways:

1. *Mixing in too large a detector volume*: If the volume of the detector is much larger than the volume of mobile phase carrying a small fraction of the eluting band, the resulting peak will begin to broaden as the differing concentrations eluting from the band mix together in it. In an extreme case, two bands just resolved on a column could completely mix in a large volume detector cell and all separation be lost. A classic example of this is using a capillary GC column with very high resolution and low flow rate with a large volume detector designed to handle the much higher flows and broader peaks associated with packed column GC. This mixing can be suppressed by supplying a flow of **makeup gas** to the detector to sweep the column effluent through it rapidly before it has time to mix significantly. The resulting dilution of the effluent may lower the sensitivity of detection relative to that which could be obtained with a detector specifically designed to handle low capillary flows.

2. *Open space within a packed column*: While this is not strictly "extra-column" the effect is very similar to what happens when a bands enter a large detector and lose resolution by mixing. This will happen when voids or openings in the bed of particles in a packed column are present. The problem occurs most frequently in HPLC separations, where the packing particles are of micron dimension and the bed is subject to large fluctuations in pressure during pump startup and shutdown. If the bed is not tightly packed between supporting frits, **voids** can develop at either end or elsewhere. **Channels** may form along the walls of the column if the packing bed is not of uniform density across the entire diameter. Dissolved gases (air) may form bubbles as pressure drops along the column. Recall what happens when a pressurized carbonated drink container is opened. Even a few small **bubbles** can disrupt the packed bed. Initially these could impede mobile phase flow, but if the gas in the bubbles redissolves in the liquid mobile phase, a void which acts as a remixing chamber within the column may remain.

3. *Injector spreading and tailing*: Mixtures deposited on the head of a column prior to elution and separation exist as a band of finite thickness. If this band is broadened beyond its ideal possible minimum when it is put on the column, maximum resolution performance from the column will not be attainable. In GC the sample is vaporized in a hot injection port, and then swept onto the column by the mobile phase flow. The analyte mixture is often dissolved in a much lower boiling solvent. If the vapor volume of solvent and analytes resulting from flash vaporization in the hot injector exceeds the capacity of the injector port, some sample may be blown back into unheated portions of the gas lines where it will continue to seep slowly back into the system. If bands of some components deposited at the head of the column move significantly down the column while more sample is still entering the column from the injector, they will start with a larger than ideal width. Many chromatographic procedures load the sample from the injector under conditions where k' is very high.

Little movement of the bands takes place until all the sample is on the head of the column and an ideal minimum initial bandwidth is achieved. Then the temperature is raised in GC, or the mobile phase polarity in LC is changed to more closely match the analytes, and movement and separation of these initially narrow bands begins.

11.10. QUALITATIVE CHROMATOGRAPHY—ANALYTE IDENTIFICATION

Each compound has its own characteristic **retention time**, t_r, in a chromatographic system. This is useful in identifying compounds, but in general it is not as specific as the compound's mass, IR, UV/VIS absorption, or NMR spectra. The t_r depends on the compound, the stationary and mobile phase compositions, the mobile phase flow rate, and column dimensions. It is generally not practical to predict t_r for an analyte precisely from all these parameters, nor to maintain reference data which will enable an analyst to identify a compound from its t_r the way we do from interpretation or comparison of spectra. Further, within the range of $k' = 2-10$, where resolution is optimal and t_r is not unreasonably long, there is space for only several hundred resolved peaks at best. There are thousands of compounds which could elute in this region, so coelutions are highly possible and a t_r is not a unique identifier of only one compound.

We can, however, use t_r to indicate the absence of a detectable level of a compound if there is no peak at the retention time we have determined for the compound on our system by separately running a standard. How closely must the t_r of an unknown and a standard peak match for us to say that they could be the same compound? A slight difference in t_r between the two runs might be due to a shift in flow, temperature, or mobile phase composition between the runs. Spiking the standard into the unknown solution and chromatographing them together at the same time and observing a single undistorted peak eliminates this uncertainty. Another way to "factor out" instrumental variations from run to run is to coinject a **retention time internal standard (IS)** with both unknowns and standards. This compound should have a t_r different from any analytes to be measured or contaminants that might elute closely enough to be mistaken for it. **Relative retention times (RRT)** can be calculated by ratioing the t_r of the standards to that of this IS. These RRTs are more reproducible and may even be compared between columns with the same stationary and mobile phases and operating conditions but having slightly different dimensions. Use of RRT values avoids the need to confirm identity by coelution with spiked standards when there are many peaks to be identified in a chromatogram.

Additional certainty in using t_r values to assign peak identity is obtained when retention times match on two columns with significantly different stationary phases, or a single planar stationary phase eluted by orthogonal, sequential application of two mobile phases. This is the principle and application of 2D chromatography discussed earlier in Section 11.3. Many analyses of complex environmental mixtures specify splitting an injection to two different capillary GC columns operating in parallel to separate detectors. An analyte is considered to be measureable only if it elutes with the appropriate RRT on both columns and the measured quantities of it on each column differ by less than a specified amount. The latter condition ensures that there is no significant coelution with yet another compound on one of the columns. The second column in such a pair is called the *confirmation column*.

Even more certainty of peak identification is obtained by using detectors that are **selective** for certain classes of compounds (e.g., only those with particular heteroatoms, such as halogens, N, P, S, etc). These will be discussed in greater detail in Chapters 12 and 13. Even more selectivity can be obtained by so-called hyphenated detection techniques. Here the effluent is delivered to a spectrometer (e.g., mass spectrometer, IR spectrometer, UV/VIS spectrometer, NMR, etc.), and full spectra or characteristic regions from them are sampled continuously and stored in computer data files as the peaks elute from the column. This provides the best of both worlds (the instrumental analytical worlds of chromatographic separation and of unique spectral fingerprints of pure separated compounds). If their spectra are sufficiently different, it is often possible to identify and quantitate compounds which the chromatograph has failed to separate and which coelute in a single peak. Recall that there is only limited "retention space" in any chromatographic system, and there are thousands of compounds which could possibly elute and coelute there. Conversely, in MS, for example, there are isomers, such as *ortho-*, *meta-*, and *para*-xylene (dimethyl benzenes), which fragment to produce identical mass spectra, but which can be distinguished by their different t_r values in a chromatographic systems. Such instrumentation is complex and expensive, and it demands computerized data handling, but it dominates instrumental analytical chemistry in the fields of biomedical and environmental monitoring and research. The techniques for interfacing chromatographs to spectrometers and acquiring and processing the resulting flood of data are discussed at the ends of Chapters 12 and 13.

11.11. QUANTITATIVE MEASUREMENTS IN CHROMATOGRAPHY

Quantitation may be done crudely on the spots separated by planar chromatographic techniques such as TLC or slab gel electrophoresis (see Chapter 13). One might compare the optical density, the fluoresence, or the degree of stationary phase fluoresence suppression by the unknown spot to a series of standards of known concentration. In contrast, the electrical signals from the variety of detectors used in various column chromatography instruments can be precisely, reproducibly, and linearly related to the amount of analyte passing through the detector cell. If all parameters of injection, separation, and detection are carefully controlled from run to run, and especially if appropriate quantitative internal standards are incorporated in the procedure, accuracy and precision better than $\pm 1\%$ may be attained.

11.11.1. Peak Areas or Peak Heights—Which are Best?

In chromatography we measure each of the signals from a series of peaks. Unlike spectrometric measurements where a single stable signal level is proportional to the amount of analyte, the peak signal rises from the baseline (background) level to a maximum and then returns to baseline. The amount of analyte is proportional to the area of the peak. The instrument or the operator must decide when a peak starts and ends, how to estimate the position of the baseline beneath the peak signal, and then to integrate the signal level above the baseline between the start and end points.

In earlier instrumentation where the signal was displayed as a pen trace from a chart recorder, the baseline was drawn manually. The area of a symmetrical Gaussian peak could be approximated by drawing tangents to each side of the peak and calculating the area of the resulting triangle formed by these tangents and the baseline. This is illustrated in Fig. 11.1. A more accurate but laborious method involved actually cutting out the peak areas from the chart paper and comparing their weights obtained on an analytical balance. Most chromatographic instrumentation now employs computerized data systems. The analog signal level is continuously sampled and digitized, and a summation of these digital "counts" during the peak's elution, after subtraction of levels calculated from a hypothetical baseline between the peak start and end points, produces a peak area value. The operator of the data system must select a "trigger level" of baseline rise that signals the start of a peak. Another one after the descent from the peak maximum level determines either that the signal has returned to a level value (baseline) or is ascending again with the elution of a following peak. These tell the system when to start and stop peak integration. At analyte levels close to the instrumental limit of detection, baseline noise (i.e., random small variations in the signal) will increasingly corrupt the proper functioning of these triggers. If they are set high enough not to be incorrectly tripped by random noise, the baseline will be drawn too high and a significant portion of the peak will not be integrated. If they are set too low they may be tripped by noise, and integration could be either started or cutoff too early. These problems of setting parameters for computerized peak integration are often the primary source of chromatographic standard curve nonlinearity near the lower limit of quantitation.

Peak heights are easier to measure either manually or in a computerized method. The signal level at the top of the peak has the greatest signal-to-noise ratio, and if closely eluting peaks are incompletely resolved the tops are the easiest points to locate. The problems associated with proper estimation of the baseline above which the height is to be measured are similar to the case for peak areas, but heights only have to be evaluated at the single point on the baseline beneath the peak maximum. For well-separated peaks of ideal Gaussian form, the peak height is directly proportional to the peak area. However, as the peak shape becomes more and more distorted due to leading or tailing processes the relation no longer holds. This introduces nonlinearity into standard curves of peak height vs. concentration, which would not occur if properly measured peak areas were used instead of heights. Therefore peak area measurements are preferred whenever possible, despite their greater complexity. The only common exception may be when measuring a small peak eluting on the front or back of a larger one. In such cases a data system may not be capable of being programmed to accurately start and stop integration and draw an appropriate baseline for the small peak. Printing the peak pairs and manually drawing and measuring baselines and peak heights of the unknown and appropriate standards may enable a quantitation calculation.

Review the description of the two different types of detector (mass detector or concentration detector) at the end of Section 11.4. Note well that the area that will be calculated for a peak measured by a mass detector is independent of mobile phase velocity, but when a concentration detector is used it is not. When calibrating against standards using a concentration detector, it is important that the analyte peaks in both the unknown and standard runs elute at the same time and under the same mobile phase flow conditions. Conversely, if quantitation is by peak height comparison using a mass detector, the same requirements apply.

11.11.2. Calibration with an External Standard

The same principles that are outlined for calibration in Section 2.5 apply. A series of solutions containing analytes at levels covering the concentration range to be measured are injected, peak areas are measured, and **standard curves** of peak areas vs. concentrations are produced. In contrast to spectrometric measurements, a well-designed chromatographic separation will move mixture components away from the analyte's retention time, so a blank background subtraction need not be applied to peak area values. It is vital, however, to analyze a blank sample using the same injection solvent, and if possible one processed using the same sample preparation procedure from a similar matrix not containing the analyte. This should confirm the absence of interfering peaks eluting at the same times as target analytes. If these are found, the chromatographic conditions should be adjusted to separate them from the analytes. Note that the **MSA (method of standard additions)** could only be applied if there were sufficient experience with the samples to warrant the assumption that no interferences eluting at the positions of the analytes would be encountered. Such an assumption will always have some level of uncertainty, so this method of calibration is seldom applied in quantitative chromatography. Since chromatography separates the analyte from most matrix components, its detectors do not often suffer from the matrix suppression effects that this calibration method corrects for in spectroscopy.

The standard curve should be linear over the calibration range. Samples with levels above this range should be diluted to fall within it and then be reanalyzed. Advanced data handling systems can fit peak heights or areas to nonlinear (e.g., quadratic, logarithmic) standard curves, but best practice is to find the linear response region and to operate within it. The precision and accuracy of an externally calibrated assay are limited by the reproducibility of sample preparation recoveries, sample volume injections, and the stability of instrument detector response over the course of calibration and analysis runs. In chromatography we may be comparing the detector response from unknowns against those from an external standard curve run hours before.

To quantitate unknown peaks against a standard curve, one plots the response (e.g., area counts from a digital integrator) on the y-axis (ordinate) against the concentration on the x-axis (abscissa). The absolute value of the y-axis intercept of a linear, least squares fit of the data to the points should be a small fraction of the value at the lowest calibrated point on the "curve" (i.e., the response extrapolates to a value close to zero at zero concentration). If so, or if we select a least squares fitting program which forces the line through zero, then we can calculate the concentration C of an unknown from its peak area A and the slope S of the calibration curve using Eq. (11.11).

$$C = \frac{A}{S}$$ (11.11)

This is only valid for areas and concentrations within the limits of the standard curve calibration levels. While in principle one may calibrate against a standard curve which has a substantial y-axis intercept, this will yield valid results only if the the background signal level causing the large intercept is the same in both the standards and unknown sample matrices. This is a dangerous and often unverifiable assumption to make, and the use of such calibration curves should be avoided.

11.11.3.   Calibration with an Internal Standard

Some of the factors mentioned earlier which introduce error when calibrating with external standards only, can be compensated for by introducing a constant amount of an internal standard in both the unknown and the standard calibration samples. This should be a compound with similar chemical nature to the analytes, so that it will pass through the sample extraction and preparation procedure similarly. In general it should elute in the chromatographic system close to the other peaks in the system, but separated from all of them, so it can be identified and measured accurately. One calculates the **ratio** of the peaks' responses to their concentrations. Then the **ratio** for each analyte, A, peak is compared to ratio of the internal standard, IS peak, to give the **relative response factor (RRF)**, for the analyte to the internal standard, as summarized in Eq. (11.12).

$$\text{Ratio A} = \frac{\text{Area A}}{\text{Conc. A}}; \quad \text{Ratio IS} = \frac{\text{Area IS}}{\text{Conc. IS}}$$

$$\text{RRF} = \frac{\text{Ratio A}}{\text{Ratio IS}} = \frac{\text{Area A} \times \text{Conc. IS}}{\text{Area IS} \times \text{Conc. A}} \tag{11.12}$$

For a given constant concentration of IS added, if we plot the ratio of the analyte peak area to the IS peak area against the analyte concentration A we should get a standard curve whose slope is the RRF. In a fashion analogous to Eq. (11.11) for external standard calibration, we can determine the Ratio U of analyte of unknown concentration to IS peak area and calculate the unknown concentration C using Eq. (11.13).

$$C = \frac{\text{Ratio U}}{\text{RRF}} \quad \text{or} \quad C = \frac{\text{Area U/Area IS}}{\text{RRF}} \tag{11.13}$$

The criteria for acceptable linearity of least squares fit and zero intercept when plotting ratios of analyte to internal standard areas vs. concentration are similar to the case for external standard calibrations described earlier. More than one IS can be used, both for calculating RRTs to compensate for retention time variations as well as the RRFs for improving quantitation. The variations that a quantitation IS can compensate for depend upon the point at which it is introduced in the analysis. If it is put into the final extract prior to injection on the chromatograph, it can correct for concentration variations due to evaporative volume changes, variations in injection volume, and variations in detector response. This is called an **injection internal standard**. If the internal standard is put into the initial sample, and into calibration standards prepared in an equivalent matrix, it can additionally correct for variations in recovery during the sample preparation process. This is called a **method internal standard**. Combined use of separate compounds for each purpose can aid in determining the causes of peak area variability.

11.12. EXAMPLES OF CHROMATOGRAPHIC CALCULATIONS

To illustrate the operation of some of the equations and calibration principles earlier outlined, consider the data from the following hypothetical chromatogram summarized in Table 11.1. This is for a capillary GC column, 30 m long, with a diameter of 0.25 mm, operated at a constant temperature of 200°C under a helium pressure of 20 psi above atmospheric pressure, resulting in an average linear flow rate 31.3 cm/s. We measured the time for an unretained component to pass through the column as 1.6 min. We determined the typical number of theoretical plates on the column to be 30,000. Ask yourself how we could easily measure and calculate these values. If the column is 30 m, or 30,000 mm, long, it is trivial to see that the height equivalent to a theoretical plate is 1 mm. This is a very efficient column!

There are five analytes, giving peaks 1 through 5, plus an internal standard peak 6. Observe peaks 2 and 3, eluting at 4 and 4.1 min, repectively. Their widths at the base are somewhat more than 5.5 s, and their peaks are separated only by 0.1 min (=6 s), so they probably overlap a bit near the bottom. This is reflected by the calculated resolution of 1.07. A rule of thumb is that we need a resolution of 1.5 or more to achieve complete baseline resolution. We note that k' is about 1.53 for this pair; a little less than the minimum preferable value of 2 or more. We can improve resolution by lowering the temperature from 200°C to a value that will increase k' and the resolution. From the last factor

Table 11.1 Data from a Hypothetical Chromatogram

Peak	t_r (min)	k'	w_b (s)	$w(1/2)$ (s)	Resolution
1	3.70	1.31	5.13	3.02	
2	4.00	1.50	5.54	3.26	3.37
3	4.10	1.56	5.68	3.34	1.07
4	4.30	1.69	5.96	3.51	2.06
5	4.35	1.72	6.03	3.55	0.50
6 IS	8.00	4.00	11.09	6.52	5.77

			Conc. (ppm)		
	10	25	50	100	200
Peak area counts for five-level standard curve					
1	64,871	138,589	300,767	654,611	1,132,301
2	53,422	150,838	345,670	617,345	1,319,829
3	65,668	151,856	235,309	530,813	1,094,460
4	54,705	116,870	253,633	552,025	9,548,54
5	49,904	140,906	322,911	604,723	1,232,931
6 IS[a]	287,296	262,670	270,879	300,977	279,087
Peak area ratios (standard/peak 6 internal standard)					
1	0.23	0.53	1.11	2.17	4.06
2	0.19	0.57	1.28	2.15	4.73
3	0.23	0.58	0.87	1.76	3.92
4	0.19	0.44	0.94	1.83	3.42
5	0.17	0.54	1.19	2.01	4.42

Note: Capillary GC 30 m × 0.25 mm i.d. column; 20 psi helium mobile phase; 31.3 cm/s; $t_m = 1.6$ min; temperature = 200°C; $N = 30,000$.
[a]The concentration of IS was 50 ppm.

in Eq. (11.9) we can calculate that increasing k' to 3.06, corresponding to t_r of about 6 min, will raise resolution from 1.07 to 1.34. Perhaps this will be sufficient. We could double the number of plates by going to a 60 m column. This improves resolution by the square root of 2; namely, 1.4. That times 1.07 gives us a resolution of 1.5. But this is at the cost of purchasing and installing a 60 m column at nearly twice the price, requiring 40 psi pressure for the same optimum flow rate (perhaps out of our instrument's range), and IS peak 6 will now take 16 min to come off before we can run the next sample or standard. It is much easier to adjust the temperature down to get better resolution of the close pair [2,3]. Speaking of close pairs the resolution of 0.5 between peaks 4 and 5 may well cause them to overlap so much that they cannot be separately quantitated. Based on what we could do by lowering temperature or lengthening the column to increase the resolution of 1.07 for peaks 2 and 3, these remedies are unlikely to be practical for resolving 4 and 5. If we need to measure them, perhaps we had best look for a column with a different stationary phase which has a selectivity, α, for this pair, which is not so close to 1. Of course we must hope that resolution for the other analyte pairs does not decrease too much with the new column. Such are the tradeoffs in finding the best column and conditions for a mixture separation. One can easily imagine that the probability of problems and conflicts increases with the number of components we would like to measure.

The second segment of Table 11.1 shows peak area counts for each of the five analyte peaks over the range of five levels from 10 to 200 ppm, and the areas of the

50 ppm IS peak 6 in each of those five standards. The ratios of each peak area to the IS area are calculated and displayed in the third segment. We can plot each line of data for each peak against the standard concentration, and obtain the slopes and intercepts of the linear least squares fits. If you do this you will find that there is some small variation of the data points about the line, but the intercept is close to zero, and much smaller than the response from the lowest level standard. We can therefore use the slopes, and area or area ratios for unknowns, to calculate concentrations using either Eq. (11.11) or (11.13) for external or internal standard method calculations, respectively.

In the top segment of Table 11.2 we display data for peaks 2, 3, and IS 6 from the standard curve runs, and for three unknown peaks (A, B, C) in a separate chromatographic run eluting with similar RT, plus D, which is the IS 6 peak in that run. In both runs we use IS 6 as a retention internal standard to calculate RRT values displayed in the third column of the table. If we only compared RTs (i.e., t_r), we might say A is 2 and B is 3. We note, however, that IS D elutes 0.2 min later than it did in the standard runs. Something has caused a shift in retention between the runs, perhaps the flow has slowed a little, or the GC oven temperature dropped a bit. If we compare RRTs we see that B and C are better matches to 2 and 3, respectively.

Notice that not only has the retention time of the D IS shifted, but its peak area is significantly lower than the average value from the five standard runs displayed directly above it. If D were employed as a method IS, we may have a recovery difference between the unknown sample and the standards, or perhaps the unknown solution was diluted more during the final step before injection, or perhaps a smaller volume was inadvertently injected, or possibly some combination of all these or other factors. The method IS will quantitatively correct for any or all of these, even if we do not know which may have occurred. The bottom section of Table 11.2 shows the different results which will be obtained with or without correction for either or both the retention and quantitation shifts. The bold values displayed at the lower right are obtained using the IS to calculate RRTs for peak assignment and RRFs for quantitation. Note how they differ from the

Table 11.2 Use of Retention and Quantitation Internal Standard to Correctly Identify Peaks and Correct for Sample Loss or Concentration Changes

Peak	RT (min)	RRT	Area counts	Slope ext. std.	Slope int. std.	Ratio (U)
2	4.00	0.500		6635	0.0235	
A	4.01	0.489	175,000			0.831
3	4.10	0.513		5435	0.0193	
B	4.12	0.502	194,500			0.924
C	4.24	0.517	254,000			1.207
6 IS	8.00		280,182			
D IS	8.20		210,500			

	Ext. std. quant. (ppm)		Int. std. quant. (ppm)		
	Wrong ID	Right ID uncorrected	Wrong ID	Right ID corrected	Identification
A	26.4	Contam.	35.4	Contam.	Unknown
B	35.8	29.3	47.9	**39.3**	Cpd. 2
C	Contam.	46.7	Contam.	**62.5**	Cpd. 3

uncorrected and incorrect values calculated without full benefit of internal standard retention and quantitation calibration.

BIBLIOGRAPHY

Ayres, G.H. *Quantitative Chemical Analysis*, 2nd ed.; Harper and Row: New York, 1968.

Giddings, J.C. *Unified Separation Science*; Wiley: New York, 1991.

Harris, D.C. *Quantitative Chemical Analysis*, 5th ed.; W.H. Freeman: New York, 1999.

Jonsson, J.A., Ed. *Chromatographic Theory and Basic Principles*; Marcel Dekker, Inc.: New York, 1987.

Miller, J.M. *Chromatography: Concepts and Contrasts*; Wiley: New York, 1987.

Scott, R.P.W. *Techniques and Practice of Chromatography*; Marcel Dekker, Inc.: New York, 1995.

Skoog, D.A.; Holler, F.J.; Nieman, T.A. *Principles of Instrumental Analysis*, 5th ed.; Harcourt, Brace and Company: Orlando, FL, 1998.

Willard, H.H.; Merrit, L.L.; Dean, J.A.; Settle, F.A. *Instrumental Methods of Analysis*, 7th ed.; Van Nostrand: New York, 1988.

PROBLEMS

11.1 In GC what is the effect on the retention time (increase, decrease, remain the same, become zero?) of a peak of:
  (a) raising the column temperature (if head pressure kept constant)?
  (b) lengthening the column?
  (c) increasing the gas flow rate?
  (d) increasing the volume of stationary phase in the column?
  (e) increasing the column head pressure?
  (f) overloading the column with sample?
  (g) having some stationary phase slowly volatilize from column?
  (h) forming a more volatile derivative of the peak compound to analyze?

11.2 In LC what is the effect on the retention time (increase, decrease, remain the same, become zero?) of a peak of:
  (a) raising the column temperature (if head pressure kept constant)?
  (b) decreasing the particle size in packed column HPLC (at constant pressure)?
  (c) Having a bubble form in the column?
  (d) Increasing the column head pressure?
  (e) Using an open tubular column of the same length with a diameter $5\times$ the spacing between particles in a packed column at the same head pressure?
  (f) Increasing the polarity of the mobile phase in "reversed-phase" HPLC?
  (g) Using a less polar stationary phase in "normal-phase" HPLC?
  (h) Using a more efficient column with twice as many plates with the same length, mobile and stationary phase compositions, operated at the same flow rate?

11.3 Can one perform planar GC? 2D GC? How?

11.4 What controls the mobile phase flow rate in ascending TLC or paper chromatography? Do you think it is constant during the separation?

11.5 Distinguish among the terms: elution, eluent, effluent.

11.6 What form of chromatography is SPE?

11.7 Acronym Quiz: Identify—HETP, HPLC, TLC, CZE, GC-MS, IS, RRT, RRF.

11.8 Identify and define in words the following formula terms, and give their defining equations from Chapter 11:

$\alpha, H, k', N, A, B, C, L, t_m$

11.9 For a pair of peaks, what is the effect on Resolution (R_s) of increasing one of each of the terms in Problem 11.8.

11.10 What do the Van Deempter equations look like for packed column chromatography, capillary chromatography, and capillary electrophoresis (hint: in the last of these there is no stationary phase)?

11.11 Draw a Van Deempter curve for each of the three techniques in Problem 11.10—the answer should be "schematic", not necessarily quantitative, explain the differences in appearance.

11.12 (a) Calculate the phase ratio for a 30 m by 0.2 mm diameter capillary column coated with a 0.5 μm film of stationary phase.

(b) If a similar column uses a 2.0 μm film of the same stationary phase how will k' change?

(c) Will t_m differ between columns a and b if operated the same way?

(d) If t_m for the column in a is 1.5 min and a peak has a t_r on it of 3.5 min, what will be the t_r of that peak on the column in b?

11.13 Under what conditions are peak heights proportional to area? When are they not?

11.14 Calculate how many theoretical plates a column must have to achieve a resolution of 1.5 (baseline) between two peaks eluting at 8 and 8.2 min.

11.15 An HPLC column received from a manufacturer was specified to have 4000 plates measured under standard conditions with a test compound. When the customer repeated the test she found only 1500 plates, despite a symmetrical Gaussian peak shape. What might account for this?

11.16 In an experiment, an analyst injected a pure compound into a gas chromatograph in volumes increasing from 0.2 to 5.0 μL in 0.2 μL increments. At first the amount of "tailing" observed in each successive peak decreased, but then started increasing at volumes above 3.0 μL. Account for what happened (hint: two different processes dominate at each end of the volume range).

Questions Based on Example in Section 11.13, Tables 11.1 and 11.2

11.17 Using a calculator or personal computer program, use the appropriate data from the second and third segments of Table 11.1 to calculate the slopes for all five peaks used for both the external standard and internal standard calibration calculation. Check your values against those reported for peaks 2 and 3 in Table 11.2.

11.18 How might you satisfy yourself that peak C and not peak B is actually component 3? Suggest two different ways. Which is easier to implement?

11.19 Describe how using separate injection and method internal standards could allow you to separate analyte recovery variations from injection volume variations.

11.20 Suppose the five standard curve mixtures in this example and then two unknowns are injected at 9 min intervals. All six peaks in the five standard levels appear normal, and peaks A, B, and C appear in the first unknown as described in Table 11.2. In the second unknown, the IS is again at

8.2 min, and peaks are seen at the same times as A, B, and C in the previous chromatogram, but there is now a symmetrical Gaussian peak at 4.55 min. with a w_b of 19 s.

(a) Should we identify and quantitate this as component 5?

(b) What does the width of this peak tell us?

(c) Where did it come from?

(d) How can we confirm our hypothesis as to its origin?

(e) How should we modify our operating procedure to deal with peaks like this?

12
Gas Chromatography

12.1. HISTORICAL DEVELOPMENT OF GC—THE FIRST CHROMATOGRAPHIC INSTRUMENTATION

In **gas chromatography** (**GC**, originally called **vapor phase chromatography**) the *mobile phase* is a gas, usually nitrogen, helium, or hydrogen, introduced to the column through a pressure regulator from a cylinder of compressed gas. These mobile phase gases do not have significant solvation interactions with analyte molecules. Initially the *stationary phases* were high boiling liquids dispersed on a particulate solid support, which was then "packed" into a column of metal tubing 1/16 to 1/4 in. in diameter, generally made of copper or stainless steel. Hence the term "**packed column**" GC. The coated particles were secured by plugs of glass wool or porous metal frits at each end. The column (often wound in a coil to fit) was maintained in a thermostatted oven set to a controlled constant temperature. Higher molecular weight analytes, with higher boiling points, would require correspondingly higher temperatures to partition sufficiently into the gaseous mobile phase to be eluted in a reasonable time. True liquid stationary phases would eventually begin to vaporize slowly at such elevated temperatures and be carried off the column as well, a phenomenon referred to as column "**bleeding**". In time such liquid stationary phases were replaced with very high molecular weight polymers, especially long-chain, substituted silicones (cf. end of Section 11.5), which would behave like liquids to the analytes, but which were stable at higher temperatures. At high enough temperatures these would eventually bleed as well, but instead of volatilization of intact molecules of the stationary phase, this was now due to stepwise loss of small fragments from the ends of the polymer chains. Mixtures of volatile liquids (e.g., the components of a petroleum oil fraction, mixed products of an organic synthesis procedure, lipid mixtures, etc.) would be sampled with a **syringe**, and injected on the GC column for separation and detection of separated components. The mobile phase gas, called the **carrier gas**, would be introduced to the head of the column into an unpacked open space above the restraining frit or plug. This empty volume lay beneath a plastic or rubber **septum**, which sealed the top of the column and through which the syringe needle could be inserted. The liquid mixture was injected, vaporized at the elevated oven temperature, and was swept along by the carrier gas to initially pile up as a narrow band on the beginning of the packing. Its components subsequently progressed more slowly, at different rates along the packed portion of the column, and the chromatographic separation ensued. If the carrier gas were helium or hydrogen, these very low molecular weight gases had much higher thermal conductivity than almost any analyte they might carry off the column. A simple detector at the effluent end of the column produced a signal proportional to the amount of analyte in each eluting peak. This resulted from the increase in electrical resistance of a wire placed in the effluent stream and heated by passage of an electrical current. This was initially thought to be a consequence of its increase in temperature due to the decrease in thermal conductivity and thus cooling power of the carrier gas stream, which was caused by the presence of the heavier analyte vapors. So GC had a nearly universal detector which readily produced an electrical signal which could be plotted on a chart recorder to produce a **chromatogram**.

The general concept of GC was put forward in a Nobel Prize winning paper by Martin and Synge in 1941, and implemented by James and Martin in 1952 under the name of vapor phase chromatography. The reasons for GC being the initial realization of modern instrumental chromatographic techniques lie in the relative ease of achieving several of the key components:

1. Only low pressures (5–250 psi) of gas were needed to achieve flow rates suitable for optimal separations. Such pressures could be contained by simple metal or glass tubing and standard nut and ferrule fittings. Compressed gas cylinders using standard two-stage regulators easily provide gases at these pressures.

2. Because diffusion is so much more rapid and viscosity so much lower in gases than in liquids, mass transfer equilibrium between stationary and mobile phases in GC was easily achieved while using large-dimension packing particles, in the range of 100–1000 μm (i.e., 1 mm). It was not necessary to use complex pumps to achieve suitable pressures, as was required for the later realization of instrumental LC, which is performed in packed beds of much smaller particles (1–10 μm).

3. One parameter for controlling partition between the phases; namely, temperature, was easily varied and maintained by readily available thermostatted oven technology.

4. A simple, universal detector, the **katharometer**, or **thermal conductivity** detector (TCD), applicable to chromatography with analytes eluting in gas phase, was available to provide an electrical signal which could be displayed as a chromatogram.

12.2. ADVANCES IN GC LEADING TO PRESENT-DAY INSTRUMENTATION

If the dimensions of stationary phase coating thickness and diffusion distance to the film from the gas phase have been optimized, for further improvement of GC resolution, it becomes necessary to increase the length of the column. This is seen from the simple relation of Eq. (11.7), which indicates that for an optimal minimized value of H (the height equivalent to a theoretical plate) the number of plates, N, is proportional to the length of the column. From Eqs. (11.8) and (11.9) we note that the resolution is proportional to the square root of N. For columns packed with particles of optimal size, and operated at the optimal linear flow rate at the minimum of the Van Deempter curve (Fig. 11.3), the typical maximum pressure of $\sim$100 psi achievable from a regulated gas cylinder requires that most packed columns be less than 4–7 m long. More typically they are only 1–2 m in length. These considerations limit the resolution achievable in packed column GC.

In 1958 Marcel Golay demonstrated that coating the stationary phase on the inner walls of narrow bore tubing of capillary dimension (generally in the range of

0.1–0.5 mm diameter) could allow columns up to 50 or even 100 m to be operated within these pressure ranges. When the flow rate was optimized to a minimum of the open tubular column relation, similar in form to the Van Deempter equation [Eq. (11.10)], the H values were similar to or even better than those obtainable from the best packed columns, so the much longer open tubular or "capillary" columns could have much larger values of N, and consequently much greater resolution. Since they have no packing, the H values for capillary columns can be smaller because there is no contribution from a multipath A term as in Eq. (11.10).

A number of additional advances in GC instrumentation were necessary before the advantages of capillary columns over packed columns could be widely employed.

1. The actual amount of stationary phase per unit column length on a capillary column is much less than that supported on the particles of the typical packed column. Therefore the capacity for analytes of the capillary is much less than that of the packed column. When the capacity is exceeded, the analytes will spread out over the front of the column after injection, and the improved resolution of the capillary will be lost. This results in the phenomenon of **leading** or **fronting** peak shape described in Section 11.4. In capillary GC, one must decrease the concentrations of the injected analytes, often by dilution with solvent to levels which are at concentrations of only the part per thousand or ppm level. For this to be practical one must have detectors orders of magnitude more sensitive than the TCD described in Section 12.1.

2. Inventors developed a variety of "ionization detectors" based on measurement of the tiny currents of ions or electrons induced or suppressed when analyte peaks eluted. One, the FID, was almost universal, giving sensitive and generally similar responses for almost all carbon containing compounds capable of eluting through a GC system. Others were even more sensitive, and displayed selectivities for certain classes of organic compounds, such as halogen or other electronegative substituent containing compounds (ECD), nitrogen or phosphorus containing compounds (NPD), or sulfur and phosphorus containing compounds (FPD). These and several others will be discussed at length later on. The key point is that the sensitivity of these detectors was sufficient to measure the low levels of analytes which were within the column capacity of capillary columns.

3. The typical GC syringe injection volume of 0.5–2 µL of liquid solvent typically expands to 0.5–1.5 mL vapor volume, depending on injector temperature. This is a volume increase on the order of 1000×. A capillary GC column is operated at a much lower flow rate than a packed column (typically on the order of 1 mL/min instead of 20–40 mL/min). The handling of the volume of vaporized solvent and analyte(s) in capillary GC required a more complex control of gas flows in and out of the injector volume. At the typical capillary column flow rates, it may require a minute or more for all this vapor volume to be transferred to the column. In order that the analytes (if not the solvent) should not spread out after entering the column, the column oven must be maintained at a temperature much lower than that used to vaporize the sample in the injector space. This requires that a separate "injector port" volume be independently heated to a different and higher temperature than the initial oven temperature.

    If the analyte concentrations are so high that they will overload the column, an automatic diversion valve may be used to "split" the sample, allowing only a small percentage (1–5%) of the injected vapor volume to enter it. On the other hand, the analytes might be present at very low concentration, and we would wish to transfer almost all of them to the column, while setting the initial column oven temperature high enough to allow the more volatile solvent pass rapidly through. The vapor cloud of solvent exiting the injector port does not do so as a discrete plug. The last portions will mix and be diluted with incoming carrier gas in the injector volume, and bleed slowly onto the column. In order that such a persistent "solvent tail" not overwhelm the detector and interfere with the much smaller analyte peaks, we may wish to program the "**split valve**" to divert the last residue of solvent vapor from continuing to enter the column after we have transferred most of the analytes to the head of the column; a procedure designated "**splitless injection**".

4. Long capillary columns must be coiled to fit in a GC oven. This can be done easily with metal capillary tubing. Unfortunately metal at the elevated temperatures used in GC often interacts unfavorably with many analytes. Such interactions, which could lead to complete analyte destruction, or tailing, as described in Section 11.4, will more severely impact the smaller sample levels required in capillary columns. Also, all the stationary phase is supported on the metal column instead of mostly on more inert packing particles, as is the case of GC on packed metal columns. For sensitive analytes both capillary and packed column GC used glass columns, whose active sites leading to tailing could be deactivated by masking dangling surface hydroxyl groups by treatment with reagents such as trimethylchlorosilane (see the discussion of silica chemistry in Section 11.5). The glass capillary columns were difficult to coil, to coat uniformly with stationary phase, and to fit to the injector and detector ends of the GC instrument without breakage. Unless the greater resolution of capillaries was absolutely necessary, more robust packed glass columns were preferred.

    The development of the flexible fused silica capillary GC column described in Section 11.5 resulted in the domination of the GC market by open tubular columns and instruments tailored to their use. Packed columns are relegated to a few special applications. Fused silica capillaries were easy to install. They were more inert (i.e., contained less active sites causing tailing), than glass or metal capillaries, or the particle supports used in packed column GC. Even when the resolution of packed columns was sufficient for simple separations, a capillary coated with the same stationary phase could achieve equivalent separation in a shorter run time. Packed columns, especially those with diameters larger than 2 mm, will possess much higher analyte capacity than capillaries. They are still preferred for preparative separations, where the separated peaks are captured for later processing or use. The TCD is still the most universal, responding to everything but hydrogen or helium when these are used as the carrier gas. Compounds such as carbon monoxide, oxygen, nitrogen, and small molecules not containing carbon require its use, and its low sensitivity often dictates the use of high concentrations of analytes requiring the peak capacity of packed columns. Some regulatory standard assay procedures developed in the early days of GC instrumentation still specify the use of packed columns, although most of these have been updated to capillary GC equivalents. Therefore, we shall discuss GC and GC instrumentation primarily from the perspective of open tubular column GC.

12.3. GC INSTRUMENT COMPONENT DESIGN (INJECTORS)

12.3.1. Syringes

The most common way of introducing a sample into a GC instrument is with a syringe. The typical hypodermic syringe used to administer vaccines or intravenous medications has a volume on the order of 1 mL, a plastic body and plunger, and a sharply pointed, detachable needle with an internal bore of 0.2–0.5 mm, and is often designed for one-time use. In contrast, the usual GC syringe has a narrower bore, permanently attached needle, long enough (2–3 in.) for the tip to penetrate to the center of an injector port, a volume on the order of 5–25 μL, and is composed of a glass body with a very tightly fitting metal plunger. The usual volume of liquid sample or solution injected is 0.5–2.0 μL when open tubular GC columns are used. Larger volumes might expand to a vapor volume (>1–2 mL) which is larger than the capacity of a standard heated injector port, and "blow back" into the gas supply system and condense there. Such vapor volumes would require an excessive time to transfer to GC columns at typical 1–2 mL/min capillary flow rates. The GC syringe needle must be very narrow, both to minimize the volume of sample remaining in the syringe when the plunger is depressed, and to pass smoothly through the deformable **septum disc** which seals the entrance to the injector port to maintain the column **head pressure**. Too wide a needle diameter could cut out a plug of the septum, a process called "*coring*" (as in coring an apple). This results in gas leaks preventing the GC from maintaining the desired head pressure, and the **septum core** is likely to be projected into the hot injection port, where it will decompose to liberate "*septum bleed*" into the column or trap sensitive analytes. Syringes designed for very small volume injection (<1 μL) may contain the sample within the bore of the needle itself; expelling it with an internal wire plunger. Such designs place a limit on how narrow the needle may be.

Gas samples may be injected with larger syringes (1 mL or larger volume), as they do not expand as much as liquids do upon vaporization. Very large gas volumes (>30 mL) may require slow injection into the larger flow rates usable with packed columns. For regular repetitive sampling of vapor streams, a rotary multiport *gas sampling valve* like that illustrated for HPLC injectors (Fig. 13.6) can be set to cycle at intervals to accept an uncondensable gas sample into a metal sample loop of known volume and then insert this into the carrier gas stream for a GC separation. Sampling and injection techniques to measure trace concentrations of gases in air or water will be discussed in Section 12.10.3.

12.3.2. Autosamplers

Modern GC instruments can sequentially analyze up to 100 or more liquid samples without manual intervention. Racks or carrousels of vials containing liquid samples or standard calibration solution are capped with thin septa, like miniature versions of the GC injector port septum. The vial holder moves to present each vial sequentially to a GC syringe mounted in a robotic assembly. The syringe penetrates the sample vial septum, the plunger is pumped several times to pull up a set volume of sample and expel air bubbles. Then the syringe is withdrawn, moved over the injector port septum, inserted smoothly and rapidly, and the plunger depressed to complete the injection. Upon injection a signal is sent to the GC electronics to initiate any temperature control programs needed in the various GC ovens and to the detector and recording electronics to start recording the chromatogram signal. Many refinements can be programmed into these operations. For example, a plug of air may be drawn into the syringe before the sample volume is pulled up, and after sampling the plunger is further withdrawn in air again to pull all of the sample out of the needle bore. The sample is thus *sandwiched* between two plugs of air. When the needle is initially inserted into the hot injector port the solvent is not present in it to start vaporizing prematurely. When the plunger is depressed the rear air plug forces all the sample from the needle, and again there is no boil off of the sample from within the hot metal needle. The long thin narrow bore needle required for these microliter injections is the weak point of the "syringe-through-septum" technique. Bad operator technique or autosampler malfunction can easily result in a severely bent needle. Once it is bent, the one-piece GC syringe/needle assembly is beyond repair.

12.3.3. Solid Phase Microextraction (SPME)

A recently developed procedure called SPME combines the principles of the mini-column sample isolation and concentration technique of SPE, and the technique of automated GC sample injection via a syringe assembly. A small amount of a solid phase into which analytes can partition from the sample is coated on one end of a fine fused silica fiber. These solid phase materials can often be the same as GC stationary phases. The other end of the fiber is fastened to a GC syringe plunger, such that the coated portion of the fiber can be extended beyond the needle tip, or be withdrawn entirely within the needle. A sample, generally liquid, is contained in a septum-capped vial. The syringe needle pierces the sample vial septum, and the fiber is extended for several minutes—either into the sample liquid or the headspace gas above its surface. Analytes partition and concentrate into the fiber coating. The volume of this solid phase coating is very small; hence the "micro" in SPME. After equilibration, the fiber is retracted; the syringe needle is pulled from the vial, and it is then inserted through the GC septum into the heated injector port, where the fiber is immediately re-extended. Analytes are thermally desorbed and are swept by the carrier gas flow to the top of the much cooler capillary GC column, where they collect. The column temperature is raised to an appropriate value and separation of analytes proceeds in the normal fashion.

The SPME injection procedure is illustrated in Fig. 12.1. If the sample matrix is not so dirty that it coats the fiber with high boiling material that is not thermally desorbed, the fiber may be used for many repeated sampling and injection cycles. If the analytes are quite volatile, it may be possible to sample them in the confined headspace volume above a "dirty" matrix, thereby avoiding the contamination of the fiber by high boiling contaminants. This **headspace analysis** is one of the most useful modes of SPME injection. The fiber concentrates only small portions of the analytes, but all of the "extracted" material gets injected into the GC by the thermal desorption process. Contrast this with classical liquid–liquid extraction or SPE procedures, where the extract is into a large volume (e.g., 1–10 mL), which must then be concentrated (to perhaps 25–100 µL), of which perhaps only 1–2 µL may be injected. A good application of headspace SPME analysis would be for volatile organic vapors (alcohols, solvents, hydrogen cyanide, etc.) from a messy whole blood sample. The amount of analyte absorbed is based on the equilibrium concentration that accumulates in the fiber, and it is far from the total amount of analyte present in the sample. This equilibrium concentration can vary substantially with equilibration time, temperature, and other conditions. Therefore to obtain quantitative data when employing SPME, it is necessary to calibrate against an IS introduced into both unknowns and calibration solutions. The IS must behave in a fashion similar to the analytes, both in its equilibration behavior and in its response to the GC detector employed. Even when using an IS, good quantitative values are difficult to obtain, unless the IS is an isotopically labeled version of the analyte, and it is used with mass spectrometric

Figure 12.1 Diagrams of SPME (a) extraction sampling, and (b) GC despoption injection. (Used with permission from Supelco, Inc.)

detection. The fiber coating in SPME is often the same as a GC stationary phase. Analytes are moved off it into the GC column by varying the same parameter, temperature, which controls GC elution. In the capillary GC column the coating is on the inside of a long cylindrical tube. On the SPME fiber it is on the outside of a short cylindrical rod, of slightly smaller diameter. For this reason SPME sampling and injection is sometimes character-ized as "inside-out gas chromatography". It is not really chromatography, because there is no continuous partitioning to mobile phase, but otherwise many of the principles and materials are similar.

12.3.4. Split Injections

The sample capacity of capillary columns is limited. GC analysis of neat mixtures or highly concentrated solutions may require that only a portion of the sample which is vaporized in the injection port enters the column. Modern capillary GC instruments often feature a "**split–splitless injector**". This has a time programmable on–off valve controlling flow to a split vent gas line from the injector port. When this valve is open, only a fraction of the carrier gas flow (and of the vaporized injected sample) passing through the injector port enters the column. The rest flows out the split vent. The relative proportions passing to the column and the vent may be adjusted by a variable needle valve (the backpressure regulator) in the split vent line. The ratio: split vent flow/column flow can typically be varied from 5 to 100, and is called the **split ratio**. When operating the GC column isothermally (at a constant temperature chosen so analytes move down the column at a reasonable rate), the sample must enter the column over a short time, if the initial chro-matographic bandwidth is to be minimized. At capillary column flow rates of ~ 1 mL/min. the ~ 1 mL vapor volume from a 1 μL injection would take about a minute to load onto the column, and the sample would spread out during loading. With the much larger flow through the port with the split vent open and set to a high split ratio, a small fraction of the vapor cloud enters the column over a much shorter period. A diagram of the flows in the **split mode** of a split–splitless injector is illustrated in the upper section of Fig. 12.2.

12.3.5. Splitless Injections

When sample concentrations are much lower (ppm to ppb, or lower) one will wish for all the analytes to be transferred to the capillary column, and overloading will be less of a

Figure 12.2 Diagram of GC flows for split and splitless injections: (a) split mode; (b) splitless mode. (Adapted with permission of Restek Inc.)

concern. As illustrated in the lower section of Fig. 12.2, when the valve to the split vent is closed, all of the vaporized sample must flow to the column. When this **splitless mode** of injection is used the valve must be closed while most of the sample and its solvent vapor have time to enter the column. As mentioned in the split mode discussion this time may be on the order of 1–2 min. In order that the analytes not spread out on the front of the column, the initial column oven temperature must be well below that at which they begin to move down the column. The lower boiling solvent they were dissolved in will pass rapidly ahead of them. In the splitless mode of operation, the column temperature will then be raised to temperatures (constant or steadily increasing) at which the analytes begin to move down the column and separate. Traces of the much higher concentration of this solvent remaining in the injection port might continue to enter the column and obscure the signals from the lower levels of analytes. Therefore, at the end of the splitless injection period, the valve is returned to the split mode position to rapidly divert the remaining solvent. A high split ratio setting enhances this.

12.4. GC INSTRUMENT COMPONENT DESIGN (THE COLUMN)

12.4.1. Column Stationary Phase

As described in Section 11.5, silicone polymers form the most common and useful class of GC stationary phases. Figure 12.3(a) displays a two-monomer segment of the PDMS chain. This is the fundamental silicone polymer. Various polymers used for GC phases differ in their polarity. The degree to which the polarity of the analyte is similar to that of the phase is an indication of the strength with which it will dissolve into and be retained by the phase. This in turn depends on the analyte's polarity. The general rules of "like dissolves like" and its converse apply here. At one end of the polarity scale are pure long-chain hydrocarbons, whose intermolecular interactions are governed almost entirely by *Van der Waals forces* (alternatively referred to as *London's dispersion forces*). These are the *least polar* compounds and phases. Unless pure hydrocarbons have a molecular weight high enough not to volatilize or decompose under the full range of GC temperatures, they make poor stationary phases. PDMS itself is of intermediate polarity, with the oxygen atoms in its backbone contributing some **dipole** and **hydrogen bonding** character to its polarity. Substitution of the methyl groups in PDMS with longer hydrocarbon substituents (e.g., $-C_8H_{17}$ or $-C_{18}H_{35}$) yields a silicone polymer with a lower polarity more similar to hydrocarbons. The oxygen in the backbone is less accessible, and there is a higher proportion of the hydrocarbon-type Van der Waals dispersive interactive surface presented to the analyte.

Figure 12.3(b) illustrates two other types of modification to the PDMS structure. In the "y" monomer, a more polar (by virtue of **π-electron bonding interaction**) phenyl group has been added. This increases the polarity of the phase, and thereby the retention of analytes which have a similar functionality. In the "x" unit, occasionally substituting a phenyl ring for an O-atom in the polymer backbone again enhances this type of polarity, but it removes the hydrogen-bonding and dipole polarity of the replaced oxygen atom. The net overall effect on the polarity interaction is difficult to predict. It will depend on the structure of the analyte. Such an inserted "**spacer**" is in fact not present for polarity adjustment, but rather to stabilize the polymer chain against thermal decomposition at high temperatures. This suppresses the phenomenon of **column bleeding**, which results in an increased detector signal and a rising baseline signal as temperature is increased. If one wishes an even more polar column, a substituent such as $-CH_2CH_2CH_2CN$, with a very strong cyano group dipole can be substituted for a PDMS methyl. The polarity changes introduced by substitutions such as these phenyl or cyanopropyl groups can be varied by adjusting their percentage on the chain.

Occasionally other substituents with reactive end functional groups can be introduced. After the phase is coated on the interior of the column, these groups can be reacted to bond with one another (**cross-linking**) or with free silanol groups (Section 11.5) on a capillary column's interior silica surface (to form a **bonded phase column**).

Figure 12.3 Representative monomer units for several common types of GC stationary phase polymers: (a) PDMS, (b) backbone phenylene-stabilized and phenyl-substituted PDMS (more polar π-electron character), and (c) Carbowax (polyethyl ether phase).

These processes further improve the thermal stability of the stationary phase, and suppress degradation and bleeding. A bleeding signal may also result from decomposition of very involatile material deposited on the column from prior injections of dirty samples. Cross-linked phases bonded to the walls of capillary columns can withstand rinsing with solvents to remove this kind of contamination. Recall that bare or nonbonded silanol groups on the silica wall can act as active sites to introduce tailing in eluting peaks. After coating and bonding a phase to the column wall, such remaining "**active sites**" may be **deactivated** by treatment with reagents which bond methyl or trimethylsilyl groups to them. With long use, these active sites may reform, and can sometimes be reblocked by injection of the derivitizing agent through the column at elevated temperature. The commercial designations of capillary columns often reflect many of these design elements. For example a widely used J&W (now Agilent) column is designated DB5-MS; the 5 indicates 5% phenyl substituted PDMS, the MS indicates the presence of phenyl spacers in the chain to suppress bleed that interferes particularly with MS detection, and DB is "Durabond", this particular manufacturer's trade designation that indicates that the phase is chemically bonded to the capillary column. Other manufacturers employ a variety of very similar designations for their equivalent phase materials. Figure 12.3(c) displays a common nonsilicone polymer, whose monomer is characterized as a "carbowax" column. The oxygen in the chain is much less shielded than the one in silicones, so wax phases are much more polar, but they lack the high thermal stability of the silicones.

The description just given highlights a few examples from the major categories of polymer stationary phase coatings. In the modern practice of GC, these are almost always employed in open tubular or capillary columns. Any assay which might employ them in packed column mode can be implemented more efficiently, or faster, or both, using open tubular columns, and a vast selection of these are now readily available from commercial vendors. Because many very simple instructional GCs only operate with packed columns, the undergraduate student may still encounter them in the instructional laboratory. They may lack separately heated injector or detector regions and be capable only of isothermal separations. The student should be aware that in industrial and research practice, the analysis time and resolution advantages of open tubular columns makes them the overwhelmingly dominant choice for most applications. Packed column GC is relegated to large-scale **preparative GC**, where the greater sample capacity of wide-bore, packed columns is required, or to a few "niche applications".

One example of the latter is the separation of "**fixed gases**" (e.g., H_2, He, Argon, O_2, N_2, CH_4, CO, CO_2). One scarcely needs a column oven for these—a column refrigerator would seem more in order. Unfortunately when most polymer phases are frozen solid, gas molecules do not partition efficiently into them. Packed particles of a "**molecular sieve**" material can separate these small ultravolatile molecules. These materials, sometimes produced from the catalyst minerals called zeolites, have pores of the dimension of the fixed gas molecules. Larger molecules cannot enter and elute with the mobile phase gas stream. The small molecules vary in the extent to which they can diffuse in and out of the pores and through the sieve, and are thereby separated. Figure 12.4 displays such a separation. The principle of excluding the larger molecules is similar to what happens in "size exclusion chromatography" described in Chapter 13. For those molecules which can enter the openings of the sieve particles, the factors governing relative retention are more complex. It is not easy to rationalize the elution order seen in Fig. 12.4.

Another type of packed column particle useful for relatively small molecule separations is a "**porous polymer**", Trade name: **Porapack**®. These are like the molecular sieve, but made of organic polymers or resins with larger sized pores. Highly volatile molecules larger than the fixed gases, but of a size containing only a few atoms, are

Figure 12.4 Molecular sieve PLOT column separation of fixed gases with TCD detection. (Used with permission from Varian Chrompack Inc.)

retained by these materials, even at temperatures high enough to be controlled by standard GC column ovens. Even these materials have now been engineered into coatings for open tubular GC, with its inherent advantages, and are sold as **porous layer open tubular (PLOT)** Columns. A representative separation of small molecules on a PLOT column is displayed in Fig. 12.5. An advantage of these porous polymer phases is that they will even tolerate and separate (albeit with poor tailing peak shape) water. This ultrapolar and strong hydrogen-bonding liquid is usually totally incompatible with and destructive of other polymer stationary phase coatings.

Figure 12.5 Porous polymer PLOT column separation of low molecular weight gases with TCD detection: analysis of PH_3 and AsH_3 in C_1-C_3 hydrocarbons. (Used with permission from Varian Chrompack Inc.)

12.4.2. Selecting a Stationary Phase for an Application

With thousands of possible analytes, even more different combinations of these analytes, and upwards of a hundred stationary phase materials generally available, how should one go about selecting the right column for the job? In earlier times, analysts compiled tables of so-called McReynolds or Rohrschneider constants of characteristic compound types for different phase materials. If the data for the compounds of interest were available from these tables, one could employ these values with calculation algorithms to estimate the relative elution orders and resolution of the components on different columns. This approach has fallen into disuse. A modern commercial capillary column with all the features described in Section 12.4.1 is a very highly engineered product, not easily created or duplicated in the analyst's lab. The column manufacturers have amassed extensive databases and application notes for optimum separations using the most suitable columns from their stables. The best place to start when looking for a column and phase to achieve separations for a group of analytes is with the **manufacturers' applications literature**. Their catalogs, in hardcopy or on searchable CD-ROM disc, and their websites, with search programs for the analyst to enter the compound type or application, illustrate thousands of optimized separations. It is beyond the scope of this chapter to attempt to summarize even the beginning of this information. Appendix 12.1 lists some selected web resources. The student is encouraged to access them, or catalog data if available, to gain a sense of the scope and organization of this information. This is where one should now start when confronted with a GC separation problem.

Once a reasonably suitable column phase has been identified, the dimensions of the column employing it should be selected, and then the separation is optimized by varying the flow and temperature parameters. The column dimension decisions will depend on the difficulty of achieving the necessary peak resolutions, which is likely to increase with the number of analytes to be measured. The column's analyte capacity depends on these dimensions as well. The column dimensions, carrier gas flow, and column oven temperature parameters thus chosen will determine the time required to perform each GC run.

It may be useful to attach a length of uncoated capillary tubing to the front of the coated column. This so called "**retention gap**" serves two purposes. It allows space for the injected sample and solvent vapors to collect on the column without being significantly retained until they encounter the coated portion. Dirty, poorly volatile contaminant residues can accumulate in this region, and front portions of the gap section can be cut away to remove these without affecting the separating and retaining functions of the column. When the gap region becomes too short, it can be replaced by another. Figure 12.6 illustrates the processes which occur following sample and solvent injection on a capillary column using a similar procedure, **direct on-column injection**. Here an extremely fine syringe needle actually enters the capillary column, bypassing the injector port (whose heater should be turned off), and deposits the liquid sample solution directly in the gap region of the column, usually at a low starting temperature near or below the solvent's boiling point. This is a delicate process, required for delicate, thermally unstable analytes, which might be destroyed by vaporizing in a hot injector port. After vaporization, the retention gap functions as illustrated; whether the injection is on-column or through a splitless injector.

12.4.3. Effects of Mobile Phase Choice and Flow Parameters

A column with a particular stationary phase and dimensions has these as fixed parameters. Changing them means obtaining or purchasing another column and installing it. New

Figure 12.6 Illustration of on-column injection with a retention gap. (Cazes, used with permission.)

columns require a significant **conditioning** or "breaking in" period after installation. Another fixed parameter is the choice of mobile phase (i.e., the carrier gas to be used). Three are commonly employed: helium (He), hydrogen (H_2), or nitrogen (N_2). As will be discussed in Section 12.7, proper operation of the GC detector being used will sometimes dictate use of a particular carrier gas. If the detector tolerates any of these gases, then their flow and diffusion characteristics govern how well each attains optimum chromatographic resolution. This is best illustrated by their respective **Van Deempter curves** [cf. Sections 11.8 and 11.9, Eq. (11.10), Fig. 11.3]. These are displayed in Fig. 12.7. N_2

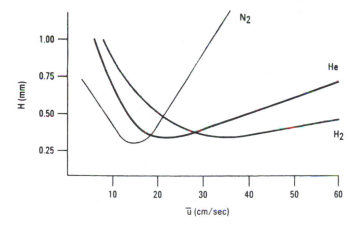

Figure 12.7 Van Deempter plots for three GC carrier gases: N_2, He, H_2: H (height equivalent to theoretical plate) vs. u (linear flow velocity, cm/s).

has the deepest **minimum H** (i.e., HETP) value (i.e., greatest efficiency), but only over a limited range of linear flow rates, u. Lighter gases He and H_2 do not quite attain such low H values, but they stay lower over a much wider range of u than does N_2. One would like to operate the column close to the flow rate of the Van Deempter minimum for the gas. Advanced chromatographs equipped with automatic flow controllers can maintain the flow at this point as other parameters change. In the absence of one of these complex and expensive devices, if one is raising the column temperature over the course of an analysis, it is better use the light gases and to start at a flow rate above the Van Deempter minimum. As column temperature goes up, gas viscosity increases, and the flow will decrease towards minimum position of H. Low values of H are more difficult to maintain over the steep-sided valley of the N_2 curve. In Section 12.7, we will see that such rising "**temperature programs**" are often required to optimize separations of mixtures of large numbers of compounds.

Helium is more expensive to purchase than N_2 or H_2 (which can be generated on site from water electrolysis). H_2 forms explosive mixtures with air. A catastrophic column rupture or leak might allow enough into the oven to contact hot heating coils and trigger an explosion. The US held a near monopoly on He, so GC users there preferred it to H_2 on grounds of safety. In Europe and elsewhere chromatographers learned to use H_2, since He was not available (recall that the Hindenburg Dirigible was forced to employ hydrogen to its ultimate demise). Its capillary GC resolution performance exceeds even that of He. As flow controllers become more available on GC instruments, and lower flow capillary systems become the norm, one could program them to recognize a sudden catastrophic H_2 leak and command the shutdown of carrier gas flow before the H_2 concentration reaches dangerous values. This facilitates the use of superior H_2 as a carrier gas.

12.5. GC INSTRUMENT OPERATION (COLUMN DIMENSIONS AND ELUTION VALUES)

The column temperature is the major instrumentally controllable variable which can be used to affect the speed and resolution of the GC separation. Much precision engineering goes into designing ovens which can precisely control and reproduce the temperature profile that the column encounters during each analysis.

Before we move to a description of how temperature affects peak elution on GC columns, let us review how some of the column dimension and stationary phase composition factors interact. In Fig. 12.8(a) we see the effect on retention time and resolution of increasing the stationary phase **film thickness** d_f, and thereby lowering the **phase ratio** [cf. Section 11.7, Eq. (11.2)]. Thicker films are especially effective at retarding faster eluting components to elute at suitable k' values [cf. Eq. (11.3)], while providing modest improvements in the resolution of close pairs eluting in the middle of the run. The primary value of the thicker film is to keep very volatile components from eluting too fast before they have time to separate well. It also confers greater column capacity against overloading, which would result in leading peaks and degrade resolution. Another way to increase resolution is to use a longer column, as illustrated in Fig. 12.8(b). Note that for the illustrated pair separation, the longest column takes a little longer and does not achieve quite as good resolution of this pair as did the thicker film column. The longer column would probably result in a much higher price and require higher pressures to achieve a flow to operate it at the Van Deempter optimum. Thus longer columns are not always the best way to achieve better separation. Going to a longer column should be the last resort. Recall from Eq. (11.8) that resolution increases only as the square root of

Figure 12.8 Effects of stationary phase film thickness or GC column length on retention time (R_t) and resolution (R). (a) film thicknesses (d_f) of 0.25, 0.50, and 1.0 μm (30 m × 0.25 mm PDMS column) and (b) column lengths (L) of 15, 30, and 60 m (0.25 mm × 0.5 μm PDMS column). (Used with permission from SGE Corporation.)

the number of plates, N, or the column length, L. Note that it is not necessarily the thicker film that provides the better resolution, it is lowering of the phase ratio. This could also be achieved with a thinner film and a correspondingly narrower bore column, to achieve the same low phase ratio. With a thinner film, equilibration in the stationary phase would be faster as well, leading to even lower values of H (plates/m) and a more efficient column. However, the narrow bore, thin film column has less capacity and is more easily overloaded, and it requires higher head pressures to be operated at or above the Van Deempter optimum. The pressure limits of the GC may limit the usable length of such columns.

In Fig. 12.9(a) we see how a shorter column, with film thickness and diameter reduced to keep phase ratio constant, can achieve nearly the same separation performance in half the time as a column of more normal dimensions. If overloading can be avoided, this is the royal road to achieving "**fast GC**". For some detectors the lower carrier gas flow is a benefit, for others designed for higher flow levels special adjustments (e.g., use of **makeup gas**, cf. Section 12.7, Item 4) may be required. If none of these resolution

NORMAL - 0.25mm ID
Chromatogram using a conventional
(30m x 0.25mm ID) BPX5 column
with a 0.25μm film.

FAST - 0.10mm ID
Chromatogram using a FAST
(10m x 0.1mm ID) BPX5
column with a 0.10μm film.

Components

1. Naphthalene	5. Phenanthrene
2. Acenaphthylene	6. Anthracene
3. Acenaphthene	7. Fluoranthene
4. Fluorene	8. Pyrene

9. Benzo(a)anthracene	13. Benzo(a)pyrene
10. Chrysene	14. Indeno(1,2,3,-c,d)pyrene
11. Benzo(b)fluoranthene	15. Dibenzo(a,h)anthracene
12. Benzo(k)fluoranthene	16. Benzo(g,h,i)perylene

(a)

Components
1. Benzene
2. Toluene
3. Ethylbenzene
4. p-Xylene
5. m-Xylene
6. o-Xylene

SolGel-WAX™
Polar

SolGel-1ms™
Non-Polar

(b)

Figure 12.9 (a) Comparison of PAH separations on a column of "normal" dimensions and on a column designed for "fast" GC. (b) Comparison of BTEX compound separations on carbowax and PDMS GC phases. (Used with permission from SGE Corporation.)

enhancing tricks and no adjustments of the column temperature will achieve resolution of critical analyte pairs, it is time to fall back and consider finding a column with a different stationary phase to do the job. A classic example is the analysis of BTEX (benzene, ethyl benzene, toluene and 3 isomeric xylenes) single ring aromatic hydrocarbons. Many silicone-based phases fail to separate the *meta*-(1,3-dimethyl) and *para*-(1,4-dimethyl) benzenes (xylenes). This is illustrated in Fig. 12.9(b), where the SolGel-1MS (a PDMS-type phase) fails, but the BTEX compounds are all separated on a SolGel-WAX (carbowax-type phase). Note that the other xylene isomer, the *ortho*-(1,2-dimethyl), is more strongly retained on either phase. In general one cannot predict relative retention for similar structures precisely enough to guess which peak is which or whether a given phase will separate them or not. This is where recourse to the manufacturer's databases on separations and phases is invaluable. Moral: database lookups trump theory for getting elution orders.

12.6. GC INSTRUMENT OPERATION (COLUMN TEMPERATURE AND ELUTION VALUES)

Finally we reach the discussion of the most easily adjusted variable controlling GC peak elution. Recall from Chapter 11 that for a given stationary phase the temperature of the column is the primary determinant of the equilibrium partition ratio between the stationary and mobile phases. The larger the percentage of time the analyte spends in the gas flow of the mobile phase the more quickly it elutes from the column. A modern GC instrument is designed to very precisely and reproducibly control the temperature of the oven compartment in which the coiled-up column resides. This will promote reproducible retention times to enable peak identification.

In a very simple GC analysis of a group of similar analytes which would elute close together, the analyst chooses an appropriate stationary phase, optimizes its film thickness, and selects the necessary column diameter and length, all as outlined in Section 12.5. Then the oven is set to a column temperature where k' for these analytes is between 2 and 10 [cf. Eq. (11.3)]. The highest value of temperature T at which adequate resolution of the closest eluting pair is observed, will yield the fastest analysis. Such operation is called **isothermal GC**. It is the simplest for the temperature controllers to maintain, and as soon as the last component has eluted, the instrument is ready to accept injection of the next sample.

Most often more complex mixtures of components varying widely in polarity and volatility must be analyzed. In Figs. 12.10 and 12.11 we will examine the problems these sorts of mixtures present for the development of a simple isothermal GC analysis, and we describe how to separate them all in a single GC run. The two figures illustrate the "**early eluting**" and "**later eluting**" peaks of a single analysis. The column parameters have been selected: a 30 m long, 0.25 mm i.d., 0.25 μm film PDMS column; the archetypical standard capillary GC column. It uses H_2 carrier flow controlled at 46.3 cm/s, the minimum of the H_2 Van Deempter curve. The column "dead time", t_m, was determined to be 1.08 min. The analytes consist of four groups:

1. Normal straight-chain hydrocarbons, labeled C9 through C15, for the number of carbons in the chain;
2. A group of three branched-chain C10 hydrocarbon isomers, labeled "**3 C10s**";
3. A group of three branched-chain C14 hydrocarbon isomers, labeled "**3 C14s**";
4. Three dimethylnaphthalene isomers, labeled D1, D2 and D3.

There are several points of caution to note when interpreting these figures:

1. These "chromatograms" were not obtained on an actual GC instrument. They are accurate simulations of retention data from a computer program using data on the column and the analytes.
2. The retention time axes *differ* for each run, A through D (i.e., the peaks do not line up with one another on a single elution time axis).
3. Only retention time and peak width data are simulated, to illustrate analysis time and resolution. The peaks are always displayed at the same height (except when they merge), while in a real chromatogram, as peaks from the same sample were eluted more slowly, their height would decrease as their width increased, thus keeping their area constant.

**Effect of Column Temperature on Retention Time and Peak Resolution
(Early Eluting Peaks)**

Figure 12.10 Separations of three branched chain C10 hydrocarbons, and C9, C10, C11 *n*-alkanes on 30 m × 0.25 mm × 0.25 μm PDMS column; at (a) 100°C, (b) 120°C, and (c) 150°C, and (d) programmed 100°C/1 min–10°C/min–140°C/4 min early eluting peaks. (Separations simulated using EZ-Chrom GC®, Restek Corp. Inc.)

Sections A, B, and C of each figure illustrate three isothermal GC analyses, at 100°C, 120°C, and 150°C, respectively. At 100°C isothermal, the three early eluting branched-chain C10s elute sufficiently far from t_m to be well resolved from each other and from C9. As the temperature is increased, they begin to elute too close to t_m (i.e., k' becomes too much less than 2) and resolution is lost. On the other hand, if we look at the later eluting sections in Fig. 12.11, we see that at 100°C isothermal, C15 requires more than 30 min to elute. If we could run the separation at 150°C isothermal, last peak C15 would be out in only 4.5 min and we would still be able to resolve all the earlier eluting peaks. Figure 12.10 (Section C) reveals that this temperature would not do for the early eluters' separation. The solution to the problem is to change the temperature as the run proceeds. Start at 100°C, and when the initial group of peaks has been resolved and eluted, increase the temperature linearly to around 150°C to speed up the elution of the less volatile components. This is what was done in the chromatograms depicted in D. The procedure is called programmed temperature GC. The "temperature program"

Effect of Column Temperature on Retention Time and Peak Resolution
(**Later** Eluting Peaks)

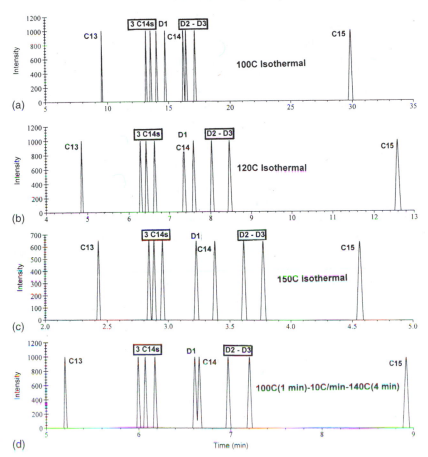

Figure 12.11 Separations of three branched chain C14 hydrocarbons, three dimethylnaphthalene isomers, and C13, C14, C15 *n*-alkanes on 30 m × 0.25 mm × 0.25 μm PDMS column; at (a) 100°C, (b) 120°C, (c) 150°C and (d) programmed 100°C/1 min–10°C/min–140°C/4 min, late eluting peaks. (Separations simulated using EZ-Chrom GC®, Restek Corp. Inc.)

illustrated is abbreviated as "100C(1 min)–10C/min–140C(4 min)". Translated, this states that the oven was held at 100°C isothermal for 1 min, then programmed to increase steadily at a rate of 10°C/min to 140°C, where it is held for 4 min. Now all peaks are resolved and the entire analysis takes only 9 min (plus enough time for the oven to be rapidly cooled back to the starting temperature to repeat the program for the next sample). The cool down time is short (a minute or two) unless the start temperature is close to room temperature, where the cooling rate becomes very slow.

The simplest and most common temperature program is the single linear ramp illustrated earlier. More complex programs, with multiple ramps interrupted by isothermal intervals may be designed to optimize critical separations and minimize overall analysis time. Environmental and biological samples often contain very slowly eluting heavier background contaminants of no interest to the analysis, which must be cleared off the column before the next injection, lest they elute with and interfere with analyte peaks from subsequent injections. Such "**late eluters**" can be recognized because their peaks appear much broader than those of the faster moving analyte peaks they coelute with. The cure for these is to rapidly program the column oven to a temperature close to the column stability limit after the last analyte of interest elutes, and to hold it there until all late eluters are cleared. Close comparison of isothermal runs A, B, and C with programmed temperature run D will reveal that in the former, peaks become broader with increasing retention time, while in the latter, the peaks have very similar widths. This can be understood because in the programmed temperature runs each peak (more precisely each band) exits the column at nearly the same rate and has been in the mobile phase for the same length of time. This may seem unobvious, but it can be rationalized if one has a clear understanding of the chromatographic process (e.g., the discussion in Section 11.5), and the explanation is left as an exercise for the student (or the instructor). Figure 12.12 illustrates another aspect of this point in a different manner. Here the same temperature program is applied to the separation of many fatty acid methyl esters on equivalent columns differing in length by a factor of two. Although the small scale of the reproduction prevents it from being seen, the resolution of closely eluting peaks on the longer column is improved by a factor of about 1.4, just the amount predicted by the square root of N relation in Eq. (11.8). Note, however, that the analysis time has NOT been extended by a factor of two, but only from 22 to a little more than 24 min! Under temperature programming conditions, the interval during which the temperature passes through the range where each component moves at a significant rate is only several minutes. This is the order of magnitude of the extra time it takes to complete the analysis on the longer, more highly resolving column.

In Sections 12.5 and 12.6 we have seen that elution and resolution behavior of analytes depend in a complex interactive way upon a number of different factors:

1. The *composition* of the stationary phase (*strongly*: effects on both absolute and relative retention) and the mobile phase gas (*less strongly*: effects on resolution via differences in the parameters which control the Van Deempter constants, but essentially no effect on retention)
2. Column *dimensions* (Length L) and phase ratio (Column i.d. and film thickness d_f)

30m vs. 60m column
Bacterial Acid Methyl Esters
130°C (hold 2 min.) to 250°C @ 4°C/min.

Figure 12.12   Effect of column length on elution time in programmed temperature GC of bacterial long-chain acid methyl esters. (Used with permission from Restek Corp.)

3. Column *gas flow* (depends on gas, pressure drop across column, and temperature)

4. Column *temperature* (the major controllable variable, can be very complex if programmed)

Most of these factors interact with one another. One could specify all the equations for these interactions and require a student to master calculations using them to predict the results on a GC separation of varying them. An even more challenging task would be to use them all to optimize a given separation. This is beyond the scope of this introductory chapter, but the need for it remains. The solution is to avail oneself of interactive computer programs which perform the linked calculations almost instantaneously. One such was used to create Figs. 12.10 and 12.11. Several are available to download from the World Wide Web at no charge. The **Agilent pressure-flow calculator** allows one to specify all the factors in 2–4, and instantly see the effect of varying any one at a time. The **Agilent method translator** allows retention information from a particular method (with a column described by factor 1) to be used to quickly predict how a separation will change with variation in any of the other three categories of factors. Appendix 12.1 provides instructions to access these programs. The student is encouraged to download them and experiment with the results of varying parameters. Thus use of free interactive GC separation optimization programs beats working through a tangle of equations whose variables interlock with one another.

12.7. GC INSTRUMENT COMPONENT DESIGN (DETECTORS)

The third and final major component analytes encounter after the injector and the column is a detector. This produces an electrical signal (usually analog, but often converted to digital) which is proportional to either the concentration or the mass flow rate of the analyte molecules in the effluent stream. The signal is displayed as a chromatogram on a chart recorder, or more often these days, on the screen of a desktop computer data system. Retention times are automatically calculated, heights of peaks are measured, or they are automatically integrated to obtain their areas, and peaks can be identified by their elution within in a retention time window, and quantitated by comparison to the areas or heights of a quantitative standard. We will not discuss the details of the operation of this signal processing equipment, but will describe only the operation and characteristics of the most useful GC detectors.

There are around a dozen GC detectors in common use. Detailed descriptions and illustrations of 16 different types, together with representative application chromatograms, can be accessed at the www.srigc.com site listed in Appendix 12.1. Spectroscopic instruments can be interfaced to the effluent of a GC and act as a form of detector which has the compound identification power of a spectroscopic measurement. This mating of a separation instrument to a spectroscopic instrument is called a **hyphenated technique**. The acronyms for the two classes of instruments are separated by a hyphen [or sometimes a slash (/)], as in GC-MS: gas chromatography-mass spectrometry. These will be discussed later in Section 12.8. Some of the general characteristics of GC detectors which need to be considered are the following:

1. *Universality vs. selectivity*: If a detector responds with similar sensitivity to a very wide variety of analytes in the effluent it is said to be universal (or at least almost so—no GC detector is absolutely universal). Such detectors are valuable when one needs to be sure that no components in the separated

sample are overlooked. In the other extreme, a selective detector may give a significant response to only a limited class of compounds: those containing only certain atoms, (e.g., atoms other than the ubiquitous C, H, and O atoms of the majority of organic compounds), or possessing certain types of functional groups or substituents which posses certain affinities or reactivities. Selective detectors can be valuable if they respond to the compounds of interest while not being subject to interference by much larger amounts of coeluting compounds for which the detector is insensitive.

2. *Destructive vs. nondestructive*: Some detectors destroy the analyte as part of the process of their operation (e.g., by burning it in a flame, fragmenting it in the vacuum of a mass spectrometer, or by reacting it with a reagent). Others leave it intact and in a state where it may be passed on to another type of detector for additional characterization.

3. *Mass flow vs. concentration response*: These two modes of detector response were described in Section 11.4. In general, destructive detectors are mass flow detectors. If the flow of analyte in effluent gas stops, the detector quickly destroys whatever is in its cell, and the signal drops to zero. A nondestructive detector does not affect the analyte, and the concentration measurement can continue for as long as the analyte continues to reside in the detector cell, without decline in the signal. Some types of nondestructive detectors (e.g., the ECD) measure the capture of an added substance (e.g., electrons). The "saturation" of this process causes a signal loss, so they are mass-flow detectors.

4. *Requirement for auxiliary gases*: Some detectors do not function well with the carrier gas composition or flow rates from a capillary column effluent. **Makeup gas**, sometimes the same as the carrier gas, may be required to increase flow rates through the detector to levels at which it responds better and/or to suppress **detector dead volume** degradation of resolution achieved on the column. Some detectors require a gas composition different from that used for the GC separation. Some detectors require both air and hydrogen supplied at different flow rates than the carrier to support an optimized flame for their operation. Makeup flow dilutes the effluent but does not change the detection mechanism from concentration to mass-flow detection.

5. *Sensitivity and linear dynamic range*: Detectors (both universal, and of course selective ones) vary in their sensitivity to analytes. Sensitivity refers to the lowest concentration of a particular analyte that can be measured with a specified **signal-to-noise ratio**. The more sensitive the detector, the lower this concentration. The range over which the detector's signal response is **linearly proportional** to the analyte's concentration is called the *linear dynamic range*. Some exquisitely sensitive detectors have limited linear dynamic ranges, so higher concentrations of analytes must be diluted to fall within this range. Another less than satisfactory solution to a limited linear dynamic range is to calibrate against a multilevel nonlinear standard curve. This requires injections of more standards and is more prone to introduce quantitative error. Dilution will not work satisfactorily if there is a wide range of concentrations in the sample. A very insensitive detector will perforce have a more limited dynamic range, and multilevel standard curves or dilution will be of no avail with it.

Figure 12.13 compares the sensitivities and dynamic ranges of several of the most common types of GC detectors. The further to the left the range bar extends, the more

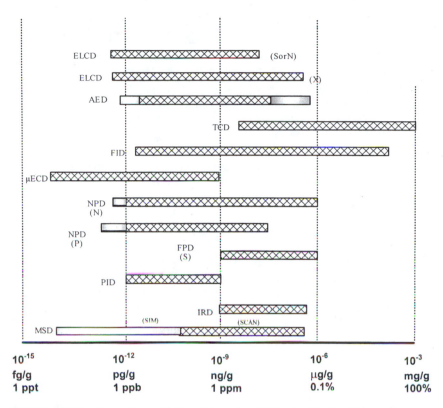

Figure 12.13 Approximate limits of detection (left end of bar) and dynamic ranges for 12 GC detectors from 1 μL sample injected.

sensitive the detector. The wider the bar, the greater the dynamic range. Each vertical dotted line denotes a span of three orders of magnitude (1000×), so the overall ranges covered are very large. The values along the x-axis assume an injected volume of 1 μL (1/1000 of an mL or cm^3) of solution. The ranges are approximate, their exact endpoints will depend on the design and model of the detector and optimization of its operating conditions. They provide a good general comparison. The IRD and MSD at the bottom are hyphenated method spectroscopic detectors, operable in several modes, and will be discussed in Section 12.8. Note that the "original" GC detector, the TCD, is the least sensitive, but also the only one suitable for handling neat (i.e., 100% pure single component) samples. The FID has the greatest single-mode dynamic range, while the "micro-ECD" is more sensitive, but with a more limited dynamic range. Let us proceed to describe the operation, characteristics, and applications for each of these nonhyphenated method detectors.

12.7.1. The Thermal Conductivity Detector (TCD)

12.7.1.1. TCD Characteristics

Universal (except for H_2 and He); non-destructive; concentration detector; no auxiliary (aux.) gas; works better with a parallel column; insensitive; limited dynamic range.

The TCD was the first widely commercially available GC detector, in the era when all the columns were packed, and samples were neat (i.e., not diluted solutions) mixtures to be separated. It measured differences in the thermal conductivity and/or specific heat of highly thermally conductive (either H_2 or He) carrier gas when diluted by small concentrations of much less conductive analyte vapor (anything else). A current through a thin resistive wire heated the wire in the detector flow cell. The thermal conductivity of the flowing carrier gas cooled the wire. When analytes were in the stream, their lower thermal conductivity produced less cooling, which caused the wire's temperature to rise and its resistance to increase. This wire resistor was in a "**Wheatstone bridge**" circuit (an arrangement of four resistors on the sides of a square). One of the other resistors was in a matching TCD cell connected to a matching column and flow with no analyte passing through. In isothermal GC, carrier flows and temperatures would remain constant, but with temperature programming of the column the temperature would increase and the flow would decrease, independently affecting the conductivity. The matching *reference cell* would compensate for these effects on the resistance changes. A fixed and a variable resistance constituted the other two legs of this bridge. A voltage sensor was connected across opposite points on the diagonal of the square array. The variable resistance was used to null (or "zero" the signal from) the voltage sensor. Once the bridge circuit was thus balanced, passage of analyte changed the resistance of the wire in one leg of the bridge, throwing the bridge out of balance and producing a voltage signal proportional to the resistance change. The designs of two TCD cells are illustrated in Fig. 12.14. This design of TCD cell was also referred to as a **katharometer** cell, a name for a gas thermal conductivity measurement device.

Note the application of the TCD for the detection of "**fixed gases**" in the chromatogram of Fig. 12.4. These are generally not seen by other detectors (except hyphenated GC-MS). Note the low sensitivity signal for H_2, whose thermal conductivity is the only one to closely match that of the He carrier gas used. If only H_2 were being measured, it would be better to use N_2 as the carrier. This would yield a peak signal in the negative direction. In Fig. 12.5, note that the detector is described as a μ-TCD. This must employ miniaturized cells to be compatible with the low carrier flows of the open-tubular

Figure 12.14   Diagrams of two types of thermal conductivity detectors (TCD). (Cazes, used with permission.)

PLOT column and the need for small detector cell volumes to avoid extra-column band broadening.

12.7.2. Flame Ionization Detector (FID)

12.7.2.1. FID Characteristics

Nearly universal (all carbon compounds except CO, CO_2, HCN, but not many inorganic gases); destructive; mass flow detector; needs H_2 and air or O_2 aux. gas; sensitive with wide dynamic range.

The FID is the most commonly employed detector, as it gives a response to almost all organic compounds. On a molecular basis the signal is roughly proportional to the number of carbon atoms in the molecule. Hydrogen and oxygen (or air) must be separately provided to fuel a flame in the detector cell. As illustrated in the diagram of Fig. 12.15, the H_2 is introduced and mixed with the carrier effluent from the GC column. Even if H_2 is used as capillary carrier gas, an additional separately controlled H_2 supply is necessary to adjust the appropriate **fuel** supply for the flame. The mixed gas enters the cell through a jet, where air or O_2 flows past to serve as the flame **oxidizer** supply. An electrical **glow plug** in the cell (not illustrated) can be pulsed to ignite the flame. The fuel and oxidizer flows are adjusted with needle valves to achieve a stable flame with optimal FID response, often by bleeding a volatile unretained hydrocarbon into the carrier stream to provide a reference signal. The **jet tip** is charged by several hundred volts positive relative to several "**collector electrodes**" or a "**collector ring**" surrounding the flame. In the absence of eluting organic analytes, no current flows in the jet–collector circuit. When a carbon-containing analyte elutes into the flame, the molecule breaks up into smaller fragments during the cascade of oxidation reactions. Some of these are positively charged ions, and they can carry current across the flame in the circuit. Although the ionization efficiency of the FID is low, its base current is also very low. Hence its signal-to-noise ratio is very high. Against such a low background, even very small ionization currents can be accurately measured using modern electronics which draw very low currents (high input impedance, voltage measurement circuits). Hence the good sensitivity and extraordinary dynamic range of this detector, often exceeding six orders of magnitude.

FID Detector for GC

Figure 12.15 Diagram of the flame ionization detector (FID). (Cazes, used with permission.)

One can detect essentially any organic compound that can survive passage through the GC to the detector. A very wide range of analyte concentrations can be measured in a single run. The FID is the forerunner of a series of more selective and sometimes even more sensitive ionization detectors; some of which will be discussed subsequently. It led the way to GC as it is currently practiced, with analytes being present in dilute concentrations in a preparation or extraction solvent solution instead of being mixtures of the neat compounds. Precise quantitation requires calibration against reference standards of the analytes being measured. The responses on a compound weight basis are sufficiently similar that for **approximate quantitation**, calibration for many components can be made against a single reference compound peak area. The wide dynamic range minimizes the number of calibration levels required for an adequate standard curve. The flame gases can be supplied from compressed gas cylinders or generated on site by **gas generators** employing hydrolysis of water, as well as compression and filtration (to avoid organic contaminants) of ambient air. If H_2 is used as carrier instead of He, all compressed gas cylinders could be dispensed with.

12.7.3. The Electron Capture Detector (ECD)

12.7.3.1.   ECD Characteristics

Very selective (for organic compounds with halogen substituents, nitro and some other oxygen-containing functional groups); non-destructive; concentration detector; needs to use N_2 or argon/CH_4 as carrier, or as makeup gas if used with H_2 or He capillary carrier flow; extreme but highly variable sensitivity but with limited dynamic range.

If the FID brought GC into the realm of characterizing dilute solutions of organics, it was the ECD that allowed it to spark a revolution in the understanding of the threat of bioaccumulation in tissue and bioconcentration up food chains, of lipophilic, persistent organic pollutants (POPs) in the environment. The classic example of this was the discovery of the threat posed by the organochlorine pesticide DDT, and its subsequent banning. J.E. Lovelock (subsequently famous as the founder of the Gaia theory of the whole earth as an organism) invented this deceptively simple but exquisitely sensitive and selective detector. It is not too much of a reach to claim that observations which were only made possible by the use of this GC detector set off the revolution in environmental consciousness in the 1960s.

How then does this little marvel function? Two versions are illustrated in Fig. 12.16. The righthand diagram displays features of the initial design (operation in DC mode). The column effluent enters from the right. A nickel foil doped with radioactive ^{63}Ni [a β particle (energetic electron)-emitter] constantly bombards the carrier gas, ionizing some of its molecules, creating an atmosphere of positive ions and the negative electrons which have been knocked loose. This radioactive source is chosen for its ability to withstand the high operating temperatures GC detectors may require to prevent condensation of high boiling analytes. A *low voltage* (several volts, instead of the hundreds of volts across the FID) between the negative inlet side of the detector and the positive outlet side is sufficient to set up a current between these two electrodes. Under this low voltage, it is the more mobile, lighter free electrons which carry the so-called "standing current" to the positive electrode.

If an analyte with highly electronegative, highly polarizable substituents enters the cell, the electrons of the standing current can be captured by these molecules. The mobility of the captured negative charge is drastically reduced, and some may be more easily neutralized by collision with the positive ions which are generated. It is the depression of

ECD for Use with Pulsed Electrode Potential   **ECD for Use with Constant Electrode Potential**

Figure 12.16 Diagrams of two types of electron capture detectors (ECD). (Cazes, used with permission.)

this standing negatively charged electron current which is displayed reversed and appears as a peak signal. The electron capture sensitivity varies dramatically (one to two orders of magnitude) with the nature and number of the substituents: I > Br > Cl ≫ F among the halogens, and 5–8 halogen atoms >4 > 3 ≫ 2 ≫ 1. Note that most electronegative F is surpassed by more polarizable Cl, Br, and I, increasing in that order, so it is the *polarizability* which dominates, although both are necessary for electron capturing effectiveness. Some other groups such as $-NO_2$, $-NO$, and so on fall somewhere on the low side of these responsiveness series. Hydrocarbons are essentially unresponsive, although in great excess they can depress the sensitivity of the ECD response of ECD active compounds if they coelute with them. So with the ECD, calibration standards for every analyte are necessary to perform quantitative work.

Note that ECDs require often special carrier or makeup gases. Most early ECD cells were designed for high packed column flow rates, and will need makeup gas at higher flow rates and of different composition when He or H_2 are employed as capillary carrier gases. Use of a radioactive ionization source in the detector requires a license from the NRC in the US, and a program of regular wipe tests to detect radioactivity leaks in and around the detector. Another problem is that the depression of the standing current soon "saturates". As the level of electron capturing analytes rises, the tiny current of electrons becomes more and more depleted. Additional increase in the analyte level is not reflected by a proportional decrease in the standing current—there just are not enough electrons left to go around (the circuit). This results in severe nonlinearity of the detector response, curvature of the standard curve (it eventually levels out as concentrations continue to increase), and a severely limited linear dynamic range. Quantitative calibration becomes a tricky problem with ECDs.

The lefthand ECD diagram shows a more modern design which attempts to alleviate some of these difficulties. The ionization is achieved by the "discharge electrode" using a special discharge gas flow, thereby replacing the radioactive foil. A special dopant gas is introduced to enhance this. Instead of maintaining a constant potential across the cell, it may be intermittently pulsed at a higher voltage. Typically a pulse at ~30 V of around 1 μs repeated at 1 ms intervals (a duty cycle of 0.1%) collects all free electrons in the

ECD cell, while during the "off period" the electrons re-establish equilibrium with the gas. The very fast collection time is enabled by admixture of 5–10% methane in the argon ECD gas, which serves to enhance the voltage-induced migration rate of the free electrons. The current during the pulsed collection is integrated to produce the standing current level. The benefit of this method of controlling the standing current is to extend the linear dynamic range, and most modern ECDs are designed to operate as pulsed-mode ECDs. Such an extended linear dynamic range is illustrated for the "μ-ECD" in Fig. 12.13. The micro-(μ) designation indicates that this model is equipped with a much smaller cell volume adapted for the effluent flow rates of capillary columns. Dispensing with the need for dilution of the effluent stream with makeup gas also increases the sensitivity even more. All other things remaining equal, pushing the sensitivity even lower extends the dynamic range since it is mainly limited by the effects of saturation of the standing current at the upper end of the range. The earlier model, large volume, DC-mode ECDs often had a linear dynamic range of only 100 or less. This made measuring multiple analytes (especially if they covered a much larger range of concentration) within a valid calibration range very complicated. Many dilutions and reassays were often required to do good quantitative work—a great contrast to operating within the huge dynamic range of the FID. In contrast the special selectivity and extraordinary sensitivity of the ECD for certain classes of compounds, such as multiply-chlorinated DDT, were indispensable for the discovery of the bioconcentration pathways of that compound and others such as PCBs, dioxins, and PBDEs (polybrominated-diphenylether flame retardants). It is an important general principle that the detection limits for given analytes in chromatographic methods applied to real samples from environmental or biological systems are more often determined by the selectivity of the detector against background coeluting interferences than by its absolute instrumental signal to noise sensitivity limit.

12.7.4. The Electrolytic Conductivity Detector (ELCD)

12.7.4.1. ELCD Characteristics

Heteroatom selective [for organic compounds with halogen (X) substituents, or N or S atom]; destructive; mass-flow detector; requires liquid reagents; good sensitivity and linearity; response is directly proportional to flow rate of the X, N, or S atoms.

The ELCD operates by consuming the analyte in a hydrogen–air flame and dissolving the resulting gases in an aqueous solution whose electrolytic conductivity is proportional to the amounts of any halogen (X), S, or N atom which were present and oxidized to the corresponding acids. It represents the first example of a detector specifically and proportionately responsive to specific atoms in organic compounds. It is complex to operate and now not commonly used. Its unique value is its proportional response only to the number of heteroatoms, not to their position in the molecule. If a mixture of compounds with only one of these elements present is analyzed, and additional information is available about the number of its atoms in each peak (obtainable from a GC-MS analysis), then a single standard peak response can serve to calibrate many other peaks, for which no standards exist. An example is the calibration of PCB congener peaks in commercial mixtures against the response of a chlorine compound standard. There are 209 of these PCB isomers, and their ECD responses vary widely. The TCD, FID, or GC-MS responses vary less, but still by too much, and individual standards were initially available for only a small subset. The exact atom-proportionate response and good dynamic range of the ELCD permits quantitation in the absence of complete standard sets for these homologous compounds.

12.7.5. The Sulfur–Phosphorous Flame Photometric Detector (SP-FPD)

12.7.5.1. SP-FPD Characteristics

Heteroatom selective (for organic compounds with S or P atoms; separately); destructive; mass-flow detector; less sensitive and shorter dynamic range than FID.

When organic compounds containing S or P atoms are burned in an FID, the flame conditions can be adjusted to produce a lower temperature which enhances the emission from S_2 fragments at 394 nm or HPO fragments at around 515 nm. The FID electrodes are omitted, and a temperature resistant window or fiber-optic light guide in the side of the detector cell passes the emitted light to filters designed to isolate these wavelengths and pass the characteristic emission to a sensitive photomultiplier tube. Operated in S-selective mode it is useful to characterize the organosulfur compounds in complex petroleum mixtures, as the much higher levels of coeluting hydrocarbon peaks give minimal response. Operated in P-selective mode it is a sensitive detector for organophosphorus pesticide trace residues in complex environmental mixtures. Only the sulfur mode response range is illustrated in Fig. 12.13.

12.7.6. The Sulfur Chemiluminescence Detector (SCD)

12.7.6.1. SCD Characteristics

Heteroatom selective (for organic compounds with S atoms only); destructive; mass-flow detector; more sensitive and greater dynamic range than S-FPD.

This detector takes the gases from a flame produced in the same manner as in an FID (again minus the electrodes), and reacts it with ozone to induce chemiluminescence from sulfur products produced from eluting organosulfur compounds burned in the flame. The intensity of this luminescence is detectable at even lower levels than the S_2 emissions produced in the S-FPD, making it useful for monitoring trace environmental pollutant sulfur organics. Its response range is not displayed in Fig. 12.13.

12.7.7. The Nitrogen–Phosphorous Detector (NPD)

12.7.7.1. NPD Characteristics

Heteroatom selective (for organic compounds with N or P atoms; separately); more sensitive for N and P compounds than FID, as well as selective for them; destructive; mass-flow detector; more sensitive than FID for N, more sensitive than P-FPD for P, but with less dynamic range than the FID.

The NPD is yet another variation on the workhorse of GC detectors, the FID. Comparison of Fig. 12.17 for an NPD to Fig. 12.15 for the FID highlights the similarities. Both are defined as "**thermionic detectors**", that is, the high temperature of a flame breaks the eluting analytes into fragments, some of them positive ions, which release electrons to carry a current under the influence of the voltage between two electrodes. The NPD is operated under **fuel (i.e., H_2)-rich** conditions. Under these conditions the normal carbon compound FID thermionic response is suppressed by orders of magnitude. The "new element" in the detector is **rubidium** in the form of a glass or ceramic **bead** doped with a rubidium salt, which is heated by immersion in the flame, but is also additionally and variably heated by passing current through thin wires which support the bead in the flame. What exactly happens near the surface of the bead while organonitrogen or organophosphorus compounds are decomposing in this not-so-hot flame is complex and poorly

Figure 12.17 Diagram of the nitrogen–phosphorus detector (NPD). (Cazes, used with permission.)

understood. In some designs the flame would not be self-supporting were it not for the independently heated bead embedded in it. Somehow C–N or C–P containing fragments interact with easily ionizable rubidium atoms to produce ions and electrons which then produce a signal like that in the FID; but one which is exquisitely sensitive and selective to either N or P atoms (depending on how the detector parameters are set). This is illustrated in Fig. 12.13. The extreme sensitivity of this detector to some organophosphorus compounds explains why it has largely supplanted the P-FPD for this application, and no range is depicted for that mode of the FPD operation in the comparison figure. The NPD produces no response to N atoms not bound to carbon in organic molecules. This is a fortunate circumstance which renders it immune from interference from ubiquitous atmospheric N_2.

12.7.8. The Photoionization Detector (PID)

12.7.8.1. PID Characteristics

Compound class selective (for organic compounds with more easily ionizable π-electrons, especially aromatic compounds); nondestructive; mass-flow detector; slightly more sensitive than FID for many compounds it detects, but with less dynamic range than the FID.

Ionization detectors like the FID and the NPD have great sensitivity. Is there any other way to selectively ionize some classes of compounds to achieve a sensitive, selective ionization detector like the NPD? Let us take the "thermo" out of thermionic by dispensing with the flame, and supply ionization energy instead by passing **high-energy UV radiation** into the cell. This is the reverse of measuring flame emission coming out as in the S-FPD. The UV radiation provides enough energy to knock electrons out of some aromatic compounds and some others which have electrons in higher energy level π-bond orbitals. Precise selection of the UV wavelength can allow some selectivity within these compound classes. Organic compounds like alkyl hydrocarbons containing only electrons in more tightly held single bonds will be unaffected. The resulting PID is selective for aromatic hydrocarbons, and other organic compounds with unsaturated bonds. The high-energy UV lamp and its power supply are much more compact and robust than the FID and its associated gas supplies. Combined with its exceptional sensitivity to a reasonably broad-based selection of organic compounds, this operational simplicity makes the PID especially suitable as a detector for field-portable GC instruments. An additional factor supporting such applications is that it is not affected by oxygen in air and can even be used with air as a carrier gas, if the column stationary phase can withstand this at the temperature conditions employed.

12.7.9. The Helium Ionization Detector (HID)

12.7.9.1. HID Characteristics

Universal detector (everything except neon); nondestructive; mass-flow detector; a little less sensitive than FID for FID-active compounds, and with less dynamic range than the FID.

The HID is like the FID, NPD, or PID in that it is a sensitive detector which measures a current of ions produced from the analyte molecules. An analogous **argon ionization detector** employing the same principle was another early GC detector. They are sometimes also referred to as "**discharge ionization detectors**" (**DIDs**). In the HID a flow of He gas atoms (as He carrier or He makeup gas) are activated to an excited metastable state by **discharge electrodes**. These have sufficient energy to ionize any analyte molecules (except Ne atoms) they come in contact with, which then produce an ionization current proportional to sample amount between two other electrodes in the detector. As with the ECD, pulsing the discharge electrode to maintain a constant current yields a design with superior dynamic range. It can be operated in series with the other universal detector, the TCD, since both are nondestructive, and their two overlapping dynamic ranges can cover the range up to 100% of the neat compound. It is often employed in simple portable GC instruments designed to measure both fixed gases and very volatile low molecular weight hydrocarbons (cf. Fig. 12.18).

1. oxygen
2. nitrogen
3. carbon monoxide
4. methane
5. carbon dioxide
6. acetylene
7. ethylene
8. ethane

ShinCarbon ST 100/120 mesh
2 meter x 1mm ID micropacked (cat.# 19808)

Sample:	permanent gases-C1+C2 hydrocarbons
Sample size:	5μL
Inj. temp.:	100°C
Carrier gas:	helium
Flow rate:	10mL/min.
Oven temp.:	40°C (hold 3 min.) to 250° @ 8°/min. (hold 10 min.)
Det.:	HID @ 200°C

Figure 12.18 Separation and detection of fixed gases and low MW hydrocarbons on high surface area molecular sieve using a helium ionization detector (HID). (Adapted with permission of Restek Inc.)

12.7.10.   The Atomic Emission Detector (AED)

12.7.10.1.   AED Characteristics

Tunable, separately or simultaneously selective for compounds containing many specific atoms (e.g., C, O, N, S, P, Si, Sn, halogens, other metals); destructive; mass-flow detector; sensitivity and dynamic range vary with element measured (cf. Fig. 12.13).

The AED, employs a **microwave-induced He plasma** to dissociate eluted analyte molecules to their component atoms and excite them to emit at characteristic wavelengths. This is very similar to the mechanism in the argon plasma ICP source (cf. Section 7.3.1). A spectrometer with a diode array detector (Fig. 7.26) isolates and measures the intensity of sensitive emission lines unique to each element. Depending on the relative sensitivity and proportion of atoms in the molecules, separate element response channels may display peaks in several element-selective chromatograms. These data may be combined with retention information to additionally characterize the peaks, to separately quantitate coeluting peaks, and to suggest identification of unknown peaks based on their elemental composition. Like the ELCD its response is atom selective and atom proportional, but each different element may be simultaneously detected, identified, and quantified. Since it interfaces a chromatographic separation instrument with a multiwavelength spectrometric detector it verges on being classified as a 2D hyphenated instrument. Such fully 2D instruments are the subject of the next section.

12.8.   HYPHENATED GC TECHNIQUES (GC-MS; GC-IR; GC-GC; OR 2D-GC)

2D analytical techniques can combine two similar (e.g., 2D-TLC, cf. Section 11.3, or 2D gel electrophoresis, cf. Chapter 13) separation procedures by sequentially varying the mobile phase compositions and developing with the second phase in a direction perpendicular to the first.

If we can

1.   find a way to join two chromatography systems with differing stationary phases,
2.   separately and repeatedly trap everything eluting from the first phase (column),
3.   then repeatedly separate each trapped fraction on a second column,
4.   and finally store and manipulate the data signals from a detector at the end of the second column,

we can produce a comprehensive 2D chromatography system (e.g., 2D-GC, cf. Section 11.3).

A final way to add a second dimension to chromatographic detection is to employ as a detector a spectroscopic instrument capable of sequentially and rapidly acquiring full spectra and storing them as a series of computer data files. In the case of GC, the two major hyphenated techniques of this type are GC-MS and GC-IR. There are four critical requirements for achieving this:

1.   A spectrometric detector capable of acquiring full spectra quickly enough to sample the eluting peak at least 5–10 times over its width. This is a challenge in capillary GC, where peak widths may be as small as several seconds.
2.   A method of **interfacing** the chromatographic effluent stream to the spectrometer's detection cell that presents the sample under the conditions required for proper detector functioning.

3. A computer data processing system fast enough to deal with the high information rate of such a detector, with sufficient data storage capacity for the many spectra to be acquired.

4. Efficient and intuitive software for viewing and manipulating these data files, extracting subsets of the information, and using it for qualitative identification and quantitative measurement.

12.8.1. Gas Chromatography-Mass Spectrometry (GC-MS)

If we consider their wide availability and capability, GC-MS instruments could have been said to provide the largest total amount of analytical power available to the instrumental analysis world. With the more recent spread of LC-MS instrumentation (Chapter 13) to serve the biochemical research market, they now must share this status. If analytes are volatile and thermally stable, capillary GC-MS can identify and quantitate hundreds (even thousands in GC-GC-TOFMS runs) of them from a single injected mixture. The various types of mass spectrometers have been described in Chapter 9. It is the **quadrupoles** and **ion-traps** which best meet the earlier-mentioned scan speed criterion. Magnetic sector instruments, with their higher mass resolution but slower scan speeds, will require that the GC peaks be broadened and slowed down. This works against the goals of faster analyses and better chromatographic resolution. On the other hand so-called "fast GC", employing short, narrow-bore, thin-film capillary columns, will require MS-scan acquisition speeds of $50-500$ s^{-1}, which are attainable only by **TOF-MS instruments**.

The major problem for GC-MS is interfacing. Mass spectrometers form and move their ions at highly reduced pressures, and the mass analyzer sections need to be operated at even lower pressures (10^{-5} or 10^{-6} torr). The effluent of a GC consists primarily of carrier gas around atmospheric pressure (760 torr). The MS vacuum pumps must remove this fast enough to maintain the necessary low pressure. With packed column flows of $10-40$ mL/min, this was impractical. With capillary GC flows around 1 mL/min, modern diffusion pumps, or even better, powerful turbomolecular pumps, could achieve this. Another reason to prefer capillary GC to packed column GC separations! Many modern GC-MS instruments simply introduce the end of the capillary column into the ion source (*direct coupling*). A separate pump evacuates the small, largely confined volume of the source, while the ions are extracted through a small orifice into the mass analyzer region, which is kept at a lower pressure by another pump. Such a MS design is called a *differentially pumped* system. To interface packed columns, a device called a **jet separator** (Fig. 12.19) preferentially removed light He or H$_2$ carrier gas

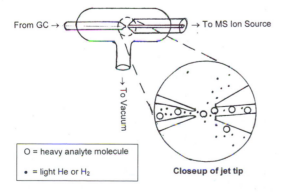

Figure 12.19 Diagram of GC-MS jet separator interface operation.

atoms or molecules from the effluent, allowing a smaller but enriched concentration of analyte molecules to enter the ion source at a lower total gas flow rate. The much higher speed of the lighter carrier gas particles causes most of them to spread more widely in the vacuum of the separator chamber, and not to pass on through the conical orifice leading to the ion source. Jet separators are not much used or needed for GC-MS now, but we will encounter a similar design feature when we come to discuss the even more challenging problem of dealing with the much denser and larger mobile phase mass in LC-MS interfacing (Section 13.1.5).

In principle one could monitor a GC effluent with a nondestructive detector like a TCD, pass it on through a suitable interface, and acquire and print out the mass spectra of peaks as they are eluting. With fast digital data converters and the speed and power of modern desktop PCs, it is better to simply acquire mass spectral data as a continuous sequence of spectra or selected mass fragment signals as the GC run proceeds. A data file in computer memory consists of a sequence of MS scans ordered by retention time. There are three main modes of acquiring and employing such data files:

1. *Full-scan*: If the file consists of a sequence of full-scan spectra, the total counts from ions of all masses in each scan may be summed, and these totals plotted against successive scan numbers (or equivalently, GC retention time). This display is called a **total ion chromatogram (TIC)**. Here the mass spectrometer acts like an ionization detector, using whatever MS ion-source ionization mode was selected (EI, CI, etc.). The TIC chromatogram will appear similar to that obtained by a FID detector. One difference is that the FID output digitized by a chromatographic data system will likely define each peak at many more points across its width than the number of scans that the MS could make (unless it is the very fast scanning TOF-MS; Section 9.2.3). For this reason integration of FID peaks yields more accurate quantitation than TIC peaks. The advantage of the TIC data file is that by selecting any scan comprising a TIC peak one can display the mass spectrum of the compound in the peak. If the peaks are sharp and narrow, the level of the analyte may vary significantly over the time of the scan, thus distorting the proportions of mass fragments in the spectrum. It would be best therefore to select a scan at the top of the peak. Even better would be to sum scans across the whole peak to get an average spectrum. Even better than that would be to select and sum an equal number of scans of background mass spectra near the peak where nothing else is obviously eluting. Then subtract that sum from the peak average sum, and thereby obtain a cleaner mass spectrum, with spurious fragments from column bleed eliminated. This illustrates the power and flexibility of manipulating a data file of continuous, contiguous MS scans. We have combined the great separation power of capillary GC to the great characterization and identification power of MS. This is particularly feasible because in GC the samples are in the vapor state, which is mandatory for the MS ionization and fragment identification process to proceed. The vast libraries of EI-MS spectra are based on this mode of operation. In many cases it is possible to automate the process of peak detection, spectral selection, library spectra searching, and separated compound identification by matching to a library spectrum. A naïve person might be forgiven for wondering whether GC-MS eliminates the need for an analyst. Not quite. But the interfacing of vapor-separation GC with vapor-requiring MS, when combined with the fluency of PC processing of digital data files is an excellent combination.

2. *Mass chromatograms (XIC)*: One may seek to locate and measure only certain categories of analytes in the chromatogram of a much larger and more complex mixture. These might have very characteristic mass fragments. An example would be mono-, di-, and trimethylnaphthalene isomers in a complex mixture of petroleum hydrocarbons such as a fuel oil. Their mass spectra consist mostly of a single intense molecular ion (M^+) at masses 142, 156, or 170, respectively. We can program a GC-MS data system to extract from the full-scan GC-MS files and plot only the ions of these masses in three separate chromatograms. Such plots are called **extracted ion chromatograms** (**XIC**), from their mode of production, or **mass chromatograms**, since they display peaks whose spectra contain the selected ion mass(es). If in the mass chromatogram of a petroleum sample at $M^+ = 142$, we observed a pair of large peaks close together, we might well suspect them of being from the two possible mono-methylnaphthalene isomers, and could confirm our suspicion by inspecting the complete spectrum of the scans at the center of each peak. Thus we may be able to determine the retention times of various compounds without injecting a standard. Unfortunately mass spectra of isomers of such compounds are indistinguishable, so we will need some sort of retention information based on standards run on the particular GC stationary phase to say which isomer is which. A method of accessing such information without actually running the standards is described in Section 12.9. Such a process could be repeated at many other characteristic masses for the different classes of hydrocarbons. The data for all masses in the range scanned, at every point in the TIC chromatogram, are present in the full-scan GC-MS data file.

3. *Selected ion monitoring (SIM)*: In full-scan mode the mass spectrometer is acquiring counts at any particular ion mass for only a brief portion of the scan. For example, if scanning from mass 50 to 550 each second, with unit mass resolution, the detector spends less than 2 ms at each mass during one scan. If we have only a few classes of analytes we wish to measure, and we know their characteristic major mass fragments, we can program the MS to acquire counts only at the selected masses, thereby increasing the dwell time at each mass, increasing the signal-to-noise ratio, and improving the sensitivity. If we wanted the three successive methyl naphthalene isomer distributions, we could monitor at only $M^+ = 142$, 156, and 170. In 1 s each of these three ions would be monitored for a little less than 330 ms instead of 2 ms, greatly improving the sensitivity over the 50–550 full-scan acquisition. In fact we could cut the cycle time from 1 to 0.2 s, still acquire for 66 ms per cycle, but now be sampling and defining the peak shape 5 times/s instead of once per second. If we know that each class of isomers elutes over a unique range of retention times, we can set MS acquisition to monitor just their most characteristic and abundant ions during this period and achieve even greater sensitivity. This mode of acquisition is called SIM. SIM improves sensitivity by collecting more counts at the masses of interest, and it improves quantitative precision by enabling the GC-MS peak to be defined and integrated using more points. The extension of the linear range of measurement to lower values with SIM vs. full-scan is represented by the two sections of the GC-MS sensitivity range in Fig. 12.13. Ultimate SIM sensitivity is a complex function of MS ionization efficiency for the particular analyte compound, number of different masses monitored, dwell time, and cycle time. Figure 12.13 indicates that in the most favorable instances GC-MS-SIM can reach detection limits below that of the FID and in the range of the ECD. It is

far more selective than any of the general GC detectors. This sort of analysis is characterized as "**target compound analysis**" since the system is tuned or programmed to select specific characteristic ions from specific target compounds expected to elute in specific retention time ranges.

To reiterate, the difference between XIC and SIM chromatograms is that in the former case the desired ion masses are extracted from a full-scan data file, while in the latter case the MS is directed to acquire the data only at those selected masses. The ability to improve sensitivity and quantitative precision by using SIM applies mainly to magnetic-sector (including use of high resolution MS) and quadrupole MS instruments. The mode of operation of **ion-trap MS** and **TOF-MS** instruments yields near-optimum sensitivity in full-scan mode. Thus there is generally no provision for SIM acquisition on these instruments, and quantitation is done on XIC files. The MS in GC-MS is a destructive, mass-flow, detector. Only with mass spectrometric detection can the analyst use the "perfect" internal standard; namely, the identical chemical species, labeled with stable isotopes of atoms with a higher mass (e.g., ^{2}H, ^{13}C, ^{15}N, ^{18}O, etc.). This procedure, **isotope-dilution mass spectrometry** (**IDMS**), can correct for differences in sample preparation recovery, derivitization efficiency, MS ionization efficiency, and so on, for which use of a different chemical species as an IS may not fully compensate. Review Section 11.12 for a discussion of chromatographic IS. However, IDMS IS materials are difficult to make, expensive to purchase, and commercial products are limited to only several thousand especially important target analytes.

12.8.2. Gas Chromatography-IR Spectrometry (GC-IR)

At the concentrations eluting from GC columns, with analytes extensively diluted as vapors in the carrier gas, IR is not nearly as sensitive a detector as MS. Unfortunately, since IR spectrometry is traditionally carried out on analytes in a liquid or solid matrix, the libraries of vapor phase IR spectra are much less extensive than those in the other phases, and the spectra differ significantly. Actually the vapor phase IR spectra are sharper, with greater interpretive structure than in the other phases, but comparable upper range concentrations are unattainable. The spectra must be acquired during a rather brief passage of the peak through a detector cell. This makes acquisition by the rapid-sampling FTIR process (Section 4.2.2) mandatory. The beam from the FTIR passes from one end to the other of a long narrow cylindrical "**light pipe**" flow cell. The GC effluent enters and exits the sides of each end of this tube, while the ends are capped by flat windows of IR transmitting material. The cell must be heated to prevent condensation of analyte vapors. The interior side walls are coated with an IR-reflective gold film, and the beam remains confined within the pipe by grazing-incidence total internal reflection.

The cell must have a volume small enough not to introduce extra-column GC band broadening. This conflicts with the need to make it as long as possible to increase the IR absorption from the dilute analyte vapor in the peak. Remember that an IR detector is a nondestructive, concentration detector, and as such its absorbance signal is proportional to both concentration and path length. Something has to give, and it is sensitivity. The FTIR acquisition process could improve sensitivity if more time could be taken to acquire more scans. One way to achieve this is to stop carrier flow (best achieved by closing shut-off valves upstream and downstream of the detector) while the peak is in the cell. For reasonable times, diffusion will not spread the band too much outside the cell. This **stopped-flow-analysis** is an awkward process. Chromatography of other peaks still on the column may slowly degrade during the pauses in flow, one has to know in advance exactly when to stop the flow, and the analysis time increases dramatically, especially if one wishes to do this for many peaks. The lower sensitivity range of GC-IR is reflected in Fig. 12.13 (the IRD). The instrumentation is as expensive, or even more so than much GC-MS.

Why then does anyone even want to do GC-IR? We have seen that MS information cannot distinguish between many isomers of the same molecular weight, such as the methyl naphthalenes mentioned earlier. Perhaps standards and retention data for the GC phase can help us distinguish them. If, however, several coelute, GC-MS cannot tell how much of each is in the peak, or which one is present or absent. An example of this is seen in an important GC assay for "BTEX" aromatic hydrocarbons [Section 12.5, Fig. 12.9(b)]. On the PDMS column phase, *para*- and *meta*-xylenes coelute. All three xylenes have indistinguishable mass spectra. The carbowax type phase will separate them [Fig. 12.9(a)], but it is not as stable as PDMS if we must also elute other analytes later at much higher temperatures. If we inspect the IR spectra for xylenes in Figs. 4.39–4.41 we can see significant differences in the positions of the large absorptions in the $1000-500$ cm^{-1} region, in particular between coeluting *p*- and *m*-xylene. In the gas phase spectra these differences would be even sharper and more distinctive. We could monitor absorbance at these distinctive wavenumbers and separately quantify each xylene, even in the presence of a coeluting one. As with GC-MS, examination of the full spectrum of resolved GC peaks could enable identification even without retention data from standards (i.e., using spectral library matching). IR spectra will distinguish among structural isomers whose MS spectra will appear identical. Since the GC-IR detector is nondestructive, after IR spectra are acquired, the peaks may be passed on to a destructive detector such as an FID, or even an MS. A GC-FTIR-MS instrument is very complex and expensive, but correspondingly powerful in its ability to characterize separated analytes!

12.8.3. Comprehensive 2D-Gas Chromatography (GC-GC or GC^2)

At the beginning of Section 12.8, we described how an additional different GC column in series could add a second dimension to GC analyses. Now the second dimension is not a spectrum from a detector, but a very fast GC column with separation capability different from the first column. The speed of the second short fast column (cf. the discussion in Section 12.5 on "fast GC") enables it to quickly separate and elute effluent collected and briefly stopped from the much longer and slower separation on the first column. Because the second column uses a phase with different polarity characteristics than the first one, it may separate components coeluting in the peak trapped from the first column. What is the trick for repeatedly trapping and then transferring sequential segments of the chromatogram in the first column for separation on the second one? A length of uncoated fused silica capillary tubing connects the two. It is alternately flash cooled to **trap** and then flash heated to **desorb** segments into the second column. In some versions the flash cooling is done by jets of liquid N$_2$, and the flash heating by electrically conductive paint on the outside of the bare capillary. This is the same material used for electrically heated automobile rear window defrosters. The very high heat conductivity of the thin-walled fused silica capillary facilitates these processes. This interfacing device is known as the "**thermal modulator**". Data from the detector at the end of this pair of serially coupled columns is acquired and stored as a sequence of fast separations (second column GC "scans") at equal retention time intervals from the first column. This is analogous to full-scan GC-MS data files. The computer system displays the data set as a planar 2D array of spots, looking like a developed 2D TLC plate or 2D gel electrophoresis plate. Each "spot" is a peak, and quantitation is by volume integrated over peak area, instead of area integrated over peak baseline as in 1D GC.

Many coelutions are thus resolved. The technique is called "**comprehensive**" because all components injected are measured, and most are resolved from one another. In this sense it differs from **heart-cutting 2D GC**, where only a selected peak is diverted by a valve to a second slow GC column for additional separation. Using comprehensive GC-GC-FID, a light fuel oil like one whose 1D separation is illustrated in Fig. 12.20, has been resolved into 3000–4000 quantified peaks! The size and resolution of the textbook page is hopelessly inadequate to display this.

An additional valuable feature of the thermal modulator is that it acts in a fashion similar to the retention gap (cf. Fig. 12.6). A slow, late eluting peak will be gradually accumulated off the first column during the cold trapping phase. Then it is quickly flashed back to vapor, and much more rapidly transferred to the head of the second column, where it is "refocused" to a very narrow band like that of an initial retention gap injection. The fast separation on the second column retains this very narrow peak width, so at the detector the peak height is correspondingly increased from what it had been exiting the first column, yielding a great improvement in signal-to-noise at the peak top. In some circumstances the S/N improvement can be an order of magnitude or more, making it worthwhile to use a thermal modulator even if the second column is not needed for or capable of resolving coelutions.

12.9. RETENTION INDICES (A GENERALIZATION OF RELATIVE R_t INFORMATION)

In the last paragraph of Section 12.8.2, we saw in the case of xylene isomers that even the great compound fingerprinting capability of full scan MS detection could not distinguish

Figure 12.20 Linear programmed temperature GC-FID analysis of hydrocarbon fuel oil. (Adapted with permission from Supelco, Inc.)

among them. The differences in elution times on a GC column could resolve this problem, if we had run authentic standards of each isomer on the system to establish where each one elutes. What if no standards are available in the lab? We know that absolute values in minutes of R_t will depend on many factors: the column phase and exact dimensions, temperature program, flow rate. Such numbers from someone else's lab will be useless. How about RRTs (Section 11.11)? This might get us closer, but the values will still vary significantly with the temperature program used. Elution orders and RRT among compounds will differ with and be specific to the stationary phase in use. The best system devised to codify data on these parameters is a library of compound **retention index (RI)** values, specific for each stationary phase. The most common such system was devised by Kovats.

Compounds of similar structure, differing by only the number of —CH_2— units in linear hydrocarbon chain substituents, or in the number of the same substituents on aromatic rings (like the methylnaphthalenes), are called homologs. The simplest example is the sequence of **n-alkanes** (methane, C1; ethane, C2; propane, C3; ⋯ octane, C8; ⋯ hexadecane, C16; etc,). The n-alkane elution sequence on most stationary phases displays a simple regularity. Under isothermal GC conditions their adjusted retention times t'_r (cf. Section 11.7) follow a **logarithmic** progression. In the **Kovats RI system** each is assigned an RI value of 100× its carbon number; thus, octane has an RI of 800 and nonane an RI of 900. By definition, these numbers are the same on any GC phase. The RI of any other compound (U) is calculated by substituting the adjusted R_t values of both the compound and the alkane pair between which it elutes into Eq. (12.1):

$$\text{RI} = 100\left[n + (N - n)\frac{\log t'_r(U) - \log t'_r(n)}{\log t'_r(N) - \log t'_r(n)}\right] \qquad (12.1)$$

where n and N are the number of carbon atoms in the smaller and larger alkanes, respectively. This formulation gives the official, most accurate, thermodynamic evaluation of a compound's RI. The measurement and calculation are fairly complex. One must measure in the appropriate isothermal temperature range, use the dead time t_m to calculate t'_r values from measured t_r, and use moderately complex Eq. (12.1).

GC runs covering a wide range of analyte boiling points will generally be done by programming temperature using a linear ramp covering the desired range of elution temperatures. Fig. 12.20 illustrates a GC-FID chromatogram of a hydrocarbon fuel oil. The n-alkanes from C10 to C24 are seen under these conditions to elute at almost regularly spaced intervals. Correspondingly, a close estimation of RI can be made much more simply:

$$\text{RI} = 100\left[n + (N - n)\frac{t_r(U) - t_r(n)}{t_r(N) - t_r(n)}\right] \qquad (12.2)$$

Note the use of actual rather than adjusted retention times. Although not as accurate or conforming to the operational definition of RI as defined in Eq. (12.1), use of Eq. (12.2) with retention times acquired over the linear ramp portion provides a useful approximate value. Essentially the RI values of unknowns are calculated by linear interpolation between the values of the flanking alkanes. If one had the PDMS phase RI values for the three dimethylnaphthalenes illustrated in Fig. 12.11, one could probably assign identity to these compounds, whose mass spectra in GC-MS would be indistinguishable. To be certain, however, one would need to have RI values for all such isomers (there are many more than these three). If there were some others whose RI values were very close to those observed in that chromatogram, it would be dangerous to rely on RI to clinch the identification. At best it would tell which were the most likely candidates, and then the confirmation standards could be selected. If RIs are within less than one unit of one another, coelution is a possibility, and resort to a different column which will resolve the candidates [as in the xylene case of Fig. 12.8(a) and (b)] is indicated. Tables of RIs for the compounds on different phases will aid in selecting such an appropriate column. In general RI values are useful for aiding in sorting out small sets of isomer elutions. There are far too many possible compounds relative to the number of reliably distinguishable RI values to make these values useful for identifying general unknowns. Kovats Indices are an adjunct to identification with hyphenated GC methods rather than a substitute for them.

12.10. THE SCOPE OF GC ANALYSES

Appropriately configured GC instrumentation can separate and identify molecules ranging from the very smallest (H_2), to those with masses on the order of 1000 Da. This upper mass range limit applies only for compounds with great thermal stability at high temperatures and very low polarity leading to minimal boiling points for a given mass. The prototype of such compounds are the n-alkanes, the reference series for the Kovats RI system, which present the characteristic "picket fence" profile of a homologous series when they elute under linear temperature programs. This is illustrated in the FID chromatogram of a light fuel oil in Fig. 12.20. Heavier petroleum crude oils might extend this pattern out to n-C60 or beyond. The limiting factors are:

1. The ability to efficiently volatilize the compound.
2. A stationary phase which will not decompose at a temperature sufficient to partition the analyte into the mobile phase for a large enough percentage of the time. This will enable it to pass through the column in a reasonable time, and come off the column quickly enough to produce a peak well above the noise background. Recall that very slowly eluting compounds with long retention times elute so gradually as they exit the column into the detector, that their peaks spread out and eventually become not much higher than and indistinguishable from background noise and baseline drift.
3. The column tubing itself must be stable at the necessary temperature. The polyimide scratch-protective outer coating of fused-silica capillaries is itself stable only to about 380°C. Some exceptional stationary phases can exceed this limit, and may require use of metal capillaries, which for some analytes at those temperatures must be deactivated—perhaps with an inner coating of silica, which will not yield a flexible column coil.

The strength of GC relative to other chromatographic techniques which combine separation, identification, and quantitation, is the number of compounds in a mixture that it can characterize in a single run. The illustration of capillary GC-FID analysis of the fuel oil in Fig. 12.20 only suggests what has been resolved. Magnification of the smaller peak regions between successive n-alkane peaks would show perhaps 15–20 resolved hydrocarbons. Use of GC-MS SIM or GC-MS XIC chromatograms could tease out many times more classes of hydrocarbon peaks which are too small and overlap too much to be distinguished in the FID chromatogram. Among these would be a majority of the methyl naphthalene isomers we discussed earlier, as well as many other such extended "families" of hydrocarbons. In the comprehensive 2D-GC-FID run of a similar fuel oil described in Section 12.8.3, 3000–4000 peaks were resolved and quantitated (approximately—certainly standards for many of these were not available!). This spectacular performance is a tribute both to the extreme resolution of multiple capillary GC separations and the sensitive, wide-dynamic-range FID.

12.10.1. GC Behavior of Organic Compound Classes

Hydrocarbons, aliphatic or aromatic, have low polarity and high stability, and perform well at molecular weights limited only by the three factors earlier listed. Organic compounds with only ketone groups (R—(C=O)—R′), or ester groups (R—(C=O)—O—R′), or ether groups (R—O—R′), or halide substituents (—F, —Cl, —Br, —I), are somewhat more polar, and will have higher boiling points for their mass, but are still relatively unreactive to components of the GC.

Compounds with strong hydrogen-bonding groups, for example, alcohols (—OH), or with acidic ionizable H^+, for example, organic acids (—COOH), or phenols (Φ—OH); or basic functionality; amines (—NH₂), amides [R—C—(NH)—R′], or aromatic ring nitrogens; all these may be reactive with one another, with themselves, or with active sites on the column, which leads to tailing peaks. They are more prone to decomposition at higher temperatures. It requires specially engineered phases and carefully deactivated columns to handle even the lighter molecules of this sort. The heavier they become and the more such features they possess, the less likely they are to make it through the GC while exhibiting good chromatographic behavior. Really active compounds, like strong acids or bases or water can destroy a column after one or two injections at elevated temperature. Ionizable compounds which form salts will also fail to elute. One must use highly specialized columns, such as the one illustrated in the chromatogram of Fig. 12.5 to handle such compounds by GC.

12.10.2. Derivitization of Difficult Analytes to Improve GC Elution Behavior

One way to deal with the recalcitrant compounds described in the preceding paragraph is to modify the troublemaking functional group(s) to a more tractable form. Their reactivity and polarity is the cause of their bad behavior. We can use this reactivity to add a substituent which converts them to a less-active, less-polar group.

1. We can **methylate** alcohols (—OH) to methyl ethers, or **esterify** organic acids (—COOH) to esters. An excellent reagent for this is **diazomethane, CH_2N_2**. This is a toxic gas which must be generated on site, dissolved in ethyl ether, and handled with great care in a fume hood. Upon addition of its ether solution to a solution (not in alcohol or water) containing analytes, their alcohol or acid functionalities are instantly converted (methylated), the byproduct is inert nitrogen gas, and excess reagent is quenched by adding water.

2. We can deactivate both acids and alcohols by replacing their end hydrogen with a **trimethylsilyl (—Si(CH₃)₃)** group, and also tame the basic amines by bonding this group to their N atom. The silyl esters, ethers, or amides thus formed present a relatively nonpolar trimethyl face to the stationary phase, where earlier an active, polar irritant was present. Despite the larger mass added by this group, such derivatives are actually more volatile, and will elute faster at lower temperatures. Derivitizing agents such as trimethylchlorosilane or hexamethyldisilazane or a host of others are used to accomplish this.

    Such agents function less instantaneously than diazomethane. They react faster at quite high temperatures. One way to achieve those has been to mix them with the injection solution. Upon flash vaporization in the hot injection port or the start of the column, the derivitization reaction is nearly instantaneous and complete. This procedure is known as **on-column derivitization.** Such a mode of introduction of this reagent also achieves the purpose of reacting with and deactivating the free —Si—OH, active site groups in the column, as described at the very end of Section 11.5. This treatment is called **silanization**.

3. We can **acylate** alcohols to esters by reacting them with active organic acids or anhydrides. For example, heptafluorobutyric anhydride derivitizes R—OH to R—O—(C=O)—C_4F_7. After quenching excess anhydride, the liberated heptafluorobutyric acid must be left behind in an aqueous phase by extracting the derivative into an organic layer under basic buffer conditions. The acid is extraordinarily strong acid, and just a trace in a single injection could utterly destroy even a robust PDMS stationary phase.

    The heavy —C_4F_7 group nevertheless is so much less polar than the —OH it replaced (think "Teflon") that the derivative will elute more easily than the parent compound. An additional advantage is that all those electronegative F-atoms yield a derivative with ECD sensitivity, making our analyte responsive to a more sensitive and selective detector. Derivitizing to add groups with heavier halides like Cl or Br, would confer even greater ECD sensitivity, albeit with perhaps less volatility. Incorporating Cl or Br atoms into the analyte derivative would produce MS mass fragments with the characteristic isotope cluster patterns of these halides (cf. Chapter 10), which would make them stand out when using GC-MS detection.

To summarize, derivatization for GC can confer greater thermal stability to an analyte. It can convert the analyte to a form which is more volatile, less polar, and with less tendency to produce tailing peaks by interaction with column active sites. It can introduce elements into it which make it detectable by more sensitive or selective detectors (remember, in real-world analyses, selectivity often confers more effective sensitivity to an assay). The scope of derivatization for GC is far greater than the several examples listed here. See the bibliography (Grob, Jennings, Scott) or product literature from specialty derivitization agent vendors (Appendix 12.1) for more examples.

12.10.3. Gas Analysis by GC

Not surprisingly, GC is the method of choice for the analysis of mixtures of gases (i.e., compounds which are vapors at room temperature). The procedures for sampling and introduction to the GC instrument sometimes differ from the standard syringe injection or SPME techniques (Section 12.3) employed for neat liquids or dilute solution samples. Gases at trace levels in air or dissolved in water may be present at levels too low to measure with the volumes one might sample and inject with a syringe; even one of several milliliter volume. Certainly one would not wish to introduce several milliliters of water into a GC column!

Trace levels of gases in air can be sampled by drawing large volumes of air (up to cubic meter volumes) with a pump through a **trap** (a tube or plate filled with a special sorbent packing) for periods up to an hour or more. One especially useful sorbent is a packed bed of **Tenax**® particles. These are beads of polyphenyl ether polymer [(··· —Φ—O—Φ—O— ···), where Φ = 6-carbon aromatic *para*-phenylene ring]. Some varieties have additional phenyl rings substituted on the phenylene units in the polymer backbone. At room temperature this will absorb small organic vapor molecules while letting fixed gases, and, most importantly, water vapor pass through. After these are concentrated from a large volume of air slowly sampled over time, a trap oven heats it rapidly to desorb the mixture almost instantaneously to a GC configured for low molecular weight gas analysis. An alternative form of trap employs **activated charcoal**. Thermal desorption is not as efficient with this, so the trap packing is emptied after sampling, and the analytes are desorbed by carbon disulfide solvent, which is then filtered, concentrated, and injected into a standard low temperature GC system. This method works best for a range of higher molecular weight vapor molecules than the range covered by thermal desorption from Tenax. Fig. 12.18 displays elution of both fixed gases and low m.w. organics from a new GC column which can handle both while operated near room temperature. Trace level sampling for this type of analysis would probably require liquid N_2 **cryotrapping** for the fixed gases after Tenax trapping for the less volatile small organics. It is remarkable that a single GC column phase can handle separation of both classes simultaneously.

Trace levels of gases or volatiles in water can be sampled by **purging** a water sample with a cloud of gas bubbles (N_2 or He) introduced through a **frit** at the bottom of the sample vessel. The purged vapors and large amounts of entrained water vapor pass through a trap packed with Tenax or porous polymer beads which does not retain the large excess of water. After the analyte vapors are completely purged from the sample and captured on the trap, they are thermally desorbed to a GC column separation and detection system in a fashion similar to the procedure for the trace gases in air. Dedicated instruments connected to a GC are designed to sequentially perform such a procedure on multiple samples of water and are called "**purge and trap samplers**".

These traps for volatiles at room temperature behave like GC columns operated at too low a temperature to have significant elution in a reasonable time. Since the sample is being loaded constantly over a long sampling time, eventually the trap may become saturated with analyte, and/or slow "elution" through the trap tube will eventually cause analytes to start coming out the downstream end of the trap. At this point the amount on the trap reaches an equilibrium value and is no longer proportional to the time-averaged concentration in the volume sampled. **Breakthrough** is said to have occurred. Often a smaller trap will be added in series behind the main trap. If it is first separately desorbed and analyzed, and no analytes are detected, and it is then removed or left at the desorption temperature, one may be confident that breakthrough has not occurred on the primary trap and proceed to desorb and analyze its contents.

12.10.4. Limitations of Gas Chromatography

GC serves well for analyzing many mixtures of compounds significant for environmental pollution monitoring, checking processes and contaminants in synthetic chemical manufacture, characterizing complex petroleum hydrocarbon mixtures, and of course it shines at the low end of the compound volatility range. The huge area of application of chromatographic instrumentation in the 21st century is in biochemical, medical, and pharmaceutical research. The chemistry of life sciences is mostly chemistry in aqueous media, and many of the compounds are correspondingly polar, to function in this very polar medium. The complexity of life processes demands large, complex molecules: enzymes, proteins, receptors, hormones, the double helix of DNA, RNA, nucleotides, and so on. These things are generally polar, full of acidic and basic functionality, and too large to be volatilized into a GC without suffering catastrophic thermal decomposition, even if resort to derivitization is attempted. Chromatography in the liquid state, at temperatures closer physiological values, will be required to separate, identify, and quantitate these complex mixtures. GC led the way in instrumentation for chromatography, but the torch is being passed to those methods which are carried out in liquids. One does not see many GC instruments in genomics or proteomics laboratories. We next focus our attention in Chapter 13 to the instruments and methods for chromatography in the liquid state.

BIBLIOGRAPHY

Cazes, J. *Encyclopedia of Chromatography*; Marcel Dekker Inc.: New York, 2001.

Grob, R.L. *Modern Practice of Gas Chromatography*, 3rd ed.; Wiley: New York, 1995.

Jennings, W.; Mittlefehlt, E.; Stremple, P. *Analytical Gas Chromatography*, 2nd ed.; Harcourt Brace: Orlando, FL, 1997.

McMaster, M. *GC/MS: A Practical Users Guide*; Wiley: New York, 1998.

McNair, H.; Miller, J.M. *Basic Gas Chromatography*; Wiley: New York, 1998.

Niessen, W.M.A., Ed. *Current Practice of Gas Chromatography–Mass Spectrometry*; Marcel Dekker, Inc.: New York, 2003.

Rood, D. *A Practical Guide to the Care, Maintenance, and Troubleshooting of Capillary Gas Chromatographic Systems*, 3rd revised Ed.; Wiley: New York, 1999.

Scott, R.P.W. *Chromatographic Detectors*; Marcel Dekker, Inc.: New York, 1996.

Scott, R.P.W. *Analytical Gas Chromatography*, 2nd Ed.; Marcel Dekker, Inc.: New York, 1998.

PROBLEMS

12.1   Indicate whether packed GC columns or open-tubular GC provide the best performance in each of the following categories:
1. analyte capacity
2. peak resolution
3. number of peaks resolvable in a given time
4. operation of detectors without need for makeup gas flow
5. ease of interfacing to mass spectrometer
6. ease of interfacing to FTIR spectrometer
7. operability without need for separately heated injection port
8. capable of using the longest columns
9. capable of producing the most inert "deactivated" column

12.2   Indicate which injection mode (split or splitless) best fits or meets the criteria below
1. more sensitive detection
2. less solvent peak tailing
3. most appropriate for injection of "neat" samples
4. not appropriate for packed column application
5. more likely to need initial temperature hold before temperature program begins
6. preferred for operation with gas sampling valve injection
7. works best when doing SPME desorption

12.3   What is the effect on retention time (longer, shorter, unchanged) of: ?
1. septum coring
2. loss of stationary phase by column bleeding
3. operating column at higher temperature
4. operating column at higher head pressure
5. clipping off front 5% of column to remove nonvolatile contaminants
6. the air peak's elution on a PDMS column when temperature is raised
7. changing carrier gas from He to N_2, if flow, temperature, and column are same
8. a small leak where the column is attached to the detector
9. a small leak where the column is attached to the injector port

12.4   Describe the sequence of events that would lead to septum bleed giving rise to spurious GC peaks

12.5   Why does column bleed not give rise to such spurious peaks? What is its GC signature? Are you more likely to observe this signature in isothermal or programmed temperature GC? Why?

12.6   Answer the questions below for each of the following analyte mixtures
1. HCN volatilized from whole blood
2. Pure gasoline
3. Organophosphorus pesticides at 1 ppm each in hexane
4. Trace volatile chlorocarbons in swimming pool water
5. Trace inert gases in air
6. Rapidly repeated sampling of major hydrocarbons in natural gas pipeline
7. Dimethylbenzothiophene ($C_{12}H_8S$) isomers in crude oil
8. <50 ppm polychlorinatedbiphenyl mixtures in transformer fluid
9. Two completely unknown peaks at 0.1% concentration each in toluene
10. BTEX vapors in refinery air

A. Which of the following sampling and injection methods would best for each sample?

Split injection, splitless injection, SPME, headspace SPME, gas valve injection, tenax trapping with thermal desorption, purge and trap sampling

B. Select one or more detectors from Section 12.7 (and only one example from hyphenated methods, Section 12.8) which would be most suitable to analyze these samples

C. Which of the following columns would be best for each sample?
a. 10 m Porapak PLOT Q
b. 20 m × 0.25 mm × 0.25 μm PDMS
c. 5 Å molecular sieve
d. 40 m × 0.18 mm × 0.1 μm PDMS
e. 10 m SolGel-Wax column

12.7   A. Draw a representative structural formula of a portion of a PDMS polymer phase deposited on fused silica

B. Draw the structure for such a crosslinked PDMS phase

C. Draw the structure for such a bonded PDMS phase

D. Draw the structure for such a bonded and crosslinked PDMS phase

12.8   Using the chromatographic system pictured in Figs. 12.10 and 12.11 to analyze two trichloronaphthalene isomers
1. Which selective detector from section 12.7 is most suitable?
2. With N_2 carrier gas what flow rate (cm/s) should be used?
3. With H_2 carrier gas what flow rate (cm/s) should be used?
4. How would you measure and set these flow rates?
5. Which compound would be most suitable to use to measure the linear flow rate (Cl_2, CH_3Cl, or chlorobenzene)?
6. What auxiliary gases (if any) will be needed to operate with H_2 carrier gas?
7. If the Kovats RI of two of the analytes are known to be 915 and 920, what temperature would be suitable for an isothermal analysis to separate this pair?
8. If chloronaphthalenes with more chlorines are present as well, what temperature program should be used?
9. If the first pair cannot be separated, describe three changes in the column that might be tried to achieve this

12.9   Which detector (from Section 12.7) is most suitable for each analyte below:
1. neon
2. krypton and xenon in air
3. nitrogen in air
4. CO_2
5. trace carbon tetrachloride in hexane
6. trace "eau de skunk", n-butyl mercaptan, C_4H_9SH
7. trace PCB (polychlorinated biphenyl) mixture in solvent
8. trace TNT (trinitrotoluene) vapor in air
9. mixture of fatty acid methyl esters
10. heavy crude oil sample
11. trace BTEX mixtures on handheld monitoring GC instrument

12.10   Which hyphenated GC system detector and mode of detection (listed in Section 12.8) would be most suitable for analysis of the following samples:
1. Identification by library searching of 100 possible different pesticides in vegetation extracts
2. Finding traces of mono-, di-, and trimethyldibenzothiophene isomers in a fuel oil sample
3. Identifying the compounds found in (2) against standards of the authentic isomers
4. Identifying the compounds in (2) against library spectra
5. Identifying the maximum possible number of components in a fuel oil sample over the widest concentration range.
6. Detection of small amounts of an unanticipated class of isomers in a hydrocarbon mixture, followed by identification of the compound class, followed by quantitation of relative amounts of the peaks for this subset of compounds assuming they all had similar relative response factors (RRFs)

12.11   Retention data for three dimethylnaphthalene isomers (D1–D3) and the n-alkanes between which they elute in Fig. 12.11 are listed in the table below:

Temperature	Compound	$R_t(n)$	$R_t(N)$	$R_t(U)$
100°C	D1	9.55	15.20	14.80
	D2	15.20	29.80	16.40
	D3	15.20	29.80	17.20
120°C	D1	4.84	7.60	7.37
	D2	7.60	12.60	8.03
	D3	7.60	12.60	8.47
150°C	D1	2.43	3.38	3.23
	D2	3.38	4.57	3.61
	D3	3.38	4.57	3.78
Program	D1	5.21	6.67	6.61
	D2	6.67	8.92	6.97
	D3	6.67	8.92	7.21

The retention times are in minutes for preceding (n) and following (N) n-alkane peaks, and for each of the three D1–D3 unknowns (U). As mentioned in the text the column dead time was 1.08 min.

Using the formulas of Eqs. (12.1) and (12.2), set up and use a spreadsheet to calculate the RI values for each of the three compounds at each of the four temperature conditions. If you cannot program this, calculate the values with a hand calculator, which will take lots more time. How well do the RI values compare? Some of the variation is due to inaccuracies in the simulation program used to generate the data, but some reflects how these values depend on the conditions under which they are acquired.

How well do the values approximated from the programmed temperature formula compare with those of the theoretically more accurate isothermal runs? Which isothermal run do you think should provide the best RI values?

Perhaps the RIs vary more with temperature because the analytes are aromatic and the reference compounds are aliphatic. If you have the spreadsheet set up, measure the retention times in the figures and put in the appropriate values for the triplets of branched hydrocarbons C10s or C14s and see whether their RIs are more consistent with variation of temperature.

There are actually 10 possible dimethylnaphthalene isomers. See if you can draw them all. Naphthalene consists of 2 hexagonal rings of 6 carbons sharing an edge (so there are 10 carbons in all). The 8 carbons to which substituents can be attached are numbered clockwise from 1 to 8 around the ring starting at the top right side. Suppose we obtained a GC-MS-SIM chromatogram of a petroleum oil at mass 156, which showed a cluster of 6 peaks, three of which matched the RIs of D1, D2, and D3, and two of these are "fatter" than the other four.

1. Why might we not be seeing all 10 isomers (more than one reason)?
2. How might we attempt to find the others?
3. How would running authentic standards of all 10 isomers help resolve the uncertainties?
4. How would having RI data on all 10 help? As much as the standards? Why or why not?
5. You find another mass 156 peak in the SIM chromatogram with an RI of 1100. Is that one of the missing isomers? What might it be? If you had acquired data with an ion-trap MS, how might you check this out?

12.12 Suppose all the compounds eluting in the chromatogram of Figs. 12.10 and 12.11 had the same detector RRF and were all present at the same concentration. How would the peak heights differ from the "phony" artificial ones in the simulated chromatogram? Sketch the appearance of the actual chromatograms for the 100°C isothermal run and the programmed temperature run using the same retention time scale. What causes the difference in peak heights, areas, and widths?

APPENDIX 12.1 GC INTERNET WEB RESOURCES

Web. URL	Site selection path	Use and contents
www.chem.agilent.com/cag/ cabu/gcapps.htm		GC applications
www.alltechWEB.com	Chromatography Access database	2,500 GC chromatograms

(continued)

Appendix   Continued

Web. URL	Site selection path	Use and contents
www.sigma-aldrich.com/supelco	Chromatogram search	Several hundred GC chromatograms
	Library/applications	Application notes
www.restekcorp.com	Chromatogram Search	Many chromatograms
www.sge.com	Chromatograms Search	Selected by compound or application
www.varianinc.com		
www.agilent.com/chem	1. GC-GC/MS systems	
	2. Tech support	
	3. User contributed software	Interactive column parameters calc.
	4. Pressure/flow Calculator	Calc. effect of gases and column geometry
	Or 4. Method Translator	Calc. how elution of compounds changes as above parameters change
www.restekcorp.com	/colsel/phases2.htm	Formulas of stationary phases (Table I)
	Retention indices	RI of model compounds on different Phases (Table II)
www.sigma-aldrich.com/supelco	Library/Apps./SPME	SPME instruction and video of process
Consumables.info.us @varianinc.com	Load prog. and CD	Select from 1,000's by col., application, cmpd.
Request CP-Scan View CD-ROM	Mailed to you	
www.srigc.com	Detectors Instructional and process control GCs	16 GC detector types, descriptions, photos, pictures of application chromatograms
www.piercenet.com	Pierce Technologies Gas chrom./HPLC	List of hundreds of GC derivitization agents
www.separationsnow.com	GC and LC courses Literature articles, (free access)	A mother lode of reading material on all forms of chromatography
www.chromatographyonline.com	Issue archive Author–subject index Useful links	Contents all issues from 1983–present Year-by-year index Dozens to other sites

13
Chromatography with Liquid Mobile Phases

13.1. HIGH-PERFORMANCE LIQUID CHROMATOGRAPHY (HPLC)

The first chromatography was performed using liquid mobile phases to separate colored plant pigments, which caused its inventor, Tswett, to give it the name of "color writing" (Section 11.1). Crude liquid chromatographic separations under gravity flow were commonly used to characterize synthetic mixtures from the organic research lab, but it was the chromatographic version of the venerable distillation process which was first instrumentalized as gas chromatography (GC). The reasons for this precedence were discussed in Section 12.1. Diffusion and equilibration in the column is much faster with gases. To attain equivalent diffusion equilibration using a liquid mobile phase requires that the dimensions for diffusion be much smaller. These dimensions will be those of pore space in a packed column or the diameter of an open tubular column. Liquids are much more viscous than gases. To force liquid to flow through such narrow pathways requires that orders of magnitude greater pressures be imposed across a column capable of withstanding these pressures. Suitably small support particles need to be coated with an extremely thin layer of stationary phase, then be packed tightly, and not develop resolution-destroying voids as high pressure pumps cycle over the immense range necessary to move the liquid through the bed of microscopic particles. Detectors for analytes in the small eluted peak volumes of the liquid mobile phases needed to be designed. The prior example of GC instrumentation was required to inspire analytical chemists to invent this whole suite of improvements, resulting in an instrumental version of LC which would match or exceed the performance of GC. The instrumental version of LC became known as HPLC. In the early days of the method this was often understood to stand for high-pressure liquid chromatography, for it was instrumentation that could control these necessary pressures that made HPLC possible.

HPLC is now a larger and more important market than GC. Why has this happened? The limitations on analytes for GC outlined in Section 12.10.4 are to compounds with relatively low molecular weight and low polarity, which are electrically neutral (i.e., uncharged) and thermally stable. Vast numbers of biochemically important molecules fall outside these categories. There are more parameters available to vary in liquid mobile phase chromatography than in GC, so it is divided into a number of techniques. Some are named for the different mechanisms of separation employed (e.g., partition, adsorption, size-exclusion, ion-exchange, immunoaffinity, etc.). Others are described by the design of the separating technique (e.g., thin layer, high-pressure column, and electrophoresis, which is not even really a chromatographic method, but has similar features such as peaks, elution times, resolution, capillary columns or 2D plates, and chromatographic type detectors).

The major method is HPLC, operating as **partition chromatography**. This is what is commonly implied when the term HPLC is used. It is the mode we will discuss initially as we cover the design and function of the pumps, columns, stationary and mobile phases, injectors, and detectors. Recall the discussion in Section 11.5, where initially, the partitioning of analytes between nonpolar liquid mobile phases and silica particles with adsorbed water became known as "**normal phase**" chromatography. For the larger population of more polar analytes encountered in biological systems, it became more advantageous to reverse this. A more polar, more aqueous mobile phase could carry such analytes past particles coated not with adsorbed water molecules, but with covalently attached nonpolar organic groups. Thus this received the name "**reversed phase**" HPLC (RP-HPC), and it became the workhorse of the field. Let us then start our discussion with RP-HPLC, and contrast the design features of the instrumentation with those of GC, which were covered in Chapter 12.

13.1.1. The HPLC Column and Stationary Phases

Compare GC with HPLC. The typical capillary GC column has an internal diameter around 0.25 mm (or 250 μm). A typical polymer stationary phase coating on the inner wall of this capillary is 0.25 μm (or 250 nm). The high diffusion rate of analytes in the gas phase permits rapid near-equilibration over such distances. Diffusion in a liquid is much slower. If we compensate by reducing the dimensions of a capillary LC system sufficiently to attain the desired near equilibration condition, they will be very small indeed, with a bore measured in a few microns. To retain a phase ratio similar to GC, the stationary coating must shrink to a thickness of several nanometers (nm). This is in the size range of single molecules. An open-tubular LC column with such dimensions would have a very minimal sample capacity before it overloaded, and it would be very difficult to build a detector that would work on such a small scale. Larger volume detectors would introduce severe extra-column band broadening or require extensive makeup flow which would further dilute the already miniscule concentrations of analyte eluting from such a column. Capillary HPLC is possible when samples are limited to very small amounts, but it is nowhere near as common as capillary GC.

The way to achieve HPLC is to return to packed columns, but to implement them with support particle and stationary phase coating dimensions much smaller than those of packed column GC. Typically one employs spherical silica particles of uniform diameters (**monodisperse**) in the range of 2–10 μm. The pore spaces between the tightly packed spheres will be of similar dimension. Unlike the much larger support particles in packed column GC, these LC particles often have a smaller scale **secondary internal porosity** of channels permeating the particle into which analytes can diffuse. The stationary phase coating in GC is like a true quasi-liquid polymer many molecular sizes thick—a true phase with a boundary across which molecules diffuse, dissolve, and partition from the mobile gas phase. In contrast, the stationary phase on these porous micro-HPLC particles is an organic monomolecular layer chemically bonded to the free silanol groups on the silica surface. The most common such group is a long-chain n-C18 hydrocarbon (**octa-decyl silyl group, ODS**). Check the discussion in Sections 11.4 and 11.5 and see Fig. 13.1. If stretched out, such a chain is on the order of several nanometers long, which is just the thickness proposed in the previous paragraph. The structure of such coated microporous particles is diagrammed in Fig. 13.2. Here are some points of contrast between HPLC and packed or capillary GC stationary phases:

1. Each LC phase molecule is *bonded* to its solid support surface, but is less often *cross-linked*.

Figure 13.1 Types of ODS-silica HPLC stationary phases: (a) i, synthesis of monomeric C18; ii, synthesis of polymeric C18; iii, endcapping process; and (b) i, monomeric C18 ligand; ii, encapped silanol; and iii, residual silanol. (Cazes, used with permission.)

2. The HPLC phase is so thin that it is not a fully randomized fluid, like a GC polymer coating.

3. Usually it is impossible, due to steric hinderance or increasing crowding, to attach a molecule [like the $—O—Si—(CH_3)_2C_{18}H_{37}$ group] to every free silanol bonding site on the silica surface.

4. These unreacted silanol sites lie closer to the mobile phase than in GC, and can easily act as active sites degrading the chromatography as described in Section 11.4.

5. The unreacted active sites on the support surface must be masked by reacting with a very short group [such as $—Si—(CH_3)_3$] which can penetrate to react and bond with most of the remaining silanol groups; a process called **endcapping**.

6. As the composition of the liquid mobile phase is varied, it can interact to cause alterations in the alignment of the stationary phase molecules, and thereby affect their partition behavior with some analytes. This dependence of stationary phase behavior upon mobile phase composition is not found in GC, and it adds more complexity to prediction of HPLC separation behavior.

7. The small interparticle pore size of HPLC columns requires stable, uniform, spherical particles to avoid their packing together in a fashion which will block uniform flow through these tiny passages. Mobile phase solvents must be **filtered** to remove micron-size particles which could plug up these interstitial pores and block or channel mobile phase flow.

8. Dissolved gases in the mobile phase may come out of solution and form bubbles as the pressure changes from the entrance to the exit of the column. Those will block and channel flow, degrading resolution or even blocking the column completely. Even if the bubbles redissolve, the resultant large cavity in the packing may introduce extra-column band broadening. Mobile phase liquids must be **degassed** prior to use, by filtering under vacuum or by **sparging** with a flow of fine bubbles of poorly soluble helium gas.

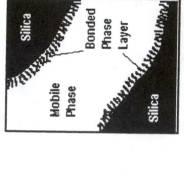

Magnified cross-section of a pore, showing a layer of alkyl chains bonded to the silica surface

(b)

A particle of column packing is honeycombed by pores. Most of the surface area is inside these pores (this drawing is *not* to scale; typical pore diameters are on the order of 1–5% of the particle diameter).

(a)

Figure 13.2   Microporous structure of bonded phase HPLC particle. (Used with permission from Supelco, Inc.)

9. As the instrument cycles the column through very large pressure changes as the pumps turn flow on and off, the packed bed may move or settle, yet again introducing the dread **void** and its extra-column band broadening.

10. Outside certain bounds of pH and/or temperature the liquid mobile phase may attack and gradually dissolve the silica support particles, and/or release the bonded stationary phase molecules.

11. Because of the smaller particle size and internal porosity of LC packings, the area covered by the molecular monolayer stationary "phase" is vastly greater than that of the thick polymer film stationary phase in GC columns of similar volume. This restores the analyte capacity of these columns to a level compatible with typical chromatographic detector designs.

12. The HPLC columns must operate at pressures, (500–6000 vs. 10–50 psi), orders of magnitude higher than GC columns. The mobile phase must be pumped instead of coming from a pressurized cylinder through a reducing valve. The columns are commonly fabricated from thick-walled stainless steel, with stainless steel mesh packing restraining frits, and use stainless steel or high impact plastic nut and ferrule fittings.

The effects on analyte resolution of particle size and column length in HPLC are analogous to those of capillary column diameter and length in GC. However, one does not vary the "film thickness" of the bonded phase molecules in HPLC. The percentage of available silanol sites to which stationary phase groups have been bonded can vary. A larger percentage does not directly translate into a larger effective volume of stationary phase. It is the surface density of the groups which varies, and with it the molecular interaction strength, not just the volume available for analyte partitioning. The percentage and completeness of endcapped silanol groups is an extra variable that also impacts the nature of the analyte interaction with the phase. The lesson here is that different preparations of a reverse phase packing category, such as the workhorse silica-ODS, can vary widely in their suitability for different analytes, depending on analyte size, polarity, acidity, or basicity, and the characteristics of the interactions with the mobile phase. All ODS columns are not created equal! Even more than with GC columns, comparison of manufacturer's specifications and application notes from catalogs or websites is the efficient first step in column selection.

13.1.1.1. *Support Particle Considerations*

Silica can be prepared as micron-sized spheres. The internal porosity can be varied. The spheres can be generated to have a narrow range of diameters, so they will pack together uniformly. The surface silanol groups make ideal attachment points for the bonded phase molecules. At acidic low pH values, the organosilyl bonds may hydrolize and the phase can degrade. More densely covered and appropriately endcapped silica surfaces may tolerate lower pHs. At pH above 8, the silica itself may begin to dissolve at the surface. Recall that glass can be etched by strongly alkaline solutions. As will be discussed subsequently under mobile phases, there could be advantages to operating the column at elevated temperatures. Unfortunately silica also becomes more soluble under these conditions. Recall that quartz crystals grow from hydrothermal solutions under pressure in the ground. HPLC column specifications need to be checked for their pH and temperature stability.

Zirconia (ZrO_2) is a new substitute for silica as an HPLC particle support, which has both greater pH range stability (1–14) and temperature stability (to 100°C or even greater—recall that we are operating at many atmospheres pressure in HPLC, and this can elevate boiling points). For some analyses, the greater cost of zirconia columns may be offset by the improvements in analysis speed and performance they can bring to HPLC.

Another new development circumvents production of the individual particles, and casts the column as a monolithic silica cylinder which possesses both the macro- and micro-scale porosities of a column "packed" with micron-scale spherical silica particles. When techniques for achieving this were invented, two major new advantages ensued. First, such monolithic HPLC columns are mechanically far more stable than a packed bed of particles. There is far less possibility for development of voids and channels. Secondly, the size of the macro pores is not constrained to the interstitial dimensions of a packed bed of spherical particles. Specifically, they can be larger, but the network of fused-together microporous support "particles" can still remain connected to one another. This additional degree of freedom in the column dimensions results in columns which can be operated at higher mobile phase flow rates, eluting analytes more quickly with lower column pressures, while still retaining the resolution performance of comparable packed columns.

For completeness we should note that spherical, porous beads of organic polymer resin, particularly cross-linked styrene–divinylbenzene copolymer, can be used as HPLC phases. Unlike silica-ODS, the whole particle is the stationary phase. These are often designated as XAD resins. Unlike hard silica particles coated with a monomolecular layer, these resin beads will take up mobile phase solvent and swell in size. Resin bead sizes cannot be as small as the spherical silica particles, lest this swelling fill and block the interstitial macro pores. The higher pressures cannot be used, and the high resolution efficiencies of true HPLC are rarely attained. These resins are very important as the substrates for attaching and supporting ion exchange groups to form the stationary phases for the technique of ion exchange chromatography. These will be discussed in Section 13.2. Pure polystyrene beads can serve as supports for covalent attachment of organic groups to make phases which are stable over the pH range of 8–12, but they also suffer from the structural shortcomings of the resin beads.

13.1.1.2.   *Stationary Phase Considerations*

Three of the most common nonpolar organic substituents employed in stationary phases for RP-HPLC with aqueous/polar mobile phases are:

Octadecyl (C18,ODS); $—(CH_2)_{17}CH_3$

Octyl(C8); $—(CH_2)_7CH_3$

Phenyl; $—(CH_2)_3C_6H_5$

The larger the substituent group, the more strongly and longer will it retain the nonpolar analyte, so retention will vary inversely to the order listed here.

Three examples of common polar organic substituents (incorporating functional groups to confer the desired polarity) for normal phase HPLC with less-polar mobile phases are:

Amino; $—(CH_2)_3\mathbf{NH_2}$

Cyano; $—(CH_2)_3—\mathbf{C \equiv N}$

Diol; $—(CH_2)_2OCH_2CH(\mathbf{OH})CH_2\mathbf{OH}$

Here it is the functional groups which dominate retention. The amino and diol phases will retain compounds with pronounced hydrogen bonding potential more strongly, while the cyano phase interacts most strongly with analytes whose polar interactions result from strong dipole moments. Complex structures may interact in a combination of these mechanisms, so relative retention is less easy to predict than with the reversed phase columns.

13.1.1.3.   Chiral Phases for Separation of Enantiomers

Even more specialized groups can be attached to supports to achieve particular separations. Complex cyclic carbohydrates called **cyclodextrins** exist in a single **enantiomeric form**. These are illustrated in Fig. 13.3. They form a truncated conical pocket, into which different optical isomers of a compound (enantiomers) will bind with different strengths. This enables HPLC separation of these isomers, which have identical chemical structures and polarity, and differ only in their stereochemical orientation of substitution at asymmetric carbons. These are described as **chiral** compounds, and use of such stationary phases enables "**chiral chromatography**" The cyclodextrin "pockets" come in a variety of sizes (α-, β-, γ- etc.) suitable for separating enantiomer pairs of differing size. This capability is especially important for determining the relative amounts of each enantiomer which may be present. Solutions of individual enantiomers rotate the plane of oscillation of plane polarized light. At high concentrations this may be measured by optical polarimetry instruments (not covered in this text). Polarimetric detectors also exist for HPLC. They are relatively insensitive (LOD ~ 5 ppm in eluent solution). For trace levels, use of even more sensitive chromatographic detectors makes resolution of enantiomers by chiral chromatography the method of choice. Often it is only one enantiomeric form of a drug compound which is active in the body. If the other form is toxic, the drug may have to be formulated to contain only the active, safe form. It then becomes critical to be able to determine whether, when, and to what extent this form may **racemize** (revert to an equilibrium mixture of both enantiomers) in the body. Chiral chromatography may be the only way to address this. A classic example is the drug Thalidomide, originally marketed as a racemic mixture of *N*-phthalylglutamic acid imide. One enantiomer had the desired pharmacological activity, but it was not realized until too late that the other was a strong teratogen, capable of causing severe fetal malformations. Cyclodextrin substituent stationary phases can be used to create **chiral GC** columns as well. HPLC is better suited than GC for many chiral separations, as the elevated temperatures used in GC may induce racemization in the column. There are other chiral substituents which can be incorporated into HPLC columns to enable them to separate enantiomer pairs, but the cyclodextrins possess exceptional discrimination ability.

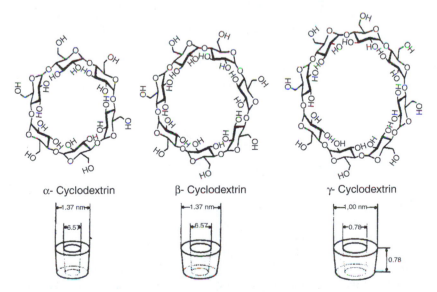

α- Cyclodextrin β- Cyclodextrin γ- Cyclodextrin

Figure 13.3   Structures of α-, β-, and γ-Cyclodextrins. (Katz et al., used with permission.)

13.1.2.   Effects on Separation of Composition of the Mobile Phase

In GC, elution temperature is the most easily controlled variable that affects elution and k'. In LC the corresponding variable is mobile phase composition. The factor which is being varied is polarity of the solvent mixture. This is more complex and less easy to define rigorously than is temperature. It is a relative value applying to the molecules to be separated, and to both the mobile phase and stationary phase materials. Polarity interactions can be of several types [e.g., dipole–dipole interactions, π-bonding interactions, and acid–base or hydrogen-bonding interactions (proton donors or acceptors)]. Both analytes and the chromatographic phases can display polarities which are combinations of these mechanisms. In GC there were only two polarities involved; those of the analyte and those of the stationary phase. In LC we add the third: the mobile phase, and it is that one that we can vary most readily. A detailed discussion of how one would evaluate the polarity and interactive components of the analyte molecule, select the most appropriate stationary phase, and then adjust the solvent mixture in the mobile phase to optimize a separation is beyond the scope of this text. What follows are a few oversimplified, general "rules of thumb".

1. The overall polarity contributions from particular functional groups increase in the order: hydrocarbons < ethers < esters < ketones < aldehydes < amides < amines < alcohols.
2. For partition chromatography, it is best to use a stationary phase of polarity opposite to that of the analytes, and to elute with a mobile phase of different polarity.
3. In the extreme of normal phase HPLC, the stationary phase is highly polar adsorbed water on silica or alumina supports, so the eluants should be nonpolar, often hydrocarbon solvents. The mechanism of normal phase HPLC is often classified as adsorption rather than partition, as the analytes are essentially binding to a highly polar surface.
4. Reverse phase (RP) partition HPLC uses less-polar bonded organic groups as the stationary phase, so the mobile phase is more polar. This is preferred for more polar biomolecules, since they have been selected by evolution to function in polar aqueous environments.
5. The most polar components of an RP-HPLC mobile phase will often be water (HOH), acetonitrile (CH_3—$C\equiv N$), or methanol (CH_3OH). The polarity of the mobile phase can be adjusted to values between these by mixing them in different proportions. Even less-polar compounds, such as ethanol, tetrahydrofuran, or diethyl ether, may be added to bring the overall eluent polarity down even more, or to introduce different interaction mechanisms of polarity which may increase the separation factor α between close pairs of analytes. Being at the extreme end of the polarity scale, water is often considered the fundamental component of the mobile phase mixture, and the other less-polar, organic solvents which may be added are termed "**organic modifiers**"

There are several methods of quantifying the relative solvent strengths or polarities of liquids used as mobile phases in HPLC. Table 13.1 displays values from two such systems for selected solvents. The first was developed to describe the relative ability of eluents in normal phase adsorption LC to solvate analytes and move compounds through the column more quickly. This "**eluent strength**", ε^0, is ordered in the so-called "**eluotropic series**". The values are obtained using alumina (Al_2O_3) particle columns. Its zero point is defined by n-pentane, and it goes to very high values for highly polar water. The corresponding values on silica are about 80% of the alumina values. Clearly

Table 13.1 Eluant Strengths, ε^0, and Polarity Indices, P', of Selected Mobile Phase Solvents

Solvent	ε^0	P'
Cyclohexane	-0.2	0.04
n-Hexane	0.01	0.1
Carbon tetrachloride	0.18	1.6
Toluene	0.29	2.4
Tetrahydrofuran	0.57	4.0
Ethanol	0.88	4.3
Methanol	0.95	5.1
Acetonitrile	0.65	5.8
Ethylene glycol	1.11	6.9
Water	$\sim\infty$	10.2

water as a mobile phase will very rapidly elute anything adsorbed to the adsorbed water of these stationary phases. The second column of Table 13.1 lists values for a "**polarity index**", P', developed by L.R. Snyder. These indices are derived from solubility of the solvent in three test solvents chosen to have high or low dipole proton acceptor-polarity or high dipole proton donor- polarity. The polarity index of mixtures of solvents can be calculated by summing the products of their respective polarity indices and volume fractions in the mixture. ($P_{AB...} = \phi_A P'_A + \phi_B P'_B + \cdots$, where ϕ is the volume fraction)

It is possible to optimize a RP-HPLC separation of several compounds by systematically varying the proportions of three appropriate solvents, usually water and two organic modifiers. By observing the relative shifts of R_t as each change is made, and using as few as seven runs with well chosen variations, one can create a "solvent polarity triangle" for the separations. Data from this will enable calculation of an optimum mobile phase composition for the separation. For HPLC instruments equipped with separate mobile phase solvent reservoirs and programmable precise mixing valves, this process can be automated, and the mobile phase optimization can even be done without operator intervention.

The major contrast with GC is that the mobile phase eluent strength (affecting k') and the relative selectivity between analytes (α) can both be readily varied. In GC the carrier gas composition has practically no effect on selectivity, and varying temperature affects only k'. The large number and range of parameters to be evaluated to optimize the eluent composition makes HPLC method development more complex. Every benefit comes with its own price!

13.1.3. Design and Operation of an HPLC Instrument

Figure 13.4 is a schematic diagram of a typical HPLC instrument. We will use this figure to discuss the features of the major components, and describe the construction of some of them. We will use this discussion to illustrate some of the major modes of operation of HPLC instruments. We proceed in order of the numbered items of the figure.

1. *Mobile phase degassing*: The figure shows a setup designed to sparge four separate reservoirs with He gas. It is distributed by the fixed splitter valve to fritted glass or metal filters which disperse the gas into a fine cloud of bubbles, driving out (sparging) other dissolved gases. Removal of dissolved gases could

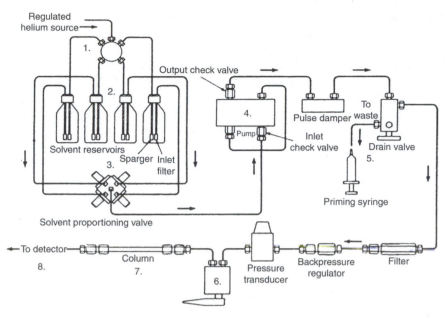

Figure 13.4 Diagram of conventional HPLC instrument. (Used with permission from Perkin Elmer Analytical Instruments.)

alternatively be done by manually filtering over vacuum prior to placing in the reservoirs.

2. *Mobile phase storage*: The figure shows four reservoirs, one for each of up to four pure solvents [e.g., water (perhaps with pH controlling buffer), acetonitrile, methanol, tetrahydrofuran, etc.]. Note the use of inlet filters. Alternatively one could prepare the mobile phase mixture to the desired composition manually, and store it in a single reservoir. Operation at a single, constant mobile phase composition is called **isocratic** HPLC elution.

3. *Mobile phase mixing*: The solvent proportioning valve can be programmed to mix solvents from the reservoirs to produce the desired mixture composition. This could be a constant proportion mixture for isocratic operation, or one could vary the solvent strength of the mobile phase with time by changing the mixture proportions. As a simple example, the mobile phase composition could be programmed to vary from 75% water:25% acetonitrile at time zero, to 25% water:75% acetonitrile at the end of the run. Such a variation of mobile phase solvent strength or polarity during a run is termed **gradient elution**. There is a "gradient" or variation in mobile phase composition and elution strength with time. A mixture of components of widely varying polarity can thus be separated and eluted more rapidly. In an RP separation, less-polar analytes will elute later during this gradient, but the run will be completed more rapidly than would be possible with isocratic operation. Gradient elution is the HPLC analog of programmed temperature GC, and it confers all the same advantages. It is especially valuable if there are "slow eluters" which must be removed before the next sample injection. If these are very slow eluters, which need not be measured, it may be better to reverse the flow through the column and *backflush* these off the front of the column and out a drain port located between the column and the injector, using backflush

valves on each end of the column (not illustrated here). It takes considerable time to re-equilibrate the stationary phase with the starting mobile phase composition after a gradient run. Unless there is a wide polarity range of components to be eluted, isocratic elution is to be preferred. It is simpler to implement, and if R_t values of separated components do not become too long, it will permit shorter analysis cycle times.

4. *HPLC pump*: A common type of HPLC pump is the reciprocating piston type illustrated in Fig. 13.5. During *fill stroke* Fig. 13.5(a) it is pulling mobile phase liquid from the solvent side, then on *exhaust stroke* Fig. 13.5(b) it is pushed through the injector and to the head of the column. This is where the high pressure that permits the high performance in HPLC is generated. The figure illustrates a "*single piston*" design. The flow from the pump will rise during the exhaust stroke, and subside during the fill stroke. The "*pulse dampener*" shown downstream attempts to smooth the flow surges to approximate a steady constant flow. The stationary phase particle bed would become unsettled by repeated pressure surges. When using mass-flow rate detectors, variation in flow would be reflected in variation of their response. Both are undesirable consequences of pulsed flow. An improved pump design effects more uniform pressure and flow using *dual pistons* in an arrangement whereby one is in the exhaust stroke while the other is in the fill stroke.

5. *Fill/drain valve:* There are many liquid transfer lines and components that the mobile phase passes through before it gets to the head of the column. The

Figure 13.5 Operation of piston and check valve reciprocating pump: (a) suction stroke and (b) exhaust stroke. (Katz et al., used with permission.)

analyst needs to fill all this with each new mobile phase, and after use one stores the system and column in a simple solvent or water rather than a complex mobile phase containing buffers, dissolved salts, and so on. A **priming syringe** pulls mobile phase through the lines of an empty system to remove air and "prime the pump". It can also be used to push mobile phase or storage solvent further downstream to clear out those lines. The **drain valve** can be opened so the pumps can quickly rinse storage solvent through the front half of the lines without having to work against the resistance of the column. Note that the solid black lines connecting the components of the system on the high-pressure side of the pump are narrow bore (e.g., 0.5 mm i.d.) stainless steel tubing.

6. *Rotary sample loop injector:* Figure 13.6 illustrates the most common design of HPLC injector. To introduce the sample as in GC with a syringe through a

Figure 13.6   (a) Diagram of operation and (b) photo of two-position rotary injection valve. (Katz et al., used with permission.)

septum into the highly pressurized mobile phase liquid flow is impractical. The top portion of the figure is a schematic end-on view of the valve, which rotates on an axis coming out of the page. There are three inlet and three outlet fittings. The valve is shown in the "*load*" position. Sample from a syringe is pushed into a **sample loop** of tubing of specified volume (e.g., 50 or 100 μL), with any excess flowing to waste. Mobile phase is flowing through one solid-line arc of the valve as shown. To *inject*, the valve is rotated 60° clockwise. Now the solid-line flow paths within the valve diagram are at the location of the dotted lines. The pump is now pushing the mobile phase and sample liquid in the loop in a direction *opposite* to that in which it was loaded, and it exits the valve to the column. All this occurs without leakage or introduction of bubbles at the high pressure maintained by the pump.

7. *The column:* HPLC columns were described in detail in Section 13.1.1. When a fused silica capillary GC column accumulates unmoving contaminants at its front end, it is a simple matter to cut off some of it and reinstall it. This is not possible with HPLC columns. Even with the most careful filtration, the fine pores may become clogged with traces of particulate matter over time. The solution is to attach in front of the "**analytical column**" a very short, inexpensive, "**guard column**", often packed with "**pellicular**" particles (i.e., without complete internal porosity, thus the bonded stationary phase layer is of much reduced area). When it becomes contaminated, it is detached, discarded, and replaced with a new guard column. If the stationary phase support will tolerate it, **operation of the column at elevated temperature** can confer advantages of both speed and improved resolution. Contrary to gases, the viscosity and resistance to flow of liquids decrease with increasing temperature. Diffusion and equilibration rates in the column are improved with increasing temperature as in GC. Unlike capillary GC, overall column temperature equilibration with an oven is not as fast in HPLC columns. To avoid radial temperature gradients across the diameter of the column which will degrade resolution, it is critical to ensure that the mobile phase is preheated to the desired elution temperature before it enters a thermostatted HPLC column. Even without operation at elevated temperature, thermostatting an HPLC column helps make elution times more reproducible.

8. *HPLC detectors:* Like GC, HPLC has a wide variety of detectors, universal or specific, destructive or nondestructive, mass flow or concentration responsive, and with even more challenging requirements for interfacing to spectrometers in hyphenated techniques. Tubing of even smaller bore than described earlier in item 5 is necessary to connect the effluent end of the column to the detector to avoid extra-column band broadening effects. Pressures are lower and high strength and density PEEK plastic may be used in place of stainless steel.

13.1.4. HPLC Detector Design and Operation

Sections 13.1.1 and 13.1.2 described the stationary phase column and the mobile phase liquid, respectively. Section 13.1.3 described how these and other HPLC instrument components are connected together and operated. The last item therein, number 8, detectors, covers such a broad and important variety of devices, that another subsection will be required to describe them. Many of the principal features of HPLC detector operation are those described for GC detectors in the introduction in Section 12.7. The student is encouraged to reread that section. Only item number 4 in Section 12.7, "Requirement for Auxiliary Gases", is peculiar to GC. An analogous feature in HPLC would be *post-column derivatization*, which will be treated later in Section 13.1.5. As in the GC chapter, hyphenated HPLC techniques are discussed separately in Section 13.1.6.

A common feature of many HPLC detectors is a small volume "**flow cell**" through which the column effluent passes and in which the detector measurement is made. When acquired from within a flow cell, even detector response data files which contain full spectral information are sometimes not classified as hyphenated HPLC techniques. Those usually require the presence of a special interface device. Typical HPLC flow rates are on the order of 1 mL/min (1000 μL/min). The analyte concentration in the effluent might be sampled on the order of once every second or two. It is necessary that the volume of effluent pass through the flow cell on this time scale. Thus flow cell measurement volumes must be on the order of 5–50 μL. Many HPLC detectors rely on emission or absorption of light, so their flow cells must have transparent entrance and exit windows. The requirement for small volumes limits the length of light paths. This is one of the factors which affect the magnitude of optical spectroscopic signals which may be obtained from a given analyte concentration [e.g., the "*b*" factor in the *abc* of Beer's Law (Section 2.4)].

13.1.4.1.   The Refractive Index (RI) Detector

RI Characteristics.   Universal; nondestructive; concentration detector; relatively insensitive; requires thermostatting and useable only with isocratic HPLC separations.

Every transparent substance will slow the speed of light passing through it, by an amount roughly proportional to its density (not mass density). As light is a wave phenomenon, this results in a bending of the light path as it passes at an angle to the interface from a less dense material to a more dense material, or vice versa. The quantity which defines this effective density is called the **refractive index (RI)**. The presence of analyte molecules in the mobile phase liquid will generally change its RI by an amount almost linearly proportional to its concentration. At low concentrations, the change is small and could easily be masked by density and RI changes arising from temperature differences or slight changes in mobile phase solvent proportions. Hence the need for careful thermostatting of an RI detector cell and the restriction to isocratic operation. Figure 13.7 diagrams one common **differential** type of RI detector. Mobile phase mixture and analyte containing eluent each pass through one of two transparent compartments separated by a tilted transparent plate. When the RI of the fluids in each compartment differ, the optical arrangement results in shifting of the source radiation beam's focus *location* on a photocell sensor. The photocell's response surface is arranged so that this shift in *position*, rather than a change in beam *intensity* (as in absorption spectrometry) results in a signal linearly proportional to the RI change produced by the analyte concentration in the cell.

The RI detector for HPLC is somewhat analogous to the TCD detector in GC. It is relatively insensitive but almost universal in its response, failing only in the unusual cases where the RI value of analyte and mobile phase are identical. Both are more readily and stably operated without "programmed" variation of mobile phase concentrations or GC oven temperatures. Response factors for most organic molecules vary over less than an order of magnitude. It is resorted to for analytes which give no response with other more sensitive or selective detectors. An example of analytes requiring HPLC with RI detection are mixtures of complex carbohydrates like sugars, which have no chromophores, fluorescence, electrochemical activity, and so on and which would be difficult, even after derivatization, to get through a GC column without decomposition. One difference between RI and TCD detectors is that RI is not very sensitive to variation in mobile phase flow rate, while the TCD is.

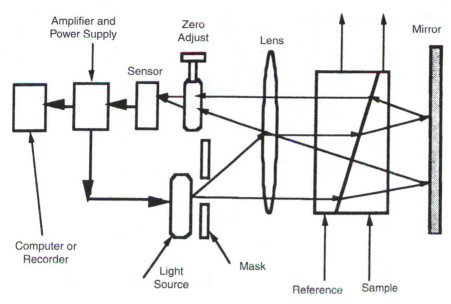

Figure 13.7 Schematic diagram of differential refractive index detector. (Katz et al., used with permission.)

13.1.4.2. The Evaporative Light Scattering Detector (ELSD)

ELSD Characteristics. Universal; destructive; mass-flow detector; sensitive; requires auxiliary gas; useable with isocratic or gradient mobile phases but with *no buffer salts*; insensitive to flow variations; more expensive than RI detector; *not* usable with mobile phase buffers.

The ELSD is the other universal HPLC detector. Instead of absorption, stimulated emission, or refraction, it uses yet another interaction of analytes with light; namely "**scattering**". Scattering occurs when light interacts with small particles of dimension comparable to its wavelength and is scattered, without change in wavelength, in a direction to the side of its original path. The particles are of a phase different from the one in which they and the light are embedded (e.g., liquid droplets or solid particles in a vapor, or solid particles in a liquid). How could we employ this principle to measure concentrations of molecules, which are *much* smaller than visible light wavelengths, when they are dissolved in liquid mobile phases? These do not scatter light. The trick is to *nebulize* the mobile phase effluent into droplets and instantly *evaporate* each of these, leaving behind a residue as a tiny solid particle of the nonvolatile analyte. The mechanism of particle formation from the droplets and the relation of scattering response to particle size is complex, but empirical calibration curves can be obtained. The response is roughly proportional to the mass of analyte eluting. As long as no buffer salts are present, so all the mobile phase can be volatilized, the ELSD is insensitive to composition variations during gradient elution. It can be orders of magnitude more sensitive than RI detection. Modern instruments illuminate the particle cloud exiting a heated droplet evaporation tube with an intense monochromatic beam from a solid-state laser. Scattered light passes through a filter tuned to pass the same wavelength to a sensitive photocell which provides the detector signal.

Figure 13.8 illustrates the three steps in the operation of an ELSD: nebulization, evaporation, and light scattering. As we have seen for ICP-OES (Section 7.1.3) and will see later for LC-MS (Section 13.1.6.1), careful design of nebulizers or spray chambers

**The ELSD detects any compound less volatile than the
mobile phase to low nanogram levels using a simple 3-Step process:**

(a) Nebulization–

Column effluent passes through
a needle and mixes with
nitrogen gas to form a
dispersion of droplets.

(b) Evaporation–

Droplets pass through a heated
'drift tube' where the mobile
phase evaporates, leaving a
fine mist of dried sample
particles in solvent vapor.

(c) Detection–

The sample particles pass
through a cell and scatter light
from a laser beam. The
scattered light is detected,
generating a signal.

Figure 13.8   Three steps of operation of evaporative light-scattering detector. (Adapted with permission from Alltech Inc.)

and thermally assisted droplet evaporation is key to many instumental interfaces designed for introducing analytes as liquid solutions. Figure 13.9 illustrates some advantages of the ELSD as a universal HPLC detector by comparing its response to the same effluent measured with other detectors.

The ELSD has a more uniform response to analytes than a UV absorption detector, whose response factors for different compounds vary widely and are wavelength dependent, and which gives no response at all to many compounds as shown in Fig. 13.9(2).

Only one of five analytes eluting here [Fig. 13.9(2)] is seen by the UV detector, and it misses the huge peak for PEG, gelatin, whereas all five are seen by the ELSD. The similarity of the mass proportionality responses enables one to immediately identify the major component.

ELSD has much better signal-to-noise response to three sugars, and better baseline stability than the RI detector. Note that the ELSD shows no solvent front peak, since the solvent droplets evaporate completely leaving no particle behind to scatter light [Fig. 13.9(3)].

Here [Fig. 13.9(4)] three derivatized amino acids or amino acid metabolites are analyzed by both LC-MS and LC-ELSD. The mass spectra in the LC-MS contain much structural information, but MS ionization efficiencies and fragmentation yields vary widely, so, TIC chromatograms convey the relative concentrations less accurately than do the ELSD peaks.

13.1.4.3.   *UV/VIS Absoption Detectors*

UV/VIS detector characteristics.   Compound-specific; nondestructive; concentration detector; compound sensitivities differ over a wide range; useable with isocratic or gradient mobile phases including buffer salts; pre-, or post-column derivatization can be used to increase number of measurable compounds.

The principles and operation of UV/VIS spectroscopic detectors are discussed at length in Section 5.2.4. The "detector" in HPLC refers to the whole spectroscopic assembly, whose design is optimized to make rapid, continuous measurements of solutions passing through a flow cell of volume much smaller than the typical sample cell or cuvette of a conventional spectrometer. The light source, optical train, dispersion device, and light detectors are similar in design to those discussed in Chapter 5, but they are often built on a

Figure 13.9 Illustrations of four advantages of ELSD vs. other HPLC detectors. (Adapted with permission from Alltech Inc.)

much smaller compressed scale, to match the size of the other HPLC components. The major differences from the large spectrometers are the much smaller measurement volume of the flow cell, the necessity to route continuous flow through it, and the need to focus light on smaller areas on the flow cell windows and extract it efficiently to send it on to the light "detector".

Figure 13.10 displays a simple UV/VIS flow cell. These are typically glass with UV transparent quartz windows bonded at each end enclosing volumes of $1-10$ µL. Note that in this design the incoming flow is directed against one window, to induce turbulent mixing which breaks up the parabolic laminar flow velocity profile across the diameter of small-bore tubes. The flow exits in a similar fashion. This "Z-design" flow cell geometry reduces laminar flow velocity variation within the cell which could contribute to band spreading.

Compounds with aromatic rings, C=C double bonds, or with functional groups containing double bonds to some heteroatoms, particularly when these structures are multiply

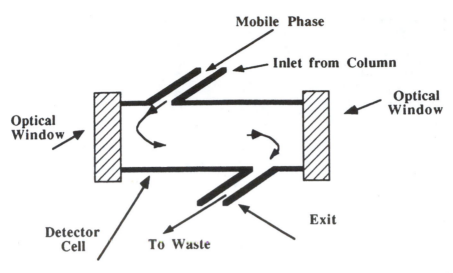

Figure 13.10 Diagram of HPLC UV-VIS detector flow cell. (Scott, used with permission.)

conjugated, may display widely varying degrees of absorption in the UV or visible regions (cf. Chapter 5). These features occur in many polar organic compounds which are separated and quantitated by RP-HPLC. In one survey, well over 70% of compounds assayed by HPLC using nonhyphenated techniques employed UV/VIS detection, and another 15% were measured with the related UV fluorescence detector. Despite not being as universal as the previously described detectors, UV/VIS and fluorescence are the workhorse detectors of HPLC. Their position is like that of the FID in GC. They possess the best combination of applicability to analytes of interest, compatibility with the widest range of mobile phases, and high sensitivity. Unlike the FID in GC, or the RI and ELSD detectors earlier described, these light absorption and emission detectors have compound response factors which vary widely, so calibration with standards of each analyte is mandatory. The optimum wavelengths for measurement will also vary with analyte structure.

    There are three main classes of UV/VIS absorbance detectors; namely, **single wavelength filter photometers**, **dispersive monochromator detectors**, and **diode array detectors**. The first of these often uses a source lamp which emits light at wavelengths specific to the excited, vaporized element employed in its design. These lines are isolated by wavelength-selective filters. The latter two use source lamps which emit over a wide wavelength range: tungsten filament for the visible and near-UV, or deuterium and xenon lamps which cover a broad emission range into the nonvacuum UV (down to around 190 nm). Many designs employ a dual beam geometry, with the reference beam and its photocell monitoring a flow cell containing only mobile phase. Sometimes the reference beam simply passes through a filter. In the first case, ratioing of the two signals can compensate for absorbance variations in the mobile phase. In both designs, variations in source intensity are compensated for. If absorbance data from a **single beam** design instrument is stored in a computer during a run, subtraction of mobile phase absorbance may be achieved post run, but stable source intensities must be maintained as source variations cannot be compensated.

    Single wavelength filter photometers. This simplest of detector designs often employs a mercury vapor lamp source, with filters between the flow cell and the photocell isolating a single line emission wavelength. The most useful such line is the intense one at

254 nm, a wavelength at which many UV-active compounds absorb strongly. Less generally useful, but sometimes more selective lines at 250, 313, 334, or 365 nm may be isolated and monitored by use of the appropriate filters. Interference filters may be used with deuterium sources to isolate specific wavelengths and measure absorbance at them.

Dispersive monochromator detector (DMD). Instead of a selected wavelength filter, a grating monochromator (Section 2.6.2.3) may be used to select a narrow wavelength band for absorption measurement. The wavelength is generally isolated *before* the beam passes through the detector flow cell. Such detectors are particularly useful when the analytes to be measured are known in advance (**target compound analysis**). The monochromator may be set to monitor the signal at the compound's wavelength of maximum absorption, and if operating under computer control, this wavelength can be varied to the optimal value for each chromatographic peak as it elutes. Conversely, if trying to use UV/VIS spectra to help identify unknown compounds, the eluent flow can be stopped while the top of the LC peak is in the flow cell, and the monochromator can be scanned across all accessible wavelengths to produce a spectrum.

Diode array detector (DAD). This detector substitutes a linear array of light sensitive solid-state diodes for the photocell. See the extensive discussion of diode array detector design and function in Section 5.2.4 under the heading of "Semiconductor Detectors". Light from the source passes through the flowcell and is then dispersed into its component wavelengths, which are directed to the different diodes of the array. The array itself is illustrated in Fig. 5.24. Each diode responds to light from a narrow wavelength range of the dispersed light, so all wavelengths are monitored simultaneously. A computer is required to read out all these parallel signals and store them in files arrayed as a function of elution time. A diagram of a diode array UV/VIS spectrometer is presented in Fig. 5.25. Figure 13.11 diagrams the components of a typical DAD designed for use in HPLC. Note that the wavelength dispersion take place *after* the beam from the source passes through the flow cell. By contrast the DMD isolates a single wavelength *before* passage through the flowcell. In either design, there is a possibility that in some cases the UV passing through

Figure 13.11 Schematic of diode array detector for HPLC. (Cazes, used with permission.)

the flowcell might excite fluorescent emission at longer wavelengths. In the DAD this would distort the observed absorption spectrum at those longer wavelengths, and in either the DAD or DMD it might add back to the absorption signal being recorded at detector and cause the absorption value to appear less. As the DAD illuminates the cell at all wavelengths, the probability of it falling victim to distortions arising by this mechanism is greater. The DAD data set can easily provide UV/VIS spectra of everything eluting during the chromatogram. If HPLC peaks are composed of a single UV-active component, the ratios of DAD responses at several different wavelengths will remain constant across the peak. If there is an impurity with a somewhat different UV spectrum within the HPLC peak, and it elutes with its peak at a slightly different retention time from that of the major component, then these wavelength response ratios will vary across the peak. This is a powerful method for detecting the presence of small impurities "buried" under the major component HPLC peak.

Infrared absorbance detectors. For completeness we should note that IR detectors for HPLC do exist. As is the case for GC-IR (Section 12.8.2) a continuous rapid response HPLC-IR detector benefits from FTIR spectral acquisition and computerized data file storage. Such spectra as may be obtained have spectral peak widths more characteristic of a liquid matrix, instead of the more information rich, sharp gas phase GC-IR peaks. The major drawback to IR detection in HPLC is that most of the commonly useful HPLC mobile phases absorb strongly in many areas of the IR spectral region. Thus HPLC-IR can be used for only a very limited set of analytes.

13.1.4.4. *The Fluorescence Detector*

Fluorescence Detector Characteristics. Compound-specific; nondestructive; concentration detector; compound sensitivities differ over a wide range; detection at right angles to and at wavelength different from excitation beam results in low background noise and thus higher signal-to-noise sensitivity than UV/VIS; useable with isocratic or gradient mobile phases including buffer salts; pre- or post-column derivatization can be used to increase number of measurable compounds.

The general principles of fluorescence spectrometry and the design of fluorescence spectrometers are discussed in Chapter 5. The simplest detectors for HPLC excite molecules with intense UV emission lines from a mercury lamp or wavelengths isolated by filters from a xenon lamp. The fluorescent emission at longer wavelength is sampled at right angles to the excitation beam, and the desired wavelength is isolated by use of a suitable filter. A more versatile design employs one tunable monochromator to select the optimum excitation wavelength from a broad band source such as a deuterium or xenon lamp, and another to select the fluorescent emission wavelength passed to the photocell or diode detector. This permits operation at optimal sensitivity for a wider range of analytes. More recent designs employ an intense monochromatic laser beam to increase the energy used for excitation. While this provides for more intense excitation, limitations of available laser wavelengths sometimes preclude excitation of analytes at their absorption maxima unless expensive tunable lasers are used. A disadvantage of the detector is that only a limited subset of analytes display substantial fluorescence emission. Its greatest advantage by comparison with the UV/VIS absorption detector is its several orders of magnitude greater sensitivity for those compounds which do display such emission. This results from the very low noise background in the measured signal. This is a consequence of measuring at right angles to the incident beam against a dark background at a wavelength longer than and isolated from that of light which might be scattered from the excitation beam. Improved sensitivity results from an improved signal-to-noise ratio.

The extreme sensitivity of fluorescence detection encourages the use of derivatization reagents to react with a nonfluorescent analyte's functional group(s) to introduce or form a fluorescent structure in the derivative. This must be done prior to injection of the sample and HPLC separation, if the derivatization reaction is slow. More often, one attempts to quickly form derivatives of separated peaks as they elute from the column prior to passing into the detector. This procedure is called **post-column derivatization**, and it will be discussed in greater detail in Section 13.1.5. One common fluorescence derivatizing reagent is dansylchloride ("**Dansyl**" = 5-dimethylaminonaphthalene-1-sulfonyl-). Dansyl derivatives of primary and secondary amines, amino acids, or phenolic compounds form readily and are highly fluorescent.

Highly sensitive detection of chromatographically (or electrophoretically) separated mixtures of fluorescent derivatives of amino acids or peptides is central to many of the instruments which support the new discipline of **proteomics**. A fluorescent derivatizing agent can derivatize the terminal nucleotide of a mixture of all fragment lengths produced by sequential removal of one nucleotide at a time from one end of a DNA polymer. Chromatography or electrophoresis can separate and elute these in inverse order of their length. The smallest will elute first. There are four different nucleotides which could be at the end of each fragment, and each such derivative has a different fluorescence emission wavelength. Use of a fluorescence detector capable of discrimination among each of these emissions allows one to identify the end nucleotide in each successively longer fragment. This then gives one the **sequence** of nucleotides (i.e., the order of the "letters" of the genetic code in the particular segment of the DNA chain being analyzed). By repeating this enough times, with hundreds of automated machines operating around the clock for years, the entire human genetic code of around 3 billion nucleotides has been sequenced! An electropherogram employing this four fluorescent wavelength detection and sequencing of a chain of close to 1000 DNA "bases" (nucleotides) is displayed in Fig. 13.36. The subdiscipline of **genomics** relies greatly upon the great sensitivity of fluorescent detection applied to small volume capillary separations.

13.1.4.5. Electrochemical Detectors (ECDs)

ECD Characteristics. Compound-specific; destructive; concentration or mass-flow detector depending on operation; compound sensitivities differ over a wide range; useable primarily with isocratic RP-HPLC including some buffer salts, pre- or post-column derivatization can be used to increase number of measurable compounds.

ECDs for HPLC are summarized in the electrochemistry chapter in Section 15.4. They are divided into two classes: **voltammetric** (subdivided into amperometric, polarographic, and coulometric), and **conductometric**. The voltammetric detectors will be described in this section under the classification of electrochemical detectors (ECDs; be careful not to confuse the acronym with the gas chromatographic ECD, where EC denotes electron capture, not electrochemical). The conductometric detectors operate on a principle different from the voltammetric ones, so they will be treated separately in the next section. One must employ RP-HPLC aqueous or polar mobile phases capable of carrying dissolved electrolytes (i.e., the analytes measured by conductometric detectors or the supporting electrolytes to suppress migration current [cf. Section 15.4]. Thus ECDs are not suitable for normal phase HPLC separations. As distinguished from conductometric detectors, ECDs respond to substances which are oxidizable or reducible at a solid electrode surface. Their signal is the current which flows through the electrode circuit when these processes occur. If flow of analyte through the detector cell and the potentials applied to the electrodes are appropriately controlled, then the rate at which the analyte is provided to the **working electrode** surface to be consumed there to liberate electrons to the electrical circuit is proportional to its amount in the cell. The usual design of electrochemical detectors employs a three-electrode configuration consisting of:

> **the working electrode**, where electrons are transferred in a redox reaction;
> **an auxiliary electrode**, to which the current flows from the working electrode; and
> **a reference electrode**, which serves as a reference against which to control and maintain the potential of the working electrode relative to the auxiliary electrode.

Many different arrangements of these three electrodes are employed in the design of ECD cells. Five are depicted schematically in Fig. 13.12.

Figure 13.12   Five HPLC electrochemical detector three-electrode arrangements. (Scott, used with permission.)

If all the analyte is consumed, integration of the current determines the total charge transferred, and such a voltammetric detector is classified as **coulometric** (from the Coulomb unit of measurement of charge). Such complete electrolysis of the analyte in the short period of peak passage through the detector cell is difficult to achieve on a standard working electrode surface. A **porous carbon electrode** can be constructed as a frit in the detector cell through which all the effluent must pass. The pressure drop through such an electrode is small, and its huge internal surface area enables complete consumption of all analytes reacting at its potential, thus enabling coulometric rather than amperometric detection. A sequence of these electrodes (each with its accompanying auxiliary and reference electrodes) can be arrayed in series along the flow path to operate at intervals of increasing potential. Such a **multielectrode array detector** will sequentially and separately produce a signal only from analytes which react in the narrow potential interval around which each electrode is set. Analytes which would react at lower potentials have already been completely consumed by prior electrodes, and those requiring higher potentials will be measured by subsequent ones. The data acquisition and multianalyte capabilities of such an array are analogous to the DAD for UV/VIS spectral responses. An advantage of coulometric detection is that if the reaction is known, that is, the number of electrons consumed or released per mole of analyte, the integrated signal quantifies the analyte without need of a calibration standard.

The most common ECD for HPLC is the **amperometric detector**. Here the analyte is completely depleted at the working electrode surface, but not in the bulk solution passing through the flow cell. An equilibrium diffusion current which is proportional to both the hydrodynamic flow parameters past the working electrode and the concentration of the oxidizable or reducible analyte is established. If steady-state effluent flow and diffusion conditions are maintained, then the amperometric current is proportional just to the analyte concentration in the cell. These conditions are more difficult to maintain during a gradient elution, so ECD detection works best with isocratic separations. There is a wide variety of modes of operation of amperometric ECDs. The working electrode may be an anode (where the analyte is oxidized). A useful material for oxidative analyses is vitreous, conductive, so-called "glassy" carbon. Reductions can also be carried out on this electrode, and various metals such as Pt, Au, and Hg are used as cathodes for reductions. These sometimes are vulnerable to build up of contamination by adsorbing material. A mercury electrode can be extruded from a capillary as a hanging drop, which can be easily replaced with a fresh, clean surface. As described in Chapter 15, the high overvoltage for H^+ reduction on Hg permits a considerable range of electrode potentials over which reductions may be induced without solvent reduction. This is the principle of polarographic analysis, and a detector with such a voltammetric working electrode is classified as a **polarographic detector**. When using such detectors, extensive degassing of the mobile phase becomes critical for the additional purpose of removing potentially interfering reduction of dissolved oxygen. A detector configured as a pair of amperometric electrode assemblies can be operated in series, with the first set to oxidize an analyte to a form suitable for subsequent reduction by the next detector in line, or vice versa. Optimization of the potentials for operation of each working electrode in such a chain can confer additional target analyte selectivity, which may also result in improved sensitivity by lowering background noise.

13.1.4.6. *The Conductometric Detector*

Conductometric Detector Characteristics.   Universal for all anions or cations in solution (i.e., a "bulk property of mobile phase" detector); nondestructive; concentration detector; compound sensitivities differ over an order of magnitude range, useable primarily with isocratic RP-HPLC without buffer salts (unless subsequent suppression column is used); premier detector for ion chromatography.

The conductometric detector is small, simple, and easy to construct. It consists of two electrically isolated electrodes immersed in or surrounding the mobile phase effluent flow, as illustrated in Fig. 13.13. All ions in the mobile phase will participate in carrying current across the cell when an *alternating potential* (AC signal) of frequency 1000–5000 Hz is imposed between the electrodes. If a constant DC potential were imposed, the ions would migrate to the electrodes opposite to their charge, concentration polarization would occur, and current through the cell would decrease to zero. With an imposed AC voltage, this polarization is suppressed, and AC current passes through the cell. The conductance is proportional to the concentration of ions, and small changes in its inverse quantity (resistance, or with AC current, more properly impedance) may be measured directly or

Figure 13.13   Three conductometric detector designs. (Scott, used with permission.)

against a reference cell in a Wheatstone bridge arrangement (cf. discussion of GC-TCD detector in Section 12.7.1 and Fig. 12.14)

The primary usefulness of the conductometric detector is for small anions or cations such as inorganic ions (e.g. Cl^-, ClO_4^-, NO_3^-, PO_4^{-3}, Na^+, NH_4^+, etc.) or small organic anions (e.g., acetate, oxalate, citrate, etc.) or cations (e.g., di-, tri-, and tetra-alkyl ammonium ions). These often possess no useful chromophore or suitable redox activity at ECDs. The difficulty with these conductometric detectors is that they are bulk property detectors responding to all ions present in the mobile phase. They cannot be made compound selective. Background ions from mobile phase buffers or the ionic eluents used in ion-exchange chromatography (Section 13.2.1) would swamp the conductance from low levels of ionic analytes. The conductometric detector first became generally useful when the ion suppressor column was developed to remove this background, giving rise to the method termed ion chromatography (discussed in Sections 13.2.2 and 15.4).

13.1.4.7.   Summary Comparison of Six Major HPLC Detectors

A comparison of the six major HPLC detectors is shown (Table 13.2).

13.1.5.   Derivatization in HPLC

In Section 12.10 we saw how in GC, analytes were reacted with derivatizing agents primarily to produce a compound with improved thermal stability and greater volatility by masking more-polar functional groups with a less-polar substituent. Sometimes this also was done to introduce atoms which rendered the derivative detectable by more sensitive and selective GC detectors. With HPLC, analyte thermal stability is not so critical, and the workhorse RP systems perform well with very polar analytes. The primary purpose for derivatization in HPLC becomes that of introducing substituents to increase sensitivity and selectivity of detection. One case where improvement of HPLC chromatography is a major goal is for analytes which exist as charged ions in solution [anions, cations, or zwitterions (containing both positively and negatively charged atoms)]. The very important group of amino acids falls into this category. Originally complete separation of complex mixtures of these required very lengthy runs on ion-exchange resin columns. Reaction of physiological amino acids with orthophthalaldehyde produces isoindole derivatives whose polarity is compatible with much more efficient RP-C18 columns, leading to much more rapid separations. Additionally these derivatives display intense fluoresence at 425 nm, which greatly improves the detection limits when monitoring with a fluorescence detector.

The second benefit in the earlier example illustrates the primary application of derivatization in HPLC. Inspection of Table 13.2 suggests that if we want to derivatize for improved selectivity and sensitivity, introduction of fluorescence should be our first choice. Table 13.3 lists a number of fluorescence derivatization agents and their target functionalities. Generally we wish to derivatize only one group at a time.

The derivatization can be carried out in solution as a last step in sample preparation. This allows for whatever temperature and reaction time may be necessary to form the derivative. This is the standard practice for forming GC derivatives, and is termed **precolumn derivatization** (or labeling). In GC we can sometimes avail ourselves of the very high temperatures in a GC injection port during flash vaporization to perform "**on-column derivatization**" (cf. Section 12.10). While this is not possible in HPLC, we have an even better option with this latter form of chromatography; namely, **post-column derivatization**. Here, the labeling agent is added to the column effluent by

Table 13.2 Comparison of Major HPLC Detectors

Detector type	RI	ELSD	UV/VIS	Fluorescent	ECD	Conductivity
Typical LOD (ng/injection)	100	1	1	0.01–0.1	0.01	1
Selectivity	No	No	Moderate	Very High	High	Low
Robustness	High	High	Excellent	High	Poor	High
Gradient elution	No	Yes	Yes	Yes	No	Yes[a]
Microsystem usable	No	No	Limited	High	High	High

[a]When used with suitable effluent suppression system.

Table 13.3 Reagents for Precolumn Fluorescense Labeling

Reagent	Analytes	λ_{ex} (nm)	λ_{em} (nm)
o-Phthalic anhydride (OPA)	Primary amines, thiols	340	455
Fluorescamine	Primary amines	390	475
Dansyl chloride (Dns-Cl)	Primary and secondary amines	350	530
9-Fluorenylmethyl chloroformate (FMOC)	Primary and secondary amines	260	305
Naphthyl Isocyanate	Hydroxy compounds	310	350
Dansylhydrazine (Dns-H)	Aldehydes and ketones	365	505
Anthryldiazomethane (ADAM)	Carboxylic acids	365	412
N-Acridinyl maleimide	Thiols	355	465

Source: Katz et al.

constant rate infusion from a separate pump, passes through a mixing chamber or coil, possibly at elevated temperature, for a precisely regulated and reproducible interval, and then passes into the detector cell. A block diagram of such a system is displayed in Fig. 13.14. When using this mode, the purpose is clearly only to convert the analyte to a better-detected form. The post-column reaction needs to be fast, as too long a mixing chamber could lead to extra-column band broadening. However, the reproducibility of the mixing and its interval yields satisfactory quantitation even if the reaction does not proceed to completion. Additionally, should the derivative be unstable, it is being quickly measured at a highly reproducible interval after formation, which would not be the case when using precolumn derivatization. If the reagent is itself fluorescent in the range being monitored, the presence of the unreacted excess will interfere with the analyte measurement. If such a reagent is used in the precolumn mode and is separable from the derivatized analyte on the HPLC column, a fluorescent reagent may be usable. When properly implemented, post-column derivatization is actually more reproducible in its yield and quantitation than is the precolumn mode. Additionally, if more than one fluorescent product may be formed in the reaction, they will all pass through the detector cell together, while the products of a precolumn procedure may elute separately in variable proportions from the HPLC column.

Derivatization can be employed to advantage with ECD detection as well. Both pre-column or post-column reactions may be employed to produce electroactive compounds analogous to the formation of fluorescent derivatives. The UV-induced reaction chamber for ECDs described in Section 15.4 and the dual electrode detector oxidation–reduction

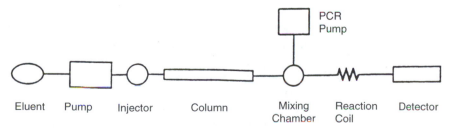

Figure 13.14 Block diagram of HPLC post-column derivatization system. (Katz et al., used with permission.)

series described at the end of Section 13.1.4.5 may be regarded as nonreagent instrumental forms of the principle of post-column derivatization.

13.1.6. Hyphenated Techniques in HPLC

Adding a second dimension to an HPLC separation lagged behind the application of this idea to GC. True HPLC columns and instrumentation were just becoming available in the early to mid-1970s while the first commercial computerized GC-MS systems came on the market. At the time of this writing, when comprehensive 2D-GC systems and their associated data handling software have just become commercially available, 2D-LC or LC-GC systems are still confined to development in individual academic and research laboratories. It is easier to interface the less-dense vapor mobile phase of GC to the vacuum conditions required for MS operation. There is no equivalent in LC of GC-GC's thermal modulator (cf. Section 12.8.3) to refocus and transfer effluent peaks from one column to another continuously with few or no moving parts. The use of a DAD detector (Section 13.1.4.3) with HPLC does enable continuous acquisition of full UV/VIS spectra of peaks eluting from an HPLC column. This can be valuable in some applications; however, its utility does not begin to match that of GC-MS or GC-IR. The range of organic compounds which present useful UV/VIS spectra is much narrower than those which yield mass spectra or IR spectra (essentially all organic compounds). The UV/VIS spectra contain much less structure than mass or IR spectra, and are much less easy to interpret for identifying unknowns or to match unambiguously with library searching programs. LC full-scan-DAD instruments are therefore often not considered true hyphenated techniques. The IR absorption of most HPLC mobile phase solvents precludes use of IR detection as a full-scan second-dimension detector.

LC-NMR: One information-rich spectral technique that is more suited to the liquid mobile phase of HPLC than to the vapor phase of GC is NMR. LC-NMR has been implemented, but it has significant limitations. To obtain interpretable spectra of unknowns, concentrations in the measurement cell must be higher than with other detectors. The cell must be smaller than the usual NMR tube, so for any but the very highest concentrations of analytes, FT-NMR acquisition is preferred, with each eluted peak being retained in the measurement cell by a stopped-flow procedure similar to that employed to increase sensitivity in GC-IR (Section 12.8.2). Expensive deuterated mobile phase solvents are required for proton NMR, which mandates the use of low mobile phase volume flow columns: narrow bore or even capillary HPLC. LC-NMR is expensive to implement and not readily available from commercial vendors at this time.

13.1.6.1. *Interfacing HPLC to Mass Spectrometry*

Given the clear superiority of HPLC over GC for separations of large polar biomolecules, analysts yearned to interface its effluent to MS instrumentation. A huge market in the pharmaceutical and biomedical research field and in clinical biomonitoring beckoned. The difficulty was the approximately three orders of magnitude greater mass of less-volatile mobile phase which must be prevented from entering the MS vacuum system (while still letting the analytes in), since no imaginable MS pumping system could remove it fast enough. Consider that a vapor is $\sim 1000 \times$ less dense than its parent liquid, and that typical mobile phase flow rates in capillary GC and packed column HPLC are both on the order of 1 mL/min. The jet separator design for GC-MS illustrated in Fig. 12.18 would only work with mobile phases of very light, fast-moving atoms or molecules like He or H_2. Chromatographers attempted to deposit HPLC effluent onto a moving conveyer belt which passed through vacuum interlocks into a pumped down chamber where the mobile phase would vaporize, leaving the analytes behind on the belt. The belt then passed through further interlocks to an even higher vacuum ion source chamber, where it was heated in hopes of volatilizing the analytes so they could be ionized and fragmented under EI-MS or CI-MS conditions as in classic GC-MS analyses. These Rube Goldberg contraptions were difficult to operate in a reproducible manner, and the prospect for volatilizing thermally sensitive large biomolecules was not much better than trying to coax them to survive passage through the high-temperature gauntlet of a GC-MS system.

The Thermospray HPLC-MS Interface. The first really successful interface did in fact look something like the GC-MS jet separator, although its mechanism was different. The HPLC effluent from a microbore column or a fraction split from a regular diameter column was coupled to a capillary tube 10–20 cm long, nestled inside a tight fitting cylindrical electrically heated tube. The exit end of the capillary pointed at a skimmer cone orifice leading to the first stage of a differentially pumped MS ion source. The arrangement was similar to the sample solution nebulizers feeding an ICP torch as described in Chapter 9 for ICP-MS instruments. If the temperature and heating rate from the heater were optimized to the mobile phase flow, it would begin to boil within the capillary close to the exit and emerge from its tip as a fine spray of droplets which would evaporate, leaving a portion of the analyte molecules to enter the ion source orifice, while the rest of the solvent vapor was pumped out of the spray chamber. Any nonvolatile salts or other material would deposit around the MS orifice entrance, requiring its frequent cleaning. Mobile phases with ions in solution often gave rise to strong analyte molecular ions, while weakly ionized solutions required an electron beam to produce EI-MS spectra. Sensitivity was not as high as could be due to the need to use low capacity HPLC columns or to split effluent flow, and because of inefficient transfer from the spray through the narrow MS entrance orifice. Nonetheless it was an effective and useful general LC-MS interface.

The Electrospray Interface for HPLC-MS (ESI). The thermospray interface retained some disadvantages, Heating the transfer capillary to the boiling temperature of the mobile phase was not as great a shock to thermally labile compounds as attempted volatilization in GC or from the moving belt interface. Still, even lower temperatures would be preferable. Larger molecules might clump together and "precipitate out" as the spray droplets evaporated. Only a small proportion might travel on the right trajectory to enter the orifice, and only a portion of these might enter as separate molecules instead of the precipitated solid particulates. The key to an improved spray interface was in the observation that analytes already existing as ions in the mobile phase or capable of having charge transferred to them from other ions in it, gave exceptionally strong MS signals.

Instead of heating the inlet capillary entering the spray chamber, in the ESI its tip is electrically charged so that as droplets emerge from it they pick up excess positive or negative charge which is transferred to them. A dry heated stream of highly purified nitrogen gas is directed at the emerging spray of charged droplets to accelerate evaporation. Two things happen as the charged droplets rapidly shrink by evaporation. Analyte and/or solvent molecules bearing charge of the excess polarity migrate to the surface under mutual repulsion. As they are forced closer together they induce a "**Coulomb explosion**" whereby the droplets fragment into multiple smaller droplets with greater surface area to support the excess charge. This process repeats in a cascade, almost instantly producing a spray of extremely small charged droplets. As the surface electric field continues to build, analyte ions begin to be individually expelled from the microdroplets as **solvated clusters** of one analyte ion embedded in a cluster of solvent molecules. This process is illustrated in Fig. 13.15(a), and it is far more efficient in converting analyte molecules to ions than is the

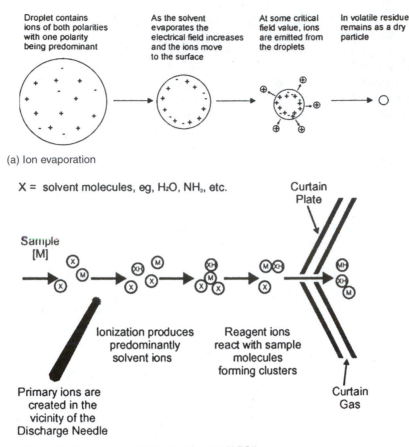

(a) Ion evaporation

(b) Atmospheric pressure chemical ionization (APCI)

Figure 13.15   LC-MS interface ionization mechanism diagrams. (a) coulomb explosion in ESI droplets and (b) reactions leading to molecular ion in APCI. (Adapted with permission from Applied Biosystems/MDS Sciex.)

thermospray interface. The spray can be directed at right angles (**orthogonal spray**) to the ion entrance orifice and skimmer assemblies instead of directly at them. By appropriately charging the entrance orifice the analyte ions can be electrically attracted to it and drawn in, while excess solvent droplets and particulate material pass by to waste. An example of such an ESI is illustrated in Fig. 13.16. The region between the entrance orifice and capillary, and the illustrated "**skimmer cone**" seen in cross-section, contains high purity nitrogen at low pressure. Collisions with this gas result in **collision induced dissociation (CID)**, which produces fragment ions from the original analyte molecular ions, in a process equivalent to the CID which takes place in the second quadrupole of a triple quadrupole MS-MS instrument (cf. Section 9.2.3.4). These solvent molecules and any others pulled in through the capillary are swept away by the CID gas flow, while the analyte molecular and fragment ions are drawn on and focused through more electric field entrance lenses into the higher vacuum region of the quadrupole mass analyzer.

Originally the electrospray interface, like the thermospray interface, was limited to use with very low flow rates of mobile phase from capillary or microbore HPLC columns or capillary electrophoretic separations. The acceleration of droplet evaporation by

Figure 13.16 Diagram of orthogonal electrospray LC-MS interface. (Adapted with permission from Agilent Inc.)

addition of heated gas flow enabled its use with higher flow rate separations. Some modern designs can even tolerate the 3–6 mL/min flow associated with high-speed separations on monolithic HPLC columns with large macropores. These versions of the electrospray interface are sometimes termed The **IonSpray**® **interface**. IonSpray or electrospray can generate multiply charged ions from compounds which have multiple charge sites, such as peptides or oligonucleotides. This enables one to observe molecular ions or high mass fragments of high molecular weight species whose masses are greater than the nominal mass limit of the MS instrument. The multiple charges produce an ion whose mass-to-charge (m/z) ratio (cf. Section 9.1) lies within the mass range of the instrument.

Note that all ionization is initiated at the entrance to the spray chamber, at the tip of the sprayer. There is no electron impact beam used to ionize and fragment analytes. Large molecules are treated gently through the whole process, and if they have natural charge centers at various pHs (as is especially the case with peptide chains comprising linked amino acids), they are ionized with great ease. This method of LC-MS interfacing, along with the one to be described next, was the long sought solution for a practical, general, and easy-to-use interface. LC-MS instruments based on these two interfaces are used by the thousands in biomedical research laboratories, and they now outsell GC-MS instrumentation. It is difficult to predict for many analytes whether negative or positive ion LC-MS operation will produce the most sensitive signal, but it is easy to shift between modes, even within a single LC-MS run, and to find out, and then to program an LC-ESI-MS instrument to shift to the optimal polarity for each analyte as it elutes.

We learned earlier in Section 13.1.4.4, that fluorescence detection of end nucleotides of capillary electrophoretic separations of DNA segments is key to the instruments which perform the DNA sequencing that supports the field of **genomics**. The LC-ESI-MS instruments, with their superb ability to characterize analogous large peptide (polymer chains of amino acids) segments of protein macromolecules, likewise are key to rapid characterization of the complex protein mixtures which reflect the various patterns of gene "expression" in biological systems. This is the foundation of the discipline of **proteomics**, whose full flowering will likely entail even more massive analytical determinations than the huge project of sequencing selected biota genomes (e.g., the 3 billion base sequence of the human genome). Achieving a full understanding of proteomics will likely occupy much of the 21st century. One of the inventors of the ESI-MS interface, Dr. John Fenn, shared the 2002 Nobel Prize in Chemistry for this contribution. He entitles his lectures on the subject: "Electrospray wings for molecular elephants".

The Atmospheric Pressure Chemical Ionization (APCI) interface. The other highly successful HPLC-MS interface design is illustrated in Fig. 13.17. This is the atmospheric pressure chemical ionization (APCI) interface. Like the ESI interface, the sample is nebulized into a spray of fine droplets, aided by and entrained in a flow of nitrogen "**sheath gas**", but without being charged by an applied potential at the spray tip. As in IonSpray, heat conducted through the sheath gas is supplied from a heated ceramic tube as the droplets are carried downstream. This gently accelerates droplet evaporation. The process occurs at atmospheric pressure. At the exit of the tube, ions are produced by the **corona discharge** of electrons emitted from the high-field gradient of a sharp metal needle charged to several thousand volts. The initial ions produced by the discharge are N_2^+, O_2^+, H_2O^+, and NO^+ from air, and through a complex cascade of charge transfer they create **reagent ions** ($[X + H]^+$ in positive mode; $[X - H]^-$ in negative mode, where X is a mobile phase solvent molecule). The reagent ions then produce stable sample ions upon collision with sample molecules. The final step in this mechanism is the same as that occurring in classical CI-MS (cf. Section 9.2.2.2) and thus gives the name for APCI. In contrast, GC-CI-MS introduces a reagent gas maintained in a confined ion source volume at pressures also near atmospheric, but its composition is different (usually methane, isobutane, ammonia, etc.) from the normal atmospheric components that serve to form the initial primary ions in LC-APCI-MS (i.e., nitrogen, oxygen, water vapor). Contrary to experience in

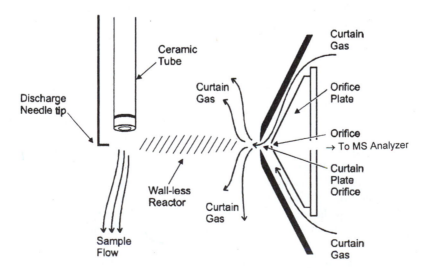

Figure 13.17 Diagram of atmospheric pressure chemical ionization LC-MS interface. (Adapted with permission from Applied Biosystems/MDS Sciex.)

GC-MS, LC-APCI-MS ion sources can be opened to the atmosphere for cleaning without venting and shutting down the rest of the MS instrument. After all, they operate under normal atmospheric conditions!

The reactions leading to the production of sample molecular ions are illustrated in Fig. 13.15(b). Note the barrier to entry of solvent vapor into the vacuum pumped portions of the MS, which is produced by outflowing N_2 **curtain gas** flowing between the curtain plate and orifice plates illustrated in Figs. 13.15(a) and 13.17. The ions are drawn through this flow by the charges applied to these plates, while uncharged molecules are swept back and away. The various charge transfer reactions forming the sample ions occur as the ions are being drawn to the entrance orifices, and this region between the corona discharge tip and the curtain plate orifice acts as an ideal "wall-less reactor". Remaining ions which do not make their way through the orifices eventually are neutralized by contact with a "wall" and play no further part in the process. While the ESI interface is especially suitable for large biomolecules with multiple charge sites, the APCI interface is more useful for smaller neutral molecules. Many analytes may be analyzed successfully using either interface. The one which produces the best sensitivity must be found by experiment. There are several modes (e.g., positive ion, negative ion, different mobile phase salts or components which affect ionization efficiency or CI ionization) which may be evaluated. Numerous ionization voltage, orifice voltage, or evaporation temperature and flow parameters may be adjusted to maximize sensitivity. Fortunately, in LC-MS one may easily continuously infuse into the MS interface a solution of a constant low-level concentration of analyte in a proposed LC mobile phase. The interface parameters may be quickly and automatically varied in sequence to find their optimal settings. Such optimization is not so quickly attained with a sequence of injections and elutions through a GC column in GC-MS. Many modern general purpose HPLC-MS instruments have interchangable ESI or APCI spray and ionization assemblies which enable rapid conversion from one mode of operation to the other without requiring venting of the MS vacuum system.

13.1.7. Applications of HPLC

In Section 12.10 we described the scope of GC. It was easy to find large important classes of compounds for which it was unsuitable, either by lack of volatility associated with size, thermal instability, polarity or acid–base functionality, or existence as an ionized form, and so on. When considering HPLC, the question is reversed to define the much smaller range of analytes for which none of its manifold variations are suitable. These would be substances which cannot be put into liquid solution or be easily retained therein. Highly volatile compounds or fixed gases are one category. GC handles these well. Speciation of forms which exist only in the solid state (e.g., metals, refractory and insoluble inorganics, cross-linked polymers of indefinitely high molecular weight), is better handled by direct spectroscopic techniques such as XRF or XRD, or by indirect techniques which employ selective dissolution prior to spectroscopic analysis, thermal analysis, and so on. We shall not attempt a comprehensive description of the range of compounds usefully characterized by the wide range of reverse phase or normal phase HPLC, adsorption chromatography, ion chromatography (Section 13.2), affinity chromatography (Section 13.3), size exclusion chromatography (Section 13.4), and the similar but nonchromatographic separations using electrophoresis (Section 13.6). The student is referred to the references in the bibliography and to the websites listed in Appendix 13.1 for discussions of thousands of applications.

Some of the variety of categories covered include:

Additives	Carbohydrates	Pesticides
Aflatoxins	Condensed aromatics	Plant pigments (cf. Tswett)
Amino Acids	Drugs	Phenols
Analgesics	Drug metabolites	Poisons
Antibiotics	Dyes	Propellants
Antioxidants	Estrogens	Sedatives
Artificial sweeteners	Herbicides	Steroids
Bile acids	Lipids	Surfactants
Blood alcohols	Narcotics	Urine extracts
Blood proteins		

13.1.7.1. Biochemical Applications in Proteomics

As an example of the utility of LC techniques in support of biochemical research we shall focus on several applications to the analysis of amino acids and peptides. **Peptides** are short polymer chains consisting of **amino acid** units. Figure 13.18 illustrates schematically a peptide structure consisting of four amino acid units. The "*backbone*" of the peptide polymer chain consists of "*peptide bonds*" between the amino end (—NH$_2$) and the carboxylic acid end (—COOH) of each of these amino acids. Even longer amino acid polymers form **proteins**, which are crucial for a vast variety of biological functions. Proteins form enzymes, which catalyze the chemical reactions in living systems. They form structural features, such as muscles, tendons, hair, and so on. They form compound selective receptors on cell surfaces which receive and "interpret" messenger compounds such as hormones which regulate biological functions. Some of these messengers are themselves amino acids or peptides. Clearly we need to characterize and measure an immense variety of these complex mixtures if we are to understand life at the molecular level. Of the thousands of possible amino acids, there are only 20 (called "**essential**" **amino acids**) which are incorporated in these biopolymers; each distinguished by a different "R-group" substituent structure. These are tabulated in Fig. 13.19. Their R-group substituents are seen to have a wide variety of polarities; some have acidic or basic functionalities, and the ability to carry both positive and negative charge. Large polymers comprising such subunits will not be separable using GC instruments. HPLC in its various

Figure 13.18 Generic structure of a four-amino acid peptide chain. (Used with permission from Phenomenex Inc.)

Figure 13.19 Structures of the 20 essential amino acids, names, three-letter and one-letter abbreviations, and molecular weights. (Used with permission from Phenomenex Inc.)

forms, perhaps using derivatization, and employing a variety of detectors to enhance measurement sensitivity and selectivity, is called for. Each essential amino acid in Fig. 13.19 is named, and three letter and one letter abbreviations for each are also listed. If amino acids 1 through 4 in the tetrapeptide in Fig. 13.18 were alanine, glycine, phenylalanine, and arginine, in that order, the peptide would be designated: "Ala-Gly-Phe-Arg" or more compactly: "AGFR" The latter nomenclature is often encountered in the output of computerized, automated instrumentation for sequencing peptide chains.

Figure 13.20 displays chromatograms from HPLC analysis of amino acids derivatized with phenylisothiocyanate to yield derivatives with strong absorption at the common Hg-lamp line of 254 nm. The single letter component abbreviations are those of Fig. 13.19, with some additional peaks such as ammonia (NH_3), phosphoserine (PH-S), phosphothreonine (PH-T), hydroxyproline (OH-P), galactosamine (Gal), and norleucine internal standard (NLE) and excess derivatizing reagent (Re). The fact that

Figure 13.20 HPLC of amino acid derivatives detected by 254 nm UV absorption: (a) 200 pmol of PTC-amino acid standard, including phosphoserine (PH-S), phosphothreonine (PH-T), hydroxy-proline (OH-P), galactosamine (Gal), norleucine (NLE, 1 nmol internal standard), excess reagent (Re), and other amino acids designated by one letter codes listed in Fig. 13.19 and (b) Analysis of a human fingerprint, taken up from a watchglass using a mixture of water and ethanol. (Courtesy of National Gallery of Art and the Andrew W. Mellon Foundation.) (Cazes, used with permission.)

Re appears as a distinct peak, indicates that derivatization took place in precolumn mode. The presence of altered or metabolized derivatives of the 20 essential amino acids is characteristic of actual biological samples. The top trace is of a standard mixture of 200 pM concentration amino acids, while the lower one is from fingerprint oils extracted by water and ethanol from a glass surface. This illustrates impressive sensitivity for HPLC with UV/VIS detection. Twenty-fold greater sensitivity is possible when using fluorescent detectors with OPA (cf. Table 13.2) derivatives of amino acids.

A major application of HPLC interfaced with MS is in protein or peptide sequencing. This can be accomplished using a wide variety of chromatographic separations followed by different kinds of mass spectrometric characterizations. We will illustrate one example of such techniques to give an impression of the power and complexity of such analyses. A very large protein might consist of a sequence of several hundred amino acids. Such a structure would be much too difficult to unambiguously characterize with a single measurement. Even if we were to know the identity and how many of each of the component amino acids were in the protein chain, there could an astronomical number of different ways they might be arranged. All these different possible proteins would have the same molecular weight. Knowing only that will not distinguish among them. Some procedures (Edman degradation) can repeatedly clip off the terminal amino acid from one end of the chain, leading to a mixture of peptide chains successively shorter by one amino acid. If we identify each of the amino acids as we repeat the procedure, we could reconstruct the order of the chain. Alternatively we might be able to use chromatography to separate the successively smaller peptide fragments, and if we could be confident of separating them all and measuring their masses, we could arrange them in order of descending mass and determine the identity of each successive amino acid by mass difference. One problem with this is that we cannot distinguish between pairs of amino acids I and L or Q and K, in Fig. 13.19, since they have identical masses. The reliability of these sequential "clipping" approaches diminishes as the length of the peptides or protein increases.

Figure 13.21 illustrates the initial steps for evaluating the sequence of amino acids in an unknown protein chain by using LC-MS data. These data might be obtained by using only the first quad of a triple quad MS instrument (cf. Section 9.2.3.4) to measure the masses of peptide molecular ions. The first step, illustrated in Fig. 3.21(a) is to break the protein into smaller peptide chains by treating it with an enzyme (which is itself a catalytic protein), which cleaves the chain only between specific pairs of amino acids occurring in a particular order from the amino to the acid end of the chain. A most commonly used enzyme is **trypsin**, which cleaves bonds on the carboxyl side of K or R (lysine or arginine) to some but not all other amino acids in the protein chain. Resulting peptide fragments are only a few dozen or less amino acids in length, and are easily separated by capillary LC. Figure 3.21(b) contrasts total ion current capillary LC-electrospray MS chromatograms of such a trypsin digest, which employ either ammonium acetate or trifluoroacetate buffers in the LC mobile phase. Note the use of buffers containing volatile salts, which are necessary to avoid buildup of salt deposits in the MS ion source. The vertical ion count scales in each chromatogram are the same, so clearly the ammonium acetate buffer gives stronger signals than trifluoroacetic acid. The separations achieved with each buffer differ. Figure 3.21(c) shows the single stage ESI-MS mass spectrum of a 12-amino acid peptide fragment from one of the chromatographic peaks in Fig. 3.21(b). This spectrum consists of ions with one through five positive charges on the peptide chain RAKMDRHVGKEK, whose molecular weight is 1453.8. Note that there are R and K amino acids within this peptide chain which were not the sites for tryptic cleavage. The number of types of peptide bonds suitable for tryptic cleavage tends to result in mixtures of peptides of the most desirable length range; neither too large nor too small to be both well separated by LC, and reliably identified by their molecular masses. The fact that the four lower m/z values in this mass spectrum are precisely $2\times$, $3\times$, $4\times$, and $5\times$ less than that of the molecular weight assures us that the spectrum arises only from one compound of this molecular weight. If the LC peak contained several unresolved peptides, we would see several such sequences—one for each peptide, and if the coeluting mixture were sufficiently simple, we could unravel the mixture mass spectrum to identify the molecular weights of each component. Note that the molecular weight alone does not uniquely identify the sequence of amino acids in the peptides. However, the tryptic digest of any particular protein will produce a well-defined series of LC-MS peptide peaks. The combination of the LC retention times or retention order of the different peptides, together with their molecular masses determined by ESI-MS, will produce a "fingerprint" which can be matched to libraries of protein trypsin digests, and will serve to identify the protein by library matching.

Adding a second stage mass spectral fragmentation by employing MS-MS (cf. Section 9.2.3.4) enables one to obtain data on the actual *sequence* of amino acids of an unknown peptide fragment. Figure 13.22 displays the collisionally induced fragmentation pattern of a molecular ion of the 14-amino acid peptide EGSFFGEENPNVAR. It has been produced by isolating the peptide by LC, isolating its molecular ion using the first stage of a triple quad MS, collisionally fragmenting this ion in the second collision cell quad, and using the third quad to scan a complete mass spectrum of all the collision fragment ions. Note that all 13 fragments produced by sequentially removing amino

Figure 13.21 LC-MS analysis of protein peptide fragments. (a) Trypsin digestion of protein to yield peptide fragments, (b) LC-MS-TIC chromatograms of pepsin digest of a protein molecule, and (c) ESI-MS spectrum of the molecular ion of separated peptide fragment. (Adapted with permission from Applied Biosystems/MDS Sciex.)

acids one at a time from the "E" end of the chain are present and identified. Their relative proportions in the collisional fragmentation spectrum vary greatly. There are also many other fragments resulting from breaking these bonds in different orders, so it would be difficult but not impossible to obtain a definite peptide sequence from this data set. However, the MS-MS pattern is fairly unique for each such peptide, and again acts as a fingerprint. Once the peptide sequence has been unraveled, its fingerprint can be filed in a database, and it can then be rapidly identified if encountered again, perhaps in a different context, by computer matching the pattern to its database entry. Thousands of such

Figure 13.22 LC-MS-MS fragmentation to sequence a 14-amino acid peptide. (Adapted with permission from Applied Biosystems/MDS Sciex.)

identified peptide fingerprints are submitted to international libraries. Evolution often selects and conserves certain peptide sequences, so one will often find a match for a peptide fragment MS-MS spectrum in the libraries, thus accelerating the process of sequencing the whole protein chain.

DNA nucleotide sequences for particular genes are often identical or very similar among species. Libraries of identified gene sequences are the foundation of genomics. Likewise the corresponding libraries of peptides resulting from particular enzymatic digests and their sequences are the building blocks of analytical proteomics. Achieving a library match of some portions of these MS-MS spectra to those of the corresponding "fingerprints" of a known protein will usually be sufficient to identify it as such without the need to completely sequence it. The particular instrumental examples described here are not necessarily the only or even the primary methods used for proteomics. They illustrate the requirements for rapid and definite identification and measurement of proteins, namely:

1. A method for cutting them specifically and reproducibly into fragments (peptides) which are small enough to be separated easily by chromatographic techniques (e.g., LC, capillary electrophoresis, 2D-TLC, electrophoresis, etc.)
2. A method for interfacing the peptides from the chromatographic separation to some form of spectroscopic detection (primarily MS, due to its superior sensitivity to underivatized peptide chains or amino acids, and the high information content of mass specta, especially when employed as multistage MS^{n}).

13.2. CHROMATOGRAPHY OF IONS DISSOLVED IN LIQUIDS

Many compounds dissolve in polar (esp. aqueous) solutions to form charged species (ions). Inorganic salts dissociate (e.g., $NaCl \rightarrow Na^{+}$ and Cl^{-}). Depending on solution pH, organic acids may dissociate (e.g., $RCOOH \rightarrow RCOO^{-}$ and H^{+}) or bases may protonate (e.g., $R_3N + HOH \rightarrow R_3NH^+$ and OH^-) as pH varies. Peptide chains containing acidic amino acid residues such as aspartic or glutamic acids, or basic residues such as lysine, arginine, and histidine (cf. Fig. 13.19) may contain both positive and/or negative charged side groups, with one or the other in excess. The net charge can be balanced at a pH corresponding to the **isoelectric point**, where the molecule has equal amounts of positive and negative charges, but these continue to interact with the solvent by charge–dipole interactions. Small molecules with single positive and negative charges at either end can exist as **zwitterions**. Ionic species generally do not partition smoothly between the classical mobile and stationary phases of reverse phase or normal phase HPLC. Special modifications, either to the mobile phase to suppress the ionic nature of the analytes, or to the stationary phase to incorporate ions of the opposite charge to attract and retain analytes, are used to facilitate the chromatography of ionic species.

Ion-pairing chromatography: One way to render ions more suitable for separation by RP-HPLC is to incorporate a **counter ion** of opposite charge into the mobile phase. If the counter ion is fairly large, with relatively nonpolar organic substituents attached, then in the polar, aqueous mobile phases characteristic of RP-HPLC, it will associate with analyte ions to form a tight "**ion-pair**" with no net charge. This entity can now equilibrate and partition between the mobile and stationary phases in a fashion not possible for the parent analyte ion. The associated ion pair will often retain or even improve the spectroscopic or other detection characteristics of the original parent ion, allowing it to be measured by the standard HPLC detectors. Positively charged organic amines can be ion paired with alkyl sulfonic acids (e.g., $C_{12}H_{25}SO_3^-$), while negatively charged, dissociated carboxylic or sulfonic acids, and various dye molecules are often paired with long chain quaternary amines [e.g., tetrabutylammonium ion $(C_4H_9)_4N^+$]. Both are **strong** acidic or basic ions, respectively (i.e., they remain dissociated in an ionic form over a wide pH range). Neither of these counterions interferes with UV detection of the analyte at wavelengths above 210 nm. **Ion-pairing chromatography (IPC)** thus extends the scope of RP-HPLC. It is less practical for use with normal phase HPLC, as the less-polar normal phase mobile phases are not as capable of maintaining the counterions in solution. The ion-exchange columns to be described in the next section have difficulties separating large ionic species, as these cannot easily penetrate the cross-linked resin networks which comprise these phases. This hinders attainment of mass-transfer equilibrium and results in loss of chromatographic efficiency. Ion-pairing RP-HPLC works better for such ions. Ionic surfactant molecules (e.g., phospholipids, phosphate detergents) have such high affinity for the ion exchange resins that they are difficult to elute from them, so ion-pairing RP-HPLC is also preferred for this class of analyte.

Ion-exchange chromatography: For many years it has been known that clays include metal ions in their mineral structure, and that these can exchange with other metal ions present in solutions in contact with the clay. **Zeolites** (sodium aluminum silicates) possess similar ion-exchange properties. Natural and artificial zeolites are employed in water softeners to remove the "hardness ions", Ca^{2+} and Mg^{2+}, by binding them while releasing a charge equivalent number of more soluble Na^+ ions. For application of ion-exchange to chromatographic separations, resin columns (cf. end of Section 13.1.1) can be modified by introducing charged functional groups into the cross-linked 3D polymeric structure. These groups will have either positive or negative charges opposite to those of the ions to be exchanged.

A particularly useful resin is that formed by the copolymerization of styrene(vinylbenzene) and divinylbenzene, and is illustrated in Fig. 13.23. By varying the proportion of divinyl benzene incorporated in the polymerization one can vary the extent of cross-linking of the polystyrene chains, which confers greater stability to the polymer structure and

Figure 13.23 Formation and structure of styrene–divinylbenzene resin polymer.

affects the porosity and tendency of the polymer to swell as it takes up mobile phase liquid. The benzene rings in the polymer structure are ideal substrates for synthetic introduction of various charged functional groups, using standard aromatic substitution reactions.

1. *Cation exchange resins.* The functional group in cation exchange resins is usually an acid. Sulfonation reactions can add the sulfonic acid group, $-SO_3H$ to essentially all the benzene rings of the resin. The weaker carboxylic acid group, $-COOH$, may be added instead. The former produces a "**strong acid**" cation exchange resin of the formula $Res-(SO_3H)_n$, where Res represents the resin polymer of matrix and $-(SO_3H)_n$ the numerous attached sulfonic acid groups. Introduction of the $-COOH$ group yields a "**weak acid**" exchange resin whose pK_a is more similar to that of benzoic acid. In each case an acidic hydrogen is attached to a functional group chemically bound to the resin. The number of such exchange sites, n, in a given weight of the ion exchange resin is a measure of its **exchange capacity**, usually expressed as milliequivalents/g (or meq/g). To maintain charge balance in the resin, the exchange and binding of multiply charged ions require interaction with multiple singly charged sites. Thus the equivalent value for such ions is calculated as m/z, where m is the number of moles, and z is the ionic charge. If the affinity of an ion such as Na^+ in solution is greater than that of H^+ for the functional group, and/or its solution concentration is sufficiently greater than the "concentration" (not to be confused with exchange capacity) of exchangable H^+ on the resin functional groups, then exchange reactions of the following type establish an equilibrium:

$$Res-COOH + Na^+ \longrightarrow Res-COONa + H^+ \tag{13.1}$$

$$Res-(SO_3H) + Na^+ \longrightarrow Res-(SO_3Na) + H^+ \tag{13.2}$$

Almost any metal ion will displace hydrogen ion from the resin under these circumstances. This provides a method for removing metal ions from aqueous solutions. The solution becomes acidic only if the metal is exchanging with a hydrogen ion. A cation exchange resin with H^+ on its functional groups is said to be in the *acid form*, and when Na^+ replaces the H^+ it is said to be in the *sodium form*. Depending on the relative affinities of the metal ions for the resin functional group, further exchanges are possible. For example Cu^{2+} will replace the sodium in a sodium form resin according to the process:

$$Res-(SO_3)_2 2Na + Cu^{2+} \longrightarrow Res-(SO_3)_2 Cu + 2Na^+ \tag{13.3}$$

Note that the cupric ion is shown to be associated with two resin functional groups, releasing two sodium ions. Half the number of moles of doubly charged copper is equivalent on the basis of charge to sodium or hydrogen. If the copper were present as singly charged cuprous ion, the equivalents would contain equal numbers of moles.

2. *Anion exchange resins*. Anions can be removed from solution using the same principles which are applied to exchange cations. The functional groups of opposite positive charge are most commonly quaternary ammonium or polyalkyl amine groups, forming either **strong** or **weak base** exchangers according to the processes illustrated in Eqs. (11.4) and (11.5), respectively

$$\text{Res}-\text{CH}_2\text{N}(\text{CH}_3)_3^+\text{OH}^- + \text{Cl}^- \longrightarrow \text{Res}-\text{CH}_2\text{N}(\text{CH}_3)_3^+\text{Cl}^- + \text{OH}^- \tag{13.4}$$

$$\text{Res}-\text{NH}(-\text{R})_2^+\text{OH}^- + \text{Cl}^- \longrightarrow \text{Res}-\text{NH}(-\text{R})_2^+\text{Cl}^- + \text{OH}^- \tag{13.5}$$

The rate of ion exchange is controlled by the law of mass action. At *equal* concentrations the greater affinity of Na^+ replaces H^+ from the cation exchange resins in the acid form. However, if a much higher concentration of strong (i.e., completely dissociated) acid is passed through the sodium form resin, it will reverse the equilibrium, displace the Na^+ ions, and convert the resin back to the acid form. In general it is usually possible to return an ion exchange resin column to a desired starting form by passing a large excess of the desired ion at very high concentration through the resin. Thus one can use an ion exchange column to retain the majority of ions of higher affinity as a less concentrated solution of them is passed through it (until most of the binding sites or functional groups have been exchanged and the exchange capacity is exceeded—a situation analogous to "overloading" in partition chromatography). Subsequent passage of a smaller volume of a very concentrated solution of the initial form, lower affinity ions restores it to that form, **regenerating** the column, and effects a *concentration* of the retained ions in the regenerating wash solution effluent.

The relative affinities (i.e., selectivities) of ions for the functional groups on the resin are governed by two rules:

1. ions with higher charge have higher affinity (e.g., $\text{Na}^+ < \text{Ca}^{2+} < \text{Al}^{3+}$ and $\text{Cl}^- < \text{SO}_4^{2-}$) and
2. the ion with the greatest size (radius) and the greatest charge has the highest affinity   (e.g., $\text{Li}^+ < \text{Na}^+ < \text{K}^+ < \text{Cs}^+ < \text{Be}^{2+} < \text{Mg}^{2+} < \text{Cu}^{2+}$ and $\text{F}^- < \text{Cl}^- < \text{Br}^- < \text{I}^-$)

By a careful choice of the ionic composition of the eluent, and the gradual adjustment of its strength (i.e., ion concentration and/or pH, which can control concentration of ions resulting from acid–base equilibrium) during elution using a controlled gradient, the components of a mixture of ions can be induced to separate just as do the components of a mixture separated by partition chromatography. The chromatographic principles discussed for the latter technique also apply. For ion-exchange chromatographic separations, the concentrations of the analyte ions being separated by a resin column need to be much less than those which would be present at values close to its exchange capacity.

The fundamental parameters controlling relative residence times of analyte or other eluent ions in the resin stationary phase or the ionic solution mobile phase are: both the relative selectivity of the resin for the ions and their relative concentrations in each phase. In contrast, in reverse phase or normal phase HPLC, selectivity is controlled by the relative polarity interactions of different analytes with both phases. Until an overload occurs this is independent of analyte concentration.

Another difference is that in ion exchange the selectivity resides in relative ion-pairing interaction strengths only in the stationary phase. The effects of mobile phase composition changes are manifested by the relative concentrations of other nonanalyte ions in the mobile phase competing for the fixed number of exchange sites on the resin.

Ion-exchange chromatography was initially developed during the Manhattan Project to preparatively separate the chemically very similar triply charged lanthanide rare earth and actinide series cations on the basis of slight differences in their ionic radii. A generally useful instrumental version suitable for quantitative analysis of small organic and inorganic ions awaited the development of a universal, sensitive detector. The conductometric detector (Section 13.1.4.6) would seem to fit the requirements, but the high background level of eluent ions present in traditional ion-exchange chromatography rendered it relatively useless.

13.2.1. Ion Chromatography

As described earlier, when attempting to use a conductivity detector for ion-exchange chromatography, the high electrolyte concentration required to elute analyte ions in a reasonable time tends to overwhelm the much lower conductivity contributed by low levels of analyte ions. This problem was eventually overcome by introducing a so-called **eluent suppressor column** to convert those ions to nonconductive, unionized molecular forms, while leaving the low concentrations of analyte ions to be measured by a conductometric detector. This mode of analytical ion-exchange chromatography goes by the name of ion chromatography. **Ion chromatography** (IC) is defined as the analysis of ionic analytes by separation on ion exchange stationary phases with **eluent suppression** of excess eluent ions. Detection in most IC systems is by the universal (for ionic analytes) conductivity detector (Section 13.1.4.6).

When *cations* are being exchanged on the analytical column to effect a separation, variable concentrations of hydrochloric acid are often employed as the eluent. In this case the suppressor column is an anion exchange column in the hydroxide form. The hydroxide on the suppressor column reacts with the hydrogen ions in the eluent exiting the analytical column to form nonconductive water, which contributes no signal to the conductivity detector. The reaction on the suppressor column is:

$$H^+(aq) + Cl^-(aq) + Res^+(OH^-) \longrightarrow Res^+(Cl^-) + HOH \ (water) \tag{13.6}$$

The cations being separated on the analytical column pass through the anion exchange suppressor column without being retained. Conversely, for anion separations the suppressor column is in the acid form of a cation exchange column. The eluent contains high concentrations of a displacing anion (e.g., OH^- or HCO_3^-) which can react with the exchangable H^+ on the suppressor column to form a largely undissociated species (e.g., H_2O or carbonic acid, H_2CO_3, respectively). Analysis of chloride anion, Cl^- by simple suppressor column IC is illustrated in Fig. 13.24(a). In this case Na^+ from NaOH eluent and NaCl analyte displaces H^+ from the head of the suppressor column, and the liberated H^+ reacts with the OH^- to form water, leaving only the chloride ion and a balancing amount of hydrogen ion to produce a conductivity signal in the detector proportional to the small amount of eluted chloride. As more NaOH in the eluent displaces H^+ in the suppressor column, the portions which are "exhausted" increase and the "suppressing" portions decrease. Eventually it becomes completely exhausted, and then must be **regenerated** by backflowing a high concentration of strong acid solution through it, to convert it back to the acid form, and then be rinsed with water to remove residual ions. The same process with a strong base is used to regenerate anion exchange suppressor columns employed in cation IC systems.

The need to periodically regenerate suppressor columns is an inconvenience, especially if one wishes to operate an ion chromatograph in an unattended, automated fashion over long periods. A **self regenerating suppressor** continually regenerates the element in an ion chromatograph which performs the function of the suppressor

Figure 13.24 Three methods of suppressing mobile phase ions in IC: (a) packed bed suppression—batch suppression, (b) self regenerating suppression—continuous regeneration, and (c) eluent generation from water. (Adapted with permission from Dionex Inc.)

column. It recycles the post-detector effluent of the ion chromatograph past an electrode which continuously electrolyzes water in the effluent to produce the desired H^+ or OH^- regenerant ions. The suppressor ion-exchange column is replaced by an array of microvolume ion-exchange membranes. Effluent from the analytical column passes over one side of these membranes, while the regeneration ions produced by the electrode flow on the other. The excess eluent ions pass through the membrane and are carried to waste in the eluent recycle stream from which the electrode-generated ions were produced. These latter ions pass through the membrane to maintain the charge balance, and they react to suppress the charge of the excess eluent ions, just as did the corresponding ions bound on a regenerated suppressor column. Figure 13.24(b) illustrates this process for the same anion IC analysis of chloride which was illustrated in Fig. 13.24(a).

The employment of an electrode system to generate H^+ or OH^- on demand in the amounts required to operate a self-regenerating membrane suppressor system led to the idea of using a similar generator to also produce the desired concentrations of these ions in the eluent streams with which analytes are eluted from analytical ion-exchange columns used for IC. For this to work effectively, it is desirable to use high-efficiency, low-capacity, ion-exchange columns for the separation. These are better suited for the low levels of analytes to be separated, and permit the use of lower eluent ion concentrations, which are easier to generate by these methods. Instead of the large porous resin beads, where diffusion in and out of the beads degrades chromatographic plate count and resolution, **pellicular packings** are used. An example of such an ion-exchange column "bead" is illustrated in Fig. 13.25. The spherical bead consists of a rigid silica sphere of 5 μm radius, coated on the surface with a 0.1 μm layer of latex to which sulfonate groups ($-SO_3^-$) are bonded. Large molecules containing multiple positively charged alkyl ammonium ion substituents bond electrostatically to these $-SO_3^-$ groups. The outer layer of positively charged ions serves as the site for exchange of anions with a mobile phase. Figure 13.24(c) displays a block diagram of an IC system which employs an eluent generating system, EG40® (Dionex Corp.), based on electrolysis of water. Such a system does away with the need for preparing and consuming eluent and regenerating solutions. The slogan used for such systems by their originator is "Just add water".

Figure 13.26 displays an IC separation of 34 small anions. Note that both inorganic halides and oxo-anions such as sulfate and bromate are measured, as well as small organic anions like formate, malonate, and phthalate. Note that pairs of ions representing different oxidation states of inorganic atoms such as nitrite/nitrate and chlorite/chlorate are readily separable. IC is particularly appropriate for speciation of inorganic elements, and is a particularly useful mode for introduction to highly specific and sensitive detectors such as ICP-OES and ICP-MS when element-specific speciation is called for.

13.2.1.1. Single-Column IC

The development of ion-exchange columns with low exchange capacity has led to a mode of IC which dispenses with the suppressor column. When using such columns, a dilute eluent solution can be sufficient to displace ions from the column. If such an eluent can

Figure 13.25 Pellicular anion exchange bead for ion chromatography. (Used with permission from Dionex Inc.)

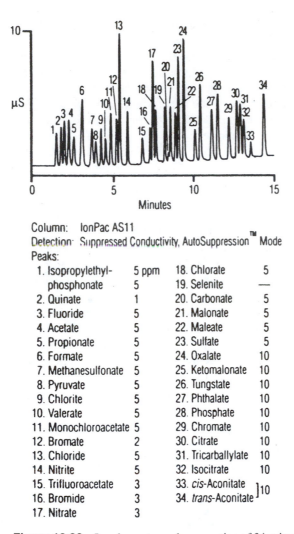

Column: IonPac AS11
Detection: Suppressed Conductivity, AutoSuppression™ Mode
Peaks:

1. Isopropylethyl-	5 ppm	18. Chlorate	5
phosphonate	5	19. Selenite	—
2. Quinate	1	20. Carbonate	5
3. Fluoride	5	21. Malonate	5
4. Acetate	5	22. Maleate	5
5. Propionate	5	23. Sulfate	5
6. Formate	5	24. Oxalate	10
7. Methanesulfonate	5	25. Ketomalonate	10
8. Pyruvate	5	26. Tungstate	10
9. Chlorite	5	27. Phthalate	10
10. Valerate	5	28. Phosphate	10
11. Monochloroacetate	5	29. Chromate	10
12. Bromate	2	30. Citrate	10
13. Chloride	5	31. Tricarballylate	10
14. Nitrite	5	32. Isocitrate	10
15. Trifluoroacetate	3	33. *cis*-Aconitate	]10
16. Bromide	3	34. *trans*-Aconitate	
17. Nitrate	3		

Figure 13.26 Ion chromatography separation of 34 anions. (Used with permission from Dionex Inc.)

be found which has a substantially lower conductivity than the analyte ions, then the conductivity of the effluent will change (increase) as the more conductive analyte ions elute through the detector. Charge balance ensures that the number of charges (or ions of the same charge) will be constant as the exchange takes place. **Single-column anion chromatography** typically employs benzoate, *p*-hydroxy benzoate, or phthalate anions as eluent anions. These are either benzene mono- or dicarboxylate anions. Adjustment of eluent pH can control average eluent anionic charge and thus the eluent strength. **Single-column cation chromatography** is performed with nitric acid as eluent for singly charged cations or doubly charged ethylenediammonium salts for doubly charged cations. While the instrumentation is much simpler, the sensitivity and linear dynamic range of single-column IC tends to be less than that of IC employing suppressor columns.

13.2.1.2. *Indirect Detection in IC*

The use of anions such as phthalate in single-column IC enables another form of detection. Many small anions will have no UV absorption in the range where the aromatic phthalate anion absorbs. If the latter is employed as an eluent, when it displaces non-UV-absorbing anions, the concentration of absorbing species passing a UV detector will decrease to the extent that the nonabsorbing analyte ions replace the eluent ions in the effluent. The effect is a drop in the previously constant level of absorbance from the eluent anions, which manifests itself as a negative peak. The peak area is proportional to the concentration of the analyte anion. A suitable UV-absorbing eluent for indirect detection in cation chromatography is cupric ion, as copper sulfate. This is called **indirect detection**, since we are not measuring the analyte directly, but instead a drop in the signal produced by absorbing eluent ions which have been replaced on an exactly equivalent basis by the "invisible" analyte ions with which they have exchanged on the column. It is the equivalence of charge exchange and balance in ion-exchange chromatography which enables this method to work quantitatively. We will see later that indirect detection can also be employed in the technique of CE, where again the analytes are ions, and can substitute for absorbing ions forming a background supporting electrolyte, which in this case is not an eluent.

13.3. AFFINITY CHROMATOGRAPHY

In Section 13.1.7 we saw how chromatography interfaced with MS could separate and characterize (even sequence) peptide chains. With sufficient information of this sort, one might even reassemble the data on fragments of a tryptic digest of a large protein molecule, and chart the sequence of amino acids in the whole molecule, thus completely characterizing and identifying this macromolecule. Complex biological systems, even at the level of particular types of cells, may contain hundreds, even thousands of proteins. How might these be separated and isolated from one another, so that this sort of analysis may proceed on a single protein? Indeed we may expect that there are chromatographic procedures or electrophoresis procedures that can separate mixtures of proteins. These may not be satisfactory for isolating one out of the hundreds that might be present, especially if they are minor components of the mixture, and we do not know exactly which peak or spot is the particular molecule of interest. Many of these biological macromolecules (not all only proteins) have been tailored by natural selection to bind very selectively to other specific biomolecules to perform their function in the organism. For example, enzymes bond selectively to their substrates, antibodies to their antigens, and receptors to their messenger hormone molecules.

If we wish to isolate a single biomolecule of this sort from a complex soup of other biomolecules, the technique of **affinity chromatography** is especially powerful. The appropriate biomolecule to which the target binds (e.g., an enzyme substrate, antigen, hormone, etc.) is isolated, and attached by a short molecular chain called a "**spacer arm**" to a chromatographic gel support material, which is permeable to solutions of the biological macromolecules. The "spacer arm" will have reactive functionalities (often carbamide groups) on either end which will attach to reactive groups on both the gel and the binding biomolecule. Dangling this target away from the gel on the spacer arm reduces the steric hinderance that the large target molecule might suffer were the binding molecule attached directly to the gel.

The steps for isolating a specific biomolecule are outlined in Fig. 13.27. The stationary phase might be a permeable cross-linked agarose polymer, or perhaps an agarose–acrylamide copolymer gel, here represented by the coiled, looping lines in the column. The binding biomolecules (termed **ligands**) attached by the spacer arm, are represented by small open circles linked to the gel polymer by a short line. A mixture containing

Figure 13.27 Four steps in performing affinity chromatography. (Katz et al., used with permission.)

three large target biomolecules is shown being applied to the column. Only one of these has a cavity of just the right shape to bind strongly to the ligand (**adsorption** step). This is done using a solvent designated as an **application buffer**, which is close to the pH, ionic strength, and polarity which the target and ligand experience in their natural environment. The other two nonbinding biomolecules are rinsed out of the column in the **wash** step using the application buffer. The next step is to remove the bound target compound in a **elution** step, using an **elution buffer**. If the target biomolecule to be eluted is very sensitive to changes in solvent condition (e.g., a protein structure that is easily denatured by changes in pH, solvent polarity, etc.) then one will not wish to break the binding by changing these buffer parameters too greatly from those of the application buffer. A more gentle elution method is **biospecific elution**, in which the elution buffer contains a similar, competing biomolecule, which *displaces* the target molecule. This can be done either by adding an agent which competes with the column ligand for the target (**normal-role elution**) or one which competes with the target for ligand-binding sites (**reversed-role elution**). While biospecific elution is very gentle, it is generally slow, and the eluted target peaks are very broad. The competing molecule may have to be removed from the eluted solution, and if analytical affinity chromatography is being performed, it must not interfere by producing a large background signal under the detection conditions being used. For analytical rather than the preparative applications for which the gentle mechanism of biospecific elution may be required, **nonspecific elution** may be employed. This employs an elution buffer of substantially different pH, ionic strength, or polarity to lower the association constant of binding to the ligand. Under such conditions, the compound of interest is eluted much more rapidly, leading to higher, narrower peaks, and consequently lower limits of detection. After the target compound is detected or isolated, the column is **regenerated** by passing the application buffer through it before injecting the next sample.

13.4. SIZE EXCLUSION CHROMATOGRAPHY (SEC)

Size exclusion chromatography (SEC) or gel permeation (or filtration) chromatography (GPC/GFC) uses a porous material as the stationary phase and a liquid as a mobile phase. The diameters of the pores of the porous material can range from 5 to 100,000 nm, which are useful for the size range of molecules having molecular weights of ~500 to 1 million. The latter penetrate the pores according to their size. Small molecules penetrate more deeply into the pores than large molecules, which frequently are excluded from fully penetrating the smaller cavities of the pores. This results in a difference in the rates at which the molecules pass down the column, the larger molecules traveling faster than the smaller molecules. Ideally there are no binding interactions between the analytes and the porous column particles, so the retention of smaller molecules is due solely to the relative ease and depth to which they can penetrate the pores by diffusion, before diffusing back again to re-enter the flow of the mobile phase. The principles of this separation are illustrated in Fig. 13.28. They are the same as those which are used for the GC separation of fixed gases using Zeolite molecular sieves or Porapak® materials described in Section 12.4.1. In the examples of Figs. 12.4 and 12.5, it is not so easy to rationalize relative retention strictly on the basis of analyte size or weight. There are some binding interactions of the analytes with the pore materials. SEC separations have less binding interactions but are not completely immune from them. Thus retention is more easily correlated with molecular size or weight.

There are generally two types of column packings: porous glasses or silicas and porous cross-linked organic gels such as dextrans, methacrylate-based gels, polyvinyl alcohol-based gels, and hydroxyethyl cellulose gels. GFC uses aqueous solvents and hydrophilic packings, while GPC employs nonpolar organic solvents and hydrophobic packings. The detectors used are based on UV fluorescence, UV absorption, or changes in refractive index. Generally a given packing material is available in a range of pore sizes. Molecules above a given molecular weight are too large to enter the pores, and therefore elute in the void volume. Any molecule above this **exclusion limit** will elute in the minimum possible time, and molecules with molecular weights above this limit cannot be separated. Molecules small enough to freely diffuse into all portions of the pores will be the most strongly retained, spending the same portion of time trapped in the pores, so they will elute at the same maximum time, called the **permeation limit**, and will likewise fail to be separated. Thus in contrast to partition chromatography, SEC or GPC have not only a lower limit for peak retention times, but also an upper limit. This is actually an advantage, since chromatographic runs will not drag on too long with late eluters. In the molecular weight range between these limits (on the order of one to two orders of magnitude) the retention time of a molecule is inversely proportional to the logarithm of the molecular weight. A packing with the appropriate pore size to cover the anticipated range of weights must be selected. Particles with different pore sizes can be mixed to accomplish separation of mixtures of a wider range of molecular weights.

If the sample is a single polymer, the chromatogram represents the molecular weight distribution. This method is very valuable for determining the molecular weight

(a)

(b)

*Schematic representation of gel-filtration chromatography. Molecules of
different size in the frame are separated according to size during migration through
the gel-filtration matrix as shown in the middle and right frames.*

Figure 13.28 Mechanism of size exclusion chromatography; (a) size selectivity as a function
of pore and analyte sizes and (b) Separation process on a Gel-Filtration Column. (Adapted with
permission from Phenomenex, Inc.)

distributions of polymers up to very high molecular weights. The actual calculation is
usually based on a comparison with a standard polymer material of known molecular
weight distribution. The procedure is also used to assess the molecular weights of biological compounds, such as proteins. GPC is also capable of separating different polymers
from each other and, under the correct conditions, mixtures of polymers can be characterized with respect to percentage and weight range of each polymer present. GPC has been
used for such important analyses as the determination of somewhat toxic aldehydes and
ketones in auto exhaust. It has also been used to separate C_{60} and C_{70} fullerenes (Bucky-
balls). Figure 13.29 illustrates a GFC separation of several large protein molecules. An
additional useful application of GFC is **desalting**; wherein salts of low molecular
weight are separated from solutions of large macromolecules in order to facilitate changing the buffer composition of their solution to one suitable for a method such as electrospray LC-MS or CE.

Care must be taken in interpreting retention data to estimate molecular weights since
the separation is based on the size of the molecule rather than on the actual molecular
weight. This means that the shape of molecule has a significant effect on the results.
For example, two molecules of the same molecular weight, one straight chained and the
other highly branched, will be retained somewhat differently because each has a different

Protein Molecular Weight Standards

Column:	**GFC-100, 30cm x 7.5mm ID, 12-15µm particles**
Mobile Phase:	50mM Tris-HCl + 100mM KCl, pH 7.5
Flow Rate:	0.5mL/min
Det.:	UV, 280nm
Inj.:	10µL

1. Albumin, 5.08mg/mL
2. Carbonic anhydrase, 2.94mg/mL
3. Cytochrome c, 1.90mg/mL
4. Aprotinin, 3.02mg/mL
5. Cytidine, 0.18mg/mL

Figure 13.29 Gel filtration separation of five protein molecules. (Used with permission from Supelco, Inc.)

penetrating power into the column particle pores. This must be taken into account when selecting standards for calibrating the molecular weights.

An advantage of SEC is that it can be carried out at room temperature and the samples are not decomposed because of exposure to high temperature. This is especially important when dealing with some very high molecular weight compounds which are easily broken into two or three fragments, resulting in great changes in apparent molecular weight. It gives a limited range of retention times which facilitates rapid sample throughput, and the peaks are narrow and therefore give good sensitivity. Since analyte interaction with the particles is minimal, sample loss and deactivation of the column is also minimized. A disadvantage of the short time range for peak elution between the exclusion and permeation limits is the limitation this imposes on the number of peaks that can be accommodated. The method generally requires differences on the order of $\pm 10\%$ in molecular weight for sample peaks to be resolved. Consequently similar sized variants of large biomolecules (e.g., isomers) are better analyzed by affinity chromatography, if the appropriate ligand can be obtained and bound in an affinity chromatography column.

13.5. SUPERCRITICAL FLUID CHROMATOGRAPHY (SFC)

A compound such as CO_2 is a gas at normal temperature and pressure (NTP) and, like all gases below a certain critical temperature, further increasing the pressure results in the formation of a liquid. Above this critical temperature, increasing the pressure increases the density of the fluid but a distinct transition to and boundary with a liquid phase never forms. At the critical temperature and pressure, the density of the gas phase and the liquid phase are the same. This state is neither a true liquid nor a true gas but is a **supercritical fluid (SCF)**. Use of a compound in this fluid state as a chromatographic mobile phase provides different properties than when it is used as either a gas or a liquid in GC or HPLC, as shown in Table 13.4. To date, most attention has been focused on CO_2, C_2H_6, and N_2O, with critical temperatures of 31°C, 32°C, and 37°C, respectively. The temperatures and the necessary pressures can be accommodated in conventional GC instruments if the effluent end of the column is maintained at a sufficiently high pressure above ambient. This can be achieved by the addition of suitable flow restrictors at the column end. An important property of SCFs is their ability to dissolve poorly volatile molecules. Certain important industrial processes are based upon the high solubility of organic species in supercritical CO_2. For example, CO_2 has been employed for extracting caffeine from coffee beans to produce decaffeinated coffee and for extracting nicotine from cigarette tobacco. Supercritical CO_2 readily dissolved n-alkanes containing up to 22 carbon atoms, di-n-alkylphthalates with the alkyl groups containing up to 16 carbon atoms, and various polycyclic aromatic hydrocarbons.

13.5.1. Operating Conditions

To adapt equipment to supercritical applications, it is only necessary to provide an independent means for controlling the internal pressure of the system. Such systems are available as adaptations of commercial GC or HPLC instrument designs.

13.5.2. Effect of Pressure

Pressure changes in supercritical chromatography affect k'. For example, increasing the average CO_2 pressure across a packed column from about 70 to 90 bars (1 bar = 0.987 atm) the elution time for hexadecane from about 25 to 5 min. This effect is general and has led to the type of gradient elution in which the column pressure is increased linearly as the elution proceeds. The results are analogous to those obtained with programmed-temperature in GC and solvent-gradient elution in LC.

Table 13.4   Typical Values for Gases, Liquids, and SCFs

Property	Gas (NTP)	SCF	Liquid
Density	1×10^{-2}	0.3	1.0
Diffusion coefficient (cm^2/s)	3×10^{-1}	10^{-3}	1×10^{-5}
Viscosity (g/cm/s)	2×10^{-4}	2×10^{-4}	2×10^{-2}

13.5.3. Stationary Phases

So far, most SFC has utilized column packings commonly used in LC. Capillary SFC has been performed with stationary phases of organic films bonded to capillary tubing. Because of the low viscosity of SCFs, long columns (50 m or more) can be used. This results in very high resolutions in a reasonable elapsed time. This is the key advantage of SFC: it combines the low resistance to flow of GC, enabling longer columns with more theoretical plates, while the mobile phase density approaches that of a liquid, and thus introduces a liquid-like solvating power that is totally lacking in GC gases like H_2, He, or N_2.

13.5.4. Mobile Phases

The most commonly used mobile phase for SFC is CO_2. It is an excellent solvent for many organic molecules, and it is transparent in the UV range. It is odorless, nontoxic, readily available, and inexpensive when compared with other chromatographic solvents. Carbon dioxide's critical temperature of $31°C$ and its pressure of 72.9 bar at the critical point permits a wide selection of temperatures and pressures without exceeding the operating limits of modern chromatographic equipment. Other substances that have served as mobile phases for SFC include ethane, pentane, dichlorodifluoromethane, diethyl ether, and tetrahydrofuran.

13.5.5. Detectors

The most widely used detectors are those found in both GC or LC (i.e., UV absorption and fluorescence, RI, flame ionization, and MS).

13.5.6. SFC vs. Other Column Methods

As shown in Table 13.4, several physical properties of SCFs are intermediate between gases and liquids. Hence, SFC combines some of the characteristics of both GC and LC. For example, like GC, SFC is inherently faster than LC because of the lower viscosity and higher diffusion rates in supercritical fluids. High diffusivity, however, leads to band spreading, a significant factor with GC but not with LC. The intermediate diffusivities and viscosities of supercritical fluids result in faster separations than are achieved with LC together with lower band spreading than encountered in GC.

Figure 13.30 compares the performance characteristics of a packed column when elution is performed with supercritical CO_2 vs. a conventional HPLC mobile phase. The rate of elution is approximately $4\times$ faster using SFC. The roles of the mobile phase in GC, LC, and SFC are somewhat different. In GC, the mobile phase serves but one function−band movement. In LC, the mobile phase provides not only transport of solute molecules but also influences selectivity factors. When a molecule dissolves in a supercritical medium, the process resembles volatilization but at a much lower temperature than under normal circumstances. Thus, at a given temperature, the vapor pressure for a large molecule in an SCF may be orders of magnitude greater than in the absence of that fluid. As a result, important compounds such as high molecular weight compounds, thermally unstable species, polymers, and biological molecules can be brought into a much more fluid state than that of a normal liquid solution of these same molecules. Interactions between solute molecules and the molecules of an SCF must occur to account for their solubility in these media. The solvent power is thus a function of the chemical composition of the

Figure 13.30   Comparison of SFC and HPLC separations (R_t in min).

fluid. Therefore, in contrast to GC, there is the possibility of varying retention substantially by changing the mobile phase (i.e., by adding modifiers to the SFC mobile phase, as is done in HPLC). The solvent power of an SCF is also related indirectly to the gas density and thus the gas pressure. One manifestation of this relationship is the general existence of a threshold density below which no solution of solute occurs. Above this critical value, the solubility increases rapidly and then levels off.

13.5.7. Applications

SFC can handle larger molecules than GC with higher efficiencies than LC. It was also considerably easier to interface with mass spectrometers than was its liquid counterpart prior to the advent of ESI and APCI-MS interfaces. The high chromatographic resolution achievable with coated capillary columns was particularly significant. Effluent collection was especially easy when using CO_2, as the mobile phase could be readily and gently removed from collected fractions by simply opening them to atmospheric pressure (bleeding the gas off slowly to avoid the CO_2 fire extinguisher effect!). This feature has made **preparative SFC** one of the major present applications of the technique, while **analytical SFC** is less often practiced.

13.6. ELECTROPHORESIS

In Section 11.4 we described the mechanism of the chromatographic separation process using an extended analogy to people stepping on and off a moving slideway. We mentioned that if different classes of people stayed always on the slideway but walked forwards or backwards at different constant rates they would separate themselves along the slideway and exit it at different times. If molecules could be induced to behave in a column in a similar fashion, the process would not be chromatographic, as they would not be transferring back and forth between a stationary and mobile phase. The separation process would, however, appear very much like chromatography, with a mixture being introduced at one end of a column with flowing fluid, and the components exiting the other end as separated, gaussian shaped peaks, to be detected and quantitated by the same type of detectors employed in HPLC.

13.6.1. Capillary Zone Electrophoresis (CZE)

If molecules in solution carry a charge (i.e., if they exist as ions in solution) application of a sufficiently high voltage between two electrodes immersed in the solution will cause positively charged cations to migrate (i.e., move through the solvent in response to the electromotive force or voltage) toward the cathode. Likewise anions will migrate toward the anode. Such a **migration** of ions in solution under the influence of an electric field is termed **electrophoresis**. In polarography (Chapter 15) such a migration current for those analytes which exist as ions is undesirable, as we wish to measure only the diffusion current, so the migration current is suppressed by swamping it with a large excess of supporting electrolyte which does not undergo a Faradaic process (oxidation or reduction) at the electrode. In the ion source of a mass spectrometer, operating under near-vacuum conditions, electric fields of several volts are sufficient to pull ions through the entrance lenses, and then somewhat higher voltages accelerate them into the analyzer section. At a constant accelerating voltage in the vacuum, the velocity of the accelerated ions is precisely inversely proportional to their masses and directly proportional to their charge, there being no other factor or force involved than the inertia of the ions.

The situation for ions immersed in solution and migrating in an electrophoresis experiment is radically different. There is intimate interaction of the ions with the surrounding solvent as they move through it, resulting in viscous frictional drag forces which act almost instantaneously to limit the motion of the ion to a steady speed through the liquid. This speed is that at which the retarding frictional force just balances the applied electromotive force (voltage). To attain useful electrophoretic migration speeds electrical fields of thousands of volts/meter must be applied. It is the *effective size* of the ion, *not its mass*, that governs the frictional force resisting the migration of the ion. The balance of the frictional force and the accelerating force at a given migration velocity is given by Eq. (13.7)

$$qE = fu_{ep} \tag{13.7}$$

where q is the charge on the ion (in C), and E is the applied field (in V/m), while f is the frictional coefficient and u_{ep} is the ion's velocity. Rearranging Eq. (13.7) to Eq. (13.8) we obtain:

$$u_{ep} = \left(\frac{q}{f}\right)E = \mu_{ep}E \tag{13.8}$$

where μ_{ep}, **the electrophoretic mobility**, is defined as the constant of proportionality between the speed of the ion and the electric field. This value is proportional to the charge on the ion and inversely proportional to the frictional coefficient. That coefficient "f" is proportional to the **effective hydrodynamic radius** so the electrophoretic mobility is seen from Eq. (13.8) to be inversely proportional to this "size" parameter of the ion. Note that this effective radius is not calculated from measurements of the molecular size, but is experimentally determined by measurements of μ_{ep}. A diprotic acid could bear either one or two negative charges depending on solution pH. Although the charge is exactly doubled, the effective size of the ion with its tightly bound, associated water molecules will vary with the degree of acid dissociation, so the mobility difference will not necessarily be exactly a factor of 2. Mobilities must be measured by experiment, as they are not subject to precise calculation. They will vary as the viscosity of the buffer solution changes, so they are specific for each buffer composition.

We can separate ions of different molecules by electrophoresis if they have different mobilities. Figure 13.31 is a schematic diagram of a CZE instrument. Two reservoirs connected by a length of capillary tubing are all filled with a buffer solution whose pH and electrolyte strength are adjusted to optimize the electrophoretic separation. A high voltage (typically 30–50 kV) is applied across the capillary between an anode and cathode in each of the reservoirs. Imagine that a narrow band of buffer solution containing a mixture of ions to be separated were somehow placed at the center of the capillary. Charge balance would require equal numbers of anions and cations. When the voltage was applied, anions would migrate to the anode, and cations to the cathode. Ions of greater μ_{ep} would move faster, and if the mobility differences among the components were great enough, they could be separated. Migration velocities are fairly slow, typically in the cm/min range, and if a long capillary were used, the experiment might take a long time. To measure both anions and cations one would need a detector at each end, and ions of each charge would have only half the capillary length to achieve separation. Note that the figure displays only one detector. If we introduced the sample mixture band at the left end of the capillary, and arranged for buffer to flow through it from left to right at a speed exceeding that of the greatest migration velocity of anions, then the cations would pass the detector first in order of their mobilities (fastest first). Following the cations, the anions migrating in a direction opposite to the faster buffer flow would pass the detector in inverse order of their mobilities. One detector gets them all, the sample is easily introduced at other end of the capillary, and we can choose the buffer flow to speed up or slow down the period of "elution" depending on how long we need the ions to remain migrating in the capillary to

Figure 13.31 Schematic of capillary zone electrophoresis apparatus. (Cazes, used with permission.)

achieve a separation. One might be able to cause the buffer to flow toward the detector simply by raising the anode reservoir. We will see that long (50–100 cm) narrow bore (50–100 μm) fused silica capillaries are generally used for CZE. To obtain a desired buffer flow rate might require pressurizing the system. Adjustment and precise control of both voltage and pressure drop across the capillary would make for a more complex instrumental design, sacrificing the elegant simplicity of the system diagrammed in Fig. 13.31. It turns out that there is a different and better way to induce buffer flow through the capillary.

Toward the end of Section 11.5 we saw that one of the many useful features of fused silica capillary tubing was the ability of its surface silanol groups to lose H^+ ions and form a chemically bound layer of negatively charged groups in aqueous solutions at pH above 2. When this happens, these excess bound negative charges are partially neutralized by tightly binding positive ions from the buffer electrolyte used in the CZE analysis. The rest of the excess negative surface charge is neutralized by the presence of an excess of cations in the **diffuse double layer** very close to the silica surface. Polarizing (adding charge to) an electrode in electrochemical techniques results in the establishment of both compact and diffuse electrical double layers at the electrode surface. The diffuse double layer cations are not tightly bound and can move in response to the applied electrophoretic voltage. Because they are in excess in this layer, more buffer cations are flowing toward the cathode than buffer anions towards the anode. This excess cationic flow in this layer drags the buffer solution as a whole towards the cathode. This motion is called the **electroosmotic flow** (EOF). Like the phenomenon of "osmotic pressure" established across semipermeable membranes, it arises from an imbalance of particles across a boundary, but the flow ends up being parallel to the boundary instead of across it, since the motion of interest is that of an excess of ions responding to a voltage gradient imposed along instead of across the surface. The thickness of the diffuse double layer which mediates this flow is inversely proportional to the concentration (more precisely the ionic strength) of the buffer. Thus changing the electrolyte composition of the buffer will affect and allow control of the velocity of the EOF. Likewise lowering the pH of the buffer will decrease the dissociated silanol surface charge density and decrease the velocity of the EOF. The origin of the EOF is illustrated in Fig. 13.32. Note that one

Figure 13.32 Diagram showing origin of electroosmotic flow (EOF). (Cazes, used with permission.)

siloxyl group lacks a tightly bound counter cation. It is the excess of mobile countering cations in the layer just above the "plane of shear" that provides the "handle" for the applied voltage to move the buffer through the column. This mechanism is sometimes referred to as **electrokinetic pumping**, in contrast to the pressure driven, **hydrodynamic flow** produced by pumps in HPLC systems. For the EOF we can write an equation similar to Eq. (13.8) relating the velocity of the electroosmotic flow (u_{eo}) to the electric field E

$$u_{eo} = \mu_{eo}E \qquad (13.9)$$

where μ_{eo} is defined as the **electroosmotic mobility**. We can then define and measure an **apparent mobility** of an ion, which is the sum of μ_{ep} and μ_{eo}:

$$\mu_{app} = \mu_{ep} + \mu_{eo} \qquad (13.10)$$

μ_{ep} is positive for cations and negative for anions, so μ_{app} will be greater or less than μ_{eo}. The apparent mobility of an ion and the electrophoretic mobility can be calculated according to:

$$\mu = \frac{L_d/t_x}{V/L_t} \qquad (13.11)$$

where L_d and L_t are the length of the column from injector to detector and the total length of the column, respectively. Note in CZE that the detector usually is before the end of the capillary, as illustrated in Fig. 13.31. V is the total voltage across the capillary. To measure μ_{app} the time for the ion to migrate from the injector end of the column to the detector is substituted for t_x. To measure μ_{eo} one substitutes in the equation the time for a neutral species capable of giving a detector response (e.g., a UV absorbing compound) to make the same journey. From these two measurements the electrophoretic mobility μ_{ep} can be calculated by taking their difference.

There is an additional important distinction between these two modes of inducing liquid flow within a narrow capillary. Figure 13.33(a) illustrates the profile of flow velocity across a capillary driven by EOF. Except at points very close to the inner wall, it is uniform. In contrast, pressure-driven hydrodynamic flow is affected by shear forces varying with distance from the center to the wall, resulting in a parabolic, laminar flow velocity profile, Fig. 13.33(b). Band spreading resulting from analyte diffusion across this hydrodynamic flow profile will be absent in electrokinetically pumped systems. The friction with solvent molecules of ions being pulled through them by the applied voltage, which creates both the EOF and the electrophoretic separation, can raise the temperature of the buffer by a process called **Joule heating**. This is closely analogous to the heating produced by a current of electrons passing through a resistor in an electrical circuit. In CZE this heat is dissipated by flowing the short distance to the wall of the capillary. If a significant temperature gradient builds up across the radius of the capillary, the higher central temperature will cause the buffer viscosity to become lower at the center than at the capillary wall, across which the heat must be dissipated. This would act to disrupt the uniform EOF flow profile, making it resemble the pressure driven laminar flow profile, and introduce band broadening. The high thermal conductivity of fused silica helps alleviate this, and placing the capillary in a cooling bath may retard the onset of the problem, but this phenomenon places a limit on the inner diameter of capillaries that can be used to full efficiency. Typical ranges are $25-100$ μm, while tubing a millimeter or more in diameter could not escape the bad effects of Joule heating.

Although we have emphasized that the fundamental principle of electrophoretic separation is different from that of chromatography, there are many similarities in the

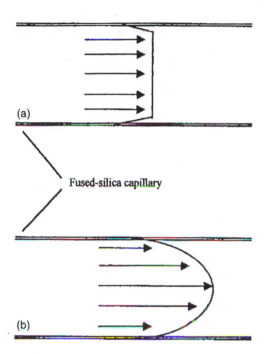

Figure 13.33 Comparison of electroosmotic and hydrodynamic flow profiles: (a) uniform electro-osmotic flow (electrically driven) and (b) laminar, parabolic, hydrodynamic flow (pressure driven). (Cazes, used with permission.)

appearance of the separation. There is sample injection, flow of a "carrier fluid" along a column, similar detectors employed near the end of the column, and separated compounds being displayed as peaks emerging at characteristic times. Similar formulas can be used to describe the resolution of separations, and to characterize the separation efficiency of the system in terms of theoretical plates. It is instructive to revisit the Van Deempter equation [Eq. (11.10)] for evaluating the impact of various processes on the height equivalent to a theoretical plate (H) described in Section 11.8. Recall that minimizing this maximizes efficiency (i.e., permits more plates for a given column length).

$$H = A + \frac{B}{u} + Cu$$

Since CZE is an open tubular technique there is no multipath A term. Since CZE is actually not a chromatographic technique, there being no stationary phase, there is no mass transfer C term. Recall that this is actually a term which combines several processes; namely, transfer in the mobile phase across the laminar flow velocity gradient, which we have shown is suppressed in a properly operated CZE system, and transfer to and from the stationary phase, of which there is none in CZE. The only remaining source of band broadening is the longitudinal diffusion term B. This is proportional to the diffusion coefficient of the analyte in the carrier fluid, and varies inversely with the size or mass of the analyte molecule. Consequently very low values of H and correspondingly high plate counts are obtained for large molecule separations. Plate counts up to 500,000 are routinely achieved for large molecule separations by CZE, which is an order of magnitude better than those of which HPLC is capable. Biological macromolecules such as large peptides or proteins, which have multiple sites for bearing charge as a result of adding or abstracting H^+ by pH variation, may often be separated with great efficiency by CZE, provided a pH can be found at which one charge predominates, and which is compatible with good electrophoretic operation. It can be demonstrated that the number of plates in a CZE separation of a given compound is related only to the voltage V across the capillary, the electrophoretic mobility, μ_{ep}, and diffusion coefficient, D, of the ion, according to Eq. (13.12):

$$N = \frac{\mu_{ep} V}{2D} \tag{13.12}$$

Contrary to our experience with chromatographic separations, N is independent of column length. Clearly we would like to operate at higher voltages, but we will be limited by the onset of Joule heating, which is proportional to the field E. For a given applied voltage V, E is proportional to V/L. Thus we must increase column length and analysis time to keep E below the value at which Joule heating degrades resolution. This is the reason for using longer columns. One can see that larger ions, which have lower values of D, will achieve greater plate counts. This is why CZE is often superior to HPLC for separations of charged biological macromolecules.

If the absolute value of μ_{ep} is less than μ_{eo} for all anions and cations in a CZE separation, then they may all be observed whether the applied voltage directs EOF either to the cathode or anode. If not, then those migrating in a direction opposite to the EOF at greater than its velocity will never reach the detector. It may then be necessary to do a second electrophoretic separation with the electrode polarities reversed to measure these. An alternative is to reverse the direction of the EOF while maintaining the same electrode polarity. To do this, one must replace the siloxyl negative charges fixed to the column wall with bound positive charges. This may be accomplished by adding a cationic **surfactant** such as cetyltrimethyl ammonium ion $[n\text{-}C_{16}H_{33}N(CH_3)_3^+]$. The long C_{16} hydrocarbon tails attract one another and form a bilayer sandwich structure with the hydrophilic quaternary ammonium cation portion on the outside and the nonpolar organic chains on the inside. One layer of positive charges is fixed by electrostatic attraction to the negative siloxyl groups on the silica wall, and the buffer is presented with the other layer of positive charges. Now excess ions in the diffuse double layer will be anions instead of cations, and the EOF will move in the opposite direction. This is illustrated in Fig. 13.34.

The ability of some organic molecules to selectively bind to the silica surface in this fashion can become a problem if they are in fact the analytes one is trying to separate and measure by CZE. Such unintended "wall effects" superpose an adsorption

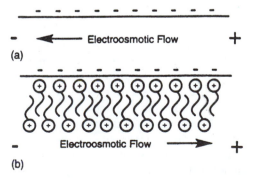

Figure 13.34 Use of cationic surfactant to reverse electroosmotic flow: (a) normal EOF toward cathode (no surfactant) and (b) reversed EOF toward anode (cationic surfactant bilayer). (Katz et al., used with permission.)

chromatographic mechanism on the electrophoretic separation. If the binding is very tight, one may initially fail to see some components migrate to the detector, only to have them begin to appear as peaks with degraded resolution and shifting migration times, as successive injections result in saturation of such "binding sites". For some applications it is necessary to pre-coat the capillary with a covalently bonded hydrophilic polymer to suppress this interference. Such complexities make developing CZE separations less simple than one might naively predict.

13.6.2. Sample Injection in CZE

In capillary GC, samples on the order of several microliters are propelled from a syringe through a septum into a complexly designed injector port where they vaporize, expanding in volume nearly a 1000-fold, and then must either be split or reconcentrated in the head of the capillary. In capillary HPLC, samples are injected with a syringe into a sample loop injector which must be designed to function at high pressures. In CZE things can be much simpler. The injection volumes are much smaller, on the order of nanoliters. One wishes to introduce a plug or zone of analytes dissolved in aqueous buffer into the injection end of the capillary. This can be done a two ways: hydrodynamic injection or electrokinetic injection. For **hydrodynamic injection** one often simply removes the injection end of the capillary from its buffer reservoir, and places it briefly in the sample buffer, which is elevated sufficiently above the detector end reservoir, and hydrostatic pressure causes sample solution to flow into the capillary. The volume of the sample flowing in will be given by:

$$\text{Volume} = \frac{\Delta P \pi d^4 t}{128 \eta L_t} \tag{13.13}$$

where ΔP is the pressure difference between the ends of the capillary; d, its inner diameter; L_t, its total length; t, the time it spends in the sample solution; and η, the viscosity of the sample solution. If the capillary is both too narrow and/or too long, it may be necessary to apply additional pressure at the injection end or suction at the detector end in order to introduce a sufficient volume in a reasonable time. For **electrokinetic injection** one places the injector end of the column into the sample solution and applies a voltage across the column (for this we will need to introduce an electrode into the sample solution). The sample ions will enter the column both by migration (at rates which will vary with their electrophoretic mobilities) and by entrainment in the EOF (at a constant rate for all ions). This difference in sample amount will cause problems in quantitative CZE work, so hydrodynamic injection is preferred in that case. In the capillary mode called capillary gel electrophoresis (discussed in Section 13.6.4) the gel in the capillary is much too viscous to employ that injection technique, so electrokinetic injection must be used.

To introduce a sufficient sample volume into a narrow CZE capillary, one might produce a band of sample much longer than the optimum bandwidth which would be implied by the number of plates calculated from Eq. (13.12) operating on an ideally narrow injected band. This is an example of an "extra column band-broadening effect". It would be desirable to have a mechanism for "focusing" the injected sample mixture into a very narrow band, as happens in chromatographic injections when the analytes riding the mobile phase into the column are arrested at its head by the stationary phase, when k' is sufficiently high. Just such a focusing effect can be achieved in CZE if one employs a buffer electrolyte concentration in the sample which is much *lower* than that in the separation buffer. The conductivity of this more dilute sample plug is much lower (thus its resistance is higher), and since the ion current through the system must be constant, by Ohm's Law ($E = IR$), the field across the sample plug must become proportionally larger. The increased field causes ions in the sample plug to migrate more rapidly to the end nearest to the electrode of opposite charge. When they reach this interface, they slow down and continue to move with the velocities characteristic of the separation buffer. This process is called **sample stacking**, and it is illustrated in Fig. 13.35 for anions in a sample plug. Cations would "stack" in a similar fashion at the opposite end of the plug. The result is to quickly produce two narrow bands of either cations or anions, separated by the width of the sample injection plug, which can then experience the full separation capability of the CZE process.

13.6.3. Detection in CZE

In contrast to detectors used in GC and HPLC, those in CZE usually are placed at some point before the end dipping into the capillary reservoir, as indicated in Fig. 13.31. Optical detectors are similar in design to HPLC detectors, but with modifications of the detector cell portion to fit the much smaller peak volumes encountered in CZE. For UV absorption or fluorescence detectors, the detection cell is often a portion of the capillary column itself. The polyimide protective coating of a short section of the fused silica column is removed, and the beams of the spectrometric detectors are focused with condensing optics on this small volume. This has the advantage of using a detector cell volume which is similar to the very small band volume of highly resolved CZE peaks. The very short path length for the beam across the diameter of the CZE capillary decreases sensitivity. This may be improved by bending the capillary at right angles or fitting a capillary length between two 90° fittings for a short distance approximating the bandwidth of a narrow peak, and introducing the illuminating beam lengthwise down the axis of this segment through optically flat windows. Another approach to increase the light path length is to coat the outside of a short length of the capillary with reflective silver

Figure 13.35   Sample stacking of anions in sample plug toward anode. (Katz et al., used with permission.)

beneath the protective polyimide coating. Light entrance windows are cleared at either end on opposite sides of this segment. Light enters one window at a slight angle to perpendicular to the capillary such that it is reflected multiple times back and forth across the capillary before exiting the other window the detector.

Miniature amperometric or conductometric detectors can be introduced into the capillary, by plating the appropriate electrodes in the inside of the capillary walls. It is often necessary to employ a porous glass or graphite joint between the capillary containing the detector electrodes and the end of the capillary in the detector end electrode reservoir, in order to isolate the detector from the high voltage used to power the electrophoretic separation.

The extremely low flow rates of buffer in a CZE capillary separation make it especially easy to interface to a mass spectrometer by using an ESI interface (cf. Section 13.1.6.1) for detection of large multiply charged biological macromolecules. Instead of a CZE detector end reservoir with its electrode, the voltage at that end is applied through a metallized coating on the inside and outside of the capillary end. This voltage also functions to charge the ions and droplets that exit the tip of the capillary end, and to form the "electrospray" with its process of coulomb explosion and charge transfer, yielding molecular ions suitable for MS analysis.

Because ions in CZE migrate past spectrometric or electrochemical detectors at different rates according to their electrophoretic mobilities, they pass through the detector zone at different rates, unlike chromatographic analytes, which enter the detector cell from the column at the constant rate of the mobile phase velocity. For concentration detectors (cf. Section 11.4), the peak areas are therefore independent of retention or elution time in chromatographic systems. However, this is **not** true in CZE. The same concentration of material giving the same response will pass more slowly through the detection region of the capillary if it has a lower migration velocity, and therefore will integrate over time to a larger peak area value. If one is quantitating against an IS of different mobility, and separation conditions and "apparent migration times" shift between the time a calibration run is made and an unknown sample quantitation run is made, calibration based on peak area ratios will become invalid.

Because CZE is a method for separating and measuring ions, like IC, considerations of charge balance make it particularly suitable for use of the method of indirect detection, as described at the end of Section 13.2.1.

13.6.4. Modes of CE

There are several different modes of conducting electrophoresis in capillary columns. We have just discussed at length the basic one, CZE, which is conceptually the simplest. It is conducted with a liquid buffer of uniform composition of electrolyte concentration and pH level. There are more complex versions, in which the buffer liquid is enmeshed in a porous gel of hydrophilic polymer, or the concentration or the pH of the separation buffer is not constant along the length of the column, leading to variations in the migration velocities of individual ions as the separation proceeds. Several of these will be discussed subsequently.

13.6.4.1. Capillary Zone Electrophoresis (CZE)

CZE has already been discussed in Sections 13.6.1–13.6.3. To summarize for purposes of comparison, it is performed using a separation buffer of constant composition. The only exception is the sample plug, which is usually introduced in a buffer of lower concentration to enable faster migration velocities within it to achieve the desirable effect of sample stacking at either end of the plug. Once migration of the sample mixture ions begins in the uniform separation buffer, the different components separate into "zones", similar conceptually to the *bands* in a chromatographic column separation. These zones migrate at an apparent constant rate, which is the sum of their electrophoretic mobility and any electroosmotic mobility which may be induced by the particular buffer pH and its electrolyte concentration. The zones are separated by spaces of buffer. With separation buffer flow toward the cathode, cation zones will lead a wide zone (of the width of the injected sample plug) of uncharged molecules moving at the EOF velocity, followed by anion zones (if $\mu_{ep} < \mu_{eo}$ for all anions). Hence the name CZE.

CZE is useful for a wide range of ion sizes, and is particularly useful for small volumes of small organic and inorganic ions. Its inherent smaller sample size requirements (nL instead of μL) make it particularly useful for sampling small volumes, even ones as small as the fluids inside a single biological cell. In its simplest form, the equipment is much less expensive than comparable IC or HPLC instruments. The interaction of so many variables such as pH, electrolyte concentration, adsorption effects, and so on, on both the electrophoretic mobility and the electroosmotic mobility make development and optimization of a CZE method for a particular separation less intuitive than for IC or HPLC. Once the CZE procedure is refined, it may be more cost effective to run than the corresponding chromatographic technique. As we shall see it is even possible to separate and quantitate electrically neutral species using a CZE apparatus by combining it with a chromatographic partitioning process.

13.6.4.2. *Capillary Gel Electrophoresis (CGE)*

As will be discussed in Section 13.7, the initial application of electrophoresis was to large biological molecule separations on a planar, unconfined surface. The separation buffer was applied to a solid planar substrate (a glass plate) mixed into a porous hydrophilic gel matrix. This mixture retards dispersion of the analyte by diffusion and convection, and such a cast **slab gel** is easier to manipulate in the instrument and to place its opposite edges into the electrode reservoirs. The gels are commonly produced by polymerizing acrylamide ($CH_2{=}CH{-}CO{-}NH_2$) in the presence of a cross-linking agent, or using other hydrophilic polymers such as agarose or polyethylene glycol. Increasing the degree of cross-linking reduces the size of the pores in the gel through which the buffer solution diffuses and the charged macromolecules and electrolyte migrate. It was found that the separation mechanism for macromolecules in such gels is actually a combination of differing migration velocities due to the charge and hydrodynamic radius of the ions, and a size discrimination effect due to the **sieving action** of the pores on the ions. This sieving action is more effective than the differences in electrophoretic mobility in achieving separations of macromolecules such as oligonucleotides, DNA fragments, and proteins having similar total charges. Such gels can be formed within capillaries to confer the same advantages for certain CZE separations. This modification of CZE is called **capillary gel electrophoresis (CGE)**. If the gel is cross-linked in the fashion employed for 2D slab gels, the capillary gel can be used only until clogging or other problems renders it unusable, as it is nearly impossible to remove such a so-called **chemical gel** from the capillary by flushing. If the capillary is filled with a **physical gel**, formed by filling it with long linear polymer chains which produce pores by entangling randomly, this may be easily flushed and replaced by fresh physical gel material as needed.

The use of CGE separations with multiple reagent, precolumn derivatization of end nucleotides on DNA segments, and laser-induced fluorescence measurement of the four possible end nucleotides was described in Section 13.1.4.4. Figure 13.36 displays a CGE electropherogram of such a separation of a mixture of DNA fragments ranging

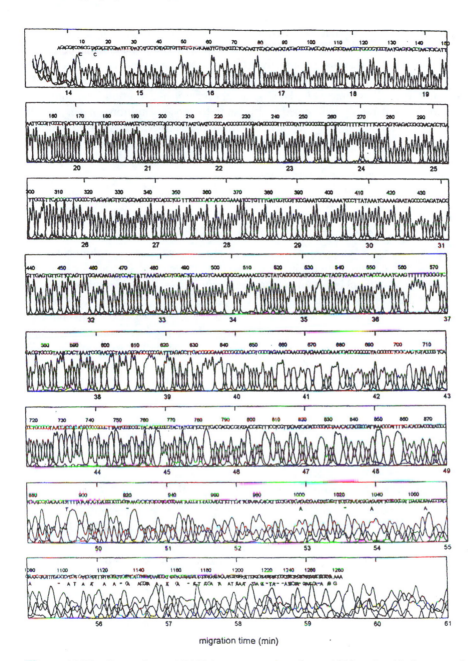

migration time (min)

Figure 13.36 Separation and DNA base sequencing of over 1000 nucleotide fragments by capillary gel electrophoresis employing detection by fluorescence derivatization. (Cazes, used with permission.)

from 3 to over 1000 nucleotides in length. These come off the column in order of their length, so the listing of each sequential end nucleotide (represented by A, G, C, T above each peak) directly yields the sequence of the nucleotides (also designated "bases") in the fragment. Automated instrumentation performs such measurements on many capillaries operating in parallel, with fluorescence excitation and emission being rapidly repeatedly scanned across the detection regions of the bank of capillaries by a single fluorescence detector spectrometer. DADs can be used to acquire multiple channels of spectral data simultaneously. Use of automatically refilled physical gels between runs, enables the system to cycle repeatedly through electrokinetic injections of banks of parallel sample vials or samples in small volume well plates, without operator intervention. The ability of large numbers of such highly automated instruments to process millions of very small volume samples of short length DNA segments is what makes projects such as the complete assembly of the 3 billion base long human genome sequence possible. Similar massively parallel automated instrumentation for sequencing protein chains forms an important resource supporting the discipline of proteomics.

13.6.4.3. *Capillary Isoelectric Focusing (CIEF)*

Proteins, peptides, and their component amino acids are categorized as **amphoteric** compounds; namely, they contain both acidic (—COOH) and basic (—NH$_2$) functionalities in the same molecule. As the pH of their surrounding solution varies from higher to lower values, they can lose protons from the carboxylic acid groups and gain protons on the amino groups. Their net charge will be negative (excess —COO$^-$) on the high pH (basic) side, and positive (excess —NH$_3^+$) on the low pH (acidic) side, of a compound-specific **isoelectric point** (pH value), at which the compound will exist as a zwitterion with equal amounts of positive and negative charges, and hence no overall net charges. Taking the simplest amino acid, glycine, as an example, we have the equilibria shown below:

$$\underline{\text{Glycine}}$$
$$\text{NH}_2\text{CH}_2\text{COOH}$$
$$\downarrow\uparrow$$
$$\underline{\text{Zwitterion}}$$

$$\text{OH}^- + \text{NH}_3{}^+\text{CH}_2\text{COOH} \Leftrightarrow \text{H}_2\text{O} + \text{NH}_3{}^+\text{CH}_2\text{COO}^- + \text{H}_2\text{O} \Leftrightarrow \text{NH}_2\text{CH}_2\text{COO}^- + \text{H}_3\text{O}^+$$

$$\underline{\text{pH}<\text{isoelectric point}} \qquad \underline{\text{pH @ isoelectic point}} \qquad \underline{\text{pH}>\text{isoelectric point}}$$

If we perform electrophoresis on such a compound, at pH below the isoelectric point it will migrate toward the cathode, at pH above that point, it will migrate toward the anode, and if it is in a buffer whose pH matches its isoelectric point value, it will be net neutral and undergo no migration at all. If we were to place a band or zone containing a mixture of such compounds in a capillary containing a separation buffer whose pH varied with a constant linear gradient along the length of the capillary, each amphoteric compound would migrate in one direction or the other, depending on whether the band started at a point whose pH was above or below the compound's isoelectric point. As it moved up or down the gradient, its excess charge would progressively become neutralized and its migration rate would slow until it would stop completely when it reached the point in the pH gradient which matched its isoelectric point. The mixture is separated into very sharply resolved motionless bands whose relative positions correspond to their isoelectric point values, which are characteristic for each compound, and may help serve to identify it. The process should be operated without the disruptive presence of EOF, so the capillaries are generally coated with an adsorbed organic compound which masks the surface siloxyl groups. This separation is called **capillary isoelectric focusing (CIEF)**.

Two big questions remain: How do we establish and maintain a pH gradient across the column while the amphoteric analytes are migrating to their focused halt points, and how then do we move the separated and focused bands past a detector while maintaining the separation which was achieved in the first step?

To form the gradient we employ a separating buffer which contains several other amphoteric compounds (called ampholytes). These **ampholyte** mixtures are composed of compounds whose isoelectric points cover the range of the target analytes. The cathodic end of the capillary is dipped in a reservoir solution of strong base such as NaOH, and the anodic end into a reservoir containing a strong acid. Under applied electrophoretic potential, H^+ from the anode migrates toward the cathode and OH^- from the cathode toward the anode. The buffer ampholytes will migrate toward the electrode of charge opposite their excess charge, which will be progressively neutralized by the strong acid or base ions migrating from the reservoirs in the opposite direction. The net effect of all these countering migrations and neutralizations is the establishment of an equilibrium state in which there is a uniform pH gradient along the column between the pH values of the highest and lowest isoelectic points of the separation buffer ampholytes. Note that when the migrating ampholyte of highest or lowest isoelectric point reaches that pH value it will stop migrating, thus setting the bounds of the gradient.

Once all the amphoteric analyte bands have completed migration to and have been focused at their isoelectric points along the gradient, all net motion stops. The pattern of focused bands can then be **mobilized** along the column so that they pass the detector to be identified by their isoelectric point values and quantitated. This can be done in a straightforward manner by applying a pressure difference across the column, using the same mechanism that brings a sample plug into the column by hydrodynamic injection. Pressure driven laminar flow profiles may cause some loss of the high resolution achieved by the isoelectric focusing step. Alternatively, either the H^+ or OH^- solutions in one of the reservoirs may be diluted with equivalent concentrations of another ion of the same charge. This will change the pH gradient, as there is a lesser concentration of one strong acid or base to neutralize the other. The gradient moves toward the electrode reservoir in which the acid–base dilution (or ion replacement) took place, and the focused bands move with it as they follow the shift in position of their isoelectric point along the length of the column as the gradient shifts.

13.6.5. Capillary Electrochromatography (CEC)

One shortcoming of the very "simple" technique of CZE is that it acts only to separate ions. While it separates ions from any neutrals left behind in the sample plug, these neutrals cannot be separated from one another. A crude way of achieving this would be to introduce a chromatographic separation mechanism by packing the capillary with microparticles of RP-HPLC packing, while retaining the liquid mobile phase flow produced by a separation buffer mobilized by EOF. This becomes simply HPLC with mobile phase flow induced by EOF instead of high-pressure pumps. The multipath A term returns to the Van Deempter equation, as well as the mass transfer C terms. Many of the resolution enhancements seen for CZE of ions, especially for biological macromolecules, are lost. The opportunity exists to achieve both neutral separations based on chromatographic partitioning to a stationary phase, as well as ion separations based on differences in electrophoretic mobility, in the same sample mixture. The number of variables to be managed becomes even more daunting than in either HPLC or CZE alone.

13.6.5.1. *Micellar Electrokinetic Capillary Chromatography (MEKC)*

Instead of a packing composed of an organic coating on a spherical solid support acting as the stationary phase of the chromatographic component of a CEC system, we may avail ourselves of something that performs the same function without blocking buffer/mobile phase flow as dramatically as a stationary HPLC particle packing. Recall that we employed a cationic surfactant bilayer to reverse electroosmotic flow, as illustrated in Fig. 13.34. If we prepare a buffer solution of an anionic surfactant such as **sodium dodecylsulfate (SDS)** $(n\text{-}C_{12}H_{25}OSO_3^- Na^+)$ we can form a variety of structures, which are illustrated in Fig. 13.37. In Fig. 13.37(a) the hydrocarbon tails coalesce together to form a small spherical **micelle**, with a charged polar exterior and an hydrophobic interior. This structure is small enough to remain in colloidal suspension in the buffer solution, and it acts in an electrophoretic system like a very large, multiply charged macromolecule. As the concentration of the surfactant in the buffer increases, it reaches a **critical micelle concentration (CMC)**, at which an equilibrium forms between a *constant* surfactant concentration and a rapidly increasing micelle concentration. If the total amount of surfactant becomes too high, an **inverted micelle** may form, with the aqueous buffer solution trapped in a multitude of spherical inverted micelles [Fig. 13.37(b)]. This would be unsuitable for electrophoresis. A spherical bilayer may form Fig. 13.37(c) comprising a surface whose inside is nonpolar organic and surrounds a pocket of the aqueous buffer. This structure is called a **bilayer vesicle**. Finally an extended **planar bilayer** [Fig. 13.37(d)] may form, especially if there is a suitable surface to which it may adhere, as was the case for cationic surfactants used to reverse EOF.

Figure 13.37 Formation of micelles, vesicles, and bilayers by surfactant molecules: (a) micelle, (b) inverted micelle, (c) vesicle, and (d) flat bilayer. (Cazes, used with permission.)

Figure 13.38 illustrates the behavior of the anionic surfactant micelles formed in a standard CE experiment. The surfactant being anionic, it does not form a planar bilayer on the capillary wall as would a cationic surfactant. The micelles have a net negative charge and migrate slowly to the left (anode). This migration speed is slower than the induced EOF toward the cathode, so the micelles will be carried toward the detector, but at a net speed less than that of the sample plug where neutrals would ordinarily remain unseparated. The neutrals (represented as **"solubilizate"** in the figure) can partition between the polar separation buffer and the less-polar interior of the micelles, as in RP-HPLC. They will move away from the sample plug in a direction opposite the flow by that percentage of the distance to the retarded back-migrating micelles, which is simply the percentage of the time they spend in the micelle. This variable retention in elution time relative to the speed at which buffer is moved through the capillary by the EOF, results from the same mechanism that causes variable peak retention in chromatography. The micelles are acting as a "pseudostationary phase" for separating the neutrals based on differences in their partition coefficient between the moving buffer and the differently moving micelle interiors. A chromatographic process has been overlaid on the electrophoretic separation of any analyte ions present. Thus the name of the procedure emphasizes that it is **miceller electrokinetic chromatography (MEKC)**. The focus of the technique is on the introduction of a chromatographic separation of neutrals in a capillary electrophoresis system. The ions themselves may interact electrostatically to some extent with the charged micelle surface, so their migration times are altered as well. The imposition of a true chromatographic process means that the mass transfer term C in the Van Deempter equation reappears, but mass transfer in and out of micelles is so efficient that this adds little to band broadening.

Surfactants other than SDS, which have different critical micelle concentrations and form different size micelles with different interior polarity, can be used. If a cationic surfactant such as cetyltrimethylammonium bromide is used, the direction of the EOF will be reversed by the formation of a capillary surface bilayer, as illustrated in Fig. 13.34. The charge on the micelle surface is also opposite that of an anionic surfactant micelle, so again the neutrals will lag the solvent plug as they are separated chromatographically by partitioning with the pseudostationary phase. The term **solubilizate** for the separated analytes comes from the original use of surfactant detergents to bring nonpolar greasy material on surfaces into polar aqueous solutions in washing processes. Introduction of cyclodextrins (cf. Fig. 13.3) to an MEKC separation can alter the times that enantiomers of optically active compounds spend in each phase, and effect chiral separations.

= Surfactant   = Electroosmotic flow
= Solubilizate   = Electrophoresis

Figure 13.38 Mechanism of micellar electrokinetic chromatography. (Cazes, used with permission.)

13.7. PLANAR CHROMATOGRAPHY AND PLANAR ELECTROPHORESIS

If the stationary phase for a chromatographic separation, or a gel for an electrophoretic separation, is bound to a surface, it is possible to spot a sample on to one corner of such a flat plat or slab, and perform a separation directed along one edge, spreading out the bands separated by chromatographic partitioning or electrophoretic migration along that edge (without eluting them off the surface or through a detector). The linear array of separated bands can then be subject to another separation process using different chromatographic mobile phase or electrophoretic buffer parameters directed at right angles to the first separation. Bands which remained unresolved in the first step may be further separated in this new direction, and the final product is a surface covered with many separated spots, which may be scanned by some detector, or reacted to visualize the separated components. These basic principles were outlined in Section 11.3.

13.7.1. Thin Layer Chromatography (TLC)

13.7.1.1. Parallel 1D TLC Separations

In TLC particles of stationary phase [e.g., silica or alumina for a normal phase separation, or large diameter particle $(10-100 \ \mu m)$ RP-HPLC type packing] are coated and bound as a thin layer to a square planar supporting **plate**, usually of glass. A liquid mobile phase is placed in the bottom of a TLC **tank** or **chamber**, a rectangular glass tank with a tightly sealing glass plate top, and with dimensions that will contain the TLC plate when it is propped up in the tank at an angle inclined a little from the vertical. The tank contains an internal lip above the solvent reservoir or separating off an unfilled part, so that initially the plate can be set on it out of contact with the liquid to equilibrate with the solvent vapor. The mixtures for separation are **spotted** on the plate along an origin line parallel to one edge, at a distance sufficient for the spots to lie above the level of the mobile phase solvent when the edge with the sample spots is lowered into the reservoir. This can be done crudely by using a glass capillary to pick up the sample and transfer it to the plate. For quantitative work a measured volume is slowly put on the plate using a micro-liter syringe, allowing time for solvent to evaporate during the application in order to minimize spot spreading in the thin layer. Figure 13.39 displays a side view of a TLC plate in a sealed tank. Solvent vapors from the liquid in the bottom equilibrate with the headspace above it [Fig. 13.39(1)]. Capillary attraction to the closely spaced stationary phase particles comprising the thin layer causes the mobile phase solvent to move up the plate. The leading edge of this flow is called the solvent front. Above the solvent front the thin layer remains conditioned by solvent vapor diffusing into it [Fig. 13.39(2)]. There will be no net evaporation from the solvent flowing up the plate [Fig. 13.39(3)], because the closed system establishes an equilibrium vapor pressure. Control of these variables ensures reproducible mobile phase flow in the system.

Multiple spots (e.g., standard mixtures, a blank, and several unknown samples) can be chromatographed in parallel. This facilitates direct comparisons of retention relative to the advance of the solvent front, and comparison of separated component spot densities for estimating relative quantities. The same separation principles apply as in HPLC column chromatography. Analytes must not have significant volatility, as there is no column to contain them. Figure 13.40 displays three stages in the process of obtaining a TLC chromatogram. The left-hand side view shows the initial, somewhat ragged line of spotted

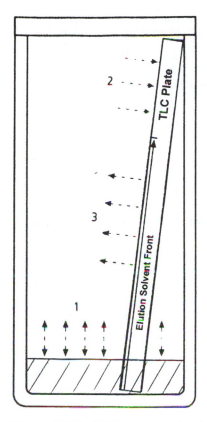

Figure 13.39   TLC plate in enclosed developing tank.

samples. If a strong elution solvent is used, initial ascent of its solvent front through the sample spots moves them to a line precisely at the solvent front, and compacts the spots into narrow elongated bands. This is called **beginning development** (center panel). As **development** continues, either with the beginning solvent or a weaker one which can discriminate better among the analytes, the bands separate until the solvent

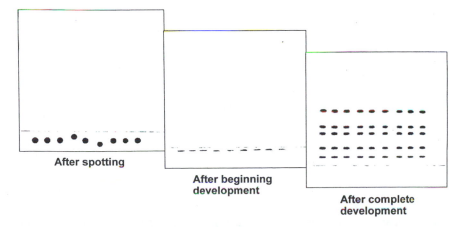

Figure 13.40   Steps in development of a TLC chromatogram.

front has run up most of the plate (right-hand side panel). This is the TLC chromatogram. The various distances of the separated bands between the start point origin and the solvent front (which the analyst can see and mark when removing the plate before drying it to remove solvent) and the bandwidths are measured. The R_f values (corresponding to relative retention time in column chromatography), and the theoretical plate number, N, are calculated as outlined in Fig. 13.41.

13.7.1.2. Detection in TLC

The separated spots or bands may be detected in a variety of ways. In contrast to column chromatography they remain on the stationary phase and are not eluted off it into a separate detection region. Detection is by scanning or imaging the "developed" plate. It is critical to understand that in the TLC context, development refers to the chromatographic separation process, and not the subsequent process of making the separated spots visible, as is done when one "develops" a latent image on a planar photographic film. If the spots are colored or opaque, imaging may be as simple as looking at them by eye. If the analytes are colorless, but are fluorescent under UV illumination, the dried, developed plate may be place in a darkened box, illuminated with UV, and the fluorescent spots imaged by eye. This can be done more precisely by using a **plate position scanning** (as opposed to wavelength scanning) fluorescence spectrometer. Fluorescence can be used as an **indirect detection method** if the TLC plate contains an "**F layer**" incorporating a bound phosphor or fluorescent indicator such as zinc sulfide. Analyte spots containing compounds (such as most organic aromatic compounds and those with conjugated double bonds) which are colorless in the visible range, but which absorb Hg-lamp UV at around 254 nm, will absorb the UV and **quench** the fluorescence of the plate at the position of their developed spots.

Some reagents or reactions will visualize almost all separated analytes. Examples of such **universal detection reagents** are: (a) iodine vapor, which will reversibly stain almost all analyte spots except some saturated alkanes, forming a dark brown color and (b) irreversible charring with heating in the presence of sulfuric acid or a mixture of 8% phosphoric acid and 3% copper sulfate. At the other end of the reagent spectrum are **selective detection reagents**, such as: (a) ninhydrin for α-amino acids, (b) 2,4-dinitrophenylhydrazone $\cdot$ HCl to produce orange zones by reaction with aldehydes, and (c) enzyme reactions with a substrate to form a colored product to generate uniform color over the

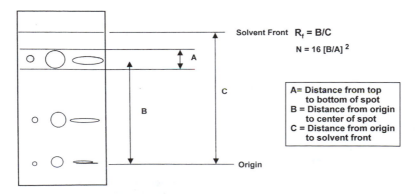

Figure 13.41 Definition and calculation of retention factor (R_f) and Plate Number (N) from a developed TLC chromatogram.

plate except on bands containing biologically active compounds which act to block the enzyme function (e.g., cholinesterase inhibitors such as organophosphorus pesticides). These reagents are generally applied to the plates in one of two fashions: (a) **dipping** them for a set period of time in a solution of the reagent. Special dipping chambers can automatically perform a reproducibly timed immersion and gentle agitation of the plate. Dipping yields the best reactions for *quantitative* purposes. (b) **spraying** solutions of the reagents as an aerosol mist from a handheld atomizer. This must often be done in a contained volume TLC spray chamber vented to a fume hood, in order to avoid exposure to hazardous solvent vapors. Intermittent spraying with intervals for drying will minimize band spreading due to mobilization and expansion of the spots from accumulated spray solvent.

TLC separations rarely approach the resolution or quantitation accuracy of chromatographic instruments employing the sequence of injection, separation on a confined capillary column, and separate detection in a specially designed detector cell. For applications which do not require such precision, TLC's simplicity and the low cost of materials and equipment recommend it, especially for initial screening applications.

13.7.1.3.   2D Separations using TLC—(2D-TLC)

Additional resolution of components which are incompletely separated by 1D development may be obtained by rotating the plate 90°, and redeveloping the line of partially separated components of a single mixture sample originally spotted in one corner. For the second development a different mobile phase solution having different selectivities for the partially separated compounds is used. This is illustrated in Fig. 13.42. Only one sample can be separated on each plate. This 2D-TLC analysis is an example of a 2D separation employing a 2D stationary (i.e., planar) phase. This relatively simple trick can be contrasted with the extremely complex 2D column chromatographic separations employing modulator devices interfacing columns of two different selectivity and analysis speeds (e.g., comprehensive 2D-GC, Section 12.8.3).

13.7.2. Planar Capillary Electrophoresis on Slab Gels

13.7.2.1. 1D Planar Gel Electrophoresis

Prior to the development of CZE, the primary application of electrophoresis was for large biological macromolecule separations on planar gels. The gels were supported on glass plates not unlike the particles in the thin layer of a TLC plate. These so-called slab gels were needed to confine the separation buffer. It was easier to prepare cross-linked

| After spotting | After 1st develop. | After 2nd develop. |

Figure 13.42   Diagram of development of a 2D TLC chromatogram.

chemical gels on the open surface than in narrow capillaries, and to remove the gel from the support plate to prepare a fresh slab gel for a new analysis. Electrophoretic separations of multiple samples in 1D could be performed and easily visualized and compared. This was the standard tool for biochemists to compare distributions of peptides from enzymatic protein digests, or DNA or RNA fragments. Comparison of multiple lanes of 1D separations enabled easy illustration and identification of abnormal distributions, or the detection of the presence of unusual fragments, which might, for example, be indicators of a mutant form or a disease state. Such patterns looked similar to the diagram of a developed TLC plate shown in the right-hand panel of Figure 13.40. Thousands of photos of such multilane slab gel electropherograms grace the pages of the molecular biology literature.

13.7.2.2. 2D Planar Gel Electrophoresis

Many DNA, RNA, or protein digests can contain hundreds or even over a thousand fragments, and extracts of particular tissues or cell types may contain similar numbers of proteins. 1D planar gel electrophoresis could not possibly resolve all of these. The same principle used with 2D-TLC was applied. There are many possible combinations of separation conditions which might be applied to sequential developments of a sample on a planar gel, to produce an array of hundreds of resolved spots. One particular sequence is a standard for **protein peptide mapping**. This procedure is so well standardized, and the reagents and instrumentation so widely commercialized, that such "maps" of the distribution of peptides as a particular array of separated spots on a 2D-gel, are archived in international computer-searchable databases, enabling researchers to submit new patterns which they have characterized, or to determine if a new pattern has already been described by others. This is analogous to the huge databases of full-scan mass spectra which help to identify unknown spectra by library searching. Similar libraries of the distributions of cell proteins are compiled by very similar procedures.

The most common standard protein or peptide mapping experiment employs first a slab gel electrophoretic separation employing **isoelectric focusing** (IEF) to spread the amphoteric analytes across one axis of the plate in order of their **isoelectric point pH** (pI) values (cf. Section 13.6.4.3). These partially separated components are then subjected to **sodium-dodecylsulphate–polyacryamide gel electrophoresis** (SDS–PAGE) separation in a direction at right angles to the initial IEF separation. The proteins are **denatured** (i.e., their specific tertiary folded structure is disrupted by elevated temperature or high ionic strength solution), and the micelle-forming SDS (at a concentration well below the critical micelle concentration) will bind an amount of anionic SDS proportional to the size of a denatured protein. The resulting protein–SDS complexes have approximately constant charge-to-mass ratios and therefore identical electrophoretic mobilities. Therefore in the sieving gel medium, the complexes migrate proportionally to their effective molecular radii, and likewise to the protein molecular weight. Thus the 2D planar electrophoresis separation initially sorts the proteins (or peptides) by a charge-related parameter of their isoelectric points, and secondly by the size-related parameter of the SDS–PAGE sieving process.

The 2D-SDS–PAGE separation described results in a planar array of separated compounds. If these are unknown proteins (perhaps just one or a few which interrupt a standard pattern, and may signal a mutant form or an abnormal biological state), the spots of interest may be cut out or dissolved out of the gel. The isolated spot can be subjected to tryptic digest, the peptide digest mixture separated by capillary HPLC, and the separated peaks characterized, and if necessary, sequenced by ESI-MS and ESI-MS-MS, as was illustrated in Figs. 13.21 and 13.22. If the 2D separation is of peptides from a single protein in a peptide mapping experiment, the developed slab gel may be put on an automatically posi tioned stage which will place each spot at the laser target point of a MALDI-TOF-MS (cf. Section 10.3.3), which may obtain enough information on each peptide to identify it based on archival library lookup and matching. The plate positioning engine and the MALDI-TOF operate so quickly to acquire data, that for gel plates containing very large numbers of closely resolved components, it may be more efficient to sample the entire plate surface at points in a closely spaced grid, store all the data in a computer, and have it automatically processed. Pause for a moment to contemplate and marvel at the number of powerful separation and spectral characterization techniques that are being interfaced, automated, and having their output processed by complex computer algorithms in such an instrument. This melding of many instrumental techniques results in systems being successfully marketed today to support the enormous demands for analytical information in the disciplines of genomics and proteomics.

BIBLIOGRAPHY

Anton, K.; Berger, C., Ed. *Supercritical Fluid Chromatography*; Marcel Dekker, Inc.: New York, 1998.
Ardrey, R.E. *Liquid Chromatography–Mass Spectrometry, An Introduction*; Wiley: New York, 2003.
Berthold, *et al. Micellar Liquid Chromatography*; Marcel Dekker, Inc.: New York, 2000.
Camilleri, P. *Capillary Electrophoresis: Theory and Practice*; CRC Press: Boca Raton, FL, 1998.
Cazes, J. *Encyclopedia of Chromatography*; Marcel Dekker Inc.: New York, 2001.
Cunico, R.L. *et al. Basic HPLC and CE of Biomolecules*; Bay Analytical Laboratories, 1998.
Fritz, J.S.; Gjerde, D.T. *Ion Chromatography*; VCH Publishers: Germany, 2000.
Guzman, N.A., Ed. *Capillary Electrophoresis Technology*; Marcel Dekker, Inc.: New York, 1993.
Katz, *et al.*, Eds. *Handbook of HPLC*; Marcel Dekker, Inc.: New York, 1998.
Landers, J.P. *Handbook of Capillary Electrophoresis*; Marcel Dekker, Inc.: New York, 1997.
Pommerening, *Affinity Chromatography: Practical and Theoretical Aspects*; Marcel Dekker, Inc.: New York, 1985.
Scott, R.P.W. *Chromatographic Detectors*; Marcel Dekker, Inc.: New York, 1996.
Sherma, J.; Fried, B., Eds. *Handbook of Thin Layer Chromatography*; Marcel Dekker, Inc.: New York, 1996.
Snyder, R.L., *et al. Practical HPLC Method Development*; Wiley: New York, 1997.
Wu, C.S., Ed. *Handbook of Size Exclusion Chromatography*; Marcel Dekker, Inc.: New York, 1995.

PROBLEMS AND EXERCISES

13.1   Compare packed column LC, capillary column LC, and capillary zone elec-trophoresis (as best, intermediate, poorest, or inapplicable for the following categories)

(a)   Analyte capacity

(b)   Highest number of theoretical plates for large molecule separations

(c)   Suitability for separating charge neutral compounds

(d)   Ability to separate anions and cations in a single run

(e)   Suitability for analyzing 50 μL sample of mix of low MW alcohols

(f)   Suitability for analyzing 10 μL sample of mix of low MW alcohols

(g)   Suitability for analyzing 10 μL sample of peptides

13.2   (a)   List advantages and disadvantages of presence of free silanol groups on silica surfaces for applications in liquid phase separations.

(b)   How may the disadvantages be suppressed?

(c)   How may the advantages be enhanced?

13.3   Describe for the following HPLC procedures: (1) What is meant by? (2) What is the purpose of? (3) How is it done?:
(a)   **Degassing** the mobile phase
(b)   **Filtering** the mobile phase
(c)   **Modifying** the mobile phase
(d)   **Endcapping** the stationary phase
(e)   **Pulse-dampening** the mobile phase
(f)   Introducing a **Guard column**
(g)   Introducing a **Suppressor column**

13.4   Describe three ways in which an HPLC support particle differs from one used in packed column GC.

13.5   Describe the primary purpose and advantages of using the following columns in HPLC:
(a)   Silica spheres with bonded organic substituents
(b)   Monolithic columns
(c)   Resin polymer spheres
(d)   Polyacrylamide gels of varying pore size
(e)   Zirconia spheres with bonded organic substituents
(f)   Bare silica spheres
(g)   Resin spheres with attached —COOH groups
(h)   Silica spheres with attached latex coating with bonded —SO_3H groups
(i)   Resin spheres with bonded —$N(CH_3)_3^+$ groups
(j)   Completely endcapped —C18 bonded silica

13.6   Answer questions (i) and (ii) for the following analyte mixtures:
(a)   A mixture of styrene–vinylbenzene polymer chain lengths
(b)   Organophosphorus pesticides in urine
(c)   Polynuclear aromatic hydrocarbons in hexane
(d)   Rare earth element ions in acidic aqueous solution
(e)   Peptides from a tryptic digest
(f)   A partially racemized, optically active, drug molecule containing phenolic groups
(g)   Isolation of a single antibody protein from a preparation of disrupted cells
(h)   A mixture of hydrocarbon ethers
(i)   A mixture of polysaccharide polymers (e.g., starches or sugars)
(j)   A mixture of inorganic halides and oxyhalides
(k)   A preparative extract of a mixture of spice essences from leaves
    (i)   Select one or more suitable liquid phase chromatographic or electrophoretic separation procedures from Sections 13.1.1, 13.2, 13.2.1, 13.3, 13.4, 13.1.5, or 13.6.
    (ii)   Select one or more suitable detectors from Sections 13.1.4 and 13.1.6 and explain why.

13.7   Draw diagrams of the structures at the atomic level of the following silica surfaces:
(a)   Silica in aqueous solution at pH 1
(b)   Silica in aqueous solution at pH 9
(c)   An ODS-silica, bonded HPLC phase (indicate no. of C atoms n, by C_n)
(d)   An endcapped, octyl bonded silica HPLC phase
(e)   A silica surface with an additive which reverses electroosmotic flow
(f)   A silica surface treated to suppress electroosmotic flow

13.8 Using the style of the diagram of the six-port rotary HPLC injector valve (Fig. 13.6) draw two diagrams showing the flow through the injection loop, the valve, and the column during the load and injection positions.

13.9 Using the information in Sections 13.1.4 and 13.1.6, construct a table contrasting the properties of various LC and CE detectors, indicating relative sensitivity, compound selectivity, gradient compatibility, buffer salt and pH compatibility, ability to extract analyte structural information, and mass or concentration mode of detection.

13.10 Contrast guard and suppressor columns with respect to their purpose, their position relative to the separation column, and the reason that pellicular packings are often desirable for use in each.

13.11 For a normal phase or reversed phase separation predict the elution order of:
(a) *n*-octane, *n*-octanol, naphthalene
(b) ethyl acetate, diethyl ether, nitrobutane

13.12 In HPLC:
(a) Why is high pressure required?
(b) Why are stationary phases bonded?
(c) Why are supports endcapped?
(d) Why are the buffers used usually at pH between 2 and 8?

13.13 In GC a plot of the Van Deempter equation is shaped like a severely distorted U. In HPLC using 3 μm particles its shape is more like the letter "L". Explain.

13.14 Describe three different modes of use for SDS to change and improve separations in chromatography and electrophoresis.

13.15 In capillary electrophoresis, what is the effect on the electroosmotic flow of:
(a) Increasing buffer pH
(b) Increasing buffer concentration
(c) Increasing the applied voltage
(d) Increasing the capillary length for a constant applied voltage
(e) Endcapping silanol groups on the silica capillary surface

13.16 In ion chromatography:
(a) How does one increase the exchange capacity of the stationary phase?
(b) Is the exchange group charge on the suppressor column the same or opposite that on the separating column? Explain why.
(c) Why and how does one regenerate the separator column?
(d) Why and how does one regenerate the suppressor column?
(e) Describe how to regenerate the suppressor continuously.

13.17 Describe the instrumentation and separation media you would choose for:
(a) Separation of amino acids by:
(i) Ion exchange chromatography
(ii) Reverse phase HPLC of derivatives
(iii) Capillary isoelectric focusing
(b) Separation of proteins by:
(i) Size exclusion chromatography
(ii) Capillary gel electrophoresis
(iii) 2D - IEF - SDS–PAGE

13.18 In Fig. 13.21(c) the molecular weight of the 11 amino acid peptide is given as 1453.8, but the M^+ ion is listed as 1454.8. Explain why (Hint: this is a MS, not an LC question).

13.19   What is the function of:
        (a)   Crosslinked polyacrylamide in size exclusion chromatography?
        (b)   Crosslinked polyacrylamide in SDS-PAGE planar electrophoresis?
        (c)   Uncrosslinked polyacrylamide in rapid DNA sequencing by CGE?
        (d)   Styrene-divinylbenzene copolymer resin in Ion Exchange Chromatography?

13.20   In what ways is supercritical fluid chromatography like LC?—Like GC? What are several advantages of the use of CO_2 as eluent in SFC?

13.21   (a)   How does the LC-APCI-MS interface differ from GC-CIMS?
        (b)   In an orthogonal electrospray ionization MS interface, where does most of the LC mobile phase go?
        (c)   Which LC-MS interface of (a) and (b) above works best for:
              (i)   small molecules
              (ii)   large molecules
              (iii) charged molecules (i.e., ions in solution)
              (iv)   neutral (uncharged) molecules

13.22   Sample injection in CZE is by either hydrodynamic or electrokinetic injection
        (a)   Which is better for capillary gel electrophoresis with physical gels?
        (b)   Which is better for capillary gel electrophoresis with chemical gels?
        (c)   How should an injection buffer differ from a separation buffer in order to achieve sample stacking? What is this, how does it work, and what are its benefits? Explain.

13.23   Explain how micellar electrokinetic chromatography combines electrophoretic and chromatographic separation principles. Why is the MEKC micelle called a "pseudostationary phase"?

13.24   If cetyltrimethylammonium ion formed micelles, and one employed it for MEKC, in which direction (towards the cathode or anode?) would:
        (a)   The electroosmotic flow move?
        (b)   The cations move?
        (c)   The anions move?
        (d)   Neutrals move away from their position in the initial injection plug?

13.25   Consider the planar 2D separation illustrated in the last panel of Fig. 13.42.
        (a)   If this were on a TLC plate how do you think one might identify the separated spots by FAB-MS? By capillary LC-ESI-MS? Describe a proposed procedure.
        (b)   If it were an SDS–PAGE separation on a slab gel, how do you think you might use MALDI-TOF-MS to identify the spots? Describe a proposed procedure.

13.26   Internet World Wide Web "Literature Research" project. Choose one of the HPLC compound class applications in Section 13.1.7, and use the links in Appendix 13.1 to download manufacturer's application notes, sample chromatograms, and suggested products for their analysis. Write a detailed experimental procedure to separate your chosen compounds.

APPENDIX 13.1 LC/CE/TLC INTERNET WEB RESOURCES

Web URL	Site selection path	Use and Contents
www.chem.agilent.com/		LC applications
www.alltechWEB.com	Chromatography access database	1300 LC chromatograms
www.sigma-aldrich.com/ supelco	Chromatogram search	Several hundred LC chromatograms
	Library/applications	Application notes
www.restekcorp.com	Chromatogram search	Many chromatograms
www.sge.com	Chromatograms search	Selected by compound or application
www.varianinc.com	Chromatography	HPLC Applications
www.agilent.com/chem	1. CE systems	
	2. Tech support	
	3. LC amd LC-MS	
	4. Tech support, FAQs and videos	
www.dionex.com		Ion chromatography HPLC
www.las.perkinelmer.com	HPLC	
www.phenomenex.com	HPLC	Large selection of HPLC stationary phase info, and SPE cartridges
www.sigma-aldrich.com/ supelco	Library/Apps./CE	Capillary electrophoresis
Consumables.info. us@varianinc.com	Load prog. and CD	Select from 1000's by col., application, compound
Request CP-Scan View CD-ROM	Mailed to you	
www.camag.com	TLC instrumentation and planar separation media	TLC instruments plates applications
www.piercenet.com	Pierce Technologies Gas chrom./HPLC	List of hundreds of LC derivitization agents
www.separationsnow.com	GC and LC courses, literature articles, (free access)	A mother lode of reading material on all forms of chromatography
www. chromatographyonline. com	Issue archive Author–subject index Useful links	Contents all issues from 1983–Present year-by-year index dozens to other sites

14

Surface Analysis

14.1. INTRODUCTION

The term **surface analysis** is used to mean the characterization of the chemical and physical properties of the **surface layer** of **solid** materials. The surface layer of a solid usually differs in chemical composition and in physical properties from the bulk solid material. A common example is the thin layer of oxide that forms on the surface of many metals such as aluminum upon contact of the surface with oxygen in air. The thickness of the surface layer that can be studied depends on the instrumental method. This layer may vary from one atom deep, an atomic **monolayer**, to 100–1000 nm deep, depending on the technique used. Surface analysis has become increasingly important because our understanding of the behavior of materials has grown. The nature of the surface layer often controls important material behavior, such as resistance to corrosion. The various surface analysis methods reveal the elements present, the distribution of the elements, and sometimes the chemical forms of the elements in a surface layer. Chemical speciation is possible when multiple surface techniques are used to study a sample.

Surface analysis has found great use in understanding important fields such as materials characterization of polymers, metals, ceramics and composites, corrosion, catalysis, failure analysis, and the functioning and failure of microelectronics and magnetic storage media. In the pharmaceutical industry, it can be used to investigate the multi-layered materials and coatings used in packaging and time-release products. In medicine, it has been used to study bone structure, the surface of teeth, indicating why SnF_2 fights tooth decay, for example, and to study the biocompatibility of metallic and polymeric implantable devices.

Spectroscopic surface analysis techniques are based on bombarding the surface of a sample with a beam of X-rays, particles, electrons, or other species. The bombardment of the surface by this **primary beam** results in the emission or ejection of X-rays, electrons, particles, and the like from the sample surface. This emitted beam is the **secondary beam**. The nature of the secondary beam is what provides us with information about the surface. A number of techniques have been developed for surface analysis but only the most common will be discussed in this chapter. The names of these spectroscopic techniques and the primary and secondary beams used for each technique are listed in Table 14.1. These techniques are frequently quite different in physical approach, but all provide information about solid surfaces. Applications of these surface analysis techniques are presented in Tables 14.2 and 14.3.

The student should be aware that there is another class of surface analysis instruments based on analytical microscopy, including scanning electron microscopy,

Table 14.1   Selected Spectroscopic Techniques for Surface Analysis

Abbreviated name	Full name	Primary beam	Secondary beam
ESCA, XPS	Electron spectroscopy for chemical analysis, X-ray photoelectron spectroscopy	X-rays	Electrons
AES	Auger electron spectroscopy	Electrons	Electrons
ISS	Ion scattering spectroscopy	Ions	Ions
SIMS	Secondary ion mass spectrometry	Ions	Ions
EM, EPMA	Electron microprobe, Electron probe microanalysis	Electrons	X-rays

transmission electron microscopy, atomic force microscopy, and scanning tunneling microscopy. A discussion of these microscopy techniques is beyond the scope of this chapter. Most industrial materials characterization laboratories will have some combination of electron spectroscopy, X-ray analysis, surface mass spectrometry, and analytical microscopy instrumentation available, depending on the needs of the industry.

Table 14.2   Surface Analysis Applications by Industry

Industry	Application
Microelectronics	Composition of deposited layers
	Thickness of deposited layers
	Defect characterization
	Particle identification
	Process residue identification
Magnetic data storage media	Lubricant type and thickness
	Carbon overcoat composition
	Magnetic layer composition
	Magnetic layer thickness
	Defect/contamination identification
	Failure analysis
Optical coatings	Coating composition and thickness
	Adhesion layers
	Composition of surface defects
Automobile industry	Paint adhesion
	Paint weathering
	Catalyst poisoning
	Lubricant chemistry
Pharmaceutical industry	Tablet coating composition and uniformity
	Distribution of active ingredients in tablet
	Patent infringement monitoring
Polymer industry	Surface coatings
	Surface chemistry
	Multilayer composition and thickness
Biotechnology	Surface chemistry of metal and polymer medical implants
	Biocompatibility of implanted devices
	Implanted device performance

Source: Information in table provided by Physical Electronics USA, Inc. (www.phi.com).

Table 14.3 Specific Surface Analysis Applications

Type of analysis	Applications
Failure analysis	Surface contamination
	Particle identification
	Fracture due to grain boundary impurities
Adhesion failure	Adhesive failure
	Cohesive failure
	Silicone surface contamination
Corrosion	Thickness and composition of surface oxides
	Identification of corrosive elements (e.g., Cl)
	Passivation layer composition and thickness
Surface cleanliness	Detergent residue identification
	Solvent residue identification
	Type and amount of surface impurities
Reverse engineering	Composition and identification of competitors' materials
Semiconductor wafers	Analysis of complete surface of wafer (up to 200 mm)
Hot sample stage analysis	Volatility of surface components
(up to 700°C)	Temperature dependence of surface chemistry
	Migration of bulk components to surface
Cold sample stage	Surface species that are volatile at RT in UHV
analysis (to −100°C)	Hydrated surfaces
	Samples that outgas at RT in UHV

Note: RT, room temperature; UHV, ultrahigh vacuum.
Source: Information in Table provided by Physical Electronics USA, Inc. (www.phi.com).

14.2. ELECTRON SPECTROSCOPY TECHNIQUES

In the discussion on X-rays (Chapter 8) it was mentioned that the source of an X-ray photon is an atom that is bombarded by high-energy electrons or photons. This can displace an inner shell electron, which is ejected from the atom, leaving an ion with a vacancy in an inner shell (Fig. 14.1). An electron from an outer shell then drops into the inner shell and an X-ray

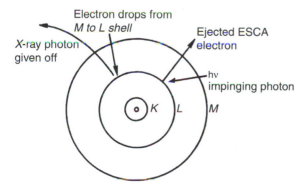

Figure 14.1 An impinging photon ejecting an inner shell ESCA electron. The kinetic energy of the ejected ESCA electron, E_k, is related to the binding energy of the electron in the atom, E_b. The binding energy is characteristic of the element. An X-ray photon of energy $E_M − E_L$ is emitted in the process. (An alternative process to the emission of the X-ray photon is emission of an Auger electron, which is shown in the next figure.)

photon is emitted simultaneously. The energy of the photon is equal to the difference between the energy of the orbital the electron was in originally and that of the one to which it descends. The energy levels of these two inner orbitals are almost independent of the chemical form of the atom, combined or otherwise. However, we know from Chapter 5 that the energy of a valence electron varies with the chemical form and chemical environment of the combined atom and provides the basis for UV absorption analysis. This variation is reflected as well in the energies of the inner shell electrons, but the changes in energy involved are very small compared with the energy of the emitted X-rays themselves. The slight differences in X-ray wavelength are extremely difficult to measure, since they are such a small fraction of the nominal wavelength of the X-rays generated. It is therefore normally accepted that the energies of the emitted X-rays are independent of the chemical form of the generating atoms and differences in energy cannot be observed except under very high resolution. XRF, as was discussed in Chapter 8, is therefore an elemental analysis technique. It tells us what elements are present, but does not give any information on the oxidation state of the element or the chemical species present.

However, the energy E of the original electron ejected from the atom is the difference between the energy E_1 of the impinging electron (or photon) and the energy E_b required to remove the electron from the atom, that is, $E = E_1 - E_b$. The energy E_b will be slightly different depending on the chemical environment of the atom. We can determine this small difference by making the energy E_1 of the impinging electron just slightly greater than the energy E_b required to eject the electron. The residual energy E of the emitted electron will then be small, but any variation in E_b will produce a larger relative variation in the energy E of the emitted electron. In this way small differences in E_b can be measured. For example, if the energy of the K_α line for Al is 1487 eV, the effect of chemical environment may change this by 2 eV, resulting in Al K_α at 1485 eV. The relative shift (2/1487) is slight, and the 1485 eV X-ray line would be difficult to distinguish from the original 1487 eV line in X-ray emission and XRF. If the energy of the *impinging electrons* generating X-rays is 1497 eV, then the energy of the ejected electron is $1497 - 1487 = 10$ eV. But if the chemical environment changes the energy needed to eject the electron to 1485 eV, then the energy of the emitted electron is $1497 - 1485 = 12$ eV. It is easier to distinguish electrons with 10 eV energy from electrons with 12 eV energy than to distinguish between photons with energies of 1487 and 1485 eV. These slight changes in the energy of the ejected electron can provide information about the chemical species present and the oxidation state of the atoms present. Based on this phenomenon, the field of **electron spectroscopy for chemical analysis** (ESCA) was developed in the 1960s by Swedish physicist Kai Siegbahn and his coworkers. ESCA is also called **X-ray photoelectron spectroscopy (XPS)**.

A companion field, **Auger electron spectroscopy** (AES), was developed simultaneously. AES does not provide chemical species information, only elemental analysis, as we will see. Since the electrons ejected in these two techniques are of low energy and the probability of electron interaction with matter is very high, the electrons cannot escape from any significant depth in the sample. Typical escape depths for XPS and AES electrons range from 0.5 to 5 nm for materials. The phenomenon is therefore confined to a few atomic layers, combined or otherwise, which are at the surface of the sample and provides a method of surface analysis.

14.2.1. Electron Spectroscopy for Chemical Analysis (ESCA) or X-ray Photoelectron Spectroscopy (XPS)

When an X-ray beam of precisely known energy impinges on a sample surface held under an ultrahigh vacuum (UHV), inner shell electrons are ejected and the energy of the ejected photoelectrons is measured. This is the phenomenon on which ESCA, also called XPS, is based. ESCA was the original name for the technique, but the name XPS is now preferred by surface scientists; both terms will be used interchangeably. Figure 14.2 shows the emission of both a photoelectron and an Auger electron for a model atom.

The kinetic energy of the escaping electron is designated as E_k. The binding energy of this electron is given by the equation

$$E_k = h\nu - E_b - \phi \tag{14.1}$$

where E_b is the binding energy of the electron; $h\nu$, the energy of the photon (either X-ray or vacuum UV); E_k, the kinetic energy of the escaping electron; and ϕ, an instrumental constant called **the work function of the spectrometer**.

The XPS spectrum is a plot of the number of emitted electrons per energy interval vs. their kinetic energy. The work function of the spectrometer can be measured and is constant for a given instrument, allowing the binding energies of the electrons to be determined.

Since E_b, the original binding energy of the emitted electron, depends on the energy of the electronic orbit and the element from which the electron is emitted, it can be used to identify the element present. In addition, the chemical form or environment of the atom affects the energy E_b to a much smaller but measurable extent. These minor variations give rise to the **chemical shift** and can be used to identify the valence of the atom and sometimes its exact chemical form. Databases of binding energies for elements and compounds are available, such as the NIST electronic XPS database which can be found at www.nist.gov/srd/surface or tables in reference books such as that by Moulder et al. listed in the bibliography. Quantitative measurements can be made by determining the intensity of the ESCA lines of each element.

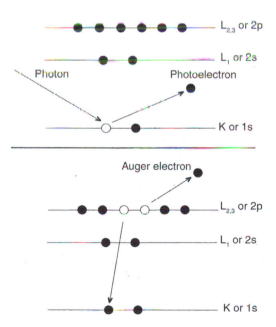

Figure 14.2 The XPS process for a model atom. The top diagram shows an incoming photon causing the ejection of the XPS (ESCA) photoelectron. The bottom diagram shows one possible relaxation process that follows the ejection of the photoelectron, resulting in the emission of an Auger electron. [From Moulder et al., courtesy of Physical Electronics USA, Inc. (www.phi.com).]

14.2.1.1.   Instrumentation for ESCA/XPS

A commercial ESCA/XPS instrument consists of four major components housed in a UHV system with magnetic shielding: (1) the radiation source, consisting of an X-ray source and a means of providing highly monochromatic X-rays; (2) the sample holder; (3) the energy analyzer, which resolves the electrons generated from the sample by energy; and (4) an electron detector. Modern instruments have computerized data recording and processing systems. The pressure required for ESCA must be very low, often less than 10^{-9} Pa in order to prevent adsorbed residual gas from interfering with the surface analysis. This requires a UHV system. As will be seen, many of the instrument components are also used for Auger spectroscopy instruments. The sample holder will be discussed in Section 14.2.3.

Radiation Source.   The radiation source used in ESCA is a standard X-ray anode tube, as described in Chapter 8. Soft X-rays are used, with Al and Mg being the most common anodes. Many commercial systems offer a dual anode X-ray tube so that the analyst can switch between excitation wavelengths. It is very important that the X-ray source be monochromatic, with a linewidth extending over as narrow an energy range as possible. Al and Mg have narrow K emission lines. The Mg K_α line has an energy of 1253.6 eV and a linewidth of 0.7 eV, while the Al K_α line has an energy of 1486.6 eV and a linewidth of 0.85 eV. Linewidths for other elements are generally about 1–4 eV. It can be seen from Eq. (14.1) that any variation in the energy of the impinging X-rays will produce a similar variation in the energy of the ejected ESCA electron. There are several ways to reduce the bandwidth of the source, especially if anodes other than Al are used. One of the most accurate is to use a crystal monochromator, such as the Rowland circle system shown in Fig. 14.3. (If necessary, the student should review diffraction of X-rays by crystals in Chapter 8.) Although this instrument gives very monochromatic radiation and narrow linewidths, which increase the spatial resolution of the technique, the intensity of the radiation is reduced in the process. Resolution achievable by using a crystal monochromator is on the order of 100 μm. It should be noted that Mg cannot be used as the source with an X-ray monochromator; there are no crystals available with the proper spacing.

If an aluminum target is used, an aluminum window placed after the target removes much of the K_β and background radiation (**Bremsstrahlung**) from the aluminum source. A magnesium filter can be used in a similar fashion with a magnesium source.

One problem with using X-ray beams as an energy source is that they cannot be focused; therefore, relatively large surface areas are examined, which limits the

Figure 14.3   Schematic diagram of an ESCA instrument using an X-ray monochromator and a multichannel detector.

spatial resolution of XPS. This problem can be overcome by attaching or depositing the sample on one side of a thin aluminum foil. The Al foil is then bombarded with an electron beam. The electron beam can be focused to a small point and excites the Al foil at that point. The excited Al emits Al K_α lines with a narrow wavelength range (narrow energy range). Only the sample in the immediate vicinity is exposed to this radiation. The net result is that samples in a small region, on the order of a few mm, can be excited with narrow band radiation, thus increasing the spatial resolution of the method. The disadvantage to this approach is that samples must be extremely thin, <1 μm, for the Al X-rays to reach the sample surface.

An alternative method for improving resolution, shown schematically in Fig. 14.4, selects the photoelectrons from a given area of the sample surface using an aperture and electron-focusing lens. Only the photoelectrons from a given small area on the sample surface are passed into the energy analyzer. The spatial resolution of this approach is on the order of 25 μm. The sample surface must be uniformly irradiated with a conventional X-ray source.

Electron Energy Analyzers. Electron energies can be filtered with energy discriminators as shown in Fig. 14.5. Figure 14.5(a) shows two parallel charge plates with two openings. The electrons with the required energy enter and leave the holes as in a slit system. Electrons with other energies do not exit the second hole. Figure 14.5(b) shows a second type of energy discriminator, which is a simple grid discriminator. Electrons with insufficient energy are repelled by the second grid and do not penetrate. This system only discriminates against electrons with low energies. Any electrons that have sufficient energy penetrate the second grid. A third system uses cylindrical plates [Fig. 14.5(c)]. If the angle between the planes of the entry and exit slits is 127.17° [i.e., $\pi/(2)^{1/2}$ rad], a double focusing effect is obtained and the intensity of the electron beam is maintained as high as possible. Electrons with the incorrect energy are lost either to the sides of the cylindrical plates or on the sides of the exit slit. Similar modifications have led to the development of a 180° cylindrical system rather than parallel plates [Fig. 14.5(d)]. This system does not provide such fine-tuning of the energy bandwidth, but it increases the intensity of the electron beam. This is another example of gaining in beam intensity but losing in energy discrimination, or gaining in power and losing in resolution.

Figure 14.4 Schematic diagram of an ESCA instrument with a single channel electron multiplier detector.

Parallel plates

(a)

(b)

(c)

Cylindrical plates

Electron focusing

e- beams

Exit

Entrance

(d)

Electron beam

Exit

Entrance

Figure 14.5   Schematic diagrams of several electron energy discriminators.

Electron energy analyzers are equivalent to the monochromators used in spectroscopy. Their function is to disperse the emitted photoelectrons based on their energies. The most commonly used electron energy analyzers incorporate an electrostatic field that is either symmetrical or hemispherical. These systems are in essence an extension of the electron energy filters shown in Fig. 14.5. All electron energy analyzers require shielding from stray magnetic fields as described subsequently. One system is shown schematically in Fig. 14.6; it is based on the system of Fig. 14.5(a) and is called a cylindrical mirror analyzer (CMA). The "plates" are now two coaxial cylinders, thus providing an efficient electron-trapping system while maintaining resolution. The electron source labeled on the figure is of course the sample surface, emitting photoelectrons. A voltage (negative potential) is applied to the outer cylinder; the inner cylinder is grounded. Only photoelectrons with the appropriate energy pass through the apertures and are focused onto the detector. Photoelectrons of energy E will be focused when:

$$E = \frac{KeV}{\ln(r_o/r_i)} \tag{14.2}$$

Cylindrical Analyzer

Electron source

Electron multiplier

Figure 14.6   Cross-section of an electrostatic cylindrical electron energy analyzer. The electron source is the excited sample surface that is emitting photoelectrons.

where K is a constant; e, the charge on the electron, V, the applied voltage, r_1, the radius of the inner cylinder; and r_o, the radius of the outer cylinder.

The CMA as shown is used for Auger electron spectroscopy. For XPS, two CMAs in series are used to obtain the required energy resolution. This design is called a double pass CMA. The transmission of electrons through a double pass CMA is good, but the resolution is poorer than that obtained using the concentric hemispherical analyzer described subsequently.

The concentric hemispherical analyzer (CHA), shown in Fig. 14.4, is widely used in both XPS and Auger instruments. The CHA consists of an input lens assembly, two concentric hemispherical shells of differing radii, and an electron detector. The input lens focuses the photoelectrons from the sample and lowers (or retards) their energy. This permits better energy resolution. The electrons then pass through a slit into the hemispherical path. The two shells are maintained at a potential difference, ΔV, with the outer sphere negatively charged and the inner one positively charged, as shown in Fig. 14.4. Only electrons with a very small energy range will pass through the hemispheres along a path of radius r_o and through the exit slit to the detector. The potential difference is varied to scan the electron energies and a spectrum of the energies of the ejected photoelectrons is recorded. Resolution of this analyzer, $\Delta E/E$, depends on the entrance angle of the electrons into the analyzer, the slit width and r_o, as follows:

$$\frac{\Delta E}{E} = \frac{w}{r_o} + \alpha^2 \tag{14.3}$$

where $\Delta E/E$ is the resolution; w, the slit width, r_o, the radius of the equipotential surface within the analyzer, and α, the angle at which the electrons enter the analyzer.

Two approaches to retardation (deceleration) of the electrons are used. One mode of operation is called **fixed-analyzer transmission (FAT)** or **constant-analyzer transmission (CAT)**. In this mode the electrons are retarded to a constant energy. This results in the absolute resolution of the system being independent of the kinetic energy of the incoming electrons and provides better quantitative analysis, especially for XPS. However, the signal-to-noise ratio for low kinetic energy peaks is poor, so identification of peaks in this part of the spectrum is difficult. The other operating mode is called **fixed retardation ratio (FRR)** or **constant relative ratio (CRR)**. In this mode the incoming electrons are decelerated by a constant ratio from their initial kinetic energies. This results in a constant relative resolution and better signal to noise ratio for low kinetic energy peaks. The FRR mode is better able to detect peaks with low kinetic energies (i.e., AES peaks) but quantitative analysis is poor.

Other analyzers have been designed based on the magnetic deflection of electrons, but in general these have not been very successful because of the difficulty of maintaining a uniform magnetic field. Electrostatic systems are used in all commercial instrumentation.

The instrument is calibrated regularly with known conductive standards such as gold or copper to establish the linearity of the energy scale and its position.

Detectors. Both single channel and multichannel detectors are used in ESCA and Auger spectroscopy. The most common single channel detector is the channel electron multiplier. The channel electron multiplier functions much like the PMT used in optical spectroscopy (discussed in Chapter 5). The major difference is that electrons constitute the signal that is being amplified, not photons. The channel electron multiplier consists of a continuous dynode surface inside a tube as depicted in Fig. 14.7(a). Channels may be straight tubes as shown or curved. The surface is a thin film conductor with a high resistance. When a voltage is supplied across the input and output, a potential gradient

Figure 14.7 (a) A schematic channel electron multiplier. A thin film conductive layer inside the tube serves as a continuous dynode surface. (b) A schematic microchannel plate configuration. [Both figures courtesy of Hamamatsu Corporation, Bridgewater, NJ (www.usa.hamamatsu.com).]

exists along the channel direction. The dynode surface and potential gradient permit electron amplification throughout the channel. An incoming electron strikes the inner wall and secondary electrons are emitted. These are accelerated by the potential gradient and travel a parabolic path until they strike the opposite wall, releasing more electrons. An ejected photoelectron from the sample enters at one end and an amplified pulse of electrons exits at the other end. Gains of up to 10^5 are possible with a 1 kV supply voltage. This detector is very efficient at counting electrons, even those with very low energy. Like the PMT, the detector can be saturated at high electron intensity.

Multichannel electron detectors of various types can be placed to cover the exit plane of the analyzer. CCDs (discussed in Chapter 7), phosphor-coated screens and position-sensitive detectors are some of the multichannel devices in use. One type of position-sensitive detector consists of a **microchannel plate (MCP)** electron multiplier. The MCP [Fig. 14.7(b)] consists of a large number of very thin conductive glass capillaries, each $6-25$ μm in diameter. The capillaries are fused together and sliced into a thin plate. Each capillary or channel works as an independent electron multiplier, exactly as the single channel electron multiplier described earlier, thereby forming a 2D electron multiplier. Used in conjunction with a phosphor screen, 2D imaging of surfaces is possible.

Electrons are detected as discrete events and the number of electrons for a given time and energy is stored and then displayed as a spectrum.

Magnetic Shielding. The ejected electrons have low energy. They are affected significantly by local magnetic fields, including the earth's and those of any stray electrical impulses as generated by wiring to lights, equipment, elevators, and so on. These stray fields must be neutralized in the critical parts of the instrument in order to obtain useful data. One method is to enclose the critical regions with high-permeability magnetic alloy, which shields the sample from stray magnetic fields. Another method is to use Helmholtz coils, which produce within themselves a homogeneous field. This field may be made exactly equal and opposite to the earth's magnetic field. A feedback system to the coils can also be used; this senses variations in local magnetic fields and varies the current in the Helmholtz coils, neutralizing transient magnetic fields as they arise. This system not only neutralizes the earth's magnetic field but also local fields as they are generated.

Electron Flood Gun. In practice, when the sample is irradiated, electrons are ejected. An electron takes with it a negative charge, leaving the sample positively charged. Depending on the conductivity of the sample, this positive charge builds up at a steady but unpredictable rate on the surface of the sample, changing the work function of the sample itself and therefore the net energy of the ejected electrons. In practice, this would lead to a variation in the observed energy of the ejected electrons and erroneous results would be derived. It is therefore highly desirable to eliminate this variable charge.

This problem can be overcome by flooding the surface of the sample with low-energy electrons, which neutralize the positive charge built up on the surface. This is done by using an electron flood gun. Of course, these electrons in turn affect the work function of the surface to some extent, but the effect is constant and reproducible data are obtainable over an extended period of time.

Ion Gun. An ion gun is used in XPS and Auger instruments for two purposes: (1) to clean the sample surface of any external contamination layer and (2) to sputter atoms from the surface in order to obtain a depth profile analysis, discussed under applications. One type of ion gun uses a heated filament to ionize inert gas atoms. The ions are accelerated by a potential placed on the ionization chamber and are focused to strike the sample surface. The ions remove atoms from the sample surface by collision. The rate of removal of surface atoms is controlled by the kinetic energy of the ions and by the nature of the surface atoms; sputtering rates may be as high as 10 nm/min.

UHV System. Since the electrons involved in ESCA are of low energy, they must not collide with other atoms or molecules in the system or they will not be detected. This is best handled by using a very high vacuum system. Pressures of $10^{-8}-10^{-9}$ Pa are typical in surface analysis systems. (One pascal, Pa, is equal to 7.50×10^{-3} torr.) In order to achieve this UHV, surface analysis instruments are fabricated of stainless steel. Stainless steel structures can be evacuated to the required low pressure with the appropriate combination of vacuum pumps. The stainless steel can also be heated ("baked out") to remove adsorbed gases. Other required components such as windows, gaskets, and the like must be made of materials that do not outgas and can withstand required elevated temperatures. The vacuum pumps used in surface analysis instruments to achieve the UHV pressures required include turbomolecular pumps, ion pumps, cryopumps, and oil diffusion pumps. Oil diffusion pumps must include a liquid nitrogen trap to prevent oil vapor from entering the instrument. Details of materials used in surface analysis systems and the vacuum pumps used can be found in the handbook by Ewing listed in the bibliography and references therein.

As we will see, surface analysis equipment is very similar for several techniques, so most surface analysis instruments are configured for multiple surface analysis techniques. A schematic diagram of a commercial instrument showing the placement of the ion gun, energy analyzer, source, and so on is shown in Fig. 14.8.

14.2.1.2. Sample Introduction and Handling for Surface Analysis

The sample must be kept under high vacuum during analysis. In older instruments and some modern lower cost instruments, the main analysis chamber must be brought to atmospheric pressure in order to insert or change the sample. The system is vented with nitrogen, the sample inserted, and the entire system pumped down to UHV pressures again before analysis can begin. This presents a problem when numerous samples are to be analyzed. Many instruments utilize sample prechambers, small volume compartments that will accommodate a number of samples and can be evacuated rapidly. The samples are then transferred into the main chamber without causing an excessive rise in pressure.

Sample holders range from those that accommodate samples of 1–2 cm diameter and about 1 cm in thickness to custom-designed sample chambers that can hold much bigger samples and those that are of irregular size and shape. Chambers designed for 8 in. silicon wafers in the semiconductor industry are common, for example. The sample

Figure 14.8 A schematic diagram of a commercial surface analysis instrument, the PHI Model 5600 MultiTechnique system. The monochromatic X-ray source is located perpendicular to the sample surface (not shown), the standard X-ray source is at 57.4° relative to the analyzer axis and the specimens are analyzed with an electron take-off angle of 70° relative to the surface place. [From Moulder et al., courtesy of Physical Electronics, USA, Inc., Eden Prairie, MN (www.phi.com).]

in its holder is placed on a stage that can be moved in all directions and allows the sample to be rotated and tilted for optimal positioning. The stage in a modern surface analysis instrument is under computer-control. Some sample chambers allow the sample to be heated or cooled as necessary. Cooling, for example, is required to analyze materials with a high vapor pressure, when volatile surface components are to be studied and for samples such as plastics that outgas at room temperature under UHV conditions.

XPS performed with commercial instruments as described is used for the analysis of the surfaces of solid samples. The solid must be in electrical contact with the spectrometer, so conducting solids are easily analyzed. Nonconductive samples can be analyzed by using a low current electron flood gun, or combination of electron flood gun and ion gun to charge the surface of an insulator. While solid samples can be analyzed "as is", they often must be cut to fit the sample holder and the surface is usually cleaned to remove layers of surface oxide and adsorbed gas so that the true sample surface is analyzed. Samples must be as clean as possible before being placed into the UHV system. Samples are often handled in a laminar flow clean hood. Sample surfaces, sample holders, and tools may be cleaned with high-purity solvents in an ultrasonic bath to remove traces of cutting fluids, oils, or grease. A layer of adsorbed gas or oxide from the environment will often contaminate the sample surface. It is common for samples to have adsorbed oxygen and nitrogen from the earth's atmosphere on their surface. But these are not really part of the surface the analyst wants to examine, in most cases. The contamination can lead to erroneous results if not recognized and corrected for. Surfaces are cleaned of this layer by baking the sample at high temperature or sputtering the surface of the sample with an ion gun, or subjecting the sample to oxygen plasma, among other techniques. Once cleaned, it is imperative that the sample be kept under high vacuum and be analyzed as soon as possible to avoid having the surface recontaminated. Even under vacuum, gas molecules will reabsorb onto a clean sample surface. Hercules and Hercules describe that at pressures between 10^{-8} and 10^{-10} torr, a surface can be covered by a monolayer of adsorbed gas molecules in from 1 to 10 h, while at 10^{-6} torr, the same process occurs in a few seconds. This rapid recontamination of a cleaned surface makes clear the need for UHV conditions for accurate surface analysis.

Some surface analysis instruments have the equipment to allow samples to be fractured within the UHV chamber, thereby exposing a clean surface for analysis. One problem with this approach is that fracturing may occur along grain boundaries. This may result in a "surface" that is not representative of the bulk material because impurities and additives in materials often segregate to the grain boundaries between crystals. Grinding the samples appropriately and then dusting the ground powder onto adhesive tape or pressing the powder into a pellet can be used to obtain bulk analysis of solid materials.

14.2.1.3. Analytical Applications of ESCA/XPS

Fundamental research in physical chemistry using XPS has included the study of gases and liquids to explore the energies of valence electrons, radicals, excited states, vibrational states, and more, but a discussion of this is beyond the scope of this text. Only the analysis of solid surfaces will be considered. For qualitative elemental analysis and especially for a "first pass" at an unknown sample, a survey scan of the energy range should be run. In order to obtain chemical state information, and for quantitative analysis of minor elements, a higher resolution detail scan is taken of the energy region of interest. Detail scans must be collected so that the exact peak position can be determined, for a sufficient period of time to give good counting statistics (high peak intensity to background intensity). Sample depth for most materials is a few atomic layers.

Tables 14.2 and 14.3 provide examples of applications of surface analysis techniques. XPS can be used for many of these applications.

Qualitative Analysis. ESCA is readily applied to the qualitative identification of elements present in a sample surface, to depths of 25 Å for metals and up to 100 Å for polymers. All elements except hydrogen and helium can be determined with the typical X-ray source. The XPS spectra for sulfur obtained with monochromated Al K_α radiation and MgK$_\alpha$ radiation are presented in Figs. 14.9 and 14.10, respectively. Each spectrum

Figure 14.9 XPS spectrum of sulfur collected with monochromated Al K$_\alpha$ radiation. [From Moulder et al., courtesy of Physical Electronics, USA, Inc. (www.phi.com).]

Figure 14.10   XPS spectrum of sulfur collected with Mg K$_\alpha$ radiation. [From Moulder et al., courtesy of Physical Electronics USA, Inc. (www.phi.com).]

consists of a limited number of photoelectron peaks in the range of 0–1400 eV in binding energy for Al (0–1200 eV for Mg). XPS photoelectron peaks from the sulfur 2s (K shell) and 2p (L shell) orbitals are seen, as well as an Auger peak (marked LMM). Some small contaminant peaks from Ar and oxygen can be seen in the spectra. Auger peaks can be distinguished from photoelectron (XPS) peaks. Because Auger electron kinetic energies are independent of the ionizing radiation energy, on a plot of binding energies, the Auger lines appear to be at different positions when different X-ray sources are used. The Auger peak is shown as an inset in each of the two spectra; as you can see, the Auger peak appears at a higher binding energy when Al X-radiation is used than when Mg X-radiation is used. The XPS peaks do not change position. Figures 14.9 and 14.10 are survey spectra, the type of scan that would be used for qualitative elemental analysis of an unknown.

For low atomic weight elements ($Z < 30$), energy peaks are observable corresponding to the K and L shells (the s and p electrons). For elements with atomic numbers between 35 and 70, the d electrons result in peaks in the XPS spectrum. For elements with atomic numbers greater than 70, the pattern includes f electrons. It can be seen that qualitative identification of the surface elements present is possible based on identification of the binding energies. Binding energies for all elements are tabulated and may be found in the NIST XPS database at http://www.nist.gov/srd/surface and in tables in references such as Moulder et al. listed in the bibliography. Typical data for S and Cu are shown in Fig. 14.11. In addition, the spin doublet patterns and ratios for the lines from the p, d, and f electrons are characteristic and also tabulated. The copper XPS spectrum is shown in Fig. 14.12. The spin doublet for the 2p electrons is seen ($2p_{1/2}$ and $2p_{3/2}$) and the ratio of the lines is about 1:2, which is the expected intensity ratio for these lines. XPS lines from the 2s, 3s, 3p, and 3d orbitals are also seen, along with the Auger LMM transition.

XPS spectra also exhibit some minor peaks due to a number of other processes. Small peaks called X-ray satellite peaks appear at lower binding energies due to the non-monochromatic nature of the X-ray source. X-ray "ghost peaks" occur due to other elements present in the X-ray source; Cu lines in spectra are an example of these ghost peaks. Copper is the base material used in the X-ray anode, so Cu lines may appear in the spectra of samples that contain no Cu. A complex photoelectric process results in

Line Positions (eV)

Photoelectron Lines

2s	2p$_{1/2}$	2p$_{3/2}$	3s
228	165	164	18

Auger Lines

L$_{23}$M$_{23}$M$_{23}$
1336 (Al)
1103 (Mg)

Line Positions (eV)

Photoelectron Lines

2s	2p$_{1/2}$	2p$_{3/2}$	3s	3p$_{1/2}$	3p$_{3/2}$
1097	953	933	123	77	75

Auger Lines

L$_3$M$_{23}$M$_{23}$	L$_2$M$_{23}$M$_{23}$	L$_3$M$_{23}$M$_{45}$ (^{1}P)	
719	712		
486	479˙	648	(Al)
		415	(Mg)

L$_3$M$_{23}$M$_{45}$ (^{3}P)	L$_2$M$_{23}$M$_{45}$ (^{1}P)	L$_3$M$_{45}$M$_{45}$	L$_2$M$_{45}$M$_{45}$	
640	628	568	548	(Al)
407	395	335	315	(Mg)

Figure 14.11 Typical XPS data tables from literature references. (Top) Data for sulfur; (bottom) data for copper. [From Moulder et al., courtesy of Physical Electronics USA, Inc. (www.phi.com).]

"shake-up" peaks, which are peaks that occur slightly higher in binding energy than the main XPS photoelectron peak. Figure 14.13 shows the shake-up peaks in a copper oxide sample; they are the peaks (a singlet and an irregular doublet) at higher binding energy than the marked photoelectron peaks.

Figure 14.12 The XPS spectrum of copper collected with Mg K$_\alpha$ radiation. [From Moulder et al., courtesy of Physical Electronics USA, Inc. (www.phi.com).]

Figure 14.13 "Shake-up" peaks in the copper XPS spectrum of a copper oxide sample. These peaks are one example of several processes that result in lines other than XPS and Auger lines in an XPS spectrum. [Modified from Moulder et al., courtesy of Physical Electronics USA, Inc. (www.phi.com).]

ESCA chemical shift. There is a shift in the binding energy of the core levels of atoms brought about by the chemical environment of the atom. The shift is called a chemical shift and is on the order of 0.1–10 eV. The shift occurs as a result of changes in the electrostatic screening felt by the core electrons as a result of valence electron position. As an example, electronegative elements such as fluorine attached to the atom of interest (e.g., sulfur in SF_6) will pull valence electrons away from the sulfur atom, resulting in a higher binding energy for the core sulfur electrons as a result of the decreased negative charge on the sulfur atom. That is, the sulfur core electrons in SF_6 are held more tightly by the positively charged sulfur nucleus due to less screening by the valence electrons than are the core sulfur electrons in elemental sulfur. Figure 14.14(a) shows how changes in chemical environment cause the binding energy of the sulfur $2p_{3/2}$ electrons to shift. Elemental sulfur has a binding energy of 164.0 eV for the $2p_{3/2}$ line. Sulfur in compounds varies from this value by a few eV; some compounds are shifted to lower binding energies and some to higher binding energies. The attachment of highly electronegative fluorine atoms to the sulfur atom results in a shift in the binding energy to a higher value as just described. Figure 14.14(b) presents similar data for copper in a variety of compounds. The three chemically different nitrogen atoms in a cobalt complex can be identified by the nitrogen 1s XPS spectrum (Fig. 14.15). In a similar fashion, the chemical environments of carbon and oxygen in a compound influence the carbon and oxygen ESCA spectra as shown in Fig. 14.16. This example is from Siegbahn, who with his coworkers, developed the ESCA technique and first reported the chemical shift in copper 1s binding energies in 1957. Siegbahn was awarded the Nobel Prize in physics in 1981 for his pioneering work. Tables of chemical shifts for the oxidation states of elements are also tabulated in the literature and at the NIST XPS database site.

An interesting example of the use of ESCA is the nondestructive identification of elements on the lunar surface. The Surveyor was an unmanned probe that landed on the surface of the moon. The lunar soil was exposed to magnesium K_α radiation and the ESCA spectrum collected. The elemental components of the soil can be clearly identified

(a)

(b)

Figure 14.14 (a) Chemical shift in the sulfur $2p_{3/2}$ binding energy in various sulfur compounds. (b) Chemical shift in the copper $2p_{3/2}$ binding energy in various copper compounds. [From Moulder et al., courtesy of Physical Electronics USA, Inc. (www.phi.com).]

(Fig. 14.17). The advantage of a surface analysis technique for this application is obvious. It is easier to examine the surface with a remote probe (X-rays) than to devise a way of having an unmanned spacecraft take a sample for an analytical method that requires elaborate sample preparation.

A high-resolution (detail) scan of the Pb 4f electron region (Fig. 14.18) for lead bromide and lead chloride using magnesium K_α radiation was obtained. In this case, 0.07 eV was the difference between the maxima for these two compounds. This resolution approaches the practical limitations of the technique.

A sample of pure lead was evaporated onto a surface, which was then exposed to air. The ESCA spectra were obtained at various time intervals as indicated in Fig. 14.19. The spectra show the formation of lead oxide on the surface, going through the intermediate of PbO and finally, after a number of hours, to PbO_2. This demonstrates the capability of ESCA/XPS to study surface chemistry.

Quantitative Analysis.   Quantitative analysis by XPS can be used to determine the relative concentrations of components of a sample. Either peak area or peak height sensitivity factors can be used, but the latter approach is more accurate.

Figure 14.15 Nitrogen 1s XPS spectrum for *trans*-[Co(NH₂CH₂CH₂NH₂)₂(NO₂)₂]NO₃. The three chemically different nitrogen atoms exhibit slightly different chemical shifts. The peak intensities are an indication of the number of each type of nitrogen atom. (Reprinted with permission from Hendrickson et al., copyright 1969 American Chemical Society.)

Figure 14.16 ESCA spectrum showing the chemical shifts of carbon and oxygen atoms in methyl acetate. (From Siegbahn, copyright 1973 with permission from Elsevier.)

Figure 14.17 The ESCA spectrum of lunar soil, taken by the Surveyor. (*Am. Lab.* 1972.)

For a homogeneous sample (one that is homogeneous in the analysis volume), the following equation can be written for the intensity of a given spectral peak:

$$I = nf\sigma\theta\lambda yAT \tag{14.4}$$

where I is the intensity (photoelectrons/s); n, the number of atoms of the element/cm^3 sample; f, the X-ray flux (photons/s cm^2); σ, the photoelectric cross-section for the

Figure 14.18 High-resolution XPS spectrum of pure PbBr$_2$ overlaid on a similarly acquired spectrum of pure PbCl$_2$. A chemical shift of 0.07 eV between the two Pb lines would enable these two compounds to be distinguished by XPS. Mg K$_\alpha$ radiation was the excitation source.

Figure 14.19 The use of ESCA to study changes in the surface composition of Pb on exposure to air and heat. (From Siegbahn, copyright 1973 with permission from Elsevier.)

transition (cm^2/atom); θ, the angular efficiency factor based on the angle between the photon path and detected electron; y, the efficiency of production of primary photoelectrons; λ, the mean free path of the photoelectrons in the sample (cm); A, the area of the sample from which photoelectrons are detected (cm^2); and T, the detection efficiency for emitted photoelectrons.

An atomic sensitivity factor, S, can be defined as:

$$S = f\sigma\theta\lambda yAT \tag{14.5}$$

To determine two elements in a sample, an intense line from each element is measured. The relationship between the amount of each element, the intensity of the line, and the factor S for each element is:

$$\frac{n_1}{n_2} = \frac{I_1/S_1}{I_2/S_2} \tag{14.6}$$

This equation can be used for all homogeneous samples because the ratio S_1/S_2 is for all purposes matrix-independent. For any spectrometer it is possible to develop a set of relative values for S for every element. (The instrument manufacturer usually provides such values.) To determine the atomic concentration of elements on the surface, the net peak intensity of the most sensitive line for each element detected must be measured. The fractional atom concentration, C_A, of element A is given by:

$$C_A = \frac{n_A}{\sum n_i} = \frac{I_A/S_A}{\sum I_i/S_i} \tag{14.7}$$

This use of atomic sensitivity factors will give semiquantitative results, within 10–20% of the value, for most homogeneous materials. While the absolute sensitivity of XPS is high, the actual sample volume analyzed is small, so the amount of an element that can be detected is in the % range with sensitivity of about 0.5 atom%.

Depth Profiling and Element Location. By using an ion source to sputter or remove the outermost surface atoms, layers below the surface can be studied. Examining sample composition as a function of depth below the surface is called **depth profiling**. The sputtering of layers must be done in a controlled manner so that the depth of the layer being examined is known. This approach is very useful in determining the composition of thin multilayered materials. One limitation of ion sputtering for depth profiling is that changes in chemical state may occur as a result of the ion bombardment. The rate of removal of atoms is material dependent; gold is removed from a surface much faster than an inorganic oxide like SiO_2, for example. This can pose a problem for inhomogeneous materials.

One simple way of distinguishing an element on the surface from an element in the bulk of the material using XPS is to change the angle at which the sample is mounted. A shallow angle between the plane of the sample surface and the analyzer axis results in a shallow electron take-off angle; this will accentuate signals from the surface of the material. This shallow angle is also called a grazing angle. Samples mounted at 90° (normal) to the analyzer axis with respect to the surface plane will have the surface signal minimized with respect to the bulk signal. An example of the effect of changing the sample angle is presented in Fig. 14.20.

Modern XPS systems have the ability to analyze areas with a diameter of $<10\ \mu m$. This enables a surface to be mapped at high resolution and a plot of element distribution and chemical state information to be made.

14.2.2. Auger Electron Spectroscopy (AES)

The process on which AES is based is similar to that of ESCA, but it is a two-step process instead of one step. As with ESCA, the sample is irradiated with either accelerated electrons or X-ray photons. An inner shell electron is ejected, leaving a vacancy in the inner shell. An electron from an outer shell falls into the inner shell as in the ESCA process.

Figure 14.20 An example of the enhanced surface sensitivity achieved by varying the electron take-off angle. A thin SiO_2 oxide layer on silicon (Si) is enhanced at the shallow grazing angle. [From Moulder et al., courtesy of Physical Electronics USA, Inc. (www.phi.com).]

The ion then either emits an X-ray photon or an Auger electron. In X-ray emission the energy balance is maintained by the emission of an X-ray photon with energy equal to the difference between the two energy levels involved (Fig. 14.1). This is XRF, discussed in Chapter 8. In the Auger process the released energy is transferred to a second electron in an outer shell, which is then emitted. This is the Auger electron. The actual process by which the energy is transferred is not clearly understood but can be represented schematically as in Figs. 14.2 and 14.21.

In the examples in both figures, an electron from the K shell was ejected by bombardment. An electron from the L shell descended to the K shell, simultaneously transferring energy to a second electron in the L shell. This second electron was ejected as an Auger electron. This Auger electron is termed a KLL Auger electron. This nomenclature arises from the fact that an electron was ejected from a K shell followed by a transition from an L to a K shell, simultaneously liberating an L Auger electron. It can be deduced that other Auger transitions, such as LMM and MNN, are also possible.

The Auger electron is ejected by the energy released by relaxation, that is, by the neighboring electron that drops into the inner orbital. The energy released on relaxation is a function of the particular atom and not of the energy of the source used to eject the initial electron from the inner orbital. Therefore the kinetic energy of the Auger electron is independent of the energy range of the radiation source falling on the sample. This means that the source does not have to be monoenergetic (or monochromatic). Because the Auger peaks are independent of the source energy, unlike the XPS peaks, they can be distinguished readily from the XPS peaks. When spectra are collected with two different source energies, as shown in Figs. 14.9 and 14.10, the binding energy of the Auger electron apparently changes with source energy while the binding energies of the XPS photoelectrons do not.

The Auger process and XRF are competitive processes for the liberation of energy from bombarded atoms. In practice, it is found that the Auger process is more likely to occur with low atomic number elements; this probability decreases with increasing atomic number. In contrast, XRF is unlikely with elements of low atomic number but increases in probability with increasing atomic number. This is illustrated in Fig. 14.22.

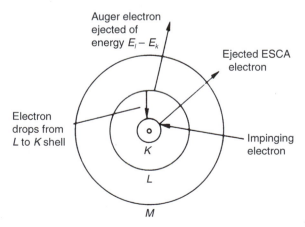

Figure 14.21 A simplified view of the process of Auger electron emission. An impinging electron ejects an inner shell electron, leaving an incomplete K shell. An electron from the L shell drops into the K shell and simultaneously an Auger electron is emitted from the L shell with energy $E_L - E_K$. This process competes with the process that produces X-ray photons (X-ray fluorescence) shown in Fig. 14.1.

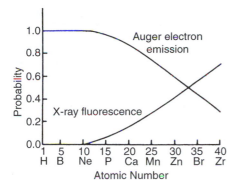

Figure 14.22 Relationship between atomic number and the probabilities of X-ray fluorescence and Auger electron emission. (Reprinted with permission from Hercules, copyright 1970 American Chemical Society.)

The energies involved in Auger spectroscopy are similar in all respects to those of ESCA, since the same atomic shells are involved. A graphical plot of Auger electron energies is shown in Fig. 14.23 and tabulated values of the prominent lines are available in the literature.

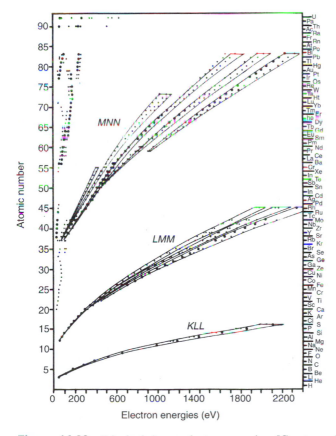

Figure 14.23 Principal Auger electron energies. [Courtesy of Physical Electronics USA, Inc. (www.phi.com).]

AES is an elemental surface analysis technique that can detect elements from lithium to uranium with a sensitivity of about 0.5 atom%. Auger spectra consist of a few peaks for each element in the same energy region, 0–1000 eV, as ESCA peaks. Auger peaks are less intense than ESCA peaks as seen in Figs. 14.9 and 14.12. Therefore, Auger spectra are generally plotted as the first derivative of the signal, $dN(E)/dE$, vs. the Auger electron energy (Fig. 14.24). The use of the first derivative of the signal is a common practice to enhance small signals and to minimize the high and sloping background from scattered electrons.

Auger Chemical Shift. The chemical environment of the atom also shifts the Auger lines in the same way the ESCA lines are shifted. The chemical shift magnitude may be greater for the Auger lines than for the ESCA chemical shift. The Auger chemical shift is very useful for the identification of chemical states. A plot of Auger electron kinetic energy vs. the XPS photoelectron binding energy is a useful tool for identifying the chemicals states of an element. We can define a quantity called the Auger parameter as the difference between the kinetic energies of the Auger electron and the photoelectron:

$$AP = KE_{(A)} - KE_{(P)} = BE_{(P)} - BE_{(A)} \tag{14.8}$$

where AP is the Auger parameter; KE, the kinetic energy of the Auger (A) and photoelectron (P) respectively; and BE, the binding energy of the respective electrons. It can be shown that

$$KE_{(P)} = h\nu - BE_{(A)}$$

which leads to

$$KE_{(A)} + BE_{(P)} = AP + h\nu \tag{14.9}$$

Plots of $KE_{(A)}$ vs. $BE_{(P)}$, with $AP + h\nu$ shown on the other ordinate (called Wagner plots) are found in the literature and in the NIST XPS database. An example for tin is presented in Fig. 14.25. The Sn $3d_{5/2}$ photoelectron and the Sn MNN Auger electron binding energies are tabulated and plotted for the element Sn and a variety of tin compounds, showing how the chemical environment of the Sn affects the electron energies.

Figure 14.24 The direct Auger spectrum of silver (bottom), the direct spectrum with a tenfold enhancement of the signal (middle), and the derivative or differential spectrum (top). The derivative spectrum significantly improves the ability to measure a small signal against a high and sloping background signal. (From Weber.)

Tin		
Compound	Sn 3d$_{5/2}$ Binding Energy (eV)	Sn MNN Kinetic Energy (eV)
Sn	484.9	437.4
Sn	485.0	437.4
Sn	485.1	437.4
SnS	485.6	435.7
Na$_2$SnO$_3$	486.2	431.7
SnO$_2$	486.7	432.7
Na$_2$SnO$_3$	486.7	431.7
Na$_2$SnO$_3$	487.2	431.7
NaSnF$_3$	487.4	430.8

Figure 14.25 Table of values and plot of the Auger MNN electron kinetic energy for tin and Auger parameter vs. the photoelectron binding energy for the tin 3d$_{5/2}$ electron in the element and a variety of compounds. The metal and different chemical states of tin in compounds can be readily distinguished. [From Moulder et al., courtesy of Physical Electronics USA, Inc. (www.phi.com).]

14.2.2.1. Instrumentation for AES

The instrumentation used in AES is very similar to that used in ESCA. The major difference is that the source used is a focused beam of electrons from an electron gun or a field emission source, not X-ray photons. A schematic diagram of an AES instrument is shown in Fig. 14.26. Many instrument manufacturers provide instruments that permit both XPS and Auger spectra to be collected on one instrument.

Figure 14.26 Schematic diagram of an Auger spectrometer, showing the electron gun source, an electron flood gun and an ion gun, along with a carousel for multiple samples. The vacuum system is not shown. [Courtesy of Physical Electronics USA, Inc. Eden Prairie, MN (www.phi.com).]

Figure 14.27   A schematic cross-section of an electron gun with a tungsten filament and an Einsel lens focusing arrangement. S is the sample, V_f is the focus voltage, V_e is the emission voltage, and V_a is the acceleration voltage. (From Turner, used with permission.)

The advantage of an electron beam is that it can be focused and deflected, unlike X-rays. The electron beam can be focused, depending on the source, from a spot size of 10 nm to a spot size of several hundred micrometers. The focused beam can be scanned over the surface. This permits compositional mapping of a surface with very high resolution and the ability to study very small features. Electron guns may use a heated tungsten filament or lanthanum hexaboride rods as the cathode. Figure 14.27 shows a schematic cross-section of an electron gun with a tungsten filament.

A field emission source uses a needle-like tungsten or carbon tip as the cathode, shown in Fig. 14.28. The tip is only nanometers wide, resulting in a very high electric field at the tip. Electrons can tunnel out of the tip with no input of thermal energy, resulting in an extremely narrow beam of electrons. Electron beams from heated filaments have a focal (cross-over) diameter of about 50 μm while a field emission source has a cross-over diameter of only about 10 nm. Field emission sources can serve as probes of surfaces at the nanometer scale (an Auger nanoprobe).

14.2.2.2.   Applications of AES

AES is primarily a surface elemental analysis technique. It is used to identify the elemental composition of solid surfaces, and can be used to quantify surface components, although quantitative analysis is not straightforward. AES is a true surface analysis technique, because the low-energy Auger electrons can only escape from the first few (three to five) atomic layers or from depths of 0.2–2.0 nm.

The electron beam can be deflected in a line across the surface of a sample to analyze multiple points or it can be rastered (moved in two dimensions) to produce a compositional map of the surface. Very small spot size and large magnifications (up to 20,000×) enable AES to study extremely small particles and intricate structures such as those used in microelectronics. Depth profiling with ion beam sputtering of the surface permits the analysis of layers and the identification of components in the grain boundaries of crystalline materials. The use of AES to study segregation of impurities in the grain boundaries of

Figure 14.28   Field emitter of single crystal tungsten is shown (left) on its supporting hairpin and (right) in a close up view of the tip. (From Stinger.)

metals and alloys has played a pivotal role in understanding embrittlement of steels, corrosion and stress corrosion cracking of alloys, and the behavior of ductile tungsten used in light bulb filaments. Very often, sulfur, phosphorous, or carbon impurities in the grain boundaries of metals and alloys are the sites at which corrosion, cracking, and failure of the alloy are initiated. Alternatively, impurities in the grain boundaries may have a beneficial effect. Potassium in the grain boundaries of ductile tungsten acts as a lubricant, allowing the tungsten wire filament to expand and contract when a light bulb is switched on and off without breaking along the grain boundaries.

Some application examples are presented. Figure 14.29 shows the Auger spectrum of stainless steel before and after heating at 750°C for 10 min. It can be seen that the major components, iron, nickel, and chromium, are unchanged by the heating. However, the carbon that was on the surface of the stainless steel before heating disappeared after heating. This surface carbon could be from traces of lubricating oil or cutting fluid, for example. Sulfur is present on the surface after heating and the phosphorus signal has increased. This may be due to diffusion from the bulk stainless steel to the surface during heating.

The electron beam can scan a surface area systematically in what is called a raster scan. By monitoring the intensity of the Auger spectrum of a particular element during scanning, it is possible to map its distribution on the surface examined. An example of this technique is shown in Fig. 14.30, which reveals the concentrations and location of sulfur and nickel on a spent catalyst surface. In another example, Fig. 14.31 shows the distribution of molybdenum, gold, and oxygen on an integrated circuit. Using this technique it is possible to map the distributions of these three elements and their positions relative to the circuit. The use of a scanning mode and small spot size is called scanning Auger microscopy (or scanning Auger microprobe), known by the initials SAM.

Figure 14.29 Auger spectra of stainless steel before and after heating in vacuum at 750°C for 10 min. As a result of heating, carbon has been lost from the surface while sulfur and phosphorus have diffused from the bulk steel to the surface.

Figure 14.30 Auger element distribution maps of Ni and S on the surface of a spent catalyst, obtained by rastering the electron beam across the surface of the catalyst.

Depth Profiling. As can be done with ESCA, another valuable application of Auger electron spectroscopy has been developed by using an ion beam to progressively strip off the surface of a sample under controlled conditions from a sample. Spectra can be collected and the distribution of elements recorded as surface material is removed.

Figure 14.31 Auger element maps of an integrated circuit, showing the distribution of oxygen, gold, and molybdenum on the surface. [Courtesy of Physical Electronics USA, Inc. Eden Prairie, MN (www.phi.com).]

Figure 14.32 Use of ion sputtering for depth profile analysis. An ion gun is used to remove surface layers during collection of Auger spectra. [Courtesy of Physical Electronics USA, Inc. Eden Prairie, MN (www.phi.com).]

The results show changes in distribution of different elements with depth, called a depth profile.

A schematic diagram of a sample undergoing depth profiling is shown in Fig. 14.32. The ion beam sputters the surface and etches a crater into the material, while the electron beam is used as the source for AES. Figure 14.33 shows the results of such a depth profile investigation of Nichrome film on a silicon substrate. It can be seen from the plot that the Nichrome film and its outer oxide layer are about 150–175 Å thick. That can be deduced from the appearance of the Si substrate after about 175 Å of material has been sputtered off. It is also evident that the chromium at the outer surface has formed an oxide layer about 75 Å thick. Chromium readily forms a protective oxide layer when exposed to air. The oxygen signal decreases after the oxide layer has been sputtered off, revealing the Ni and Cr film itself. Information about surface oxide layers, layer thickness and composition is critical to understanding corrosion chemistry, surface reactions, material behavior, and device fabrication.

Another example of use of depth profiling is shown in Fig. 14.34; AES clearly shows the very localized region containing Cl at a depth of about 1100 Å from the surface of the

Figure 14.33 Depth profile analysis of a Nichrome film on a silicon substrate. The outermost layer shows both Cr and oxygen, probably as a stable chromium oxide layer. The Ni and Cr (the Nichrome film) signals drop off at about 175 Å while the Si substrate signal starts to appear, indicating the approximate film thickness. (From Weber.)

Figure 14.34 Depth profile of material with a highly localized narrow region containing Cl. The high lateral resolution of Auger electron spectroscopy allows the characterization of thin layers and small features, such as this 100 Å thick Cl-containing region. (From Weber.)

material. The Cl-containing region is only about 100 Å wide. This extremely fine lateral resolution is one of the strengths of AES over other analytical techniques.

14.3. ION SCATTERING SPECTROSCOPY

In **ion scattering spectroscopy** (ISS) a beam of ions is directed at the sample. On collision with the surface, the ions are scattered by the sample atoms at the surface. Some of the bombarding ions are scattered after a single binary elastic collision with the surface atoms. For such ions scattered at a given angle, conservation of momentum results in the energy of these ions being dependent only on the mass of the surface atom and the energy of the bombarding (primary) ion. By measuring the energy of the *scattered ions*, one can determine the mass of the *scattering atoms* on the surface.

For a scattering angle of 90°, the relationship between the energy of the scattered ions and the masses of the bombarding ion and scattering atom are given by the equation

$$E_S = E_0 \frac{M_S - M_{\text{ion}}}{M_S + M_{\text{ion}}} \tag{14.10}$$

or

$$\frac{E_S}{E_0} = \frac{M_S - M_{\text{ion}}}{M_S + M_{\text{ion}}}$$

where E_s is the energy of the scattered ion (after collision); E_0, the energy of the bombarding ion (before collision); M_s, the mass of the scattering (surface) atom; and M_{ion}, the mass of the bombarding ion.

The process is illustrated in Fig. 14.35. In this system the bombarding ion is $^3\text{He}^+$; the scattering atoms were ^{16}O and ^{18}O. Based on Eq. (14.10), the energy of the scattered helium ions can be calculated if the primary beam energy is known. The energy of the scattered helium ion will differ depending on whether it was scattered by an atom of ^{16}O or ^{18}O. By measuring the relative abundance (signal intensity) of the energies of scattered helium ions, the relative amounts of ^{16}O and ^{18}O on the surface of the sample can be determined.

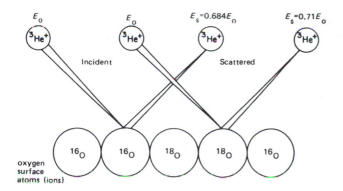

Figure 14.35 The process of ion scattering spectroscopy. The incoming helium ions are scattered with different energies from ^{16}O and ^{18}O. (From Czanderna et al.)

Only exposed atoms on the surface, that is, the top monolayer of atoms, contributes to the signal in ISS. It is therefore the most surface sensitive of the surface analysis techniques. ISS is one technique that provides isotopic information on surface atoms. ISS can be used to determine all elements with an atomic number greater than that of the bombarding ion. The elemental and isotopic compositions of the surface can be determined both qualitatively and quantitatively. Sensitivity is about 1% of a monolayer for most elements.

A schematic diagram of instrumentation used for ISS is shown in Fig. 14.36. This instrument is based on a CMA, discussed in Section 14.2.1.1. The instrument shown consists of an ion source, a vacuum system (not shown), an energy analyzer, and a detector. The major instrument components including the sample are under vacuum.

The most common ion guns use inert (noble) gases, particularly helium, argon, and neon, to produce monoenergetic ions. Ion guns are capable of focusing the ion beam from 100 μm to 1 mm diameter spot size and are able to control the current density of the ion beam. Ions are produced in the ion gun by electron bombardment of the gas atoms. A stream of the inert gas passes through an electron grid, which ionizes the gas by electron impact. These ions can then be accelerated to a known energy using an accelerating

Figure 14.36 Ion scattering spectrometer based on a cylindrical mirror analyzer.

electrode. The accelerated ion beam is then focused onto the surface of the sample. Here the ions are scattered, and those scattered at some predetermined angle enter an energy-analyzing system, where they are separated before reaching an ion detector.

The scattering efficiencies of different rare gas ions are different for different surfaces. Heavy surface atoms respond better to bombardment with a heavy ion such as argon rather than He. The ability of ISS to resolve atoms close in mass decreases as the difference between the bombarding ion and the target atom increases. Spatial resolution of ISS is limited to about 100 μm, so it is not as good as AES in this respect.

Applications of ISS are similar to the applications of static SIMS, described subsequently. ISS is used to study surface reaction mechanisms, catalyst behavior, and adsorption–desorption processes at surfaces. Because of its ability to discriminate among isotopes of an element, ISS can be used to study diffusion or any other reactions involving the replacement of one isotope for another.

14.4. SECONDARY ION MASS SPECTROMETRY (SIMS)

A companion method of analysis to ISS is secondary ion mass spectrometry (SIMS). This process is slightly different from ISS, as shown in Fig. 14.37. In this instance the surface is again bombarded with a beam of an ionized inert gas (the **primary ion beam**). In addition to scattering, the primary ions displace (sputter) ions from the sample surface by setting up a "collision cascade". In this collision cascade, chemical bonds are ruptured in the solid by direct impact or by indirect energy transfer. Ions as well as neutral molecules are sputtered from the surface as a result of this process. The sputtered ions, called *secondary* ions, may be a complete molecular ion or an ionized fragment of a molecule. In addition, the sputtered ions may be positively or negatively charged; in practice, both types of ions are utilized in the analytical process. The mass-to-charge ratios of the sputtered ions are measured, as in other forms of MS (Chapters 9 and 10), and a spectrum of the intensity of the secondary ions vs. mass-to-charge ratio is generated. The detected secondary ions are sputtered from the top two or three monolayers of the surface, making SIMS a very surface-sensitive technique.

SIMS can be performed in two modes: **dynamic SIMS**, which uses a high primary ion beam intensity; and **static SIMS**, which uses a very low primary ion beam intensity.

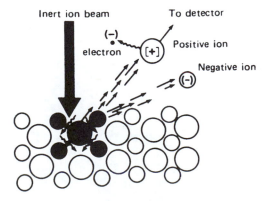

Figure 14.37 The SIMS process. Positive and negative ions are ejected from the surface by bombardment with inert ions (He^+).

Dynamic SIMS results in high sputter rates, high secondary ion production and therefore low detection limits. The sensitivity for surface analysis by dynamic SIMS is about 1 ppm. Dynamic SIMS is a destructive technique because of the high rate of removal of surface layers. The sputtering rate depends on the ion beam intensity, the energy of the ions, and the angle of incidence, as well as the mass of the bombarding ion and the nature of the sample itself. The sample mass, crystal orientation, specific chemical composition, and other sample properties also affect the rate at which surface atoms or molecules are removed. Static SIMS can be used to study samples that are extremely thin or need to be analyzed nondestructively. The low primary beam intensity removes only a few mono-layers of the surface. The sensitivity is not as good as that of dynamic SIMS because the secondary ion production is low. Detection limits are generally in the ppm–0.1% range for static SIMS. All elements including hydrogen can be determined, and molecular species can also be measured.

14.4.1. Instrumentation for SIMS

In its most elementary form, a SIMS system consists of a source of primary ions, a sample holder, secondary ion extraction optics, a mass spectrometer, and an ion detector, all housed in a UHV compartment. Systems are also equipped with data processing and output systems. A schematic diagram of part of a SIMS system with a quadrupole mass analyzer is shown in Fig. 14.38. The design and operation of mass spectrometers is covered in Chapter 9 and will be only briefly reviewed here.

14.4.1.1. Primary Ion Sources

SIMS instruments frequently are equipped with two ion sources, one source designed to generate positive secondary ions and the other to generate negative secondary ions. The use of an electronegative primary ion such as O_2^+ results in enhanced generation of positive secondary ions, while the use of an electropositive primary ion such as Cs^+ gives enhanced generation of negative secondary ions. The oxygen-based ion beam is generated by passing the gas through a low-voltage, low-pressure hot cathode arc, which produces a plasma of negatively and positively charged gas ions. The desired ions are extracted from the plasma, accelerated, and focused by electrostatic and magnetic fields onto the

Figure 14.38 Partial schematic diagram of an SIMS instrument equipped with an ion gun source and a quadrupole mass analyzer.

sample surface. Such ion beams can be focused to spot sizes from about 50 μm to as little as 0.5 μm. The second source is normally a cesium ion gun for the generation of negative secondary ions. A primary ion mass filter can be used to remove doubly charged ions and neutrals (Fig. 14.39).

TOF-SIMS instruments, described subsequently, may use two ion guns, a nanosecond pulsed Ga^+ ion gun to acquire the spectrum and a second Cs^+ ion gun for sputtering the surface and depth profiling. The gallium and cesium ion guns have a reservoir of the metal, which is heated to form gas phase atoms. The gas phase atoms are ionized at a charged tungsten grid, extracted, and focused to form the ion beam.

14.4.1.2. *Mass Spectrometers*

The three most common types of mass analyzers in SIMS systems are: (1) double focusing magnetic sector instruments, (2) time of flight (TOF) mass spectrometers, and (3) quadrupole mass spectrometers. The choice of mass analyzer depends on whether dynamic or static SIMS is needed, on the requirements of mass range and resolution, and on transmission efficiency, among other factors. The mass analyzers have been discussed in Chapter 9 in detail and this chapter should be reviewed as necessary.

Double-focusing magnetic sector instruments use a magnetic field to separate the ions by mass-to-charge ratio. The mass range covered by this type of analyzer is up to 500 Da with a mass resolution, $M/\delta M$, of 10,000 and high transmission efficiency at low resolution. The advantages of this type of mass analyzer are high mass resolution, direct ion imaging capability with an imaging detector in place, and high sensitivity over the mass range. Sensitivity is lower when high mass resolution is required. When a magnetic sector instrument is used, only a single mass is measured at a time. Secondary ions are generated constantly by the continuous primary ion beam bombardment, but most of these ions are not detected. Figure 14.39 shows a schematic double-focusing ion microprobe mass spectrometer with a magnetic sector used as a primary ion mass filter.

Figure 14.39   A double-focusing ion microprobe mass spectrometer with a magnetic sector to filter the primary ion beam.

Quadrupole mass analyzers operate by varying the applied RF and DC voltages to the four rods that make up the quadrupole. Only ions of a given m/z ratio can pass through the rods at a given set of conditions; all other ions collide with the rods and are not detected. Direct ion imaging cannot be done with a quadrupole, although surface mapping is possible. The mass range that can be covered is up to 1000 Da, but the mass resolution is only about 200 with a transmission efficiency of about 10%, or 30% that of a magnetic sector instrument. The advantages of a quadrupole mass analyzer are its low cost, simple operation, its ability to switch rapidly from positive to negative ion mode, and its flexibility in handling nonconducting samples. Like the magnetic sector analyzers, only one mass is measured at a time with a quadrupole.

The TOF mass analyzer has an almost unlimited mass range, with a mass resolution of 10,000 and a transmission efficiency midway between a quadrupole and a magnetic sector. The primary ion beam is pulsed, and the pulse of secondary ions generated is mass separated in the drift tube of the TOF analyzer. Ions of light m/z reach the detector before heavy m/z ions. Unlike the other two mass analyzers, the TOF analyzer collects and analyzes all of the secondary ions. This makes the TOF system very useful for static SIMS. Advantages of the TOF analyzer are the ability to determine very large molecules such as polymers, high mass resolution with no loss of sensitivity (unlike magnetic sector analyzers), as well as high and constant sensitivity for all masses. Imaging of the surface is done by rastering the primary ion beam across the surface and collecting a mass spectrum at each pixel in a 256×256 array, for example.

The detectors used for SIMS are electron multipliers (discussed in Section 14.2.1.1 and in Chapter 9); array detectors are used for imaging.

14.4.2. Analytical Applications of SIMS

SIMS is used for the identification of the surface composition of solid samples by the mass spectra obtained from the surface species. Compositional information comes from the top two to three monolayers of the surface. Elements and molecular species can be determined. Surface distribution, isotope ratios, and depth profiles can be measured. SIMS, like ISS, is used to determine residues on surfaces from cleaning processes. These residues may be organic or inorganic and can be potentially damaging by causing adhesion failure of coatings, contamination of high purity materials for semiconductor manufacture, or undesired chemical reactions. Adsorbed gases on surfaces can be studied. Element distribution in materials can provide information about homogeneity, corrosion, failure mechanisms, and the strength of materials. Distribution of deliberately added trace elements (dopants, implanted ions) in materials can be evaluated.

Both positive and negative secondary ions can be generated and analyzed. In practice, positively charged ions are most valuable for examining the elements on the left side of the periodic table, whereas the negative ions are most useful for elements on the right side of the periodic table (nonmetals).

In both SIMS and ISS, insulating samples can be analyzed if a flood gun or combination of flood gun and ion gun is used to charge the surface and to maintain charge balance. This can be seen in the SIMS spectrum of a glass surface (Fig. 14.40) collected without a flood gun (top spectrum) and with a flood gun (bottom spectrum). It can be seen that the spectrum has a flatter baseline and improved resolution when the flood gun is used.

The principles of ISS and SIMS are similar, in the method of bombarding the sample, the utilization of high vacuum, and the measurement of ions. In ISS the scattered ion beam is examined and the energies of the ions measured; in SIMS the ions ejected from the surface are examined and their mass-to-charge ratio measured. Both techniques are

Figure 14.40 Effect of charge neutralization on the SIMS spectrum of soda lime glass. The bottom spectrum, taken with the use of a flood gun, shows improved resolution and a flat baseline compared with the top spectrum.

very surface sensitive, with ISS obtaining information from only the top monolayer of a surface and SIMS obtaining information from two to three monolayers of the surface. Generally, the scattered and secondary ions vary considerably in mass and do not interfere with each other. The two systems can therefore be run simultaneously (ISS-SIMS), and this provides a wealth of information on surface elemental analysis, isotope ratios, and surface molecular analysis, depending on the fragmentation and ionization of the molecules involved.

SIMS can be used to obtain elemental information (so it complements other atomic mass spectrometry techniques). Examples are shown in Fig. 14.41. SIMS can also be used to obtain molecular information. As an example of the type of molecular information available from a TOF-SIMS instrument, the positive ion mass spectrum of a pharmaceutical compound is shown in Fig. 14.42. The M+1 ion, that is, molecular ion + H, and a variety fragment ions are seen, permitting the molecular weight and structure of the compound to be worked out using the principles presented earlier in the text for interpreting mass spectra (Chapter 10). Using the TOF-SIMS, chemical images can be generated by collecting a mass spectrum at every pixel of a 256 × 256 array as the primary ion beam is rastered across the sample surface. Both molecular and elemental image maps can be produced. For example, a time-release drug tablet or coated tablet can be imaged for both elements present and the active drug compound, to identify coating thickness and evaluate homogeneity.

Figure 14.41 (a) Positive SIMS spectrum of molybdenum. All peaks are isotopes of Mo. (b) Negative SIMS spectrum showing both elements and polyatomic fragments of molecules. [Courtesy of Physical Electronics USA, Inc. Eden Prairie, MN (www.phi.com).]

14.4.2.1. Quantitative Analysis

The analytical signal obtained in SIMS depends on a number of factors, namely, the abundance of the isotope examined on the surface, the properties of the surface, and the bombarding ion. The relationship is complex; calibration curves may or may not be continuous. Certified calibration standards that are matrix-matched and cover the analytical range are required. For bulk analysis using dynamic SIMS, certified standards can be obtained from sources such as NIST or commercial standards firms for many metals and alloys and for some ceramics and glasses, but such standards are rarely available for surfaces, thin layers, and multilayered materials. The excellent sensitivity of SIMS allows it to distinguish between ppb and ppm concentrations of analyte in solids. SIMS is not used for quantitative analysis at the % level; XPS or methods such as XRF are better suited to this purpose.

Figure 14.42 Positive SIMS spectrum of metoprolol, showing molecular and fragment peaks. [Courtesy of Physical Electronics USA, Inc. Eden Prairie, MN (www.phi.com).]

SIMS is used for quantitative depth profile determinations of trace elements in solids. These traces can be impurities or deliberately added elements, such as dopants in semiconductors. Accurate depth profiles require uniform bombardment of the analyzed area and the sputter rate in the material must be determined. The sputter rate is usually determined by physical measurement of the crater depth; for multilayered materials, each layer may have a unique sputter rate that must be determined. Depth profile standards are required. Government standards agencies like NIST have such standard reference materials available for a limited number of applications. For example, SRM depth profile standards of phosphorus in silicon, boron in silicon, and arsenic in silicon are available from NIST for calibration of SIMS instruments. P, As, and B are common dopants in the semiconductor industry and their accurate determination is critical to semiconductor manufacture and quality control.

14.4.2.2. Ion Microprobe Mass Spectrometry

Ion microprobe mass spectrometry (IMMS) is an extension of SIMS. In this system an ion beam is focused to a very small spot size onto the sample so that micron-sized points on the surface can be examined. The ion beam can be constricted by electrostatic lenses to a diameter of $3-10$ μm, although spot sizes as small as $0.2-0.5$ μm are attainable. A schematic diagram of such a system is shown in Fig. 14.39. The secondary ions produced from the sample surface pass through an electrostatic field, which brings most of the ions to the same velocity. The ions then pass through a magnetic sector, which resolves them by mass-to-charge ratio. This is the same type of double-focusing mass spectrometer described in Chapter 9.

This ion beam may be used in any one of three modes: (1) it may be held stationary on the sample and the analysis at that particular spot obtained; (2) it may be rastered across the sample surface to obtain a distribution map of the elements on that surface; or (3) it may be focused on a single spot for an extended period of time, producing a crater that can be used to obtain a depth profile of the elements of interest. With this system multi-element analysis of the surface is possible. Mapping of the surface by element is also possible. In addition, the focusing potential of IMMS permits analysis of very small points on the sample surface. For example, IMMS can be used in the determination of elements on the surface of collected airborne particles and dust. Solid airborne particulates are collected by pulling a known volume of air through a small diameter filter. The particles, often 0.5 μm or so in diameter, are trapped on the filter for analysis. Elements adsorbed on the surface of the particulates can be examined qualitatively and quantitatively. Lead, thallium, manganese, and chromium on the surface of airborne particles, all metals of environmental and toxicological interest, have been studied using this approach.

14.5. ELECTRON MICROPROBE (ELECTRON PROBE MICROANALYSIS)

The electron probe microanalyzer (EM or EPMA) uses a beam of high-energy electrons to bombard the surface of a solid sample. This results, as we have already seen, in the removal of an inner shell electron. As discussed in Section 14.2, this can result in the ejection of a photoelectron (the basis of ESCA) and the emission of an X-ray photon. The X-ray photons emitted have wavelengths characteristic of the elements present. The EPMA uses either a wavelength dispersive (WD) or energy dispersive (ED) X-ray spectrometer to detect and identify the emitted X-rays. This is very much analogous to XRF spectrometry

Figure 14.43 Approximate depth of sample surface probed by different surface analysis techniques.

(Chapter 8), where the primary beam was of X-rays, not electrons. This instrument was discussed at the end of Chapter 8, which should be reviewed if necessary.

The instrumentation for EM uses the same type of X-ray spectrometers discussed in detail in Chapter 8, with an electron beam as the source and a UHV system that includes the sample compartment. An ED X-ray spectrometer allows the simultaneous collection and display of the X-ray spectrum of all elements from boron to uranium. The ED spectrometer is used for rapid qualitative survey scans of sample surfaces. The wavelength dispersive spectrometer has much better resolution and is used for quantitative analysis of elements. The WD spectrometer is usually equipped with several diffracting crystals to optimize resolution and to cover the entire spectral range. The electron beam, sample stage, spectrometer, data collection, and processing are all under computer control.

Applications of EPMA include elemental analysis of surfaces and of micron-sized features at a surface. The sensitivity of the method is about 0.2 atom%. It provides a rapid, accurate method for compositional analysis of microscopic features. Elemental mapping of the elements present at the surface can also be done, and the composition correlated to topographical maps obtained from an analytical microscopy method, allowing correlation of topographical features of a surface with elemental composition. Like XRF, EPMA is strictly an elemental analysis technique; no information on chemical speciation or oxidation state is obtained. All elements from boron to uranium can be determined. Given that X-rays can escape from depths of 1000 Å or so, EPMA has the "deepest" definition of "surface" of the techniques discussed in this chapter. In fact, like XRF, it can be considered to be a bulk analysis technique assuming the sample is homogeneous.

The relative depth of penetration for the surface analysis techniques discussed is shown in Fig. 14.43 for comparison. ISS is the most surface sensitive, while EPMA is the least of the techniques that have been covered.

BIBLIOGRAPHY

Am. Lab. **1972**, *4* (2), 7.

Benninghoven, A.; Rudenauer, F.G.; Werner, H.W. *Secondary Ion Mass Spectrometry: Basic Concepts, Instrumental Aspects, Applications and Trends*; John Wiley and Sons: New York, 1987.

Briant, C.L.; Messmer, R.P., Eds. Auger electron spectroscopy. In *Treatise on Materials Science and Technology*; Academic Press, Inc.: San Diego, CA, 1988; Vol. 30.

Carlson, T.A. *Photoelectric Auger Spectroscopy*; Plenum Press: New York, 1975.

Czarderna, A.W.; Metler, A.C.; Jellinek, H.H.G.; Kochi, H. *Ind. Res.* **1978**, *January*.

Davis, L.E.; MacDonald, N.C.; Palmberg, P.W.; Riach, G.E.; Weber, R.E. *Handbook of Auger Electron Spectroscopy*, 2nd ed.; Physical Electronics Inc.: Eden Prairie, MN, 1976. (Now Physical Electronics USA, www.phi.com.)

Ewing, G.W., Ed. *Analytical Instrumentation Handbook*, 2nd ed.; Marcel Dekker, Inc.: New York, 1997.

Hendrickson, D.N.; Hollander, J.M.; Jolly, W.L. *Inorg. Chem.* **1969**, *8*, 2642.

Hercules, D.M. *Anal. Chem.*, **1970**, *42*(1), 20A.

Hercules, D.M. *Anal. Chem.*, **1986**, *58*, 1177A.

Hercules, D.M.; Hercules, S.H. *J. Chem. Ed.*, **1984**, *61*, 403.

Lee, L.H., Ed. *Characterization of Metal and Polymer Surfaces*; Academic Press Inc.: San Diego, CA, 1974; Vol. 1.

Lifshin, E., Ed. Characterization of materials, parts I and II. In *Materials Science and Technology: A Comprehensive Treatment*; Cahn, R.W., Haasen, P., Kramer, E.J., Eds.; VCH Publishers, Inc.: New York, 1992; Vols. 2A and 2B.

Moulder, J.F.; Stickle, W.F.; Sobol, P.E.; Bomben, K.D. In *Handbook of X-ray Photoelectron Spectroscopy*; Chastian, J., King, R.C., Eds.; Physical Electronics, Inc.: Eden Prairie, MN, 1995 (Now Physical Electronics USA, www.phi.com.)

NIST, XPS Database, www.nist.gov/srd/surface, copyright 2003 by the US Secretary of Commerce on behalf of the United States of America. All rights reserved.

Perry, S.S.; Somorjai, G.A. Characterization of organic surfaces. *Anal. Chem.* **1994**, *66* (7), 403A.

Settle, F., Ed. *Handbook of Instrumental Techniques for Analytical Chemistry*; Prentice Hall PTR: Upper Saddle River, NJ, 1997.

Siegbahn, E. *Endeavor* **1973**, *32*, 51.

Stinger, K. *Res. Dev.* **1977**, *28* (9), 40.

Thompson, M.; Baker, M.D.; Christie, A.; Tyson, J.F. *Auger Electron Spectroscopy*; John Wiley and Sons, Inc.: New York, 1985.

Turner, N.H. Electron spectroscopy. In *Analytical Instrumentation Handbook*, 2nd ed.; Ewing, G.W., Ed.; Marcel Dekker, Inc.: New York, 1997.

Wilson, R.G.; Stevie, F.A.; Magee, C.W. *Secondary Ion Mass Spectrometry*; John Wiley and Sons Inc., New York, 1989.

Weber, R.E. *Res. Dev.*, **1972**, *10*, 22.

PROBLEMS

14.1   The wavelength of XRF radiation of an element is virtually independent of the chemical form of the element, but in ESCA the photoelectron energy is not independent of the chemical environment of the element. Explain.

14.2   Why is an electron flood gun used in ESCA, Auger, and other surface analysis instruments?

14.3   ESCA and Auger instruments are useful for the analysis of elements present in concentrations greater than 1% although the absolute sensitivity of the method is about 10^{-10} g. Explain.

14.4   Draw a schematic diagram of an ESCA instrument.

14.5   Show the relationship between the processes that give rise to XRF, ESCA, and AES using a schematic diagram of an atom.

14.6   Why is the ESCA spectrum plotted directly as signal intensity vs. energy, but an Auger spectrum presented as the first derivative of the signal?

14.7   XRF is most sensitive for elements with a high atomic number. What is the relationship between Auger sensitivity and atomic number?

14.8 The concentration of nickel in a sample varies with depth from the surface. How may this variation be measured?

14.9 Why is it necessary to analyze surface samples under vacuum? How may several samples be examined consecutively?

14.10 Describe ISS. What depth of surface can be measured by this method? How does that compare with ESCA?

14.11 How can the concentration and distribution of a selected metal on a plane surface be measured and displayed?

14.12 What sources are commonly used for generating ESCA and Auger spectra?

14.13 How can the effects of rapid variation in local magnetic fields on ESCA signals be eliminated? Why is it necessary to do that?

14.14 You want to identify the location, distribution, concentration, and chemical form (molecular, elemental, oxidation state) of ppm levels of potassium in tungsten wire. What techniques will you use? Explain your reasoning. Can surface analysis techniques provide all of the information needed? If not, what else will you use?

14.15 What is SIMS? Describe its analytical uses.

14.16 What are the advantages of (a) ISS and (b) SIMS over XRF?

14.17 Describe the major components of an IMMS.

14.18 Describe the detectors used in ESCA.

14.19 Describe two types of energy discriminators. What are their functions?

14.20 Describe two types of energy analyzers. What is their function?

14.21 What is meant by the ESCA chemical shift? What is meant by the Auger chemical shift?

15

Electroanalytical Chemistry

Electrochemistry is the area of chemistry that studies the interconversion of chemical energy and electrical energy. Electroanalytical chemistry is the use of electrochemical techniques to characterize a sample. The original analytical applications of electrochemistry, electrogravimetry and polarography, were for the quantitative determination of trace metals in aqueous solutions. The latter method was reliable and sensitive enough to detect concentrations as low as 1 ppm of many metals. Since that time, many different types of electrochemical techniques have evolved, each useful for particular applications in organic, inorganic, and biochemical analyses.

A species that undergoes reduction or oxidation is known as an **electroactive** species. Electroactive species in general may be solvated or complexed, ions or molecules, in aqueous or nonaqueous solvents. Electrochemical methods are now used not only for trace metal ion analyses, but also for the analysis of organic compounds, for continuous process analysis, and for studying the chemical reactions within a single living cell. Applications have been developed that are suited for quality control of product streams in industry, *in vivo* monitoring, materials characterization, and pharmaceutical and biochemical studies, to mention a few of the myriad applications. Under normal conditions, concentrations as low as 1 ppm can be determined without much difficulty. By using electrodeposition and then reversing the current, it is possible to extend the sensitivity limits for many electroactive species by three or four orders of magnitude, thus providing a means of analysis at the ppb level.

In practice, electrochemistry not only provides a means of elemental and molecular analysis, but also can be used to acquire information about equilibria, kinetics, and reaction mechanisms from research using polarography, amperometry, conductometric analysis, and potentiometry. The analytical calculation is usually based on the determination of current or voltage or on the resistance developed in a cell under conditions such that these are dependent on the concentration of the species under study. Electrochemical measurements are easy to automate because they are electrical signals. The equipment is often far less expensive than spectroscopy instrumentation. Electrochemical techniques are also commonly used as detectors for LC, as discussed in Chapter 13.

Coauthor: R.J. Gale, Department of Chemistry, Louisiana State University, Baton Rouge, LA

15.1. FUNDAMENTALS OF ELECTROCHEMISTRY

Electrochemistry is the study of **reduction–oxidation** reactions (called **redox** reactions) in which electrons are transferred from one reactant to another. A chemical species that loses electrons in a redox reaction is **oxidized**. A species that gains electrons is **reduced**. A species that oxidizes is also called a **reducing agent** because it causes the other species to be reduced; likewise, an **oxidizing agent** is a species that is itself reduced in a reaction. An oxidation–reduction reaction requires that one reactant gain electrons (be reduced) from the reactant which is oxidized. We can write the reduction and the oxidation reactions separately, as half-reactions; the sum of the half-reactions equals the net oxidation–reduction reaction. Examples of oxidation half-reactions include:

$$Fe^{2+} \longrightarrow Fe^{3+} + e^-$$
$$Cu(s) \longrightarrow Cu^{2+} + 2e^-$$
$$AsH_3(g) \longrightarrow As(s) + 3H^+ + 3e^-$$
$$H_2C_2O_4 \longrightarrow 2CO_2(g) + 2H^+ + 2e^-$$

Examples of reduction half-reactions include:

$$Co^{3+} + e^- \longrightarrow Co^{2+}$$
$$(IO_3)^- + 6H^+ + 5e^- \longrightarrow \frac{1}{2}I_2(s) + 3H_2O$$
$$Cl_2(g) + 2e^- \longrightarrow 2Cl^-$$
$$Ag^+ + e^- \longrightarrow Ag(s)$$

If the direction of an oxidation reaction is reversed, it becomes a reduction reaction; that is, if Al^{3+} accepts 3 electrons, it is reduced to $Al(s)$. All of the reduction reactions are oxidation reactions if they are written in the opposite direction. Many of these reactions are reversible in practice, as we shall see.

A net oxidation–reduction reaction is the sum of the appropriate reduction and oxidation half-reactions. If necessary, the half-reactions must be multiplied by a factor so that no electrons appear in the net reaction. For example, the reaction between $Cu(s)$, Cu^{2+}, $Ag(s)$, and Ag^+ is:

$$Cu(s) + 2Ag^+ \longrightarrow Cu^{2+} + 2Ag(s)$$

We shall see why the reaction proceeds in this direction shortly. The net reaction is obtained from the half-reactions as follows:

Oxidation reaction: $Cu(s) \longrightarrow Cu^{2+} + 2e^-$
Reduction reaction: $Ag^+ + e^- \longrightarrow Ag(s)$

Each mole of copper gives up 2 moles of electrons, while each mole of silver ion accepts only 1 mole of electrons. Therefore the entire reduction reaction must be multiplied by 2, so that there are no electrons in the net reaction after summing the half-reactions:

Oxidation reaction: $Cu(s) \longrightarrow Cu^{2+} + 2e^-$
Reduction reaction: $2(Ag^+ + e^- \longrightarrow Ag(s))$
Net reaction: $Cu(s) + 2Ag^+ \longrightarrow Cu^{2+}2Ag(s)$

The equal numbers of electrons on both sides of the arrow cancel out.

Electrochemical redox reactions can be carried out in an **electrochemical cell** as part of an electrical circuit so that we can measure the electrons transferred, the current, and the voltage. Each of these parameters provides us with information about the redox reaction, so it is important to understand the relationship between charge, voltage, and current. The absolute value of the charge of one electron is 1.602×10^{-19} coulombs (C); this is the fundamental unit of electric charge. Since 1.602×10^{-19} C is the charge of one electron, the charge of one mole of electrons is:

$$(1.602 \times 10^{-19}\,\text{C/e}^-)(6.022 \times 10^{23}\,\text{e}^-/\text{mol}) = 96{,}485\,\text{C/mol} \tag{15.1}$$

This value 96,485 C/mol is called the Faraday constant (F), and provides the relationship between the total charge, q, transferred in a redox reaction and the number of moles, n, involved in the reaction.

$$q = n \times F \tag{15.2}$$

In an electric circuit, the quantity of charge flowing per second is called the **current**, i. The unit of current is the ampere, A; 1 A equals 1 C/s. The potential difference, E, between two points in the cell is the amount of energy required to move the charged electrons between the two points. If the electrons are attracted from the first point to the second point, the electrons can do work. If the second point repels the electrons, work must be done to force them to move. Work is expressed in joules, J, and the potential difference, E, is measured in volts. The relationship between work and potential difference is:

$$w \text{ (in joules)} = E \text{ (in volts)} \times q \text{ (in coulombs)} \tag{15.3}$$

Since the unit of charge is the coulomb, 1 V equals 1 J/C.

The relationship between current and potential difference in a circuit is expressed by Ohm's Law:

$$i = \frac{E}{R} \tag{15.4}$$

where i is the current; E, the potential difference, and R, the resistance in the circuit. The units of resistance are V/A or ohms, Ω.

15.2. ELECTROCHEMICAL CELLS

At the heart of electrochemistry is the electrochemical cell. We will consider the creation of an electrochemical cell from the joining of two half-cells. When an electrical conductor such as a metal strip is immersed in a suitable ionic solution, such as a solution of its own ions, a potential difference (voltage) is created between the conductor and the solution. This system constitutes a half-cell or electrode (Fig. 15.1). The metal strip in the solution is called an **electrode** and the ionic solution is called an **electrolyte**. We use the term electrode to mean both the solid electrical conductor in a half-cell (e.g., the metal strip) and the complete half-cell in many cases, for example, the standard hydrogen electrode, the calomel electrode. Each half-cell has its own characteristic potential difference or *electrode potential*. The electrode potential measures the ability of the half-cell to do work, or the driving force for the half-cell reaction. The reaction between the metal strip and the ionic solution can be represented as

$$M^0 \longrightarrow M^{n+} + ne^- \tag{15.5}$$

Figure 15.1 A half-cell composed of a metal electrode M^0 in contact with its ions, M^+, in solution. The salt bridge or porous membrane is shown on the lower right side.

where M^0 is an uncharged metal atom, M^{n+} is a positive ion, and e^- is an electron. The number of electrons lost by each metal atom is equal to n, where n is a whole number. This is an oxidation reaction, because the metal has lost electrons. It has been oxidized from an uncharged atom to a positively charged ion. In the reaction, the metal ions enter the solution (dissolve). By definition, the electrode at which oxidation occurs is called the **anode**. We say that at the anode, oxidation of the metal occurs according to the reaction shown in Eq. (15.5).

Some examples of this type of half-cell are:

$$Cd(s) \longrightarrow Cd^{2+} + 2e^-$$
$$Ag(s) \longrightarrow Ag^+ + e^-$$
$$Cr(s) \longrightarrow Cr^{3+} + 3e^-$$

Note that in normal usage, the zero oxidation state of the solid metal is understood, not shown with a zero superscript. It has been found that with some metals the spontaneous reaction is in the opposite direction and the metal ions tend to become metal atoms, taking up electrons in the process. This reaction can be represented as

$$M^{n+} + ne^- \longrightarrow M^0 \tag{15.6}$$

This is a reduction reaction because the positively charged metal ions have gained electrons, lost their charge, and become neutral atoms. The neutral atoms deposit on the electrode, a process called *electrodeposition*. This electrode is termed a **cathode**. At the cathode, reduction of an **electroactive species** takes place. An electroactive species is one that is oxidized or reduced during reaction. Electrochemical cells also contain nonelectroactive (or inert) species such as counterions to balance the charge, or electrically conductive electrodes that do not take part in the reaction. Often these inert electrodes are made of Pt or graphite, and serve only to conduct electrons into or out of the half-cell.

It is not possible to measure directly the potential difference of a single half-cell. However, we can join two half-cells to form a complete cell as shown in Fig. 15.2. In this example, one half-cell consists of a solid copper electrode immersed in an aqueous solution of $CuSO_4$; the other has a solid zinc electrode immersed in an aqueous solution

Figure 15.2 A complete Zn/Cu galvanic cell with a salt bridge separating the half-cells.

of $ZnSO_4$. The two half-cell reactions and the net **spontaneous** reaction are shown:

$$\text{Anode (oxidation) reaction: } Zn(s) \longrightarrow Zn^{2+} + 2e^-$$

$$\text{Cathode (reduction) reaction: } Cu^{2+} + 2e^- \longrightarrow Cu(s)$$

$$\text{Net reaction: } Zn(s) + Cu^{2+} \longrightarrow Zn^{2+} + Cu(s)$$

No reaction will take place, and no current will flow, unless the electrical circuit is complete. As shown in Fig. 15.2, a conductive wire connects the electrodes externally through a voltmeter (potentiometer). A salt bridge, a glass tube filled with saturated KCl in agar gel, physically separates the two electrolyte solutions. The salt bridge permits ionic motion to complete the circuit while not permitting the electrolytes to mix. The reason we need to prevent the mixing of the electrolytes is that we want to obtain information about the electrochemical system by measuring the current flow through the external wire. If we had both electrodes and both ionic solutions in the same beaker, the copper ions would react directly at the Zn electrode, giving the same net reaction but no current flow in the external circuit. In the electrolyte solution and the salt bridge the current flow is ionic (ion motion), and in the external circuit the current flow is electronic (electron motion). This cell and the one in Fig. 15.3 show the components needed for an electrochemical cell: two electrical conductors (electrodes), suitable electrolyte solutions and a means of allowing the movement of ions between the solutions (salt bridge in Fig. 15.2, a semipermeable glass frit or membrane in Fig. 15.3), external connection of the electrodes by a conductive wire and the ability for an oxidation reaction to occur at one electrode, and a reduction reaction at the other. Of course there are counterions present in each solution (e.g., sulfate ions) to balance the charge; these ions are not electroactive and do not take part in the redox reaction. They do flow (ionic motion) to keep charge balanced in the cell. A cell that uses a spontaneous redox reaction to generate electricity is called a **galvanic cell**. Batteries are examples of galvanic cells. A cell set up to cause a nonspontaneous reaction to occur by putting electricity into the cell is called an **electrolytic cell**.

   The complete cell has a potential difference, a cell potential, which can be measured by the voltmeter. The potential difference for the cell, E_{cell}, can be considered to be equal

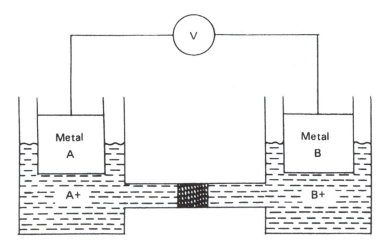

Figure 15.3 A schematic complete cell with a porous frit separating the half-cells.

to the difference between the two electrode potentials, *when both half-cell reactions are written as reductions.* That is,

$$E_{cell} = E_{cathode} - E_{anode} \qquad\qquad (15.7)$$

when the potentials are **reduction potentials**. This convention is necessary to calculate the sign of E_{cell} correctly, even though the anode reaction is an oxidation. The cell potential is also called the **electromotive force** or **emf**.

While we cannot measure a given single electrode potential directly, we can easily measure the cell potential for two half-cells joined as described. So now we have a means of measuring the **relative electrode potential** for any half-cell by joining it to a designated **reference electrode (reference half-cell)**. In real cells, there is another potential difference that contributes to E_{cell}, called a junction potential. If there is a difference in the concentration or types of ions of the two half-cells, a small potential is created at the junction of the membrane or salt bridge and the solution. Junction potentials can be sources of error. When a KCl salt bridge is used the junction potential is very small because the rates of diffusion of K^+ and Cl^- ions are similar, so the error in measuring a given electrode potential is small.

15.2.1. Line Notation for Cells and Half-Cells

Writing all the equations and arrows for half-cells and cells is time consuming and takes up space, so a shorthand or line notation is often used. For example, the half-cell composed of a silver electrode and aqueous 0.0001 M Ag^+ ion (from dissolution of silver nitrate in water) is written in line notation as

$$Ag(s) \mid Ag^+ \ (0.0001 \ M)$$

The *vertical stroke or line* between Ag and Ag^+ indicates a *phase boundary*, that is, a difference in phases (e.g., solid | liquid, solid 1 | solid 2, or liquid | gas) in the constituents of the half-cells that are in contact with each other. The complete cell in Fig. 15.2 can be represented as

$$Zn(s) \mid Zn^{2+}(0.01 \ M) \parallel Cu^{2+}(0.01 \ M) \mid Cu(s)$$

The *double vertical stroke* after Zn^{2+} indicates a membrane junction or salt bridge. The double stroke shows the termination of one half-cell and the beginning of the second. This cell could also be written to show the salts used, as shown:

$$Zn(s) \mid ZnSO_4(aq) \parallel CuSO_4(aq) \mid Cu(s)$$

It is conventional to write the electrode that serves as the anode on the left in an electrochemical cell. The other components in the cell are listed as they would be encountered moving from the anode to the cathode.

15.2.2. Standard Reduction Potentials

15.2.2.1. The Standard Hydrogen Electrode

In order to compile a table of relative electrode potentials, chemists must agree upon the half-cell that will serve as the reference electrode. The composition and construction of the half-cell must be carefully defined. The value of the electrode potential for this reference half-cell could be set equal to any value, but zero is a convenient reference point. In practice, it has been arbitrarily decided and agreed upon that the **standard hydrogen electrode (SHE)** has an assigned electrode potential of exactly zero volts at all temperatures. The SHE, shown in Fig. 15.4, consists of a platinum electrode with a surface coating of finely divided platinum (called a *platinized* Pt electrode) immersed in a solution of 1 M hydrochloric acid, which dissociates to give H^+. Hydrogen gas, H_2, is bubbled into the acid solution over the Pt electrode. The finely divided platinum on the electrode surface provides a large surface area for the reaction

$$2H^+ + 2e^- \longrightarrow H_2(g) \quad E^0 = 0.000\,V \tag{15.8}$$

Figure 15.4 The standard hydrogen electrode (SHE). This design is shown with a presaturator containing the same 1 M HCl solution as in the electrode to prevent concentration changes by evaporation. (Aikens et al., by permission, Waveland Press Inc., Long Grove, IL, 1984. All rights reserved.)

In addition, the Pt serves as the electrical conductor to the external circuit. Under standard state conditions, that is, when the H_2 pressure equals 1 atm and the *ideal* concentration of the HCl is 1 M, and the system is at $25°C$, the reduction potential for the reaction given in Eq. (15.8) is exactly 0 V. (The potential actually depends on the chemical activity of the HCl, not on its concentration. The relationship between activity and concentration is discussed subsequently. For an ideal solution, concentration and activity are equal.) The potential is symbolized by E^0, where the superscript zero means standard state conditions. The term **standard reduction potential** means that the ideal concentrations of all solutes are 1 M and all gases are at 1 atm; other solids or liquids present are pure (e.g., pure Pt solid). By connecting the SHE half-cell with any other standard half-cell and measuring the voltage difference developed, we can determine the standard reduction potential developed by the second half-cell.

Consider, for example, a cell at $25°C$ made up of the two half-cells:

$$Zn(s) \mid Zn^{2+}(aq, 1\ M) \parallel H^+(aq, 1\ M) \mid H_2(g, 1\ atm) \mid Pt(s)$$

This cell has a Zn half-cell as the anode and the SHE as the cathode. All solutes are present at *ideal* 1 M concentrations, gases at 1 atm and the other species are pure solids, and so both half-cells are at standard conditions. The measured cell emf is $+0.76$ V and this is the standard cell potential, E^0, because both half-cells are in their standard states. From Eq. (15.7), we can write:

$$E^0_{cell} = E^0_{cathode} - E^0_{anode} \tag{15.9}$$

The total voltage developed under standard conditions is $+0.76$ V. But the voltage of the SHE is 0 by definition; therefore the standard reduction potential of the Zn half-cell is:

$$+0.76\ V = 0.000 - E^0_{Zn}$$
$$E^0_{Zn} = -0.76\ V$$

Therefore we can write:

$$Zn^{2+} + 2e^- \longrightarrow Zn(s) \quad E^0_{Zn} = -0.76\ V$$

We have determined the Zn standard reduction potential even though the galvanic cell we set up has Zn being oxidized. By substituting other half-cells, we can determine their electrode potentials (actually, their relative potentials) and build a table of standard reduction potentials. If we set up a galvanic cell with the SHE and Cu, we have to make the SHE the anode in order for a spontaneous reaction to occur. This cell,

$$Pt(s) \mid H_2(g, 1\ atm) \mid H^+(aq, 1\ M) \parallel Cu^{2+}(aq, 1\ M) \mid Cu(s)$$

has a measured cell emf $= +0.34$ V. Therefore the standard reduction potential for Cu is

$$+0.34\ V = E^0_{Cu} - 0.000\ V$$
$$E^0_{Cu} = +0.34\ V$$

The quantity E^0 is the emf of a half-cell under standard conditions. A half-cell is said to be under standard conditions when the following conditions exist at a temperature of $25°C$:

1. All solids and liquids are pure (e.g., a metal electrode in the standard state)
2. All gases at a pressure of 1 atm (760 mmHg)
3. All solutes are at 1 M concentration (more accurately, at **unit activity**)

The true electrode potential is related to the activity of the species in solution, not the concentration. For a pure substance, its mole fraction and its activity $= 1$. For a pure substance that is not present in the system, its mole fraction and its activity $= 0$. From general chemistry, you should remember that Raoult's Law predicts that for an ideal solution, the mole fraction of the solute and its activity are equal. However, most solutions deviate from linearity because of interactions between the solute and solvent molecules. These deviations can be positive or negative, because the species may attract or repel each other. The amount of attraction or repulsion affects the activity of the solute. For dilute solutions, the activity is proportional to concentration (Henry's Law). Activity is equal to the concentration times the activity coefficient for the species in solution. That is

$$a_{ion} = [M_{ion}]\gamma_{ion} \tag{15.10}$$

where a_{ion} is the activity of given ion in solution; $[M_{ion}]$, the molar concentration of the ion; and γ_{ion}, the activity coefficient of the ion.

Activity depends on the ionic strength of the solution. If we compare a 1 M solution and a 0.01 M solution, the more concentrated solution may act as though it is less than $100\times$ more concentrated than the dilute solution. It is then said that the activity of the 1 M solution is less than unity or the activity coefficient is less than 1. For solutions with positive deviations from Henry's Law, the activity coefficient will be greater than 1. For very dilute solutions (low ionic strength) the activity coefficient γ approaches 1, so concentration is approximately equal to activity for very dilute solutions. We will use concentrations in the calculations in this text instead of activities, but the approximation is only accurate for dilute solutions (<0.005 M) and for ions with single charges. Details on activity corrections can be found in most analytical chemistry texts, such as the ones by Harris or Enke listed in the bibliography.

Looking back at our cell in Fig. 15.2, with the Zn half-cell as the anode and the Cu half-cell as the cathode, we can calculate the standard cell potential for this galvanic cell. In the spontaneous reaction, Zn is oxidized and Cu^{2+} is reduced, therefore Zn is the anode and Cu is the cathode.

$$Zn^{2+} + 2e^- \longrightarrow Zn^0 \quad E^0 = -0.76 \text{ V}$$
$$Cu^{2+} + 2e^- \longrightarrow Cu^0 \quad E^0 = +0.34 \text{ V}$$

The standard cell potential developed is calculated from Eq. (15.9):

$$+0.34 - (-0.76 \text{ V}) = +1.10 \text{ V}$$

In tables of standard potentials, all of the half-cell reactions are expressed as reductions. The sign is reversed if the reaction is reversed to become an oxidation. In a spontaneous reaction, when both half-cells are written as reductions, the half-cell with the more negative potential will be the one that oxidizes. The negative sign in Eq. (15.9) reverses one of the reduction processes to an oxidation. Some standard reduction potentials for common half-cells are given in Appendix 15.1. These can be used to calculate E^0 for other electrochemical cell combinations as we have done for the Zn/Cu cell. More complete lists of half-cell potentials can be found in references such as Bard et al., listed in the bibliography.

In the reaction of our example, zinc metal dissolves, forming zinc ions and liberating electrons. Meanwhile, an equal number of electrons are consumed by copper ions, which plate out as copper metal. The net reaction is summarized as

$$Zn^0 + Cu^{2+} \longrightarrow Cu^0 + Zn^{2+} \quad +1.10 \text{ V}$$

Zinc metal is oxidized to zinc ions, and copper ions are reduced to copper metal. The copper cathode becomes depleted of electrons because these are taken up by the copper ions in solution. At the same time the zinc anode has an excess of electrons because the neutral zinc atoms are becoming ionic and liberating electrons in the process. The excess electrons from the anode flow to the cathode. The flow of electrons is the source of external current; the buildup of electrons at the anode and the depletion at the cathode constitute a potential difference that persists until the reaction ceases. The reaction comes to an end when either all the copper ions are exhausted from the system or all the zinc metal is dissolved, or an equilibrium situation is reached when both half-cell potentials are equal. If the process is used as a battery, such as a flashlight battery, the battery becomes dead when the reaction ceases.

When a cell spontaneously generates a voltage, the electrode that is negatively charged is the anode and the positively charged electrode is the cathode.

15.2.3. Sign Conventions

The sign of half-cell potentials has been defined in a number of different ways, and this variety of definitions has led to considerable confusion. The convention used here is in accord with the recommendations of the IUPAC meeting in Stockholm in 1953. In this convention the standard half-cell reactions are written as reductions. Elements that are more powerful reducing agents than hydrogen show negative potentials, and elements that are less powerful reducing agents show positive potentials. For example, the $Zn \mid Zn^{2+}$ half-cell is negative and the $Cu \mid Cu^{2+}$ half-cell is positive.

15.2.4. The Nernst Equation

We have seen how to use standard reduction potentials to calculate E^0 for cells. Real cells are usually not constructed at standard state conditions. In fact, it is almost impossible to make measurements at standard conditions because it is not reasonable to adjust concentrations and ionic strengths to give unit activity for solutes. We need to relate standard potentials to those measured for real cells. It has been found experimentally that certain variables affect the measured cell potential. These variables include the temperature, concentrations of the species in solution, and the number of electrons transferred. The relationship between these variables and the measured cell emf can be derived from simple thermodynamics (see any introductory general chemistry text). The relationship between the potential of an electrochemical cell and the concentration of reactants and products in a general redox reaction

$$aA + bB \longrightarrow cC + dD$$

is given by the Nernst equation:

$$E = E^0 - \frac{RT}{nF} \ln \frac{[C]^c [D]^d}{[A]^a [B]^b} \tag{15.11}$$

where E is the measured potential (emf) of the cell; E^0, the emf of the cell under standard conditions; R, the gas constant (8.314 J/K mol = 8.314 V C/K mol); T, the temperature (K); n, the number of moles of electrons transferred during reaction (from the balanced half-reactions); F, the Faraday constant (96,485 C/mol e^-); and ln = natural logarithm to base e.

The logarithmic term has the same form as the equilibrium constant for the reaction. The term is called Q, the reaction quotient, when the concentrations (rigorously, the activities) are not the equilibrium values for the reaction. As in any equilibrium constant expression, pure liquids and pure solids have activities equal to 1, so they are omitted from the expression. If the values of R, T ($25°C = 298$ K), and F are inserted into the equation and the natural logarithm is converted to log to the base 10, the Nernst equation reduces to

$$E = E^0 - \frac{0.05916}{n} \log \frac{[C]^c [D]^d}{[A]^a [B]^b} \qquad (15.12)$$

It should be noted that the square brackets literally mean "the molar concentration of". For example, $[Fe^{2+}]$ means "the molar concentration of ferrous ion" or moles of ferrous ion per liter of solution.

Concentrations in molarity should really be the activities of the species, but we often do not know the activities. For that reason, we define the **formal potential**, $E^{0'}$, as the measured potential of the cell when the species being oxidized and reduced are present at concentrations such that the ratio of the concentrations of oxidized to reduced species is unity and other components of the cell are present at designated concentrations. The use of the formal potential allows us to avoid activity coefficients, which are often unknown. This gives us:

$$E = E^{0'} - \frac{0.05916}{n} \log \frac{[C]^c [D]^d}{[A]^a [B]^b} \qquad (15.13)$$

The Nernst equation is also used to calculate the electrode potential for a given half-cell at nonstandard conditions. For example, for the half-cell $Fe^{3+} + e^- \rightarrow Fe^{2+}$ which has an $E^0 = 0.77$ V and $n = 1$, the Nernst equation would be:

$$E = 0.77 - \frac{0.05916}{1} \log \frac{[Fe^{2+}]}{[Fe^{3+}]}$$

To calculate the electrode potential, the molar concentrations of ferrous and ferric ion used to construct the half-cell would be inserted. The general form of the Nernst equation for an electrode written as a reduction is:

$$E = E^0 - \frac{0.0591}{n} \log \left(\frac{[red]}{[ox]} \right) \qquad (15.14)$$

where [red] means the molarity of the reduced form of the electroactive species and [ox] means the molarity of the oxidized form of the electroactive species. In cells where the reduced form is the metal (e.g., in the Cu/Cu^{2+} half-cell) [red] $= 1$ because pure metals such as Cu have unit activity. There is the equivalent definition of formal potential for each half-cell.

The Nernst equation gives us the very important relationship between E, the emf of the half-cell, and the concentration of the oxidized and reduced forms of the components of the solution. Measurements using pH electrodes and ion-selective electrodes are based on this relationship. If the two electrode reactions are written as reductions, the potentials can be calculated for the cathode and anode (or the electrode inserted into the positive terminal of the potentiometer, E_+ and the electrode inserted into the negative terminal of the potentiometer, E_-) The cell voltage is the difference

$$E_{cell} = E_+ - E_- \tag{15.15}$$

15.2.5. Activity Series

The tendency for a species to become oxidized or reduced determines the sign and potential of the half-cell. The tendency is strongly related to the chemical reactivity of the species concerned in aqueous systems. Based on the potential developed in a half-cell under controlled conditions, the elements may be arranged in an order known as the activity series or electromotive series (Table 15.1). In general, the metals at the top of the activity series are most chemically reactive and tend to give up electrons easily, following the reaction of $M^0 \rightarrow M^{n+} + ne^-$. The metals at the bottom of the series are more "noble" and therefore less active. They do not give up electrons easily; in fact, their cations will accept electrons from metals above them in the activity series. In the process, the cations become neutral metal atoms and plate out of solution, while the more active metals become ionic and dissolve. This is illustrated as follows:

$$\text{metal A (active)} \longrightarrow \text{metal ion A}^+ + e^-$$
$$\text{metal ion B}^+\text{(noble)} + e^- \longrightarrow \text{metal B}$$
$$\text{net result: A} + \text{B}^+ \longrightarrow \text{A}^+ + \text{B}$$

Table 15.1　The Activity Series of Metals

Metal	E^0 (V)	Chemical reactivity
Li	−3.05	These metals displace hydrogen from acids and dissolve in all acids, including water
K	−2.93	
Ba	−2.91	
Sr	−2.89	
Ca	−2.87	
Na	−2.71	
Mg	−2.37	These metals react with acids or steam
Al	−1.66	
Mn	−1.18	
Zn	−0.76	
Cr	−0.74	
Fe	−0.44	
Cd	−0.40	
Co	−0.28	These metals react slowly with all acids
Ni	−0.26	
Sn	−0.14	
Pb	−0.13	
H_2	0.00	
Bi	+0.32	These metals react with oxidizing acids (e.g., HNO_3)
Cu	+0.34	
Ag	+0.80	
Hg	+0.85	
Pd	+0.92	
Pt	+1.12	These metals react with aqua regia (3:1 v/v HCl/HNO_3)
Au	+1.50	

In short, the more active metals displace the less active metals from solution. As an example, if an iron strip is immersed in a solution of copper sulfate, some of the iron dissolves, forming iron ions, while the copper ions become metallic and copper metal plates out on the remaining iron strip. The activity series can be used to predict displacement reactions between atoms and ions in compounds of the type $A + BC \rightarrow AC + B$, where A and B are atoms. Using the activity series, any atom A will displace from a compound any element B listed below it, but will not displace any element listed above it.

This reaction is particularly important when B^+ in BC refers to a hydrogen ion solution (acid). Based on this principle, all elements above hydrogen in the activity series are capable of displacing hydrogen from solution; that is, it is possible for them to be dissolved in an acid solution by reducing the H^+ ion. The metal ionizes (oxidizes) and enters the solution while the H^+ (acid) reduces to H_2 and usually bubbles off. We would predict that Al and Zn would dissolve in HCl, but that Cu will not. In contrast, a simple mineral acid such as HCl cannot dissolve noble metals, such as platinum and gold, because the metals will not displace hydrogen ion. These metals require an oxidizing acid such as nitric acid plus a complexing ion such as the chloride ion from HCl, as is found in aqua regia, a mixture of HCl and HNO_3, to force them into solution.

15.2.6. Reference Electrodes

In order to measure the emf of a given half-cell, it is necessary to connect it with a second half-cell and measure the voltage produced by the complete cell. In general, the second half-cell serves as a reference cell and should be one with a known, stable electrode potential. Although the standard hydrogen electrode serves to define the standard reduction potential, in practice it is not always convenient to use an SHE as a reference electrode. It is difficult to set up and control. Other, more convenient reference electrodes have been developed. In principle, any metal-ion half-cell could be used under controlled conditions as a reference electrode, but in practice, many metals are unsatisfactory materials. Active metals, such as sodium and potassium, are subject to chemical attack by the electrolyte. Other metals, such as iron, are difficult to obtain in the pure form. With some metals, the ionic forms are unstable to heat or to exposure to the air. Also, it is frequently difficult to control the concentration of the electrolytes accurately. As a result, only a few systems provide satisfactory stable potentials.

The characteristics of an ideal reference electrode are that it should have a fixed potential over time and temperature, long term stability, the ability to return to the initial potential after exposure to small currents (i.e., it should be reversible), and it should follow the Nernst equation. Two common reference electrodes that come close to behaving ideally are the saturated calomel electrode and the silver/silver chloride electrode.

15.2.6.1. The Saturated Calomel Electrode

The **saturated calomel electrode (SCE)** is composed of metallic mercury in contact with a saturated solution of mercurous chloride, or *calomel* (Hg_2Cl_2). A Pt wire in contact with the metallic Hg conducts electrons to the external circuit. The mercurous ion concentration of the solution is controlled through the solubility product by placing the calomel in contact with a saturated potassium chloride solution. It is the saturated KCl solution that gives this electrode the "saturated" name; there are other calomel reference electrodes used that differ in the concentration of KCl solution, but all contain saturated mercurous chloride solution. A typical calomel electrode is shown in Fig. 15.5. The half-cell

Figure 15.5 The saturated calomel electrode (SCE). (Aikens et al., by permission, Waveland Press Inc., Long Grove, IL, 1984. All rights reserved.)

reaction is

$$Hg_2Cl_2(s) + 2e^- \longleftrightarrow 2Hg(l) + 2Cl^- \quad E^0 = +0.268 \text{ V}$$

Applying Eq. (15.12), we obtain

$$E = E^0 - \frac{0.05916}{n} \log \frac{[C]^c [D]^d}{[A]^a [B]^b}$$

$$E = E^0 - \frac{0.0591}{2} \log \left(\frac{[Hg^0]^2 [Cl^-]^2}{[Hg_2Cl_2]} \right)$$

But $Hg_2Cl_2(s)$ and $Hg(l)$ are in standard states at unit activity; therefore

$$E = E^0 - \frac{0.0591}{2} \log \left(\frac{1 \times [Cl^-]^2}{1} \right)$$

$$= E^0 - \frac{0.0591}{2} \log ([Cl^-]^2)$$

When the KCl solution is saturated and its temperature is 25°C, the concentration of chloride ion, $[Cl^-]$, is known and

$$E = +0.244 \text{ V for the SCE}$$

For a 1 N solution of KCl, the electrode is called a normal calomel electrode (NCE) and $E(\text{NCE}) = +0.280$ V. The advantage of the SCE is that the potential does not change if some of the liquid evaporates from the electrode, because the Cl^- concentration does not change. Remember that "saturated" for the SCE refers to the KCl concentration, not to the mercurous chloride (calomel).

15.2.6.2. The Silver/Silver Chloride Electrode

Another common reference cell found in the chemical-processing industry and useful in organic electrochemistry is the silver/silver chloride half-cell. The cell consists of silver metal coated with a paste of silver chloride immersed in an aqueous solution saturated with KCl and AgCl. The half-cell reaction is

$$AgCl(s) + e^- \longleftrightarrow Ag(s) + Cl^- \quad E^0 = +0.222 \text{ V}$$

The same principle applies as in the SCE, but in this case the silver ion activity depends on the solubility product of AgCl, which is in contact with a solution of known chloride ion activity. The measured E for an Ag/AgCl electrode saturated in KCl is $+0.199$ V at $25°C$. A saturated KCl solution is about 4.5 M KCl. This reference electrode can also be filled with 3.5 M KCl (saturated with AgCl); the potential in this case is $+0.205$ V at $25°C$. The silver/silver chloride electrode provides a reproducible and reliable reference electrode free from mercury salts.

Both reference electrode designs can permit KCl to leak through the porous plug junction and contaminate the analyte solution. In addition, analytes that react with Ag or Hg ions, such as sulfides, can precipitate insoluble salts that clog the junction and impair the performance of the electrode. Electrode manufacturers have designed double-junction electrodes with an outer chamber that can be filled with an electrolyte that does not interfere with a given measurement and electrodes with large, flushable sleeve junctions that can be cleaned easily and are not prone to clogging.

15.2.7. The Electrical Double Layer

While it is easy to say that we can measure the potential difference in an electrochemical cell, the reality of what we are measuring is not obvious. The potential difference between a single electrode and the solution cannot be measured. A second electrode immersed in the solution is required. The second electrode has its own potential difference between itself and the solution; this difference cannot be measured either. All we can measure are relative potential differences. The potential difference between an electrode and the solution actually exists only in a very small region immediately adjacent to the electrode surface. This region of potential difference is often called the "**electrical double layer**" and is only nanometers wide. A very simplistic view of this interface is presented. A negatively charged electrode surface will have a "layer" of excess positive ions in solution adjacent to (or adsorbed to) the surface of the electrode; hence the term "double layer". Immediately adjacent to the positive ion layer there is a complex region of ions in solution whose distribution is determined by several factors. Outside of that, and only nanometers away from the electrode surface, there is the bulk solution. A positively charged electrode surface will have a similar interfacial region with the charges reversed. The exact nature of the charged interface and the ion distribution is very complex and is not yet completely understood. Outside of this very narrow region of high potential gradient next to each electrode, there is no potential gradient in the bulk of the solution. As a consequence of the placement of electrodes in solution and the creation of charged interfaces between the electrode surface and the solution, electrochemical measurements are made in a very nonhomogeneous environment.

Any reactions that occur must occur at the electrode–solution interface and the reacting species must be brought to the electrode surface by diffusion or mass transport through stirring of the solution (convection). The ions in the bulk of the solution are not attracted to the electrodes by potential difference; there is no potential gradient in the bulk of the solution. Even when we are interested in the bulk properties of the solution, we are only analyzing an extremely small amount of the solution that is no longer homogeneous because of the formation of the complex layers of charged species at the electrode interface.

15.3. ELECTROANALYTICAL METHODS

Diverse analytical techniques, some highly sensitive, have been developed based on measurements of current, voltage, charge, and resistance in electrochemical systems. One variable is measured while the others are controlled. Electroanalytical methods can be classified according to the variable being measured. Table 15.2 provides a summary of the more important methods. The methods are briefly defined below and then discussed at length in subsequent sections.

The Nernst equation indicates the relationship between the activity of species in solution and the emf produced by a cell involving those species. The electrochemical technique called **potentiometry** measures the potential developed by a cell consisting of an **indicator electrode** and a **reference electrode**. In theory, the indicator electrode response is a measure of the activity of a single component of the solution, the analyte. In practice, the indicator electrode is calibrated to respond to concentration rather than activity and is usually not specific for a single analyte. The potential, E_{cell}, is measured under negligible current flow to avoid significant changes in the concentration of the component being measured. Modern potentiometers permit measurement of potential with currents <1 pA, so potentiometry can be considered to be a nondestructive technique. Potentiometry is the basis for measurement of pH, ion-selective electrode measurements, and potentiometric titration.

Coulometry is the term used for a group of methods based on electrolytic oxidation or reduction of an analyte. The electrolysis is performed either by controlling the potential or the current, and is carried out to quantitatively convert the analyte to a new oxidation state. One form of coulometry is **electrogravimetry**, in which metallic elements are reduced, plated out onto an electrode, and weighed. The weight of the metal deposited is a measure of the concentration of the metal originally in solution. Coulometry is based on Faraday's Law, which states that the extent of reaction at an electrode is proportional to the current. It is known that 1 F (96,485 C) of electricity is required to reduce (or oxidize) 1 gram-equivalent weight of an electroactive analyte. By measuring the quantity of electricity required to reduce (or oxidize) a given sample exhaustively, the quantity of analyte can be determined, provided the reaction is 100% efficient (or of known efficiency). Mass, or charge $q = i$ (A) $\times t$ (s), can be used as a measure of the extent of the electrochemical reaction.

Table 15.2 Electroanalytical Techniques

Technique	Controlled parameter	Parameter measured
Conductometry	Voltage, V (AC)	Conductance $G = 1/R$
Conductometric titration	Voltage, V (AC)	Titrant volume vs. conductance
Potentiometry	Current, $i = 0$	Potential, E
Potentiometric titration	Current, $i = 0$	Titrant volume vs. E
Cyclic voltammetry	Potential, E	i vs. E
Polarography	Potential, E	i vs. E
Anodic stripping voltammetry	Potential, E	i vs. E
Coulometry	E or i	Charge, q (integrated current)
Electrogravimetry	E or i	Weight deposited
Coulometric titration	i	Time, t

In *voltammetry* a controlled potential is applied to one electrode and the current flowing through the cell is monitored over time. A powerful family of techniques is available using voltammetry. *Polarography* is a technique that requires three electrodes. Polarography uses as the **working electrode** a dropping mercury electrode or a static (hanging) mercury drop. The auxiliary electrode (or counterelectrode) is normally a Pt wire or foil. A third (reference) electrode is used as a basis for control of the potential at the working electrode. The current of analytical interest flows between the working and auxiliary electrodes, and the reference acts only as a high-impedance probe. As the voltage is progressively increased or decreased with time (sweep voltammetry), resultant changes in the anodic or cathodic currents occur whenever an electroactive species is oxidized or reduced, respectively. Polarography is a special case of *linear sweep voltammetry* because the electrode area increases with time as each drop grows and falls every 4 s or so; that is, the voltage is changing as the electrode area increases. Polarography is especially useful for analyzing and studying metal ion and metal complex reductions and solution equilibria. Another technique for studying electrochemical reaction rates and mechanisms involves reversing the potential sweep direction to reveal the electroactive products formed in the forward sweep. In this way it is possible to see if the products undergo reaction with other species present or with the solvent. This technique is called *cyclic voltammetry* and is usually carried out at a solid electrode. In a special case of voltammetry called *anodic stripping voltammetry* the electrolyzed product is preconcentrated at an electrode by deposition for a reasonable period at a fixed potential. The product is then stripped off with a rapid reverse potential sweep in the positive direction. Peak currents on the reverse sweep are used to determine the analyte concentration from a standard additions calibration. In principle, this method is applicable to both anodic and cathodic stripping.

Amperometry is the measurement of current at a fixed potential. An analyte undergoes oxidation or reduction at an electrode with a known, applied potential. The amount of analyte is calculated from Faraday's law. Amperometry is used to detect titration endpoints, as a detector for liquid chromatography and forms the basis of many new sensors for biomonitoring and environmental monitoring applications.

In *conductometry* an alternating (AC) voltage is applied across two electrodes immersed in the same solution. The applied voltage causes a current to flow. The magnitude of the current depends on the electrolytic conductivity of the solution. This method makes it possible to detect changes of composition in a sample during chemical reactions (e.g., during a titration) although the measurement itself cannot identify the species carrying the current. Conductivity or conductance measurements are used routinely to monitor water quality and process streams. Conductivity detectors are used for measuring ion concentrations in commercial ion chromatography instruments.

A variety of electroanalytical methods are used as detectors for liquid chromatography. Detectors based on conductometry, amperometry, coulometry, and polarography are commercially available.

15.3.1. Potentiometry

Potentiometry involves measurement of the potential, or voltage, of an electrochemical cell. Accurate determination of the potential developed by a cell requires a negligible current flow during measurement. A flow of current would mean that a faradaic reaction is taking place, which would change the potential from that existing when no current is flowing.

Measurement of the potential of a cell can be useful in itself, but it is particularly valuable if it can be used to measure the potential of a half-cell or **indicator electrode** (also called a sensing electrode), which responds to the concentration of the species to be determined. This can be accomplished by connecting the indicator electrode to a reference half-cell to form a complete cell as shown in Fig. 15.6. When the total voltage of the cell is measured, the difference between this value and the voltage of the reference electrode is the voltage of the indicator electrode. This can be expressed as

$$E(\text{total}) = E(\text{indicator}) - E(\text{reference})$$

Compare this equation with Eqs. (15.7) and (15.15). By convention, the reference electrode is connected to the negative terminal of the potentiometer (the readout device). The common reference electrodes used in potentiometry are the SCE and the silver/silver chloride electrode, which have been described. Their potentials are fixed and known over a wide temperature range. Some values for these electrode potentials are given in Table 15.3. The total cell potential is measured experimentally, the reference potential is known, and therefore the variable indicator electrode potential can be calculated and related to the concentration of the analyte through the Nernst equation. In practice, the concentration of the unknown analyte is determined after calibration of the potentiometer with suitable standard solutions. The choice of reference electrode depends on the application. For example, the Ag/AgCl electrode cannot be used in solutions containing species such as halides or sulfides that will precipitate or otherwise react with silver.

15.3.1.1. Indicator Electrodes

The indicator electrode is the electrode that responds to the change in analyte activity. An ideal indicator electrode should be specific for the analyte of interest, respond rapidly to

Figure 15.6 An electrochemical measuring system for potentiometry. The indicator (sensing, working) electrode responds to the activity of the analyte of interest. The potential difference developed between the reference electrode and the indicator electrode is "read out" on the potentiometer (voltmeter). [Courtesy of Thermo Orion, Beverly, MA (www.thermoorion.com).]

Table 15.3 Potentials of Reference Electrodes at Various Temperatures

Temperature (°C)	Potential (V) vs. SHE	
	Saturated calomel (SCE)	Saturated Ag/AgCl
15	0.251	0.209
20	0.248	0.204
25	0.244	0.199
30	0.241	0.194
35	0.238	0.189

changes in activity, and follow the Nernst equation. There are no specific indicator electrodes, but there are some that show a high degree of selectivity for certain analytes. Indicator electrodes fall into two classes, metallic electrodes and membrane electrodes.

Metallic Electrodes. A metal electrode of the **first kind** is just a metal wire, mesh, or solid strip that responds to its own cation in solution. Cu/Cu^{2+}, Ag/Ag^+, Hg/Hg^+, and Pb/Pb^{2+} are examples of this type of electrode. There are significant problems encountered with these electrodes. They have poor selectivity, responding not only to their own cation but also to any other more easily reduced cation. Some metal surfaces are easily oxidized, giving erratic or inaccurate response unless the solution has been purged of air. Some metals can only be used in limited pH ranges because they will dissolve in acids or bases. Silver and mercury are the most commonly used electrodes of the first kind.

A metal electrode of the **second kind** consists of a metal coated with one of its sparingly soluble salts (or immersed in a saturated solution of its sparingly soluble salt). This electrode responds to the anion of the salt. For example, a silver wire coated with AgCl will respond to changes in chloride activity because the silver ion activity is controlled by the solubility of AgCl. The electrode reaction is $AgCl(s) + e^- \longleftrightarrow Ag(s) + Cl^-$, with a potential $E^0 = 0.222$ V. The Nernst equation expression for the electrode potential at 25°C is $E = 0.222 - 0.05916 \log[Cl^-]$.

A metal electrode of the **third kind** uses two equilibrium reactions to respond to a cation other than that of the metal electrode. EDTA complexes many metal cations, with different stabilities for the complexes but a common anion (the EDTA anion) involved in the equilibria. A mercury electrode in a solution containing EDTA and Ca will respond to the Ca ion activity, for example. The complexity of the equilibria make this type of electrode unsuitable for complex sample matrices.

The last type of metallic electrode is the **redox indicator electrode**. This electrode is made of Pt, Pd, Au, or other inert metals, and serves to measure redox reactions for species in solution (e.g., Fe^{2+}/Fe^{3+}, Ce^{3+}/Ce^{4+}). These electrodes are often used to detect the endpoint in potentiometric titrations. Electron transfer at inert electrodes is often not reversible, leading to nonreproducible potentials. Although not a metal electrode, it should be remembered that carbon electrodes are also used as redox indicator electrodes, because carbon is also not electroactive at low applied potentials.

Membrane Electrodes. Membrane electrodes are a class of electrodes that respond *selectively* to ions by the development of a potential difference (a type of junction potential) across a membrane that separates the analyte solution from a reference solution. The potential difference is related to the concentration difference in the specific ion measured on either side of the membrane. Remember that electrodes really respond to activity differences, but we will think in terms of concentration because we are interested (usually) in the concentration of the analyte in our sample. These electrodes do not involve a redox reaction at the surface of the electrode as do metallic electrodes. Because these electrodes respond to ions, they are often referred to as **ion-selective electrodes (ISEs)**. The ideal membrane allows the transport of only one kind of ion across it; that is, it would be specific for the measurement of one ionic species only. As of this writing, there are no specific ISEs, but there are some highly selective ones. Each electrode is more or less selective for one ion; therefore a separate electrode is needed for each species to be measured. In recent years, a large number of different types of membrane electrodes has been developed for a wide variety of measurements.

ISEs are relatively sensitive, as we will see. They are capable of detecting concentrations as low as 10^{-12} M for some electrodes. To avoid writing small concentrations in exponential form, the term **pIon** has been defined, where pIon equals the negative logarithm (base 10) of the molar concentration of the ion. That is,

$$pIon = -\log[Ion] \tag{15.16}$$

For example, pH is the term used for the negative logarithm of the hydrogen ion concentration, where H^+ is expressed as (moles of H^+)/L of solution. Concentrations of other ions can be expressed similarly: pCa for calcium ion concentration, pF for fluoride ion concentration, pOH for hydroxide ion concentration, and pCl for chloride ion concentration. If the concentration of H^+ ion in solution equals 3.00×10^{-5} M, the pH $= -\log(3.00 \times 10^{-5}) = 4.522$, for example. If the concentration is 1.00×10^{-7} M, the pH $= -\log(1.0 \times 10^{-7}) = 7.00$.

Glass Membrane Electrodes. The first membrane electrode to be developed was the glass electrode for measurement of pH, the concentration of hydrogen ion, H^+, in solution. The pH electrode consists of a thin hydrogen ion-sensitive glass membrane, often shaped like a bulb, sealed onto a glass or polymer tube. The solution inside the electrode contains a known concentration of H^+ ion, either as dilute HCl or a buffer solution. The solution is saturated in AgCl. The activity of H^+ inside the electrode is constant and keeps the internal potential fixed. An internal reference electrode is sealed inside the tube and is attached to one terminal of the potentiometer. The glass pH electrode is shown schematically in Fig. 15.7(a). The glass pH electrode is used in combination with a reference electrode, either a separate Ag/AgCl electrode or an SCE, as shown in Fig. 15.7(b). Both electrodes are immersed in a solution of unknown pH and the cell potential developed is a measure of the hydrogen ion concentration in the solution on the outside of the glass membrane of the pH electrode, since all other potentials are fixed.

In a standard pH electrode, the glass membrane is composed of SiO_2, Na_2O, and CaO. The response of this glass to changes in hydrogen ion activity outside the membrane is complex. The glass electrode surfaces must be hydrated in order for the glass to function as a pH sensor. It is thought that a hydrated gel layer containing adsorbed H^+ ions exists on the inner and outer glass surfaces. It is necessary for charge to move across the membrane in order for a potential difference to be measured, but studies have proved that H^+ ion does not move through the glass membrane. The singly charged Na^+ cation is mobile within the 3D silicate lattice. It is the movement of sodium ions in the lattice that is responsible for electrical conduction within the membrane. The inner and outer glass membrane surfaces contain negatively charged oxygen atoms. Hydrogen ions from the solution inside the electrode equilibrate with the inner glass surface, neutralizing some of the negative charge. Hydrogen ions from the solution outside the electrode (our sample) equilibrate with the oxygen anions on the outer glass surface and neutralize some of the charge.

Figure 15.7 (a) A schematic glass electrode for pH measurement. (b) A complete pH measurement cell, with the glass indicator electrode and an external saturated calomel reference electrode. (c) A commercial combination pH electrode, with built-in internal Ag/AgCl reference electrode.

The side of the glass exposed to the higher H^+ concentration has the more positive charge. Na^+ ions in the glass then migrate across the membrane from the positive side to the negative side, which results in a change in the potential difference across the membrane. The only variable is the hydrogen ion concentration in the analyte solution. Because the electrode behavior obeys the Nernst equation, the potential changes by 0.05916 V for every 10-fold change in [H^+] or for every unit change in pH. The glass pH electrode is highly selective for hydrogen ions. The major interfering ion is Na^+. When Na^+ concentration is high and H^+ is very low, as occurs in very basic solutions of sodium hydroxide, the electrode responds to sodium ion as if it were hydrogen ion. This results in a negative error; the measured pH is lower than the true pH. This is called the *alkaline error*. The use of pH electrodes to measure pH is discussed below under applications of potentiometry.

Commercial glass pH electrodes come in all sorts of shapes and sizes to fit into every imaginable container, including into NMR tubes and to measure pH in volumes of solution as small as a few microliters. Some electrodes come as complete cells, with a second reference electrode built into the body of the pH electrode. These are called **combination electrodes**, and eliminate the need for a second external reference electrode. Combination electrodes are required if you want to measure small volumes in small containers and have no room for two separate electrodes. A combination electrode is shown schematically in Fig. 15.7(c). Many glass pH electrodes have polymer bodies and polymer shields around the fragile glass bulb to help prevent breakage. A microelectrode for measuring microliter volumes in well plates is shown in Fig. 15.8.

Other glass compositions incorporating aluminum and boron oxides have been used in membrane electrodes to make the membrane selective for other ions instead of hydrogen ion. For example, a glass whose composition is 11% Na_2O, 18% Al_2O_3, 71% SiO_2 is highly selective toward sodium, even in the presence of other alkali metals. For example, at pH 11 this electrode is approximately 3000× more sensitive to Na^+ than to K^+. The ratio of the response of the electrode to a solution of potassium to its response to a solution of sodium, the analyte, for solutions of equal concentration, is called the *selectivity coefficient*. The selectivity coefficient should be a very small number ($\ll 1$) for high selectivity for the analyte. An ISE has different selectivity coefficients for each ion that responds. Commercial glass ISEs are available for all of the alkali metal ions (Li, Na, K, Rb, and Cs), and for ammonium ion (NH_4^+), Ag^+, Fe^{3+}, Pb^{2+}, and Cu^{2+}.

Crystalline Solid-State Electrodes. Crystalline solid-state electrodes have membranes that are single crystal ionic solids or pellets pressed from ionic salts under high pressure. The ionic solid must contain the target analyte ion and must not be soluble in the solution to be measured (usually aqueous solution). These membranes are generally about 1–2 mm thick and 10 mm in diameter. Sealing the solid into the end of a polymer tube forms the electrode. Like the pH electrode, the interior of the polymer tube contains an internal electrode to permit connection to the potentiometer and a filling solution containing a fixed concentration of the analyte ion. A concentration difference in the analyte ion on the outside of the crystalline membrane causes the migration of charged species across the membrane. These electrodes generally respond to concentrations as low as 10^{-6} M of the analyte ion.

The fluoride ion selective electrode is the most commonly used single crystal ISE. It is shown schematically in Fig. 15.9. The membrane is a single crystal of LaF_3 doped with EuF_2. The term "**doped**" means that a small amount of another substance (in this case, EuF_2) has been added intentionally into the LaF_3 crystal. (If the addition were not intentional, we would call the europium an impurity or contaminant!) Note that the two salts do not have the same stoichiometry. Addition of the europium fluoride creates fluoride ion vacancies in the lanthanum fluoride lattice. When exposed to a variable concentration of F^- ion outside the membrane, the fluoride ions in the crystal can migrate. Unlike the pH electrode, it is the F^- ions that actually move across the membrane and result in the electrode response. The F^- ISE is extremely selective for fluoride ion. The only ion

Figure 15.8   A commercial microelectrode for pH measurements in volumes as small as 10 μL. The electrode has a solid-state pH sensor and a flexible polypropylene stem, making it extremely rugged and suitable for use in the field. This electrode can be used to determine pH in single droplets of water on leaves of plants or blades of grass to study acid rain deposition. [Courtesy of Lazar Research Laboratories, Inc., Los Angeles, CA (www.lazarlab.com).]

that interferes is OH^-, but the response of the electrode to fluoride ion is more than $100\times$ greater than the response of the electrode to hydroxide. The hydroxide interference is only significant when the OH^- concentration is 0.1 M or higher. The electrode only responds to fluoride ion, so the pH of the solution must be kept high enough so that HF does not form.

Other crystalline solid-state electrodes are commercially available to measure chloride, bromide, iodide, cyanide, and sulfide anions. Most of these electrode membranes are made from the corresponding silver salt mixed with silver sulfide, due to the low solubility of most silver salts in water. In addition, mixtures of silver sulfide and the sulfides of copper, lead, and cadmium make solid-state electrodes for Cu^{2+}, Pb^{2+}, and Cd^{2+} available. An advantage of the silver sulfide-based electrodes is that a direct connection can be made to the membrane by a silver wire, eliminating the need for electrolyte filling

Figure 15.9 A commercial solid-state fluoride ISE. The ion-sensitive area is a solid crystal of LaF_3 doped with EuF_2. The filling solution contains NaF and NaCl. The electrical connection is made by an Ag/AgCl electrode. [Courtesy of Thermo Orion, Beverly, MA (www.thermoorion.com).]

solutions. These electrodes will also respond to both silver ion and sulfide ion as well as the intended analyte.

Liquid Membrane ISEs. Liquid membrane electrodes are based on the principle of ion exchange. Older electrodes, such as the one diagrammed in Fig. 15.10(a), consisted of a hydrophilic, porous membrane fixed at the base of two concentric tubes. The inner tube contains an aqueous solution of the analyte ion and the internal reference electrode. The outer tube contains an ion-exchanger in organic solvent. The ion-exchanger may be a cation exchanger, an anion exchanger, or a neutral complexing agent. Some of the organic phase is absorbed into the hydrophilic membrane, forming an organic ion-exchange phase separating the aqueous sample solution and the aqueous internal solution with a known concentration of analyte ion. An ion-exchange equilibrium is established at both surfaces of the membrane and the concentration difference results in a potential difference. Modern liquid membrane electrodes have the liquid ion-exchanger immobilized in or covalently bound to a polymer film, shown schematically in Fig. 15.10(b). The selectivity of these electrodes is determined by the degree of selectivity of the ion-exchanger for the analyte ion. Selective lipophilic complexing agents have been developed and this area is still an important area of research.

Commercial liquid membrane electrodes are available for calcium, calcium plus magnesium to measure water hardness, potassium, the divalent cations of Zn, Cu, Fe, Ni, Ba, and Sr, and anions BF_4^- and nitrate, NO_3^-, among others. In general, these electrodes can be used only in aqueous solution, to avoid attack on the membrane. Lifetimes are limited by leaching of the ion-exchanger from the membrane, although newer technology (such as covalent attachment of the ion-exchanger) is improving this.

The polymer film membrane is being used to coat metal wire electrodes to make miniature ISEs for *in vivo* analysis. These coated wire ISEs require no internal reference solution and have been made with electrode tip diameters of about 0.1 μm. Electrodes have been made small enough to measure ions inside a single cell, as seen in Fig. 15.11.

Gas-Sensing Electrodes. Gas-sensing electrodes are really entire electrochemical cells that respond to dissolved gas analytes or to analytes whose conjugate acid or base is a

Internal aqueous filling solution

Ion exchange reservoir

Liquid ion exchange layer

Ag/AgCl
(a) reference electrode

Porous membrane

Electrical Contact

Module Housing

Aqueous Reference
Solution

Reference Element
(AgCl)

Ion Selective
Membrane

(b)

Figure 15.10 (a) A liquid membrane ISE. (b) An ISE with ion exchange material in a solid polymer membrane. The sensed ion is exchanged across the membrane, creating the potential. This type of ISE is available for Ca, K, nitrate ion and nitrite ion determinations. [Part (b) courtesy of Thermo Orion, Beverly, MA (www.thermoorion.com).]

gaseous species. Because they are complete cells, the term "probe" is often used instead of electrode. A typical gas-sensing probe consists of an ISE and a reference electrode sealed behind a gas-permeable membrane, through which the analyte gas diffuses. The ISE used can be selective for the species, for example an Ag_2S electrode to measure hydrogen sulfide that effuses through the gas permeable membrane. Several gas-sensing probes use a glass pH electrode as the indicator electrode for gas species that change the pH of the internal solution. Examples include the determination of CO_2, which dissolves in the aqueous internal solution to form H^+ and HCO_3^-, and the determination of ammonia, which dissolves to form NH_4^+ and OH^-. The potential that develops in all of these cells depends on the concentration of the analyte gas in the external sample solution. Gas-sensing electrodes typically have long response times, on the order of a few minutes, because the gas must cross the membrane. These electrodes are highly selective and free from interference from nonvolatile species; only gases can cross the gas-permeable membrane. If a pH electrode is used as the ISE, obviously any gas that dissolves and

Figure 15.11   Microscopic carbon-fiber electrodes for measurements in a single cell. (From Michael and Wightman, used with permission.)

changes the internal solution will be detected; therefore some gases may interfere with other gas analytes.

  Immobilized Enzyme Membrane Electrodes.   Enzymes are biocatalysts that are highly selective for complex organic molecules of biochemical interest, such as glucose. Many enzymes catalyze reactions that produce ammonia, carbon dioxide, and other simple species for which ISEs are available as detectors. Coating an ISE with a thin layer of an enzyme holds the promise of making highly selective biosensors for complex molecules with the advantages of potentiometry—speed, low cost, simplicity. The enzyme is generally immobilized on the electrode by incorporation into a gel, by covalent bonding to a polymer support or by direct adsorption onto the electrode surface, so that a small amount of enzyme may be used for many measurements. Very few species interfere with enzyme-catalyzed reactions, although the enzyme can be inhibited. No commercial potentiometric enzyme-based electrodes are available; however, considerable research in this area is in progress.

  Ion-Selective Field Effect Transistors.   Ion-selective field effect transistors (ISFETs) are semiconductor devices related to the solid-state detectors used in spectroscopy (discussed in Chapter 5). In this case, the surface of the transistor is covered with silicon nitride, which adsorbs H^+ ions from the sample solution. The degree of adsorption is a function of the pH of the sample solution and the adsorption of H^+ ions results in a change in the conductivity of the ISFET channel. The cell requires an external reference electrode. ISFET pH sensors can be made extremely small (about 2 mm^2) and are extremely rugged, unlike the fragile glass bulb pH electrode. They have rapid response times and can operate in corrosive samples, slurries, and even wet solids such as food products. The sensor can be scrubbed clean with a toothbrush, stored in a dry condition, and does not require hydrating before use. The price of an ISFET pH electrode is about half that of a standard glass electrode. A schematic of an ISFET pH electrode is shown in Fig. 15.12, with the ISFET sensor embedded in a polymer body electrode.

15.3.1.2. Instrumentation for Measuring Potential

Understanding electrochemical instrumentation requires a basic knowledge of electricity and basic electronics. Coverage of these fundamentals is impossible in a text of this size. The student is advised to review the concepts of electricity learned in general physics, and to understand the definitions of current, voltage, resistance, and similar basic terms. The texts by Kissinger and Heineman, Malmstadt et al., or Diefenderfer and Holton, listed in the bibliography, are excellent sources of information on electronics used in instrumentation. The electrochemical cell is one circuit element with specific electrical properties in the complete instrumentation circuit.

A **potentiometric cell** is a galvanic cell connected to an external potential source that is exactly equal but opposite to that of the galvanic cell. Negligible current flows in this type of cell. The potential of a complete cell can be measured with a solid-state circuit that requires negligible input current but can output the voltage to a meter, digital readout, recorder, or computer. The circuit should have high input impedance, high gain, low output impedance and ideally, no output signal when there is no input signal. **Operational amplifiers** have these properties. Figure 15.13 illustrates an operational amplifier (op-amp) set up as a voltage follower. The triangle represents the amplifier with two inputs on the left and an output on the right. It may contain a large number of discrete components, for example, transistors. Usually such devices require ± 15 V power lines, which are omitted in circuit diagrams. The voltage input marked $+$ is called a *noninverting input*,

Fast response silicon
chip sensor

Built-in reference and medical grade
temperature sensor

Figure 15.12   A commercial ISFET pH probe, with a silicon chip pH sensor, a built in reference electrode, and a built-in temperature sensor. This pH probe is housed in a stainless steel body with a slanted tip for easy insertion into meat or other soft solid samples. (Courtesy of IQ Scientific Instruments, Inc. San Diego, CA.)

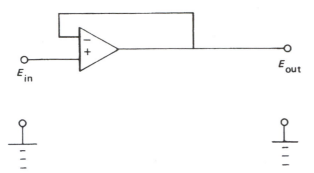

Figure 15.13   Schematic of an operational amplifier (called an op-amp) with voltage feedback (the voltage follower).

and the one marked $-$ is the *inverting input*. If A is the gain (typically 10^5), the output voltage E_{out} is

$$E_{out} = -A(E^- - E^+) \tag{15.17}$$

where E^- and E^+ are the voltage values at the inverting and noninverting inputs, respectively. The minus sign means that the polarity of the voltage difference is reversed. Rearranging, we have

$$E^- = E^+ - \frac{E_{out}}{A}$$

and because A is very large,

$$E^- \approx E^+$$

Notice that E_{out} is fed back into the noninverting input (E^-). This is termed voltage feedback.

The function of this type of circuit is to offer a high-input impedance which can be typically 10^5–10^{13} ohms (Ω). This means that the $+$ and $-$ inputs require negligible current and can be used to monitor the voltage of a highly resistive source such as a glass electrode. However, the output current and voltage capabilities are considerable, depending on the device properties and the power supply limitations. It seems odd, at first, that the circuit just reproduces the voltage of the cell but does so in such a way as to make available much larger currents than are available from the electrochemical cell. The electrical resistance of the glass electrode for pH measurement is enormous (50–10,000 MΩ). An ordinary potentiometer or voltmeter would indicate practically no observable voltage, which is why a high-impedance solid-state measuring circuit is required. Modern instruments drain no more than 10^{-10}–10^{-12} A from the cell being measured.

Commercial solid-state potential measuring devices based on the type of op-amp described are often called **pH or pIon meters** and are designed to work with glass pH electrodes, ion selective electrodes, and other indicator electrodes described earlier. Research quality pIon meters have built-in temperature measurement and compensation, autocalibration routines for a three-point (or more) calibration curve, recognition of electrodes (so you do not try measuring fluoride ion with your pH electrode!), and the ability to download data to computer data collection programs. The relative accuracy of pH measurements with such a meter is about ± 0.005 pH units. Meters are available as handheld portable devices for field or plant use; these meters have less accuracy, on the order of ± 0.02 pH units, but are much less expensive. The field portable meters are battery-powered and waterproof. Commercial meters will read out in mV, pH, concentration of ion for an ISE, and in mS for conductivity measurements and many will also read out the temperature.

A second important use of operational amplifiers and feedback circuits is to control the potential at a working electrode. The circuit that performs this task is called a **potentiostat.** Potentiostats and modern polarography equipment are called **three-electrode devices** because they attach to a working electrode, a reference electrode, and an auxiliary electrode. also called a counterelectrode (Fig. 15.14). The principle of operation is that the potential between the working electrode (WE) and the reference electrode (REF) is maintained by a feedback circuit. The applied voltage E_{appl} may be a constant voltage or some signal generator voltage, such as a linear ramp. To understand how this circuit works, consider starting with no potential at the amplifier output. The output voltage E_{out} will be given, as before, by

$$E_{out} = -A(E^- - E^+)$$

Because at time $t = 0$, $E^- = E_{appl}$, the voltage E_{out} will rise, with the same sign, until $E_{out} = E_{appl}$ at the point REF. Assuming that the uncompensated resistance R_u between WE and REF can be essentially ignored, this ensures that WE is held at E_{appl} and the potential of the counterelectrode can "float" to maintain zero differential input at the $-$ and $+$ inputs. Note that negligible current flows to the inverting ($-$) input through the reference electrode. The measured cell current flows between the WE and the counterelectrode (CE). In practice, the uncompensated resistance R_u is kept small by placing the reference electrode close to WE, by using an excess of supporting electrolyte, and by keeping the net cell current small by using microelectrodes, for example, a dropping mercury electrode. Then the voltage drop in the electrolyte between WE and REF, iR_u, need not be compensated for.

On modern potentiostats, the applied voltage is given as the *actual* sign of the voltage on the WE, and there is no need to assume a potential reversal. In other words, if you set the potentiostat at -0.500 V, WE is at -0.500 V vs. whatever reference electrode you have in the actual cell. If, for example, this electrode was a SCE, then on the

Figure 15.14 Schematic of a simple potentiostat and representation of the cell as an electrical (equivalent) circuit.

SHE scale at 25°C,

$$E_{WE} = -0.500 + 0.241$$
$$= -0.259 \text{ V vs. SHE}$$

If the cell had a silver/silver chloride reference (3.5 M) and the equipment were set at -0.500 V vs. REF, then

$$E_{WE} = -0.500 + 0.205$$
$$= -0.295 \text{ V vs. SHE}$$

In each of these cases, when we state our values on the hydrogen scale, the potential at the working electrode is numerically less. On the other hand, for an applied potential of $+0.500$ V vs. SCE on the potentiostat,

$$E_{WE} = +0.500 + 0.241$$
$$= 0.742 \text{ V vs. SHE}$$

This raises an important point. It does not matter which reference electrode we use; it is best to use the one most suited for the chemical system being studied. We should always, however, quote our potentials vs. the particular reference and be sure to label the axes of current–voltage curves with the appropriate reference electrode. Electrode potentials are relative values.

15.3.1.3. *Analytical Applications of Potentiometry*

Potentiometry is widely used in industrial, environmental, agricultural, clinical, and pharmaceutical laboratories to measure ions, acids, bases, and gases. Measurements may be made in a standard laboratory setting, but potentiometry is ideal for online process monitoring, *in vivo* monitoring, and field measurements using flow-through or portable instruments. Quantitative measurement of pH is extremely important and will be discussed in detail below. Quantitative measurement of inorganic and organic ions, acidic and basic ions, and gases, and determination of ions in specific oxidations states are commonly performed by potentiometry. Potentiometry is used in research to determine stability constants of complexes, solubility product constants, to determine reaction rates and elucidate reaction mechanisms, and to study enzyme and other biochemical reactions.

Samples must be in the form of liquids or gases for analysis. Sample preparation may include buffering the sample to an appropriate pH for some ISE and gas measurements or adding ionic strength "buffer" to make all samples and standards of equivalent ionic strength. The detection limit for most common electrodes is about 10^{-6} M, while gas probes have detection limits in the low ppm range.

An understanding of stoichiometry, acid–base theory, and simple equilibrium calculations is required for the following quantitative applications. Review your introductory chemistry text or a basic quantitative analysis text such as those by Harris or Enke listed in the bibliography.

Direct Measurement of an Ion Concentration. The concentration of an ion in solution may be measured directly via the potential developed by the half-cell involved and by applying the Nernst equation. For example, a silver/silver ion half-cell plus an SCE gives the following relationship:

$$E(\text{cell}) = E(\text{Ag}) - E(\text{SCE})$$

where $E(Ag)$ is the emf of the silver half-cell. But

$$E(Ag) = E^0(Ag) - \frac{RT}{nF}\ln\frac{1}{[Ag^+]}$$

Therefore

$$E(cell) = E^0(Ag) - (0.05916)\log\frac{1}{[Ag^+]} - E(SCE)$$

and at 25°C, $E(SCE) = 0.244$ V. From Appendix 15.1, $E^0(Ag) = 0.799$ V. Substituting, we get:

$$E(cell) = 0.799 - (0.05916)\log\frac{1}{[Ag^+]} - 0.244$$

This gives us $E = 0.555 + 0.05916\log[Ag^+]$. In an experiment conducted at 25°C, the cell potential for such a cell was measured and found to be $= 0.400$ V. Therefore:

$$\log[Ag^+] = \frac{0.400 - 0.555}{0.05916}$$

$$= -\frac{0.155}{0.05916}$$

$$\log[Ag^+] = -2.62$$

$$\text{The concentration of } Ag^+ = \text{antilog}(-2.62)$$

$$= 2.4 \times 10^{-3} \text{ M}$$

Potentiometric Titrations. Potentiometry is a useful way to determine the end-point in many titrations. For example, the concentration of Ag^+ ion in solution can be used to determine the **equivalence point** in the titration of Ag^+ with Cl^-. In this titration the following reaction takes place:

$$Ag^+ + Cl^- \longrightarrow AgCl(s) \downarrow (\text{precipitation})$$

The concentration of the Ag^+ in solution steadily decreases as Cl^- is added. At the equivalence point $[Ag^+] = [Cl^-]$. But, for the sparingly soluble salt silver chloride, $[Ag^+][Cl^-]$ is a constant called the *solubility product*, K_{sp}. In the case of the AgCl precipitation reaction, $K_{sp} = 1 \times 10^{-10}$ at 25°C. Therefore,

$$[Ag^+][Cl^-] = 1 \times 10^{-10}$$

But, as we have seen, $[Ag^+] = [Cl^-]$ at the equivalence point; therefore

$$[Ag^+]^2 = 1 \times 10^{-10}$$

$$[Ag^+] = \sqrt{1 \times 10^{-10}} = 1 \times 10^{-5} \text{ M}$$

From this calculation, at the equivalence point $[Ag]^+ = 1 \times 10^{-5}$ M. By applying the Nernst equation to this solution, we get

$$E(Ag) = E^0 + \frac{0.05916}{1}\log[Ag^+]$$

and

$$E^0(Ag) = +0.799$$

therefore

$$E(Ag) = +0.799 + \frac{0.05916}{1} \times (-5)$$

$$E(Ag) = +0.799 - 0.295$$

$$E(Ag) = +0.504$$

At the equivalence point the emf of the half-cell is calculated to be 0.504 V vs. SHE. If we make a complete cell by using the silver half-cell and an SCE half-cell (+0.244 V), the observed voltage at the equivalence point will be given by the relationship

$$E(observed) = E(Ag) - E(reference)$$

$$= +0.504 - 0.244$$

$$= +0.260 \, V \text{ vs. SCE}$$

At the equivalence point the measured cell emf is +0.260 V vs. SCE. In the titration of silver ion, the chloride solution should be added in small portions and the emf of the cell measured after each addition. By plotting the relationship between the volume of Cl⁻ solution added and the voltage of the cell, we can determine the volume of Cl⁻ solution necessary to reach the equivalence point. A typical curve relating the volume of chloride to the voltage for titration of silver ion with 0.100 N Cl⁻ solution is shown in Fig. 15.15. In a real titration, the endpoint is determined from the inflection point in the titration curve, not from the calculated voltage.

In the following example, the equivalence point is reached after adding 2.00 mL of 0.100 Cl⁻ solution to 5.00 mL of silver solution. The normality of the silver can be

Figure 15.15 Relationship between Ag^+ concentration, the cell potential and the volume of Cl⁻ solution added.

determined from the equation:

$$\text{(volume of } Cl^-) \times \text{(normality of } Cl^-) = \text{(volume of } Ag^+)$$
$$\times \text{(normality of } Ag^+)$$

We have $(2.00 \text{ mL})(0.100 \text{ N}) = (5.00 \text{ mL})(\text{N } Ag^+)$ or

$$\text{normality of silver solution} = \frac{2.00 \text{ mL} \times 0.100 \text{ N}}{5.00 \text{ mL}} = 0.0400 \text{ N}$$

From this, the weight of Ag^+ present in the solution can be calculated as follows:

1 L of 1 N Ag^+ contains 108 g Ag^+, thus

g Ag = (108 g Ag/eq)(0.0400 eq/L)(1 L/1000 mL)(5.00 mL)

g Ag = 0.0216 g

In the example just discussed, we used the known K_{sp} to calculate the potential at the equivalence point. Clearly, it is possible to work backwards and to use the measured potential at the equivalence point to determine the K_{sp} for a sparingly soluble salt.

In Fig. 15.15 the titration curve was simple and the equivalence point easily detected; however, in dilute solutions, nonaqueous solvents and titrations involving slower reactions, the titration curve may be flatter (the inflection point hard to find) and difficult to interpret. The problem can be simplified by using equipment that records the **first and/or second differential** of the titration curve. The curves obtained for a relationship such as that in Fig. 15.15 are shown in Fig. 15.16. When the first differential of the curve is taken, the slope is measured and plotted against the volume added. As the potential changes, the slope changes. At the endpoint, the slope of the curve ceases to increase and begins to decrease, that is, the slope goes through a maximum. This can be seen in the curve of Fig. 15.16(b). The second differential measures the slope of the first differential curve and plots this against the volume added. At the endpoint the curve in Fig. 15.16(b) goes through a maximum. At the maximum, the slope of the curve is horizontal; that is, it has a zero slope. The second differential curve is shown in Fig. 15.16(c). As can be seen, there is a rapid change in the second differential curve, whose values go from positive to negative, and at the equivalence point the value of the second differential equals zero. There are commercial automatic potentiometric titrators available that accurately and rapidly deliver titrant, perform the potentiometric measurements, calculate the derivatives, stop the titrant after the equivalence point is reached, and accurately calculate the concentration of the analyte. Many of these come with autosamplers and computer data handling systems so that multiple titrations can be performed unattended.

pH Measurements. One of the most important applications of potentiometry is the determination of $[H^+]$ or the pH of a solution, where pH is defined as the negative logarithm to the base 10 of the hydrogen ion activity (and is approximately equal to the negative logarithm to the base 10 of the hydrogen ion concentration under certain conditions discussed subsequently),

$$\text{pH} = -\log(H^+) \approx -\log[H^+] \tag{15.18}$$

where (H^+) stands for the activity of the hydrogen ion and $[H^+]$ is the molar concentration of the hydrogen ion.

The procedure can be used for the determination of the pH of most solutions, which is important if the solution is drinking water, the water in a swimming pool, your favorite soda, or a chemical in an industrial process. Frequently, organic and inorganic synthetic reactions,

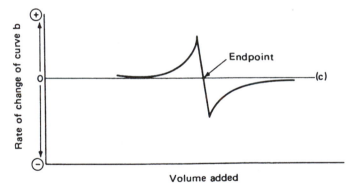

Figure 15.16   Relationship between (a) a simple potentiometric titration curve and its (b) first, and (c) second differentials.

biochemical reactions, and other chemical reactions carried out by research or industrial chemists are very sensitive to pH, which therefore must be measured and controlled.

The determination of $[H^+]$ in a solution is a special case of the determination of ion concentration (activity). The importance of this determination has merited special equipment designed for the specific application—field-portable, online analysis, and a wide variety of benchtop units. Hydrogen ion concentration can be determined directly by employing the Nernst equation, an SHE and a REF such as the SCE.

The potential developed by the cell is given by (if we ignore the liquid junction potential)

$$E(\text{cell}) = E(H^+) - E(\text{reference}) \tag{15.19}$$

Also,

$$E(\mathrm{H}^+) = E^0(\mathrm{H}) + \frac{RT}{nF}\ln[\mathrm{H}^+]$$

$$E(\mathrm{cell}) = -E(\mathrm{reference}) + E^0(\mathrm{H}) + \frac{RT}{nF}\ln[\mathrm{H}^+]$$

$$= -E(\mathrm{reference}) + E^0(\mathrm{H}) + \frac{0.05916}{1}\log[\mathrm{H}^+] \text{ at } 25^\circ\mathrm{C}$$

But $E(\mathrm{H})$, the emf of a standard hydrogen cell, is zero by definition. Hence

$$E(\mathrm{cell}) = -E(\mathrm{reference}) + \frac{0.0591}{1}\log[\mathrm{H}^+]$$

The emf of an SCE $= +0.241$ V vs. SHE; therefore

$$E(\mathrm{cell}) = -0.241 + 0.0591\log[\mathrm{H}^+]$$

By rearranging we get

$$-\log[\mathrm{H}^+] = \frac{[E(\mathrm{cell}) + E(\mathrm{reference})]}{0.0591}$$

$$= \frac{[E(\mathrm{cell}) + 0.241]}{0.0591}$$

If in a particular experiment the observed voltage of the cell is -0.6549 V vs. SCE, then

$$-\log[\mathrm{H}^+] = -\frac{(-0.6549 + 0.2412)}{0.05916} = 6.99$$

therefore,

$$[\mathrm{H}^+] = 1.0 + 10^{-7}\,\mathrm{M}$$

SHEs are not used for routine pH determinations due to the difficulty of construction and operation of the SHE. The most common hydrogen half-cell in practical use is the glass electrode. An SCE is commonly used as the reference electrode. The cell may be formulated as

SCE $\|$ $[\mathrm{H_3O}^+]_{\text{(unknown)}}$ $|$ glass membrane $|$ $[\mathrm{H_3O}^+]_{\text{(internal reference)}}$
$\|$ 0.1 M Cl$^-$, AgCl(sat.) $|$ AgCl(s), Ag(s)

The SCE is the external reference electrode and the Ag/AgCl electrode is the internal reference electrode. Theoretically, the only unknown is the activity (concentration) of the H^+ in the sample or unknown solution. The Nernst equation tells us that the voltage of the pH electrode should change by 59.16 mV for every unit change in pH at 25°C. However, we cannot in practice ignore the junction potential. The real equation for a cell is given by

$$E_{\mathrm{cell}} = E_{\mathrm{ind}} - E_{\mathrm{ref}} + E_{\mathrm{j}} \tag{15.20}$$

In addition to the junction potential E_{j}, membrane electrodes have asymmetry potentials of unknown magnitude. These potentials cannot be computed theoretically or easily measured individually, so direct pH measurements require that the pH electrode system be calibrated. In fact, the definition of pH used by NIST and IUPAC is an *operational definition* based on calibration with a series of standard buffer solutions.

Buffer solutions are those in which the pH remains relatively constant (1) over wide ranges of concentration and (2) after the addition of small quantities of acid or base. Standard buffer solutions to calibrate pH electrode systems are commercially available from most major science supply companies, from electrode manufacturers, or may be formulated from mixtures described in chemical handbooks such as the one by Dean listed in the bibliography. Certified buffers can be purchased from NIST, but these are only needed for extremely accurate work.

The pH meter is calibrated (to compensate for asymmetry and other potentials and variations in the reference electrode) by immersing the glass electrode and the reference electrode in a buffer solution of known pH, which we will call pH_{std}. The potential of the cell generated by immersing the electrodes in the buffer is measured, and is designated E_{std}. The electrodes are then placed in the sample, pH_{unk}, and the potential E_{unk} is measured. The pH of the sample is calculated from:

$$pH_{unk} = pH_{std} - \frac{(E_{unk} - E_{std})}{0.05916} \tag{15.21}$$

Equation (15.21) is the operational definition of pH at $25°C$. This is our best experimental estimation of the activity of the hydrogen ion, (H^+), from Eq. (15.18).

It is best to calibrate the instrument at a pH near to that of the sample or at two pH values that bracket the sample pH. If necessary, calibration can be performed at multiple pH levels, such as pH 4.00, 7.00, and 10.00 to check the linearity of the system. The use of a calibration curve gives the best empirical relationship between voltage and concentration, but the accuracy depends on matching the ionic strength of the sample to that of the standards to avoid differences in activity coefficients. This can be difficult to do in high ionic strength samples and complex matrices. In these cases, the MSA, discussed subsequently, may be used to advantage.

Errors in pH Measurement with Glass Electrodes.   There are several sources of error in the routine measurement of pH. One source of error that may occur with any pH probe, not just glass electrodes, is in the preparation of the calibration buffer or buffers. Any error in making the buffer or any change in composition on storage of the buffer will result in error in the pH measured. Common problems with buffers are bacterial growth or mold growth in organic buffers, and absorption of CO_2 from air by very basic buffers (thereby making them less basic). Guidelines on storage of buffers, preservatives that can be added, and other practical advice can be found in the reference by Dean. The accuracy of the measured pH can be no more accurate than the pH of the calibration buffer.

Glass electrodes become sensitive to alkali metal ions in basic solution (pH > 11) and respond to H^+ *and* Na^+, K^+, and so on. This results in the measured pH being *lower* than the true pH. The magnitude of the alkaline error depends on the composition of the glass membrane and the cation interfering. This error is called the **alkaline error**. It results from ion-exchange equilibria at the glass membrane surface between the alkali metal ions in the glass and the alkali metal ions in solutions. Special glass compositions are made for electrodes that are used in highly alkaline solutions to minimize the response to non-H^+ ions.

Glass electrodes also show an error in extremely acidic solutions (pH < 0.5). The **acid error** is in the opposite direction to the alkaline error; the measured pH values are too high. The cause of the acid error is not understood.

Variations in junction potential between calibration standards and samples lead to errors in pH measurement. Absolute accuracy of 0.01 pH units is difficult to obtain, but relative differences in pH as small as 0.001 pH units can be measured. The operational definition of pH is only valid in dilute solutions, with ionic strengths <0.1 and in the pH range 2–12. Both very high ionic strength solutions and very low ionic strength solutions (e.g., unpolluted lake water) can have serious errors in pH measurement caused by nonreproducible junction potentials.

pH Titrations. Direct measurement of pH can be used for quantitative acid–base titrations. For example, if NaOH is added slowly to HCl, the following reaction takes place:

$$\text{Original solution: HCl} \longrightarrow \text{H}^+ + \text{Cl}^-$$
$$\text{Addition of NaOH: H}^+ + \text{Cl}^- + \text{Na}^+ + \text{OH}^- \longrightarrow \text{H}_2\text{O} + \text{Na}^+ + \text{Cl}^-$$

In the process $[\text{H}^+]$ decreases and the potential of the pH electrode changes 59.16 mV for every unit change in pH, as we have seen. Near the neutralization point the pH changes very rapidly with the addition of small amounts of NaOH. A plot of cell potential vs. the volume (in mL) of NaOH added gives the relationship in Fig. 15.17. The equivalence point of the titration is the inflection point of the curve.

Use of potentiometry for pH titration allows analyses to be carried out in colored or turbid solutions. Also, it solves the problem of selecting the correct indicator for a particular acid–base titration. The endpoint can be determined more accurately by using a first or second differential curve as described earlier. It also permits pH titrations in nonaqueous solvents for the determination of organic acids and bases as described subsequently. In addition, it can be readily automated for unattended operation.

Titrations of Weak Acids and Bases. Most organic acids and bases and some inorganic weak acids and bases are too weak to be titrated in aqueous solution. This is a result of the **leveling effect**; the strongest acid that can exist in water is H^+ and the strongest base that can exist in water is OH^-. All acids that dissociate to give a proton are leveled to the same acid strength in water. For very weak acids and bases, the equilibrium constant for the titration reaction in water may not be large enough to give a distinct endpoint. It is common practice, therefore, to choose a nonaqueous solution for these determinations.

Figure 15.17   Schematic plot of the potentiometric titration of a strong acid with NaOH. A pH electrode can be used as the indicator electrode.

In a nonaqueous solvent, different acids are not leveled to the same extent. The nonaqueous solvents are chosen based on their acidity, dielectric constant, and the solubility of the sample in the solvent. It is important that the acidity of the solvent not be too great; otherwise the titrant is used up in titrating the solvent rather than the sample. The same principles apply to the titration of bases, in that the solvent must not be too basic; otherwise reaction between the sample base and the titrant is not observed. The dielectric constant affects the relative strength of dissolved acids. For example, negatively charged acids become relatively stronger compared to uncharged acids (such as formic acid) as the dielectric constant of the solvent is decreased. It is also important that the sample be soluble in the solvent, otherwise the reactions will be incomplete and unobservable. An advantage of using an organic solvent is that substances that are not soluble in water may be determined.

The types of cells used vary with the solvent employed. Glass and SCE electrodes may be used if the solvent is an alcohol, ketone, or acetonitrile but special modification of the salt bridge in the SCE is necessary. Platinum electrodes may be used for titrations with tetrabutylammonium hydroxide (TBAH), $(C_4H_9)_4NOH$, dissolved in a benzene–methanol mixture or in MIBK. When TBAH is added to a sample that contains more than one acid functionality, it first reacts with the strongest acid in solution until reaction is virtually complete. It then commences reaction with the second strongest acid until the second reaction is complete, and so on. In a solution containing two different acid groups, if the K_a values differ enough, two equivalence points can be observed in a plot of emf (in mV) vs. mL of titrant. These two acid groups may be on the same compound, that is, a diprotic acid, or they could be two different monoprotic acids (a mixture). A schematic illustration of a titration of a mixture of three weak organic acids with TBAH is shown in Fig. 15.18. **Nonaqueous titrations** have been used to determine amino acids, amines, alkaloids, carboxylic acids, many drugs such as barbiturates and antihistamines, and alcohols, aldehydes, and ketones.

The experimental setup for titrations in nonaqueous solvents differs from those in aqueous solution and specialized texts (e.g., Huber) should be consulted.

Quantitative Determination of Ions Other than Hydrogen.   ISEs can be used for the analysis of aqueous and nonaqueous solutions alike and have therefore found increasing applications in organic chemistry, biochemistry, and medicine. Furthermore, in many circumstances no sample preparation is necessary. As a result, the use of ISEs is particularly valuable for obtaining rapid results with no loss of volatile analytes. ISE potentiometry can be considered nondestructive, since the amount of contamination of the sample by leaking electrolyte is usually negligible. These electrodes can even be used for continuous analysis of waters and industrial plant streams simply by inserting the electrodes into the sample (such as the river or plant stream) and measuring the voltages generated.

It must be remembered that the cell potential is proportional to the ln(activity) of the ion rather than its concentration. The activity is a measure of the extent of thermodynamic nonideality in the solution. The activity coefficient is usually less than unity, so the activity of a solution is generally lower than the total concentration, but the values of activity and concentration approach each other with increasing dilution. If a compound is not completely ionized, the activity is further decreased. Decreased ionization can be brought about because of a weak ionization constant, chemical complexation, or a high salt concentration in the solution. Any of these factors will cause a change in the potential of the cell, even if the ion concentration is constant. In practice, it is better to determine the relationship between the cell potential and ion *concentrations* experimentally.

ISEs are calibrated in a manner similar to pH electrodes. Standard solutions of known concentrations of the ion to be measured are used and a plot of the cell emf vs.

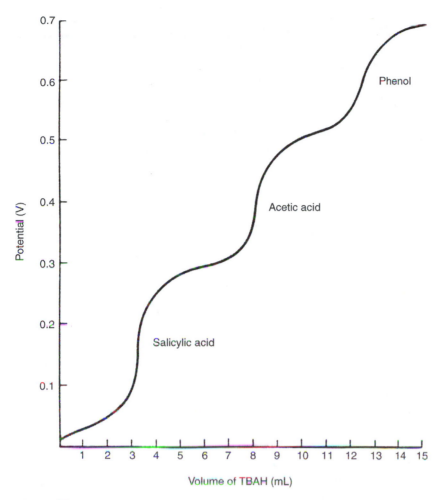

Figure 15.18 Schematic plot of the potentiometric titration of three weak acids with tetrabutylammonium hydroxide. Note the equivalence points after the addition of 3.5, 8.5, and 13.0 mL of titrant.

concentration is made. It is important to keep the ionic strength and other matrix components of the samples and standards the same. This is often done by adding the same amount of a high ionic strength buffer to all samples and calibration standards. In some cases, the buffer may also adjust the pH if the form of the ion to be measured is pH-dependent. The determination of fluoride ion is an example of this approach. All samples and standards have the same amount of a commercial buffer called **TISAB (total ionic strength adjusting buffer)** that also controls pH to insure that fluoride is present as F^-, not HF, because the fluoride electrode does not respond to HF.

Electrodes that behave in a Nernstian manner will have slopes equal to 59.16 mV per pIon unit for monovalent cations, (59.16/2) mV per pIon unit for divalent cations, -59.16 mV per pIon unit for monovalent anions, and so on, at 25°C. [Where pIon is defined exactly as is pH, as the negative logarithm to the base 10 of the ion concentration (activity) in M. So we have pF, pCa, pAg, and so on.] Many plots of concentration vs. voltage for ISEs deviate from linearity due to activity coefficient effects, so an alternate approach is to use the MSA for calibration, especially for complex matrices.

Calibration by MSA (Method of Standard Additions).   In the construction of standard addition curves for calibration we assume that the potential is proportional to the log of the concentration. Let the original ion concentration be C_0 and the solution volume be V_0 (in L). The initial potential

$$E_1 = \text{const.} + 2.303 \frac{RT}{nF} \log C_0 \tag{15.22}$$

It is usual to add a small volume V_s of a standard solution of concentration C_s from a microliter pipette such that the concentration becomes $(C_0 V_0 + C_s V_s)/(V_0 + V_s)$ and the new potential E_2 is given by

$$E_2 = \text{const.} + 2.303 \frac{RT}{nF} \log\left(\frac{C_0 V_0 + C_s V_s}{V_0 + V_s}\right) \tag{15.23}$$

Subtracting Eq. (15.22) from Eq. (15.23), we have

$$\Delta E = 2.303 \frac{RT}{nF} \log\left(\frac{C_0 V_0 + C_s V_s}{C_0 (V_0 + V_s)}\right)$$

If we note that $V_0 \gg V_s$,

$$\Delta E \approx 2.303 \frac{RT}{nF} \log\left(\frac{C_0 V_0 + C_s V_s}{C_0 V_0}\right) \approx 2.303 \frac{RT}{nF} \log\left(1 + \frac{C_s V_s}{C_0 V_0}\right)$$

This expression, in turn, can be rearranged as

$$1 + \frac{C_s V_s}{C_0 V_0} = 10^{nF\Delta E/2.303RT} \tag{15.24}$$

from which we can obtain the concentration of the unknown by plotting $(10^{nF\Delta E/2.303RT} - 1)$ vs. V_s for several additions of the standard. Then the slope m will have the value $C_s/C_0 V_0$ and the unknown concentration is calculated from

$$C_0 = \frac{C_s}{m V_0} \tag{15.25}$$

The method is usually very sensitive but depends upon the particular ISE. Also, special buffers may be required to avoid problems with interfering ions. In general, two additions should be made where possible, with the concentration doubled from the original concentration. It is assumed that the additions do not change the ionic strength or the junction potentials significantly.

ISEs have been used for the determination of sodium and potassium in bile, nerve and muscle tissue, kidneys, blood plasma, urine, and other body fluids. ISEs are used for the analysis of ions in sea water, river water, and industrial water and wastewater, as well as in a wide variety of commercial products, such as personal care and cosmetic products. The advantages of ISEs are that they are fast, with response times <1 min for most ISEs; they are nondestructive, have a linear range of about six orders of magnitude in concentration, usually over the 10^{-6} to 1 M range, and they can be used in turbid or highly colored solutions. The disadvantages are that a different electrode is needed for each ionic species, the electrodes are selective but not specific, so interferences can occur, and the electrodes can become plugged or contaminated by components of the sample. The ionic species must be in solution and in the proper oxidation state to be detected by the electrode.

Oxidation–Reduction Titrations. Potentiometry can be used to follow reduction–oxidation (redox) titrations. For example, the oxidation of stannous ions by ceric ions follows the chemical reaction

$$Sn^{2+} + 2Ce^{4+} \longrightarrow Sn^{4+} + 2Ce^{3+}$$

It is usually best to first identify the two half-cell reactions and to find the total number of electrons n in the reaction (n must be the same for each half-cell reaction, so it may be necessary to balance the half-reactions) and to look up the standard reduction potentials in Appendix 15.1:

$$Sn^{2+} \longrightarrow Sn^{4+} + 2e^- \quad E^0 = +0.13 \, V$$
$$2Ce^{4+} + 2e^- \longrightarrow 2Ce^{3+} \quad E^0 = +1.74 \, V$$

The net $E^0_{cell} = 1.74 \, V - 0.13 \, V = +1.61 \, V$. The emf $E(A)$ of the half-cell is given by the Nernst equation for a stannous/stannic half-cell, with the stannous ion oxidizing to stannic ion:

$$E(A) = E^0(Sn^{2+}/Sn^{4+}) + \frac{0.05916}{2} \log\left(\frac{[Sn^{4+}]}{[Sn^{2+}]}\right) \tag{15.26}$$

For the ceric/cerous half-cell, with ceric ion reduced

$$E(B) = E^0(Ce^{3+}/Ce^{4+}) - \frac{0.0591}{2} \log\left(\frac{[Ce^{3+}]^2}{[Ce^{4+}]^2}\right) \tag{15.27}$$

At all times the emf produced by each half-cell must be equal to that of the other because the mixture cannot exist at two potentials. Therefore, from the Nernst equation [Eq. (15.12)], we have

$$E_{cell} = E^0_{cell} - \frac{0.05916}{2} \log \frac{[Sn^{4+}][Ce^{3+}]^2}{[Sn^{2+}][Ce^{4+}]^2}$$

Since the equilibrium constant for this reaction is:

$$K_{eq} = \frac{[Sn^{4+}][Ce^{3+}]^2}{[Sn^{2+}][Ce^{4+}]^2}$$

at equilibrium,

$$E^0_{cell} = \frac{0.0591}{2} \log K_{eq} \tag{15.28}$$

A similar relationship is true for all equilibrium redox reactions. With knowledge of $E^0(Sn^{2+}/Sn^{4+})$ and $E^0(Ce^{3+}/Ce^{4+})$ it is possible to calculate the equilibrium constant for the reaction.

It is possible to calculate the *equivalence point potential* by noting that we cannot ignore the small concentrations of Sn^{2+} and Ce^{4+} remaining, even though the overall reaction may be close to completion. The stoichiometry at the equivalence point demands that

$$2[Sn^{4+}] = [Ce^{3+}]$$

and

$$2[Sn^{2+}] = [Ce^{4+}]$$

If we let the potential at equivalence be E_{eq},

$$E_{eq} = E^0(Ce^{3+}/Ce^{4+}) + \frac{0.0591}{2} \log\left(\frac{[Ce^{4+}]^2}{[Ce^{3+}]^2}\right)$$

$$2E_{eq} = 2E^0(Sn^{2+}/Sn^{4+}) + \frac{0.0591}{2} \log\left(\frac{[Sn^{4+}]^2}{[Sn^{2+}]^2}\right)$$

Adding these two expressions and substituting the values for the Ce ion species in terms of Sn, the log term becomes equal to zero:

$$\log\left(\frac{2[Sn^{2+}]^2[Sn^{4+}]^2}{2[Sn^{4+}]^2[Sn^{2+}]^2}\right) = \log 1 = 0$$

Thus $3E_{eq} = E^0(Ce^{3+}/Ce^{4+}) + 2E^0(Sn^{2+}/Sn^{4+})$. In practice, it is often easier to observe the equivalence point than to calculate it! If you must calculate the shape of a redox titration curve, the use of a spreadsheet program such as Excel is invaluable because of the multiple equilibrium equations that must be solved. (See the text by Harris for excellent examples of spreadsheet calculations for redox titrations.)

The relationship between the emf of the cell and the number of milliliters of ceric salt added to the stannous solution is of the same form as Fig. 15.15. The equivalence point is denoted by a rapid change in potential as the ceric salt is added. If necessary, the first and second differentials of the curve may be taken to detect the endpoint as shown in Fig. 15.16. From the data obtained, calculations of solution concentrations and other variables are done in the same manner as in conventional volumetric analysis.

The same principle can be used for many redox reactions. The potential of the cell depends on the concentration of the oxidized and reduced forms of ions present. During a titration these concentrations vary as the chemical reaction proceeds:

$$red(A) + ox(B) \longrightarrow ox(A) + red(B)$$

where red(A) is the reduced form of ion A, ox(B) is the oxidized form of ion B, and so on. At the equivalence point there is a rapid change in potential with the addition of the titrating solution. This change makes it possible to detect the equivalence point of the reaction. Typical inorganic oxidation–reduction reactions include

$$2Fe^{2+} + Sn^{4+} \longrightarrow 2Fe^{3+} + Sn^{2+}$$
$$5V^{2+} + 3MnO_4^- + 4H^+ \longrightarrow 5VO_2^+ + 3Mn^{2+} + 2H_2O$$
$$IO_3^- + 8I^- + 6H^+ \longrightarrow 3H_2O + 3I_3^-$$
$$I_3^- + 2S_2O_3^{2-} \longrightarrow 3I^- + S_4O_6^{2-}$$

Organic compounds may be determined by redox titrimetry, including glucose and other reducing sugars, vitamin C, thiols such as cysteine, and many other organic chemicals of biological importance. There are thousands of potentiometric redox titration methods published in the chemical literature.

It may be necessary to correct for interferences to a method depending on your sample. For example, the presence of traces of other oxidizing or reducing compounds may displace the equilibrium and give an incorrect voltage for the equivalence points. Many equilibria are pH sensitive, and this too may generate inaccurate results. Care must be taken to correct for these interference effects.

15.3.2. Coulometry

It was shown earlier that when a standard cell is made from a Zn/Zn^{2+} half-cell and a Cu/Cu^{2+} half-cell, Zn dissolves, Cu^{2+} is deposited as Cu metal, and a potential of 1.100 V is developed from the spontaneous cell thermodynamics. A cell operating spontaneously is called a **galvanic cell**. It is possible to cause a cell to react in the nonspontaneous direction, by application of a sufficient potential. A cell operated in this manner is called an **electrolytic cell**. The process of causing a thermodynamically nonspontaneous oxidation or reduction reaction to occur by application of potential or current is called **electrolysis**. Electrolysis is carried out for a sufficient length of time to convert a species quantitatively to a new oxidation state. There are three primary electroanalytical methods based on electrolysis; they are constant-current coulometry (coulometric titrimetry), constant-potential coulometry, and **electrogravimetry**. In electrogravimetry, the product of the electrolysis (usually a metal) is plated out on one of the electrodes and the amount determined by weighing the electrode. In the **coulometric methods**, the charge is measured. In contrast to potentiometry, where we did not want to change the concentration of the species in solution, electrolytic methods are designed to completely consume the species being measured by converting it quantitatively to a new species.

If we consider the Zn/Cu galvanic cell, we might consider how to reverse the reaction. If a sufficient voltage is applied in the opposite direction, from a power source or battery, there will be a tendency for these reactions to reverse, provided that no new reactions occur. In this case, if we try to electrodeposit Zn from aqueous solution by forcing the reaction

$$Zn^{2+} + 2e^- \longrightarrow Zn$$

the evolution of hydrogen from the water will take precedence by the reaction

$$2H^+ + 2e^- \longrightarrow H_2$$

There is much more water in the cell than Zn^{2+}, so the applied voltage goes to form H_2 and we get no deposition of Zn(s). We say that the Zn/Zn^{2+}(aq) half-cell is *chemically irreversible*. On the other hand, the copper will dissolve if made an anode in an acidic solution, so the reaction

$$Cu \longrightarrow Cu^{2+} + 2e^-$$

is said to be *chemically reversible*.

The differences and similarities between an electrochemical cell and an electrolysis cell are summarized in Fig. 15.19. Oxidation always occurs at the anode and reduction always occurs at the cathode; notice, however, that in a spontaneous cell the positive electrode is the one at which reduction takes place, whereas in an electrolysis cell the negative electrode is the one at which reduction takes place. Negatively charged anions, such as Cl^-, NO_3^-, and SO_4^{2-}, are attracted to a positive electrode. Positively charged cations, such as Na^+, Ca^{2+}, and Mg^{2+}, are attracted to a negative electrode. Under the influence of the applied voltage in an electrolysis cell the H^+ and Zn^{2+} ions are attracted to the negative electrode, the cathode; the SO_4^{2-} and OH^- ions are attracted to the positive electrode, the anode. Electrons are consumed at **the cathode** by H^+ ions to evolve hydrogen gas in the electrolysis cell, but are consumed by Cu^{2+} ions in the spontaneous cell to deposit Cu atoms. For continuous operation of either type of cell, it is necessary for both **the anode** and cathode reactions to proceed with equal numbers of electrons.

Frequently the reaction that proceeds at either the anode or cathode is the decomposition of the solvent; if water is the solvent, it may electrolyze. Note that water electrolysis

Spontaneous reactions:

Zn → Zn²⁺ + 2e⁻ (anodic oxidation)

Cu²⁺ + 2e⁻ → Cu (cathodic reduction)

(a)

Electrolysis:

2H+ + 2e⁻ → H₂ (cathodic reduction)

Cu → Cu²⁺ + 2e⁻ (anodic oxidation)

(b)

Figure 15.19 Comparison of (a) a galvanic cell and (b) an electrolysis cell.

depends on the pH:

$$\text{For reductions:} \quad 2H^+ + 2e^- \longrightarrow H_2$$
$$2H_2O + 2e^- \longrightarrow H_2 + 2OH^-$$
$$\text{For oxidations:} \quad 2H_2O - 4e^- \longrightarrow O_2 + 4H^+$$
$$4OH^- - 4e^- \longrightarrow O_2 + 2H_2O$$

15.3.2.1.   *Electrogravimetry*

If a solution of a metallic ion, such as copper, is electrolyzed between electrodes of the same metal (i.e., copper), the following reaction takes place:

$$Cu^0 \longrightarrow Cu^{2+} + 2e^- \quad \text{(anode)}$$
$$Cu^{2+} + 2e^- \longrightarrow Cu^0 \qquad \text{(cathode)}$$

The net result is that metal dissolves from the anode and deposits on the cathode. The phenomenon is the basis of electroplating (e.g., chromium plating of steel), electrowinning, and electrorefining. Also, it is the analytical basis of an electrodeposition method known as *electrogravimetry*. This involves the separation and weighing of selected components of a sample. Most metal elements can be determined in this manner, usually deposited as the M^0 species, although some metal elements can be deposited as oxides. The halides can be determined by deposition as the silver halide. Metals commonly determined include Ag, Bi, Cd, Co, Cu, In, Ni, Sb, Sn, and Zn.

15.3.2.2. Instrumentation for Electrogravimetry and Coulometry

The basic apparatus required is a power supply (potentiostat with a DC output voltage), an inert cathode and anode (usually platinum foil, gauze, or mesh), and arrangements for stirring. Sometimes a heater is used to facilitate the processes.

Electrogravimetry is usually carried out under conditions of controlled potential, so an auxiliary reference electrode is used together with the working electrode. (The working electrode can be either the anode or the cathode.) The auxiliary electrode–working electrode pair is connected to a potentiostat that fixes the potential of the working electrode by automatically adjusting the applied emf throughout the course of the electrolysis.

The working electrode should have a large surface area. Platinum is used because it is inert; Pt gauze or mesh electrodes are used to provide a large surface area. The electrode "mesh" is like wire window-screening material, with the wire diameter about 0.2 mm. The mesh is welded into an open cylinder to make the electrode. A standard size is a cylinder about 5 mm high and 5 mm in diameter, but Pt mesh electrodes of many sizes are available.

Because the deposition needs to be quantitative, the solution must be stirred during electrolysis. A magnetic stirrer is commonly used, often as a combination stirrer–hot plate. For electrogravimetry, an analytical balance of the appropriate capacity and sensitivity is needed. The electrode on which deposition will occur is weighed before the experiment is started. The electrode with the deposit is removed from solution upon completion of the electrolysis, washed, dried, and then reweighed. For coulometric methods, the charge is measured. The charge is obtained by integration of the current using an electronic integrator.

It is critical for these techniques that the current efficiency be as close to 100% as possible. The percent current efficiency tells us how much of the applied current results in the reaction of interest. The percent current efficiency is defined as being equal to $100 \times (i_{applied} - i_{residual})/i_{applied}$, where $i_{applied}$ is the current applied to the cell and $i_{residual}$ is the "background" current. All real cells have some residual current. To achieve a 99.9% current efficiency, the applied current must be approximately $1000\times$ the residual current. This is readily achieved for currents in the μA to 100 mA range used in constant-current coulometry.

15.3.2.3. The Applied Potential

Electrogravimetry is generally performed under a constant, controlled potential but can also be performed under conditions of controlled current. For example, a metal alloy may contain nickel and copper. The alloy can be dissolved and the solution electrolyzed, with the result that the copper is selectively and exhaustively deposited on the cathode. The electrodeposited copper should form an adherent coating so that it can be washed, dried, and weighed. (The formation of appropriate forms of deposit has been studied thoroughly and is critical to obtaining accurate results. See the handbook by Dean for detailed information and references.) Instead of dissolution of the platinum metal taking place at the platinum anode, oxygen gas is liberated from the aqueous solution. As a consequence, platinum is a useful electrode material for redox purposes, since it is not consumed in the process. The initial reactions are as follows:

$$\text{cathode: } Cu^{2+} + 2e^- \longrightarrow Cu \quad E^0 = +0.337 \text{ V}$$
$$\text{anode: } 2H_2O - 4e^- \longrightarrow O_2 + 4H^+ \quad E^0 = -1.229 \text{ V}$$
$$\text{Net reaction: } 2H_2O + 2Cu^{2+} \longrightarrow 2Cu + O_2 + 4H^+ \quad E_{net} = -0.892 \text{ V}$$

We may now consider on a more quantitative basis what influence the applied potential has. If the cell could behave as a thermodynamic cell and the reactants were under standard conditions, we might expect that the voltage needed would be that given by the net reaction above, $E_{net} = -0.892$ V. But this would be for an infinitesimal, reversible change and, similar to any other chemical reaction, an electrochemical reaction has an *activation energy* that must be overcome for the reaction to occur. In fact, an additional voltage termed the **overpotential** η is required to drive the electrode reactions. About 2 V has to be applied to reach the current onset of the cell described, and this threshold is sometimes called the *decomposition voltage*. Thus the total potential that we must apply across the cell is

$$E_{decomp} = E_{net} + IR + \eta \qquad (15.29)$$

An overpotential contribution is required for both the anodic and the cathodic reaction and, in this case, it is mainly to oxidize water at the Pt anode. As copper is deposited and the concentration of Cu^{2+} ions falls, both the ohmic potential drop IR across the electrolyte and the overpotential decrease. If we assume that the solution resistance R remains fairly constant, the ohmic potential drop is directly proportional to the net cell current. The overpotential increases exponentially with the rate of the electrode reaction. Once the Cu^{2+} ions cannot reach the electrode fast enough, we say that *concentration polarization* has set in. An ideally polarized electrode is one at which no faradaic reactions ensue; that is, there is no flow of electrons in either direction across the electrode–solution interface. When the potential of the cathode falls sufficiently to reduce the next available species in the solution (H^+ ions, or nitrate depolarizer), the copper deposition reaction is no longer 100% efficient.

Vigorous stirring, therefore, is necessary for several purposes. It helps to dislodge the gas bubbles that form at the electrodes. More importantly, it provides convective transport of the Cu^{2+} ions to the cathode as the solution becomes increasingly dilute in Cu^{2+} ion. At this point, if the applied voltage is raised to increase the current density, the potential could become sufficiently negative to cause the reduction of the next easiest reduced species present, in this case H^+ ions:

$$2H^+ + 2e^- \longrightarrow H_2 \quad E^0 = 0.00 \text{ V}$$

Such evolution of H_2 gas can produce unsatisfactory copper deposits, which may flake off and lead to a poor analytical determination. To prevent this, it is usual to add a cathodic depolarizer of NO_3^- ion to the solution. A *depolarizer* is a species more readily reducible than H^+ ion but one that does not complicate the cathodic deposit by occlusion or gas entrapment. In this instance, the product is ammonium ions:

$$NO_3^- + 10H^+ + 8e^- \longrightarrow NH_4^+ + 3H_2O$$

If the potential becomes too negative at the cathode, the Ni^{2+} ion from our alloy sample might commence to deposit and form an alloy deposit. Note that the order of selectivity is not always that predicted by the standard potentials in the activity series. Control of the pH and the chelates or complexing agents in a solution may alter the reduction–oxidation potentials from those of an uncomplexed species.

Perhaps the most common example of overvoltage encountered in electrochemistry is that needed to reduce H^+ ions at a mercury electrode. On a catalytic Pt surface (platinized Pt, which is a large surface-area, black Pt deposit on Pt metal) the H_2/H^+ ion couple is said to behave reversibly. This means that one can oxidize hydrogen gas, or reduce H^+ ions, at the standard reduction potential, 0.00 V, under standard conditions. At a mercury cathode, however, it takes about -1 V vs. SHE to reduce H^+ ions, because hydrogen evolution is kinetically slow. This phenomenon has great importance in polarography, because it enables the analysis of many of the metals whose standard potentials are negative of the SHE.

Controlled potential electrolysis (potentiostatic control) requires a three-electrode cell, so as not to polarize the reference electrode. Controlled potential methods enable one to be very selective in depositing one metal from a mixture of metals. If two components have electrochemical potentials that differ by no more than several hundred millivolts, it may still be possible to shift these potentials by complexing one of the species. One disadvantage of exhaustive electrolysis is the time required for analysis, and faster methods of electrochemical analysis are described.

In summary, important practical considerations in precise electrodeposition are (1) rapid stirring, (2) optimum temperature, (3) correct controlled potential or correct current density (usually expressed in A/dm^2), and (4) control of pH. The proper conditions have all been worked out for hundreds of systems; when these conditions prevail, deposits are bright and adherent and no unusual precautions are necessary in handling (washing, drying, weighing, etc.) the electrode. In commercial electroplating, additives (glue, gelatin, thiourea, etc.) are used as brighteners and for good adherence. They are less useful in analytical deposition, because they would cause the weight of deposit to be too high and would create an interference in the method. Electrodeposits improve the appearance, wear, and corrosion resistance of many fabricated metal parts. Electrolysis is also used to recharge batteries, study electrode reactions, extract pure metals from solutions and create shapes that cannot be machined (electroforming), and eliminate metallic impurities from solutions.

15.3.2.4. Analytical Determinations Using Faraday's Law

Electrogravimetry depends on weighing the WE before and after plating out the element under test. Therefore, it is limited to the determination of electroactive species where the product of electrolysis is a solid that forms a suitable deposit. As an alternative to weighing the deposit, the quantity of electricity used to deposit the metal can be measured. From this measurement the quantity of metal deposited or the amount of ions reduced or oxidized can be calculated. This is an advantage if the product of electrolysis is a gas or another soluble ionic species; electrogravimetry will not work for these analytes. The calculation is based on Faraday's Law, which states that equal quantities of electricity cause chemical changes of equivalent amounts of the various substances electrolyzed. The advantage of measuring the charge is that coulometry can measure electroactive species that do not form solid deposits, so it has much wider application. The accuracy is as good as gravimetric methods and coulometry is much faster than gravimetry and classical volumetric methods of analysis.

Charge is measured in either coulombs or faradays. One coulomb (C) is the amount of charge transported in one second by a current of 1 ampere. One faraday (F) is the charge in coulombs of one mole of electrons, so $1\ F = 96,485\ C/mol\ e^-$.

The relationship between charge and amount of analyte for a constant current, i, can be stated as:

$$q = nFVM = \frac{nFw}{M_w} = i \times t \tag{15.30}$$

where q is the charge(in coulombs); n, the number of equivalents per mole of analyte; F, the Faraday constant; V, the volume (L); M, the molar concentration of analyte (mol/L);

w, the weight of analyte (g); M_w is the molecular weight of the analyte (g/mol); i, the current (A); and t, time (s). This relationship involves only fundamental quantities; there are no empirically determined "calibration factors", so standardization and calibration are not required. This makes coulometry an **absolute method**; it does not require calibration with external standards. The current efficiency must be 100%. For a variable current, the charge is given by:

$$q = \int_0^t i\,dt$$

Example A. A sample of stannic chloride was reduced completely to stannous chloride according to the reaction

$$Sn^{4+} + 2e^- + Sn^{2+}$$

The applied current was 9.65 A and the time taken for reduction was 16.0 min 40 s. What was the initial weight of stannic ion present? First convert 16.0 min 40 s to s. 16.0 min $\times$ 60 s/min $= 960$ s. Adding the additional 40 s gives a total time (in s) of 1.00×10^3 s.

$$\text{number of coulombs} = i(A) \times t(s)$$
$$= 9.65\,A \times 1.00 \times 10^3\,s$$
$$= 9650\,C$$
$$\text{number of faradays} = \frac{9650}{96{,}500} = 0.100\,F$$

But 1 F will reduce 1 g-eq weight of stannic ion:

$$1\,\text{g-eq wt.} = \frac{\text{atomic wt. of tin}}{\text{valence change}} = \frac{118.69\,g/mol}{2\,eq/mol}$$

that is, 1 F will reduce 59.35 g of stannic ion, or 0.100 F will reduce 5.935 g of stannic ion.

Example B. If the original volume of the solution was 250.0 mL, what was the molarity of the solution?

Molarity means moles of solute contained in exactly 1 L of solution. The atomic weight of tin, the solute, is 118.69 g/mol. It was shown in example (A) that 5.935 g of tin were reduced and example (A) stated that the reduction was complete. The molarity of the solution is calculated from:

$$\text{mol Sn} = 5.935\,g\,Sn \times \frac{1\,\text{mol Sn}}{118.69\,g}$$
$$\text{mol Sn} = 0.05000\,\text{mol}$$

and

$$L_{\text{solution}} = 250.0\,mL \times \frac{1.000\,L}{1000\,mL} = 0.2500\,L$$

Therefore the molarity of the solution was 0.05000 mol Sn/0.2500 L $= 0.200$ M stannic chloride to three significant figures.

Equation (15.30) allows us to calculate the amount of an electroactive species in either molarity or weight of material electrolyzed.

Controlled Potential Coulometry. Coulometry is frequently carried out under conditions of controlled potential. This is achieved by using a third electrode in the system, as described earlier. The three electrode system maintains the potential at the working electrode at a constant value or permits an applied voltage pulse or ramp to be added to the working electrode. The reason for using controlled potential can be seen when we examine the Nernst equation. As a metal is oxidized or reduced under experimental conditions, the concentration of the remaining metal ions in the original oxidation state in solution steadily decreases. Therefore in order to continue the deposition the potential applied to the system must be steadily increased. When the potential is increased, different elements may begin to react or deposit and interfere with the results. Their deposition results in an increased weight of metal deposited (in electrogravimetry) and an increased number of coulombs passing through the cell in all coulometric methods. Controlled potential is therefore used to eliminate interferences from other reactions that take place at different potentials.

In theory, the equivalence point is never reached, because the current decays exponentially. Since a small amount of material always remains in solution, a correction must be made for it by measuring the current flowing at the end of the analysis. This quantity should be subtracted from the integrated signal in order to give an accurate measure of the material that has deposited or reacted during the experiment.

Controlled potential coulometry is usually used to determine the number of electrons involved in a reaction when studies are being carried out on new inorganic or organic compounds. Using a coulometer, we measure q. If we know M_w, from mass spectrometric measurements, for example, the number of moles of electrons can be calculated using Eq. (15.30). This indicates how the species oxidizes or reduces; it gives us information about the reaction chemistry of new compounds. Controlled potential coulometry is also valuable in generating unstable or highly reactive substances *in situ* with good quantitative control.

Coulometric Titrations. An aqueous iodide sample may be titrated with mercurous ion by anodizing a mercury pool electrode. When metallic mercury is oxidized to mercurous ion by a current passing through the system, the mercurous ion reacts directly with the iodide ion to precipitate yellow Hg_2I_2:

$$2Hg - 2e^- \longrightarrow (Hg_2^{2+})$$
$$Hg_2^{2+} + 2I^- \longrightarrow Hg_2I_2$$

The reaction continues and current passes until all the iodide is used up. At this point some means of endpoint detection is needed. Two methods are commonly adopted. The first uses an amperometric circuit with a small imposed voltage that is insufficient to electrolyze any of the solutes. When the mercury ion concentration suddenly increases, the current will rise because of the increase in the concentration of the conducting species. The second method involves using a suitable indicator electrode. An indicator electrode may be a metal electrode in contact with its own ions or an inert electrode in contact with a redox couple in solution. The signal recorded is potentiometric (a cell voltage vs. a stable reference electrode). For mercury or silver we may use the elemental electrodes, because they are at positive standard reduction potentials to the hydrogen/hydrogen ion couple.

Figure 15.20 illustrates an apparatus suitable for the coulometric titration described. The anode and cathode compartments are separated with a fine porous glass membrane to prevent the anode products from reacting at the cathode, and vice versa. The porosity should be such as to allow minimal loss of titrant in the course of the experiment. The

Figure 15.20 Apparatus for coulometric titration with potentiometric endpoint detection.

electrolysis circuit is distinct from the endpoint detector circuit. A constant current source may be used, but it is not mandatory, provided that the current is integrated over the time of the reaction, that is

$$q = it \text{ (constant current)}$$

or

$$q = \int_0^t i(t)\, dt \text{ (integrator)}$$

The potentiometric cell can be written as:

$$\text{SCE} \parallel 0.1\,\text{M HClO}_4 \mid \text{Hg}_2^{2+}, 0.2\,\text{M NaNO}_3 \mid \text{Hg}$$

The endpoint is sensed by recording the voltage across the potentiometric cell.

The **Luggin probe** used to monitor the potential of the Hg indicator electrode is spaced close (~1 mm) to the mercury droplet to reduce ohmic resistance. The top barrel is cleaned, wetted with concentrated NaNO$_3$ solution, and closed. Sufficient conductivity exists in this solution layer to permit the small currents necessary for potentiometric determination. When an excess of mercurous ion is generated, the potential of the Hg/Hg$_2^{2+}$ couple varies in accordance with the Nernst equation, and a voltage follower may be used to output the voltage curve to a Y-time recorder or computer data system. For precise work, dissolved oxygen should be removed from the supporting electrolyte and efficient stirring is needed.

Coulometric titrimetry has been used for the determination of mercaptan, halide, and phosphorus compounds by using a silver electrode. The principle has also been used for many other types of titrations. Another example is the determination of ferrous ion, Fe^{2+}, in the presence of ferric ion, Fe^{3+}. This reaction can be controlled directly by coulometric analysis, but cannot be carried out to completion. As the concentration of Fe^{2+} decreases, the Nernst equation indicates that the potential necessary to continue oxidation will steadily increase until water is oxidized and oxygen is evolved. This will take place before completion of the oxidation of ferrous to ferric ions. The problem can be overcome by adding an excess of cerous ions, Ce^{3+}, to the solution. The Ce^{3+} ion acts as a *mediator* and itself undergoes no net reaction. With coulometry, the cerous ions can be oxidized to ceric ions, which are then immediately reduced by the ferrous ions present back to cerous ions, generating ferric ions. The reaction continues until all the ferrous ions have been oxidized. At this point any new ceric ions that are formed are not reduced but remain stable in solution, and there is a change in the current flow that signals the endpoint. The endpoint can also be detected using an indicator sensitive to ceric ions.

Analytical Applications of Coulometry. The major advantage of coulometry is its high accuracy, because the "reagent" is electrical current, which can be well controlled and accurately measured. Coulometry is used for analysis, for generation of both unstable and stable titrants "on demand", and for studies of redox reactions and evaluation of fundamental constants. With careful experimental technique, it is possible to evaluate the Faraday constant to seven significant figures, for example.

Some typical important industrial applications of coulometry include the continuous monitoring of mercaptan concentration in the materials used in rubber manufacture. The sample continuously reacts with bromine, which is reduced to bromide. A third electrode measures the potential of Br_2 vs. Br^- and, based on the measurement, automatically regulates the coulometric generation of the bromine. Coulometry is used in commercial instruments for the continuous analysis and process control of the production of chlorinated hydrocarbons. The chlorinated hydrocarbons are passed through a hot furnace, which converts the organic chloride to HCl. The latter is dissolved in water and the Cl^- titrated with Ag^+. The Ag^+ is generated by coulometry from a silver electrode, Ag^0. It is necessary for the sample flow rate to be constant at all times. Integration of the coulometric current needed to oxidize the silver to silver ion results in a measurement of the Cl^- concentration.

Another interesting application of coulometry is in the generation of a chemical reagent in solution. Such reagents as acids, bases, reducing agents, and oxidizing agents can all be generated in solution and allowed to react in the manners used in volumetric analysis. In effect, volumetric titrations can be carried out by generating the titrant electrochemically; the amount of titrant used is then measured directly by the amount of electricity used to generate it. This method is very exact, frequently more so than weighing the reagent. Reagents generated electrochemically are extremely pure. With coulometry we can obtain results that are more accurate than those possible with the purest commercially available reagents.

15.3.3. Conductometric Analysis

The ability of a solution to conduct electricity can provide analytical information about the solution. The property measured is electrical conductivity between two electrodes by ions in solution. All ions in solution contribute to the electrical conductivity, so this is not a specific method of analysis. Electrical conductivity is used to provide qualitative analysis, such as the purity of an organic solvent and relative quantitative analysis for quality control of materials, or comparison of drinking water quality in terms of total ionic contaminants.

Ohm's Law states that the resistance of metal wire is given by the equation

$$R = \frac{E}{i} \tag{15.31}$$

where E is the voltage applied to the wire (V); i, the current of electrons flowing through the wire (A); and R, the resistance of the wire (Ω).

The resistance R depends on the dimensions of the conductor:

$$R = \frac{\rho L}{A} \tag{15.32}$$

where ρ, is the resistivity; L, the length; and A, the cross-sectional area.

Another valuable parameter, especially when we consider the mechanisms of current flow in solutions, is electrolytic conductivity, κ, where

$$\kappa - \frac{1}{\rho} = \frac{(L/A)}{R} \tag{15.33}$$

The units of electrolytic conductivity are $\Omega^{-1}\,m^{-1}$ (reciprocal ohm m^{-1} also called mho/ m, mho m^{-1}, or S m^{-1}, where S is the siemen). The SI unit is the S/m, but practical measurement units are usually in μS/cm. **Electrolytic conductivity** is also called *specific conductance*, not to be confused with conductance. The electrolytic conductivity of a solution is a measure of how well it carries a current, in this instance by ionic carriers rather than electron transfer, and it is an intrinsic property of the solution. A related property, the conductance, G, is also used and defined as $G = 1/R$. The **conductance** is a property of the solution *in a specific cell*, at a specific temperature and concentration. The conductance depends on the cell in which the solution is measured; the units of G are siemens (S).

The charge carriers are ions in electrolyte solutions, fused salts, and colloid systems. The positive ions M^+ **migrate** through the solution toward the cathode, where they may or may not react faradaically to pick up electrons. Anions, symbolized as A^-, migrate toward the anode, where they may or may not deliver electrons. The net result is a flow of electrons across the solution, but the electron flow itself stops at each electrode. Faradaic reaction of the easiest reduced and oxidized species present may occur, and hence compositional changes (reduction and oxidation) may accompany ionic conductance.

Electrolyte conductivity depends on three factors: the ion charges, mobilities, and concentrations of ionic species present. First, the number of electrons each ion carries is important, because A^{2-}, for example, carries twice as much charge as A^-. Second, the speed with which each ion can travel is termed its *mobility*. The mobility of an ion is the limiting velocity of the ion in an electric field of unit strength. Factors that affect the mobility of the ion include (1) the solvent (e.g., water or organic), (2) the applied voltage, (3) the size of the ion (the larger it is, the less mobile it will be), and (4) the nature of the ion (if it becomes hydrated, its effective size is increased). The mobility is also affected by the viscosity and temperature of the solvent. Under standard conditions the mobility is a reproducible physical property of the ion. Because in electrolytes the ion concentration is an important variable, it is usual to relate the electrolytic conductivity to **equivalent conductivity**. This is defined by

$$\Lambda = \frac{\kappa}{C_{eq}} \tag{15.34}$$

where, the Λ, is the equivalent conductivity ($\Omega^{-1}\,cm^2$/equivalent or S cm^2/equivalent); κ, electrolytic conductivity (mho/m); and C_{eq}, the equivalent concentration (i.e., normality of solution, where N = M × charge on ion).

Electrolyte solutions only behave ideally as infinite dilution is approached. This is because of the electrostatic interactions between ions, which increase with increasing concentration. As infinite dilution is approached, the equivalent conductance of the electrolyte Λ approaches Λ^0, where:

$$\Lambda^0 = F(U_+^0 + U_-^0) = \lambda_+^0 + \lambda_-^0 \qquad (15.35)$$

where U_+^0 and U_-^0 are the cation and anion mobilities, respectively, and λ_+^0 and λ_-^0 are the cation and anion equivalent limiting ion conductivities, respectively, at infinite dilution.

The λ^0 values are not accessible to direct measurement, but they may be calculated from transport numbers. Kohlrausch's law of independent ionic conductivities states that at low electrolyte concentrations the conductivity is directly proportional to the sum of the n individual ion contributions, that is,

$$\Lambda^0 = \sum_{i=1}^{n} \lambda_i^0 \qquad (15.36)$$

Table 15.4 shows some typical limiting ionic equivalent conductivities. From this we can deduce that the limiting equivalent conductivity of potassium nitrate is $(K^+) + (NO_3^-) = 74 + 71 = 145$, and that of nitric acid is $(H^+) + (NO_3^-) = 350 + 71 = 421$, assuming 100% dissociation into ions. The change in conductivity of a solution upon dilution or replacement of one ion by another by titration is the basis of conductometric analysis.

The electrolytes KNO_3 and HNO_3 are strong and dissociate completely in water. Weak electrolytes do not dissociate completely and many exist in a combined form as non-ionized molecules. Now the effects of interionic forces are less important because the ion concentrations are lower and, in fact, the degree of ionization (α) is readily obtainable from conductance measurements:

$$\alpha \approx \frac{\Lambda}{\Lambda^0} \qquad (15.37)$$

For example, the weak acid HA partially dissociates into H^+ and A^-:

$$HA \longleftrightarrow H^+ + A^- \quad \text{with} \quad K' = \frac{[H^+][A^-]}{[HA]}$$

If the initial total molar concentration of HA $= c$, at equilibrium, the molar concentration of HA $= (1 - \alpha)c$, and the molar concentration of $H^+ = A^- = \alpha c$.

Table 15.4 Equivalent Conductances of Various Ions at Infinite Dilution at 25°C

Anions	$(\Omega^{-1}\,cm^2/eq)$	Cations	$(\Omega^{-1}\,cm^2/eq)$
OH^-	198	H^+	350
Cl^-	76	Na^+	50
Br^-	78	K^+	74
NO_3^-	71	NH_4^+	74
ClO_4^-	67	Ag^+	62
$HCOO^-$	55	Cu^{2+}	55
CH_3COO^-	41	Zn^{2+}	53
SO_4^{2-}	80	Fe^{3+}	68

We can express an apparent dissociation constant in terms of conductivities by substitution.

$$K' = \frac{\alpha^2 c^2}{(1-\alpha)c} \approx \frac{\Lambda^2 c}{\Lambda^0(\Lambda^0 - \Lambda)} \tag{15.38}$$

This is an expression of Ostwald's dilution law, and the equation can be used to determine K', α, or Λ^0. For example, from Table 15.4, we can calculate Λ^0 for formic acid, HCOOH, to be equal to $350 + 55 = 405$ S cm^2/eq. If we measure the equivalent conductance of a 0.020 N solution of formic acid, we find it equal to 36.6 S cm^2/eq at 25°C. The degree of dissociation, α, is calculated to be:

$$\alpha = \frac{\Lambda}{\Lambda^0} = \frac{36.6}{405} = 9.04 \times 10^{-2}$$

and from this we can calculate the dissociation constant, K',

$$K' = \frac{(9.04 \times 10^{-2})^2(0.020)}{1 - 9.04 \times 10^{-2}} = 1.8 \times 10^{-4}$$

This agrees well with the literature value for the K_a for HCOOH. Conductance measurements may also be used to find the solubility of sparingly soluble salts and complexation equilibrium constants.

15.3.3.1. Instrumentation for Conductivity Measurements

The equipment is basically a Wheatstone bridge and conductivity cell, as illustrated in Fig. 15.21. Resistance A is made up of the cell containing the sample; B is a variable resistance; resistances D and E are fixed. Resistor B and variable capacitor C may be adjusted so that the balance point can be reached. At this point

$$\frac{R_A}{R_B} = \frac{R_D}{R_E}$$

The resistances B, D, and E can be measured, and from these measurements the resistance and hence the conductance of the cell can be calculated. A *small* superimposed AC voltage

Figure 15.21   Wheatstone bridge arrangement for conductometric analysis.

($\sim$20 mV peak to peak) at 1000 Hz is best as a signal, because then faradaic polarization at the electrodes is minimized. The null detector may be a sensitive oscilloscope or a tuned amplifier and meter. Stirring is often used to minimize polarization.

A wide variety of cell geometries and sizes are available for conductivity measurements, designed with two, three, or four electrodes, depending on the use. A typical dip cell (so called because it is dipped into a beaker containing the sample) usually is constructed with two parallel, platinized Pt foil electrodes, each about 1 cm^2 in area. Because the dimensions of a constructed cell are not exact, one must calibrate a particular cell. The recommended method of calibration is to use primary standard KCl solutions of known strength. For example, at 25°C, 7.419 g of KCl in 1000 g of solution (or 0.1 molal) has a specific conductivity of 0.01286 Ω^{-1}/cm. Primary standard solutions are not always practical to make: one has to include the conductivity of the water used to prepare the standard, for example. NIST has a set of SRMs covering the range 5 μS/cm to 100,000 μS/cm available for purchase that are satisfactory for most calibration needs. (Platinizing the Pt electrodes is achieved by cleaning the Pt in hot concentrated HNO$_3$ and electrodepositing a thin film of Pt black from a 2% solution of platinic chloride in 2 N HCl. The Pt black is a porous Pt film, which increases the surface area of the electrodes and further reduces faradaic effects.)

Cells are available for sample volumes as small as 2 mL, while standard sample size is 25–50 mL. Special cells for highly accurate conductivity measurements are available for research purposes. Flow-through cells are available for online monitoring of process streams. The reference by Berezanski depicts several cell types.

Very accurate measurements of conductivity require the use of a thermostatted bath with temperature controlled to within 0.005°C.

In addition to the laboratory meter setup described, there are handheld devices available for making conductivity measurements in the field. These are generally battery-powered and have relatively poor accuracy (1% RSD) compared with laboratory meters, which may have RSDs of less than 0.05%.

15.3.3.2. *Analytical Applications of Conductometric Measurements*

Conductivity measurements can be performed on many types of solutions with no sample preparation required. The method is used to monitor solutions for their ionic content. Examples include drinking water, natural water, high purity (deionized) water, high purity solvents, and potable beverages. Conductivity is common as a detector in ion chromatography, HPLC and other chromatographic techniques where charged species are produced (Chapter 13). Conductivity is a powerful tool for endpoint detection in titrimetry in aqueous and nonaqueous solvents.

Conductimetric Titrations in Aqueous Solution. When one ion is replaced in solution by a different ion with a significantly different equivalent conductivity, a change in total conductivity occurs. As seen in Table 15.4, hydrogen ion and hydroxide ion have the highest equivalent conductivities; replacing them with less conductive ions can form the basis of conductimetric titrations for acids and bases. For example, when NaOH is added to HCl, the following reaction occurs:

$$\text{Chemical reaction: } HCl + NaOH \longrightarrow NaCl + H_2O$$

$$\text{Ionic reaction: } H^+ + Cl^- + Na^+ + OH^- \longrightarrow Na^+ + Cl^- + H_2O$$

The original solution contains $H^+ + Cl^-$ in water; the final solution contains $Na^+ + Cl^-$ in water. It can be seen that Na^+ replaces H^+. If we plot the conductivity measured while NaOH is being added, we observe the relationship depicted in Fig. 15.22. In part A of

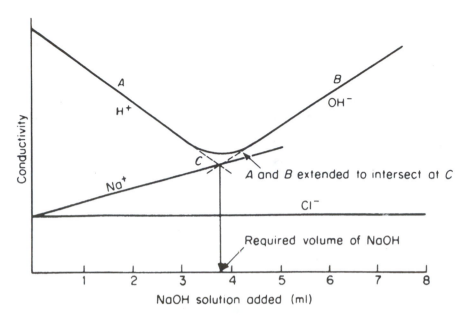

Figure 15.22 Solution conductivity during an HCl–NaOH titration. Point C is the neutralization point.

the curve, H^+ ions are being removed by the OH^- of the NaOH added to the solution. The conductivity slowly decreases until the neutralization point C is reached. In part B of the curve, the H^+ ions have been effectively removed from solution. Further addition of NaOH merely adds Na^+ and OH^- to the solution. An increase in conductivity results. The contribution of the Cl^-, Na^+, H^+, and OH^- ions to the total conductivity of the solution can be seen from Fig. 15.22. The volume of NaOH required to titrate the HCl can be measured by projecting lines A and B (extrapolation from points away from the endpoint) to the intersection C. Point C indicates the required volume of NaOH to neutralize the acid present.

Weak acids can also be titrated with NaOH and the endpoint detected by conductivity. A typical curve is shown in Fig. 15.23. As with all weak acids, the H^+ concentration is low and in equilibrium with the acetate ion. At the equivalence point, however, all the H^+ has been neutralized. Any further addition of NaOH has the effect of adding Na^+ and OH^- to the solution. A sharp increase in conductivity occurs, as shown by point C in Fig.15.23. Mixtures of weak and strong acids can be titrated with NaOH and the endpoints detected by conductivity changes. In such mixtures the strong acid is neutralized before the weak acid. Titrations involving mixtures of acids such as these are difficult to perform using indicators because of the problem involved in detecting both endpoints separately. With conductimetric titration the problem is simplified, as shown in Fig. 15.24. It can be seen that two abrupt changes in conductivity take place (C' and C''); these correspond to the titration endpoints for the strong and weak acid, respectively. For known acids, we can determine the concentrations present in the sample quantitatively. The contributions from the Na^+, Cl^-, acetate$^-$, OH^-, and H^+ to the total conductivity are noted in the figure. But remember that the conductivity measurement itself cannot distinguish individual ion contributions in a mixture. The method can be used to show how many acids are in a mixture and to indicate whether they are weak or strong acids, but not to identify them if the sample is an unknown. Conductimetric titrations can also be used to find the

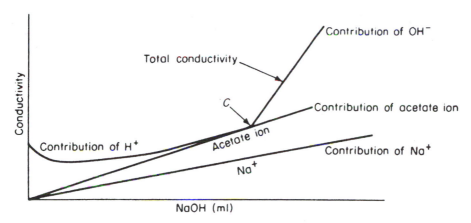

Figure 15.23 Solution conductivity during an acetic acid–NaOH titration. Point C is the neutralization point.

endpoints of reactions where precipitation takes place, as in the titration of silver solution with chloride solution. Again the endpoint is manifested by a sharp change in conductivity.

Conductimetric Titrations in Nonaqueous Solvents. In nonaqueous solvents, such as alcohol, toluene, or pyridine, it is possible to titrate Lewis acids or bases that cannot be titrated in aqueous solutions. For example, it is possible to titrate phenols dissolved in an organic solvent. Phenols act as Lewis acids, releasing hydrogen ion, which can be titrated with a suitable basic material. Since the titration is carried out in a nonaqueous medium, it is necessary to use a base that is soluble in the solvent. A base commonly used is tetramethylammonium hydroxide (TMAH). This material is basic but soluble in organic solvents and will neutralize Lewis acids. Nonaqueous solvents can be used to advantage to investigate molecular species that are not soluble in water. However, interpretation of the titration curve may not be straightforward. Solvent effects, viscosity, temperature, intermolecular attractions between solvent and solute (e.g., ion-pairing, complexation) are some of the factors that must be considered. Methods and more detail may be found in the references by Huber and Kucharsky listed in the bibliography.

Figure 15.24 Solution conductivity during titration of a mixture of HCl and CH_3COOH with NaOH. Here C′ and C″ designate the respective endpoints. HCl is titrated first, then the weaker acetic acid is neutralized.

Other Applications of Conductivity Measurements. Conductivity is used to determine the purity of drinking water and other natural water and wastewater. Dedicated instruments for this purpose, including handheld, portable meters, are often calibrated to read in "total dissolved solids" or TDS, or in "salinity". These tell the environmental scientist, water treatment plant operator, or geochemist about the amount of ionized species (e.g., sodium chloride is assumed to be the species for salinity of seawater) in solution. It is important to remember that the conductivity measurement cannot tell if the conductivity really is due to sodium chloride. It measures any ion that contributes to conductivity. The chemist or engineer must know something about the sample, or make good scientific judgments based on other information before drawing conclusions from a salinity or TDS reading on an unknown sample. Methods for conductivity measurements in potable water and wastewater may be found in "Standard Methods for Examination of Water and Wastewater" and for all types of water, deionized to brackish, in the ASTM standards Volumes 11.01 and 11.02. The references are listed in the bibliography.

In addition to drinking water and environmental applications, water purity is critical to many industries. Conductivity detectors are used in semiconductor and chip fabrication plants, to monitor cleanliness of pipelines in the food and beverage industry, to monitor incoming water for boilers to prevent scale buildup and corrosion. Any process stream with ions in it can be analyzed by conductometry. Conductivity detectors are part of commercial laboratory deionized water systems, to indicate the purity of the water produced and to alert the chemist when the ion-exchange cartridges are exhausted. The detector usually reads out in resistivity; theoretically, completely pure water has a resistivity of 18 MΩ cm.

Conductivity detectors are widely used in ion chromatography instruments using eluant suppression detection. The detectors are inexpensive, simple, rugged, and easy to miniaturize. The same type of detectors can be used for any chromatographic process to detect charged species in a nonionic eluant.

15.3.4. Polarography

When a potential difference is applied across two electrodes immersed in a solution, even in the absence of an electroactive species of interest, a small current arises due to background reactions and dissolved impurities. If the solution contains various metal ions, these do not electrolyze until the applied negative potential exceeds the reduction potential of the metal ion, that is, becomes more negative than the reduction potential. The difference in the current flowing through a solution under two conditions—(1) with the potential less negative than the metal ion reduction potential and (2) with the potential more negative than the metal ion reduction potential—is the basis of polarography. Polarography is not restricted to reductions (negative potential sweep), although these are more common. It is also possible to sweep to positive potentials and obtain oxidation curves. Metal cations, anions, complexes, and organic compounds all can be analyzed using polarography. The plot of current against applied potential for a sample solution is called a polarogram.

Polarography is the study of the relationship between the current flowing through a conducting solution and the voltage applied to a **dropping mercury electrode** (DME). It was discovered by Jaroslav Heyrovsky more than 60 years ago and has since resulted in tens of thousands of research studies. Heyrovsky's pioneering work in the field earned him the Nobel Prize in 1959. In the past two decades many variations of Heyrovsky's classical polarographic method have evolved, principally to improve the sensitivity and resolution of this analytical method. Two important variations are described below, namely, normal pulse polarography and differential pulse polarography.

Classical DC polarography uses a linear potential ramp (i.e., a linearly increasing voltage). It is, in fact, one subdivision of a broader class of electrochemical methods called **voltammetry**. Voltammetric methods measure current as a function of applied potential where the working electrode is *polarized*. This polarization is usually accomplished by using microelectrodes as working electrodes; electrode surface areas are only in the μm^2 to mm^2 range. The term *polarography* is usually restricted to electrochemical analyses at the dropping mercury electrode. Figure 15.25 illustrates the potential excitation used in DC polarography and the net current waveform relationship, usually shown as an $i-E$ rather than an $i-t$ curve. The net S-shaped, or sigmoidal, curve obtained when a substance is reduced (or oxidized) is called a **polarographic wave** and is the basis for both qualitative and quantitative analyses. The *half-wave potential* $E_{1/2}$ is the potential value at a current one-half of the limiting current i_L of the species being reduced (or oxidized). The magnitude of the limiting **diffusion current** i_L, obtained once the half-wave potential is passed, is a

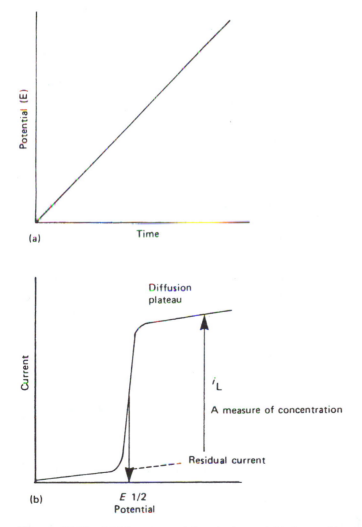

Figure 15.25 (a) Linear potential excitation and (b) the resulting polarographic wave.

measure of the species concentration. It is standard to extrapolate the background residual current to correct for impurities that are reduced (or oxidized) before the $E_{1/2}$ potential of the species of interest is reached. Although DC polarography is no longer commonly used for analytical purposes, it provides the basis of the newer, more powerful variants of classical polarography.

15.3.4.1. Classical or DC Polarography

A modern apparatus for DC polarography is shown in Fig. 15.26. The three-electrode system for aqueous work consists of the DME as the working electrode, a Pt wire or foil auxiliary electrode, and an SCE for reference. The potential of the working electrode (DME) is changed by imposition of a slow voltage ramp vs. a stable reference (Fig. 15.25). At the heart of polarography is the DME. A very narrow capillary is connected to a mercury column, which has a pressure head that can be raised and lowered. A drop of mercury forms at the tip of the capillary, grows, and finally falls off when it becomes too large. Typically, the level of the mercury column above the tip of the capillary is about 60 cm and the natural interval between drops is 2–6 s. The choice of mercury for the electrode is important for several reasons:

1. Each fresh drop exposes a new Hg surface to the solution. The resulting behavior is more reproducible than that with a solid surface, because the liquid drop surface does not become contaminated in the way solid electrodes can be contaminated. Organic contaminants or adsorbants must undergo reequilibration with each new drop and are less likely to interfere.

2. As mentioned earlier, there is a high overpotential for H^+ ion reduction at mercury. This means that it is possible to analyze many of the metal ions whose standard reduction potentials are more negative than that of the H_2/H^+ ion couple. It is easier, too, to reduce most metals to their mercury amalgam than to a solid deposit. Conversely, however, mercury is easily oxidizable, which severely restricts the use of the DME for the study of oxidation processes.

3. Solid electrodes have surface irregularities because of their crystalline nature. Liquid mercury provides a smooth, reproducible surface that does not depend on any pretreatment (polishing or etching) or on substrate inhomogeneity (epitaxy, grain boundaries, imperfections, etc).

In electroanalysis, diffusion currents are quite small ($<100 \mu A$), which means that the aqueous solution IR drop between the reference electrode and the DME can be neglected in all but the most accurate work. Electrolytes prepared with organic solvents, however, may have fairly large resistances, and in some instances IR corrections must be made.

As each new drop commences to grow and expand in radius, the resulting current is influenced by two important factors. The first is the depletion by electrolysis of the electroactive substance at the mercury drop surface. This gives rise to a diffusion layer in which the concentration of the reactant at the surface is reduced. As one travels radially outward from the drop surface, the concentration increases and reaches that of the bulk homogeneous concentration. The second factor is the outward growth of the drop itself, which tends to counteract the formation of a diffusion layer. The net current waveform for a single drop is illustrated in Fig. 15.27.

A mathematical description of the diffusion current, in which the current i_L is measured at the top of each oscillation just before the drop dislodges, is given by the following equation:

$$i_L = 708nCD^{1/2}m^{2/3}t^{1/6} \tag{15.39}$$

Figure 15.26 Modern polarographic cell and three-electrode circuit.

where i_L is the maximal current of drop (μA); n, the number of electrons per electroactive species; C, the concentration of electroactive species (mM); D, the diffusion coefficient of electroactive species; m, the mercury flow rate (mg/s); and t, the drop time (s)

This is called the **Ilkovic equation**. For a particular capillary and pressure head of mercury, $m^{2/3}t^{1/6}$ is a constant. Also, the value of n and that of the diffusion coefficient for a particular species and solvent conditions are constants. Thus i_L is proportional to the concentration C of the electroactive species, and this is the basis for quantitative analysis. The Ilkovic equation is accurate in practice to within several percent, and routinely $\pm 1\%$

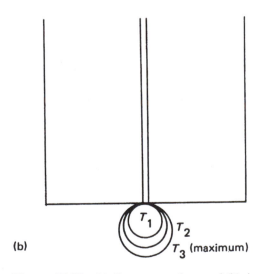

Figure 15.27   (a) Current waveform and (b) time frame of an expanding drop.

precision is possible. It is commonplace to use standard additions to obtain a calibration curve, or an internal standard. Internal standards are useful when chemical sampling and preparation procedures involve the possibility of losses. The principle is that the ratio of the diffusion currents due to the sample and the added standard should be a constant for a particular electrolyte.

Before a polarogram can be obtained from a sample, two important steps are necessary: (1) the sample must be dissolved in a suitable **supporting electrolyte** and (2) the solution must be adequately **degassed**. To understand the purpose of the supporting electrolyte, we should first review the three principal forms of transport that occur in ionic solutions: (1) diffusion, (2) convection, and (3) migration.

Diffusion is the means by which an electroactive species reaches the electrode when a concentration gradient is created by the electrode reaction. The electron transfer process can decrease the concentration of an electroactive species or produce a new species (not originally present in the bulk solution) that diffuses away from the electrode surface. *Convection*, that is, forced motion of the electrolyte, can arise from natural thermal currents always present within solutions, by density gradients within the solution or be produced deliberately by stirring the electrolyte or rotating the electrode. In general, natural convection must be minimized by electrode design and by making short time scale measurements (on the order of milliseconds). The third form of transport is migration of the charged species. *Migration* refers to the motion of ions in an electric field and must be suppressed if the species is to obey diffusion theory. This is done by adding an excess of inert *supporting electrolyte* to the solvent. The role of the supporting electrolyte is twofold: it ensures that the electroactive species reaches the electrode by diffusion and it lowers the resistance of the electrolyte.

Typically, supporting electrolyte concentrations are 0.1–1 M. An example is KCl solution. The K^+ ions in this solution are not easily reduced and have the added advantage that the K^+ and Cl^- ions migrate at about equal velocities in solution. In a solution of $ZnCl_2$ (0.0001 M) and KCl (0.1 M), the migration current of the zinc is reduced to a negligible amount, and we are therefore able to measure the current response due to zinc ion reduction under diffusion conditions.

Actually, in polarography, even though measurements are made in a quiescent solution (no stirring by either gas bubbling or magnetic bar is permissible), the transient currents arising with each drop depend on both diffusion and the convective motion due to the expanding mercury drop. These effects combine to produce a current that is proportional to $t^{1/6}$.

A further important step is the adequate degassing of the electrolyte by bubbling purified nitrogen or argon. This is necessary because dissolved oxygen from the air is present in the electrolyte and, unless removed, would complicate interpretation of the polarogram. Figure 15.28 shows an actual polarogram of a supporting electrolyte that is air saturated and without added electroactive sample. The abrupt wave at -0.1 V vs. SCE is due to the reduction of molecular oxygen, and the drawn-out wave at -0.8 V vs. SCE is assigned to reduction of a product from the oxygen reduction, namely, hydrogen peroxide. Once the potential exceeds the threshold for peroxide reduction, both of these reactions can occur at each new drop as it grows, and the limiting current is approximately twice as large. The first reaction is sharp and can be used for the determination of dissolved oxygen in solutions. The addition of a few drops of a dilute Triton X-100 solution is necessary to prevent the formation of maxima in the current at the diffusion plateau. Such maxima distort the waveforms and complicate the measurement of the diffusion currents. Maximum suppressants, such as gelatin and Triton X-100, are capillary-active substances, which presumably damp the streaming currents around the drop that cause the current distortions.

The relationship between current and voltage for a well-degassed solution of $Pb(NO_3)_2$ is shown in Fig. 15.29. (This polarogram is actually a normal pulse polarogram, not a DC polarogram. The difference is discussed.) Within the voltage range of 0 to -0.3 V vs. SCE only a small *residual current* flows. It is conventional in polarography to represent cathodic (reduction) currents as positive and anodic (oxidation) currents as negative. The residual current actually comprises two components, a faradaic contribution and a charging or double-layer current contribution:

$$i_{res} = i_F + i_{ch}$$

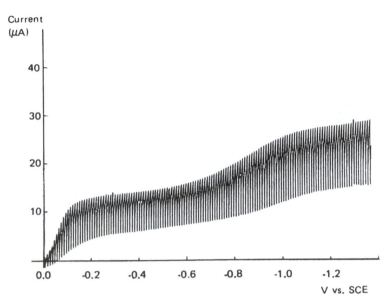

Figure 15.28 Polarogram of dissolved oxygen in an electrolyte, 1 M KNO$_3$ (5 mV/s, 2 s drop time, 0.1 mA, y-scale = 10 V, x = 0.2 V/in., y = 0.1 V/in.).

The faradaic current i_F arises from any residual electroactive species that can be electrolyzed in this potential range, such as traces of metals or organic contaminants. Such impurities may be introduced by the supporting electrolyte, which will have to be purified in some trace analyses. The charging current i_{ch} is nonfaradaic; in other words, no electron flow occurs across the metal–solution interface, and neither redox nor permanent chemical changes result from its presence. The mercury–solution interface acts, to a

Figure 15.29 Normal pulse polarogram of Pb^{2+} ions (5 ppm) in 1 M KNO$_3$ (scan rate 5 mV/s, 2 s drop time).

first approximation, as a small capacitor, and charge flows to the interface to create an electrical double layer. At negative potentials this can be thought of as a surplus of electrons at the surface of the metal and a surplus of cations at the electrode surface. In reality, the electrical double layer capacity varies somewhat with the potential; so the fixed capacitor analogy is not strictly accurate. An important consequence of charging currents is that they limit the analytical sensitivity of polarography. This is why pulse polarographic methods, which discriminate against charging currents, can be used to determine much lower concentrations of analyte.

Referring again to Fig. 15.29, we can see that as the potential becomes increasingly negative, it reaches a point where it is sufficient to cause the electroreduction

$$Pb^{2+} + 2e^- \longrightarrow Pb(Hg)$$

The Pb^{2+} concentration in the immediate vicinity of the electrode decreases as these ions are reduced to lead amalgam. With a further increase in negative potential, the Pb^{2+} concentration at the surface of the electrode becomes zero, even though the lead ion concentration in the bulk solution remains unchanged. Under these conditions, the electrode is said to be polarized. The use of an electrode of small area, such as the DME, means that there is no appreciable change of Pb^{2+} ions in the bulk of the solution from a polarographic analysis.

When the electrode has become polarized, a fresh supply of M^+ ions to its surface is controlled by the diffusion of such ions from the bulk of the solution, through the zone depleted of M^+. In other words, the current flowing is dependent on the diffusion of ions from the bulk liquid. This current is called the diffusion current and the plateau region of the curve can be used to measure the limiting diffusion current i_L. It is usual to extrapolate the residual current background and to construct a parallel line through the diffusion current plateau to correct for the residual current contribution to i_L.

15.3.4.2. Half-Wave Potential

An important point on the curve is that at which the diffusion current is equal to one-half of the total diffusion current; the voltage at which this current is reached is the *half-wave potential $E_{\frac{1}{2}}$*. The half-wave potential is used to characterize the current waveforms of particular reactants. Whether a process is termed reversible or not depends on whether equilibrium is reached at the surface of the electrode in the time frame of the measurements. In other words, a process is reversible when the electron transfer reactions are sufficiently fast so that the equilibrium

$$ox + ne^- \longleftrightarrow red$$

is established and the Nernst equation describes the ratio [ox]/[red] as a function of potential, that is,

$$E = E^0 + 2.303 \frac{RT}{nF} \log\left(\frac{[ox]}{[red]}\right)$$

When this is the case, it can be shown that there is a relation between the potential, the current, and the diffusion current i_L, which holds true throughout the polarographic wave:

$$E = E_{\frac{1}{2}} + \frac{0.0591}{n} \log\left(\frac{i_L - i}{i}\right) \quad \text{at } 25°C \tag{15.40}$$

This equation, in fact, may be practically used to test the reversibility (or Nernstian behavior) of an electroactive species. The graph of E vs. $\log[(i_L - i)/i]$ will be linear with a slope of

$0.0591/n$ and intercept $E_{1/2}$. Determination of the slope enables us to determine n, the number of electrons involved in the process. Substitution of $i_L/2$ in Eq. (15.40) reveals that $E = E_{1/2}$ when the surface concentrations [ox] and [red] are equal.

Note, however, that the half-wave potential $E_{1/2}$ is usually similar but not exactly equivalent to the thermodynamic standard potential E^0. First, the product of reduction may be stabilized by amalgam formation in metal ion reductions; second, there will always be a small liquid junction potential in electrochemical cells of this type that should be corrected for; and finally, it can be shown that the potential $E_{1/2}$ is the sum of two terms:

$$E_{1/2} = E_0 + \frac{0.0591}{n} \log\left(\frac{\gamma_{ox}}{\gamma_{red}}\right)\left(\frac{D_{red}}{D_{ox}}\right)^{1/2}$$

where the γ terms are the activity coefficients and the D terms are the diffusion coefficients of the oxidized and reduced species. In most analytical work the activity corrections are ignored and $D_{ox} \approx D_{red}$. This is because the size of the electroactive product and reactant is not greatly affected by the gain (or loss) of an electron. The value of the half-wave potential is that it can be used to characterize a particular electroactive species qualitatively. It is not affected by the analyte concentration or by the capillary constant. It can, however, be severely affected by changes in the supporting electrolyte medium. Table 15.5 lists some half-wave potentials of diverse species that may be analyzed by polarography.

15.3.4.3. Normal Pulse Polarography

Unlike classical DC polarography, normal pulse polarography does not use a linear voltage ramp; instead, it synchronizes the application of a square-wave voltage pulse of progressively increasing amplitude with the last 60 ms of the life of each drop. This is shown in Fig. 15.30. It is necessary to use an electronically controlled solenoid to knock the drop from the capillary at a preset time, for example, every 2 s. Each drop has the same lifetime, which is shorter than its natural (undisturbed) span. The initial voltage E_{init} is chosen such that no faradaic reactions occur during most of the growth of the drop. Then, when the rate of

Table 15.5 Half-Wave Potentials of Common Metal Ions

		$E_{1/2}$ vs. SCE
Ag(I)	1 M NH_3/0.1 M NH_4Cl	-0.24
Cd(II)	1 M HCl	-0.64
	0.1 M CH_3COONa/0.1 M CH_3COOH	-0.65
Cu(II)	1 M HCl	-0.22
	0.1 M CH_3COONa/0.1 M CH_3COOH	-0.07
O_2	0.1 M KNO_3	-0.05
Pb(II)	1 M HCl	-0.44
	0.1 M CH_3COONa/0.1 M CH_3COOH	-0.50
Tl(I)	1 M HCl	-0.48
Zn(II)	1 M NH_3/0.1 M NH_4Cl	-1.35
	1 M CH_3COONa/0.1 M CH_3COOH	-1.1

Note: More complete data are available from textbooks (e.g., Bard, 1980) and from polarographic equipment suppliers (e.g., EG&G Princeton Applied Research Application Briefs and Application Notes).

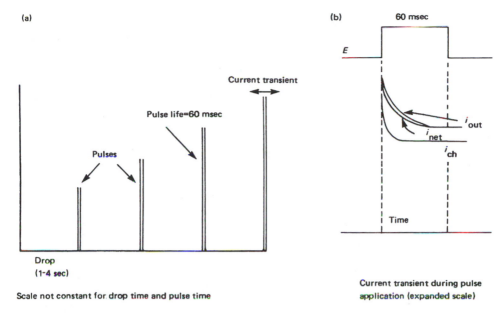

(a)

Current transient

Pulse life=60 msec

Pulses

Drop
(1-4 sec)

Scale not constant for drop time and pulse time

(b)   60 msec

E

i_{out}

i_{net}

i_{ch}

Time

Current transient during pulse
application (expanded scale)

Figure 15.30   (a) Waveform for normal pulse polarography and (b) current transient response (not to scale).

change of the drop surface area is less than during the drop's early formation stages, the pulse is applied. As a first approximation the drop surface area can be considered to be constant during the pulse application. If the potential is such that faradaic reaction can take place at the pulse potential, the resultant current decay transient has both a faradaic and a charging current component, as shown in Fig. 15.30. The measurement of current may be made during the last 17 ms of the pulse by a sample-and-hold circuit. This outputs the average current and holds this value until the next current is ready for output. Thus the output current is a sequence of steady-state values lasting the drop life, for example, 2 s. The $i–E$ curve does not have the large oscillations reminiscent of classical polarography (compare Figs. 15.28 and 15.29), but may provide a Nernstian waveform directly.

It is readily seen from Fig. 15.30 why the current is sampled in this manner. A faradaic current i_F at an electrode of constant area will decay as a function of $t^{-1/2}$ if the process is controlled by diffusion; on the other hand, the charging current transient i_{ch} will be exponential, with a far more rapid decay. This permits better discrimination of the faradaic current and charging current contributions.

A further important advantage of this technique is that during the first 1.94 s of the life of each drop no current flows (remember, the pulse is applied in the last 0.060 s of a 2 s drop life). To understand how this is advantageous, we must reconsider what happens in classical DC polarography. At the start of a drop life, the potential is at a given value and increases slowly as the drop grows. If faradaic reaction can occur, this means that the electroactive species is being depleted during the whole life of the drop, such that when it has grown for about 1.94 s there is a large depletion region extending from the surface. In normal pulse polarography no reaction occurs until the drop area is quite large, and this enhances the current response considerably. The analytical implication is that it is possible to detect concentration levels of about 1×10^{-6} M with the normal pulse technique, which is much better than the useful range of about 1×10^{-4}–1×10^{-2} M in classical DC polarography.

15.3.4.4.   Differential Pulse Polarography

Differential pulse polarography has the most complex waveforms of the polarographic methods discussed, but it is the easiest to interpret for analytical purposes. The applied voltage is a linear ramp with imposed pulses added during the last 60 ms of the life of each drop (Fig. 15.31). In this technique, however, the pulse height is maintained constant above the ramp and is termed the *modulation amplitude*. The modulation amplitude may be varied between 10 and 100 mV. As in normal pulse polarography, the current is not measured continuously, but in this technique it is sampled twice during the mercury drop lifetime: once just prior to the imposition of a pulse and once just before each drop is mechanically dislodged. As in classical DC polarography the presence of a potential at the start of the life of each drop reduces the analyte concentration by creating a depleted diffusion layer. However, the sampled current output is a differential measurement,

$$\Delta i_{out} = i_{after\ pulse} - i_{before\ pulse}$$

and, as in normal pulse polarography, the time delay permits the charging current to decay to a very small value. As the measured signal is the difference in current, a peak-shaped $i-E$ curve is obtained, with the peak maximum close to $E_{1/2}$ if the modulation amplitude is sufficiently small. A differential pulse waveform is shown in Fig. 15.32.

To understand why the waveform is a peak it is necessary to consider what is being measured. The output is a current difference arising from the same potential difference, that is, $\Delta i / \Delta E$. If ΔE becomes very small ($\Delta E \rightarrow dE$), we should obtain the first derivative of the DC polarogram, and the peak and $E_{1/2}$ would coincide. However, because the modulation amplitude is finite and, in fact, has to be reasonably large in order to produce an adequate current signal, the E_{peak} value of the differential pulse polarogram is shifted positive at the half-wave potential for reduction processes:

$$E_{peak} = E_{1/2} - \Delta E/2$$

where ΔE is the modulation amplitude. Increasing ΔE increases the current response but reduces the resolution of the waveform, and experimentally it is best to vary the instrument

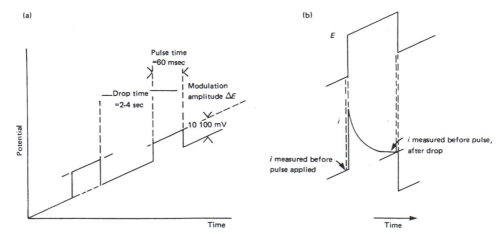

Figure 15.31   Differential pulse polarography (not to scale).

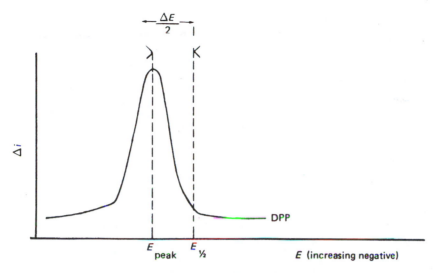

Figure 15.32 Waveform obtained in differential pulse polarography.

parameters until a suitable combination is found for the ramp scan rate and the modulation amplitude.

The simplicity of the differential pulse polarographic method lies in the fact that the peak height is proportional to the analyte concentration (Fig. 15.33). If adequate resolution exists, it is possible to analyze several ionic species simultaneously with polarography. Differential pulse polarography is especially useful for trace analyses when working close to the electrolyte background reduction or oxidation wave. Its limit of detection is typically 1×10^{-7} M or better, depending on the sample and conditions.

Figure 15.33 Effect of standard additions on a differential pulse polarogram.

The addition of surfactants is not always recommended in normal and differential pulse polarography, because their presence may reduce the sensitivity of these methods in certain circumstances. Many other forms of polarography have been suggested and tested. Prominent among these are AC polarographic methods, which use sinusoidal and other periodic waveforms. Most modern instruments offer a range of techniques to the electroanalytical chemist.

The solvent used plays a very important role in polarography. It must be able to dissolve a supporting electrolyte and, if necessary, be buffered. In many cases the negative potential limit involves the reduction of a hydrogen ion or a hydrogen on a molecule. It is therefore vital to control the pH of the solution with a buffer. The buffer normally serves as the supporting electrolyte. Furthermore, it must conduct electricity. These requirements eliminate the use of many organic liquids such as benzene. A polar solvent is necessary, the most popular being water. To dissolve organic compounds in water, a second solvent, such as ethanol, acetone, or dioxane, may first be added to the water. The mixture of solvents dissolves many organic compounds and can be conditioned for polarography. As an alternative to the aqueous/nonaqueous systems, a pure polar solvent may be used, especially if water is to be avoided in the electrolysis. Commonly used solvents are acetonitrile, dimethylformamide, dimethylsulfoxide, and propylene carbonate. Tetralkylammonium perchlorates and tetrafluoroborates are useful supporting electrolytes because their cations are not readily reduced. A list of functional groups that can be determined by polarography is shown in Table 15.6.

In general, simple saturated hydrocarbons, alcohols, and amines are not readily analyzed at the DME. However, aldehydes and quinones are reducible, as well as ketones.

Similarly, olefins (in aqueous solutions) may be reduced according to the equilibrium

$$R{-}C{=}C{-}R' + 2e^- + 2H_2O \longleftrightarrow R{-}CH_2CH_2{-}R' + 2OH^-$$

Table 15.6 Typical Functional Groups that can be Determined by Polarography

Functional group	Name	$E_{1/2}$ (V)
RCHO	Aldehyde	-1.6
RCOOH	Carboxylic acid	-1.8
RR'C=O	Ketone	-2.5
R—O—N=O	Nitrite	-0.9
R—N=O	Nitroso	-0.2
R—NH$_2$	Amine	-0.5
R—SH	Mercaptan	-0.5

Note: Electrochemical data on a large number of organic compounds are compiled in the CRC Handbook Series in Organic Electrochemistry.

In short, polarography can be used for the analysis of C—N, C—O, N—O, O—O, S—S, and C—S groups and for the analysis of heterocyclic compounds. Also, many important biochemical species are electroactive, such as vitamin C (ascorbic acid), fumaric acid, vitamin B factors (riboflavin, thiamine, niacin), antioxidants such as tocopherols (vitamin E), N-nitrosamines, ketose sugars (fructose and sorbose), and the steroid aldosterone.

15.3.5. Voltammetry

We mentioned that polarography is a form of voltammetry in which the electrode area does not remain constant during electrolysis. It is possible, however, to use electrode materials other than mercury for electroanalyses, provided that the potential "window" available is suitable for the analyte in question. Table 15.7 summarizes the accessible potential ranges for liquid mercury and for solid platinum electrodes. The precious metals and various forms of carbon are the most common electrodes in use, although a great many materials, both metallic and semiconducting, find use as analytical electrode substrates. Voltammetry is conducted using a microelectrode as the working electrode under conditions where polarization at the working electrode is enhanced. This is in sharp contrast to both potentiometry and coulometry where polarization is absent or minimized by experimental conditions. In voltammetry, very little analyte is used up in the measurement process unlike coulometry where complete consumption of the analyte is desired.

In polarography the DME is renewed regularly during the voltage sweep, with the consequence that the bulk concentration is restored at the electrode surface at the start of each new drop at some slightly higher (or lower) potential. At a solid electrode, the electrolysis process initiated by a voltage sweep proceeds to deplete the bulk concentration of analyte at the surface without interruption. This constitutes a major difference between classical polarography and sweep voltammetry at a solid electrode. Normal pulse polarography is a useful technique at solid electrodes, because the bulk concentration is restored by convection at the surface during the off-pulse, provided that this is at least 10× longer than the electrolysis period. It is possible to use a mercury electrode in a so-called "static" or "**hanging**" mode. Such a stationary Hg droplet may be suspended from a micrometer syringe capillary. Alternatively, commercially available electrodes have been developed

Table 15.7 Potential Windows for Commonly Used Electrodes and Solvents

		Range (V vs. SCE[a])
Mercury		
Aqueous	1 M $HClO_4$	+0.05 to −1.0
	1 M NaOH	−0.02 to −2.5
Nonaqueous	0.1 M TEAP/CH_3CN[b]	+0.6 to −2.8
Platinum		
Aqueous	1 M H_2SO_4	+1.2 to −0.2
	1 M NaOH	+0.6 to −0.8
Nonaqueous	0.1 M TBABF$_4$/CH_3CN[c]	+2.5 to −2.5

[a]The SCE may only be used in nonaqueous systems if traces of water are acceptable.
[b]TEAP, tetraethylammonium perchlorate.
[c]TBABF$_4$, tetrabutylammonium tetrafluoroborate.

that inject a Hg drop of varying size to the opening of a capillary. These electrodes find application for stripping voltammetry (see subsequently).

A powerful group of electrochemical methods uses *reversal* techniques. Foremost among these is *cyclic voltammetry*, which is invaluable for diagnosing reaction mechanisms and for studying reactive species in unusual oxidation states. In cyclic voltammetry, a microelectrode is used as the WE. The potential is increased linearly and the current is measured. The current increases as the potential of the electroactive material is reached. The area of the working electrode and the rate at which the analyte can diffuse to the electrode surface limits the current. A single voltage ramp is reversed at some time after the electroactive species reacts and the reverse sweep is able to detect any electroactive products generated by the forward sweep. This is **cyclic voltammetry**. A cyclic voltammogram of a typical reversible oxidation–reduction reaction is shown diagrammatically in Fig. 15.34. Usually, an *XY* recorder or computer data system is used to track the voltage on the time axis so that the reverse current appears below the peak obtained in the forward sweep, but with opposite polarity. The shapes of the waves and their responses at different scan rates are used for diagnostic purposes. Substances are generally examined at concentrations around the millimolar level, and the electrode potentials at which the species undergo reduction and oxidation may be rapidly determined.

The height of the current peak of the first voltage sweep can be calculated from the Randles–Sevcik equation:

$$i_p = 2.69 \times 10^5 n^{2/3} A D^{1/2} C v^{1/2} \tag{15.41}$$

where n is the number of electrons transferred; A, the electrode area (cm^2); D, the diffusion coefficient of the electroactive species (cm^2/s); C, the concentration of the electroactive species (mol/cm^3); and v, the potential scan rate (V/s). The standard potential E^0 is related to the anodic and cathodic peak potentials, E_{pa} and E_{pc}:

$$E^0 = \frac{1}{2}(E_{pc} + E_{pa}) \tag{15.42}$$

The peak separation is related to the number of electrons involved in the reaction:

$$E_{pa} - E_{pc} = \frac{0.059}{n} \tag{15.43}$$

assuming that the *IR* drop (resistance) in the cell is not too large, which will increase the peak separation.

Cyclic voltammetry is not primarily a quantitative analytical technique. The references at the end of this chapter provide additional guidance to its applications and interpretation. Its real value lies in the ability to establish the nature of the electron transfer reactions—for example, fast and reversible at one extreme, slow and irreversible at the other—and to explore the subsequent reactivity of unstable products formed by the forward sweep. Suffice it to say that such studies are valuable for learning the fate and degradation of such compounds as drugs, insecticides, herbicides, foodstuff contaminants or additives, and pollutants.

15.3.5.1. Instrumentation for Voltammetry

A cyclic voltammetry mode of operation is featured on many modern polarographs, or a suitable voltage ramp generator may be used in combination with a potentiostat. The three-electrode configuration is required. Pretreatment of the working electrode is necessary for

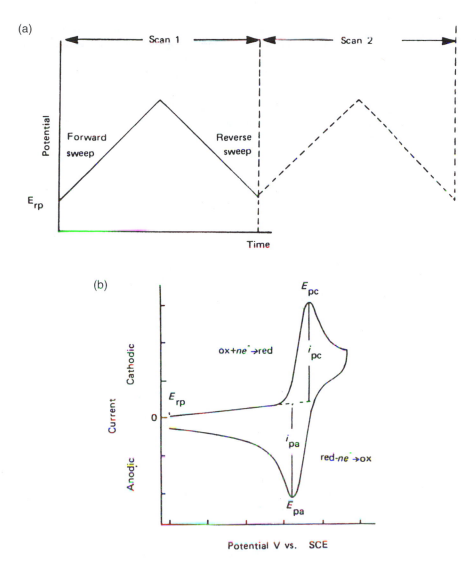

Figure 15.34 (a) Excitation waveform and (b) current response for a reversible couple obtained in cyclic voltammetry.

reproducibility (unless a hanging MDE is used), and normally this entails polishing the electrode mechanically with successively finer grades of abrasives.

Some confusion results in choosing an initial voltage at which to start the voltammogram. It is best to measure the open-circuit voltage between the working and reference electrodes with a high-impedance digital voltmeter. This emf is known as the **rest potential** E_{rp}. Scans may then be made in the negative and/or positive directions starting at the rest potential set on the potentiostat. By this means the potentials for reductions and oxidations, respectively, and the electrolyte limits (the "windows") are established. In other words, it is generally necessary to commence at a potential at which no faradaic reaction is possible, or else voltammograms will be distorted and irreproducible.

Scan rates typically vary from 10 mV/s to 1 V/s if an *XY* recorder is used for current output. The lower limit is due to thermal convection, which is always present in an electrolyte. To record faster responses an oscilloscope or fast transient recorder is required. The currents in successive scans differ from those recorded in the first scan. For quantitation purposes the first scan should always be used.

15.3.5.2.   Stripping Voltammetry

The technique of *stripping voltammetry* may be used in many areas requiring trace analyses to the ppb level. It is especially useful for determining heavy metal contaminants in natural water samples or biochemical studies. Either anodic or cathodic stripping is possible in principle, but analyses by anodic stripping voltammetry are more often used. Stripping analysis is a two-step technique involving (1) the preconcentration of one or several analytes by reduction (in anodic stripping) or oxidation (in cathodic stripping) followed by (2) a rapid oxidation or reduction, respectively, to strip the products back into the electrolyte. Analysis time is on the order of a few minutes. The overall determination involves three phases:

1.   preconcentration
2.   quiescent (or rest) period
3.   stripping process, for example, by sweep voltammetry or differential pulse

It is the preconcentration period that enhances the sensitivity of this technique. In the preconcentration phase precise potential control permits the selection of species whose decomposition potentials are exceeded. The products should form an insoluble solid deposit or an alloy with the substrate. At Hg electrodes the electroreduced metal ions form an amalgam. Usually the potential is set 100–200 mV in excess of the decomposition potential of the analyte of interest. Moreover, electrolysis may be carried out at a sufficiently negative potential to reduce all of the metal ions possible below hydrogen ion reduction at Hg, for example. Concurrent H^+ ion reduction is not a problem, because the objective is to separate the reactants from the bulk electrolyte. In fact, methods have been devised to determine the group I metals and NH_4^+ ion at Hg in neutral or alkaline solutions of the tetraalkylammonium salts. Exhaustive electrolysis is not mandatory and 2–3% removal suffices. Additionally, the processes of interest need not be 100% faradaically efficient, provided that the preconcentration stage is reproducible for calibration purposes, which is usually ensured by standard addition.

Typical solid substrate electrodes are wax-impregnated graphite, glassy (vitreous) carbon, platinum, and gold; however, mercury electrodes are more prevalent in the form of either a **hanging mercury drop electrode (HMDE)** or thin-film mercury electrode (TFME). TFMEs may be electro-deposited on glassy carbon electrodes from freshly prepared $Hg(NO_3)_2$ dissolved in an acetate buffer (pH 7). There is an art to obtaining good thin films, and usually some practice is necessary to get uniform, reproducible coverage on the carbon substrates. It is recommended that literature procedures be adhered to carefully. Some procedures recommend simultaneously pre-electrolyzing the analyte and a dilute mercury ion solution ($\sim 10^{-5}$ M) so that the amalgam is formed in a single step.

It is important to stir the solution or rotate the electrode during the preconcentration stage. The purpose of this is to increase the analyte mass transport to the electrode by convective means, thereby enhancing preconcentration. In general, in electroanalysis one seeks to obtain proper conditions for diffusion alone to permit mathematical expression of the process rate (the current). In this and controlled flow or rotation cases it is advantageous to purposely increase the quantity of material reaching the electrode surface. Pre-electrolysis times are typically 3 min or longer.

In the rest period or quiescent stage, the stirrer is switched off for perhaps 30 s but the electrolysis potential is held. This permits the concentration gradient of material within the Hg to become more uniform. The rest period is not obligatory for films produced on a solid substrate.

A variety of techniques have been proposed for the stripping stage. Two important methods are discussed here. In the first, in anodic stripping, for example, the potential is scanned at a constant rate to more positive values. With this single sweep voltammetry the resolution of a TFME is better than that of an HMDE, because stripping of the former leads to a more complete depletion of the thin film. As illustrated in Fig. 15.35, it is possible to analyze for many metal species simultaneously. The height of the stripping peak is taken to be directly proportional to concentration. Linearity should be established in the working range with a calibration curve. The second method of stripping involves the application of a differential pulse scan to the electrode. As in polarographic methods, an increase in sensitivity is obtained when the differential pulse waveform is used.

Extremely high sensitivities in trace analyses require good analytical practice, especially in the preparation and choice of reagents, solvents, and labware. Glass cells and volumetric glassware must be soaked 24 h in trace-metal purity 6 M HNO_3. Plastic electrochemical cells are recommended if loss of sample by adsorption to the vessel walls is a likely problem. The inert gas (N_2 or Ar) used to remove dissolved oxygen should be purified so as not to introduce additional contaminants. Solid catalysts and drying agents are recommended for oxygen and water removal. A presaturator is recommended to reduce electrolyte losses by volatilization, especially when low vapor pressure organic solvents are used. A major source of contamination is the supporting electrode itself. This may be purified in part by recrystallization, but the use of sustained preelectrolysis at a mercury pool electrode may ultimately be required. Dilute standards and samples ought to be prepared daily because of the risk of chemical (e.g., hydrolysis) or physical (e.g., adsorption) losses.

15.3.5.3. Applications of Anodic Stripping Voltammetry

Anodic stripping voltammetry is readily applicable for those metals that form an amalgam with mercury, for example, Ag, As, Au, Bi, Cd, Cu, Ga, In, Mn, Pb, Sb, Sn, Tl, and Zn. One important cause of interferences is intermetallic compound formation of insoluble alloys

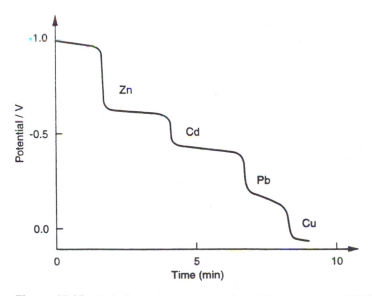

Figure 15.35 Stripping voltammogram of metal-ion species at a TFME.

between the metals within the amalgam. For example, In/Au and Cu/Ni can form in the Hg drop and then not respond to the stripping stage, so their measured concentrations will be too low. It is imperative to carefully select the electrolyte such that possibly interfering compounds are complexed and electroinactive. This is also a means of improving resolution when there are two overlapping peaks. Some elements can be analyzed in aqueous electrolytes only with difficulty (groups I and II), but, fortunately, their analysis by flame or atomic absorption methods are sensitive and easy to carry out. Anions of carbon compounds can also be stripped by either (1) anodic preconcentration as sparingly soluble Hg salts, (2) adsorption and decomposition, or (3) indirect methods, such as displacement of a metal complex. Examples of complexes used include thiourea, succinate, and dithizone. Anions may be determined as mercurous or silver(I) salts if these are sparingly soluble. Based on solubility determinations, it is possible to estimate some theoretical values for the minimum determinable molar concentrations of anions at mercury: Cl^-, 5×10^{-6}; Br^-, 1×10^{-6}; I^-, 5×10^{-8}; S^{2-}, 5×10^{-8}; CrO_4^{2-}, 3×10^{-9}; WO_4^{2-}, 4×10^{-7}; MoO_4^{2-}, 1×10^{-6}; and $C_2O_4^{2-}$, 1×10^{-6}. With the exception of Cl^-, the mercurous salts are less soluble than the silver salts.

In conclusion, stripping voltammetry is an inexpensive, highly sensitive analytical tool applicable to multicomponent systems; in fact, it is not recommended for metal-ion samples whose concentrations are greater than 1 ppm. Careful selection of operating conditions and especially the electrolyte buffer is necessary. The sensitivity of stripping voltammetry is less for nonmetallic and anionic species than for metals. More recently, flow-through systems have been devised for continuous monitoring purposes. Stripping voltammetry has been applied to numerous trace metal analyses and environmental studies, for example, to determine impurities in oceans, rivers, lakes, and effluents; to analyze body fluids, foodstuffs, and soil samples; and to characterize airborne particulates and industrial chemicals.

15.4. LC DETECTORS

Electrochemical detectors for the determination of trace amounts of ionic and molecular components in liquid chromatographic effluents are sensitive, selective, and inexpensive. For many separations they provide the best means of detection, outperforming, for example, UV/VIS and fluorescence spectroscopy and refractive index methods. Two major categories of electrochemical detectors will be described subsequently: voltammetric and conductometric. An efficient redox center in the ion or molecule to be detected is necessary for voltammetric methods, but this is not mandatory in the species actually undergoing separation, because many organic compounds can be derivatized with an electroactive constituent before or after column separation. Alternatively, certain classes of organic compounds can be photolytically decomposed by a post-column online UV lamp into electroactive products, which can be detected. For example, organic nitrocompounds can be photolyzed to nitrite and nitrate anions, which undergo electrolysis at suitable working electrode potentials. In the case of conductivity detectors the method does not require species that are redox active, or chromophoric; rather, changes in the resistivity of the electrolyte eluent are monitored. Examples of species that may be detected are simple ions such as halides, SO_4^{2-}, PO_4^{3-}, NO^{3-}, metal cations, NH_4^+, as well as organic ionic moieties.

15.4.1. Voltammetric Detection

Various waveforms and signal presentation techniques are available, such as steady-state voltammetry with amperometric or coulometric (integrated current) outputs and sweep

Figure 15.36 Schematic of an LC system with electrochemical detection.

procedures such as differential pulse. In the simplest mode a sufficient steady-state potential (constant E) is applied to a working electrode to reduce or oxidize the component(s) of interest. The output signal is a transient current peak collected on a Y-time recorder (Fig. 15.36) or for modern chromatographs, collected and processed by a computer data system. To obtain reasonable chromatographic resolution, the electrochemical cell, positioned at the outlet of the column, must be carefully designed. The volume of the detector cell must be small to permit maximum band resolution. As the reduction or oxidation is performed under hydrodynamic flow conditions, the kinetic response of the electrode/redox species should be as rapid as possible. Glassy (vitreous) carbon is a popular electrode material, and it is possible to chemically alter the electrode surface to improve the response performance. However, other substrates are available, such as Pt, Hg, and Au; these prove useful in cases where they are not passivated by the adsorption of organic reaction products. Platinum and gold electrodes are particularly prone to blockage from the oxidation products of amino acids, carbohydrates, and polyalcohols, for example. Typically, a three-electrode thin-layer cell is used, with the reference and auxiliary electrodes downstream from the working electrode. Dissolved oxygen may interfere in some assays and may therefore have to be removed by prior degassing.

15.4.2. Conductometric Detection

Many important inorganic and organic ions at the trace level are not easily detected by reduction–oxidation or spectroscopic methods. This has led to the development of sophisticated separation and conductometric detectors capable of ultrasensitive analyses. It is difficult to measure small changes in conductivity due to a trace analyte ion when the background level is orders of magnitude larger; consequently procedures have been developed that are capable of suppressing high ionic strength backgrounds, allowing the net signal to be measured more easily. Combined with miniature conductance detectors, these procedures have led to a technique known as *ion chromatography* with eluent suppression (cf 13.2.1; Fig. 13.24). Figure 15.37 shows the setup for a typical application: the analysis of the constituents of acid rain samples.

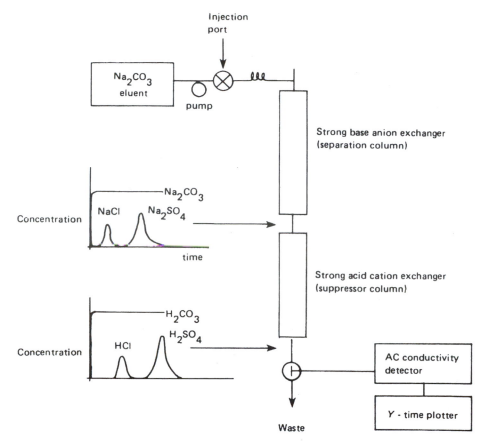

Figure 15.37 Schematic of the eluent suppression approach for ion chromatography that permits conductometric detection of analytes.

In the first ion exchange column, whose resin exists in its carbonate form ($R^{2+}CO_3^{2-}$), the sample species, such as Cl^- and SO_4^{2-}, are selectively separated and eluted as sodium salts in a background of sodium carbonate. Detection of the chloride and sulfate by conductivity at this stage would not be possible because of the large excess of conducting ions Na^+ and CO_3^{2-}. In the second column, the cations undergo ion exchange to produce highly conducting acids (such as the strong acids HCl and H_2SO_4, which are fully ionized) with a background of slightly ionized carbonic acid:

$$H_2CO_3 \longleftrightarrow 2H^+ + CO_3^{2-}$$

Now the highly ionized chloride and sulfate can be measured because the carbonic acid has a low conductance.

A major problem is that the capacity of the second column needs frequent regeneration if large amounts of Na_2CO_3 eluent are used. Various approaches have been tested to improve the practical operation of this form of separation. The first objective is to shorten the time needed to resolve the dilute sample constituents. This is done with a low-capacity separator, which decreases the resolution time and thereby the exchange consumption of the suppressor column. The next objective is to provide continuous regeneration of the suppressor column by automation or by a novel ion exchange membrane methodology that continuously replenishes the column. For cation analyses, lightly sulfonated beads of cross-linked polystyrene have been employed; for anion analysis composite beads are formed by electrostatically coating these beads with anion-exchanging spheres.

Liquid chromatographic electrochemical detection has been widely used for metabolite studies in complex matrices and has general applicability in many fields, for example, the pharmaceutical industry, forensic science, medicine, the explosives industry, and agriculture.

BIBLIOGRAPHY

Adams, R. *Electrochemistry at Solid Electrodes*; Marcel Dekker, Inc.: New York, 1969.

Aikens, D.; Bailey, R.; Moore, J.; Giachino, G.; Torukins, R. *Principles and Techniques for an Integrated Chemistry Laboratory*; Waveland Press, Inc.: Prospect Heights, IL, 1985.

ASTM. *Annual Book of ASTM Standards, Water and Environmental Technology*; ASTM: West Conshohocken, PA, 2000; Vols. 11.01 and 11.02.

Bard, A.J., Ed. *Electroanalytical Chemistry*; Marcel Dekker, Inc.: New York, 1993.

Bard, A.J.; Faulkner, L.R. *Electrochemical Methods—Fundamentals and Applications*; John Wiley and Sons, Inc.: New York, 1980.

Bard, A.J.; Parsons, R.; Jordan, J., Eds. *Standard Potentials in Aqueous Solution*; Marcel Dekker, Inc.: New York, 1985.

Baizer, M.M.; Lund, H. *Organic Electrochemistry*; Marcel Dekker, Inc.: New York, 1983.

Bates, R.G. *Determination of pH*; John Wiley and Sons, Inc.: New York, 1973.

Berezanski, P. In *Handbook of Instrumental Techniques for Analytical Chemistry*; Settle, F.A., Ed.; Prentice-Hall PTR: NJ, 1997.

Bockris, J.O. *Modern Electrochemistry*; Plenum Press: New York, 1970; Vols. 1 and 2.

Bond, A.M. *Modern Polarographic Methods in Analytical Chemistry*; Marcel Dekker, Inc.: New York, 1980.

Dean, J.A. *Analytical Chemistry Handbook*; McGraw-Hill, Inc.: New York, 1995.

Diefenderfer, A.J.; Holton, B.E. *Principles of Electronic Instrumentation*, 3rd ed.; Saunders College Publishing: Philadelphia, PA, 1994.

Dryhurst, G. *Electrochemistry of Biological Molecules*; Academic Press: New York, 1977.

Enke, C.G. *The Art and Science of Chemical Analysis*; John Wiley and Sons, Inc.: New York, 2000.

Evans, A. *Potentiometry and Ion Selective Electrodes*; John Wiley and Sons, Inc.: New York, 1987.

Ewing, G.W., Ed. *Analytical Instrumentation Handbook*, 2nd ed.; Marcel Dekker, Inc.: New York, 1997.

Fritz, J.S. *Acid–Base Titrations in Non-Aqueous Solvents*; G. Frederick Smith Chemical Company: Columbus, OH, 1952 (This publication is still available from GFS Chemicals upon request.)

Fry, A.J. *Synthetic Organic Electrochemistry*; Harper and Row: New York, 1972.

Greenberg, A.; Clesceri, L.; Eaton, A., Eds. *Standard Methods for the Examination of Water and Wastewater*, 18th ed.; American Public Health Association: Washington, DC, 1992.

Harris, D.C. *Quantitative Chemical Analysis*, 5th ed.; W.H. Freeman and Co.: New York, 1999.

Hart, J.P. *Electroanalysis of Biologically Important Compounds*; Ellis Horwood: New York, 1990.

Huber, W. *Titrations in Nonaqueous Solvents*; Academic Press: New York, 1967.

Ives, D.J.G.; Janz, G.J., Eds. *Reference Electrodes—Theory and Practice*; Academic Press: New York, 1961.

Kalvoda, R. *Operational Amplifiers in Chemical Instrumentation*; Ellis Horwood: Chichester, 1975.

Kissinger, P.T.; Heineman, W.R. *Laboratory Techniques in Electroanalytical Chemistry*, 2nd Ed.; Marcel Dekker, Inc.: New York, 1996.

Koryta, J.; Stulik, K. *Ion-Selective Electrodes*, 2nd Ed.; Cambridge University Press: Cambridge, 1983.

Kucharsky, J.; Safarik, L. *Titrations in Nonaqueous Solvents*; Elsevier: Amsterdam, 1965.

Malmstadt, H.V.; Enke, C.G.; Crouch, S.R. *Microcomputers and Electronic Instrumentation: Making the Right Connections*; American Chemical Society: Washington, DC, 1994.

Meites, L.; Zuman, P.; Rupp, E.B., Eds. *CRC Handbook Series in Organic Electrochemistry*; CRC Press, Inc.: Boca Raton, FL, 1982; Vols. 1–5.

Michael, A.; Wightman, R.M. In *Laboratory Techniques in Electroanalytical Chemistry*, 2nd Ed.; Kissinger, P., Heineman, W., Eds.; Marcel Dekker, Inc.: New York, 1996.

Settle, F.A., Ed. *Handbook of Instrumental Techniques for Analytical Chemistry*; Prentice-Hall PTR: NJ, 1997.

Streuli, C.A. Titrimetry: acid–base titrations in non-aqueous solvents. In *Treatise on Analytical Chemistry, Part I*; Kolthoff, I.M., Elving, P.J., Eds.; Wiley-Interscience: New York, 1975; Vol. II.

Vydra, F.; Stulik, K.; Julakova, E. *Electrochemical Stripping Analysis*; Ellis Horwood, 1976.

Wang, J. *Stripping Analysis*; VCH Publishers, Inc.: Deerfield Beach, FL, 1985.

Warner, M. *Anal. Chem.* **1994**, *66*, 601A.

SUGGESTED EXPERIMENTS

15.1   Run a DC polarogram with a known volume of 1 M KNO_3 electrolyte that has not been deaerated. Observe and identify the two reduction waves of dissolved oxygen.

15.2   Purge the electrolyte in problem 15.1 with purified nitrogen gas for 20 min or so. Add sufficient lead nitrate solution to make the electrolyte 1.0×10^{-4} M in Pb^{2+} ion. Record the DC polarogram and measure the limiting current and the half-wave potential ($E_{1/2}$) for the lead ion reduction wave.

15.3   Measure the mercury flow rate with the column height set as for problem 15.2. Collect about 20 droplets under the electrolyte (why?) and determine the drop lifetime with a stopwatch. Use the Ilkovic equation to calculate the diffusion equation for Pb^{2+} ion and compare your derived value with a literature value.

15.4   Record a normal pulse polarogram of (a) 1.00×10^{-5} M Zn^{2+}, (b) Cd ion and, (c) Cu ion in a degassed 1 M KNO_3 electrolyte. Construct a graph of potential (E) vs. $\log[(i_L - i)/i]$. Determine the value of n from the slope and comment on the reversibility of this reaction. How does the electrolyte temperature affect the slope of this curve?

15.5   Record a differential pulse polarogram of a background electrolyte without and with added tapwater. Which metal species are present? Use a standard additions calibration to quantify one of the metal impurities. Compare E_{peak} with $E_{1/2i}$.

15.6   Use a F^- ion selective electrode per manufacturer's instructions to find the concentration of F^- ion in your drinking water and in your brand of (a) toothpaste and (b) mouthwash. Use a standard additions calibration procedure. Some sample preparation may be needed for the toothpaste. Several extraction procedures are available in the literature.

15.7   Collect some samples for pH analyses, for example, rainfall, snow, lake water, pool water, soft drinks, fruit juice. Standardize the pH meter with a standard buffer of pH close to the particular sample of interest. Design some experiments to change the temperature of these samples. What is the influence of temperature on pH measurements?

15.8   (a)   Use a glass pH electrode combination and titrate unknown strength weak acid(s) with 0.2000 N NaOH. Stir the electrolyte continuously.

         (b)   Calculate the strength(s) of the acid(s) (i.e., the pK_a) and explain the shape of the curve.

15.9   Copper sulfate may be analyzed by gravimetry by exhaustive electrolysis at a weighed platinum electrode. Add 2 mL H_2SO_4, 1 mL HNO_3, and 1 g urea to 25 mL of the copper solution. Suggest a stoichiometry for your copper sulfate based on a variable water of hydration, for example, $CuSO_4 \cdot 2H_2O$.

15.10   Tap waters, as well as a variety of biological fluids and natural waters, can be analyzed for heavy metals by anodic stripping voltammetry at a hanging mercury drop electrode. Clean glassware and sample bottles by leaching for about 5 h in 6 N HNO_3. First, analyze for Cd^{2+} and Pb^{2+} ions at acidic pH (2–4). Deposition is made with gentle stirring for 2 min exactly. After a 15 s quiescent period, scan from -0.7 V SCE in the positive direction. Zn^{2+} and Cu^{2+} ions can be analyzed by raising the pH to 8–9 with 1 M NH_4Cl/1 M NH_4OH buffer. Deposit at -1.2 V SCE and strip as before. It will be necessary to adjust the deposition time and/or current output sensitivity to achieve best results.

15.11 Cyclic voltammetry can be used to study the reversible reduction–oxidation couple of $[Fe(CN)_6]^{3-} + e^- \leftrightarrow [Fe(CN)_6]^{4-}$. You will need solutions of 1 M KNO_3, 5×10^{-3} M $K_3[Fe(CN)_6]$ in 1 M KNO_3, and 5×10^{-3} M $VOSO_4$ in 1 M KNO_3. A CV equipped with a three-electrode cell (Pt microelectrode as the working electrode, Pt foil electrode as the auxiliary electrode and a calomel reference electrode) and an X–Y recorder is required. The potassium nitrate solution serves as the baseline. All solutions must be purged with nitrogen to eliminate oxygen before scanning the potential. A sweep from 0.80 to -0.12 V and back vs. SCE is suitable for the Fe system. The range may need to be changed for the vanadium system. Can you evaluate n for each system? Is the value what you expect for each reaction? Evaluate E^0 for each system. Do these values agree with those in the literature? (Experiment courtesy of Professor R.A. Bailey, Department of Chemistry, Rensselaer Polytechnic Institute.)

PROBLEMS

15.1 Give an example of a half-cell. The absolute potential of a half-cell cannot be measured directly. How can the potential be measured?

15.2 Using the Nernst equation, complete the following table:

pH of solution	Concentration of H^+	E of hydrogen half-cell
		0
1.2		
3		
	10^{-5}	
	10^{-10}	
7		
12		
14		

15.3 Describe and illustrate an SCE. Write the half-cell reaction.

15.4 How is pH measured with a glass electrode? Why does a glass electrode give pH readings lower than the actual pH in strongly basic solutions? What other errors can occur in pH measurement with a glass electrode?

15.5 Calculate the theoretical potential of the following cell. Is the cell as written galvanic or electrolytic?

$$Pt|Cr^{3+}(3.00 \times 10^{-2} M), Cr^{2+}(4.00 \times 10^{-5} M) \| Sn^{2+}(2.00 \times 10^{-2} M),$$
$$Sn^{4+}(2.00 \times 10^{-4} M)|Pt$$

15.6   A salt of two monovalent ions M and A is sparingly soluble. The E^0 for the metal is $+0.621$ V vs. SHE. The observed emf of a saturated solution of the salt is 0.149 V vs. SHE. What is the solubility product K_{sp} of the salt? (Neglect the junction potential.)

15.7   Describe the process of electrodeposition and how it is used for electrogravimetric analysis.

15.8   Can two metals such as iron and nickel be separated completely by electrogravimetry? Explain your answer and state any assumptions you make.

15.9   State Faraday's Law. A solution of Fe^{3+} has a volume of 1.00 L, and 48,246 C are required to reduce the Fe^{3+} to Fe^{2+}. What was the original molar concentration of iron? What is the molar concentration of Fe^{2+} after the passage of 12,061 C?

15.10   What are the three major forms of polarography? State the reasons why pulse polarographic methods are more sensitive than classical DC polarography.

15.11   What are the advantages of mercury electrodes for electrochemical measurements? What are the advantages of the dropping mercury electrode vs. a Pt microelectrode for polarography? What are the disadvantages of the DME?

15.12   Describe the method of anodic stripping voltammetry. What analytes can it be used to determine? Why is stripping voltammetry more sensitive than other voltammetric methods?

15.13   Describe the principle of an ISE. Why is the term ion-specific electrode not used?

15.14   How can conductometric measurements be used in analytical chemistry? Give two examples.

15.15   Briefly outline two types of electrochemical detectors used for chromatography.

15.16   The fluoride ISE is used routinely for measuring fluoridated water and fluoride ion in dental products such as mouthwash. A 50 mL aliquot of water containing sodium fluoride is analyzed using a fluoride ion electrode and the method of standard additions. The pH and ionic strength are adjusted so that all fluoride ion is present as free F^- ion. The potential of the ISE/reference electrode combination in a 50 mL aliquot of the water was -0.1805 V. Addition of 0.5 mL of a 100 mg/L F^- ion standard solution to the beaker changed the potential to -0.3490 V. Calculate the concentration of (1) fluoride ion and (2) sodium fluoride in the water sample.

15.17   Copper is deposited as the element on a weighed Pt cathode from a solution of copper sulfate in an electrolytic cell. If a constant current of 0.600 A is used, how much Cu can be deposited in 10.0 min? (Assume no other reductions occur and that the reaction at the anode is the electrolysis of water to produce oxygen.)

15.18   Why does coulometry not require external calibration standard solutions?

15.19   Explain why a silver electrode can be an indicator electrode for chloride ion.

15.20   What are the three processes by which an analyte in solution is transported to an electrode surface? What single transport process is desired in polarography? Explain how the other transport processes are minimized in polarography.

15.21   Would you expect the half-wave potential for the reduction of copper ion to copper metal to be different at a Hg electrode from that at a platinum electrode? Explain your answer.

15.22   Sketch a schematic cyclic voltammogram for a nonreversible reduction reaction. (See Fig. 15.34 for a CV of a reversible reaction.)

APPENDIX 15.1.   SELECTED STANDARD REDUCTION POTENTIALS AT 25°C

Half-reaction	E^0 (V)
$F_2(g) + 2e^- \longrightarrow 2F^-$	+2.87
$Co^{3+} + e^- \longrightarrow Co^{2+}$	+1.82
$H_2O_2 + 2H^+ + 2e^- \longrightarrow 2H_2O$	+1.77
$Ce^{4+} + e^- \longrightarrow Ce^{3+}$	+1.74
$PbO_2(s) + 4H^+ + SO_4^{2-} + 2e^- \longrightarrow PbSO_4(s) + 2H_2O$	+1.70
$MnO_4^- + 8H^+ + 5e^- \longrightarrow Mn^{2+} + 4H_2O$	+1.51
$Au^{3+} + 3e^- \longrightarrow Au(s)$	+1.50
$Cl_2(g) + 2e^- \longrightarrow 2Cl^-$	+1.36
$Cr_2O_7^{2-} + 14H^+ + 6e^- \longrightarrow 2Cr^{3+} + 7H_2O$	+1.33
$MnO_2(s) + 4H^+ + 2e^- \longrightarrow Mn^{2+} + 2H_2O$	+1.23
$O_2(g) + 4H^+ + 4e^- \longrightarrow 2H_2O$	+1.23
$Br_2(l) + 2e^- \longrightarrow 2Br^-$	+1.07
$NO_3^- + 4H^+ + 3e^- \longrightarrow NO(g) + 2H_2O$	+0.96
$2Hg^{2+} + 2e^- \longrightarrow Hg_2^{2+}$	+0.92
$Hg_2^{2+} + 2e^- \longrightarrow 2Hg(l)$	+0.85
$Ag^+ + e^- \longrightarrow Ag(s)$	+0.80
$Fe^{3+} + e^- \longrightarrow Fe^{2+}$	+0.77
$O_2(g) + 2H^+ + 2e^- \longrightarrow H_2O_2$	+0.68
$MnO_4^- + 2H_2O + 3e^- \longrightarrow MnO_2(s) + 4OH^-$	+0.59
$I_2(s) + 2e^- \longrightarrow 2I^-$	+0.53
$O_2(g) + 2H_2O + 4e^- \longrightarrow 4OH^-$	+0.40
$Cu^{2+} + 2e^- \longrightarrow Cu(s)$	+0.34
$AgCl(s) + e^- \longrightarrow Ag(s) + Cl^-$	+0.22
$Cu^{2+} + e^- \longrightarrow Cu^+$	+0.15
$Sn^{4+} + 2e^- \longrightarrow Sn^{2+}$	+0.13
$2H^+ + 2e^- \longrightarrow H_2(g)$	0.00
$Pb^{2+} + 2e^- \longrightarrow Pb(s)$	−0.13
$Sn^{2+} + 2e^- \longrightarrow Sn(s)$	−0.14
$Ni^{2+} + 2e^- \longrightarrow Ni(s)$	−0.25
$Co^{2+} + 2e^- \longrightarrow Co(s)$	−0.28
$Cd^{2+} + 2e^- \longrightarrow Cd(s)$	−0.40
$Fe^{2+} + 2e^- \longrightarrow Fe(s)$	−0.44
$Cr^{3+} + 3e^- \longrightarrow Cr(s)$	−0.74
$Zn^{2+} + 2e^- \longrightarrow Zn(s)$	−0.76
$2H_2O + 2e^- \longrightarrow H_2(g) + 2OH^-$	−0.83
$Mn^{2+} + 2e^- \longrightarrow Mn(s)$	−1.18
$Al^{3+} + 3e^- \longrightarrow Al(s)$	−1.66
$Na^+ + e^- \longrightarrow Na(s)$	−2.71
$Ca^{2+} + 2e^- \longrightarrow Ca(s)$	−2.87
$Ba^{2+} + 2e^- \longrightarrow Ba(s)$	−2.90
$K^+ + e^- \longrightarrow K(s)$	−2.93
$Li^+ + e^- \longrightarrow Li(s)$	−3.05

Note: All species in aqueous solution (aq) unless otherwise indicated. All dissolved species = 1 M; all gases = 1 atm.

16

Thermal Analysis

When matter is heated, it undergoes certain physical and chemical changes. Physical changes include phase changes such as melting, vaporization, crystallization, transitions between crystal structures, changes in microstructure in metal alloys and polymers, volume changes (expansion and contraction), and changes in mechanical behavior. Chemical changes include reactions to form new products, oxidation, corrosion, decomposition, dehydration, chemisorption, and the like. These physical and chemical changes take place over a wide temperature range. The rates of chemical reactions vary with temperature and the properties of some materials, such as semicrystalline polymers and metal alloys, depend on the rate at which they are cooled. Materials are used over a wide range of temperatures, from Arctic cold to tropical heat, in corrosive environments, variable humidity, and under load (stress). It is necessary to characterize materials and their behavior over a range of temperatures to determine what materials are suitable for specific uses and to determine what temperature range materials or chemicals can withstand without changing. This sort of information is used to predict safe operating conditions for products, such as which type of tire material is best for vehicles in extremely cold or extremely hot climates, the average expected lifetime of materials such as paints and polymers exposed to temperature changes, processing conditions for materials, and the curing times and temperatures for dental filling material, among other uses.

The physical changes and chemical reactions a sample undergoes when heated are characteristic of the material being examined and the atmosphere in which the heating occurs. By measuring the temperature at which such reactions occur and the heat involved in the reaction, we can determine a great deal about the material. Composition of pure compounds and mixtures, including polymer blends, can be determined. The purity of pharmaceutical compounds can be determined. Rate of reaction, rate of crystallization, glass transition temperatures, decomposition temperatures, and catalysis can be studied. The effectiveness of additives and stabilizers in materials can be evaluated. Enthalpy changes (ΔH) of reactions can be measured. Percent crystallinity of polymers, which greatly affects the mechanical properties of polymers, can be determined. By heating materials under load (stress), mechanical properties such as modulus, ductility, yield point (the point at which nonpermanent elastic deformation changes to permanent plastic deformation), and volume change as a function of the load can be measured. These parameters are very important in engineering design to ensure safe and functional products.

The analytical techniques used to study changes in physical properties with temperature are called **thermal analysis** techniques. They include thermogravimetric analysis (TGA), differential thermal analysis (DTA), differential scanning calorimetry (DSC), thermometric titration (TT), and direct injection enthalpimetry, dynamic mechanical analysis (DMA), and thermomechanical analysis (TMA). Thermal analysis techniques are used in the characterization of inorganic and organic compounds, polymers, cosmetics, pharmaceuticals, metals, alloys, geological samples, ceramics, glasses, and many manufactured products. Table 16.1 presents a brief overview of some applications for various thermal analysis methods. Table 16.2 lists some typical sample types and some of the properties that have been measured by thermal analysis methods.

16.1. THERMOGRAVIMETRY

Thermogravimetry or thermogravimetric analysis (TGA) measures the mass (weight) of a sample in a specified atmosphere as the temperature of the sample is programmed. The most common temperature program is a linear increase in temperature with time, although isothermal programs, stepped temperature programs, and so on can be used. In the most common TGA experiment, the sample temperature is increased linearly over a time period and the mass of the sample is constantly recorded. The output from a TGA experiment is a plot of mass (or mass%) vs. temperature. The TGA plot is called a **thermal curve**. An example of a TGA thermal curve for the decomposition of calcium carbonate is shown in Fig. 16.1. Weight or weight% is plotted along the y-axis and temperature (or time for a linear temperature ramp) along the x-axis. The change in weight of a sample as the temperature changes tells us several things. First, this determines the temperature at which the material loses (or gains) weight. Loss of weight indicates decomposition or evaporation of the sample. A gain in weight can indicate adsorption by the sample of a component in the atmosphere or a chemical reaction with the atmosphere. Second, the temperatures at which no weight change takes place are determined, which indicate the temperature stability of the material. These weight changes at certain temperatures are physical properties of chemical compounds under the conditions of the experiment (atmosphere, heating rate). This information can be used to determine if a sample is the same as a "standard" or "good" material in a production process, for example. Knowledge of the temperatures at which a sample is unstable and subject to decomposition or chemical change is important to the engineer because it reveals the temperature above which such materials as polymers, alloys, and building materials may *not* be used, as well as the temperatures at which they may be used safely.

Table 16.1   Some Applications of Thermal Analysis Methods

Applications	TGA	DSC, DTA	TMA	DMA
Compositional analysis	×	×		
Curing studies		×		×
Glass transition		×	×	×
Heat of reaction		×		
Oxidative stability	×	×		
Corrosion	×	×		
Creep			×	×
Stress relaxation			×	×
Thermal stability	×	×		
Viscoelastic properties			×	×
Protein denaturation		×		
Shrinkage			×	×

Table 16.2 Partial List of Sample Types and Properties Examined by Thermal Analysis

Properties	Samples						
	Chemicals	Elastomers	Explosives	Soils	Plastics	Textiles	Metals
Identification	×	×	×	×	×	×	×
Quantitative composition	×	×	×	×	×	×	
Phase diagram	×	×	×		×		×
Thermal stability	×	×	×		×	×	
Polymerization	×	×			×	×	
Catalytic activity	×						×
Reactivity	×	×			×	×	×
Thermochemical constants	×	×	×	×	×		×
Reaction kinetics	×	×	×		×	×	×

Note: × = Sample type that has been characterized for this property.

The *weight lost* by a sample heated to a given temperature helps the inorganic or analytical chemist to determine the composition of a compound and follow the reactions involved in its decomposition. It also enables the analytical chemist to identify crystals of unknown composition or determine the percentage of a given compound in a mixture of compounds. For example, if pure calcium carbonate ($CaCO_3$) is heated to 850°C, it loses 44% of its weight (Fig. 16.1). Also, the gas evolved can be collected and identified as CO_2. This observation virtually confirms that the reaction

$$CaCO_3(s) \longrightarrow CaO(s) + CO_2(g)$$

takes place at this temperature. How do we know this? If we know our sample was pure $CaCO_3$ and CO_2 is identified as a product of the reaction, the need for mass balance tells us

Figure 16.1 TGA thermal curve for pure anhydrous calcium carbonate, $CaCO_3$. The loss in mass is due to the loss of CO_2 (g) and the compound remaining is CaO.

that CaO is a reasonable "other" product. If we had 50.0 mg of $CaCO_3$ to start with, the mass loss due to CO_2 can be calculated:

$$mg\ CO_2 = \frac{(50.0\,mg\ CaCO_3) \times (1\,mmol\ CaCO_3) \times (1\,mmol\ CO_2) \times (44.00\,mg\ CO_2)}{(100.00\,mg)\ \ (1\,mmol\ CaCO_3)\ \ (mmol\ CO_2)}$$

$$mg\ CO_2 = 22.0\,mg$$

The loss of CO_2 equals a loss in weight of $(22.0\ mg/50.0\ mg) \times 100\% = 44\%$. From the stoichiometry, it is expected that 1 mole of CO_2 is lost for every mole of $CaCO_3$ present, which corresponds to a loss of 44% of the mass of 1 mole of $CaCO_3$. The experimental mass loss supports the theoretical loss if the decomposition of $CaCO_3$ proceeds according to the reaction proposed. The formation of CO_2 can be verified by having the evolved gas analyzed by MS or by online IR spectroscopy. The CaO can be confirmed by analysis of the residue by XRD or other techniques.

The TGA technique was developed to solve problems encountered in gravimetric analysis. For example, a common precipitating and weighing form for the determination of calcium was calcium oxalate. In practice, the calcium oxalate was precipitated, filtered, and the filtrate dried and weighed. The drying step was difficult to reproduce. A particular procedure might recommend drying at 110°C for 20 min and cooling prior to weighing. The analyst would then obtain very reproducible results derived from his or her own work. A second procedure might recommend drying at 125°C for 10 min, cooling, and weighing. As before, the analyst using this method would obtain very reproducible results, but these results might not agree with those of the first research worker. The thermal curve of hydrated calcium oxalate is shown in Fig. 16.2. When we examine the

Figure 16.2 TGA thermal curve of hydrated calcium oxalate, $Ca(COO)_2 \cdot xH_2O$, with both adsorbed water from the precipitation process and water of crystallization. There is a loss of adsorbed water starting at about 90°C, and loss of bound water at about 150°C. The stable compound above 225°C is anhydrous calcium oxalate, $Ca(COO)_2$. This loses CO at about 450°C to form $CaCO_3$. The calcium carbonate is stable until approximately 600°C, when it loses CO_2 to form CaO. (Compare this step with Fig. 16.1.)

thermal curve, we find that calcium oxalate loses weight over a range of temperatures from 100°C up to 225°C. The most logical explanation for this weight loss is that water is evaporating from the sample. The water that is driven off includes not only absorbed or adsorbed water from the precipitating solution, but also water of crystallization (also called water of hydration) that is bound to the calcium oxalate as $CaC_2O_4 \cdot nH_2O$. It is only when both types of water are driven off that reproducible results are achievable. If the sample were heated to 110°C, the absorbed water would be driven off, but a small amount of bound water (i.e., water of crystallization) would also be lost. Similarly, if the drying temperature were 125°C, the absorbed water would be driven off and a different small amount of bound water would be lost. As long as a drying temperature was rigidly adhered to, the results would be reproducible, but results using different drying temperatures would show differences because the amount of bound water lost would vary. The TGA experiment shows that drying to a temperature above 225°C but below 400°C will result in a stable form suitable for gravimetric analysis.

16.1.1. TGA Instrumentation

Modern TGA equipment has a sensitive balance, usually a microbalance, for continuously measuring sample weight, a furnace surrounding a sample holder, and a purge gas system for providing inert or reactive atmospheres. A computer generally controls the furnace and the data (weight vs. sample temperature) is collected and processed by computer. Intelligent autosamplers are available for most instruments that permit the unattended analysis of samples.

Modern analytical microbalances of several different designs are commercially available—torsion balances, spring balances, and electrobalances have been used in TGA instruments. In general, the balance is designed so that a change in sample weight generates an electrical signal proportional to the weight change. The electrical signal is transformed into weight or weight loss by the data processing system and plotted on the y-axis of the thermal curve. TGA balances are available for sample masses from 1 to 1000 mg, with the usual sample weighing between 5 and 20 mg. There are specialized high-capacity TGA systems available that can accommodate samples of up to 100 g and systems that can handle microgram quantities of sample. Figures 16.3 and 16.4 show the schematic of a TGA electrobalance and a cutaway of a commercial high-capacity TGA based on the electrobalance. The usual sample size for TGA is very small, so care must be taken to obtain a homogeneous or representative sample. The balance itself must be thermally isolated from the furnace, although the sample holder and sample must be in the furnace. There are two possible configurations of the balance and furnace, a horizontal furnace or a vertical furnace. Both types of configuration suffer from drift as the temperature increases. Vertical configurations suffer from buoyancy effects due to the change in gas density with temperature. The horizontal configuration was designed to minimize buoyancy effects, but horizontal configurations experience changes in the length of the quartz rod connecting the sample to the balance as the temperature changes. Buoyancy effects and changes in the quartz rod result in error in determining the mass of the sample.

The furnace surrounds the sample and sample holder. It must be capable of being programmed for a linear heating rate. Modern instruments can be heated and cooled rapidly, which increases sample throughput. Instruments that heat at rates of up to 1000°C/min are available. One commercial instrument can heat at rates up to 200°C/min from room temperature to about 1200°C; cooling by forced air can be done at ~50°C/min. There are furnaces available with upper temperatures of 1500°C,

(a)

(h)

Figure 16.3 (a) Block diagram of a TGA system. S represents the sample pan hanging from the balance arm in position in the furnace (represented by the solid bars on each side). (b) Schematic of a commercial TGA, showing the purge gas inlet and outlet and the thermocouple position beneath the sample pan. [Figure 16.3(b) courtesy of PerkinElmer, Inc. Shelton, CT, www.perkinelmer.com.]

1700°C, or 2400°C; these higher temperature instruments are useful for studying refractory materials and engineering materials. The furnace must be able to be purged with a desired gas, to provide the correct atmosphere for the experiment and to remove gaseous products from the sample compartment. Argon or nitrogen is used when an inert atmosphere is desired. Reactive gas atmospheres can be used for certain studies. Air is often used for oxidation and combustion studies. Hydrogen gas may be used to provide a reducing atmosphere, with the appropriate precautions to prevent explosions. Modern instruments permit the purge gas to be switched automatically, so that the sample can start heating in an inert atmosphere and be switched to air or other reactive gas at high temperatures, for example.

The sample holder and any instrument parts inside the furnace, such as the thermocouple for measuring the temperature, must be able to withstand high temperature and be

Figure 16.4 (a) Electrobalance for a TGA. As the sample weight changes, the balance arm tips, resulting in a change in the amount of light reaching the photocell. This generates a current to restore the arm position; the current is proportional to the change in weight of the sample. (b) Cutaway view of a high-capacity TGA capable of holding 100 g samples. Typical types of large samples in quartz sample holders are shown in the photographs on the right. (Courtesy of ThermoCahn, Madison, WI, www.thermo.com.)

inert at these high temperatures. Quartz, platinum, and various ceramics are used for the sample holder and other parts. The sample is placed in a small pan or crucible made of Pt, quartz, or ceramic.

Ideally, the temperature recorded is the exact temperature of the sample. This entails measuring the temperature of the sample while the analysis is carried out. It is particularly important to measure the temperature of the sample rather than that of the furnace. This is difficult because the temperature is measured with a thermocouple that is near but not in the sample. The temperature of the sample inside the furnace is measured with a thermocouple, either a chromel/alumel thermocouple or one made of Pt alloy. The thermocouple is never inserted directly into the sample because of possible sample contamination (or contamination of the thermocouple resulting in errors in temperature), inadvertent initiation of a catalytic reaction, particle size effects, sample packing effects, and possible weighing errors. The thermocouple is made as small as possible and placed close to the sample holder, sometimes in contact with the bottom of the sample pan. The temperature actually recorded may be slightly different from the sample temperature; the sample temperature generally is lower than the temperature recorded by the thermocouple. This is due to factors such as rate of heating, gas flow, thermal conductivity of the sample, and the sample holder. Modern instruments have a temperature–voltage control program in the computer software that permits reproducible heating of the furnace. With precise, reproducible heating, the thermocouple can be calibrated to provide accurate furnace temperatures, but the relationship between actual sample temperature and recorded temperature is complex. The problem is compounded by the fact that at temperatures below 500°C, most of the heat transferred from the furnace to the sample takes place by convection and conduction, but at temperatures above 500°C, where the furnace is red-hot, most of the energy is transferred by radiation. The switch from conduction–convection to radiative energy transfer makes choosing the position of the thermocouple to obtain accurate temperature measurements of the sample quite a complicated problem.

Temperature calibration of TGA instruments with samples of pure materials with well-characterized weight losses can be done, but often is not satisfactory. For example, one problem with this approach is that black samples such as coal behave differently from white samples such as calcium phosphate under the influence of radiant energy, and the sample temperatures will therefore be different under the same furnace conditions. A more accurate calibration method uses the Curie temperature of various ferromagnetic standard materials. The materials undergo specific and reversible changes in magnetic behavior at their Curie temperature. Standards are available covering the temperature range of 242–771°C. A **ferromagnetic material** is magnetic under normal conditions, but at a characteristic temperature (**the Curie temperature**) its atoms become disoriented and paramagnetic and the material loses its magnetism. To take advantage of this phenomenon, we weigh the ferromagnetic material continuously, with a small magnet placed above the balance pan. The standard's apparent weight is then its gravitational weight minus the magnetic force it experiences due to the magnet immediately above. At the Curie temperature, the standard loses its magnetism and the effect of the magnet is lost. There is a change in the apparent weight of the standard, and this can be recorded. The temperatures of the Curie transitions of ferromagnetic materials are well known and therefore can be used for calibration purposes. An example of calibration using the Curie temperatures of alumel and nickel is shown in Fig. 16.5.

16.1.2. Analytical Applications of Thermogravimetry

One of the first important applications of thermogravimetry was the determination of correct drying temperatures for precipitates used in gravimetric analysis. This knowledge was of vital importance if accurate and reproducible results were to be obtained from gravimetric analysis. A second important application was the identification of the gases given off while a sample's temperature is increased. In addition, the composition of the residue can be determined using techniques such as XRD, XRF, and other techniques.

Figure 16.5 The TGA Curie point method records for each standard an apparent sharp weight change at a well-defined temperature, which corresponds to a known transformation in the standard's ferromagnetic properties at that temperature. The figure shows the relative temperature precision from three replicate calibration runs using alumel and nickel Curie point standards. (Courtesy of TA Instruments, New Castle, DE, www.tainst.com.)

This information reveals the chemical decomposition process occurring when materials are heated and permits identification of the formulas of the starting materials. TGA is very important in determining the upper use temperatures of materials such as polymers by identifying the temperature at which oxidative degradation occurs on heating in air.

From Fig. 16.6 it can be determined that when pure calcium oxalate monohydrate, $Ca(COO)_2 \cdot H_2O(s)$, is heated, it first loses water of crystallization and forms $Ca(COO)_2(s)$.

Figure 16.6 TGA thermal curve of calcium oxalate monohydrate, $Ca(COO)_2 \cdot H_2O$.

The fact that the compound contains only 1 mole of water of hydration can be determined from the mass loss. The reaction is:

$$Ca(COO)_2 \cdot H_2O(s) \longrightarrow Ca(COO)_2(s) + H_2O(g) \uparrow$$

Molecular weights (g/mol)146   128   18

Therefore the $\%W$ lost $= (18/146) \times 100 = 12.3\%$. If our initial sample mass was 20.00 mg, the mass loss at the first step would be 2.46 mg if there is 1 mole of water of crystallization. If the mass loss actually measured was 4.92 mg, that would mean there were 2 moles of water of crystallization and the formula would be $Ca(COO)_2 \cdot 2H_2O(s)$, a dihydrate.

Upon further heating up to about 400°C, CO(g) is given off and the reaction that occurs is:

$$Ca(COO)_2(s) \longrightarrow CaCO_3(s) + CO(g) \uparrow$$

Molecular weights (g/mol)   28

The $\%W$ lost from the initial compound $= (28/146) \times 100 = 19.2\%$. The total $\%W$ lost at this point is the sum of the two steps, $12.3 + 19.2 = 31.5\%$.

Finally, at even higher temperatures (about 800°C), the $CaCO_3$ formed at 400°C decomposes:

$$CaCO_3(s) \longrightarrow CaO(s) + CO_2(g) \uparrow$$

Molecular weights (g/mol)   44

The $\%W$ lost from the original compound $= (44/146) \times 100 = 30.1\%$. The total mass loss is the sum of all three steps: $12.3 + 19.2 + 30.1 = 61.6\%$. The losses correspond to what is seen in the decomposition of calcium oxalate monohydrate in Fig. 16.6. The $\uparrow$ symbol indicates gas evolved from the sample and swept out of the TGA system. The gases may be identified if the TGA analyzer is connected to an IR spectrometer or a mass spectrometer.

TGA can be used for the identification of compounds present in mixtures of materials. When such mixtures are heated using a thermogravimetric analyzer, the thermal curve produced consists of all possible weight losses from all components superimposed on each other. Interpretation of the complete thermal curve requires that the individual thermal events be separated and identified. In many cases, the components of the mixture can be identified and a quantitative determination of each is possible from the thermal curve. An example of how this can be done is shown in Fig. 16.7. The uppermost curve is the weight loss curve for pure compound A. The next lower curve is the weight loss curve for pure compound B. The bottom curve is the weight loss curve for a mixture of A and B. The amount of A present in the mixture can be determined from the weight loss between points δ and ε, while the amount of B is determined from the weight loss between points β and γ.

An illustration of the application of TGA to quantitative analysis of mixtures is the determination of the magnesium oxalate content of a mixture of magnesium oxalate and magnesium oxide, MgO. Magnesium oxalate is less stable than calcium carbonate; it decomposes to magnesium oxide, MgO, at ~500°C. Pure MgO is stable at room temperature and to well above 500°C; MgO does not lose any weight. The TGA curve for this mixture shows two mass losses, one at about 200°C. The second mass loss occurs in the 397–478°C range. Thinking about what happened to calcium oxalate, it is reasonable to suppose that the first mass loss is due to loss of water, both adsorbed and water of crystallization. From the formula for $Mg(COO)_2 \cdot 2H_2O$, we see that there

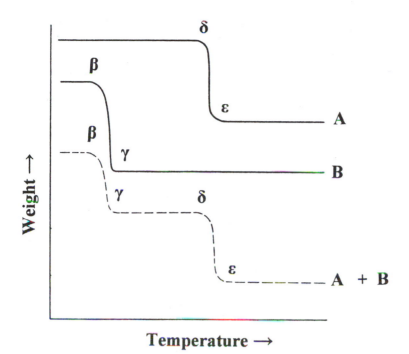

Figure 16.7 TGA thermal curves for (top) pure A, (middle) pure B, and (bottom dotted line) a mixture of A and B. Because A and B have unique temperatures at which mass is lost, the composition of the mixture may be determined.

are 2 moles of water of crystallization for every mole of magnesium oxalate. Only one other mass loss is seen, and we know MgO is stable at these temperatures. We can assume that when $Mg(COO)_2 \cdot 2H_2O$ is heated to a temperature above 500°C, it forms MgO according to the reaction:

$$Mg(COO)_2 \cdot 2H_2O \longrightarrow MgO(s) + CO + CO_2 + 2H_2O$$

Molecular
weights (g/mol)   148   40   28   44   18

The final weight loss is therefore: $\%W = ([28 + 44 + 2(18)]/148) \times 100 = 73\%$

We would lose 73 mg for every 100 mg of pure magnesium oxalate dihydrate we had in the mixture on heating the mixture to 500°C. Note that the decomposition of magnesium oxalate does not follow the same process as the decomposition of calcium oxalate. Calcium oxalate first forms calcium carbonate, which only loses carbon dioxide to form CaO at temperatures above 600°C. Magnesium oxalate appears to decompose directly to MgO in one step at about 500°C. The weight remaining from the magnesium oxalate is equal to:

$$\frac{\text{mol wt MgO}}{\text{mol wt Mg(COO)}_2 \cdot 2H_2O} \text{ per gram of } Mg(COO)_2 \cdot 2H_2O$$

$$= \left(\frac{40}{148}\right) = 0.27 \text{ g MgO/g Mg(COO)}_2 \cdot 2H_2O$$

The following data were obtained from our TGA curve:

Original weight of sample	25.00 mg
Weight of sample after heating to 500°C	10.40 mg
Loss in weight	14.60 mg

But we have already calculated that $Mg(COO)_2 \cdot 2H_2O$ loses 73% of its mass upon heating to 500°C; therefore the weight of $Mg(COO)_2 \cdot 2H_2O$ in the original sample was:

$$\text{mg } Mg(COO)_2 \cdot 2H_2O = (14.60 \text{ mg lost})\frac{100 \text{ mg } Mg(COO)_2 \cdot 2H_2O}{73 \text{ mg lost}} = 20 \text{ mg}$$

Then the concentration of $Mg(COO)_2 \cdot 2H_2O$ in the original sample was:

$$\%Mg(COO)_2 \cdot 2H_2O = \frac{\text{mg } Mg(COO)_2 \cdot 2H_2O}{\text{total sample weight in mg}} \times 100$$

$$\%Mg(COO)_2 \cdot 2H_2O = \frac{20 \text{ mg}}{25.00 \text{ mg}} \times 100 = 80\%$$

Thus the other 20% of the starting mixture is MgO.

  Similarly, we can determine the magnesium oxalate content and the calcium oxalate content in a mixture of the two compounds. The TGA curve of the mixture is shown in Fig. 16.8. We have already seen in Fig. 16.6 that calcium carbonate decomposes to calcium oxide above 600°C. We can deduce from the thermal curve for magnesium oxalate that magnesium carbonate is not stable because the oxalate decomposes directly to MgO. So at temperatures above 500°C, but below 600°C we should have a mixture of MgO and $CaCO_3$. Above 850°C the $CaCO_3$ will have decomposed to CaO, so the

Figure 16.8 TGA thermal curve of a mixture of calcium oxalate monohydrate and magnesium oxalate dihydrate. The last weight loss is due to the loss of CO_2 only from calcium oxalate monohydrate. Therefore, the composition of the mixture can be determined, even though the other steps have combined mass losses from both compounds.

residue is a mixture of CaO and MgO. By examining Fig. 16.6 and the information we have above on the magnesium salt, it is evident that the weight loss in Fig. 16.8 above 600°C is due to the evolution of CO_2 from the $CaCO_3$ that came from the $Ca(COO)_2 \cdot H_2O$. We know one molecule of $Ca(COO)_2 \cdot H_2O$ generates one molecule of CO_2. If the weight loss above 600°C (due to CO_2 loss) was 18%, then the weight% of $Ca(COO)_2 \cdot H_2O$ in the mixture is

$$\frac{\text{mol wt}\,Ca(COO)_2 \cdot H_2O}{\text{mol wt}\,CO_2} \times 18\% = \frac{146}{44} \times 18\% = 60\%$$

Assuming that the weights of calcium oxalate and magnesium oxalate total 100%, then it follows that the percentage of $Mg(COO)_2 \cdot 2H_2O = 100 - 60 = 40\%$. As you can see from the examples given, TGA can be used for quantitative analysis, but not without some knowledge of the sample. If there were other components in our mixture of oxalates that we did not know about, our assumption that the two oxalates composed 100% of the sample would be wrong. If there is another component that loses weight above 600°C we have another error. Without some knowledge of the sample, our calculated value for one (or both) of the components could be wrong.

TGA also provides quantitative information on organic compound decompositions, and is particularly useful for studying polymers. An example is the use of TGA to determine the amount of vinyl acetate in copolymers of vinyl acetate and polyethylene. On heating, each mole of vinyl acetate present loses 1 mole of acetic acid. A TGA study of several vinyl acetate-polyethylene copolymers is presented in Fig. 16.9.

TGA is very useful for providing qualitative information about samples of many types. TGA can provide qualitative information on the stability of polymers when they are heated in air or under inert atmospheres. From the decomposition temperatures of various polymers heated in a TGA in an air atmosphere, the upper use temperatures of

Figure 16.9 TGA thermal curves of vinyl acetate copolymers. The weight loss at 340°C is due to loss of acetic acid and gives a quantitative measure of the amount of vinyl acetate in the polymer. (Courtesy of TA Instruments, New Castle, DE, www.tainst.com.)

polymer materials can be determined. Figure 16.10 presents the decomposition temperatures for a variety of common polymeric materials; such a TGA comparison can be used to choose a polymer that will be stable below a certain temperature for a given application.

Another example of the application of thermogravimetry is in the characterization of coal. A TGA thermal curve for a coal sample heated in nitrogen or some other inert atmosphere indicates, in a single analysis, the percentage of volatiles present and the "fixed carbon"; if the atmosphere in the TGA is then automatically switched to air, the fixed carbon will burn and the amount of ash in the coal is determined as the residue left. This is very valuable information in the characterization of the quality of the coal and for its handling and subsequent use. Coals vary in their volatiles content and in their carbon content. Coals often contain a considerable amount of volatile material. The composition of this volatile material is variable and may contain valuable chemicals that can be used in the pharmaceutical industry, the dyestuff industry, and the chemical industry. Of course it can be used as a fuel, "coal gas". The newest form of carbon to be discovered, C_{60}, called buckminsterfullerene, and related fullerene compounds are found in soot. TGA can be used to characterize soot for its fullerene content, as the fullerenes are more volatile than graphite. Similarly, nonvolatile additives in polymers can be measured quantitatively by TGA. The pigment additive "carbon black", which is not volatile, is measured in a sample of nylon polymer, shown in Fig. 16.11. The polymer is burned off and the nonvolatile pigment determined by the mass remaining. Other polymer additives such as silica, titanium dioxide, and inorganic pigments such as cadmium red, and nonvolatile flame-retardants can be measured in a similar manner.

A high-precision TGA may be used under isothermal or nonisothermal conditions as a mass detector to study kinetics. Such an approach has been used to study oxidation of metals, corrosion, rates of reaction involving mass changes, and phase changes involving small changes in oxygen content for materials like high-temperature superconductors. Many applications of TGA can be found in the scientific literature and on the websites of thermal analysis instrument manufacturers. Websites with applications include the TA Instruments site at www.tainst.com, the Mettler–Toledo site at www.mt.com, and the PerkinElmer site at www.perkinelmer.com.

Figure 16.10 TGA thermal curves showing the decomposition temperatures of some common polymers: PVC, polyvinylchloride; PMMA, polymethylmethacrylate; HDPE, high-density polyethylene; PTFE, polytetrafluoroethylene; PI, polyimide. (Courtesy of TA Instruments, New Castle, DE, www.tainst.com.)

Figure 16.11   TGA determination of carbon black pigment in nylon. The sample was heated under nitrogen up to 650°C, and then the purge gas was automatically switched to air. The polymer decomposes between 350°C and 450°C, leaving the carbon black pigment, which will not decompose in nitrogen, as the 3.05% by weight residue. On switching to air, the entire residue decomposes (final mass of 0.00%). This shows that the entire sample including the pigment was organic in nature. The solid line is the thermal curve, while the dotted line is the first derivative (DTG) plot. (Courtesy of TA Instruments, New Castle, DE, from Thermal Analysis Applications Brief TA-122.)

16.1.3.   Derivative Thermogravimetry

Examination of a TGA curve will show that a sample's weight loss associated with a particular decomposition occurs over a considerable temperature range, not at a single temperature. When TGA is used to identify an unknown compound, this wide range is a handicap because the uncertainty of identification is increased. This problem can be partially overcome by derivative thermogravimetry (DTG). In DTG, the first derivative of the TGA curve is plotted with respect to temperature. The plot that results has the change in weight with time, $d(w)/dt$, which is the rate of weight change, on the y-axis. Figure 16.12(a) shows the TGA curve and its DTG curve for a hydrated sodium silicate, general formula $Na_xSi_yO_z \cdot nH_2O$. From the TGA curve, the temperature range over which the loss of water occurs is broad and does not have a smooth slope. The DTG curve shows that three separate steps occur in the range of 50–200°C, all probably due to loss of water bound in different ways to the sodium silicate. Another example of the power of the derivative plot is shown in Fig. 16.12(b). This is a TGA and its DTG curve for a mixture of hydrated salts of barium, strontium, and calcium. From the TGA, there appears to be a single weight loss occurring between 130°C and 210°C. This might be misinterpreted to be a single pure compound if all we had was the TGA thermal curve. But from the very sensitive DTG, it is clear that there are three different losses of water, occurring at 140°C, 180°C, and 205°C, respectively. These peaks are in fact loss of water first from the barium salt, then from the strontium salt and finally from the calcium salt. The DTG gave us a clue that more than one event was taking place, a clue that the TGA did not provide. Consequently, DTG is a valuable method of data presentation for thermal analysis. A similar example of the power of the DTG plot is shown in Fig. 16.13, which shows desorption of chemisorbed basic compounds from the acidic sites of a zeolite catalyst. **Zeolites**, an important class of catalysts, are porous crystalline aluminosilicates with acidic sites. The number and relative strength

Figure 16.12   (a) TGA (solid line) and DTG (dotted line) thermal curves of a pure hydrated sodium silicate. The TGA mass loss from 50–150°C suggests loss of water in more than one form because of the change in slope seen during the step. The first derivative DTG plot clearly shows three separate mass losses due to water bound in different forms. (Courtesy of PerkinElmer, Inc. Shelton, CT, www.perkinelmer.com.) (b) Partial TGA (bottom curve) and DTG (top curve) thermal curves for a mixture of hydrated barium, strontium, and calcium oxalates. L. Erdey et al. (*Talanta* **1962**, *9*, 489–493.) showed that the three hydrated oxalates lost water at three different temperatures. The TGA seems to show only one step, but the DTG clearly shows the three separate losses, one for each salt.

of the acidic sites can be estimated by chemisorption of a basic compound, such as ammonia, and then studying the desorption by TGA/DTG. As seen in Fig. 16.13, the weight change is small, about 1% total spread over two broad steps, but the two steps are clearly evident from the DTG plot.

The decomposition of polyvinyl chloride polymer (PVC) also demonstrates the power of the DTG plot. The TGA curve (solid line in Fig. 16.14) seems to show two weight loss events, but the DTG plot (dotted line in Fig. 16.14) clearly shows three steps. The loss at 280°C is due to volatilatization of HCl, while the mass losses at 320°C and 460°C are due to loss of hydrocarbons. We will learn how we know this later in the chapter.

Figure 16.13 TGA and DTG thermal curves showing desorption of basic compounds from the acidic sites of a zeolite catalyst. Losses from weakly acidic sites can be distinguished from strongly acidic sites using the DTG curve, despite the fact that the overall mass loss is only about 1%. The mass scale is on the left-handside y-axis. (Courtesy of TA Instruments, New Castle, DE, from Thermal Analysis Applications Brief TA-231.)

16.1.4. Sources of Error in Thermogravimetry

Errors in thermogravimetry can lead to inaccuracy in temperature and weight data. Proper placement of the TGA instrument in the laboratory, away from sources of vibration and heat, is essential to minimize fluctuations in the balance mechanism. Older instruments suffered from an apparent gain in weight of a sample container when heated, known as the buoyancy effect. This effect, due to the decreased buoyancy of the atmosphere,

Figure 16.14 TGA (solid line) and DTG (dotted line) thermal curves of polyvinylchloride polymer heated under nitrogen. The first mass loss at 280°C is due to loss of HCl. (Courtesy of PerkinElmer, Inc. Shelton, CT, www.perkinelmer.com.)

changes in convection on heating and other complex factors, has been to a large extent eliminated in modern TGA instruments. The buoyancy effect can be evaluated and compensated for by running a blank (empty sample container) under the same conditions of heating and gas flow used for the samples. Errors can arise due to turbulence caused by the gas flow and due to convection on heating. Gas flow rates and heating rates should be kept as low as possible to minimize these effects.

Placement of the thermocouple is critical to accurate temperature measurement. Ideally, having the thermocouple in the sample itself would give the most accurate reading of the sample temperature. However, there are problems associated with putting the thermocouple into the sample. These include reaction with the sample, reproducible sample packing, sample mass, and thermal conductivity, among others. Modern instruments generally have the thermocouple in contact with the sample pan or close to the sample pan. The sample temperature is generally lower than the recorded temperature due to several factors including the finite heating rate, thermal conductivity of both the sample itself and the sample container, the gas flow rate and similar factors. There is also the heat of reaction to take into account. An endothermic reaction will cause self-cooling of a sample and therefore an even greater lag in the sample temperature than would otherwise occur, while an exothermic reaction will decrease the lag in sample temperature.

16.2. DIFFERENTIAL THERMAL ANALYSIS

Differential thermal analysis (DTA) is a technique in which the difference in temperature, ΔT, between the sample and an inert reference material is measured as a function of temperature. Both sample and reference material must be heated under carefully controlled conditions. If the sample undergoes a physical change or a chemical reaction, its temperature will change while the temperature of the reference material remains the same. That is because physical changes in a material such as phase changes and chemical reactions usually involve changes in **enthalpy**, the heat content of the material. Some changes result in heat being absorbed by the sample. These types of changes are called **endothermic**. Examples of endothermic changes include phase changes such as melting (fusion), vaporization, sublimation, and some transitions between two different crystal structures for a material. Chemical reactions can be endothermic, including dehydration, decomposition, oxidation–reduction, and solid–state reactions. Other changes result in heat being given off by the sample. Such changes are termed **exothermic**. Exothermic changes include phase changes such as freezing (crystallization), some transitions between different crystal structures and chemical reactions; decomposition, oxidation–reduction, and chemisorption can be exothermic. There are also physical changes that are not simple phase changes that still cause the sample temperature to change. Examples of such physical changes include adsorption and desorption of gases from surfaces and **glass transitions** in amorphous glasses and some polymers. The glass transition is a change in an amorphous material from a brittle, vitreous state to a plastic state. Glass transitions are second order phase transitions.

DTA and the related technique of DSC to be discussed later in the chapter are capable of measuring many types of physical and chemical changes that result in enthalpy changes. It is not necessary that the sample's weight change in order to produce a DTA response. However, if a weight change does take place, as occurs on loss of water, the enthalpy of the sample invariably changes, and a DTA response will be observed. So DTA is capable of measuring the same changes measured by TGA, plus many additional changes that

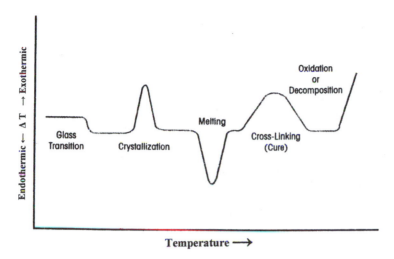

Figure 16.15 Hypothetical DTA thermal curve for a semicrystalline polymer with the ability to cross-link. The plot shows the baseline shift that occurs at the glass transition temperature, T_g, exothermic peaks for crystallization and cross-linking (or curing), an exothermic peak (offscale) for oxidative decomposition, and an endothermic peak for melting of the polymer. A similar thermal plot would be obtained by DSC analysis. (Courtesy of TA Instruments, New Castle, DE, www.tainst.com.)

TGA cannot measure because no mass change occurs. A DTA plot or thermal curve has ΔT on the y-axis and T (or time) on the x-axis, as shown schematically in Fig. 16.15. The x-axis temperature can be the temperature of the heating block, the temperature of the sample or the temperature of the reference, or it can be time. By convention, exothermic changes are plotted as positive, and the peaks point up, while endothermic changes are plotted as negative, and the peaks point down. (The same convention may be used for DSC; however, some instrument software uses the opposite convention. We will see examples of both conventions.) Some changes, such as the glass transition shown in Fig. 16.15, do not result in a peak, but only a step change in the baseline. The reason will be discussed later in the chapter.

16.2.1. DTA Instrumentation

The equipment used in DTA studies is shown schematically in Fig. 16.16. The sample is loaded into a crucible, which is then inserted into the sample well (marked S). A reference sample is made by placing a similar quantity of inert material (such as Al_2O_3) in a second crucible. This crucible is inserted in the reference well, marked R. The dimensions of the two crucibles and of the cell wells are as nearly identical as possible; furthermore, the weights of the sample and the reference should be virtually equal. The sample and reference should be matched thermally and arranged symmetrically with the furnace so that they are both heated or cooled in an identical manner. The metal block surrounding the wells acts as a heat sink. The temperature of the heat sink is slowly increased using an internal heater. The sink in turn simultaneously heats the sample and reference material. A pair of matched thermocouples is used. One pair is in contact with the sample or the sample container (as shown); the other pair is in contact with the reference. The output of the differential thermocouple, $T_s - T_r$ or ΔT, is amplified and sent to the data acquisition system. This allows the difference in temperature between the sample and

Single heat source

Figure 16.16 Schematic DTA instrument. (Courtesy of PerkinElmer, Inc. Shelton, CT, www. perkinelmer.com.)

the reference to be recorded as a function of either the sample temperature, the reference temperature or time. If there is no difference in temperature, no signal is generated, even though the actual temperatures of the sample and reference are both increasing. Operating temperatures for DTA instruments are generally room temperature to about 1600°C, although one manufacturer makes a DTA capable of operating from $-150°C$ to 2400°C. To reach the very low subambient temperatures, a liquid nitrogen cooling accessory is needed. Some low temperatures (but not $-150°C$) may be reached with electrical cooling devices or with forced air-cooling.

When a physical change takes place in the sample, heat is absorbed or generated. For example, when a metal carbonate decomposes, CO_2 is evolved. This is an endothermic reaction; heat is absorbed and the sample temperature decreases. The sample is now at a lower temperature than the reference. The temperature difference between the sample and reference generates a net signal, which is recorded. A typical example of a DTA thermal curve is shown in Fig. 16.15. If, in the course of heating, the sample undergoes a phase transition or a reaction that results in the generation of heat, that is, an exothermic reaction, the sample becomes hotter than the reference material. In this case, the sample heats up to a temperature higher than that of the reference material until the reaction is completed. The sample then cools down or the temperature of the reference cell catches up until its temperature and that of the heat cell once again become equal. Such an effect is shown on the thermal curve as a peak that moves in a positive (upward) direction rather than in the negative direction, which allows us to distinguish between exothermic and endothermic reactions. The DTA experiment is performed under conditions of constant pressure (usually atmospheric pressure). Under constant pressure, the change in heat content of a sample (the change in *enthalpy*) is equal to the heat of reaction, ΔH. Any chemical or physical change that results in a change in ΔH gives a peak in the DTA thermal curve. There are some types of changes that do not result in a peak in the thermal curve but only a change in the baseline, as shown schematically in Fig. 16.15. These types of changes do not undergo a change in ΔH, but a change in their *heat capacity*, C_p. Heat capacity is sometimes referred to as specific heat, and is the amount of heat required to raise the temperature of a given amount of material by 1 K. If the amount of material is specified to be 1 mole, the heat capacity is therefore the molar heat capacity, with units of J/mol K. The most common process that gives rise to a change in baseline but not a peak in the DTA is a "glass transition" in materials such as polymers or glasses. The glass transition is discussed briefly under applications of DTA.

Modern DTA instruments have the ability to change atmospheres from inert to reactive gases, as is done in TGA. As is the case with TGA, the appearance of the DTA thermal curve depends on the particle size of the sample, sample packing, the heating rate, flow characteristics inside the furnace, and other factors. Thermal matching between the sample and the reference is often improved by diluting the sample with the inert reference, keeping the total masses in each crucible as close to each other as possible.

The peak area in a DTA thermal curve is related to the enthalpy change for the process generating the peak, so DTA instruments must be calibrated for both temperature and for peak area. The National Institute for Standards and Technology (NIST), a US government agency, has certified high purity metals like indium, tin, and lead with melting points known to six significant figures and enthalpies known to three or four significant figures, and a series of high purity salts with solid state transition enthalpies which are accurately known. These materials can be used to calibrate both temperature and peak area.

Sample crucibles are generally metallic (Al, Pt) or ceramic (silica) and may or may not have a lid. Many metal pans with lids have the lid crimped on using a special tool. Best results are obtained when the area of contact between the sample and the pan or crucible is maximized. Samples are generally in the 1–10 mg range for analytical applications.

The peak area in DTA is related to the enthalpy change, ΔH, to the mass of sample used, m, and to a large number of factors like sample geometry and thermal conductivity. These other factors result in the area, A, being related to the mass and ΔH by an empirically determined calibration constant, K:

$$A = K(m)(\Delta H) \tag{16.1}$$

Unfortunately, K is highly temperature-dependent in the DTA experiment, so it is necessary to calibrate the peak area in the same temperature region as the peak of interest. This may require multiple calibration standards and can be time consuming. As we shall see, the calibration constant K for DSC is not temperature dependent; therefore DTA is usually used for qualitative analysis, while DSC is used for quantitative measurements of ΔH and heat capacity.

16.2.2. Analytical Applications of DTA

DTA is based on changes of heat flow into the sample. Using DTA, we can detect the decomposition or volatilization of the sample, just as we can with TGA. In addition, however, physical changes that do not involve weight changes can be detected by DTA. Such changes include crystallization, melting, changes in solid crystal phases, and homogeneous reactions in the solid state. In each of these changes there is a flow of heat between the sample and its surroundings caused by endothermic or exothermic transitions or by changes in the heat capacity. The main use of DTA is to detect thermal processes and characterize them as exothermic or endothermic, reversible or irreversible, but only qualitatively. DTA thermal curves can be used to determine the order of a reaction (kinetics), and can provide the information required to construct phase diagrams for materials. DTA can be used for characterization of engineering materials, for the determination of the structural and chemical changes occurring during sintering, fusing, and heat treatments of alloys to change microstructure, identification of different types of synthetic rubbers, and determination of structural changes in polymers.

An instance of the use of transitions where no change in weight occurs is the DTA characterization of polymers. The physical properties, such as strength, flexibility, and solubility, of a polymer depend (among other things) on its degree of crystallinity. Crystalline materials are those materials that exhibit a high degree of long-range and short-range order in the arrangement of their molecules or atoms. No polymers are 100% crystalline, but some polymers can partially crystallize; these are called **semicrystalline** polymers. A polymer is a gigantic organic molecule, or **macromolecule**, with a high molecular weight, typically 5000–40,000 g/mol. Polymer molecules generally exist as long and flexible chains. The chains are capable of bending and twisting, as shown in Fig. 16.17(a). A bulk polymer consists of large numbers of these chains, all intertwined like a bowl of cooked spaghetti noodles or rumpled pieces of string. A polymer in this state, with no short-range or long-range order, is said to be **amorphous**. With some types of polymers and proper treatment, such as slow cooling from the molten state, some of these long chains can form crystal-like zones that are regularly oriented [Fig. 16.17(b)]. The DTA curves of the two samples in Fig. 16.17 would appear as in Fig. 16.18 for samples that had been rapidly cooled (quenched) from the liquid state. At the glass transition temperature, T_g, a polymer goes from a glassy, rigid state to a rubbery state as the temperature now permits large-scale molecular motion. At the glass transition, there is a change in the heat capacity, C_p, of the polymer, which is seen as a step-change in the baseline of the thermal curve. There is no change in the enthalpy at T_g, so no peak occurs in the thermal curve. There is also a change in the rate of volume expansion of the polymer at T_g, which can be measured by thermomechanical analysis, TMA. On further heating, crystallization of the semicrystalline polymer occurs as the viscosity drops, molecular mobility increases, and the chains align themselves into ordered regions, resulting in an exothermic peak. The amorphous polymer shows no such exothermic peak. Additional heating will generally result in the melting of semicrystalline polymer, seen as an endothermic peak in the thermal curve. None of these transitions involves a change in mass of the sample, so none of these changes would be seen in a TGA thermal curve. Amorphous polymers do not melt, but may show an endothermic decomposition peak on continued heating. This decomposition would be seen by both DTA and TGA, since decomposition does involve weight loss. The degree of crystallinity in the semicrystalline polymer can be calculated from the heat of fusion (the area under the crystallization exotherm), but quantitative measurements of this sort are usually performed by DSC because DSC is more accurate than DTA for quantitative analysis.

Qualitative identification of materials is done by comparing the DTA of the sample to DTA thermal curves of known materials. DTA thermal curves serve as fingerprints for materials.

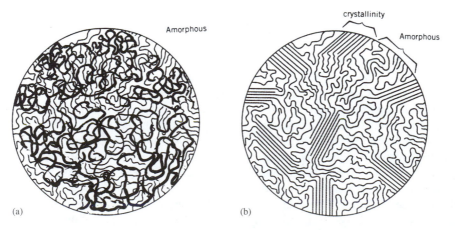

Figure 16.17 (a) Amorphous polymer structure (b) semicrystalline polymer structure with aligned chains in the crystalline regions and random structure in the amorphous region.

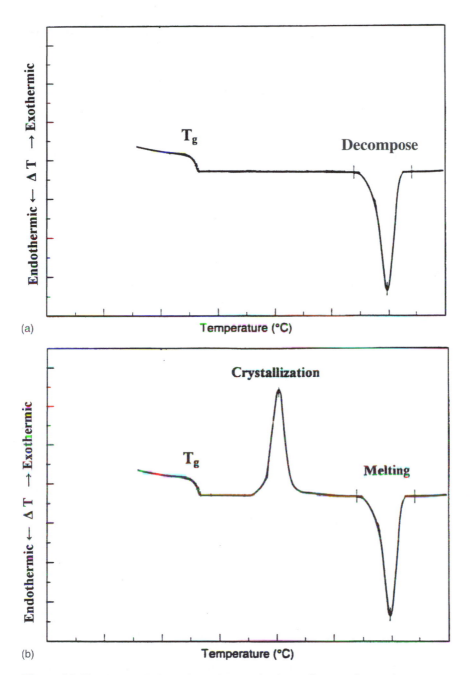

Figure 16.18 Schematic DTA thermal curves for the totally amorphous polymer structure and the semicrystalline polymer structure shown in Fig. 16.17. Both show T_g; only the semicrystalline polymer has a crystallization exotherm.

Theoretically, the area under each DTA peak should be proportional to the enthalpy change for the process that gave rise to the peak. Unfortunately, in a traditional simple DTA, there are many factors that are not compensated for. For example, DTA response generally decreases as the temperature increases. This makes the area under the peaks unreliable for enthalpy measurements unless the DTA has been calibrated in the temperature range of interest. Semiquantitative results for enthalpies can be obtained, but for quantitative enthalpy measurements, we turn to the DSC.

16.3. DIFFERENTIAL SCANNING CALORIMETRY

In DSC, differences in heat flow into a reference and sample are measured vs. the temperature of the sample. The difference in heat flow is a difference in energy; DSC is a calorimetric technique, and results in more accurate measurement of changes in enthalpy and heat capacity than that obtained by DTA.

16.3.1. DSC Instrumentation

The DSC measurement requires a sample and a reference, as does DTA. Modern DSC sample and reference pans are small and usually made of aluminum. They may or may not have lids. Sample size is generally $1-10$ mg. Often the reference pan is left empty, but an inert reference material such as is used in DTA may be used. Commercial DSC equipment can operate at temperatures from $-180°C$ to $700°C$, with specialized instruments capable of maximum temperatures of $1600°C$. The DSC must be able to be heated and cooled in a controlled manner. To achieve the very low end of the temperature range, a special liquid nitrogen cooling accessory is needed; for other cooling applications, electrical cooling or forced air cooling is used. Modern DSC instruments are available with automatic intelligent sample changers that permit the unattended analysis of as many as 50 samples or more in any order specified by the analyst.

There are two main types of DSC instrumentation, heat flux DSC and power compensated DSC. A schematic of a commercial **heat flux DSC** is presented in Fig. 16.19. In a heat flux instrument, the same furnace heats both the sample and the reference. In heat flux DSC, the temperature is changed in a linear manner while the differential heat flow into the sample and reference is measured. The sample and reference pans sit on the heated thermoelectric disk, made of a Cu/Ni alloy (constantan). The differential heat flow to the sample and reference is monitored by area thermocouples attached to the bottom of the sample and reference positions on the thermoelectric disk. The differential heat flow into the pans is directly proportional to the difference in the thermocouple signals. The sample temperature is measured by the alumel/chromel thermocouple under the sample position. This temperature is an estimated sample temperature because the thermocouple is not inserted into the sample itself. The accuracy of this temperature will depend on the thermal conductivity of the sample and its container, the heating rate, and other factors. As shown in Fig. 16.19, the sample and reference pans both have lids and the reference pan is an empty pan. A schematic of a **power compensated DSC** is presented in Fig. 16.20. The major difference in power compensated DSC instruments is that two separate heating elements are used for the sample (marked P in Fig. 16.20) and the reference. A change in temperature between the sample and the reference serves as the signal to "turn on" one of the heaters so that the sample and the reference stay at the same temperature. When a phase change, reaction, glass transition, or similar event occurs in the sample, the sample and reference temperatures become different. This causes extra power to be directed to the cell at the lower temperature in order to heat it. In this manner, the temperatures of the reference and sample cells are kept virtually equal ($\Delta T = 0$) throughout the experiment. The power and temperature are measured accurately and continuously. The temperatures of the sample and reference are measured using Pt resistance sensors, shown in Fig. 16.20. The difference in power input is plotted vs. the average temperature

Figure 16.19 Schematic of a heat-flux DSC. (Courtesy of TA Instruments, New Castle, DE, www.tainst.com.)

of the sample and reference. Power compensation provides high calorimetric accuracy, high precision, and high sensitivity. This permits analysis of very small samples as demonstrated in Fig. 16.21, showing the determination of the heat of fusion of a 6 μg sample of indium metal. Note that this figure is plotted opposite to the convention normally used for endothermic peaks.

The DSC peak area must be calibrated for enthalpy measurements. The same types of high purity metals and salts from NIST discussed for calibration of DTA equipment are also used to calibrate DSC instruments. As an example, NIST SRM 2232 is a 1 g piece of high purity indium metal for calibration of DSC and DTA equipment. The indium SRM is certified to have a temperature of fusion equal to 156.5985°C ± 0.00034°C and a certified enthalpy of fusion equal to 28.51 ± 0.19 J/g. NIST offers a range of similar standards. These materials and their certified values can be found on the NIST website at www.nist.gov. Government standards organizations in other countries offer similar reference materials.

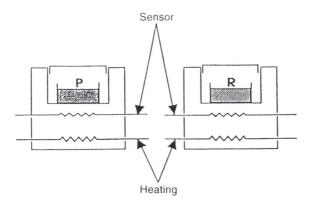

Figure 16.20   Schematic of a power-compensated DSC. (Courtesy of PerkinElmer, Inc. Shelton, CT, www.perkinelmer.com.)

The resultant thermal curve is similar in appearance to a DTA thermal curve, but the peak areas are accurate measures of the enthalpy changes. Differences in heat capacity can also be accurately measured and are observed as shifts in the baseline before and after an endothermic or exothermic event or as isolated baseline shifts due to a glass transition. Because DSC provides accurate quantitative analytical results, it is now the most used of the thermal analysis techniques. A typical DSC thermal curve for polyethylene tereph-thalate, the polymer used in many soft drink bottles, is shown in Fig. 16.22.

16.3.2. Applications of DSC

DSC is used to study all of the types of reactions and transitions that can be studied using DTA, with the added advantage of accurate quantitative measurements of ΔH and ΔC_p.

Figure 16.21   DSC thermal curve showing the determination of the enthalpy of vaporization of 6 µg of indium metal. (Courtesy of PerkinElmer, Inc. Shelton, CT, www.perkinelmer.com.)

Figure 16.22 DSC thermal curve for a sample of polyethylene terephthalate (PET). T_g is observed at 78.9°C. Crystallization begins at 137°C and the area under the exothermic peak is equivalent to 36.36 J/g PET. Melting begins at about 323°C; the area under the endothermic peak is equivalent to 40.13 J/g PET. (Courtesy of TA Instruments, New Castle, DE, www.tainst.com.)

Polymer chemists use DSC extensively to study percent crystallinity, crystallization rate, polymerization reaction kinetics, polymer degradation, and the effect of composition on the glass transition temperature, heat capacity determinations, and characterization of polymer blends. Materials scientists, physical chemists, and analytical chemists use DSC to study corrosion, oxidation, reduction, phase changes, catalysts, surface reactions, chemical adsorption and desorption (chemisorption), physical adsorption and desorption (physisorption), fundamental physical properties such as enthalpy, boiling point, and equilibrium vapor pressure. DSC instruments permit the purge gas to be changed automatically, so sample interactions with reactive gas atmospheres can be studied.

For example, from Fig. 16.22, the heat of crystallization is measured to be 36.36 J/g and the heat of melting (or fusion) is measured to be 40.13 J/g by integration of the respective peaks by the instrument data analysis software. For this polymer sample, the measured heat of crystallization is slightly lower than the measured heat of melting, indicating that the polymer was partly crystalline at the start of the experiment. The T_g and the specific heat can also be accurately measured from this thermal curve.

The **heat of fusion** is a useful measure of the **percent crystallinity** of polymers. Table 16.3 presents heat of fusion data and the calculated percent crystallinity for three

Table 16.3 DSC Results for Three Samples of Polyethylene

Sample	Melt onset (°C)	Melt peak (°C)	Enthalpy (J/g)	%crystallinity
1	121.9	132.9	195.9	67.6
2	121.3	132.6	194.5	67.1
3	122.3	131.6	180.1	62.1

Source: From TA Instruments Applications Brief TA-123, courtesy of TA Instruments, Inc., New Castle, DE, USA.

different samples of polyethylene. The percent crystallinity is calculated from calibration with either a 100% crystalline sample or a sample of known crystallinity. The first two samples in the table are identical in percent crystallinity and in their melting behavior; the third sample, however, has a lower percent crystallinity and a sharper melting profile (compare the melt onset and melt peak temperature columns). This tells the polymer chemist that sample 3 has been processed differently and will have different physical and mechanical properties than samples 1 and 2. The sharp melting of sample 3 is shown in Fig. 16.23.

16.3.2.1. Pressure DSC

Normally, DSC experiments are run at atmospheric pressure. Running samples under different pressures can extend the usefulness of DSC. DSC instruments are available with high-pressure cells capable of operating at up to 1000 psi (47 kPa). Changing the pressure will affect any reactions involving gases, while not having any significant effect on condensed phases. Boiling is the phase change of a material from liquid to gas. The boiling point of a material increases as the pressure increases, while the melting point (a phase change involving only solid and liquid) does not change significantly with pressure. Therefore, changing the pressure facilitates interpretation of peaks in a DSC thermal curve. If the peak shifts in temperature as a function of pressure, a gas is involved in the reaction that gave rise to the peak. Because the pressure affects the boiling point of a liquid, a series of experiments in a pressure DSC cell at known pressures yields boiling point shifts. These shifts can be used to obtain quantitative vapor pressure information on liquids.

Increasing the pressure will often increase the rate of a reaction involving gases. Adsorption of hydrogen is often used as a means of characterizing precious metal catalysts, such as the platinum catalysts used in automobile catalytic converters and the Pt and Pd catalysts used in large-scale organic chemical synthesis. A typical adsorption study may require large samples and take 6 h or more. The same type of study can be run in a pressure DSC cell in ∼15 min. The high pressure of hydrogen accelerates the reaction. The adsorption and

Figure 16.23   Melting point of a polyethylene sample by DSC. (Courtesy of TA Instruments, New Castle, DE. From Thermal Analysis Applications Brief TA-123.)

desorption of hydrogen on palladium as studied by DSC is shown in Fig. 16.24. Many sample–atmosphere reactions such as oxidation of oils and greases can be accelerated by increasing the pressure of air or oxygen and can also be studied effectively by pressure DSC. Other pressure sensitive reactions, such as volatilization, are slowed down by increased pressure. Suppression of these reactions can be a valuable tool when pressure DSC is used to study a complex system.

16.4. HYPHENATED TECHNIQUES

16.4.1. Hyphenated Thermal Methods

While TGA provides useful data when a mass change is involved in a reaction, we have seen that many reactions do not have a change in mass associated with them. The use of both TGA and DTA or TGA and DSC provides much more information about a sample than either technique alone provides. There are several commercial thermal analysis instrument manufacturers who offer simultaneous combination systems. Simultaneous TGA-DTA and simultaneous TGA-DSC instruments are available. Instrument combinations cover a wide temperature range and come in both "analytical sample" size (1–20 mg) and high-capacity sample size. A schematic of a simultaneous TGA-DTA instrument is presented in Fig. 16.25, showing the dual sample and reference pans. The instrument monitors both the weight change and the temperature difference between the sample and reference, and plots both the TGA and DTA thermal curves simultaneously. A hypothetical sample might have plots like those in Fig. 16.26, where two of the DTA peaks are clearly associated with the mass losses in the TGA, and there are two other "events" with enthalpy changes but no mass changes.

16.4.2. Evolved Gas Analysis

The evolution of gas from a thermal analyzer such as a TGA, DTA, or DSC may be determined using *evolved gas detection* (EGD) or, if qualitative or quantitative analysis of the gas is required, *evolved gas analysis* (EGA). These techniques are essentially a combination of thermal analysis and MS, tandem mass spectrometry (MS-MS), GC-MS or other

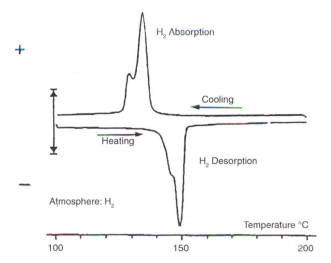

Figure 16.24 DSC thermal curve of adsorption and desorption of hydrogen from a precious metal catalyst under constant pressure.

TG/DTA Schematic

Furnace

Sample Holder   **Driving Coil**   **Deflection Sensor**

Temp.Signal

DTA Signal

Heater Power

Balance Circuit

TG Signal

Figure 16.25 A schematic simultaneous TGA-DTA instrument. A weight change on the sample side displaces the beam and a drive coil current returns the beam to zero. The current is proportional to the weight change and serves as the TGA signal. The temperature change between the sample and reference pans is measured by thermocouples attached to the pans. The temperature differential is the DTA signal. (Courtesy of ThermoHaake, Inc., Paramus, NJ, www.thermo.com, and Seiko Instruments, Japan.)

TGA

DTA

A   B   C   D

Figure 16.26 The TGA and DTA curves of a hypothetical sample. The two steps in the TGA curve result from weight losses. They correspond to the endothermic peaks A and C in the DTA plot. The DTA thermal curve also shows an exothermic event, B, and an endothermic event, D, that do not involve a change in mass.

species-selective detectors, such as FTIR. The evolved gases from the furnace are carried through a heated transfer line to the mass spectrometer or FTIR or other gas analyzer. This permits real-time identification of the gases given off by the sample during the thermal program. Figure 16.27(a) presents a commercial interface for capturing the

Purge Gas Flow
(from balance)

Transfer Line
To FTIR or Mass
Spectrometer

Sniffer Tube

Condensation Baffle
Protects balance
& hang-down wire

Gas Seperation
Zone

Evolved gases
drawn into Sniffer
tube

(a)

Reaction Gas Flow

Synergy System—TGA-FTIR/MS/GC

GC
Interface

FTIR
Spectrometer

TG
Interface

TGA

Thermal
Desorption

GC

MS

(b)

Figure 16.27 (a) The Sniffer® interface for evolved gas analysis. The Sniffer tube sits directly above the sample pan and draws evolved gases to both an FTIR and a GC-MS with minimal dilution. The interface is shown in a high-capacity TGA. (Courtesy of ThermoCahn, Madison, WI, www.thermo.com.) (b) The high-capacity TGA with Sniffer coupled simultaneously to an FTIR spectrometer and a GC-MS system. (Courtesy of ThermoCahn, Madison, WI, www.thermo.com.)

evolved gases from a TGA and sending evolved gas to both an FTIR and a GC-MS simultaneously [Fig. 16.27(b)]. This particular interface is used with a high-capacity TGA (up to 100 g sample size), so whole electronic components and fabricated parts can be analyzed. This can make failure analysis and production control very simple.

The alternative to real-time detection is to have a system for trapping the evolved gases and storing the trapped gases for analysis at a later time. Sorbents for collecting the evolved gases can be liquid or solid. The sorbents can be selective for a specific gas or a class of gaseous compounds. If no adsorbent material is desired, cryogenic trapping may be used.

The two most common instruments interfaced to thermal analyzers are FTIR instruments, discussed in Chapter 4 and MS instruments, discussed in Chapters 9 and 10. FTIR is less sensitive and less versatile than MS, but is simpler and cheaper. For hyphenated thermal analysis-FTIR, a heated transfer line from the thermal analyzer is connected to a heated FTIR gas cell, as shown in Fig. 16.28. The interface is relatively simple, because FTIR normally operates at ambient pressure as does the thermal analyzer. IR spectra are simple to interpret and reference libraries of gas phase IR spectra are available for the common gases and volatile organic compounds. This makes positive identification of the evolved gas straightforward if the gas is one of the gases that often accompany the decomposition of a material, such as water vapor, CO, or HCl, for example. The decomposition of PVC polymer can be studied by TGA-FTIR. Looking back at Fig. 16.14, we have the TGA thermal curve for PVC heated in nitrogen atmosphere. If the TGA is connected to an FTIR, the spectrum corresponding to the gas evolved at the first mass loss is the gas phase spectrum of HCl, Fig. 16.29, indicating that the first step in the decomposition of PVC in an inert atmosphere is loss of HCl from the polymer.

MS has more analytical flexibility than FTIR, but interfacing a thermal analyzer is difficult because of the low operating pressure required for MS. MS instruments typically operate at approximately 10^{-5} torr, while thermal analyzers are usually at atmospheric pressure. One approach is to evacuate the thermal analyzer, but the common method

Figure 16.28 Hyphenated TGA-FTIR, showing the heated transfer line and heated IR gas cell. (Courtesy of Netzsch Instruments, Inc., Paoli, PA.)

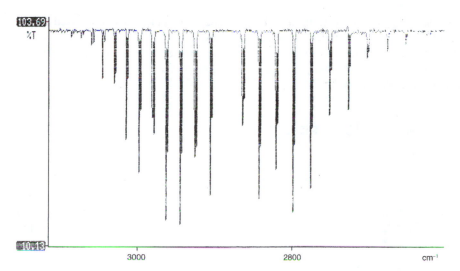

Figure 16.29 Gas phase FTIR spectrum of HCl. Spectrum collected on a PerkinElmer Paragon 1000 FTIR by Mr. Mark Bielaska, Rensselaer Polytechnic Institute, Troy, NY. Used with permission.

used is a differential pumping system such as that used for GC-MS. This reduces the pressure from the thermal analyzer in several stages prior to allowing the gas flow into the mass analyzer. A commercial interface for a thermal analyzer-MS system, shown in Fig. 16.30, uses a supersonic jet to skim analyte molecules into the MS, in a manner

Figure 16.30 Commercial "supersonic" interface between a thermal analyzer and a mass spectrometer. The mass spectrum of evolved gases (up to molecular weights of 1024 Da) can be obtained to provide identification of the structure with no condensation during transfer between the thermal analyzer and the mass spectrometer. (Courtesy of Setaram, SA, Caluire, France.)

similar to the jet separators used in GC-MS. Jet separator operation is discussed in the chapter on GC (Chapter 12) under hyphenated techniques.

Evolved gas analysis is used for materials characterization, polymer analysis, characterization of oil shale, oxidation and reduction studies, evaluation of catalysts, and many other applications.

16.5. THERMOMETRIC TITRIMETRY

Thermometric titration depends on measuring the heat generated during a chemical reaction; therefore it is another **calorimetric technique**. Usually the reactions take place at room temperature under conditions such that no heat enters or leaves the titration cell except that brought in by introduction of the titrant. In this procedure a titrant of known concentration is added to a known volume of sample. The titration reaction follows the generalized chemical equation

$$A \text{ (sample)} + B \text{ (titrant)} \longrightarrow C \text{ (products)} + \text{heat}$$

For any particular reaction, a mole of A titrated with a mole of B will generate a fixed quantity of heat, which is the heat of reaction. If there is half a mole of A present, half as much heat is generated, even if there is an excess of B. Heat is generated only while A and B react with each other; an excess of either one does not cause generation of heat. The reaction is usually set up in an insulated beaker, a Dewar flask, or a Christiansen adiabatic cell. This ensures a minimum of heat loss from the system during titration.

In practice, the temperature of the system is measured as the titrant is added to the sample. A thermometric titration plot generally has ΔT on the y-axis and volume of titrant on the x-axis. A typical plot for an exothermic reaction is shown in Fig. 16.31. We will

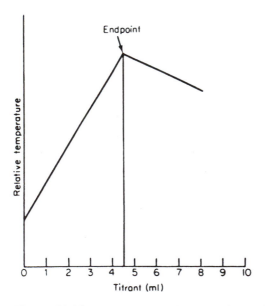

Figure 16.31 Representative thermometric titration plot. The endpoint is denoted by an abrupt change in the slope of the curve. The thermometric titration plot for HCl (a strong acid) with NaOH would look virtually identical to the thermometric titration plot for boric acid (a very weak acid) with NaOH because the ΔH of reaction is almost the same for both acids.

assume that the sample solution and the titrant are initially at room temperature. As titrant is added and the exothermic reaction occurs, the temperature of the solution increases. It can be seen that the temperature of the mixture increases until approximately 4.5 mL of titrant have been added to the sample. This is the equivalence point of the titration. Further addition of titrant causes no further reaction, because the entire sample A has been consumed. The temperature of the mixture steadily decreases as the cooler titrant is added to the warm mixture. The endpoint can therefore be determined as the abrupt change in slope that occurs at the equivalence point. In these circumstances it is not necessary to add exactly the stoichiometric amount of titrant. Any excess added does not result in an increase in temperature, and is therefore not a source of error as it is in conventional volumetric analysis.

For simple titrations the equipment necessary is not particularly sophisticated; a thermometer, a buret, and an insulated beaker (or even two Styrofoam cups, one sitting in the other) will do. However, if one wants to measure specific heats or the heat of reaction, better control is required. Modern automated thermometric titrators consist of a constant delivery pump for the titrant, a temperature control system for the titrant, an insulated cell, calibration circuitry, electronic temperature sensing, and a data processing system. Most modern instruments are totally computerized, so different methods can be programmed and run unattended.

The titrant is delivered with a motorized syringe pump. This permits the volume of titrant to be calculated from the rate of delivery. These pumps are able to deliver rates of flow down to microliters per minute with high precision. The temperature control system, usually a thermostatted bath for the titrant and a heater for the sample cell, is used to bring the titrant and the sample to exactly the same temperature. This is required for high precision measurements of heat capacity and enthalpy. Modern thermostats can maintain the temperature to within $0.001°C$.

For precise measurements of reaction parameters such as rate constants, equilibrium constants, and enthalpies, a well-designed insulated titration cell is required. The cell should have minimum heat loss to the surroundings. It should have as small a mass as possible to minimize the contribution of the cell to the total heat capacity and to maximize the response of the system to temperature changes. The calibration circuitry is used to determine the heat capacity of the system, which must be known if accurate heats of reaction are to be measured. Thermistors, which are temperature-sensitive semiconductors, are used as the temperature sensors in these systems due to their small size, fast response, and sensitivity.

16.5.1. Applications of Thermometric Titrimetry

A major use of thermometric titrimetry is for the titration of very weak acids or bases. The pH for the titration of a weak acid with a strong base (or vice versa) is easily calculated and such calculations can be found in standard analytical chemistry texts. Strong acids such as HCl, when titrated with strong base, give a large change in pH as the equivalence point is reached; therefore the quantitative determination of a strong acid using a potentiometric titration with a glass pH electrode is very simple. As the acid becomes weaker, the change in pH near the equivalence point decreases until the equivalence point becomes too shallow to detect by potentiometric titration. For example, the weak acid boric acid, with $pK_1 = 9.24$ and $pK_2 = 12.7$, does not give a good end point in a potentiometric titration with NaOH because the inflection is too small. However, in a thermometric titration with NaOH, boric acid gives a sharply defined end point, and is easily determined, because the ΔH for the boric acid neutralization is large. It is very similar in magnitude to that for neutralization of HCl by NaOH. Thermometric titrations can be used for quantitative analysis using neutralization reactions, oxidation-reduction reactions, and complexation reactions, as well as to determine heats of mixing and to determine equilibrium constants.

In thermometric titration it is assumed that the rate of reaction is relatively fast and that the endpoint will occur as soon as a small excess of titrant is added. This assumption is not valid if the reaction is slow. The position of the endpoint will be distorted and the results will be inaccurate.

16.6. DIRECT INJECTION ENTHALPIMETRY

When a chemical reaction occurs at constant pressure, heat is liberated or absorbed. The heat flow into or out of a system at constant pressure is quantified using a quantity called enthalpy, H. We usually measure the change in enthalpy, ΔH, called the enthalpy of reaction. It is a reproducible physical property for a given reaction $A + B \rightarrow C + \Delta H$. Therefore the magnitude of ΔH depends on the quantity of reactants involved in the reactions (e.g., multiplying all of the reaction coefficients by 2 means we have to multiply ΔH by 2). Excess amounts of any of the reactants do not take part in the generation or absorption of heat.

Direct injection enthalpimetry (DIE) is similar in many respects to thermometric titrimetry. One essential difference is that an excess of titrant is added very rapidly to the sample and the reactants mixed vigorously. The temperature is then measured against time following the injection of the titrant, as shown in Fig. 16.32. We may suppose that the following exothermic reaction takes place:

$$A \text{ (analyte)} + B \text{ (titrant)} \longrightarrow C \text{ (products)} + \text{heat}$$

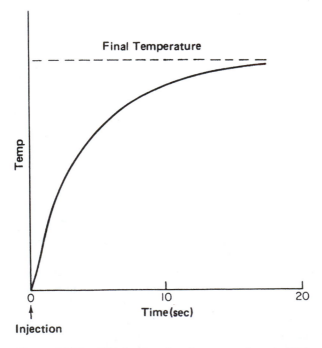

Figure 16.32   DIE titration plot. An excess of reactant B is added rapidly. The amount of analyte A is calculated from the final temperature.

The quantity of heat generated is a function of the number of moles that take part in the reaction. This, in turn, is controlled by the amounts of A and B. Any excess of either A or B does not react. In this instance, we have added an excess of B; therefore the amount of heat generated must be a function of the amount of A present in the mixture. If the amount of A present were halved, then the amount of heat generated would be halved. Since there is an excess of B present in both cases, only the amount of A affects the amount of heat generated.

The method is quite useful, particularly if the rate of chemical reaction between A and B is slow. The final quantity of heat evolved is not a function of time, but a function of the concentration of sample. This is a distinct advantage over conventional volumetric analysis and, in some instances, thermometric titrations. As stated earlier, slow reactions give rise to errors in the endpoint determination in thermometric titrations.

The equipment used for DIE is identical to that for thermometric titrimetry. The titrant must be at the same temperature as the sample at the start of the experiment, and the syringe is emptied rapidly into the cell to deliver the titrant "instantaneously".

DIE may be used for the same applications as discussed for thermometric titrations, for example, for the volumetric analysis of materials, such as boric acid, which are virtually impossible to titrate using endpoint indicators or pH indicators. DIE can also be used in biological studies where the reaction rates may be slow. For example, proteins have been titrated with acid or base, antibodies have been titrated with antigen, and enzyme–coenzyme systems have been studied. DIE is used to determine kinetic parameters for slow reactions. The use of a large excess of one reactant (the titrant) favors the forward reaction (according to Le Chatelier's principle) even if the equilibrium constant is small, so equilibria may be studied using DIE that cannot be studied using other titrimetric methods.

A word of caution should be mentioned about possible interferences, particularly if two simultaneous reactions take place. Since both reactions may generate heat, a direct error will be involved with injection enthalpimetry methods. However, with thermometric titrations, frequently consecutive reactions take place and can be determined consecutively without prior separation. This has been demonstrated in the titration of calcium and magnesium with oxalate.

16.7. THERMOMECHANICAL ANALYSIS AND DYNAMIC MECHANICAL ANALYSIS

Many materials are used under conditions that subject them to forces or loads. Examples include aluminum alloys used in airplanes, glass in windows and doors, polymers used in molded automobile bumpers, and polymer fibers in "stretch" clothing. The *mechanical* properties of these materials are very important; mechanical properties include ductility, stiffness, hardness and strength. (For an overview of mechanical properties of materials, see an introductory materials science text, such as the one by Callister listed in the bibliography.) The mechanical properties of a material are a function of the applied load, the duration of the load, and environmental conditions, including temperature. The thermal properties of materials are important to understand for the proper selection of materials that will function at elevated or subambient temperatures or for materials exposed to temperature gradients or thermal cycling. The response of mechanical behavior as a function of temperature is studied using **thermomechanical methods of analysis**.

Two important classes of thermomechanical methods of analysis are **thermomechanical analysis (TMA)** and **dynamic mechanical analysis (DMA)**. TMA measures changes in the physical dimensions of a sample as a function of temperature under an applied external load or stress. The dimensional change is generally measured in one direction (the length of the sample) with respect to the applied load; 1D changes are easier to measure than volume changes. The applied load is a static load in TMA. If the applied load is zero (i.e., no external load is applied), the TMA technique is called *thermodilatometry*. In the absence of an external load, changes in dimension as a function of temperature still can be measured. These changes are expressed as the thermal expansion coefficients of a material. TMA is used to measure expansion, compression, softening point, bending properties and extension (elastic and plastic deformation) for materials, especially polymers, composites, ceramics, and glasses. DMA applies an oscillating load to the sample. Deformation of the sample under changes in temperature and an alternating (dynamic) stress is measured as a function of time, temperature, load, and frequency. A great deal of data about the elastic and plastic behavior of materials is obtained from DMA. Table 16.4 presents a brief overview of some parameters measured by TMA and DMA and the information obtained from these measurements.

16.7.1. Instrumentation

16.7.1.1. TMA Equipment

A schematic of a commercial TMA instrument is shown in Fig. 16.33. The instrument consists of a dimensionally stable (with temperature) sample holder and measuring probe, a programmable furnace, a **linear variable displacement transducer (LVDT)** to measure the change in length, a means of applying force (load) to the sample via the probe (core rod, push rod), and a temperature sensor (usually a thermocouple).

The sample holder is generally made of quartz or steel. A sample is placed on the holder and the selected probe is placed in contact with the sample surface. A measured load is applied and the temperature program started. As the sample expands, softens, or

Table 16.4   Information Obtained from TMA and DMA experiments

Type of measurement	Information	Results
TMA Static load	Change in length Expansion Compression	Expansion coefficient Glass transition temperature Shrinkage
Thermodilatometry No external load	Penetration Extension Bending	Softening Penetration with given force, T Strain Static elastic modulus (Young's Modulus, i.e., the slope of the stress–strain curve in the elastic region) Crystalline morphological transitions

DMA
  Dynamic
  (alternating) load
	Deformation
	(cyclical change
	in length)
	Penetration
	Bending
	Extension
		Complex dynamic modulus
		Storage and loss modulus
		Mechanical loss ("tan delta")
		Glass transition temperature
		Polymerization, hardening, curing
		Post-polymerization cross-linking
		Decomposition

Source: Table courtesy of PerkinElmer, Inc, Shelton, CT (www.perkinelmer.com).

Figure 16.33 Schematic of a TMA. Three types of sample probes are shown with samples in place (dark colored areas are the samples). The expansion probe and sample are shown as they would be located in the furnace. (Courtesy of Netzsch Instruments, Inc., Paoli, PA.)

contracts, the position of the probe changes. The position change is measured by the LVDT, which provides a voltage proportional to the deflection of the probe.

The probe is usually also made of quartz or steel. Probes come in a variety of shapes, with the shape determining the type of dimensional change measured. Schematic diagrams of common TMA and DMA probes are presented in Fig. 16.34. A flat end on the probe is used for expansion measurements and the determination of linear expansion coefficients. Compression and penetration probes have a variety of rounded or pointed ends; the hemispherical probe is used to determine the modulus of soft, rubbery materials, for example. The three-point bend apparatus is a standard mechanical test for materials. A picture of a steel three-point bending sample holder is shown in Fig. 16.35. The sample holder has two knife-edge supports on which the sample is placed and a knife-edge probe through which the load is applied to the upper side of the sample. Tension or extension probes are used for measuring extension or shrinkage in thin films and fibers. The tension holder consists of a pair of grips, clamps or hooks, one fixed and one moveable, to pull the sample in tension under load. Figure 16.33 shows in the upper left corner the placement of a sample in expansion, tension, and penetration modes.

Loads of 0.1–200 g, or forces of 0.1–100 N may be selected and furnace temperatures may be programmed from −150°C to 1000°C, depending on the particular instrument. Samples of various sizes and shapes may be measured; lengths of 1 μm to 50 mm are common, with diameters in the 10 mm range, depending on the particular probe and instrument design.

TMA instruments must be calibrated for both temperature and dimensional motion. The temperature is calibrated using the same type of melting point standards used for DSC. High-purity In, Sn, Zn, and Al are often used. The LVDT is calibrated by using a standard with a well-known linear coefficient of thermal expansion, such as Al. The initial length of

Expansion

Compression

Bending

Extension

Figure 16.34   Representative sample probe shapes for TMA and DMA measurements. (Courtesy of PerkinElmer, Inc. Shelton, CT, www.perkinelmer.com.)

the standard is determined at room temperature and then the change in length of the standard with temperature is measured. The linear coefficient of thermal expansion, α, is defined by

$$\frac{L_f - L_0}{L_0} = \alpha(T_f - T_0) \tag{16.2}$$

Figure 16.35   On the right is a schematic of a three-point bend setup for TMA or DMA. On the far left is a photograph of a three-point bend probe made of quartz. The center photograph is of a similar probe made of stainless steel. (Courtesy of PerkinElmer, Inc. Shelton, CT, www.perkinelmer.com.)

where L_f is the final sample length, L_0 is the initial sample length, and T_f and T_0 are the final and initial temperatures (in °C). The calculated value of α is compared to the literature value for the material and a correction factor is applied to the data if needed.

16.7.1.2. DMA Equipment

Commercial DMA instruments vary in their design. One commercial instrument is shown in Fig. 16.36, set up for a three-point bend test under dynamic load. A different commercial instrument schematic, Fig. 16.37 shows a sample clamped between two arms that are free to move about the pivot points [Fig. 16.37(a)]; the electromagnetic drive and arm/sample assembly are shown in Fig. 16.37(b). The electromagnetic motor oscillates the arm/sample system and drives the arm/sample system to a preselected amplitude (strain). The sample undergoes a flexural deformation as seen in Fig. 16.37(a). An LVDT on the driver arm measures the sample's response to the applied stress, calculates the modulus (stiffness) and the damping properties (energy dissipation) of the material.

    DMA instruments have furnaces programmable from -170°C to 600°C. Frequency ranges can be varied from 0.001 to 1000 Hz, depending on the manufacturer. Probe types are the same as those for TMA.

16.7.2. Applications of TMA and DMA

TMA is used to measure expansion, compression, elastic relaxation, penetration, softening point, glass transitions, shrinkage, swelling, and elastic modulus (the stiffness, or resistance of a material to elastic deformation—the higher the modulus, the stiffer the material). TMA is widely used to characterize polymers, metals, coatings, ceramics, glasses, and composite materials. TMA has been used to measure the expansion behavior of printed circuit boards. As seen in Fig. 16.38, three different circuit board samples were measured using an expansion probe. Glass transition temperatures for all three can be determined from the change in slope, marked on the curves. One sample shows the melting of the tin solder used on this board, at about 180°C. At temperatures at or above 300°C, each

Figure 16.36   Schematic of a DMA analyzer. (Courtesy of Netzsch Instruments, Inc., Paoli, PA.)

SAMPLE DEFORMATION DMA

(a)

DMA

ELECTROMECHANICAL SYSTEM

(b)

Figure 16.37   Schematic of (a) a sample holder for DMA and (b) the DMA electromechanical system. (Courtesy of TA Instruments, New Castle, DE, www.tainst.com.)

board shows an abrupt expansion. This is the point at which the boards delaminate, caused by the explosive release of internal stress in the composite materials.

TMA is used to characterize polymer fibers. The characterization of polymer fibers is important in the textile industry. Processing (or texturing) the fibers by heating, spinning, twisting, and so on results in different types of stresses in the fibers. These can affect textile properties such as wrinkle-resistance and stiffness, so textile chemists and polymer chemists need to evaluate the effect of processing on spun and processed fibers. Figure 16.39 compares the fiber tension in an untextured (solid line) and textured (dotted line) PET fiber. These samples were 4–5 mm long and were clamped in a fiber extension probe for testing. Fiber tension is clearly lost during the particular texturing process for this sample. Figure 16.40 shows the behavior of three different polymer fiber samples. Two of the samples (#2 and #3) show severe shrinkage prior to melting,

Figure 16.38 TMA of three printed circuit board samples showing the glass transitions, the melting of solder on the middle sample, and the large change in dimension (expansion) due to delamination of the composite boards at high temperature. (Courtesy of Mettler-Toledo GmbH, Analytical, Schwerzenbach, Switerzerland, www.mt.com.)

due to relaxation of stresses induced during spinning. Fiber sample #1 is free of induced stress (no change in length until the melting point is reached).

Penetration mode can be used to measure coating thickness, such as the polymer coating inside beverage cans and the silicone release coating on the paper backing of "peel-off" labels. Penetration can be used to measure softening points of waxes and fats used in food and cosmetics. TMA in expansion mode is used to measure shrinkage of solid-state materials, and is useful in optimizing sintering conditions for materials

Figure 16.39 TMA curves of textured and untextured fibers measured in tension, showing the dramatic difference in behavior due to processing of the fiber. (Courtesy of TA Instruments, New Castle, DE.)

Figure 16.40 TMA curves of three types of polymer fiber, showing shrinkage due to stress relaxation before melting in polymers #2 and #3. (Courtesy of Netzsch Instruments, Inc., Paoli, PA.)

made from compacted powders, such as ceramics and UO_2 for nuclear fuel rods. TMA can study expansion and contraction of solid-state materials such as high-temperature superconductors to help understand the behavior of these materials. Figure 16.41 shows the difference in expansion behavior for a glass fiber-reinforced composite. The expansion is very different (much larger) in the direction perpendicular to fiber alignment than in the direction along the fibers, an important behavior for the end user of this composite to know.

DMA provides materials scientists, polymer chemists, and design engineers with detailed information on the elastic and inelastic (plastic) deformation of materials, modulus, and damping (energy dissipation) properties of materials. The damping behavior of synthetic and natural rubbers is important in vibration and acoustic application, for example.

The mechanical modulus can be measured under different types of load-shear, flexure, and so on. The measured modulus can be expressed as the storage modulus (related to the

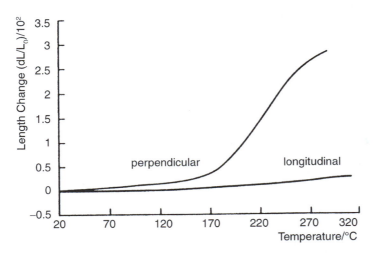

Figure 16.41 TMA curve of a fiber composite material. Thermally induced expansion differs depending on the direction of measurement because the strength of the material is related to the orientation of the fiber layers. (Courtesy of Netzsch Instruments, Inc., Paoli, PA.)

stiffness of the material) and the loss modulus (related to the viscous behavior or softness of the material) under the various types of loads. DMA is used to follow the curing of polymers, crosslinking of polymers, and to measure glass transition temperatures. A recent application of DMA is in the characterization of food. DMA is sensitive and can handle larger samples

(a)

(b)

Figure 16.42 DMA plots of (a) a sample of fresh bread and (b) a sample of "stale" bread. The changes in the modulus and tan δ in the fresh bread sample are associated with water in the sample, while the overnight exposure to air of the stale sample has caused it to dry out. The room temperature modulus for the stale sample is greater than that for the fresh sample, indicating that the stale bread is stiffer than the fresh bread. (Courtesy of TA Instruments, New Castle, DE, from Thermal Analysis Applications Brief TA-119.)

than DSC, so its use for inhomogeneous materials like food should have advantages. Samples of commercial white bread were studied using DMA. A fresh sample was compared with a sample exposed to the air overnight. (You know what happens to bread left "out"—it gets dry and stiff, that is, brittle.) The DMA plots of storage modulus, loss modulus, and tan δ (tan delta) for both samples are shown in Fig. 16.42(a) and (b). For the fresh sample, the changes in moduli and tan δ are postulated to be associated with changes in the physical state of water in the bread. For example, the tan δ "doublet" at about 0°C in the fresh sample could be melting of two different forms of bound water or it could be a surface/interior water phenomenon. In any case, the "stale" bread is clearly lacking many of the features seen in the fresh sample, as we might expect if the stale sample has lost the water responsible for the features. The room temperature modulus for the fresh sample was measured to be 2 MPa, while the stale bread was 16 MPa; the increase in modulus correlates with the increased stiffness of stale bread. In a related experiment, the moduli of cooked pasta were calculated from a series of DMA experiments. The plot of the storage modulus, Fig. 16.43, shows a sharp drop at about 8 min, indicating that the pasta has become "limp". The DMA data could be used by pasta manufacturers to suggest optimum cooking times to their customers. (Details of these food experiments are from TA Instruments Thermal Analysis Applications Brief Number TA-119, available for downloading from the website at www.tainst.com.)

16.8. SUMMARY

Thermometric methods of analysis have wide analytical applications. They can be used not only for the identification of chemical compounds, but also for the quantitative analysis of mixtures of compounds and for characterizing many types of engineering materials such as

Figure 16.43 The storage modulus, G', of cooked pasta measured by DMA. The modulus drops sharply at about 8 min, indicating that the pasta has lost stiffness (or become limp). (Courtesy of TA Instruments, New Castle, DE, from Thermal Analysis Applications Brief TA-119.)

polymers, glasses, ceramics, alloys, and composites. Physical and chemical properties of samples can be measured, including fundamental properties like equilibrium constants, rate constants, phase diagrams, thermal expansion coefficients, enthalpies, and specific heats. The fields of thermometric titrations and direct injection enthalpimetry have distinct advantages over conventional volumetric analysis and can be used to advantage in the analysis of biological systems. Thermomechanical testing complements room temperature tensile testing for materials and is extremely important in the evaluation of the physical behavior of a wide variety of engineering materials.

BIBLIOGRAPHY

Brown, M.E. *Introduction to Thermal Analysis: Techniques and Applications*; Chapman and Hall: New York, 1988.

Callister, W.D. *Materials Science and Engineering: An Introduction*, 5th ed.; John Wiley and Sons: New York, 2000.

Connolly, M.; Tobias, B. *Am. Lab.* **1992**, *1*, 38.

Duval, C. *Inorganic Thermogravimetric Analysis*, 2nd ed.; American Elsevier: New York, 1963.

Earnest, C. Modern thermogravimetry. *Anal. Chem.* **1984**, *November*, 1471A.

Gallagher, P.K. Thermoanalytical methods. In *Materials Science and Technology, A Comprehensive Treatment, Part 1* Cahn, R.W.; Haasen, P.; Kramer, E.J.; Eds.; VCH: Weinheim, 1992; Vol. 2A.

Groves, I.; Lever, T.; Hawkins, N. *TA Instruments Applications Brief TA-123*; TA Instruments, Ltd: UK.

Pazstor, A.J. Thermal analysis techniques. In *Handbook of Instrumental Techniques for Analytical Chemistry*; Settle, F.A. Ed.; Prentice-Hall, Inc.: NJ, 1997.

Speyer, R.F. *Thermal Analysis of Materials*; Marcel Dekker, Inc.: New York, 1994.

Turi, E.A., Ed. *Thermal Characterization of Polymeric Materials*, 2nd Ed.; Academic Press: New York, 1996.

Tyrell, H.V.; Beezer, A.E. *Thermometric Titrimetry*; Chapman and Hall: London, 1968.

Utschick, H. *Methods of Thermal Analysis: Applications from Inorganic Chemistry, Organic Chemistry, Polymer Chemistry and Technology*; Ecomed Verlagsgesellschaft AG and Co KG: Landsberg, Germany, 1999.

Washall, J.W.; Wampler, T.P. Direct-pyrolysis Fourier transform-infrared spectroscopy for polymer analysis. *Spectroscopy* **1992**, *6* (4), 38.

Wendlandt, W.W. *Thermal Methods of Analysis*, 3rd ed.; John Wiley and Sons, Inc.: New York, 1986.

Wunderlich, B. *Thermal Analysis*; Academic Press: Boston, MA, 1990.

SUGGESTED EXPERIMENTS

16.1 Using a TGA instrument determine the TGA thermal curve for copper sulfate pentahydrate. A sample size of approximately 10 mg should be heated in nitrogen from room temperature until at least 300°C. Explain the thermal plot in terms of loss of the water of hydration.

16.2 Repeat experiment 16.1 using KCl. Note the initial loss of free (adsorbed) water and then the loss of bound interstitial water up to temperatures above 200°C. Could temperatures below 200°C be used to dry KCl crystals successfully?

16.3 Using a solution of $AlCl_3$ (in acid), neutralize and precipitate the Al as $Al(OH)_3$ by the addition of NH_4OH. Filter the precipitate. Obtain the TGA thermal curve of the precipitate. Heat the precipitate until no change in mass occurs (to at least 1000°C). Could this precipitate be dried to a constant weight at any temperature below 1000°C?

16.4   Take a fresh piece of ham (about 10 mg for an analytical TGA). Obtain the TGA thermal plot in air up to 150°C. Note the loss in weight due to (a) water evaporation. Do any other losses occur? What is being lost, do you think? Run six additional samples of the same ham. Evaluate the accuracy and precision of the water loss measurement and explain your observations. Many commercial hams (and other meats) contain added water or broth. State regulations may specify how "water-added" products are labeled. Would there be an advantage to a high-capacity TGA if you had to determine if the amount of water in hams met the labeling regulations on a routine basis?

16.5   Obtain the TGA curve of butanol. Heat slowly. Note the slow evaporation of the butanol throughout and then the rapid loss at the boiling point.

16.6   Using about 10 mg samples of polyvinylchloride (PVC) powder, obtain the TGA thermal curves (to at least 700°C) in air and in nitrogen. Compare the curves. The first loss in mass was described in the text. What is it due to? Explain any other mass losses and note any differences in the two curves.

16.7   Using DTA or DSC equipment, obtain the thermal curve of benzoic acid. In the first instance, heat the acid rapidly; in the second, heat it slowly. Note the difference in the apparent melting and boiling ranges of the benzoic acid when the different heating rates are used. Which rate gives the more accurate data?

16.8   Obtain the DTA (or DSC) and TGA curves of benzoic acid. Note that the TGA curve does not indicate the melting point but does indicate the boiling point.

16.9   Obtain the DSC curves for samples of commercial butter, margarines, lard, and solid butter substitutes. The temperature range for the DSC should be from $-50°C$ to $150°C$. Compare the melting behavior and explain your observations. You can also run this experiment with various types of solid chocolate.

16.10   Obtain DSC thermal curves of several semicrystalline polymers such as polymethylmethacrylate (PMMA), polystyrene, polycarbonate, high-density polyethylene, low-density polyethylene and look for the glass transition in these polymers. The DSC run may need to be repeated twice with rapid cooling between runs. Many "as received" polymers will show a small peak on top of the glass transition on the first run due to "relaxation effects" in the polymer. The second run should not show this "peak", but only a step change in the baseline. Compare your values of T_g to literature values. Deviations may indicate the presence of plasticizers or other additives in the polymer.

PROBLEMS

16.1   Describe the major components of a TGA instrument. Draw a schematic for a simple TGA instrument. What is measured by the TGA experiment?

16.2   A sample of mixed calcium oxalate monohydrate and anhydrous silica weighed 7.020 mg. After heating to 600°C, the weight of the mixture was 6.560 mg. What was the weight of calcium oxalate in the original sample?

16.3   What information can be obtained by DTA that cannot be obtained by TGA?

16.4   Describe the components of a DTA instrument and discuss the differences between a DTA and a TGA instrument.

16.5   What are the advantages of using a combination of DTA and TGA to characterize a sample?

16.6   Describe the components of a DSC instrument. Draw a schematic of a heat-flux DSC. What is measured in DSC?

16.7 Make a list of the types of changes that can give exothermic peaks in DSC and DTA. Make a list of the types of changes that can give endothermic peaks in DSC and DTA.

16.8 What applications can you find for DSC that cannot be performed by TGA?

16.9 Why are Pt and quartz commonly used for sample holders in thermal analysis?

16.10 The following figure shows the TGA thermal curves of a glass-fiber filled polymer (solid line) and the nonfilled polymer (dotted line). (a) If the original sample weight for the glass-fiber filled polymer was 23.6 mg, what is the weight% glass fiber in the sample? (b) What would you recommend as the upper temperature limit for the use of this material, assuming you want to build in a safety factor of 20% (i.e., you want to be no higher than 80% of the decomposition temperature)?

16.11 The following figure shows the TGA thermal curves for copper sulfate pentahydrate, $CuSO_4 \cdot 5H_2O$. The solid line is the TGA of the crystalline salt. (a) Assuming that all of the losses are due to water, how many different types of water are present? (b) How many moles of water are lost in each step? (c) Does the total mass loss confirm that this is $CuSO_4 \cdot 5H_2O$? The dotted line is the TGA of the crystalline salt that has been stored in a desiccator with magnesium perchlorate as the desiccant. (d) Comment on the differences observed between the two thermal curves. (e) How much water has the desiccant removed from the compound?

16.12 Shown in the following figure are the TGA (solid line) and the DTG (dotted line) thermal curves for a sample of hydrated citric acid, $C_6H_8O_7 \cdot nH_2O$. The formula for anhydrous citric acid is $C_6H_8O_7$. The mass of the sample used is 20.09 mg. (a) Assuming the first mass loss is the loss of water, how many moles of water of hydration are present? (What is the value of n in the formula $C_6H_8O_7 \cdot nH_2O$?) (b) What decomposition temperature would you report for anhydrous citric acid? (c) What drying temperature would you recommend for producing anhydrous citric acid from the hydrated form? Which thermal curve is more useful for assigning the temperatures in (b) and (c)?

16.13   The DTA thermal curves for two polyolefins are shown in the following figure. Curve 1 is a sample of low-density polyethylene. Curve 2 is a sample of stabilized polypropylene. The melting points, T_m, are shown. (a) What are the melting points for each polymer? (b) Which polymer would be better to use for hot water pipes in houses? Why? (c) What process is occurring at the high temperature end of each thermal curve?

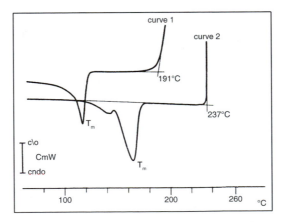

16.14   The DSC thermal curve of a sample of "light" margarine is shown. (*Note*: This plot has the endothermic peaks pointing up!) (A) What process and what substance do you think gives rise to the endothermic peak at $4.60°C$? (Think about what might be added to margarine to reduce the calories.) (B) Is the endothermic peak at $-24.6°C$ due to the melting of the oil used to make the margarine? Explain your answer.

(*Note*: The figures for Problems 16.10, 16.11, 16.12, and 16.14 are courtesy of PerkinElmer, Inc, Shelton, CT. The figure for Problem 16.13 is courtesy of Mettler-Toledo GmbH, Switzerland.)

Acronyms Index

Instrumentation for Analytical Chemistry gives rise to many abbreviations forming "acronyms". These are often more encountered than the terms they abbreviate, and they appear extensively in the text. A reader new to the field may become lost or disoriented in this thicket of initials. To aid the student in reading the text, the **Acronym Index** below translates these and indicates the chapter and page where they are best defined or characterized. These acronyms are frequently compounded, as in UV/VIS (ultraviolet/visible) or LC-CI-TOFMS (Interfaced Liquid Chromatograph to Time-of-Flight Mass Spectrometer operating in Chemical Ionization mode). The components of such compounded acronyms are listed individually in the index, but not all the possible combinations.

Acronym	Chap.	Page	Abbreviated Term
AAS	6	385	Atomic Absorption Spectrometry
AC	15	935	Alternating Current [or voltage]
AED	12	780	Atomic Emission Detector
AES	14	897	Auger Electron Spectroscopy
AES	7	449	Atomic Emission Spectroscopy
AFS	7	516	Atomic Fluorescence Spectroscopy
AOAC	1	14	Association of Official Analytical Chemists
APCI	9	625	Atmospheric Pressure Chemical Ionization
APDC	6	428	Ammonium Pyrrolidine Dithiocarbamate
APHA	1	14	American Public Health Association
API	5	357	American Petroleum Institute
APT	3	162	Attached Proton Test
ASE	1	46	Accelerated Solvent Extraction
ASTM	1	14	American Society for Testing and Materials
ATR	4	250	Attenuated Total Reflectance
BTEX	12	764	Benzene, Toluene, Ethylbenzene, Xylenes
CAT	14	885	Constant-Analyzer Transmission
CAT	8	535	Computed Axial Tomography
CCD	4	295	Charge Coupled Device [or Detector]
CE	13	859	Capillary Electrophoresis
CE	15	979	Counter Electrode [in Electrochem.]
CEC	13	863	Capillary Electrochromatography

Acronym	Chap.	Page	Abbreviated Term
CEM	9	645	Channel Electron Multiplier
CFFAB	9	631	Continuous Flow-FAB
CGE	13	860	Capillary Gel Electrophoresis
CI	9	624	Chemical Ionization
CID	4	295	Charge Injection Device [or Detector]
CID	13	826	Collision Induced Dissociation
CIEF	13	862	Capillary Isoelectric Focusing
CL	1	33	Confidence Level
cm	2	65	centimeter [length]
CMA	14	884	Cylindrical Mirror Analyzer
CMC	13	864	Critical Micelle Concentration
COSY	3	162	Correlated Spectroscopy
CP-MAS	3	189	Cross-Polarization-MAS
CRR	14	885	Constant Relative Ratio
CT	8	535	Computed Tomography
CTD	7	474	Charge Transfer Device
CVAAS	6	431	Cold Vapor-AAS
CZE	13	851	Capillary Zone Electrophoresis
DAD	13	815	Diode Array Detector
DCP	7	486	Direct Current Plasma
DEPT	3	162	Distortionless Enhancement by Polarization Transfer
DID	12	779	Discharge Ionization Detector
DIE	16	1038	Direct Injection Enthalpimetry
DIN	7	493	Direct Insertion Nebulizer
DMA	16	1039	Dynamic Mechanical analysis
DMD	13	815	Dispersive Monochromator Detector
DME	15	976	Dropping Mercury Electrode
DMSO	4	287	Dimethylsulfoxide
DNA	10	692	Deoxyribonucleic Acid
DRC	10	708	Dynamic Reaction Cell (tm)
DRIFTS	4	252	Diffuse-Reflectance-IR-FT-Spectroscopy
DSC	16	1026	Differential Scanning Calorimetry
DTA	16	1020	Differential Thermal Analysis
DTG	16	1017	Derivative Thermogravimetry
DTGS	4	240	Deuterated Triglycine Sulfate
ECD	12	774	Electron Capture Detector [in GC]
ECD	13	817	Electrochemical Detector [in HPLC]
ED	8	547	Energy Dispersive [with XRF]
EDL	6	392	Electrodeless Discharge Lamp
EDTA	15	937	Ethylenediaminetetraacetic Acid
EE^+	10	667	Even Electron [ion in MS]
EGA	16	1031	Evolved Gas Analysis
EGD	16	1031	Evolved Gas Detection
EI	9	622	Electron Ionization
EIE	7	499	Easily Ionized Element
ELCD	12	776	Electrolytic Conductivity Detector

Acronym	Chap.	Page	Abbreviated Term
ELSD	13	811	Evaporative Light Scattering Detector
EM	9	645	Electron Multiplier
EM	14	914	Electron Microprobe
EMA	8	593	Electron Microprobe Analyzer
EMF	15	924	Electromotive Force [voltage]
EOF	13	853	Electroosmotic Flow
EPA	1	14	Environmental Protection Agency
EPMA	14	914	Electron Probe Microanalyzer
ESCA	14	881	Electron Spectroscopy for Chemical Analysis
ESI	9	625	Electrospray Ionization
ETA	6	396	Electrothermal Atomizer
ETV	10	699	Electrothermal Vaporization
FAAS	6	402	Flame-AAS
FAB	9	629	Fast Atom Bombardment
FAT	14	885	Fixed-Analyzer Transmission
FIA	6	433	Flow Injection Analysis
FID	3	130	Free Induction Decay [in NMR]
FID	12	773	Flame Ionization Detector [in GC]
FPA	4	306	Focal Plane Array
FPC	9	647	Focal Plane Camera
FRR	14	885	Fixed Retardation Ratio
FT	1	55	Fourier Transform
FTICR	9	644	=FT-MS
FTIR	4	230	Fourier Transform IR
FTMS	9	644	=FT-ICR-MS
FTNMR	3	117	Fourier Transform-NMR
FWHM	8	562	Full-Width at Half-Maximum
GC	11	723	Gas Chromatography
GD	7	506	Glow Discharge
GDMS	10	698	—GD-MS
GFAAS	6	396	Graphite Furnace-AAS
GFC	13	845	Gel Filtration Chromatography
GF-LEAFS	7	520	Graphite Furnace Laser-Excited AFS
GPC	13	845	Gel Permeation Chromatography
HCL	6	390	Hollow Cathode Lamp
HEPS	7	468	High Energy Prespark
HETCOR	3	162	Heteronuclear Chemical Shift Correlation
HETP	11	723	Height Equivalent to a Theoretical Plate
HGAAS	6	432	Hydride Generation-AAS
HID	12	779	Helium Ionization Detector
HMDE	15	992	Hanging Mercury Drop Electrode
HPLC	13	797	High Performance [or Pressure] Liquid Chromatography
Hz	2	66	Hertz [frequency]
IC	13	839	Ion Chromatography
ICP	7	483	Inductively Coupled Plasma
ICR	9	642	Ion Cyclotron Resonance
IDMS	12	784	Isotope-Dilution Mass Spectrometry

Acronym	Chap.	Page	Abbreviated Term
IEF	13	870	Isoelectric Focusing
IMMS	14	914	Ion Microprobe Mass Spectrometry
INADEQUATE	3	162	Incredible Natural-Abundance Double-Quantum Transfer
IPC	13	836	Ion-Pairing Chromatography
IR	4	213	Infrared
IRE	4	250	Internal Reflection Element
IS	11	739	Internal Standard
ISE	15	938	Ion Selective Electrode
ISFET	15	944	Ion-Selective Field-Effect Transistor
ISS	14	906	Ion Scattering Spectroscopy
IUPAC	15	928	International Union of Pure and Applied Chemistry
JEOL	10	691	Japan Electron Optics Laboratory
KLL	14	898	Auger Peak-Energy Level Designator
LC	11	723	Liquid Chromatography
LIMS	1	21	Laboratory Information Management System
LMM	14	890	Auger Peak-Energy Level Designator
LOD	1	58	Limit of Detection
LOQ	1	59	Limit of Quantitation
LPDA	5	339	Linear Photodiode Array
LVDT	16	1040	Linear Variable Displacement Transducer
m	2	65	meter [length]
MALDI	9	626	Matrix-Assisted Laser-Desorption Ionization
MAS	3	178	Magic Angle Spinning
MCA	8	568	Multichannel Analyzer
MCP	14	886	Microchannel Plate
MCT	4	240	Mercury Cadmium Telluride
MEKC	13	863	Micellar Electrokinetic Chromatography
MIBK	1	45	Methyl Isobutyl Ketone
MIP	7	488	Microwave Induced Plasma
MIT	7	478	Massachusetts Institute of Technology
MNN	14	898	Auger Peak-Energy Level Designator
MRI	3	117	Magnetic Resonance Imaging
MS	9	613	Mass Spectrometry
MSA	2	84	Method of Standard Additions
MW	10	660	Molecular Weight
NIR	4	285	Near Infrared
NIRIM	4	307	NIR-Raman Imaging Microscope
NIST	16	1027	National Institute for Standards and Technology
nm	2	65	nanometer [length]
NMR	3	117	Nuclear Magnetic Resonance
NOE	3	176	Nuclear Overhauser Effect
NOESY	3	162	NOE-Spectroscopy
NPD	12	777	Nitrogen-Phosphorous Detector
NRC	12	775	Nuclear Regulatory Commission
ODS	13	798	Octadecylsilyl
$OE^{\cdot+}$	10	667	Odd Electron [ion in MS]

Acronym	Chap.	Page	Abbreviated Term
OES	7	449	Optical Emission Spectroscopy
PAGE	13	870	Polyacrylamide Gel Electrophoresis
PBDE	10	693	Polybrominated Diphenyl Ether
PCB	1	29	Polychlorinated Biphenyl
PDA	4	295	Photodiode Array
PDMS	12	757	Polydimethylsiloxane
PET	16	1045	Polyethylene Terephthalate
PID	12	777	Photoionization Detector
PIXE	8	548	Particle Induced X-ray Emission
PLOT	12	759	Porous Layer Open Tubular
PMT	5	334	Photomultiplier Tube
POP	12	774	Persistant Organic Pollutant
PVC	16	1018	Polyvinyl Chloride
QIT	9	642	Quadrupole Ion Trap
REE	10	702	Rare Earth Elements
REF	15	947	Reference Electrode
RF	3	117	Radiofrequency
RGA	10	693	Residual Gas Analyzer
RI	12	787	Retention Index [in GC]
RI	13	810	Refractive Index [as in LC detector]
RNA	10	692	Ribonucleic Acid
RP	13	798	Reverse Phase
RRF	11	742	Relative Response Factor
RRT	11	739	Relative Retention Time
RU	7	480	raies ultimes
S/N	1	53	Signal-to-Noise [ratio]
SAM	14	903	Scanning Auger Spectroscopy
SCD	12	777	Sulfur Chemiluminescence Detector
SCD	7	490	Segmented Charge Detector
SCE	15	931	Saturated Calomel Electrode
SCF	13	848	Supercritical Fluid
SDS	13	864	Sodium Dodecyl Sulfate
SEC	13	845	Size Exclusion Chromatography
SEM	8	593	Scanning Electron Microscope
SERS	4	302	Surface-Enhanced Raman Spectroscopy
SFC	13	848	Supercritical Fluid Chromatography
SHE	15	925	Standard Hydrogen Electrode
SI	1	3	Système International d'Unités
SIM	12	783	Selected Ion Monitoring
SIMS	14	908	Secondary Ion Mass Spectrometry
SPADNS	5	361	see end of Table 5.11
SPE	1	48	Solid Phase Extraction
SP-FPD	12	777	Sulfur-Phosphorous Flame Photometric Detector
SPME	12	754	Solid Phase Microextraction
SRM	10	704	Standard Reference Material
TBAH	15	956	Tetrabutylammonium Hydroxide

Acronym	*Chap.*	*Page*	*Abbreviated Term*
TCD	12	771	Thermal Conductivity Detector
TFME	15	992	Thin Film Mercury Electrode
TGA	16	1004	Thermogravimetric Analysis
TIC	12	782	Total Ion Chromatogram
TISAB	15	957	Total Ionic Strength Adjusting Buffer
TLC	13	866	Thin Layer Chromatography
TMAH	15	975	Tetramethylammonium Hydroxide
TMA	16	1039	Thermomechanical Analysis
TMS	3	131	Tetramethylsilane
TOF	9	635	Time-of-Flight [in mass spectrometry]
UHV	14	887	Ultra High Vacuum
μm	2	65	micrometer [length]
USN	7	493	Ultrasonic Nebulizer
UV	5	317	Ultraviolet
VIS	5	317	Visible [light wavelength range]
WD	8	547	Wavelength Dispersive [with XRF]
WE	15	947	Working Electrode
XIC	12	783	Extracted Ion Chromatogram
XPS	14	881	X-ray Photoelectron Spectroscopy
XRD	8	546	X-ray Diffraction
XRF	8	544	X-ray Fluorescence
YAG	5	373	Yttrium Aluminum Garnet
ZPD	4	231	Zero Path Difference

Index

2D techniques
  gas chromatography, 724–5, 780, 785–6
  nuclear magnetic resonance, 152, 180–8
  planar capillary gel electrophoresis, 870–1
  thin layer chromatography, 724–5, 869
AAS *see* Atomic absorption spectrometry
Absorbance, 77–81
Absorption, 69
  edges, 544
  laws, 76–81
  NMR line width, 125–8
  spectra, 71
Absorptivity, 78–9, 325–6
Accelerated solvent extraction (ASE), 46–7
Accuracy, 24–5, 364
Acetic acid, 170–1, 172
Acetone, 274
Acetylene, 672, 674
Acid anhydrides, 274–5
Acid dissolution, 40–3
Acid errors, 954
Activity series, 930–1
Adenosine triphosphate (ATP), 195–6
AED *see* Atomic emission detectors
AES *see* Auger electron spectroscopy;
      Optical emission spectroscopy
Affinity chromatography, 843–5
AFS *see* Atomic fluorescence spectrometry
Agilent method translator, 769
Agilent pressure-flow calculator, 769
Air, 249, 670
Air-acetylene flames, 403–8
Alcohols, 169, 269–74, 676–80
Aldehydes, 169–70, 274–5, 353, 680–2
Aliphatic hydrocarbons, 162–6, 263–6,
      670–3
Aliquots, 17
Alizarin garnet R, 374–5
Alkaline errors, 954
Alkanes, 162–6, 263–6, 670–2

Alkenes, 162–6, 263–6, 672
Alkyl halides, 162–6
Alkynes, 162–6, 263–6, 672, 674
Alpha cleavage, 678–9
Alumina, 729
Aluminium, 499–502
Aluminon, 360
Amides, 172–3, 685
Amines, 171–2, 275–6, 684–5
Amino acids, 277, 279, 307, 830–3
4-Aminophenazone, 360
Amorphous polymers, 585, 1024, 1025
Amperometric detectors, 817, 819, 935
Amplitude, 65–7
Analyst errors, 28
Analytes, 3, 4
Analyzing crystals, 556–9
Anion exchange resins, 838–9
Anisotropic effect, 127–8, 133
Anodic stripping voltammetry, 935,
      993–4
Anti-Stokes direct-line fluorescence, 517
Anti-Stokes scattering, 291–3
Antibonding orbitals, 322–4
APCI *see* Atmospheric pressure chemical
      ionization
API *see* Atmospheric pressure ionization
Arc emission spectroscopy *see* Atomic optical
      emission spectroscopy
Argon ionization detectors, 779
Aromatic compounds, 166–9, 266–9, 355–6,
      357, 674–6
Array detectors, 647
ASE *see* Accelerated solvent extraction
Assessment of data, 57–9
Asymmetrical stretching, 217
Atmospheric pressure chemical ionization
      (APCI), 625–6, 828–9
Atmospheric pressure ionization (API),
      625–6

Atomic absorption spectrometry (AAS), 385–448
  absorption wavelengths, 438–45
  analytical range, 425–6
  applications, 8, 11, 424–33
  atomizers, 389–90, 393–9, 402–9
  calibration, 426–8
  degree of absorption, 389
  detection limits, 445–7
  detectors, 401
  flame AAS, 393–6, 402–8, 532–3
  gas samples, 431
  graphite furnace AAS, 520, 531, 533
  hydride generation AAS, 398–9, 432–3
  instrumentation, 389–402
  liquid samples, 428
  modulation, 401–2
  monochromators, 399–400
  nonspectral interferences, 406, 409–17
  optical systems, 399–401
  principles, 385–9
  qualitative analysis, 424–5
  quantitative analysis, 385, 425–8
  radiation sources, 390–3
  sensitivity, 447–8
  solid samples, 429–31
  spectral bandwidth, 388
  spectral interferences, 417–24
Atomic emission detectors (AED), 780
Atomic emission spectroscopy (AES) see
    Optical emission spectroscopy
Atomic fluorescence spectrometry (AFS), 450, 516–21, 522
Atomic mass spectrometry, 694–710
  applications, 697–704
  ICP-MS, 695–7
  interferences, 704–9
  limitations, 709–10
Atomic optical emission spectroscopy, 462–83
  applications, 479–83
  calibration, 478–9
  commercial systems, 521–2
  detectors, 472–6
  Echelle monochromators, 471–2
  excitation sources, 463–8
  instrumentation, 463–76
  interferences, 476–9
  polychromators, 470–1
  preburn time, 481–3
  sample holders, 468–70
Atomic spectral interferences, 417
Atomizers, 389–90, 393–9, 402–9
ATP see Adenosine triphosphate

Attenuated total reflectance (ATR), 250–1, 283–5
Auger electron spectroscopy (AES), 880, 897–906
  applications, 902–4
  chemical shift, 900–1
  depth profiling, 904–6
  instrumentation, 901–2
  principles, 897–900
Auger electrons, 536, 880–1, 890, 897–906
Autosamplers, 753–4
Auxochromes, 345
Average mass, 659–60
Averaging signals, 54–5

Babington nebulizers, 493
Background absorption, 417–23
Bandwidth, 101, 104
Barium, 413
Barrier layer cells, 333
Base peaks, 617, 652
Baselines, 53
Bathochromic shift, 345–6
Beam-splitters, 106–7, 231, 236
Beer–Lambert–Bouguer Law, 79–81, 82–4, 90–3
  atomic absorption spectrometry, 389
  infrared, 247, 253, 260, 282–3
  X-ray spectroscopy, 543, 573
Benchtop NMR spectrometers, 154, 193–4
Bending modes, 217–24
Benzaldehyde, 179, 182
Benzene
  infrared, 267
  mass spectrometry, 618–19, 652–3, 657, 659
  NMR, 133, 140–1, 179–80
  UV/VIS, 328–9
Benzoic acid, 170–1, 173
Benzophenone, 681
Benzylic protons, 166–9
Benzylmercaptan, 278, 283
Biospecific elution, 844
Blanks, 15
Bleeding, 749, 757
Blue shift, 345, 346–8
Bolometers, 236–7
Boltzmann distribution, 387
Boltzmann excess, 123
Bonding orbitals, 322–4
Boron, 512–15
Bragg equation, 547, 578
Bremsstrahlung, 540–1, 548–50
1-Bromo-4-chlorobenzene, 168
Bromoethane, 163–5

Bromomethane, 663
BTEX analysis, 764, 785
Bulk analysis, 507–8
Burner assemblies, 451–2
Burning velocity, 395–6
1-Butanal, 136–9, 169–70
Butane, 141, 143, 147
Butanol, 677–9
Butylamine, 174
Butyronitrile, 276–7, 278

^{13}C nuclear magnetic resonance, 173–80
Cadmium, 414–16, 501–2
Calcium, 411
Calcium carbonate, 1005
Calcium oxalate, 1006, 1011–15
Calibration, 52, 81–93
  atomic absorption spectrometry, 426–8
  atomic OES, 478–9
  Beer's Law errors, 90–3
  chromatography, 739, 741–3
  curves, 56–7
  flame photometry, 459–61
  internal standards, 87–90
  method of standard additions, 84–7
  plasma emission spectroscopy, 497–9
  potentiometry, 958
  standard solutions, 81–4
  standards, 15, 56, 497–9
  X-ray spectroscopy, 591
Capillary electrochromatography (CEC), 863–5
Capillary electrophoresis (CE), 8, 11, 702
Capillary gel electrophoresis (CGE), 860–2
Capillary isoelectric focusing (CIEF), 862–3
Capillary zone electrophoresis (CZE), 727, 851–6
Carbide formation, 413–14
Carbon dioxide, 669
Carbon disulfide, 278, 282
Carbonic anhydrase, 627
Carboxylic acids, 170–1, 269–75, 682–4
CAT *see* Computed axial tomography
Cation exchange resins, 837
CCD *see* Charge-coupled devices
CE *see* Capillary electrophoresis
CEC *see* Capillary electrochromatography
Ceramics, 191, 592–3
Cesium iodide, 574–5
CFFAB *see* Continuous flow FAB
CGE *see* Capillary gel electrophoresis
Characterization, 1
Charge-coupled devices (CCD), 105, 473, 475–6

Charge-injection devices (CID), 105, 473, 474, 475–6
Charge transfer devices (CTD), 105, 474–5, 476
Chemical exchange, 156–8
Chemical imaging, 306–8
Chemical interferences
  atomic absorption spectrometry, 406, 410–11
  atomic fluorescence spectrometry, 519
  flame photometry, 455–6
  plasma emission spectroscopy, 499
Chemical ionization (CI), 624–5
Chemical modification, 414–17
Chemical shifts
  Auger electrons, 900–1
  nuclear magnetic resonance, 127–8, 130–5
  X-ray photoelectron spectroscopy, 881, 892–3
Chemical speciation, 505–6, 521, 702–3
Chiral chromatography, 803
Chloranilic acid, 360
Chloroethane, 140, 146, 165
Chloroform, 278, 280
2-Chloropropane, 144
Chromatography, 721–48
  see also individual techniques
  calculation examples, 743–6
  calibration, 739, 741–3
  column variables, 734–7
  definitions, 722–3
  detectors, 728, 738
  development, 722–5
  efficiency, 734–7
  equations describing separations, 731–4
  extra-column band broadening effects, 737–9
  molecular processes, 725–8
  optimization, 736–7
  peak area/height, 740–1
  qualitative analysis, 739–40
  quantitative analysis, 740–3
  resolution, 733
  reversed phase, 729–30
  stationary phases, 723, 725–6, 728–31
Chromium, 404–5
Chromophores, 324–5, 345
CI *see* Chemical ionization
CID *see* Charge-injection devices; Collision induced dissociation
CIEF *see* Capillary isoelectric focusing
Classical polarography, 978–83
CMA *see* Cylindrical mirror analyzers
CMC *see* Critical micellar concentration
Coals, 1016
Cocaine, 297
Coelution, 724–5

Cold vapor-AAS (CVAAS), 398–9, 431–2
Collection of samples, 15–20
Collision cells, 708–9
Collision induced dissociation (CID), 826
Colored compounds, 358–9
Column bleeding, 749, 757
Combination bands, 218
Combination electrodes, 940
Combination spectrometers, 489
Combustion, 44
Composite samples, 18
Computed axial tomography (CAT), 535
Computed tomography (CT), 535
Concentration units, 3
Concentric nebulizers, 493
Conductometric analysis, 935, 969–76, 995 7
    detectors for LC, 817, 819–21
Confidence, 24, 33, 36–8
Confirmation columns, 739
Conjugated aldehydes, 353
Conjugated dienes, 348–52
Conjugated ketones, 352–5
Connes' advantage, 235
Constant-current coulometry, 961
Constant-potential coulometry, 961
Contamination, 29
Continuous flow FAB (CFFAB), 631–2
Continuous wave NMR, 148
Continuous-dynode electron multipliers, 645
Continuum radiation, 540–1, 548–50
Continuum source background correction,
    418–20
Controlled potential coulometry, 967
Corrected absorbances, 81
Correlation coefficients, 57
COSY, 181–7
Coulometry, 934, 961–9
Coupling, IR, 218
Crack cocaine, 297
Critical micellar concentration (CMC), 864
Cross-flow nebulizers, 493, 494
Cross-polarization, 179
Crystal analyzers, 556–9
Crystalline solid-state electrodes, 940–2
Crystallinity, 1003, 1029–30
Crystallography, 546–7, 576–85
CT see Computed tomography
CTD see Charge transfer devices
Cumene, 143
Curie temperature, 1010, 1011
Cuvettes, 77, 341–3
CVAAS see Cold vapor-AAS
Cyclic voltammetry, 935, 990
Cycloalkanes, 672

Cyclodextrins, 803
Cyclohexane, 265, 673
Cyclohexanol, 179, 181
Cyclohexanone, 170–1
Cylindrical mirror analyzers (CMA),
    884–5, 907
CZE see Capillary zone electrophoresis

DAD see Diode array detectors
Daltons, 613–14
Dark current, 335–6
Data handling
    accuracy, 24–5
    assessment of data, 57–9
    calibration, 52, 56–7
    confidence, 24, 33, 36–8
    errors, 25–31, 32–9
    noise, 52–6
    performing measurements, 51–7
    precision, 24–5, 36
    rejection of results, 39–40
    signals, 52–6
    significant figures, 21–4
    spreadsheets, 33–6
    statistics, 21–40
    variance, 32, 38–9
DC arcs, 464–7
DC GD sources, 506–7, 522
DC polarography, 978–83
DCP see Direct current plasma
Deactivation, 128
Deceleration methods, 885
1-Decene, 266
Degenerate vibrations, 218
Degrees of freedom, 33
Depolarizers, 964
Depth profile analysis, 508–9, 897, 904–6
Derivative thermogravimetry (DTG), 1017–19
Derivatization, 789–90, 817, 821–4
Design of methodology, 14–15
Desorption chemical ionization, 625
Desorption ionization, 626–32
Destructive analysis, 5–6, 12
Detection limits see Limit of detection
Detectors, 52, 104–5
    atomic absorption spectrometry, 401
    capillary zone electrophoresis, 858–9
    chromatography, 728, 738
    electroanalytical chemistry, 994–7
    gas chromatography, 750, 769–80
    HPLC, 809–21, 822
    infrared, 236–42
    liquid chromatography, 994–7
    thin layer chromatography, 868–9

Detectors (*contd.*)
  UV/VIS, 333–41, 373, 812–16
  X-ray photoelectron spectroscopy, 885–6
  X-ray spectroscopy, 559–61
Determinate errors, 25–9
Deuterium arc lamps, 331–2
Diamond anvil cells, 244
Diazomethane, 789
1,2-Dichlorobenzene, 167–8
1,1-Dichloroethane, 140, 165
1,3-Dichloropropane, 166
DID *see* Discharge ionization detectors
DIE *see* Direct injection enthalpimetry
Difference bands, 218
Differential pulse polarography, 986–9
Differential scanning calorimetry (DSC), 1003,
  1026–31
Differential thermal analysis (DTA), 1003,
  1020–6, 1031–2
Diffraction gratings, 96–102, 471–2
Diffraction patterns, 578
Diffuse reflectance (DRIFTS), 252–3
Diffusion, 981
Digestion, 40–3
DIN *see* Direct insertion nebulizers
Diode array detectors (DAD), 336–41, 815–16
Diodes, 336–41
Dipole moments, 214–16
Dipropylamine, 275, 277
Direct current plasma (DCP), 449, 486–8,
  489–91, 503–5, 522
Direct exposure probes, 621
Direct injection enthalpimetry (DIE), 1038–9
Direct insertion nebulizers (DIN), 493
Direct insertion probes, 621
Direct on-column injection, 760–1
Discharge ionization detectors (DID), 779
Discrete-dynode electron multipliers, 645, 646
Dispersive . . .
  infrared spectroscopy, 230, 235
  optical layouts, 107–8
  power, 99–102
  spectrometer systems, 295
Dispersive monochromator detectors (DMD), 815
Dissolution, 40–3
Dissolved oxygen, 981–2
Distillation, 721
DMA *see* Dynamic mechanical analysis
DMD *see* Dispersive monochromator detectors
DME *see* Dropping mercury electrodes
DNA sequencing, 860–2, 869–70
Documentation, 4
l-Dopa, 127
Double resonance, 158–61

Double-beam
  AAS, 400–1
  optics, 105–7
  spectrophotomers, 230
  UV/VIS, 329–30
DR *see* Diffuse reflectance
Drift, 54, 105, 131
Drift tubes, 635–6
DRIFTS *see* Diffuse reflectance
Dropping mercury electrodes (DME), 976, 978
Dry ashing, 44
DSC *see* Differential scanning calorimetry
DTA *see* Differential thermal analysis
DTG *see* Derivative thermogravimetry
Duane–Hunt Law, 541–2
Dynamic mechanical analysis (DMA), 1003,
  1039–48
Dynamic SIMS, 908–9
Dynodes, 334–5

ECD *see* Electrochemical detectors; Electron
  capture detectors
Echelle gratings, 101–2, 471–2
EDL *see* Electrodeless discharge lamps
Effluents, 722
EGA *see* Evolved gas analysis
EGD *see* Evolved gas detection
EI *see* Electron ionization
ELCD *see* Electrolytic conductivity detectors
Electric sector mass spectrometers, 633–5
Electrical double layers, 933–4
Electroanalytical chemistry, 919–1001
  activity series, 930–1
  background, 919
  conductometric analysis, 935, 969–76,
    995–7
  coulometry, 934, 961–9
  definitions, 920–4
  detectors, 994–7
  electrical double layers, 933–4
  electrochemical cells, 921–34
  line notation, 924–5
  Nernst equation, 928–30, 934
  polarography, 935, 976–89
  potentiometry, 934, 935–60
  reference electrodes, 925–8, 931–3, 934
  sign conventions, 928
  standard reduction potentials, 925–8, 1001
  voltammetry, 935, 977, 989–95
Electrobalances, 1007, 1008, 1009
Electrochemical cells, 921–34
Electrochemical detectors (ECD), 817–19
Electrodeless discharge lamps (EDL), 392–3
Electrodeposition, 922

Electrogravimetry, 934, 961, 962–6
Electrokinetic injection, 857–8
Electrolytic cells, 923, 961
Electrolytic conductivity detectors
        (ELCD), 776
Electromagnetic radiation, 65–71
Electromotive force, 924
Electron capture detectors (ECD), 774–6
Electron flood guns, 886–7
Electron ionization (EI), 614, 622–4
Electron multipliers (EM), 644–5, 646
Electron probe microanalysis (EPMA),
        593–4, 914–15
Electron spectroscopy, 879–906
Electron spectroscopy for chemical analysis
        (ESCA) see X-ray photoelectron
        spectroscopy
Electronic excitation, 319–23, 326–8, 385–6
Electronic transitions, 71–2, 76, 535–47
Electroosmotic flow (EOF), 853–7, 865
Electrophoresis, 722, 850–71
Electrospray ionization (ESI), 625–6, 627,
        628, 825–8
Electrothermal atomizers (ETA), 396–8, 531
Electrothermal vaporization (ETV), 496–7
Elemental analysis, 8, 11–12
ELSD see Evaporative light scattering detectors
Eluents, 722, 839–40
    see also Mobile phase
Eluotropic series, 804–5
Elution, 722, 762–9
EM see Electron multipliers
EMA (electron microprobe analysis) see
        Electron probe microanalysis
Emission, 69
    measurements, 249–50, 253–4
    spectra, 71, 73–4
Empirical formulae, 6
Enantiomers, 7–8, 803
Endcapping, 799
Endothermic processes, 1020
Energy-dispersive XRF, 547, 568–72, 587–90
Environmental applications
    concentration units, 3
    conductometric analysis, 976
    mass spectrometry, 693, 703–4
    plasma emission spectroscopy, 504
EOF see Electroosmotic flow
EPMA see Electron probe microanalysis
Equivalence point, 949–50, 955, 959
Equivalent hydrogen atoms, 139
Errors, 25–31, 32–9
ESCA (electron spectroscopy for chemical
        analysis) see X-ray photoelectron
        spectroscopy

Escape peaks, 561–2, 572
ESI see Electrospray ionization
Essential amino acids, 830–3
Esters, 274–5, 682–4
ETA see Electrothermal atomizers
Ethanol, 131–2, 144–6, 156–7, 272
Ethers, 680
Ethylbenzene, 159, 168–9
ETV see Electrothermal vaporization
Evaporative light scattering detectors (ELSD),
        811–12, 813
Even electron ions, 655
Evolved gas analysis (EGA), 1031–6
Evolved gas detection (EGD), 1031–6
Exact mass, 659
EXAFS see Extended X-ray absorption fine
        structure spectroscopy
Excitation, 70–2, 73–4
    interferences, 456
    sources
        atomic OES, 463–8
        flame photometry, 453–5
        plasma emission spectroscopy, 449,
        483–9, 503–5
Exocyclic double bonds, 351–2
Exothermic processes, 1020
Expanding drop method, 978–80
Extended X-ray absorption fine structure
        spectroscopy (EXAFS), 576
Extra-column band broadening effects, 737–9
Extracted ion chromatograms (XIC), 783–4
Extraction, 44–51, 721

FAAS see Flame atomic absorption
        spectrometry
FAB see Fast atom bombardment
Far-IR, 213, 229–30, 238
Faraday cups, 644, 645–7
Faraday's Law, 965–6
Fast atom bombardment (FAB), 629–32
Fast Fourier transform (FFT), 55–6
Fast gas chromatography, 763–4
Fatigue, 584–5
Fellgett's advantage, 110, 235
Fermi resonance, 275
Ferromagnetism, 1010
FFT see Fast Fourier transform
FIA see Flow injection analysis
Fiber optic probes, 344
Fiber-reinforced plastics (FRP), 189
FID see Flame ionization detectors
Field free drift tubes, 635–6
Filtered wavelengths, 95–6
Filters, 452–3, 554–5

Fingerprint region, 262–3
First order spectra, 147–8
Flame atomic absorption spectrometry (FAAS), 393–6, 402–8, 532–3
Flame atomizers, 393–6, 402–8
Flame ionization detectors (FID), 773–4
Flame photometry, 450–62
  applications, 8, 11, 458–62
  calibration, 459–61
  detection limits, 462, 529–31
  excitation sources, 453–5
  gas chromatography detectors, 777
  instrumentation, 451–5
  interferences, 455–8
Flicker noise, 54
Flow cells, 810, 813–14
Flow injection analysis (FIA), 433
Flow proportional counters, 563
Flow-through samplers, 343–4
Fluorescence, UV/VIS, 366–76
Fluorescence detectors, 816–17
Fluorides, 193–4
Fluorophores, 367
Focal plane cameras (FPC), 647
Food chemistry, 703–4
Forensic science, 257–8, 260
Formal potential, 929
Forward bias, 338
Fourier transform (FT), 54–6, 108–10
  infrared, 213, 230–6, 254–8, 260, 283–5, 1033–6
  mass spectrometry, 643–4
  microscopy, 254–8, 303–8
  nuclear magnetic resonance, 128–30, 148–9
  Raman, 295–7
FPC see Focal plane cameras
Fragmentation patterns, 654–6
Frequency (wave property), 66–7
FRP see Fiber-reinforced plastics
FT see Fourier transform
Full width at half maximum (FWHM) method, 620
Functional group region, 262
Fundamental parameters method, 591
Fundamental transitions, 218
Fused silica capillary columns, 752
Fusions, 43–4
FWHM see Full width at half maximum

Gadolinium, 196
Gallium arsenide, 512, 514
Galvanic cells, 923, 945–6, 961

Gas chromatography (GC), 749–96
  see also High performance liquid chromatography
  applications, 8, 10, 788–91
  column design, 757–69
  derivatization, 789–90
  detection limits, 771
  detectors, 750, 769–80
  development, 723, 749–52
  dimensions of column, 762–5
  elution values, 762–9
  gas analysis, 790–1
  hyphenated techniques, 10, 621–2, 702–3, 769, 780–6, 1031–6
  injectors, 753–6
  internet resources, 795–6
  limitations, 791
  mobile phases, 760–2
  retention indices, 786–8
  selectivity, 769–70
  stationary phases, 757–60
  temperature of column, 765–9
Gas expansion systems, 621
Gas phase infrared, 214–15
Gas samples, 18–19
  atomic absorption spectrometry, 431
  infrared, 248, 288
  mass spectrometry, 692–3
  UV/VIS, 341–3
  GC, 754, 758–9
Gas-filled detectors, 559–61
Gas-sensing electrodes, 942–4
Gaussian distributions, 30, 725
GC see Gas chromatography
GD see Glow discharge
Geiger-Müller tubes, 561
Gel boat holders, 344
Gel filtration chromatography (GFC), 845–8
Gel permeation chromatography (GPC), 845–8
Genomics, 722, 817, 827–8
Geological applications, 700–2
GF see Graphite furnace
GF-LEAFS see Graphite furnace laser-excited atomic fluorescence spectrometry
GFAAS see Graphite furnace atomic absorption spectrometry
GFC see Gel filtration chromatography
Ghost peaks, 890
Glass electrodes, 944–5
Glass membrane electrodes, 938–40
Glass transitions, 1020–1, 1022, 1043–4
Globars, 227
Glow discharge (GD), 506–9, 522, 632
Glucopyranoside, 183, 187
Golay detectors, 239

Gold, 501, 503

Goniometers, 556–7

GPC *see* Gel permeation chromatography

Gradient elution, 806–7

Graphite furnace atomic absorption
   spectrometry (GFAAS), 520,
   531, 533

Graphite furnace (GF) atomizers, 396–8,
   408–9, 413–14, 423–4

Graphite furnace laser-excited atomic
   fluorescence spectrometry (GF-LEAFS),
   520–1

Grotrian diagrams, 386–7

Ground state, 70–2, 73–4

Half-cells, 921–5

Half-wave potential, 977, 983–4

Halogen isotopic clusters, 663–6

Hanging mercury drop electrode (HMDE), 989,
   992–3

HCL *see* Hollow cathode lamps

Headspace analysis, 754

Heat detectors, 105

Heat flux differential scanning calorimetry,
   1026–7

Heat of vaporization, 1028

Heated wires, 227

Height equivalent to a theoretical plate (HETP),
   723, 733

Heisenberg uncertainty principle, 125–6

Helium ionization detectors (HID), 779

Helium MIP systems, 509–16

Helium-neon laser sources, 234

HEPS *see* High-energy prespark

Heptanoic acid, 273–4, 275

3-Heptanone, 681

HETCOR, 181, 183–4

Heterogeneous samples, 4

Heteronuclear decoupling, 161, 176

HETP *see* Height equivalent to a theoretical
   plate

Hexane, 264, 672

1-Hexene, 674

1-Hexyne, 266–7

HGAAS *see* Hydride generation-AAS

HID *see* Helium ionization detectors

High performance liquid chromatography
   (HPLC), 797–835
   see also Liquid chromatography
   applications, 8, 10–11, 829–35
   chiral phases, 803
   derivatization, 817, 821–4
   detectors, 809–21, 822
   development, 797–8

   hyphenated techniques, 194–5, 506, 621–2,
      702–3, 816, 824–9, 832–5
   instrument design and operation, 805–9
   limitations, 829
   mobile phases, 804–7
   resolution, 801
   reversed phase, 798, 802
   stationary phases, 798–803
   support particles, 801–2

High-energy prespark (HEPS), 468, 481

High-resolution mass spectrometry, 687–9, 708

HMDE *see* Hanging mercury drop electrode

Hollow cathode lamps (HCL), 390–2

Homoannular dienes, 349–51

Homogeneous field, 125

Homogeneous samples, 4

Homonuclear decoupling, 158–61

Hooke's Law, 219

HPLC *see* High performance liquid
   chromatography

Hydride generation-AAS (HGAAS), 398–9,
   432–3

Hydrides, 521

Hydrodynamic injection, 857–8

Hydrogen chloride, 215, 662–3

Hydrogen cyanide, 670

Hydrogen electrodes, 925–8

Hydrogen fluoride, 214–16

Hyperchromism, 345

Hyphenated techniques
   see also Mass spectrometry
   gas chromatography, 10, 622–3, 701–2, 769,
      780–6, 1031–6
   HPLC, 506, 621–2, 702–3, 816, 824–9,
      832–5
   liquid chromatography, 622, 702
   nuclear magnetic resonance, 194–5, 824
   optical emission spectroscopy, 505–6, 534
   thermal analysis, 1031–6
   thermogravimetric analysis, 1031–2

Hypochromism, 345

Hypsochromic shift, 345, 346–8

IC *see* Ion chromatography

ICP *see* Inductively coupled plasma

ICR *see* Ion cyclotron resonance trap

IDMS *see* Isotope-dilution mass spectrometry

IEC *see* Interelement correction

Ilkovic equation, 978–80

Immiscible liquids, 19

Immobilized enzyme membrane electrodes, 944

IMMS *see* Ion microprobe mass spectrometry

In-plane bending, 217

INADEQUATE, 184

Indeterminate errors, 30–1, 32–9
Indicator electrodes, 934, 935–45
Indirect detection, 842–3
Indium, 1028
Inductively coupled plasma (ICP)
  atomic MS, 695–7
  mass spectrometry, 632–3
  optical emission spectroscopy, 483–6,
    489–91, 503–5, 522, 531, 533–4
Infrared (IR) spectroscopy, 213–316
  see also Raman spectroscopy
  absorption frequencies, 221–4, 287
  absorption types, 214–24
  applications, 8, 9, 259–85, 288–90
  background correction, 248–9
  chemical imaging, 306–8
  detectors, 236–42
  dipole moments, 214–16
  emission measurements, 249–50, 253–4
  Fourier transform, 213, 230–6, 254–8, 260,
    283–5, 1033–6
  gas samples, 248, 288
  hyphenated techniques, 780, 784–5, 816,
    824, 1033–6
  instrumentation, 225–42, 286
  interferometers, 230–6
  liquid samples, 245–8, 287–8
  microscopy, 254–8, 303–8
  monochromators, 230–6
  near-IR spectroscopy, 213, 228–9, 238, 285–
    90, 306–8
  nondispersive systems, 258–9
  optical materials, 225, 226, 236
  qualitative analysis, 261–81
    aliphatic hydrocarbons, 263–6
    aromatic hydrocarbons, 266–9
    halogenated organic compounds, 277–8
    heteroatom (S, P, Si) compounds,
      278–81
    nitrogen-containing organic compounds,
      275–7
    oxygen-containing organic compounds,
      269–75
  quantitative analysis, 281–5
  radiation sources, 225–30
  reflectance measurements, 249–53, 260
  response time, 241–2
  sampling techniques, 242–54
  solid samples, 243–5, 288
  spectral databases, 262, 309
  transmission measurements, 242–9
  vibrational modes, 217–24
Injection internal standards, 743
Injectors, 738–9, 753–6

Inorganic compounds
  infrared, 300
  mass spectrometry, 632–3, 685, 686–7
  UV/VIS, 320, 358
Instrumentation errors, 28
Integer mass, 659
Intensity of incident beam, 76–7
Interelement correction (IEC), 502
Interferences, definition, 4
Interferometers, 109–10, 230–6
Internal reflection elements (IRE), 250–1
Internal standards, 87–90
  atomic OES, 478–9
  chromatography, 739
  plasma emission spectroscopy, 499
Intersystem crossings, 367
Ion chromatography (IC), 839–43
Ion cyclotron resonance trap (ICR), 642, 643–4
Ion-exchange chromatography, 836–9
Ion guns, 887
Ion microprobe mass spectrometry (IMMS), 914
Ion-pairing chromatography (IPC), 836
Ion scattering spectroscopy (ISS), 906–8
Ion-selective electrodes (ISE), 938, 942, 943
Ion-selective field effect transistors (ISFET),
  944–5
Ion trap see Quadrupole ion trap
Ionization, 450
Ionization interferences, 412–13, 456, 499
IPC see Ion-pairing chromatography
IR see Infrared spectroscopy
IRE see Internal reflection elements
Iron, 587–90
ISE see Ion-selective electrodes
ISFET see Ion-selective field effect transistors
Isobutane, 141, 143
Isocratic HPLC elution, 806
Isoelectric point, 836, 862–3
Isomers, qualitative analysis, 6
Isothermal gas chromatography, 765
Isotope-dilution mass spectrometry (IDMS),
  784
Isotopes
  abundances, 658–66, 697–9, 719–20
  clusters, 663–6, 697–9
  dilution, 689
  labeling, 689, 704
  masses, 614, 618
ISS see Ion scattering spectroscopy
IUPAC notation, 606–7

Jacquinot's advantage, 110, 235
Jet separators, 781–2
Joule heating, 854

K lines, 539, 543–4, 602–6, 608–11
Katharometers, 750, 771–3
Ketones, 170, 274–5, 352–5, 680–2
Kohlrausch's law, 971
Kovats retention indices, 787–8
Kubelka–Munk equation, 253

L lines, 539, 543–4, 604–6, 608–11
Laminar flow, 726–7, 734–6
Laminates, 255–7
Larmor equation, 122
Laser ablation, 497, 498
Laser desorption, 626
    see also Matrix-assisted laser desorption
        ionization
Laser light sources, 372–3
Laser sources, 230, 234
Laue photographs, 582
LC *see* High performance liquid
        chromatography; Liquid
        chromatography
Least squares method, 56–7
Light scattering, 69, 364–6
Light sources, Raman, 294–5
Limit of detection (LOD), 58–9
Limit of quantitation (LOQ), 59
Linear dynamic range, 770
Linear photodiode arrays (LPDA), 338–41
Linear sweep voltammetry, 935
Linear variable displacement transducers
        (LVDT), 1040–3
Linear working ranges, 58
Liquid chromatography (LC)
    see also High performance liquid
        chromatography
    2D separations, 866–71
    affinity chromatography, 843–5
    applications, 8, 10–11
    detectors, 994–7
    electrochemical detectors, 994–7
    electrophoresis, 850–71
    hyphenated techniques, 622, 702
    internet resources, 875
    ions dissolved in liquids, 835–43
    size exclusion chromatography, 845–8
    supercritical fluid chromatography,
        848–50
    thin layer chromatography, 866–9
Liquid membrane ion-selective electrodes,
        942, 943
Liquid samples, 19, 245–8, 287–8
    atomic absorption spectrometry, 428
    atomic OES, 469–70
    plasma emission spectroscopy, 491–5

UV/VIS, 341–3
    X-ray spectroscopy, 566–7
Liquid–liquid extraction, 45–6
Lithium, 459–61
LOD *see* Limit of detection
Long-term precision, 36
Longitudinal diffusion term, 734, 735–6
Longitudinal relaxation, 126
LOQ *see* Limit of quantitation
Lorentz broadening, 388
Loss modulus, 1047–8
LPDA *see* Linear photodiode arrays
Luggin probes, 968
Luminescence, 68, 366–76
LVDT *see* Linear variable displacement
        transducers
L'vov platforms, 409, 415

McLafferty rearrangement, 680–1
Magic angle spinning (MAS), 127–8, 178–9, 189
Magnesium, 404
Magnesium oxalate, 1012–15
Magnet specifications, 151–2
Magnetic field strength, 123–5
Magnetic moment, 118–19
Magnetic resonance imaging (MRI), 117,
        195–200
Magnetic sector mass spectrometers, 615–17,
        633–5
Magnetic shielding, 886
Magnetogyric ratio, 119, 121
MALDI *see* Matrix-assisted laser desorption
        ionization
MAS *see* Magic angle spinning
Mass analyzers, 633–44
Mass chromatograms (XIC), 783–4
Mass spectrometry (MS), 613–49, 651–720
    applications, 8, 10, 11, 687–94, 697–704
    atomic MS, 694–710
    detectors, 644–7
    development, 651
    environmental applications, 693, 703–4
    Fourier transform, 643–4
    gas analysis, 692–3
    high-resolution, 687–9, 708
    hyphenated techniques
        gas chromatography, 10, 621–2, 702–3,
            769, 780–4
        HPLC, 621–2, 702–3, 824–9, 832–5
        nuclear magnetic resonance, 195
        optical emission spectroscopy, 505–6, 534
        thermal analysis, 1031–6
    inorganic MS, 632–3
    instrumentation, 620–47

Mass spectrometry (*contd.*)
  interpretation of mass spectra, 652–87
    alcohols, 676–80
    carbonyl compounds, 680–2
    carboxylic acids/esters, 682–4
    common mass losses, 667
    ethers, 680
    examples, 667–87
    fragmentation patterns, 654–6
    halogen isotopic clusters, 663–6
    heteroatoms, 662–6, 684–7
    hydrocarbons, 670–6
    isotopic abundances, 658–66
    molecular formulae, 658–62
    neutral fragments, 667–8
    nitrogen rule, 656–7
    rings plus double bonds, 666–7
  ionization sources, 614, 622–33
  limitations, 694, 709–10
  mass analyzers, 633–44
  principles, 613–20, 652–6
  proteomics, 691–2
  quantitative analysis, 689–91, 699–700
  resolution, 619–20
  sample input systems, 621–2
  SIMS, 908–14
  spectral databases, 653–4
  surface analysis, 908–14
  tandem MS, 640–1
Mass transfer term, 734, 736
Matched cells, 343
Matrix
  blanks, 15
  definition, 4
  interferences, 411–12, 477, 704–6
  modification, 414–17
Matrix-assisted laser desorption ionization
    (MALDI), 626–30
Maxwell–Boltzmann equation, 387
Mechanical slit width, 103–4
Medical applications
  infrared, 304
  mass spectrometry, 692–3, 703–4
  nuclear magnetic resonance, 195–200
  X-ray spectroscopy, 573–5, 593
MEKC *see* Micellar electrokinetic capillary
    chromatography
Membrane electrodes, 937–40, 942–4
Membrane junctions, 923, 925
Mercury, 521
Mercury arc lamps, 372–3
Metallic electrodes, 937
Metallurgical applications, 575, 584–5,
    587–90, 592, 594

Metals, acid dissolution, 42–3
Metastable ions, 676
Methane, 655–6
Methanol, 157–8, 656–7, 660–1
Method internal standards, 743
Method of least squares, 56–7
Method of standard additions (MSA), 84–7,
    412, 742, 958, 987
Method validation, 4
Methodology, 14–15, 29
2-Methylpropane, 141, 143
Micellar electrokinetic capillary
    chromatography (MEKC), 863–5
Microconcentric nebulizers, 493
Microscopy, 254–8, 302–8
Microwave assisted extraction, 46–7
Microwave digestion, 41
Microwave-induced plasma (MIP), 488–9
Mid-IR, 213, 225–8, 238, 261–81
Migration, 981
Miller indices, 576–8
MIP *see* Microwave-induced plasma
Mixed crystal lattices, 583–4
Mobile phases
  eluents, 722, 839–40
  gas chromatography, 760–2
  HPLC, 804–7
  liquid chromatography, 849
  principles, 723, 725–6
Modulation, 401–2
Molar absorptivity, 79, 325–6
Molecular analysis, 8, 9–11
Molecular emission spectrometry, 8–9, 10,
    366–70
Molecular formulae, 6, 658–62
Molecular ions, 614, 654–6
Molecular sieves, 758, 759
Molybdenum, 410
Monochromatic light, 66
Monochromators
  atomic absorption spectrometry, 399–400
  atomic OES, 471–2
  diffraction gratings, 96–102, 471–2
  flame photometry, 451–2
  infrared spectroscopy, 230–6
  UV/VIS, 332, 371
  X-ray spectroscopy, 556–9
Monolayers, 877
Moseley's Law, 542
MRI *see* Magnetic resonance imaging
MS *see* Mass spectrometry
MS-MS/MS^n *see* Tandem MS
MSA *see* Method of standard additions
Mulling, 243

Multicomponent determinations, 361–2
Multipath term, 734–5
Multiphase materials, 583–4
Multiplex instruments, 108–10
Multiplicity, 139

Naphthalene, 140, 142, 674–5
NCE *see* Normal calomel electrodes
Near-IR (NIR) spectroscopy, 213, 228–9, 238,
    285–90, 306–8
Nebulizers, 393–6, 483–6, 492–6
Nephelometry, 364–6
Nernst equation, 928–30, 934
Nernst glowers, 225–6
NIR *see* Near-IR
Nitrate ions, 358
Nitro groups, 173
Nitrobenzene, 278
Nitrogen fluoride, 685
Nitrogen rule, 656–7
Nitrogen-containing organic compounds,
    171–3, 275–7, 684–6
Nitrogen–phosphorus detectors (NPD), 777–8
Nitropropane, 175
Nitrous oxide-acetylene flames, 403–8
NMR *see* Nuclear magnetic resonance
    spectroscopy
NOE *see* Nuclear Overhauser effect
Noise, 52–6, 104, 235, 770
Nominal mass, 659
Nonaqueous titrations, 956
Nonbonding electrons, 321–4
Noncrystalline materials, 585
Nondestructive analysis, 2, 5, 12
Nondispersive IR systems, 258–9
Nonspecific elution, 844–5
Normal calomel electrodes (NCE), 932
Normal distributions, 30
Normal phase chromatography, 729
Normal pulse polarography, 984–5
NPD *see* Nitrogen–phosphorus detectors
Nuclear magnetic resonance (NMR)
    spectroscopy, 117–212
  ^{13}C NMR, 173–80
  2D NMR, 152, 180–8
  absorption line width, 125–8
  active nuclei, 175
  applications, 8, 9, 154–94
  chemical exchange, 156–8
  chemical shifts, 127–8, 130–5
  computer specifications, 153–4
  development, 117–18
  double resonance, 158–61
  Fourier transform, 128–30, 148–9

hyphenated techniques, 194–5, 824
instrumentation, 148–54
interpretation of ^{1}H spectra, 161–73
  aliphatic hydrocarbons, 162–6
  alkyl halides, 162–6
  aromatic compounds, 166–9
  nitrogen-containing organic compounds,
    171–3
  oxygen-containing organic compounds,
    169–71
limitations, 200
magic angle spinning, 127–8, 178–9, 189
magnet specifications, 151–2
magnetic field strength, 123–5
NMR imaging, 195–200
nuclear Overhauser effect, 161, 176–8, 191
nuclear properties, 118–19
^{31}P NMR, 191, 198–9
peak area, 155–6, 177–8
qualitative analysis, 155–90
quantitative analysis, 190–4
quantization of ^{1}H nuclei, 119–25
RF generation and detection, 152–3
sample holders, 149–50
sample preparation, 154–5
sample probes, 150–1
saturation factor, 123–5
signal integrator, 153–4
solid samples, 178–9, 190
spectral databases, 185–7, 201
spin–spin coupling, 135–48, 161, 174–6
standards, 131–2
tacticity of polymers, 187–9
wide-line spectrometers, 154
Nuclear Overhauser effect (NOE), 161,
    176–8, 191
Nuclear spin, 118–19
Nujol mulls, 243

Octane, 163, 191–2
Octane number, 193
Octene, 166–7, 191–2
Odd electron ions, 655
OES *see* Optical emission spectroscopy
On-column derivatization, 789–90, 821–3
On-column injection, 760–1
Operational amplifiers, 945–6
Optical emission spectroscopy (OES), 449–534
  applications, 8, 11
  atomic fluorescence spectrometry, 516–21
  atomic OES, 462–83
  commercial systems, 521–2
  excitation process, 450
  flame photometry, 450–62

Optical emission spectroscopy (OES) (*contd.*)
glow discharge emission spectrometry, 506–9
helium MIP systems, 509–16
hyphenated techniques, 505–6, 534
mass spectrometry, 695
particle characterization, 509–16
plasma emission spectroscopy, 483–506
resources, 522–3
Optical systems, 93–110
atomic absorption spectrometry, 399–401
detectors, 104–5
dispersive optical layouts, 107–8
double-beam optics, 105–7
infrared, 225, 226, 236
optical slits, 103–4
radiation sources, 95
single-beam optics, 105–7
UV/VIS, 329–30
wavelength selection, 95–102
Organic modifiers, 804
Organometallic compounds, 300
Orthogonal spray, 826–7
Ostwald's dilution law, 972
Out-of-plane bending, 217
Outliers, 39
Overtones, 218, 267–8
Oxidizing agents, 920
Oxygen-containing organic compounds,
169–71, 269–75

^{31}P nuclear magnetic resonance, 191, 198–9
Packed columns, 752, 758–9, 798–801
Paper chromatography, 724
Particle characterization, 509–16
Pascal's triangle, 137
Paul ion trap *see* Quadrupole ion trap
PDMS *see* Polydimethylsiloxanes
Peak area, NMR, 155–6, 177–8
Pelleting, 243–4
Pellicular packings, 841
PET *see* Polyethylene teraphthalate
Petroleum industry, 592, 693, 703
PH electrodes, 944–5, 946–7, 951–6
Pharmaceuticals, 260, 290, 300
Phase transitions, 584
Phenolphthalein, 363
Phenols, 169–70, 273, 679–80
3-Phenylpropionaldehyde, 274–5
Phosphorescence, 366–8, 376
Phosphorus compounds, 278–81
Photodiodes and photodiode arrays, 105,
336–41
Photoelectric transducers, 333
Photoelectrons, 536, 880–1

Photoionization detectors (PID), 778–9
Photoluminescence, 366–76
Photolysis, 20
Photometers, definition, 111
Photomultiplier tubes (PMT), 105, 334–6,
470–3
Photons, 66–7
detectors, 105, 240–1, 242
Pi bonds, 321–4
PID *see* Photoionization detectors
Planar capillary gel electrophoresis, 869–71
Plane-polarized light, 65–6
Plasma emission spectroscopy, 483–506
applications, 503–6
calibration, 497–9
chemical speciation, 505–6
excitation sources, 449, 483–91, 503–5
hyphenated techniques, 505–6, 534
instrumentation, 483–97
interferences, 497–503
sample introduction systems, 491–7
Plasma sources, 449
Plate beam splitters, 107
Plate theory, 723
PLOT *see* Porous layer open tubular
PMT *see* Photomultiplier tubes
Polarity indices, 805
Polarography, 935, 976–89
classical (DC) polarography, 978–83
detectors, 817, 819
differential pulse polarography, 986–9
half-wave potential, 983–4
normal pulse polarography, 984–5
principles, 976–8
Polyatomic species, 706–8
Polychromatic light, 66
Polychromators, 470–1, 489
Polydimethylsiloxanes (PDMS), 188–9,
730–1, 757
Polyethylene, 1029–30
Polyethylene teraphthalate (PET), 1029, 1045
Polyglycol, 283–4
Polyimide coating, 730
Polymers
FTIR microscopy, 255–7
mass spectrometry, 629, 694
near-IR, 288–9
plasma emission spectroscopy, 505
quantitative NMR, 191
Raman, 299, 307
tacticity, 187–9
thermal analysis, 1003, 1015–19, 1023–4,
1029–30, 1044–8
X-ray spectroscopy, 567

Polymethylvinylsilane, 189–90

Poly(1,4-phenylene ether sulfone), 178

Polystyrene, 215

Polyvinyl chloride (PVC), 1018–19

Porous layer open tubular (PLOT) columns, 759

Porous polymers, 758–9

Portable NMR spectrometers, 154

Post-column derivatization, 817, 821–3

Potassium bromide pellets, 243–4

Potentiometry, 934, 935–60

    applications, 948–60

        calibration by MSA, 958

        ion concentration, 948–51

        pH measurements, 951–6

        potentiometric titrations, 949–51

        quantitative analysis, 956–7

        redox titrations, 959–60

        weak acids and bases, 955–6

    indicator electrodes, 934, 935–45

    measurement of potential, 945–8

Potentiostats, 947, 965

Powdered crystalline diffractometers, 581–2

Power compensated differential scanning
        calorimetry, 1026–8

Preburn time, 481–3

Precession, 120–1

Precipitation, 721

Precision, 24–5, 36, 364

Precolumn derivatization, 821–3

Precursor ions, 641

Preparation of samples, 40–51

Preparative gas chromatography, 758

Preparative supercritical fluid chromatography,
        850

Pressure differential scanning calorimetry,
        1030–1

Pressurized fluid extraction, 46

Primary ion sources, 909–10

Prisms, 96–7, 99–100, 225

2-Propenoic acid, 683

Propionamide, 180, 183

Proportional counters, 560–1, 563–5

Proportional errors, 26–7

Propylamine, 275–6

Proteins

    infrared, 289

    liquid chromatography, 843–8, 862–3, 870–1

    mass spectrometry, 628, 644, 691–2

Proteomics, 691–2, 722, 817, 827–8, 830–5

Pulse height, 562–3

Pulse polarography, 984–9

Pulsed NMR, 128–30

Purge and trap samplers, 791

PVC *see* Polyvinyl chloride

Pyridine, 319

Pyroelectric devices, 236–7, 240

p–n junctions, 337–40

Quadrupole ion trap (QIT) mass spectrometers,
        640, 642–3

Quadrupole mass analyzers, 638–40

Qualitative analysis, 5–8

    atomic absorption spectrometry, 424–5

    atomic OES, 479–80

    chromatography, 739–40

    definition, 1

    elemental analysis, 8, 11–12

    enantiomers, 7–8

    flame photometry, 458

    infrared, 261–81

    isomers, 6

    molecular analysis, 8, 9–11

    nuclear magnetic resonance, 155–90

    Raman, 299–300

    sensitivity, 12

    thermal analysis, 1024–5

    UV/VIS, 318–19, 356–7

    X-ray fluorescence, 586–90

    X-ray photoelectron spectroscopy, 889–92

Quantitation limits *see* Limit of quantitation

Quantitative analysis, 8–14

    atomic absorption, 385, 425–8

    atomic OES, 481

    chromatography, 740–3

    definition, 1

    elemental analysis, 8, 11–12

    flame photometry, 458–62

    infrared, 281–5

    mass spectrometry, 689–91, 699–700

    molecular analysis, 8, 9–11

    nuclear magnetic resonance, 190–4

    potentiometry, 956–7

    Raman, 299

    sensitivity, 12

    SIMS, 913–14

    UV/VIS, 318–19, 357–61

    X-ray fluorescence, 590–3

    X-ray photoelectron spectroscopy, 893–6

Quantization of ^{1}H nuclei, 119–25

Quartz, 729

Quartz tungsten-halogen lamps, 228–9

Quinalizarin, 361

Radiant power, 76–7

Radiation sources, 95

    atomic absorption spectrometry, 390–3

    infrared, 225–30

    UV/VIS, 330–2, 371–3

Radiofrequency (RF) radiation, 117–18, 152–3, 483–6, 506–7, 522, 638–40
Radioisotope sources, 551–2
Raies Ultimes (RU) lines, 480–1
Raman spectroscopy, 290–308
    applications, 298–301
    chemical imaging, 306–8
    Fourier transform, 295–7
    instrumentation, 293–8
    light scattering, 290–3
    microscopy, 302–8
    resonance effect, 298, 301–2
    sample holders, 297–8
    surface enhanced, 302
Randles–Sevcik equation, 990
Random diffusion, 725–6
Random errors, 30–1, 32–9
Rayleigh scattering, 291–3
Reaction cells, 708–9
Reaction kinetics, 362, 1023
Reagent blanks, 15
Reagent errors, 28
Red shift, 345–6
Redox indicator electrodes, 937
Redox reactions, 920
Redox titrations, 959–60
Reducing agents, 920
Reduction potentials, 924, 925–8, 1001
Reference beams, 106–7
Reference electrodes, 925–8, 931–3, 934
Reference standards, 15
Reflectance measurements, 249–53, 260
Reflection, 69
Reflectrons, 637
Refractive index (RI) detectors, 810–11, 813
Rejection of results, 39–40
Relative electrode potential, 924
Relative response factors (RRF), 742–3
Relative retention times (RRT), 739
Relative uncertainty, 91–2
Relaxation time, 125–6
Repeatability, 36
Replicate analysis, 23–4, 25
Representative samples, 14, 15–16
Reproducibility, 36
Resolution, 99–102, 619–20, 733, 801
Resonance fluorescence, 517
Resonance lines, 386
Resonance Raman spectroscopy, 298, 301–2
Response time, 241–2
Retardation methods, 885
Retention gap, 760
Retention indices (RI), 786–8
Reverse bias, 338

Reverse electroosmotic flow, 856
Reverse engineering, 260
Reversed phase chromatography, 729–30, 798, 802
RF see Radiofrequency
Rhodium, 540
RI see Refractive index; Retention indices
Ring current, 133
Ringbom plots, 91–3
Rings plus double bonds rule, 666–7
Robustness, 36
Rock salt region, 225
Rocking, 217
Rotary sample loop injectors, 808–9
Rotating disk beam splitters, 107
Rotating disk electrodes, 469–70
Rotational transitions, 71–2, 74–5, 326–8
Rowland circle polychromators, 471
RRF see Relative response factors
RRT see Relative retention times
RU see Raies Ultimes
Ruggedness, 36

Salinity, 976
Salt bridges, 923, 925
Sample beams, 106–7
Samples
    see also Gas samples; Liquid samples; Solid samples
    collection, 15–20
    holders
        atomic OES, 468–70
        cuvettes, 77
        dynamic mechanical analysis, 1044
        UV/VIS, 341–5, 373
        X-ray photoelectron spectroscopy, 887–9
        X-ray spectroscopy, 565–7
    infrared, 242–54, 287–8
    introduction systems, 491–7, 621–2
    nuclear magnetic resonance, 149–51, 154–5
    preparation, 40–51
    Raman, 297–8
    stacking, 858
    storage, 20–1
Saturated calomel electrodes (SCE), 931–2
Saturation factor, 123–5
Scanning electron microscopes (SEM), 593
Scattering, 69
SCD see Sulfur chemiluminescence detectors
SCE see Saturated calomel electrodes
Scintillation counters, 564–5
Scissoring, 217
Scrubbing gas samples, 19
Sealed proportional counters, 563–4

Sealed vessel digestion, 41–3
SEC *see* Size exclusion chromatography
Secondary ion mass spectrometry (SIMS),
     626, 908–14
Selected ion monitoring (SIM), 624, 783–4
Selective detection reagents, 868
Selectivity, 769–70
Selenium, 416–17
Self regenerating suppressors, 839–40
SEM *see* Scanning electron microscopes
Semiconductors, 240–1, 242, 336–41,
     511–15, 569–72
Semicrystalline polymers, 1024, 1025
Semipermeable frits, 923, 925
Sensitivity, definition, 58
Sensors, 52
Sequential spectrometers, 489
SERS *see* Surface enhanced Raman
     spectroscopy
SFC *see* Supercritical fluid chromatography
SFE *see* Supercritical fluid extraction
Shake-up peaks, 890–1, 892
SHE *see* Standard hydrogen electrodes
Shielding, 131–2
Shim coils, 150, 151–2
Short-term precision, 36
Shot noise, 91–2
SI system, 3
Siegbahn notation, 606–7
Sigma bonds, 321–4
Signal integrators, 153–4
Signal-to-noise ratio, 53–4
Signals, 52–6
Significant figures, 21–4
Silanization, 789–90
Silica, 511–12, 513, 728–31, 798–803
Silicon, 584
Silicon compounds, 278–82
Silicones, 730–1
Silver/silver chloride electrodes, 933
SIM *see* Selected ion monitoring
SIMS *see* Secondary ion mass spectrometry
Simultaneous spectrometers, 489
Simultaneous wavelength-dispersive XRF, 568
Single wavelength filter photometers, 813–14
Single-beam atomic absorption spectrometry,
     400–1
Single-beam optics, 105–7
Single-column ion chromatography, 841–2
Single-crystal diffractometers, 579–81
Size exclusion chromatography (SEC), 845–8
Slab gels, 869–71
Slurry nebulizers, 495–6
Smith–Hieftje background correction, 422–3

Sodium, 512–15
Sodium silicate, 1017–18
Solid phase extraction (SPE), 48–9, 723
Solid phase microextraction (SPME), 49–51,
     52, 754–5
Solid samples, 19–20
  atomic absorption spectrometry, 429–31
  atomic OES, 469
  infrared, 243–5, 288
  nuclear magnetic resonance, 178–9, 190
  plasma emission spectroscopy, 495–7
  UV/VIS, 344
  X-ray spectroscopy, 565–7
Solvents
  absorption, 249
  extraction, 45–8
  UV/VIS, 328–9, 345–8
SOP *see* Standard operating procedures
Soxhlet extraction, 46–7
SP-FPD *see* Sulfur–phosphorus flame
     photometric detectors
SPADNS, 361
Spark ablation, 497
Spark emission spectroscopy *see* Atomic optical
     emission spectroscopy
Spark sources, 467–8, 632
SPE *see* Solid phase extraction
Speciation, 505–6, 521, 702–3
Specificity, 58
Spectral bandwidth, 101, 104
Spectral databases, 185–7, 201, 262, 309,
     653–4
Spectral interferences
  AFS, 519
  atomic MS, 706–8
  atomic OES, 478
  flame photometry, 456–8
  plasma emission spectroscopy, 499–503
Spectroelectrochemistry, 363
Spectrographs, 111, 470
Spectrometry, definition, 111
Spectrophotometry, 111, 358, 359–61
Spectroscopy, 65–116
  see also individual techniques
  absorption laws, 76–81
  atomic spectroscopy, 72–4
  Beer–Lambert–Bouguer Law, 79–81, 82–4,
     90–3
  calibration, 81–93
  detectors, 104–5
  dispersive optical layouts, 107–8
  double-beam optics, 105–7
  electromagnetic radiation, 65–71
  Fourier transform, 108–10

Spectroscopy (*contd.*)
  instrument nomenclature, 111
  optical slits, 103–4
  optical systems, 93–110
  radiation sources, 95
  single-beam optics, 105–7
  technique, 111
  wavelength selection, 95–102
Specular reflectance, 251–2
Spin angular momentum, 118–19
Spin decoupling, 158–61
Spin quantum number, 118–19
Spin–spin coupling, 135–48, 161, 174–6
Split injections, 755–6
Splitless injections, 752, 755–6
SPME *see* Solid phase microextraction
Spray chambers, 495
Spreadsheets, 33–6
Sputtering, 391, 507, 508
Stack plots, 198–9
Standard deviation, 24, 32
Standard hydrogen electrodes (SHE), 925–8
Standard operating procedures (SOP), 28–9
Standard reduction potentials, 925–8, 1001
Standards
  see also Calibration
  atomic absorption spectrometry, 427
  internal, 87–90, 478–9, 499, 739
  methodology, 14–15
  nuclear magnetic resonance, 131–2
Stark broadening, 388
Static SIMS, 908–9
Stationary phases
  gas chromatography, 757–60
  HPLC, 798–803
  liquid chromatography, 849
  principles, 723, 725–6, 728–31
  selection, 760
Statistics
  accuracy, 24–5
  confidence, 24, 33, 36–8
  definitions, 31–2
  errors, 25–31, 32–9
  precision, 24–5, 36
  quantifying random error, 32–9
  rejection of results, 39–40
  significant figures, 21–4
  spreadsheets, 33–6
  variance, 32, 38–9
Stokes direct-line fluorescence, 517
Stokes scattering, 291–3
Stopped flow analysis, 784
Storage modulus, 1046–8
Storage of samples, 20–1

Stretching modes, 217–24
Stripping voltammetry, 992–4
Student's *t* values, 37
Substituted benzene rings, 355–6, 357
Sucrose, 159–60, 177, 183–4, 186, 188
Sulfur chemiluminescence detectors (SCD), 777
Sulfur compounds, 278–81, 686–7
Sulfur dioxide, 686
Sulfur–phosphorus flame photometric detectors (SP-FPD), 777
Sum peaks, 572
Supercritical fluid chromatography (SFC), 848–50
Supercritical fluid extraction (SFE), 47–8
Supporting electrolytes, 980–1
Surface analysis, 877–917
  applications, 878–9, 889–97, 902–4, 908, 911–15
  Auger electron spectroscopy, 880, 897–906
  electron probe microanalysis, 914–15
  electron spectroscopy, 879–906
  ion scattering spectroscopy, 906–8
  principles, 877–9, 880–1, 897–900
  secondary ion mass spectrometry, 908–14
  X-ray photoelectron spectroscopy, 880–97
Surface coatings, 508–9, 513–16, 587–90, 593, 702
Surface enhanced Raman spectroscopy (SERS), 302
Surfactants, 856–7, 863–5, 988
Symmetrical stretching, 217
Syringe injectors, 753
Systematic errors, 25–9
Système International d'Unités *see* SI system

TA *see* Thermal analysis
Tacticity of polymers, 187–9
Tandem MS, 640–1, 833–5, 1031
Target compound analysis, 784
TCD *see* Thermal conductivity detectors
TDS *see* Total dissolved solids
Tetramethylsilane (TMS), 131–2
TFME *see* Thin-film mercury electrode
TGA *see* Thermogravimetric analysis
Thermal analysis (TA), 1003–53
  applications, 8, 10, 1004
  differential scanning calorimetry, 1003, 1026–31
  differential thermal analysis, 1003, 1020–6
  direct injection enthalpimetry, 1038–9
  dynamic mechanical analysis, 1003, 1039–48
  hyphenated techniques, 1031–6
  thermogravimetric analysis, 1003, 1004–20
  thermomechanical analysis, 1003, 1039–48
  thermometric titrimetry, 1003, 1036–8

Thermal conductivity detectors (TCD), 750,
    771–3
Thermal detectors, 105
Thermionic detectors, 777
Thermistors, 236–7, 239
Thermocouples, 236–7, 239
Thermogravimetric analysis (TGA), 1003,
    1004–20
  applications, 1010–17
  derivative, 1017–19
  error sources, 1019–20
  hyphenated techniques, 1031–2
  instrumentation, 1007–10
Thermomechanical analysis (TMA), 1003,
    1039–48
Thermometric titrimetry (TT), 1003, 1036–8
Thermospray ionization (TSI), HPLC, 825
Thin film deposition, 243–4
Thin layer chromatography (TLC), 724,
    866–9
Thin-film mercury electrode (TFME), 992–3
Thiols, 686
Thiourea, 360
Three-electrode devices, 947
Throughput advantage, 110
TIC see Total ion chromatograms
Time of flight (TOF), 635–8, 781, 910–11
Titanium nitride, 508, 509
Titrations, 362–3
TLC see Thin layer chromatography
TMA see Thermomechanical analysis
TMS see Tetramethylsilane
TOF see Time of flight
Toluene, 124, 268–9
Total dissolved solids (TDS), 976
Total ion chromatograms (TIC), 782–3
Trace analysis, 632
Transducers, 52
Transitions, 68
Transmission measurements, 242–9
Transmission quadrupole mass analyzers, 639,
    640
Transmittance, 77
Transverse relaxation, 126
Trichloroethylene, 665–6
Tripropylamine, 276, 277
Tropylium ions, 676
Trypsin, 833
TSI see Thermospray ionization
TT see Thermometric titrimetry
Tunable laser sources, 230
Tungsten, 500–1
Tungsten-halogen lamps, 330–1, 332
Turbidimetry, 364–6

Twisting, 217
Two-line background correction, 418

UHV system, 887
Ultrasonic nebulizers (USN), 493–4
Ultraviolet and visible (UV/VIS) spectroscopy,
    317–84
  absorption by molecules, 323–5
  accuracy, 364
  advantages, 376
  applications, 8–9, 10, 356–63, 374–6
  conjugated dienes, 348–52
  conjugated ketones, 352–5
  detectors, 333–41, 373, 812–16
  disadvantages, 376
  electronic excitation in molecules, 319–23
  instrumentation, 329–45, 370–3
  luminescence, 366–76
  molar absorptivity, 325–6
  molecular emission spectrometry, 8–9, 10,
    366–70
  monochromators, 332, 371
  multicomponent determinations, 361–2
  nephelometry, 364–6
  optical system, 329–30
  precision, 364
  qualitative analysis, 318–19, 356–7
  quantitative analysis, 318–19, 357–61
  radiation sources, 330–2, 371–3
  reaction kinetics, 362
  sample holders, 341–5, 373
  shape of absorption curves, 319, 326–8
  solvents, 328–9, 345–8
  spectroelectrochemistry, 363
  spectrophotometric reagents, 359–61
  substituted benzene rings, 355–6, 357
  titrations, 362–3
  turbidimetry, 364–6
Uncertainty, 91–2, 125–6
Unified atomic mass units, 613–14
Universal detection reagents, 868
USN see Ultrasonic nebulizers
UV/VIS see Ultraviolet and visible
    spectrophotometry

V-groove nebulizers, 493, 494
Vacuum phototubes, 334
Vacuum UV radiation, 317–18
Valine, 279
Van Deempter equation, 734–7, 751, 761
Vapor phase IR, 214–15
Variance, 32, 38–9
Vibrational modes, 217–24
Vibrational transitions, 71–2, 75–6, 326–8

Vinyl acetate copolymers, 1015–17
Visible spectrophotometry *see* Ultraviolet and
    visible spectrophotometry
Voltammetry, 935, 977, 989–95

Wagging, 217
Water, 216, 661–2
Wavelength, 65–7
    selection, 95–102
Wavelength-dispersive XRF, 547, 552–3,
    555–65, 568
Weak acids and bases, 955–6, 1037–8
Wet ashing, 40–3
White noise, 54
White radiation, 540–1, 548–50
Wide-line NMR spectrometers, 154, 193–4
Woodward–Fieser Rules, 348
Working electrodes, 935

X-ray diffractometry (XRD), 8, 10, 535, 546–7,
    576–85
X-ray fluorescence (XRF) spectroscopy, 535
    applications, 8, 11, 585–93
    energy-dispersive, 547, 568–72, 587–90
    instrumentation, 550–72
    principles, 544–6
    wavelength-dispersive, 547, 552–3, 555–65,
        568
X-ray photoelectron spectroscopy (XPS),
    880–97
    applications, 889–97
    chemical shift, 881, 892–3
    depth profiling, 897
    detectors, 885–6
    electron energy analyzers, 883–5
    electron flood guns, 886–7
    element location, 897
    instrumentation, 882–7
    ion guns, 887
    magnetic shielding, 886
    principles, 880–1
    qualitative analysis, 889–92

quantitative analysis, 893–6
    radiation source, 882–3
    sample holders, 887–9
    UHV system, 887
X-ray spectroscopy, 535–611
    absorption process, 543–4, 573–6,
        608–11
    analyzing crystals, 556–9
    applications, 8, 10, 572–94
    calibration, 591
    characteristic wavelengths and energies,
        601–7
    collimators, 552–4
    detectors, 559–61, 569–72
    development, 535
    electron probe microanalysis, 593–4
    electronic transitions, 535–47
    escape peaks, 561–2, 572
    EXAFS, 576
    filters, 554–5
    goniometers, 556–7
    instrumentation, 547–72
    IUPAC notation, 606–7
    proportional counters, 560–1, 563–5
    pulse height, 562–3
    sample holders, 565–7
    Siegbahn notation, 606–7
    sources, 548–52
Xenon arc lamps, 332, 372
XIC *see* Mass chromatograms
XPS *see* X-ray photoelectron spectroscopy
XRD *see* X-ray diffractometry
XRF *see* X-ray fluorescence spectroscopy
Xylenes, 268–9, 270

Zeeman effect, 120–1, 388, 420–2
Zeolites, 836, 1017–18, 1019
Zero path length difference (ZPD), 231–3
Zinc, 587–90
Zirconia, 730, 801–2
ZPD *see* Zero path length difference
Zwitterions, 727, 836, 862–3

SI Base Units

Quantity	Name	Symbol
Length	meter	m
Mass	kilogram	kg
Time	second	s
Electric current	ampere	A
Thermodynamic temperature	kelvin	K
Amount of substance	mole	mol
Luminous intensity	candela	cd

Common Scientific Units and Symbols

Symbol	Unit	Unit expressed in other SI units	Symbol	Unit	Unit expressed in other SI units
A	ampere		K	kelvin	
Å	ångstrom		L	liter	
atm	atmosphere		M	mol/L	
C	coulomb	s A	m	meter	
°C	degree Celsius		min	minute	
cal	calorie		N	newton	$m\,kg/s^2$
eV	electron volt		Pa	pascal	N/m^2
F	farad	C/V	s	second	
G	gauss		T	tesla	kg/s^2A
g	gram		V	volt	W/A
h	hour		W	watt	J/s
Hz	hertz	s^{-1}	Ω	ohm	V/A
J	joule	N m			

SI Prefixes for Common Multiplication Factors

Prefix	Symbol	Factor
tera	T	10^{12}
giga	G	10^{9}
mega	M	10^{6}
kilo	k	10^{3}
deci	d	10^{-1}
centi	c	10^{-2}
milli	m	10^{-3}
micro	μ	10^{-6}
nano	n	10^{-9}
pico	p	10^{-12}